MW00452938

Reveal MATH®

Integrated I

mheducation.com/prek-12

Cover: (t to b, l to r) whitestone/123RF, Ingram Publishing

Send all inquiries to:
McGraw-Hill Education
8787 Orion Place
Columbus, OH 43240

ISBN: 978-0-07-700687-7
MHID: 0-07-700687-9

Printed in the United States of America.

4 5 6 7 8 9 10 11 12 13 14 LWI 28 27 26 25 24 23 22 21

Contents in Brief

Reveal AGA® Makes Math Meaningful...

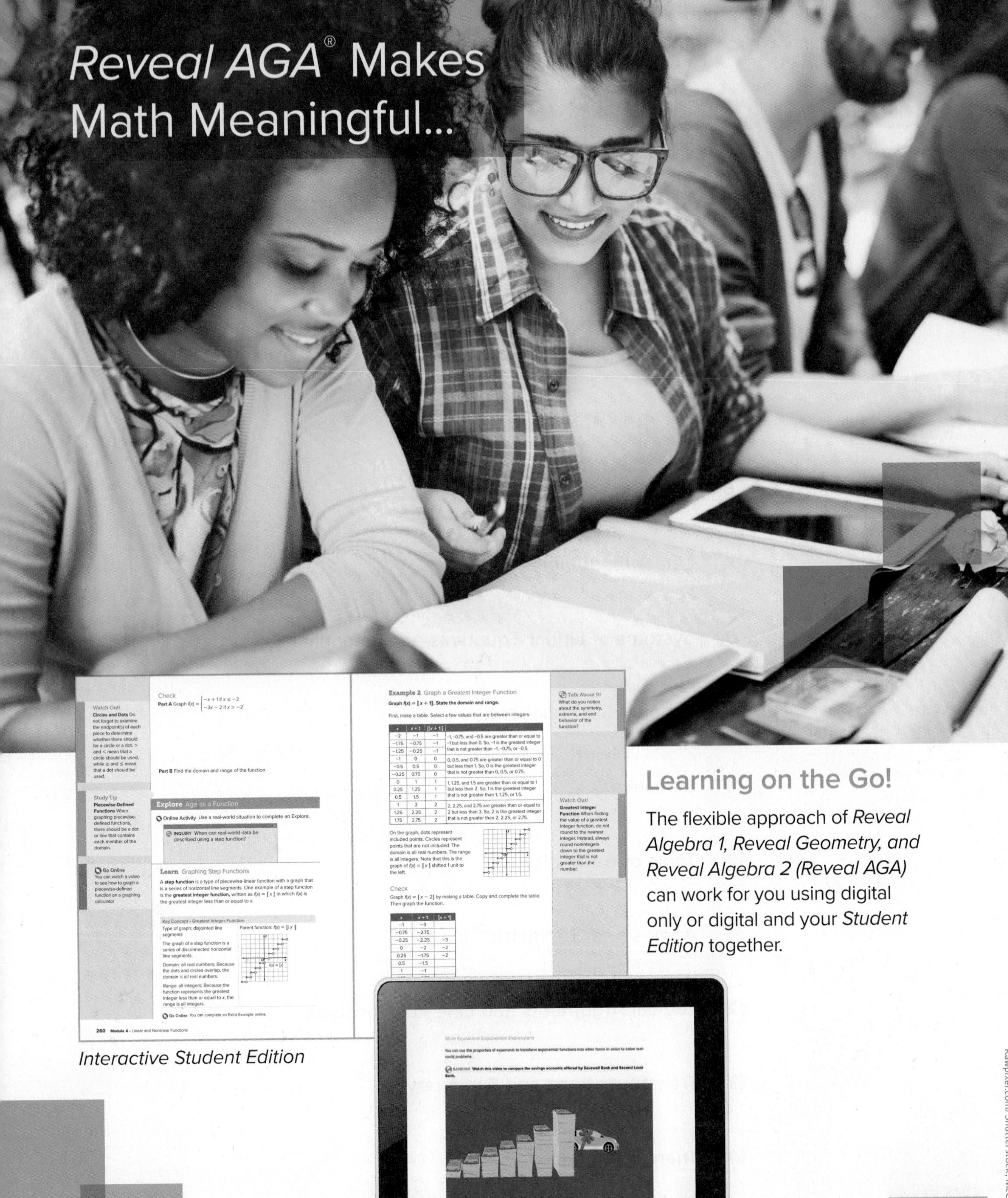

Watch Out!
Circles and Dots Do not forget to examine the endpoint(s) of each piece to determine whether there should be a circle or a dot. > and < mean that a circle should be used, while ≥ and ≤ mean that a dot should be used.

Check

Part A Graph $f(x) = \begin{cases} -x + 1 \text{ if } x \leq -2 \\ -3x - 2 \text{ if } x > -2 \end{cases}$

Part B Find the domain and range of the function.

Study Tip
Piecewise-Defined Functions When graphing piecewise-defined functions, there should be a dot or line that contains each member of the domain.

Explore Age as a Function

Online Activity Use a real-world situation to complete an Explore.

INQUIRY When can real-world data be described using a step function?

Go Online
You can watch a video to see how to graph a piecewise-defined function on a graphing calculator

Learn Graphing Step Functions

A **step function** is a type of piecewise-linear function with a graph that is a series of horizontal line segments. One example of a step function is the **greatest integer function,** written as $f(x) = [\![x]\!]$ in which $f(x)$ is the greatest integer less than or equal to x.

Key Concept • Greatest Integer Function

Type of graph: disjointed line segments

The graph of a step function is a series of disconnected horizontal line segments.

Domain: all real numbers; Because the dots and circles overlap, the domain is all real numbers.

Range: all integers; Because the function represents the greatest integer less than or equal to x, the range is all integers.

Parent function: $f(x) = [\![x]\!]$

Go Online You can complete an Extra Example online.

260 Module 4 • Linear and Nonlinear Functions

Example 2 Graph a Greatest Integer Function

Graph $f(x) = [\![x + 1]\!]$. State the domain and range.

First, make a table. Select a few values that are between integers.

x	$x + 1$	$[\![x + 1]\!]$	
−2	−1	−1	−1, −0.75, and −0.5 are greater than or equal to −1 but less than 0. So, −1 is the greatest integer that is not greater than −1, −0.75, or −0.5.
−1.75	−0.75	−1	
−1.25	−0.25	−1	
−1	0	0	0, 0.5, and 0.75 are greater than or equal to 0 but less than 1. So, 0 is the greatest integer that is not greater than 0, 0.5, or 0.75.
−0.5	0.5	0	
−0.25	0.75	0	
0	1	1	1, 1.25, and 1.5 are greater than or equal to 1 but less than 2. So, 1 is the greatest integer that is not greater than 1, 1.25, or 1.5.
0.25	1.25	1	
0.5	1.5	1	
1	2	2	2, 2.25, and 2.75 are greater than or equal to 2 but less than 3. So, 2 is the greatest integer that is not greater than 2, 2.25, or 2.75.
1.25	2.25	2	
1.75	2.75	2	

On the graph, dots represent included points. Circles represent points that are not included. The domain is all real numbers. The range is all integers. Note that this is the graph of $f(x) = [\![x]\!]$ shifted 1 unit to the left.

Check

Graph $f(x) = [\![x - 2]\!]$ by making a table. Copy and complete the table. Then graph the function.

x	$x + 1$	$[\![x + 1]\!]$
−1	−3	
−0.75	−2.75	
−0.25	−2.25	−3
0	−2	−2
0.25	−1.75	−2
0.5	−1.5	
1	−1	

Talk About It!
What do you notice about the symmetry, extrema, and end behavior of the function?

Watch Out!
Greatest Integer Function When finding the value of a greatest integer function, do not round to the nearest integer. Instead, always round nonintegers down to the greatest integer that is not greater than the number.

Interactive Student Edition

Learning on the Go!

The flexible approach of *Reveal Algebra 1, Reveal Geometry, and Reveal Algebra 2 (Reveal AGA)* can work for you using digital only or digital and your *Student Edition* together.

Student Digital Center

...to Reveal YOUR Full Potential!

Reveal AGA® Brings Math to Life in Every Lesson

Reveal AGA is a blended print and digital program that supports access on the go. Use your student edition as a reference as you work through assignments in the Student Digital Center, access interactive content, animations, videos, eTools, and technology-enhanced practice questions.

Go Online! my.mheducation.com

Web Sketchpad® Powered by The Geometer's Sketchpad®- Dynamic, exploratory, visual activities embedded at point of use within the lesson.

Animations and Videos – Learn by seeing mathematics in action.

Interactive Tools – Get involved in the content by dragging and dropping, selecting, highlighting, and completing tables.

Personal Tutors – See and hear a teacher explain how to solve problems.

eTools – Math tools are available to help you solve problems and develop concepts.

TABLE OF CONTENTS

Module 1

Expressions

Module 2

Equations in One Variable

TABLE OF CONTENTS

Module 3

Relations and Functions

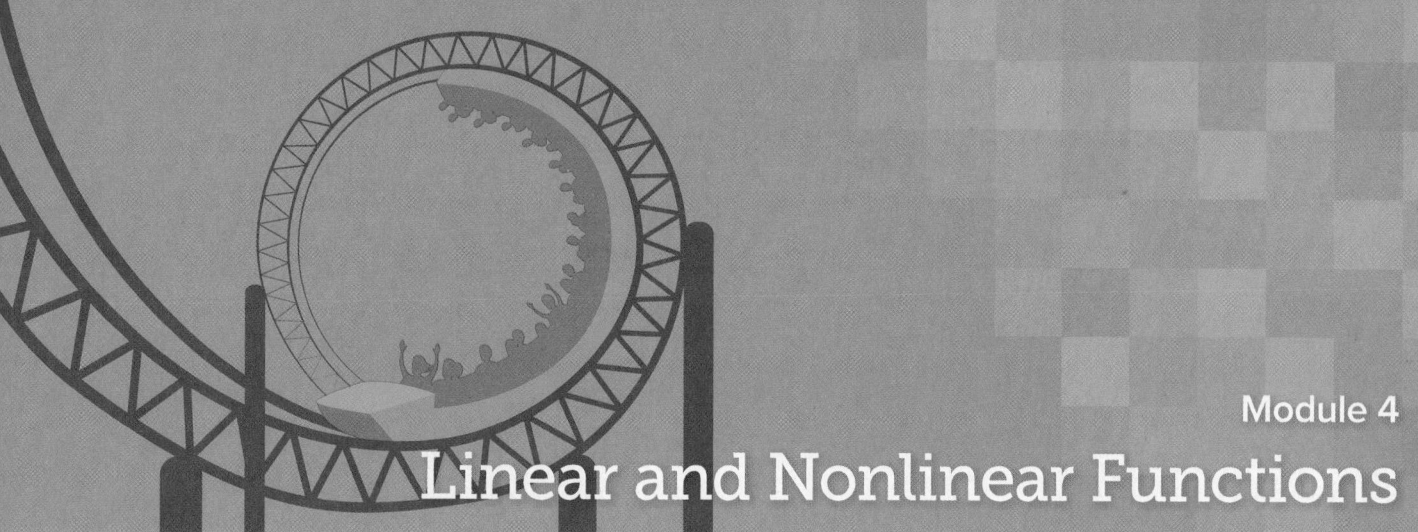

Module 4

Linear and Nonlinear Functions

TABLE OF CONTENTS

Module 5

Creating Linear Equations

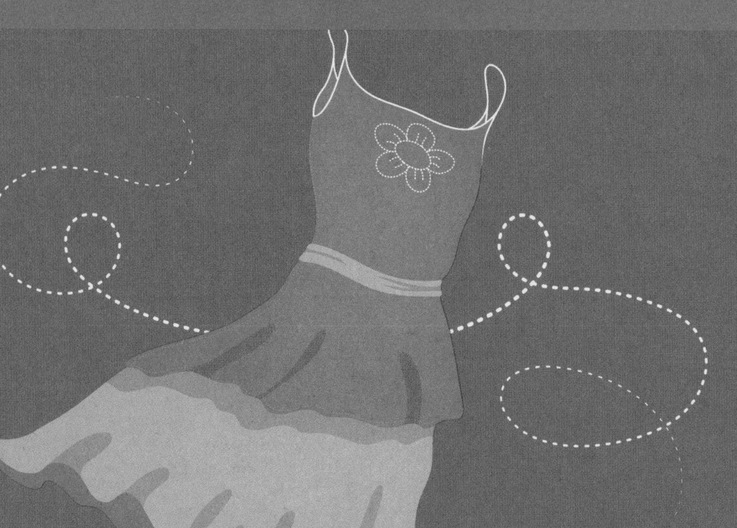

Module 6

Linear Inequalities

TABLE OF CONTENTS

Module 7

Systems of Linear Equations and Inequalities

Module 8

Exponential Functions

TABLE OF CONTENTS

Module 9
Statistics

Module 10

Tools of Geometry

TABLE OF CONTENTS

Module 11

Angles and Geometric Figures

Module 12

Logical Arguments and Line Relationships

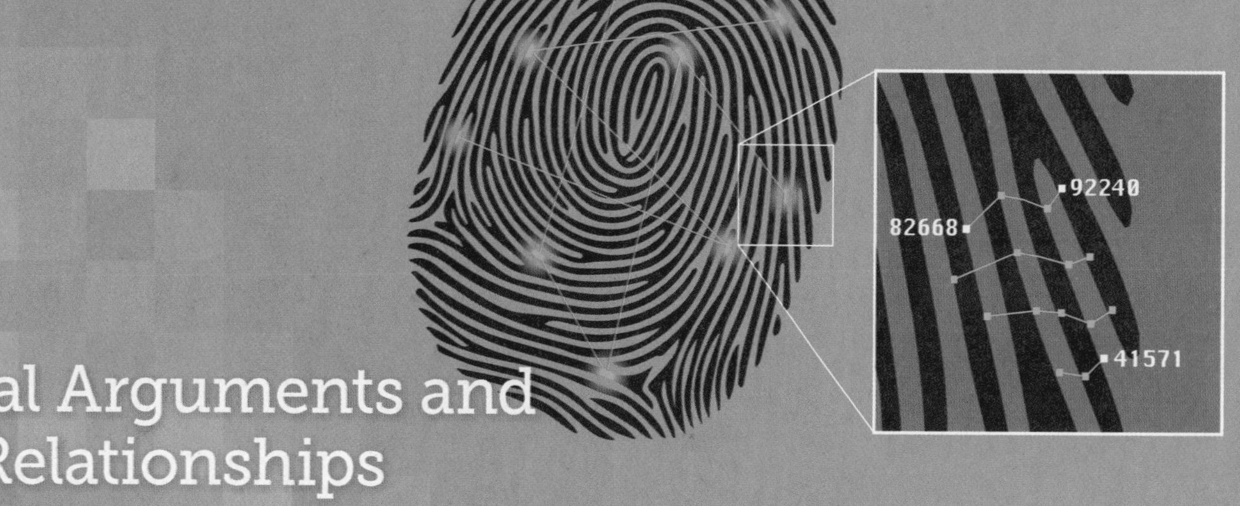

TABLE OF CONTENTS

Module 13

Transformations and Symmetry

Module 14

Triangles and Congruence

Module 1

Expressions

Essential Question

How can mathematical expressions be represented and evaluated?

What Will You Learn?

How much do you already know about each topic **before** starting this module?

KEY

— I don't know. — I've heard of it. — I know it!

	Before			After		
	I don't know.	I've heard of it.	I know it!	I don't know.	I've heard of it.	I know it!
write numerical expressions						
evaluate numerical expressions						
use the order of operations						
write algebraic expressions						
evaluate algebraic expressions						
identify properties of equality						
apply the Identity and Inverse Properties to evaluate expressions						
apply the Commutative, Associative, and Distributive Properties to evaluate expressions						
write and evaluate absolute value expressions						
use descriptive modeling to describe real-world situations						
choose a level of accuracy appropriate to limitations on measurements						

Foldables Make this Foldable to help you organize your notes about expressions. Begin with three sheets of notebook paper.

1. **Fold** three sheets of paper in half along the width. Then cut along the crease.
2. **Staple** the six half-sheets together to form a booklet.
3. **Cut** five centimeters from the bottom of the top sheet, four centimeters from the second sheet, and so on.
4. **Label** each tab with a lesson number.

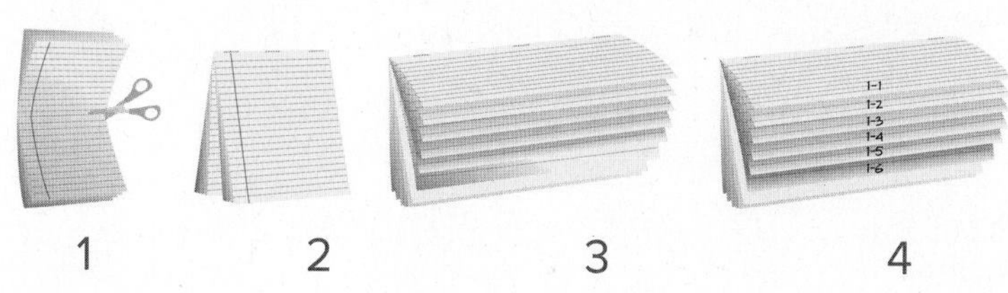

What Vocabulary Will You Learn?

- absolute value
- accuracy
- additive identity
- additive inverses
- algebraic expression
- base
- closed
- coefficient
- constant term
- define a variable
- descriptive modeling
- equivalent expressions
- evaluate
- exponent
- like terms
- metric
- multiplicative identity
- multiplicative inverses
- numerical expression
- reciprocals
- simplest form
- term
- variable
- variable term

Are You Ready?

Complete the Quick Review to see if you are ready to start this module.
Then complete the Quick Check.

Quick Review

Example 1

Write $\frac{24}{40}$ in simplest form.

Find the greatest common factor (GCF) of 24 and 40.

factors of 24: 1, 2, 3, 4, 6, 8, 12, 24

factors of 40: 1, 2, 4, 5, 8, 10, 20, 40

The GCF of 24 and 40 is 8.

$\frac{24 \div 8}{40 \div 8} = \frac{3}{5}$ Divide the numerator and denominator by their GCF, 8.

Example 2

Find $2\frac{1}{4} \div 1\frac{1}{2}$.

$2\frac{1}{4} \div 1\frac{1}{2} = \frac{9}{4} \div \frac{3}{2}$ Write mixed numbers as improper fractions.

$= \frac{9}{4} \times \frac{2}{3}$ Multiply by the reciprocal.

$= \frac{18}{12} = \frac{3}{2}$ or $1\frac{1}{2}$ Simplify.

Quick Check

Write each fraction in simplest form.

1. $\frac{24}{36}$
2. $\frac{34}{85}$
3. $\frac{5}{65}$
4. $\frac{64}{88}$

Evaluate.

5. $6 \times \frac{2}{3}$
6. $\frac{7}{8} - \frac{1}{6}$
7. $\frac{3}{8} \div \frac{1}{4}$
8. $\frac{1}{3} + \frac{3}{4}$

How did you do?

Which exercises did you answer correctly in the Quick Check?

Numerical Expressions

Explore Order of Operations

Online Activity Use a real-world situation to complete the Explore.

INQUIRY How can you evaluate a numerical expression?

Today's Goals
- Write numerical expressions for verbal expressions.
- Evaluate numerical expressions.

Today's Vocabulary
numerical expression
exponent
base

Learn Writing Numerical Expressions

A **numerical expression** is a mathematical phrase that contains only numbers and mathematical operations. For example, $6 + 2 \div 1$ is a numerical expression.

Some numerical expressions contain multiplication. Multiplication can be represented in several ways, including a raised dot or parentheses. Here are some ways to represent the product of 2 and 3.

$2 \cdot 3$ 2(3) (2)3 (2)(3)

Example 1 Translate a Verbal Expression

Translate *one plus eight divided by three* into a numerical expression.

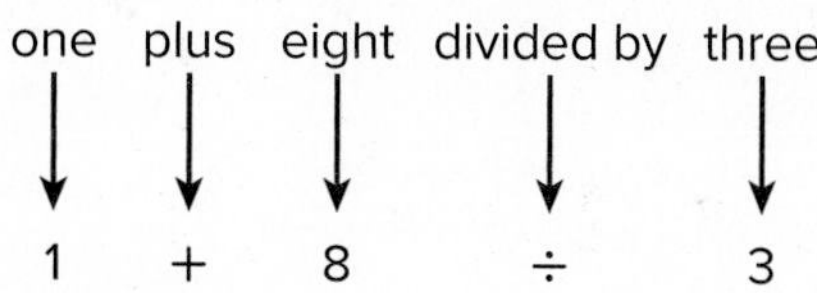

The translation is $1 + 8 \div 3$.

Think About It!
Identify the factors of the numerical expression 6(11).

Check

Translate the verbal expression into a numerical expression.
seven times fifteen minus two times nine

Think About It!
What verbal phrase represents addition? subtraction? multiplication? division?

Go Online You can complete an Extra Example online.

Example 2 Translate a Verbal Expression with Grouping Symbols

Translate *the sum of five and nine divided by two* into a numerical expression.

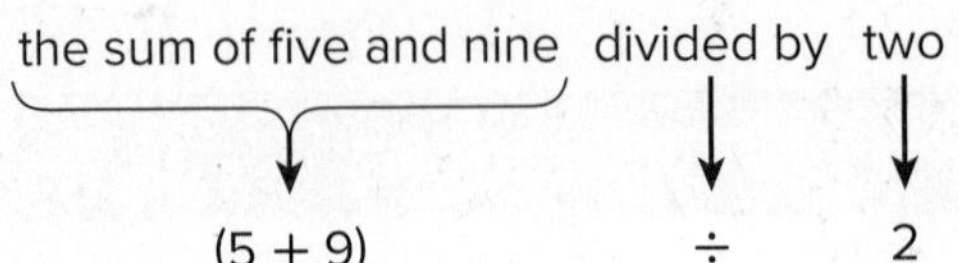

The translation is $\frac{5+9}{2}$.

Example 3 Write a Numerical Expression

BOWLING **A person's handicap in bowling is usually found by subtracting the person's average from 200, multiplying by 2, and dividing by 3. Marley's bowling average is 170. Write a numerical expression for Marley's handicap in bowling.**

The first step to find the handicap is to subtract the average from 200.

Marley's average is 170. $200 - 170$

Next, multiply the difference by 2. $2(200 - 170)$

Finally, divide the result by 3. $\frac{2(200-170)}{3}$

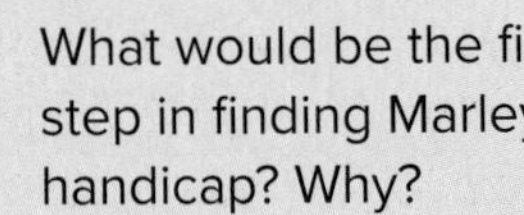

Think About It!

What would be the first step in finding Marley's handicap? Why?

Check

TEMPERATURE To convert a temperature in degrees Celsius to degrees Fahrenheit, multiply the temperature in degrees Celsius by 9, divide by 5, and add 32. Which numerical expression below converts 37° Celsius to Fahrenheit?

A. $\frac{37 \cdot 9}{5} + 32$

B. $\frac{37 \cdot 9}{5 + 32}$

C. $\frac{37 \cdot 9}{5} \cdot 32$

D. $\frac{37 \cdot 5}{9} + 32$

Watch Out!

Verbal Expressions The first step mentioned in the verbal expression does not always mean it is the first step in writing the numerical expression.

Go Online You can complete an Extra Example online.

Learn Evaluating Numerical Expressions

An expression of the form x^n is read "*x* to the *n*th power." The word *power* is used to refer to the expression, the value, or the exponent of the expression.

When *n* is a positive integer, the **exponent** indicates the number of times a number is multiplied by itself. In a power, the **base** is the number being multiplied by itself.

To **evaluate** an expression means to find its value. If a numerical expression contains more than one operation, then the rule that lets you know which operation to perform first is called the **order of operations**.

Key Concept • Order of Operations
Step 1 Evaluate expressions inside grouping symbols.
Step 2 Evaluate all powers.
Step 3 Multiply and/or divide from left to right.
Step 4 Add and/or subtract from left to right.

Talk About It!

Explain how the order of operations applies when using the formula $\frac{1}{2}h(b_1 + b_2)$ to find the area of a trapezoid.

Example 4 Evaluate Expressions

Evaluate each expression.

a. 2^4

$2^4 = 2 \cdot 2 \cdot 2 \cdot 2$ — Use 2 as a factor 4 times.

$= 16$ — Multiply.

b. 4^5

$4^5 = 4 \cdot 4 \cdot 4 \cdot 4 \cdot 4$ — Use 4 as a factor 5 times.

$= 1024$ — Multiply.

Example 5 Order of Operations

Evaluate $20 - 7 + 8^2 - 7 \cdot 11$.

$20 - 7 + 8^2 - 7 \cdot 11 = 20 - 7 + 64 - 7 \cdot 11$ — Evaluate 8^2.

$= 20 - 7 + 64 - 77$ — Multiply 7 and 11.

$= 13 + 64 - 77$ — Subtract 7 from 20.

$= 77 - 77$ — Add 13 and 64.

$= 0$ — Subtract 77 from 77.

Think About It!

Write an expression that uses exponents and at least three different operations. Explain the steps you would take to evaluate the expression.

 **Go Online** You can complete an Extra Example online.

Check

Write the steps of the order of operations in the correct order.

Step 1 ?

Step 2 ?

Step 3 ?

Step 4 ?

Use a Source

Find data about the scoring in a game of interest to you where you can score different numbers of points for different plays. Write and evaluate an expression to represent a possible score.

Example 6 Write and Evaluate a Numerical Expression

ARCADE **Mellie is playing a bowling game at an arcade. She rolls two balls into the 30-point hole, four balls into the 20-point hole, and three balls into the 50-point hole. Write and evaluate an expression to find Mellie's total score.**

Part A Complete the table to write an expression for Mellie's total score.

To find Mellie's total score, find the number of points scored from each hole and add the products.

Words	two balls rolled into the 30-point hole	plus	four balls rolled into the 20-point hole	plus	three balls rolled into the 50-point hole
Expression	$2 \cdot 30$	$+$	$4 \cdot 20$	$+$	$3 \cdot 50$

Part B Evaluate the expression.

$2 \cdot 30 + 4 \cdot 20 + 3 \cdot 50 = 60 + 80 + 150$ Multiply.

$= 290$ Add.

Mellie scored 290 points.

Check

COMPUTERS A computer technician charges a flat fee of $50 plus $25 per hour. On Monday, he worked on Aika's computer for 2 hours. On Tuesday, he worked on Aika's computer for 3 hours.

Part A Which expression(s) represents Aika's bill? Select all that apply.

?

A. 50 + 25(2) + 25(3)

B. 50 + 25(2 + 3)

C. 25(2 + 3)

D. 50 + 25(5)

E. 25 + 50(2 + 3)

Part B How much money does Aika owe the technician?

?

Go Online You can complete an Extra Example online.

Example 7 Expressions with Grouping Symbols

Evaluate each expression.

a. $\frac{(4 + 5)^2}{3\,(7 - 4)}$

$\frac{(4 + 5)^2}{3\,(7 - 4)} = \frac{(9)^2}{3(3)}$ Evaluate inside parentheses.

$= \frac{81}{3(3)}$ Evaluate the power in the numerator.

$= \frac{81}{9}$ Multiply 3 and 3 in the denominator.

$= 9$ Divide 81 by 9.

b. $15 - [10 + (3 - 2)^2] + 6$

$15 - [10 + (3 - 2)^2] + 6 = 15 - [10 + (1)^2] + 6$ Evaluate innermost parentheses.

$= 15 - [10 + 1] + 6$ Evaluate power.

$= 15 - [11] + 6$ Add.

$= 4 + 6$ Subtract.

$= 10$ Add.

Check

Evaluate $\frac{2^3 - 5}{15 + 9}$.

Learn Plan for Problem Solving

Using a **four-step problem-solving plan** can help you make sense of problems and persevere in solving them.

Key Concept • Four-Step Problem-Solving Plan
Step 1 Identify and understand the task. Read the task carefully, and make sure you understand what question to answer or problem to solve.
Step 2 Plan your approach. Choose a strategy. Plan the steps you will use to complete the task.
Step 3 Solve the problem. Use the strategy you chose in Step 2 to solve the problem.
Step 4 Check the solution. Make sure that your solution is reasonable and completes the task.

Go Online You can complete an Extra Example online.

Think About It!
Equivalent expressions have the same value. Are the expressions $(30 + 17) \times 10$ and $10 \times 30 + 10 \times 17$ equivalent? Why or why not?

Study Tip
Grouping Symbols such as parentheses ( ), brackets [ ], braces { }, and fraction bars are used to clarify or change the order of operations. So, evaluate expressions inside grouping symbols first. For fraction bars, evaluate the numerator and denominator before completing the division.

Think About It!
How can using this four-step problem-solving plan help you effectively solve problems?

Math History Minute

Hungarian mathematician **George Pólya (1887–1985)** is known as the father of mathematical problem solving. One of his books, *How to Solve It*, was a bestseller that was translated into 17 languages. In it, Pólya describes a four-step plan for problem solving.

Example 8 Write and Evaluate Expressions

MONEY **Thursdays are Student Days at LSC Theaters. Student tickets are $5 and popcorn refills are free. You will buy a ticket and a 20-ounce bottle of water, and you will split the cost of a large tub of popcorn and large candy with a friend. How much money should you bring?**

Popcorn		Drinks		Snacks	
Mini Tub	$3.00	20 oz	$5.00	Fruit Snacks	$3.00
Medium Tub	$5.00	32 oz	$5.50	Hot Dog	$5.00
Large Tub	$7.50	44 oz	$6.00	Candy	$3.50

UNDERSTAND **We are given what you will buy, the cost of each item, and the cost of the popcorn and candy that will be split with a friend. We are asked to find the amount of money that will be spent.**

PLAN

Step 1 Write an expression to represent the total cost of the ticket, drink, and food.

Step 2 Evaluate the expression to find the cost.

SOLVE

Step 1 Write the expression for the total cost in dollars.

ticket		drink		popcorn		candy
5	+	5	$+\frac{1}{2}$	7.50	$+\frac{1}{2}$	3.50

Step 2 Use the order of operations to evaluate the expression.

Original expression

$5 + 5 + \frac{1}{2} \cdot 7.50 + \frac{1}{2} \cdot 3.50 = 5 + 5 + 3.75 + 1.75$

$= \mathbf{15.50}$

Check

The cost of the ticket and water is $__?__. The total cost of the popcorn and candy is $__?__. So half of that cost is $__?__. The total is $10 + $5.50 or $ __?__.

Check

PETS While she was on a 7-day vacation, Ms. Hernandez boarded her dog at a kennel. If her boarding budget is $250, can Ms. Hernandez afford a daily extra walk and one wash and nail trimming? Explain.

Service	Cost
Board	$24 per day
Wash	$45
Extra walk	$4 per day
Nail trimming	$12
Vitamins	$1 per day

Go Online You can complete an Extra Example online.

Practice

Go Online You can complete your homework online.

Example 1

Write a numerical expression for each verbal expression.

1. two plus twelve divided by four
2. eighteen more than five
3. seven more than three times eleven
4. twenty-five less than one hundred
5. six minus three minus one
6. fourteen decreased by three times four
7. twenty-four divided by six plus seven
8. eight times six divided by two minus nine
9. one hundred sixteen divided by four plus twenty-eight minus thirty-three
10. two hundred fifty-nine minus eighty-five plus sixty-two divided by two

Example 2

Write a numerical expression for each verbal expression.

11. the sum of three and seven divided by two
12. the difference of six and two divided by four
13. the sum of four and nine times three
14. eighteen divided by the sum of two and seven
15. ten divided by the product of four and five
16. the difference of eleven and four times five
17. the sum of one and two divided by twenty
18. the sum of two and four and six times eight
19. the sum of twelve and sixteen divided by the sum of three and four
20. the difference of twenty-two and six divided by the sum of five and three
21. the sum of thirty-six and fourteen divided by the product of two and five
22. the quotient of thirty-two and four divided by the sum of one and three
23. the sum of six and fifteen divided by the difference of thirteen and nine
24. the difference of thirty-one and seventeen divided by the product of ten and four

Example 3

25. **SOLAR SYSTEM** It takes Earth about 365 days to orbit the Sun. It takes Uranus about 85 times as long. Write a numerical expression to describe the number of days it takes Uranus to orbit the Sun.

26. **TEST SCORES** To find the average of a student's test scores, add the scores and divide by the number of tests. Suppose Ryan scored 85, 92, 88, and 98 on four tests. Write a numerical expression to describe Ryan's average test score.

27. **HOMEWORK** It took Carrie five less minutes than twice the amount of time as Hua to complete her homework. It took Hua thirty-five minutes to complete her homework. Write a numerical expression to describe the amount of time it took Carrie to complete her homework.

28. **PERIMETER** The perimeter of Stephanie's triangle is half the perimeter of Juan's triangle. Juan's triangle is shown. Write a numerical expression to describe the perimeter of Stephanie's triangle.

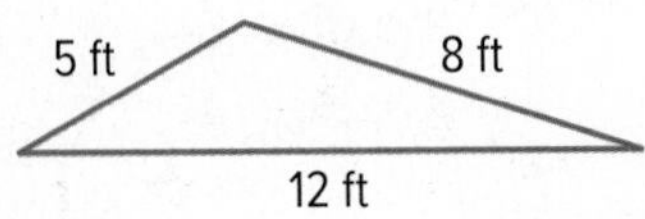

29. **BEDROOM** Shenandoah's rectangular bedroom is 12 feet long and 7 feet wide. Write a numerical expression to describe the area of Shenandoah's bedroom.

Examples 4, 5, and 7

Evaluate each expression.

30. 7^2

31. 14^3

32. 2^6

33. $35 - 3 \cdot 8$

34. $18 \div 9 + 2 \cdot 6$

35. $10 + 8^3 \div 16$

36. $[(6^3 - 9) \div 23]4$

37. $\frac{8 + 3^3}{12 - 7}$

38. $\frac{(1 + 6)9}{5^2 - 4}$

39. $4(16 \div 2 + 6)$

40. $13 - \frac{1}{3}(11 - 5)$

41. $(5 \cdot 2 - 9) + 2 \cdot \frac{1}{2}$

42. $62 - 3^2 \cdot 8 + 11$

43. $4^3 \div 8$

44. $20 + 3(8 - 5)$

45. $3[4 - 8 + 4^2(2 + 5)]$

46. $\frac{2 \cdot 8^2 - 2^2 \cdot 8}{2 \cdot 8}$

47. $25 + \left[(16 - 3 \cdot 5) + \frac{12 + 3}{5}\right]$

48. $7^3 - \frac{2}{3}(13 \cdot 6 + 9)4$

Example 6

49. BIOLOGY Lavania is studying the growth of a population of fruit flies in her laboratory. After 6 days, she had nine more than five times as many fruit flies as when she began the study. If she observes 20 fruit flies on the first day of the study, write and evaluate an expression to find the population of fruit flies Lavania observed after 6 days.

a. Write an expression for the population of fruit flies Lavania observed after 6 days.

b. Find the population of fruit flies Lavania observed after 6 days.

50. PRECISION The table shows how scores are calculated at diving competitions. Each of the five judges scores each dive from 1 to 10 in 0.5-point increments. Tyrell performs a dive with a degree of difficulty of 2.5. His scores from the judges are 8.0, 7.5, 6.5, 7.5, and 7.0.

Calculating a Diving Score	
Step 1	Drop the highest and lowest of the five judges' scores.
Step 2	Add the remaining scores to find the raw score.
Step 3	Multiply the raw score by the degree of difficulty.

a. Write an expression to find Tyrell's score for the dive.

b. What was Tyrell's score for the dive?

51. RAMP The side panel of a skateboard ramp is a trapezoid, as shown.

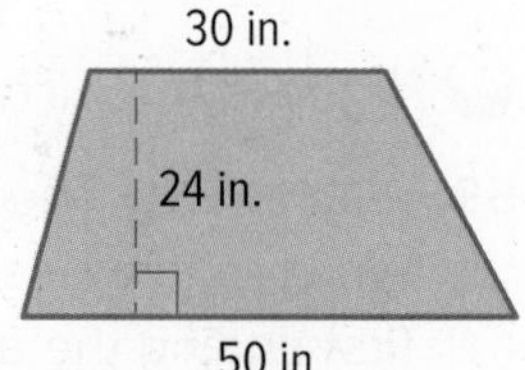

a. Write an expression to find the amount of wood needed to build the two side panels of a skateboard ramp.

b. How much wood is needed to build the two side panels of a skateboard ramp?

Example 8

52. SKIING The cost of a ski tip is shown. The Sanchez family wants to purchase lift tickets for 2 adults and 3 children. They also need to rent 2 complete pairs of skis. If they also buy a 16-ounce hot chocolate for each person, find the total cost of the ski trip.

Lift Tickets		Rentals		Hot Chocolate	
Children	\$34	Skis	\$32	12 oz	\$3
Seniors	\$36	Poles Only	\$10	16 oz	\$4
Adults	\$42	Snowboards	\$29	20 oz	\$5

Mixed Exercises

Write a numerical expression for each verbal expression.

53. eight to the fourth power increased by six

54. the sum of three and five to the third power times five plus one

55. CONSTRUCT ARGUMENTS Isabel wrote the expression $6 + 3 \times 5 - 6 + 8 \div 2$ and asked Tamara to evaluate it. When Tamara evaluated it, she got a value of 19. Isabel told Tamara that her value was incorrect and said that the value should have been 38. Who is correct? Justify your argument.

56. REASONING Write an expression that includes the numbers 2, 4, and 5, has a value of 50, and includes one set of parentheses.

57. CONSTRUCT ARGUMENTS Kelly buys 3 video games that cost \$18.96 each. She also buys 2 pairs of earbuds that cost \$11.50 each. She has a coupon for \$2 off the price of each video game. Kelly uses a calculator, as shown, to find that the total cost of the items is \$77.85. The cashier tells her that the total cost is \$73.85. Who is correct? Justify your argument.

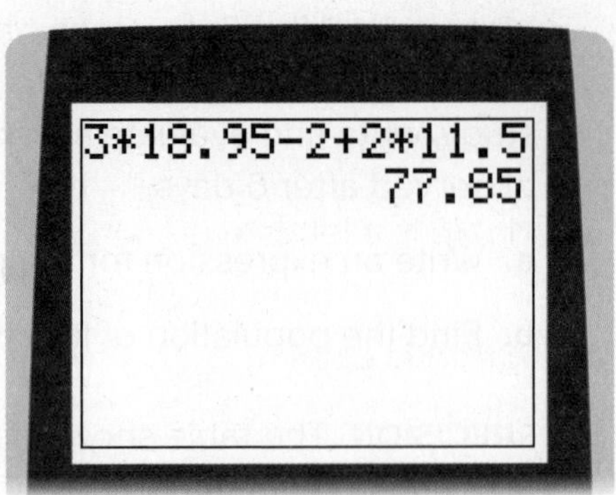

58. REASONING The expression $13.25 \times 5 + 6.5$ gives the total cost in dollars of renting a bicycle and helmet for 5 days. The fee for the helmet does not depend upon the number of days.

a. What does 13.25 represent?

b. How would the expression be different if the cost of the helmet were doubled?

Higher-Order Thinking Skills

59. PERSEVERE The figure shows a floor plan for a two-room apartment. Write an expression for the area of the apartment, in square feet, by first finding the area of each room and then adding. Then describe how you can write the expression in a different way.

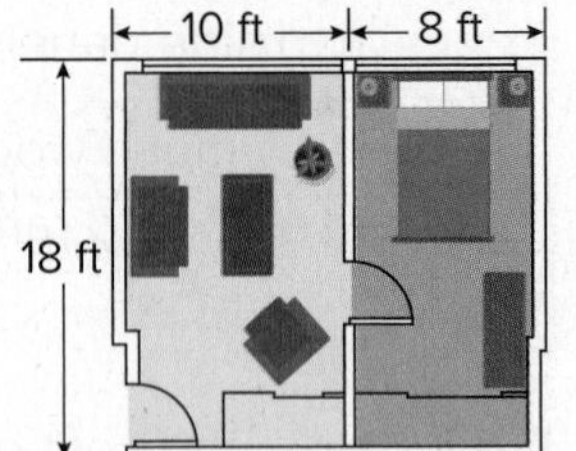

60. CREATE Describe a situation that could be represented by each expression.

a. $9.95 + 0.75 \times 3$

b. $15 \times 6 - 5 \times 6$

c. $59 \times 5 - 25 \times 5 - 30$

61. FIND THE ERROR A student was asked to evaluate an expression. The student's work is shown. Describe any errors and find the correct value of the expression. Explain your reasoning.

$6^2 - 5 \times 2 + 2(9 - 7)$
$= 6^2 - 5 \times 2 + 2(2)$
$= 36 - 5 \times 2 + 2(2)$
$= 36 - 10 + 4$
$= 36 - 14$
$= 22$

62. PERSEVERE When is $4x < 4$? Use a drawing to justify your reasoning.

Algebraic Expressions

Learn Writing Algebraic Expressions

A **variable** is a letter used to represent an unspecified number or value.

An **algebraic expression** is an expression that contains at least one variable.

A **term** of an expression is a number, a variable, or a product or quotient of numbers and variables.

A **variable term** is a term that contains a variable.

A **constant term** is a term that does not contain a variable.

Today's Goals

- Write algebraic expressions for verbal expressions.
- Evaluate algebraic expressions.

Today's Vocabulary

variable

algebraic expression

term

variable term

constant term

define a variable

Example 1 Write a Verbal Expression

There are common verbal phrases associated with operations. These phrases can be used to interpret an algebraic expression.

Interpreting an Algebraic Expression	
Operation	Verbal Phrases
addition	more than, sum, plus, increased by, added to
subtraction	less than, subtracted from, difference, decreased by, minus
multiplication	product of, multiplied by, times
division	quotient of, divided by

Write a verbal expression for $5x^3 + 2$.

five times x to the third power plus two

Think About It!

Write an algebraic expression that uses at least two variables, one constant, one product, and one power. Identify the terms, variables, constant, product, factors, and powers.

Check

Which verbal expression represents $\frac{2}{5}m^2$?

A. two fifths times the product of m and two

B. the quotient of two and five times the product of m and two

C. two fifths of m squared

D. two fifths times m cubed

Go Online You can complete an Extra Example online.

Example 2 Write a Verbal Expression with Grouping Symbols

Write a verbal expression for $a^4 + \frac{6b}{7}$.

a^4	$+$	$\frac{6b}{7}$
a to the fourth power	plus	the quotient of 6 times *b* and 7

a to the fourth power plus the quotient of 6 times *b* and 7

Try an alternate method.

a^4	$+$	$\frac{6b}{7}$
a to the fourth power	plus	6 times *b* divided by 7

a to the fourth power plus 6 times *b* divided by 7

Compare and contrast the methods.
Quotient implies division, so you can replace quotient with divided by. Thus, the two methods have the same meaning but use a different word to describe division.

Go Online An alternate method is available for this example.

Watch Out!
When a verbal expression refers to a quotient, the first term mentioned is the numerator and the second term mentioned is the denominator. For example, *the quotient of 6 and t* is written as $\frac{6}{t}$, while *the quotient of t and 6* is written as $\frac{t}{6}$.

Example 3 Write an Algebraic Expression

Write an algebraic expression for each verbal expression.

a. 2 times the quantity *y* plus 11

The word *times* implies multiplication, *the quantity* implies parentheses, and *plus* implies addition.

So, the expression is written as $2(y + 11)$.

b. *n* cubed increased by 5

The word *cubed* implies a power of 3, and *increased by* suggests addition.

The expression could be written as $n^3 + 5$ or $5 + n^3$.

Study Tip
Exponents The exponents 2 and 3 are frequently used in math and have special verbal phrases associated with them: x^2 is read "*x* squared," and x^3 is read "*x* cubed."

Check

Use the verbal phrase *18 increased by* the *product of* 3 and *d*.

Part A The key verbal phrases in the expression are italicized. Identify which operation each key phrase implies.

increased by implies ____?____

product of implies ____?____

Part B Write an algebraic expression for the verbal phrase.

Go Online You can complete an Extra Example online.

When writing an expression to represent a situation, choose a variable to represent each unknown value in the problem. This is called **defining a variable.**

Example 4 Write an Expression

SOCCER In the group play stage of the FIFA World Cup, teams are placed in groups of 4, and they play each other. A team is awarded 3 points for a win, 1 point for a tie, and no points for a loss. Write an algebraic expression that represents the number of points accumulated by one team in the group play stage of the World Cup.

Define variables for the unknown values.

Let w be the number of wins, t be the number of ties, and z be the number of losses for one team.

So, the number of points awarded for wins is $3w$, the number of points awarded for ties is t, and the number of points awarded for losses is $0z$. The number of points accumulated is $3w + t$.

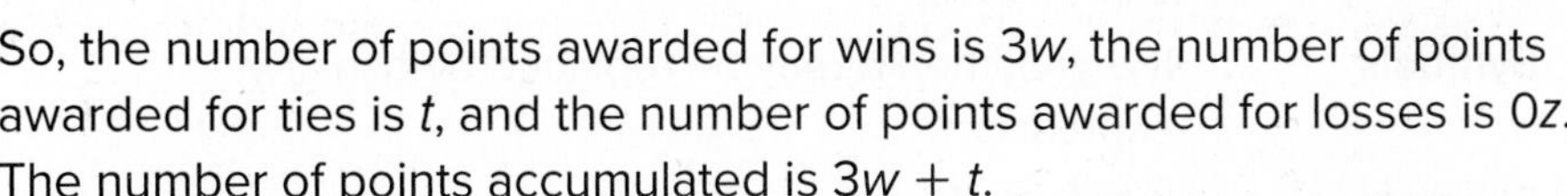

a. Write a verbal expression for the number of points accumulated and interpret the meaning of the variables in the context of the problem.

$3w + t$; 3 points times the number of wins plus the number of ties

b. What units are associated with the variables, the coefficients, and the expression?

The variables represent numbers of games, the coefficients represent points per game, and the expression represents the total number of points.

c. How would the expression change if a point were deducted for each loss?

You would subtract z from the original expression; $3w + t - z$.

Check

MUSIC A music festival offers one-day and three-day passes. A one-day pass costs $100, and a three-day pass costs $250. Write an expression for the total ticket sales if n one-day passes and t three-day passes are sold.

Go Online You can complete an Extra Example online.

> **Study Tip**
>
> **Modeling** When writing an expression to model a situation, begin by identifying the important quantities and relationships.

Red Fred/Alamy

Explore Using Algebraic Expressions in the Real World

Online Activity Use a real-world situation to complete the Explore.

INQUIRY How are algebraic expressions useful in the real world?

Learn Evaluating Algebraic Expressions

Algebraic expressions can be evaluated for given values of the variables. The Substitution Property allows us to evaluate an algebraic expression by replacing the variables with their values.

Key Concept • Substitution Property	
Words	A quantity may be substituted for its equal in any expression.
Symbols	If $a = b$, then a may be replaced by b in any expression.
Example	If $m = 11$, then $4m = 4 \cdot 11$.

Think About It!

Use the Substitution Property to replace each variable in the expressions with the appropriate value. Let $p = 6$, $q = 0.5$, and $r = 10$.

$3p = 3 \cdot 6$

$q^2 = 0.5^2$

$4q + r = 4 \cdot 0.5 + 10$

$pr^3 = 6 \cdot 10^3$

Example 5 Evaluate an Algebraic Expression

After applying the Substitution Property to an algebraic expression, you can find the value of the numerical expression by using the order of operations.

Evaluate $a^2(3b - a + 5) \div c$ if $a = 2$, $b = 6$, and $c = 4$.

$a^2(3b - a + 5) \div c =$

$2^2(3 \cdot 6 - 2 + 5) \div 4$	$a = 2, b = 6, c = 4$
$= 2^2(18 - 2 + 5) \div 4$	Multiply 3 by 6.
$= 2^2(21) \div 4$	Subtract 2 from 18, add 5.
$= 4(21) \div 4$	Evaluate 2^2.
$= 84 \div 4$	Multiply 4 by 21.
$= 21$	Divide 84 by 4.

Talk About It!

How would you evaluate $a[(b - c) \div d] - f$ if you were given values of a, b, c, d, and f? How would you evaluate the expression differently if the expression were $a \cdot b - c \div d - f$?

Check

Evaluate $\frac{b(9 - c)}{a^2}$ if $a = 4$, $b = 6$, $c = 8$.

Go Online You can complete an Extra Example online.

Example 6 Write and Evaluate an Algebraic Expression

WORLD RECORDS **In 2004, Chad Fell set the record for the largest bubblegum bubble blown. Assume that the bubble was spherical. The surface area of a sphere is four times π multiplied by the radius squared.**

Part A **Complete the table to write an expression that represents the surface area of a sphere.**

Words	four times π multiplied by radius squared
Variable	Let r = radius
Expression	$4 \times \pi r^2$ or $4\pi r^2$

Part B **The record-setting bubble had a radius of 25.4 centimeters. Find the surface area of this bubble.**

$A = 4\pi r^2$	Surface area of a sphere
$= 4\pi(25.4)^2$	Replace r with 25.4.
$= 4\pi(645.16)$	Evaluate $25.4^2 = 645.16$.
$= 2580.64\pi$	Multiply 4 by 645.16.
≈ 8107.32	Simplify.

The surface area of the bubble is approximately 8107.32 cm^2.

Study Tip

Assumptions Assuming that the bubble was spherical allows us to use the formula for the surface area of a sphere to estimate the surface area of the bubble. Although the bubble was not perfectly spherical, using the formula for a sphere allows for a reasonable estimate.

Check

FOOTBALL The seating capacities of team stadiums in the AFC East Division of the National Football League are shown in the table.

Team	Number of Seats
Miami Dolphins	65,326
New England Patriots	66,829
Buffalo Bills	71,608
New York Jets	82,500

Part A Write an expression that represents the maximum number of attendees at Jets, Dolphins, and Patriots home games during the season. Let j be the number of Jets home games, d be the number of Dolphins home games, and p be the number of Patriots home games.

Part B Suppose that after the sixth week of the season, the Jets had played 4 home games, the Dolphins had played 3 home games, and the Patriots had played 2 home games. Based on your expression from Part A, find the maximum number of attendees at the Jets, Dolphins, and Patriots games after the sixth week of the season.

Go Online You can complete an Extra Example online.

Pause and Reflect

Did you struggle with anything in this lesson? If so, how did you deal with it?

Practice

Go Online You can complete your homework online.

Examples 1 and 2

Write a verbal expression for each algebraic expression.

1. $4q$
2. $\frac{1}{8}y$
3. $15 + r$
4. $w - 24$
5. $3x^2$
6. $\frac{r^4}{9}$
7. $2a + 6$
8. $r^4 \cdot t^3$
9. $25 + 6x^2$
10. $6f^2 + 5f$
11. $\frac{3a^5}{2}$
12. $9(a^2 - 1)$
13. $5g^6$
14. $(c - 2)d$
15. $4 - 5h$
16. $2b^2$
17. $7x^3 - 1$
18. $p^4 + 6r$
19. $3n^2 - x$
20. $(2 + 5)p$
21. $18(p + 5)$

Example 3

Write an algebraic expression for each verbal expression.

22. x more than 7
23. a number less 35
24. 5 times a number
25. one third of a number
26. f divided by 10
27. the quotient of 45 and r
28. three times a number plus 16
29. 18 decreased by 3 times d
30. k squared minus 11
31. 20 divided by t to the fifth power
32. the sum of a number and 10
33. 15 less than the sum of k and 2
34. the product of 18 and q
35. 6 more than twice m

Example 4

36. TECHNOLOGY There are 1024 bytes in a kilobyte. Write an expression that describes the number of bytes in a computer chip with n kilobytes.

37. THEATER H. Howard Hughes, Professor Emeritus of Texas Wesleyan College, and his wife Erin Connor Hughes attended a record 6136 theatrical shows. Write an expression for the average number of shows they attended per year if they accumulated the record over y years.

38. TIDES The difference between high and low tides along the Maine coast one week is 19 feet on Monday and x feet on Tuesday. Write an expression to show the average difference between the tides for Monday and Tuesday.

39. SALE The cost of a T-shirt is shown. Monica has a $10-off coupon. Write an expression that describes the cost of t T-shirts, not including sales tax.

40. GYM MEMBERSHIP Juliana wants to join a gym. The cost of a gym membership is a one-time $100 fee plus $30 per month. Write an expression that describes the cost of a gym membership after m months.

41. BOWLING The cost for bowling is $5 per player for shoe rentals and $45 per hour to book a lane. Suppose a group of f friends go bowling for h hours. Write an expression for the total cost for the group of friends to go bowling.

Example 5

Evaluate each expression if $g = 2$, $r = 3$, and $t = 11$.

42. $g + 6t$

43. $7 - gr$

44. $r^2 + (g^3 - 8)^5$

45. $(2t + 3g) \div 4$

46. $t^2 + 8rt + r^2$

47. $3g(g + r)^2 - 1$

Evaluate each expression if $a = 8$, $b = 4$, and $c = 16$.

48. $a^2bc - b^2$

49. $\frac{c^2}{b^2} + \frac{b^2}{a^2}$

50. $\frac{2b + 3c^2}{4a^2 - 2b}$

51. $\frac{3ab + c^2}{a}$

52. $\left(\frac{a}{b}\right)^2 - \frac{c}{a - b}$

53. $\frac{2a - b^2}{ab} + \frac{c - a}{b^2}$

Evaluate each expression if $x = 6$, $y = 8$, and $z = 3$.

54. $xy + z$

55. $yz - x$

56. $2x + 3y - z$

57. $2(x + z) - y$

58. $5z + (y - x)$

59. $5x - (y + 2z)$

60. $z^3 + (y^2 - 4x)$

61. $\frac{y + xz}{2}$

62. $\frac{3y + x^2}{z}$

Example 6

63. SCHOOLS Jefferson High School has 100 less than 5 times as many students as Taft High School.

a. Write an expression to find the number of students at Jefferson High School if Taft High School has t students.

b. How many students are at Jefferson High School if Taft High School has 300 students?

64. GEOGRAPHY Guadalupe Peak in Texas has an altitude that is 671 feet more than double the altitude of Mount Sunflower in Kansas.

a. Write an expression for the altitude of Guadalupe Peak if Mount Sunflower has an altitude of n feet.

b. What is the altitude of Guadalupe Peak if Mount Sunflower has an altitude of 4039 feet?

65. TRANSPORTATION The Plaid Taxi Cab Company charges a $1.75 base fee plus $3.45 per mile. Deangelo plans to take a Plaid taxi to the airport.

a. Write an expression to find the cost for Deangelo to take a Plaid taxi m miles to the airport.

b. How much will it cost for Deangelo to take a Plaid taxi 8 miles to the airport?

66. GEOMETRY The area of a circle is given by the product of π and the square of the radius.

a. Write an expression for the area of a circle with radius r.

b. What is the area of the circle shown at the right? Use 3.14 for π.

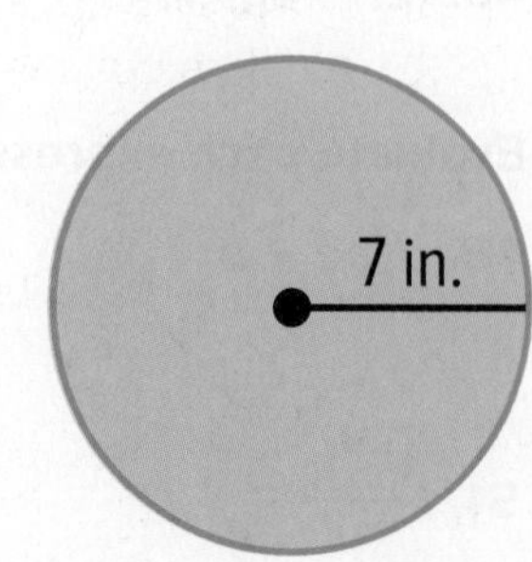

Mixed Exercises

67. Consider the expression $\frac{5x}{2} + y^3$.

a. Write two verbal expressions for $\frac{5x}{2} + y^3$.

b. Evaluate the expression if $x = 4$ and $y = 2$.

68. Evaluate $\frac{7a + b}{b + c}$, if $a = 2$, $b = 6$, and $c = 4$.

69. Evaluate $x^2 + y^2 + z$, if $x = 7$, $y = 6$, and $z = 4$.

70. Evaluate $\frac{2b + c^2}{a}$, if $a = 2$, $b = 4$, and $c = 6$.

71. Evaluate $2 + x(2y + z)$, if $x = 5$, $y = 3$, and $z = 4$.

72. **STRUCTURE** Write an algebraic expression that includes a sum and a product. Write a verbal expression for your algebraic expression.

73. **STRUCTURE** Write a verbal expression that includes a difference and a quotient. Write an algebraic expression for your verbal expression.

74. **USE A MODEL** A toy manufacturer produces a set of blocks, with edge b, that can be used by children to build play structures. The production team is analyzing the amount of paint they need for a block.

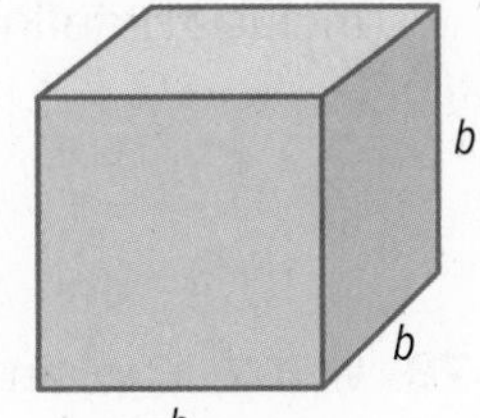

a. The production team decides to use one coat of paint for each block. Write an expression representing the minimum amount of paint needed for one block with edge b.

b. The production team decides one coat of paint is not enough, so they want to use two coats of paint for each block. Write an expression representing the minimum amount of paint needed for one block with edge b.

c. The production team purchases cans of paint that will cover 60 in^2. Write an inequality representing the maximum length of edge b, in inches, when the block is covered with the minimum amount of paint needed for two coats of paint.

75. **REASONING** During a long weekend, Devon paid a total of x dollars for a rental car so he could visit his family. He rented the car for 4 days at a rate of $36 per day. There was an additional charge of $0.20 per mile after the first 200 miles driven.

a. Write an expression to represent the amount Devon paid for additional mileage.

b. Write an expression to represent the number of miles over 200 miles that Devon drove.

c. How many miles did Devon drive overall if he paid a total of $174 for the car rental?

Higher-Order Thinking Skills

76. **PERSEVERE** For the cube, x represents a positive whole number. Find the value for x such that the volume of the cube and 6 times the area of one of its faces have the same value.

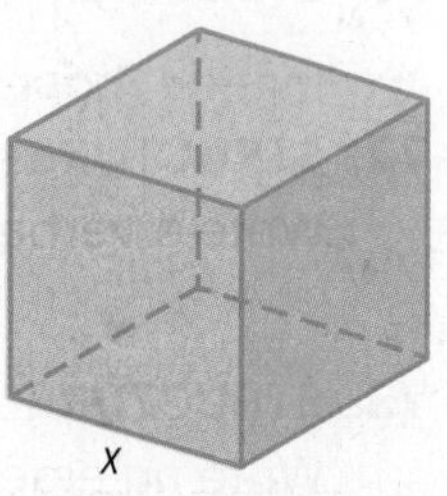

77. **WRITE** Describe how to write an algebraic expression from a real-world situation. Include a definition of algebraic expression in your own words.

78. **WRITE** Explain the difference between an algebraic expression and a verbal expression.

79. **CREATE** Write a real-world situation that can be modeled by the inequality $10t + 5.25 \leq 50$.

Properties of Real Numbers

Learn Properties of Equality

There are several properties of equality that apply to the addition and multiplication of real numbers.

Key Concept • Properties of Equality	
Reflexive Property	
Words	Any quantity is equal to itself.
Symbols	For any number a, $a = a$.
Examples	$3 = 3$ $9 + 2 = 9 + 2$
Symmetric Property	
Words	If one quantity equals a second quantity, then the second quantity equals the first.
Symbols	For any numbers a and b, if $a = b$, then $b = a$.
Example	If $7 = 3 + 4$, then $3 + 4 = 7$.
Transitive Property	
Words	If one quantity equals a second quantity and the second quantity equals a third quantity, then the first quantity equals the third quantity.
Symbols	For any real numbers a, b, and c, if $a = b$ and $b = c$, then $a = c$.
Example	If $5 + 1 = 2 + 4$ and $2 + 4 = 6$, then $5 + 1 = 6$.

Today's Goals

- Recognize the properties of equality and identiy.
- Evaluate numerical expressions by applying the Inverse and Identity Properties.
- Evaluate numerical expressions by applying the Commutative and Associative Properties.

Today's Vocabulary

additive identity

additive inverses

multiplicative identity

multiplicative inverses

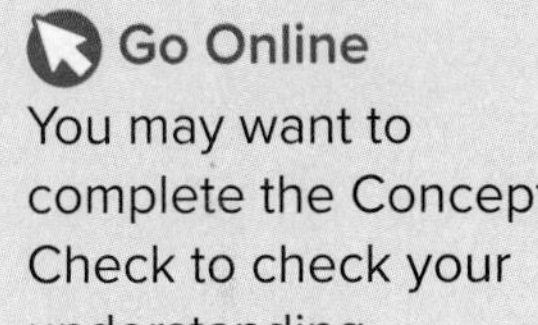

Go Online
You may want to complete the Concept Check to check your understanding.

Example 1 Identify Properties of Equality

Identify the property of equality used to justify each statement. Explain your reasoning.

a. If $13 + 25 = 38$, then $38 = 13 + 25$.

Symmetric Property of Equality; $13 + 25 = 38$ and $38 = 13 + 25$

b. $y + 4 = y + 4$

Reflexive Property of Equality; $y + 4$ is equal to itself.

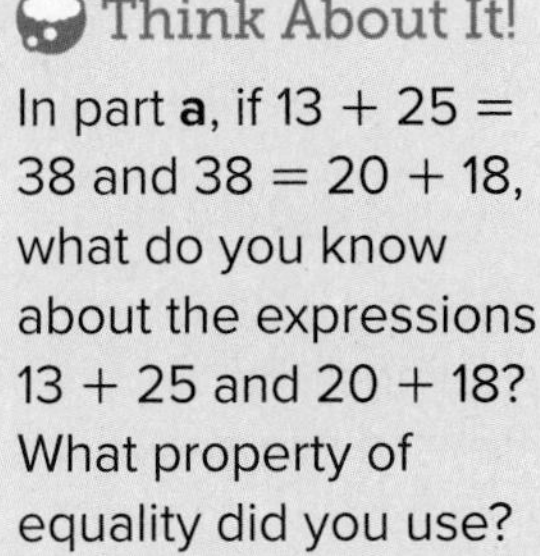

Think About It!
In part **a**, if $13 + 25 = 38$ and $38 = 20 + 18$, what do you know about the expressions $13 + 25$ and $20 + 18$? What property of equality did you use?

Check

Identify the property of equality used to justify each statement.

a. $22 + 7 = 22 + 7$

b. If $36 = 17 + 19$, then $17 + 19 = 36$.

Go Online You can complete an Extra Example online.

Example 2 Interpret Properties of Equality

COOKING **If the amount of sugar in a recipe is equal to the amount of flour plus 2 tablespoons, and the amount of flour plus 2 tablespoons is equal to the amount of milk, then the amount of sugar is equal to the amount of milk.**

a. Write a verbal expression in the spaces below.

sugar = flour and 2 tablespoons, and flour and 2 tablespoons = milk, then

sugar = milk.

Let a = the amount of sugar in the recipe, b = the amount of flour in the recipe plus 2 tablespoons, and c = the amount of milk in the recipe.

If we substitute a, b, and c for the verbal expressions, we can write the following algebraic expression.

If $a = b$ and $b = c$, then $a = c$.

b. This statement is an example of which property of equality?

This is an example of the Transitive Property of Equality.

Talk About It!

Suppose the example stated "If the amount of milk in a recipe is equal to the amount of sugar and the amount of sugar is equal to the amount of flour plus 2 tablespoons, then the amount of milk is equal to the amount of flour plus 2 tablespoons." Would this example still describe the Transitive Property of Equality? Explain your reasoning.

Check

WEIGHT The weight of a bag of oranges plus 13 ounces is equal to 5 pounds.

Part A Which quantity is equivalent to 5 pounds?

Part B This statement is an example of which property of equality?

Watch Out!

Choosing a Variable Remember that any letter can be used to represent an algebraic expression, not just a, b, and c.

 **Go Online** You can complete an Extra Example online.

Example 3 Use Properties of Equality

Use the given property of equality to complete each statement.

a. $y - 21 = \underline{\quad ? \quad}$; **Reflexive Property of Equality**

The Reflexive Property of Equality states that any quantity equals itself, so $y - 21 = y - 21$.

b. If $24 + 11 = 9 + 26$ and $9 + 26 = z$, then $24 + 11 = \underline{\quad ? \quad}$; Transitive Property of Equality

The Transitive Property of Equality states that if one quantity $(24 + 11)$ equals a second quantity $(9 + 26)$ and the second quantity $(9 + 26)$ equals a third quantity (z), then the first quantity $(24 + 11)$ equals the third quantity (z). So, $24 + 11 = z$.

Think About It!

Avery says that in part **b**, $z = 35$. Do you agree or disagree? Explain your reasoning.

Check

Use the given property of equality to complete each statement.

a. If $43 + 9 = 10 + 42$ and $10 + 42 = 52$, then $43 + 9 = \underline{\quad ? \quad}$; Transitive Property of Equality

b. $2m - 1 = \underline{\quad ? \quad}$; Reflexive Property of Equality

Learn Identities and Inverses

The sum of any number a and 0 is equal to a. Thus, 0 is called the **additive identity**. If the sum of two numbers is equal to the additive identity, like $4 + (-4) = 0$, then the two numbers are **additive inverses**.

Key Concept • Addition Properties	
Additive Identity Property	
Words	For any real number a, the sum of a and 0 is a.
Symbols	$a + 0 = 0 + a = a$
Examples	$5 + 0 = 5$ $0 + 5 = 5$
Additive Inverse Property	
Words	A real number and its opposite are additive inverses of each other.
Symbols	$a + (-a) = 0$
Examples	$2 + (-2) = 0$ $7 - 7 = 0$

Go Online You can complete an Extra Example online.

The product of any number a and 1 is equal to a. Thus, 1 is called the **multiplicative identity**.

The product of any number a and 0 is equal to 0. This is called the Multiplicative Property of Zero.

Two numbers with a product of 1 are called **multiplicative inverses** or **reciprocals**.

Key Concept • Multiplication Properties	
Multiplicative Identity Property	
Words	For any real number a, the product of a and 1 is a.
Symbols	$a \cdot 1 = a$ $1 \cdot a = a$
Examples	$4 \cdot 1 = 4$ $1 \cdot 4 = 4$
Multiplicative Property of Zero	
Words	For any real number a, the product of a and 0 is 0.
Symbols	$a \cdot 0 = 0$ $0 \cdot a = 0$
Examples	$12 \cdot 0 = 0$ $0 \cdot 12 = 0$
Multiplicative Inverse Property	
Words	For every real number $\frac{a}{b}$ where $a, b \neq 0$, there is exactly one number $\frac{b}{a}$ such that $\frac{a}{b} \cdot \frac{b}{a}$ is 1.
Symbols	$\frac{a}{b} \cdot \frac{b}{a} = 1$ $\frac{b}{a} \cdot \frac{a}{b} = 1$
Examples	$\frac{2}{3} \cdot \frac{3}{2} = 1$ $\frac{3}{2} \cdot \frac{2}{3} = 1$

Study Tip

Decimals as Fractions When multiplying decimal values, consider the values as fractions to see if you can evaluate the expression using multiplicative inverses. For example, the value of $0.\overline{6} \cdot 1.5$ is not immediately clear, but if you rewrite the expression using fractions, then you can easily see that $\frac{2}{3} \cdot \frac{3}{2} = 1$.

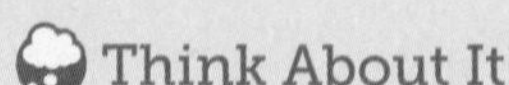

Think About It!

If $x - y$ is an example of the Additive Inverse Property, then what must be true about the relationship between x and y?

Example 4 Evaluate Using the Addition Properties

Evaluate $4 - 2^2 + 8(2)$.

$4 - 2^2 + 8(2) = 4 - 4 + 8(2)$	Simplify 2^2.
$= 4 - 4 + 16$	Multiply 8 by 2.
$= 0 + 16$	Additive inverses: $4 - 4 = 0$
$= 16$	Additive identity: $0 + 16 = 16$

Check

Evaluate $13 + 2^2 + 0 =$ ___?___.

Go Online You can complete an Extra Example online.

Example 5 Evaluate Using the Multiplicative Identity and Multiplicative Inverse

Evaluate $\frac{3}{2} \cdot \frac{2}{3} \cdot 7$

$\frac{3}{2} \cdot \frac{2}{3} \cdot 7 = 1 \cdot 7$ Multiplicative inverses: $\frac{3}{2} \cdot \frac{2}{3} = 1$

$= 7$ Multiplicative identity: $1 \cdot 7 = 7$

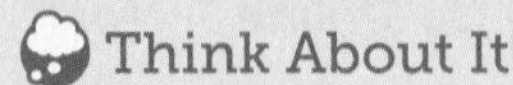

Why is the multiplicative identity 1 and not the same value as the additive identity, 0?

Check

Identify the property used in each step of the evaluation process.

$4 \cdot 1 + 0 - 4 + 3 = 4 + 0 - 4 + 3$

$= 4 - 4 + 3$

$= 0 + 3$

$= 3$

Example 6 Evaluate Using the Multiplicative Property of Zero

Evaluate $[4 - 3(2) + 7] \cdot 0$.

Notice that this expression is the product of an expression and 0.

According to the Multiplicative Property of Zero, the product of any number and 0 is 0.

Therefore, $[4 - 3(2) + 7] \cdot 0 = 0$.

Check

Evaluate $(13 - 1) \cdot 0 + \frac{3}{4} \cdot \frac{4}{3} =$ ___?___.

Watch Out!

Parentheses
Before evaluating the expression using the Multiplicative Property of Zero, pay close attention to parentheses to see if the entire expression or only part of it is being multiplied by 0. For example, $(5 + 9 - 2) \cdot 0 = 0$, but $5 + (9 - 2) \cdot 0 = 5 + 0 = 5$.

Go Online You can complete an Extra Example online.

Go Online to practice what you've learned about the properties of real numbers in the Put It All Together over Lessons 1-1 through 1-3.

Explore Testing the Associative Property

Online Activity Use a table to complete the Explore.

INQUIRY For what operations does the Associative Property hold true? For what operations does it not?

Learn Commutative and Associative Properties

An easy way to find the sum or product of numbers is to group, or associate, the numbers using the Associative Property.

For the addition and multiplication of real numbers, the order does not change their sum or product. This is called the Commutative Property.

Key Concept • Associative and Commutative Properties	
Associative Property	
Words	The way you group three or more numbers when adding or multiplying does not change their sum or product.
Symbols	For any numbers a, b, and c, $(a + b) + c = a + (b + c)$ and $(ab)c = a(bc)$.
Examples	$(2 + 9) + 4 = 2 + (9 + 4)$ $(3 \cdot 6) \cdot 5 = 3 \cdot (6 \cdot 5)$
Commutative Property	
Words	The order in which you add or multiply numbers does not change their sum or product.
Symbols	For any numbers a and b, $a + b = b + a$ and $a \cdot b = b \cdot a$.
Examples	$8 + 12 = 12 + 8$ $4 \cdot 9 = 9 \cdot 4$

Example 7 Evaluate Using the Associative Property

PERSONAL FINANCE **Jalen wants to add up how much money he spent on gasoline in July. He grabs the four receipts he has for July and writes the expression that represents the total amount he spent. Evaluate the expression to determine how much money Jalen spent on gasoline in July.**

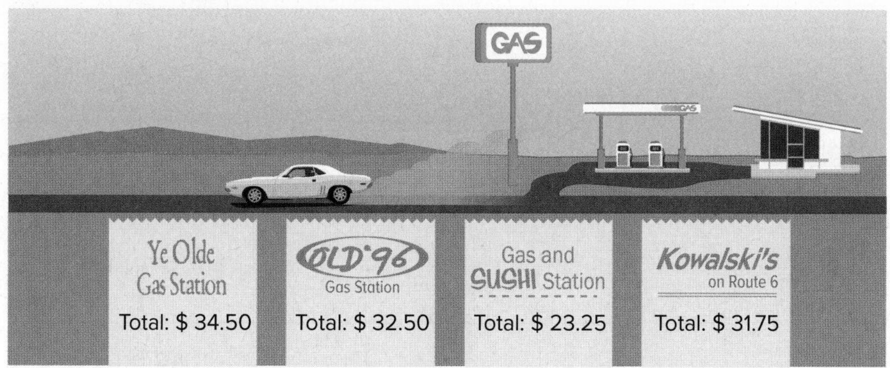

$34.50 + 32.50 + 23.25 + 31.75$	Original expression
$= (34.50 + 32.50) + (23.25 + 31.75)$	Associative (+)
$= 67.00 + 55.00$	Simplify.
$= 122.00$	Add.

Jalen spent $122.00 on gasoline in July.

Think About It!

Why did you use the Associative Property to evaluate this expression?

Check

Determine the property used for each step of the evaluation process.

$114 + 71 + 19 + 26 = 114 + (71 + 19) + 26$	
$= 114 + 90 + 26$	Simplify.
$= 114 + 26 + 90$	
$= 140 + 90$	Simplify.
$= 230$	Simplify.

Go Online You can complete an Extra Example online.

Watch Out!

Associative and Commutative Properties

Before you apply the Associative or Commutative Property, make sure that only one operation is involved. For example, in $3 + 5 \cdot 2$, you cannot group $3 + 5$ or switch the 3 and 5 because that does not adhere to the order of operations.

Think About It!

For which step(s) did Jade use the Commutative Property? For which step(s) did Jade use the Associative Property?

Go Online to learn about operations with rational numbers in Expand 1-3.

Example 8 Evaluate Using the Commutative Property

Evaluate $\frac{5}{6} \cdot 9 \cdot \frac{6}{5}$.

$\frac{5}{6} \cdot 9 \cdot \frac{6}{5} = \frac{5}{6} \cdot \frac{6}{5} \cdot 9$ Commutative (×)

$= 1 \cdot 9$ Multiplicative inverses: $\frac{5}{6} \cdot \frac{6}{5} = 1$

$= 9$ Multiplicative Identity

Check

Josefina needs to evaluate $\frac{7}{3} \cdot 2 \cdot 3$. She wants to use the Commutative Property to more easily evaluate the expression. What should she do first?

Example 9 Evaluate Using the Associative and Commutative Properties

GROCERIES **Jade buys groceries once each week, and she wants to calculate how much she spent on groceries over a 4-week period. Evaluate the expression to find how much Jade spent at the grocery store.**

Groceries	
Week	Amount Spent ($)
1	32
2	27
3	28
4	33

$32 + 27 + 28 + 33 = 32 + 28 + 27 + 33$ Step 1

$= (32 + 28) + (27 + 33)$ Step 2

$= 60 + 60$ Step 3

$= 120$ Step 4

Jade spent $120 on groceries over the 4-week period.

Check

Use the Commutative and Associative Properties to evaluate $\frac{5}{9} \cdot 9 \cdot \frac{9}{5} \cdot 4$.

Go Online You can complete an Extra Example online.

Practice

Go Online You can complete your homework online.

Examples 1–3

Identify the property of equality used to justify each statement.

1. If $4 + 17 = 21$, then $21 = 4 + 17$.

2. $x + 3 = x + 3$

3. If $16 = 9 + 7$, then $9 + 7 = 16$.

4. If $6 + 2 = 4 + 4$ and $4 + 4 = 8$, then $6 + 2 = 8$.

Use the given property of equality to complete each statement.

5. If $23 + 14 = 37$, then $37 = 23 +$ __?__; Symmetric Property of Equality

6. If $a + 5 = b + 3$ and $a + 5 = 12$, then $b + 3 =$ __?__; Transitive Property of Equality

7. If $34 = 19 + 15$, then $19 + 15 =$ __?__; Symmetric Property of Equality

8. $b + 5 + 12 =$ __?__; Reflexive Property of Equality

9. TOLL ROADS Some toll highways assess tolls based on where a car entered and exited. The table shows the highway tolls for a car entering and exiting at a variety of exits. Assume that the toll for the reverse direction is the same.

Entered	Exited	Toll
Exit 8	Exit 10	\$0.25
Exit 10	Exit 15	\$1.00
Exit 15	Exit 18	\$0.50
Exit 18	Exit 22	\$0.75

a. Julio travels from Exit 8 to Exit 15. Which quantity is equivalent to Exit 8 to Exit 15?

b. What property would you use to determine the toll?

Examples 4–6

Evaluate each expression. Name the property used in each step.

10. $3(22 - 3 \cdot 7)$

11. $[3 \div (2 \cdot 1)]\frac{2}{3}$

12. $2(3 \cdot 2 - 5) + 3 \cdot \frac{1}{3}$

13. $2[5 - (15 \div 3)]$

14. $6 + 9[10 - 2(2 + 3)]$

15. $2(6 \div 3 - 1) \cdot \frac{1}{2}$

Examples 7–9

Evaluate each expression using properties of numbers. Name the property used in each step.

16. $25 + 14 + 15 + 36$

17. $4\frac{4}{9} + 7\frac{2}{9}$

18. $4.3 + 2.4 + 3.6 + 9.7$

19. $2 \cdot 8 \cdot 10 \cdot 2$

20. $1\frac{5}{6} \cdot 24 \cdot 3\frac{1}{11}$

21. $2\frac{3}{4} \cdot 1\frac{1}{8} \cdot 32$

22. $16 + 8 + 14 + 12$

23. $2 \cdot 4 \cdot 5 \cdot 3$

24. $6.4 + 2.7 + 1.6 + 5.3$

25. $\frac{4}{3} \cdot 7 \cdot 3 \cdot 10$

Evaluate each expression if $a = -1$, $b = 4$, and $c = 6$.

26. $4a + 9b - 2c$

27. $-10c + 3a + a$

28. $a - b + 5a - 2b$

29. $8a + 5b - 11a - 7b$

30. $3c^2 + 2c + 2c^2$

31. $3a - 4a^2 + 2a$

32. Name the property that is used in $5 \cdot n \cdot 2 = 0$. Then find the value of n.

33. Name two properties used to evaluate $7 \cdot 1 - 4 \cdot \frac{1}{4}$.

34. Evaluate $7 \cdot 2 \cdot 7 \cdot 5$ using properties of numbers. Name the property used in each step.

Mixed Exercises

Find the value of x. Then name the property used.

35. $8 = 8 + x$

36. $3.2 + x = 3.2$

37. $10x = 10$

38. $\frac{1}{2} \cdot x = \frac{1}{2} \cdot 7$

39. $x + 0 = 5$

40. $1 \cdot x = 3$

41. $\frac{4}{3} \cdot \frac{3}{4} = x$

42. $2 + 8 = 8 + x$

43. $x + \frac{3}{4} = 3 + \frac{3}{4}$

44. $\frac{1}{3} \cdot x = 1$

45. MENTAL MATH The triangular banner has a base of 9 centimeters and a height of 6 centimeters. Using the formula for area of a triangle, the banner's area can be expressed as $\frac{1}{2} \times 9 \times 6$. Gabrielle finds it easier to write and evaluate $\left(\frac{1}{2} \times 6\right) \times 9$ to find the area. Is Gabrielle's expression equivalent to the area formula? Explain.

46. FINANCE Felicity put down $800 on a used car. She took out a loan to pay off the balance of the cost of the car. Her monthly payment will be $175. After 9 months, how much will she have paid for the car?

47. ANATOMY The human body has 126 bones in the upper and lower extremities, 28 bones in the head, and 52 bones in the torso. Use the Associative Property to write and evaluate an expression that represents the total number of bones in the human body.

48. SCHOOL SUPPLIES At a local school supply store, a highlighter costs $1.25, a ballpoint pen costs $0.80, and a spiral notebook costs $2.75. Use mental math and the Associative Property of Addition to find the total cost if one of each item is purchased.

Higher-Order Thinking Skills

49. PERSEVERE Write two equations showing the Transitive Property of Equality. Justify your reasoning.

50. ANALYZE Determine whether the following statement is *sometimes*, *always*, or *never* true. Justify your argument. The Commutative Property holds for subtraction.

51. ANALYZE Provide examples to show that there is no Commutative Property or Associative Property for division. What is the relationship between the results when the order of division of two numbers is switched?

52. WRITE Explain why 0 has no multiplicative inverse.

53. ANALYZE The sum of any two whole numbers is always a whole number. So, the set of whole numbers {0, 1, 2, 3, 4, ...} is said to be closed under addition. This is an example of the **Closure Property**. State whether each statement is *true* or *false*. If false, justify your reasoning.

a. The set of whole numbers is closed under subtraction.

b. The set of whole numbers is closed under multiplication.

c. The set of whole numbers is closed under division.

54. ANALYZE Explain whether 1 can be an additive identity. Give an example to justify your reasoning.

55. WHICH ONE DOESN'T BELONG? Identify the equation that does not belong with the other three. Justify your conclusion.

$x + 12 = 12 + x$	$7h = h \cdot 7$	$1 + a = a + 1$	$(2j)k = 2(jk)$

56. CREATE Write an expression that simplifies to 160 using the Commutative and Associative Properties.

Distributive Property

Today's Goals
- Use the Distributive Property to evaluate expressions.
- Use the Distributive Property to simplify expressions.

Today's Vocabulary
coefficient
like terms
simplest form
equivalent expressions

Explore Using Rectangles with the Distributive Property

Online Activity Use dynamic geometry software to complete the Explore.

INQUIRY What is the product of a and $(b + c)$?

Explore Modeling the Distributive Property

Online Activity Use algebra tiles to complete the Explore.

INQUIRY How can you use algebra tiles to find the product of two expressions?

Learn Distributive Property with Numerical Expressions

The expressions $3(4 + 2)$ and $3 \cdot 4 + 3 \cdot 2$ are equivalent expressions because they have the same value, 18. This concept shows how the Distributive Property combines addition and multiplication. Multiplying a number by a sum of numbers is the same as doing each multiplication separately and then adding the products.

Key Concept • Distributive Property	
Symbols	For any numbers a, b, and c, $a(b + c) = ab + ac$ and $(b + c)a = ba + ca$ and $a(b - c) = ab - ac$ and $(b - c)a = ba - ca$.
Examples	$4(9 + 3) = 4 \cdot 9 + 4 \cdot 3$ $4(12) = 36 + 12$ $48 = 48$ $5(8 - 2) = 5 \cdot 8 - 5 \cdot 2$ $5(6) = 40 - 10$ $30 = 30$

The Symmetric Property of Equality allows the Distributive Property to also be written in the reverse order $ab + ac = a(b + c)$.

Go Online You can complete an Extra Example online.

Think About It!
If $a(b + c) = ab + ac$, then what does $ab + ac$ equal?

Apply Example 1 Use the Distributive Property

Think About It!
Write an expression for the amount spent on back-to-school shopping in June and September. Then evaluate the expression.

SHOPPING **Assume that high school students spend an average of $180 on back-to-school shopping during the months of June to September. Write an expression to represent the amount of money spent in August if the amount of money spent in August is equal to the amount of money spent in July minus the amount of money spent in June. Evaluate your expression using the Distributive Property.**

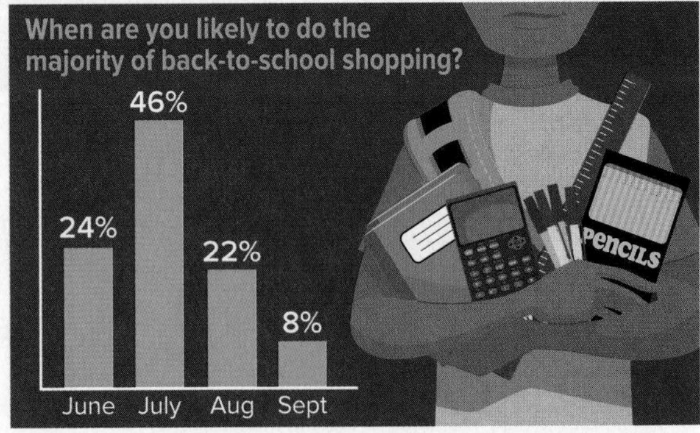

1. What is the task?

Describe the task in your own words. Then list any questions that you may have. How can you find answers to your questions?

Sample answer: We are given the total amount of money spent on back-to-school shopping, as well as percentages of money spent during each month of the back-to-school shopping season. We are asked to find the amount of money that will be spent in August.

Watch Out!
Decimals and Percents Make sure to convert percents to decimals before performing calculations.

2. How will you approach the task? What have you learned that you can use to help you complete the task?

Sample answer: I will determine the percent of the total spent in June and July. Then I will estimate the amount spent in August. I will write and evaluate an expression for the amount spent in August. Finally, I will use my estimate to check my solution.

Study Tip
Making Sense The four-step problem solving plan is a tool for making sense of any problem. When making and executing your plan, continually ask yourself, "Does this make sense?" Monitor and evaluate your progress and change course if necessary.

3. What is your solution?

Use your strategy to solve the problem.
What expression represents the amount spent in August?

180(0.46 – 0.24)

What is the total amount spent in August?

$39.60

4. How can you know that your solution is reasonable?

Write About It! Write an argument that can be used to defend your solution.

The total percentage for August is 22%, and the total amount spent on back-to-school shopping is $180. Multiply 0.22 and 180 to get 39.6. Therefore, the total amount spent in August is $39.60. This amount also makes sense with our estimate of a total spending of more than $36.

Go Online You can complete an Extra Example online.

Check

SWIMMING Verdell's swim team practices 5 days a week. Each day they spend 15 minutes stretching, 45 minutes swimming laps, and 30 minutes lifting weights.

Part A Which expression(s) represent the number of minutes Verdell's team spends in practice each week? Select all that apply.

A. $5(15 + 45 + 30)$ B. $5(15) + 45 + 30$

C. $5(15) + 5(45) + 5(30)$ D. $5(15) + 5(45) + 30$

E. $5 + 15 + 45 + 30$

Part B How much time does Verdell's team spend in practice each week?

A. 90 minutes B. 150 minutes

C. 330 minutes D. 450 minutes

Example 2 Mental Math

Use the Distributive Property to rewrite and evaluate each expression.

a. 5 · 99

$5 \cdot 99 = 5(100 - 1)$	Think: $99 = 100 - 1$
$= 5(100) - 5(1)$	Distributive Property
$= 500 - 5$	Multiply.
$= 495$	Subtract.

b. 4 · 1002

$4 \cdot 1002 = 4(1000 + 2)$	Think: $1002 = 1000 + 2$
$= 4(1000) + 4(2)$	Distributive Property
$= 4000 + 8$	Multiply.
$= 4008$	Add.

Think About It!

How can you use the Distributive Property to rewrite and evaluate the expression 8(1100)?

Check

Part A Estimate the value of the expression 7(51).

7(__?__) = 350, so 7(51) will be a little __?__ than 350.

Part B Which expression(s) use(s) the Distributive Property to rewrite and find the exact value of the expression 7(51)?

A. $51(7 - 3)$; 204 B. $7(50 + 1)$; 357 C. $51(7 + 3)$; 510

D. $51(7) - 51(3)$; 204 E. $7(50) + 7(1)$; 357 F. $51(7) + 51(3)$; 510

Go Online You can complete an Extra Example online.

Learn Distributive Property with Algebraic Expressions

The **coefficient** is the numerical factor of a term.

Like terms are terms with the same variables, with corresponding variables having the same exponent.

An expression is in **simplest form** when it is replaced by an equivalent expression having no like terms or parentheses.

The Distributive Property and the properties of equality can be used to show that $6x + 2x = 8x$. In this expression, $6x$ and $2x$ are like terms.

$6x + 2x = (6 + 2)x$	Distributive Property
$= 8x$	Substitution

The expressions $6x + 2x$ and $8x$ are called **equivalent expressions** because they represent the same value for any value of the variable.

Example 3 Distribute an Algebraic Expression from the Left

Rewrite $4(5x - 7)$ using the Distributive Property. Then simplify.

$4(5x - 7) = 4 \cdot 5x - 4 \cdot 7$	Distributive Property
$= 20x - 28$	Multiply.

Problem-Solving Tip

Make a Model
It can be helpful to visualize a problem using algebra tiles or folded paper.

Check

Simplify the expression.

$-6(r + 3g - t)$

Go Online You can complete an Extra Example online.

Example 4 Distribute an Algebraic Expression from the Right

Rewrite $(3y^2 + y - 8)6$ using the Distributive Property. Then simplify.

$(3y^2 + y - 8)6 = 6(3y^2) + 6(y) + 6(-8)$ Distributive Property

$= 18y^2 + 6y - 48$ Multiply.

Talk About It!

Emilio says you can add $18y^2$ and $6y$ to get $24y^3$. Do you agree or disagree? Justify your answer.

Example 5 Combine Like Terms

Simplify each expression.

a. $14a + 18a$

$14a + 18a = (14 + 18)a$ Distributive Property

$= 32a$ Substitution

b. $4b^2 + 9b - 3b$

$4b^2 + 9b - 3b = 4b^2 + (9 - 3)b$ Distributive Property

$= 4b^2 + 6b$ Substitution

 **Think About It!**

What are the like terms in part **a** and part **b**?

Check

Simplify the expression. If not possible, choose *simplified*.

$b^2 + 13b + 13$

Watch Out!

Like Terms
$4b^2$ and $6b$ are not like terms because they have different exponents.

Go Online You can complete an Extra Example online.

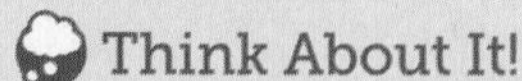

Caitlyn is using the Distributive Property to simplify $-4(3k + 7p)$. She got $-12k + 28p$. Do you agree or disagree? Justify your answer.

Example 6 Write and Simplify Expressions

Part A Complete the table to write an algebraic expression for *three times the sum of 2x and 3y decreased by twice the difference of 4x and y.*

Words	three times the sum of $2x$ and $3y$	decreased by	twice the difference of $4x$ and y
Expression	$3(2x + 3y)$	$-$	$2(4x - y)$

Part B Simplify the expression and indicate the properties used.

$3(2x + 3y) - 2(4x - y)$

$= 3(2x) + 3(3y) - 2(4x) - 2(-y)$	Distributive Property
$= 6x + 9y - 8x + 2y$	Multiply.
$= 6x - 8x + 9y + 2y$	Commutative (+)
$= (6 - 8)x + (9 + 2)y$	Distributive Property
$= -2x + 11y$	Simplify.

Study Tip

Distributive Property Remember, when simplifying the expression $3(2x + 3y)$, you must distribute the 3 to both terms inside the parentheses.

Check

Which expressions are equivalent to *4 times the sum of 2 times x and 6*?

- $8(x - 3)$
- $(4 + 2x)6$
- $8x + 24$
- $4(2x + 6)$
- $4(2 + x + 6)$
- $4(2x) + 4(6)$

Equivalent	Not Equivalent

Go Online You can complete an Extra Example online.

Practice

Go Online You can complete your homework online.

Example 1

Use the Distributive Property to rewrite each expression. Then evaluate.

1. $(4 + 5)6$

2. $7(13 + 12)$

3. $6(6 - 1)$

4. $(3 + 8)15$

5. $14(8 - 5)$

6. $(9 - 4)19$

7. OPERA Aran's drama class is planning a field trip to see Mozart's famous opera *Don Giovanni*. Tickets cost $39 each, and there are 23 students and 2 teachers going on the field trip.

a. Write an expression to find the group's total ticket cost.

b. What is the group's total ticket cost?

8. SALARY In a recent year, the median salary for an engineer in the United States was $55,000 and the median salary for a computer programmer was $52,000.

a. Write an expression to estimate the total cost for a business to employ an engineer and a programmer for 5 years.

b. Estimate the total cost for a business to employ an engineer and a programmer for 5 years.

9. COSTUMES Isabella's ballet class is performing a spring recital for which they need butterfly costumes. Each butterfly costume is made from $3\frac{3}{5}$ yards of fabric.

a. Write an expression to find the number of yards of fabric needed for 10 costumes.

b. Use the Distributive Property to find the number of yards of fabric needed for 10 costumes. Show your work. (Hint: A mixed number can be written as the sum of an integer and a fraction.)

10. REASONING Letisha and Noelle each opened a checking account, a savings account, and a college fund. The chart shows the amounts that they deposit into each account every month.

	Checking	Savings	College
Letisha	$125	$75	$50
Noelle	$250	$50	$50

a. Write an expression to find the amount in Letisha's checking, savings, and college accounts after 12 months.

b. How much is in Letisha's checking, savings, and college accounts after 12 months?

Example 2

Use the Distributive Property to rewrite and evaluate each expression.

11. $7 \cdot 497$

12. $6(525)$

13. $36 \cdot 3\frac{1}{4}$

14. $\left(4\frac{2}{7}\right)21$

15. $5 \cdot 89$

16. $9 \cdot 99$

17. $15 \cdot 104$

18. $15\left(2\frac{1}{3}\right)$

19. $12 \cdot 98$

20. $8 \cdot 1.5$

21. $3 \cdot 10.2$

22. $5\left(4\frac{1}{5}\right)$

Examples 3 and 4

Rewrite each expression using the Distributive Property. Then simplify.

23. $2(x + 4)$

24. $(5 + n)3$

25. $(4 - 3m)8$

26. $-3(2x - 6)$

27. $(2 - 4n)17$

28. $11(4d + 6)$

29. $\left(\frac{1}{3} - 2b\right)27$

30. $4(8p + 16q - 7r)$

31. $6(2c - cd^2 + d)$

32. $7(h - 10)$

33. $3(m + n)$

34. $2(x - y + 1)$

35. $\left(\frac{1}{2} + 6a\right)14$

36. $-2(7m - 8n - 5p)$

37. $(0.3 - 6x)9$

38. $-4(4a + 2b - \frac{1}{2}c)$

Example 5

Simplify each expression. If not possible, write *simplified*.

39. $13r + 5r$

40. $3x^3 - 2x^2$

41. $7m + 7 - 5m$

42. $5z^2 + 3z + 8z^2$

43. $7m + 2m + 5p + 4m$

44. $6x + 4y + 5x$

45. $3m + 5g + 6g + 11m$

46. $4a + 5a^2 + 2a^2 + a^2$

47. $5k + 3k^3 + 7k + 9k^3$

48. $6x^2 + 14x - 9x$

49. $17g + g$

50. $2x^2 + 6x^2$

51. $7a^2 - 2a^2$

52. $3y^2 - 2y + 9$

53. $3q^2 + q - q^2$

Example 6

Consider each verbal expression.

a. Write an algebraic expression to represent the verbal expression.

b. Simplify the expression and indicate the properties used.

54. *The product of 9 and t squared, increased by 3 times the sum of 2 and t squared*

55. *The product of 3 and a, plus 5 times the difference of a and b*

56. *3 times the sum of r and d squared increased by 2 times the sum of r and d squared*

Mixed Exercises

Simplify each expression.

57. $3x + 7(3x + 4)$

58. $4(fg + 3g) + 5g$

59. $6d + 4(3d + 5)$

60. $2(6x + 4) + 7x$

61. $4y^3 + 3y^3 + y^4$

62. $a + \frac{a}{5} + \frac{2}{5}a$

63. $4(2b - b)$

64. $2(n + 2n - 5)$

65. $7(2x + y) + 6(x + 5y)$

66. REASONING A theater has m seats per row on the left side of the aisle and n seats per row on the right side of the aisle. There are r rows of seats.

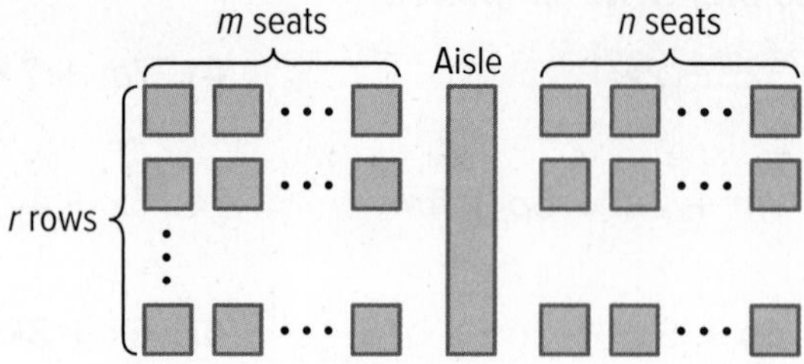

a. Explain how you can use the Distributive Property to write two different expressions that represent the total number of seats in the theater.

b. Suppose you double the number of seats in each row on the left side of the aisle. Does this double the number of seats in the theater? Use one of the expressions you wrote in part **a** to justify your answer.

67. FENCES Demonstrate the Distributive Property by writing two equivalent expressions to represent the perimeter of the fenced dog park.

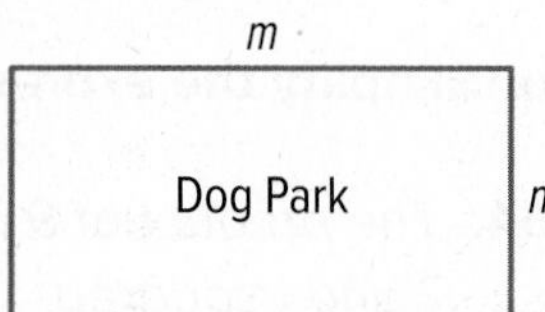

68. PERSEVERE Use the Distributive Property to simplify $6x^2[(3x - 4) + (4x + 2)]$.

69. FIND THE ERROR Ariana is shipping 8 bags of granola to a customer. Each bag weighs 22 ounces and the maximum weight she can ship in one box is 10 pounds 5 ounces. She makes the calculations at the right and decides that she can ship the bags in one box. Do you agree? Explain your reasoning.

$$\begin{aligned} 80(22) &= 8(20 + 2) \\ &= 8(20) + 2 \\ &= 160 + 2 \\ &= 162 \text{ ounces} \end{aligned}$$

70. ANALYZE Determine whether the following statement is *true or false*. Justify your argument. The Distributive Property is a property of both addition and multiplication.

71. WRITE Why is it helpful to represent verbal expressions algebraically?

72. CREATE Write an expression that simplifies to $2a + 14$ using the Distributive Property.

Expressions Involving Absolute Value

Today's Goal
- Evaluate absolute value expressions.

Today's Vocabulary
absolute value

Explore Distance Between Points on a Number Line

Online Activity Use graphing technology to complete the Explore.

INQUIRY How can you find the distance between any two values *x* and *y* on a number line?

Learn Evaluating Expressions Involving Absolute Value

The **absolute value** of a number is its distance from 0 on a number line.

Key Concept • Absolute Value of Variables	
Words	For any real number *x*, if *x* is positive or zero, then the absolute value of *x* is *x*. If *x* is negative, then the absolute value of *x* is the opposite of *x*.
Symbols	For any real number *x*, $\|x\| = x$ if $x \geq 0$, and $\|x\| = -x$ if $x < 0$.
Examples	If $x = 12$, then $\|x\| = 12$. If $x = -7$, then $\|x\| = -(-7) = 7$.

Study Tip

Interpreting $-x$ Do not read $-x$ as *negative x* but as *the opposite of x*. Because *x* represents a negative number whenever $x < 0$, the expression $-x$ represents a positive number.

Example 1 Write an Absolute Value Expression

THERMOMETERS **The accuracy of a meat thermometer is the positive difference between the temperature reading on the thermometer *t* and the actual temperature of the meat *m*. Complete the statements about which expressions are and are not equivalent to the accuracy of a meat thermometer.**

$|m - t|$ This expression is equivalent to the accuracy of a meat thermometer because it represents the difference between the measurements, and it is always positive because it is the absolute value of the difference.

$|m| - |t|$ This expression is not equivalent to the accuracy of a meat thermometer in all cases because if $t > m$, then $|m| - |t|$ is negative.

$|t - m|$ This expression is equivalent to the accuracy of a meat thermometer because it represents the difference between the measurements, and it is always nonnegative because it is the absolute value of the difference.

$|t| - |m|$ This expression is not equivalent to the accuracy of a meat thermometer in all cases because if $m > t$, then $|t| - |m|$ is negative.

Think About It!

Is the expression $-x$ always a negative value? If not, for what values of *x* is $-x$ positive?

Think About It!

Give three pairs of values for *c* and *d* for which $|c + d|$ is not equal to $|c| + |d|$.

Check

AMUSEMENT PARKS At an amusement park, a vendor attempts to guess a person's weight. If they are not within 3 pounds, the person will win a prize. If a person weighs 168 pounds and the vendor makes a guess of x pounds, which expression represents the number of pounds a winning guess is away from the actual weight?

A. $|3 - x|$ **B.** $|x + 3|$

C. $|168 - x|$ **D.** $|3 - 168|$

Study Tip

Opposites If a and b are real numbers and $a \neq b$, the difference $a - b$ is the opposite of $b - a$. The absolute values of opposites are always equal.

 Think About It!

How would you evaluate the expression

$|-2(xy + 5y)|$?

Example 2 Evaluate the Absolute Value of an Algebraic Expression

Evaluate $|-2xy + 5y|$ if $x = 6$ and $y = -3$.

$|-2xy + 5y| = |-2(6)(-3) + 5(-3)|$	Replace x with 6 and y with -3.
$= |36 - 15|$	Multiply.
$= |21|$	Subtract 15 from 36.
$= 21$	$|21| = 21$

Check

Evaluate $-|5a + 3(2ab - 1)|$ if $a = -2$ and $b = -3$.

Study Tip

Order of Operations Apply the order of operations when evaluating an expression inside grouping symbols. Perform all the multiplication from left to right and then do the addition and subtraction from left to right.

Example 3 Evaluate an Expression Involving Absolute Value

When evaluating algebraic expressions, absolute value bars act as a grouping symbol. Perform any operations inside the absolute bars first.

Evaluate $23 - |3 + 4x|$ if $x = 2$. Select a statement below to justify each step.

Replace x with 2. Multiply. $3 + 8 = 11$

$|11| = 11$ Simplify.

$23 - |3 + 4x| = 23 - |3 + 4(2)|$	Replace x with 2.
$= 23 - |3 + 8|$	Multiply.
$= 23 - |11|$	$3 + 8 = 11$
$= 23 - 11$	$|11| = 11$
$= 12$	Simplify.

Watch Out!

Additive Inverses Although $|-2| = |2|$, replacing x with -2 will not give the same value as replacing x with 2. The evaluation of the expression will be the same only if $3 + 4x$ equals -11 for some other value of x.

Check

Evaluate $1.4 - 2.5|5y + 0.6|$ if $y = -3$.

Go Online You can complete an Extra Example online.

Practice

Go Online You can complete your homework online.

Example 1

1. POOLS The accuracy of a pool thermometer is the positive difference between the temperature reading on the thermometer t and the actual temperature of the pool p. Write two absolute value expressions equivalent to the accuracy of a pool thermometer.

2. ROLLERCOASTER At a theme park, a person must be a certain height to ride a rollercoaster. A person must be h inches tall, plus or minus 1.5 inches. Anoki says the absolute value expression $|1.5 - h|$ represents an acceptable height. David says the absolute value expression $|h - 1.5|$ represents an acceptable height. Who is correct? Explain.

3. WATER DEPTH An *echo sounder* is a device used to determine the depth of water by measuring the time it takes a sound produced just below the water surface to return, or echo, from the bottom of the body of water. The accuracy of an echo sounder is the positive difference between the depth of water reading on the echo sounder r and the actual depth of water w. Write two absolute value expressions equivalent to the accuracy of an echo sounder.

4. GOLF A certain company designs and ships boxes of golf balls. Each box must weigh 540 grams. Write two absolute value expressions that represent the number of grams a box weighing g grams is away from the desired weight.

Example 2

Evaluate each expression if $m = -4$, $n = 1$, $p = 2$, $q = -6$, $r = 5$, and $t = -2$.

5. $|-n - 2mp|$

6. $|12 + 2t|$

7. $|q - 2mt|$

8. $|3r + 6m|$

9. $|p + 4q - 3r|$

10. $|16 + 4(3q + p)|$

11. $|2m + 6(q - t)|$

12. $-|10 - 7r + 8m + 2p|$

13. $-|14 - 6n + 7(q + 2t)|$

Example 3

Evaluate each expression if $a = 2$, $b = -3$, $c = -4$, $h = 6$, $y = 4$, and $z = -1$.

14. $|2b - 3y| + 5z$

15. $15 - |2 - 3a|$

16. $|a - 5| - 1$

17. $|b + 1| + 8$

18. $5 - |c + 1|$

19. $|a + b| - c$

20. $5 + |2b|$

21. $|4 - h| - b$

22. $|2 - b - h| - h$

Evaluate each expression if $a = -2$, $b = -3$, $c = 2$, $x = 2.1$, $y = 3$, and $z = -4.2$.

23. $|2x + z| + 2y$

24. $4a - |3b + 2c|$

25. $-|5a + c| + |3y + 2z|$

26. $-a + |2x - a|$

27. $|y - 2z| - 3$

28. $3|3b - 8c| - 3$

Evaluate each expression if $a = -\frac{1}{2}$, $b = \frac{3}{4}$, and $c = -\frac{2}{3}$.

29. $-|6c - 16b| + 1$

30. $14 + 2|3c + 10a|$

31. $|-2a - 20b| - 12c$

32. $12a - |-16b|$

33. $|5 - 15c| + a$

34. $|2 - (a - 6b)| + 18c$

Mixed Exercises

35. PRECISION A golf GPS is a device that can be used to determine the distance a golf ball is from a pin. The accuracy of a golf GPS is the positive difference between the distance a golf ball is from a pin on the golf GPS g and the actual distance a golf ball is from a pin d.

a. Write two absolute value expressions equivalent to the accuracy of a golf GPS.

b. Evaluate the expression if the actual distance from the golf ball to the pin is 70 meters and the distance the golf ball is from the pin on the golf GPS is 75 meters.

36. STRUCTURE The students in Mrs. Mangione's class attempt to guess the number of marbles in a jar to earn 2 extra credit points on their next exam. Suppose there are 1206 marbles in a jar and a student makes a guess of m marbles.

a. Write two absolute value expressions that represent the difference between the guess and the actual number of marbles in the jar.

b. Evaluate the expression if a student guesses there are 1100 marbles in the jar.

37. CREATE Describe a real-world situation that could be represented by the absolute value expression $|x - 89|$.

38. FIND THE ERROR The accuracy of a rain gauge is the positive difference between the amount of rain in the rain gauge g and the actual amount of rain r. Sam says the absolute value expression $|g| - |r|$ is equivalent to the accuracy of a rain gauge. Is Sam correct? Explain your reasoning.

39. ANALYZE Diaz claims that if a and b are real numbers, then $|a + b|$ is always equal to $|a| + |b|$. Determine whether his claim is *true* or *false*. Justify your argument.

Descriptive Modeling and Accuracy

Learn Descriptive Modeling

Descriptive modeling is a way to mathematically describe real-world situations and the factors that cause them. A **metric** is a rule for assigning a number to some characteristic or attribute.

Metrics can be used to make comparisons. In sports, earned run average is used to compare baseball pitchers, and the quarterback rating compares the performance of football quarterbacks. In banking, a person's debt-to-income ratio can determine whether the person qualifies for a loan. In a good metric, factors that are important in the situation are considered and included in the metric. For example, the quarterback rating includes measures of passing, running, penalties, and other factors that make a quarterback effective.

Today's Goals

- Define and use appropriate quantities for the purpose of descriptive modeling.
- Choose a level of accuracy appropriate to limitations on measurements when reporting quantities.

Today's Vocabulary

descriptive modeling

metric

accuracy

Example 1 Use Descriptive Modeling

COLLEGE ATHLETICS **Some universities use a metric called the Academic Index to qualify high school athletic recruits. A student athlete with a 3.1 G.P.A. received scores of 610 in reading, 640 in writing, and 700 in math on the SAT. Use the expression to determine whether the student athlete qualifies at a university that requires an Academic Index of 186 or greater.**

G.P.A.	G.P.A. Value
4.0	80
3.9	79
3.8	78
3.7	77
3.6	75
3.5	73
3.4	71
3.3	70
3.2	69
3.1	68
3.0	67

$$2\left[\frac{\left(\frac{\text{Reading Score} + \text{Writing Score}}{2}\right) + \text{Math Score}}{20}\right] + \text{G.P.A. Value}$$

Step 1 Find all values for the metric.

Use the table to determine the G.P.A. value. For a 3.1 G.P.A., the value is 68.

(continued on the next page)

Think About It!

Some universities require student athletes to take the SAT twice and then use both scores when determining their Academic Index. How do you think this could affect the score?

Step 2 Substitute values in the metric.

$$2\left[\frac{\left(\frac{\text{Reading Score} + \text{Writing Score}}{2}\right) + \text{Math Score}}{20}\right] + \text{G.P.A. Value}$$

$$= 2\left[\frac{\left(\frac{610 + 640}{2}\right) + 700}{20}\right] + 68$$ Reading 610, writing 640, math 700, and G.P.A. 68

$$= 2\left[\frac{\left(\frac{1250}{2}\right) + 700}{20}\right] + 68$$ Add 610 and 640.

$$= 2\left[\frac{625 + 700}{20}\right] + 68$$ Divide by 2.

$$= 2\left[\frac{1325}{20}\right] + 68$$ Add 625 and 700.

$$= 2(66.25) + 68$$ Divide by 20.

$$= 132.5 + 68$$ Multiply by 2.

$$= 200.5$$ Simplify.

The Academic Index is 200.5.

Step 3 Evaluate by using the metric.

Because the Academic Index is greater than 186, this student would be qualified to attend the university as an athlete.

Think About It!

What other attributes of high school recruits do you think universities might consider when creating metrics to determine qualification?

Check

PARKS Rachelle wants to determine the best state park for hiking and fishing.

Part A Use the metric to calculate a score for each park. Round to the nearest tenth.

$$\text{Park Score} = 100\left[0.2\left(\frac{\text{online rating}}{5}\right) + 0.4\left(\frac{\text{miles of trails}}{25}\right) + 0.4\left(\frac{\text{fish weight}}{10}\right)\right]$$

State Park	Online Star Rating	Hiking Trails (mi)	Best Fish (lb)	Park Score
Gooseberry Falls	4.8	20	9.8	
Lake Maria	4.3	14	8.2	
Maplewood	4.9	25	7.3	
Camden	4.3	15.8	10.1	

Part B How might someone who enjoys hiking much more than fishing change this metric?

Go Online You can complete an Extra Example online.

Example 2 Compare Metrics

HEIGHT **A child's adult height can be predicted using several metrics. Given the height of the mother, father, and their son at 2 years old, use the metrics to predict the son's height as an adult.**

Method 1 The Gray Method

The Gray Method uses the average heights of the parents, adjusted by the gender of the child. For a boy, the mother's height is multiplied by $\frac{13}{12}$, and for a girl, the father's height is multiplied by $\frac{12}{13}$.

$$\frac{\frac{13}{12} \cdot \text{mother's height} + \text{father's height}}{2}$$

$= \frac{\frac{13}{12} \cdot 65 + 72}{2}$ mother's height 65, father's height 72

$= \frac{70.42 + 72}{2}$ Multiply $\frac{13}{12}$ and 65.

$= \frac{142.42}{2}$ Add 70.42 and 72.

$= 71.21$ Simplify.

Using the Gray Method, the boy will be about 71 inches tall as an adult.

Method 2 The Doubling Method

The Doubling Method multiplies the height of a child by 2 at a specific age to predict the child's height as an adult. Height at 24 months is used for boys, and height at 18 months is used for girls.

$2 \cdot$ height of child

$= 2 \cdot 35$ Boy's height of 35 inches

$= 70$ Simplify.

Using the Doubling Method, the boy will be 70 inches tall as an adult.

Use a Source

Find information to create a metric to measure something that is important to you. Explain how your metric includes the factors that you think are important to measure.

Go Online You can complete an Extra Example online.

Check

LOANS Elan is applying for a home loan. At National Road Bank, Elan's debt-to-income ratio must be 0.36 or less to qualify for a loan, and at New Savings Bank his mortgage-to-income ratio must be 0.28 or less.

Part A Use the two metrics and the information provided to determine whether Elan qualifies for a home loan at each bank. Round to the nearest hundredth.

Monthly Income	Monthly Debt	Monthly Mortgage
$3650	$1165	$1068

National Road Bank: Debt-to-Income Ratio

$\frac{\text{Monthly Debt}}{\text{Monthly Income}} = \underline{\quad ? \quad}$

New Savings Bank: Mortgage-to-Income Ratio

$\frac{\text{Monthly Mortgage}}{\text{Monthly Income}} = \underline{\quad ? \quad}$

Part B Compare the results of the two metrics. How effective are each of the metrics as measures of whether Elan can afford to buy a house?

Learn Accuracy

All measurements are approximations. When you measure something, you are limited by the measurement tool that you are using. **Accuracy** is the nearness of a measurement to the true value of the measure.

The accuracy needed for baking cookies, timing the final seconds of a basketball game, and determining the gold medalist of a 100-meter dash are very different.

Whether measurements should be rounded depends on how the measurement will be used and the limitations of the units in which the measurement is taken.

Study Tip

Fractions Fractions may be more accurate than rounded decimals. For example, the sum of $\frac{3}{7} + \frac{2}{3}$ is more accurately reported as $\frac{23}{21}$ than 1.095.

Go Online You can complete an Extra Example online.

Example 3 Decide Where to Round

ROAD TRIP **Damien and two of his friends are taking a road trip. They plan to share the responsibility of driving and will each drive an equal distance. Damien's GPS shows that the total distance is 172 miles. Determine the exact distance that each person should drive. Then determine a more appropriate driving distance for each person given the limitations of the situation.**

To determine the exact distance each person should drive, divide the total distance by 3.

$$172 \text{ miles} \div 3 = 57.\overline{3} \text{ miles}$$

Because the distance given by the GPS is accurate to the nearest mile, the distance each driver will drive should be rounded to the nearest mile. Each driver will drive about 57 miles.

Check

VACATION Inchiro has saved $400 to spend on his 7-day vacation. He plans to budget his $400 by spending the same amount each day of the vacation. Determine the appropriate amount he should spend each day.

$__?__

> **Think About It!**
> The total cost of fuel for the trip was $20. If they split the cost equally, how much should each person pay? What unit of measure limits the accuracy of the solution?

Example 4 Find an Appropriate Level of Accuracy

SPACE SHUTTLE **In 2012, NASA's space shuttle *Endeavor* traveled approximately 897 miles from the Kennedy Space Center to Houston, Texas, by a shuttle carrier aircraft. Then it traveled about 1381 miles to the Los Angeles International Airport. Finally, a truck pulled *Endeavor* 12 miles through the streets of Los Angeles to the California Science Center.**

If the shuttle carrier aircraft flew at an average speed of 287 miles per hour and the truck pulled *Endeavor* at an average speed of 1.3 miles per hour, determine the total amount of time it took *Endeavor* to travel from the Kennedy Space Center to the California Science Center with a reasonable level of accuracy.

Because the parts of the space shuttle's journey from the Kennedy Space Center to Houston and then from Houston to Los Angeles are at the same speed, add those two distances, 897 + 1381 or 2278.

$$\frac{2278 \text{ miles}}{287\frac{\text{miles}}{\text{hours}}} + \frac{12 \text{ miles}}{1.3\frac{\text{miles}}{\text{hours}}} \approx 7.937 \text{ hours} + 9.231 \text{ hours}$$
$$\approx 17.168 \text{ hours}$$

The total travel time for *Endeavor* from the Kennedy Space Center to the California Science Center was about 17 hours.

 **Go Online** You can complete an Extra Example online.

> **Talk About It!**
> Why is it unreasonable to say that it took 17.168 hours for Endeavor to reach the science center?

Check

POSTAGE A school is hosting a marching band competition and plans to mail postcards, fliers, and large information packets to other schools. The school will mail 150 postcards, which cost $0.34 in postage, and between 75 to 100 fliers, which cost $0.49 in postage. Forty-three information packets will be mailed. The cost of mailing the information packets varies, but the average is $1.59. Determine the total mailing cost that represents the most reasonable level of accuracy.

Example 5 Determine Accuracy

POPULATION The U.S. Census Bureau Web site shows a counter that displays the population of the United States on a certain day as 329,158,023. How accurate is the reported population? Explain your reasoning.

United States Population

329,158,023

COMPONENTS OF POPULATION CHANGE

One birth every **7 seconds**

One death every **13 seconds**

One international migrant (net) every **29 seconds**

Net gain of one person every **11 seconds**

Because there is no way to count every person in the United States at any given moment, giving an exact population does not make sense. The number of births, deaths, and immigrations varies, so the population does not increase at a steady rate. The Web site uses averages to estimate the population at a specific time.

It would be more appropriate for the Web site to report the population as 329.2 million.

Check

BIOLOGY A science magazine reported that there are, on average, 37 trillion cells that make up the human body. Select the option that best describes the accuracy of the magazine.

A. The magazine is accurate because scientists can count every cell.

B. The magazine is probably accurate because the number is not very specific.

C. The magazine is not accurate because there is no way to count all of the cells of a person.

D. The magazine is not accurate because the number of cells is always changing.

Go Online You can complete an Extra Example online.

Practice

Go Online You can complete your homework online.

Example 1

1. TEST SCORES A teacher compares the ratio of the number of questions answered correctly to the total number of questions on a test as a metric. For a student to earn an A or B on a test, the ratio must be greater than or equal to 0.8. The last test given by the teacher had a total of 40 questions. Using this metric, what is the least number of questions a student can answer correctly to earn an A or B on the test?

$$\frac{\text{number of questions answered correctly}}{\text{total number of questions}}$$

2. DRIVER'S TEST The Department of Motor Vehicles, DMV, uses a metric to determine whether a person earns a driver's license. In one state, the total number of possible points on the written portion of the driver's exam is 46. A person will pass the written portion of the driver's exam by scoring 84% or greater. The table shows the number of points different people earned on the written portion of the driver's exam. How many people passed the written portion of the driver's exam?

38	40	41	45	39
35	40	46	43	37
41	42	44	41	46
40	38	32	44	45
39	40	30	43	45

3. TRACK A college track coach compares the ratio of time it takes a runner to run 100 meters to 12 seconds. For a runner to be on the team, the ratio must be less than or equal to 0.95. What is the slowest time 100 meters can be run to make the team?

Example 2

4. DEBT-TO-INCOME RATIO Find Jada's debt-to-income ratio if her monthly expenses are $1850 and her monthly salary is $2500.

5. DEBT-TO-INCOME RATIO Find Victoria's debt-to-income ratio if her monthly expenses are $1280 and her monthly salary is $2500.

6. PLUMBING Raven is deciding between two plumbing services. Service Provider A multiplies the average number of hours spent at a residence by $50, where the average number of hours spent at a residence is 1.5 hours for a new house and 3.75 hours for an old house. Service Provider B multiplies the exact number of hours spent at a residence by $60. Suppose Raven has a new house and needs a plumber for 1.75 hours. Find the cost of service charge using both methods.

7. INVESTING Hector is deciding how much he should invest each year. The Automatic Method multiplies the average income by 10%, where the average income is $50,000 for an employee that has been at the same company for 10 years or less and $60,000 for an employee that has been at the same company for more than 10 years. The Exact Method multiplies the exact income by 7.5%. Suppose Hector has been at the same company for 12 years and his income last year was $75,000. Find the amount Hector should invest using both methods.

Example 3

8. MONEY Jordan has $20 to share among 3 people. Jordan types 20 ÷ 3 into his calculator and gets 6.666666667. How much should he give to each person?

9. SNACKS Ms. Miller has 14 snack bars to share among 6 students. Ms. Miller types 14 ÷ 6 into her calculator and gets 2.333333333. How many snack bars should she give each student?

10. EVENT PLANNING Max is planning a banquet for the National Honors Society. Approximately 60 people will be attending the banquet. If 8 people can fit comfortably at a table, how many tables should he have?

11. GARDEN Emily wants to plant flowers in a narrow rectangular plot that is 1 foot by 4.5 feet. The flowers she wants to plant need to be spaced at least 8 inches apart. How many plants should she buy for the garden?

12. LEMONADE Justin has 64 ounces of lemonade to divide among 9 people. When he types 64 ÷ 9 into his calculator, the number that appears is 7.1111111. How much lemonade should he give to each person?

13. MONEY Darnell has $1000 he wants to divide among his 3 children. How much should Darnell give each of his children?

Example 4

14. FUNDRAISING At a bake sale, the golf team sold all 50 cupcakes for $1.50 each. They sold almost all of the 100 cookies for $1.00 each. The team also received donations from 7 people averaging $4.25 per donation. Determine the total amount of money the golf team collected from the bake sale with a reasonable level of accuracy.

15. EXPENSES Santino spends an average of $125 per month on clothing, not including sales tax. He also spends about $180 per month going out to eat and an average of $130 per week on groceries, including sales tax. If the tax rate is 7.25%, find the total amount he spends on food and clothing in a year, including sales tax, with a reasonable level of accuracy.

16. TIME MANAGEMENT Ava spends about 45 minutes studying each day. She also practices the piano for an average of 20 minutes per day, and she practices soccer 1.5 hours three times a week. If Ava decides to reduce the time she does each activity by $\frac{1}{6}$, find the total number of hours she spends studying and practicing in a year with a reasonable level of accuracy.

Example 5

17. SCHOOLS The superintendent at Hartgrove High School says there are 3103 students enrolled at the school. How accurate is the reported enrollment? Explain your reasoning.

18. POPULATION The U.S. Census Bureau Web site shows that the population of Texas on July 1, 2016 was 27,862,596. How accurate is the reported population? Explain your reasoning.

19. TRAFFIC LIGHTS A map maker reported that there were about 12,000 traffic lights in New York City. How accurate is the report? Explain your reasoning.

20. SAND A mathematician reported that there are 1,578,932 grains of sand in one cubic foot. How accurate is the report? Explain your reasoning.

Mixed Exercises

21. USE A SOURCE A coach compares the ratio of the number of free throws made to the total number of attempted free throws as a metric. For a player to be selected as a free throw shooter when the other team is given a technical foul, the ratio must be greater than or equal to 0.82. Find the number of free throws made and the number of free throws attempted for three former NBA players. Using the metric, which players would and would not be selected as a free throw shooter when the other team is given a technical foul?

22. REASONING A carpenter is measuring the length of a living room. Should the carpenter measure the length in feet, inches, meters, or kilometers to be most accurate? Explain.

23. SPACE Which unit of measure is the most appropriate for measuring the distance from Earth to the star Polaris: feet, kilometers, or light-years? Explain.

24. POPULATION Juanita and Trevor are doing research about the deer population in Ohio. Juanita says there are over 750,000 deer in Ohio. Trevor says there are 734,928 deer in Ohio. Who is more accurate? Explain your reasoning.

The graph shows how the number of visitors at a local zoo is related to the average daily temperature. The line shown is the line that most closely approximates the data in the scatter plot. Use the graph for Exercises 25–27.

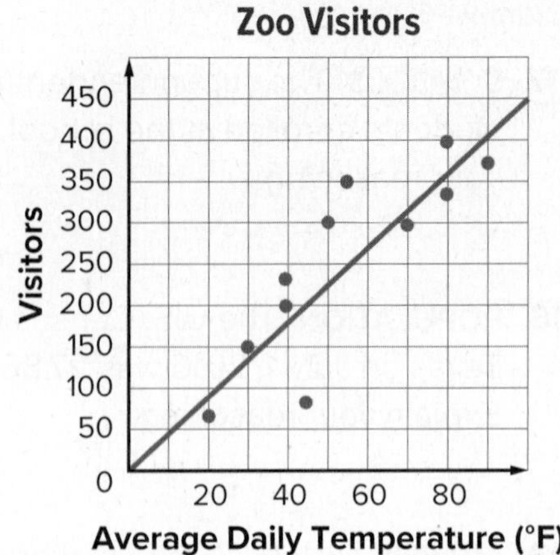

25. STRUCTURE Describe the line in terms of accuracy.

26. USE ESTIMATION Use the line to approximate the number of visitors at the zoo for an average daily temperature of 50°F. Compare this to the actual number of visitors given by the point on the graph for an average daily temperature of 50°F.

27. REASONING Explain why some points are above the line and some points are below the line.

28. METRICS Suppose two mortgage companies compare the ratio of the monthly mortgage payment to the total monthly income as their metric. Suppose 0.3 is the ideal metric for the debt-to-income ratio for Company A, and 0.28 is the ideal metric for the debt-to-income ratio for the Company B. Provided the target mortgage payment is the same for either company, then which of these mortgage companies requires a greater monthly income? Explain.

Higher-Order Thinking Skills

29. WRITE Suppose you start your own company. When hiring employees, you want to set certain metrics, such as typing speed. What other attributes of employees do you think you might consider when creating metrics to determine hiring qualifications?

30. FIND THE ERROR Mr. Moreno's students are weighing materials for a chemistry experiment. Four students weigh the same sample using different scales:

100 g   104 g   105 g   103.5 g

Mr. Moreno tells the students that they each weighed the amount correctly. Explain how this is possible.

31. WRITE Lamont stops at a gas station that sells gasoline at $\$3.29\frac{9}{10}$ per gallon. He pumps 8.618 gallons of gasoline into the tank. How much will Lamont pay for gas? How much accuracy is possible? How much accuracy is necessary? Explain.

Review

Essential Question

How can mathematical expressions be represented and evaluated?

You can represent mathematical expressions verbally, numerically, and algebraically. They can be evaluated by applying properties and rules. For example, you can translate a sentence to a numerical or algebraic expression and use the order of operations to simplify or evaluate the expression.

Module Summary

Lessons 1-1 and 1-2

Numerical and Algebraic Expressions

- A numerical expression contains only numbers and mathematical operations.
- To evaluate an expression means to find its value. If a numerical expression contains more than one operation, the rule that lets you know which operation to perform first is called the *order of operations*.
- The Substitution Property allows you to evaluate an algebraic expression by replacing the variables with their values.

Lessons 1-3 and 1-4

Properties of Real Numbers and Distributive Property

- The Reflexive Property states that any quantity is equal to itself.
- The Symmetric Property states that if one quantity equals a second quantity, then the second quantity equals the first.
- The Transitive Property states that if one quantity equals a second quantity and the second quantity equals a third quantity, then the first quantity equals the third quantity.
- The Associative Property states that the way you group three or more numbers when adding or multiplying does not change their sum or product.
- The Commutative Property states that the order in which you add or multiply numbers does not change their sum or product.
- The Distributive Property states that multiplying a number by a sum of numbers is the same as doing each multiplication separately and then adding the products.

Lesson 1-5

Expressions Involving Absolute Value

- The absolute value of a number is its distance from 0 on the number line.
- Absolute value is always greater than or equal to zero.

Lesson 1-6

Descriptive Modeling and Accuracy

- Descriptive modeling is a way to mathematically describe real-world situations and the factors that cause them.
- A metric is a rule for assigning a number to a characteristic or attribute.
- Accuracy is the nearness of a measurement to the true value of the measure.

Study Organizer

Foldables

Use your Foldable to review the module. Working with a partner can be helpful. Ask for clarification of concepts as needed.

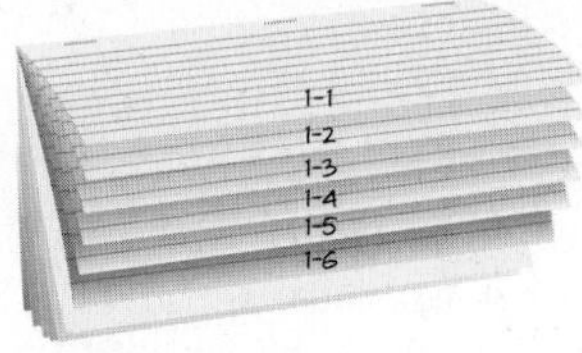

Test Practice

1. **MULTIPLE CHOICE** Which is equivalent to 2^5? (Lesson 1-1)

   A. 10

   B. 16

   C. 24

   D. 32

2. **MULTI-SELECT** The table shows the prices of several items at a movie theater. Which expressions represent the total cost of 4 movie tickets, 2 popcorns, and 1 bottled water? Select all that apply. (Lesson 1-1)

Item	Cost
Ticket	$9.75
Popcorn	$6.25
Soda	$5.50
Water	$4.75
Box of Candy	$3.50

   A. 4(9.75) + 2(6.25) + 4.75

   B. 4(9.75) + 2(5.50) + 4.75

   C. 39.00 + 12.50 + 4.75

   D. 54.75

   E. 56.25

3. **OPEN RESPONSE** Write an algebraic expression that represents *five times the quantity x increased by seven, minus four cubed.* (Lesson 1-2)

4. **MULTIPLE CHOICE** What is the value of the expression $9x^2 + 4x - 11$ when $x = 3.2$? Express your answer as a decimal, rounded to the nearest hundredth. (Lesson 1-2)

   A. 20.232

   B. 30.6

   C. 59.4

   D. 93.96

5. **MULTIPLE CHOICE** Which algebraic expression represents the verbal expression *the product of five and a number, decreased by eleven*? (Lesson 1-2)

   A. $5n - 11$

   B. $11 - 5n$

   C. $5(n - 11)$

   D. $11 - (n + 5)$

6. **OPEN RESPONSE** Evaluate the expression $5[13 - (3^2 + 2^2)]$. (Lesson 1-3)

7. **MULTIPLE CHOICE** Which expression is NOT a way to represent $2 \cdot 3\frac{5}{6} + 2 \cdot 12 + 2 \cdot 1\frac{1}{6}$? (Lesson 1-3)

   A. $2\left(3\frac{5}{6} + 12 + 1\frac{1}{6}\right)$

   B. $2 \cdot 5 + 12$

   C. $2(5 + 12)$

   D. 34

8. OPEN RESPONSE Ayumi, a chef, wants to determine how many meals she cooked in one evening. The table shows the four meals she made and the number of people that were served each meal. (Lesson 1-3)

Meal	Number of People
Lasagna	27
Spaghetti & Meatballs	21
Steak & Potatoes	19
Shrimp Scampi	13

She uses the following steps to determine how many total meals she cooked.

Step 1: $27 + 21 + 19 + 13$
Step 2: $27 + 13 + 21 + 19$
Step 3: $(27 + 13) + (21 + 19)$
Step 4: $40 + 40$
Step 5: 80
Which property did Ayumi use in Step 3?

9. OPEN RESPONSE Indicate whether each of the statements is *true* or *false*. (Lesson 1-3)

A. $4(6 - 2 \times 3) = 0$

B. $11(3^2 - 9) + 2\left(\frac{1}{2}\right) = 0$

C. $4 \cdot 0 + 4^2 - 2^3 - (2 + 2 \cdot 3) = 0$

10. MULTI-SELECT Which expressions could be used to evaluate 418(27)? (Lesson 1-4)

A. $418(20 - 7)$

B. $(420 - 2)(27)$

C. $(400 - 18)(27)$

D. $(418)(20 + 7)$

E. $(418)(30 - 3)$

11. MULTI-SELECT Indicate whether each expression represents the verbal expression *negative seven times the quantity triple m minus eleven.* (Lesson 1-4)

A. $-7(m^3 - 11)$

B. $-7(3m) - 7(-11)$

C. $-21m - 77$

D. $-21m + 77$

E. $-21m - 11$

F. $-7m^3 + 77$

12. MULTIPLE CHOICE Which is the simplified expression of $-8(2m + 9k - 13)$? (Lesson 1-4)

A. $-16m + 9k - 13$

B. $-16m - 72k + 104$

C. $-16m - 72k - 104$

D. $16m - 72k - 104$

13. **MULTI-SELECT** A group of 8 artists plans to attend a quilting class and purchase lunch. (Lesson 1-4)

Which expression(s) represents the total cost for all 8 artists?

A. 8(10) + 14

B. 10(8 + 14)

C. 8(10 + 14)

D. 80 + 140

E. 80 + 112

14. **MULTI-SELECT** If x and y are both integers, which expression(s) are equivalent to $|x - y|$? (Lesson 1-5)

A. $|y - x|$

B. $|y + x|$

C. $|y| - |x|$ if $x \leq y$ and $x \geq 0$

D. $|x| - |y|$ if $y > x$

E. $|y| + |x|$ if $y > x$

15. **OPEN RESPONSE** What is the value of the expression $4^2 + 3|4x - 9|$ when $x = -2$? (Lesson 1-5)

16. **OPEN RESPONSE** A player's secondary average (SecA) is a way to look at the extra bases gained without regard to batting average. The formula for SecA is $\text{SecA} = \frac{T - H + B + S - C}{A}$, where T is total bases, H is hits, B is bases from balls or walks, S is stolen bases, C is number of times caught stealing, and A is times at bat.

Player	*T*	*H*	*B*	*S*	*C*	*A*
Altuve	186	116	39	22	3	330
Murphy	183	110	17	2	3	315
Ortiz	187	94	45	2	0	279
Ramos	139	84	23	0	0	251

Find each player's SecA. Round to the nearest hundredth if necessary. (Lesson 1-6)

17. **OPEN RESPONSE** A marketing manager bought 4 advertisements for $1345 each. She reported to her supervisor that she spent about $4000 out of her budget. Did the marketing manager report her spending to a reasonable level of accuracy? Explain. (Lesson 1-6)

Module 2

Equations in One Variable

Essential Question

How can writing and solving equations help you solve problems in the real world?

What Will You Learn?

How much do you already know about each topic **before** starting this module?

KEY

— I don't know. — I've heard of it. — I know it!

	Before			After		
	I don't know.	I've heard of it.	I know it!	I don't know.	I've heard of it.	I know it!
write equations to represent relationships						
interpret equations that represent relationships						
solve one-step equations by using addition and subtraction						
solve one-step equations by using multiplication and division						
solve multi-step equations						
solve equations with variables on each side						
solve equations by applying the Distributive Property						
solve equations that involve absolute value						
solve proportions						
solve an equation with more than one variable for a specific variable						
convert units of measure by using dimensional analysis						

Foldables Make this Foldable to help you organize your notes about equations. Begin with four sheets of grid paper.

1. **Fold** four sheets of grid paper in half along the width.
2. **Unfold** each sheet and tape to form one long piece.
3. **Label** each piece with the lesson number as shown. Label the last piece for vocabulary. Refold to form a booklet.

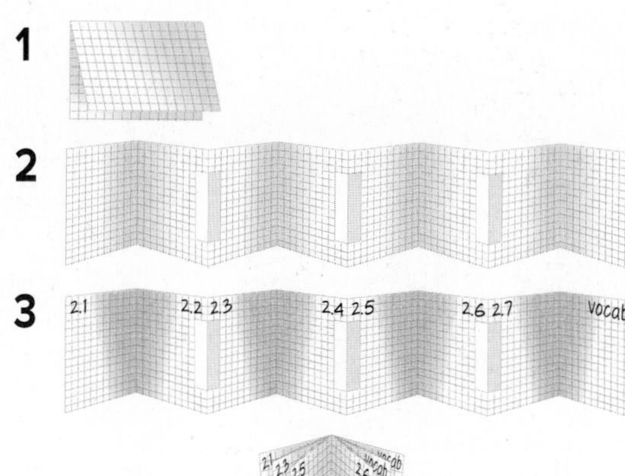

What Vocabulary Will You Learn?

- constraint
- dimensional analysis
- equation
- equivalent equations
- formula
- identity
- literal equation
- multi-step equation
- proportion
- solution
- solve an equation

Are You Ready?

Complete the Quick Review to see if you are ready to start this module.
Then complete the Quick Check.

Quick Review

Example 1

Write an algebraic expression for the phrase *the quotient of five and w decreased by eight.*

the quotient of five and w decreased by eight

$\frac{5}{w}$ $-$ 8

The expression is $\frac{5}{w} - 8$.

Example 2

Evaluate $9 + \frac{4^2}{2} - 2(5 \times 2 - 8)$.

$9 + \frac{4^2}{2} - 2(5 \times 2 - 8)$	Original expression
$= 9 + \frac{4^2}{2} - 2(2)$	Evaluate inside the parentheses.
$= 9 + 8 - 2(2)$	Evaluate the power and divide.
$= 9 + 8 - 4$	Multiply.
$= 13$	Add and subtract.

Quick Check

Write an algebraic expression for each verbal expression.

1. six times a number n increased by two
2. a number d squared minus three
3. the sum of four times b and nine

Evaluate each expression.

4. $(7 + 3)^2 - 4$
5. $4(11 - 5) \div 3$
6. $\frac{1}{3}(21) + \frac{1}{8}(32)$
7. $3 \cdot 2^3 + 64 \div 8$
8. $\frac{11 - 3}{2} + 9$
9. $6[(5 - 3)^2 + 8] \div 2$

How did you do?

Which exercises did you answer correctly in the Quick Check?

Writing and Interpreting Equations

Today's Goals
- Translate sentences into equations.
- Translate equations into sentences.

Today's Vocabulary
equation
constraint

Explore Writing Equations by Modeling a Real-World Situation

Online Activity Use a real-world situation to complete the Explore.

INQUIRY What steps can you use to write equations to represent a real-world situation?

Learn Writing Equations

A mathematical statement that contains two expressions and an equal sign, =, is an **equation**.

Key Concept • Writing Equations	
Step 1	Identify each unknown and assign a variable to it.
Step 2	Identify the givens and their relationships.
Step 3	Write the sentence as an equation.

Think About It!
What distinguishes an expression from an equation?

Example 1 Write an Equation for a Sentence

Write an equation for the sentence.

Twenty minus the quotient of 7 and *x* is the same as twice *x*.

Recall that a quotient is the result of division.

Twenty minus the quotient of 7 and x is the same as ***twice x.***

$$20 - \quad \frac{7}{x} \quad = \quad 2x$$

The equation is $20 - \frac{7}{x} = 2x$.

Study Tip
Verbal Phrases When writing the verbal form of an equation, *is* and *equals* can be used interchangeably.

Check

Write an equation for the sentence.

Four times a number less 10 is equal to 16.

A. $4x - 10 = 16$ **B.** $4(x - 10) = 16$

C. $10 - 4x = 16$ **D.** $4x + 10 = 16$

Go Online You can complete an Extra Example online.

Example 2 Write an Equation

LIFE ONLINE **Of 799 teens surveyed about what they do online, some use a social network. Of those on a social network, 430 say people their age are "mostly kind" online and the remaining 193 do not. Write an equation to find the number of teens surveyed who are not on a social network.**

Social Teens Mostly Kind Online

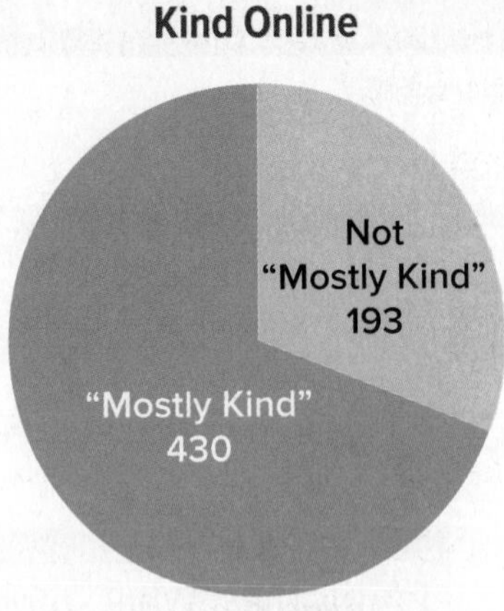

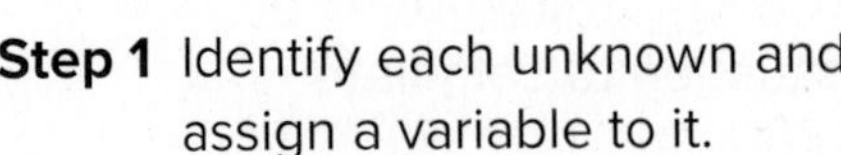

Step 1 Identify each unknown and assign a variable to it.

Let n = the number of teens surveyed who are not on a social network.

Step 2 Identify the givens and their relationship.

The givens are:

- 799 teens were surveyed.
- Some number of the teens use a social network.
- 430 of those on a social network say people their age are "mostly kind" online. The other 193 do not.

The 430 and 193 make up the group on a social network. The rest of the 799 surveyed are not on a social network.

Step 3 Write the sentence as an equation.

The sum of the teens on a social network and those not on a social network is 799.

$(430 + 193) + n = 799$

Think About It!

Is there only one equation that represents the situation? Justify your argument.

Check

READING Etu has read 12 of the 32 chapters in his assigned book. He plans to finish the book by reading c chapters each for 8 days until the book is due. Which equation best represents the situation?

A. $12 - 8c = 32$

B. $12 + \frac{c}{8} = 32$

C. $12 + 8c = 32$

D. $12 - \frac{c}{8} = 32$

Go Online You can complete an Extra Example online.

Example 3 Write an Equation with Multiple Variables

GEOMETRY **Translate the sentence into a formula.**

The perimeter of a rectangle is twice the sum of the length and the width.

Step 1 Identify unknowns.

perimeter, the length, and the width

Step 2 Assign variables.

Let P = perimeter, ℓ = length, and w = width.

Step 3 Identify the givens and their relationships.

Twice *means two times.*

Twice the sum *means you add first, then multiply.*

Step 4 Write an equation.

The formula for the perimeter of a rectangle is $P = 2(\ell + w)$.

Think About It!

Why is it helpful to identify all the unknowns before writing an equation?

Check

Translate the sentence into a formula.

MOTORS The horsepower of a motor is the product of the motor speed and the torque divided by 5252.

A. $H = \frac{M}{5252T}$

B. $H = \frac{MT}{5252}$

C. $H = \frac{5252}{MT}$

D. $H = \frac{5252M}{T}$

BAGELS Plain and cinnamon raisin bagels are the most popular flavors. Each year, 24 million more than twice as many packages of plain bagels are sold as cinnamon raisin. There were 136 million packages of plain bagels sold last year. Create an equation that can be used to find the number of millions of packages of cinnamon raisin bagels, c, sold last year.

$136 = 2c + 24$ OR $136 = 24 + 2c$ OR $2c + 24 = 136$ OR $24 + 2c = 136$

Talk About It!

What is an example of a real-life constraint? Explain.

Learn Interpreting Equations

Look for the relationships in an equation by interpreting each part of the expressions in the equation.

As you interpret an equation that represents a real-life situation, consider that the equation may be viewed as a constraint in the situation. In mathematics, a **constraint** is a condition that a solution must satisfy. These conditions limit the number of possible solutions. The solutions of the equation meet the constraints of the problem.

Go Online You can complete an Extra Example online.

Example 4 Write a Sentence for an Equation

Write a sentence for the equation.

$$2z - 1 = 5$$

Sample answer: Two z minus one equals five.

Check

Write a sentence for the equation.

a. $6z - 15 = 45$

A. Six times a number z, minus fifteen equals forty-five.

B. Six times the quantity of a number z minus fifteen equals forty-five.

C. Six times the difference of a number z and fifteen equals forty-five.

D. Six plus z minus fifteen equals forty-five.

b. $(y + 3)^2 = 25$

A. y plus 3 squared is 25.

B. y squared plus 3 is 25.

C. The quantity y plus 3 squared is 25.

D. y plus 3 is 25.

Example 5 Write a Sentence for an Equation with Grouping Symbols

Write a sentence for the equation.

$$3(y + 1) = 12$$

The parentheses tell us that 3 is *three times* the expression in the parentheses.

The parentheses can be written as *the quantity*.

Write $y + 1$ as y plus 1 or *the sum of y and one*.

Write $= 12$ as *equals twelve* or *is twelve*.

One of the sentences that represents the equation $3(y + 1) = 12$ is:

Three times the quantity y plus one equals twelve.

Watch Out!

Parentheses When writing a sentence for an equation with parentheses, the phrase *the quantity* should be written immediately before the terms that are contained in the parentheses, not at the beginning of the sentence.

Go Online

An alternate method is available for this example.

Go Online You can complete an Extra Example online.

Check

Select the sentence(s) that represent(s) the equation.

$7(p + 23) = 102$

A. Seven times the sum of *p* and twenty-three is the same as one hundred two.

B. Seven times *p* plus twenty-three equals one hundred two.

C. Seven times the quantity *p* plus twenty-three equals one hundred two.

D. The quantity seven times *p* plus twenty-three is the same as one hundred two.

E. Seven times the sum of *p* and twenty-three is one hundred two.

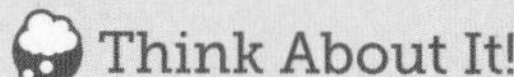

Think About It!

What words communicate grouping symbols in an expression or equation?

Example 6 Interpret an Equation

GEOMETRY **Write a sentence for the formula for the surface area of a rectangular prism $S = 2\ell w + 2\ell h + 2wh$. Then interpret the equation in the context of the situation.**

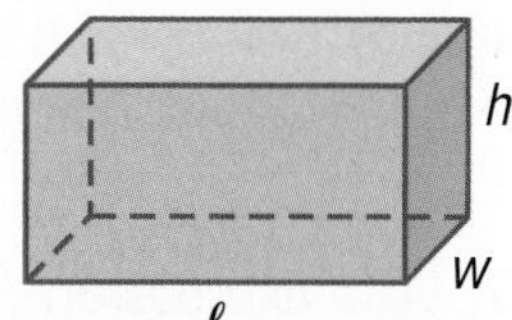

From the equation, we see that the surface area of a rectangular prism depends on the length, width, and height. Complete the table.

$S =$	$2\ell w +$	$2\ell h +$	$2wh$
Surface area equals	two times length times width plus	Two times length times height plus	two times width times height

- The first term, $2\ell w$, is two times the area of a rectangle . In the prism above, the area of the bottom face is ℓw. The top is the same shape, so it has the same area. This term represents the sum of the areas of the bottom and top faces.

- The second term, $2\ell h$, is the sum of the areas of the front and back faces.

- The third term, $2wh$, is the sum of the areas of the left and right faces.

So, the surface area of the rectangular prism is the sum of the areas of the faces.

Think About It!

What did you already know that helped you interpret the formula?

Go Online

You can watch a video to see how to use algebra tiles with this example.

Go Online You can complete an Extra Example online.

Check

FINANCE The formula for a loan balance is $b = p\left(1 + \frac{r}{12}\right) - d$ if the previous balance is p, the annual interest rate is r, and a payment of d is made.

Part A Write a sentence for the formula.

A. The balance equals the previous balance multiplied by the quantity one plus the annual interest rate divided by 12 minus the payment.

B. The balance equals the previous balance multiplied by the annual interest rate divided by 12 minus the payment.

C. The balance equals the previous balance multiplied by the sum of one and the annual interest rate minus the payment.

D. The balance equals the previous balance minus the payment.

Part B Select each sentence that is a correct interpretation of the equation in the context of the situation. Select all that apply.

A. The expression $\frac{r}{12}$ represents the monthly interest rate.

B. The expression $p\left(1 + \frac{r}{12}\right)$ represents the previous balance plus interest.

C. If no payments are made, then $d = 0$ and the balance is the same as the previous balance.

D. The expression $\left(1 + \frac{r}{12}\right)$ represents the monthly interest rate.

E. If no payment is made, then $d = 0$ and the balance is the previous balance plus interest.

Pause and Reflect

Did you struggle with anything in this lesson? If so, how did you deal with it?

Go Online You can complete an Extra Example online.

Practice

Go Online You can complete your homework online.

Example 1

Write an equation for each sentence.

1. Two added to three times a number m is the same as 18.

2. The product of five and the sum of a number x and three is twelve.

3. The quotient of 24 and x equals 14 minus 2 times x.

4. Nine times a number y subtracted from 85 is seven times the sum of four and y.

Example 2

5. **WALKING** Lily has walked 2 miles. Her goal is to walk 6 miles. Lily plans to reach her goal by walking 3 miles each hour h for the rest of her walk. Write an equation to find the number of hours it will take Lily to reach her goal.

6. **MATH** Paulina has completed 24 of the 42 math problems she was assigned for homework. She plans to finish her homework by completing 9 math problems each hour h. Write an equation to find the number of hours it will take Paulina to complete her math homework assignment.

7. **ATHLETICS** Of 107 athletes surveyed about what sport they play, some play basketball. Of those that play basketball, 48 play baseball and the remaining 33 do not play baseball. Write an equation to find the number of athletes surveyed who do not play basketball.

8. **SALES** Cars and trucks are the most popular vehicles. Last year, the number of cars sold was 39,000 more than three times the number of trucks sold. There were 216,000 cars sold last year. Write an equation that can be used to find the number of trucks, t, sold last year.

Example 3

Translate each sentence into an equation or formula.

9. Twice a increased by the cube of a equals b.

10. Seven less than the sum of p and t is as much as 6.

11. The sum of x and its square is equal to y times z.

12. Four times the sum of f and g is identical to six times g.

13. The area A of a square is the length of a side ℓ squared.

14. The perimeter P of a triangle is equal to the sum of the lengths of sides a, b, and c.

15. The perimeter of a rectangle is equal to 2 times the length plus twice the width.

16. The density of an object is the quotient of its mass and its volume.

17. Simple interest is computed by finding the product of the principal amount p, the interest rate r, and the time t.

18. The surface area of a rectangular prism is 2 times the sum of the width, w, times height, h, and length, l, times width and length times height.

Examples 4 and 5

Write a sentence for each equation.

19. $j + 16 = 35$

20. $4m = 52$

21. $7(p + 23) = 102$

22. $r^2 - 15 = t + 19$

23. $\frac{2}{5}v + \frac{3}{4} = \frac{2}{3}x^2$

24. $\frac{1}{3} - \frac{4}{5}z = \frac{4}{3}y^3$

25. $g + 10 = 3g$

26. $2(t + 4q) = 2q + 4t$

27. $4(a + b) = 9a$

28. $8(2y - 6x) = 4 + 2x$

29. $\frac{1}{2}(f + y) = f - 5$

30. $k^2 - n^2 = 2b$

Example 6

Write a sentence for each formula. Then interpret the equation in the context of the situation.

31. GEOMETRY The formula for the volume of a cylinder is $V = \pi r^2 h$, where V is the volume, r is the length of the radius of the base, and h is the height of the cylinder.

32. GEOMETRY The formula for the volume of a cube is $V = s^3$, where V is the volume and s is the side length.

33. FINANCE The simple interest formula is given by $I = Prt$, where I = interest, P = principal, r = rate, and t = time.

34. FINANCE The compound interest formula is given by $A = P\left(1 + \frac{r}{n}\right)^{nt}$, where A is the amount, P is the principal, r is the rate, n is the number of times interest is compounded per year, and t is the time in years.

35. SCIENCE Newton's second law of motion is $F = ma$, where F is the force acting on an object, m is the mass of the object and a is the acceleration of the object.

36. SCIENCE The formula $d = rt$ relates the distance traveled d, the rate of travel r, and the time spent traveling t.

Mixed Exercises

For Exercises 37–40, match each sentence with an equation.

A. $g^2 = 2(g - 10)$ **B.** $\frac{1}{2}g + 32 = 15 + 6g$ **C.** $g^3 = 24g + 4$ **D.** $3g^2 = 30 + 9g$

37. One half of g plus thirty-two is as much as the sum of fifteen and six times g.

38. A number g to the third power is the same as the product of 24 and g plus 4.

39. The square of g is the same as two times the difference of g and 10.

40. The product of 3 and the square of g equals the sum of thirty and the product of nine and g.

Translate each sentence into an equation.

41. The difference of the square of y and twelve is the same as the product of five and x.

42. The difference of f and five times g is the same as 25 minus f.

43. Three times b less than 100 is equal to the product of 6 and b.

44. Four times the sum of 14 and c is a squared.

Translate each equation into a sentence.

45. $4n = x(5 - n)$ **46.** $2b - 10 = 4$ **47.** $y + 3x^2 = 5x$

Translate each sentence into a formula.

48. The area A of a circle is pi times the radius r squared.

49. The volume V of a rectangular prism equals the product of the length ℓ, the width w, and the height h.

50. REASONING The area of a kitchen is 182 square feet. This is 20% of the area of the first floor of the house. Let F represent the area of the first floor. Write an equation to represent the situation.

51. REASONING Katie is twice as old as her sister Mara. The sum of their ages is 24. Write a one-variable equation to represent the situation.

52. GEOMETRY The formula $F + V = E + 2$ shows the relationship between the number of faces F, edges E, and vertices V of a polyhedron, such as a pyramid. Write the formula in words.

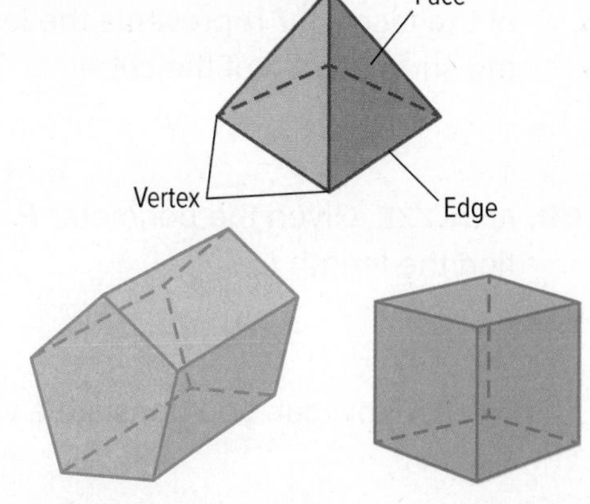

53. STRUCTURE A recycling company charges business owners \$10 for each cubic yard of waste removed from their facility plus a 10% fuel charge based on the total monthly bill. Let w represent the number of cubic yards of waste removed during the month. Write an equation to describe the total cost c of the recycling service per month.

Higher Order Thinking Skills

54. WRITE Determine whether the two sentences describe the same equation. Explain.

The product of x and y plus z equals w.

The product of x and the sum of y and z equals w.

55. ANALYZE Determine whether the equation is an accurate translation of the sentence. Explain.

a. The square of the product of 4 and a number is equal to 8 times the sum of the number and 6. $(4n)^2 = 8(n + 6)$

b. Three more than one-half a number is equal to 2 less than the number.

$\frac{n}{\frac{1}{2}} + 3 = n - 2$

56. PERSEVERE Translate the formula $A = \frac{b_1 + b_2}{2} \cdot h$ into words. Let A represent the area. List any constraints on the variables.

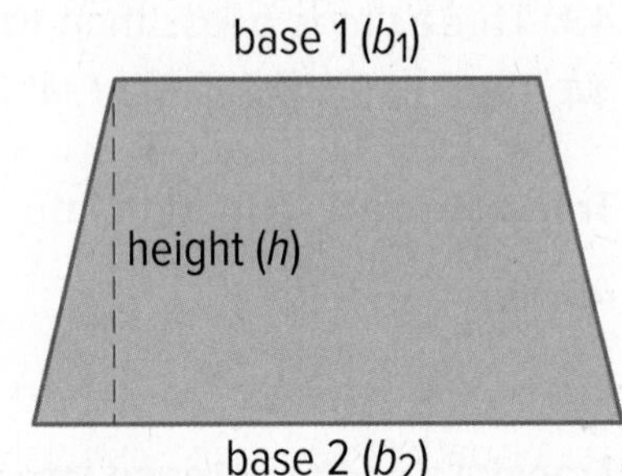

57. CREATE Write a scenario for the equation $12a + 10(a - 1) = 188$.

58. CREATE Write a problem about your favorite television show that uses the equation $x + 8 = 30$.

59. ANALYZE The surface area of a three-dimensional object is the sum of the area of the faces. If ℓ represents the length of the side of a cube, write a formula for the surface area of the cube.

60. ANALYZE Given the perimeter P and width w of a rectangle, write a formula to find the length ℓ.

61. WRITE How can you translate a verbal sentence into an algebraic equation? Explain.

Solving One-Step Equations

Explore Using Algebra Tiles to Solve One-Step Equations Involving Addition or Subtraction

Online Activity Use algebra tiles to complete the Explore.

> **INQUIRY** How can you model and solve addition and subtraction equations?

Explore Using Algebra Tiles to Solve One-Step Equations Involving Multiplication

Online Activity Use algebra tiles to complete the Explore.

> **INQUIRY** How can you use algebra tiles to solve multiplication equations?

Learn Solving One-Step Equations Involving Addition or Subtraction

To **solve an equation** means to find all values of the variable that make the equation true. Each value that makes an equation true is a **solution**. **Equivalent equations** have the same solution.

Key Concept • Addition Property of Equality	
Words	If a number is added to each side of a true equation, the resulting equivalent equation is also true.
Symbols	For any real numbers a, b, and c, if $a = b$, then $a + c = b + c$.

Key Concept • Subtraction Property of Equality	
Words	If a number is subtracted from each side of a true equation, the resulting equivalent equation is also true.
Symbols	For any real numbers a, b, and c, if $a = b$, then $a - c = b - c$.

Today's Goals
- Solve equations by using addition and subtraction.
- Solve equations by using multiplication and division.

Today's Vocabulary
solve an equation

solution

equivalent equations

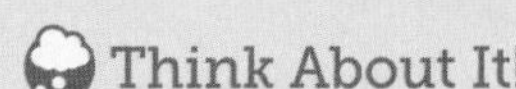

Think About It!

What happens if you add 5 to each side of $x - 5 = 15$? Which Property of Equality are you using?

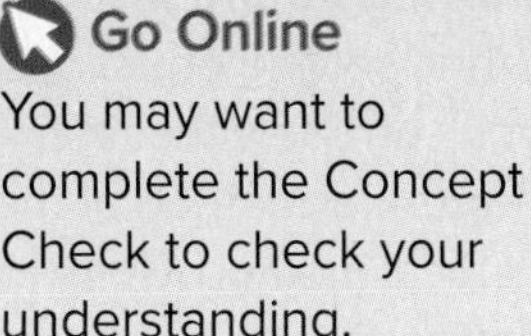

Go Online

You may want to complete the Concept Check to check your understanding.

Example 1 Solve by Adding

Use the Addition Property of Equality to solve $g - 25 = 113$.

Horizontal Method		*Vertical Method*
$g - 25 = 113$	Original equation	$g - 25 = 113$
$g - 25 + 25 = 113 + 25$	Add 25 to each side.	$+ 25 + 25$
$g = 138$	Simplify.	$g = 138$

CHECK

$g - 25 = 113$	Original equation
$138 - 25 \stackrel{?}{=} 113$	Substitute 138 for g.
$113 = 113$	True

Check

Solve $\frac{2}{3} + w = 1\frac{1}{2}$. State which property of equality you used.

A. $\frac{5}{6}$; Subtraction Property of Equality

B. $\frac{5}{6}$; Addition Property of Equality

C. $\frac{13}{6}$; Addition Property of Equality

D. $\frac{13}{6}$; Subtraction Property of Equality

Example 2 Solve by Subtracting

Use the Subtraction Property of Equality to solve $27 + k = 30$.

Horizontal Method		*Vertical Method*
$27 + k = 30$	Original equation	$27 + k = 30$
$27 - 27 + k = 30 - 27$	Subtract 27 from each side.	$-27 \quad -27$
$k = 3$	Simplify.	$k = 3$

CHECK

$27 + k = 30$	Original equation
$27 + 3 \stackrel{?}{=} 30$	Substitute 3 for k.
$30 = 30$	True

Check

Solve $a + 26 = 35$.

$a =$ ___?___

Go Online You can complete an Extra Example online.

Watch Out!

Equivalent Equations
Do not forget to add the same number to each side of the equation so the result is an equivalent equation.

Talk About It!

Ann says that you could also solve $27 + k = 30$ by using the Addition Property of Equality. Explain Ann's reasoning.

Study Tip

Solving Equations
When solving equations, you can use either the horizontal method or the vertical method. Both methods will produce the same answer.

Example 3 Write a One-Step Equation

TENNIS **In tennis, the Grand Slam tournaments are the four most prestigious annual events. At one point in his career, Roger Federer had won three more Grand Slam singles titles than Rafael Nadal. If at that time Roger Federer held the record for the most Grand Slam singles titles won with 17, how many Grand Slam singles titles had Rafael Nadal won?**

Complete the table to write an equation that represents the number of Grand Slam singles titles Rafael Nadal won.

Words	Roger Federer	won	three	more than	Rafael Nadal
Variable	Let n = the number of singles Rafael Nadal won.				
Equation	17	=	n	+	3

$17 = n + 3$ Original equation

$17 - 3 = n + 3 - 3$ Subtract 3 from each side.

$14 = n$ Simplify.

Rafael Nadal had won 14 Grand Slam singles titles.

Check

DOGS On average, a male bulldog weighs 15 pounds less than a male golden retriever. If the average male bulldog weighs 50 pounds, write and solve an equation to find the average weight of a male golden retriever.

A. $50 = w - 15$; 65 pounds

B. $50 = w + 15$; 35 pounds

C. $50 = w - 15$; 35 pounds

D. $50 = 15 - w$; 65 pounds

Go Online You can complete an Extra Example online.

Use a Source

Choose another men's singles tennis player and research the number of Grand Slam singles titles he has won. Write your own equation relating the number of Grand Slam singles titles he has won to the 17 titles of Roger Federer.

Think About It!

What happens if you divide each side of $8x = 32$ by 8? Which property of equality does this demonstrate?

Learn Solving One-Step Equations Involving Multiplication or Division

You can also use the Multiplication Property of Equality and the Division Property of Equality to solve equations.

Key Concept • Multiplication Property of Equality	
Words	If an equation is true and each side is multiplied by the same nonzero number, then the resulting equation is equivalent.
Symbols	For any real numbers a, b, and c, $c \neq 0$, if $a = b$, then $ac = bc$.
Example	If $x = 3$, then $8x = 24$.

Think About It!

How could you use the Division Property of Equality to simplify $ax = 32$ to $x = \frac{32}{a}$? How does this relate to using the Division Property of Equality to solve $8x = 32$?

Key Concept • Division Property of Equality	
Words	If an equation is true and each side is divided by the same nonzero number, the resulting equation is equivalent.
Symbols	For any real numbers a, b, and c, $c \neq 0$, if $a = b$, then $\frac{a}{c} = \frac{b}{c}$.
Example	If $x = -35$, then $\frac{x}{7} = \frac{-35}{7}$ or -5.

Example 4 Solve Equations by Multiplying or Dividing

Solve each equation.

a. $\frac{3}{8}x = \frac{9}{4}$

$\frac{3}{8}x = \frac{9}{4}$ Original equation

$\left(\frac{8}{3}\right)\frac{3}{8}x = \left(\frac{8}{3}\right)\frac{9}{4}$ Multiply each side by $\frac{8}{3}$, the reciprocal of $\frac{3}{8}$.

$x = 6$ Simplify.

Think About It!

Describe a method you could use to check your solution for **part a.**

b. $42 = -14y$

$42 = -14y$ Original equation

$\frac{42}{-14} = \frac{-14y}{-14}$ Divide each side by -14.

$-3 = y$ Simplify.

Check

Solve the equation $6y = 54$.

$y =$ __?__

Go Online You can complete an Extra Example online.

Apply Example 5 Solve by Multiplying

SURVEY **Kenji took a survey of the sophomore class. If 96 sophomores, or two-thirds of the class, said they were going to the football game on Saturday, how many sophomores were in the survey?**

1. What is the task?

Describe the task in your own words. Then list any questions that you may have. How can you find the answers to your questions?

96 sophomores are going to the game on Saturday. They make up $\frac{2}{3}$ of the class. How many sophomores were surveyed? I can find the answer to my question by writing and solving an equation to represent the situation.

2. How will you approach the task? What have you learned that you can use to help you complete the task?

I will write an equation to represent the situation and then solve it. I have learned how to translate a sentence to a mathematical equation. I have learned how to solve equations.

3. What is your solution?

What equation represents the number of sophomores surveyed?

$\frac{2}{3}n = 96$ where $n =$ the number of sophomores surveyed

How many sophomores were in the survey?

144

4. How can you know that your solution is reasonable?

Write About It! Write an argument that can be used to defend your solution.

Sample answer: It makes sense that the number of sophomores surveyed is greater than the number of sophomores attending the football game, because the number of sophomores attending the football game is only part of the total number of sophomores surveyed. So, the whole (144) should be greater than the part (96).

Go Online You can complete an Extra Example online.

Think About It!

How would your equation change if the 96 sophomores planning to attend the game represented three-fourths of the class? What would you multiply each side of the equation by to solve the new equation?

Check

FASHION Imani is making costumes for a play. She spent $146.58 on 21 yards of fabric. Write and solve an equation to find how much Imani paid for each yard of fabric.

A. $21p = 146.58$; $6.98 per yard

B. $146.58p = 21$; $0.14 per yard

C. $146.58(21) = p$; $3078.18 per yard

D. $21p = 146.58$; $3078.18 per yard

Pause and Reflect

Did you struggle with anything in this lesson? If so, how did you deal with it?

Go Online You can complete an Extra Example online.

Practice

Go Online You can complete your homework online.

Examples 1, 2, and 4

Solve each equation.

1. $v - 9 = 14$

2. $44 = t - 72$

3. $-61 = d + (-18)$

4. $18 + z = 40$

5. $-4a = 48$

6. $12t = -132$

7. $18 - (-f) = 91$

8. $-16 - (-t) = -45$

9. $\frac{1}{3}v = -5$

10. $\frac{u}{8} = -4$

11. $\frac{a}{6} = -9$

12. $-\frac{k}{5} = \frac{7}{5}$

13. $\frac{3}{4} = w + \frac{2}{5}$

14. $-\frac{1}{2} + a = \frac{5}{8}$

15. $-\frac{t}{7} = \frac{1}{15}$

16. $-\frac{5}{7} = y - 2$

17. $v + 914 = -23$

18. $447 + x = -261$

19. $-\frac{1}{7}c = 21$

20. $-\frac{2}{3}v = -22$

21. $\frac{3}{5}q = -15$

22. $\frac{n}{8} = -\frac{1}{4}$

23. $\frac{c}{4} = -\frac{9}{8}$

24. $\frac{2}{3} + r = -\frac{4}{9}$

25. $y - 7 = 8$

26. $w + 14 = -8$

27. $p - 4 = 6$

28. $-13 = 5 + x$

29. $98 = b + 34$

30. $y - 32 = -1$

31. $n + (-28) = 0$

32. $y + (-10) = 6$

33. $-1 = t + (-19)$

34. $j - (-17) = 36$

35. $14 = d + (-10)$

36. $u + (-5) = -15$

37. $11 = -16 + y$

38. $c - (-3) = 100$

39. $47 = w - (-8)$

40. $x - (-74) = -22$

41. $4 - (-h) = 68$

42. $-56 = 20 - (-j)$

43. $12z = 108$

44. $-7t = 49$

45. $18f = -216$

46. $-22 = 11v$

47. $-6d = -42$

48. $96 = -24a$

49. $\frac{c}{4} = 16$

50. $\frac{a}{16} = 9$

51. $-84 = \frac{d}{3}$

52. $-\frac{d}{7} = -13$

53. $\frac{t}{4} = -13$

54. $31 = -\frac{1}{6}n$

Examples 3 and 5

55. SUPREME COURT Chief Justice William Rehnquist served on the Supreme Court for 33 years until his death in 2005. Write and solve an equation to determine the year he was confirmed as a justice on the Supreme Court.

56. SALARY In a recent year, the annual salary of the Governor of New York was \$179, 000. During the same year, the annual salary of the Governor of Tennessee was \$94,000 less than that. Write and solve an equation to find the annual salary of the Governor of Tennessee in that year.

57. WEATHER On a cold January day, Kiara noticed that the temperature dropped 21 degrees over the course of the day to −9°C. Write and solve an equation to determine what the temperature was at the beginning of the day.

58. FARMING The Rolling Hills Farm is 126 acres. This is $\frac{1}{4}$ the size of the Briarwood Farm. Write and solve an equation to determine the number of acres of the Briarwood Farm.

59. SOCCER During the season, 13% of the players who signed up for the soccer league dropped out. A total of 174 players finished the season.

a. Assign a variable. Write an expression for the number of players who finished the season. Explain your reasoning.

b. Write an equation to find the number of players who signed up for the soccer league.

c. Solve the equation to find the number of players who signed up for the soccer league.

Mixed Exercises

Write an equation for each sentence. Then solve the equation.

60. Six times a number is 132.

61. Two thirds equals negative eight times a number.

62. Five elevenths times a number is 55.

63. Four fifths is equal to ten sixteenths of a number.

64. Three and two thirds times a number equals two ninths.

65. Four and four fifths times a number is one and one fifth.

Solve each equation. Check your solution.

66. $\frac{x}{9} = 10$

67. $\frac{b}{7} = -11$

68. $\frac{3}{4} = \frac{c}{24}$

69. $\frac{2}{3} = \frac{1}{8}y$

70. $\frac{2}{3}n = 14$

71. $\frac{3}{5}g = -6$

72. $4\frac{1}{5} = 3p$

73. $-5 = 3\frac{1}{2}x$

74. $6 = -\frac{1}{2}n$

75. $-\frac{2}{5} = -\frac{z}{45}$

76. $-\frac{g}{24} = \frac{5}{12}$

77. $-\frac{v}{5} = -45$

78. $-6 = \frac{2}{3}z$

79. $\frac{2}{7}q = -4$

80. $\frac{5}{9}p = -10$

81. $\frac{a}{10} = \frac{2}{5}$

82. $d - 8 = 6$

83. $-28 = p + 21$

84. $-7x = 63$

85. $-\frac{t}{5} = -8$

86. $y + (-16) = -12$

87. $\frac{3}{5}y = -9$

88. $-8d = -64$

89. $-\frac{3}{4}y = \frac{8}{20}$

90. VACATION The Lopez family is on vacation in Tennessee. They drove 210 miles from Memphis to Nashville and continued driving to Knoxville. By the time they reached Knoxville, they had travelled a total of 390 miles.

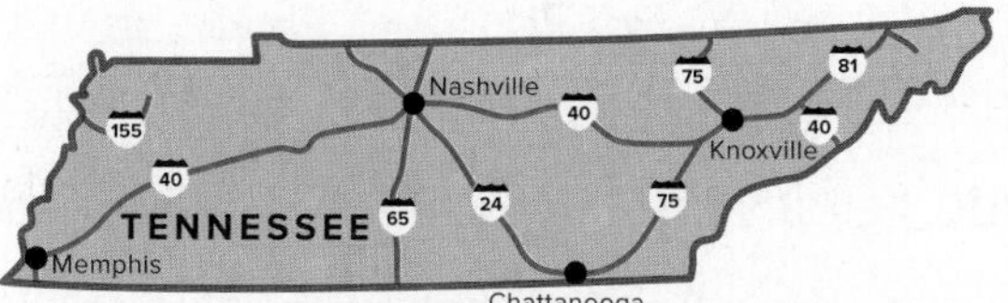

a. Define the variable and write an equation that represents the distance from Nashville to Knoxville.

b. Which property of equality could you use to isolate the variable in your equation? Explain your reasoning.

c. How far is Knoxville from Nashville? How can you verify that your solution is accurate?

d. If the Lopez family drives from Nashville to Chattanooga instead of Nashville to Knoxville, they will drive 47 fewer miles. Write an equation that represents the distance from Memphis to Chattanooga through Nashville. How far is Chattanooga from Nashville?

91. TICKETS Julian and Makayla order season tickets for the local soccer team. The ticket package they choose costs $780 and includes tickets to 12 games.

a. Write and solve an equation that represents the cost per game.

b. Single game tickets cost $85. How much do they save per game by using season tickets?

92. TACOS Orlando spent $18 at a taco truck. He ordered 4 tacos and a drink. If the drink cost $2, write and solve an equation to find the cost of each taco.

STRUCTURE Solve each equation. State the Property of Equality used.

93. $\frac{x}{9} = 24$

94. $m - 183 = -79$

95. $972 + y = 748$

96. $-\frac{4}{5}p = 32$

97. $135 = 9b$

98. $45 = \frac{3}{2}z$

Higher Order Thinking Skills

99. WHICH ONE DOESN'T BELONG Identify the equation that does not belong with the other three. Justify your conclusion.

$n + 14 = 27$	$12 + n = 25$	$n - 16 = 29$	$n - 4 = 9$

100. PERSEVERE Determine the value for each statement below.

a. If $x - 9 = 12$, what is the value of $x + 1$?

b. If $n + 7 = -4$, what is the value of $n + 1$?

101. CREATE Write an equation that you would use the Addition Property of Equality to solve.

102. ANALYZE Determine whether each sentence is *sometimes*, *always*, or *never* true. Justify your argument.

a. $x + x = x$

b. $x + 0 = x$

103. ANALYZE How would you solve $5x = 35$? How would you solve $5 + x = 35$? How are the methods similar and how are they different?

104. WRITE Consider the Multiplication Property of Equality and the Division Property of Equality. Explain why they can be considered the same property. Which one do you think is easier to use?

Solving Multi-Step Equations

Today's Goal
- Solve equations involving more than one operation.

Today's Vocabulary
multi-step equation

Explore Using Algebra Tiles to Model Multi-Step Equations

Online Activity Use algebra tiles to complete the Explore.

INQUIRY How can you model and solve a multi-step equation?

Learn Solving Multi-Step Equations

A **multi-step equation** is an equation that uses more than one property of equality to solve it. To solve this type of equation, you can undo each operation using properties of equality. Working backward in the order of operations makes this process simpler. Each step in this process results in equivalent equations.

Operation	Opposite Operation
Addition	Subtraction
Subtraction	Addition
Multiplication	Division
Division	Multiplication

Think About It!
In $4x - 2 = 5$, which two operations are being used? Which operations would you use to work backward in the order of operations to solve the equation?

Example 1 Solve Multi-Step Equations

Use properties of equality to solve each equation. Check your solutions.

a. $\mathbf{2a - 6 = 4}$

$2a - 6 = 4$	Original equation.
$2a - 6 + 6 = 4 + 6$	Add 6 to each side.
$2a = 10$	Simplify.
$\frac{2a}{2} = \frac{10}{2}$	Divide each side by 2.
$a = 5$	Simplify.

(continued on the next page)

CHECK

Check your solution by substituting the result back into the original equation.

$2a - 6 = 4$	Original equation
$2(5) - 6 \stackrel{?}{=} 4$	Substitute 5 for a.
$10 - 6 \stackrel{?}{=} 4$	Simplify.
$4 = 4$	True

b. $\frac{n+1}{-2} = 15$

$\frac{n+1}{-2} = 15$	Original equation.
$-2\left(\frac{n+1}{-2}\right) = -2(15)$	Multiply each side by −2.
$n + 1 = -30$	Simplify.
$n + 1 - 1 = -30 - 1$	Subtract 1 from each side.
$n = -31$	Simplify.

Check your solution by substituting the result back into the original equation.

Study Tip

Assumptions To solve an equation, you must assume that the original equation has a solution.

Check

Solve $3m + 4 = -11$.

$m =$ ___?___

Solve $8 = \frac{x-5}{7}$.

$x =$ ___?___

Go Online

You can watch a video to see how to use a graphing calculator with this example.

Example 2 Write and Solve a Multi-Step Equation

Study Tip

Multiplicative Inverse A number multiplied by its reciprocal is 1.

Talk About It!

Would it work to first multiply each side by $\frac{5}{2}$? Explain your reasoning.

FUNDRAISING **The student council raised $\frac{2}{5}$ of the money they need to cover the cost of the school dance with a bake sale. They raised an additional $150 selling raffle tickets. If the student council has raised $630, what is the cost of the dance? Write an equation for the problem. Then solve the equation.**

Jupiterimages/liquidlibrary/Getty Images

Go Online You can complete an Extra Example online.

Complete the table to write an equation for the cost of the dance.

Words	Two fifths of the cost	plus 150	is 630.
Variable	Let c = the cost of the dance.		
Equation	$\frac{2}{5} \cdot c$	$+ 150$	$= 630$

$$\frac{2}{5}c + 150 = 630$$ Original equation

$$\frac{2}{5}c + 150 - 150 = 630 - 150$$ Subtract 150 from each side.

$$\frac{2}{5}c = 480$$ Simplify.

$$\frac{5}{2}\left(\frac{2}{5}\right)c = \frac{5}{2}(480)$$ Multiply each side by $\frac{5}{2}$.

$$c = 1200$$ Simplify.

The dance costs \$ 1200.

Check

BASKETBALL A sporting goods store sold $\frac{2}{3}$ of its basketballs, but 8 were returned. Now the store has 38 basketballs. How many were there originally? Write an equation for the problem. Then solve the equation.

A. $\frac{2}{3}b + 8 = 38$; 45

B. $\frac{2}{3}b - 8 = 38$; 69

C. $\frac{1}{3}b + 8 = 38$; 90

D. $\frac{1}{3}b - 8 = 38$; 138

Example 3 Solve Multi-Step Equations with Letter Coefficients

Some equations have coefficients that are represented by letters. To solve these equations, apply the process of solving equations to isolate the variable.

 **Think About It!**

Why do you have to assume that $a \neq 0$ when solving the equation?

Solve $ax + 7 = 5$ for x. Assume that $a \neq 0$.

$$ax + 7 = 5$$ Original equation

$$ax + 7 - 7 = 5 - 7$$ Subtract 7 from each side.

$$ax = -2$$ Simplify.

$$\frac{ax}{a} = \frac{-2}{a}$$ Divide each side by a.

$$x = \frac{-2}{a}$$ Simplify.

Go Online You can complete an Extra Example online.

Watch Out!

Isolate the Variable
Make sure that you are isolating x in the equation. Remember that a represents a coefficient, not the variable.

Check

Solve $2 - ax = -8$ for x. Assume $a \neq 0$.

A. $x = -\frac{10}{a}$

B. $x = -\frac{6}{a}$

C. $x = \frac{6}{a}$

D. $x = \frac{10}{a}$

Pause and Reflect

Did you struggle with anything in this lesson? If so, how did you deal with it?

Practice

Go Online You can complete your homework online.

Examples 1

Use properties of equality to solve each equation. Check your solution.

1. $3t + 7 = -8$

2. $8 = 16 + 8n$

3. $-34 = 6m - 4$

4. $9x + 27 = -72$

5. $\frac{y}{5} - 6 = 8$

6. $\frac{f}{-7} - 8 = 2$

7. $1 + \frac{r}{9} = 4$

8. $\frac{k}{3} + 4 = -16$

9. $\frac{n - 2}{7} = 2$

10. $14 = \frac{6 + z}{-2}$

11. $-11 = \frac{a - 5}{6}$

12. $\frac{22 - w}{3} = -7$

Example 2

13. SHOPPING Ricardo spent half of his allowance on school supplies. Then he bought a snack for \$5.25. When he arrived home, he had \$22.50 left. Write and solve an equation to find the amount of Ricardo's allowance a.

14. SHOPPING Liza earned some money by taking care of her neighbor's pet. She bought a drink for \$1.95, and a concert ticket for \$30. She bought a ring for \$7.20, and then spent two-thirds of the remaining money on a wireless speaker. If Liza has \$38.50 left, write and solve an equation to find the amount of money m Lisa earned by taking care of her neighbor's pet.

15. PET SHELTERS Henry works at a pet shelter after school. He purchases a large package of dog treats. He sets aside 10 treats and distributes the rest equally among the 15 dogs in the shelter. If each dog received 4 treats, write and solve an equation to find the number of treats t that were in the original package.

16. BASKETBALL The average number of points a basketball team scored for three games was 63 points. In the first two games, they scored the same number of points, which was 6 points more than they scored in the third game. Write and solve an equation to find the number of points the team scored in each game.

17. HUMAN HEIGHT Micah's adult height is one less than twice his height at age 2. Micah's adult height is 71 inches. Write and solve an equation to find Micah's height h at age 2.

Example 3

Solve each equation for x. Assume $a \neq 0$.

18. $ax + 3 = 23$

19. $4 = ax - 14$

20. $ax - 5 = 19$

21. $6 + ax = -29$

22. $\frac{8}{ax} - 5 = -3$

23. $18 - ax = 42$

24. $5 = \frac{5}{ax} + 1$

25. $-3 = ax + 11$

26. $-7 = -ax - 16$

Mixed Exercises

Solve each equation. Check your solution.

27. $3x + 8 = 29$

28. $\frac{a}{6} - 5 = 9$

29. $\frac{5r}{2} - 6 = 19$

30. $\frac{n}{3} - 8 = -2$

31. $5 + \frac{x}{4} = 1$

32. $-\frac{h}{3} - 4 = 13$

33. $5(1 + n) = -5$

34. $-27 = -6 - 3p$

35. $-\frac{a}{6} + 5 = 2$

REASONING **Write and solve an equation to find each number.**

36. A number is divided by 2, and then the quotient is increased by 8. The result is 33.

37. Two is subtracted from a number, and then the difference is divided by 3. The result is 30.

Higher Order Thinking Skills

38. PERSEVERE The sum of 4 consecutive odd integers is equal to zero.

a. Write an equation to model the sentence.

b. Solve the equation to find the numbers. Check your solution.

39. FIND THE ERROR Kadija and Jorge are solving $\frac{1}{2}n + 5 = \frac{17}{2}$. Jorge uses the Subtraction Property of Equality followed by the Multiplication Property of Equality. Kadija also uses the Subtraction Property of Equality, but because n is multiplied by $\frac{1}{2}$, Kadija claims that the Division Property of Equality can be used to isolate the variable. Which student is correct? Explain your reasoning.

40. CREATE Write a problem that can be represented by the equation $11.9p + 23.1 = 273$. Define the variable and solve the equation.

41. ANALYZE Solve each equation for x. Assume that $a \neq 0$.

a. $ax + 7 = 5$ **b.** $\frac{1}{a}x - 4 = 9$ **c.** $2 - ax = -8$

42. ANALYZE Determine whether each equation has a solution. Justify your answer.

a. $\frac{a + 4}{5 + a} = 5$ **b.** $\frac{1 + b}{1 - b} = 1$ **c.** $\frac{c - 5}{5 - c} = 1$

43. ANALYZE Determine whether the following statement is *sometimes*, *always*, or *never* true. Justify your argument.

The sum of three consecutive odd integers equals an even integer.

44. WRITE Write a paragraph explaining the order of the steps that you would take to solve a multi-step equation.

Lesson 2-4

Solving Equations with the Variable on Each Side

Explore Modeling Equations with the Variable on Each Side

 **Online Activity** Use graphing technology to complete the Explore.

INQUIRY How can you solve an equation with the variable on each side?

Today's Goals

- Solve equations with the variable on each side.
- Solve equations by applying the Distributive Property.
- Prove that equations are identities or have no solution.

Today's Vocabulary

identity

Learn Solving Equations with the Variable on Each Side

Sometimes, the variable will appear on each side of an equation. To solve these equations, use the Addition or Subtraction Property of Equality to write an equivalent equation with the variable terms on one side and the numbers without variables, or constants, on the other side.

Example 1 Solve an Equation with the Variable on Each Side

Solve $5 + 7a = 4a - 13$. Check your solution.

$5 + 7a = 4a - 13$	Original equation
$-4a = -4a$	Subtract $4a$ from each side.
$5 + 3a = -13$	Simplify.
$-5 \quad = -5$	Subtract 5 from each side.
$3a = -18$	Simplify.
$\frac{3a}{3} = \frac{-18}{3}$	Divide each side by 3.
$a = -6$	Simplify.

CHECK $5 + 7a = 4a - 13$	Original equation
$5 + 7(-6) \stackrel{?}{=} 4(-6) - 13$	Substitution, $a = -6$
$5 + -42 \stackrel{?}{=} -24 - 13$	Multiply.
$-37 = -37$	True

Go Online You can complete an Extra Example online.

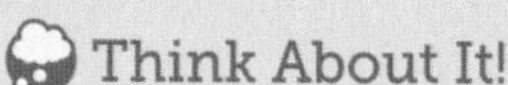

Think About It!

Leon says that when you solve $5 + 7a = 4a - 13$, you can just combine $7a$ and $4a$ because they are like terms. Explain whether Leon is correct.

Example 2 Write an Equation with the Variable on Each Side

Study Tip

Solving an Equation You may want to combine the terms with a variable on one side before isolating a constant.

CONTEST **The results of the 2015 Nathan's Hot Dog Eating Contest are shown.**

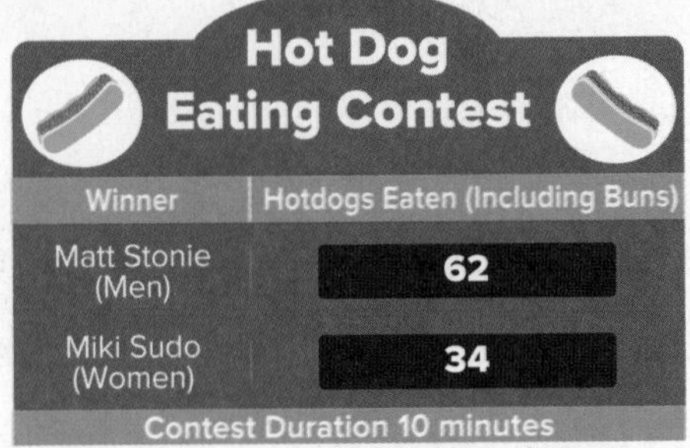

Hot Dog
Eating Contest
Winner
Hotdogs Eaten (Including Buns)
Matt Stonie (Men)
62
Miki Sudo (Women)
34
Contest Duration 10 minutes

Suppose the men's and women's winners, Matt and Miki, decide to compete against each other. To make the competition more interesting, Matt will not start until Miki has eaten 20 hot dogs. Assume that Matt and Miki eat at a constant rate throughout the competition. Based on the number of hot dogs eaten in 10 minutes by Matt and Miki, how many minutes after Matt starts eating will they have eaten the same number of hot dogs?

Read the Problem

We want to find the number of minutes m for which Matt and Miki have eaten the same number of hot dogs.

Matt eats 62 hot dogs in 10 minutes, which is a rate of 6.2 hot dogs per minute. The number of hot dogs he has eaten m minutes after starting is $6.2m$.

Miki eats 38 hot dogs in 10 minutes, which is a rate of 3.8 hot dogs per minute. She is given a 20-hot dog head start, so the number of hot dogs she has eaten m minutes after Matt starts is $3.8m + 20$.

The equation $6.2m = 3.8m + 20$ represents this situation.

Solve the Problem

$6.2m = 3.8m + 20$	Original equation
$6.2m - 3.8m = 3.8m + 20 - 3.8m$	Subtract $3.8m$ from each side.
$2.4m = 20$	Simplify.
$\frac{2.4m}{2.4} = \frac{20}{2.4}$	Divide each side by 2.4.
$m \approx 8.3$	Simplify.

After approximately $8\frac{1}{3}$ minutes, Matt and Miki will have eaten the same number of hot dogs.

Go Online You can complete an Extra Example online.

Check

BASKETBALL Nolan and Victor were two of the top scoring freshman players in a college basketball conference last season. The table shows how many points Nolan and Victor scored and how many games they played last season. The points Nolan and Victor score this season will be combined with their points from last season to give their total career points. This season, Nolan is hoping to catch up to Victor and have the same number of career points. Assume that Nolan and Victor play every game and score at the same constant rate as last season.

Player	Games Played	Points
Nolan	30	750
Victor	34	782

Part A

Based on each player's average scoring rate, write an equation that represents the number of games it will take Nolan to accumulate the same number of career points as Victor.

$$25p + \underline{\ ?\ } = \underline{\ ?\ }\, p + \underline{\ ?\ }$$

Part B

Based on your equation in Part A, after how many games this season will Nolan and Victor have scored the same number of career points?

___?___ games

Learn Solving Equations Involving the Distributive Property

Some equations contain grouping symbols. Grouping symbols can include parentheses ( ), brackets [ ], and fraction bars.

The steps for solving an equation can be summarized as follows.

Step 1 Simplify the expressions on each side. Remove any grouping symbols. Use the Distributive Property as needed.

Step 2 Use the Addition and/or Subtraction Properties of Equality to get the variable terms on one side of the equation and the constant terms on the other side. Simplify.

Step 3 Use the Multiplication and Division Properties of Equality to solve.

Go Online You can complete an Extra Example online.

Think About It!

Describe the steps you would take to solve $2(1 + t) = 8t$.

Study Tip

Grouping Symbols Some expressions, like $2 - [11 + 5(p - 8)]$, contain grouping symbols inside of grouping symbols. To simplify these expressions, work from the inside out by first simplifying the expression within the innermost grouping symbol.

Example 3 Solve an Equation with Grouping Symbols

Solve $7(n-1) = -2(3+n)$.

$7(n-1) = -2(3+n)$	Original equation
$7n - 7 = -6 - 2n$	Distributive Property
$7n - 7 + 2n = -6 - 2n + 2n$	Add $2n$ to each side.
$9n - 7 = -6$	Simplify.
$9n - 7 + 7 = -6 + 7$	Add 7 to each side.
$9n = 1$	Simplify.
$\frac{9n}{9} = \frac{1}{9}$	Divide each side by 9.
$n = \frac{1}{9}$	Simplify.

Check

Solve $7(n-2) + 8 = 3(n-4) - 2$.

$n =$ ___?___

Example 4 Solve an Equation with a Fraction Bar

Solve $5y = \frac{12y+16}{4}$.

$5y = \frac{12y+16}{4}$	Original equation
$4(5y) = 4\left(\frac{12y+16}{4}\right)$	Multiply each side by 4.
$20y = 12y + 16$	Simplify.
$20y - 12y = 12y + 16 - 12y$	Subtract $12y$ from each side.
$8y = 16$	Simplify.
$\frac{8y}{8} = \frac{16}{8}$	Divide each side by 8.
$y = 2$	Simplify.

Go Online You can complete an Extra Example online.

Watch Out!

Keep the Negative
Do not forget to distribute −2 to each term on the right side of the equation.

Math History Minute

During the short life of Indian mathematician **Srinivasa Ramanujan (1887–1920)**, he compiled nearly 3900 results, which included proofs of theorems, equations, and identities, nearly all of which have been proven correct. Ramanujan was known as a genius and an *autodidact*, which is a person who is self-taught.

Example 5 Write an Equation with Grouping Symbols

GEOMETRY **Find the value of x so that the figures have the same area.**

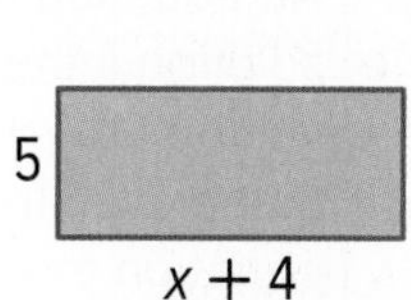

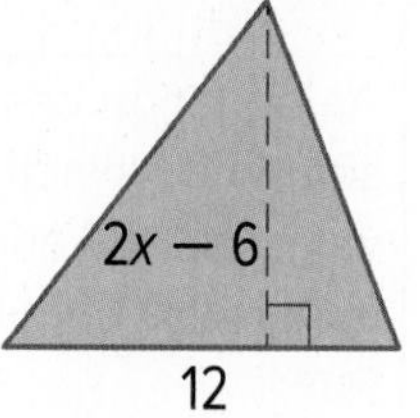

The area of the rectangle is $5(x + 4)$, and the area of the triangle is $\frac{1}{2}(12)(2x - 6)$. The equation $5(x + 4) = \frac{1}{2}(12)(2x - 6)$ represents the situation where the areas of the figures are the same.

$5(x + 4) = \frac{1}{2}(12)(2x - 6)$	Original equation
$5(x + 4) = 6(2x - 6)$	Multiply $\frac{1}{2}$ and 12.
$5x + 20 = 12x - 36$	Distributive Property
$5x + 20 + 36 = 12x - 36 + 36$	Add 36 to each side.
$5x + 56 = 12x$	Simplify.
$5x + 56 - 5x = 12x - 5x$	Subtract $5x$ from each side.
$56 = 7x$	Simplify.
$\frac{56}{7} = \frac{7x}{7}$	Divide each side by 7.
$8 = x$	Simplify.

Check

GEOMETRY Find the value of x so that the figures have the same area.

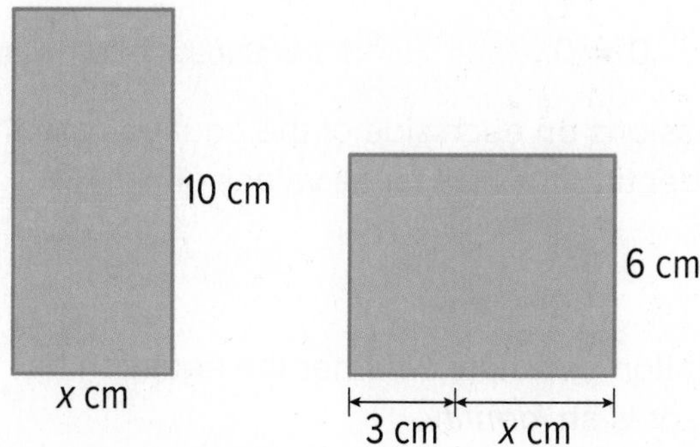

$x =$ ___?___

Go Online You can complete an Extra Example online.

Go Online to practice what you've learned about solving linear equations in the Put It All Together over Lessons 2-1 through 2-4.

Learn Identities and Equations with No Solutions

One solution	No solution	Identity
Words		
An equation has one solution if exactly one value of the variable makes the equation true.	An equation has no solution if there is no value of the variable that makes the equation true.	An **identity** is an equation that is true for all values of its variables.

Example 6 Solve an Equation with No Solution

Solve $6(y - 5) = 2(10 + 3y)$.

$6(y - 5) = 2(10 + 3y)$	Original equation
$6y - 30 = 20 + 6y$	Distributive Property
$6y - 30 - 6y = 20 + 6y - 6y$	Subtract $6y$ from each side.
$-30 \neq 20$	Simplify.

Since $-30 \neq 20$, this equation has no solution.

Example 7 Solve an Identity

Solve $7x + 5(x - 1) = 12x - 5$.

$7x + 5(x - 1) = 12x - 5$	Original equation
$7x + 5x - 5 = 12x - 5$	Distributive Property
$12x - 5 = 12x - 5$	Simplify.
$0 = 0$	Subtract $12x - 5$ from each side.

Since the expressions on each side of the equation are the same, this equation is an identity. It is true for all values of x.

Talk About It!
Could you tell that the equation was an identity before the final step? Explain your reasoning.

Check

Solve each equation and state whether the equation has *one solution*, has *no solution*, or is an *identity*.

A. $8(g + 6) = 5g + 3(g + 16)$

B. $5x + 5 = 3(5x - 4) - 10x$

C. $3w + 2 = 7w$

D. $3(2b - 1) - 7 = 6b - 10$

Go Online
to practice what you've learned in Lessons 2-1 through 2-4.

Go Online You can complete an Extra Example online.

Practice

Go Online You can complete your homework online.

Examples 1, 3, and 4

Solve each equation. Check your solution.

1. $7c + 12 = -4c + 78$

2. $2m - 13 = -8m + 27$

3. $9x - 4 = 2x + 3$

4. $6 + 3t = 8t - 14$

5. $\frac{b-4}{6} = \frac{b}{2}$

6. $\frac{3v+12}{6} = \frac{4v}{3}$

7. $2(r + 6) = 4(r + 4)$

8. $6(n + 5) = 3(n + 16)$

9. $5(g + 8) - 7 = 117 - g$

10. $12 - \frac{4}{5}(x + 15) = \left(\frac{2}{5}x + 6\right)$

11. $3(3m - 2) = 2(3m + 3)$

12. $6(3a + 1) - 30 = 3(2a - 4)$

13. $7n + 6 = 4n - 9$

14. $-6(2r + 8) = -10(r - 3)$

15. $5 - 3(w + 4) = w - 7$

16. $2x - 5(x - 3) = 2(x - 10)$

Example 2

17. OLYMPICS In the 2010 Winter Olympic Games, the United States won 1 more than 4 times the number of gold medals France won. The United States won 7 more gold metals than France. Write and solve an equation to find the number of gold medals each country won.

18. REASONING Diego's sister is twice his age minus 9 years. She is also as old as half the sum of the ages of Diego and both of his 12-year-old twin brothers. Write and solve an equation to find the ages of Diego and his sister.

19. NATURE The table shows the current heights and average growth rates of two different species of trees. Write and solve an equation to find how long it will take for the two trees to be the same height.

Tree Species	Current Height	Annual growth
A	38 inches	4 inches
B	45.5 inches	2.5 inches

20. WEIGHT A dog weighs two pounds less than three times the weight of a cat. The dog also weighs twenty-two more pounds than the cat. Write and solve an equation to find the weights of the dog and the cat.

Example 5

21. GEOMETRY Supplementary angles are two angles with measures that have a sum of 180°. Complementary angles are two angles with measures that have a sum of 90°. The measure of the supplement of an angle is 10° more than twice the measure of the complement of the angle. Let $90 - x$ equal the degree measure of the complement angle and $180 - x$ equal the degree measure of the supplement angle. Write and solve an equation to find the measure of the angle.

22. GEOMETRY Write and solve an equation to find the value of x so that the figures have the same area.

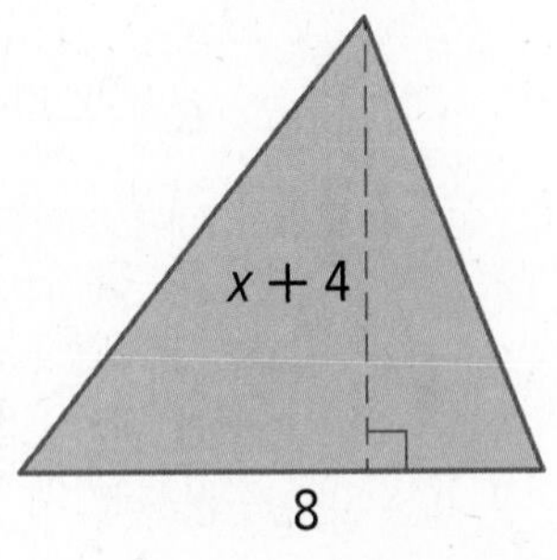

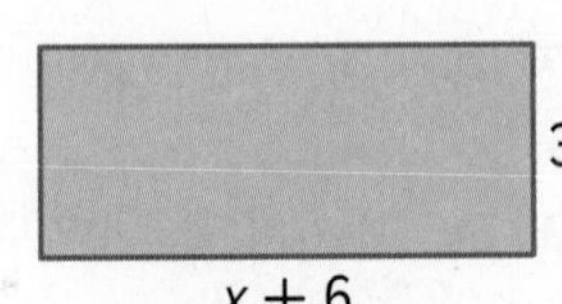

23. GEOMETRY Write and solve an equation to find the value of x so that the figures have the same area.

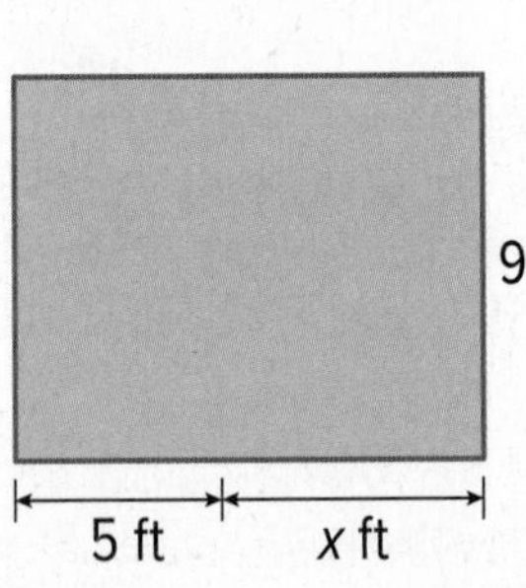

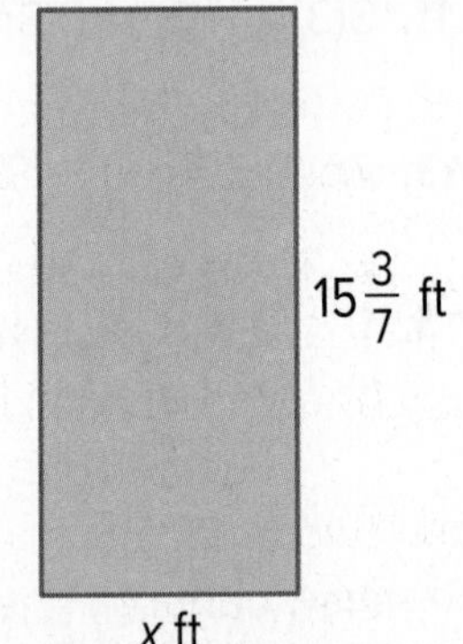

24. GEOMETRY Write and solve an equation to find the value of x so that the figures have the same area. The area of a trapezoid is $\frac{1}{2}h(b_1 + b_2)$.

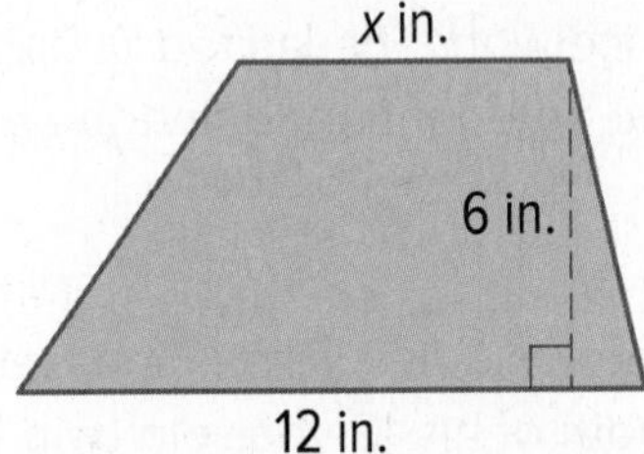

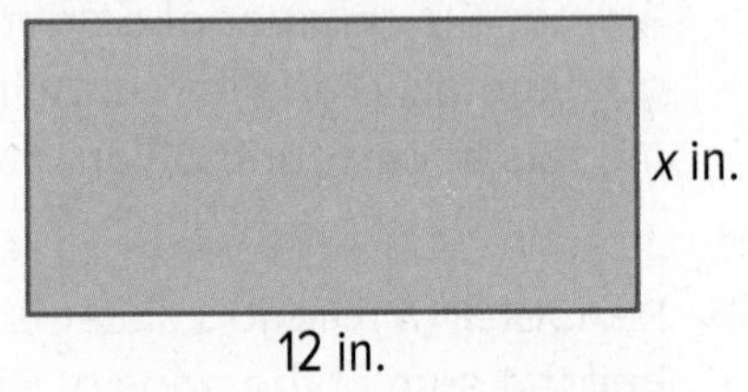

Examples 6 and 7

Solve each equation and state whether the equation has *one solution*, *no solution*, or is an *identity*.

25. $-6y - 3 = 3 - 6y$

26. $\frac{1}{2}(x + 6) = \frac{1}{2}x - 9$

27. $8q + 12 = 4(3 + 2q)$

28. $21(x + 1) - 6x = 15x + 21$

29. $12y + 48 - 4y = 8(y - 6)$

30. $8(z + 6) = 4(2z + 12)$

31. $2a + 2 = 3(a + 2)$

32. $\frac{1}{4}x + 5 = \frac{1}{4}x$

33. $7(c + 9) = 7c + 63$

34. $4k + 3 = \frac{1}{4}(8k + 16)$

35. $3b - 13 + 4b = 7b + 1$

36. $\frac{1}{2}(\frac{1}{2}m - 8) = \frac{1}{4}(m - 16)$

Mixed Exercises

Solve each equation. Check your solution.

37. $2x = 2(x - 3)$

38. $\frac{2}{5}h - 7 = \frac{12}{5}h - 2h + 3$

39. $-5(3 - q) + 4 = 5q - 11$

40. $2(4r + 6) = \frac{2}{3}(12r + 18)$

41. $\frac{3}{5}f + 24 = 4 - \frac{1}{5}f$

42. $\frac{1}{12} + \frac{3}{8}y = \frac{5}{12} + \frac{5}{8}y$

43. $6.78j - 5.2 = 4.33j + 2.15$

44. $14.2t - 25.2 = 3.8t + 26.8$

45. $3.2k - 4.3 = 12.6k + 14.5$

46. $5[2p - 4(p + 5)] = 25$

47. $m - 9 = \frac{2m - 12}{3}$

48. $\frac{3d - 2}{8} = -d + 16\frac{1}{4}$

49. Twice the greater integer of two consecutive odd integers is 13 less than three times the lesser integer.

a. Write an equation to find the two consecutive odd integers.

b. What are the integers in ascending order?

50. Two times the quantity of eight times a number plus two is equal to three times the quantity of two times the same number minus seven.

a. Write an equation to find the number.

b. Solve the equation to find the number.

51. USE A MODEL The perimeter of Figure 1 is four times a number minus three. The perimeter of Figure 2 is two times the same number plus five. The perimeters of Figure 1 and Figure 2 are the same.

a. Write an equation to find the number.

b. Solve the equation to find the number.

c. What is the perimeter of Figure 2? Explain.

d. Find the perimeter of Figure 1. Compare the perimeter of Figure 1 to the perimeter you found for Figure 2 to justify the value of k is correct. Show your work.

52. STRUCTURE Find two consecutive even integers such that twice the lesser of two integers is 4 less than two times the greater integer.

a. Write and solve an equation to find the integers.

b. Does the equation have one solution, no solution, or is it an identity? Explain.

Higher Order Thinking Skills

53. FIND THE ERROR Anthony and Patty are solving the equation $y - m = m - y + 1$ for y. Is either correct? Explain why or why not.

Anthony	Patty
$y - m = m - y + 1$	$y - m = m - y + 1$
$2y - m = m + 1$	$2y - m = m + 1$
$2y = 2m + 1$	$2y = 1$
$y = m + \frac{1}{2}$	$y = \frac{1}{2}$

54. PERSEVERE Write an equation with variables on each side of the equal sign, at least one fractional coefficient, and a solution of −6. Discuss the steps you used.

55. CREATE Create an equation with at least two grouping symbols for which there is no solution.

56. WRITE Compare and contrast solving equations with variables on both sides of the equation to solving one-step or multi-step equations with a variable on one side of the equation.

57. ANALYZE Determine whether each solution is correct. If it is incorrect; find the correct solution. Justify your argument.

a.
$$\begin{aligned} 2(g + 5) &= 22 \\ 2g + 5 &= 22 \\ 2g + 5 - 5 &= 22 \\ 2g &= 17 \\ g &= 8.5 \end{aligned}$$

b.
$$\begin{aligned} 5d &= 2d - 18 \\ 5d - 2d &= 2d - 18 - 2d \\ 3d &= -18 \\ d &= -6 \end{aligned}$$

c.
$$\begin{aligned} -6z + 13 &= 7z \\ -6z + 13 - 6z &= 7z - 6z \\ 13 &= z \end{aligned}$$

58. PERSEVERE Find the value of k for which each equation is an identity.

a. $k(3x - 2) = 4 - 6x$ **b.** $15y - 10 + k = 2(ky - 1) - y$

59. CREATE Write an equivalent equation to $x = 8$ that has the variable x on both sides.

60. ANALYZE Solve $5x + 2 = ax - 1$ for x. Assume $a \neq 5$. Describe each step.

Lesson 2-5

Solving Equations Involving Absolute Value

Explore Modeling Absolute Value

Online Activity Use an infographic to complete the Explore.

INQUIRY How is margin of error related to absolute value?

Learn Solving Equations Involving Absolute Value

Absolute value equations contain at least one absolute value expression. The simplest form of an absolute value equation is $|x| = n$. Since absolute value represents distance, you must consider the case where the solution is x units from zero in the negative direction and the case where the solution is x units from zero in the positive direction.

Key Concept • Solving Absolute Value Equations	
Words	When solving equations that involve absolute values, there are two cases to consider. **Case 1** The expression inside the absolute value symbol is positive or zero. **Case 2** The expression inside the absolute value symbol is negative.
Symbols	For any real numbers a and b, if $\|a\| = b$ and $b \geq 0$, then $a = b$ or $a = -b$.
Examples	$\|d\| = 3$, so $d = 3$ or $d = -3$.

Consider the equation $|x| = 3$. This means that the two points on the number line where the distance between 0 and x is 3 are solutions to the equation. The distance between 0 and -3 is 3, so -3 is a solution to the equation. The distance between 0 and 3 is 3, so 3 is also a solution to the equation.

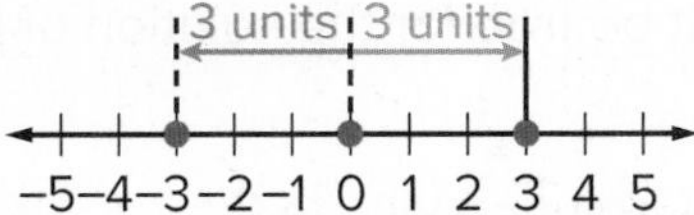

If $|x| = 3$, then $x = -3$ or $x = 3$. Thus the solution set is $\{-3, 3\}$. You can graph the solution set by graphing each solution on the number line.

For each absolute value equation, you must consider both cases. To solve an absolute value equation, first isolate the absolute value on one side of the equals sign if it is not already by itself.

Today's Goal
- Solve absolute value expressions.

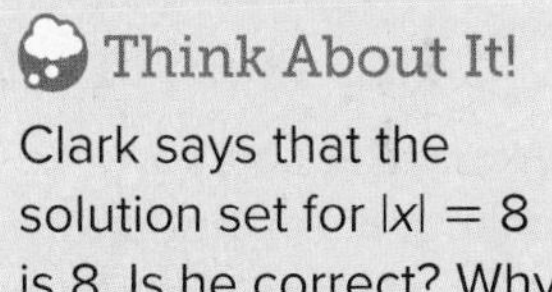

Think About It!

Clark says that the solution set for $|x| = 8$ is 8. Is he correct? Why or why not?

Example 1 Solve an Absolute Value Equation When $n > 0$

Solve $|y + 2| = 4$. Then graph the solution set.

Case 1

If y is nonnegative, then $|y| = y$.

$y + 2 = 4$ Original equation

$y + 2 - 2 = 4 - 2$ Subtract 2 from each side.

$y = 2$ Simplify.

Case 2

If y is negative, then $|y| = -y$.

$y + 2 = -4$

$y + 2 - 2 = -4 - 2$

$y = -6$

The solution set is $\{-6, 2\}$.

Graph points at -6 and 2 on the number line.

CHECK

Substitute -6 and 2 into the original equation.

$\lvert y + 2\rvert = 4$	Original equation	$\lvert y + 2\rvert = 4$
$\lvert -6 + 2\rvert \stackrel{?}{=} 4$	Substitute.	$\lvert 2 + 2\rvert \stackrel{?}{=} 4$
$\lvert -4\rvert \stackrel{?}{=} 4$	Simplify.	$\lvert 4\rvert \stackrel{?}{=} 4$
$4 = 4$ ✓	Take the absolute value.	$4 = 4$ ✓

Study Tip

Scale Keep in mind the values that you will need for graphing. If the values are very large, use a number line and scale that are reasonable for the situation.

Check

Graph the solution set of $|2t - 4| = 8$.

−5 −4 −3 −2 −1 0 1 2 3 4 5 6 7 8 9 10 11 12 13 14 15

Example 2 Solve an Absolute Value Equation When $n < 0$

Solve $|3x - 4| = -1$.

$|3x - 4| = -1$ means that the distance between $3x$ and 4 is -1. Since distance cannot be negative, the solution is the empty set $\varnothing$. The solution set is $\varnothing$.

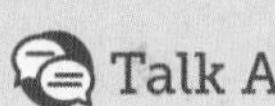

Talk About It!

Would the solution change if the equation were changed to $|ax - 4| = -n$, where $n > 0$? Explain your reasoning.

Check

Which statement must be true for the solution of $|ax + b| = c$ to be $\varnothing$?

A. If a is negative, the solution will be $\varnothing$.

B. If b is negative, the solution will be $\varnothing$.

C. If c is negative, the solution will be $\varnothing$.

D. If c is positive, the solution will be $\varnothing$.

Go Online You can complete an Extra Example online.

Example 3 Solve an Absolute Value Equation

MUSIC **Depending on the size of each song, Luna's phone holds an average of 2000 songs, give or take 250 songs. Write and solve an equation involving absolute value to find the maximum and minimum number of songs that Luna's phone can hold.**

The maximum and minimum number of songs will differ from the average by 250 songs. Complete the table to write an equation that represents the maximum and minimum number of songs.

Words	The difference between the number of songs and 2000 is 250.
Variable	Let x = the number of songs on Luna's phone.
Equation	$\|x - 2000\| = 250$

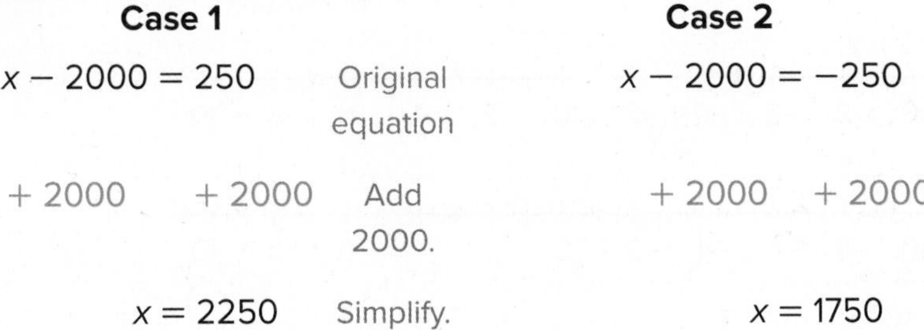

Case 1		Case 2
$x - 2000 = 250$	Original equation	$x - 2000 = -250$
$+2000 \quad +2000$	Add 2000.	$+2000 \quad +2000$
$x = 2250$	Simplify.	$x = 1750$

The solution set is {1750, 2250}. The maximum and minimum number of songs are 2250 and 1750, respectively.

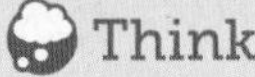

Think About It!

Does the solution set mean that Luna's phone can hold only 1750 or 2250 songs? If not, how many songs can it hold?

Check

SKYDIVING It takes approximately 6 minutes for a skydiver to land after she jumps out of a plane, give or take 30 seconds. What is the range of time, in seconds, it could take the skydiver to land?

[? , ?]

Go Online You can complete an Extra Example online.

Example 4 Write an Absolute Value Equation

Write an equation involving absolute value for the graph.

16 17 18 19 20 21 22 23 24 25 26 27 28

Find the point that is the same distance from 17 and from 27 on the number line. This is the midpoint between 17 and 27, which is 22.

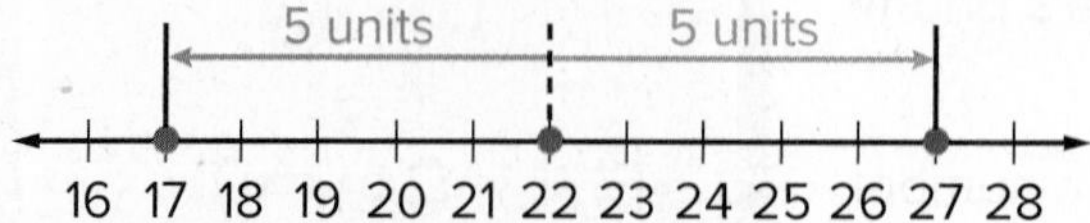

So an equation is $|x - 22| = 5$.

Check

Label each graph with the correct equation.

$|x + 3| = 6$ $|x - 1| = 3$ $|x - 1| = 6$ $|x - 3| = 5$

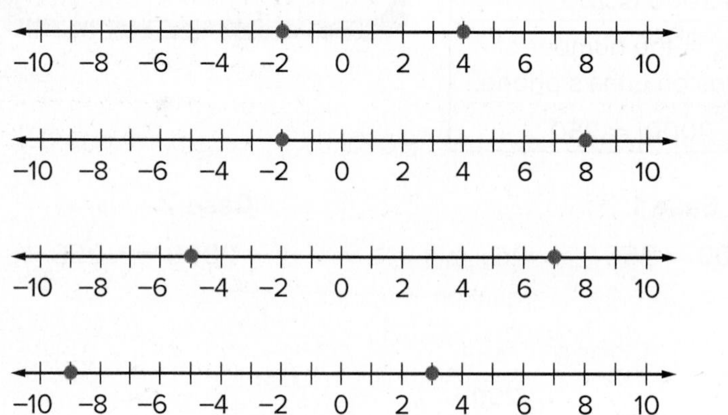

Study Tip

Find the Midpoint To find the point midway between two points, add the values together and divide by 2. For the example, $17 + 27 = 44$, and $44 \div 2 = 22$. So 22 is the point halfway between 17 and 27.

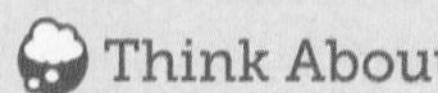

Think About It!

Write a general rule that can be used for exercises similar to Example 4.

Pause and Reflect

Did you struggle with anything in this lesson? If so, how did you deal with it?

Go Online You can complete an Extra Example online.

Practice

Go Online You can complete your homework online.

Examples 1 and 2

Solve each equation. Then graph the solution set.

1. $|n - 3| = 5$

-3 -2 -1 0 1 2 3 4 5 6 7 8 9

2. $|f + 10| = 1$

3. $|v - 2| = -5$

4. $|4t - 8| = 20$

5. $|8w + 5| = 21$

6. $|6y - 7| = -1$

7. $|x + 5| = -3$

8. $|-2y + 6| = 6$

9. $\left|\frac{3}{4}a - 3\right| = 9$

10. $|2x - 3| = 7$

Solve each equation.

11. $|7 - 2q| = 3$

12. $|4x - 2| = 26$

13. $|w + 1| = 5$

14. $|n + 2| = -1$

15. $|m - 2| = 2$

16. $|5c - 3| = 1$

17. $|2t + 6| = 4$

18. $|8k - 5| = -4$

Example 3

19. ENGINEERING *Tolerance* is an allowance made for imperfections in a manufactured object. The manufacturer of an oven specifies a temperature tolerance of ±15°F. This means that the temperature inside the oven will be within 15°F of the temperature to which it is set. Write and solve an absolute value equation to find the maximum and minimum temperatures inside the oven when the thermostat is set to 400°F.

20. POLLS Candidate A and Candidate B are running for mayor. A poll was taken to determine which candidate would likely win the election. The poll is accurate within ±5%. Write and solve an absolute value equation to find the maximum and minimum percent of voters who will vote for Candidate A if 38% of the voters in the poll voted for Candidate A.

21. STATISTICS The most familiar statistical measure is the arithmetic mean, or average. A second important statistical measure is the standard deviation, which is a measure of how far the data are from the mean. For example, the mean score on the Wechsler IQ test is 100 and the standard deviation is 15. This means that people within one standard deviation of the mean have IQ scores that are 15 points higher or lower than the mean.

   a. One year, the mean mathematics score on the ACT test was 20.9 with a standard deviation of 5.3. Write an absolute value equation to find the maximum and minimum scores within one standard deviation of the mean.

   b. What is the range of ACT mathematics scores within one standard deviation of the mean? within two standard deviations of the mean?

22. AVIATION The graph shows the results of a survey that asked 4300 students ages 7 to 18 what they thought would be the most important benefit of air travel in the future. There are about 40 million students in the United States. If the margin of error is ±3%, what is the range of the number of students ages 7 to 18 who would likely say that "finding new resources for Earth" is the most important benefit of future flight?

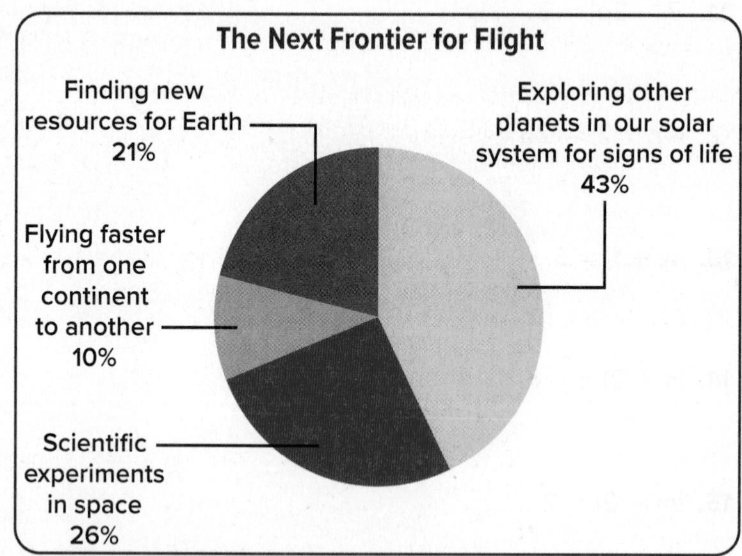

Source: *The World Almanac*

23. MANUFACTURING A hardware store sells bags of rock salt that are labeled as weighing 35 pounds. The equipment used to package the salt produces bags with a weight that is within 8 ounces of the label weight. Write and solve an absolute value equation to determine the maximum and minimum weights for the bag of rock salt. Justify each step in the solution.

Example 4

Write an equation involving absolute value for each graph.

24. 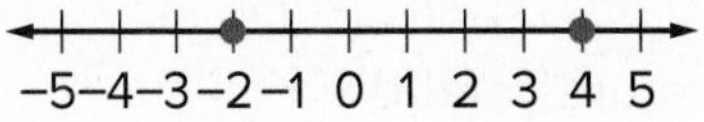

25. 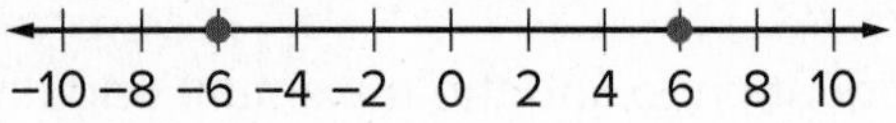

26. 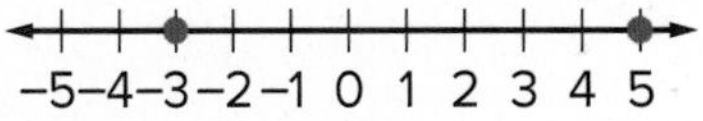

27. 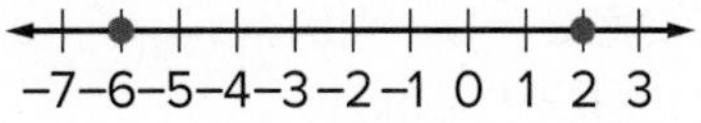

28. 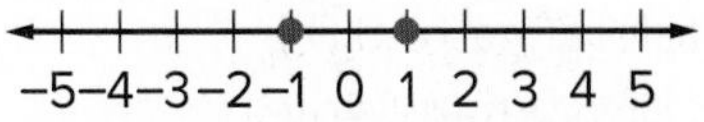

29. 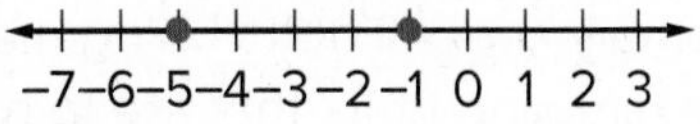

30. 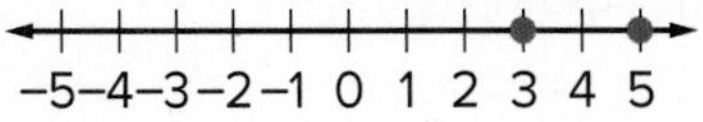

31. 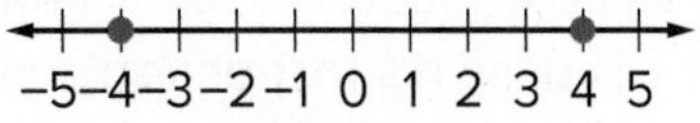

Mixed Exercises

Solve each equation. Then graph the solution set.

32. $\left|-\frac{1}{2}b - 2\right| = 10$

33. $|-4d + 6| = 12$

34. $|5f - 3| = 12$

35. $2|h| - 3 = 8$

36. $4 - 3|q| = 10$

37. $\frac{4}{|p|} + 12 = 14$

Write an equation involving absolute value for each graph.

38. 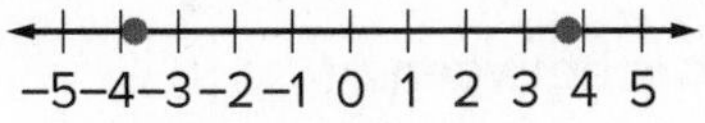

39. 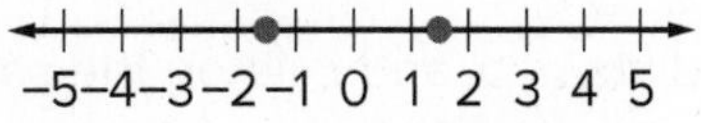

40. 

41. 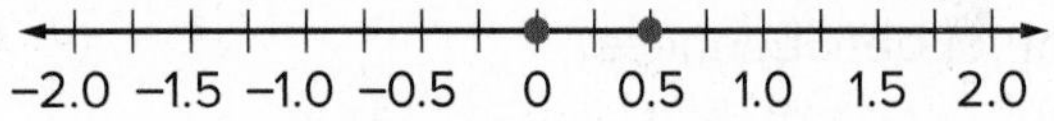

42. 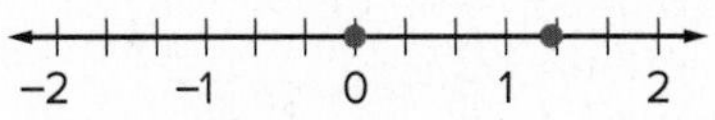

43. 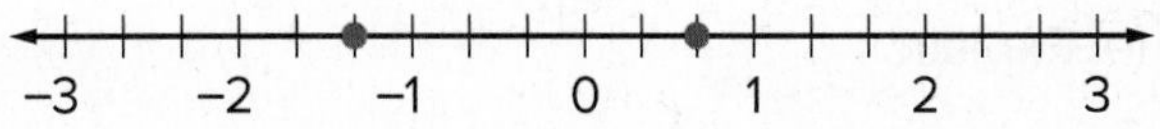

44. REGULARITY For tropical fish, aquarium water should be set to 76°F with an allowance of 4°.

a. Explain how to write an absolute value equation to represent this situation.

b. Explain the steps to solve the absolute value equation. What do the solutions represent?

45. **REASONING** The temperature of a refrigerator is 38°F give or take 2°.

a. Write an equation to find the maximum and minimum temperatures of the refrigerator.

b. Solve the equation to find the maximum and minimum temperatures of the refrigerator.

46. **STRUCTURE** A quality control inspector at a bolt factory examines random bolts that come off the assembly line. All bolts being made must be a tolerance of 0.04 mm. The inspector is examining bolts that are to have a diameter of 6.5 mm. Write and solve an absolute value equation to find the maximum and minimum diameters of bolts that will pass his inspection.

47. **SWIMMING POOL** Chlorine is added to a swimming pool to sanitize the water and make it safe for swimming. The chlorine should be in the range of 2–4 ppm (parts per million). Write an equation that represents the maximum and minimum chlorine concentration.

48. **FISH TANK** Tom has a 10 gallon fish tank that he wants to fill with neon tetra fish. Tom calculates the number of fish that will fit in the tank using three different methods. Write an equation to represent the maximum and minimum number of fish that will fit in the tank.

Method	Number of Fish
Method 1	5
Method 2	9
Method 3	8

Higher Order Thinking Skills

49. **PERSEVERE** If three points *a*, *b*, and *c* lie on the same line, then *b* is between *a* and *c* if and only if the distance from *a* to *c* is equal to the sum of the distances from *a* to *b* and from *b* to *c*. Write an absolute value equation to represent the definition of betweenness.

50. **ANALYZE** Translate the sentence $x = 5 \pm 2.3$ into an equation involving absolute value. Explain.

51. **FIND THE ERROR** Chris and Cami are solving $|x + 3| = -6$. Is either of them correct? Explain your reasoning.

Chris

$|x + 3| = 6$ or $|x + 3| = -6$

$x + 3 = 6$ or $x + 3 = -6$

$x = 3$ or $x = -9$

Cami

$|x + 3| = -6$

The solution is Ø.

52. **WRITE** Explain why an absolute value can never be negative.

53. **CREATE** Describe a real-world situation that could be represented by the absolute value equation $|x - 4| = 10$.

Solving Proportions

Today's Goal
- Solve proportions.

Today's Vocabulary
proportion

Explore Comparing Two Quantities

Online Activity Use graphing technology to complete the Explore.

INQUIRY How can you solve for an unknown value if two quantities have a proportional relationship?

Learn Solving Proportions

A **proportion** is an equation stating that two ratios are equivalent.

Example 1 Solve a Proportion

Solve the proportion. If necessary, round to the nearest hundredth.

$\frac{x}{45} = \frac{15}{25}$

$\frac{x}{45} = \frac{15}{25}$	Original proportion
$45\left(\frac{x}{45}\right) = 45\left(\frac{15}{25}\right)$	Multiply each side by 45.
$x = \frac{45(15)}{25}$	Simplify.
$x = \frac{675}{25}$	Multiply.
$x = 27$	Divide.

CHECK

Check your solution by substituting into the original proportion and check to see if the fractions are equal.

$\frac{x}{45} = \frac{15}{25}$	Original proportion
$\frac{27}{45} \stackrel{?}{=} \frac{15}{25}$	Substitute.
$\frac{3}{5} = \frac{3}{5}$	True.

Think About It!
What is another equation that you could write to solve the proportion? Explain your reasoning.

Go Online You can complete an Extra Example online.

Check

Solve $\frac{n-4}{8} = \frac{3}{2}$. If necessary, round to the nearest hundredth.

A. 16

B. 9.33

C. 8

D. 4.75

Example 2 Solve a Proportion with Two Missing Quantities

Solve $\frac{x}{9} = \frac{2x-3}{24}$. If necessary, round to the nearest tenth.

$\frac{x}{9} = \frac{2x-3}{24}$	Original proportion
$9\left(\frac{x}{9}\right) = 9\left(\frac{2x-3}{24}\right)$	Multiply each side by 9.
$x = \frac{9(2x-3)}{24}$	Simplify.
$x = \frac{18x-27}{24}$	Distributive Property.
$24x = 24\left(\frac{18x-27}{24}\right)$	Multiply each side by 24.
$24x = 18x - 27$	Simplify.
$24x - 18x = 18x - 18x - 27$	Subtract $18x$ from each side.
$6x = -27$	Simplify.
$\frac{6x}{6} = \frac{-27}{6}$	Divide each side by 6.
$x = -4.5$	Simplify.

Think About It!
How would the problem differ if the second ratio were 24 over $2x - 3$?

Check

Solve $\frac{x}{12} = \frac{2x-5}{18}$. If necessary, round to the nearest hundredth.

A. −10

B. −3.75

C. 0.83

D. 10

Go Online You can complete an Extra Example online.

Example 3 Solve a Proportion by Using a Constant Rate

GEOGRAPHY **Parts of Mexico City are sinking at a rate of 140 centimeters every 5 years. If this rate remains constant, how many centimeters will the city sink in the next 12 years?**

Step 1 Estimate the solution.

In 10 years, Mexico City will sink 140(2) or 280 centimeters. Because 10 years is slightly less than 12 years, Mexico City will sink more than 280 centimeters in 12 years.

Step 2 Write a proportion.

Let c represent the number of centimeters.

$$\frac{\text{city sinks 140 cm}}{\text{in 5 years}} = \frac{\text{city sinks } c \text{ cm}}{\text{in 12 years}}$$

Step 3 Solve the proportion.

$\frac{140}{5} = \frac{c}{12}$	Original proportion
$12\left(\frac{140}{5}\right) = 12\left(\frac{c}{12}\right)$	Multiply each side by 12.
$\frac{12(140)}{5} = c$	Simplify.
$\frac{1680}{5} = c$	Simplify.
$336 = c$	Divide.

CHECK

How do you know your solution is reasonable?

Sample answer: A rate of 140 centimeters every 5 years is a unit rate of 28 centimeters per year. In 12 years Mexico City would sink 12 years$\cdot \frac{\text{28 centimeters}}{\text{1 year}}$ or 336 centimeters. This makes sense with our estimate of more than 280 centimeters.

Talk About It!

Would you really expect the rate of sinking to remain constant over the entire time period? Explain.

Check

MIXTURE Oscar makes fruit punch to sell from his food truck by mixing 8 parts cranberry juice to 3 parts pineapple juice. How many cups of pineapple juice would Oscar need to mix with 48 cups of cranberry juice to make his punch? ___?___ cups

Go Online You can complete an Extra Example online.

Jess Kraft/Shutterstock

Example 4 Solve a Percent Problem by Using a Proportion

Study Tip

Setting Up Ratios It is a good idea to write the ratio in words to start the problem. Then read the problem to find the numbers or expressions to write each of the two ratios in the proportion.

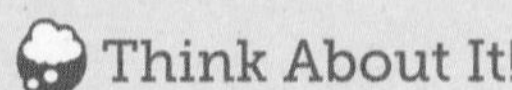

Think About It!

After multiplying each side by $r + 14$ in Method 1, Raja's resulting equation was $r = \frac{3}{10}r + 14$.

What error did Raja make?

MIXTURES **A guide company makes a trail mix of raisins and mixed nuts. How many pounds of raisins does the guide company need to mix with 14 pounds of mixed nuts to make the trail mix 30% raisins?**

METHOD 1 : raisins : trail mix

Let r represent the number of pounds of raisins.
Let $r + 14$ represent the number of pounds of the trail mix.
Write and solve a proportion.

$\frac{\text{raisins}}{\text{trail mix}} = \frac{30}{100}$ — 30% of the trail mix is raisins.

$\frac{r}{r + 14} = \frac{30}{100}$ — Substitute $r + 14$ for the amount of the trail mix.

$(r + 14)\left(\frac{r}{r + 14}\right) = (r + 14)\frac{30}{100}$ — Multiply each side by $r + 14$.

$r = (r + 14)\frac{3}{10}$ — Simplify.

$r = \frac{3}{10}r + \frac{14 \cdot 3}{10}$ — Distributive Property

$r - \frac{3}{10}r = \frac{3}{10}r - \frac{3}{10}r + \frac{42}{10}$ — Subtract $\frac{3}{10}r$ from each side.

$\frac{7}{10}r = \frac{42}{10}$ — Simplify.

$\frac{10}{7}\left(\frac{7}{10}r\right) = \frac{10}{7}\left(\frac{42}{10}\right)$ — Multiply each side by $\frac{10}{7}$.

$r = \frac{42}{7}$ or 6 — Simplify.

METHOD 2 : raisins : mixed nuts

Let r represent the number of pounds of raisins, when the number of pounds of mixed nuts is 14.
Write and solve a proportion.

$\frac{\text{raisins}}{\text{mixed nuts}} = \frac{30}{70}$ — The ratio of raisins to mixed nuts is 30 : 70.

$\frac{r}{14} = \frac{30}{70}$ — Substitute 14 for the amount of mixed nuts.

$14\left(\frac{r}{14}\right) = 14\left(\frac{30}{70}\right)$ — Multiply each side by 14.

$r = \frac{14 \cdot 30}{70}$ — Simplify.

$r = \frac{420}{70}$ or 6 — Simplify.

Check

MIXTURE Ayita is making a plant food mixture to use in her garden. The mixture is to be 20% plant food and 80% water. She needs to make 12 gallons of the mixture to cover her entire garden. Which proportions can be used to find the amount of plant food p she will need? Select all that apply.

A. $\frac{20}{80} = \frac{p}{12}$ **B.** $\frac{20}{100} = \frac{p}{12}$ **C.** $\frac{80}{100} = \frac{p}{12}$

D. $\frac{12 - p}{12} = \frac{80}{100}$ **E.** $\frac{20}{p} = \frac{80}{12}$

Go Online You can complete an Extra Example online.

Practice

Go Online You can complete your homework online.

Example 1

Solve each proportion. If necessary, round to the nearest hundredth.

1. $\frac{3}{8} = \frac{15}{a}$
2. $\frac{t}{2} = \frac{6}{12}$
3. $\frac{4}{9} = \frac{13}{q}$
4. $\frac{15}{35} = \frac{g}{7}$
5. $\frac{7}{10} = \frac{m}{14}$
6. $\frac{8}{13} = \frac{v}{21}$
7. $\frac{w}{2} = \frac{4.5}{6.8}$
8. $\frac{1}{0.19} = \frac{12}{n}$
9. $\frac{2}{0.21} = \frac{8}{n}$
10. $\frac{2.4}{3.6} = \frac{k}{1.8}$
11. $\frac{t}{0.3} = \frac{1.7}{0.9}$
12. $\frac{7}{1.066} = \frac{z}{9.65}$
13. $\frac{x-3}{5} = \frac{6}{10}$
14. $\frac{7}{x+9} = \frac{21}{36}$
15. $\frac{10}{15} = \frac{4}{x-5}$
16. $\frac{6}{14} = \frac{7}{x-3}$
17. $\frac{7}{4} = \frac{f-4}{8}$
18. $\frac{3-y}{4} = \frac{1}{9}$

Example 2

Solve each proportion. If necessary, round to the nearest hundredth.

19. $\frac{4v+7}{15} = \frac{6v+2}{10}$
20. $\frac{9b-3}{9} = \frac{5b+5}{3}$
21. $\frac{2n-4}{5} = \frac{3n+3}{10}$
22. $\frac{2}{g+6} = \frac{4}{5g+10}$
23. $\frac{x}{3} = \frac{3x+2}{6}$
24. $\frac{w+3}{7} = \frac{w-1}{8}$
25. $\frac{4q-3}{5} = \frac{2q+1}{7}$
26. $\frac{5}{7k+4} = \frac{2}{2k-3}$
27. $\frac{m+1}{9} = \frac{m+2}{2}$
28. $\frac{j-5}{2} = \frac{j+8}{7}$
29. $\frac{9f+3}{10} = \frac{2f-4}{5}$
30. $\frac{2c-1}{3} = \frac{c+2}{4}$
31. $\frac{5n-2}{8} = \frac{n+8}{3}$
32. $\frac{h-7}{4} = \frac{2h+1}{3}$
33. $\frac{14}{3y+5} = \frac{3}{y}$
34. $\frac{p+10}{8} = \frac{2p-7}{4}$
35. $\frac{7}{14-d} = \frac{3}{18+d}$
36. $\frac{2z-4}{5} = \frac{3z+3}{10}$

Example 3

37. **BOATING** Dedra's boat used 5 gallons of gasoline in 4 hours. At this rate, how many gallons of gasoline will the boat use in 10 hours?

38. **WATER** A dripping faucet wastes 3 cups of water every 24 hours. How much water is wasted in a week?

39. **PRECISION** In November 2010 the average cost of 5 gallons of regular unleaded gasoline in the United States was $14.46. What was the average cost for 16 gallons of gasoline?

40. **SHOPPING** Stevenson's Market is selling 3 packs of stylus pens for $5.00. How much will 10 packs of stylus pens cost at this price?

41. **STATE YOUR ASSUMPTION** During basketball practice, Brent made 36 free throws in 3 minutes.

    a. How many free throws will Brent make in 5 minutes?

    b. What assumption did you make in part **a**? Explain.

42. **NAILS** Human fingernails grow at an average rate of 3.47 millimeters per month. How much will they grow in 20 months?

43. **PICTURE** Jasmine enlarged the size of a picture to a height of 15 inches. What is the new width of the picture if it was originally 6 inches wide by 4 inches tall?

44. **TRAVEL** Roscoe is exchanging $121 for Euros for his upcoming trip to Germany. If $2 can be exchanged for 1.78 Euros, how many Euros will Roscoe have?

Example 4

45. **FUNDRAISER** Owen is organizing a fundraiser. The proceeds will be split between a charity and the expenses from the fundraiser. Owen would like the cost of the fundraiser to be 15% of the proceeds. If the fundraiser will cost $500, how much money do they need to raise at the fundraiser?

46. **COFFEE** A barista is mixing a house blend of coffee that is 25% light roast. If there are 8 pounds of the light roast available, how much of the blend can the barista make?

47. **CHEMISTRY** A chemistry teacher needs to mix an acid solution for an experiment. How much hydrochloric acid needs to be mixed with 1500 milliliters of water to make a solution that is 12% acid?

48. **LEMONADE** Laronda wants to make fresh lemonade. The recipe she finds online recommends that the fresh lemon juice should be 20% of the total volume. She has 18 ounces of fresh lemon juice. How much water should she mix with the lemon juice?

Mixed Exercises

Solve each proportion. If necessary, round to the nearest hundredth.

49. $\frac{9}{g} = \frac{15}{10}$

50. $\frac{3}{a} = \frac{1}{6}$

51. $\frac{6}{z} = \frac{3}{5}$

52. $\frac{5}{f} = \frac{35}{21}$

53. $\frac{12}{7} = \frac{36}{m}$

54. $\frac{6}{23} = \frac{y}{69}$

55. $\frac{42}{56} = \frac{6}{f}$

56. $\frac{7}{b} = \frac{1}{9}$

57. $\frac{10}{14} = \frac{30}{m}$

58. $\frac{3}{4} = \frac{n}{20}$

59. $\frac{6}{4} = \frac{x}{18}$

60. $\frac{33}{b} = \frac{15}{45}$

61. $\frac{m-2}{4} = \frac{5}{20}$

62. $\frac{9}{5} = \frac{3}{x+7}$

63. $\frac{5}{b} = \frac{3}{b-6}$

64. $\frac{2p+3}{3} = \frac{4p-7}{2}$

65. $\frac{3y+4}{5} = \frac{y-1}{4}$

66. $\frac{2}{w} = \frac{7}{w+5}$

67. $\frac{7n-2}{6} = \frac{3n-2}{4}$

68. $\frac{-a-8}{10} = \frac{-a+3}{2}$

69. $\frac{c+2}{c-2} = \frac{4}{8}$

70. USE A SOURCE Find the heights of Willis Tower and the John Hancock Center in Chicago including the tip. Suppose you build a scale model of each building. If you make the model of the Willis Tower 3 meters tall, what would be the approximate height of the John Hancock Center model? Round to the nearest hundredth.

71. USE TOOLS A map of Waco, Texas and neighboring towns is shown.

a. Use a metric ruler to measure the distances between Robinson and Neale on the map.

b. If the scale on the map is 1 cm = 3 mi, find the actual distance between Robinson and Neale.

c. How many square miles are shown on this map?

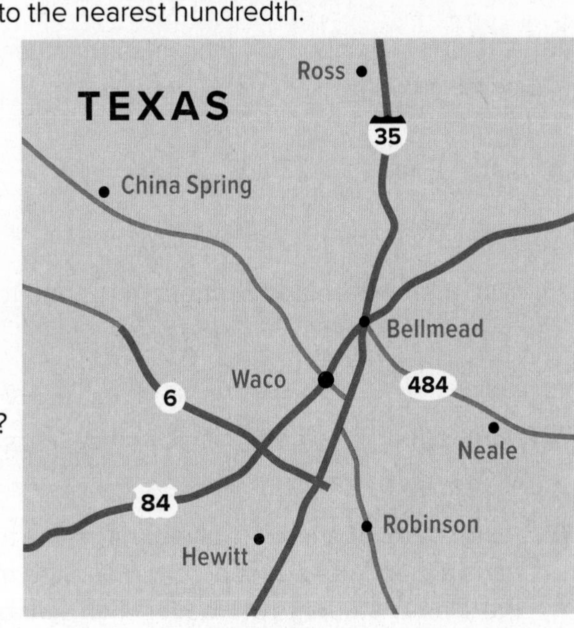

72. BUDGET Shawnda spent \$259.20 on a scooter. This was 80% of her budget for a new scooter. How much was the total budget?

73. USE TOOLS On average, 8 potatoes cost $1.50 at a farmer's market. Fernando needs to buy 22 potatoes.

a. Estimate the cost of 22 potatoes. Explain.

b. How much will it cost Fernando to buy 22 potatoes? How does this compare to your estimate?

c. How many potatoes could Fernando buy with $7?

d. What is the unit cost per potato?

74. USE A MODEL Kina and Aiesha started walking from the same location at the same time. Kina walked 6 miles. Aiesha walked 8 miles and walked 1 mile per hour faster than Kina. They each walked for the same amount of time.

a. Describe how a proportion could be used to find the rate that each person walked.

b. Solve the proportion. Explain the meaning of the solution.

Higher Order Thinking Skills

75. PERSEVERE If $\frac{a+1}{b-1} = \frac{5}{1}$ and $\frac{a-1}{b+1} = \frac{1}{1}$, find the value of $\frac{b}{a}$. (Hint: Choose values of a and b for which the proportions are true and evaluate $\frac{b}{a}$.).

76. CREATE Describe how a business can use ratios. Include a real-world situation in which a business would use a ratio.

77. WRITE Compare and contrast ratios and rates.

78. PERSEVERE Find b if $\frac{17}{34} = \frac{a}{32}$ and $\frac{a}{40} = \frac{b}{60}$.

79. PERSEVERE A survey showed that x% of the students at Hoover High school have a job. Write a proportion to find the number of students that have a job, z, if there are y students at Hoover High school.

Using Formulas

Explore Centripetal Force

Online Activity Use a video to complete the Explore.

INQUIRY Why might you want to solve a formula for a specified value?

Today's Goals
- Solve equations for specific variables.
- Convert units of measure.

Today's Vocabulary
formula
literal equation
dimensional analysis

Learn Solving Equations for Given Variables

A **formula** is an equation that expresses a relationship between certain quantities. A formula or equation that involves several variables is called a **literal equation**.

Think About It!
How do you solve an equation for a specific variable?

Example 1 Solve for a Specific Variable

Solve $5a - 2b = 15$ for a.

$5a - 2b = 15$ Original equation

$5a - 2b + 2b = 15 + 2b$ Add $2b$ to each side.

$5a = 15 + 2b$ Simplify.

$\frac{5a}{5} = \frac{15 + 2b}{5}$ Divide each side by 5.

$a = \frac{15}{5} + \frac{2b}{5}$ Simplify.

$a = 3 + \frac{2b}{5}$ Simplify.

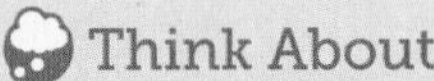

Think About It!
Describe how you would solve for b instead of a. How would the answer change?

Go Online You can complete an Extra Example online.

Example 2 Solve for a Specific Variable When the Variable Is on Each Side

Solve $4p - 7r = pq + 16$ for p.

$4p - 7r = pq + 16$	Original equation
$4p - 7r + 7r = pq + 16 + 7r$	Add $7r$ to each side.
$4p = pq + 16 + 7r$	Simplify.
$4p - pq = pq + 16 + 7r - pq$	Subtract pq from each side.
$4p - pq = 16 + 7r$	Simplify.
$p(4 - q) = 16 + 7r$	Distributive Property
$\frac{p(4-q)}{4-q} = \frac{16+7r}{4-q}$	Divide each side by $4 - q$.
$p = \frac{16+7r}{4-q}$	Simplify.

Talk About It!

Would the solution be the same if the variable p were isolated on the right side of the equation instead of the left? Explain your reasoning.

Check

Solve $2v = \frac{w+v}{t}$ for v.

$v =$ ___?___

Example 3 Solve Literal Equations for a Given Variable

GEOMETRY The area of a trapezoid is $A = \frac{h(b_1 + b_2)}{2}$. A represents the area, h represents the height, b_1 represents the length of one base, and b_2 represents the length of the other base.

Part A

Solve the formula for h.

$A = \frac{h(b_1 + b_2)}{2}$	Area Formula
$2A = 2\frac{h(b_1 + b_2)}{2}$	Multiply each side by 2.
$2A = h(b_1 + b_2)$	Simplify.
$\frac{2A}{b_1 + b_2} = \frac{h(b_1 + b_2)}{b_1 + b_2}$	Divide each side by $b_1 + b_2$.
$\frac{2A}{b_1 + b_2} = h$	Simplify.

Part B

Find the height of a trapezoid with an area of 70 square feet and bases that are 22 feet and 18 feet.

$h = \frac{2A}{(b_1 + b_2)}$ Formula for height

$h = \frac{2(70)}{(22 + 18)}$ $A = 70, b_1 = 22, b_2 = 18$

$h = \frac{140}{40}$ Simplify.

$h = 3.5$ Divide.

The height of the trapezoid is 3.5 feet.

Check

BUSINESS A Mason jar company wants to increase the volume of its cylindrical jars by 6 cubic inches. The company's designer wants a formula that states the height h of a jar given its volume V and radius r. The volume of a cylindrical jar is modeled by the equation $V = \pi r^2 h$.

Part A

What formula should be used to find the height h?

A. $h = \pi \cdot r^2$

B. $h = \sqrt{\frac{V}{\pi r}}$

C. $h = \frac{V}{\pi r^2}$

D. $h = \frac{V}{r^2}$

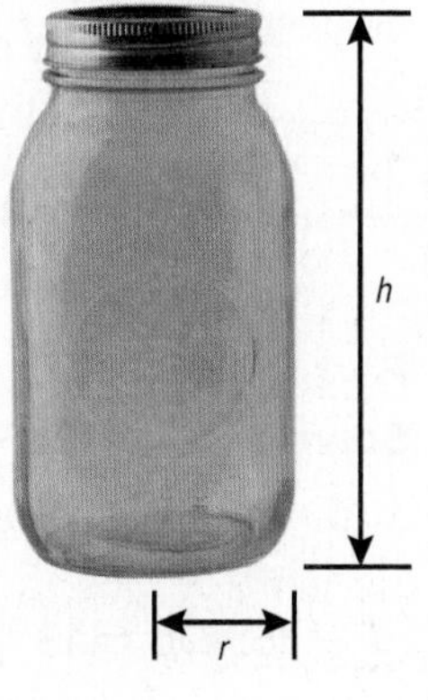

Part B

If the radius of the jar is 1.5 inches and the original volume is 30 cubic inches, then what height should the company make the height of its new jar to increase the volume by 6 cubic inches?

Go Online You can complete an Extra Example online.

Watch Out!

Dividing by a Quantity
Do not forget to divide each side by the entire quantity $b_1 + b_2$.

Think About It!

How would the height of the trapezoid change if the area were doubled and all other measures remained the same?

Study Tip

Solving for a Specific Variable
When an equation has more than one variable, it can be helpful to highlight the variable for which you are solving on a piece of paper.

Example 4 Use Literal Equations

PARTY The total amount of money Kishi spends on pizza for a party is $T = 13.49c + 15.49p$. T represents the total amount of money she spends, c represents the number of cheese pizzas she buys, and p represents the number of pepperoni pizzas she buys.

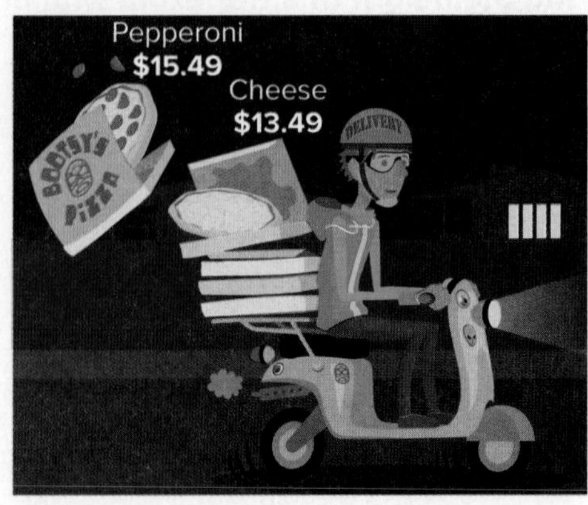

Part A

If Kishi has $85 to spend on pizza, describe the constraints on $T = 13.49c + 15.49p$.

- The maximum number of cheese pizzas Kishi can buy is 6, because she can buy 6 cheese pizzas and no pepperoni pizzas without exceeding $85.
- The maximum number of pepperoni pizzas Kishi can buy is 5, because she can buy 5 pepperoni and no cheese pizzas without going over her budget.
- The minimum number of each type of pizza she can buy is 0, because you cannot buy a negative number of pizzas.

Part B

Solve $T = 13.49c + 15.49p$ for c.

$T = 13.49c + 15.49p$	Original equation
$T - 15.49p = 13.49c + 15.49p - 15.49p$	Subtract 15.49p.
$T - 15.49p = 13.49c$	Simplify.
$\frac{T - 15.49p}{13.49} = \frac{13.49c}{13.49}$	Divide by 13.49.
$\frac{T - 15.49p}{13.49} = c$	Simplify.

Think About It!

Why did you round your answer in Part C to 2 instead of 3?

Part C

If Kishi has $85 to spend on pizza and she needs to buy 3 pepperoni pizzas, find the maximum number of cheese pizzas she can buy.

$c = \frac{T - 15.49p}{13.49}$	Original equation solved for c
$c = \frac{85 - 15.49(3)}{13.49}$	$T = 85, p = 3$
$c = \frac{38.53}{13.49}$	Simplify.
$c \approx 2.86$	Divide.

Kishi can buy a maximum of 2 cheese pizzas.

Go Online You can complete an Extra Example online.

Check

VIDEO GAMES Ella makes a video game that becomes very popular. She creates a formula, $P = 40c - 300$, to model her profit P given the number of copies sold c, taking into account the $300 fee that she has paid a retailer to sell her game. Which equation would model how many copies c must be sold to yield a specific amount of profit?

A. $c = \frac{P + 40}{300}$

B. $c = \frac{P}{40} + 300$

C. $c = \frac{P}{40} - 300$

D. $c = \frac{P + 300}{40}$

Think About It!
Why might you want to convert units?

Explore Using Dimensional Analysis

Online Activity Use a real-world situation to complete the Explore.

INQUIRY Why might you want to convert the units for a given quantity or measurement?

Learn Dimensional Analysis

When using formulas, you may want to use dimensional analysis. **Dimensional analysis** or **unit analysis** is the process of performing operations with units.

As you plan your solution method, think about

- what units were given,
- what units you need for the solution, and
- the step(s) you need to take to convert your units from what you are given to what you will need for the solution.

Example 5 Multiply by a Conversion Factor

POOLS **Mark is purchasing an above-ground swimming pool. The salesperson says that the pool will hold 97,285 liters of water. If 1 gallon = 3.785 liters, determine approximately how many gallons of water Mark's pool will hold.**

Study Tip
Precision
Notice that the question asks for an estimate, not an exact answer.

Two ratios can be used to compare liters and gallons, $\frac{1 \text{ gallon}}{3.785 \text{ liters}}$ and $\frac{3.785 \text{ liters}}{1 \text{ gallon}}$. To convert from liters to gallons, multiply by $\frac{1 \text{ gallon}}{3.785 \text{ liters}}$.

Number of liters the pool will hold × gallons to liters

$$97{,}285 \text{ liters} \times \frac{1 \text{ gallon}}{3.785 \text{ liters}}$$

$$97{,}285 \cancel{\text{liters}} \times \frac{1 \text{ gallon}}{3.785 \cancel{\text{liters}}} \approx 25{,}703 \text{ gallons}$$

Mark's pool will hold approximately 25,703 gallons of water.

Check

COOKING For the chefs to prepare the dishes for the next hour, Adelina needs to provide them with 96 cloves of garlic. However, she can only purchase bags of whole heads of garlic. If each bag of garlic contains 3 heads and each head has about 8 cloves, then how many bags of garlic should she purchase?

Part A

What assumption must Adelina make when calculating how many bags of garlic to buy?

A. Each head of garlic has exactly 8 cloves.

B. The chefs need 96 heads of garlic.

C. The chefs need 96 cloves of garlic.

D. Each bag has 3 heads of garlic.

Part B

How many bags of garlic should Adelina purchase?

A. 4 **B.** 12

C. 32 **D.** 96

Avoid a Common Error

Remember that a unit will only cancel when you divide it by itself. What error does this solution make?

$97{,}285 \text{ liters} \times \frac{3.785 \text{ liters}}{1 \text{ gallon}}$

$\approx 368{,}224$ gallons

Example 6 Use Dimensional Analysis to Convert Units

Think About It!

When you are converting, how do you know which unit goes in the numerator and which unit goes in the denominator?

COOKING **A recipe calls for 20 fluid ounces of milk. If Nita buys a half gallon of milk, how many batches of that recipe can she make? (Hint: 8 fluid ounces = 1 cup)**

RECIPE
BREAD PUDDING
1.5 tbsp. butter
12 oz. bread
20 fl. oz. milk
3 eggs
1 tbsp. vanilla
2 cups sugar
Milk

First, convert gallons to fluid ounces.

total amount of milk × gallons to quarts × quarts to pints × pint to cups × cups to fluid ounces

$$0.5 \text{ gallon} \times \frac{4 \text{ quarts}}{1 \text{ gallon}} \times \frac{2 \text{ pints}}{1 \text{ quart}} \times \frac{2 \text{ cups}}{1 \text{ pint}} \times \frac{8 \text{ fl. oz.}}{1 \text{ cup}} = 64 \text{ fl. oz.}$$

Use the following conversion factors to change gallons to ounces.

1 gallon = 4 quarts 1 quart = 2 pints
1 pint = 2 cups 1 cup = 8 fluid ounces

Nita has 64 ounces of milk. Each batch calls for 20 fluid ounces, so to find the number of batches she can make, divide by 20 fluid ounces.

$$64 \text{ fl. oz.} \times \frac{1 \text{ batch}}{20 \text{ fl. oz.}} = 3.2 \text{ batches}$$

Nita has enough milk to make 3 batches with some milk left over.

Go Online You can complete an Extra Example online.

Check

AGRICULTURE On average, a dairy cow produces 832 ounces of milk a day. About how many gallons of milk does a dairy cow produce each year? (Hint: 1 cup = 8 ounces, 1 quart = 4 cups, and 1 gallon = 4 quarts)

A. 6.5 gallons per year

B. 2372.5 gallons per year

C. 4357.2 gallons per year

D. 303,680 gallons per year

Example 7 Use Dimensional Analysis to Convert Rates

SPEED **In a novel, the main character, Aiko, can run long distances at 16.5 *kanejaku* per second. Carla knows that the Olympic record for running a marathon distance of 26.2 miles is about 126.5 minutes. She wonders if Aiko could beat that record. If 1 *kanejaku* = $\frac{10}{33}$ meters, find how far Aiko could run, in miles, in that amount of time. (Hint: 1 mile ≈ 1609.344 meters)**

Use the formula $d = rt$ that relates distance d, rate r, and time t to find the distance Aiko could run in 126.5 minutes.

$d = rt$ Distance equation

$d = \left(\frac{16.5 \text{ kanejaku}}{1 \text{ second}}\right) \cdot t$ Substitute Aiko's rate.

In order to compare Aiko to the Olympic runner, convert Aiko's rate in *kanejaku* to miles per minute.

Step 1 Convert distance.

You want distance in miles, but Aiko's distance is in *kanejaku*. Use the given conversion rates that relate to distance to convert Aiko's rate in *kanejaku* per second to miles per second.

$$\frac{16.5 \text{ kanejaku}}{1 \text{ second}} \times \frac{\frac{10}{33} \text{ meters}}{1 \text{ kanejaku}} \times \frac{1 \text{ mile}}{1609.344 \text{ meters}} = \frac{5 \text{ miles}}{1609.344 \text{ seconds}}$$

Step 2 Convert time.

You want time in minutes, but Aiko's time is in seconds. Use the resulting rate from step 2 to convert seconds to minutes.

$$\frac{5 \text{ miles}}{1609.344 \text{ seconds}} \times \frac{60 \text{ seconds}}{1 \text{ minute}} = \frac{300 \text{ miles}}{1609.344 \text{ minutes}}$$

(*continued on the next page*)

Go Online You can complete an Extra Example online.

Problem-Solving Tip

Make a Plan
Before you solve a problem, think about what the question is asking and what information will apply to the solution.

Watch Out!

Canceling Units
Do not forget to cancel your units as you multiply so that you can see what units are left. The units that are left are the units of your final answer.

Go Online An alternate method is available for this example.

Step 3 Substitute.

Substitute Aiko's rate in miles per minute and the given time into the formula and simplify.

$$d = \frac{300 \text{ miles}}{1609.344 \cancel{\text{minutes}}} \times 126.5 \cancel{\text{minutes}} \approx 23.6 \text{ miles}$$

Aiko would run approximately 23.6 miles in the time it took the Olympic runner to complete 26.2 miles. So, Aiko would not beat the Olympic marathon record time.

Pause and Reflect

Did you struggle with anything in this lesson? If so, how did you deal with it?

Practice

Examples 1

Solve each equation or formula for the variable indicated.

1. $x - 2y = 1$, for y

2. $d + 3n = 1$, for n

3. $7f + g = 5$, for f

4. $3c - 8d = 12$, for c

5. $7t = x$, for t

6. $r = wp$, for p

7. $q - r = r$, for r

8. $4m - t = m$, for m

9. $7a - b = 15a$, for a

10. $-5c + d = 2c$, for c

Go Online You can complete an Extra Example online.

Go Online You can complete your homework online.

Example 2

Solve each equation or formula for the variable indicated.

11. $u = vw + z$, for v

12. $x = b - cd$, for c

13. $fg - 9h = 10j$, for g

14. $10m - p = -n$, for m

15. $r = \frac{2}{3}t + v$, for t

16. $\frac{5}{9}v + w = z$, for v

17. $\frac{10ac - x}{11} = -3$, for a

18. $\frac{df + 10}{6} = g$, for f

Example 3

19. RECTANGLES The formula $P = 2\ell + 2w$ represents the perimeter of a rectangle. In this formula, ℓ is the length of the rectangle and w is the width.

a. Solve the formula for ℓ.

b. Find the length when the width is 4 meters and the perimeter is 36 meters.

20. BASEBALL The formula $a = \frac{h}{b}$ can be used to find the batting average a of a batter who has h hits in b times at bat.

a. Solve the formula for b.

b. If a batter has a batting average of 0.325 and has 39 hits, how many times has the player been at bat?

21. SHOPPING Thomas went to the store to buy videogames for $13.50 each and controllers. The total amount Thomas spent can be represented by $c = 13.50g + p$, where c is the total cost, g is the number of games he bought, and p is the cost of the controllers. The controllers cost $55 and Thomas spent $136 total.

a. Solve the equation for g.

b. Find how many games Thomas bought.

22. GEOMETRY The volume of a box V is given by the formula $V = \ell wh$, where ℓ is the length, w is the width, and h is the height.

a. Solve the formula for h.

b. What is the height of a box with a volume of 50 cubic meters, length of 10 meters, and width of 2 meters?

Example 4

23. COFFEE SHOP Consuelo is buying flavored coffee and plain coffee. The total amount of money she spends on coffee is $T = 5.50p + 7f$, where p represents the cost of a package of plain coffee and f represents the cost of a package of flavored coffee.

a. If Consuelo has $40 to spend on coffee, describe the constraints on the formula.

b. Solve for f.

c. If Consuelo needs to buy 3 packages of plain coffee, what is the maximum number of packages of flavored coffee she can buy?

24. SHOPPING Kimberly is ordering bath towels and washcloths for the inn where she works. The total cost of the order is $T = 6b + 2w$, where T is the total cost, b is the number of bath towels, and w is the number of washcloths.

a. Her budget is $85. Describe the constraints.

b. Solve for b.

c. She needs to order at least 20 washcloths. How many bath towels can she order and stay under budget?

Examples 5–7

25. ENVIRONMENT The United States released 5.877 billion metric tons of carbon dioxide into the environment through the burning of fossil fuels in a recent year. If 1 trillion pounds = 0.4536 billion metric tons, how many trillion pounds of carbon dioxide did the United States release in that year?

26. EUROS Trent purchases 44 euros worth of souvenirs while on vacation in France. If $1 U.S. = 0.678 euros, find the cost of the souvenirs in United States dollars.

27. LENGTH A pencil is 13.5 centimeters long. If 1 centimeter = 0.39 inch, what is the length of the pencil in feet, to the nearest hundredth?

28. TRACK If a track is 400 meters around, how many laps around the track would it take to run 3.1 miles? Round to the nearest tenth. (*Hint:* 1 foot = 0.3048 meter)

29. **BIKING** Imelda rode her bicycle 39 kilometers. If 1 meter = 1.094 yards, find the distance Imelda rode her bicycle to the nearest mile. (*Hint:* 1 mi = 1760 yd)

30. **MANUFACTURING** Aluminum, Inc. produces cans at a rate of 0.04 per hundredth of a second. How many cans can be produced in a 7-hour day?

31. **WATER USAGE** Each minute, 8.8 quarts of water flow from a shower. If the average person spends 8.2 minutes in the shower, how many gallons of water will the average person have used after taking five showers?

32. **TRAVEL** The swim team is going to finals. If the meet is in 85 days, determine how many seconds there are until the meet.

33. **PRECISION** The chemistry teacher set out a 5-pound jar of salt at the beginning of the day. If each student needs 27.6 grams of salt for an experiment, how many students can perform the experiment before the jar is empty? (*Hint:* 1 lb = 454 g)

Mixed Exercises

Solve each equation for the variable indicated.

34. $rt - 2n = y$, for t

35. $bc + 3g = 2k$, for c

36. $kn + 4f = 9v$, for n

37. $8c + 6j = 5p$, for c

38. $\frac{x - c}{2} = d$, for x

39. $\frac{x - c}{2} = d$, for c

40. $-14n + q = rt - 4n$, for n

41. $18t + 11v = w - 13t$, for t

42. $ax + z = aw - y$, for a

43. $10c - f = -13 + cd$, for c

44. **STRUCTURE** Jethro used dimensional analysis to convert from one rate of speed to another. Two of the conversion factors he used are $\frac{5280 \text{ ft}}{1 \text{ mi}}$ and $\frac{1 \text{ hr}}{60 \text{min}}$. What could be the units of the initial rate of speed and final rate of speed? Justify your answer.

45. **REASONING** The formula $A = P(1 + r)$ represents the amount of money A in an account after 1 year, where P is the amount initially deposited and r is the interest rate. Note that the interest rate is written as a decimal. Dennis deposits \$2150 into a savings account. After 1 year, he has \$2182.25 in his account. Solve the equation for r, and determine the interest rate. Show your work.

46. REASONING The regular octagon shown is divided into 8 congruent triangles. Each triangle has an area of 21.7 square centimeters. The perimeter of the octagon is 48 centimeters.

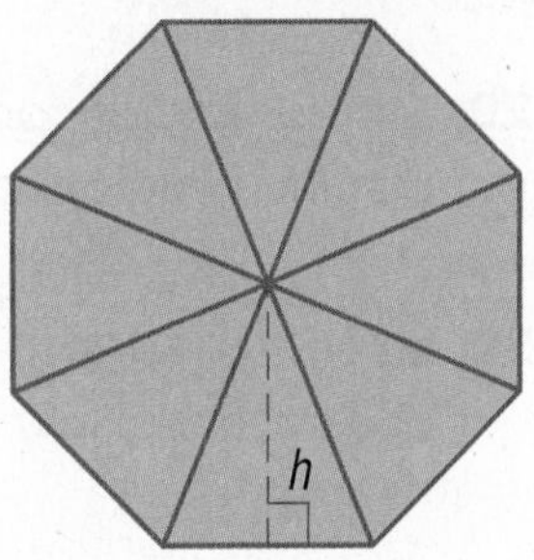

a. What is the length of each side of the octagon?

b. Solve the formula for the area of a triangle for h.

c. What is the height of each triangle? Round to the nearest tenth.

47. Consider the equation $\frac{ry + z}{m} - t = x$.

a. Solve the equation for y.

b. Would there be any restrictions on the value of each variable? If so, explain the restrictions.

Higher Order Thinking Skills

48. CREATE Think about the area formula of some geometric figures that involve a fraction.

a. Write a formula for A, the area of a geometric figure that includes a fraction.

b. Solve the formula for a variable other than A.

49. FIND THE ERROR The formula represents the relationship between temperatures in degrees Fahrenheit F and degrees Celsius C. Sasha solves the formula for C. Her solution is shown at the right. Is Sasha's solution correct? Explain your reasoning.

Sasha

$F = \frac{9}{5}C + 32$

$\frac{5}{9}F = C + 32$

$C = \frac{9}{5}F - 32$

50. PERSEVERE The formula $A = P(1 + r)^t$ represents the amount of money A in an account after t years, where P is the amount initially deposited and r is the interest rate. Patricia currently has $1839.79 in an account that has an interest rate of 2.5%. She opened the account 8 years ago and has made no additional deposits since then.

a. Solve the formula for P and find the amount of Patricia's initial deposit.

b. Nia says that for the formula in **part a**, A is always greater than P when r is positive and t is a positive integer. Do you agree? Why or why not?

Review

 Essential Question

How can writing and solving equations help you solve problems in the real world?

Equations can be written to describe the relationship between quantities in the real world. Solving these equations provides information about unknown quantities.

Module Summary

Lessons 2-1 and 2-2

One-Step Equations

- To write an equation, first identify each unknown and assign a variable to it. Then identify the givens and their relationships. Finally, write the sentence as an equation.
- Solving an equation is the process of finding all values of the variable that make the equation true.
- If a number is added to or subtracted from each side of a true equation, the resulting equivalent equation is also true.
- If an equation is true and each side is multiplied or divided by the same nonzero number, the resulting equation is equivalent.

Lessons 2-3 and 2-4

Multi-Step Equations

- To solve a multi-step equation, you can undo each operation using properties of equality. Working backward in the order of operations makes this process simpler. Each step in this process results in equivalent equations.
- To solve an equation with the variable on each side, write an equivalent expression with all of the variable terms on one side and the constants on the other side.
- When a grouping symbol appears in an equation, it must first be removed before continuing to solve the equation.

Lessons 2-5 and 2-6

Absolute Value Equations and Proportions

- When solving equations that involve absolute values, there are two cases to consider.

  Case 1: The expression inside the absolute value symbol is positive or zero.

  Case 2: The expression inside the absolute value symbol is negative.
- A proportion is a statement that two ratios are equivalent.

Lesson 2-7

Formulas

- A formula is an equation that expresses a relationship between certain quantities.
- A formula or equation that involves several variables is called a literal equation. To solve a literal equation, solve for a specific variable.
- Dimensional analysis is the process of performing operations with units.

Study Organizer

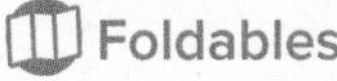

Use your Foldable to review the module. Working with a partner can be helpful. Ask for clarification of concepts as needed.

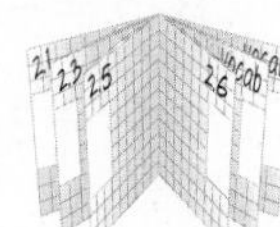

Test Practice

1. **MULTIPLE CHOICE** Which equation represents this sentence? (Lesson 2-1)

   The sum of 5 times a number m and 12 is equal to 27.

   A. $5m + 12 = 27$

   B. $5(m + 12) = 27$

   C. $5 + 12m = 27$

   D. $5(12m) = 27$

2. **OPEN RESPONSE** A concert venue surveyed 680 concert attendees about the concession stand. Of those that visited the concession stand, 527 said the concession stand prices are excessive, and the remaining 44 did not. Write an equation to find the number of attendees *a* who did not visit the concession stand. (Lesson 2-1)

3. **MULTIPLE CHOICE** Write a verbal sentence for the algebraic equation $5x^2 + 2 = 22$. (Lesson 2-1)

   A. 5 times *x* squared less 2 is 22.

   B. Five plus *x* squared plus 2 is 22.

   C. The product of 5 times *x* squared and 2 is 22.

   D. Five times *x* squared plus 2 is 22.

4. **MULTIPLE CHOICE** Solve $n - 8 = 5$. (Lesson 2-2)

   A. $n = -13$

   B. $n = -3$

   C. $n = 3$

   D. $n = 13$

5. **MULTIPLE CHOICE** Solve $\frac{1}{5}t = 10$ for *t*. (Lesson 2-2)

   A. 2

   B. 5

   C. 15

   D. 50

6. **OPEN RESPONSE** Solve $z + 12 = -3$ for *z*. Explain. (Lesson 2-2)

7. **MULTIPLE CHOICE** If $3x - 6 = 42$, what is the value of *x*? (Lesson 2-3)

   A. 8

   B. 12

   C. 16

   D. 20

8. **OPEN RESPONSE** Solve $8 = 11 - 3v$. (Lesson 2-3)

   $v =$ ___?___

9. **MULTI-SELECT** Jaime bought a notebook and a box of pencils for \$5.00. The notebook cost \$3.00 and there are 10 pencils in a box. The equation $3 + 10p = 5$ can be used to find the cost of one pencil. Select all equations that are equivalent. (Lesson 2-3)

   A. $10p = 8$

   B. $10p = 2$

   C. $p = 0.80$

   D. $p = 0.20$

   E. $10p = 5$

10. MULTIPLE CHOICE Solve the equation $3x + 1 = 4x - 8$ for x. (Lesson 2-4)

A. $x = -9$

B. $x = -1$

C. $x = 1$

D. $x = 9$

11. MULTIPLE CHOICE Solve the equation $-2(x + 4) + 3x = x - 8$. (Lesson 2-4)

A. $x = 2$

B. $x = 4$

C. all real numbers

D. no solution

12. MULTIPLE CHOICE Solve the equation $5(2y + 1) = 4y + 10$. (Lesson 2-4)

A. $y = \frac{5}{6}$

B. $y = \frac{3}{2}$

C. $y = \frac{15}{14}$

D. $y = \frac{5}{2}$

13. OPEN RESPONSE A park has a ginkgo tree, a dogwood tree, and 2 blue spruce trees. The blue spruce trees are 8 years old. The ginkgo tree is 2 years less than three times the age of the dogwood tree. The ginkgo tree is also half the sum of the ages of the dogwood tree and both of the blue spruce trees. Write and solve an equation to find the ages of the ginkgo and dogwood trees. (Lesson 2-4)

14. MULTI-SELECT Select all the values of x that are solutions of $|3x - 6| = 12$. (Lesson 2-5)

A. -18

B. -6

C. -2

D. 2

E. 6

F. 18

15. MULTIPLE CHOICE A thermometer is accurate to $\pm 2°F$. Which absolute value equation can be used to find the greatest and least possible temperatures if the thermometer reading is 17°F? (Lesson 2-5)

A. $|t - 17| = 2$

B. $|t + 17| = 2$

C. $|t - 2| = 17$

D. $|t + 2| = 17$

16. OPEN RESPONSE Solve the absolute value equation $-5|x + 1| + 2 = 12$. If there is no solution, state no solution. (Lesson 2-5)

17. OPEN RESPONSE Solve $\frac{b}{12} = \frac{10}{15}$. (Lesson 2-6)

18. MULTIPLE CHOICE Solve $\frac{3}{x+4} = \frac{2}{x-4}$. (Lesson 2-6)

A. $x = -4$

B. $x = 4$

C. $x = 8$

D. $x = 20$

19. MULTIPLE CHOICE A biologist estimated that 5% of the seagulls in a flock have been banded. There were 22 seagulls that have been banded. Which equation and solution represent the approximate number of seagulls, g, that were in the flock? (Lesson 2-6)

A. $0.005g = 22$; $g = 4400$

B. $0.05g = 22$; $g = 440$

C. $22g = 500$; $g = 22$

D. $\frac{5}{g} = 22$; $g = 227$

20. MULTIPLE CHOICE On a map of Texas, the distance between Dallas and Houston is 4.8 inches. If 1 inch = 50 miles, what is the distance, in miles, between the two cities? (Lesson 2-6)

A. 96 miles

B. 104 miles

C. 240 miles

D. 250 miles

21. MULTIPLE CHOICE If 1 foot $\approx$ 0.305 meter, approximately how many feet are in 8 meters? (Lesson 2-6)

A. 2.44

B. 2.62

C. 24.4

D. 26.2

22. OPEN RESPONSE Solve the formula for the circumference of a circle, $C = 2\pi r$, for r. (Lesson 2-7)

23. OPEN RESPONSE The volume of a right pyramid is given by the formula $V = \frac{1}{3}Bh$, where B is the area of the base and h is the height. (Lesson 2-7)

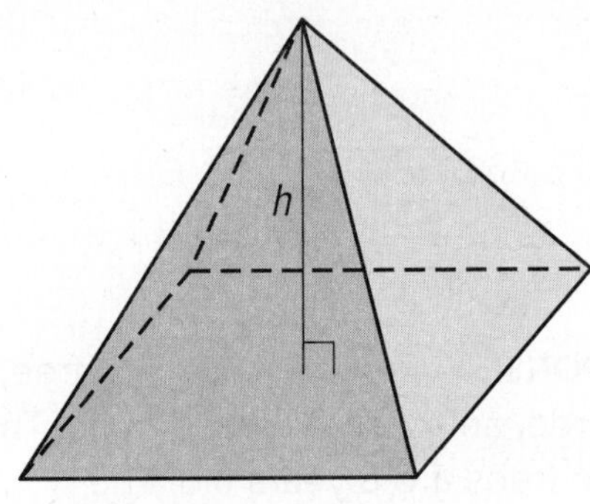

Part A Solve the formula for h.

Part B Find the height, in inches, of a right pyramid with a volume of 900 cubic inches and a base area of 225 square inches.

Module 3

Relations and Functions

Essential Question

Why are representations of relations and functions useful?

What Will You Learn?

How much do you already know about each topic **before** starting this module?

KEY

— I don't know. — I've heard of it. — I know it!

	Before			After		
	I don't know	I've heard of it	I know it	I don't know	I've heard of it	I know it
represent relations using ordered pairs, tables, graphs, and mappings						
analyze graphs of relations						
choose and interpret the scale on a coordinate graph						
determine whether relations are functions						
use function notation						
determine whether a graph is discrete, continuous, or neither						
determine whether a function is linear or nonlinear						
write linear functions in standard form						
find *x*- and *y*-intercepts of graphs						
interpret intercepts of graphs of functions						
determine whether a graph has line symmetry						
identify where a graph is increasing and where it is decreasing						
find extrema of a function						
describe the end behavior of a function						
sketch graphs of functions						
solve equations by graphing						

Foldables Make this Foldable to help you organize your notes about expressions. Begin with one sheet of 11″ × 17″ paper.

1. **Fold** the short sides to meet in the middle.
2. **Fold** the booklet in thirds lengthwise.
3. **Open and cut** the booklet in thirds lengthwise.
4. **Label** the tabs as shown.

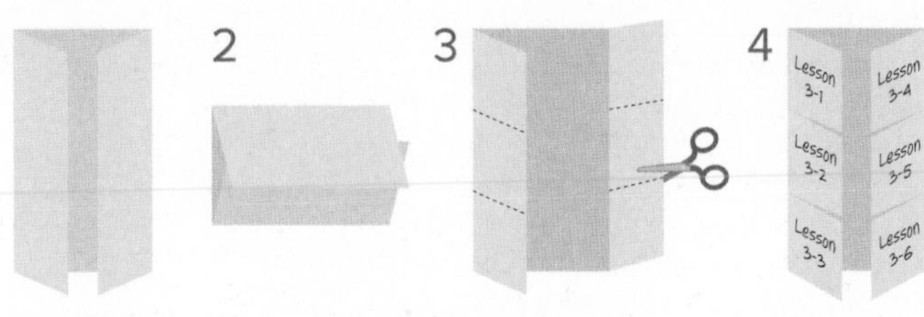

What Vocabulary Will You Learn?

- continuous function
- decreasing
- dependent variable
- discrete function
- domain
- end behavior
- extrema
- function
- function notation
- increasing
- independent variable
- line symmetry
- linear equation
- linear function
- mapping
- negative
- nonlinear function
- positive
- range
- relation
- relative maximum
- relative minimum
- root
- scale
- *x*-intercept
- *y*-intercept
- zero

Are You Ready?

Complete the Quick Review to see if you are ready to start this module. Then complete the Quick Check.

Quick Review

Example 1

Graph and label the point $A(3, 5)$ on the coordinate plane.

Start at the origin. Since the *x*-coordinate is positive, move 3 units to the right. Then move 5 units up since the *y*-coordinate is positive. Draw a dot and label it *A*.

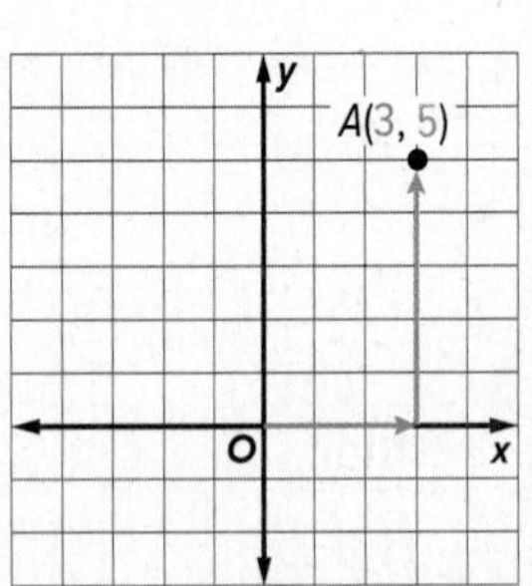

Example 2

Evaluate $5x + 13$ for $x = 4$.

Substitute the known value for *x*. Then follow the order of operations.

$5x + 13$	Original expression
$= 5(4) + 13$	Substitute 4 for *x*.
$= 20 + 13$	Multiply.
$= 33$	Add.

Quick Check

Graph and label each point on the coordinate plane.

1. $B(4, 2)$
2. $C(0, 3)$
3. $G(3, 1)$
4. $H(2, 0)$

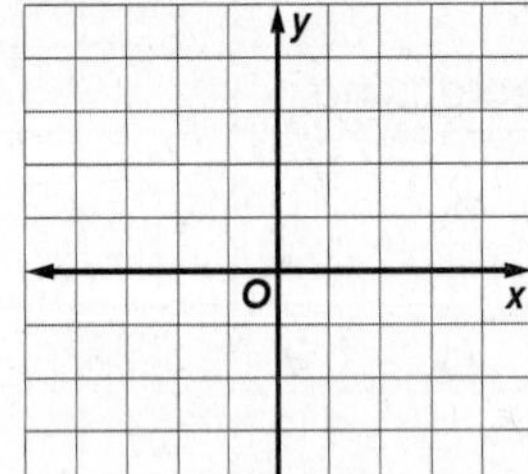

Evaluate each expression for the given value.

5. $3x - 1$ for $x = 7$
6. $\frac{1}{2}x$ for $x = 12$
7. $5x + 2$ for $x = -1$
8. $2x - 9$ for $x = -3$

How Did You Do?

Which exercises did you answer correctly in the Quick Check?

Representing Relations

Learn Relations

You can use math to represent the relationship between two sets of numbers. For example, suppose you recorded the number of minutes you spent driving for your driver's training course each day for a week. You may use the set {1, 2, 3, 4, 5, 6, 7} to represent the days and the set {30, 45, 20, 40, 45, 90, 60} to represent the minutes. The relationship between the sets pairs each day with the time driven that day. This pairing of the numbers is called a **relation**. The set of days is the **domain** of the relation and the set of times is the **range**.

A relation is a set of ordered pairs. The set of the first numbers of the ordered pairs in a relation is called the domain. The set of second numbers of the ordered pairs in a relation is called the range.

A relation can be represented in multiple ways. A **mapping** illustrates the relationship between the domain and range by showing how each element of the domain is paired with an element in the range. An equation shows the relationship between the domain and range of a relation where substituting each value in the domain for x results in the corresponding y-value of the range. The x-values in the domain and the resulting y-values can be written as ordered pairs and are called solutions of the equation because they make the equation true. Below is a mapping, table, graph, and equation of the relation {(3, 5), (−4, −2), (0, 2)}.

mapping

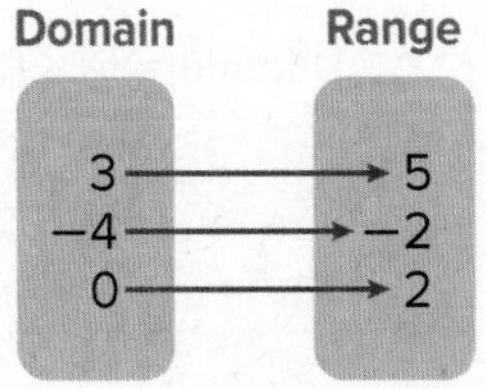

Domain	Range
3 →	5
−4 →	−2
0 →	2

table

x	y
3	5
−4	−2
0	2

graph

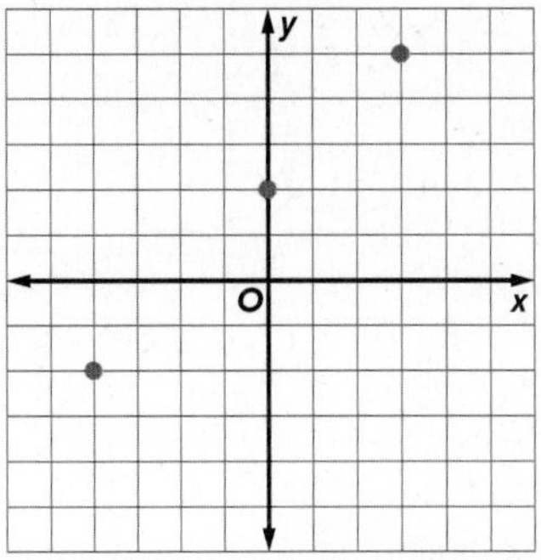

equation

$y = x + 2$

Today's Goals

- Represent relations.
- Interpret graphs of relations.
- Choose and interpret appropriate scales for the axes and origins of graphs.

Today's Vocabulary

relation
domain
range
mapping
independent variable
dependent variable
scale

Think About It!

Compare the table and mapping of the relation. What conclusions can you draw about the x- and y-coordinates and the domain and range?

Example 1 Representations of a Relation

Use {(2, 0), (0, 4), (3, −5), (−3, −5)}.

Part A Express the relation as a table, a graph, and a mapping.

table

x	y
−3	−5
0	4
2	0
3	−5

graph

mapping

Domain: 2, 0, 3, −3 → Range: 0, 4, −5

Part B Determine the domain and the range of the relation.

The domain of the relation is {−3, 0, 2, 3}.

The range of the relation is {−5, 0, 4}.

Check

List the ordered pairs in each relation.

x	y
−1	−5
4	7
−2	3

Domain: 1, 5, −6; Range: 5, −1

Table: ____?____

Graph: ____?____

Mapping: ____?____

 **Go Online** You can complete an Extra Example online.

Learn Analyzing Graphs of Relations

Graphing the total time driven during your driving course can help you visualize the progress.

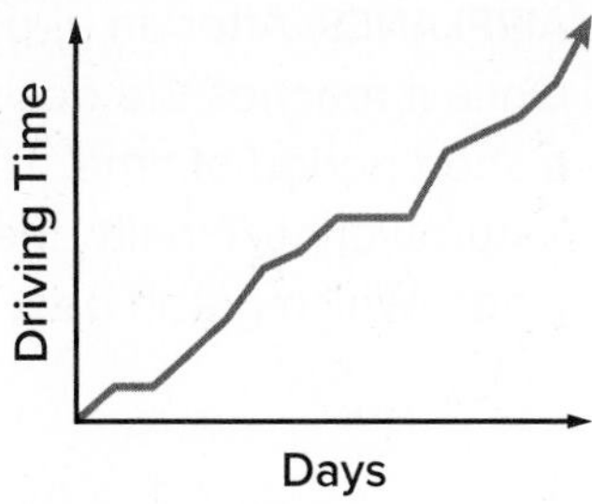

A relation can be graphed without a scale on either axis to show the relationship between the independent and dependent variables. These graphs can be interpreted by analyzing their shapes.

The values in the domain correspond to the independent variable in a relation. The **independent variable**, usually x, has a value that is subject to choice. In the graph above, the independent variable is the days.

The values in the range correspond to the dependent variable of the relation. The **dependent variable** is the variable in a relation, usually y, with values that depend on x. In the graph above, the dependent variable is the driving time.

Example 2 Analyze Graphs

TEXTING **The graph represents the number of text messages sent by Nora throughout the day.**

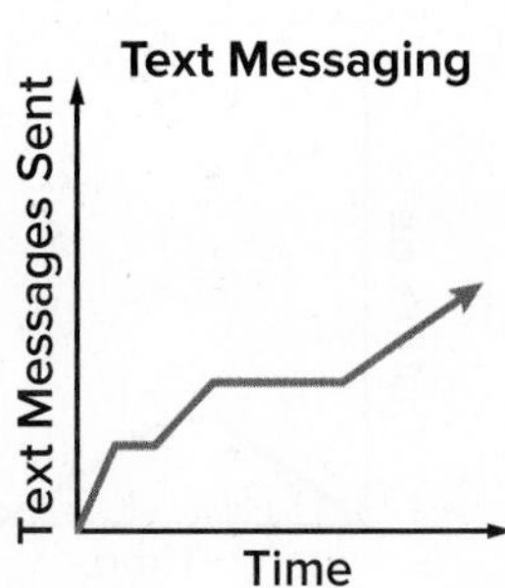

Part A Identify the independent and dependent variables of the relation. independent variable: time dependent variable: number of text messages sent

Part B Describe what happens in the graph.

As you move from left to right along the graph, time increases and the number of text messages sent increases until the graph becomes a horizontal line.

The horizontal line means that time is increasing, but the number of text messages sent remains constant. During this time, Nora stopped sending text messages.

Then she continued to send text messages until she stopped again for a period of time.

Finally, Nora began sending text messages again.

Check

Identify the independent and dependent variables of each relation.

a. The average price of a ticket to an amusement park has steadily increased over time.

The average price of a ticket is the ____?____ variable.
Time is the ____?____ variable.

b. The air pressure inside a soccer ball decreases with time.

Time is the ____?____ variable.
Air pressure is the ____?____ variable.

Go Online You can complete an Extra Example online.

Think About It!
What can you conclude from the graph about the rates at which Nora sent text messages throughout the day?

Check

AIRPLANES After an airplane takes off, its altitude increases rapidly. Once it reaches the desired altitude, it continues to fly at that level for a short period of time. Then the plane's altitude fluctuates slightly due to turbulence. Finally, the altitude decreases quickly as the plane lands. Which graph best represents this situation?

A.
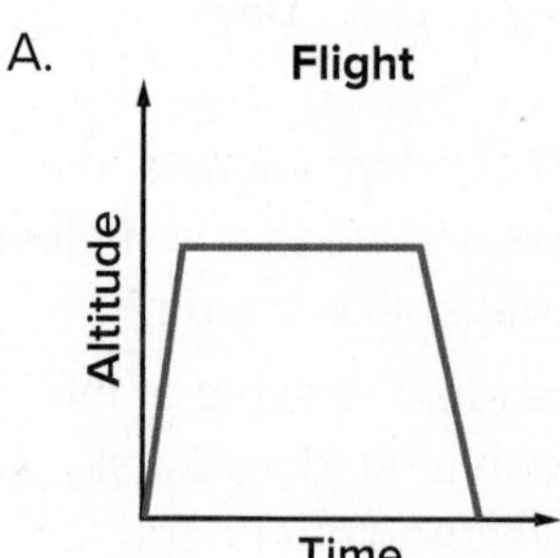

B.
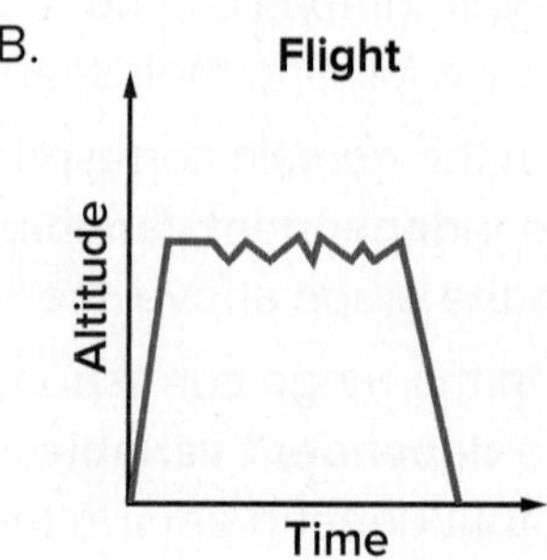

C.
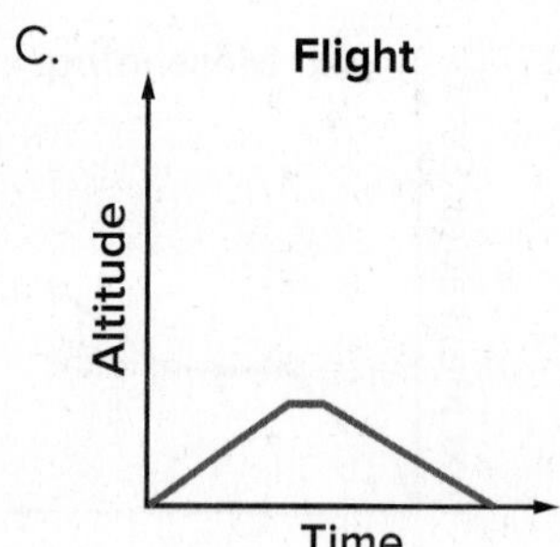

D.
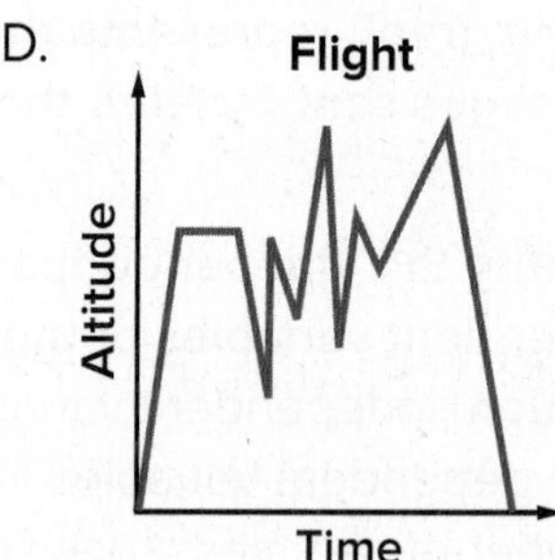

TRAFFIC Kane notices that traffic congestion on his street on weekday mornings depends on the time of day. He draws a graph to represent the relationship between the time of day and the volume of traffic congestion on his street on a weekday morning. Analyze the orange segment. Select the statement that describes what is happening during this time.

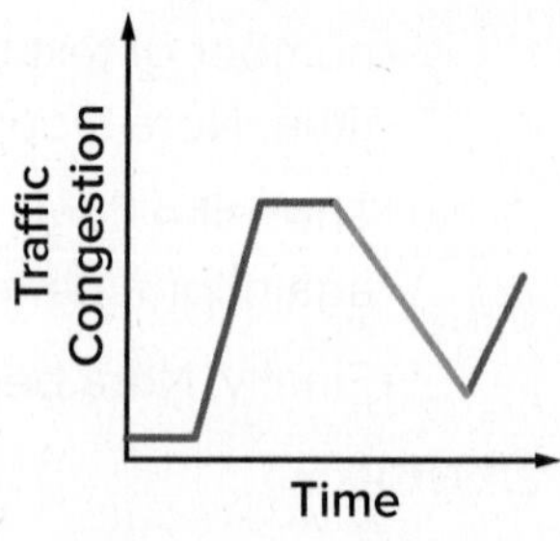

A. Traffic congestion is steadily increasing at a very fast rate.

B. Traffic congestion is constant and very light.

C. Traffic congestion is constant and very heavy.

D. Traffic congestion is decreasing at a steady rate.

Go Online You can complete an Extra Example online.

Explore Choosing Scales

Online Activity Use a real-world situation to complete the Explore.

> INQUIRY How can you tell if an appropriate scale is being used to represent a relationship?

Learn The Coordinate System

When graphing on the coordinate system, the **scale** of a graph refers to the distance, or interval, between tick marks on the *x*- and *y*-axes. For example, if one tick mark represents 5 units, then the scale of the graph is 5. Each axis may have a different scale.

A scale of 1 tick mark = 1 unit is frequently used in mathematics. However, using a different scale may make it easier to graph a given set of ordered pairs.

Example 3 Use Appropriate Scales

Graph (5, 80), (−36, 48), (25, −91), (38, 95), (−10, −50), and (1, 22).

Step 1 The *x*-coordinates are between −36 and 38.
The *y*-coordinates are between −91 and 95.

Step 2 The *x*-axis should include values from about −40 to 40.
The *y*-axis should include values from about −100 to 100.

A scale of 5 on the *x*-axis is appropriate.
A scale of 10 on the *y*-axis is appropriate.

Step 3 Graph.

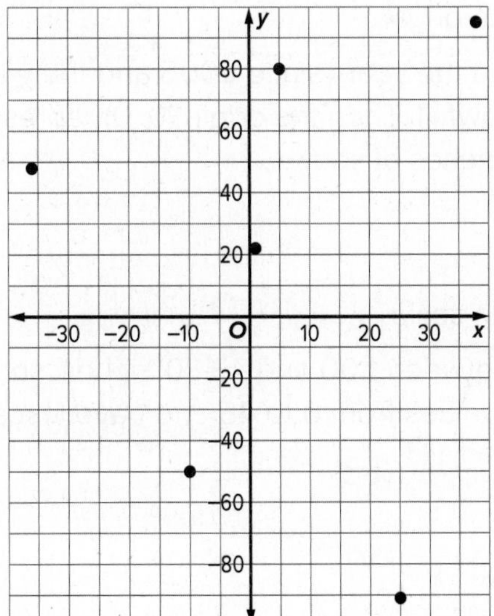

Go Online You can complete an Extra Example online.

Talk About It!

Describe a real-world situation in which it might be easier to count by a value other than 1.

Watch Out!

Large Scales Be careful not to choose scales that are too large. If a scale is too large, it can be difficult to accurately graph points.

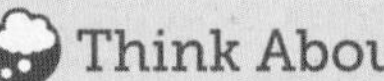

Think About It!

Gwen used a scale of 4 on the *x*-axis. Is her scale appropriate? Explain your reasoning.

Study Tip

Notation Scales of graphs are sometimes written as [−40, 40] scl: 5 or −40 to 40; scale: 5, where −40 and 40 are the minimum and maximum values and 5 is the scale.

Check

When graphing (16, 32), (−10, 11), (4, −27), and (−7, −5), select the most appropriate scale for

a. the *x*-axis	**b.** the *y*-axis
A. −20 to 20; scale: 1	**A.** −30 to 35; scale: 1
B. −12 to 18; scale: 2	**B.** −30 to 30; scale: 5
C. −12 to 18; scale: 6	**C.** −30 to 35; scale: 5
D. −20 to 20; scale: 10	**D.** −30 to 40; scale: 10

Example 4 Choose an Appropriate Origin

WEATHER **The table shows the total snowfall in January for Boston.**

Year	Total Snowfall (inches)
2005	43.30
2006	8.10
2007	1.00
2008	8.30
2009	23.70
2010	13.20
2011	38.30
2012	6.80
2013	5.00
2014	21.80
2015	34.30

Study Tip

Appropriate Origins When choosing an appropriate origin, you are still using the point (0, 0) as the origin, but are changing what it represents.

Part A Choose an appropriate origin.

Let the *x*-axis represent the years since 2005 and the *y*-axis represent the total snowfall. Then the origin (0, 0) represents the year 2005 and 0 inches of snow.

Part B Choose an appropriate scale.

The total snowfall is between 1.00 and 43.30 inches, so the *y*-axis should include values from 0 to 45 and have a scale of 5 inches.

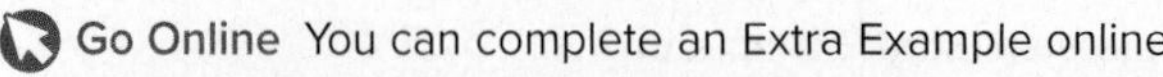

Part C Graph the data points on the coordinate plane.

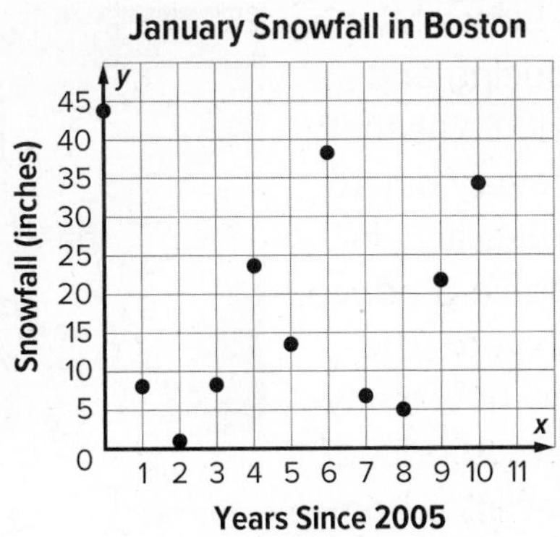

Example 5 Interpret Scales and Origins

SOCIAL MEDIA **The average number of posts per day on a social media site each year is given in the table. Interpret the meaning of the axes, scale, and origin of the corresponding graph of the data.**

Year	Average Posts Per Day (millions)
2008	15
2009	20
2010	35
2011	50
2012	100
2013	200
2014	340
2015	500
2016	560
2017	625

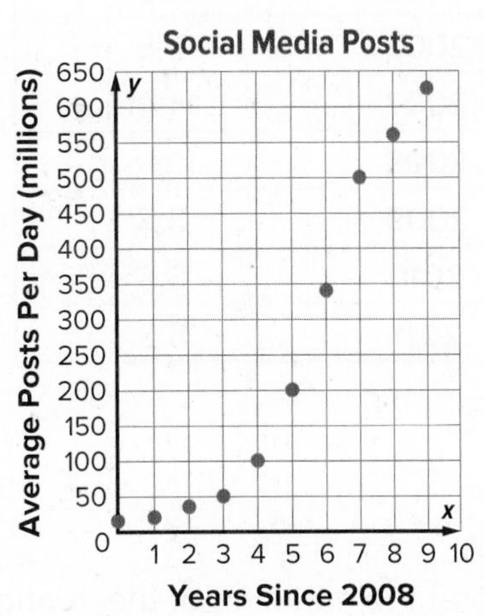

x- and *y*-axes

The *x*-axis represents the number of years since 2008, because it is not appropriate to start at the year 0.

The *y*-axis represents the average number of posts per day (in millions) on the social media site.

Scale

The *x*-axis has a scale of 1 mark = 1 year.
The *y*-axis has a scale of 1 mark = 50 million posts.

Origin

The origin (0, 0) represents the year 2008 and 0 posts per day.

Think About It!

Describe another situation where it might be necessary to choose a different meaning for the axes and origin.

Use a Source

Find data about the number of participants in a sport or other activity of interest to you in recent years. If a graph is provided, interpret the scales of the axes. If no graph is provided, create one with appropriate scales.

Check

MONEY The United States Mint is responsible for producing and distributing circulating coins that are used by people every day to buy and sell goods. Facilities in Denver and Philadelphia produce billions of coins each year. The table and graph show how many circulating coins were produced each year at these facilities from 2001 to 2011.

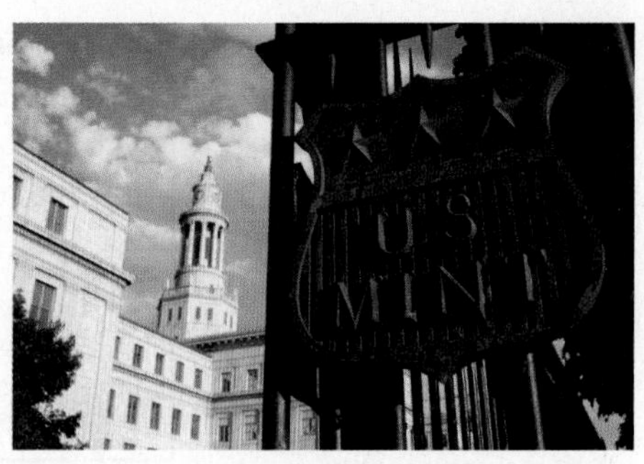

Year	Circulating Coins Produced (billions)
2001	19.4
2002	14.4
2003	12
2004	13.2
2005	15.3
2006	15.5
2007	14.4
2008	10.1
2009	3.5
2010	6.4
2011	8.2

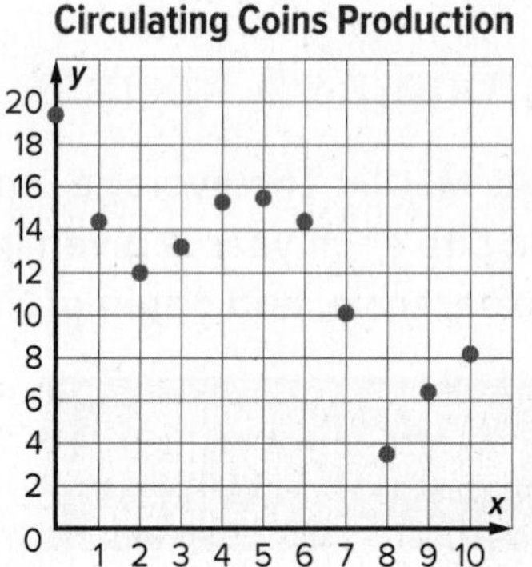

Part A

Interpret the meaning of the x- and y-axes in the context of the situation.

The x-axis represents ____?____ and has a scale of 1 mark = __?__.

The y-axis represents ____?____ and has a scale of 1 mark = __?__.

Part B

Interpret the meaning of the origin in the context of the situation.

The origin (0, 0) represents the year __?__ and __?__ coins produced.

 **Go Online** You can complete an Extra Example online.

Practice

Go Online You can complete your homework online.

Example 1

Express each relation as a table, a graph, and a mapping. Then determine the domain and range.

1. {(–1, –1), (1, 1), (2, 1), (3, 2)}

2. {(0, 4), (– 4, – 4), (–2, 3), (4, 0)}

3. {(3, –2), (1, 0), (–2, 4), (3, 1)}

Example 2

4. PAYCHECK The graph represents the amount of Seth's paycheck for different numbers of hours he works.

a. Identify the independent and dependent variables of the relation.

b. Describe what happens in the graph.

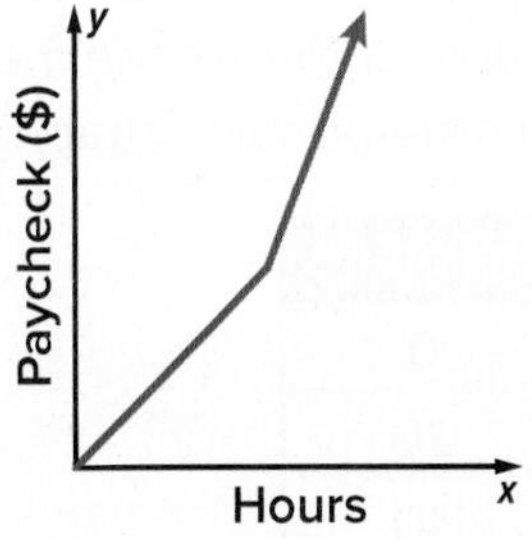

5. DEMAND The graph represents the price of an item and the number of items purchased.

a. Identify the independent and dependent variables of the relation.

b. Describe what happens in the graph.

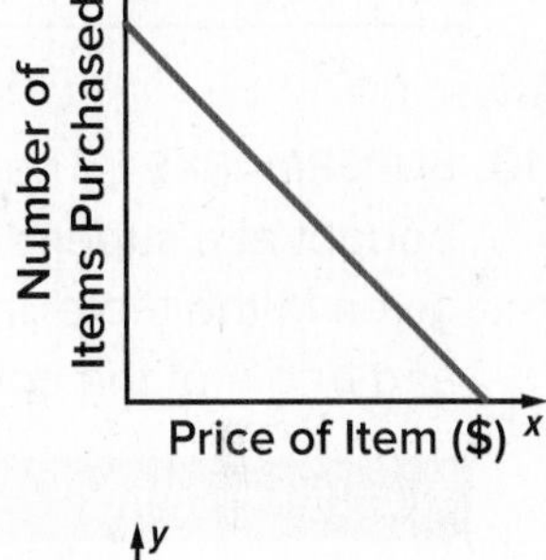

6. AIRPLANES The graph represents the number of hours of a flight and the distance an airplane is from the ground.

a. Identify the independent and dependent variables of the relation.

b. Describe what happens in the graph.

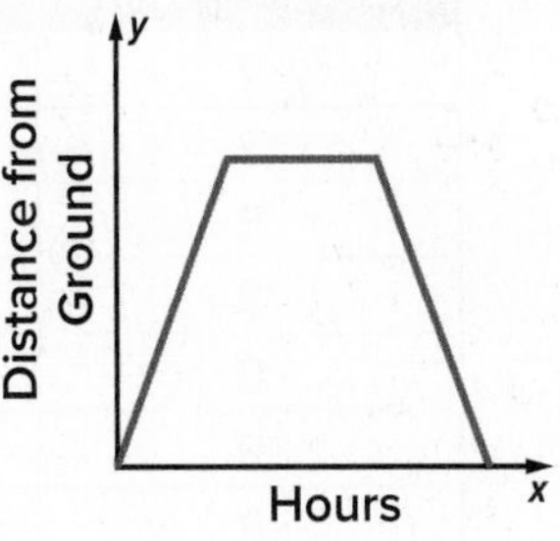

Examples 3 and 4

7. HEALTH The American Heart Association recommends that your target heart rate during exercise should be between 50% and 75% of your maximum heart rate. Use the data in the table below to graph the approximate maximum heart rates for people of given ages.
Source: American Heart Association

Age (years)	20	25	30	35	40
Maximum Heart Rate (beats per minute)	200	195	190	185	180

8. USE TOOLS The following ordered pairs give the length in feet and the weight in pounds of five snakes at the reptile house at a zoo. Graph the data. {(5.5, 4.5), (3, 0.5), (3, 2), (8, 4.5), (2, 0.5)}

Example 5

9. ELEVATOR The height of an elevator above the ground is given in the table. Interpret the meaning of the axes, scale, and origin of the corresponding graph of the data.

Time (s)	Height (ft)
0	0
1	20
2	40
3	60
4	80

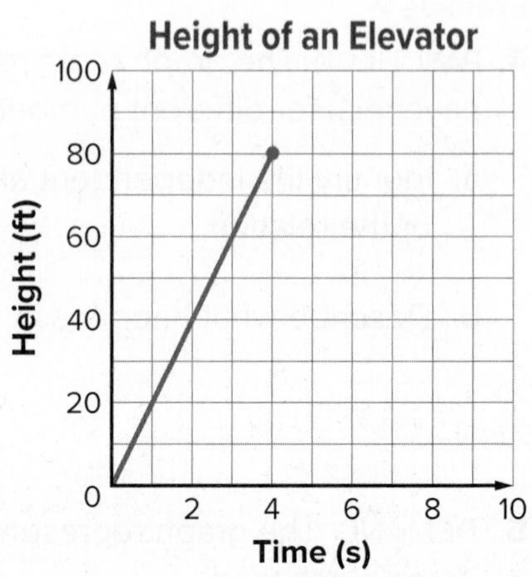

10. SUPERMARKET The number of items that eight customers bought at a supermarket and the total cost of the items is given in the table. Interpret the meaning of the axes, scale, and origin of the corresponding graph of the data.

Number of Items	Total Cost ($)
2	2
3	6
5	8
5	10
6	16
8	12
8	18
9	14

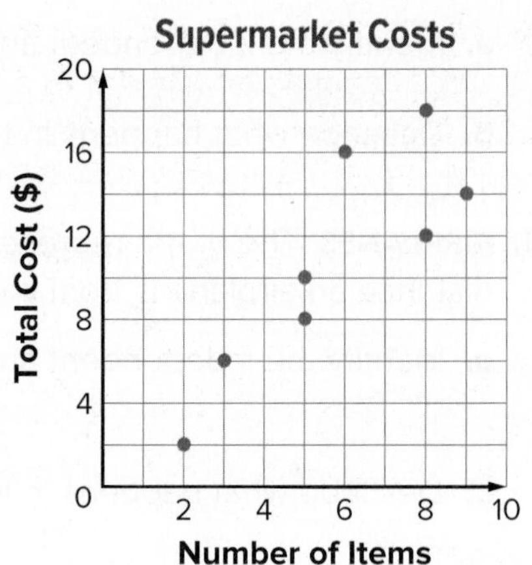

Mixed Exercises

For Exercises 11–14, express the relation in each table, graph, or mapping as a set of ordered pairs.

11.

x	y
1	7
3	45
5	11
13	15

12.

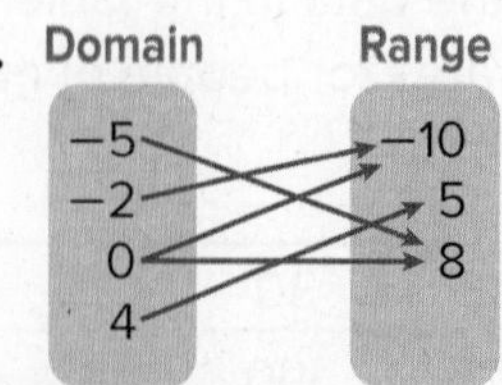

13. 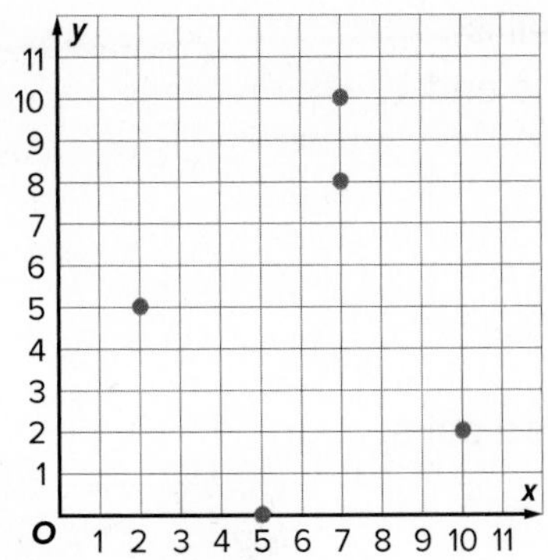

14.

x	y
4	2
7	−7
15	9
12	0

15. NATURE Maple syrup is made by collecting sap from sugar maple trees and boiling it down to remove excess water. The graph shows the number of gallons of tree sap required to make different quantities of maple syrup. Express the relation as a set of ordered pairs. Interpret the meaning of the axes, scale, and origin of the corresponding graph of the data.
Source: Vermont Maple Sugar Makers' Association

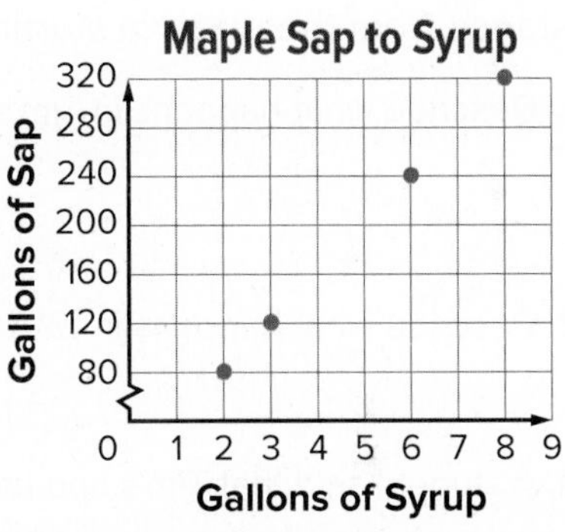

16. TRAVEL Omari drives a car that gets 18 miles per gallon of gasoline. The car's gasoline tank holds 15 gallons. The distance Omari drives before refueling is a function of the number of gallons of gasoline in the tank. Identify a reasonable domain for this situation.

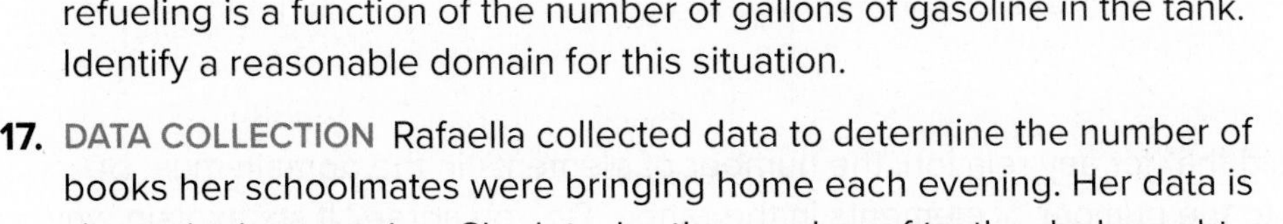

17. DATA COLLECTION Rafaella collected data to determine the number of books her schoolmates were bringing home each evening. Her data is shown in the mapping. She let x be the number of textbooks brought home after school, and y be the number of students with x textbooks.

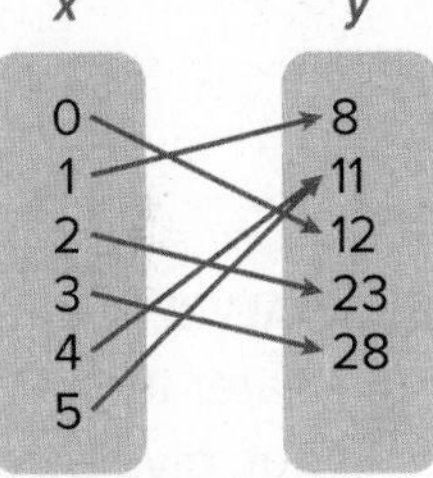

a. Express the relation as a set of ordered pairs.

b. What is the domain of the relation?

c. What is the range of the relation?

18. COOKIES Identify the graph that best represents the relationship between the number of cookies and the equivalent number of dozens.

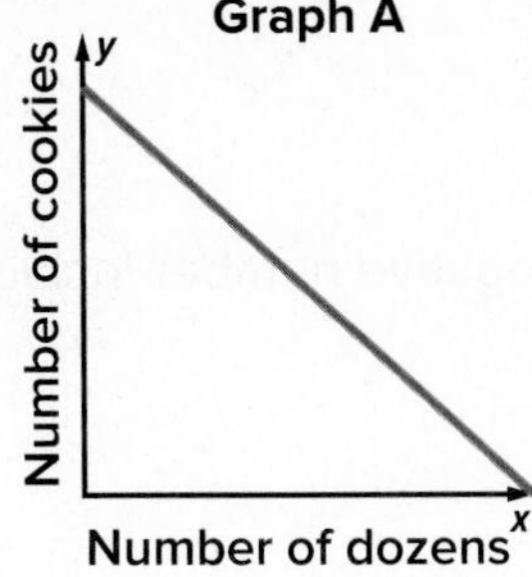

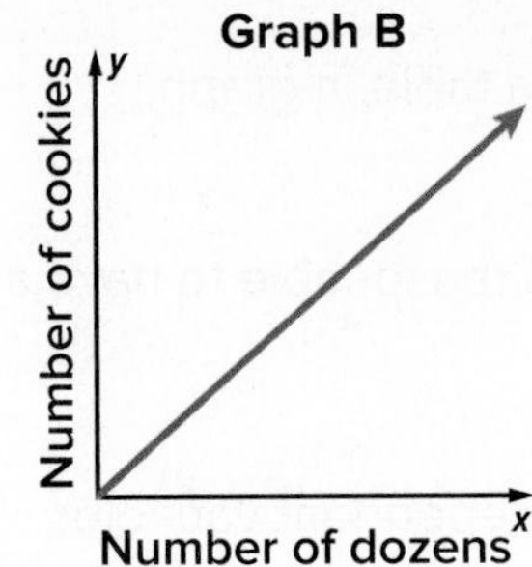

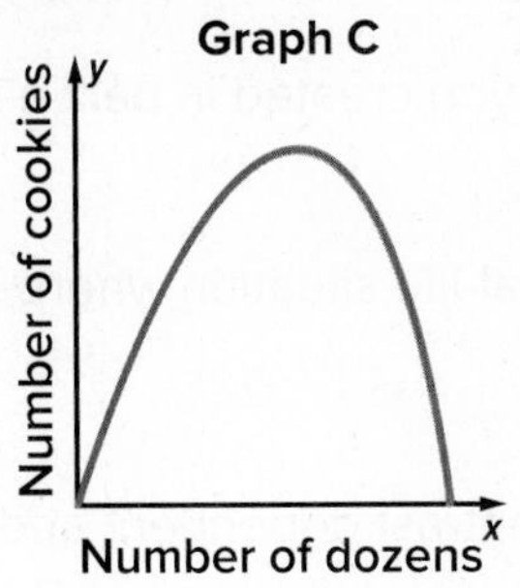

19. **MOWING** Cordell is mowing his front lawn. His mailbox is on the edge of the lawn. Draw a reasonable graph that shows the distance Cordell is from the mailbox as he mows. Let the horizontal axis show the time and the vertical axis show the distance from the mailbox.

20. **STRUCTURE** Express the relation in the mapping as a set of ordered pairs.

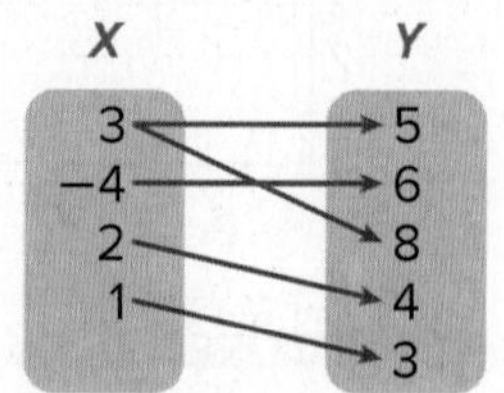

Tim and Lauren use their cars to deliver pizzas. The graph represents their distance from the pizzeria starting at 6 P.M. Use the graph for Exercises 21–24.

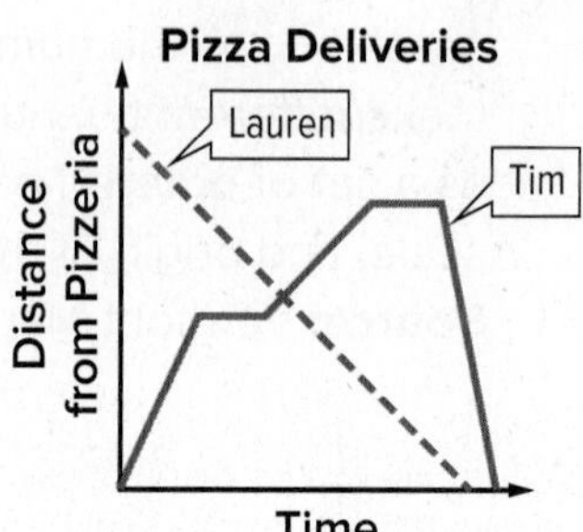

21. Describe what happens in Tim's graph.

22. Describe what happens in Lauren's graph.

23. A student said that Tim's and Lauren's graphs intersect, so their cars must have crashed at some time after 6 p.m. Do you agree or disagree? Explain.

24. After 6 p.m., which delivery person was the first to return to the pizzeria? How do you know?

25. **ANALYZE** Cameron said that for any relation, the number of elements in the domain must be greater than or equal to the number of elements in the range. Do you agree? If so, explain why. If not, give a counterexample.

26. **CREATE** Think of a situation that could be modeled by this graph. Then label the axes of the graph and write several sentences describing the situation.

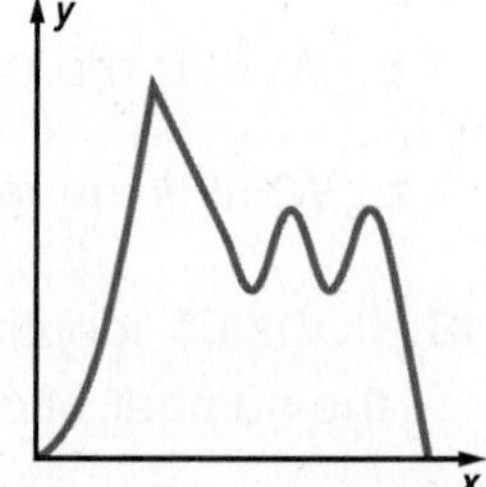

27. **CREATE** **Use the set {−1, 0, 1, 2} as a domain and the set {−3, −1, 4, 5} as a range.**
    a. Create a relation. Express the relation as a set of ordered pairs.
    b. Express the relation you created in **part a** as a table, a graph, and a mapping.

28. **ANALYZE** Describe a real-life situation where it is reasonable to have a negative number included in the domain or range.

29. **WRITE** Compare and contrast dependent and independent variables.

Functions

Today's Goals
- Determine whether relations are functions.
- Evaluate functions in function notation for given values.

Today's Vocabulary
function

function notation

Explore Vertical Line Test

 **Online Activity** Use graphing technology to complete the Explore.

INQUIRY How can you tell whether a relation is a function?

Learn Functions

A **function** is a relationship between input and output. In a function, there is exactly one output for each input. The relation shown is a function because each element of the domain is paired with *exactly* one element in the range.

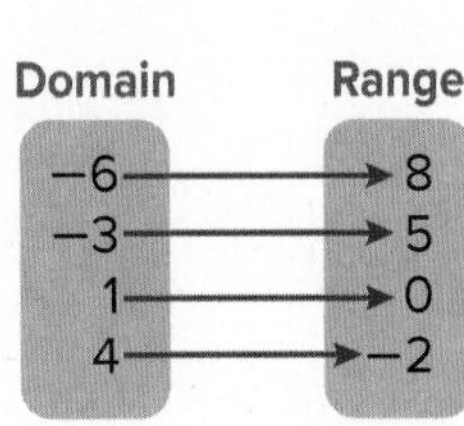

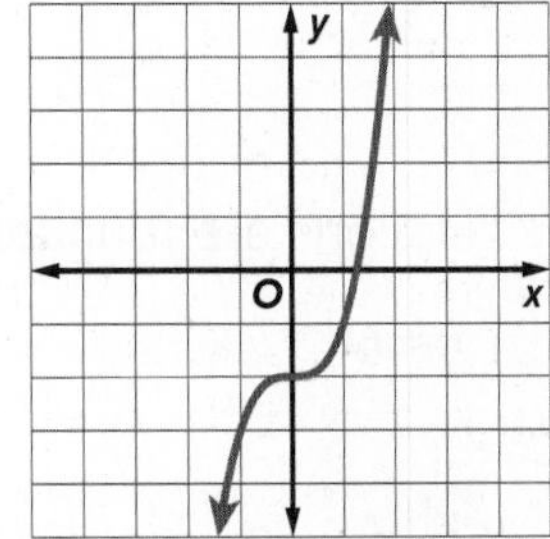

Talk About It!
Describe a way, other than using the vertical line test, that you could use to determine that a relation is not a function.

You can use the vertical line test to see if a graph represents a function.

Key Concept • Vertical Line Test	
function	**not a function**
A relation is a function if it passes the vertical line test, meaning a vertical line intersects the graph no more than once.	A relation is not a function if it fails the vertical line test, meaning that a vertical line intersects the graph more than once.
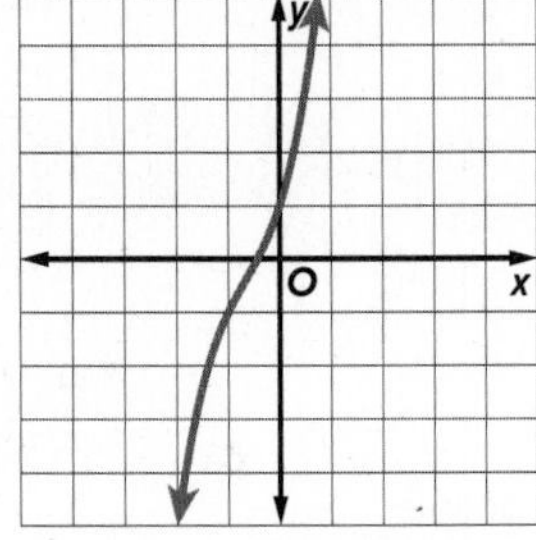	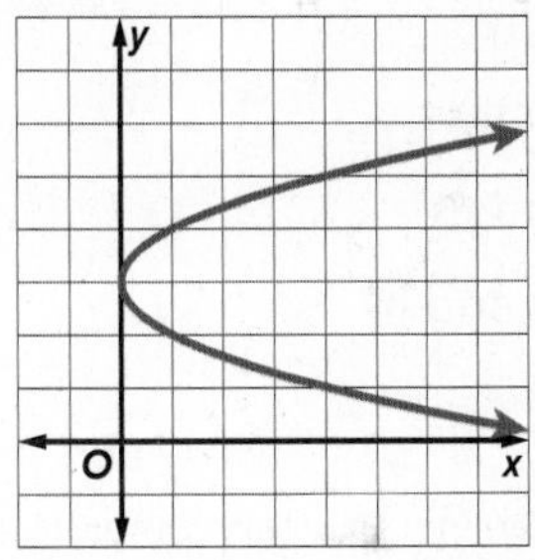

Example 1 Identify Functions

Determine whether each relation is a function. Explain.

a. **Domain** **Range**

−4 → −1
−2 → 3
2 → 3
5 → 6

For each element of the domain, there is only one element of the range. So this mapping represents a function.

b.

Domain	2	6	8	2
Range	3	5	5	−3

The element 2 in the domain is paired with both 3 and –3 in the range. This relation is not a function.

c. (2, 5), (4, 7), (8, 11), (4, 13)

The element 4 in the domain is paired with both 7 and 13 in the range. This relation is not a function.

Think About It!

Suppose the last element in the domain in part **b** was 4 as shown in the table below. Would this relation be a function? Justify your argument.

Domain	2	6	8	4
Range	3	5	5	−3

Check

Which of these relations are functions?

I.

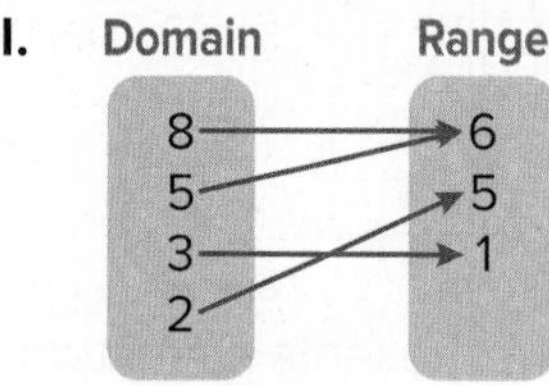

II.

Domain	0	2	4	0	−2
Range	8	6	3	−1	−3

III. {(−4, 5), (−2, 1), (1, −5), (−2, −7), (5, −13)}

A. I only
B. III only
C. I and II
D. I, II, and III

Go Online You can complete an Extra Example online.

Example 2 Analyze Data

LONG JUMP **Five schools are competing in the long jump portion of a track meet. The distances of the players with the best jump on each team are as follows: Team 1, 20.6 feet; Team 2, 21.5 feet; Team 3, 20.9 feet; Team 4, 19.4 feet; Team 5, 20.2 feet.**

Part A Make a table.

Team Number	1	2	3	4	5
Best Jump (ft)	20.6	21.5	20.9	19.4	20.2

Part B Determine the domain and range of the relation.

Domain: {1, 2, 3, 4, 5}

Range: {20.6, 21.5, 20.9, 19.4, 20.2}

Part C Determine whether the relation is a function.

For each element of the domain, there is only one element of the range. So, this relation is a function.

Think About It!

Explain why the domain and range of the function are not all real numbers.

Study Tip

Remember the domain is related to the independent variable, and the range is related to the dependent variable.

Check

HEIGHT Bailey recorded the heights of five of her friends. The heights, in inches, are as follows: Hunter, 62; Ling, 66; Ela, 65; Omar, 66; Alma, 67.

Part A Find the domain and range of the relation.

A. D: {Hunter, 62, Ling, 66, Ela}; R: {65, Omar, 66, Alma, 67}

B. D: {65, Omar, 66, Alma, 67}; R: {Hunter, 62, Ling, 66, Ela}

C. D: {Hunter, Ling, Ela, Omar, Alma}; R: {62, 65, 66, 67}

D. D: {62, 65, 66, 67}; R: {Hunter, Ling, Ela, Omar, Alma}

Part B Is the relation a function? Explain your reasoning.

A. Yes; for each element of the domain, there is only one element of the range.

B. No; it fails the vertical line test.

C. No; an element in the domain is paired with more than one element in the range.

D. No; the domain and range are all real numbers.

Go Online You can complete an Extra Example online.

Think About It!

Find the domain and range of the function.

Study Tip

Vertical Line Test You can also perform the vertical line test by placing a pencil at the left of the graph and slowly moving it across the graph to represent a vertical line.

Example 3 Equations as Functions

Determine whether $6x + 2y = 14$ is a function. Explain.

For any value x, the vertical line passes through no more than one point on the graph. So, the graph and the equation represent a function.

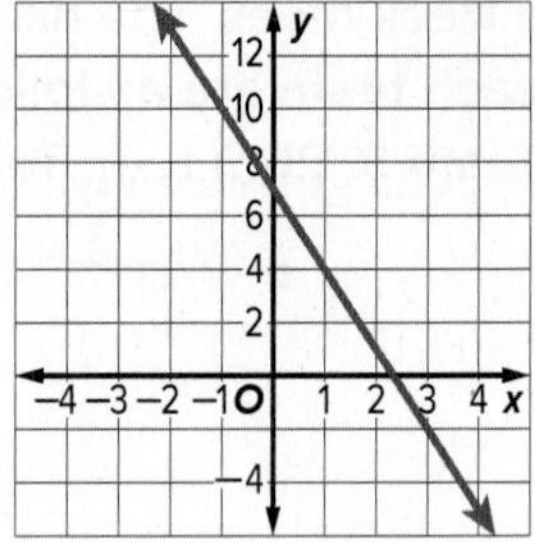

Check

Which relations described in the chart represent functions? Write Yes or No.

	Yes	No
A. The relation passes the vertical line test.		
B. An element of the domain is paired with only two elements of the range.		
C. A vertical line passes through more than one point on the graph.		
D. The domain represents each student in class, while the range represents the age of each student.		
E. The domain represents the number of a day in May, while the range represents the high temperature for the day.		
F. The relation is of the form $y = x^2 + b$.		

Learn Function Values

Equations that are functions can be written in a form called **function notation**. Function notation is a way of writing an equation so that $y = f(x)$.

In a function, x represents the elements of the domain, and $f(x)$ represents the elements of the range. The graph of $f(x)$ is the graph of the equation $y = f(x)$.

Go Online You can complete an Extra Example online.

Example 4 Find Function Values

For $f(x) = -2x + 9$, find each value.

$f(4) + 3$

$f(4) + 3 = [-2(4) + 9] + 3$ $x = 4$

$= (-8 + 9) + 3$ Multiply.

$= 1 + 3$ Add.

$= 4$ Add.

$f(5) - f(1)$

$f(5) - f(1) = [-2(5) + 9] - [-2(1) + 9]$ Substitute for x.

$= (-10 + 9) - (-2 + 9)$ Multiply.

$= -1 - 7$ Add.

$= -8$ Subtract.

Check

For $f(x) = -3x + 2$, find each value.

a. $f(-4) =$ _?_   **b.** $f(5) + 8 =$ _?_   **c.** $f(3) - f(6) =$ _?_

Example 5 Evaluate Functions

For $h(x) = -6x^2 + 18x + 36$, find each value.

a. $h(2)$

$h(2) = -6(2)^2 + 18(2) + 36$ $x = 2$

$= -24 + 36 + 36$ Multiply.

$= 48$ Add.

b. $h(4) - h(1)$

$h(4) - h(1) = [-6(4)^2 + 18(4) + 36] - [-6(1)^2 + 18(1) + 36]$

$= (-96 + 72 + 36) - (-6 + 18 + 36)$ Multiply.

$= 12 - 48$ Add.

$= -36$ Subtract.

c. $h(5) - 7$

$h(5) - 7 = [-6(5)^2 + 18(5) + 36] - 7$ $x = 5$

$= (-150 + 90 + 36) - 7$ Multiply.

$= -24 - 7$ Add.

$= -31$ Subtract.

Go Online You can complete an Extra Example online.

Think About It!
Is $f(3) - f(-2)$ the same as $f(3) + f(2)$? Justify your argument.

Think About It!
Find $h(3)$. Then find $h(-3)$. What was similar and what was different about finding these values?

Check

For $g(x) = 2x^2 - 8x - 10$, find each value.

a. $g(2) =$ __?__

b. $g(5) + 12 =$ __?__

c. $g(8) - 18 =$ __?__

d. $g(-3) - g(4) =$ __?__

Example 6 Interpret Function Values

EMPLOYMENT **Mason works at the movie theater after school. The function $g(x) = 9.25x$ represents the amount of money he makes before taxes for each hour that he works. Evaluate and interpret the function for each value.**

a. $g(8)$

$g(x) = 9.25(8) = 74$

This means that if Mason works for 8 hours, he will make $74 before taxes.

b. $g(14)$

$g(x) = 9.25(14) = 129.5$

This means that if Mason works for 14 hours, he will make $129.50 before taxes.

c. $g(27.5)$

$g(x) = 9.25(27.5) = 254.375$

This means that if Mason works for 27.5 hours, he will make $254.38 before taxes.

Think About It!

Find $g(-4)$. What does it mean in the context of the situation?

Check

POOLS A swimming pool that holds 10,000 gallons of water is being filled at a rate of 600 gallons per hour. The function $h(x) = 600x$ represents the amount of water in the pool after x hours.

Part A Find $h(3.25)$.

$h(3.25) =$ __?__

Part B Describe the meaning of $h(3.25)$ in the context of the situation.

After __?__ the total amount of water in the pool will be __?__.

Watch Out!

Remember, it is possible for Mason to work for zero hours and make no money, but it is not possible for him to work a negative number of hours.

Go Online You can complete an Extra Example online.

Practice

Go Online You can complete your homework online.

Examples 1 and 3

Determine whether each relation is a function. Explain.

1.

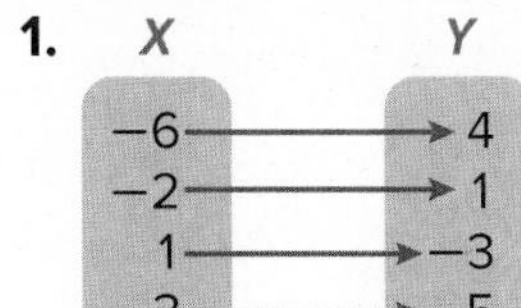

2.

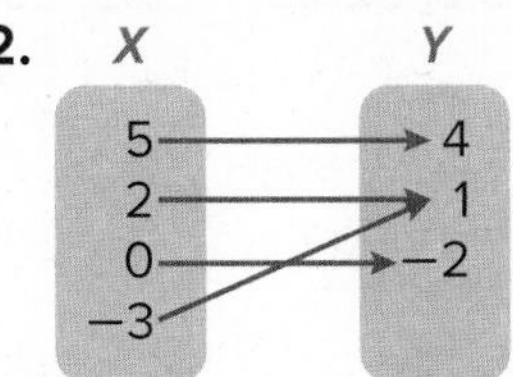

3.

X Y
4 → 2
4 → 5
6 → −1
7 → 3

4.

x	y
4	−5
−1	−10
0	−9
1	−7
9	1

5.

x	y
2	7
5	−3
3	5
−4	−2
5	2

6.

x	y
3	7
−1	1
1	0
3	5
7	3

7. {(2, 5), (4, −2), (3, 3), (5, 4), (−2, 5)}

8. {(6, −1), (−4, 2), (5, 2), (4, 6), (6, 5)}

9. $y = 2x - 5$

10. $y = 11$

11.

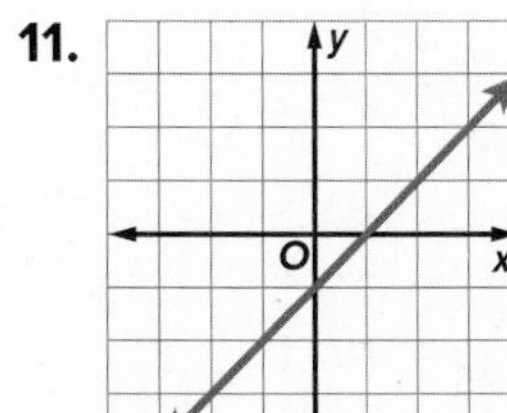

12.

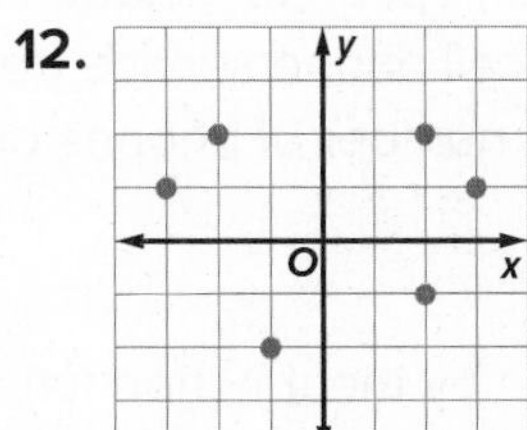

13.

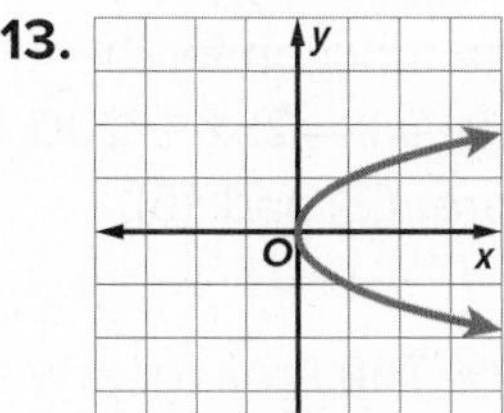

Example 2

14. TURNPIKE The total cost for cars entering President George Bush Turnpike at Beltline Road is related to the number of cars entering the turnpike at Beltline Road. The costs of different numbers of cars entering the turnpike at Beltline Road are as follows: 1 car, \$0.75; 2 cars, \$1.50; 3 cars, \$2.25; 4 cars, \$3.00; 5 cars, \$3.75.

a. Make a table.

b. Determine the domain and range of the relation.

c. Determine whether the relation is a function.

15. HOME VALUE The average value of a house changes over time. The average values of a house in January in Denver, Colorado, for different years are as follows: 2014, \$254,000; 2015, \$293,000; 2016, \$338,000; 2017, \$372,000.

a. Make a table.

b. Determine the domain and range of the relation.

c. Determine whether the relation is a function.

Examples 4 and 5

If $f(x) = 3x + 2$ and $g(x) = x^2 - x$, find each value.

16. $f(4)$

17. $f(8)$

18. $f(-2)$

19. $g(2)$

20. $g(-3)$

21. $g(-6)$

22. $f(2) + 1$

23. $f(1) - 1$

24. $g(2) - 2$

25. $g(-1) + 4$

26. $f(x) + 1$

27. $g(3b)$

Example 6

28. CELL PHONES Many cell phone plans have an option to include more than one phone. The function for the monthly cost of cell phone service from a wireless company is $f(x) = 25x + 200$, where x is the number of phones on the plan. Find and interpret $f(3)$ and $f(5)$.

29. GEOMETRY The area for any square is given by the function $f(x) = x^2$, where x is the length of a side of the square. Find and interpret $f(3.5)$.

30. TRANSPORTATION The cost of riding in a cab is \$3.00 plus \$0.75 per mile. The function that represents this relation is $f(x) = 0.75x + 3$, where x is the number of miles traveled and $f(x)$ is the cost of the trip. Find and interpret $f(17)$.

31. GYM MEMBERSHIP The cost of a gym membership is \$75 plus \$30 per month. The function that represents this relation is $f(x) = 30x + 75$, where x is the number of months and $f(x)$ is the cost. Find and interpret $f(12)$.

Mixed Exercises

32. USE A MODEL Mario collected data about some of the players on a women's basketball team. The data are shown in the table, mapping, and graph. Is each relation a function? Why or why not?

Team History	
Years on Team	Games Played
1	24
2	45
3	82
3	88
5	120

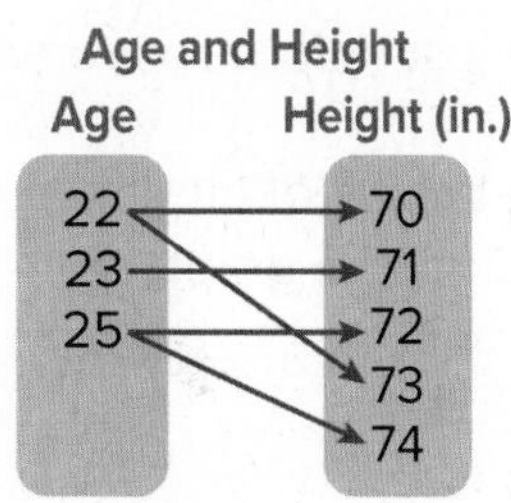

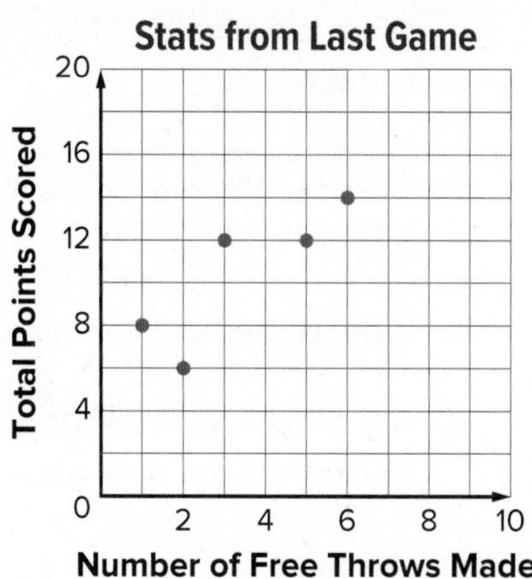

If $f(x) = -2x - 3$ and $g(x) = x^2 + 5x$, find each value.

33. $f(-1)$ **34.** $f(6)$ **35.** $g(2)$

36. $g(-3)$ **37.** $g(-2) + 2$ **38.** $f(0) - 7$

39. $f(4y)$ **40.** $g(-6m)$ **41.** $f(c) - 5$

42. $f(r) + 2$ **43.** $5[f(d)]$ **44.** $3[g(n)]$

45. FINANCIAL LITERACY Aisha has $40 to spend for her ornithology club. She spends some of it buying birdseed and saves the rest. Her savings is given by $f(x) = 40 - 1.25x$, where x is the number of pounds of birdseed she buys at $1.25 per pound.

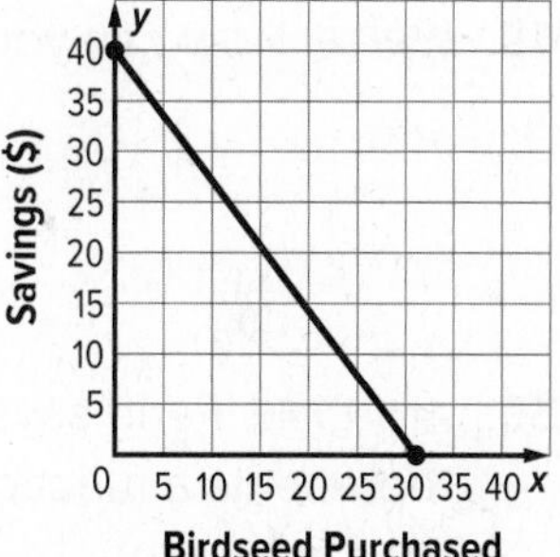

a. Graph the equation.

b. Is the relation a function? Explain.

c. Find $f(3)$, $f(18)$, and $f(36)$. What do these values represent?

d. How many pounds of birdseed can Aisha buy if she wants to save $30?

46. USE A MODEL A recipe for homemade pasta dough says that the number of eggs you need is always one more than the number of servings you are making.

a. Make a graph that shows the relationship between the number of servings and the number of eggs.

b. Is the relation a function? Explain.

c. What is the domain of the relation? Describe its meaning in the context of the situation.

47. **REASONING** The height h of a balloon, in feet, t seconds after it is released is given by the function $h(t) = 2t + 6$.

a. What is the value of $h(20)$, and what does it mean in the context of the situation?

b. Explain how to use the function to find the height of the balloon 2 minutes after it is released.

c. What is the height of the balloon just before it is released? How do you know?

d. Are there any restrictions on the values of t that can be used as inputs for the function? If so, how would this affect the graph of the function? Explain.

Higher-Order Thinking Skills

48. **CONSTRUCT ARGUMENTS** The following set of ordered pairs represents a function, but one of the values is missing: {(–4, –1), (–3, –1), (3, 2), (5, 2), (?, 2)}. What conclusions can you make about the missing value? Justify your arguments.

49. **WRITE** How can you determine whether a relation represents a function?

50. **ANALYZE** Feng says that the set of ordered pairs {(0, 1), (3, 2), (3, −5), (5, 4)} represents a function. Determine whether his statement is *true* or *false*. Justify your argument.

51. **PERSEVERE** Consider $f(x) = -4.3x - 2$. Write $f(g + 3.5)$ and simplify by combining like terms.

52. **REASONING** For the function $y = 15x - 4$, assume the domain is only values of x from 0 to 5. What is the range of the function?

53. **PERSEVERE** If $f(3b - 1) = 9b - 1$, find one possible expression for $f(x)$.

Lesson 3-3

Linearity and Continuity of Graphs

Explore Representing Discrete and Continuous Functions

Online Activity Use a real-world situation to complete the Explore.

INQUIRY How can you use the graph of a function to determine whether it is discrete?

Learn Discrete and Continuous Functions

Discrete Function	Continuous Function	Neither Discrete nor Continuous
Points are not connected.	Points are connected to form a line or curve.	Some points are connected by a line or curve, but it is not continuous everywhere.
Domain: set of individual values	Domain: all real numbers	Domain: varies
Range: set of individual values	Range: one interval of real numbers	Range: varies

Today's Goals

- Determine whether functions are continuous, discrete, or neither.
- Determine whether functions are linear or nonlinear.

Today's Vocabulary

discrete function

continuous function

linear function

linear equation

nonlinear function

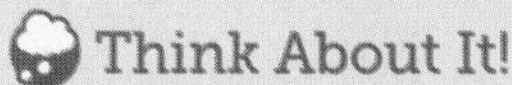

Think About It!

Can a function be both discrete and continuous? Justify your answer.

Example 1 Determine Continuity

BOOKS The Bargain Book Barn sells young adult novels on a sliding scale. That is, the more books you buy, the cheaper they are. Let $f(x)$ model the store's prices for given quantities. Is $f(x)$ *discrete* or *continuous*? Explain your reasoning.

The Bargain Book Barn

Number of Books	Price ($)
1	1.50
2	2.90
3	4.30
4	5.70
5	7.10

Use the table and context of the situation. The quantity and price correspond to ordered pairs like (1, 1.50) which can be graphed. Because books are not sold in fractional quantities, the number of books and their corresponding prices cannot be between the points given. So the points should not be connected and the function is discrete. The domain and range are sets of individual values.

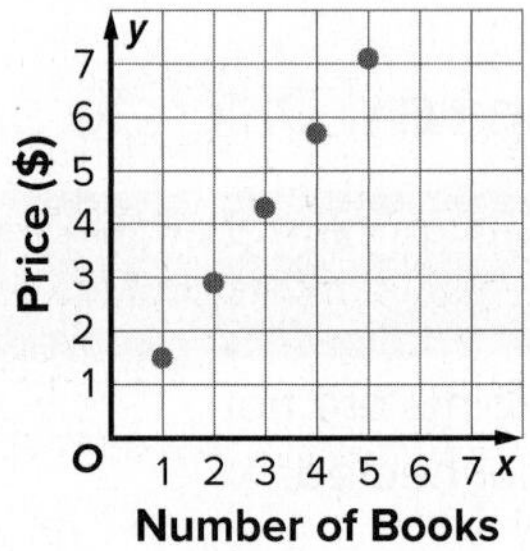

Think About It!

If Bargain Book Barn did not use a sliding scale, that is, 1 book costs $1.50 and 2 books cost $3.00, then would the function still be discrete? Explain your reasoning.

Example 2 Determine Continuity by Using Graphs

Determine whether $f(x)$ and $g(x)$ are *continuous, discrete,* or *neither.* Explain your reasoning.

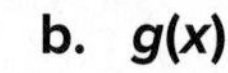

a. $f(x)$	b. $g(x)$
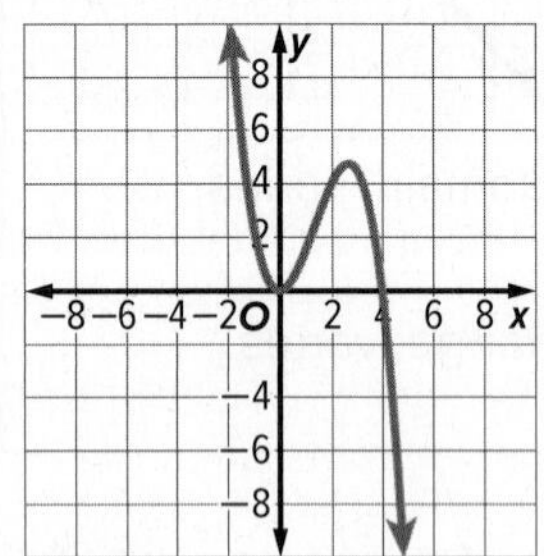	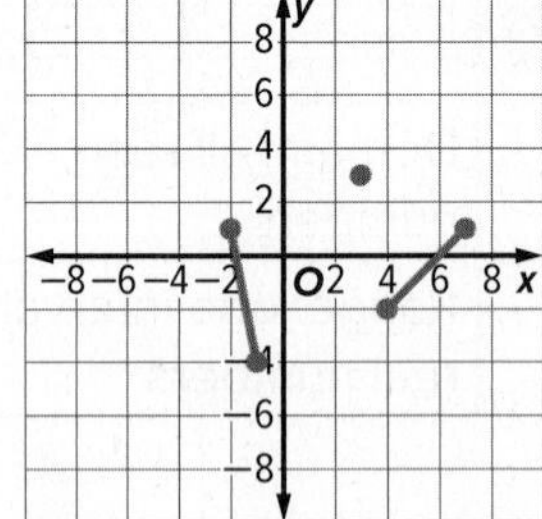
Because $f(x)$ is graphed with a single curve, it is a continuous function. The domain and range are both all real numbers.	Because $g(x)$ has continuous sections, but is not a single line or curve, it is neither discrete nor continuous. The domain and range are intervals of values.

Go Online You can complete an Extra Example online.

Check

Determine whether *f*(*x*) is *discrete*, *continuous*, or *neither*.

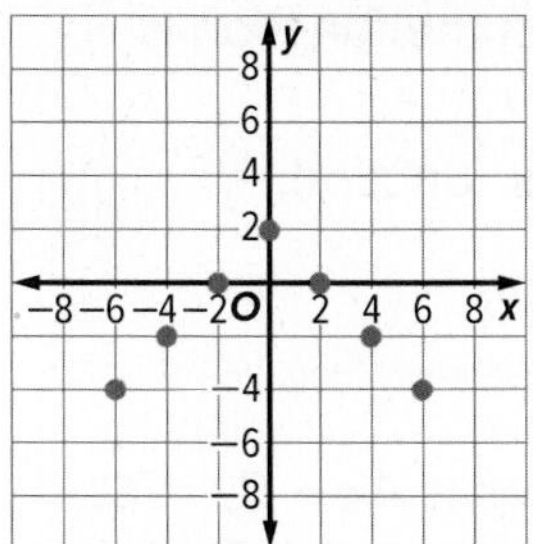

The function is ___?___.

Example 3 Apply Discrete and Continuous Functions

DETECTIVE **As a private investigator, Tia charges $25 per hour for any amount of time up to eight hours and then a flat rate of $250 per day. Use the graph to determine if the function that models this situation is *discrete*, *continuous*, or *neither*.**

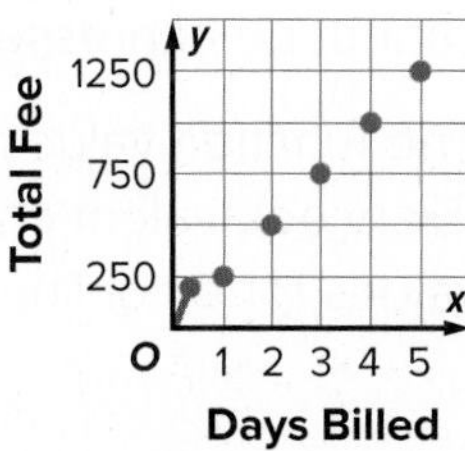

Discrete?

No; the function is not made up entirely of individual points.

Continuous?

No; the function cannot be drawn with a straight line or smooth curve.

Neither?

Yes; the function is neither continuous nor discrete.

Check

PLANTS Circadian rhythms are cycles of behavior that occur over twenty-four hours, based on a day-night cycle. One aspect of a plant's circadian rhythm is the percentage of its flowers that are open or closed.

If *f*(*x*) is represented by the curve, then is *f*(*x*) *discrete*, *continuous*, or *neither*?

The function is ___?___.

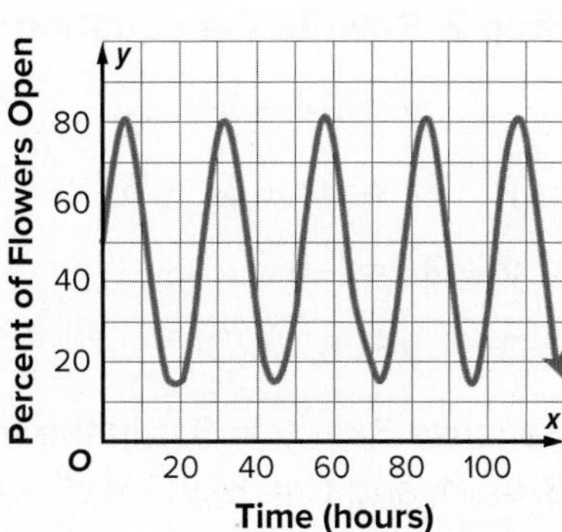

Go Online You can complete an Extra Example online.

Watch Out!

Continuous Intervals
Recall that a function can have a continuous interval, but the function itself is not continuous unless it is continuous over its entire domain.

Think About It!

How could Tia alter her pricing model to make it a discrete function? a continuous function?

Learn Linear and Nonlinear Functions

A **linear function** is a function that has a graph that is a line. If the domain of the function is all real numbers, then the function is continuous.

A **linear equation** can be used to describe a linear function.

Linear equations are often written in standard form.

Key Concept • Standard Form of a Linear Equation	
Words	The standard form of a linear equation is $Ax + By = C$, where $A \geq 0$, A and B are not both zero, and A, B, and C are integers with a greatest common factor of 1.
Examples	In $2x + 5y = 7$, $A = 2$, $B = 5$, and $C = 7$. In $x = -3$, $A = 1$, $B = 0$, and $C = -3$.

A **nonlinear function** has a graph with a set of points that cannot all lie on the same line. An equation that represents a nonlinear function cannot be expressed in the form $Ax + By = C$.

The function values of a linear function change at a constant rate for every equivalent change in the x-values. The change in the function values for a nonlinear function will vary for equivalent changes in x.

Think About It!

Can a function be both linear and nonlinear? Explain your reasoning.

Example 4 Linear and Nonlinear Functions

Determine whether $y = 4x^2 - (2x)^2 + 3x - 5$ is an equation for a *linear* or *nonlinear* function.

Step 1 Simplify the equation.

$y = 4x^2 - (2x)^2 + 3x - 5$	Original equation
$= 4x^2 - 4x^2 + 3x - 5$	Simplify exponents.
$= 0 + 3x - 5$	Subtract.
$= 3x - 5$	Simplify.

Step 2 Rewrite the equation.

$y = 3x - 5$	Simplified equation
$y - 3x = 3x - 3x - 5$	Subtract $3x$ from each side.
$y - 3x = -5$	Simplify.
$3x - y = 5$	Multiply each side by -1.

Because $3x - y = 5$ is in the form $Ax + By = C$, where $A = 3$, $B = -1$, and $C = 5$, $y = 4x^2 - (2x)^2 + 3x - 5$ is linear.

Think About It!

Can $y = -5x^2$ be written in standard form? If so, write the standard form of the equation.

Check

The function $3x^2 - \sqrt{9}x^2 - y = 17$ is ____?____.

Go Online You can complete an Extra Example online.

Example 5 Identify Linear and Nonlinear Functions

Determine whether $y = 3x^3 - x^3 + 3x + 6$ is an equation for a *linear* or *nonlinear* function.

Step 1 Simplify the equation.

$y = 3x^3 - x^3 + 3x + 6$ Original equation

$= 2x^3 + 3x + 6$ Subtract.

Step 2 Rewrite the equation.

$y = 2x^3 + 3x + 6$ Simplified equation

$y - 2x^3 - 3x = 2x^3 + 3x + 6 - 2x^3 - 3x$

Subtract $2x^3$ and $3x$ from each side.

$y - 2x^3 - 3x = 6$ Simplify.

$2x^3 + 3x - y = -6$ Multiply each side by –1.

Because $2x^3 + 3x - y = -6$ is not in the form $Ax + By = C$, $2x^3 + 3x - y = -6$ is nonlinear.

Check

The function $4x - (2y)^2 = 3$ is ____?____.

Example 6 Functions in Table Form

SOCCER **Salina kicks a soccer ball. The height of the ball after each half second is recorded in the table. Is the function that models the height of the ball a *linear* or *nonlinear* function?**

Time (s)	Height (ft)
0	2
0.5	28
1	46
1.5	56
2	58
2.5	52
3	38
3.5	16

First Half-Second Interval

During the first half-second interval, the ball goes from a height of 2 feet to a height of 28 feet. That is an increase of 26 feet.

Second Half-Second Interval

During the second half-second interval, the ball goes from 28 feet to 46 feet. That is an increase of 18 feet.

Because the change in the height varies over the two equivalent intervals, the height of the soccer ball must be modeled by a nonlinear function.

Think About It!

Is the function represented by $2x + 9 = 5y - 3xy$ a linear or nonlinear function? Justify your argument.

Think About It!

What do you notice about the height of the soccer ball over time that might indicate that the function is not linear?

Go Online You can complete an Extra Example online.

Check

Determine whether the values in each table are best modeled by a *linear* or *nonlinear* function.

x	y
−5	−15
−1	−3
0	0
2	6
3	9
7	21

x	y
0	0
1	−1
2	−8
3	−27
5	−125
7	−343

Example 7 Identify Linear Functions by Graphing

POOL Fernando uses a garden hose to fill his empty pool. The table shows the amount of water in the pool after every five minutes.

Time (min)	Water (gal)
5	60
10	120
15	180
20	240
25	300

Part A Determine linearity.

The amount of water in the pool increases by 60 gallons during the first 5 minutes. It also increases by 60 gallons during each following 5-minute intervals up to 25 minutes. Because the change in the amount of water is constant for every equivalent interval of 5 minutes, the numbers in the table can be modeled by a linear function.

Part B Graph.

The points on the graph can be connected by a straight line.

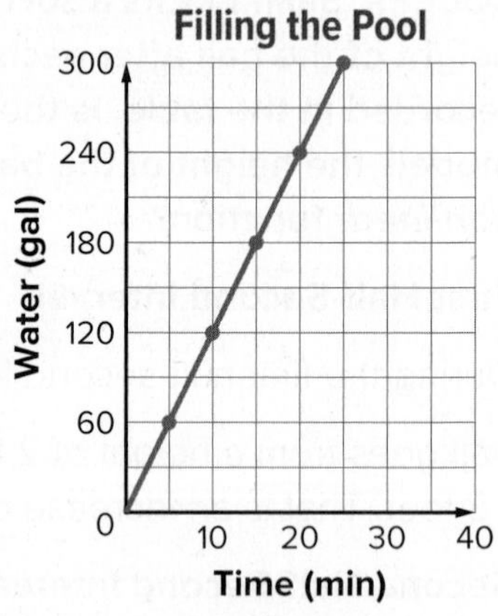

Talk About It!

After 25 minutes, Fernando turns off the hose. He takes 5 minutes to remove any wrinkles from the pool liner. Then he returns to filling the pool at the same rate. Would the function that represents filling the pool from 0 to 40 minutes be a linear function? Explain your reasoning.

Check

MINIMUM WAGE The table shows the federal minimum wage rates during years in which the wage increased. Which statement best describes the function that models the wages over time?

Year	1990	1991	1996	1997	2007	2008	2009
Wage (dollars)	3.80	4.25	4.75	5.15	5.85	6.55	7.25

A. The function is linear because the increase is \$0.70 between 1997 and 2007 and \$0.70 between 2008 and 2009.

B. The function is nonlinear because the increase is \$0.45 between 1990 and 1991 and \$0.70 between 2007 and 2008.

C. The function is linear because it is constantly increasing.

D. The function is nonlinear because it is discrete.

Go Online You can complete an Extra Example online.

Practice

Go Online You can complete your homework online.

Examples 1 and 2

Determine whether the function is *discrete*, *continuous*, or *neither*. Explain.

1. 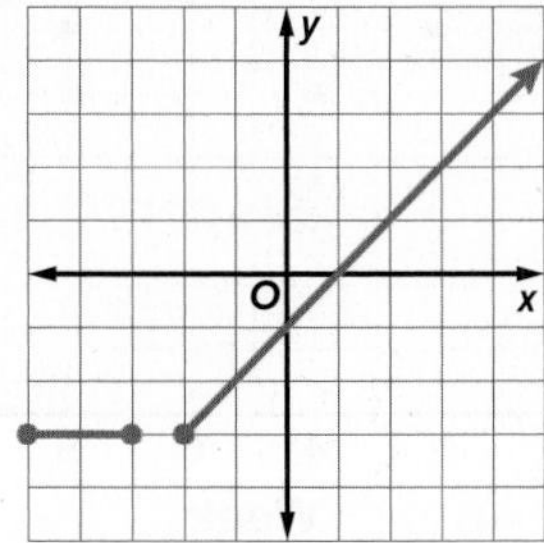

2. 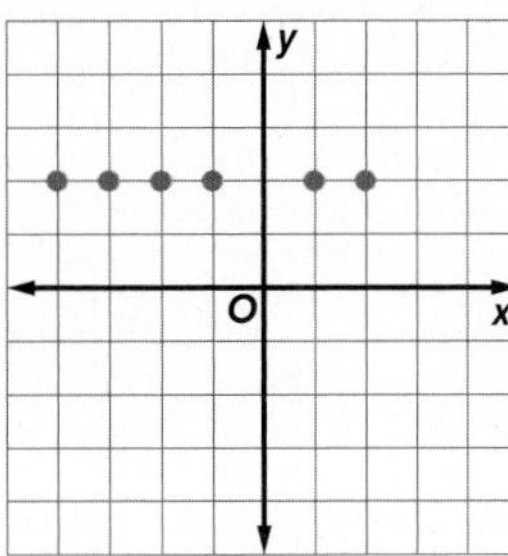

3. 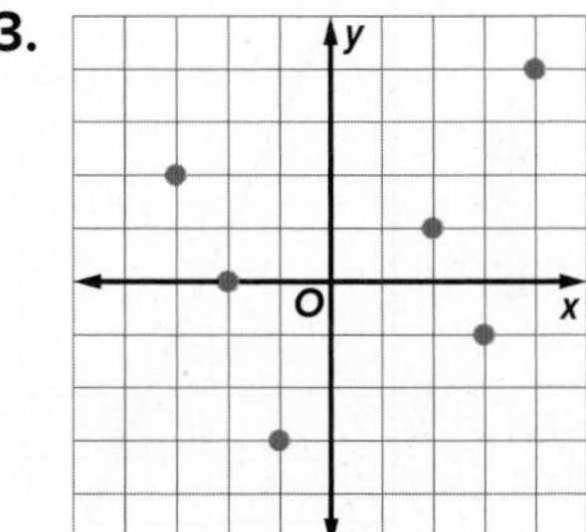

4. 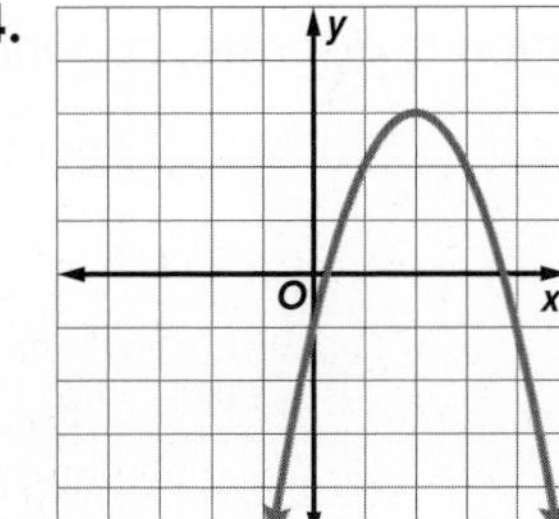

5. 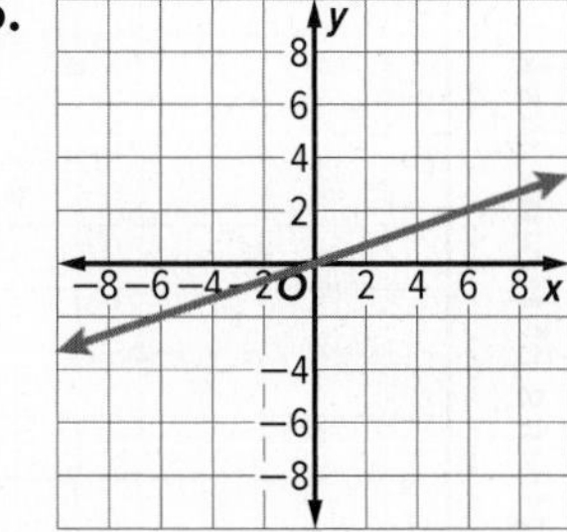

6. 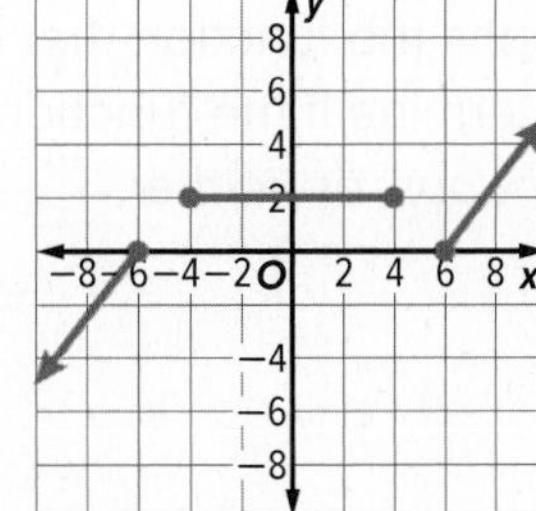

7. 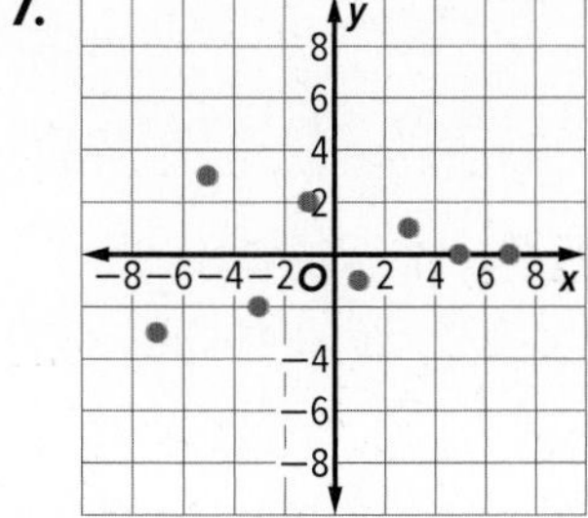

8. 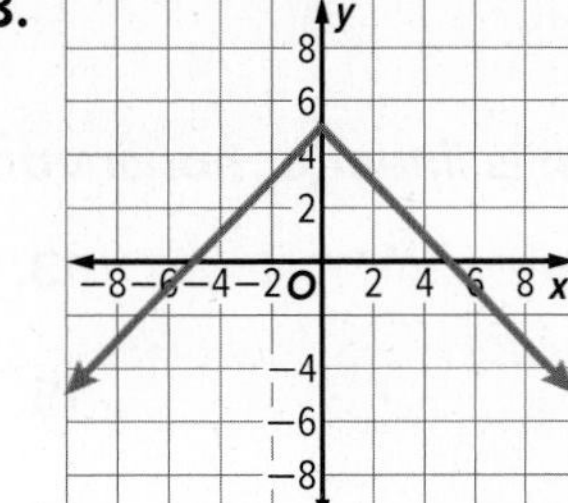

Example 3

9. EMPLOYMENT Kylie records the number of hours she works in her after-school job each week and then graphs the function that models the situation. Use the graph to determine if the function that models this situation is *discrete*, *continuous*, or *neither*.

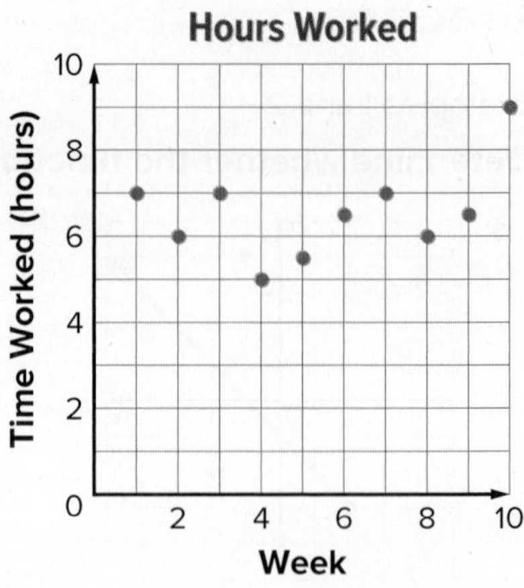

10. BAKING Kalynda keeps track of the number of cups of flour she has after baking batches of cookies and then graphs the function that models the situation. Use the graph to determine if the function that models this situation is *discrete*, *continuous*, or *neither*.

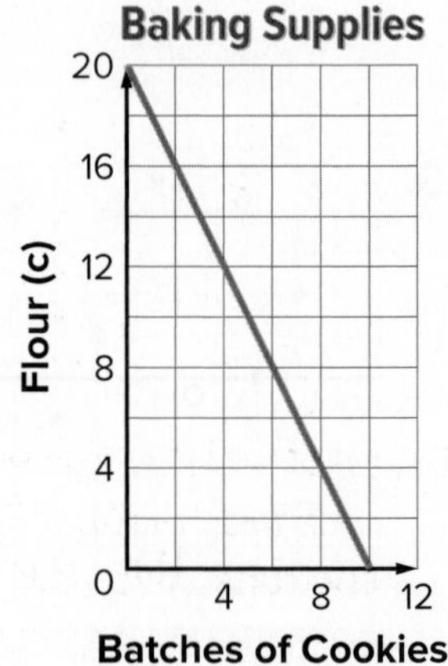

11. HOMEWORK Nayati records the number of hours of homework he has each day and then graphs the function that models the situation. Use the graph to determine if the function that models this situation is *discrete*, *continuous*, or *neither*.

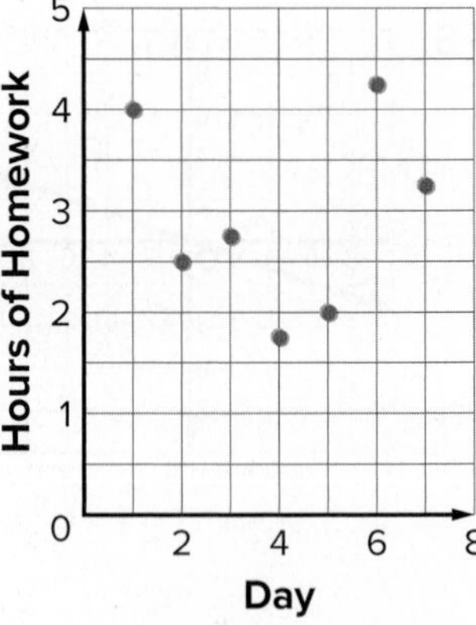

Examples 4, 5, and 6

Determine whether each function is *linear* or *nonlinear*.

12. $y - \frac{1}{x} = 11$

13. $x + (\sqrt{2}x)^2 + y - 2x^2 = 11$

14. $y = 2x + 5$

15. $y = 3x^2 - x + 5$

16. $2y + \frac{x}{3} = -6$

17. $9x - (9x)^2 = 19 - y$

18. $-9x + (\sqrt{4}x)^2 + y = -4 + 7x^2$

19. $y = -x^3 + 2x$

20.

x	y
1	100
2	125
3	150
4	175
5	200

21.

x	y
1	3
2	5
3	11
4	21
5	35

22.

x	y
−5	−10
−2	−4
0	0
2	4
3	6

23.

x	y
5	35
6	41
7	47
8	53
9	59

Example 7

24. ENDURANCE Tamika wants to improve her running endurance and thus tries to run for longer periods of time each day. The distance she runs, given the day, is provided in the table.

Day	1	2	3	4	5	6	7
Distance	2	2.25	2.75	3.5	4	4.5	5

a. Determine linearity.

b. Graph.

25. SCUBA DIVING Marco wants to know his elevation compared to sea level for periods of time each minute while scuba diving. His elevation is provided in the table.

Minute	1	2	3	4	5	6	7
Elevation	−3	−4.5	−6	−7.5	−9	−10.5	−12

a. Determine linearity.

b. Graph.

Mixed Exercises

26. EXERCISE Rashona keeps track of the total number of minutes she exercises and then graphs the function that models the situation. Use the graph to determine if the function that models this situation is *discrete, continuous,* or *neither.*

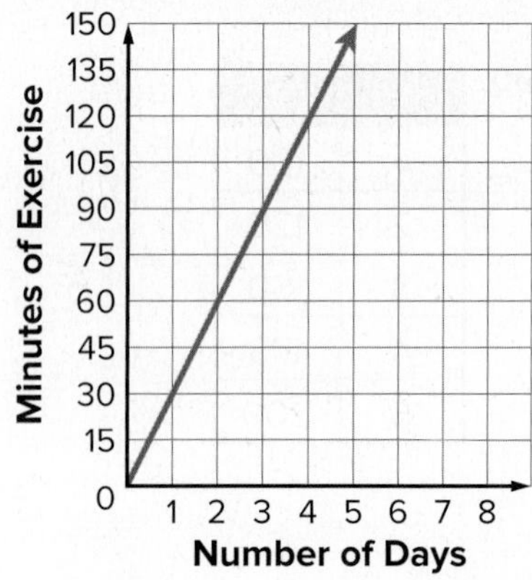

27. HOT AIR BALLOON The elevation of a hot air balloon, in feet, compared to sea level for periods of time each minute is provided in the table.

Minute	1	2	3	4	5	6	7	8
Elevation	10	15	12	18	25	45	40	42

a. Determine linearity.

b. Graph.

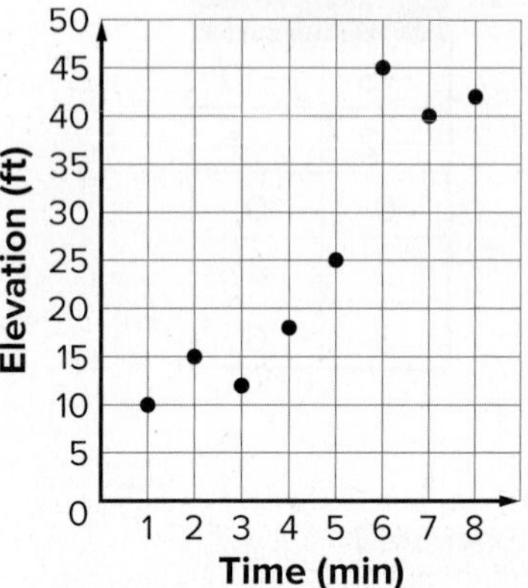

Determine whether the function is *discrete*, *continuous*, or *neither*. Then determine whether each function is *linear* or *nonlinear*.

28. $y - 14x = 3$

29. $(2x)^2 - 4y = 2$

30.

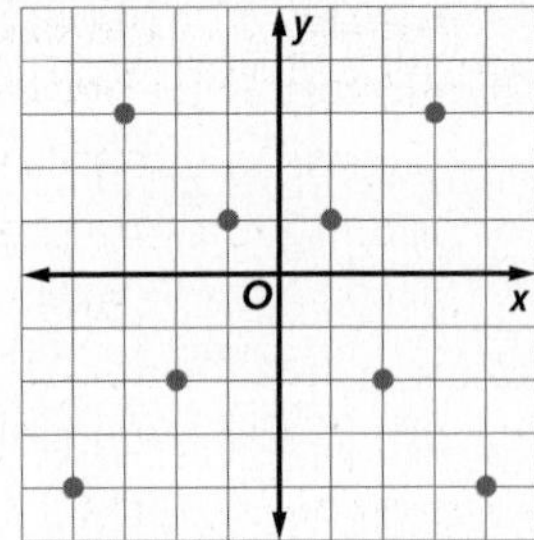

31.

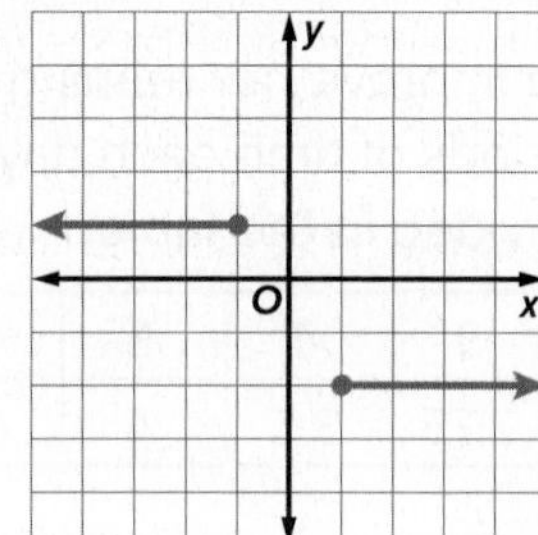

32.

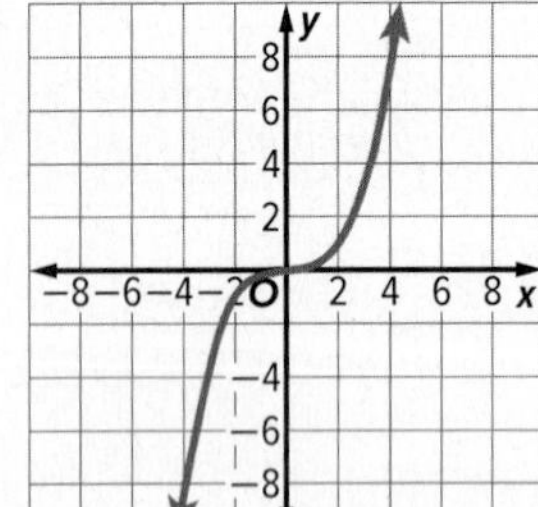

33.

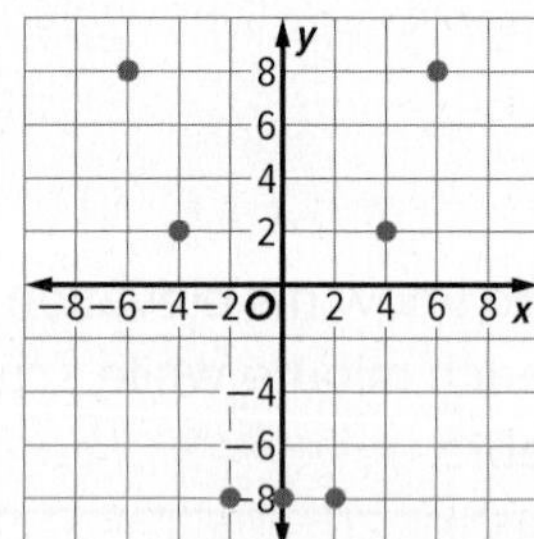

Higher-Order Thinking Skills

34. ANALYZE If $f(x)$ is a linear function, then is $f(x + a)$, where a is a real number, *sometimes, always,* or *never* a continuous function? Justify your argument.

35. WRITE Describe a real-world situation where the function that models the situation is neither discrete nor continuous.

36. ANALYZE A function that consists of a finite set of ordered pairs is *sometimes, always,* or *never* continuous? Justify your argument.

Lesson 3-4

Intercepts of Graphs

Learn Intercepts of Graphs of Functions

The intercepts of graphs are points where the graph intersects an axis.

The **x-intercept** is the x-coordinate of a point where a graph crosses the x-axis.

The **y-intercept** is the y-coordinate of a point where a graph crosses the y-axis.

A function is **positive** when its graph lies *above* the x-axis.

A function is **negative** when its graph lies *below* the x-axis.

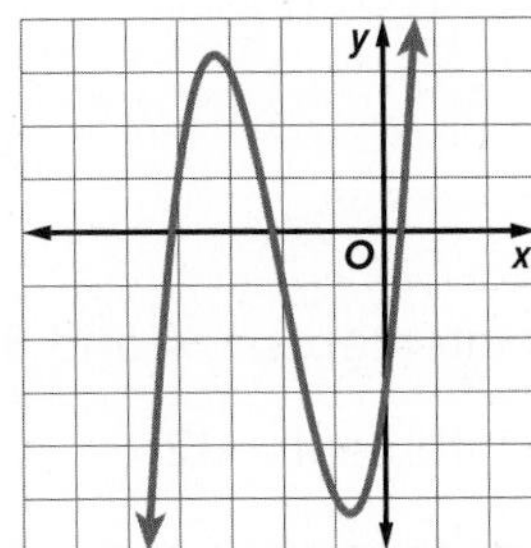

Today's Goals

- Identify the intercepts of functions and intervals where functions are positive and negative.
- Solve equations by graphing.

Today's Vocabulary

x-intercept
y-intercept
positive
negative
root
zero

Example 1 Intercepts of the Graph of a Linear Function

Use the graph to estimate the x- and y-intercepts of the function and describe where the function is positive and negative.

The x-intercept is the point where the graph crosses the x-axis, (3, 0).

The y-intercept is the point where the graph crosses the y-axis, (0, 6).

A function is positive when its graph lies above the x-axis, or when $x < 3$.

A function is negative when its graph lies below the x-axis, or when $x > 3$.

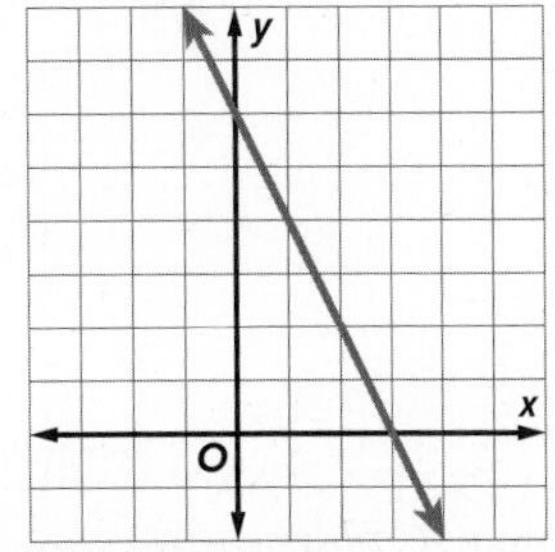

Study Tip

Intercepts Notice that *intercept* can be used to refer to either the point where the graph intersects the axis or the nonzero coordinate of the point where the graph intersects the axis.

 Think About It!

Explain why this function is linear.

Check

Use the graph to estimate the x- and y-intercepts of the function and describe where the function is positive and negative.

A. x-intercept: $(-2, 0)$; y-intercept: $(0, -6)$; positive: $x > -2$; negative: $x < -2$

B. x-intercept: $(0, -6)$; y-intercept: $(-2, 0)$; positive: $x < -2$; negative: $x > -2$

C. x-intercept: $(-2, 0)$; y-intercept: $(0, -6)$; positive: $x < -2$; negative: $x > -2$

D. x-intercept: $(0, -6)$; y-intercept: $(-2, 0)$; positive: $x > -2$; negative: $x < -2$

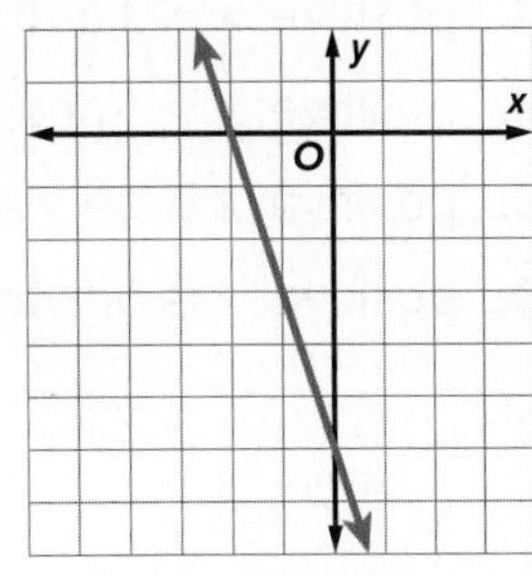

Go Online You can complete an Extra Example online.

Study Tip

x- and y-intercepts To help remember the difference between the x- and y-intercepts, remember that the x-intercept is where the graph intersects the x-axis, and the y-intercept is where the graph intersects the y-axis.

Think About It!

Explain why this function is nonlinear.

Watch Out!

Intercepts of Nonlinear Functions The graphs of nonlinear functions can have more than one x-intercept.

Example 2 Intercepts of the Graph of a Nonlinear Function

Use the graph to estimate the x- and y-intercepts of the function and describe where the function is positive and negative.

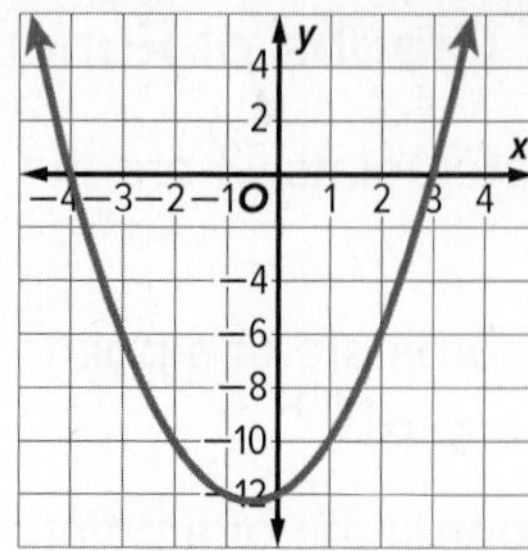

x-intercepts: -4 and 3.

y-intercept: -12.

positive: when $x < -4$ and when $x > 3$.

negative: x is between -4 and 3.

Check

Use the graph of the function to determine key features.

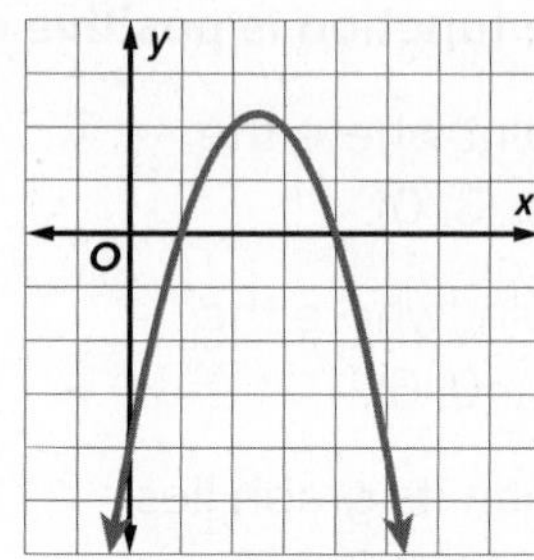

Part A Determine whether each ordered pair represents an *x-intercept,* a *y-intercept,* or *neither.*

(1, 0)

(0, 1)

Part B Describe where the function is positive and negative.

A. positive: $x < 1$ and $x > 4$; negative: x is between 1 and 4

B. positive: $x > 1$ and $x < 4$; negative: x is between 1 and 4

C. positive: x is between 1 and 4; negative: $x > 1$ and $x < 4$

D. positive: x is between 1 and 4; negative: $x < 1$ and $x > 4$

Go Online You can complete an Extra Example online.

Example 3 Find Intercepts from a Graph

SPORTS The graph shows the height of a ball for each second *x* that it is airborne. Use the graph to estimate the *x*- and *y*-intercepts of the function, where the function is positive and negative, and interpret the meanings in the context of the situation.

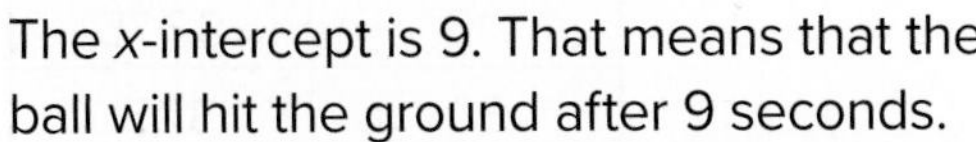

The *x*-intercept is 9. That means that the ball will hit the ground after 9 seconds.

The *y*-intercept is 4. This means that at time 4, the ball was at a height of 0 feet.

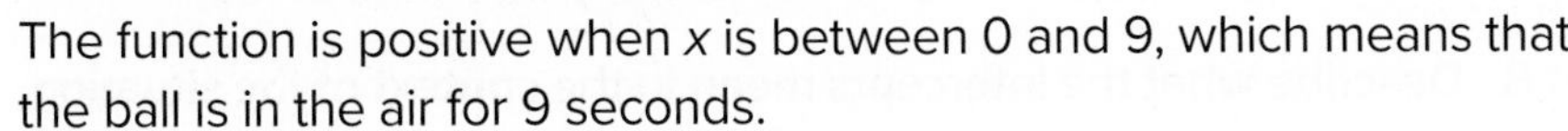

The function is positive when *x* is between 0 and 9, which means that the ball is in the air for 9 seconds.

No portion of the graph shows that the function is negative.

Think About It!

The function is only graphed from 0 to 9 seconds. What can you assume about the function when $x > 9$? Interpret this meaning. Does it make sense in the context of the situation?

Check

FITNESS The graph shows the number of people *y* at a gym *x* hours after the gym opens.

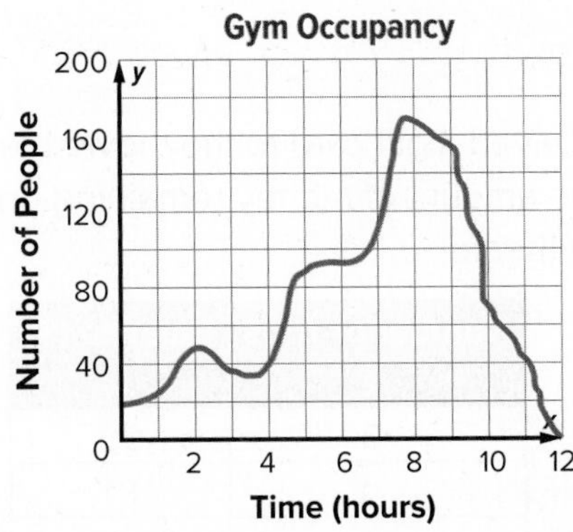

Part A Use the graph to estimate the *x*- and *y*-intercepts.

x-intercept: (__?__, __?__)

y-intercept: (__?__, __?__)

Part B Which statements describe the meaning of the *x*- and *y*-intercepts in the context of the situation? Select all that apply.

A. There were 20 people at the gym when it opened.

B. The gym closed after 20 hours.

C. The gym closed after 12 hours.

D. There were 12 people at the gym when it opened.

E. There was no one at the gym when it opened.

Go Online You can complete an Extra Example online.

Example 4 Find Intercepts from a Table

LUNCH **Violet starts the semester with \$150 in her student lunch account. Each day she spends \$3.75 on lunch. The table shows the function relating the amount of money remaining in her lunch account to the number of days Violet has purchased lunch.**

Time (Days)	Balance (\$)
x	y
0	150
2	142.50
5	131.25
10	112.50
15	93.75
30	37.50
40	0

Part A Find the intercepts.

The x-intercept is where $y = 0$, so the x-intercept is 40.

The y-intercept is where $x = 0$, so the y-intercept is 150.

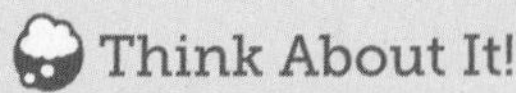

Think About It!

Explain why the x-coordinate of the y-intercept is always 0.

Part B Describe what the intercepts mean in the context of the situation.

The x-intercept means that after buying lunch for 40 days, Violet will have \$0 left in her lunch account, or it will take Violet 40 days to use all of the money in her lunch account. The y-intercept means that Violet's lunch account has \$150 after buying lunch for 0 days, or the beginning balance of her lunch account is \$150.

Check

MOVIES Ashley received a gift card to the movie theater for her birthday. The table shows the amount of money remaining on her gift card y after x trips to the movie theater.

Number of Trips	Balance (\$)
x	y
0	90
1	81
2	72
3	63
5	45
7	27
10	0

Watch Out!

Intercepts The y-coordinate of the x-intercept will always be 0, not the x-coordinate. The x-coordinate of the y-intercept will always be 0, not the y-coordinate.

Part A Find the y-intercept. (__?__, __?__)

Part B Find the x-intercept and describe what it means in the context of the situation.

A. (10, 0); The initial balance on the gift card was \$10.

B. (90, 0); The initial balance on the gift card was \$90.

C. (10, 0); After 10 trips to the movies, there will be no money left on the gift card.

D. (90, 0); After 90 trips to the movies, there will be no money left on the gift card.

Go Online You can complete an Extra Example online.

Learn Solving Equations by Graphing

The solution, or **root**, of an equation is any value that makes the equation true. A **zero** is an x-intercept of the graph of the function.

For example, the root of $3x = 6$ is 2. A linear equation, like $3x = 6$, has at most one root, while a nonlinear equation, like $x^2 + 4x - 5 = 0$, may have more than one.

Equation	Related Function
$3x = 6$	$f(x) = 3x - 6$ or $y = 3x - 6$
$x^2 + 4x - 5 = 0$	$f(x) = x^2 + 4x - 5$ or $y = x^2 + 4x - 5$

The graph of the related function can be used to find the solutions of an equation. The related function is formed by solving the equation for 0 and then replacing 0 with $f(x)$ or y.

Values of x for which $f(x) = 0$ are located at the x-intercepts of the graph of a function and are called the zeros of the function f. The roots of an equation are the same as the zeros of its related function. The solutions and roots of an equation are the same value as the zeros and x-intercepts of its related function. For the equation $3x = 6$:

- 2 is the solution of $3x = 6$.
- 2 is the root of $3x = 6$.
- 2 is the zero of $f(x) = 3x - 6$.
- 2 is the x-intercept of $f(x) = 3x - 6$.

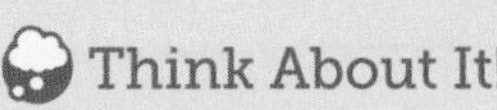

Think About It!

What is the difference between a *root* and a *zero*?

Example 5 Solve a Linear Equation by Graphing

Solve $-2x + 7 = 1$ by graphing. Check your solution.

Find the related function.

$-2x + 7 = 1$	Original equation
$-2x + 7 - 1 = 1 - 1$	Subtract 1 from each side.
$-2x + 6 = 0$	Simplify.

Graph the left side of the equation. The related function is $f(x) = -2x + 6$, which can be graphed.

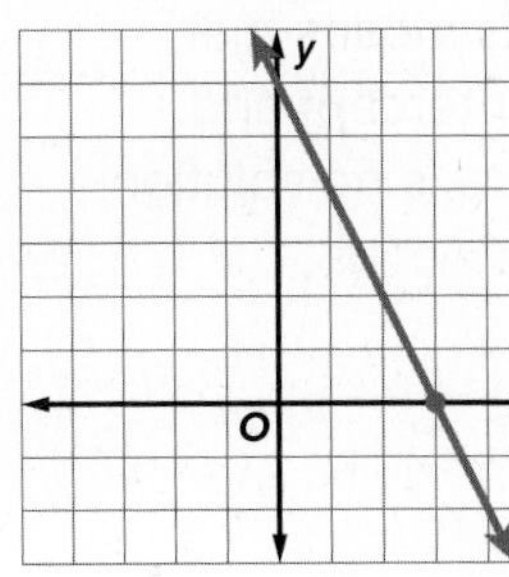

The graph intersects the x-axis at 3. This is the x-intercept, or zero, which is also the root of the equation. So, the solution of the equation is 3.

Check your solution by solving the equation algebraically.

Go Online You can complete an Extra Example online.

Go Online

An alternate method is available for this example.

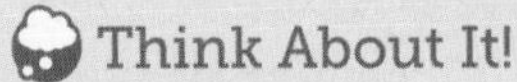

Think About It!

Suppose you first solved the equation algebraically. How could you use your solution to graph the zero of the related function?

Example 6 Solve a Nonlinear Equation by Graphing

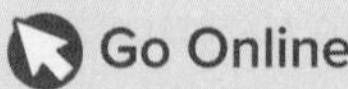

Go Online
You can watch a video to see how to use a graphing calculator with this example.

Solve $x^2 - 4x = -3$ by graphing. Check your solution.

Find the related function.

$x^2 - 4x = -3$	Original equation
$x^2 - 4x + 3 = -3 + 3$	Add 3 to each side.
$x^2 - 4x + 3 = 0$	Simplify.

Graph the left side of the equation. The related function is $f(x) = x^2 - 4x = 3$, which can be graphed.

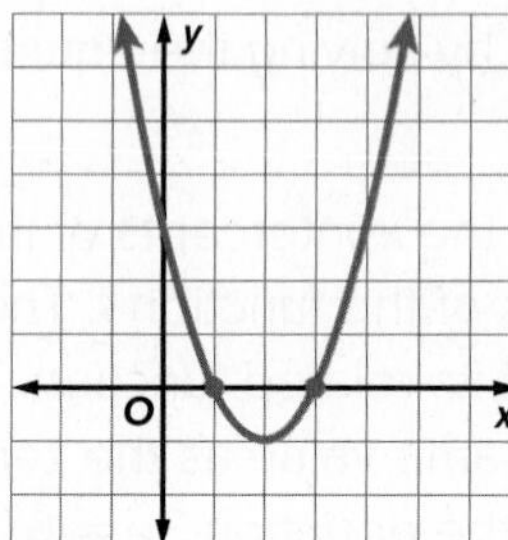

The graph intersects the x-axis at 1 and 3. These are the x-intercepts, or zeros, which are also the roots of the equation. So, the solutions of the equation are 1 and 3.

Think About It!
Name two ways that you can tell that this is a nonlinear function.

Example 7 Solve an Equation of a Horizontal Line by Graphing

Solve $4x + 3 = 4x - 5$ by graphing. Check your solution.

Find the related function.

$4x + 3 = 4x - 5$	Original equation
$4x + 3 + 5 = 4x - 5 + 5$	Add 5 to each side.
$4x + 8 = 4x$	Simplify.
$4x - 4x + 8 = 4x - 4x$	Subtract $4x$ from each side.
$8 = 0$	Simplify.

Talk About It!
Does solving the equation algebraically give a different solution? Explain your reasoning.

Graph the left side of the equation. The related function is $f(x) = 8$, which can be graphed.

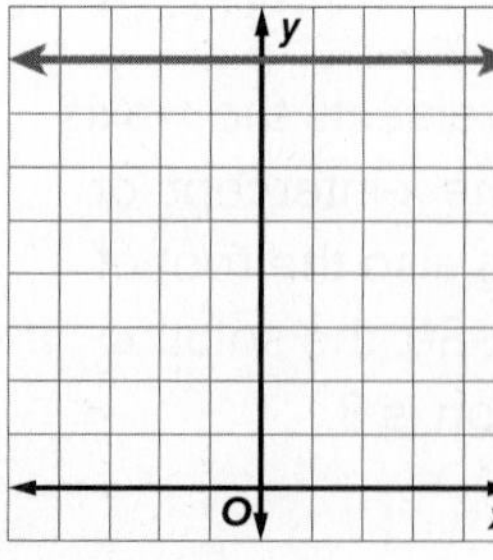

The graph does not intersect the x-axis. This means that there is no x-intercept and, therefore, there is no solution.

Go Online You can complete an Extra Example online.

Check

Equations and the graphs of their related functions are shown. Write the related function and its zero(s) under the appropriate graph.

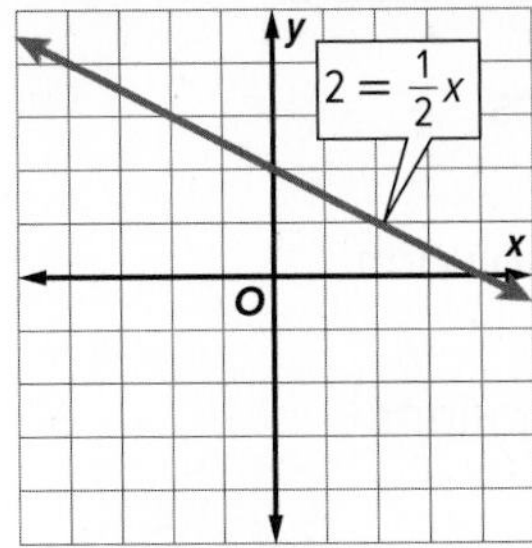

related function:

zeros:

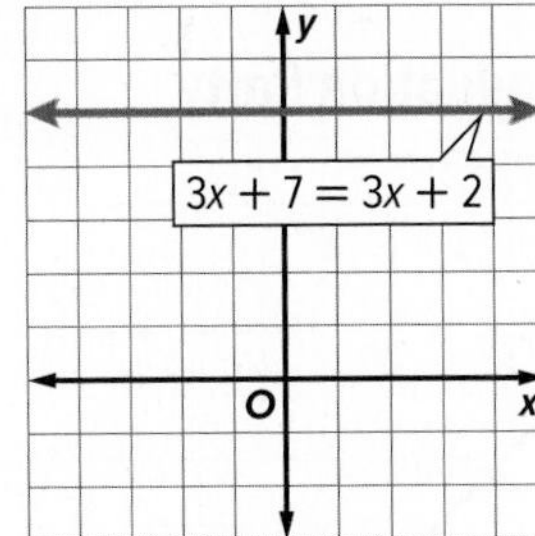

related function:

zeros:

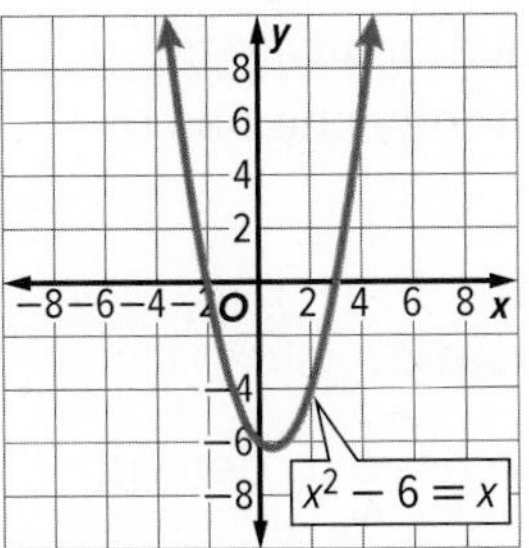

related function:

zeros:

Apply Example 8 Estimate Solutions by Graphing

PARTY Haley is ordering invitations for her graduation party. She has \$40 to spend and each invitation costs \$0.96. The function $m = 40 - 0.96p$ represents the amount of money m Haley has left after ordering p party invitations. Find the zero of the function. Describe what this value means in the context of this situation.

1 What is the task?

Describe the task in your own words. Then list any questions that you may have. How can you find answers to your questions?

Sample answer: I need to find the zero of the function and describe what it means. How can I determine the meaning of the zero from a graph of the function? I can review graphing linear functions and labeling axes.

2 How will you approach the task? What have you learned that you can use to help you complete the task?

Sample answer: I will graph the function by making a table of values. I will estimate the x-intercept of the graph to find the zero. I will then check my solution by solving the equation algebraically. I will use the axes labels to help me interpret my solution.

(continued on the next page)

Go Online You can complete an Extra Example online.

3 What is your solution?

Use your strategy to solve the problem.

Graph the function.

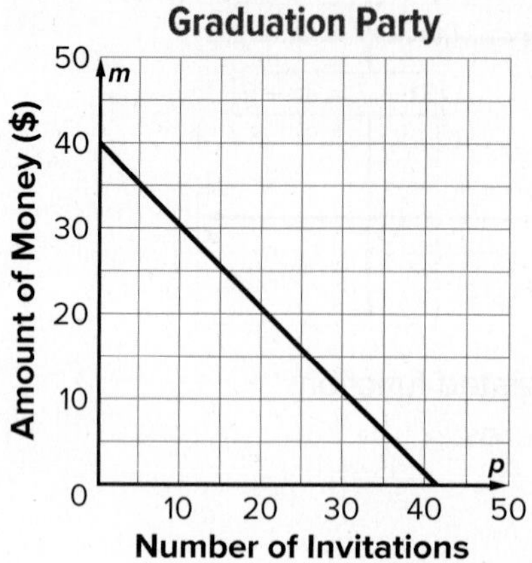

Estimate the solution.

42 invitations

Check the solution.

$\approx$ 41.67 invitations

What does your solution mean in the context of the situation?

Sample answer: Haley can order 41 invitations with the amount of money she has to spend.

4 How can you know that your solution is reasonable?

Write About It! Write an argument that can be used to defend your solution.

Sample answer: This amount is close to the estimated zero of 42 invitations from the graph.

Check

DATA Blair's cell phone plan allows her to use 3 GB of data, and she uses approximately 0.14 GB of data each day. The function $g = 3 - 0.14d$ represents the amount of data g in GB she has left after d days.

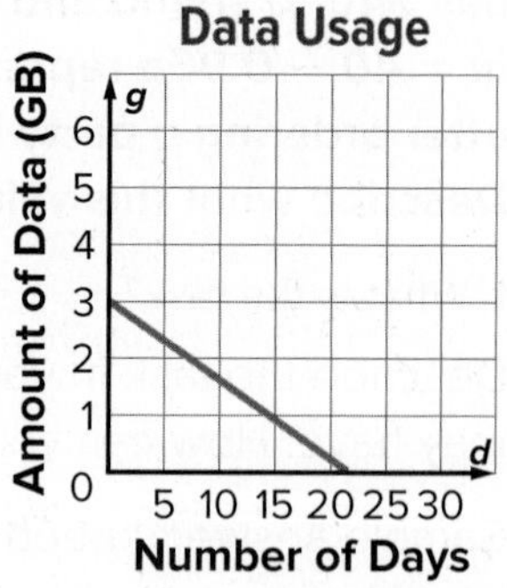

Part A Examine the graph of the function to estimate its zero to the nearest day.

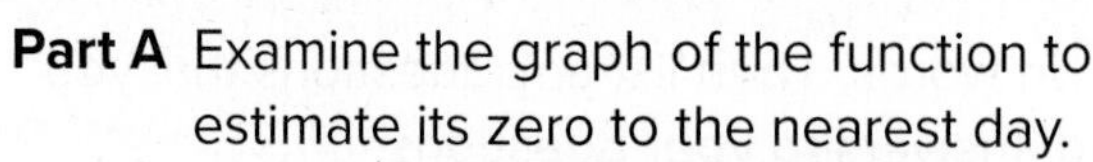

The graph appears to intersect the x-axis at __?__.

Part B Solve algebraically to check your answer. Round to the nearest tenth.

$x =$ __?__

Part C Describe what your answer to Part B means in this context.

After __?__ days, Blair has __?__ GB left.

 **Go Online** You can complete an Extra Example online.

Practice

Go Online You can complete your homework online.

Examples 1 and 2

Use the graph to estimate the *x*- and *y*-intercepts of the function and describe where the function is positive and negative.

1.

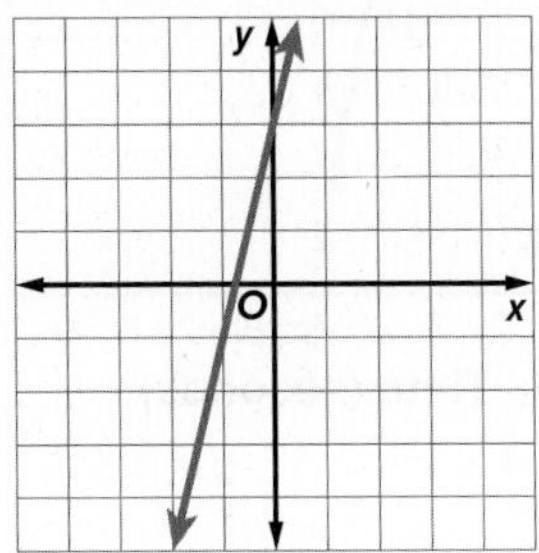

2.

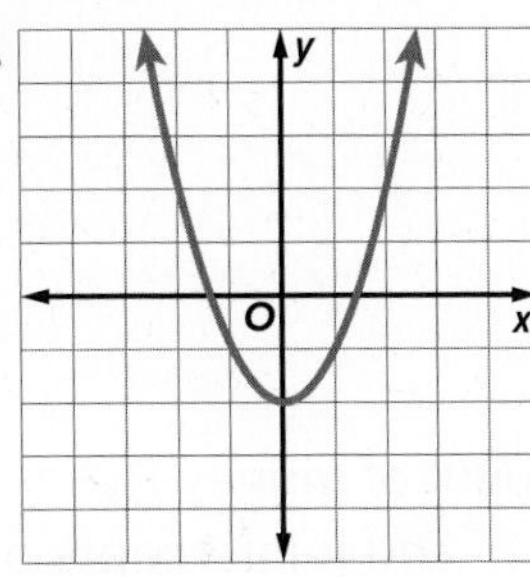

3.

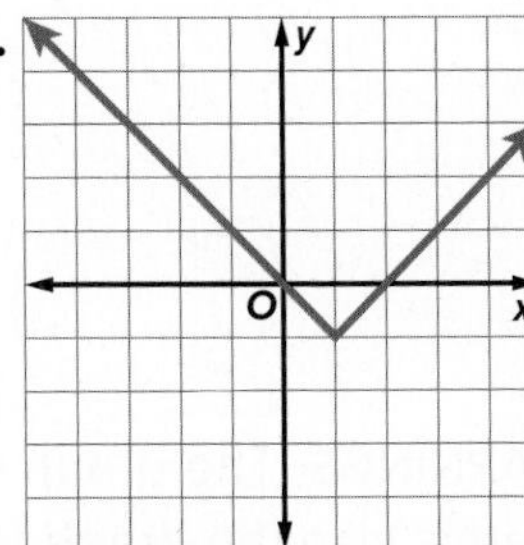

4.

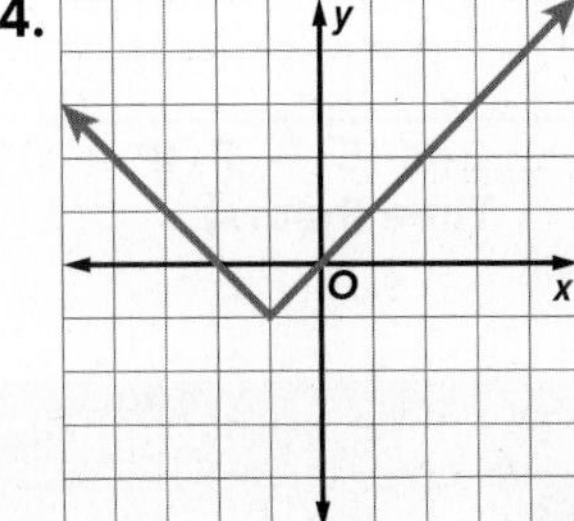

5.

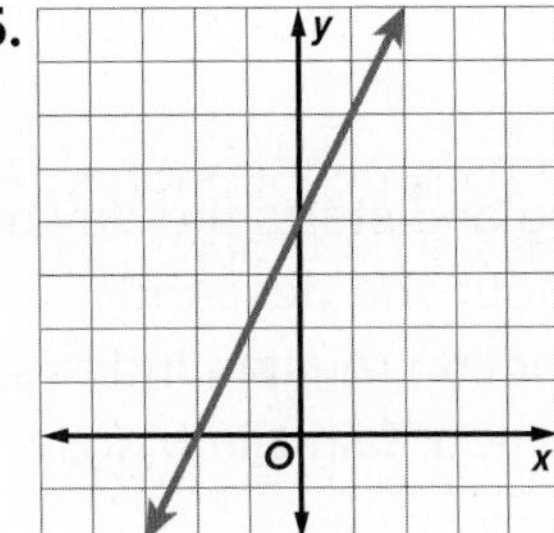

6.

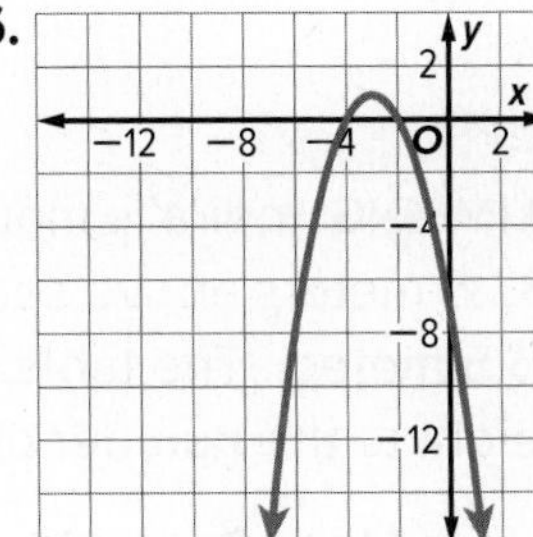

7.

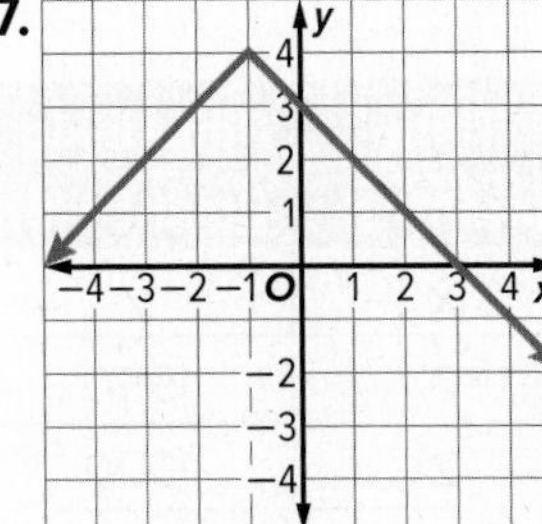

8.

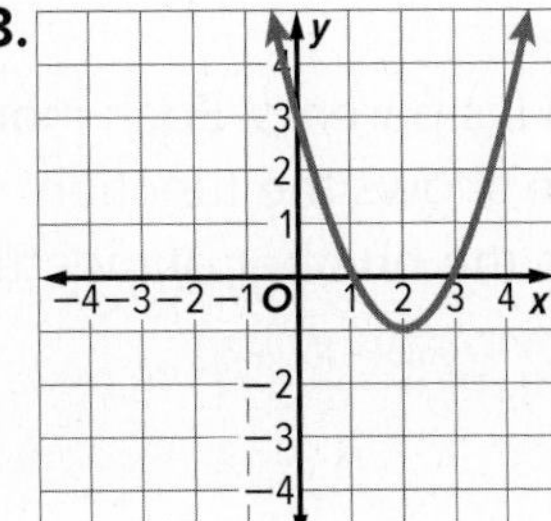

9.

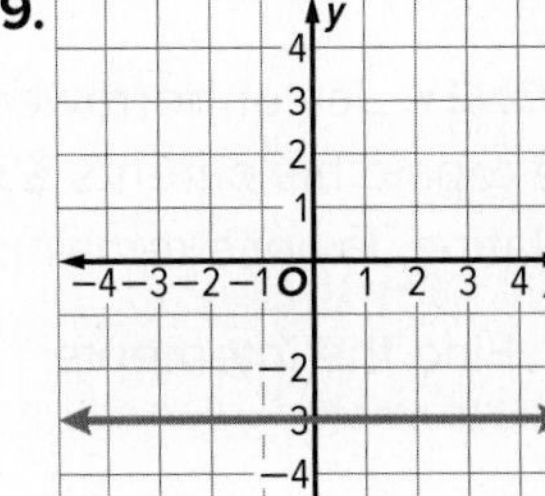

Example 3

10. FOOTBALL The graph shows the height of a football after being thrown. Use the graph to estimate the x- and y-intercepts of the function, where the function is positive and negative, and interpret the meanings in the context of the situation.

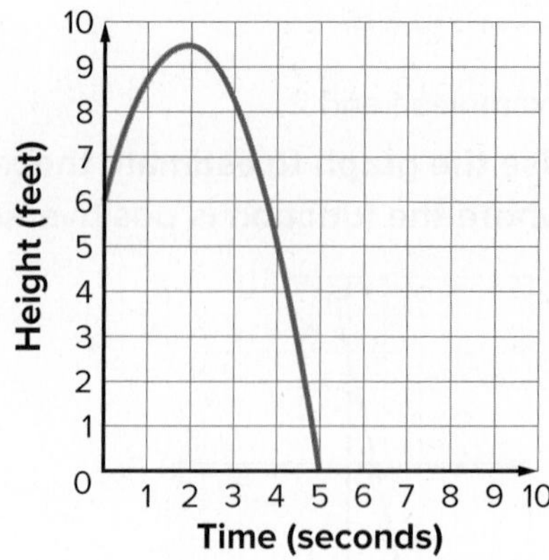

11. EARNINGS The graph shows the amount of money Ryan earns. Use the graph to estimate the x- and y-intercepts of the function, where the function is positive and negative, and interpret the meanings in the context of the situation.

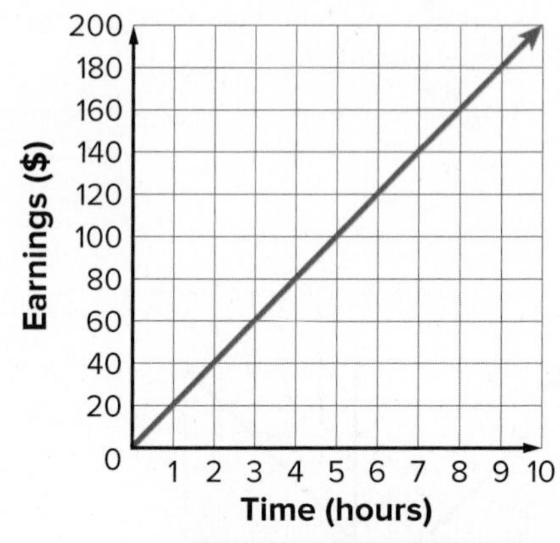

Example 4

12. CLIMBING Indira is mountain climbing and starts the day at 182.5 meters above sea level. Each hour she descends 36.5 meters. The table shows the function relating Indira's height to the number of hours she is mountain climbing.

a. Find the intercepts.

b. Describe what the intercepts mean in the context of the situation.

Time (hours)	Height (meters)
x	y
0	182.5
2	109.5
4	36.5
5	0

13. MONEY Javier borrowed $1950 from his parents. Each month he repaid his parents $325. The table shows the function relating Javier's remaining balance to the number of months.

a. Find the intercepts.

b. Describe what the intercepts mean in the context of the situation.

Time (months)	Remaining Balance ($)
x	y
0	1950
2	1300
5	325
6	0

Examples 5, 6, and 7

Solve each equation by graphing. Check your solution.

14. $2x - 3 = 3$

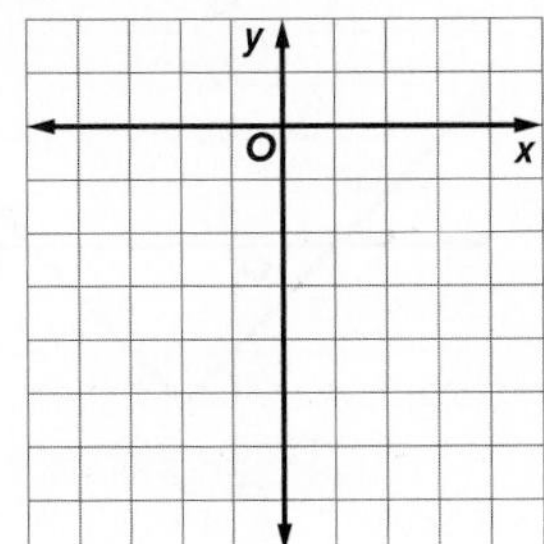

15. $-4x + 2 = -4x + 1$

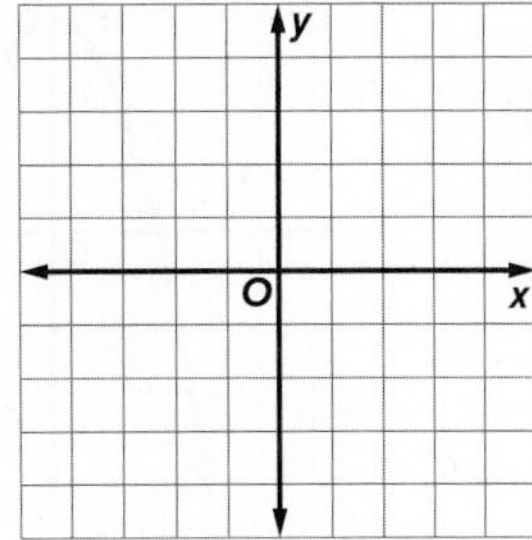

16. $4 = \frac{1}{2}x + 5$

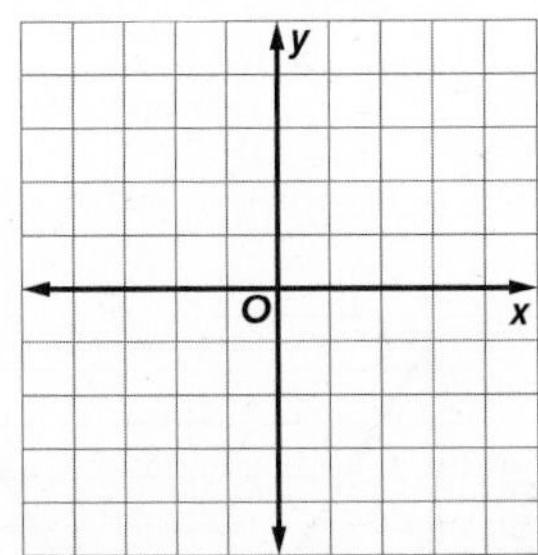

17. $3x + 1 = -5$

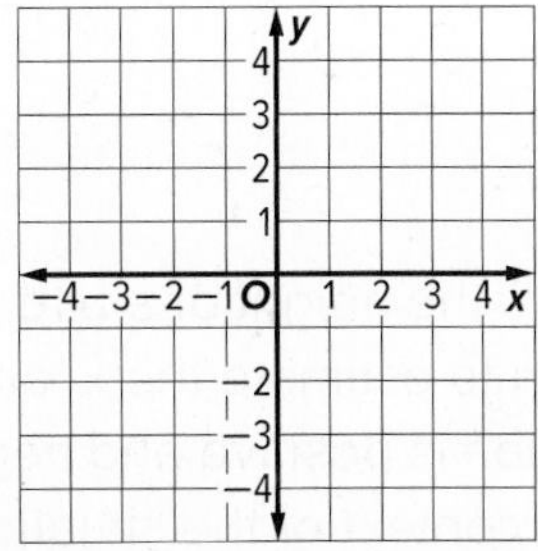

18. $3x - 5 = 3x - 3$

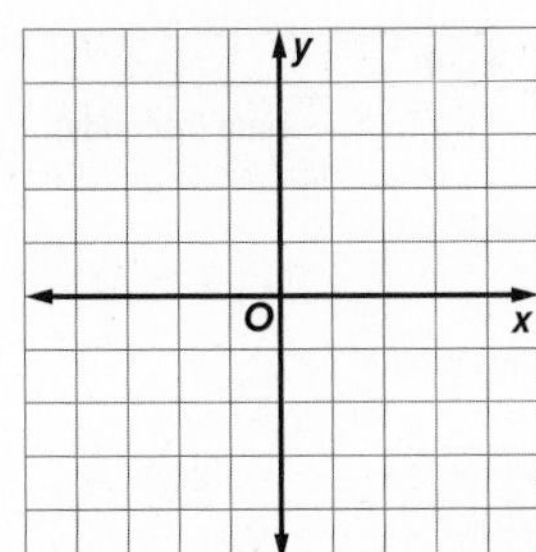

19. $9x - 7 = -6x + 8$

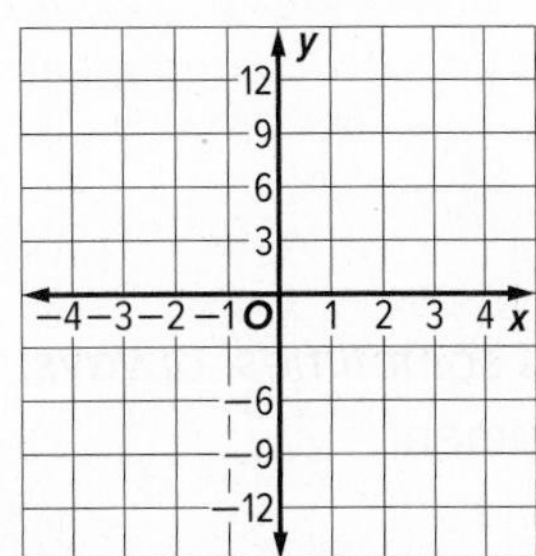

Example 8

20. INVITATIONS Moesha and Keyon are mailing invitations for their wedding. They have \$50 to spend and each invitation costs \$1.25 to mail. The function $m = 50 - 1.25w$ represents the amount of money m Moesha and Keyon have left after mailing w wedding invitations. Find the zero of the function. Describe what this value means in the context of this situation.

21. GIFT BAGS Juanita is tying ribbon on gift bags. She has 24 feet of ribbon and each gift bag uses 0.75 foot of ribbon. The function $r = 24 - 0.75g$ represents the amount of ribbon r Juanita has left after tying ribbon on g gift bags. Find the zero of the function. Describe what this value means in the context of this situation.

Mixed Exercises

Use the graph to estimate the *x*- and *y*-intercepts of the function and describe where the function is positive and negative.

22. 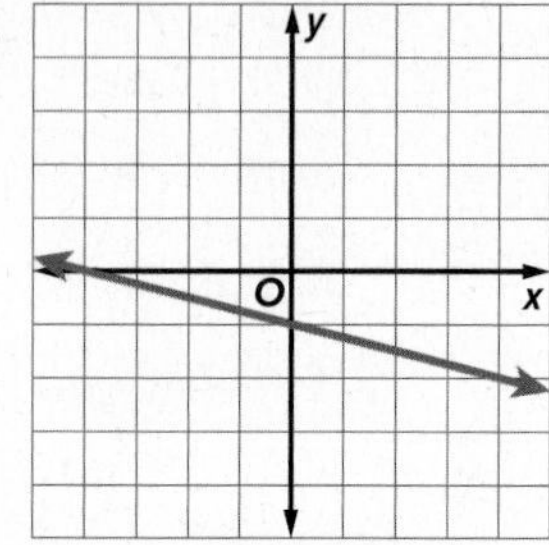

23. 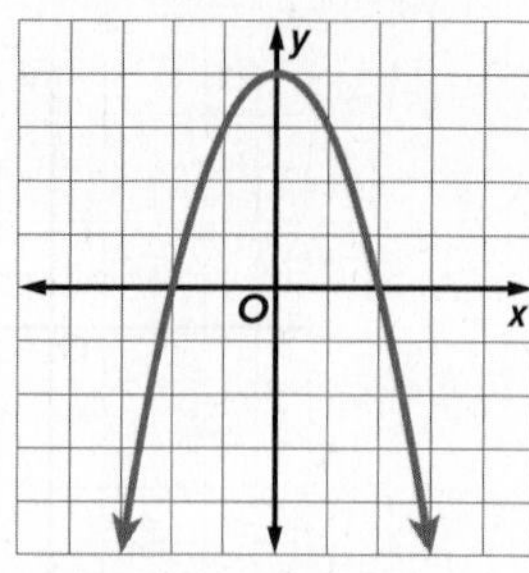

24. 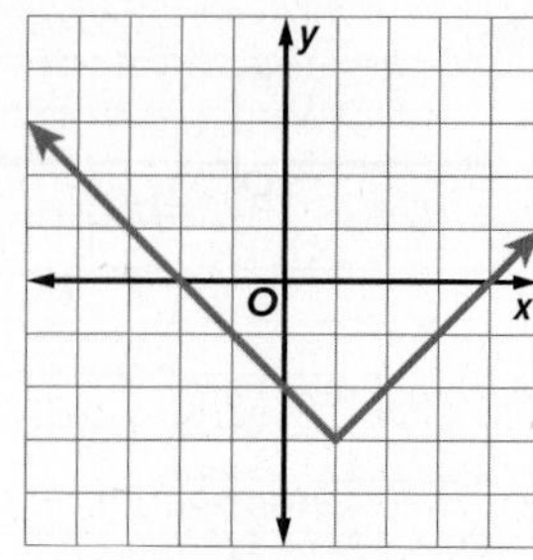

25. **REASONING** The graph shows the height of a bird compared to sea level over time. Use the graph to estimate the *x*- and *y*-intercepts of the function, where the function is positive and negative, and interpret the meanings in the context of the situation.

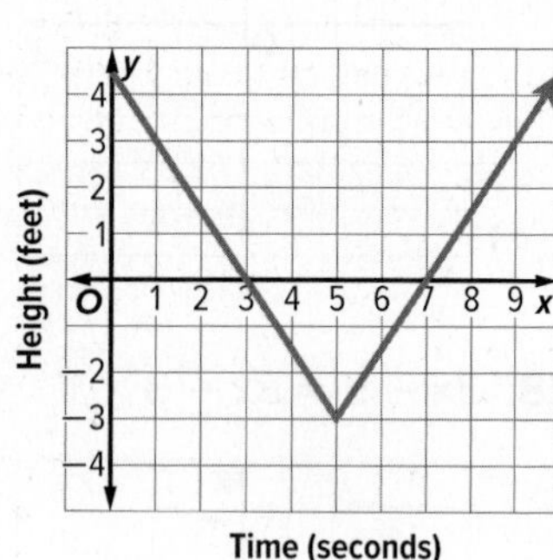

Higher-Order Thinking Skills

26. **ANALYZE** Do linear equations *sometimes, always,* or *never* have *x*- and *y*-intercepts? Justify your argument.

27. **WRITE** Describe how to find *x*- and *y*-intercepts by graphing and by using tables.

28. **CREATE** Write a word problem for the function $y = 60 - 2.5x$. Find the zero of the function. Describe what this value means in the context of your situation.

29. **PERSEVERE** Describe the steps you use to solve the equation $16 = x + 4 + (2^4 - 6)$ by graphing. Then explain how you can check your solution.

Shapes of Graphs

Explore Line Symmetry

Online Activity Use graphing technology to complete the Explore.

INQUIRY How can you use the graph of a function to determine whether it is symmetric?

Learn Symmetry and Graphs of Functions

The graphs of some functions exhibit a key feature called symmetry. A figure has **line symmetry** if each half of the figure matches the other side exactly.

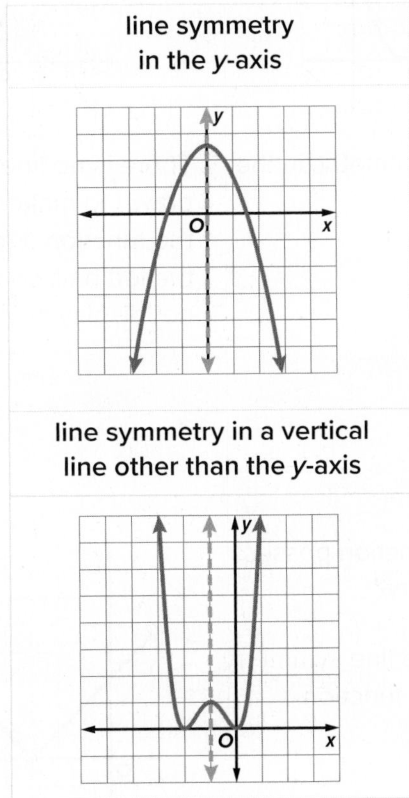

Today's Goals

- Determine whether functions have line symmetry and, if so, find the line of symmetry.
- Identify extrema and where functions are increasing and decreasing.
- Determine the end behaviors of graphs of functions.

Today's Vocabulary

line symmetry
increasing
decreasing
extrema
relative minimum
relative maximum
end behavior

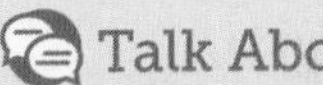

Talk About It!

Can the graph of a function be symmetric about the x-axis? Justify your argument.

Example 1 Line Symmetry

Determine whether each function has line symmetry. Explain.

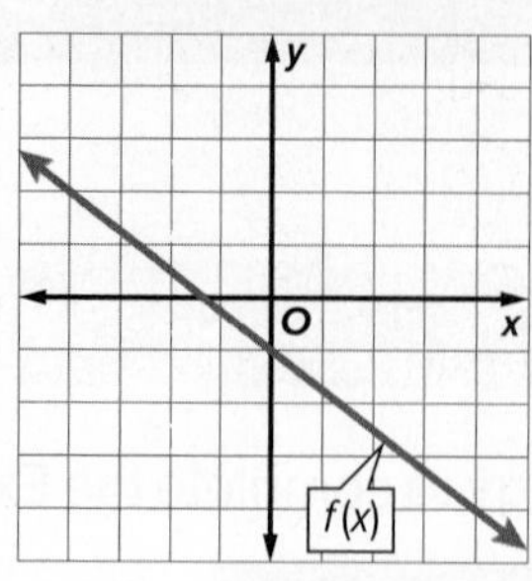	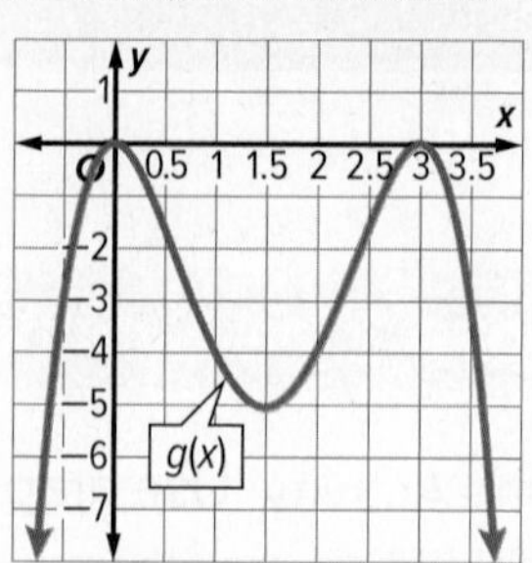
There is no line that can be drawn to make the right half a mirror image of the left half, so the function does not display any line symmetry.	This function is symmetric in the line $x = 1.5$.
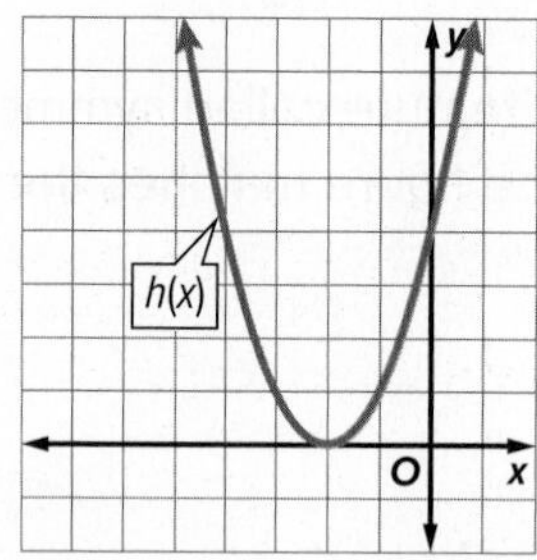	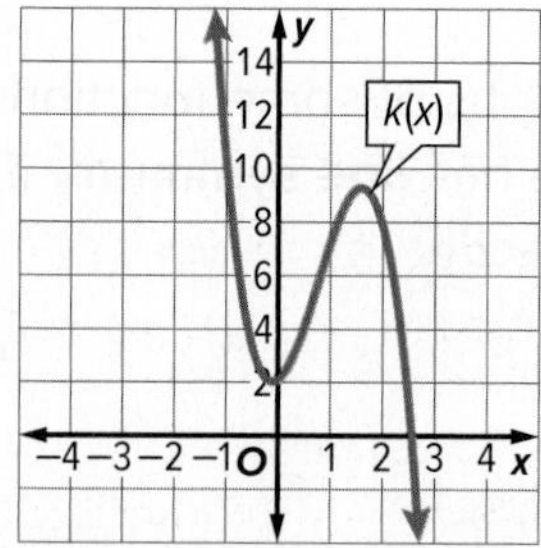
This function is symmetric in the line $x = -2$.	There is no line that can be drawn to make the right half of the function a mirror image of the left half, so the function does not display any line symmetry.

Think About It!

Find the *y*-intercepts of the functions.

Check

Examine the function.

Part A Does the function possess line symmetry?

Part B Describe the line symmetry, if any, of the function.

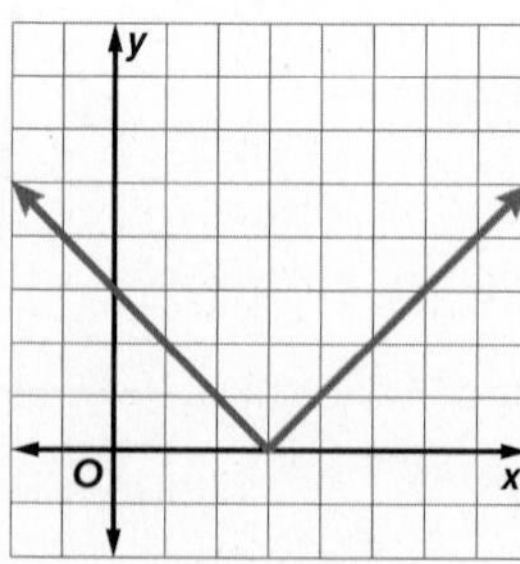

Study Tip

Symmetry Remember a graph can be symmetric in the *y*-axis or any other vertical line.

Go Online You can complete an Extra Example online.

Example 2 Interpret Symmetry

FOUNTAINS **A fountain is spraying a stream of water into the air. The solid portion of the graph represents the path of the water, where *x* is the distance in feet from the fountain and *y* is the height in feet of the stream. Find and interpret any symmetry in the graph of the function.**

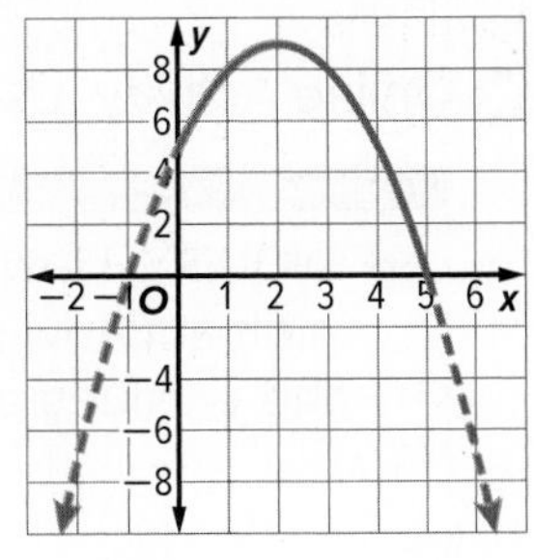

The right half of the graph is the mirror image of the left half in the line $x = 2$.

In the context of the situation, the symmetry of the graph tells you that the height of the stream of water when it is from 0 to 2 feet away from the fountain is the same as the height of the stream of water when it is from 2 to 4 feet away from the fountain.

Check

GOLF The solid portion of the graph represents the path of a golf ball after it is hit off of a platform, where *x* is the distance in feet a golf ball travels and *y* is the height in feet of the golf ball.

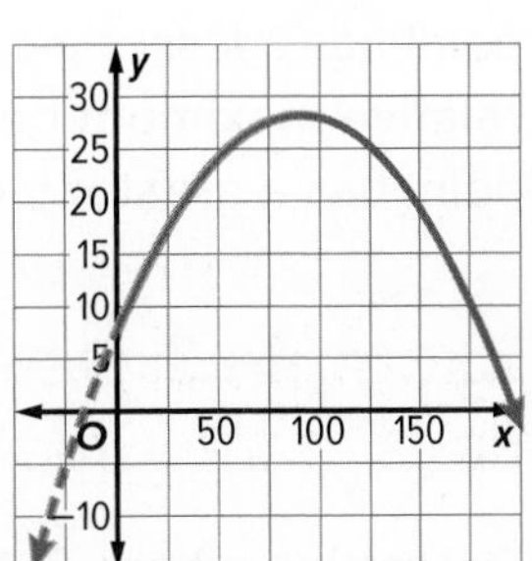

Part A Use the graph to describe any symmetry of the graph of the function.

A. symmetric in the *y*-axis

B. symmetric in the line $x = 8$

C. symmetric in the line $x = 28.25$

D. symmetric in the line $x = 90$

Part B Interpret the symmetry in the context of the situation.

A. The height of the golf ball when it has traveled a distance of 0 to 8 feet is the same as the height of the golf ball when it has traveled a distance of 8 to 28.25 feet.

B. The height of the golf ball when it has traveled a distance of 0 to 90 feet is the same as the height of the golf ball when it has traveled a distance of 90 to 180 feet.

C. The distance the golf ball has traveled when it is 0 to 8 feet in the air is the same as the distance the golf ball has traveled when it is 8 to 28.25 feet in the air.

D. The distance the golf ball has traveled when it is 0 to 90 feet in the air is the same as the distance the golf ball has traveled when it is 90 to 180 feet in the air.

Go Online You can complete an Extra Example online.

Think About It!

Find the maximum height of the stream of water.

Problem-Solving Tip

To help visualize a line of symmetry, imagine folding the graph in half. If the graph lines up perfectly, the graph is symmetric about the line you have created with the fold.

Explore Relative High and Low Points

Online Activity Use graphing technology to complete the Explore.

INQUIRY How do the y-values of relatively high and low points on the graph compare to the y-values of nearby points?

Study Tip

Reading Math *Extrema* in this context is the plural form of *extreme point*. The plural forms of *maximum* and *minimum* are *maxima* and *minima*, respectively.

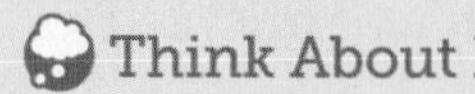

Think About It!

If $f(x)$ has a relative maximum when $x = 2$, then is $f(x)$ increasing or decreasing as x approaches 2? as x moves past 2?

Learn Extrema of Graphs of Functions

A function is **increasing** where the graph goes up and **decreasing** where the graph goes down when viewed from left to right.

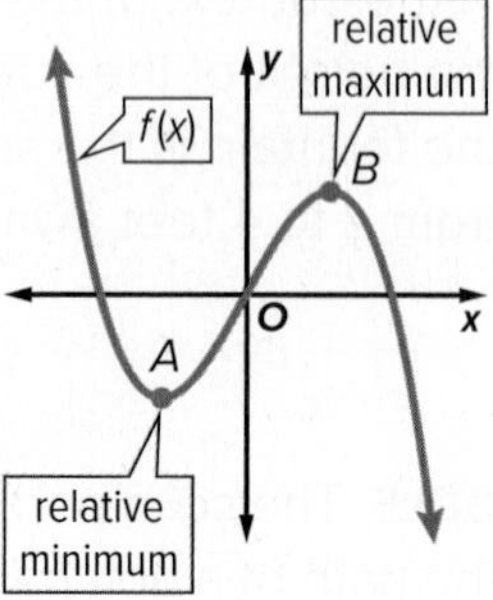

Points that are the locations of relatively high or low function values are called **extrema**. Point A is a **relative minimum** because no other nearby point has a lesser y-coordinate. Point B is a **relative maximum** because no other nearby point has a greater y-coordinate.

Example 3 Determine Increasing and Decreasing Parts of the Graph of a Function

Determine where $f(x)$ is increasing and/or decreasing.

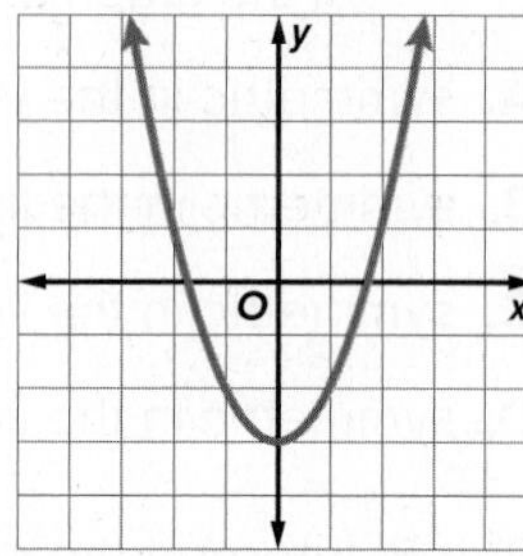

When $x > 0$, the graph goes up when viewed from left to right. So, the function is increasing for $x > 0$.

When $x < 0$, the graph goes down when viewed from left to right. So, the function is decreasing for $x < 0$.

Check

For $x > 1$, $f(x)$ is ______?______.

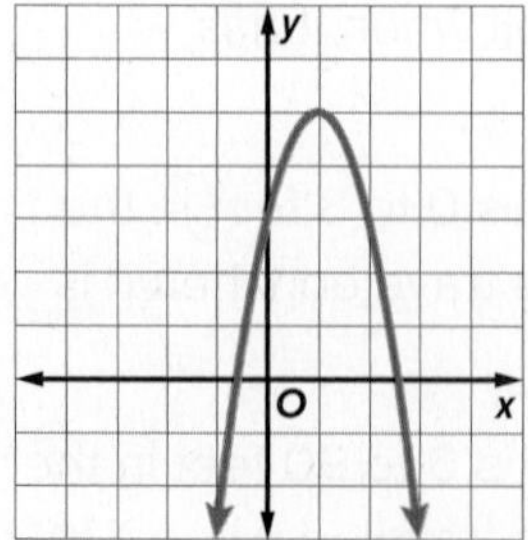

 Go Online You can complete an Extra Example online.

Example 4 Determine Extrema of the Graph of a Function

Determine the extrema of $f(x)$. Then identify each point as a relative maximum or relative minimum.

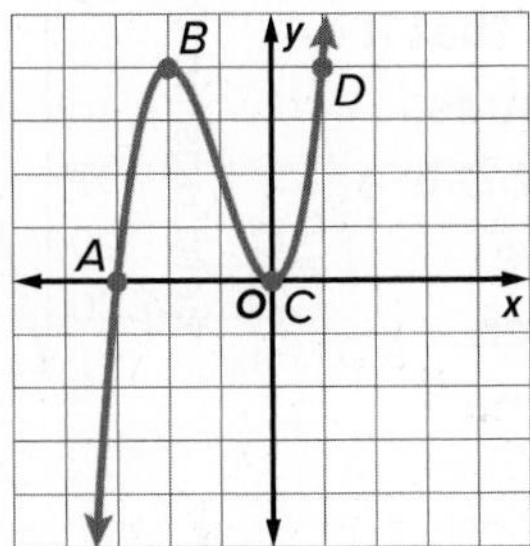

Extrema: Point B and point C are the locations of relatively high or low function values. So, they are the extrema of the function.

Relative Minimum: No other points nearby point C have a lesser y-coordinate. So, point C is a relative minimum.

Relative Maximum: No other points nearby point B have a greater y-coordinate. So, point B is a relative maximum.

Check

Which point(s) is(are) a relative minimum? Select all that apply.

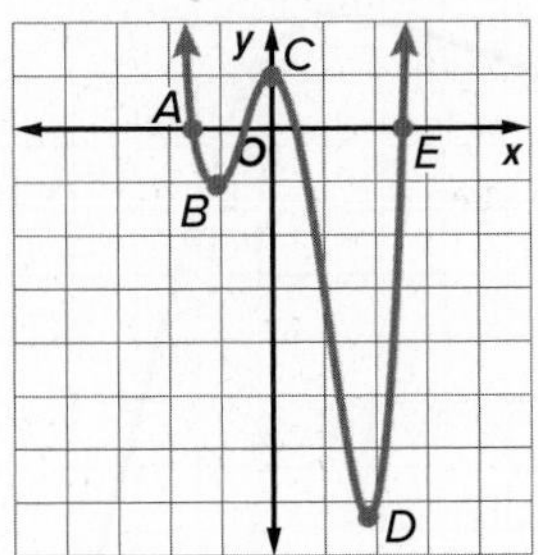

A. A

B. B

C. C

D. D

E. E

Go Online You can complete an Extra Example online.

Think About It!

Can a point be a relative minimum or relative maximum and not be an extreme point? Explain.

Go Online You can watch a video to see more about the comic book store.

Think About It! Does this function have a relative minimum? Explain your reasoning.

Example 5 Interpret Extrema of the Graph of a Function

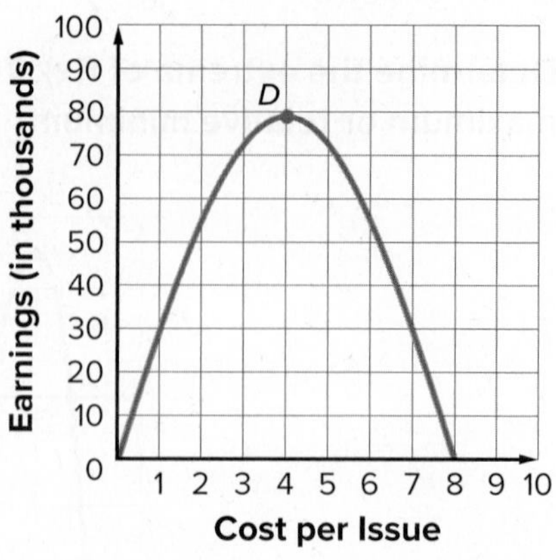

COMIC BOOKS **A comic book store uses a function to model its profit in thousands of dollars given the price in dollars that it charges for individual issues. Determine whether point *D* is a relative minimum, relative maximum, or neither. Then interpret its meaning in the context of the situation.**

Point *D* is a relative maximum because all nearby points have a lesser *y*-coordinate.

Point *D* represents the greatest profit that the comic book store can earn given the price it charges per issue.

Check

AEROBATICS Aerobatics, or stunt flying, is the practice of intentional maneuvers of an aircraft that are not necessary for normal flight. Lincoln Beachey, an inventor of aerobatics, was known for his stunt called the "Dip of Death" in which his plane would plummet toward the ground from 5000 feet until he leveled the plane. His distance from the ground during the stunt can be approximately modeled by the function $f(x)$. Identify any extrema in the context of the situation.

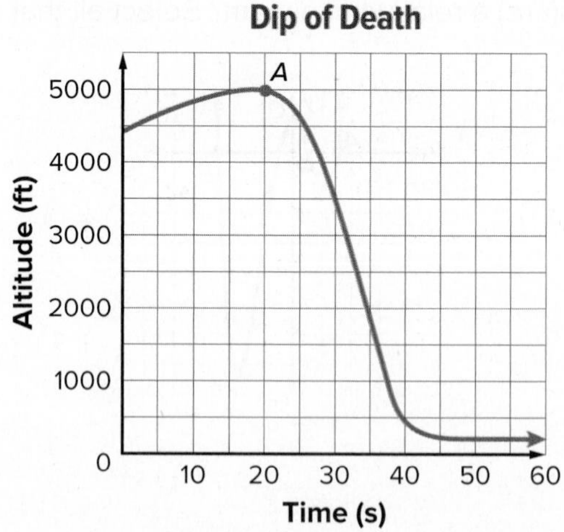

Point A represents that at about 20 seconds, Lincoln Beachey reached a ___?___ relative maximum in height at 5000 feet.

After point A, his height is ___?___ decreasing until he levels out the plane.

Go Online You can complete an Extra Example online.

Learn End Behavior of Graphs of Functions

End behavior describes the values of a function at the positive and negative extremes in its domain.

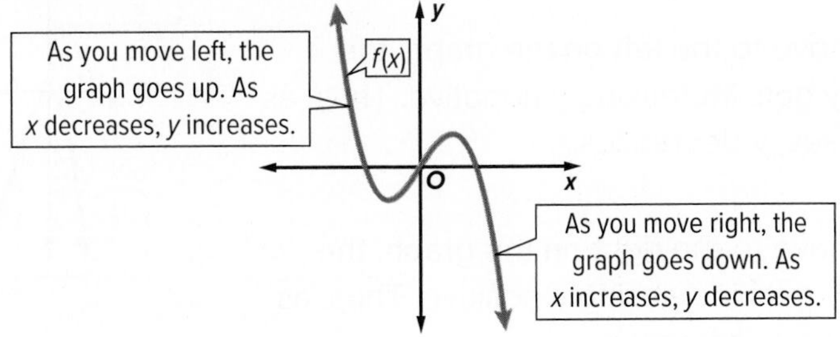
As you move left, the graph goes up. As x decreases, y increases.
f(x)
y
O
x
As you move right, the graph goes down. As x increases, y decreases.

Example 6 Determine End Behavior of the Graph of a Linear Function

Determine the end behavior of $f(x)$.

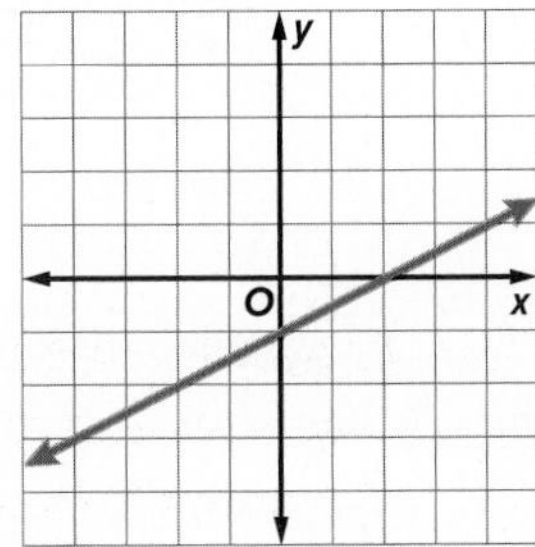
y
O
x

As you move to the left on the graph, the value of y gets increasingly negative. Thus, as x decreases, y decreases.

As you move to the right on the graph, the value of y gets increasingly positive. Thus, as x increases, y increases.

Check

Determine the end behavior of $f(x)$.

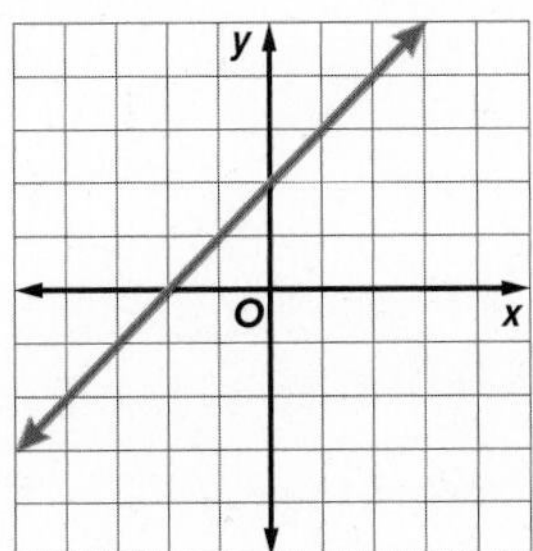
y
O
x

As x increases, y ____?____.

As x decreases, y ____?____.

Go Online You can complete an Extra Example online.

Go Online
You may want to complete the Concept Check to check your understanding.

Watch Out!
End Behavior The end behavior of some graphs can be described as approaching a specific value.

Think About It!
Make a conjecture, or educated guess, about the end behavior of a linear function when the slope is positive or negative.

Study Tip
Conjecture A conjecture is an educated guess based on known information.

Example 7 Determine End Behavior of the Graph of a Nonlinear Function

Determine the end behavior of *f*(*x*).

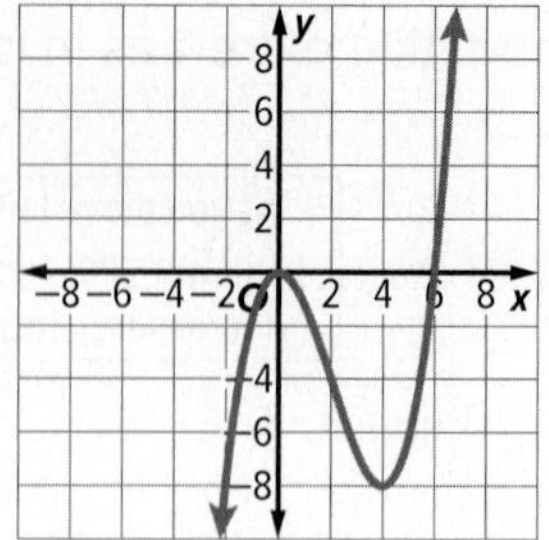

As you move to the left on the graph, the value of *y* gets increasingly negative. Thus, as *x* decreases, *y* decreases.

As you move to the right on the graph, the value of *y* gets increasingly positive. Thus, as *x* increases, *y* increases.

Go Online to practice what you've learned about interpreting graphs in the Put It All Together over Lessons 3-1 through 3-5.

Check

Determine the end behavior of each function.

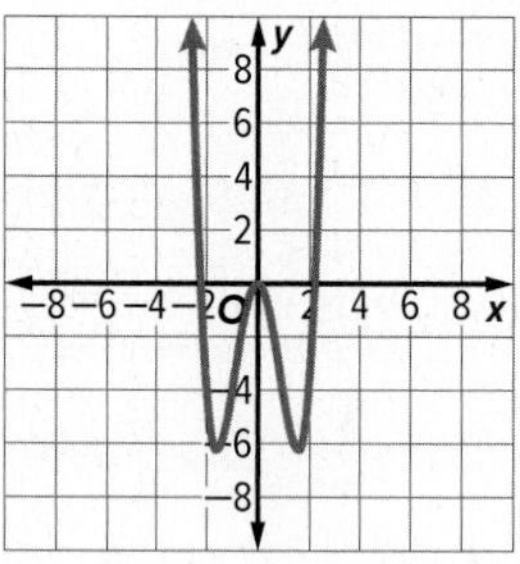

As *x* increases, *y* ___?___.

As *x* decreases, *y* ___?___.

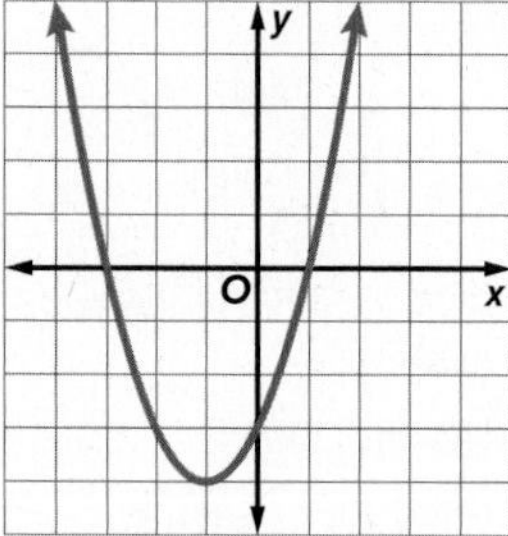

As *x* increases, *y* ___?___.

As *x* decreases, *y* ___?___.

Go Online You can complete an Extra Example online.

Practice

Go Online You can complete your homework online.

Example 1

Determine whether each function has line symmetry. Explain.

1. 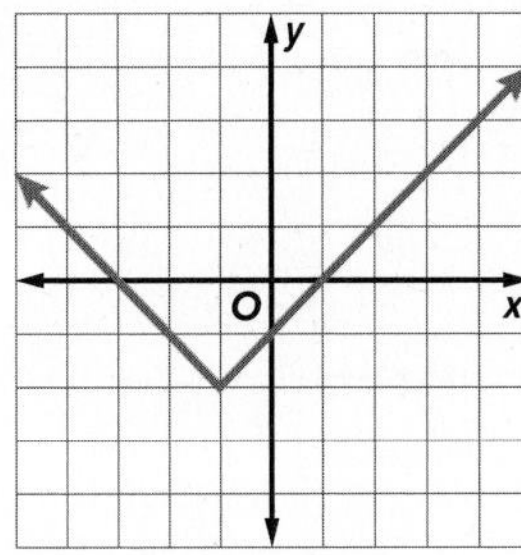

2. 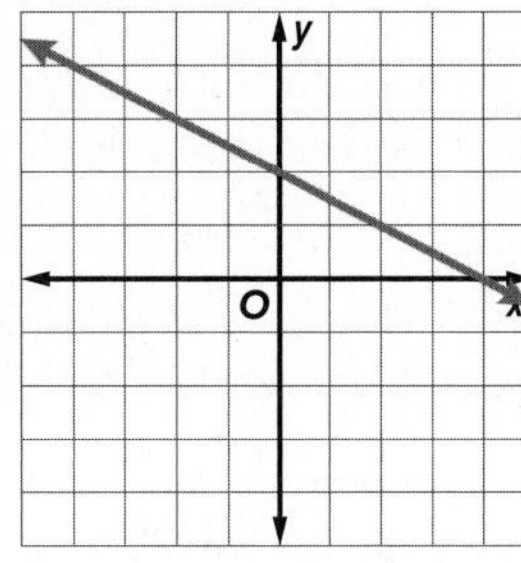

3. 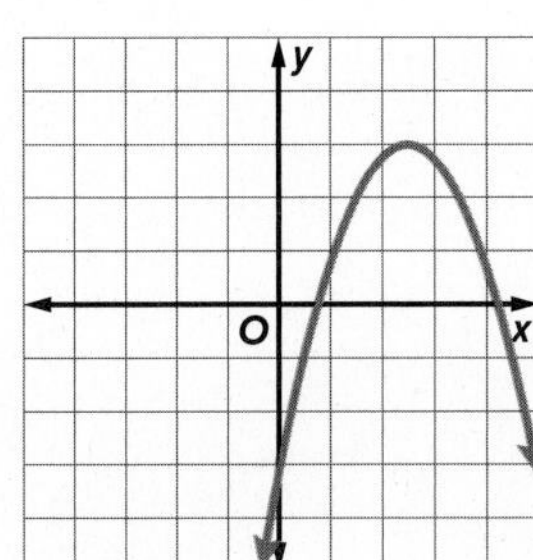

4. 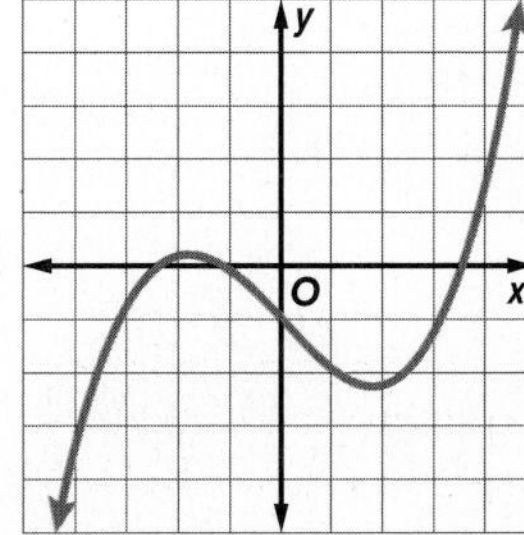

Example 2

5. **GEOMETRY** The solid portion of the graph represents the relationship between the width of a rectangle in centimeters x and the area of the rectangle in centimeters squared y. Find and interpret any symmetry in the graph of the function.

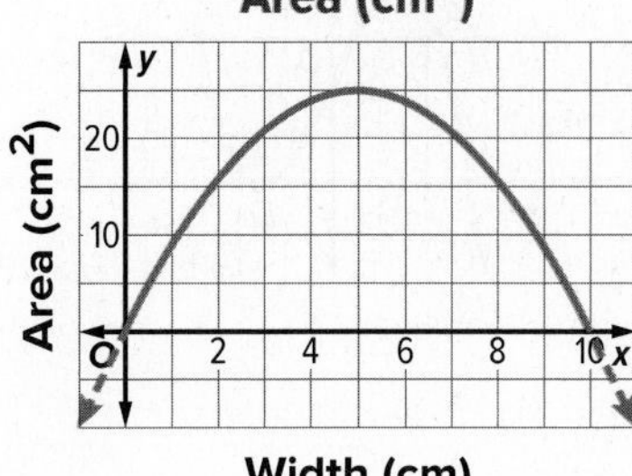

6. **SPRINKLERS** A sprinkler is spraying a stream of water into the air. The solid portion of the graph represents the path of the water, where x represents the distance in feet from the sprinkler and y represents the height in feet of the water. Find and interpret any symmetry in the graph of the function.

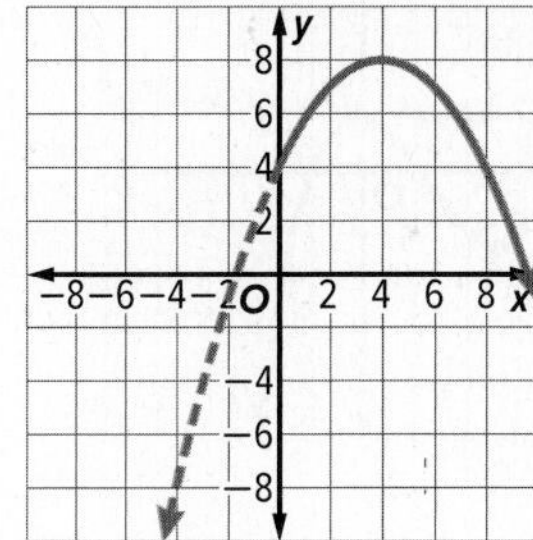

Example 3

Determine where *f*(*x*) is increasing and/or decreasing.

7.

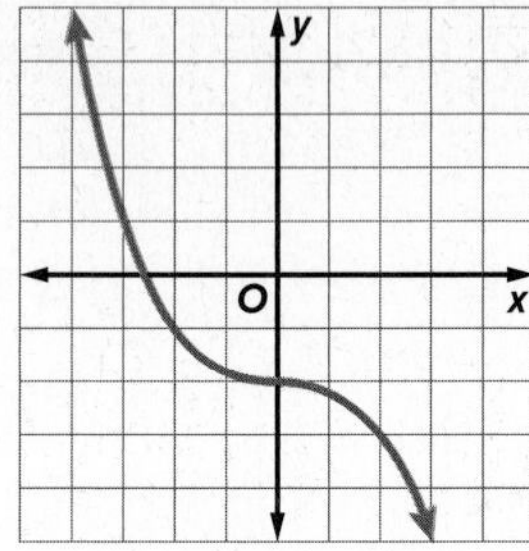

8.

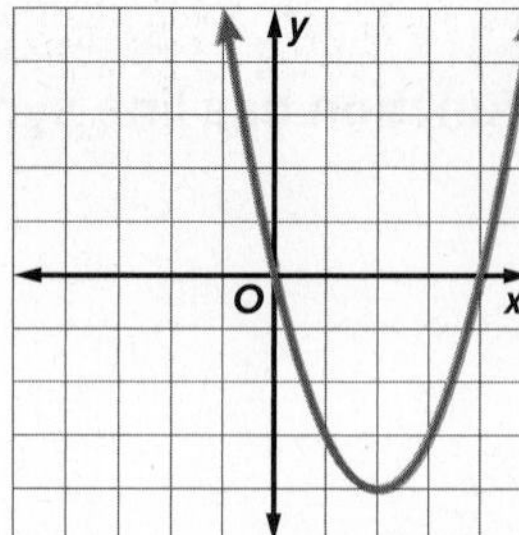

9.

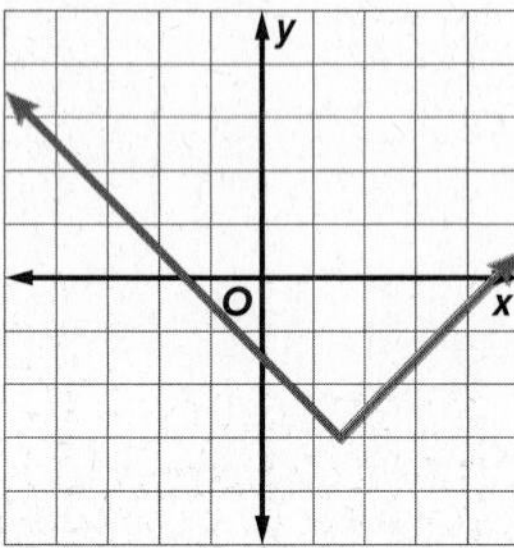

10.

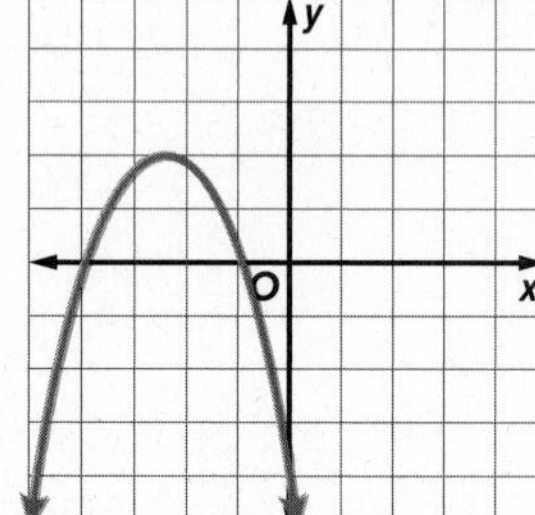

Example 4

Determine the extrema of *f*(*x*). Then identify each point as a relative maximum or relative minimum.

11.

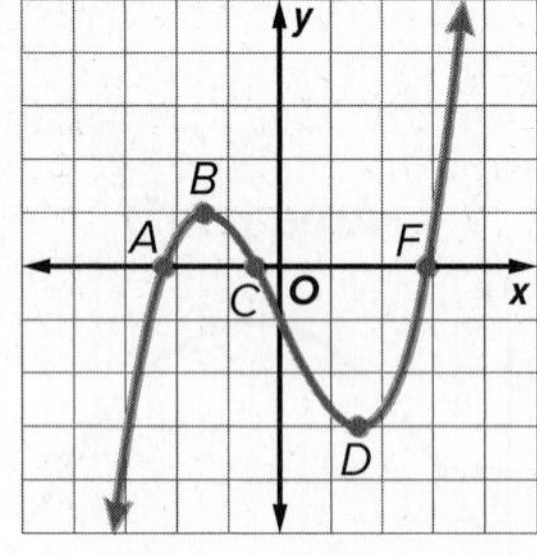

12.

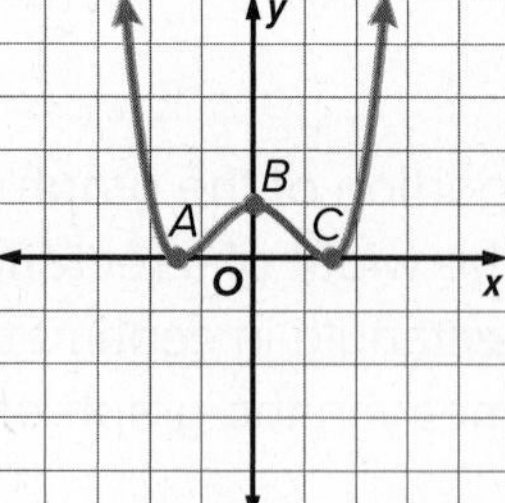

13.

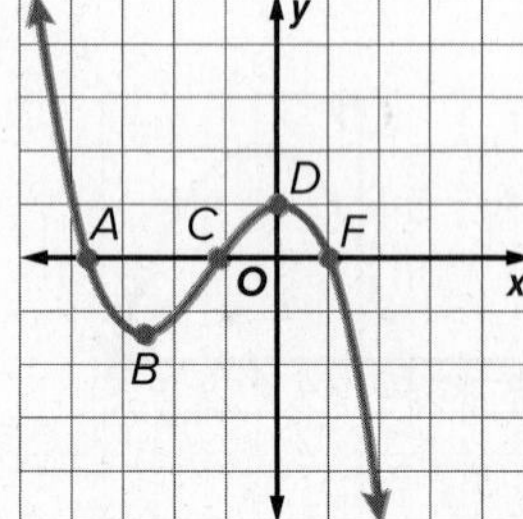

14.

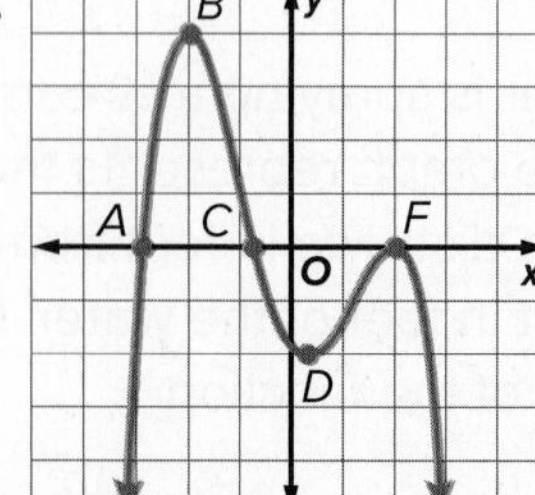

Example 5

15. GOLF The height of a golf ball compared to the distance the golf ball is from the tee is shown in the graph. Determine whether point A is a *relative minimum, relative maximum,* or *neither*. Describe what this value means in the context of this situation.

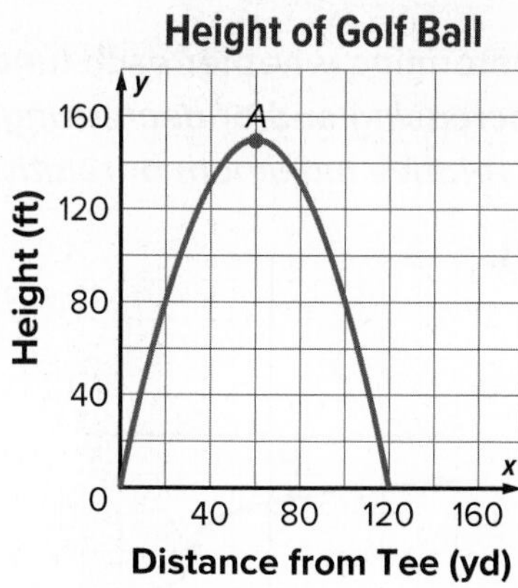

16. ROLLERCOASTER The height of a rollercoaster compared to the distance from start is shown in the graph. Determine whether points B, C, D, and F are *relative minima, relative maxima,* or *neither*. Describe what each value means in the context of this situation.

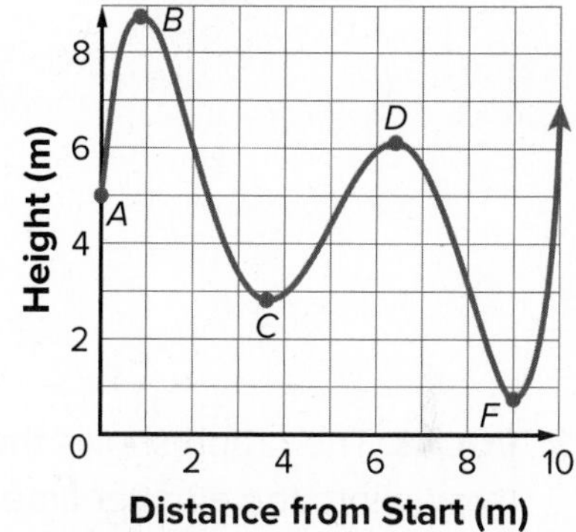

Examples 6 and 7

Determine the end behavior of $f(x)$.

17.

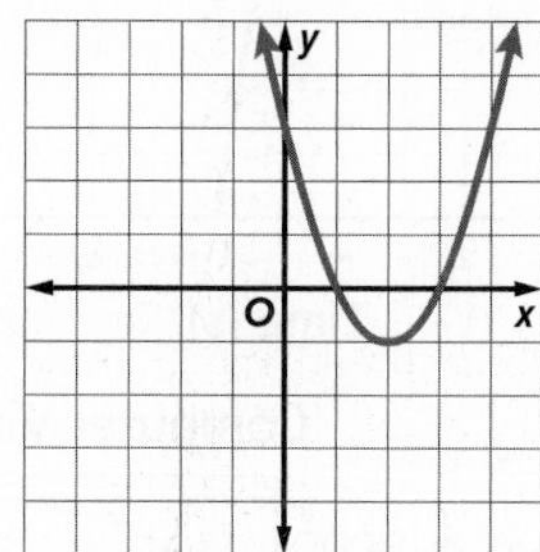

18.

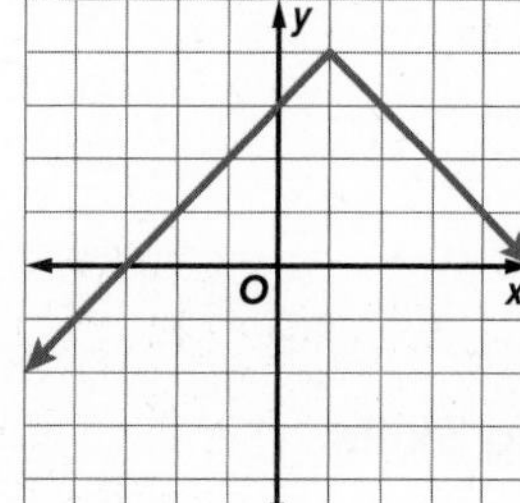

19.

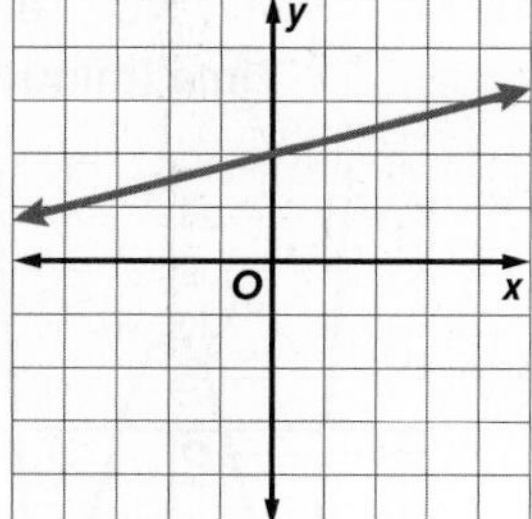

20.

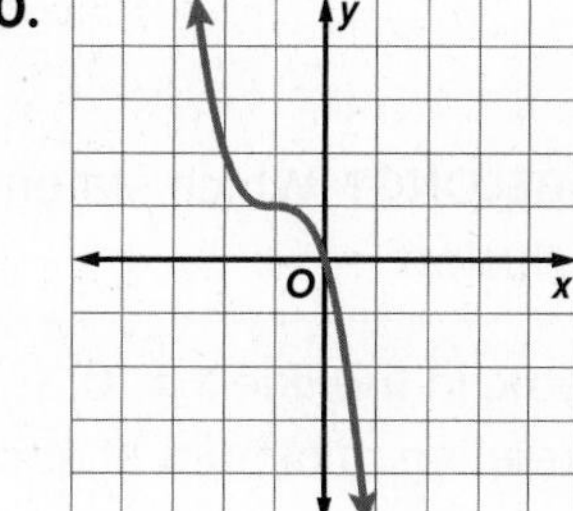

Mixed Exercises

Determine whether each function has line symmetry and where the function is *increasing* and/or *decreasing*. Determine the extrema. Then identify each point as a *relative maximum* or *relative minimum*. Determine the end behavior.

21.

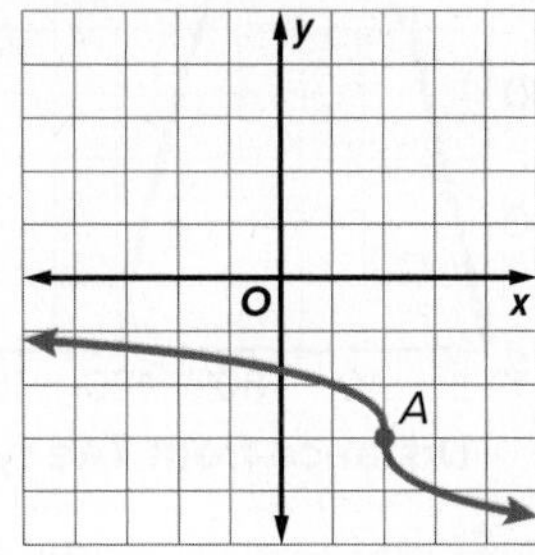

22.

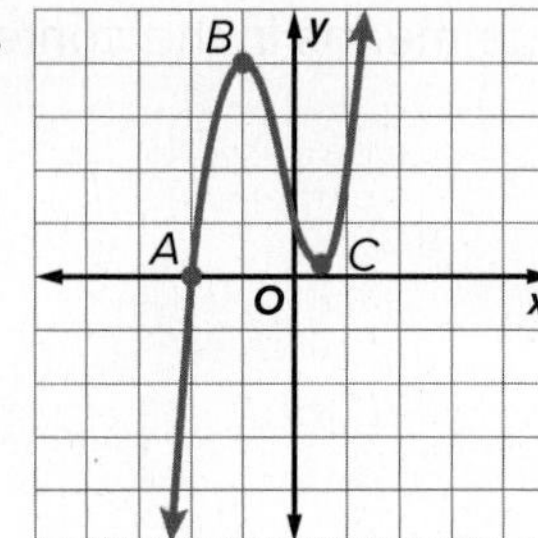

23. ROCKS The graph shows the height of a rock after it is thrown into the air over time. Determine the extrema. Then identify each point as a *relative minimum, relative maximum,* or *neither*. Describe what each value means in the context of this situation.

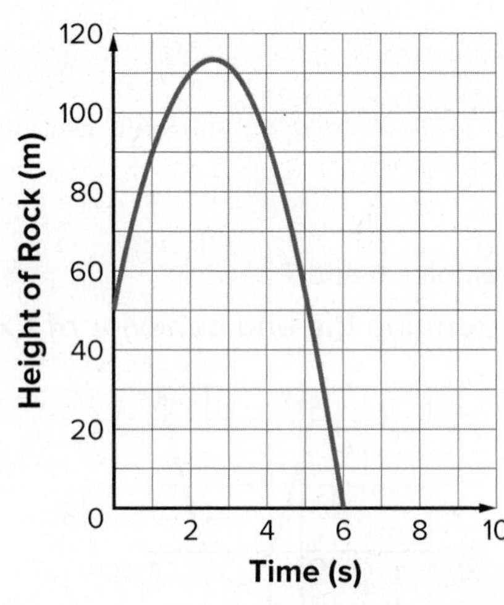

Higher-Order Thinking Skills

24. WRITE The graph shows the number of computers that are affected by a virus over time. Determine and interpret the end behavior.

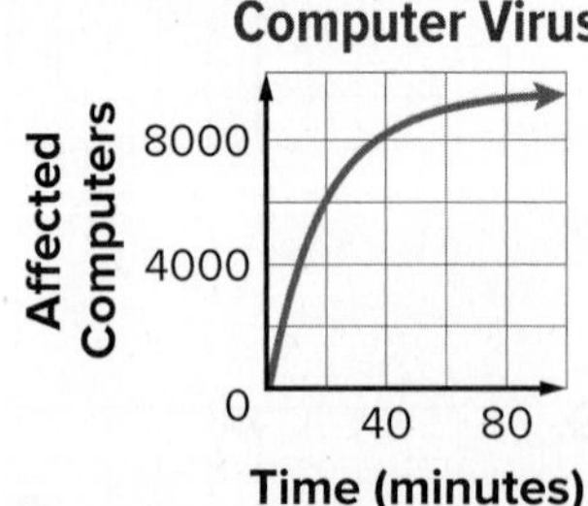

25. WHICH ONE DOESN'T BELONG? Which statement about the graph is not true? Justify your conclusion.

> The graph is symmetric in the line $x = 0$.
> The graph has one relative minimum at about $(-2.25, -16)$.
> The graph has one relative maximum at about $(0, 9)$.
> As x decreases, y increases. As x increases, y increases.

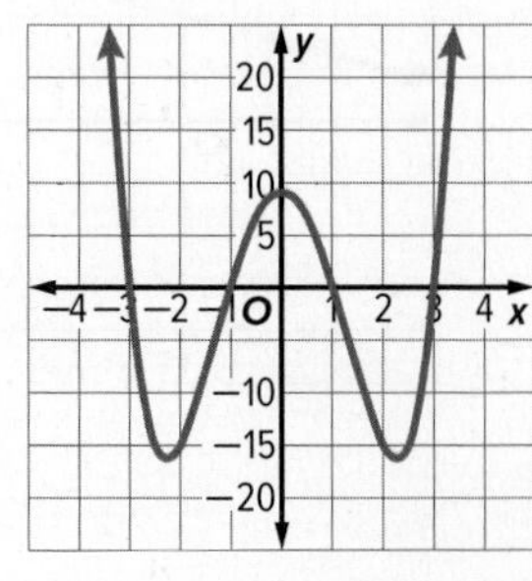

Sketching Graphs and Comparing Functions

Explore Modeling Relationships by Using Functions

 **Online Activity** Use the infographic to complete the Explore.

INQUIRY How can you use key features to approximate the graphs of functions?

Learn Sketching Graphs of Functions

You can sketch the graph of a function using its key features. Knowing the domain, range, intercepts, symmetry, end behavior, and extrema of a function as well as intervals where the function is increasing, decreasing, positive, or negative provides a clear idea of what the graph of the function looks like.

Label the graph to identify its key features

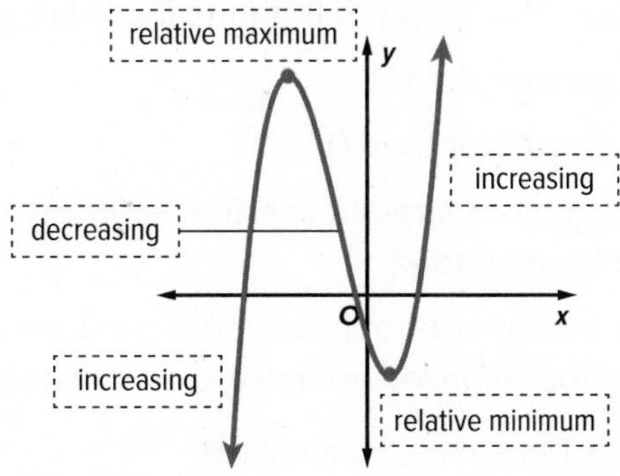

Today's Goal
- Sketch graphs of functions.

 **Think About It!**

What other key features do you see in the graph of the function? Give two examples.

Example 1 Sketch the Graph of a Linear Function

CYCLING **In 2015, Christoph Strasser set a new 24-hour cycling record by riding 556 miles in a 24-hour period. The distance he rode over the 24 hours can be represented by a function. Sketch a graph that shows the distance traveled *y* as a function of time *x*.**

Sam Edwards/age fotostock

Before sketching, consider any possible constraints of the situation. It is not possible for Christoph Strasser to ride for a negative amount of time or ride negative miles. Therefore, the domain and range are restricted to nonnegative *x*- and *y*-values, and the graph exists only in the first quadrant.

***y*-Intercept:** No distance traveled when he has ridden for 0 hours.

The point represents 0 hours ridden and 0 miles traveled. Graph this on the coordinate plane.

Linear or Nonlinear: The graph of the function is linear.

Positive: for time greater than 0

Increasing: for time greater than 0

The graph is a straight line that is positive and increasing for all hours ridden.

End Behavior: As the number of hours he has ridden increases, the number of miles he has traveled increases.

As *x* increases, *y* increases.

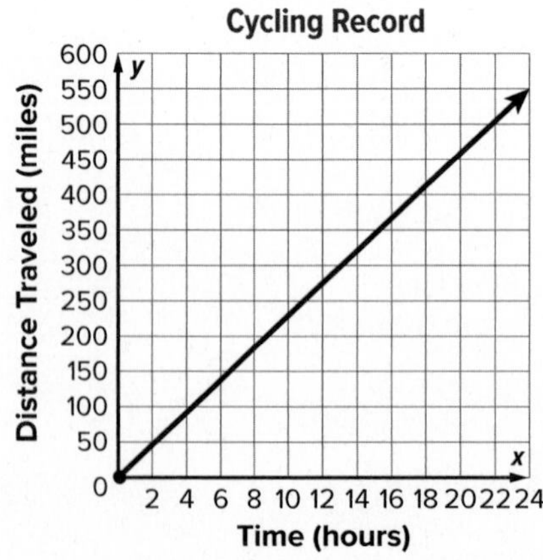

Go Online You can complete an Extra Example online.

Think About It!

What assumption is made when graphing Christoph Strasser's record-setting bike ride? Why is it necessary to make this assumption?

Watch Out!

If information about some key features is not provided, do not assume that it is not important. In this example, the missing information about where the function is negative was not necessary because it did not apply in the context of the situation. However, this is not always the case.

Example 2 Sketch the Graph of a Symmetric Function

WEATHER **A person's happiness can be affected by temperature. Sketch a nonlinear graph that shows the happiness of a person *y* as a function of temperature *x*. Interpret the key features.**

Positive: between about 25°F and 89°F

Negative: for temperatures less than 25°F and greater than 89°F

Increasing: for temperatures less than about 57°F

Decreasing: for temperatures greater than about 57°F

Relative Maximum: at about 57°F, when a person's happiness is about 85

A relative maximum occurs at 57°F, or $x = 57$, and a happiness of 85, or $y = 85$. This is represented by the point (57, 85), which we can graph on the coordinate plane.

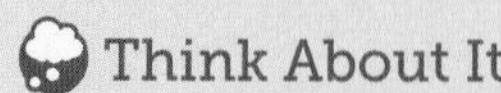

Think About It!
Describe the location of the points where the function changes from positive to negative or negative to positive.

End Behavior: As temperature increases or decreases, a person's happiness decreases.

For temperatures less than 25°F or $x < 25$, the graph is negative. For these temperatures, the graph is also increasing. For temperatures between 25°F and 57°F, the graph is positive and increasing. The graph is positive and decreasing for temperatures between 57°F and 89°F. This interval of the graph is also symmetric to the graph from 25°F to 57°F. This means that the right half of the graph is the mirror image of the left half. For temperatures greater than 89°F, or $x > 89$, the graph is negative, decreasing, and symmetric to the interval of the graph that is less than 25°F. As temperature increases or decreases, a person's happiness decreases. This means that happiness will get increasingly negative to move right and left on the graph.

Think About It!
What does a negative *y*-value represent in the context of this situation?

Symmetry: A person's happiness for temperatures less than 57°F is the same as their happiness for temperatures greater than 57°F.

A person is happiest when it is 57°F. As the temperature gets increasingly cold or hot, a person becomes less happy. When the temperature is below about 25°F or above about 89°F a person is unhappy.

(*continued on the next page*)

Go Online You can complete an Extra Example online.

One method for sketching a graph given key features is to first graph any given points. Then analyze the key features of small intervals from left to right to make sketching the graph of the function easier.

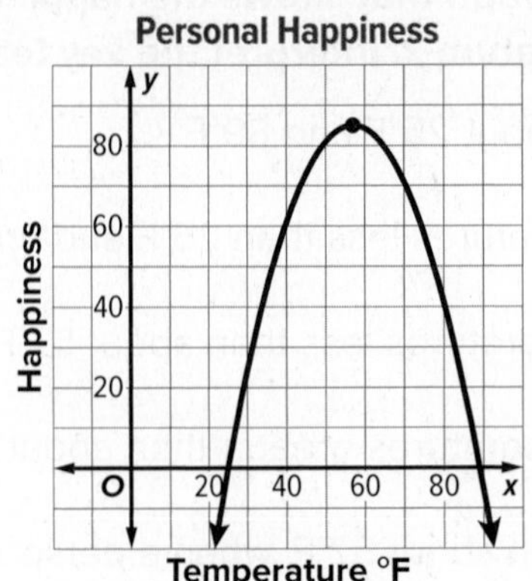

Check

Mariana used the key features to sketch the graph of x as a function of y. Examine the key features and graph to identify which key features Mariana graphed correctly and incorrectly.

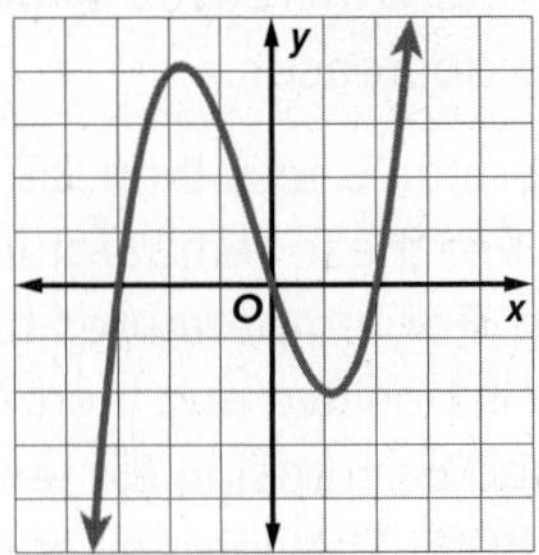

A. Positive: between $x = -3$ and $x = 0$ and for $x > 2$

B. Negative: for $x < -3$ and between $x = 0$ and $x = 2$

C. Increasing: for about $x < -2$ and between about $x = 1$ and $x = 3$

D. Decreasing: for between about $x = -2$ and $x = 1$ and for $x > 3$

E. Intercepts: The graph intersects the x-axis at $(-3, 0)$, $(0, 0)$, and $(2, 0)$.

F. Relative Minimum: $(1, -2)$

G. Relative Maximum: $(-2, 4)$ and $(3, 2)$

H. End Behavior: As x increases and decreases, the value of y decreases.

Correct	Incorrect

Example 3 Sketch the Graph of a Nonlinear Function

AMUSEMENT PARK **The number of people in line for a rollercoaster throughout the day can be modeled by a function. Use the key features to sketch a graph of the function. Then interpret the key features if x represents the time in hours since the ride opened at 10:00 A.M. and y represents the number of people in line.**

Positive: between $x = -0.5$ and $x = 12$

Negative: for $x < -0.5$ and $x > 12$

Increasing: for $x < 1.4$ and between $x = 5.3$ and $x = 9.9$

Decreasing: for between $x = 1.4$ and $x = 5.3$ and for $x > 9.9$

Intercepts: The graph intersects the x-axis at $(-0.5, 0)$ and $(12, 0)$ and intersects the y-axis at $(0, 220)$.

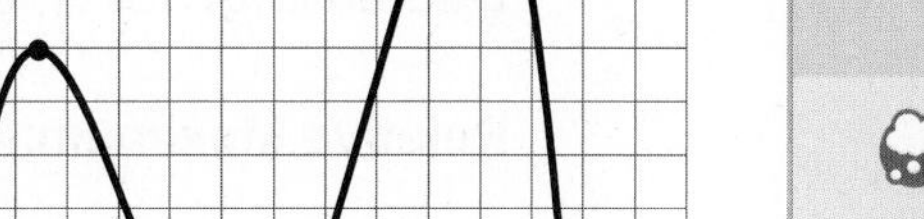

Relative Minimum: at $(5.3, 133)$

Relative Maximum: at $(1.4, 448)$ and $(9.9, 643)$

End Behavior: As x increases or decreases, the value of y decreases.

The x-intercepts mean that the number of people in line is zero a half hour before the ride opened and 12 hours after it opened. The y-intercept means that 220 people were in line when the ride opened.

The ride experienced a relative low in the number of people in line 5.3 hours after the ride opened and two relative peaks in the number of people in line 1.4 hours and 9.9 hours after it opened.

The number of people in line was negative but increasing until a half hour before the ride opened, positive and increasing from a half hour before the ride opened until 1.4 hours after it opened and again from 5.3 hours after the ride opened until 9.9 hours after it opened, negative and decreasing after the ride had been open for 12 hours, and positive but decreasing from 1.4 hours after the ride opened until 5.3 hours after it opened and again from 9.9 hours after the ride opened until 12 hours after the ride opened.

The graph indicates a period where there is a negative number of people in line. Because it is not possible to have a negative number of people, this graph appears to only model the number of people in line for the ride from a half hour before the ride opened until 12 hours after it opened.

Go Online You can complete an Extra Example online.

Think About It!

Why might there be a relative minimum in the number of people in line around 3:00 P.M., 5.3 hours after the ride opened? Why might there be zero people in line 12 hours after the ride opened at 10:00 P.M.?

Math History Minute

Sometimes the coordinate plane is referred to as the *Cartesian plane*. It is named for French mathematician **Rene Descartes (1596–1650)**, who is known as the father of analytical geometry, the bridge between algebra and geometry.

Check

MARINE LIFE The path of a dolphin jumping out of the ocean can be modeled with a symmetric function.

Part A Which graph(s) could be used to show the height of a dolphin above the water *y* as a function of time since it emerged from the water *x*?

Positive: between 0 and 8 seconds

Negative: for time less than 0 seconds and greater than 8 seconds

Decreasing: for time greater than 4 seconds

Relative Maximum: at 4 seconds

End Behavior: As *x* increases and decreases, the value of *y* decreases.

Symmetry: The right half of the graph is the mirror image of the left half in approximately the line $x = 4$.

A.

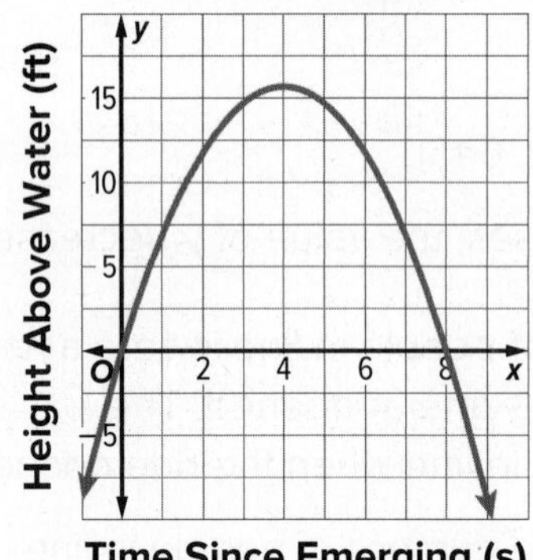

B.

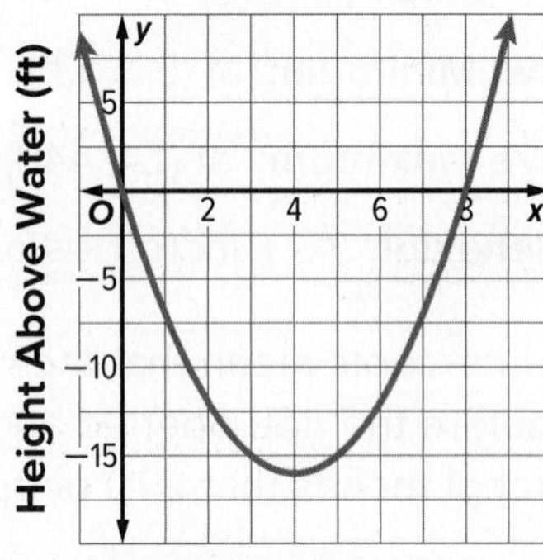

C.

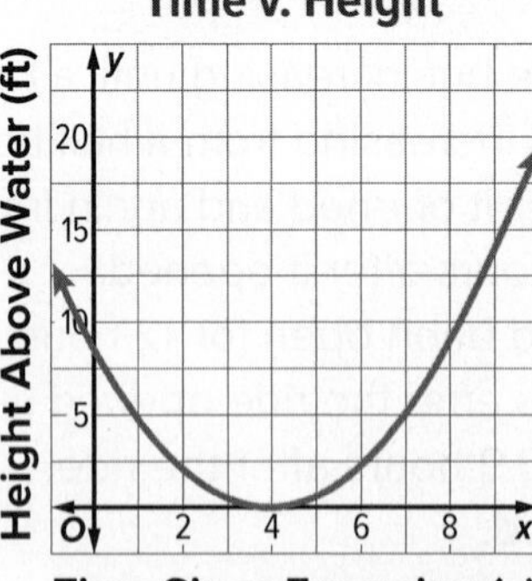

D.

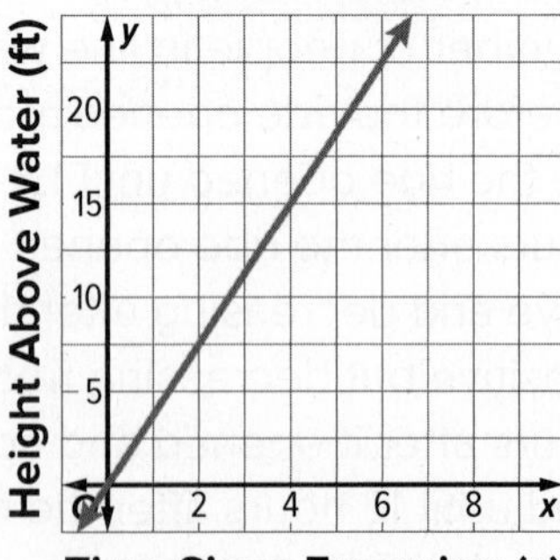

Part B Negative *x*-values represent ___?___ and negative *y*-values represent ___?___.

Example 4 Compare Properties of Functions

TENNIS **Hawk-Eye is a computer system used in tennis to track the path of the ball. It is used as an officiating aid to locate the landing spot of a tennis ball when players challenge a call. Use the description and graph of a player's forehand and backhand shots to compare the paths of the two shots if y is the vertical height and x is the distance.**

Forehand

During the forehand, the ball leaves the player's racquet at a height of 2.8 feet and travels 29 feet, when it reaches a height of about 10 feet. Then, the height of the ball decreases until it hits the ground 58 feet from where it was hit.

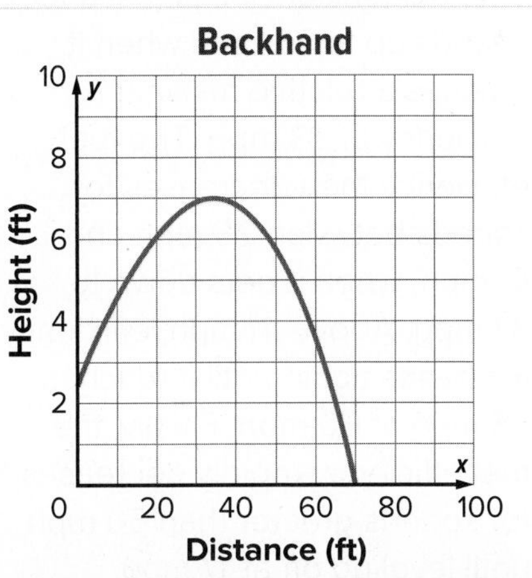

	Forehand	Backhand
x-intercept	58	70
y-intercept	2.8	2.5
Extrema	maximum height of 10 feet when $x = 29$.	maximum height of 7 feet when $x = 35$.
Increasing and Decreasing	increases to a height of 10 feet from $x = 0$ to $x = 29$ and then decreases from $x = 29$ to $x = 58$ to a height of 0 feet.	increases to a height of 7 feet from $x = 0$ to $x = 35$ and then decreases from $x = 35$ to $x = 70$ to a height of 0 feet.

x-intercept

The tennis ball travels 12 feet farther during the backhand shot.

y-intercept

The y-intercepts of the two functions mean that the tennis ball is about 0.3 foot higher at the beginning of the forehand shot.

Extrema

The maximum height of the tennis ball is 3 feet higher during the forehand shot.

Increasing and Decreasing

The height of the tennis ball increases over a shorter interval during the forehand shot, but it reaches a higher maximum height. This means that the tennis ball increases at a faster rate during the forehand.

 **Go Online** You can complete an Extra Example online.

Talk About It!

Compare the intervals over which each shot is positive and/or negative. Does this make sense in the context of the situation? Explain your reasoning.

Check

CARS Use the description and graph to compare the fuel economy of two cars, where y is the fuel efficiency in miles per gallon and x is the speed in miles per hour.

Car A	**Car B**
The fuel efficiency increases for speeds up to 25 mph when it reaches a relative maximum efficiency of 53 mpg. The fuel efficiency then decreases for speeds between 25 mph and 41 mph, when it gets down to 37 mpg. Above 41 mph, efficiency increases again until it reaches 48 mpg at 60 mph. Finally, the fuel efficiency rapidly decreases for speeds greater than 60 mph until leveling off at 17 mpg.	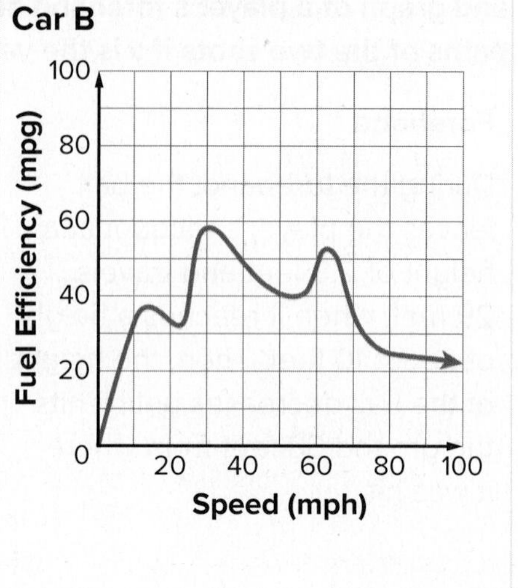

Which statements about the fuel efficiencies of the two cars are true? Select all that apply.

A. Car A has the greatest maximum fuel efficiency.

B. Car B has more relative maximum fuel efficiencies than Car A.

C. As speed increases, Car A levels off at a greater fuel efficiency than Car B.

D. Both cars get 0 mpg when they are traveling at 0 mph.

E. Car A has the least relative minimum fuel efficiency.

F. Both cars increase in fuel efficiency for speeds between 0 mph and about 13 mph.

G. Neither car reaches fuel efficiency below 0 mpg.

Pause and Reflect

Did you struggle with anything in this lesson? If so, how did you deal with it?

Practice

Go Online You can complete your homework online.

Example 1

1. SAVINGS David is saving money to buy a new car. The amount he saves can be represented by a function. Sketch a graph that shows the amount in savings y, in dollars, as a function of time x, in weeks.

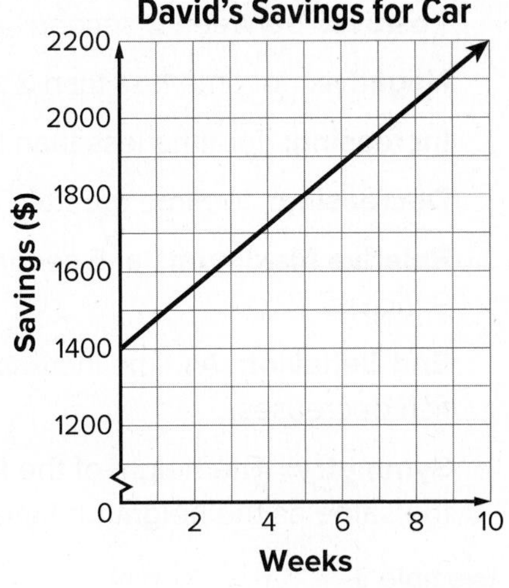

x-Intercept: none

y-Intercept: $1400

Linear or Nonlinear: The graph of the function is linear.

Positive: for time greater than 0

Increasing: for time greater than 0

End Behavior: As the number of weeks he has saved increases, the amount saved increases.

2. SWIMMING Yukio is keeping track of the number of calories she burns while swimming freestyle laps. The number of calories she burns can be represented by a function. Sketch a graph that shows the number of calories burned y as a function of time x, in hours.

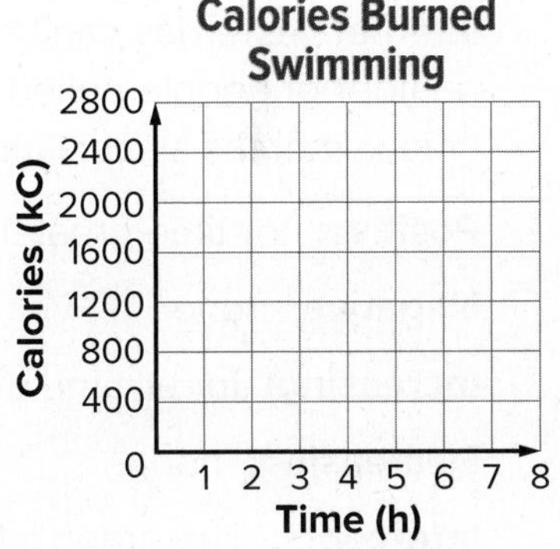

y-Intercept: No calories burned when she has swum for 0 hours.

Linear or Nonlinear: The graph of the function is linear.

Positive: for time greater than 0

Increasing: for time greater than 0

End Behavior: As the number of hours she has swum increases, the number of calories burned increases.

Example 2

3. FOOTBALL The flight of a football thrown by a quarterback can be modeled by an interval of a function. Sketch a nonlinear graph that shows the height of a football y, in feet, as a function of time x, in seconds.

Height of Football (ft) vs. Time (s)

Positive: between 0 seconds and 5 seconds

Negative: for time greater than 5 seconds (*represents time after the ball hits the ground*)

Increasing: for time less than 2 seconds

Decreasing: for time greater than 2 seconds

Relative Maximum: at 2 seconds, when the height of the football is 9 feet

End Behavior: As time increases, the height of the football decreases.

Symmetry: The height of the football for time between 0 seconds and 2 seconds is the same as the height for time between 2 seconds and 4 seconds.

4. FISH The height of a fish compared to sea level as it jumps out of the ocean water can be represented by a function. Sketch a nonlinear graph that shows the height of a fish y, in inches, as a function of time x, in seconds.

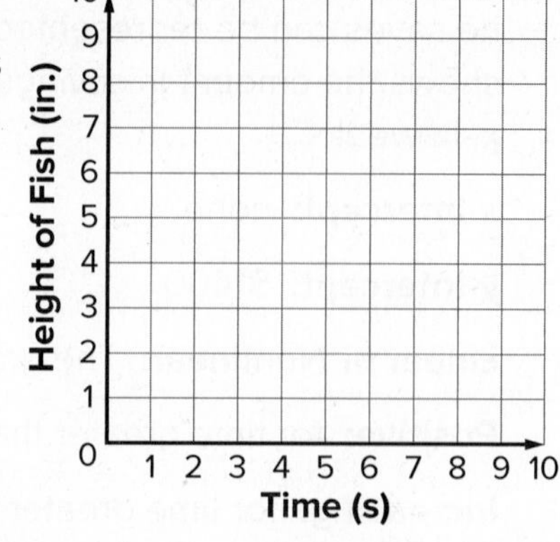

Positive: between 2 seconds and 8 seconds

Negative: for time less than 2 seconds and greater than 8 seconds

Increasing: for time less than 5 seconds

Decreasing: for time greater 5 seconds

Relative Maximum: at 5 seconds, when the height of the fish is 9 inches

End Behavior: As time increases or decreases, the height of the fish decreases.

Symmetry: The height of the fish for time less than 5 seconds is the same as the height for time greater than 5 seconds.

Example 3

5. TECHNOLOGY The results of a poll that asks Americans whether they used the Internet yesterday can be modeled as a function. Sketch a graph that shows the number of people polled that responded yes to the survey y as a function of time x, months since January 2005.

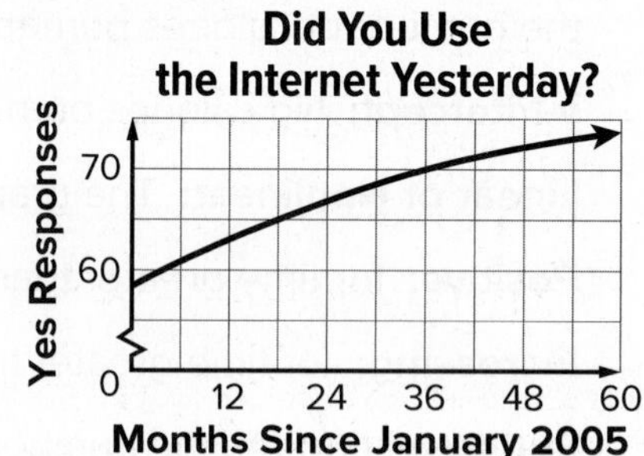

Positive: for time greater than 0 months

Negative: none

Increasing: for all time greater than 0 months

Decreasing: none

Intercepts: The graph intersects the y-axis at about (0, 58).

Extrema: none

End Behavior: As x increases, y increases. The data represent people, so the maximum it could ever reach is the maximum number of people surveyed.

6. MUSIC The results of a poll that asks Americans whether they have listened to online music can be modeled as a function. Sketch a graph that shows the number of people polled that have listened online y as a function of time x, months since August 2000.

Online Music

Have Listened Online

60

40

20

0

40 80 120

Months Since August 2000

Positive: for time greater than 0 months

Negative: none

Increasing: between 0 months and 10 months, for time greater than about 65 months

Decreasing: between 10 months and about 65 months

Intercepts: The graph intersects the y-axis at about (0, 37).

Relative Minimum: at about (65, 31)

Relative Maximum: at about (10, 39)

End Behavior: As x decreases, y decreases. As x increases, y increases. The data represent people, so the maximum it could ever reach is the maximum number of people surveyed.

Example 4

7. INTERNET Use the description and graph to compare Internet use at home and Internet use away from home, where y is the number of people polled, in thousands, that use the Internet several times a day and x is the number of months since March 2004. Use the description and graph to compare Internet use at home and Internet use away from home since March 2004.

Internet Use at Home

About 10,000 of those polled used the Internet at home in March 2004. The number of users decreased to 7000 at 36 months since March 2004. The number of users continued to increase after 36 months since March 2004.

Internet Use Away from Home

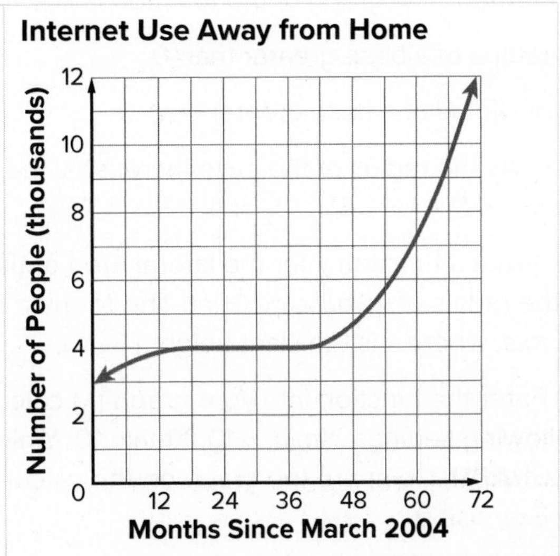

8. SPENDING Use the description and graph to compare the amount of U.S. spending on electronics and education, where y is the amount spent in billions and x is the number of years since 1949. Write statements to compare U.S. spending on electronics and education since 1949.

U.S. Electronic Spending

In 1949, the U.S. spent $0 on electronics. Twenty years after 1949, the U.S. spent about $4 billion on electronics. Thirty years after 1949, the U.S. spent about $3 billion on electronics. Seventy years after 1949, the U.S. spent about $7.5 billion on electronics and spending continues to increase.

U.S. Education Spending

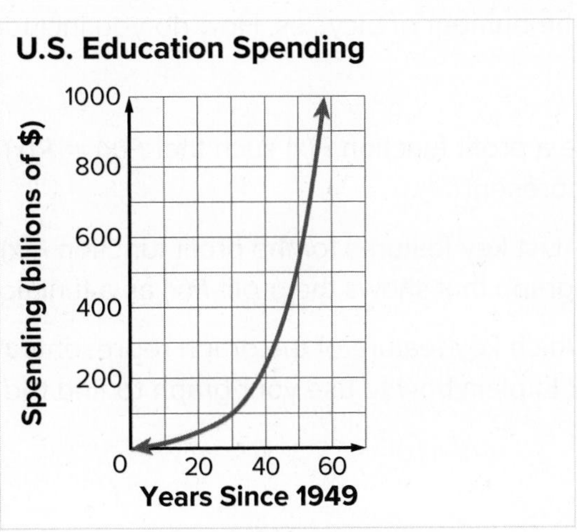

Mixed Exercises

The costume department of a theatre company is making cone-shaped hats for a play set in medieval times. Each hat will be covered with satin over its entire lateral surface area. The slant height of each hat will remain constant at 20 inches, but the radius of the base will vary to accommodate different head sizes. Using 3.14 for π, the lateral area of the hat can be expressed as a function. Use this information for Exercises 9–11.

9. USE A MODEL Sketch a graph that shows the lateral area of the hat y, in square inches, as a function of the radius of its base x, in inches.

y-Intercept: No lateral area when the radius of its base is 0 inches.

Linear or Nonlinear: The graph of the function is linear.

Positive: for radius of a base greater than 0

Increasing: for radius of a base greater than 0

End Behavior: As the radius of the base increases, the lateral area increases.

10. REASONING Write a function y for the lateral area of the hat as a function of the radius of its base, x. (Hint: The formula for the lateral area of a cone is $y = \pi xs$, where s is the slant height.)

11. USE TOOLS Enter the function into your graphing calculator. Press **WINDOW** and enter the following settings: Xmin: −10; Xmax: 10; Ymin: −1000; Ymax: 1000. Then press **GRAPH**. Compare the graph on the calculator to the graph you sketched in Exercise 9.

Higher-Order Thinking Skills

Aidan buys used bicycles, fixes them up, and sells them. His average cost to buy and fix each bicycle is $47. He also incurred a one-time cost of $840 to purchase tools and a small shed to use as his workshop. He sells bikes for $75 each. Use this information for Exercises 12–15.

12. WRITE Write revenue and cost functions $R(x)$ and $C(x)$ for Aidan's situation, where x is the number of bicycles. How do you include the one-time cost in $C(x)$?

13. WRITE Write a profit function $P(x)$ such that $P(x) = R(x) - C(x)$. In words, what does $P(x)$ represent?

14. PERSEVERE List key features for the profit function $P(x)$. Then use the key features to sketch a graph that shows the profit $P(x)$ as a function of x bicycles.

15. ANALYZE Which key feature of the graph represents Aidan's break-even point (profit = 0)? Explain how to use your graph to find the most accurate value for this feature.

16. CREATE Research the population in your state over a 10-year period. Sketch a graph to model the data. Then list the key feature of the graph.

Review

 Essential Question

Why are representations of relations and functions useful?

Relations and functions can help you visualize relationships between quantities. They can also be used to display data, identify trends, and make predictions.

Module Summary

Lesson 3-1

Representing Relations

- A relation is a set of ordered pairs.
- Relations can be shown with ordered pairs, with a table, with a graph, or with a mapping.

Lesson 3-2

Functions

- A function is a relationship between input and output. In a function, there is exactly one output for each input.
- If a vertical line intersects the graph of a relation more than once, then the relation is not a function.
- Function notation is a way of writing an equation so that $y = f(x)$.

Lessons 3-3 through 3-5

Interpreting Graphs

- A discrete function is a set of points that are not connected. A continuous function has points that connect to form a line or curve.
- An x-intercept of a graph is a point where the graph intersects the x-axis. The y-intercept of a graph is the point where the graph intersects the y-axis.
- Equations can be solved by graphing related functions.
- A figure has line symmetry if each half of the figure matches the other side exactly.
- A function is increasing where the graph goes up and decreasing where the graph goes down when viewed from left to right.
- Points that are the locations of relatively high or low function values are called extrema.
- A point is a relative minimum when no other nearby point has a lesser y-coordinate.
- A point is a relative maximum when no other nearby point has a greater y-coordinate.

Lesson 3-6

Sketching Graphs and Comparing Functions

- Knowing the intercepts, symmetry, end behavior, and extrema of a function, as well as intervals where the function is increasing, decreasing, positive, or negative, provides a clear idea of what the graph of the function looks like.

Study Organizer

 **Foldables**

Use your Foldable to review this module. Working with a partner can be helpful. Ask for clarification of concepts as needed.

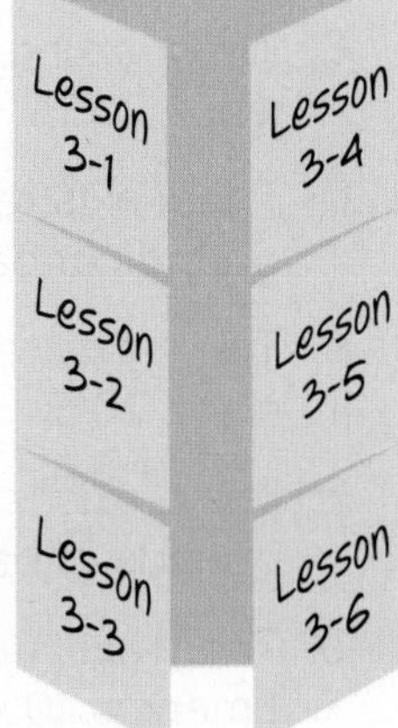

Test Practice

1. MULTI-SELECT The table and graph show the number of pounds of bananas sold at a local grocery store from 2013 to 2018. (Lesson 3-1)

x	y
0	40
1	60
2	55
3	25
4	30
5	50

Number of Pounds of Bananas Sold

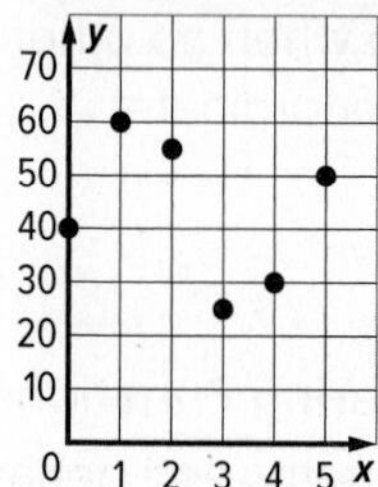

Which of the following statements correctly describes the relation? Select all that apply.

A. The x-axis has a scale mark of 1 mark = 1 pound of bananas.

B. The x-axis has a scale mark of 1 mark = 1 year.

C. The x-axis represents the years since 2013.

D. The x-axis represents the number of pounds of bananas sold.

E. The y-axis represents the years since 2013.

F. The y-axis represents the number of pounds of bananas sold.

G. The y-axis has a scale mark of 1 mark = 10 years.

2. MULTIPLE CHOICE The Hillsborough State Park in Thonotosassa, Florida, charges an admission fee of $4 plus a camping fee of $20 per night. This can be represented by the function $f(x) = 20x + 4$, where $f(x)$ is the total cost and x is the number of nights spent camping. What is the value of $f(5)$, which is the cost of 5 nights camping? (Lesson 3-2)

A. 80

B. 100

C. 104

D. 120

3. MULTIPLE CHOICE If $f(x) = -9x + 8$, then find $f(-2)$. (Lesson 3-2)

A. −84

B. −8

C. 26

D. 100

4. MULTIPLE CHOICE Which function includes the data set below?

$\{(2, -2), (6, 10), (13, 31)\}$ (Lesson 3-2)

A. $f(x) = \frac{1}{2}x - 3$

B. $f(x) = -2x + 2$

C. $f(x) = 3x - 8$

D. $f(x) = 4x - 10$

5. OPEN RESPONSE Determine whether the relation shown in the table is a function. Explain. (Lesson 3-2)

Domain	−4	2	−4	5
Range	8	11	13	13

6. OPEN RESPONSE Indicate whether each of the following relations are functions. (Lesson 3-2)

A. 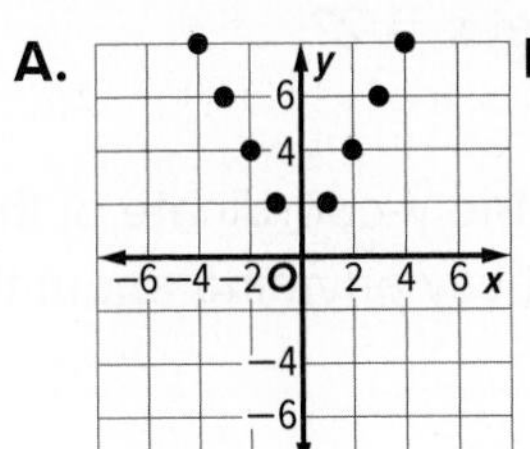

B. 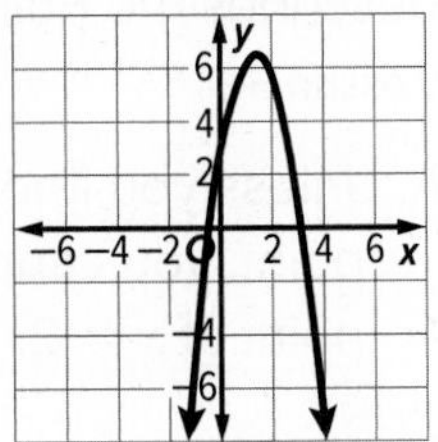

C. 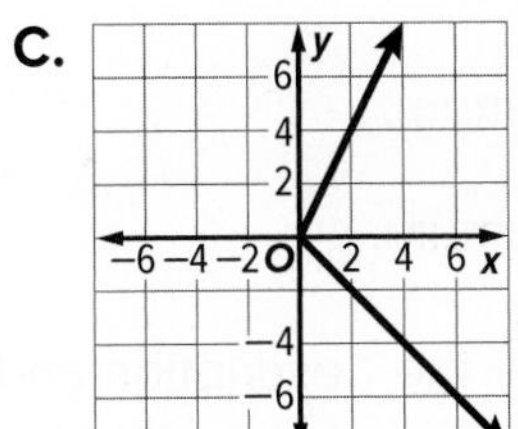

D. 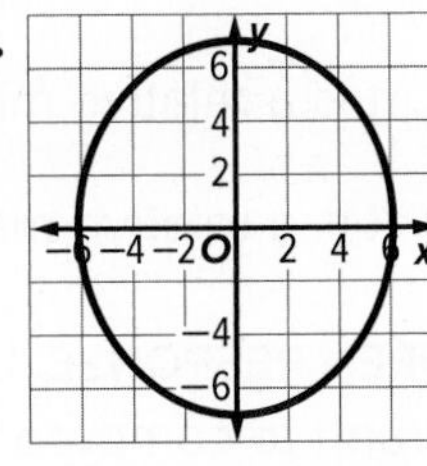

7. MULTIPLE CHOICE What is the domain of this function? (Lesson 3-3)

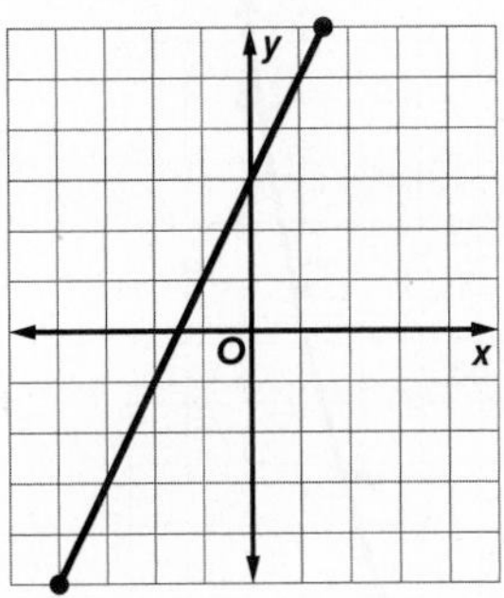

A. $-4 < x < 1.5$

B. $-4 \leq x \leq 1.5$

C. $-6 < y < 6$

D. $-6 \leq y \leq 6$

8. OPEN RESPONSE The graph shows the relationship between the number of Fun Pass tickets sold and the total value of the sales. Use the graph to estimate the *x*- and *y*-intercepts of the function, where the function is positive and negative, and interpret the meanings in the context of the situation. (Lesson 3-4)

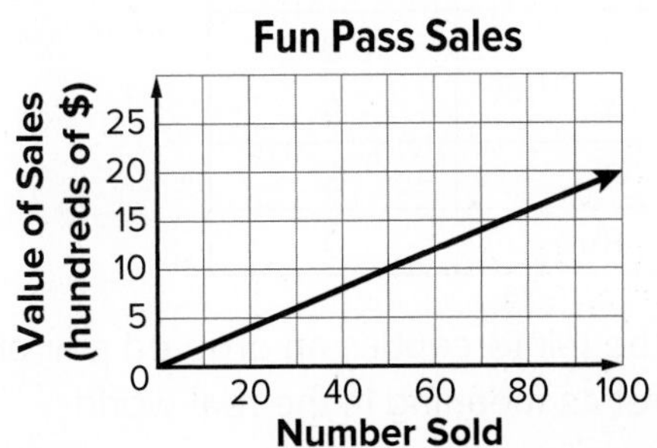

9. OPEN RESPONSE What are the intercepts for the function $y = -x - 1$ graphed below? (Lesson 3-4)

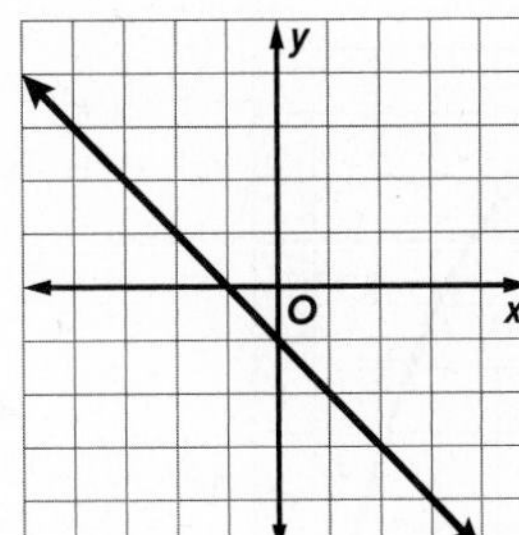

10. **OPEN RESPONSE** A baker bought a bag of flour to make banana bread. The recipe calls for 1.5 cups of flour per loaf. The number of cups of flour left in the bag y after making x loaves of bread is shown in the table. (Lesson 3-4)

x	y
0	12
2	9
4	6
6	3
8	0

Write the y-intercept as an ordered pair and interpret its meaning in the real-world context.

11. **OPEN RESPONSE** A garden supply store manager found that if she used a certain function she could determine the best price to charge for the shovels she sells to maximize the revenue. The graph represents the revenue ($) y of the store at x price ($) per shovel. Use the graph to find and interpret the symmetry of the function in the context of the situation. (Lesson 3-5)

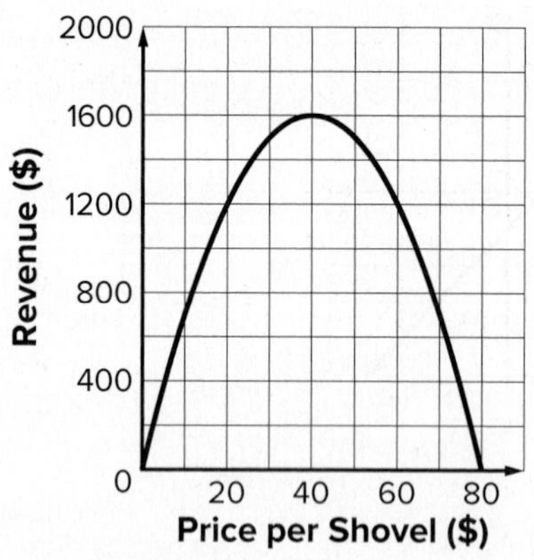

Symmetry: The graph is symmetric in the line $x =$ __?__.

Interpret symmetry: The revenue gained when a shovel is sold for $20 is the same as it is when a shovel is sold for $ __?__.

12. **MULTIPLE CHOICE** Suppose the graph of a function is increasing to the left of $x = 2$ and decreasing to the right of $x = 2$. Which describes the point at $x = 2$? (Lesson 3-5)

A. Unless you know the y-coordinate of the point, you cannot say anything about the point at $x = 2$.

B. It is an x-intercept.

C. It is a relative minimum.

D. It is a relative maximum.

13. **OPEN RESPONSE** Use the description and graph to compare the population data for Ohio and Florida, where y is the population in millions and x is the number of decades since 1900. Write statements about the populations of Ohio and Florida since 1900. (Lesson 3-6)

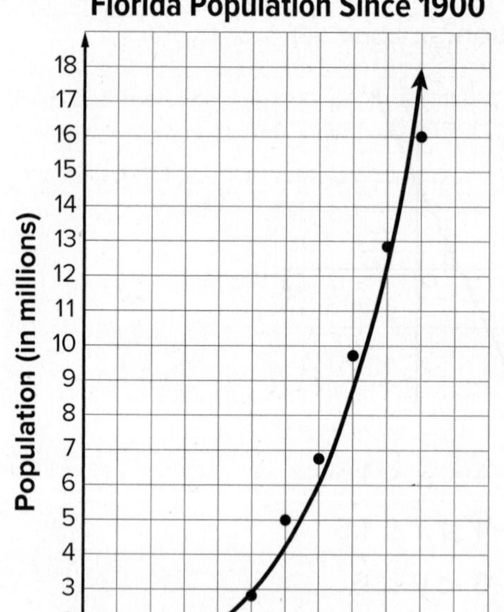

Ohio Population Since 1900

In 1900 the population of Ohio was about 4.2 million. Between 1900 and 1950, the population of Ohio nearly doubled to about 8 million. Then between 1950 and 2000, the population of Ohio grew to approximately 11.4 million. Beyond 2000, the population of Ohio continues to gradually increase.

Module 4

Linear and Nonlinear Functions

Essential Question

What can a function tell you about the relationship that it represents?

What Will You Learn?

How much do you already know about each topic **before** starting this module?

KEY

— I don't know. — I've heard of it. — I know it!

	Before			After		
	I don't know.	I've heard of it.	I know it!	I don't know.	I've heard of it.	I know it!
graph linear equations by using a table						
graph linear equations by using intercepts						
find rates of change						
determine slopes of linear equations						
write linear equations in slope-intercept form						
graph linear functions in slope-intercept form						
translate, dilate, and reflect linear functions						
identify and find missing terms in arithmetic sequences						
write arithmetic sequences as linear functions						
model and use piecewise functions, step functions, and absolute value functions						
translate absolute value functions						

Foldables Make this Foldable to help you organize your notes about functions. Begin with five sheets of grid paper.

1. **Fold** five sheets of grid paper in half from top to bottom.
2. **Cut** along fold. Staple the eight half-sheets together to form a booklet.
3. **Cut** tabs into margin. The top tab is 4 lines wide, the next tab is 8 lines wide, and so on. When you reach the bottom of a sheet, start the next tab at the top of the page.
4. **Label** each tab with a lesson number. Use the extra pages for vocabulary.

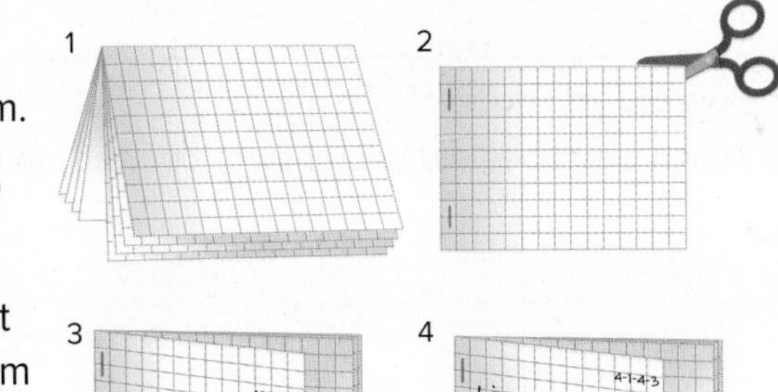

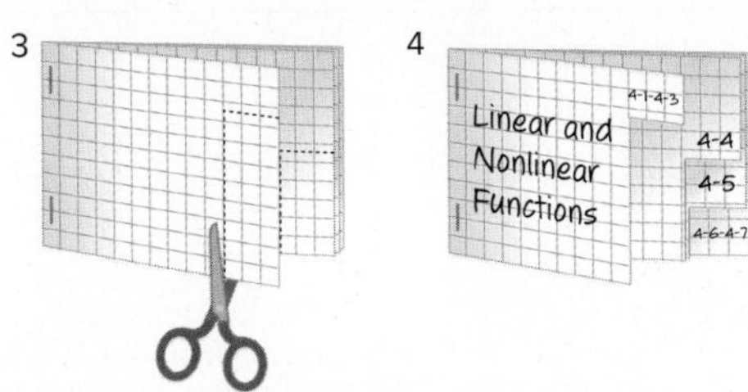

What Vocabulary Will You Learn?

- absolute value function
- arithmetic sequence
- common difference
- constant function
- dilation
- family of graphs
- greatest integer function
- interval
- identity function
- *n*th term of an arithmetic sequence
- parameter
- parent function
- piecewise-defined function
- piecewise-linear function
- rate of change
- reflection
- sequence
- slope
- step function
- term of a sequence
- transformation
- translation
- vertex

Are You Ready?

Complete the Quick Review to see if you are ready to start this module. Then complete the Quick Check.

Quick Review

Example 1

Graph *A*(3, −2) on a coordinate grid.

Start at the origin. Since the *x*-coordinate is positive, move 3 units to the right. Then move 2 units down since the *y*-coordinate is negative. Draw a dot and label it *A*.

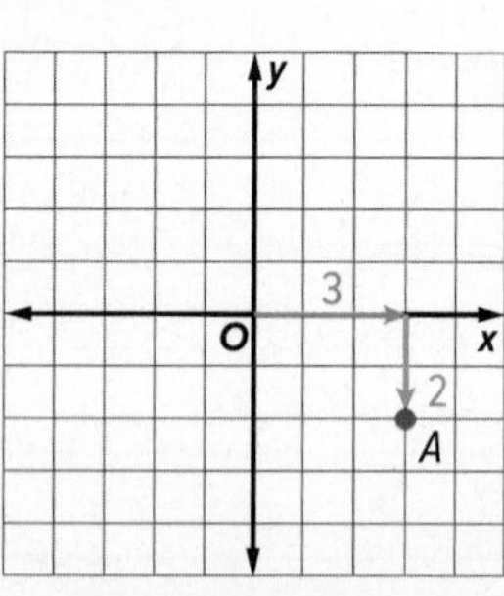

Example 2

Solve $x - 2y = 8$ for *y*.

$x - 2y = 8$	Original expression
$x - x - 2y = 8 - x$	Subtract *x* from each side.
$-2y = 8 - x$	Simplify.
$\frac{-2y}{-2} = \frac{8 - x}{-2}$	Divide each side by −2.
$y = \frac{1}{2}x - 4$	Simplify.

Quick Check

Graph and label each point on the coordinate plane.

1. *B*(−3, 3) **2.** *C*(−2, 1) **3.** *D*(3, 0)

4. *E*(−5, −4) **5.** *F*(0, −3) **6.** *G*(2, −1)

Solve each equation for *y*.

7. $3x + y = 1$ **8.** $8 - y = x$

9. $5x - 2y = 12$ **10.** $3x + 4y = 10$

11. $3 - \frac{1}{2}y = 5x$ **12.** $\frac{y+1}{3} = x + 2$

How did you do?

Which exercises did you answer correctly in the Quick Check?

Graphing Linear Functions

Explore Points on a Line

Online Activity Use an interactive tool to complete an Explore.

INQUIRY How is the graph of a linear equation related to its solutions?

Learn Graphing Linear Functions by Using Tables

A table of values can be used to graph a linear function. Every ordered pair that makes the equation true represents a point on its graph. So, a graph represents all the solutions of an equation.

Linear functions can be represented by equations in two variables.

Example 1 Graph by Making a Table

Graph $-2x - 3 = y$ by making a table.

Step 1 Choose any values of x from the domain and make a table.

Step 2 Substitute each x-value into the equation to find the corresponding y-value. Then, write the x- and y-values as an ordered pair.

x	$-2x - 3$	y	(x, y)
−4	−2(−4) −3	5	(−4, 5)
−2	−2(−2) −3	1	(−2, 1)
0	−2(0) −3	−3	(0, −3)
1	−2(1) −3	−5	(1, −5)
3	−2(3) −3	−9	(3, −9)

Step 3 Graph the ordered pairs in the table and connect them with a line.

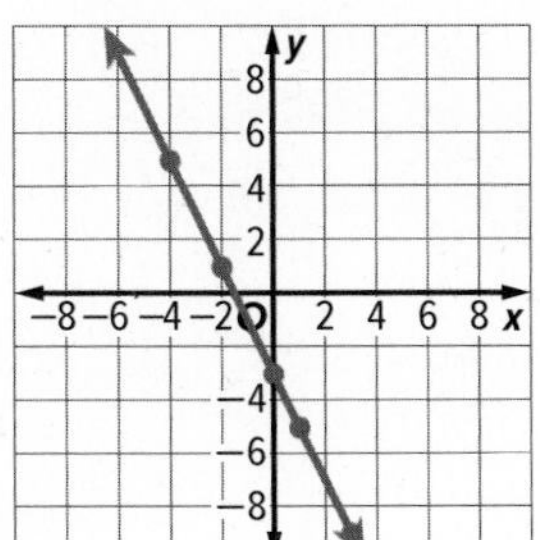

Go Online You can complete an Extra Example online.

Today's Goals

- Graph linear functions by making tables of values.
- Graph linear functions by using the x- and y-intercepts.

Talk About It!

What values of x might be easiest to use when graphing a linear equation when the x-coefficient is a whole number? Justify your argument.

Study Tip

Exactness Although only two points are needed to graph a linear function, choosing three to five x-values that are spaced out can verify that your graph is correct.

Check

Graph $y = 2x + 5$ by using a table. Copy and complete the table. Then graph the function.

x	y
−5	
−3	
−1	
0	
2	

Think About It!

What are some values of x that you might choose in order to graph $y = \frac{1}{7}x - 12$?

Example 2 Choose Appropriate Domain Values

Graph $y = \frac{1}{4}x + 3$ by making a table.

Step 1 Make a table.

Step 2 Find the y-values.

Step 3 Graph the ordered pairs in the table and connect them with a line.

x	$\frac{1}{4}x + 3$	y	(x, y)
−8	$\frac{1}{4}(-8) + 3$	1	(−8, 1)
−4	$\frac{1}{4}(-4) + 3$	2	(−4, 2)
0	$\frac{1}{4}(0) + 3$	3	(0, 3)
4	$\frac{1}{4}(4) + 3$	4	(4, 4)
8	$\frac{1}{4}(8) + 3$	5	(8, 5)

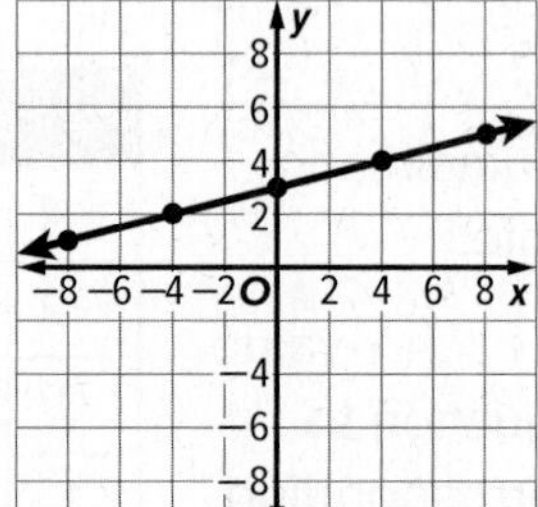

Watch Out!

Equivalent Equations Sometimes, the variables are on the same side of the equal sign. Rewrite these equations by solving for y to make it easier to find values for y.

Check

Graph $y = \frac{3}{5}x - 2$ by making a table. Copy and complete the table. Then graph the function.

x	y
−10	
−5	
0	
5	
10	

Go Online You can complete an Extra Example online.

Example 3 Graph $y = a$

Graph $y = 5$ by making a table.

Step 1 Rewrite the equation.
$y = 0x + 5$

Step 2 Make a table.

x	$0x + 5$	y	(x, y)
−2	$0(-2) + 5$	5	(−2, 5)
−1	$0(-1) + 5$	5	(−1, 5)
0	$0(0) + 5$	5	(0, 5)
1	$0(1) + 5$	5	(1, 5)
2	$0(2) + 5$	5	(2, 5)

Step 3 Graph the line.

The graph of $y = 5$ is a horizontal line through $(x, 5)$ for all values of x in the domain.

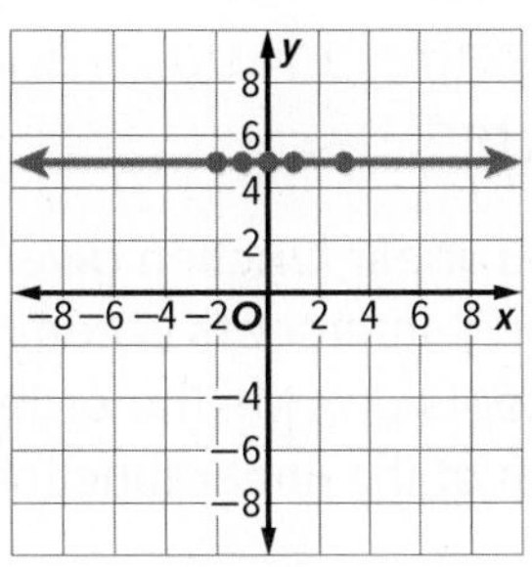

Think About It!

In general, what does the graph of an equation of the form $y = a$, where a is any real number, look like?

Example 4 Graph $x = a$

Graph $x = -2$.

You learned in the previous example that equations of the form $y = a$ have graphs that are horizontal lines. Equations of the form $x = a$ have graphs that are vertical lines.

The graph of $x = -2$ is a vertical line through $(-2, y)$ for all real values of y. Graph ordered pairs that have x-coordinates of −2 and connect them with a vertical line.

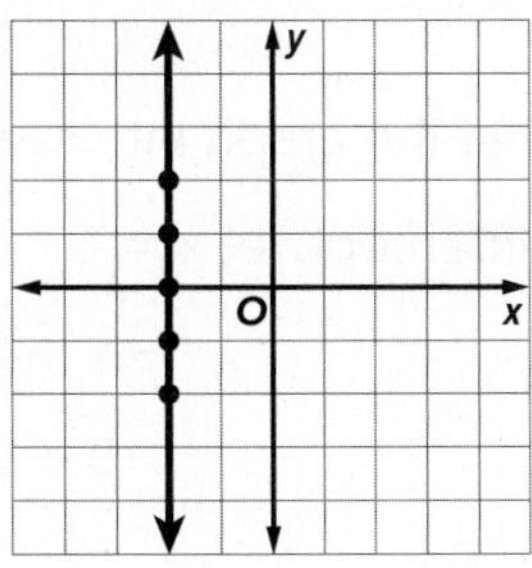

Think About It!

Is the graph of $x = a$ a function? Why or why not?

Check

Graph $x = 6$.

Go Online You can complete an Extra Example online.

Explore Lines Through Two Points

Online Activity Use graphing technology to complete an Explore.

INQUIRY How many lines can be formed with two given points?

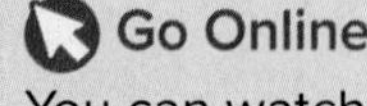

Go Online You can watch a video to see how to graph linear functions.

Think About It! Why are the x- and y-intercepts easy to find?

Learn Graphing Linear Functions by Using the Intercepts

You can graph a linear function given only two points on the line. Using the x- and y-intercepts is common because they are easy to find. The intercepts provide the ordered pairs of two points through which the graph of the linear function passes.

Example 5 Graph by Using Intercepts

Graph $-x + 2y = 8$ by using the x- and y-intercepts.

To find the x-intercept, let $y = 0$.

$-x + 2y = 8$	Original equation
$-x + 2(0) = 8$	Replace y with 0.
$-x = 8$	Simplify.
$x = -8$	Divide.

This means that the graph intersects the x-axis at $(-8, 0)$.

To find the y-intercept, let $x = 0$.

$-x + 2y = 8$	Original equation
$-0 + 2y = 8$	Replace x with 0.
$2y = 8$	Simplify.
$y = 4$	Divide.

This means that the graph intersects the y-axis at $(0, 4)$.

Graph the equation.

Step 1 Graph the x-intercept.

Step 2 Graph the y-intercept.

Step 3 Draw a line through the points.

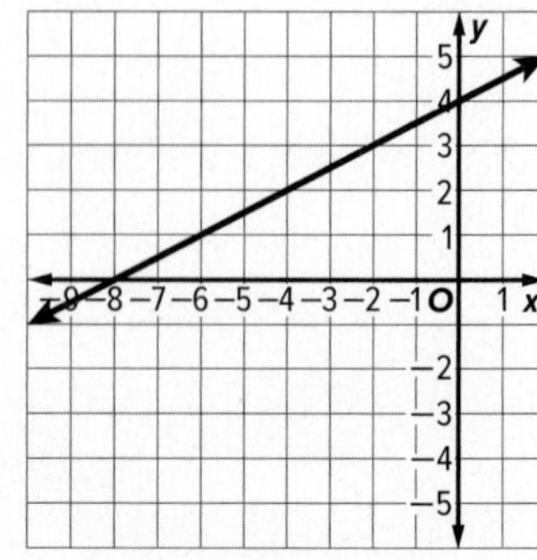

Think About It! What does a line that only has an x-intercept look like? a line that only has a y-intercept?

Study Tip

Tools When drawing lines by hand, it is helpful to use a straightedge or a ruler.

Go Online You can complete an Extra Example online.

Check

Graph $4y = -12x + 36$ by using the *x*- and *y*- intercepts.

x-intercept: __?__

y-intercept: __?__

Example 6 Use Intercepts

PETS Angelina bought a 15-pound bag of food for her dog. The bag contains about 60 cups of food, and she feeds her dog $2\frac{1}{2}$ or $\frac{5}{2}$ cups of food per day. The function $y + \frac{5}{2}x = 60$ represents the amount of food left in the bag *y* after *x* days. Graph the amount of dog food left in the bag as a function of time.

Part A

Find the *x*- and *y*-intercepts and interpret their meaning in the context of the situation.

To find the *x*-intercept, let $y = 0$.

$y + \frac{5}{2}x = 60$	Original equation
$0 + \frac{5}{2}x = 60$	Replace *y* with 0.
$\frac{5}{2}x = 60$	Simplify.
$x = 24$	Multiply each side by $\frac{2}{5}$.

The *x*-intercept is 24. This means that the graph intersects the *x*-axis at (24, 0). So, after 24 days, there is no dog food left in the bag.

To find the *y*-intercept, let $x = 0$.

$y + \frac{5}{2}x = 60$	Original equation
$y + \frac{5}{2}(0) = 60$	Replace *x* with 0.
$y = 60$	Simplify.

The *y*-intercept is 60. This means that the graph intersects the *y*-axis at (0, 60). So, after 0 days, there are 60 cups of food in the bag.

(continued on the next page)

Go Online
You can watch a video to see how to use a graphing calculator with this example.

Think About It!
Find another point on the graph. What does it mean in the context of the problem?

Think About It!

What assumptions did you make about the amount of food Angelina feeds her dog each day?

Part B

Graph the equation by using the intercepts.

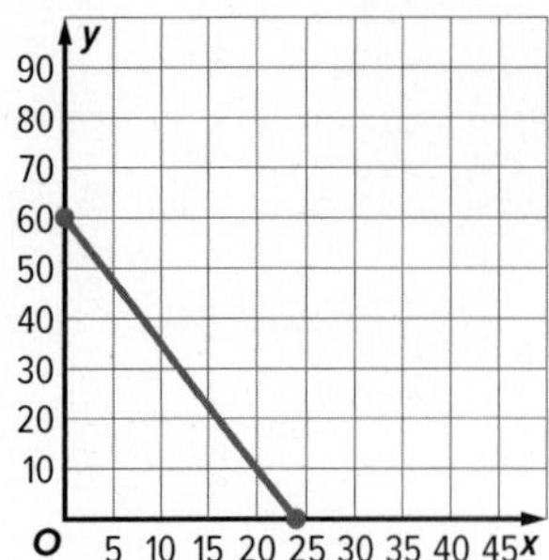

Check

PEANUTS A farm produces about 4362 pounds of peanuts per acre. One cup of peanut butter requires about $\frac{2}{3}$ pound of peanuts. If one acre of peanuts is harvested to make peanut butter, the function $y = -\frac{2}{3}x + 4362$ represents the pounds of peanuts remaining y after x cups of peanut butter are made.

x-intercept: ___?___

y-intercept: ___?___

Which graph uses the x- and y-intercepts to correctly graph the equation?

Go Online You can watch a video to see how to graph a linear function using a graphing calculator.

A.

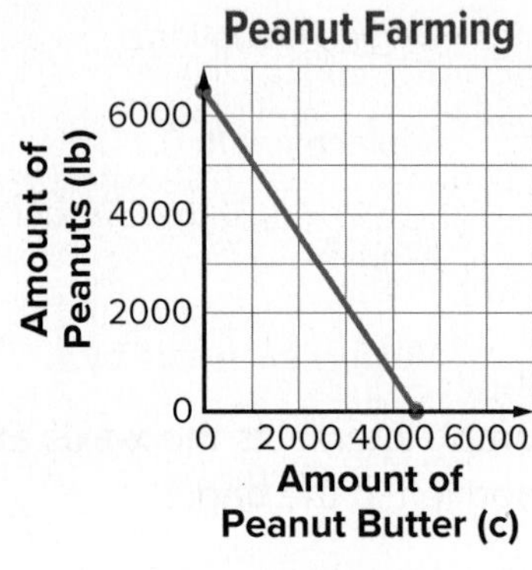

B.

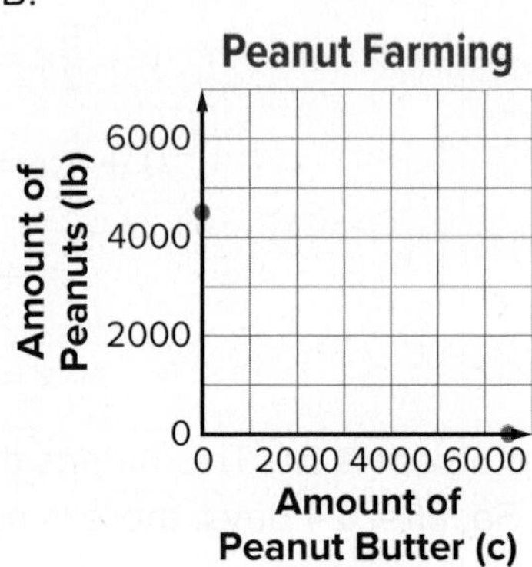

C.

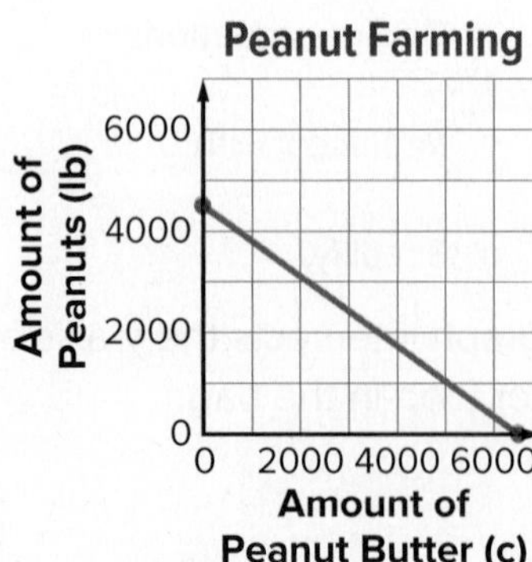

D.

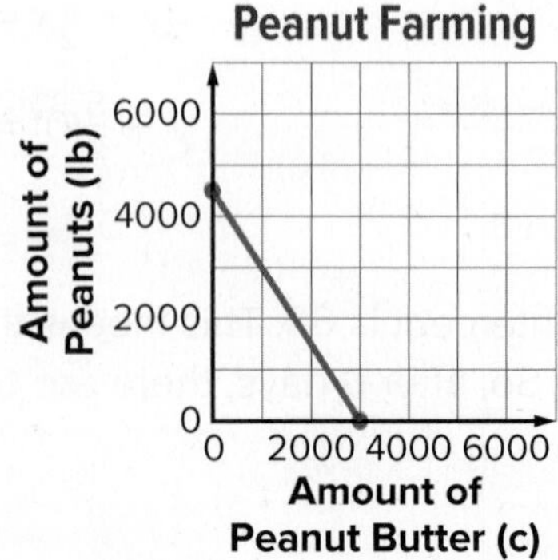

 **Go Online** You can complete an Extra Example online.

Practice

Go Online You can complete your homework online.

Examples 1 through 4

Graph each equation by making a table.

1. $x = -2$

2. $y = -4$

3. $y = -8x$

4. $3x = y$

5. $y - 8 = -x$

6. $x = 10 - y$

7. $y = \frac{1}{2}x + 1$

8. $y + 2 = \frac{1}{4}x$

Example 5

Graph each equation by using the *x*-and *y*-intercepts.

9. $y = 4 + 2x$

10. $5 - y = -3x$

11. $x = 5y + 5$

12. $x + y = 4$

13. $x - y = -3$

14. $y = 8 - 6x$

Example 6

15. SCHOOL LUNCH Amanda has \$210 in her school lunch account. She spends \$35 each week on school lunches. The equation $y = 210 - 35x$ represents the total amount in Amanda's school lunch account y for x weeks of purchasing lunches.

a. Find the *x*- and *y*-intercepts and interpret their meaning in the context of the situation.

b. Graph the equation by using the intercepts.

16. SHIPPING The *OOCL Shenzhen,* one of the world's largest container ships, carries 8063 TEUs (1280-cubic-feet containers). Workers can unload a ship at a rate of 1 TEU every minute. The equation $y = 8063 - 60x$ represents the number of TEUs on the ship y after x hours of the workers unloading the containers from the *Shenzhen.*

a. Find the *x*- and *y*-intercepts and interpret their meaning in the context of the situation.

b. Graph the equation by using the intercepts.

Mixed Exercises

Graph each equation.

17. $1.25x + 7.5 = y$

18. $2x - 3 = 4y + 6$

19. $3y - 7 = 4x + 1$

Find the *x*-intercept and *y*-intercept of the graph of each equation.

20. $5x + 3y = 15$

21. $2x - 7y = 14$

22. $2x - 3y = 5$

23. $6x + 2y = 8$

24. $y = \frac{1}{4}x - 3$

25. $y = \frac{2}{3}x + 1$

26. HEIGHT The height of a woman can be predicted by the equation $h = 81.2 + 3.34r$, where h is her height in centimeters and r is the length of her radius bone in centimeters.

a. What are the r- and h-intercepts of the equation? Do they make sense in the situation? Explain.

b. Graph the equation by using the intercepts.

c. Use the graph to find the approximate height of a woman whose radius bone is 25 centimeters long.

27. TOWING Pick-M-Up Towing Company charges \$40 to hook a car and \$1.70 for each mile that it is towed. Write an equation that represents the total cost y for x miles towed. Graph the equation. Find the y-intercept, and interpret its meaning in the context of the situation.

28. USE A MODEL Elias has \$18 to spend on peanuts and pretzels for a party. Peanuts cost \$3 per pound and pretzels cost \$2 per pound. Write an equation that relates the number of pounds of pretzels y and the number of pounds of peanuts x. Graph the equation. Find the x- and y-intercepts. What does each intercept represent in terms of context?

29. **REASONING** One football season, a football team won 4 more games than they lost. The function $y = x + 4$ represents the number of games won y and the number of games lost x. Find the x- and y-intercepts. Are the x- and y-intercepts reasonable in this situation? Explain.

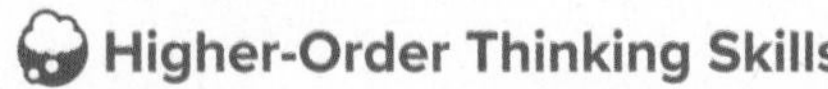

Higher-Order Thinking Skills

30. **WRITE** Consider real-world situations that can be modeled by linear functions.
    - **a.** Write a real-world situation that can be modeled by a linear function.
    - **b.** Write an equation to model your real-world situation. Be sure to define variables. Then find the x- and y-intercepts. What does each intercept represent in your context?
    - **c.** Graph your equation by making a table. Include a title for the graph as well as labels and titles for each axis. Explain how you labeled the x- and y-axes. State a reasonable domain for this situation. What does the domain represent?

31. **FIND THE ERROR** Geroy claims that every line has both an x- and a y-intercept. Is he correct? Explain your reasoning.

32. **WHICH ONE DOESN'T BELONG?** Which equation does not belong with the other equations? Justify your conclusion.

$y = 2 - 3x$	$5x = y - 4$	$y = 2x + 5$	$y - 4 = 0$

33. **ANALYZE** Robert sketched a graph of a linear equation $2x + y = 4$. What are the x- and y-intercepts of the graph? Explain how Robert could have graphed this equation using the x- and y-intercepts.

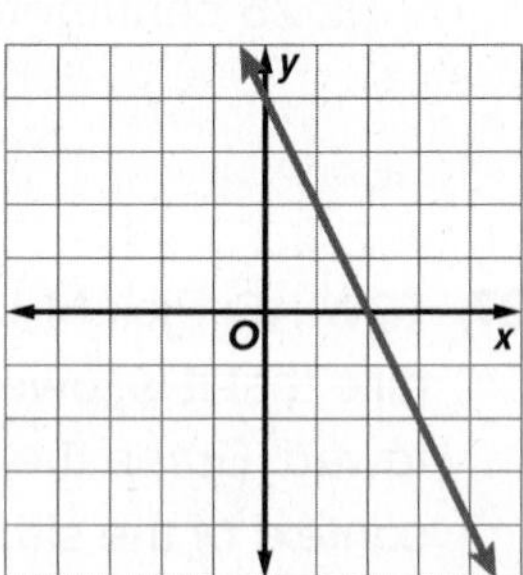

34. **ANALYZE** Compare and contrast the graph of $y = 2x + 1$ with the domain $\{1, 2, 3, 4\}$ and $y = 2x + 1$ with the domain all real numbers.

CREATE **Give an example of a linear equation in the form $Ax + By = C$ for each condition. Then describe the graph of the equation.**

35. $A = 0$

36. $B = 0$

37. $C = 0$

Rate of Change and Slope

Learn Rate of Change of a Linear Function

The **rate of change** is how a quantity is changing with respect to a change in another quantity.

If x is the independent variable and y is the dependent variable, then rate of change $= \frac{\text{change in } y}{\text{change in } x}$.

Example 1 Find the Rate of Change

COOKING **Find the rate of change of the function by using two points from the table.**

Amount of Flour x (cups)	Pancakes y
2	12
4	24
6	36

$$\begin{aligned}\text{rate of change} &= \frac{\text{change in } y}{\text{change in } x}\\ &= \frac{\text{change in pancakes}}{\text{change in flour}}\\ &= \frac{24-12}{4-2}\\ &= \frac{12}{2} \text{ or } \frac{6}{1}\end{aligned}$$

The rate is $\frac{6}{1}$ or 6. This means that you could make 6 pancakes for each cup of flour.

Check

Find the rate of change.

$\underline{\quad ? \quad}$ $\frac{\text{dollars}}{\text{gallons}}$

Amount of Gasoline Purchased (Gallons)	Cost (Dollars)
4.75	15.77
6	19.92
7.25	24.07
8.5	28.22

Go Online You can complete an Extra Example online.

Today's Goals
- Calculate and interpret rate of change.
- Calculate and interpret slope.

Today's Vocabulary
rate of change
slope

Think About It!
Suppose you found a new recipe that makes 6 pancakes when using 2 cups of flour, 12 pancakes when using 4 cups of flour, and 18 pancakes when using 6 cups of flour. How does this change the rate you found for the original recipe?

Study Tip
Placement Be sure that the dependent variable is in the numerator and the independent variable is in the denominator. In this example, the number of pancakes you can make *depends* on the amount of flour you can use.

Example 2 Compare Rates of Change

STUDENT COUNCIL **The Jackson High School Student Council budget varies based on the fundraising of the previous year.**

Part A Find the rate of change for 2000–2005 and describe its meaning in the context of the situation.

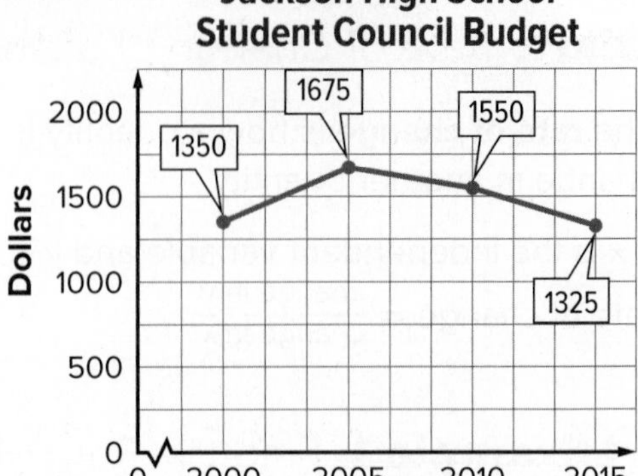

$\frac{\text{change in budget}}{\text{change in time}} =$

$\frac{1675 - 1350}{2005 - 2000} = \frac{325}{5}$, or 65

This means that the student council's budget increased by \$325 over the 5-year period, with a rate of change of \$65 per year.

Part B Find the rate of change for 2010–2015 and describe its meaning in the context of the situation.

$$\frac{\text{change in budget}}{\text{change in time}} = \frac{1325 - 1550}{2015 - 2010} = \frac{-225}{5}, \text{ or } = -45$$

This means that the student council's budget was reduced by \$225 over the 5-year period, with a rate of change of −\$45 per year.

Think About It!

How is a greater increase or decrease of funds represented graphically?

Study Tip

Assumptions In this example, we assumed that the rate of change for the budget was constant between each 5-year period. Although the budget might have varied from year to year, analyzing in larger periods of time allows us to see trends within data.

Check

TICKETS The graph shows the average ticket prices for the Miami Dolphins football team.

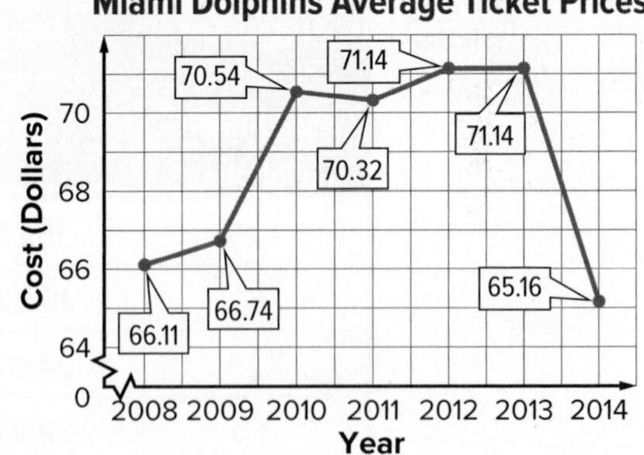

Part A Find the rate of change in ticket prices between 2009–2010.

___?___ $\frac{\text{dollars}}{\text{year}}$

Part B The ticket prices have the greatest rate of change between ___?___

Part C Between ___?___ and ___?___, the rate of change is negative.

Go Online You can complete an Extra Example online.

Example 3 Constant Rate of Change

Determine whether the function is linear. If it is, state the rate of change.

x	y
11	−5
8	−3
5	−1
2	1
−1	3

Find the changes in the x-values and the changes in the y-values.

Notice that the rate of change for each pair of points shown is $-\frac{2}{3}$.

The rates of change are constant, so the function is linear. The rate of change is $-\frac{2}{3}$.

Example 4 Rate of Change

Determine whether the function is linear. If it is, state the rate of change.

x	y
22	−4
29	−1
36	1
43	4
50	6

Find the changes in the x-values and the changes in the y-values.

The rates of change are not constant. Between some pairs of points the rate of change is $\frac{3}{7}$, and between the other pairs it is $\frac{2}{7}$. Therefore, this is not a linear function.

Study Tip

Linear Versus Not Linear Remember that the word *linear* means that the graph of the function is a straight line. For the graph of a function to be a line, it has to be increasing or decreasing at a constant rate.

Check

Copy and complete the table so that the function is linear.

x	y
	−2.25
	1
11	
10.5	7.5
10	10.75
9.5	

Go Online You can complete an Extra Example online.

Go Online You can watch a video to see how to find the slope of a nonvertical line.

Think About It!
If the point (1, 3) is on a line, what other point could be on the line to make the slope positive? negative? zero? undefined?

Explore Investigating Slope

Online Activity Use graphing technology to complete an Explore.

 **INQUIRY** How does slope help to describe a line?

Learn Slope of a Line

The **slope** of a line is the rate of change in the *y*-coordinates (rise) for the corresponding change in the *x*-coordinates (run) for points on the line.

Key Concept • Slope	
Words	The slope of a nonvertical line is the ratio of the rise to the run.
Symbols	The slope *m* of a nonvertical line through any two points (x_1, y_1) and (x_2, y_2) can be found as follows. $m = \frac{y_2 - y_1}{x_2 - x_1}$
Example	

Think About It!
Can a line that passes through two specific points, such as the origin and (2, 4), have more than one slope? Explain your reasoning.

The slope of a line can show how a quantity changes over time. When finding the slope of a line that represents a real-world situation, it is often referred to as the *rate of change*.

Example 5 Positive Slope

Find the slope of a line that passes through (−3, 4) and (1, 7).

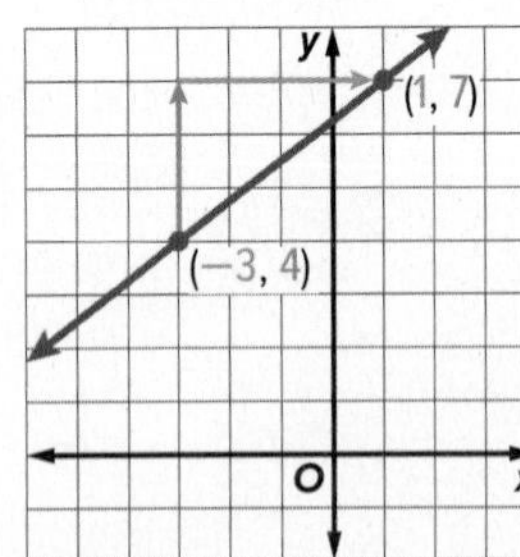

$$m = \frac{y_2 - y_1}{x_2 - x_1}$$
$$= \frac{7 - 4}{1 - (-3)}$$
$$= \frac{3}{4}$$

Think About It!
How would lines with slopes of $m = \frac{1}{8}$ and $m = 80$ compare on the same coordinate plane?

Check

Determine the slope of a line passing through the given points. If the slope is undefined, write *undefined*. Write your answer as a decimal if necessary.

(−1, 8) and (7, 10)

Go Online You can complete an Extra Example online.

Example 6 Negative Slope

Find the slope of a line that passes through (–1, 3) and (4, 1).

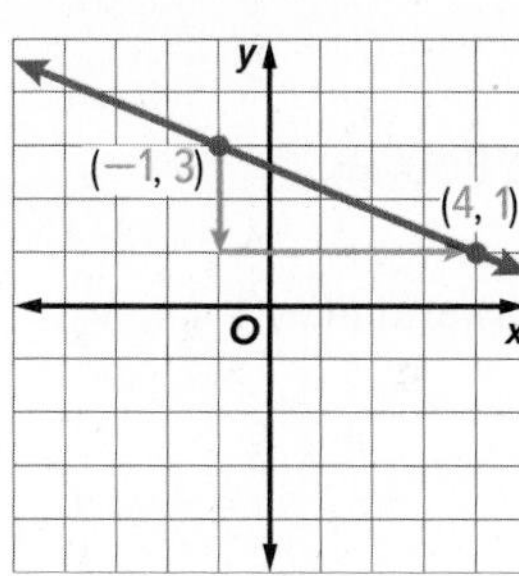

$$m = \frac{y_2 - y_1}{x_2 - x_1}$$

$$= \frac{1 - 3}{4 - (-1)}$$

$$= -\frac{2}{5}$$

Check

Determine the slope of a line passing through the given points. If the slope is undefined, write *undefined*. Write your answer as a decimal if necessary.

a. (5, −4) and (0, 1)

Example 7 Slopes of Horizontal Lines

Find the slope of a line that passes through (–2, –5) and (4, –5).

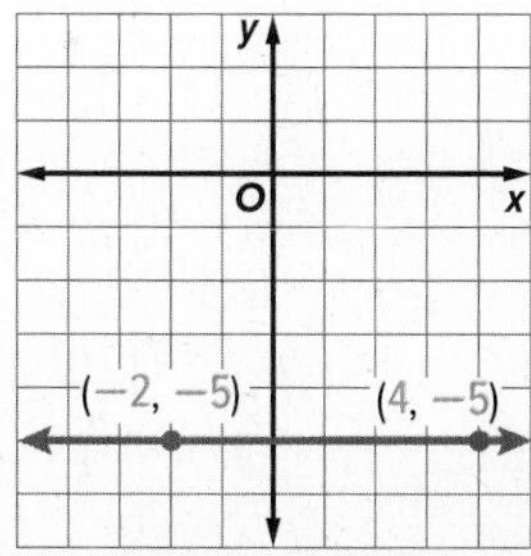

$$m = \frac{y_2 - y_1}{x_2 - x_1}$$

$$= \frac{-5 - (-5)}{4 - (-2)}$$

$$= \frac{0}{6} \text{ or } 0$$

Example 8 Slopes of Vertical Lines

Find the slope of a line that passes through (–3, 4) and (–3, –2).

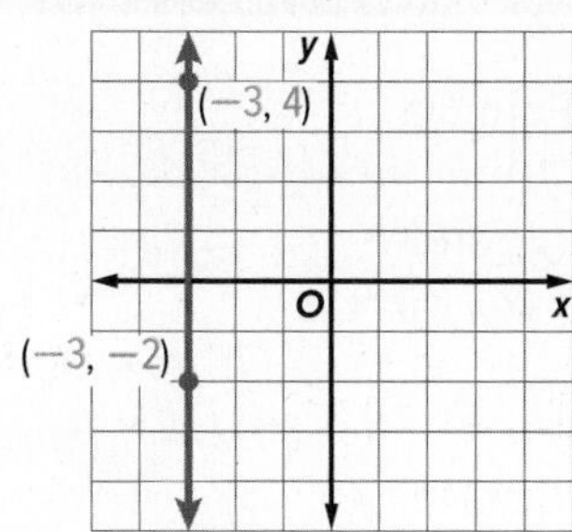

$$m = \frac{y_2 - y_1}{x_2 - x_1}$$

$$= \frac{-2 - 4}{-3 - (-3)}$$

$$= -\frac{6}{0} \text{ or undefined}$$

Go Online You can complete an Extra Example online.

Study Tip

Positive and Negative Slope To know whether a line has a positive or negative slope, read the graph of the line just like you would read a sentence, from left to right. If the line "goes uphill," then the slope is positive. If the line "goes downhill," then the slope is negative.

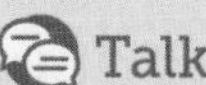

Why is the slope for vertical lines always undefined? Justify your argument.

Study Tip

Converting Slope When solving for an unknown coordinate, like the previous example, converting a slope from a decimal or mixed number to an improper fraction might make the problem easier to solve. For example, a slope of $1.\overline{333}$ can be rewritten as $\frac{4}{3}$.

Example 9 Find Coordinates Given the Slope

Find the value of r so that the line passing through (−4, 5) and (4, r) has a slope of $\frac{3}{4}$.

$m = \frac{y_2 - y_1}{x_2 - x_1}$	Use the Slope Formula.
$\frac{3}{4} = \frac{r - 5}{4 - (-4)}$	$(-4, 5) = (x_1, y_1)$ and $(4, r) = (x_2, y_2)$
$\frac{3}{4} = \frac{r - 5}{8}$	Subtract.
$8\left(\frac{3}{4}\right) = \frac{8(r - 5)}{8}$	Multiply each side by 8.
$6 = r - 5$	Simplify.
$6 + 5 = r - 5 + 5$	Add 5 to each side.
$11 = r$	Simplify.

Check

Find the value of r so that the line passing through $(-3, r)$ and $(7, -6)$ has a slope of $2\frac{2}{5}$.

$r =$ ___?___

Think About It!

If a crab is walking along the ocean floor 112 meters away from the shoreline to 114 meters away from the shoreline, how far does it descend?

Example 10 Use Slope

OCEANS

What is the slope of the continental slope at Cape Hatteras?

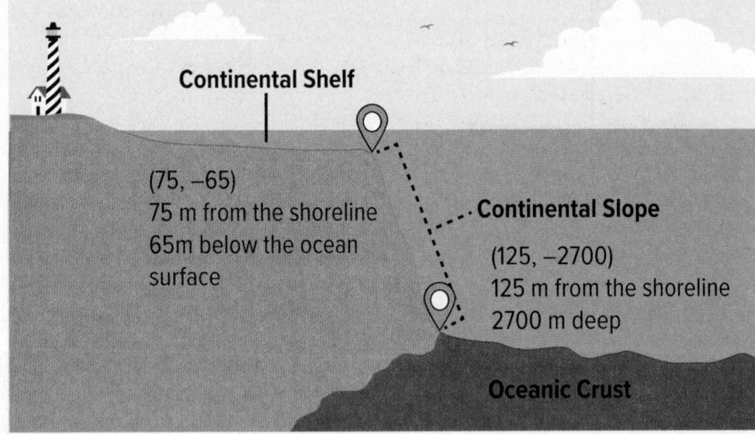

$m = \frac{y_2 - y_1}{x_2 - x_1}$	Use the Slope Formula
$= \frac{-2700 - (-65)}{125 - 75}$	$(75, -65) = (x_1, y_1)$ and $(125, -2700) = (x_2, y_2)$
$= \frac{-2635}{50}$ or -52.7	Simplify.

The continental slope at Cape Hatteras has a slope of −52.7.

Go Online You can complete an Extra Example online.

Practice

Go Online You can complete your homework online.

Example 1

Find the rate of change of the function by using two points from the table.

1.

x	y
5	2
10	3
15	4
20	5

2.

x	y
1	15
2	9
3	3
4	−3

3. POPULATION DENSITY The table shows the population density for the state of Texas in various years. Find the average annual rate of change in the population density from 2000 to 2009.

4. BAND In 2012, there were approximately 275 students in the Delaware High School band. In 2018, that number increased to 305. Find the annual rate of change in the number of students in the band.

Population Density	
Year	People Per Square Mile
1930	22.1
1960	36.4
1980	54.3
2000	79.6
2009	96.7

Source: Bureau of the Census, U.S. Dept. of Commerce

Example 2

5. TEMPERATURE The graph shows the temperature in a city during different hours of one day.

a. Find the rate of change in temperature between 6 A.M. and 7 A.M. and describe its meaning in the context of the situation.

b. Find the rate of change in temperature from 1 P.M. and 2 P.M. and describe its meaning in the context of the situation.

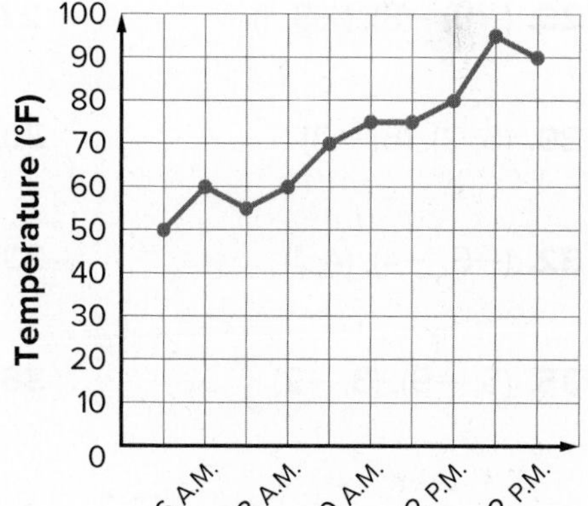

6. COAL EXPORTS The graph shows the annual coal exports from U.S. mines in millions of short tons.

a. Find the rate of change in coal exports between 2000 and 2002 and describe its meaning in the context of the situation.

b. Find the rate of change in coal exports between 2005 and 2006 and describe its meaning in the context of the situation.

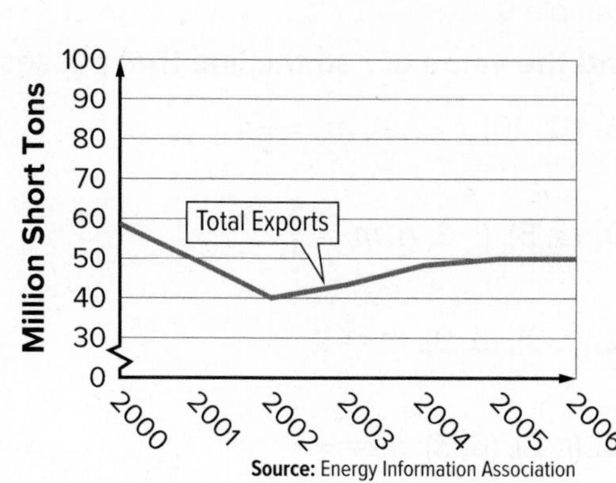

Examples 3 and 4

Determine whether the function is linear. If it is, state the rate of change.

7.

x	4	2	0	−2	−4
y	−1	1	3	5	7

8.

x	−7	−5	−3	−1	0
y	11	14	17	20	23

9.

x	−0.2	0	0.2	0.4	0.6
y	0.7	0.4	0.1	0.3	0.6

10.

x	$\frac{1}{2}$	$\frac{3}{2}$	$\frac{5}{2}$	$\frac{7}{2}$	$\frac{9}{2}$
y	$\frac{1}{2}$	1	$\frac{3}{2}$	2	$\frac{5}{2}$

Examples 5 through 8

Find the slope of the line that passes through each pair of points.

11. (4, 3), (−1, 6)

12. (8, −2), (1, 1)

13. (2, 2), (−2, −2)

14. (6, −10), (6, 14)

15. (5, −4), (9, −4)

16. (11, 7), (−6, 2)

17. (−3, 5), (3, 6)

18. (−3, 2), (7, 2)

19. (8, 10), (−4, −6)

20. (−12, 15), (18, −13)

21. (−8, 6), (−8, 4)

22. (−8, −15), (−2, 5)

23. (2, 5), (3, 6)

24. (6, 1), (−6, 1)

25. (4, 6), (4, 8)

26. (−5, −8), (−8, 1)

27. (2, 5), (−3, −5)

28. (9, 8), (7, −8)

29. (5, 2), (5, −2)

30. (10, 0), (−2, 4)

31. (17, 18), (18, 17)

32. (−6, −4), (4, 1)

33. (−3, 10), (−3, 7)

34. (2, −1), (−8, −2)

35. (5, −9), (3, −2)

36. (12, 6), (3, −5)

37. (−4, 5), (−8, −5)

Example 9

Find the value of r so the line that passes through each pair of points has the given slope.

38. (12, 10), (−2, r), $m = -4$

39. (r, −5), (3, 13), $m = 8$

40. (3, 5), (−3, r), $m = \frac{3}{4}$

41. (−2, 8), (r, 4), $m = -\frac{1}{2}$

42. (r, 3), (5, 9), $m = 2$

43. (5, 9), (r, −3), $m = -4$

44. (r, 2), (6, 3), $m = \frac{1}{2}$

45. (r, 4), (7, 1), $m = \frac{3}{4}$

Example 10

46. ROAD SIGNS Roadway signs such as the one shown are used to warn drivers of an upcoming steep down grade. What is the grade, or slope, of the hill described on the sign?

47. HOME MAINTENANCE Grading the soil around the foundation of a house can reduce interior home damage from water runoff. For every 6 inches in height, the soil should extend 10 feet from the foundation. What is the slope of the soil grade?

48. USE A SOURCE Research the Americans with Disabilities Act (ADA) regulation for the slope of a wheelchair ramp. What is the maximum slope of an ADA regulation ramp? Use the slope to determine the length and height of an ADA regulation ramp.

49. DIVERS A boat is located at sea level. A scuba diver is 80 feet along the surface of the water from the boat and 30 feet below the water surface. A fish is 20 feet along the horizontal plane from the scuba diver and 10 feet below the scuba diver. What is the slope between the scuba diver and fish?

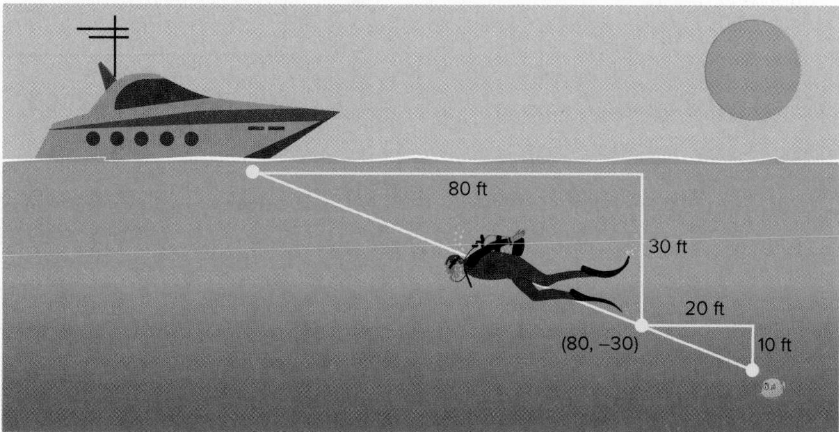

Mixed Exercises

STRUCTURE **Find the slope of the line that passes through each pair of points.**

50.

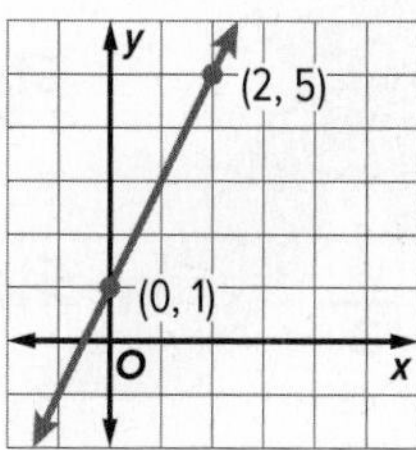

51.

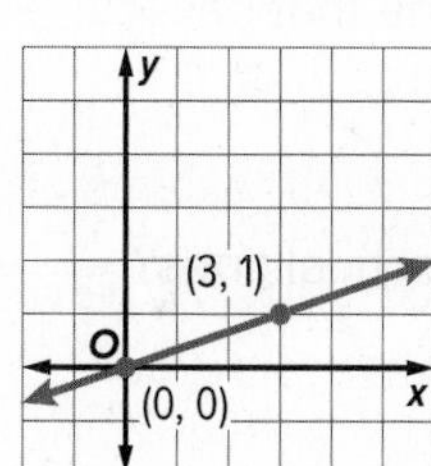

52.

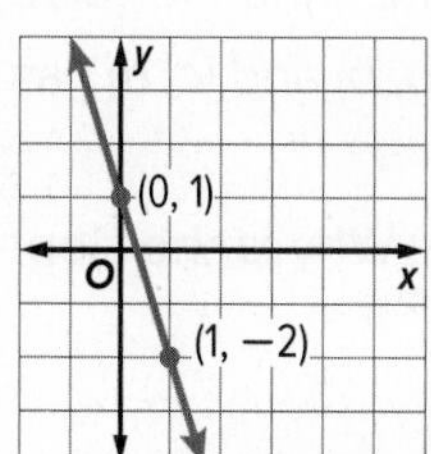

53. (6, −7), (4, −8)

54. (0, 5), (5, 5)

55. (−2, 6), (−5, 9)

56. (5, 8), (−4, 6)

57. (9, 4), (5, −3)

58. (1, 4), (3, −1)

59. **REASONING** Find the value of r that gives the line passing through (3, 2) and (r, −4) a slope that is undefined.

60. **REASONING** Find the value of r that gives the line that passing through (−5, 2) and (3, r) a slope of 0.

61. **CREATE** Draw a line on a coordinate plane so that you can determine at least two points on the graph. Describe how you would determine the slope of the graph and justify the slope you found.

62. **ARGUMENTS** The graph shows median prices for small cottages on a lake since 2005. A real estate agent says that since 2005, the rate of change for house prices is $10,000 each year. Do you agree? Use the graph to justify your answer.

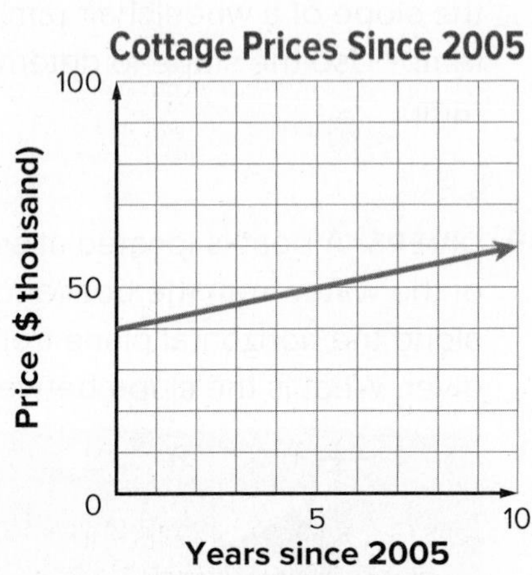

Higher-Order Thinking Skills

63. **CREATE** Use what you know about rate of change to describe the function represented by the table.

Time (wk)	Height of Plant (in.)
4	9.0
6	13.5
8	18.0

64. **WRITE** Explain how the rate of change and slope are related and how to find the slope of a line.

65. **FIND THE ERROR** Fern is finding the slope of the line that passes through (−2, 8) and (4, 6). Determine in which step she made an error. Explain your reasoning.

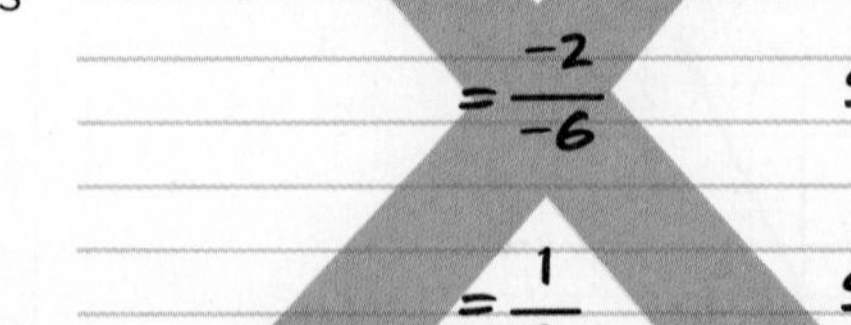

66. **PERSEVERE** Find the value of d so that the line that passes through (a, b) and (c, d) has a slope of $\frac{1}{2}$.

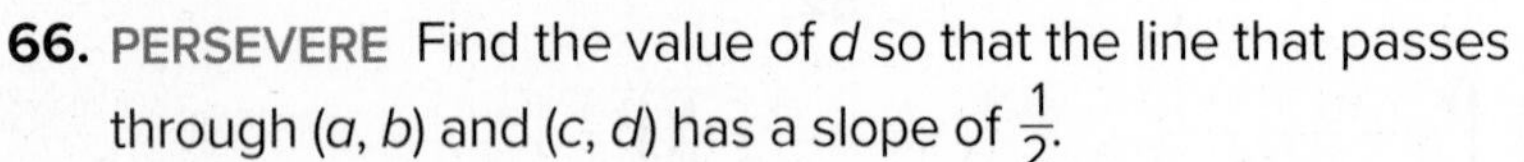

67. **ANALYZE** Why is the slope undefined for vertical lines? Explain.

68. **WRITE** Tarak wants to find the value of a so that the line that passes through (10, a) and (−2, 8) has a slope of $\frac{1}{4}$. Explain how Tarak can find the value of a.

Slope-Intercept Form

Learn Writing Linear Equations in Slope-Intercept Form

An equation of the form $y = mx + b$, where m is the slope and b is the y-intercept, is written in slope-intercept form. When an equation is not in slope-intercept form, it might be easier to rewrite it before graphing. An equation can be rewritten in slope-intercept form by using the properties of equality.

Key Concept • Slope Intercept Form	
Words	The slope-intercept form of a linear equation is $y = mx + b$, where m is the slope and b is the y-intercept.
Example	$y = mx + b$ $y = 3x + 2$

Today's Goals

- Rewrite linear equations in slope-intercept form.
- Graph and interpret linear functions.

Today's Vocabulary

parameter

constant function

Think About It!

Explain why the y-intercept of a linear equation can be written as $(0, b)$, where b is the y-intercept.

Example 1 Write Linear Equations in Slope-Intercept Form

Write an equation in slope-intercept form for the line with a slope of $\frac{4}{7}$ and a y-intercept of 5.

Write the equation in slope-intercept form.

$y = mx + b$ Slope-intercept form.

$y = \left(\frac{4}{7}\right)x + 5$ $m = \frac{4}{7}, b = 5$

$y = \frac{4}{7}x + 5$ Simplify.

Check

Write an equation for the line with a slope of −5 and a y-intercept of 12.

Go Online You can complete an Extra Example online.

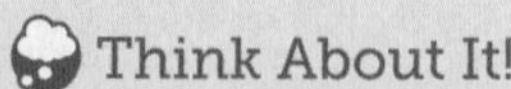

Think About It!

Can $x = 5$ be rewritten in slope-intercept form? Justify your argument.

Example 2 Rewrite Linear Equations in Slope-Intercept Form

Write $-22x + 8y = 4$ in slope-intercept form.

$-22x + 8y = 4$	Original equation
$-22x + 8y + 22x = 4 + 22x$	Add $22x$ to each side.
$8y = 22x + 4$	Simplify.
$\frac{8y}{8} = \frac{22x + 4}{8}$	Divide each side by 8.
$y = 2.75x + 0.5$	Simplify.

Check

What is the slope intercept form of $-16x - 4y = -56$?

Example 3 Write Linear Equations

JOBS **The number of job openings in the United States during a recent year increased by an average of 0.06 million per month since May. In May, there were about 4.61 million job openings in the United States. Write an equation in slope-intercept form to represent the number of job openings in the United States in the months since May.**

Use the given information to write an equation in slope-intercept form.

- You are given that there were 4.61 million job openings in May.
- Let x = the number of months since May and y = the number of job openings in millions.
- Because the number of job openings is 4.61 million when $x = 0$, $b = 4.61$, and because the number of job openings has increased by 0.06 million each month, $m = 0.06$.
- So, the equation $y = 0.06x + 4.61$ represents the number of job openings in the United States since May.

Think About It!

When $x = 2$, describe the meaning of the equation in the context of the situation.

Check

SOCIAL MEDIA In the first quarter of 2012, there were 183 million users of a popular social media site in North America. The number of users increased by an average of 9 million per year since 2012. Write an equation that represents the number of users in millions of the social media site in North America after 2012.

Go Online You can complete an Extra Example online.

Explore Graphing Linear Equations by Using the Slope-Intercept Form

Online Activity Use graphing technology to complete an Explore.

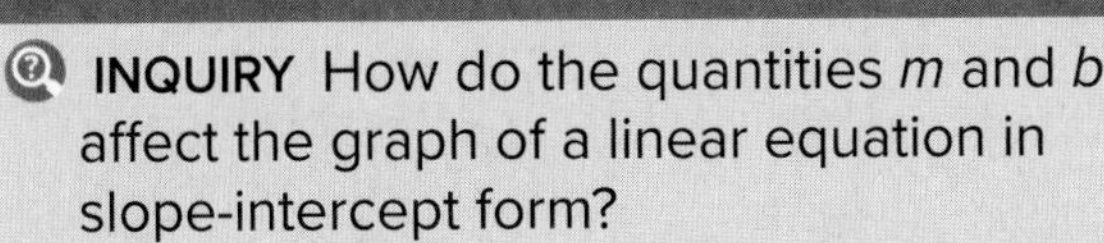

INQUIRY How do the quantities m and b affect the graph of a linear equation in slope-intercept form?

Learn Graphing Linear Functions in Slope-Intercept Form

The slope-intercept form of a linear equation is $y = mx + b$ where m is the slope and b is the y-intercept. The variables m and b are called **parameters** of the equation because changing either value changes the graph.

A **constant function** is a linear function of the form $y = b$. Constant functions where $b \neq 0$ do not cross the x-axis. The graphs of constant functions have a slope of 0. The domain of a constant function is all real numbers, and the range is b.

Example 4 Graph Linear Equations in Slope-Intercept Form

Graph a linear equation with a slope of $-\frac{3}{2}$ and a y-intercept of 4.

Write the equation in slope-intercept form and graph the equation.

$y = mx + b$

$y = \left(-\frac{3}{2}\right)x + 4$

$y = -\frac{3}{2}x + 4$

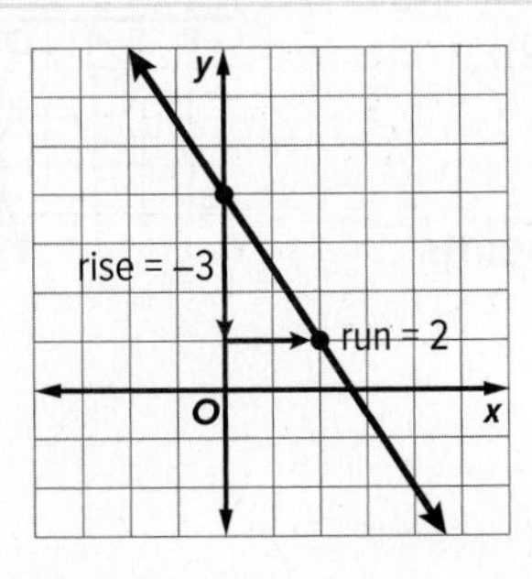

Study Tip

Negative Slope When counting rise and run, a negative sign may be associated with the value in the numerator or denominator. In this case, we associated the negative sign with the numerator. If we had associated it with the denominator, we would have moved up 3 and left 2 to the point $(-2, 7)$. Notice that this point is also on the line. The resulting line will be the same whether the negative sign is associated with the numerator or denominator.

Think About It!

Use the slope to find another point on the graph. Explain how you found the point.

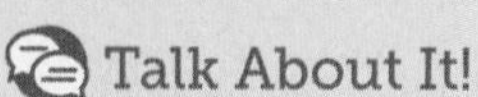

Talk About It!

Why is it useful to write an equation in slope-intercept form before graphing it?

Check

Graph a linear function with a slope of -2 and a y-intercept of 7.

Example 5 Graph Linear Functions

Graph $12x - 3y = 18$.

Rewrite the equation in slope-intercept form.

$12x - 3y = 18$	Original equation
$12x - 3y - 12x = 18 - 12x$	Subtract $12x$ from each side.
$-3y = -12x + 18$	Simplify.
$\frac{-3y}{-3} = \frac{-12x + 18}{-3}$	Divide each side by -3.
$y = 4x - 6$	Simplify.

Graph the equation.

Plot the y-intercept $(0, -6)$.

The slope is $\frac{rise}{run} = 4$. From $(0, -6)$, move up 4 units and right 1 unit. Plot the point $(1, -2)$.

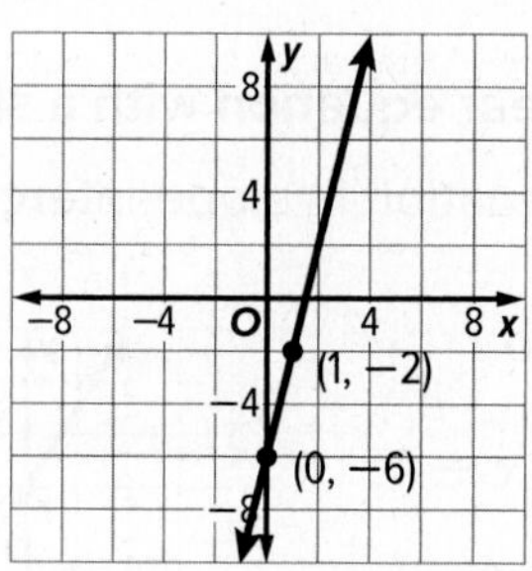

Draw a line through the points $(0, -6)$ and $(1, -2)$.

Go Online You can complete an Extra Example online.

Example 6 Graph Constant Functions

Graph $y = 2$.

Step 1 Plot (0, 2).

Step 2 The slope of $y = 2$ is 0.

Step 3 Draw a line through all the points that have a y-coordinate of 2.

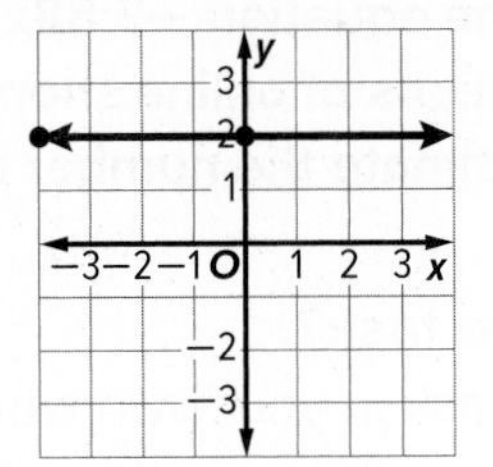

Check

Graph $y = 1$.

Match each graph with its equation.

? $y = 8$	? $3x + 7y = -28$	? $y = \frac{3}{7}x - 4$
? $y = -4$	? $y = -3x + 8$	? $3x - y = 8$

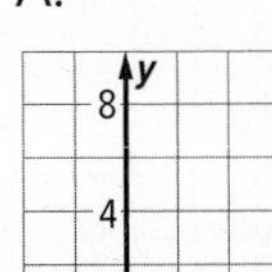

A.

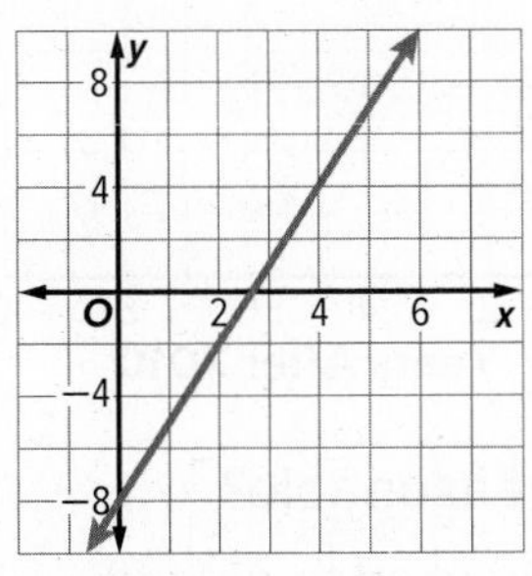

B.

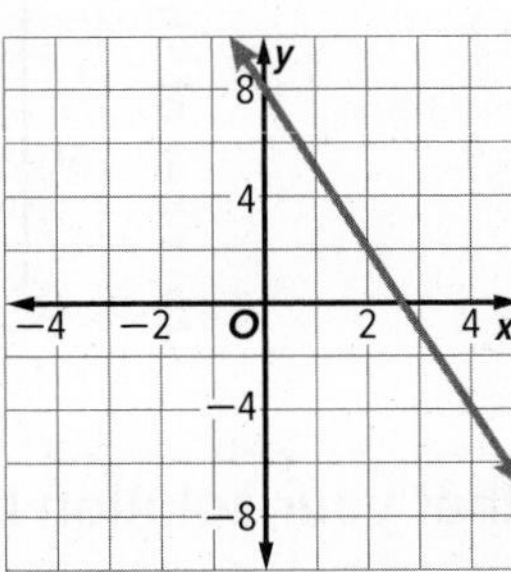

C.

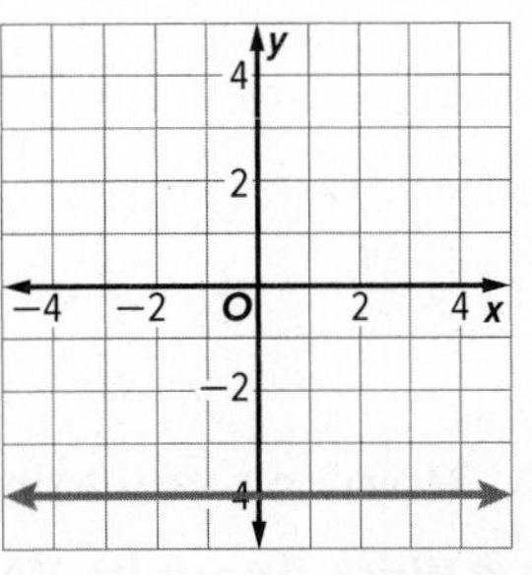

D.

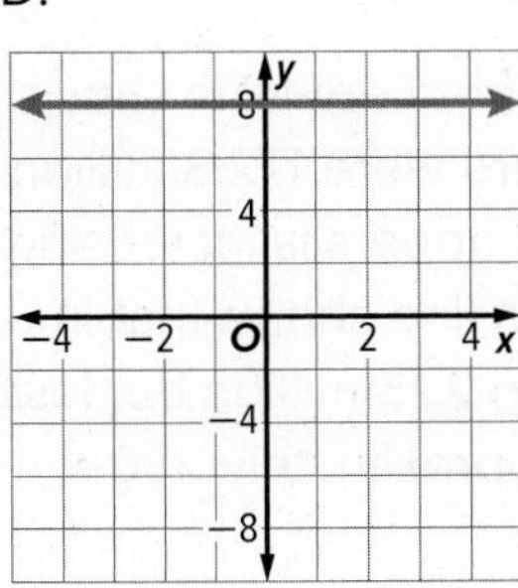

E.

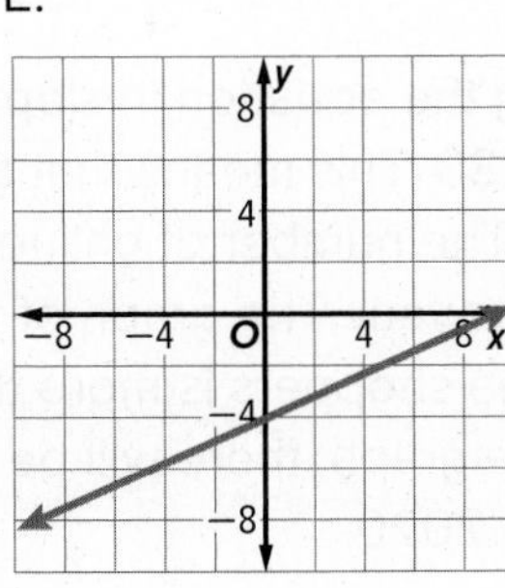

F.

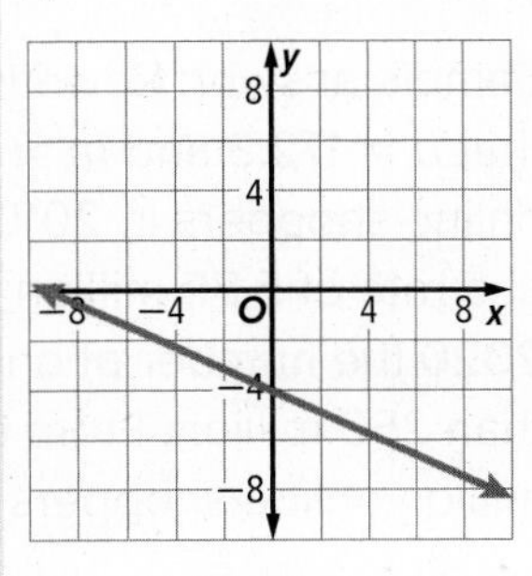

Go Online You can complete an Extra Example online.

Think About It!

How do you know that $y = 2$ has a slope of 0?

Watch Out!

Slope A line with zero slope is not the same as a line with no slope. A line with zero slope is horizontal, and a line with no slope is vertical.

Apply Example 7 Use Graphs of Linear Functions

SHOPPING **The number of online shoppers in the United States can be modeled by the equation $-5.88x + y = 172.3$, where y represents the number of millions of online shoppers in the United States x years after 2010. Estimate the number of people shopping online in 2020.**

1. What is the task?

Describe the task in your own words. Then list any questions that you may have. How can you find answers to your questions?

Sample answer: I need to find the number of people who shop online in 2020.

2. How will you approach the task? What have you learned that you can use to help you complete the task?

Sample answer: I will graph the given equation. Then I can figure out from the graph how many people will be shopping online in 2020.

3. What is your solution?

Use your strategy to solve the problem. Graph the equation.

In 2020, there will be approximately 230 million online shoppers in the United States.

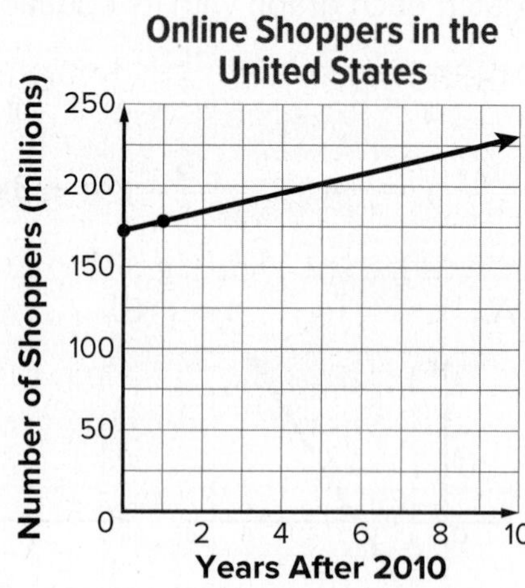

4. How can you know that your solution is reasonable?

Write About It! Write an argument that can be used to defend your solution.

Sample answer: Rewriting the equation in slope-intercept form shows that $b = 172.3$ and $m = 5.88$. This means that there were 172.3 million online shoppers in 2010. The number of online shoppers is increasing at a rate of 5.88 million per year. The graph of this line shows that in 2020 the number of online shoppers is more than 225 million but less than 250 million. From the graph, there will be approximately 230 million online shoppers in 2020.

Think About It!

Estimate the year when the number of online shoppers in the United States will reach 271 million.

Go Online

to learn about intervals in linear growth patterns in Expand 4-3.

Go Online You can complete an Extra Example online.

Practice

Go Online You can complete your homework online.

Example 1

Write an equation of a line in slope-intercept form with the given slope and *y*-intercept.

1. slope: 5, y-intercept: -3

2. slope: -2, y-intercept: 7

3. slope: -6, y-intercept: -2

4. slope: 7, y-intercept: 1

5. slope: 3, y-intercept: 2

6. slope: -4, y-intercept: -9

7. slope: 1, y-intercept: -12

8. slope: 0, y-intercept: 8

Example 2

Write each equation in slope-intercept form.

9. $-10x + 2y = 12$

10. $4y + 12x = 16$

11. $-5x + 15y = -30$

12. $6x - 3y = -18$

13. $-2x - 8y = 24$

14. $-4x - 10y = -7$

Example 3

15. SAVINGS Wade's grandmother gave him \$100 for his birthday. Wade wants to save his money to buy a portable game console that costs \$275. Each month, he adds \$25 to his savings. Write an equation in slope-intercept form to represent Wade's savings y after x months.

16. FITNESS CLASSES Toshelle wants to take strength training classes at the community center. She has to pay a one-time enrollment fee of \$25 to join the community center, and then \$45 for each class she wants to take. Write an equation in slope-intercept form for the cost of taking x classes.

17. EARNINGS Macario works part time at a clothing store in the mall. He is paid \$9 per hour plus 12% commission on the items he sells in the store. Write an equation in slope-intercept form to represent Macario's hourly wage y.

18. ENERGY From 2002 to 2005, U.S. consumption of renewable energy increased an average of 0.17 quadrillion BTUs per year. About 6.07 quadrillion BTUs of renewable power were produced in the year 2002. Write an equation in slope-intercept form to find the amount of renewable power P in quadrillion BTUs produced in year y between 2002 and 2005.

Example 4

Graph a linear equation with the given slope and *y*-intercept.

19. slope: 5, y-intercept: 8

20. slope: 3, y-intercept: 10

21. slope: -4, y-intercept: 6

22. slope: -2, y-intercept: 8

Examples 5 and 6

Graph each equation.

23. $5x + 2y = 8$

24. $4x + 9y = 27$

25. $y = 7$

26. $y = -\frac{2}{3}$

27. $21 = 7y$

28. $3y - 6 = 2x$

Example 7

29. STREAMING An online company charges $13 per month for the basic plan. They offer premium channels for an additional $8 per month.

a. Write an equation in slope-intercept form for the total cost c of the basic plan with p premium channels in one month.

b. Graph the equation.

c. What would the monthly cost be for a basic plan plus 3 premium channels?

30. CAR CARE Suppose regular gasoline costs $2.76 per gallon. You can purchase a car wash at the gas station for $3.

a. Write an equation in slope-intercept form for the total cost y of purchasing a car wash and x gallons of gasoline.

b. Graph the equation.

c. Find the cost of purchasing a car wash and 8 gallons of gasoline.

Mixed Exercises

Write an equation of a line in slope-intercept form with the given slope and y-intercept.

31. slope: $\frac{1}{2}$, y-intercept: -3

32. slope: $\frac{2}{3}$, y-intercept: -5

Graph an equation of a line with the given slope and y-intercept.

33. slope: 3, y-intercept: -4

34. slope: 4, y-intercept: -6

Graph each equation.

35. $-3x + y = 6$

36. $-5x + y = 1$

Write an equation in slope-intercept form for each graph shown.

37.

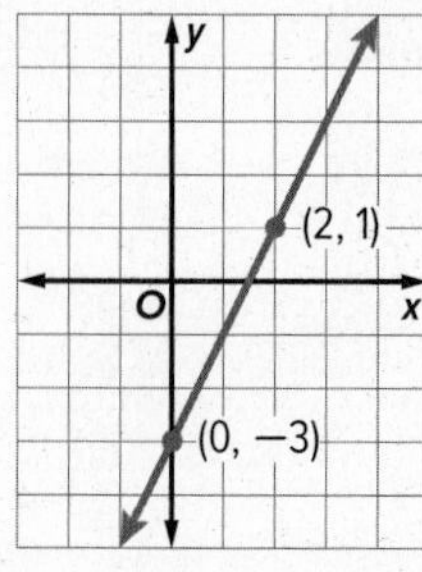

38.

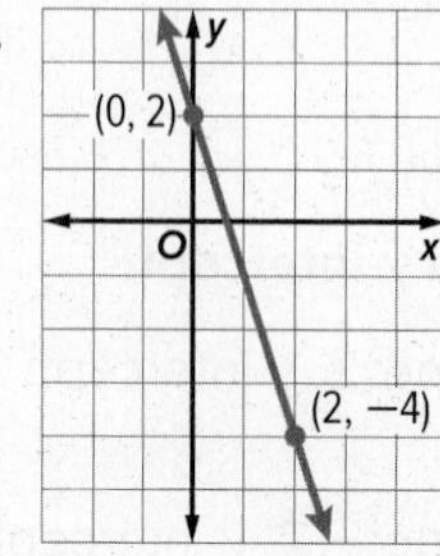

39.

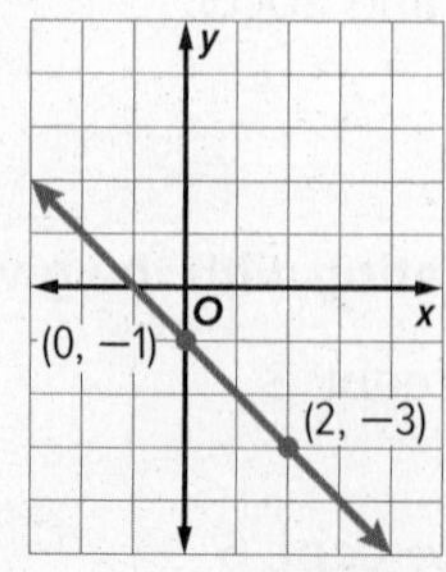

40. MOVIES MovieMania, an online movie rental Web site charges a one-time fee of \$6.85 and \$2.99 per movie rental. Let m represent the number of movies you watch and let C represent the total cost to watch the movies.

a. Write an equation that relates the total cost to the number of movies you watch from MovieMania.

b. Graph the equation.

c. Explain how to use the graph to estimate the cost of watching 13 movies at MovieMania.

d. SuperFlix has no sign-up fee, just a flat rate per movie. If renting 13 movies at MovieMania costs the same as renting 9 movies at SuperFlix, what does SuperFlix charge per movie? Explain your reasoning.

e. Write an equation that relates the total cost to the number of movies you watch from SuperFlix. Round to the nearest whole number.

41. FACTORY A factory uses a heater in part of its manufacturing process. The product cannot be heated too quickly, nor can it be cooled too quickly after the heating portion of the process is complete.

a. The heater is digitally controlled to raise the temperature inside the chamber by 10°F each minute until it reaches the set temperature. Write a function to represent the temperature, T, inside the chamber after x minutes if the starting temperature is 80°F.

b. Graph the equation.

c. The heating process takes 22 minutes. Use your graph to find the temperature in the chamber at this point.

d. After the heater reaches the temperature determined in **part c**, the temperature is kept constant for 20 minutes before cooling begins. Fans within the heater control the cooling so that the temperature inside the chamber decreases by 5°F each minute. Write a function to represent the temperature, T, inside the chamber x minutes after the cooling begins.

42. SAVINGS When Santo was born, his uncle started saving money to help pay for a car when Santo became a teenager. Santo's uncle initially saved \$2000. Each year, his uncle saved an additional \$200.

a. Write an equation that represents the amount, in dollars, Santo's uncle saved y after x years.

b. Graph the equation.

c. Santo starts shopping for a car when he turns 16. The car he wants to buy costs \$6000. Does he have enough money in the account to buy the car? Explain.

43. **STRUCTURE** Jazmin is participating in a 25.5-kilometer charity walk. She walks at a rate of 4.25 km per hour. Jazmin walks at the same pace for the entire event.

a. Write an equation in slope-intercept form for the remaining distance, *y*, in kilometers of walking for *x* hours.

b. Graph the equation.

c. What do the *x*- and *y*-intercepts represent in this situation?

d. After Jazmin has walked 17 kilometers, how much longer will it take her to complete the walk? Explain how you can use your graph to answer the question.

Higher-Order Thinking Skills

For Exercises 44 and 45, refer to the equation $y = -\frac{4}{5}x + \frac{2}{5}$ where $-2 \le x \le 5$.

44. **ANALYZE** Copy and complete the table to help you graph the equation $y = -\frac{4}{5}x + \frac{2}{5}$ over the interval. Identify any values of *x* where maximum or minimum values of *y* occur.

x	$-\frac{4}{5}x + \frac{2}{5}$	y	(x, y)
−2			
0			
5			

45. **WRITE** A student says you can find the solution to $-\frac{4}{5}x + \frac{2}{5} = 0$ using the graph. Do you agree? Explain your reasoning. Include the solution to the equation in your response.

46. **PERSEVERE** Consider three points that lie on the same line, (3, 7), (−6, 1), and (9, *p*). Find the value of *p* and explain your reasoning.

47. **CREATE** Linear equations are useful in predicting future events. Create a linear equation that models a real-world situation. Make a prediction from your equation.

Transformations of Linear Functions

Explore Transforming Linear Functions

Online Activity Use graphing technology to complete an Explore.

INQUIRY How does performing an operation on a linear function change its graph?

Today's Goals
- Apply translations to linear functions.
- Apply dilations to linear functions.
- Apply reflections to linear functions.

Today's Vocabulary
family of graphs
parent function
identity function
transformation
translation
dilation
reflection

Learn Translations of Linear Functions

A **family of graphs** includes graphs and equations of graphs that have at least one characteristic in common. The **parent function** is the simplest of the functions in a family.

The family of linear functions includes all lines, with the parent function $f(x) = x$, also called the **identity function**. A **transformation** moves the graph on the coordinate plane, which can create new linear functions.

One type of transformation is a translation. A **translation** is a transformation in which a figure is slid from one position to another without being turned. A linear function can be slid up, down, left, right, or in two directions.

Study Tip

Slope When translating a linear function, the graph of the function moves from one location to another, but the slope remains the same.

Vertical Translations

When a constant k is added to a linear function $f(x)$, the result is a vertical translation. The y-intercept of $f(x)$ is translated up or down.

Key Concept • Vertical Translations of Linear Functions

The graph of $g(x) = x + k$ is the graph of $f(x) = x$ translated vertically.

If $k > 0$, the graph of $f(x)$ is translated k units up.	If $k < 0$, the graph of $f(x)$ is translated $\lvert k \rvert$ units down.
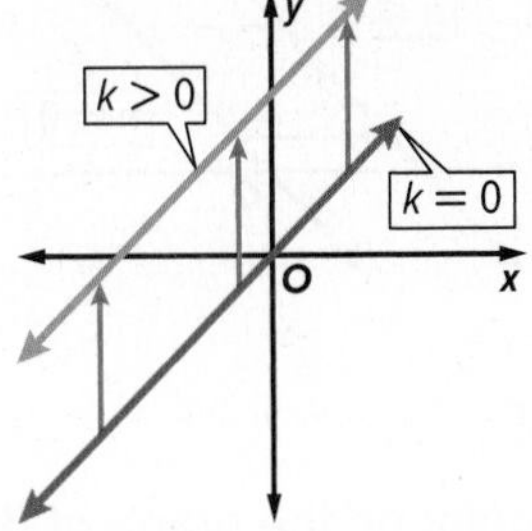	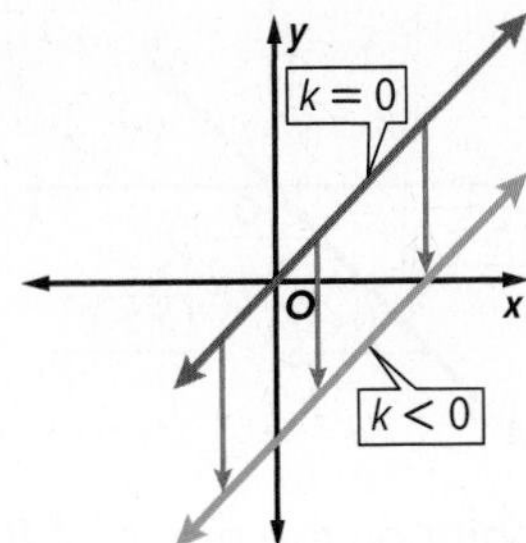
Every point on the graph of $f(x)$ moves k units up.	Every point on the graph of $f(x)$ moves $\lvert k \rvert$ units down.

Watch Out!

Translations of $f(x)$ When a translation is the only transformation performed on the identity function, adding a constant before or after evaluating the function has the same effect on the graph. However, when more than one type of transformation is applied, this will not be the case.

Example 1 Vertical Translations of Linear Functions

Describe the translation in $g(x) = x - 2$ as it relates to the graph of the parent function.

Graph the parent graph for linear functions.

x	$f(x)$	$f(x) - 2$	$(x, g(x))$
-2	-2	-4	$(-2, -4)$
0	0	-2	$(0, -2)$
1	1	-1	$(1, -1)$

Because $f(x) = x$, $g(x) = f(x) + k$ where $k = -2$.

$g(x) = x - 2 \rightarrow f(x) + (-2)$

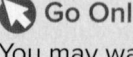

The constant k is not grouped with x, so k affects the output, or y-values. The value of k is less than 0, so the graph of $f(x) = x$ is translated $|-2|$ units down, or 2 units down.

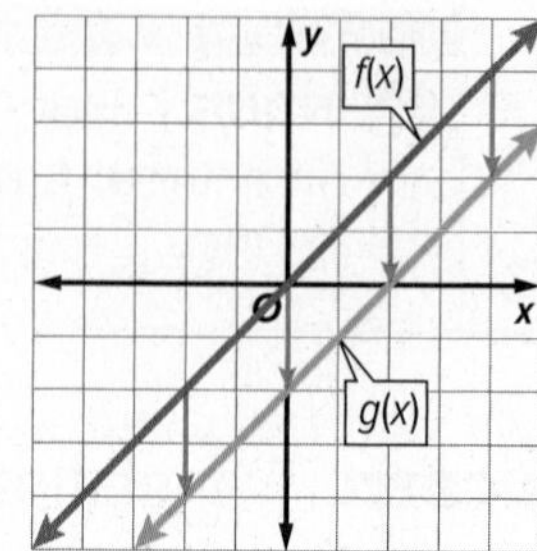

$g(x) = x - 2$ is the translation of the graph of the parent function 2 units down.

Check

Describe the translation in $g(x) = x - 1$ as it relates to the graph of the parent function.

The graph of $g(x) = x - 1$ is a translation of the graph of the parent function 1 unit ___?___.

Think About It!
What do you notice about the y-intercepts of vertically translated functions compared to the y-intercept of the parent function?

Horizontal Translations

When a constant h is subtracted from the x-value before the function $f(x)$ is performed, the result is a horizontal translation. The x-intercept of $f(x)$ is translated right or left.

Key Concept • Horizontal Translations of Linear Functions

The graph of $g(x) = (x - h)$ is the graph of $f(x) = x$ translated horizontally.

If $h > 0$, the graph of $f(x)$ is translated h units right.

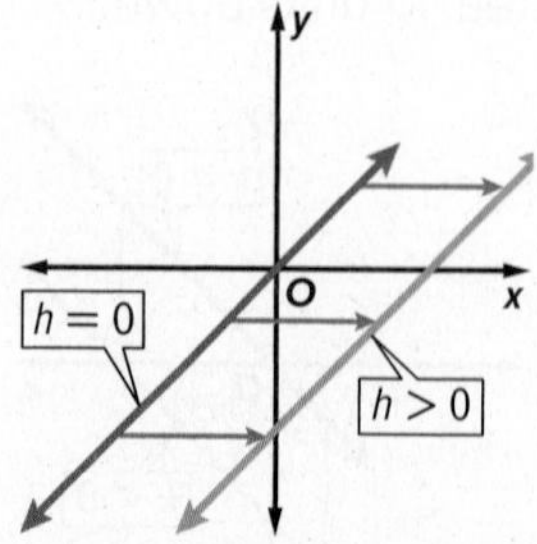

Every point on the graph of $f(x)$ moves h units right.

If $h < 0$, the graph of $f(x)$ is translated $|h|$ units left.

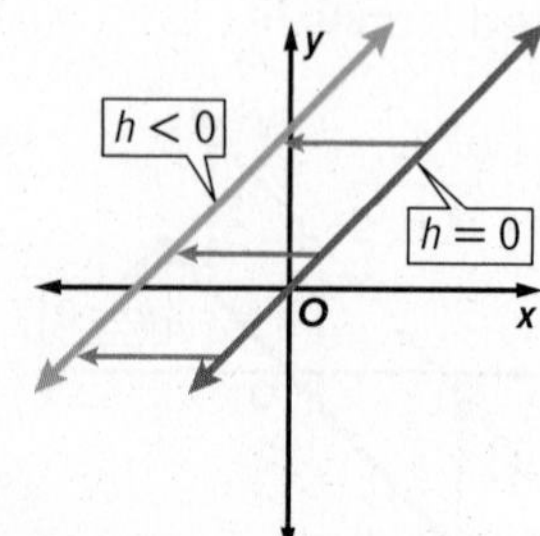

Every point on the graph of $f(x)$ moves $|h|$ units left.

Go Online You can complete an Extra Example online.

Go Online
You can watch a video to see how to describe translations of functions.

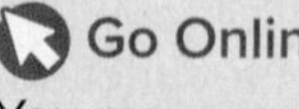

Go Online
You may want to complete the Concept Check to check your understanding.

Example 2 Horizontal Translations of Linear Functions

Describe the translation in $g(x) = (x + 5)$ as it relates to the graph of the parent function.

Graph the parent graph for linear functions.

x	$x + 5$	$f(x + 5)$	$(x, g(x))$
−2	3	3	(−2, 3)
0	5	5	(0, 5)
1	6	6	(1, 6)

Because $f(x) = x$, $g(x) = f(x-h)$ where $h = -5$.

$g(x) = (x + 5) \rightarrow g(x) = f(x - (-5))$

The constant h is grouped with x, so k affects the input, or x-values. The value of h is less than 0, so the graph of $f(x) = x$ is translated $|-5|$ units left, or 5 units left.

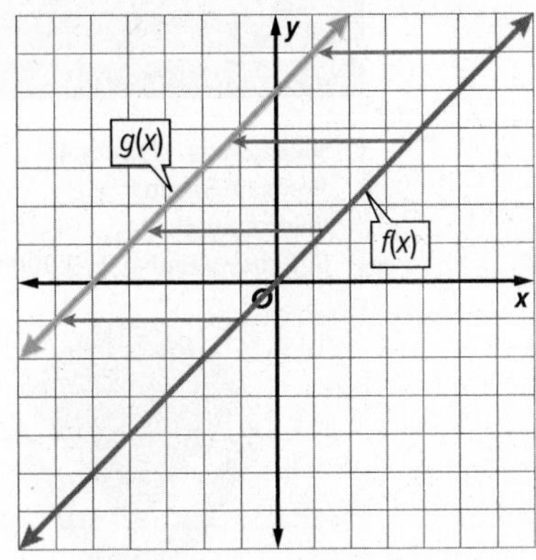

$g(x) = (x + 5)$ is the translation of the graph of the parent function 5 units left.

Check

Describe the translation in $g(x) = (x + 12)$ as it relates to the graph of the parent function.

The graph of $g(x) = (x + 12)$ is a translation of the graph of the parent function 12 units ___?___.

> **Think About It!**
> What do you notice about the x-intercepts of horizontally translated functions compared to the x-intercept of the parent function?

Example 3 Multiple Translations of Linear Functions

Describe the translation in $g(x) = (x - 6) + 3$ as it relates to the graph of the parent function.

Graph the parent graph for linear functions.

x	$x - 6$	$f(x - 6)$	$f(x - 6) + 3$	$(x, g(x))$
−2	−8	−8	−5	(−2, −5)
0	−6	−6	−3	(0, −3)
1	−5	−5	−2	(1, −2)

Because $f(x) = x$, $g(x) = f(x - h) + k$ where $h = 6$ and $k = 3$.

$g(x) = (x - 6) + 3 \rightarrow g(x) = f(x - 6) + 3$

The value of h is grouped with x and is greater than 0, so the graph of $f(x) = x$ is translated 6 units right.

The value of k is not grouped with x and is greater than 0, so the graph of $f(x) = x$ is translated 3 units up.

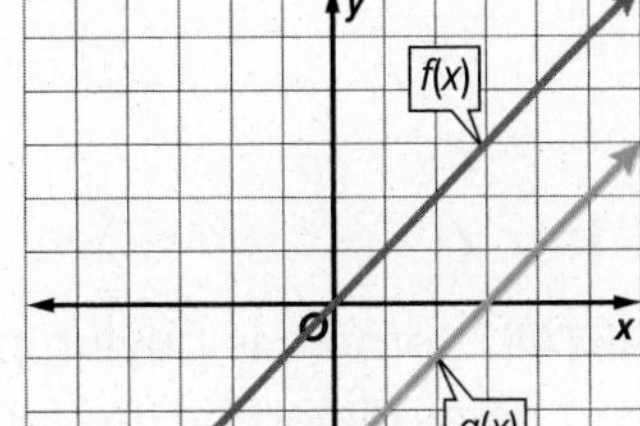

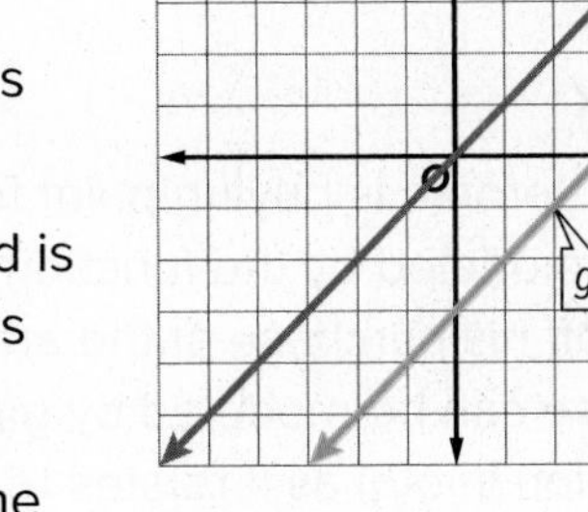

$g(x) = (x - 6) + 3$ is the translation of the graph of the parent function 6 units right and 3 units up.

> **Think About It!**
> Eleni described the graph of $g(x) = (x - 6) + 3$ as the graph of the parent function translated down 3 units. Is she correct? Explain your reasoning.

Go Online You can complete an Extra Example online.

Example 4 Translations of Linear Functions

TICKETS **A Web site sells tickets to concerts and sporting events. The total price of the tickets to a certain game can be modeled by $f(t) = 12t$, where t represents the number of tickets purchased. The Web site then charges a standard service fee of $4 per order. The total price of an order can be modeled by $g(t) = 12t + 4$. Describe the translation of $g(t)$ as it relates to $f(t)$.**

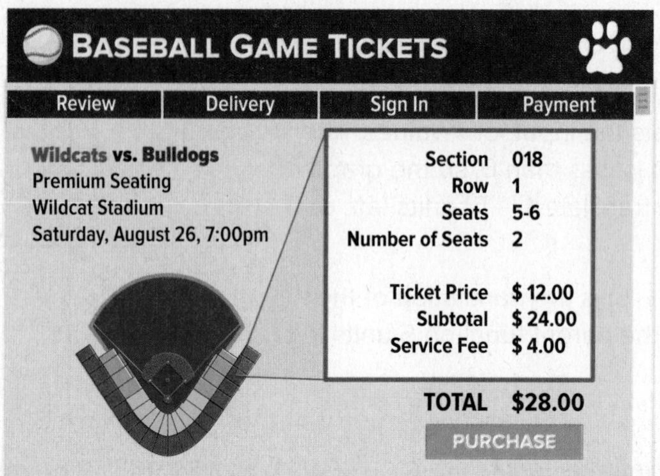

Complete the steps to describe the translation of $g(t)$ as it relates to $f(t)$.
Because $f(t) = 12t$, $g(t) = f(t) + k$, where $k = 4$. $g(t) = 12t + 4 \rightarrow f(t) + 4$

The constant k is added to $f(t)$ after the total price of the tickets has been evaluated and is greater than 0, so the graph of will be shifted 4 units up. $g(t) = 12t + 4$ is the translation of the graph of $f(t)$ 4 units up.

Graph the parent function and the translated function.

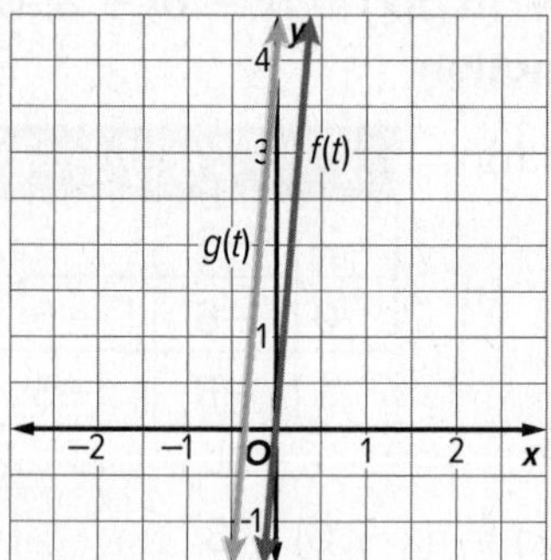

Check

RETAIL Jerome is buying paint for a mural. The total cost of the paint can be modeled by the function $f(p) = 6.99p$. He has a coupon for $5.95 off his purchase at the art supply store, so the final cost of his purchase can be modeled by $g(p) = 6.99p - 5.95$. Describe the translation in $g(p)$ as it relates to $f(p)$.

Go Online You can complete an Extra Example online.

Learn Dilations of Linear Functions

A **dilation** stretches or compresses the graph of a function.

When a linear function $f(x)$ is multiplied by a positive constant a, the result $a \cdot f(x)$ is a vertical dilation.

Key Concept • Vertical Dilations of Linear Functions

The graph of $g(x) = ax$ is the graph of $f(x) = x$ stretched or compressed vertically.

If $\|a\| > 1$, the graph of $f(x)$ is stretched vertically away from the x-axis.	If $0 < \|a\| < 1$, the graph of $f(x)$ is compressed vertically toward the x-axis.
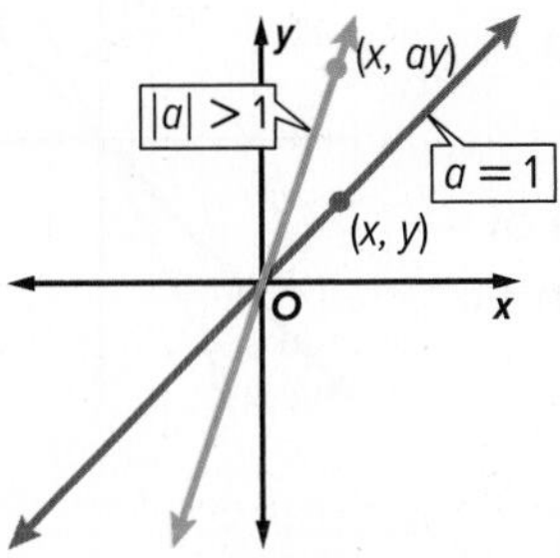	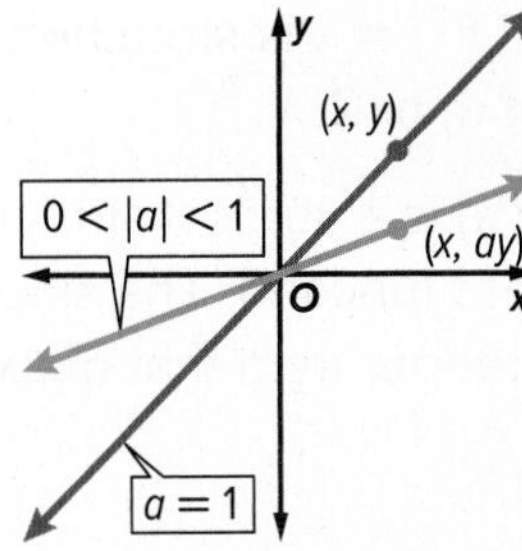
The slope of the graph of $a \cdot f(x)$ is steeper than that of the graph of $f(x)$.	The slope of the graph of $a \cdot f(x)$ is less steep than that of the graph of $f(x)$.

When x is multiplied by a positive constant a before a linear function $f(a \cdot x)$ is evaluated, the result is a horizontal dilation.

Key Concept • Horizontal Dilations of Linear Functions

The graph of $g(x) = (a \cdot x)$ is the graph of $f(x) = x$ stretched or compressed horizontally.

If $\|a\| > 1$, the graph of $f(x)$ is compressed horizontally toward the y-axis.	If $0 < \|a\| < 1$, the graph of $f(x)$ is stretched horizontally away from the y-axis.
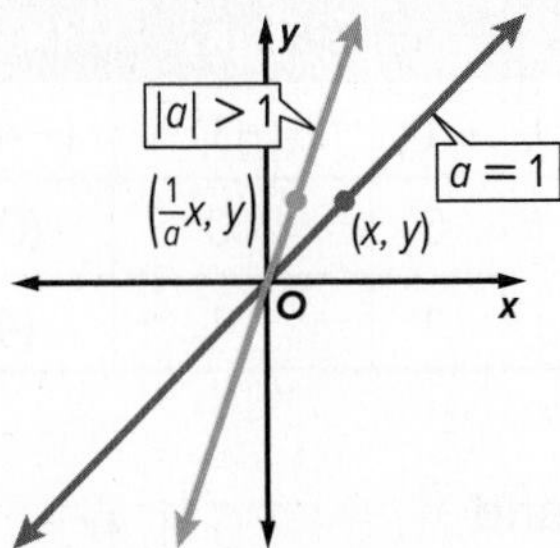	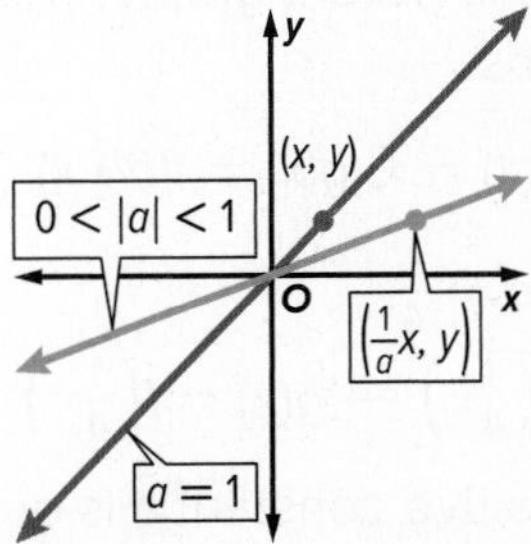
The slope of the graph of $f(a \cdot x)$ is steeper than that of the graph of $f(x)$.	The slope of the graph of $f(a \cdot x)$ is less steep than that of the graph of $f(x)$.

Go Online You can complete an Extra Example online.

Watch Out!

Dilations of $f(x) = x$ When a dilation is the only transformation performed on the identity function, multiplying by a constant before or after evaluating the function has the same effect on the graph. However, when more than one type of transformation is applied, this will not be the case.

Go Online

You can watch a video to see how to describe dilations of functions.

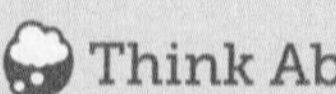

Think About It!

What do you notice about the slope of the vertical dilation $g(x)$ compared to the slope of $f(x)$?

How does this relate to the constant a in the vertical dilation?

Example 5 Vertical Dilations of Linear Functions

Describe the dilation in $g(x) = 2(x)$ as it relates to the graph of the parent function.

Graph the parent graph for linear functions.

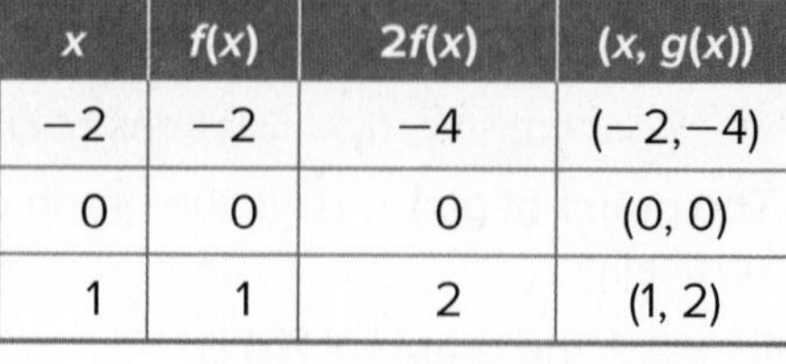

x	$f(x)$	$2f(x)$	$(x, g(x))$
−2	−2	−4	(−2,−4)
0	0	0	(0, 0)
1	1	2	(1, 2)

Since $f(x) = x$, $g(x) = a \cdot f(x)$ where $a = 2$.

$g(x) = 2(x) \rightarrow g(x) = 2f(x)$

The positive constant a is not grouped with x, and $|a|$ is greater than 1, so the graph of $f(x) = x$ is stretched vertically by a factor of a, or 2.

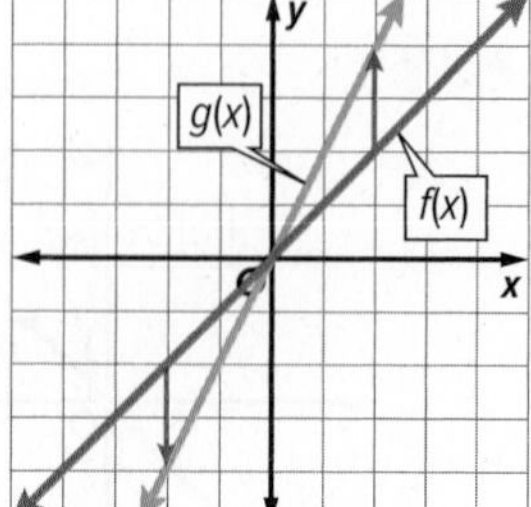

$g(x) = 2(x)$ is a vertical stretch of the graph of the parent function. The slope of the graph of $g(x)$ is steeper than that of $f(x)$.

Check

Describe the transformation in $g(x) = 6(x)$ as it relates to the graph of the parent function.

The graph of $g(x) = 6(x)$ is a ____?____ of the graph of the parent function.

The slope of the graph $g(x)$ is ____?____ than that of the parent function.

Think About It!

What do you notice about the slope of the horizontal dilation $g(x)$ compared to the slope of $f(x)$?

How does this relate to the constant a in the horizontal dilation?

Example 6 Horizontal Dilations of Linear Functions

Describe the dilation in $g(x) = \left(\frac{1}{4}x\right)$ as it relates to the graph of the parent function.

Graph the parent graph for linear functions.

x	$\frac{1}{4}x$	$f\left(\frac{1}{4}x\right)$	$(x, g(x))$
−4	−1	−1	(−4, −1)
0	0	0	(0, 0)
4	1	1	(4, 1)

Since $f(x) = x$, $g(x) = f(a \cdot x)$

where $a = \frac{1}{4}$.

$g(x) = \left(\frac{1}{4}x\right) \rightarrow g(x) = f\left(\frac{1}{4}x\right)$

The positive constant a is grouped with x, and $|a|$ is between 0 and 1, so the graph of $f(x) = x$ is stretched horizontally by a factor of $\frac{1}{a}$, or 4.

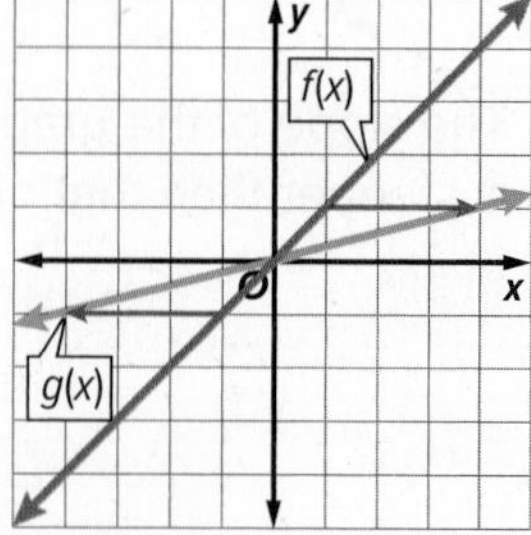

$g(x) = \left(\frac{1}{4}x\right)$ is a horizontal stretch of the graph of the parent function. The slope of the graph of $g(x)$ is less steep than that of $f(x)$.

Go Online You can complete an Extra Example online.

Learn Reflections of Linear Functions

A **reflection** is a transformation in which a figure, line, or curve, is flipped across a line. When a linear function $f(x)$ is multiplied by -1 before or after the function has been evaluated, the result is a reflection across the x- or y-axis. Every x- or y-coordinate of $f(x)$ is multiplied by -1.

Key Concept • Reflections of Linear Functions

The graph of $-f(x)$ is the reflection of the graph of $f(x) = x$ across the x-axis.	The graph of $f(-x)$ is the reflection of the graph of $f(x) = x$ across the y-axis.
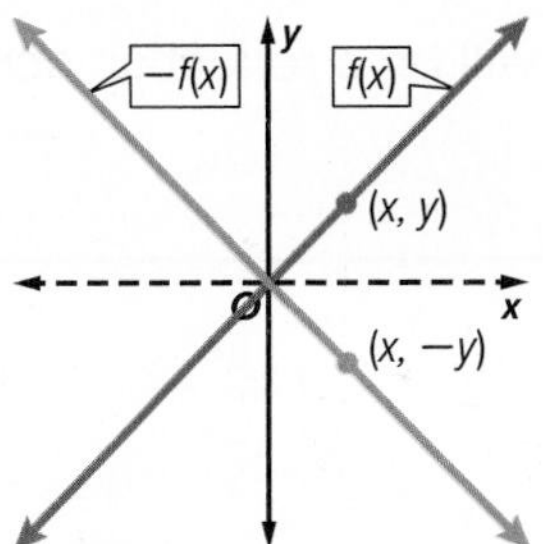	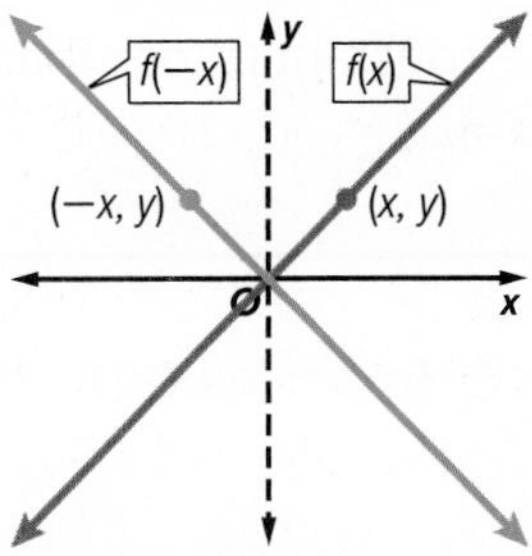
Every y-coordinate of $-f(x)$ is the corresponding y-coordinate of $f(x)$ multiplied by -1.	Every x-coordinate of $f(-x)$ is the corresponding x-coordinate of $f(x)$ multiplied by -1.

Example 7 Reflections of Linear Functions Across the x-Axis

Describe the dilation in $g(x) = -\frac{1}{2}(x)$ as it relates to the graph of the parent function.

Graph the parent graph for linear functions.

Since $f(x) = x$, $g(x) = -1 \cdot a \cdot f(x)$

where $a = \frac{1}{2}$.

x	$f(x)$	$-\frac{1}{2}f(x)$	$(x, g(x))$
−2	−2	1	(−2, 1)
0	0	0	(0, 0)
4	4	−2	(4, −2)

$g(x) = -\frac{1}{2}(x) \rightarrow g(x) = -\frac{1}{2}f(x)$

The constant a is not grouped with x, and $|a|$ is less than 1, so the graph of $f(x) = x$ is vertically compressed.

The negative is not grouped with x, so the graph is also reflected across the x-axis.

The graph of $g(x) = -\frac{1}{2}(x)$ is the graph of the parent function vertically compressed and reflected across the x-axis.

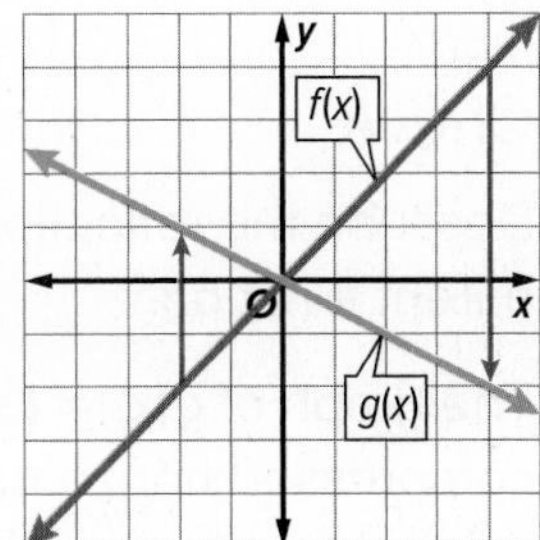

Go Online You can complete an Extra Example online.

Go Online
You can watch a video to see how to describe reflections of functions.

Watch Out!

Reflections of $f(x) = x$
When a reflection is the only transformation performed on the identity function, multiplying by -1 before or after evaluating the function appears to have the same effect on the graph. However, when more than one type of transformation is applied, this will not be the case.

Talk About It!

In the example, the slope of $g(x)$ is negative. Will this always be the case when multiplying a linear function by -1? Justify your argument.

Check

How can you tell whether multiplying −1 by the parent function will result in a reflection across the *x*-axis?

A. If the constant is not grouped with *x*, the result will be a reflection across the *x*-axis.

B. If the constant is grouped with *x*, the result will be a reflection across the *x*-axis.

C. If the constant is greater than 0, the result will be a reflection across the *x*-axis.

D. If the constant is less than 0, the result will be a reflection across the *x*-axis.

Example 8 Reflections of Linear Functions Across the *y*-Axis

Describe the dilation in $g(x) = (-3x)$ as it relates to the graph of the parent function.

Graph the parent graph for linear functions.

x	$-3x$	$f(-3x)$	$(x, g(x))$
−1	3	3	(−1, 3)
0	0	0	(0, 0)
1	−3	−3	(1, −3)

Since $f(x) = x$, $g(x) = f(-1 \cdot a \cdot x)$ where $a = 3$.

$g(x) = -3x \rightarrow g(x) = f(-3x)$

The constant a is grouped with x, and $|a|$ is greater than 1, so the graph of $f(x) = x$ is horizontally compressed.

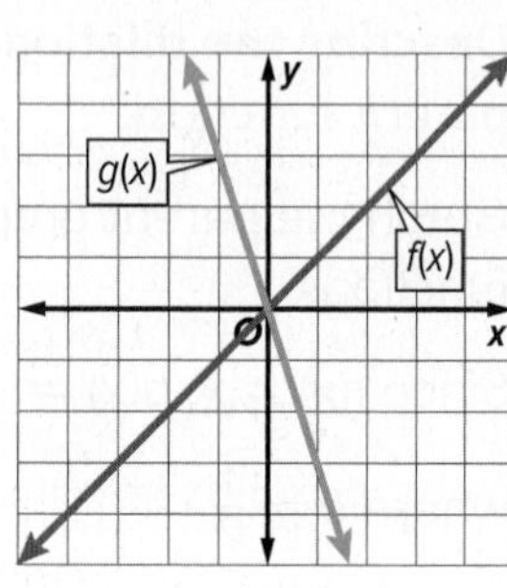

The negative is grouped with x, so the graph is also reflected across the y-axis.

The graph of $g(x) = (-3x)$ is the graph of the parent function horizontally compressed and reflected across the *y*-axis.

Go Online You can watch a video to see how to graph transformations of a linear function using a graphing calculator.

Check

Describe the reflection in $g(x) = (-10x)$ as it relates to the graph of the parent function.

The graph of $g(x) = (-10x)$ is the graph of the parent function compressed horizontally and reflected across the ___?___.

Go Online You can complete an Extra Example online.

Practice

Go Online You can complete your homework online.

Examples 1 through 3

Describe the translation in each function as it relates to the graph of the parent function.

1. $g(x) = x + 11$

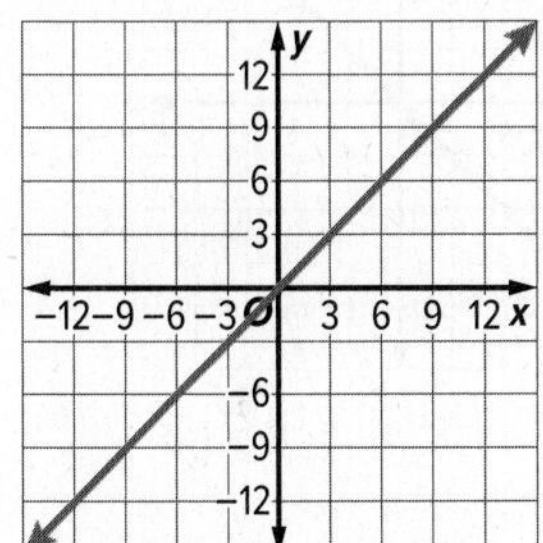

2. $g(x) = x - 8$

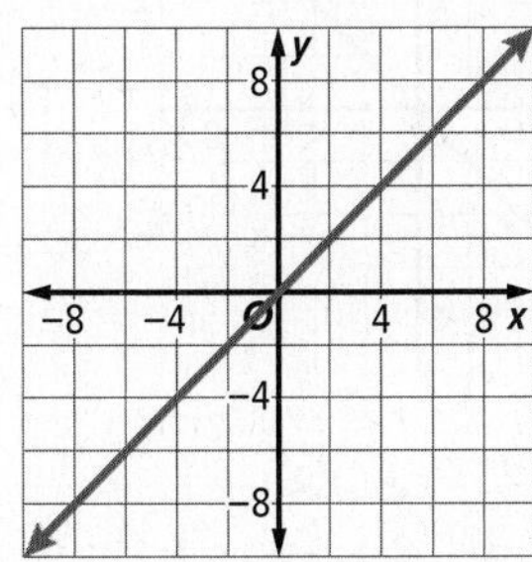

3. $g(x) = (x - 7)$

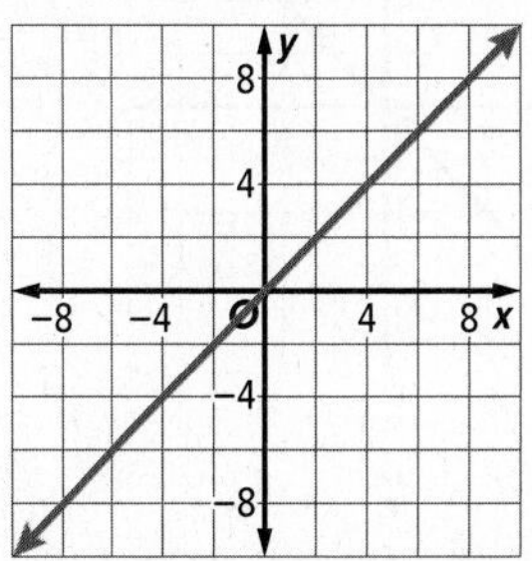

4. $g(x) = (x + 12)$

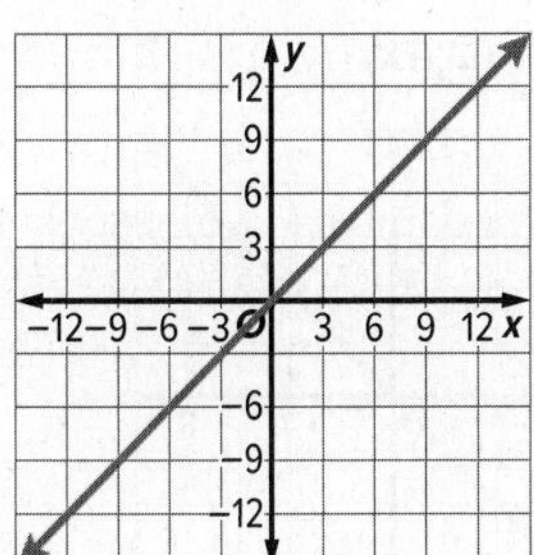

5. $g(x) = (x + 10) - 1$

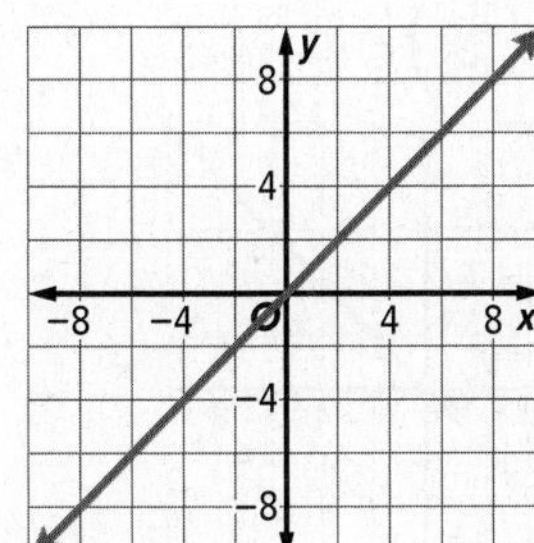

6. $g(x) = (x - 9) + 5$

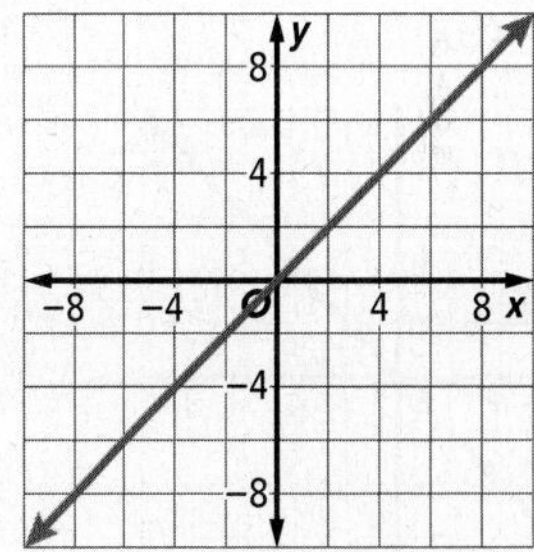

Example 4

7. BOWLING The cost for Nobu to go bowling is \$4 per game plus an additional flat fee of \$3.50 for the rental of bowling shoes. The cost can be modeled by the function $f(x) = 4x + 3.5$, where x represents the number of games bowled. Describe the graph of $g(x)$ as it relates to $f(x)$ if Nobu does not rent bowling shoes.

8. SAVINGS Natalie has \$250 in her savings account, into which she deposits \$10 of her allowance each week. The balance of her savings account can be modeled by the function $f(w) = 250 + 10w$, where w represents the number of weeks. Write a function $g(w)$ to represent the balance of Natalie's savings account if she withdraws \$40 to purchase a new pair of shoes. Describe the translation of $f(w)$ that results in $g(w)$.

9. BOAT RENTAL The cost to rent a paddle boat at the county park is \$8 per hour plus a nonrefundable deposit of \$10. The cost can be modeled by the function $f(h) = 8h + 10$, where h represents the number of hours the boat is rented. Describe the graph of $g(h)$ as it relates to $f(h)$ if the nonrefundable deposit increases to \$15.

Examples 5 and 6

Describe the dilation in each function as it relates to the graph of the parent function.

10. $g(x) = 5(x)$

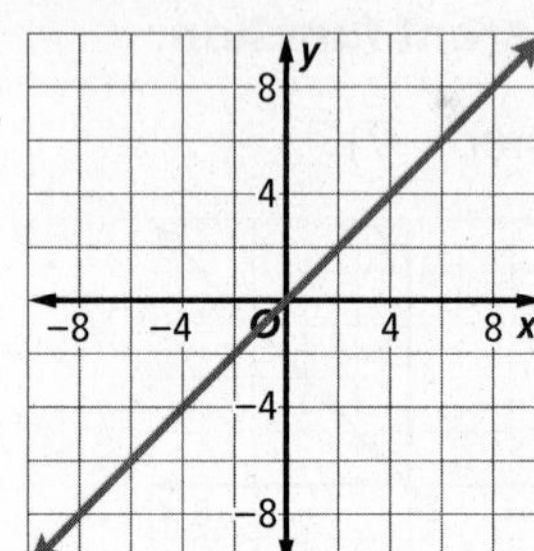

11. $g(x) = \frac{1}{3}(x)$

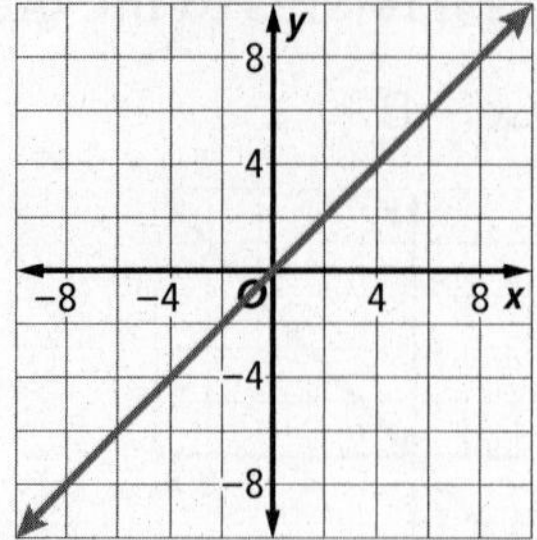

12. $g(x) = 1.5(x)$

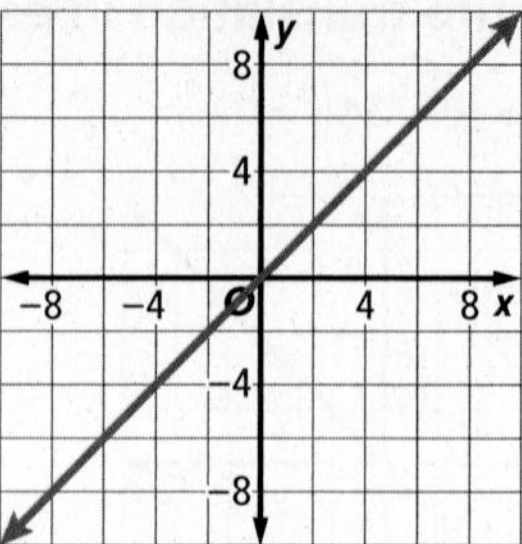

13. $g(x) = (3x)$

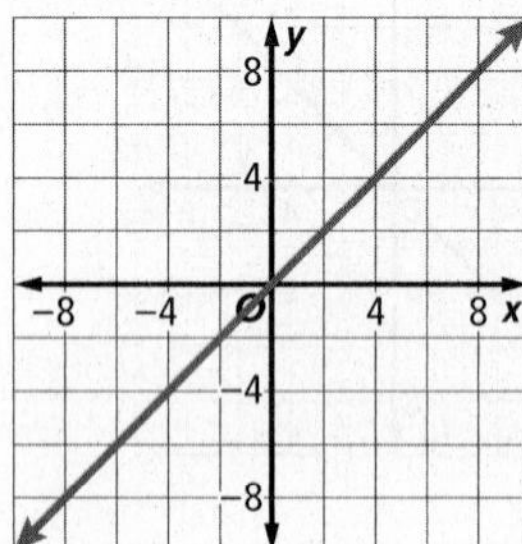

14. $g(x) = \left(\frac{3}{4}x\right)$

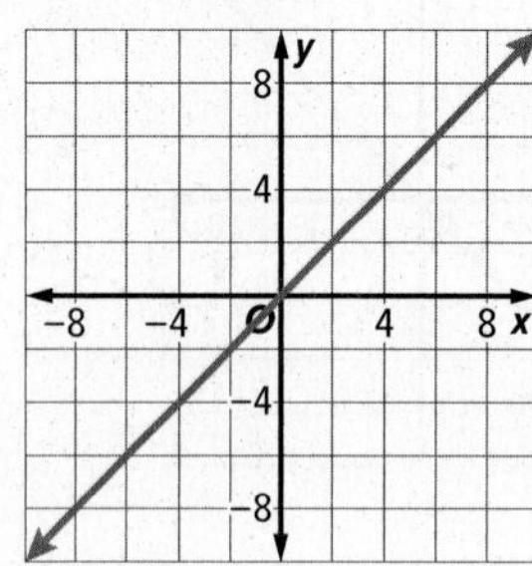

15. $g(x) = (0.4x)$

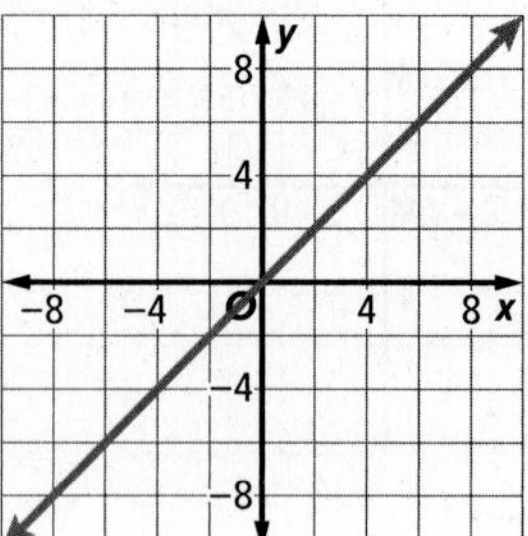

Example 7

Describe the dilation in each function as it relates to the graph of the parent function.

16. $g(x) = -4(x)$

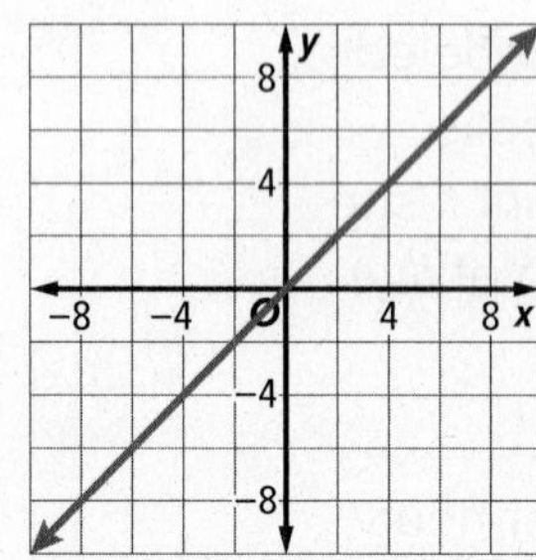

17. $g(x) = -8(x)$

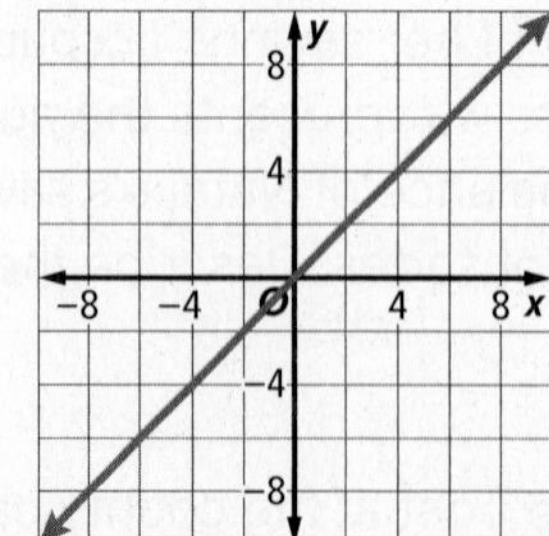

18. $g(x) = -\frac{2}{3}(x)$

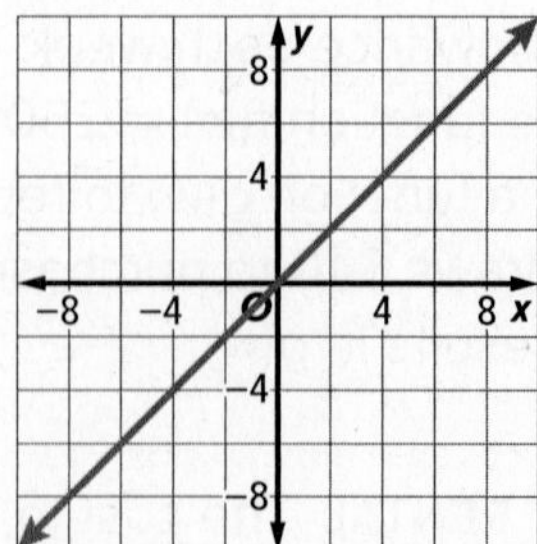

Example 8

Describe the dilation in each function as it relates to the graph of the parent function.

19. $g(x) = \left(-\frac{4}{5}x\right)$

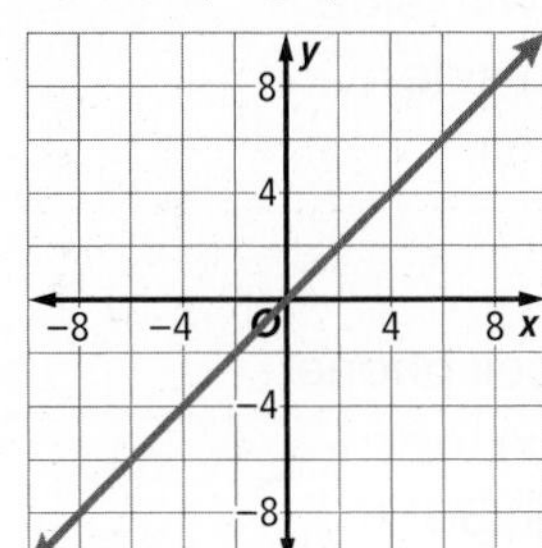

20. $g(x) = (-6x)$

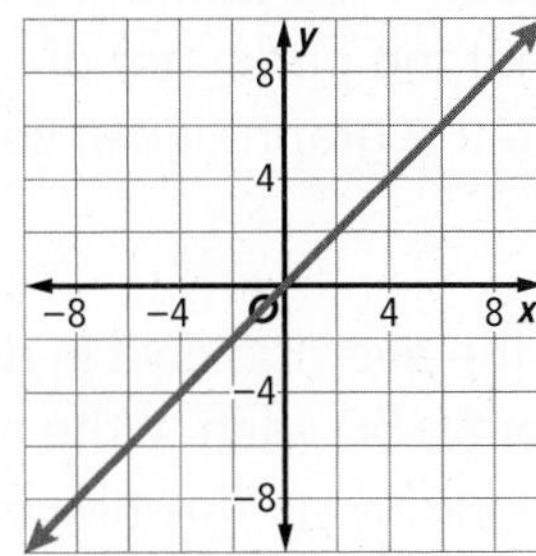

21. $g(x) = (-1.5x)$

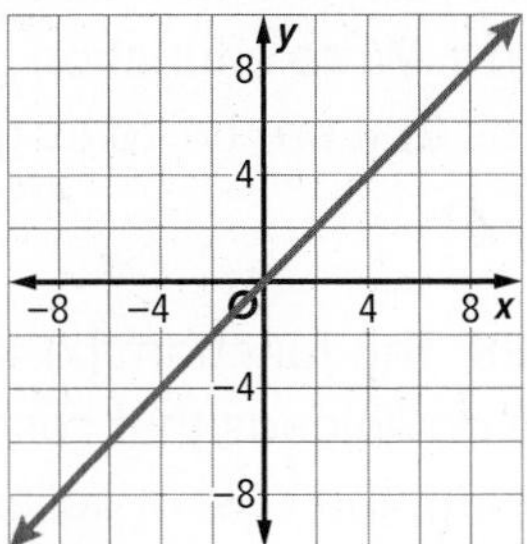

Mixed Exercises

Describe the transformation in each function as it relates to the graph of the parent function.

22. $g(x) = x + 4$

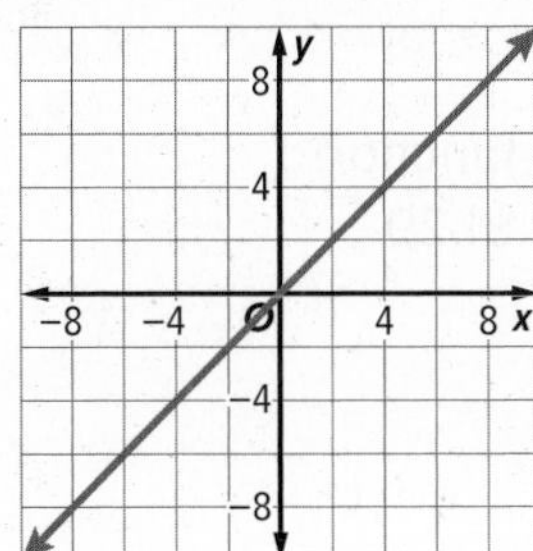

23. $g(x) = (x - 2) - 8$

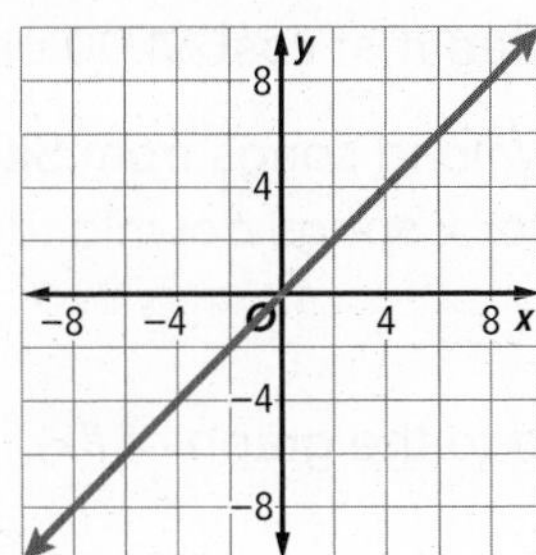

24. $g(x) = \left(-\frac{5}{8}x\right)$

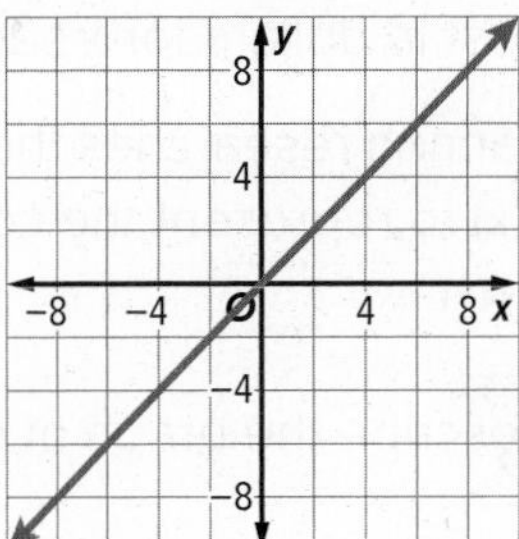

25. $g(x) = \frac{1}{5}(x)$

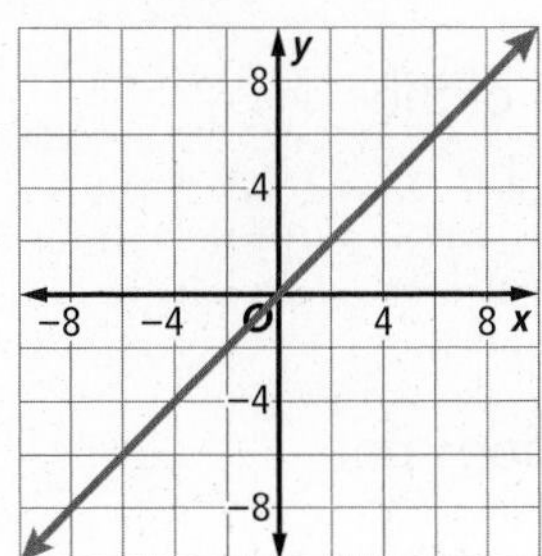

26. $g(x) = -3(x)$

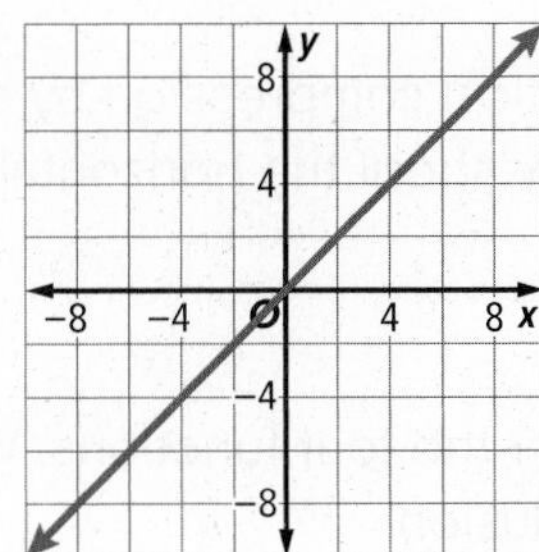

27. $g(x) = (2.5x)$

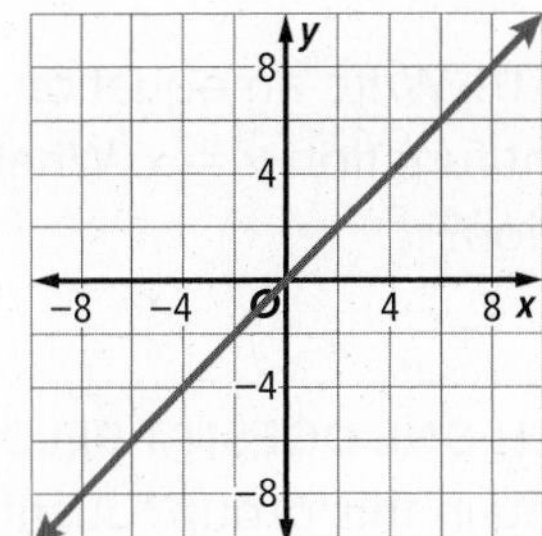

REASONING Write a function $g(x)$ to represent the translated graph.

28. $f(x) = 3x + 7$ translated 4 units up.

29. $f(x) = x - 5$ translated 2 units down.

30. PERIMETER The function $f(s) = 4s$ represents the perimeter of a square with side length s. Write a function $g(s)$ to represent the perimeter of a square with side lengths that are twice as great. Describe the graph of $g(s)$ compared to $f(s)$.

31. GAMES The function $f(x) = 0.50x$ gives the average cost in dollars for x cell phone game downloads that cost an average of \$0.50 each. Write a function $g(x)$ to represent the cost in dollars for x cell phone game downloads that cost \$1.50 each. Describe the graph of $g(x)$ compared to $f(x)$.

32. TRAINER The function $f(x) = 90x$ gives the cost of working out with a personal trainer, where \$90 is the trainer's hourly rate, and x represents the number of hours spent working out with the trainer. Describe the dilation, $g(x)$ of the function $f(x)$, if the trainer increases her hourly rate to \$100.

33. DOWNLOADS Hannah wants to download songs. She researches the price to download songs from Site F. Hannah wrote the function $f(x) = x$, which represents the cost in dollars for x songs downloaded that cost \$1.00 each.

a. Hannah researches the price to download songs from Site G. Write a function $g(x)$ to represent the cost in dollars for x songs downloaded that cost \$1.29 each.

b. Describe the graph of $g(x)$ compared to the graph of $f(x)$.

34. PERSEVERE For any linear function, replacing $f(x)$ with $f(x + k)$ results in the graph of $f(x)$ being shifted k units to the right for $k < 0$ and shifted k units to the left for $k > 0$. Does shifting the graph horizontally k units have the same effect as shifting the graph vertically $-k$ units? Justify your answer. Include graphs in your response.

35. CREATE Write an equation that is a vertical compression by a factor of a of the parent function $y = x$. What can you say about the horizontal dilation of the function?

36. WHICH ONE DOESN'T BELONG Consider the four functions. Which one does not belong in this group? Justify your conclusion.

$f(x) = 2(x + 1) - 3$	$f(x) = \frac{1}{2}x - 4$	$f(x) = -3x + 10$	$f(x) = 5(x - 7) + 3$

Arithmetic Sequences

Learn Arithmetic Sequences

A **sequence** is a set of numbers that are ordered in a specific way. Each number within a sequence is called a **term of a sequence**.

In an **arithmetic sequence**, each term after the first is found by adding a constant, the **common difference** d, to the previous term.

Words	An arithmetic sequence is a numerical pattern that increases or decreases at a constant rate called the common difference.
Examples	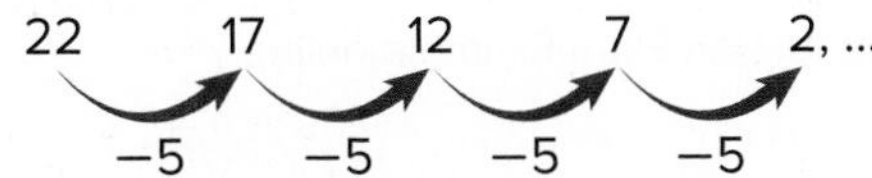 The common difference is −5. 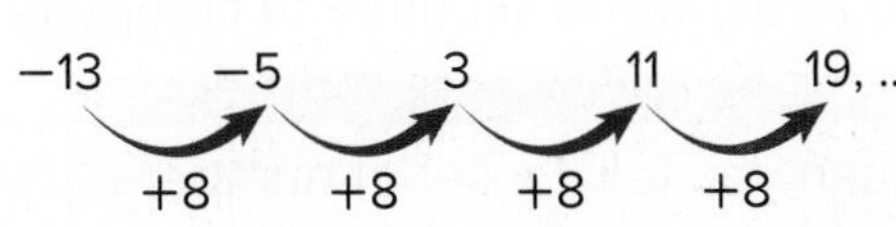 The common difference is 8.

Example 1 Identify Arithmetic Sequences

Determine whether the sequence is an arithmetic sequence. Justify your reasoning.

17, 14, 10, 7, 3

Check the difference between terms.

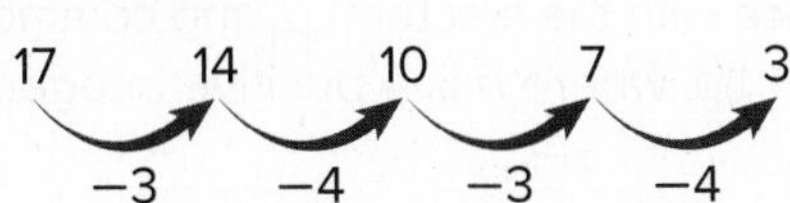

This sequence does not have a common difference between its terms. This is not an arithmetic sequence.

Check

Determine whether the sequence is an arithmetic sequence. Justify your reasoning.

82, 73, 64, 55, . . .

Go Online You can complete an Extra Example online.

Today's Goals

- Construct arithmetic sequences.
- Apply the arithmetic sequence formula.

Today's Vocabulary

sequence

term of sequence

arithmetic sequence

common difference

*n*th term of an arithmetic function

Think About It!

How are arithmetic sequences and number patterns alike and different?

Example 2 Find the Next Term

Talk About It!

Why would it be useful to develop a rule to find terms of a sequence? Explain.

Determine the next three terms in the sequence.

11, 7, 3, –1

Find the common difference between terms. -4

Add the common difference to the last term of the sequence to find the next terms.

$-1 + (-4) = -5$ $-5 + (-4) = -9$ $-9 + (-4) = -13$

Check

Determine the next three terms in the sequence.

31, 18, 5, __?__, __?__, __?__

Go Online You can complete an Extra Example online.

Explore Common Differences

Online Activity Use a real-world situation to complete the Explore.

 **INQUIRY** How can you tell if a set of numbers models a linear function?

Watch Out!

Subscripts Subscripts are used to indicate a specific term. For example, a_8 means the 8th term of the sequence. It does not mean $a \times 8$.

Learn Arithmetic Sequences as Linear Functions

Each term of an arithmetic sequence can be expressed in terms of the first term a_1 and the common difference d.

Key Concept • *n*th Term of an Arithmetic Sequence

The ***n*th term of an arithmetic sequence** with the first term a_1 and common difference d is given by $a_n = a_1 + (n - 1)d$, where n is a positive integer.

Think About It!

Why is the domain of a sequence counting numbers instead of all real numbers?

The graph of an arithmetic sequence includes points that lie along a line. Because there is a constant difference between each pair of points, the function is linear. For the equation of an arithmetic sequence,

$a_n = a_1 + (n - 1)d$

- n is the independent variable,
- a_n is the dependent variable, and
- d is the slope.

The function of an arithmetic sequence is written as $f(n) = a_1 + (n - 1)d$, where n is a counting number.

Go Online You can complete an Extra Example online.

Example 3 Find the *n*th Term

Use the arithmetic sequence −4, −1, 2, 5, . . . to complete the following.

Part A Write an equation.

$a_n = 3n - 7$

Part B Find the 16th term of the sequence.

Use the equation from Part A to find the 16th term in the arithmetic sequence.

$a_n = 3n - 7$	Equation from Part *A*
$a_{16} = 3(16) - 7$	Substitute 16 for *n*.
$a_{16} = 48 - 7$	Multiply.
$a_{16} = 41$	Simplify.

Check

RUNNING Randi has been training for a marathon, and it is important for her to keep a constant pace. She recorded her time each mile for the first several miles that she ran.

- At 1 mile, her time was 10 minutes and 30 seconds.
- At 2 miles, her time was 21 minutes.
- At 3 miles, her time was 31 minutes and 30 seconds.
- At 4 miles, her time was 42 minutes.

Part A Write a function to represent her sequence of data. Use *n* as the variable.

Part B How long will it take her to run a whole marathon? Round your answer to the nearest thousandth if necessary. (Hint: a marathon is 26.2 miles.)

Example 4 Apply Arithmetic Sequences as Linear Functions

MONEY **Laniqua opened a savings account to save for a trip to Spain. With the cost of plane tickets, food, hotel, and other expenses, she needs to save $1600. She opened the account with $525. Every month, she adds the same amount to her account using the money she earns at her after school job. From her bank statement, Laniqua can write a function that represents the balance of her savings account.**

(continued on the next page)

Go Online You can complete an Extra Example online.

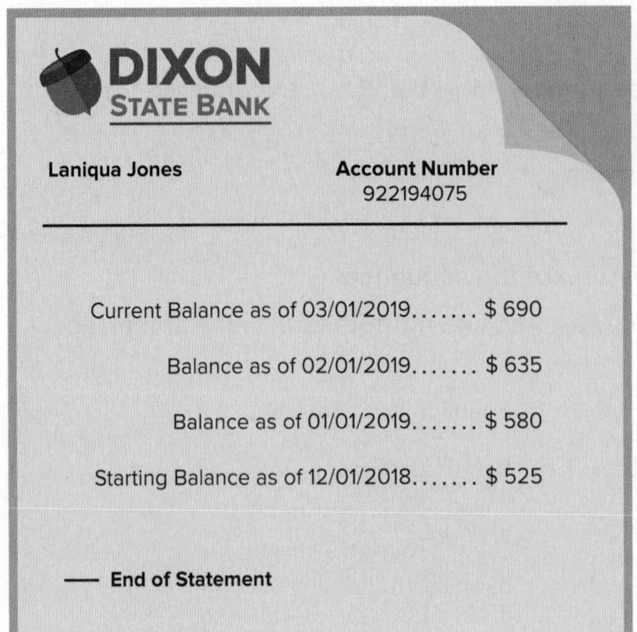
DIXON
STATE BANK

Laniqua Jones

Account Number
922194075

Current Balance as of 03/01/2019.......	$ 690
Balance as of 02/01/2019.......	$ 635
Balance as of 01/01/2019.......	$ 580
Starting Balance as of 12/01/2018.......	$ 525

— End of Statement

Part A Create a function to represent the sequence.

First, find the common difference.

525 580 635 690

+55 +55 +55

The balance after 1 month is \$580, so let $a_1 = 580$. Notice that the starting balance is \$525. You can think of this starting point as $a_0 = 525$.

$$f(n) = a_1 + (n - 1)d$$ Formula for the nth term.

$$= 580 + (n - 1)(55)$$ $a_1 = 580$ and $d = 55$

$$= 580 + 55n - 55$$ Simplify.

$$= 55n + 525$$

Part B Graph the function and determine its domain.

n	$f(n)$
0	525
1	580
2	635
3	690
4	745
5	800
6	855

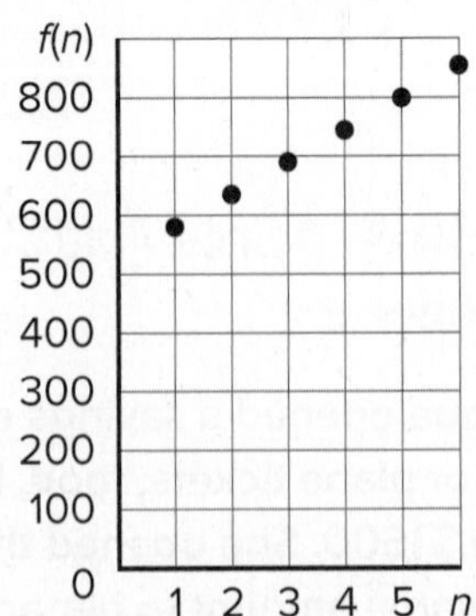

The domain is the number of months since Laniqua opened her savings account. The domain is {0, 1, 2, 3, 4, 5, ...}

Use a Source

Find the cost of a flight from the airport closest to you to Madrid, the capital of Spain. How many months would Laniqua need to save to afford the ticket?

Study Tip

Graphing You might not need to create a table of the sequence first. However, it might serve as a reminder that an arithmetic sequence is a series of points, not a line.

Go Online You can complete an Extra Example online.

Practice

Go Online You can complete your homework online.

Example 1

ARGUMENTS **Determine whether each sequence is an arithmetic sequence. Justify your reasoning.**

1. −3, 1, 5, 9, ...

2. $\frac{1}{2}, \frac{3}{4}, \frac{5}{8}, \frac{7}{16}, \ldots$

3. −10, −7, −4, 1, ...

4. −12.3, −9.7, −7.1, −4.5, ...

5. 4, 7, 9, 12, ...

6. 15, 13, 11, 9, ...

7. 7, 10, 13, 16, ...

8. −6, −5, −3, −1, ...

Example 2

Find the common difference of each arithmetic sequence. Then find the next three terms.

9. 0.02, 1.08, 2.14, 3.2, ...

10. 6, 12, 18, 24, ...

11. 21, 19, 17, 15, ...

12. $-\frac{1}{2}, 0, \frac{1}{2}, 1, \ldots$

13. $2\frac{1}{3}, 2\frac{2}{3}, 3, 3\frac{1}{3}, \ldots$

14. $\frac{7}{12}, 1\frac{1}{3}, 2\frac{1}{12}, 2\frac{5}{6}, \ldots$

15. 3, 7, 11, 15, ...

16. 22, 19.5, 17, 14.5, ...

17. −13, −11, −9, −7, ...

18. −2, −5, −8, −11, ...

Example 3

Use the given arithmetic sequence to write an equation and then find the 7th term of the sequence.

19. −3, −8, −13, −18, ...

20. −2, 3, 8, 13, ...

21. −11, −15, −19, −23, ...

22. −0.75, −0.5, −0.25, 0, ...

Example 4

23. SPORTS Wanda is the manager for the soccer team. One of her duties is to hand out cups of water at practice. Each cup of water is 4 ounces. She begins practice with a 128-ounce cooler of water.

a. Create a function to represent the arithmetic sequence.

b. Graph the function.

c. How much water is remaining after Wanda hands out the 14th cup?

24. THEATER A theater has 20 seats in the first row, 22 in the second row, 24 in the third row, and so on for 25 rows.

a. Create a function to represent the arithmetic sequence.

b. Graph the function.

c. How many seats are in the last row?

25. POSTAGE The price to send a large envelope first class mail is 88 cents for the first ounce and 17 cents for each additional ounce. The table shows the cost for weights up to 5 ounces.

Weight (ounces)	1	2	3	4	5
Postage (dollars)	0.88	1.05	1.22	1.39	1.56

Source: United States Postal Service

a. Create a function to represent the arithmetic sequence.

b. Graph the function.

c. How much did a large envelope weigh that cost $2.07 to send?

26. VIDEO DOWNLOADING Brian is downloading episodes of his favorite TV show to play on his personal media device. The cost to download 1 episode is $1.99. The cost to download 2 episodes is $3.98. The cost to download 3 episodes is $5.97.

a. Create a function to represent the arithmetic sequence.

b. Graph the function.

c. What is the cost to download 9 episodes?

27. USE A MODEL Chapa is beginning an exercise program that calls for 30 push-ups each day for the first week. Each week thereafter, she has to increase her push-ups by 2.

a. Write a function to represent the arithmetic sequence.

b. Graph the function.

c. Which week of her program will be the first one in which she will do at least 50 push-ups a day?

Mixed Exercises

CONSTRUCT ARGUMENTS **Determine whether each sequence is an arithmetic sequence. Justify your argument.**

28. −9, −12, −15, −18, ...

29. 10, 15, 25, 40, ...

30. −10, −5, 0, 5, ...

31. −5, −3, −1, 1, ...

Write an equation for the *n*th term of each arithmetic sequence. Then graph the first five terms of the sequence.

32. 7, 13, 19, 25, ...

33. 30, 26, 22, 18, ...

34. −7, −4, −1, 2, ...

35. SAVINGS Fabiana decides to save the money she's earning from her after-school job for college. She makes an initial contribution of $3000 and each month deposits an additional $500. After one month, she will have contributed $3500.

a. Write an equation for the *n*th term of the sequence.

b. How much money will Fabiana have contributed after 24 months?

36. NUMBER THEORY One of the most famous sequences in mathematics is the Fibonacci sequence. It is named after Leonardo de Pisa (1170–1250) or Filius Bonacci, alias Leonardo Fibonacci. The first several numbers in the Fibonacci sequence are shown.

1, 1, 2, 3, 5, 8, 13, 21, 34, 55, 89, . . .

Does this represent an arithmetic sequence? Why or why not?

37. STRUCTURE Use the arithmetic sequence 2, 5, 8, 11, ...

a. Write an equation for the *n*th term of the sequence.

b. What is the 20th term in the sequence?

Higher-Order Thinking Skills

38. CREATE Write a sequence that is an arithmetic sequence. State the common difference, and find a_6.

39. CREATE Write a sequence that is not an arithmetic sequence. Determine whether the sequence has a pattern, and if so describe the pattern.

40. REASONING Determine if the sequence 1, 1, 1, 1, . . . is an arithmetic sequence. Explain your reasoning.

41. CREATE Create an arithmetic sequence with a common difference of −10.

42. PERSEVERE Find the value of x that makes $x + 8$, $4x + 6$, and $3x$ the first three terms of an arithmetic sequence.

43. CREATE For each arithmetic sequence described, write a formula for the nth term of a sequence that satisfies the description.

a. first term is negative, common difference is negative

b. second term is −5, common difference is 7

c. $a_2 = 8$, $a_3 = 6$

Andre and Sam are both reading the same novel. Andre reads 30 pages each day. Sam created the table at the right. Refer to this information for Exercises 44–46.

Sam's Reading Progress	
Day	Pages Left to Read
1	430
2	410
3	390
4	370

44. ANALYZE Write arithmetic sequences to represent each boy's daily progress. Then write the function for the nth term of each sequence.

45. PERSEVERE Enter both functions from Exercise 44 into your calculator. Use the table to determine if there is a day when the number of pages Andre has read is equal to the number of pages Sam has left to read. If so, which day is it? Explain how you used the table feature to help you solve the problem.

46. ANALYZE Graph both functions on your calculator, then sketch the graph. How can you use the graph to answer the question from Exercise 45?

Piecewise and Step Functions

Learn Graphing Piecewise-Defined Functions

Some functions cannot be described by a single expression because they are defined differently depending on the interval of x. These functions are **piecewise-defined functions.** A **piecewise-linear function** has a graph that is composed of some number of linear pieces.

Example 1 Graph a Piecewise-Defined Function

To graph a piecewise-defined function, graph each "piece" separately.

Graph $f(x) = \begin{cases} 2x + 4 \text{ if } x \leq 1 \\ -x + 3 \text{ if } x > 1 \end{cases}$. State the domain and range.

First, graph $f(x) = 2x + 4$ if $x \leq 1$.

- Create a table for $f(x) = 2x + 4$ using values of $x \leq 1$.
- Because x is *less than or equal to* 1, place a dot at (1, 6) to indicate that the endpoint is included in the graph.
- Then, plot the points and draw the graph beginning at (1, 6).

x	y
1	6
0	4
−1	2
−2	0
−3	−2

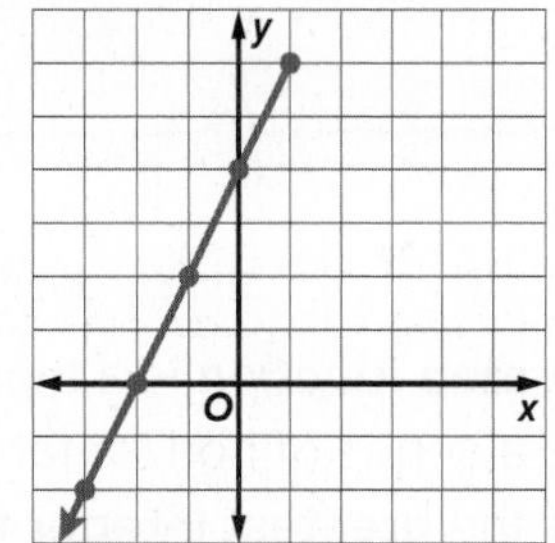

Next, graph $f(x) = -x + 3$ if $x > 1$.

- Create a table for $f(x) = -x + 3$ using values of $x > 1$.
- Because x is *greater than but not equal to* 1, place a circle at (1, 2) to indicate that the endpoint is not included in the graph.
- Then, plot the points and draw the graph beginning at (1, 2).

x	y
1	2
2	1
3	0
4	−1
5	−2

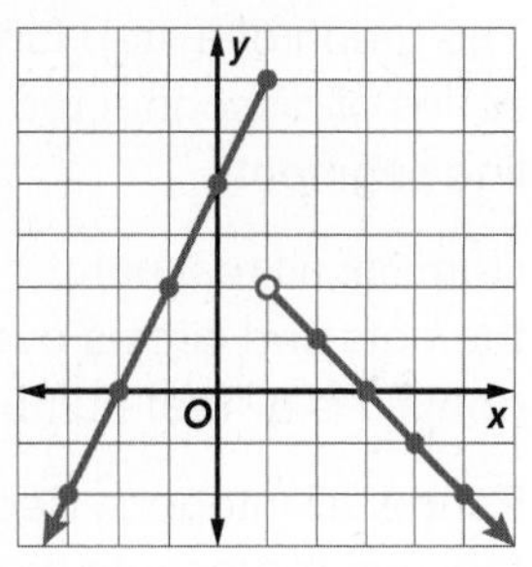

The domain is all real numbers. The range is $y \leq 6$.

Go Online You can complete an Extra Example online.

Today's Goals

- Identify and graph piecewise-defined functions.
- Identify and graph step functions.

Today's Vocabulary

piecewise-defined function
piecewise-linear function
step function
greatest integer function

Think About It!

What would be an advantage of graphing the entire expression and removing the portion that is not in the interval?

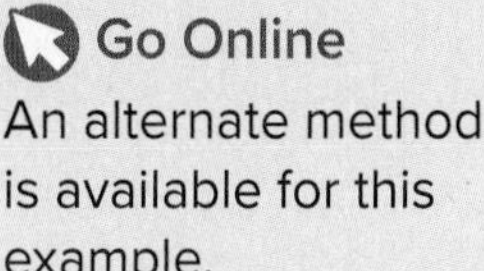

Go Online

An alternate method is available for this example.

Check

Part A Graph $f(x) = \begin{cases} -x + 1 \text{ if } x \leq -2 \\ -3x - 2 \text{ if } x > -2 \end{cases}$.

Part B Find the domain and range of the function.

Watch Out!

Circles and Dots Do not forget to examine the endpoint(s) of each piece to determine whether there should be a circle or a dot. $>$ and $<$ mean that a circle should be used, while $\geq$ and $\leq$ mean that a dot should be used.

Study Tip

Piecewise-Defined Functions When graphing piecewise-defined functions, there should be a dot or line that contains each member of the domain.

Go Online

You can watch a video to see how to graph a piecewise-defined function on a graphing calculator

Explore Age as a Function

Online Activity Use a real-world situation to complete an Explore.

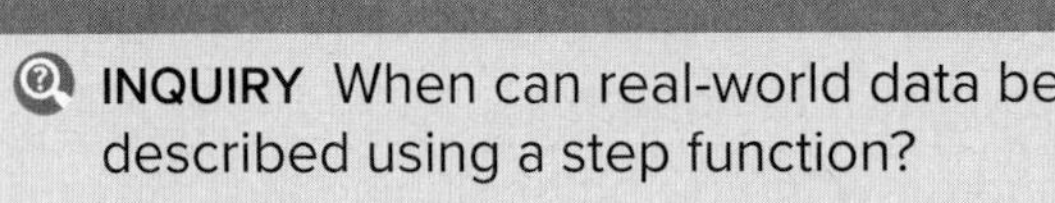

INQUIRY When can real-world data be described using a step function?

Learn Graphing Step Functions

A **step function** is a type of piecewise-linear function with a graph that is a series of horizontal line segments. One example of a step function is the **greatest integer function,** written as $f(x) = [\![x]\!]$ in which $f(x)$ is the greatest integer less than or equal to x.

Key Concept • Greatest Integer Function	
Type of graph: disjointed line segments The graph of a step function is a series of disconnected horizontal line segments. Domain: all real numbers; Because the dots and circles overlap, the domain is all real numbers. Range: all integers; Because the function represents the greatest integer less than or equal to x, the range is all integers.	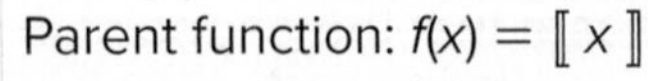 Parent function: $f(x) = [\![x]\!]$ 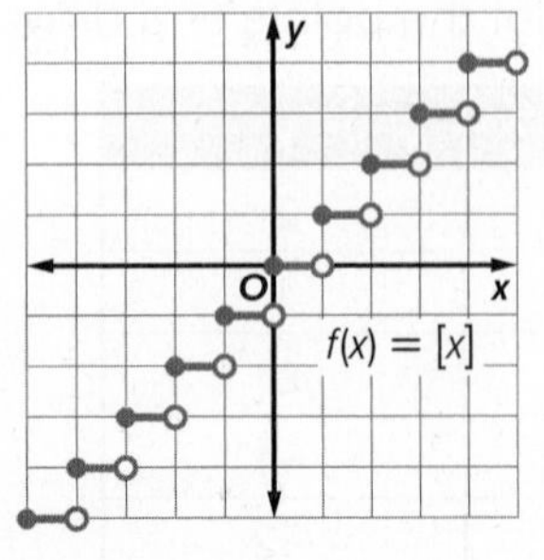

Go Online You can complete an Extra Example online.

Example 2 Graph a Greatest Integer Function

Graph $f(x) = [\![x + 1]\!]$. State the domain and range.

First, make a table. Select a few values that are between integers.

x	$x+1$	$[\![x+1]\!]$	
−2	−1	−1	−1, −0.75, and 0.25 are greater than or equal to −1 but less than 0. So, −1 is the greatest integer that is not greater than −1, −0.75, or 0.25.
−1.75	−0.75	−1	
−1.25	−0.25	−1	
−1	0	0	0, 0.5, and 0.75 are greater than or equal to 0 but less than 1. So, 0 is the greatest integer that is not greater than 0, 0.5, or 0.75.
−0.5	0.5	0	
−0.25	0.75	0	
0	1	1	1, 1.25, and 1.5 are greater than or equal to 1 but less than 2. So, 1 is the greatest integer that is not greater than 1, 1.25, or 1.5.
0.25	1.25	1	
0.5	1.5	1	
1	2	2	2, 2.25, and 2.75 are greater than or equal to 2 but less than 3. So, 2 is the greatest integer that is not greater than 2, 2.25, or 2.75.
1.25	2.25	2	
1.75	2.75	2	

On the graph, dots represent included points. Circles represent points that are not included. The domain is all real numbers. The range is all integers. Note that this is the graph of $f(x) = [\![x]\!]$ shifted 1 unit to the left.

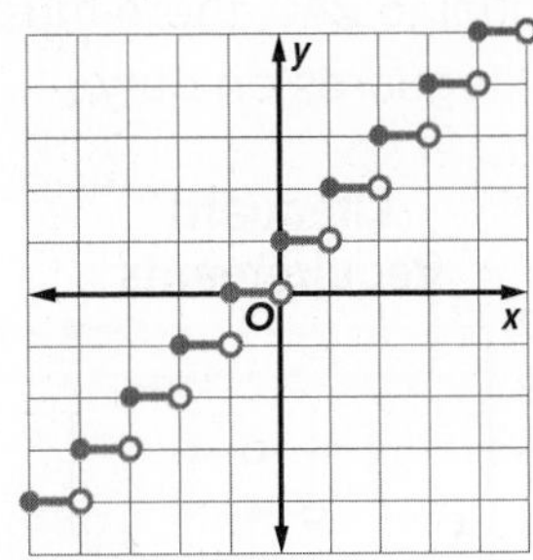

Check

Graph $f(x) = [\![x - 2]\!]$ by making a table. Copy and complete the table. Then graph the function.

x	$x-2$	$[\![x-2]\!]$
−1	−3	
−0.75	−2.75	
−0.25	−2.25	−3
0	−2	−2
0.25	−1.75	−2
0.5	−1.5	
1	−1	
1.25	−0.75	−1
1.5	−0.5	
2	0	
2.25	0.25	

Go Online You can complete an Extra Example online.

Talk About It!

What do you notice about the symmetry, extrema, and end behavior of the function?

Watch Out!

Greatest Integer Function When finding the value of a greatest integer function, do not round to the nearest integer. Instead, always round nonintegers down to the greatest integer that is not greater than the number.

Example 3 Graph a Step Function

SAFETY A state requires a ratio of 1 lifeguard to 60 swimmers in a swimming pool. This means that 1 lifeguard can watch up to and including 60 swimmers. Make a table and draw a graph that shows the number of lifeguards that must be on duty $f(x)$ based on the number of swimmers in the pool x.

The number of lifeguards that must be on duty can be represented by a step function.

- If the number of swimmers is greater than 0 but fewer than or equal to 60, only 1 lifeguard must be on duty.
- If the number of swimmers is greater than 60 but fewer than or equal to 120, there must be 2 lifeguards on duty.
- If the number of swimmers is greater than 180 but fewer than or equal to 240, there must be 4 lifeguards on duty.

x	$f(x)$
$0 < x \leq 60$	1
$60 < x \leq 120$	2
$120 < x \leq 180$	3
$180 < x \leq 240$	4
$240 < x \leq 300$	5
$300 < x \leq 360$	6
$360 < x \leq 420$	7

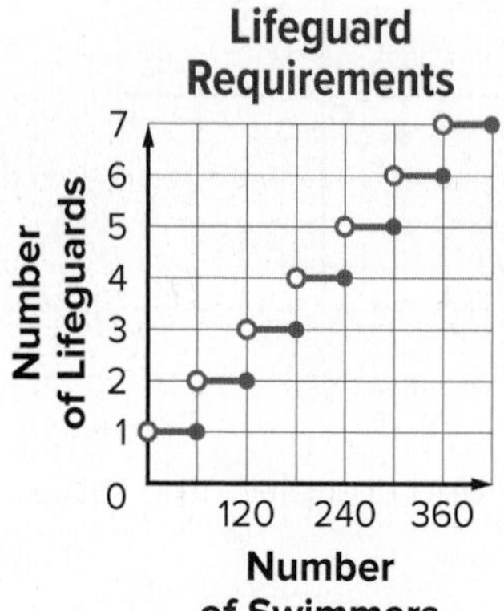

The circles mean that when there are more than a multiple of 60 swimmers, another lifeguard is required.

The dots represent the maximum number of swimmers that can be in the pool for that particular number of lifeguards on duty.

Think About It!

How would the graph change if 1 certified lifeguard could watch up to 59 swimmers? For example, if there are greater than or equal to 60, but fewer than 120 swimmers, there must be 2 lifeguards on duty.

Math History Minute

Oliver Heaviside (1850–1925) was a self-taught electrical engineer, mathematician, and physicist who laid much of the groundwork for telecommunications in the 21st century. Heaviside invented the Heaviside step function,

$$f(x) = \begin{cases} 0 & \text{if } x < 0 \\ \frac{1}{2} & \text{if } x = 0, \\ 1 & \text{if } x > 0 \end{cases}$$

which he used to model the current in an electric circuit.

Check

PETS At Luciana's pet boarding facility, it costs $35 per day to board a dog. Every fraction of a day is rounded up to the next day. Copy and complete the table. Then graph the function.

Days	Cost ($)
$0 < x \leq 1$	
$1 < x \leq 2$	
$2 < x \leq 3$	
$3 < x \leq 4$	
$4 < x \leq 5$	
$5 < x \leq 6$	

Go Online You can complete an Extra Example online.

Practice

Go Online You can complete your homework online.

Example 1

Graph each function. State the domain and range.

1. $f(x) = \begin{cases} \frac{1}{2}x - 1 \text{ if } x > 3 \\ -2x + 3 \text{ if } x \leq 3 \end{cases}$

2. $f(x) = \begin{cases} 2x - 5 \text{ if } x > 1 \\ 4x - 3 \text{ if } x \leq 1 \end{cases}$

3. $f(x) = \begin{cases} 2x + 3 \text{ if } x \geq -3 \\ -\frac{1}{3}x + 1 \text{ if } x < -3 \end{cases}$

4. $f(x) = \begin{cases} 3x + 4 \text{ if } x \geq 1 \\ x + 3 \text{ if } x < 1 \end{cases}$

5. $f(x) = \begin{cases} 3x + 2 \text{ if } x > -1 \\ -\frac{1}{2}x - 3 \text{ if } x \leq -1 \end{cases}$

6. $f(x) = \begin{cases} 2x + 1 \text{ if } x < -2 \\ -3x - 1 \text{ if } x \geq -2 \end{cases}$

Example 2

Graph each function. State the domain and range.

7. $f(x) = 3 [\![x]\!]$

8. $f(x) = [\![-x]\!]$

9. $g(x) = -2 [\![x]\!]$

10. $g(x) = [\![x]\!] + 3$

11. $h(x) = [\![x]\!] - 1$

12. $h(x) = \frac{1}{2}[\![x]\!] + 1$

Example 3

13. BABYSITTING Ariel charges $8 per hour as a babysitter. She rounds every fraction of an hour up to the next half-hour. Draw a graph to represent Ariel's total earnings y after x hours.

14 FUNDRAISING Students are selling boxes of cookies at a fundraiser. The boxes of cookies can only be ordered by the case, with 12 boxes per case. Draw a graph to represent the number of cases needed y when x boxes of cookies are sold.

Mixed Exercises

15. PRECISION A package delivery service determines rates for express shipping by the weight of a package, with every fraction of a pound rounded up to the next pound. The table shows the cost of express shipping packages that weigh no more than 5 pounds. Write a piecewise-linear function representing the cost to ship a package that weighs no more than 5 pounds. State the domain and range.

Weight (pounds)	Rate (dollars)
1	16.20
2	19.30
3	22.40
4	25.50
5	28.60

16. EARNINGS Kelly works in a hospital as a medical assistant. She earns $8 per hour the first 8 hours she works in a day and $11.50 per hour each hour thereafter.

   a. Organize the information into a table. Include a row for hours worked x, and a row for daily earnings $f(x)$.

   b. Write the piecewise equation describing Kelly's daily earnings $f(x)$ for x hours.

   c. Draw a graph to represent Kelly's daily earnings.

17. REASONING Write a piecewise function that represents the graph.

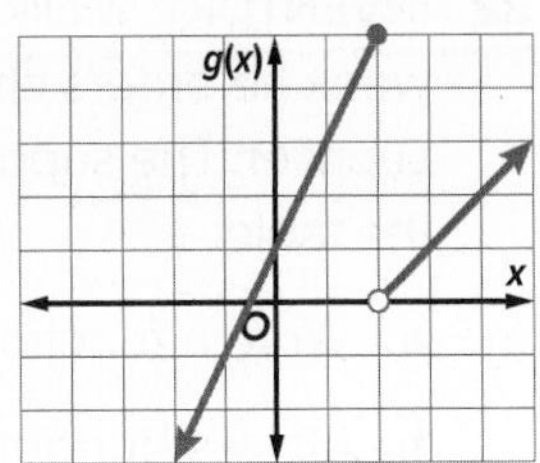

18. STRUCTURE Suppose $f(x) = 2[\![x - 1]\!]$.

a. Find $f(1.5)$.

b. Find $f(2.2)$.

c. Find $f(9.7)$.

d. Find $f(-1.25)$.

19. RENTAL CARS Mr. Aronsohn wants to rent a car on vacation. The rate the car rental company charges is $19 per day. If any fraction of a day is counted as a whole day, how much would it cost for Mr. Aronsohn to rent a car for 6.4 days?

20. USE A MODEL A roadside fruit and vegetable stand determines rates for selling produce, with every fraction of a pound rounded up to the next pound. The table shows the cost of tomatoes by weight in pounds.

a. Write a piecewise-linear function representing the cost of purchasing tomatoes that weigh no more than 5 pounds, where C is the cost in dollars and p is the number of pounds.

Weight (pounds)	Rate (dollars)
1	3.50
2	7.00
3	10.50
4	14.00
5	17.50

b. Graph the function.

c. State the domain and range.

d. What would be the cost of purchasing 8.3 pounds of tomatoes at the roadside stand?

21. ELECTRONIC REPAIRS Tech Repairs charges $25 for an electronic device repair that takes up to one hour. For each additional hour of labor, there is a charge of $50. The repair shop charges for the next full hour for any part of an hour.

a. Copy and complete the table to organize the information. Include a row for hours of repair x, and a row for total cost $f(x)$.

x	0	2	4	6	8
$f(x)$					

b. Write a step function to represent the total cost for every hour h of repair.

c. Graph the function.

d. Devesh was charged $125 to repair his tablet. How long did the repair take to complete?

22. INVENTORY Malik owns a bakery. Every week he orders chocolate chips from a supplier. The supplier's pricing is shown in the table.

Chocolate Chip Pricing	
$4 per pound	Up to 3 pounds
$1.50 per pound	For each pound over 3 pounds

a. Write a function to represent the cost of chocolate chips.

b. Malik's budget for chocolate chips for the week is $25. How many whole pounds of chocolate chips can he order?

Higher-Order Thinking Skills

23. CREATE Write a piecewise-defined function with three linear pieces. Then graph the function.

24. FIND THE ERROR Amy graphed a function that gives the height of a car on a roller coaster as a function of time. She said her graph is the graph of a step function. Is this possible? Explain your reasoning.

25. WRITE What is the difference between a step function and a piecewise-defined function?

26. ANALYZE Does the piecewise relation $y = \begin{cases} -2x + 4 \text{ if } x \geq 2 \\ -\frac{1}{2}x - 1 \text{ if } x \leq 4 \end{cases}$ represent a function? Justify your argument.

ANALYZE **Refer to the graph for Exercises 27–31.**

27. Write a piecewise function to represent the graph.

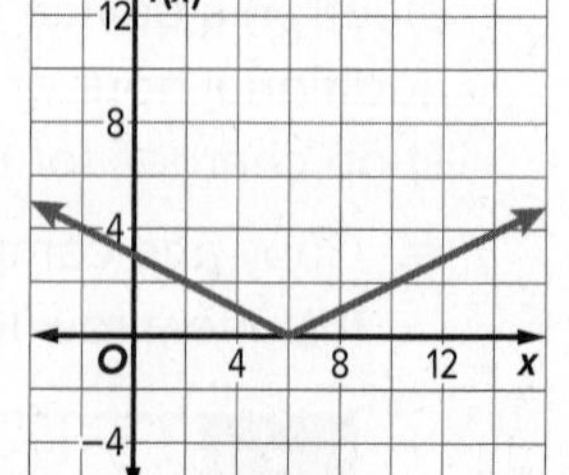

28. What is the domain?

29. What is the range?

30. Find $f(8.5)$.

31. Find $f(1.2)$.

Absolute Value Functions

Explore Parameters of an Absolute Value Function

Online Activity Use graphing technology to complete the Explore.

INQUIRY How does performing an operation on an absolute value function change its graph?

Learn Graphing Absolute Value Functions

The **absolute value function** is a type of piecewise-linear function. An absolute value function is written as $f(x) = a|x - h| + k$, where a, h, and k are constants and $f(x) \geq 0$ for all values of x.

The **vertex** is either the lowest point or the highest point of a function. For the parent function, $y = |x|$, the vertex is at the origin.

Key Concept • Absolute Value Function	
Parent Function	$f(x) = |x|$, defined as $f(x) = \begin{cases} x \text{ if } x \geq 0 \\ -x \text{ if } x < 0 \end{cases}$
Type of Graph	V-Shaped
Domain:	all real numbers
Range:	all nonnegative real numbers

Learn Translations of Absolute Value Functions

Key Concept • Vertical Translations of Absolute Value Functions

If $k > 0$, the graph of $f(x) = |x|$ is translated k units up.
If $k < 0$, the graph of $f(x) = |x|$ is translated $|k|$ units down.

Key Concept • Horizontal Translations of Linear Functions

If $h > 0$, the graph of $f(x) = |x|$ is translated h units right.
If $h < 0$, the graph of $f(x) = |x|$ is translated $|h|$ units left.

Example 1 Vertical Translations of Absolute Value Functions

Describe the translation in $g(x) = |x| - 3$ as it relates to the graph of the parent function.

Graph the parent function, $f(x) = |x|$, for absolute values.

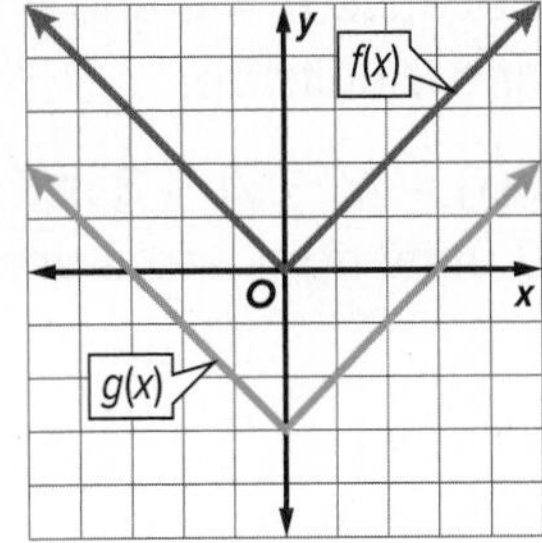

The constant, k, is outside the absolute value signs, so k affects the y-values. The graph will be a vertical translation.

(continued on the next page)

Today's Goals
- Graph absolute value functions.
- Apply translations to absolute value functions.
- Apply dilations to absolute value functions.
- Apply reflections to absolute value functions.
- Interpret constants within equations of absolute value functions.

Today's Vocabulary
absolute value function

vertex

Think About It!
Why does adding a positive value of k shift the graph k units up?

Go Online
You can watch a video to see how to describe translations of functions.

Study Tip
Horizontal Shifts Remember that the general form of an absolute value function is $y = a|x - h| + k$. So, $y = |x + 7|$ is actually $y = |x - (-7)|$ in the function's general form.

Since $f(x) = |x|$, $g(x) = f(x) + k$ where $k = -3$.
$g(x) = |x| - 3 \longrightarrow g(x) = f(x) + (-3)$

The value of k is less than 0, so the graph will be translated $|k|$ units down, or 3 units down.

$g(x) = |x| - 3$ is a translation of the graph of the parent function 3 units down.

Example 2 Horizontal Translations of Absolute Value Functions

Describe the translation in $j(x) = |x - 4|$ as it relates to the parent function.

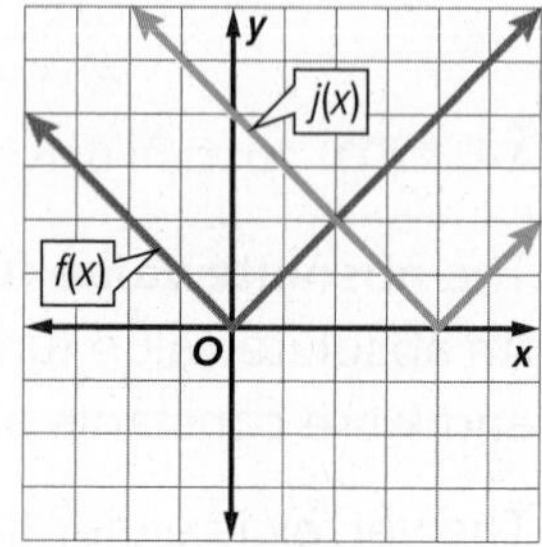

Graph the parent function, $f(x) = |x|$, for absolute values.

The constant, h, is inside the absolute value signs, so h affects the input or, x-values. The graph will be a horizontal translation.

Since $f(x) = |x|$, $j(x) = f(x - h)$, where $h = 4$.
$j(x) = |x - 4| \longrightarrow j(x) = f(x - 4)$

The value of h is greater than 0, so the graph will be translated h units right, or 4 units right.

$j(x) = |x - 4|$ is the translation of the graph of the parent function 4 units right.

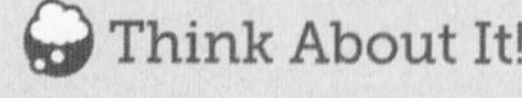

Think About It!

Since the vertex of the parent function is at the origin, what is a quick way to determine where the vertex is of $q(x) = |x - h| + k$?

Example 3 Multiple Translations of Absolute Value Functions

Describe the translation in $g(x) = |x - 2| + 3$ as it relates to the graph of the parent function.

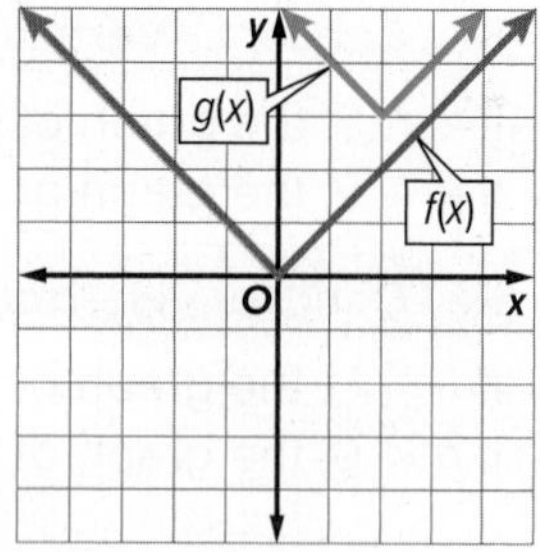

The equation has both h and k values. The input and output will be affected by the constants. The graph of $f(x) = |x|$ is vertically and horizontally translated.

Since $f(x) = |x|$, $g(x) = f(x - h) + k$ where $h = 2$ and $k = 3$.

Because $h = 2$ and $k = 3$, the graph is translated 2 units right and 3 units up.

$g(x) = |x - 2| + 3$ is the translation of the graph of the parent function 2 units right and 3 units up.

Emilio says that the graph of $g(x) = |x + 1| - 1$ is the same graph as $f(x) = |x|$. Is he correct? Why or why not?

Go Online You can complete an Extra Example online.

Example 4 Identify Absolute Value Functions from Graphs

Use the graph of the function to write its equation.

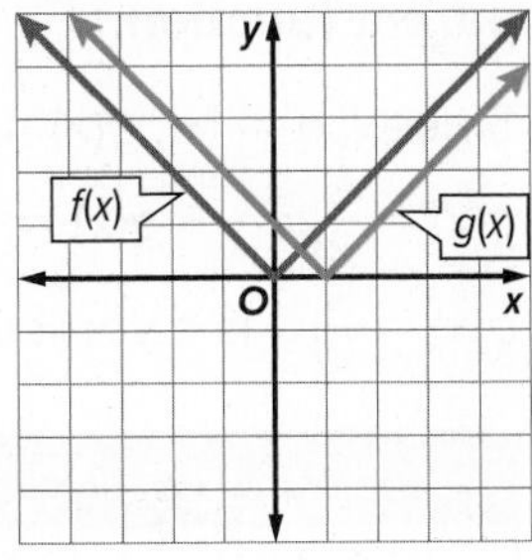

The graph is the translation of the parent graph 1 unit to the right.

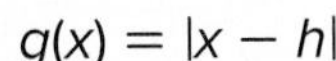

$g(x) = \lvert x - h \rvert$	General equation for a horizontal translation
$g(x) = \lvert x - 1 \rvert$	The vertex is 1 unit to the right of the origin.

Example 5 Identify Absolute Value Functions from Graphs (Multiple Translations)

Use the graph of the function to write its equation.

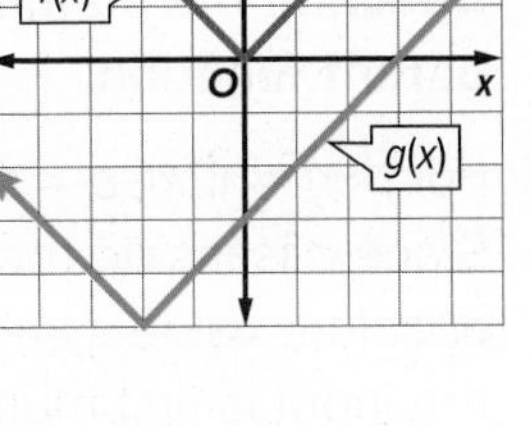

The graph is a translation of the parent graph 2 units to the left and 5 units down.

$g(x) = \lvert x - h \rvert + k$	General equation for a translations
$g(x) = \lvert x - (-2) \rvert + k$	The vertex is 2 units left of the origin.
$g(x) = \lvert x - (-2) \rvert + (-5)$	The vertex is 5 units down from the origin.
$g(x) = \lvert x + 2 \rvert - 5$	Simplify.

Learn Dilations of Absolute Value Functions

Multiplying by a constant a after evaluating an absolute value function creates a vertical change, either a stretch or compression.

Key Concept • Vertical Dilations of Absolute Value Functions

If $|a| > 1$, the graph of $f(x) = |x|$ is stretched vertically.
If $0 < |a| < 1$, the graph of $f(x) = |x|$ is compressed vertically.

When an input is multiplied by a constant a before for the absolute value is evaluated, a horizontal change occurs.

Key Concept • Horizontal Dilations of Absolute Value Functions

If $|a| > 1$, the graph of $f(x) = |x|$ is compressed horizontally.
If $0 < |a| < 1$, the graph of $f(x) = |x|$ is stretched horizontally.

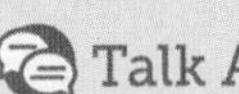

Talk About It!

How is the value of a in an absolute value function related to slope? Explain.

Go Online You can complete an Extra Example online.

Example 6 Dilations of the Form $a|x|$ When $a > 1$

Describe the dilation in $g(x) = \frac{5}{2}|x|$ as it relates to the graph of the parent function.

Since $f(x) = |x|$, $g(x) = a \cdot f(x)$, where $a = \frac{5}{2}$.

$g(x) = \frac{5}{2}|x| \longrightarrow g(x) = \frac{5}{2} \cdot f(x)$

$g(x) = \frac{5}{2}|x|$ is a vertical stretch of the graph of the parent graph.

x	$\|x\|$	$\frac{5}{2}\|x\|$	$(x, g(x))$
−4	$\|-4\| = 4$	10	(−4, 10)
−2	$\|-2\| = 2$	5	(−2, 5)
0	$\|0\| = 0$	0	(0, 0)
2	$\|2\| = 2$	5	(2, 5)
4	$\|4\| = 4$	10	(4, 10)

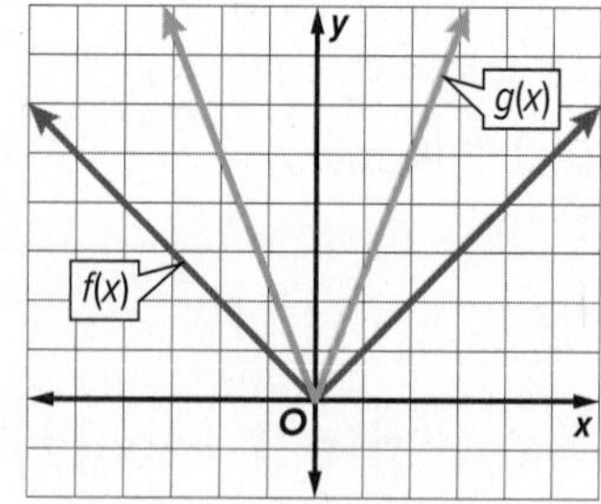

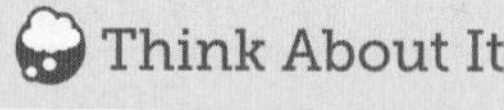

Go Online
You can watch a video to see how to describe dilations of functions.

Think About It!
How are $a|x|$ and $|ax|$ evaluated differently?

Example 7 Dilations of the Form $|ax|$

Describe the dilation in $p(x) = |2x|$ as it relates to the graph of the parent function.

For $p(x) = |2x|$, $a = 2$.
Since a is inside the absolute value symbols, the input is first multiplied by a. Then, the absolute value of ax is evaluated.

x	$\|2x\|$	$p(x)$	$(x, p(x))$
−4	$\|2(-4)\| = \|-8\|$	8	(−4, 8)
−2	$\|2(-2)\| = \|-4\|$	4	(−2, 4)
0	$\|2(0)\| = \|0\|$	0	(0, 0)
2	$\|2(2)\| = \|4\|$	4	(2, 4)
4	$\|2(4)\| = \|8\|$	8	(4, 8)

Plot the points from the table.

Since $f(x) = |x|$, $p(x) = f(ax)$ where $a = 2$.

$p(x) = |2x| \rightarrow p(x) = f(2x)$

$p(x) = |2x|$ is a horizontal compression of the graph of the parent graph.

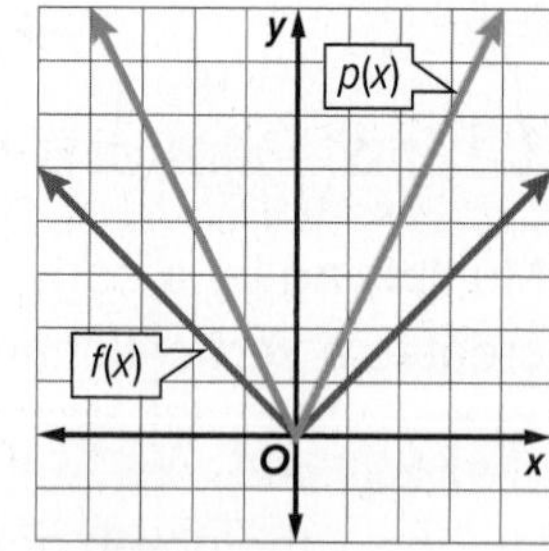

Watch Out!

Differences in Dilations
Although $a|x|$ and $|ax|$ appear to have the same effect on a function, they are evaluated differently and that difference is more apparent when a function is dilated and translated horizontally. For a function with multiple transformations, it is best to first create a table.

Check

Tell whether each equation is an example of a *vertical stretch, vertical compression, horizontal stretch, or horizontal compression.*

A. $j(x) = \left|\frac{4}{3}x\right|$

B. $q(x) = \left|\frac{1}{5}x\right|$

C. $p(x) = 6|x|$

D. $g(x) = \frac{5}{7}|x|$

Go Online You can complete an Extra Example online.

Example 8 Dilations When $0 < a < 1$

Describe how the graph of $j(x) = \frac{1}{3}|x|$ as it relates to the graph of the parent function.

For $j(x) = \frac{1}{3}|x|$, $a = \frac{1}{3}$. Because a is outside the absolute value signs, the absolute value of the input is evaluated first. Then, the function is multiplied by a.

x	$\lvert x\rvert$	$\frac{1}{3}\lvert x\rvert$	$(x, j(x))$
−6	$\lvert -6\rvert = 6$	2	(−6, 2)
−3	$\lvert -3\rvert = 3$	1	(−3, 1)
0	$\lvert 0\rvert = 0$	0	(0, 0)
3	$\lvert 3\rvert = 3$	1	(3, 1)
6	$\lvert 6\rvert = 6$	2	(6, 2)

Plot the points from the table.

Because $f(x) = |x|$, $j(x) = a \cdot f(x)$ where $a = \frac{1}{3}$.

$j(x) = \frac{1}{3}|x| \rightarrow j(x) = \frac{1}{3} \cdot f(x)$

$j(x) = \frac{1}{3}|x|$ is a vertical compression of the graph of the parent function.

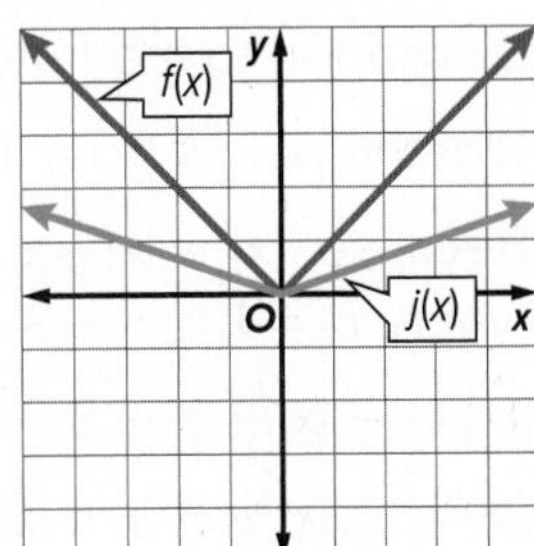

Check

Write an equation for each graph shown.

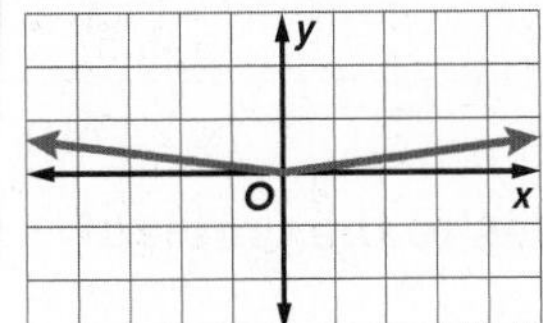

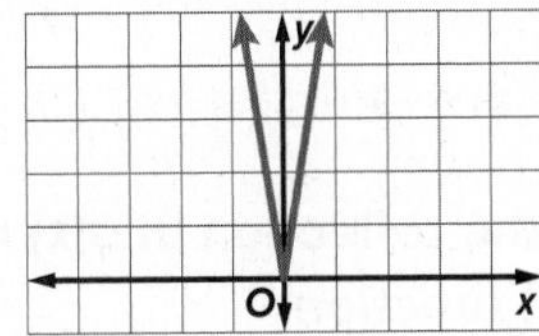

Learn Reflections of Absolute Value Functions

The graph of $-a|x|$ appears to be flipped upside down compared to $a|x|$, and they are symmetric about the x-axis.

Key Concept • Reflections of Absolute Value Functions Across the x-axis

The graph of $-af(x)$ is the reflection of the graph of $af(x) = a|x|$ across the x-axis.

When the only transformation occurring is a reflection or a dilation and reflection, the graphs of $f(ax)$ and $f(-ax)$ appear the same.

Key Concept • Reflections of Absolute Value Functions Across the y-axis

The graph of $f(-ax)$ is the reflection of the graph of $f(ax) = |ax|$ across the y-axis.

Go Online You can watch a video to see how to describe reflections of functions.

Think About It!

Why would $g(x) = |-2x|$ and $j(x) = |2x|$ appear to be the same graphs?

Go Online You can complete an Extra Example online.

Example 9 Graphs of Reflections with Transformations

Describe the reflection in $j(x) = -|x + 3| + 5$ as it relates to the graph of the parent function.

x	$\lvert x + 3\rvert$	$-\lvert x + 3\rvert$	$-\lvert x + 3\rvert + 5$	$(x, j(x))$
−5	$\lvert -5 + 3\rvert = \lvert -2\rvert = 2$	−2	$-2 + 5 = 3$	(−5, 3)
−4	$\lvert -4 + 3\rvert = \lvert -1\rvert = 1$	−1	$-1 + 5 = 4$	(−4, 4)
−3	$\lvert -3 + 3\rvert = \lvert 0\rvert = 0$	0	$0 + 5 = 5$	(−3, 5)
−2	$\lvert -2 + 3\rvert = \lvert 1\rvert = 1$	−1	$-1 + 5 = 4$	(−2, 4)
−1	$\lvert -1 + 3\rvert = \lvert 2\rvert = 2$	−2	$-2 + 5 = 3$	(−1, 3)

First, the absolute value of $x + 3$ is evaluated. Then, the function is multiplied by $1 \cdot a$. Finally, 5 is added to the function.

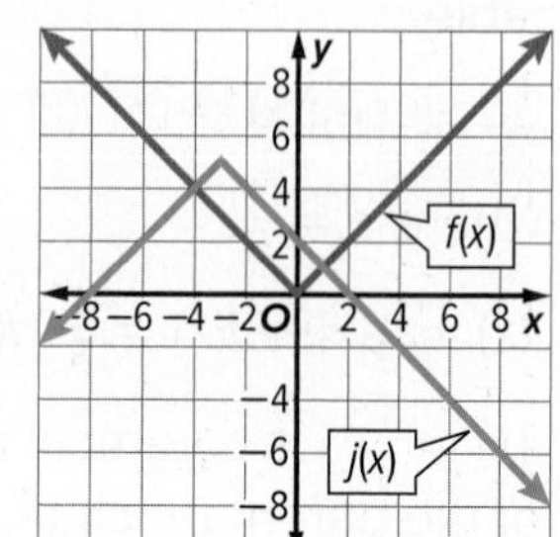

Plot the points from the table.

Because $f(x) = |x|$, $j(x) = -1 \cdot a \cdot f(x)$ where $a = 1$, $h = -3$, and $k = 5$.

$j(x) = -|x + 3| + 5 \rightarrow j(x) = -1 \cdot f(x + 3) + 5$

$j(x) = -|x + 3| + 5$ is the graph of the parent function reflected across the x-axis, and translated 3 units left and 5 units up.

Example 10 Graphs of $y = -a|x|$

Describe the reflection in $q(x) = -\frac{3}{4}|x|$ as it relates to the graph of the parent function.

First, the absolute value of x is evaluated. Then, the function is multiplied by $1 \cdot a$.

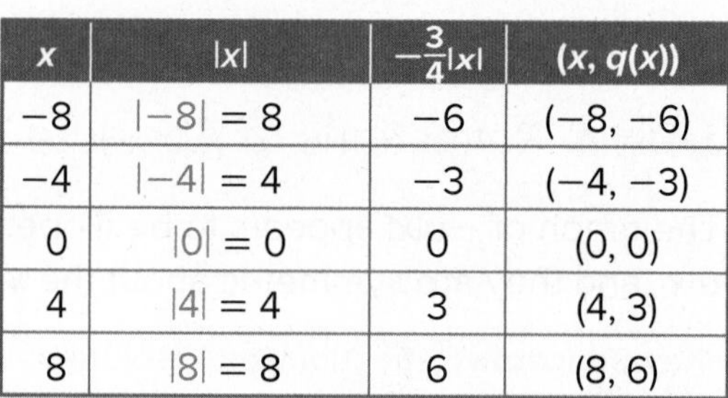

x	$\lvert x\rvert$	$-\frac{3}{4}\lvert x\rvert$	$(x, q(x))$
−8	$\lvert -8\rvert = 8$	−6	(−8, −6)
−4	$\lvert -4\rvert = 4$	−3	(−4, −3)
0	$\lvert 0\rvert = 0$	0	(0, 0)
4	$\lvert 4\rvert = 4$	3	(4, 3)
8	$\lvert 8\rvert = 8$	6	(8, 6)

Plot the points from the table.

Because $f(x) = |x|$, $q(x) = -1 \cdot a \cdot f(x)$ where $a = \frac{3}{4}$.

$q(x) = -\frac{3}{4}|x| \rightarrow q(x) = -\frac{3}{4}f(x)$

$q(x) = -\frac{3}{4}|x|$ is the graph of the parent function reflected across the x-axis and vertically compressed.

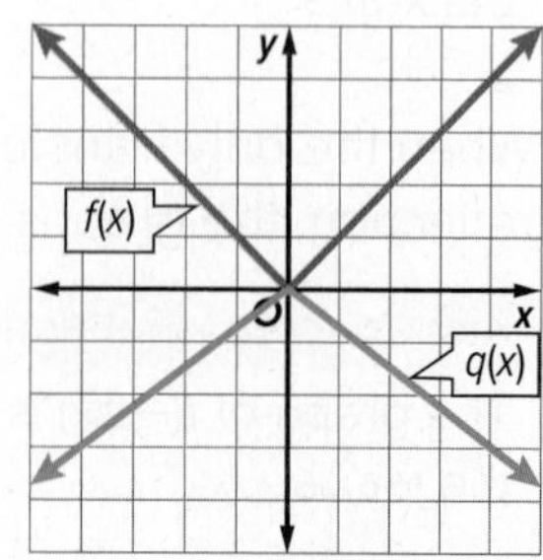

Go Online You can complete an Extra Example online.

Example 11 Graphs of $y = |-ax|$

Describe the reflection in $g(x) = |-4x|$ as it relates to the graph of the parent function.

First, the input is multiplied by $1 \cdot a$. Then the absolute value of $-ax$ is evaluated.

Because $f(x) = |x|$, $g(x) = f(-1 \cdot a \cdot x)$

where $a = 4$.

$g(x) = |-4x| \rightarrow g(x) = f(-1 \cdot 4 \cdot x)$

$g(x) = |-4x|$ is the graph of the parent function reflected across the y-axis and horizontally compressed.

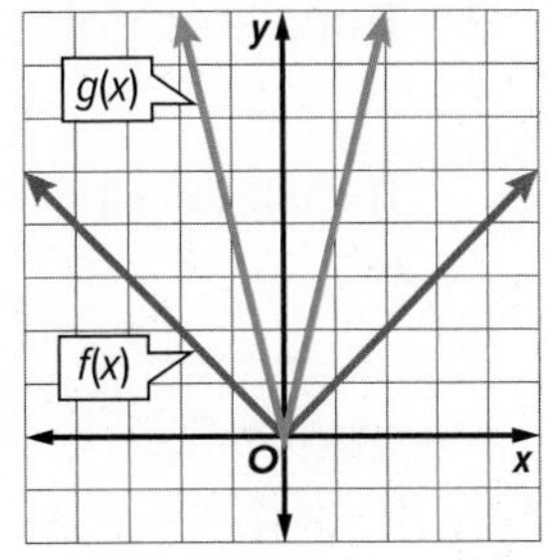

Think About It!

Describe how the graph of $y = |-ax|$ is related to the parent function.

Learn Transformations of Absolute Value Functions

You can use the equation of a function to understand the behavior of the function. Because the constants a, h, and k affect the function in different ways, they can help develop an accurate graph of the function.

Why does there appear to be no reflection for the graph of $y = |-ax|$?

Concept Summary Transformations of Graphs of Absolute Value Functions

$g(x) = a|x - h| + k$

Horizontal Translation, h	**Vertical Translation, k**
If $h > 0$, the graph of $f(x) = \|x\|$ is translated h units right. If $h < 0$, the graph of $f(x) = \|x\|$ is translated $\|h\|$ units left. 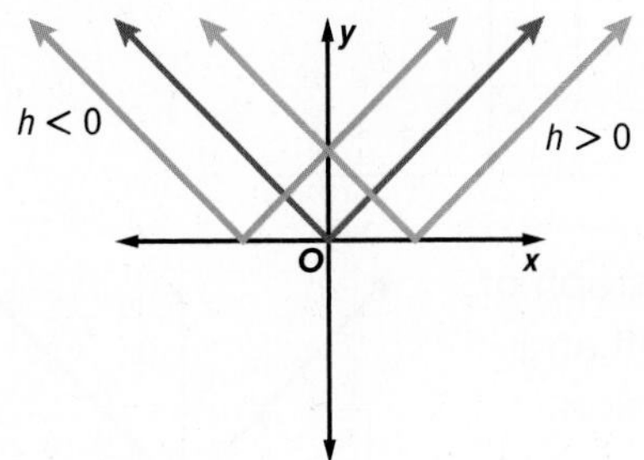	If $k > 0$, the graph of $f(x) = \|x\|$ is translated k units up. If $k < 0$, the graph of $f(x) = \|x\|$ is translated $\|k\|$ units down. 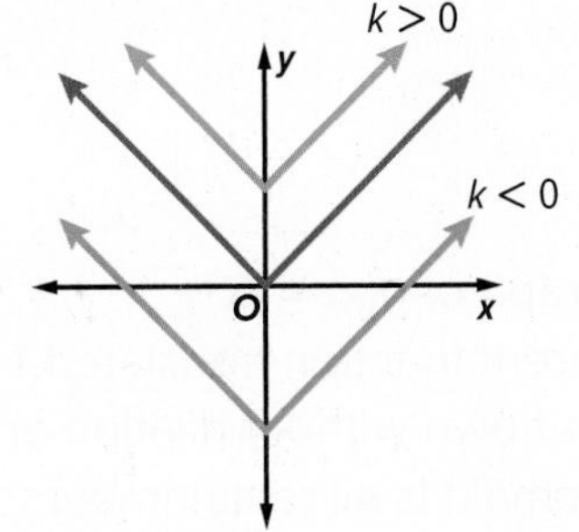
Reflection, a	**Dilation, a**
If $a > 0$, the graph opens up. If $a < 0$, the graph opens down. 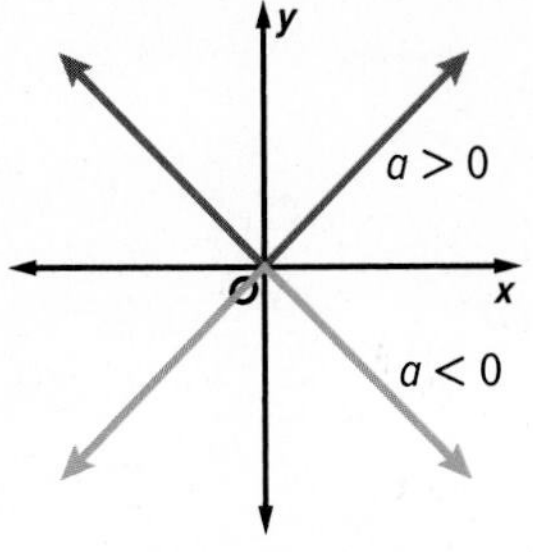	If $\|a\| > 1$, the graph of $f(x) = \|x\|$ is stretched vertically. If $0 < \|a\| < 1$, the graph is compressed vertically. 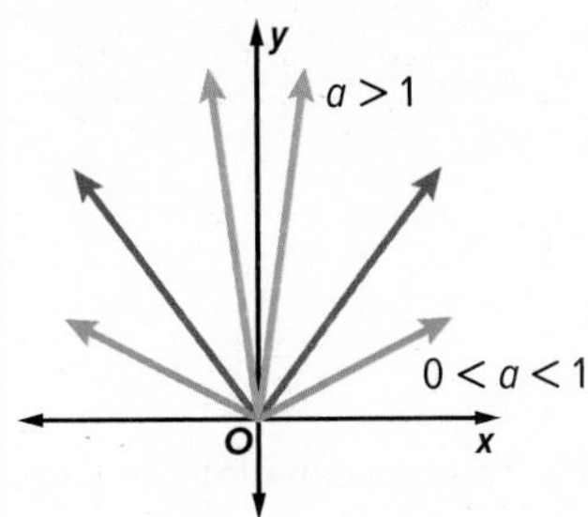

Go Online You can complete an Extra Example online.

Write the phrase that best describes how each parameter affects the graph of $g(x) = -5|x - 2| + 3$ in relation to the parent function.

-5 Reflects and stretches vertically

2 Translates right

3 Translates up

Example 12 Graph an Absolute Value Function with Multiple Translations

Graph $g(x) = |x + 1| - 4$. State the domain and range.

$a = 1$	The graph is not reflected or dilated in relation to the parent function.	
$h = -1$	The graph is translated 1 unit left from the parent function.	
$k = -4$	The graph is translated 4 units down from the parent function.	

The graph of $g(x) = |x + 1| - 4$ is the graph of the parent function translated 1 unit left and 4 units down without dilation or reflection. The domain is all real numbers. The range is all real numbers greater than or equal to -4.

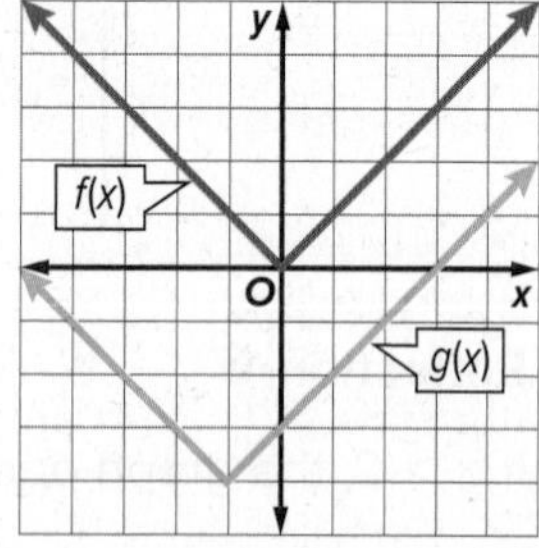

Watch Out!

Dilations and Translations

Don't assume that $j(x) = 2|x - 5| + 1$ and $p(x) = |2x - 5| + 1$ are the same graph. Functions are evaluated differently depending on whether a is inside or outside the absolute value symbols. It might be best to create a table to generate an accurate graph.

Go Online

You can watch a video to see how to graph a transformed absolute value function.

Go Online You can complete an Extra Example online.

Example 13 Graph an Absolute Value Function with Translations and Dilation

Graph $j(x) = |3x - 6|$. State the domain and range.

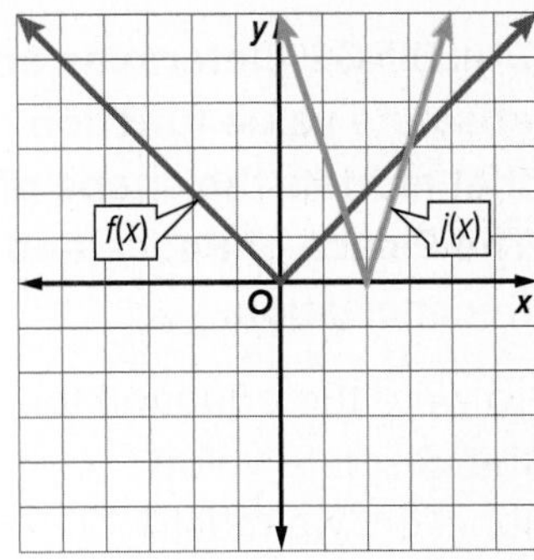

Because a is inside the absolute value symbols, the effect of h on the translation changes.

Evaluate the function for several values of x to find points on the graph.

x	$(x, j(x))$
0	(0, 6)
1	(1, 3)
2	(2, 0)
3	(3, 3)
4	(4, 6)

The graph of $j(x) = |3x - 6|$ is the graph of the parent function compressed horizontally and translated 2 units right.

The domain is all real numbers. The range is all real numbers greater than or equal to 0.

Example 14 Graph an Absolute Value Function with Translations and Reflection

Graph $p(x) = -|x - 3| + 5$. State the domain and range.

In $p(x) = -|x - 3| + 5$, the parent function is reflected across the x-axis because the absolute value is being multiplied by -1.

The function is then translated 3 units right.

Finally, the function is translated 5 units up.

$p(x) = -|x - 3| + 5$ is the graph of the parent function translated 3 units right and 5 units up and reflected across the x-axis.

The domain is all real numbers. The range is all real numbers less than or equal to 5.

Think About It!

How is the vertical translation k of an absolute value function related to its range?

Go Online You can complete an Extra Example online.

Example 15 Apply Graphs of Absolute Value Functions

BUILDINGS **Determine an absolute value function that models the shape of The Palace of Peace and Reconciliation.**

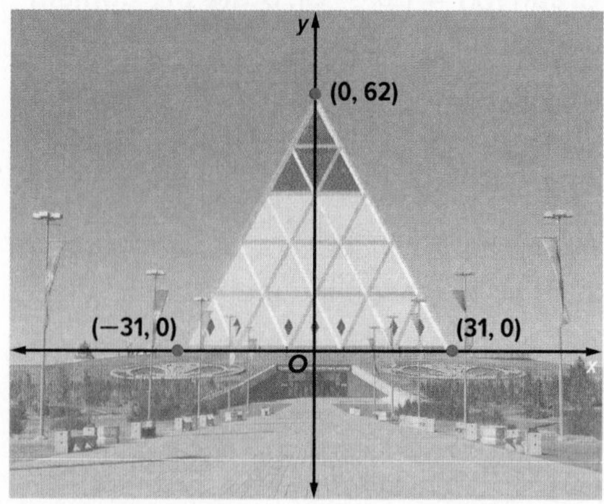

To write the equation for the absolute value function, we must determine the values of a, h, and k in $f(x) = a|x - h| + k$ from the graph.

If we consider the absolute value as a piecewise function, we can find the slope of one side of the graph to determine the value of a.

Because this function opens downward, the graph is a reflection of the parent graph across the x-axis. So we know that the a-value in the equation should be negative.

$m = \frac{y_2 - y_1}{x_2 - x_1}$ The Slope Formula

$= \frac{0 - 62}{31 - 0}$ $(0, 62) = (x_1, y_1)$ and $(31, 0) = (x_2, y_2)$

$= -\frac{62}{31}$ or -2

Next, notice that the vertex is not located at the origin. It has been translated. The absolute value function is not shifted left or right, but has been translated 62 units up from the origin.

$y = -2|x - 0| + 62$ $a = -2, h = 0, k = 62$

$y = -2|x| + 62$ Simplify.

So, $y = -2|x| + 62$ models the shape of The Palace of Peace and Reconciliation.

Check

GLASS PRODUCTION Certain types of glass heat and cool at a nearly constant rate when they are melted to create new glass products. Use the graph to determine the equation that represents this process.

$y =$ __?__ $|x -$ __?__ $| +$ __?__

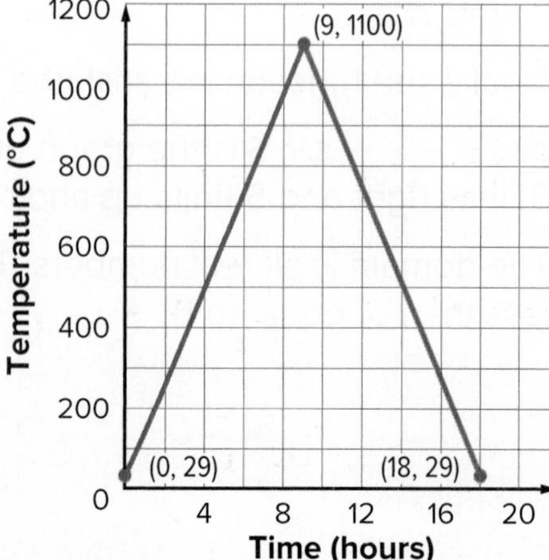

Go Online You can complete an Extra Example online.

Go Online to practice what you've learned about graphing special functions in the Put It All Together over Lessons 4–6 through 4–7.

Dmitry Chulov/123RF

Practice

Go Online You can complete your homework online.

Examples 1 through 3

Describe the translation in $g(x)$ as it relates to the graph of the parent function.

1. $g(x) = |x| - 5$

2. $g(x) = |x + 6|$

3. $g(x) = |x - 2| + 7$

4. $g(x) = |x + 1| - 3$

5. $g(x) = |x| + 1$

6. $g(x) = |x - 8|$

Examples 4 and 5

Use the graph of the function to write its equation.

7.

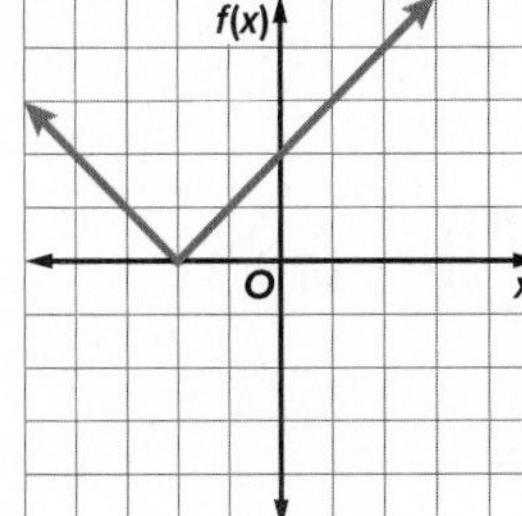

8.

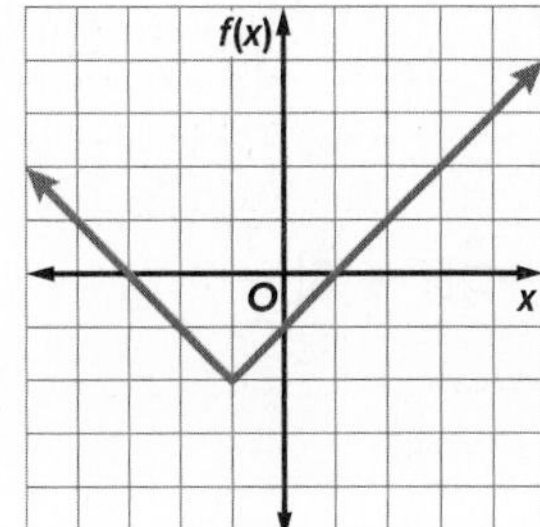

9.

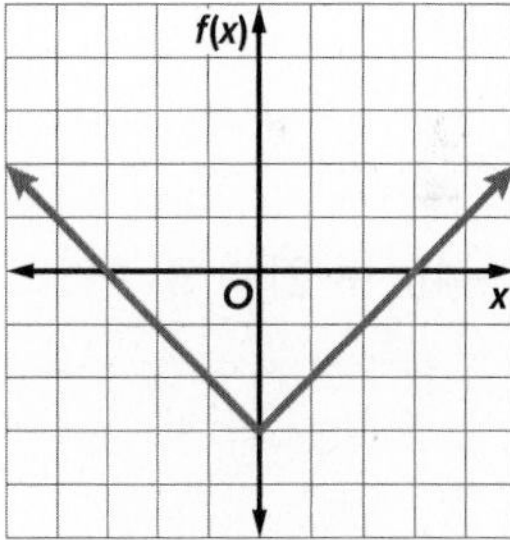

10.

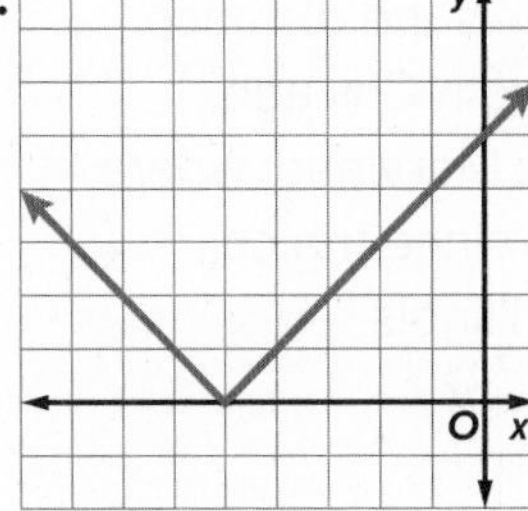

11.

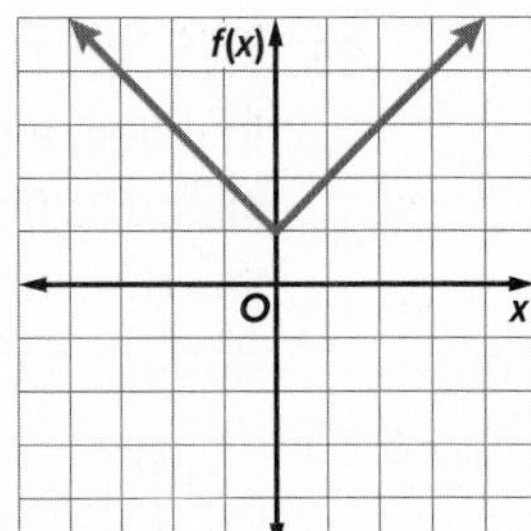

12.

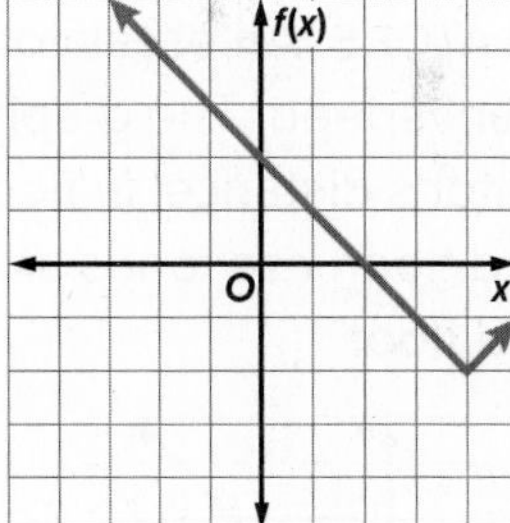

Examples 6 through 8

Describe the dilation in $g(x)$ as it relates to the graph of the parent function.

13. $g(x) = \frac{2}{5}|x|$

14. $g(x) = |0.7x|$

15. $g(x) = 1.3|x|$

16. $g(x) = |3x|$

17. $g(x) = \left|\frac{1}{6}x\right|$

18. $g(x) = \frac{5}{4}|x|$

Examples 9 through 11

Describe the reflection in $g(x)$ as it relates to the graph of the parent function.

19. $g(x) = -3|x|$

20. $g(x) = -|x| - 2$

21. $g(x) = \left|-\frac{1}{4}x\right|$

22. $g(x) = -|x - 7| + 3$

23. $g(x) = |-2x|$

24. $g(x) = -\frac{2}{3}|x|$

Examples 12 through 14

USE TOOLS **Graph each function. State the domain and range.**

25. $g(x) = |x + 2| + 3$

26. $g(x) = |2x - 2| + 1$

27. $f(x) = \left|\frac{1}{2}x - 2\right|$

28. $f(x) = |2x - 1|$

29. $f(x) = \frac{1}{2}|x| + 2$

30. $h(x) = -2|x - 3| + 2$

31. $f(x) = -4|x + 2| - 3$

32. $g(x) = -\frac{2}{3}|x + 6| - 1$

33. $h(x) = -\frac{3}{4}|x - 8| + 1$

Example 15

Determine an absolute value function that models each situation.

34. ESCALATORS An escalator travels at a constant speed. The graph models the escalator's distance, in floors, from the second floor x seconds after leaving the ground floor.

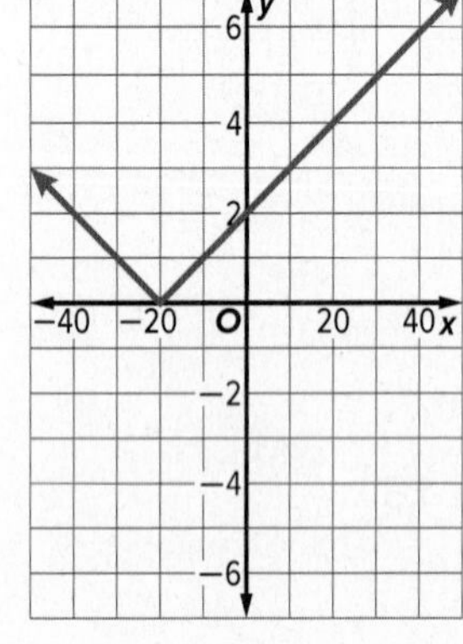

35. TRAVEL The graph models the distance, in miles, a car traveling from Chicago, Illinois is from Annapolis, Maryland, where x is the number of hours since the car departed from Chicago, Illinois.

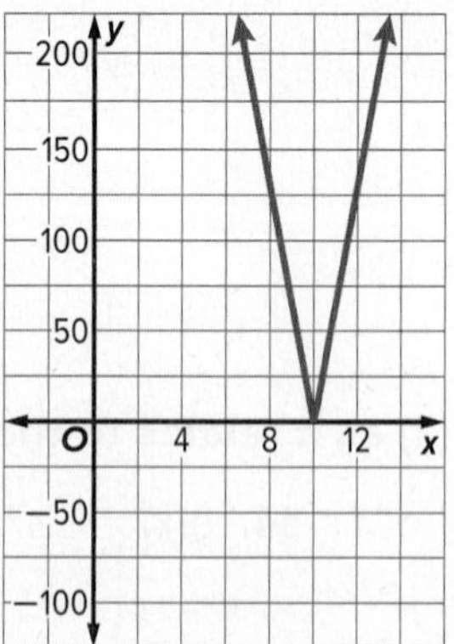

Mixed Exercises

MODELING **Graph each function. State the domain and range. Describe how each graph is related to its parent graph.**

36. $f(x) = -4|x - 2| + 3$

37. $f(x) = |2x|$

38. $f(x) = |2x + 5|$

Use the graph of the function to write its equation.

39. 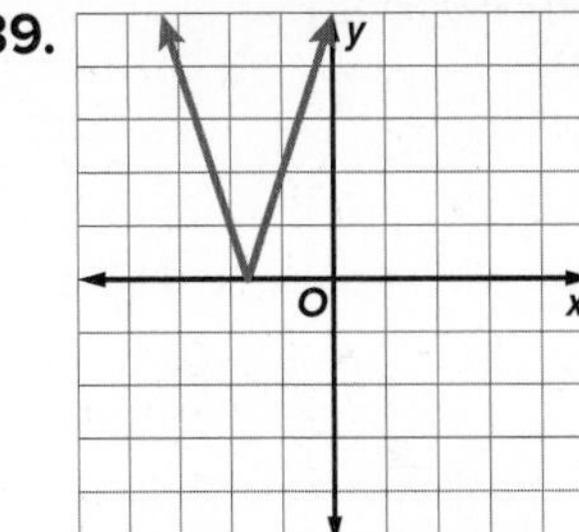

40. 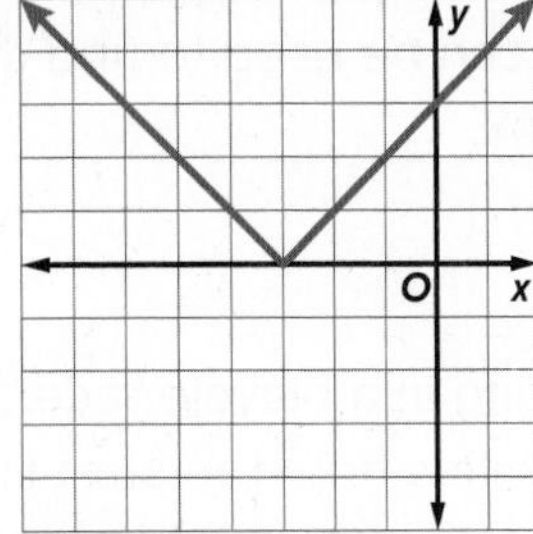

41. 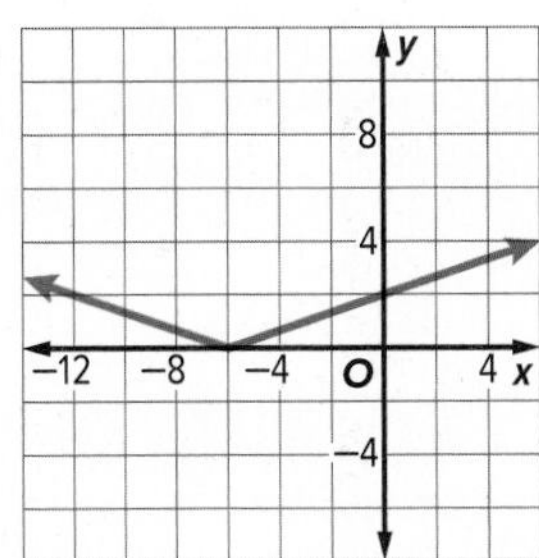

42. 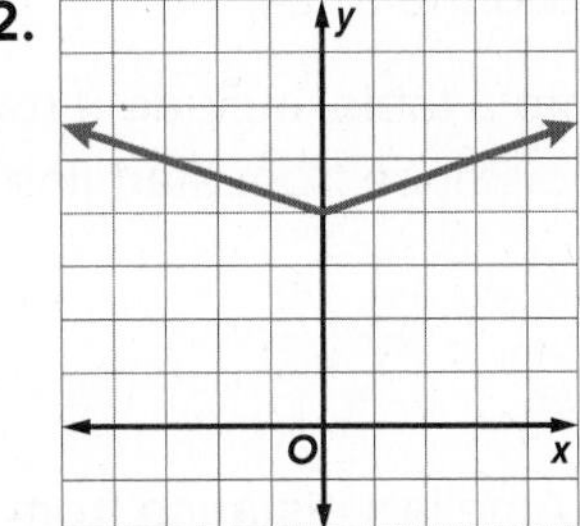

43. SUNFLOWER SEEDS A company produces and sells bags of sunflower seeds, s. A medium-sized bag of sunflower seeds must contain 16 ounces of seeds. If the amount of sunflower seeds in the medium-sized bag differs from the desired 16 ounces by more than x, the bag cannot be delivered to companies to be sold. Write an equation that can be used to find the highest and lowest amounts of sunflower seeds in a medium-sized bag.

44. REASONING The function $y = \frac{5}{4}|x - 5|$ models a car's distance in miles from a parking lot after x minutes. Graph the function. After how many minutes will the car reach the parking lot?

45. STATE YOUR ASSUMPTION A track coach set up an agility drill for members of the track team. According to the coach, 21.7 seconds is the target time to complete the agility drill. If the time differs from the desired 21.7 seconds by more than x, the track coach may require members of the track team to change their training. Write an equation that can be used to find the fastest and slowest times members of the track team can complete the agility drill so that their training does not have to change. If $x = 3.2$, what can you assume about the range of times the coach wants the members of the track team to complete the agility drill? Solve your equation for $x = 3.2$ and use the results to justify your assumption.

46. SCUBA DIVING The function $y = 3|x - 12| - 36$ models a scuba diver's elevation in feet compared to sea level after x minutes. Graph the function. How far below sea level is the scuba diver at the deepest point in their dive?

47. MANUFACTURING A manufacturing company produces boxes of cereal, b. A small box of cereal must have 12 ounces. If the amount of cereal in a small box differs from the desired 12 ounces by more than x, the box cannot be shipped for selling. Write an equation that can be used to find the highest and lowest amounts of cereal in a small box.

48. STRUCTURE Amelia is competing in a bicycle race. The race is along a circular path. She is 6 miles from the start line. She is approaching the start line at a speed of 0.2 mile per minute. After Amelia reaches the start line, she continues at the same speed, taking another lap around the track.

a. Organize the information into a table. Include a row for time in minutes x, and a row for distance from start line $f(x)$.

b. Draw a graph to represent Amelia's distance from the start line.

49. WRITE Use transformations to describe how the graph of $h(x) = -|x + 2| - 3$ is related to the graph of the parent absolute value function.

50. ANALYZE On a straight highway, the town of Garvey is located at mile marker 200. A car is located at mile marker x and is traveling at an average speed of 50 miles per hour.

a. Write a function $T(x)$ that gives the time, in hours, it will take the car to reach Garvey. Then graph the function on the coordinate plane.

b. Does the graph have a maximum or minimum? If so, name it and describe what it represents in the context of the problem.

51. PERSEVERE Write the equation $y = |x - 3| + 2$ as a piecewise-defined function. Then graph the piecewise function.

52. CREATE Write an absolute value function, $f(x)$, that has a domain of all real numbers and a range that is greater than or equal to 4. Be sure your function also includes a dilation of the parent function. Describe how your function relates to the parent absolute value graph. Then graph your function.

Review

 Essential Question

What can a function tell you about the relationship that it represents?
It can tell you about the rate of change, whether the relationship is positive or negative, the locations of the *x*- and *y*-intercepts, and what points fall on the graph.

Module Summary

Lessons 4-1 through 4-3

Graphing Linear Functions, Rate of Change, and Slope

- The graph of an equation represents all of its solutions.
- The *x*-value of the *y*-intercept is 0. The *y*-value of the *x*-intercept is 0.
- The rate of change is how a quantity is changing with respect to a change in another quantity. If *x* is the independent variable and *y* is the dependent variable, then rate of change $= \frac{\text{change in } y}{\text{change in } x}$.
- The slope *m* of a nonvertical line through any two points can be found using $m = \frac{y_2 - y_1}{x_2 - x_1}$.
- A line with positive slope slopes upward from left to right. A line with negative slope slopes downward from left to right. A horizontal line has a slope of 0. The slope of a vertical line is undefined.

Lesson 4-4

Transformations of Linear Functions

- When a constant *k* is added to a linear function *f*(*x*), the result is a vertical translation.
- When a linear function *f*(*x*) is multiplied by a constant *a*, the result $a \cdot f(x)$ is a vertical dilation.
- When a linear function *f*(*x*) is multiplied by −1 before or after the function has been evaluated, the result is a reflection across the *x*- or *y*-axis.

Lesson 4-5

Arithmetic Sequences

- An arithmetic sequence is a numerical pattern that increases or decreases at a constant rate called the common difference.
- The *n*th term of an arithmetic sequence with the first term a_1 and common difference *d* is given by $a_n = a_1 + (n - 1)d$, where *n* is a positive integer.

Lessons 4-6, 4-7

Special Functions

- A piecewise-linear function has a graph that is composed of a number of linear pieces.
- A step function is a type of piecewise-linear function with a graph that is a series of horizontal line segments.
- An absolute value function is V-shaped.

Study Organizer

 Foldables

Use your Foldable to review this module. Working with a partner can be helpful. Ask for clarification of concepts as needed.

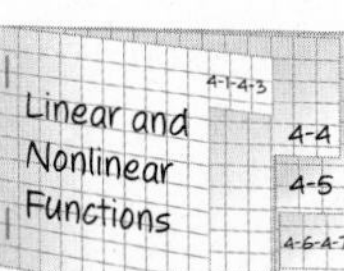

Test Practice

1. **GRAPH** Jalyn made a table of how much money she will earn from babysitting. (Lesson 4-1)

Hours Babysitting	Money Earned
1	5
2	10
3	15
4	20

Use the table to graph the function.

2. **OPEN RESPONSE** Copy and complete the table to find the missing values in the table that show the points on the graph of $f(x) = 2x - 4$. (Lesson 4-1)

x	−2	0	2	4	6
$f(x)$	−8	−4			

3. **OPEN RESPONSE** Mr. Hernandez is draining his pool to have it cleaned. At 8:00 A.M., it had 2000 gallons of water and at 11:00 A.M. it had 500 gallons left to drain. What is the rate of change in the amount of water in the pool? (Lesson 4-2)

4. **MULTIPLE CHOICE** Find the slope of the graphed line. (Lesson 4-2)

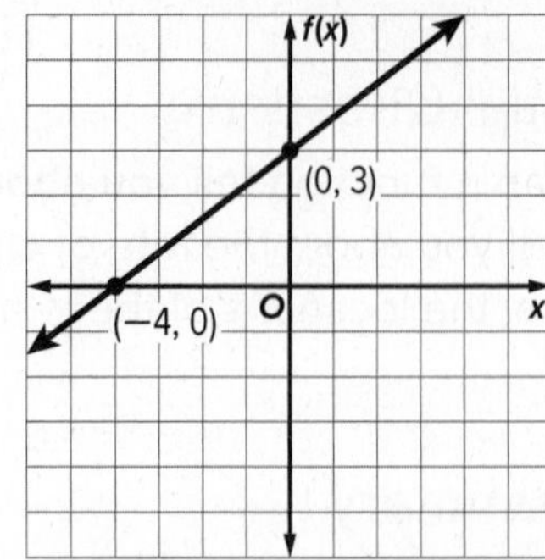

A. $-\frac{4}{3}$

B. $-\frac{3}{4}$

C. $\frac{3}{4}$

D. $\frac{4}{3}$

5. **MULTIPLE CHOICE** Determine the slope of the line that passes through the points (4, 10) and (2, 10). (Lesson 4-2)

A. −1

B. 0

C. 1

D. undefined

6. **GRAPH** Graph the equation of a line with a slope of −3 and a y-intercept of 2. (Lesson 4-3)

7. **MULTIPLE CHOICE** What is the slope of the line that passes through (3, 4) and (−7, 4)? (Lesson 4-3)

A. 0

B. undefined

C. −2

D. −10

8. MULTIPLE CHOICE A teacher buys 100 pencils to keep in her classroom at the beginning of the school year. She allows the students to borrow pencils, but they are not always returned. On average, she loses about 8 pencils a month. Write an equation in slope-intercept form that represents the number of pencils she has left, y, after a number of x months. (Lesson 4-3)

A. $y = -8x - 100$

B. $y = -8x + 100$

C. $y = 8x + 100$

D. $y = 8x - 100$

9. OPEN RESPONSE Name the transformation that changes the slope, or the steepness of, the graph of a linear function. (Lesson 4-4)

10. OPEN RESPONSE Describe the dilation of $g(x) = \frac{1}{2}(x)$ as it relates to the graph of the parent function, $f(x) = x$. (Lesson 4-4)

11. MULTIPLE CHOICE Arjun begins the calendar year with \$40 in his bank account. Each week he receives an allowance of \$20, half of which he deposits into his bank account. The situation describes an arithmetic sequence. Which function represents the amount in Arjun's account after n weeks? (Lesson 4-5)

A. $f(n) = 20n + 40$

B. $f(n) = 40n + 20$

C. $f(n) = 40 + 10n$

D. $f(n) = 10 + 40n$

12. OPEN RESPONSE What number can be used to complete the equation below that describes the nth term of the arithmetic sequence −2, −1.5, −1, −0.5, 0, 0.5, ...? (Lesson 4-5)

$a_n = 0.5n - \underline{\quad ? \quad}$

13. OPEN RESPONSE Write and graph a function to represent the sequence 1, 10, 19, 28, ... (Lesson 4-5)

14. OPEN RESPONSE Christa has a box of chocolate candies. The number of chocolates in each row forms an arithmetic sequence as shown in the table. (Lesson 4-5)

Row	1	2	3	4
Number of Chocolates	3	6	9	12

Write an arithmetic function that can be used to find the number of chocolates in each row.

15. **OPEN RESPONSE** Daniel earns \$9 per hour at his job for the first 40 hours he works each week. However, his pay rate increases to \$13.50 per hour thereafter. This situation can be represented with the function

$$f(x) = \begin{cases} 9x, \text{ if } x \leq 40 \\ 360 + 13.5(x - 40), \text{ if } x > 40 \end{cases}$$

Use this function to copy and complete the table with the correct values. (Lesson 4-6)

Hours Worked, x	Money Earned, $f(x)$
30	
35	315
40	
45	427.5
50	

16. **GRAPH** Graph the function $f(x) = 2[[x]]$. (Lesson 4-6)

17. **MULTIPLE CHOICE** Which of the following describes the effect a dilation has upon the graph of the absolute value parent function? (Lesson 4-7)

A. Flipped across axis

B. Stretch or compression

C. Rotated about the origin

D. Shifted horizontally or vertically

18. **MULTI-SELECT** Describe the transformation(s) of the function graphed below in relation to the absolute value parent function. Select all that apply. (Lesson 4-7)

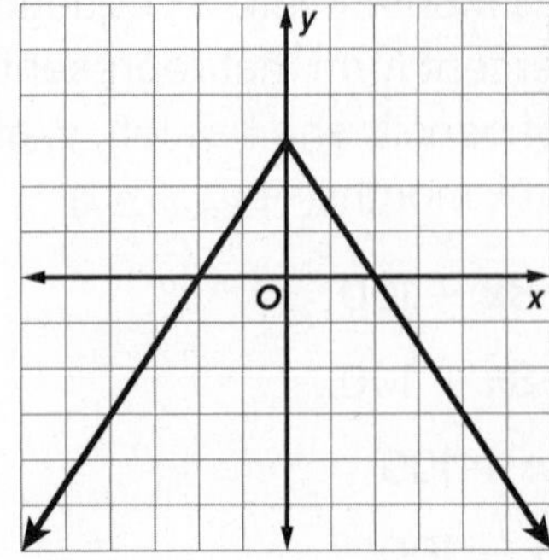

A. Reflected across x-axis

B. Vertical stretch

C. Vertical compression

D. Reflected across y-axis

E. Translated right 3

F. Translated up 3

19. **OPEN RESPONSE** Describe the graph of $g(x) = |x| + 5$ in relation to the graph of the absolute value parent function. (Lesson 4-7)

20. **OPEN RESPONSE** Across which axis is the graph of $h(x) = -5|x|$ reflected? (Lesson 4-7)

21. **OPEN RESPONSE** Use the graph of the function to write its equation. (Lesson 4-7)

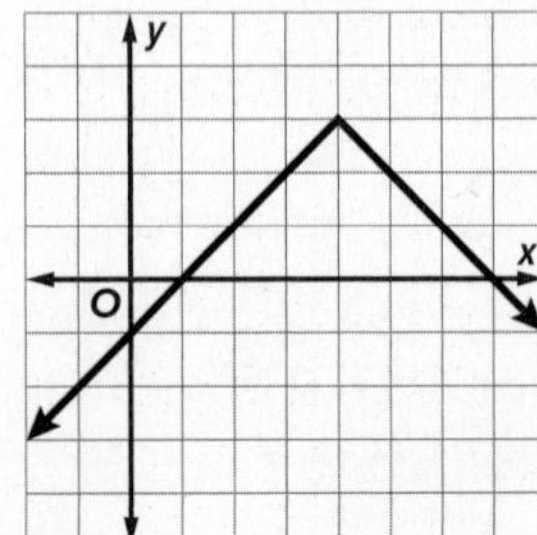

Module 5

Creating Linear Equations

Essential Question

What can a function tell you about the relationship that it represents?

What Will You Learn?

How much do you already know about each topic **before** starting this module?

KEY

— I don't know. — I've heard of it. — I know it!

	Before			After		
	I don't know	I've heard of it	I know it	I don't know	I've heard of it	I know it
write linear equations in slope-intercept form when given the slope and the coordinates of a point						
write linear equations in slope-intercept form when given the coordinates of two points on the line						
write linear equations in standard form						
write linear equations in point-slope form						
write equations of parallel and perpendicular lines						
examine scatter plots to describe relationships between quantities						
make and evaluate predictions by fitting linear functions to sets of data						
distinguish between correlation and causation						
write equations of best-fit lines						
plot and analyze residuals						
find inverses of linear relations and functions						

Foldables Make this Foldable to help you organize your notes about linear equations. Begin with one sheet of 11″ × 17″ paper.

1. **Fold** each end of the paper in about 2 inches.
2. **Fold** along the width and the length. Unfold. Cut along the fold line from the top to the center.
3. **Fold** the top flaps down. Then fold in half and turn to form a folder. Staple the flaps down to form pockets.
4. **Label** the front with the chapter title.

1

2

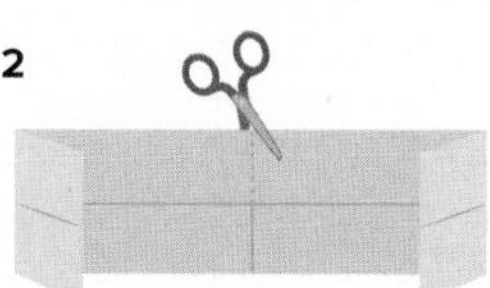

3

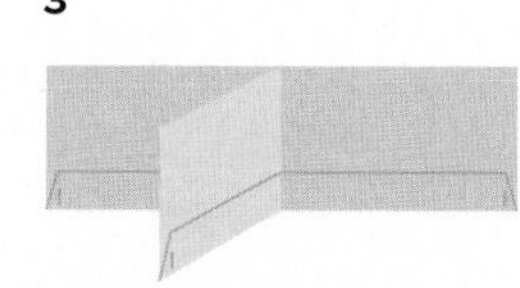

4

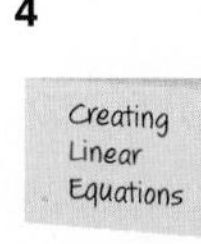

What Vocabulary Will You Learn?

- best-fit line
- bivariate data
- causation
- correlation coefficient
- inverse functions
- inverse relations
- line of fit
- linear extrapolation
- linear interpolation
- linear regression
- parallel lines
- negative correlation
- no correlation
- perpendicular lines
- positive correlation
- residual
- scatter plot
- trend

Are You Ready?

Complete the Quick Review to see if you are ready to start this module.
Then complete the Quick Check.

Quick Review

Example 1

Solve $5x + 15y = 9$ for x.

$5x + 15y = 9$	Original equation
$5x + 15y - 15y = 9 - 15y$	Subtract 15y from each side.
$5x = 9 - 15y$	Simplify.
$\frac{5x}{5} = \frac{9 - 15y}{5}$	Divide each side by 5.
$x = \frac{9}{5} - 3y$	Simplify.

Example 2

Write the ordered pair for A.

Step 1 Begin at point A.

Step 2 Follow along a vertical line to the x-axis. The x-coordinate is −4.

Step 3 Follow along a horizontal line to the y-axis. The y-coordinate is 2.

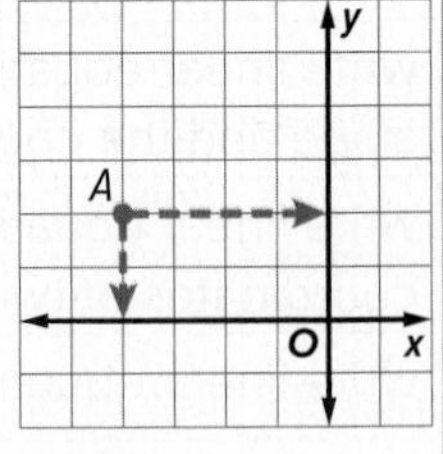

The ordered pair for point A is (−4, 2).

Quick Check

Solve each equation for the given variable.

1. $x + y = 5$ for y
2. $2x - 4y = 6$ for x
3. $y - 2 = x + 3$ for y
4. $4x - 3y = 12$ for x

Write the ordered pair for each point.

5. A
6. B
7. C
8. D
9. E
10. F

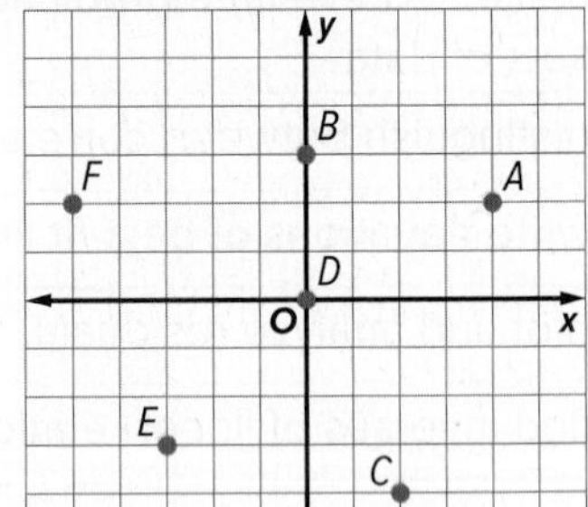

How Did You Do?

Which exercises did you answer correctly in the Quick Check?

Writing Equations in Slope-Intercept Form

Explore Slope-Intercept Form

Online Activity Use graphing technology to complete the Explore.

INQUIRY How does changing the coordinates of two points on a line affect the slope of the line?

Today's Goals

- Write an equation of a line in slope-intercept form given the slope and one point.
- Write an equation of a line in slope-intercept form given two points.

Learn Creating Linear Equations in Slope-Intercept Form Given the Slope and a Point

If you are given the slope of a line and the coordinates of any point on that line, you can create an equation for that line.

Think About It!
How can you determine whether the given point is the y-intercept of the line?

Key Concept • Creating Equations in Slope-Intercept Form Given the Slope and a Point	
Step 1	Determine whether the given point is the y-intercept. If not, substitute the given information into the slope-intercept form equation to find the y-intercept.
Step 2	Use the given slope and y-intercept you found in Step 1 to write the equation of the line in slope-intercept form.

Example 1 Write an Equation Given the Slope and a Point

Write an equation of the line that passes through (−8, 6) and has a slope of $-\frac{3}{4}$.

Step 1 Find the y-intercept.

$y = mx + b$	Slope-intercept form
$6 = -\frac{3}{4}(-8) + b$	$m = -\frac{3}{4}$, $x = -8$, and $y = 6$
$6 = 6 + b$	Simplify.
$0 = b$	Subtract 6 from each side.

Step 2 Write the equation in slope-intercept form.

$y = mx + b$	Slope-intercept form
$y = -\frac{3}{4}x + 0$	$m = -\frac{3}{4}$ and $b = 0$
$y = -\frac{3}{4}x$	Simplify.

Go Online You can complete an Extra Example online.

Think About It!
What does it mean if $b = 0$ when an equation is written in slope-intercept form?

Study Tip
Slope-Intercept Form Remember, you need two things to write an equation in slope-intercept form: the slope and the y-intercept.

Example 2 Write an Equation in Slope-Intercept Form

BAKING Marissa is baking a recipe that calls for her to turn down the temperature on her oven for part of the baking time. Write an equation to represent the situation if the temperature in her oven drops 25°F every 30 seconds, and after 2 minutes the temperature is 350°F.

Step 1 Determine a point on the line and the slope.

After 2 minutes the temperature is 350°F.

Let x = the time in minutes and y = the temperature in °F.

So, the point (2, 350) is on the line.

The temperature drops 25°F every 30 seconds.

The change in x is 30 seconds, or 0.5 minute.

"Drops" means a negative change, so the change in y is −25°F.

Slope $= \frac{\text{change in } y}{\text{change in } x} = \frac{-25}{0.5}$ or −50°F per minute

So, the slope is −50.

Study Tip

Slope When determining the slope, words like "drops" and "decreasing" represent a negative slope, and words like "growing" and "increasing" represent a positive slope.

Step 2 Find the y-intercept.

$y = mx + b$	Slope-intercept form
$350 = -50(2) + b$	$m = -50$, $x = 2$, and $y = 350$
$350 = -100 + b$	Simplify.
$450 = b$	Add 100 to each side.

This means that the temperature of the oven was 450°F when it was turned off.

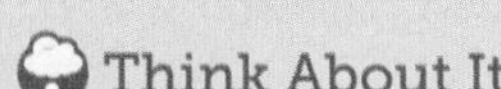

Think About It!

Find the domain of your equation, and describe the meaning in the context of the situation.

Step 3 Write the equation in slope-intercept form.

$y = mx + b$	Slope-intercept form
$y = -50x + 450$	$m = -50$ and $b = 450$

Check

MEMBERSHIP The total monthly cost of Ayzha's gym membership increases by $5 per class she attends. After signing up for 4 classes one month, her total cost is $49.99. Which equation represents Ayzha's total monthly cost y after attending x classes?

A. $y = -5x + 29.99$

B. $y = -5x + 69.99$

C. $y = 5x + 29.99$

D. $y = 5x + 69.99$

Go Online You can complete an Extra Example online.

Learn Creating Linear Equations in Slope-Intercept Form Given Two Points

If you are given the coordinates of any two points on a line, you can create an equation for that line.

Key Concept • Creating Equations in Slope-Intercept Form Given Two Points	
Step 1	Use the given points to find the slope of the line containing the points.
Step 2	Use the slope from Step 1 and either of the given points to find the y-intercept of the line.
Step 3	Use the slope you found in Step 1 and the y-intercept you found in Step 2 to write the equation of the line in slope-intercept form.

Talk About It!

Will your equation for the line be different depending on the point you choose in Step 2? Justify your argument.

Example 3 Write Equations Given Two Points

Write an equation of the line that passes through (1.2, −0.7) and (−3.4, 1.6).

Step 1 Find the slope.

$m = \frac{y_2 - y_1}{x_2 - x_1}$ Slope Formula

$m = \frac{1.6 - (-0.7)}{-3.4 - 1.2}$ $(x_1, y_1) = (1.2, -0.7), (x_2, y_2) = (-3.4, 1.6)$

$m = \frac{2.3}{-4.6}$ Simplify.

$m = -0.5$ Simplify.

Step 2 Use either point to find the y-intercept.

$y = mx + b$ Slope-intercept form

$1.6 = -0.5(-3.4) + b$ $m = -0.5$, $x = -3.4$, and $y = 1.6$

$1.6 = 1.7 + b$ Simplify.

$-0.1 = b$ Subtract 1.7 from each side.

Step 3 Write the equation in slope-intercept form.

$y = mx + b$ Slope-intercept form

$y = -0.5x - 0.1$ $m = -0.5$ and $b = -0.1$

Watch Out!

Subtraction If the (x_1, y_1) coordinates are negative, be sure to account for both the negative signs and the subtraction symbols in the Slope Formula. Remember, the result of subtracting a negative number is the same as adding its opposite.

Check

Write an equation of the line that passes through (−5, −3) and (−7, −12).

 **Go Online** You can complete an Extra Example online.

Think About It!

Use the table to make an estimate of the number of students enrolled in public high schools in 2030. Then, use the equation to predict the number of students enrolled. How does your estimate compare to the number of students that you calculated?

Apply Example 4 Write an Equation Given Real-World Data

SCHOOLS **The number of students enrolled in public high schools in the United States has risen slightly since 2010. Write an equation that could be used to predict the number of students enrolled in public high schools if enrollment continues to grow at the same rate.**

Year	Students (in thousands)
2011	14,749
2012	14,753
2013	14,754
2014	14,826
2015	14,912

1. What is the task?

Describe the task in your own words. Then list any questions that you may have. How can you find answers to your questions?

Sample answer: I need to write an equation to predict enrollment in public high schools. How can I write an equation when given a table? I can review finding rate of change and writing equations in slope-intercept form.

2. How will you approach the task? What have you learned that you can use to help you complete the task?

Sample answer: I will use what I have learned about finding the rate of change from a table to help me find the slope. I will then use the slope and one of the points to find the *y*-intercept. I will use what I have learned about writing equations in slope-intercept form to write an equation.

Problem-Solving Tip

Use a Graph You can also estimate and make predictions using a graph. Plot two points from the table, connect them with a line, and then estimate using the graph.

3. What is your solution?

Use your strategy to solve the problem.

Find the slope.

$m = 40.75$

Find the *y*-intercept.

$b = 14{,}708.25$

Write an equation to predict the number of students enrolled in public high schools if enrollment continues to grow at the same rate.

$y = 40.75x + 14{,}708.25$

Study Tip

Units The number of students enrolled is in thousands. While it is impossible to have one-quarter of a student, 14,708.25 thousand students really means 14,708,250 students. So, this solution is within the constraints of the situation.

4. How can you know that your solution is reasonable?

Write About It! Write an argument that can be used to defend your solution.

Sample answer: For the year 2015, $x = 5$. I substituted $x = 5$ into my equation to check my solution, and the result matched the number of students enrolled in 2015.

Go Online You can complete an Extra Example online.

Practice

Go Online You can complete your homework online.

Example 1

Write an equation of the line that passes through the given point and has the given slope.

1. (4, 2); slope $\frac{1}{2}$
2. (3, −2); slope $\frac{1}{3}$
3. (6, 4); slope $-\frac{3}{4}$
4. (−5, 4); slope −3
5. (4, 3); slope $\frac{1}{2}$
6. (1, −5); slope $-\frac{3}{2}$

Example 2

7. **EXERCISE** Carlos is jogging at a constant speed. He starts a timer when he is 12 feet from his starting position. After 3 seconds, Carlos is 21 feet from his starting position. Write a linear equation to represent the distance *d* of Carlos from his starting position after *t* seconds.

8. **JOBS** Mr. Kimball sells computer software. He earns a base salary of $41,250 and 8% commission on his sales. Write an equation to represent Mr. Kimball's total pay *p* after selling *d* dollars of software.

9. **USE A MODEL** In 2006, the average ticket price for a National Football League game was $62.38. Since then the cost has increased an average of $2.54 per year. Write a linear equation to represent the cost *C* of an NFL ticket *y* years after 2006.

10. **TYPING** Nebi has already typed 250 words. He then starts a timer and finds that he types 150 words in 3 minutes. If Nebi types at a constant rate, write a linear equation to represent the number of words *w* Nebi types *m* minutes after starting the timer.

Example 3

Write an equation of the line that passes through each pair of points.

11. (0, −4), (5, −4)
12. (−4, −2), (4, 0)
13. (−2, −3), (4, 5)
14. (0, 1), (5, 3)
15. (−3, 0), (1, −6)
16. (1, 0), (5, −1)
17. (9, 2), (−2, 6)
18. (−6, 5), (−6, −4)
19. (5, −2), (7, −1)
20. (5, −3), (2, 5)
21. $\left(\frac{5}{4}, 1\right), \left(-\frac{1}{4}, \frac{3}{4}\right)$
22. $\left(\frac{5}{12}, -1\right), \left(-\frac{3}{4}, \frac{1}{6}\right)$

Example 4

23. **GUITAR** Lydia wants to purchase guitar lessons. She sees a sign that gives the prices for 7 guitar lessons and 11 guitar lessons. Write a linear equation to find the total cost C for d lessons.

24. **CENSUS** The population of Laredo, Texas, was about 215,500 in 2007. It was about 123,000 in 1990. If we assume that the population growth is constant, write a linear equation with an integer slope to represent p, Laredo's population t years after 1990.

25. **WEATHER** A meteorologist finds that the temperature at the 6000-foot level of a mountain is 76°F and the temperature at the 12,000-foot level of the mountain is 49°F. Write a linear equation to represent the temperature T at an elevation of x, where x is in thousands of feet.

26. **FUNDRAISING** Natalia and her friends held a bake sale to benefit a local charity. The friends sold 15 cakes on the first day and 22 cakes on the second day of the bake sale. They collected $60 on the first day and $88 on the second day. Write an equation to represent the amount R Natalia and her friends raised after selling c cakes.

Mixed Exercises

Write an equation of each line.

27.

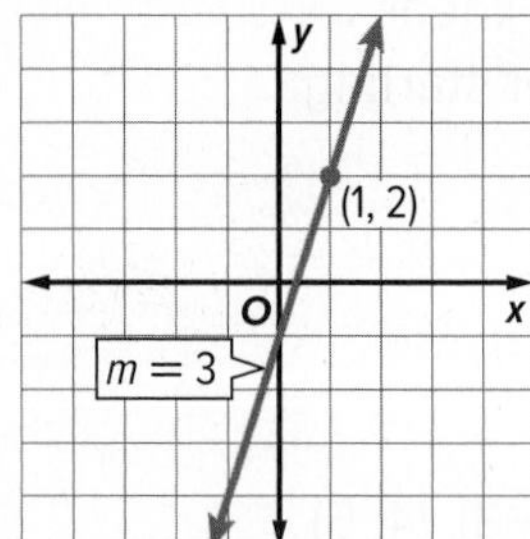

28.

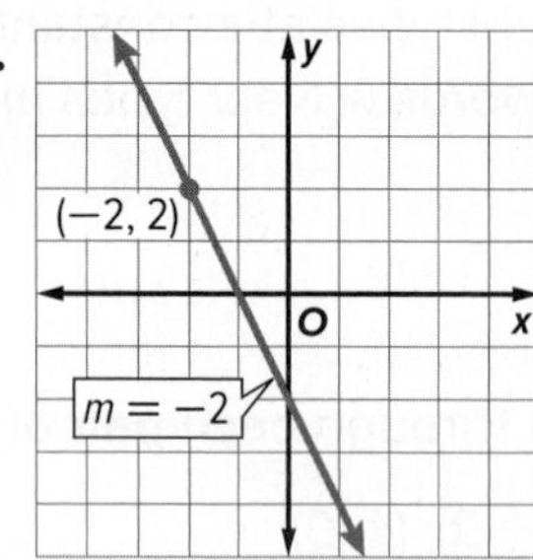

29.

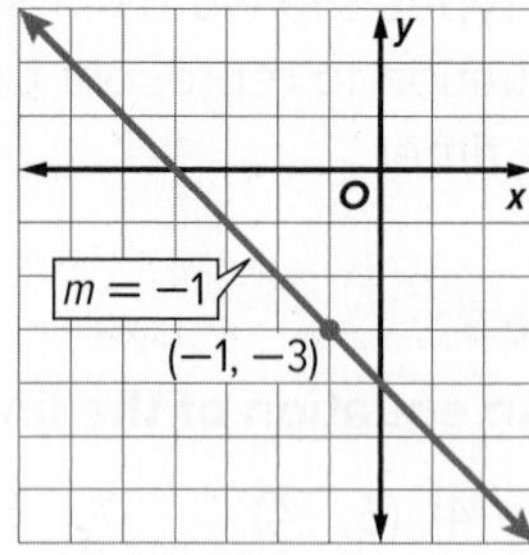

30.

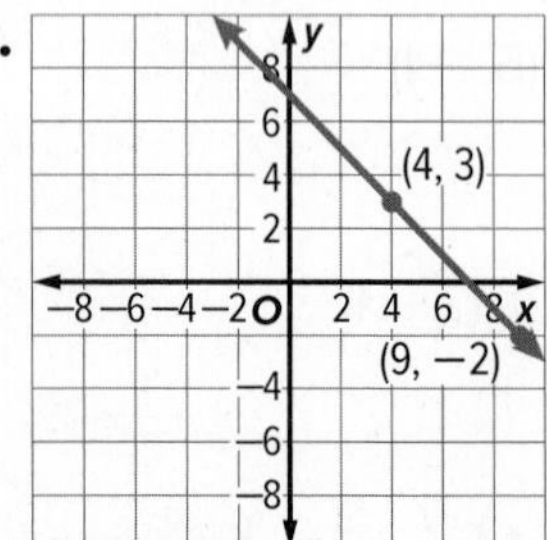

31.

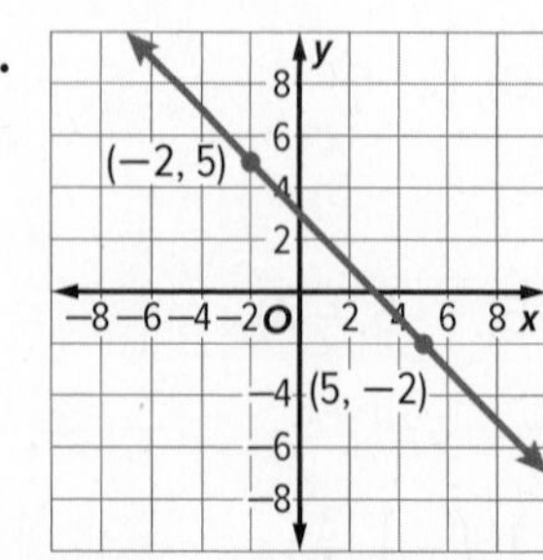

32.

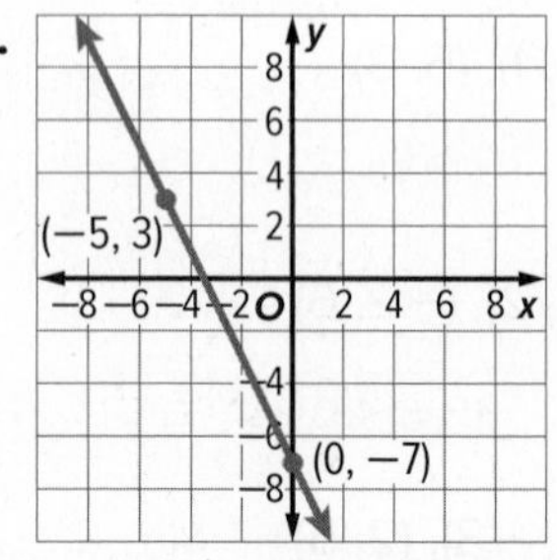

Determine whether the given point is on the line. Explain your reasoning.

33. (3, −1); $y = \frac{1}{3}x + 5$

34. (6, −2); $y = \frac{1}{2}x - 5$

35. (15, −13); $y = -\frac{1}{5}x - 10$

36. (3, 3); $y = -\frac{2}{3}x + 1$

Determine another point on a line given two points on the line.

37. (2, −4), (4, −2)

38. (0, 5), (4, 1)

39. (−3, 1), (−1, −3)

40. (0, 4), (2, 5)

41. (−2, 9), (2, −1)

42. (3, 0), (12, 3)

For Exercises 43–45, determine which equation best represents each situation. Explain the meaning of each variable.

A. $y = \frac{1}{25}x + 300$ **B.** $y = 25x + 300$ **C.** $y = 300x + 25$

43. PLANES Plane tickets cost $300 each plus a fee of $25 to select seats per order.

44. SAVINGS Larry has $300. He saves $25 each week.

45. OIL The current oil level in a tank is 25 feet. The rate that oil is being poured into the tank is $\frac{1}{25}$ inch per hour.

46. USE A MODEL The table of ordered pairs shows the coordinates of the two points on the graph of a function. Write an equation that describes the function.

x	y
−2	2
4	−1

47. USE A SOURCE The table shows how women's shoe sizes in the United Kingdom compare to women's shoe sizes in the United States.

	Women's Shoe Sizes						
U.K.	3	3.5	4	4.5	5	5.5	6
U.S.	5.5	6	6.5	7	7.5	8	8.5

Source: DanceSport UK

a. Write a linear equation to determine the U.S. size y if you are given the U.K. size x.

b. What would be the U.S. shoe size for a woman who wears a U.K. size 7.5?

c. Research women's shoe sizes in Australia compared to women's shoe sizes in the United States. Write a linear equation to determine the U.S. size y if you are given the Australia size x.

48. REASONING Shikita borrowed money from her brother and paid back a set amount each week. The table shows how much she owed in a given week.

Week	3	6	8	10	13
Amount Owed	$32.50	$25.00	$20.00	$15.00	$7.50

a. Let x represent the number of weeks and y represent the amount owed. Write an equation in slope-intercept form to model the amount Shikita owed each week.

b. Describe the graph of the equation you found in **part a.** What does its shape tell you about the problem?

49. REASONING Koby tracked the weight of his puppy for 6 months. Her growth is shown in the table where $x =$ age in months and $y =$ weight in pounds.

x	y
2	16
3	23.5
4	31
5	38.5
6	46

a. Write an equation in slope-intercept form to model the growth of Koby's puppy.

b. What is the y-intercept? What does the y-intercept mean in the context of the problem?

c. What is the slope? What does the slope mean in the context of the problem?

Higher-Order Thinking Skills

50. PERSEVERE Write the equation of each line in slope-intercept form.

a. slope: $\frac{4}{5}$, y-intercept: -8

b. $(-3, 0)$, $(3, -16)$

c. What point do the graphs of both equations have in common, and what does this tell you about their graphs?

51. FIND THE ERROR Tess and Jacinta are writing an equation of the line through $(3, -2)$ and $(6, 4)$. Is either of them correct? Explain your reasoning.

Tess	Jacinta
$m = \frac{4-(-2)}{6-3} = \frac{6}{3}$ or 2	$m = \frac{4-(-2)}{6-3} = \frac{6}{3}$ or 2
$y = mx + b$	$y = mx + b$
$6 = 2(4) + b$	$-2 = 2(3) + b$
$6 = 8 + b$	$-2 = 6 + b$
$-2 = b$	$-8 = b$
$y = 2x - 2$	$y = 2x - 8$

52. WRITE Linear equations are useful in predicting future events. Describe some factors in real-world situations that might affect the reliability of the graph in making any predictions.

53. CREATE Create a real-world situation that fits the graph at the right. Define the two quantities and describe the functional relationship between them. Write an equation to represent this relationship, and describe what the slope and y-intercept mean.

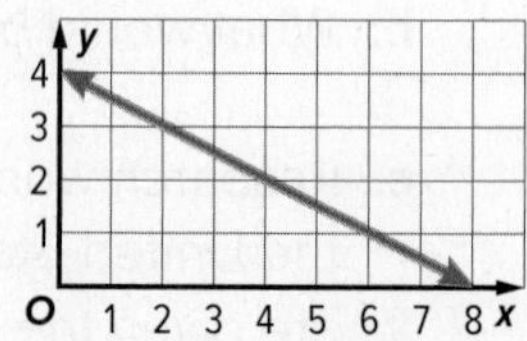

Writing Equations in Standard and Point-Slope Forms

Explore Forms of Linear Equations

Online Activity Use graphing technology to complete the Explore.

INQUIRY How are the point-slope and slope-intercept forms of a linear equation related?

Today's Goals
- Write equations of lines in point-slope form.
- Create and identify equations of parallel or perpendicular lines.

Today's Vocabulary
parallel lines
perpendicular lines

Learn Creating Linear Equations in Point-Slope Form

When the slope and the coordinates of one point of a line are known, an equation for the line can be written in point-slope form.

Key Concept • Point-Slope Form	
Words	The linear equation $y - y_1 = m(x - x_1)$ is written in point-slope form, where (x_1, y_1) is a given point on a nonvertical line and m is the slope of the line.
Symbols	$y - y_1 = m(x - x_1)$
Example	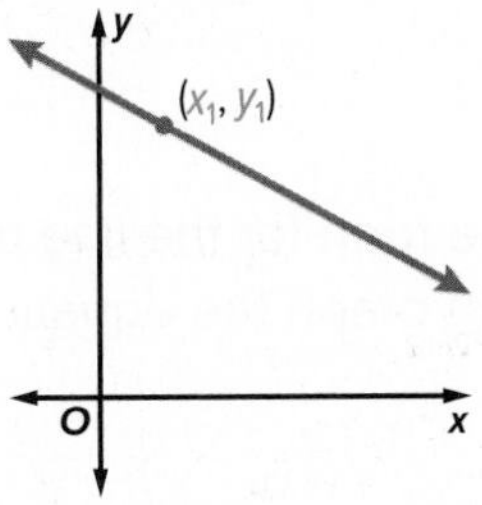

Talk About It!
Why must a line be nonvertical in order to be written in point-slope form? Explain.

If you are given two points on the line or a point on the line and its slope, you can write an equation for the line in point-slope form.

Key Concept • Writing Equations of Lines in Point-Slope Form			
Given the Slope and One Point		**Given Two Points**	
Step 1	Let the x and y coordinates be (x_1, y_1).	**Step 1**	Find the slope.
Step 2	Substitute the values of m, x_1, and y_1 into the equation of a line in point-slope form.	**Step 2**	Choose one of the two points to use.
		Step 3	Follow the steps for writing an equation given the slope and one point.

Example 1 Equation in Point-Slope Form Given Slope and a Point

Write an equation in point-slope form for the line that passes through (−2, 7) with a slope of $-\frac{3}{2}$. Then graph the equation.

$y - y_1 = m(x - x_1)$	Point-slope form
$y - 7 = -\frac{3}{2}[x - (-2)]$	$(x_1, y_1) = (-2, 7)$ and $m = -\frac{3}{2}$
$y - 7 = -\frac{3}{2}(x + 2)$	Simplify.

Step 1 Plot the given point (−2, 7).

Step 2 Use the slope, $-\frac{3}{2}$, to plot another point on the line.

Step 3 Draw a line through the points.

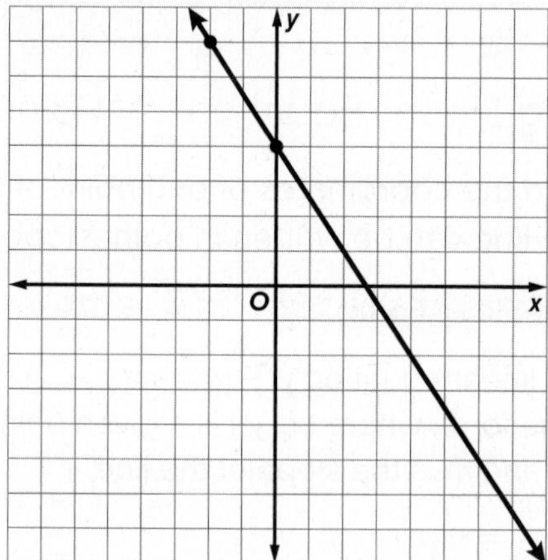

Check

Determine the equation in point-slope form for the line that passes through (7, 5) with a slope of −3. Then graph the equation.

y __?__ = __?__ $(x$ __?__ $)$

Go Online You can complete an Extra Example online.

Example 2 Equation in Point-Slope Form Given Two Points

Write an equation in point-slope form for the line that passes through the given points.

(2, −7) and (6, −3)

Step 1 Find the slope.

$m = \frac{y_2 - y_1}{x_2 - x_1}$ Slope Formula

$m = \frac{-3-(-7)}{6-2} = \frac{4}{4}$ or 1 $(x_1, y_1) = (2, -7)$ and $(x_2, y_2) = (6, -3)$

Step 2 Write an equation.

You can select either point for (x_1, y_1) in point-slope form.

$y - y_1 = m(x - x_1)$ Point-slope form

$y - (-3) = 1(x - 6)$ $(x_1, y_1) = (6, -3)$ and $m = 1$

$y + 3 = (x - 6)$ Simplify.

Check

Select an equation in point-slope form for the line that passes through (−16, 18) and (−11, −2).

A. $y + 2 = -4(x + 11)$

B. $y + 2 = -\frac{1}{4}(x + 11)$

C. $y - 2 = -4(x + 11)$

D. $y - 2 = -\frac{1}{4}(x - 11)$

E. None of these

Think About It!

Write another equation in point-slope form for the line with the points given.

Why are there multiple correct answers with the same given information?

Example 3 Change to Slope-Intercept Form

Write $y + 4 = -2(x - 6)$ in slope-intercept form.

$y + 4 = -2(x - 6)$ Original Equation

$y + 4 = -2x + 12$ Distributive Property

$y = -2x + 8$ Subtract 4 from each side.

Check

Write $y + 3 = -\frac{1}{2}(x - 8)$ in slope-intercept form.

$y =$ ___?___ $x +$ ___?___

Go Online You can complete an Extra Example online.

Study Tip

Checking Your Work To check your work, you can substitute the point from the original point-slope form of the equation, in this case (6, −4), into the slope-intercept form of the equation. If it is a true statement, the equation is correct.

$y = -2x + 8$

$-4 = -2(6) + 8$

$-4 = -12 + 8$

$-4 = -4$ ✓

Example 4 Apply Point-Slope Form

READING Nadia's book club is ordering new novels. She knows that the total cost of 5 books is $61.25, and 15 books cost $159.75. Write an equation in point-slope form to represent the total cost *y* of ordering *x* books.

Step 1 Find the slope.

$m = \frac{y_2 - y_1}{x_2 - x_1}$ — Slope Formula

$m = \frac{159.75 - 61.25}{15 - 5} = \frac{98.5}{10}$ or 9.85 — $(x_1, y_1) = (5, 61.25)$ and $(x_2, y_2) = (15, 159.75)$

Step 2 Write an equation.

$y - y_1 = m(x - x_1)$ — Point-slope form

$y - 61.25 = 9.85(x - 5)$ — $(x_1, y_1) = (5, 61.25)$ and $m = 9.85$

Think About It!
Use the equation to find the cost of purchasing 12 books.

Check

TAXIS The total cost of a taxi fare is given in the table. Determine the equation(s) in point-slope form that model(s) this situation if *x* represents the distance in miles and *y* represents the cost in dollars.

Distance (miles)	1.5	4	7.5	12.25
Cost (dollars)	6.90	13.40	22.50	34.85

A. $y - 13.4 = 2.6(x - 4)$

B. $y - 22.5 = 2.6(x - 7.5)$

C. $y = 2.6x + 3$

D. $y - 6.9 = \frac{5}{13}(x - 1.5)$

E. $y + 34.85 = 2.6(x + 12.25)$

Example 5 Change to Standard Form

Write $y - 1 = -\frac{2}{5}(x + 3)$ in standard form.

$y - 1 = -\frac{2}{5}(x + 3)$ — Original equation

$5(y - 1) = -2(x + 3)$ — Multiply each side by 5 to eliminate the fraction.

$5y - 5 = -2x - 6$ — Distributive Property

$5y = -2x - 1$ — Add 5 to each side.

$2x + 5y = -1$ — Add 2x to each side.

Study Tip

Fractional Slopes When working with an equation with a fractional slope, it is often simpler to first multiply each side of the equation by the denominator. This will eliminate distributing a fraction later in the equation.

Check

Write $y = -\frac{7}{2}x + 5$ in standard form.

Go Online You can complete an Extra Example online.

Example 6 Standard Form Given Two Points

Write an equation in standard form for the line that passes through (8, −4) and (−6, −11).

Step 1 Find the slope.

$m = \frac{y_2 - y_1}{x_2 - x_1}$ Slope Formula

$m = \frac{-11 - (-4)}{-6 - 8} = \frac{-7}{-14}$ or $\frac{1}{2}$ $(x_1, y_1) = (8, -4)$ and $(x_2, y_2) = (-6, -11)$

Step 2 Write an equation in slope-intercept form.

$y = mx + b$ Slope-intercept form

$-4 = \frac{1}{2}(8) + b$ $(x, y) = (8, -4)$ and $m = \frac{1}{2}$

$-4 = 4 + b$ Simplify.

$-8 = b$ Subtract 4 from each side.

$y = \frac{1}{2}x - 8$ Replace m with $\frac{1}{2}$ and b with −8.

Step 3 Write the equation in standard form.

$2y = 2\left(\frac{1}{2}x - 8\right)$ Multiply each side by 2.

$2y = x - 16$ Distributive Property

$-x + 2y = -16$ Subtract x from each side.

$x - 2y = 16$ Multiply each side by −1.

Check

Select the equation in standard form for the line that passes through (−9, 8) and (1, −12).

A. $x + 2y = -20$

B. $2x + y = -13$

C. $2x + y = 7$

D. $2x + y = -10$

E. $x + 2y = 20$

Go Online You can complete an Extra Example online.

Go Online
An alternate method is available for this example.

Think About It!
In Step 2, why is it possible to write an equation in either slope-intercept form or point-slope form and still get the same equation in standard form?

Learn Equations of Parallel and Perpendicular Lines

Nonvertical lines in the same plane that have the same slope are called **parallel lines**. Nonvertical lines in the same plane for which the product of the slopes is −1 are called **perpendicular lines**.

Key Concept • Slopes of Parallel and Perpendicular Lines

Parallel Lines	Perpendicular Lines
If two nonvertical lines are parallel, their slopes are the same. 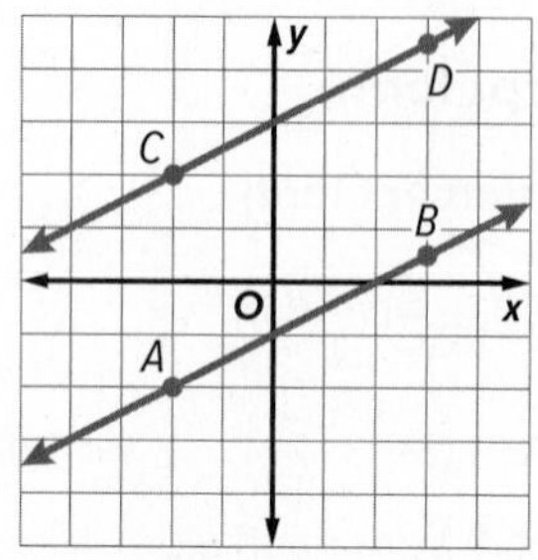 Since both lines have a slope of $\frac{1}{2}$, $\overleftrightarrow{AB} \parallel \overleftrightarrow{CD}$.	If two nonvertical lines are perpendicular, the product of their slopes is −1. 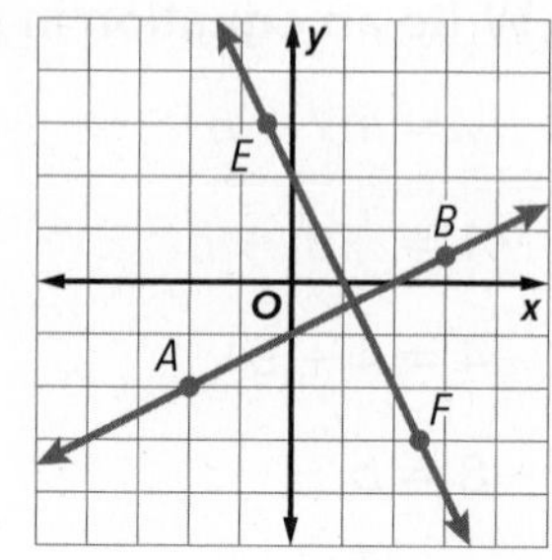 Since $\frac{1}{2}(-2) = -1$, $\overleftrightarrow{AB} \perp \overleftrightarrow{EF}$.

Think About It!

If the given line is vertical, what is the slope of any line parallel to the given line? perpendicular to the given line?

You can write an equation of a line parallel or perpendicular to a given line if you know a point on the line and an equation of the given line.

Go Online

You may want to complete the Concept Check to check your understanding.

Key Concept • Writing Equations of Lines Parallel or Perpendicular to a Given Line

Parallel Lines		Perpendicular Lines	
Step 1	Identify the slope m of the given line.	**Step 1**	Identify the slope m of the given line. The slope of the line perpendicular to the original line is $-\frac{1}{m}$.
Step 2	Use the point-slope form with slope m and the coordinates of the given point.	**Step 2**	Use the point-slope form with slope $-\frac{1}{m}$ and the coordinates of the given point.
Step 3	Rewrite the equation in the needed form.	**Step 3**	Rewrite the equation in the needed form.

Example 7 Parallel Line Through a Given Point

Write an equation in slope-intercept form for the line that passes through (−4, 2) and is parallel to the graph of $y = 3x - 5$.

Step 1 Identify the slope of the given line.

The slope of the line with equation $y = 3x - 5$ is 3. The line parallel to that line has the same slope, 3.

Go Online You can complete an Extra Example online.

Steps 2, 3 Write the equation of the parallel line.

Use the point-slope form to rewrite the equation in slope-intercept form.

$y - y_1 = m(x - x_1)$	Point-slope form
$y - 2 = 3[x - (-4)]$	$(x_1, y_1) = (-4, 2)$ and $m = 3$
$y - 2 = 3(x + 4)$	Simplify.
$y - 2 = 3x + 12$	Distributive Property
$y = 3x + 14$	Add 2 to each side.

Study Tip

Checking Your Work To check that your equation represents the correct line, graph both lines. Verify that the lines appear to be parallel and that your line passes through the given point.

Check

Write an equation for the line that passes through (8, 2) and is parallel to the graph of $y = \frac{3}{4}x + 2$.

Example 8 Perpendicular Line Through a Given Point

Write an equation in slope-intercept form for the line that passes through (1, −2) and is perpendicular to the graph of $3x + 2y = 12$.

Step 1 Identify the slope of the given line.

Write the equation in slope-intercept form.

$3x + 2y = 12$	Original equation
$3x - 3x + 2y = 12 - 3x$	Subtract $3x$ from each side.
$2y = -3x + 12$	Simplify.
$\frac{2y}{2} = \frac{-3x}{2} + \frac{12}{2}$	Divide each side by 2.
$y = -\frac{3}{2}x + 6$	Simplify.

The slope of the line with equation $3x + 2y = 12$ is $-\frac{3}{2}$. The slope of the line perpendicular to that line is the opposite reciprocal, $\frac{2}{3}$.

Steps 2, 3 Write the equation of the perpendicular line.

Use the point-slope form to rewrite the equation in slope-intercept form.

$y - y_1 = m(x - x_1)$	Point-slope form
$y - (-2) = \frac{2}{3}(x - 1)$	$(x_1, y_1) = (1, -2)$ and $m = \frac{2}{3}$
$y + 2 = \frac{2}{3}(x - 1)$	Simplify.
$y + 2 = \frac{2}{3}x - \frac{2}{3}$	Distributive Property
$y = \frac{2}{3}x - \frac{8}{3}$	Subtract 2 from each side.

Go Online You can complete an Extra Example online.

Check

Select the equation in slope-intercept form for the line that passes through (5, 0) and is perpendicular to the graph of $x - 6y = 1$.

A. $y = -6x + 30$

B. $6x + y = 30$

C. $y = -\frac{1}{6}x + 2$

D. $x - 6y = 5$

Example 9 Determine Line Relationships

Determine whether $\overleftrightarrow{AB}$ and $\overleftrightarrow{EF}$ are *parallel, perpendicular,* or *neither* for $A(6, 8)$, $B(2, 5)$, $E(-6, -3)$, and $F(0, 5)$.

Step 1 Find the slope of each line.

slope of $\overleftrightarrow{AB} = \frac{8-5}{6-2} = \frac{3}{4}$ slope of $\overleftrightarrow{EF} = \frac{-3-5}{-6-0} = \frac{-8}{-6}$ or $\frac{4}{3}$

Step 2 Determine the relationship.

parallel To determine whether the lines are parallel, compare their slopes. The two lines do not have the same slope, so they are not parallel.

perpendicular To determine whether the lines are perpendicular, find the product of their slopes. $\frac{3}{4} \cdot \frac{4}{3} = 1$

Since the product of the slopes is not -1, $\overleftrightarrow{AB}$ and $\overleftrightarrow{EF}$ are not perpendicular.

Check

Determine whether $\overleftrightarrow{CD}$ and $\overleftrightarrow{KL}$ are *parallel, perpendicular,* or *neither* for $C(4, 10)$, $D(-1, 12)$, $K(6, -5)$, and $L(1, -3)$.

$\overleftrightarrow{CD}$ and $\overleftrightarrow{KL}$ are ___?___.

Complete each sentence given $y = ax - 5$ and $y = bx + 3$.

When $a = 4$ and $b = 4$, the graphs are ___?___.

When $a = -3$ and $b = 5$, the graphs are ___?___.

When $a = -2$ and $b = \frac{1}{2}$, the graphs are ___?___.

Go Online to practice what you've learned about writing linear equations in the Put It All Together over Lessons 5-1 and 5-2.

Go Online You can complete an Extra Example online.

Practice

Go Online You can complete your homework online.

Example 1

Write an equation in point-slope form for the line that passes through each point with the given slope. Then graph the equation.

1. $(-6, -3), m = -1$

2. $(-7, 6), m = 0$

3. $(-2, 11), m = \frac{4}{3}$

Example 2

Write an equation in point-slope form for the line that passes through the given points.

4. $(-4, 6), (-2, 22)$

5. $(1, -3), (4, -15)$

6. $(4, -6), (6, -4)$

7. $(3, 3), (6, 7)$

Example 3

Write each equation in slope-intercept form.

8. $y - 1 = \frac{4}{5}(x + 5)$

9. $y + 5 = -6(x + 7)$

10. $y + 6 = -\frac{3}{4}(x + 8)$

11. $y + 2 = \frac{1}{6}(x - 4)$

Example 4

12. **NATURE** The frequency of a male cricket's chirp is related to the outdoor temperature. The relationship is expressed by the graph, where y is the temperature in degrees Fahrenheit and x is the number of chirps the cricket makes in 14 seconds. Write an equation for the line in point-slope form.

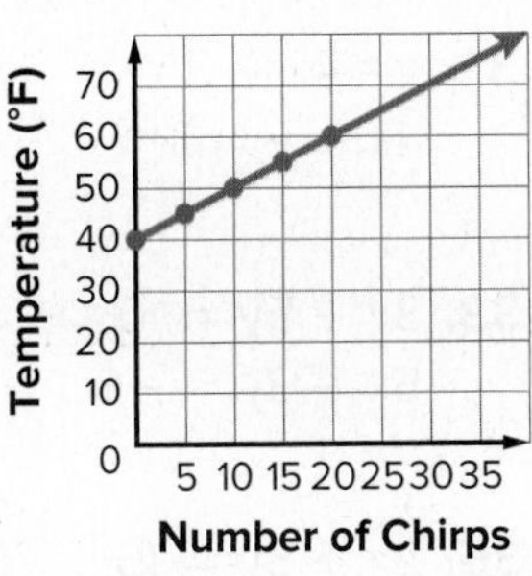

13. **CANOEING** Geoff paddles his canoe at an average speed of 3.5 miles per hour. After 5 hours of canoeing, Geoff has traveled 18 miles. Write an equation in point-slope form to find the total distance y Geoff travels after x hours.

14. **GEOMETRY** The perimeter of a square is four times the length of one side. If the side length of a square is 1 centimeter, then the perimeter of the square is 4 centimeters. Write an equation in point-slope form to find the perimeter y of a square with side length x.

Example 5

Write each equation in standard form.

15. $y - 10 = 2(x - 8)$

16. $y + 7 = -\frac{3}{2}(x + 1)$

17. $2y + 3 = -\frac{1}{3}(x - 2)$

18. $4y - 5x = 3(4x - 2y + 1)$

19. $y = x + 1$

20. $y = \frac{1}{3}x - 10$

Example 6

Write an equation in standard form for the line that passes through the given points.

21. $(-2, -3), (4, -7)$

22. $(2, 7)$ and $(-5, 2)$

23. $(-4, 9), (2, -9)$

24. $(-1, 19)$ and $(3, 35)$

Examples 7 and 8

Write an equation in slope-intercept form for the line that passes through the given point and is parallel to the graph of the equation. Then write an equation for the line that passes through the given point and is perpendicular to the graph of the equation.

25. $(3, -2); y = x + 4$

26. $(4, -3); y = 3x - 5$

27. $(0, 2); y = -5x + 8$

28. $(-4, 2); y = -\frac{1}{2}x + 6$

29. $(-2, 3); y = -\frac{3}{4}x + 4$

30. $(9, 12); y = 13x - 4$

Example 9

Determine whether the graphs of each pair of equations are *parallel*, *perpendicular*, or *neither*.

31. $y = 4x + 3$
$4x + y = 3$

32. $y = -2x$
$2x + y = 3$

33. $3x + 5y = 10$
$5x - 3y = -6$

34. $-3x + 4y = 8$
$-4x + 3y = -6$

35. $2x + 5y = 15$
$3x + 5y = 15$

36. $2x + 7y = -35$
$4x + 14y = -42$

Mixed Exercises

37. Write an equation in standard form with an *x*-intercept of 4 and a *y*-intercept of 5.

Write each equation in slope-intercept and standard forms.

38. $y + 3 = -\frac{1}{3}(2x + 6)$

39. $y + 4 = 3(3x + 3)$

40. $y - 6 = -3(x + 2)$

41. $y - 9 = -6(x + 9)$

42. $y + 4 = \frac{2}{3}(x + 7)$

43. $y + 7 = \frac{9}{10}(x + 3)$

44. Consider the graphs of the following equation.

$y = -2x$ $\qquad 2y = x$ $\qquad 4y = 2x + 4$

a. Which equations are parallel? Explain your reasoning.

b. Which equations are perpendicular? Explain your reasoning.

45. INSPECTIONS Mrs. Sanchez is inspecting a shed to determine if it is safe to use for storing football equipment. Mrs. Sanchez mapped the top view of the ceiling walls of the shed on a coordinate plane. If one of the walls lies from (−6, 11.5) to (2, 9.5) and the second wall lies from (−1, −2.5) to (2, 9.5), are the walls perpendicular? Explain your reasoning.

46. Nya mapped a quadrilateral on a coordinate plane. If she plots one segment from (−3, −3) to (3, 9) and another segment from (−5, 12) to (1, 0), are the segments parallel? Justify your reasoning.

47. STRUCTURE Immediately after take-off, a jet plane consistently climbs 20 feet for every 40 feet it moves horizontally. The graph shows the trajectory of the jet.

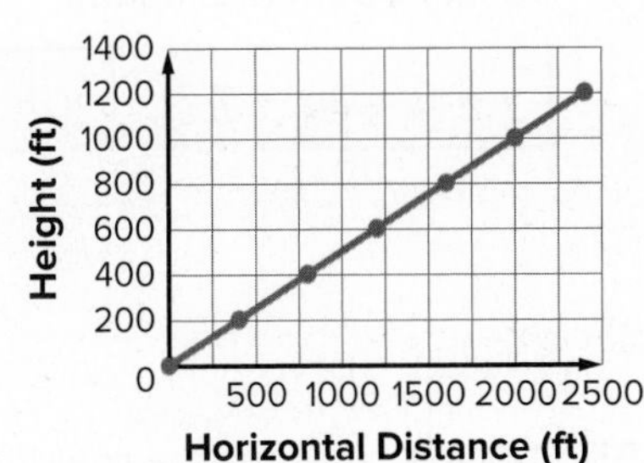

a. Write an equation in point-slope form for the line representing the jet's trajectory.

b. Write the equation from **part a** in slope-intercept form.

c. Write the equation in standard form.

48. CONSTRUCT ARGUMENTS Consider three points, (3, 7), (−6, 1), and (9, p) on the same line. Find the value of p. Justify your argument.

49. WRITE What information is needed to write the equation of a line? Explain.

50. ANALYZE Levy claims that the line through (−6, −2) and (2, 10) is perpendicular to the graph of $3x - 2y = 10$. Do you agree? Justify your argument.

51. ANALYZE Jeremiah says the line through (7, −10) and (3, −2) is parallel to $2x - y = -5$. Do you agree? Justify your argument.

52. FIND THE ERROR Alonae says that the line through (1, −4) and (5, −6) is parallel to the line through (2, −7) and (5, −6). How can you tell she is mistaken without determining the slope? Explain your reasoning.

53. PERSEVERE Write an equation in point-slope form for the line that passes through the points (f, g) and (h, j).

54. WHICH ONE DOESN'T BELONG? Identify the equation that does not belong. Justify your conclusion.

$y - 5 = 3(x - 1)$	$y + 1 = 3(x + 1)$	$y + 4 = 3(x + 1)$	$y - 8 = 3(x - 2)$

55. CREATE Describe a real-life scenario that has a constant rate of change and a value of y for a particular value of x. Represent this situation using an equation in point-slope form and an equation in slope-intercept form.

Scatter Plots and Lines of Fit

Learn Scatter Plots

Bivariate data consists of pairs of values. A **scatter plot** is a graph of bivariate data that consists of ordered pairs on a coordinate plane. Using a scatter plot can help you see the **trend,** or general pattern, in the data. Trends can be described as positive or negative correlations.

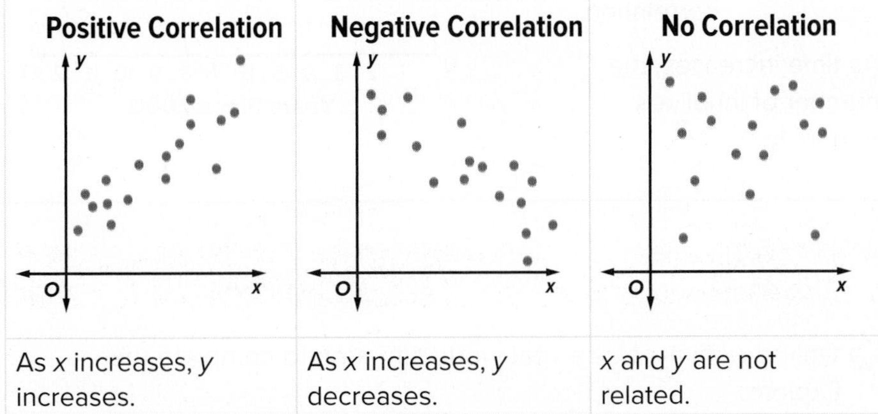

As *x* increases, *y* increases.	As *x* increases, *y* decreases.	*x* and *y* are not related.

Notice that in the graphs for positive and negative correlations, many of the points form **clusters** of points that slope upward or downward. Points outside of clusters are **outliers**.

Today's Goals

- Categorize the correlation of a set of data in a scatter plot.
- Make and evaluate predictions by fitting linear functions to sets of data.

Today's Vocabulary

bivariate data
scatter plot
trend
positive correlation
negative correlation
no correlation
line of fit
linear extrapolation
linear interpolation

Study Tip

Labeling Axes Because scatter plots display bivariate data, it is critical to label axes with their corresponding units. Otherwise, the graph may not make sense.

Example 1 Evaluate Correlation

FOOTBALL **The scatter plot displays the height and weight of New Orleans Saints football players. Determine whether the scatter plot shows a *positive, negative,* or *no* correlation. If the correlation is positive or negative, describe its meaning in the situation.**

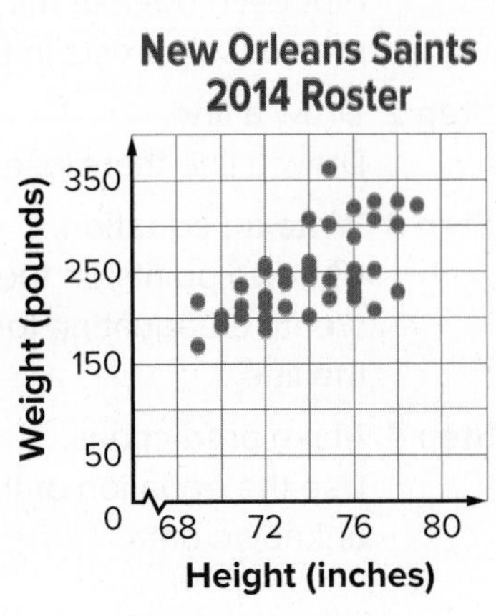

The scatter plot shows a positive correlation. As the height of the football player increases, weight usually increases.

Think About It!

What type of correlation would you expect between a player's jersey number and his birth month?

 **Go Online** You can complete an Extra Example online.

Check

TELEPHONES The scatter plot displays the number of landline telephones in the United States, in 100 millions, since 2000.

Determine whether the scatter plot shows a *positive*, *negative*, or *no correlation*. Describe the correlation's meaning in the situation.

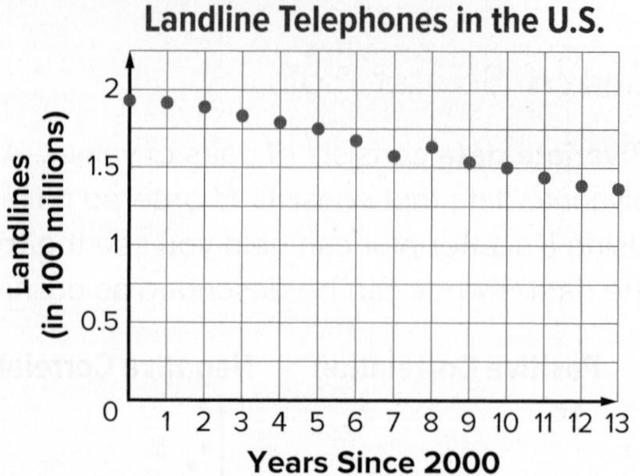

The scatter plot shows ___?___ correlation.

As time increases, the number of landlines generally ___?___.

Study Tip

Determining Correlation Similar to slope, when data points are generally increasing from left to right, there is a positive correlation. Negative correlation occurs when the data points generally decrease from left to right. If you are unable to tell if the data are increasing or decreasing, there is probably no correlation.

Explore Make Predictions by Using a Scatter Plot

Online Activity Use a real-world situation to complete the Explore.

INQUIRY How can you use a scatter plot to estimate unknown data?

Learn Lines of Fit

A **line of fit** is used to describe the trend of the data in a scatter plot.

Key Concept • Using a Linear Function to Model Data	
Step 1	**Make a scatter plot.** Plot each point of the data and determine whether any relationship exists in the data.
Step 2	**Draw a line.** Draw a line that closely follows the trend in the data.
Step 3	**Write an equation.** Use two points on the line of fit to find the slope of the line and create an equation for the line using the slope and a point on the line.
Step 4	**Make predictions.** Use the equation of the line of fit to make predictions about unknown data.

Linear extrapolation is the use of a linear equation to predict values that are outside of the range of data. **Linear interpolation** is the use of a linear equation to predict values that are inside of the data range.

Go Online You can complete an Extra Example online.

Think About It!

How can you ensure that your data predictions that are outside the range of data are as accurate as possible?

Example 2 Write an Equation for a Line of Fit

BOATS **The table shows the average cost of a jet boat in the years after 2000. Write an equation to represent the data. Then, use the equation to predict the cost of a jet boat in 2005 and 2025.**

Years Since 2000	Cost ($)
0	17,663
1	19,144
2	21,176
3	20,584
4	23,280
6	24,443
7	27,784

Years Since 2000	Cost ($)
8	28,088
9	29,774
10	32,752
11	34,082
12	35,589
13	37,618

Step 1 Make a scatter plot.

The independent variable is the number of years since 2000 and the dependent variable is the cost of the jet boats. As the years increase, the cost of the jet boats also increases. This scatter plot shows positive correlation.

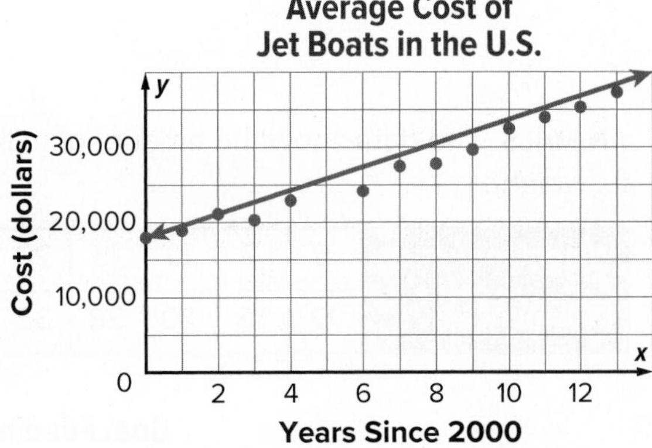

Step 2 Draw a line of fit.

A line is drawn that follows the trend of the data points and passes close to most of the points.

Step 3 Write an equation.

The line of fit passes close to the data points (5, 25,108) and (10, 32,752).

Find the slope.

$m = \frac{y_2 - y_1}{x_2 - x_1}$ Slope Formula

$= \frac{32,752 - 25,108}{10 - 5}$ $(x_1, y_1) = (5, 25,108)$ and $(x_2, y_2) = (10, 32,752)$

$= \frac{7644}{5}$ or 1528.8 Simplify.

Use $m = 1528.8$ and a point to write an equation.

$y - y_1 = m(x - x_1)$ Point-slope form

$y - 25,108 = 1528.8(x - 5)$ $(x_1, y_1) = (5, 25,108)$

$y - 25,108 = 1528.8x - 7644$ Distribute.

$y = 1528.8x + 17,464$ Simplify.

(continued on the next page)

Go Online You can complete an Extra Example online.

Watch Out!

Variations Equations for scatter plots generally do not have an exact correct solution. Equations will vary depending on how the line of fit was drawn and which points were selected when writing the equation. So, your solutions may not be exactly the same as another student's solutions or the sample answers given.

Think About It!

What do the slope and *y*-intercept mean in the context of this example?

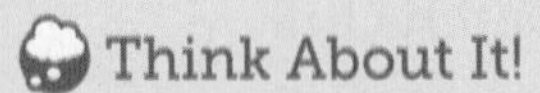

Think About It!

What assumptions are made when using a line of fit to make predictions about the cost of a jet boat in a given year?

Step 4 Predict the cost in 2005.

Use evaluation to predict the cost of a jet boat in 2005.

Since the independent variable represents the number of years after 2000, the value of x is 2005 − 2000 or 5.

$y = 1528.8x + 17{,}464$	Equation of the line of fit
$y = 1528.8(5) + 17{,}464$	$x = 5$
$y = 7644 + 17{,}464$ or 25,108	Simplify.

We can predict that the cost of a jet boat in 2005 was about $25,108.

Step 5 Predict the cost in 2025.

Extrapolate the data to determine the cost of a jet boat in 2025.

$y = 1528.8x + 17{,}464$	Equation of the line of fit
$y = 1528.8(25) + 17{,}464$	$x = 25$
$y = 38{,}220 + 17{,}464$ or 55,684	Simplify.

We can predict that the cost of a jet boat in 2025 will be about $55,684.

Check

ANIMALS The data show the amount of milk that a baby goat needs by its weight.

Weight (pounds)	5	7	10	15	20	25	30	40	50
Milk (ounces)	12	16	20	28	32	40	48	64	80

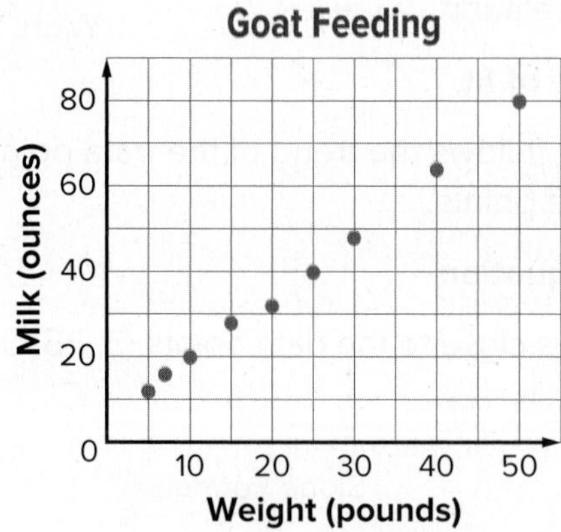

Part A Use the data points (10, 20) and (50, 80), which are contained in the line of fit, to write an equation of the line in slope-intercept form.

Part B Use the equation from Part A to predict the amount of milk needed for a 17-pound goat and a 55-pound goat.

17-pound goat: ___?___ ounces

55-pound goat: ___?___ ounces

 **Go Online** You can complete an Extra Example online.

Practice

Go Online You can complete your homework online.

Example 1

Determine whether each scatter plot shows a *positive, negative,* or *no* correlation. If the correlation is positive or negative, describe its meaning in the situation.

1.

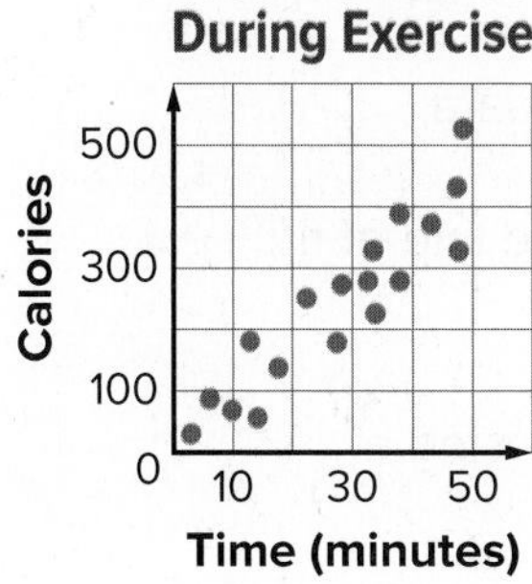

2.

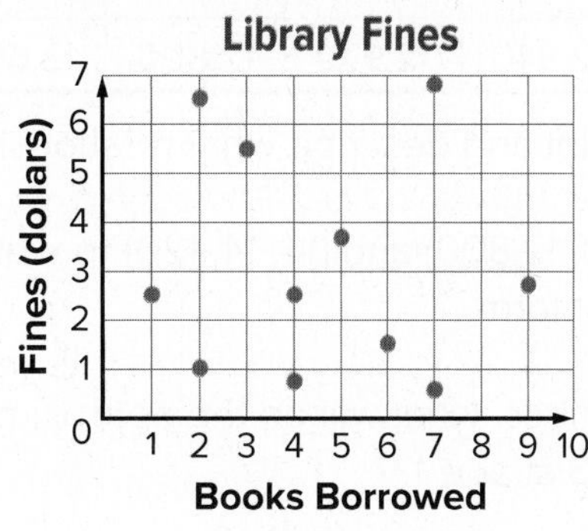

3.

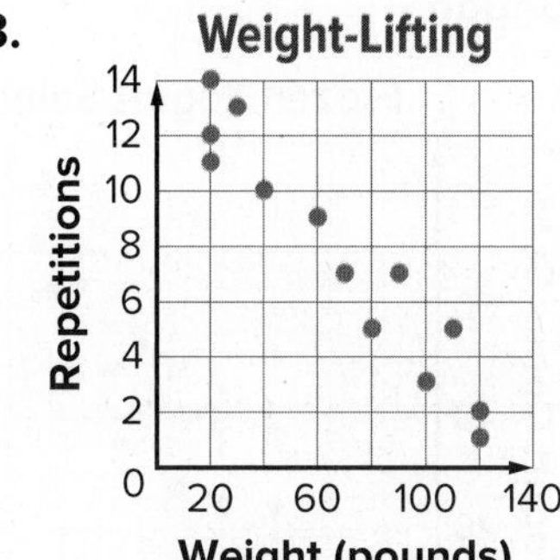

4.

Example 2

5. **MUSIC** The scatter plot shows the number of CDs in millions that were sold from 2011 to 2016.

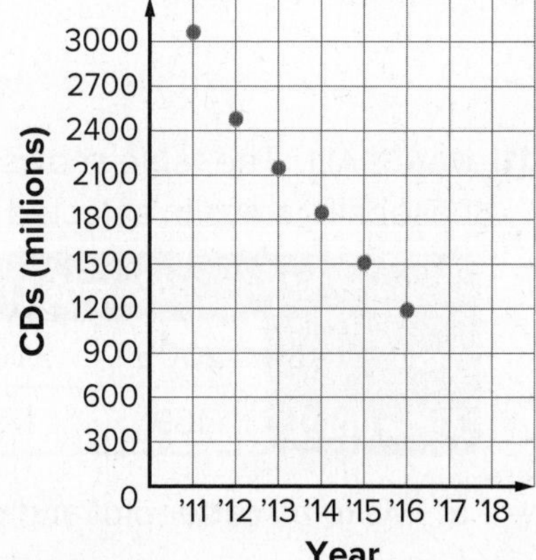

   a. Use the points (2, 2485.6) and (6, 1172.5) to write an equation of the line of fit in slope-intercept form. Let x be the years since 2010.

   b. If the trend continued, about how many CDs were sold in 2019?

6. **HOUSING** The data show the median price of an existing home from 2010 to 2015.

Year	2010	2011	2012	2013	2014	2015
Price	222,900	226,900	238,400	258,400	275,200	296,500

   a. Use the points (1, 226.9) and (4, 275.2) to write an equation for the line of fit in slope-intercept form, where x is the number of years since 2010 and y is the median price in thousands of dollars.

   b. If the trend continues, what will be the approximate median price of an existing home in 2025?

Mixed Exercises

7. **FAMILY** The table shows the predicted annual cost for a middle-income family to raise a child from birth until adulthood.

Cost of Raising a Child Born in 2013					
Child's Age	2	5	8	11	14
Annual Cost ($)	12,940	12,970	12,800	13,600	14,420

  a. Make a scatter plot and describe what relationship exists within the data.

  b. Use the points (8, 12,800) and (14, 14,420) to write the equation of the line of fit in slope-intercept form.

  c. If the trend continues, what will be the approximate annual cost of raising a child born in 2013 at age 17?

Determine whether each scatter plot shows a *positive, negative,* or *no* correlation. If the correlation is positive or negative, describe its meaning in the situation.

8. 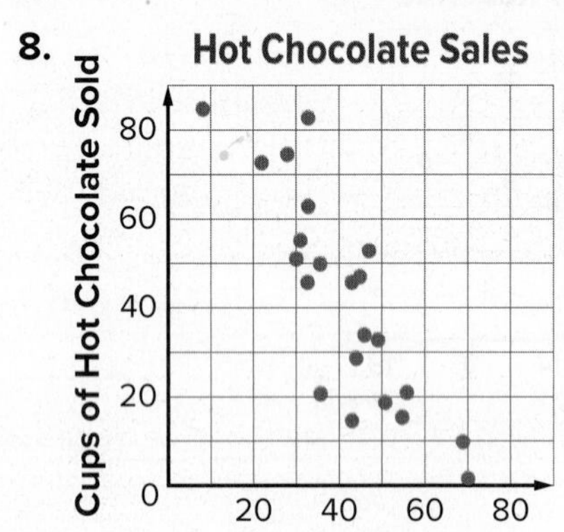

9. 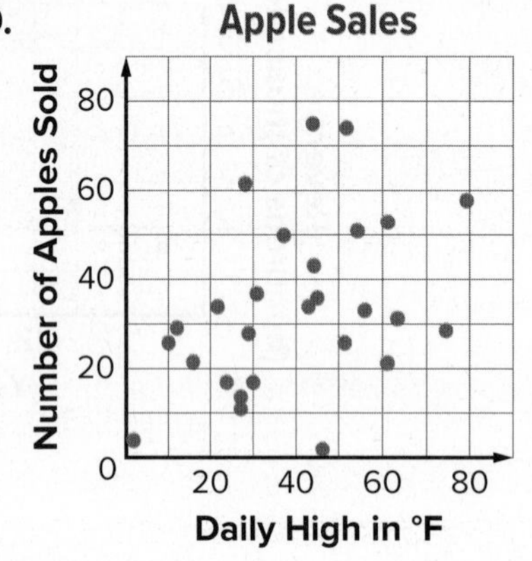

10. 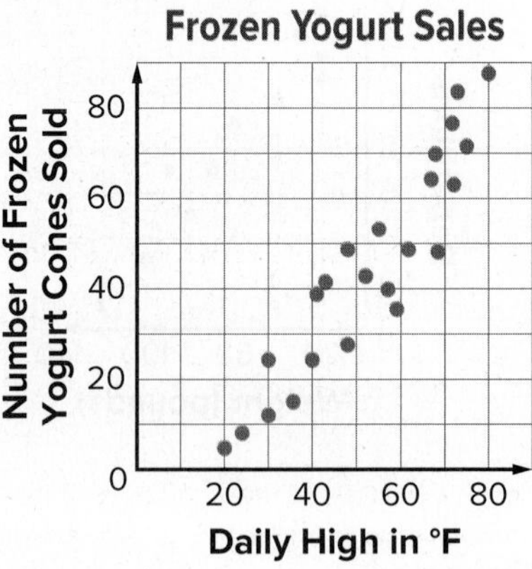

11. **BASEBALL** The table shows the average length in minutes of professional baseball games in selected years.

Average Length of Major League Baseball Games							
Year	2005	2007	2009	2011	2013	2015	2017
Time (min)	169	175	175	176	184	180	189

  a. Make a scatter plot and draw a line of best fit.

  b. Write the equation of the line of fit in slope-intercept form where x is the number of years since 2005. Explain your process.

  c. If the trend continues, what will be the approximate length of a major league baseball game in 2021?

  d. How reliable is the predicted length of a major league baseball game in 2021? Justify your argument.

12. **INCOME** The table shows the average median income for selected ages.

Age (years)	26	27	28	29	30
Median Income ($1000)	16.8	19.1	23.3	25.8	33.9

a. Make a scatter plot relating age to median income. Then draw a line of fit for the scatter plot.

b. Determine whether the graph shows a *positive, negative,* or *no correlation.* If the correlation is positive or negative, describe its meaning in the situation.

c. Use the table to write an equation of the line of fit.

d. Use the line of best fit to predict the median income for 32-year-olds.

13. **FOOTBALL** The scatter plot shows the average price of a National Football League ticket from 2007 to 2016.

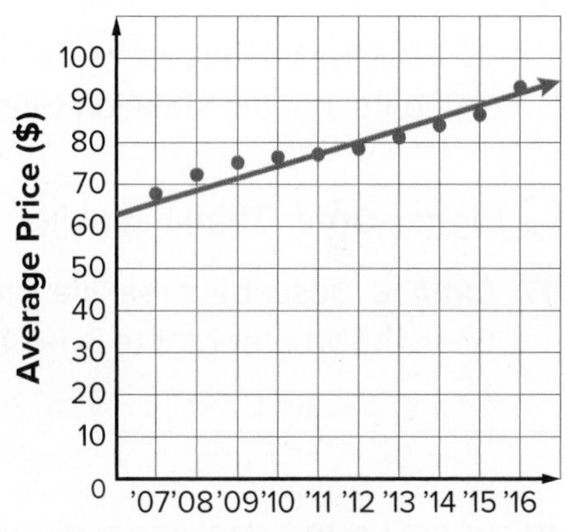

a. Determine what relationship, if any, exists in the data. Explain.

b. Use the points (2007, 67.11) and (2016, 92.98) to write the slope-intercept form of an equation for the line of fit shown, where x is the number of years since 2006. Round to the nearest hundredth.

c. Predict the price of a ticket in 2030.

14. **STRUCTURE** Refer to the scatter plot at the right.

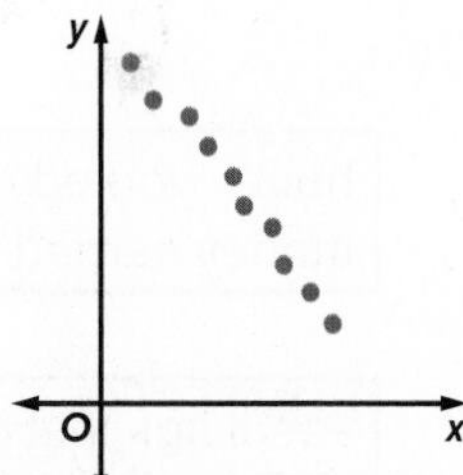

a. Describe the trend in the data shown in the scatter plot and the relationship between x and y.

b. Describe a real-life situation that could be modeled by the given scatter plot. Explain your reasoning.

15. **USE A MODEL** The table gives the life expectancy of a child born in the United States in a given year.

a. Make a scatter plot of the data. Then draw a line of fit.

b. Use the data to predict the life expectancy of a baby born in 2023. Round to the nearest tenth. Does your answer follow the trend of the data? Explain.

c. Explain any assumptions you made when using the line of fit to extrapolate and find the life expectancy of a baby born in 2023.

Years of Life Expected at Birth	
Year of Birth	Life Expectancy (years)
1930	59.7
1940	62.9
1950	68.2
1960	69.7
1970	70.8
1980	73.7
1990	75.4
2000	77.0
2010	78.7

16. **USE TOOLS** Several groups volunteered to clean up litter along a mile of the highway near their town. The table shows how many people were in each group and how long it took each group to finish the job.

Workers	9	16	18	8	15	11	9	17	9	15	11	12
Minutes	80	40	35	90	60	60	70	30	70	50	80	70

a. Graph the data on a scatter plot.

b. Draw a line of best fit to show the trend of the data.

c. Choose two points on the line of fit. Then find the equation of the line in slope-intercept form.

d. Another group wants to get done in 45 minutes. About how many workers should they have? Explain your reasoning.

e. Find the y-intercept of the line of fit. Does the y-intercept make sense in the context of the situation? Justify your argument.

Higher-Order Thinking Skills

17. **CREATE** Describe a real-life situation that can be modeled using a scatter plot. Describe whether there is *positive, negative,* or *no* correlation.

18. **WHICH ONE DOESN'T BELONG?** Analyze the following situations and determine which one does not belong. Justify your conclusion.

hours worked and amount of money earned	height of an athlete and favorite color
seedlings that grow an average of 2 centimeters each week	number of photos stored on a camera and capacity of camera

19. **ANALYZE** Determine which line of fit shown is a better fit for the data in the scatter plot. Justify your argument.

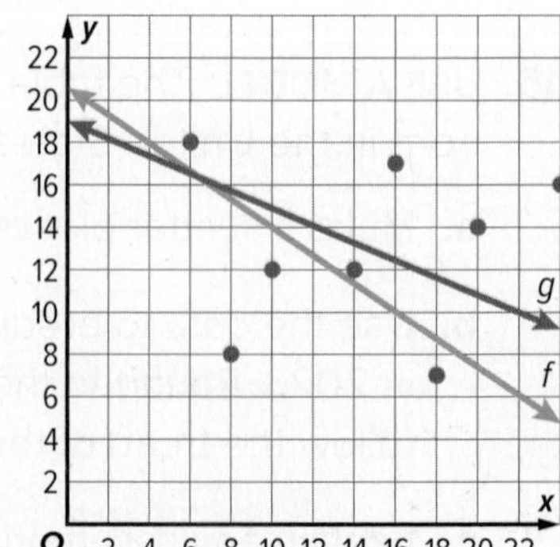

20. **WRITE** Does an accurate line of fit always predict what will happen in the future? Explain your reasoning.

21. **CREATE** Make a scatter plot that shows the height of a person and age. Explain how you could use the scatter plot to predict the age of a person given his or her height. How can the information from a scatter plot be used to identify trends and make decisions?

Correlation and Causation

Today's Goal
- Determine correlation or causation.

Today's Vocabulary
causation

Explore Collecting Data to Determine Correlation and Causation

Online Activity Use a real-wolrd situation to complete an Explore.

> **INQUIRY** What is the difference between correlation and causation?

Learn Correlation and Causation

Causation occurs when a change in one variable produces a change in another variable. It is the relationship between cause and effect. Correlation, however, can be observed between many variables.

Think About It!
Why does correlation not prove causation?

Key Concept • Correlation and Causation	
Step 1	Graph ordered pairs to create a scatter plot.
Step 2	Determine whether the scatter plot shows a positive or negative correlation.
Step 3	Determine whether the two sets of data are related. Does one variable *cause* the other? Could other factors be influencing the data results?
Step 4	Decide if the data illustrate correlation or causation.

Example 1 Correlation and Causation by Graphing

ANALYSIS **The data show the per capita consumption of mozzarella cheese and the number of civil engineering doctoral degrees awarded in the United States. Determine whether the data plotted on the graph illustrate a *correlation* or *causation*.**

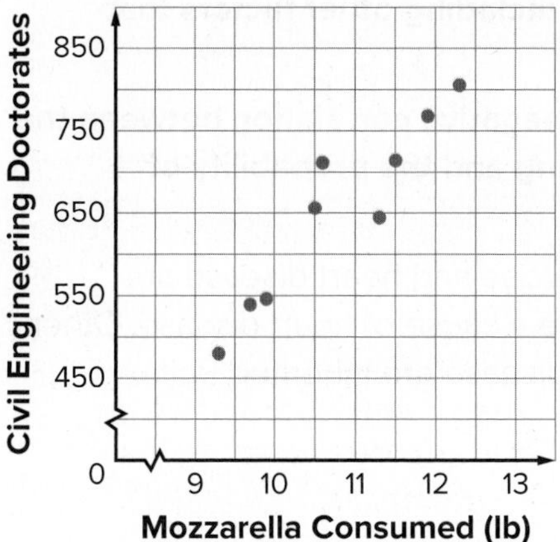

(continued on the next page)

Go Online You can complete an Extra Example online.

Step 1 Determine the correlation.

As the amount of mozzarella consumed increases, the number of civil engineering doctorates also increases. The scatter plot shows a positive correlation.

Step 2 Determine causation.

Consumption of mozzarella does not cause anyone to obtain a doctoral degree in civil engineering. These two sets of data are not related. Many factors may affect the increase in these two areas. As the demand for more roadways, airports, and water and sewage treatment plants grows, the demand for more civil engineers also increases. An increase in the per capita consumption of mozzarella may be related to increased pizza sales or dairy production. Both variables are affected by a general increase in population.

Step 3 Determine whether the data illustrate a *correlation* or *causation*.

The data exhibit a correlation, but there is no causation.

Talk About It!

Describe an experiment that could be conducted to show causation between the number of civil engineers who were awarded a doctoral degree and another factor.

Check

ANALYSIS Determine whether the data illustrate a *correlation* or *causation*.

Month	March	April	May	June	July	August
Sunscreen Sold	14	37	84	117	135	98
Sunglasses Sold	6	11	28	36	40	39

The data show a ___?___ correlation. As the number of bottles of sunscreen sold increases, the number of sunglasses sold ___?___. These data illustrate a ___?___.

Example 2 Correlation and Causation by Situation

Determine whether the situation illustrates a *correlation* or *causation*. Explain your reasoning, including other factors that might be involved.

A university experiment showed a negative correlation between the average weekly time spent exercising and the probability of developing heart disease.

This situation models causation. Exercise and heart disease are related, and lack of exercise could be a cause of heart disease. Other factors that might have led to heart disease are inherited traits, smoking, or a poor diet.

 **Go Online** You can complete an Extra Example online.

Practice

Go Online You can complete your homework online.

Example 1

1. FROZEN DESSERTS The table shows the number of pounds of frozen yogurt and the number of pounds of sherbet consumed per capita in the United States from 2009 to 2014.

Year	2009	2010	2011	2012	2013	2014	2015	2016
Pounds of Frozen Yogurt	0.9	1	1.2	1.1	1.4	1.3	1.4	1.2
Pounds of Sherbet	1	1	0.9	0.8	0.9	0.9	0.8	0.8

a. Graph the ordered pairs (pounds of frozen yogurt, pounds of sherbet) to create a scatter plot.

b. Does the scatter plot show a *positive, negative,* or *no* correlation? Explain.

c. Determine whether the data illustrate a *correlation* or *causation*. What other factors may influence the data?

2. LEISURE ACTIVITIES The table shows the average number of minutes a person reads per weekday and the average number of minutes a person watches television per weekday.

Age	15	25	35	45	55	65
Minutes Reading	7	9	12	17	30	50
Minutes Watching Television	117	115	113	127	155	236

a. Graph the ordered pairs as a scatter plot (minutes reading, minutes watching television).

b. Does the scatter plot show a *positive, negative,* or *no* correlation? Explain.

c. Determine whether the data illustrate a *correlation* or *causation*. What other factors may influence the data?

Example 2

Determine whether each situation illustrates a *correlation* or *causation*. Explain your reasoning.

3. A class experiment shows a negative correlation between the width of a person's palm and the amount of time they spend watching television each day.

4. The larger a person's shoe size, the higher a person's reading level.

5. At a grocery store, there is a negative correlation between the price of cereal and number of boxes of cereal sold.

6. Hae notices that the lower the daily temperature is, the less time she spends outside.

Mixed Exercises

7. **GARDENING** Jalen weighs each type of fruit his garden produces each week.

Week	1	2	3	4	5
Strawberries (lb)	6.5	8	12	13.5	20
Blueberries (lb)	6	5	4.5	3	2.5

a. Graph the ordered pairs (pounds of strawberries, pounds of blueberries) to create a scatter plot.

b. Does the scatter plot show a *positive, negative,* or *no* correlation? Explain.

c. Determine whether the data illustrate a *correlation* or *causation*. What other factors may influence the data?

8. **SHOES** The table shows the number of pairs of sandals and snow boots sold at a certain store during various months of the year.

Month	January	April	July	December
Sandals	12	153	215	27
Snow Boots	268	34	6	272

a. Graph the ordered pairs (sandals sold, snow boots sold) to create a scatter plot.

b. Does the scatter plot show a *positive, negative,* or *no* correlation? Explain.

c. Determine whether the data illustrate a *correlation* or *causation*. What other factors may influence the data?

Determine whether each situation illustrates a *correlation* or *causation*. If there is a correlation, describe the trend. Explain your reasoning.

9. **PIZZA** The more pizzas a restaurant sells, the more cheese it uses.

10. **BOOKS** Sam notices that as the number of words in a book increases, the number of pages in the book increases.

Higher-Order Thinking Skills

11. **CONSTRUCT ARGUMENTS** What is meant by this statement: *Correlation does not imply causation*? Justify your argument.

A study compared the average monthly amount spent on swimsuits with the average monthly amount spent on air conditioning for several months in Sunnyside. The data is shown in the scatter plot. Use this information for Exercises 12 and 13.

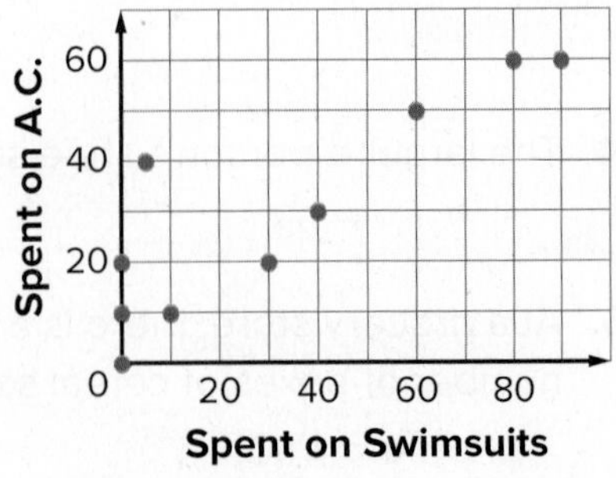

12. **ANALYZE** Explain what the scatter plot shows and describe any correlation.

13. **WRITE** Explain whether this statement is accurate: "There is a strong positive correlation between spending money on swimsuits and spending money on air conditioning. Therefore, to cut down the amount of electricity used in Sunnyside, people should buy fewer swimsuits."

Linear Regression

Learn Linear Regression and Best-Fit Lines

A calculator can find the line that most closely approximates data in a scatter plot, called the **best-fit line**. **Linear regression** is one algorithm used to find a precise line of fit for a set of data.

Calculators may also compute a number r called the **correlation coefficient**. This measure shows how well data are modeled by a linear equation. It will tell you if a correlation is positive or negative and how closely the equation is modeling the data. The closer the correlation coefficient is to 1 or –1, the more closely the equation models the data.

Weak Correlation	Moderate Correlation	Strong Correlation
$r = 0.02$	$r = 0.72$	$r = -0.97$

Example 1 Find a Best-Fit Line

BASEBALL **The table shows Jackie Robinson's total hits during each season of his major league career. Use a graphing calculator to write an equation for the best-fit line for the data. Then find and interpret the correlation coefficient.**

Year	1947	1948	1949	1950	1951	1952	1953	1954	1955	1956
Total Hits	175	170	203	140	185	157	159	120	81	98

Step 1 Enter the data.

Before you begin, make sure that your Diagnostic setting is on. You can find this under the **CATALOG** menu. Press **D** and then scroll down and click **DiagnosticOn.** Then press **enter**.

(continued on the next page)

Today's Goals

- Write equations of best-fit lines using linear regressions.
- Determine how well functions fit sets of data.

Today's Vocabulary

best-fit line
linear regression
correlation coefficient
residual

Think About It!

Write the following correlation coefficients in order from weakest to strongest.

0.85 0.3 1 –0.78

0.54 –0.06 –0.9

Study Tip

Correlation Coefficient The table shows a rule of thumb for determining how well the equation models the data based on the correlation coefficient.

Correlation Coefficient	Strength of Correlation
$\lvert r \rvert \geq 0.8$	Strong
$0.5 \leq \lvert r \rvert < 0.8$	Moderate
$\lvert r \rvert < 0.5$	Weak

Go Online to see how to use a graphing calculator with this example.

Math History Minute

One of the areas of interest of British statistician **Florence Nightingale David (1909–1993)**, who was named after family friend Florence Nightingale, was the distribution of correlation coefficients. In 1938, she released a book entitled *Tables of the Correlation Coefficient*, for which all of the calculations were done on a hand-cranked mechanical calculator.

Use a Source

Choose another baseball player and research that player's total number of hits by season. Use a graphing calculator to write an equation for the best-fit line, and decide whether the equation models the data well.

Enter the data by pressing **stat** and selecting the **Edit** option. Let the year 1947 be represented by year 0. Enter the years since 1947 into List 1 (**L1**). These will represent the **x**-values. Enter the total hits into List 2 (**L2**). These will represent the **y**-values.

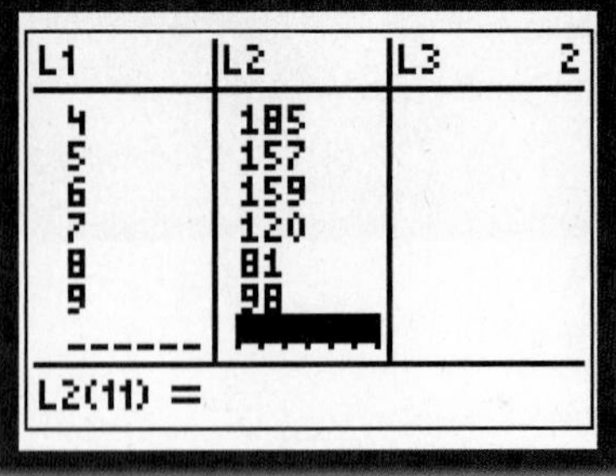

Step 2 Perform the regression.

Perform the regression by pressing **stat** and selecting the **CALC** option. Scroll down to **LinReg (ax+b)** and press **enter**. Make sure **L1** is the **Xlist** and **L2** is the **Ylist**. Then select **Calculate**.

Step 3 Interpret the results.

Write the equation of the regression line by rounding the a and b values on the screen. The form that we chose for the regression was $ax + b$, so the equation is $y = -10.32x + 195.22$. The correlation coefficient is about -0.8022, which means that the equation models the data well. Its negative value means that as the years since 1947 increase, the total number of Jackie Robinson's hits decreases.

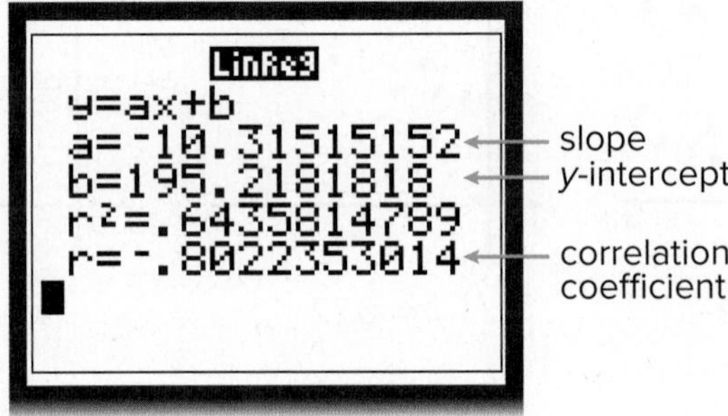

Check

TEMPERATURE The table shows the average annual temperature for the top 10 most populous states in 2014.

Rank	1	2	3	4	5	6	7	8	9	10
Temperature (°F)	59.4	64.8	70.7	45.4	51.8	48.8	50.7	63.5	59	44.4

Part A Use a graphing calculator to write an equation for the best-fit line for the data. Round to the nearest hundredth.

$y =$ ___?___ $x +$ ___?___

Part B Find the correlation coefficient r. Round to the nearest hundredth.

$r =$ ___?___

Part C Based on your answer to **part b**, does the equation model the data well? Yes or No?

 **Go Online** You can complete an Extra Example online.

Best-fit lines can be used to estimate values that are not in the data. Recall that when we estimate values that are between known values, this is called linear interpolation. When we estimate a number outside the range of data, it is called linear extrapolation.

Example 2 Use a Best-Fit Line

SHOPPING **The table shows U.S. desktop online sales on Cyber Monday since 2009. Estimate the Cyber Monday sales in 2025.**

Year	2009	2010	2011	2012	2013	2014	2015	2016
Sales (millions of dollars)	887	1028	1251	1465	1735	2038	2280	2671

Step 1 Graph the data.

Enter the data from the table into the lists. Let 2009 be represented by 0. Then the years since 2009 are the *x* -values. Let the sales be the *y*-values. Graph the scatter plot. Turn on **Plot1** under the **STAT PLOT** menu and choose the scatter plot icon. Use **L1** for the **Xlist** and **L2** for the **Ylist**.

[−0.7, 7.7] scl: 1 by [583.72, 2974.28] scl: 1

Change the viewing window so that all data are visible by pressing **zoom** and then selecting **ZoomStat**.

Step 2 Perform the regression.

Perform the regression using the data in the lists. The equation is about $y = 254.51x + 778.58$. The correlation coefficient is 0.9935, which means that the equation models the data well.

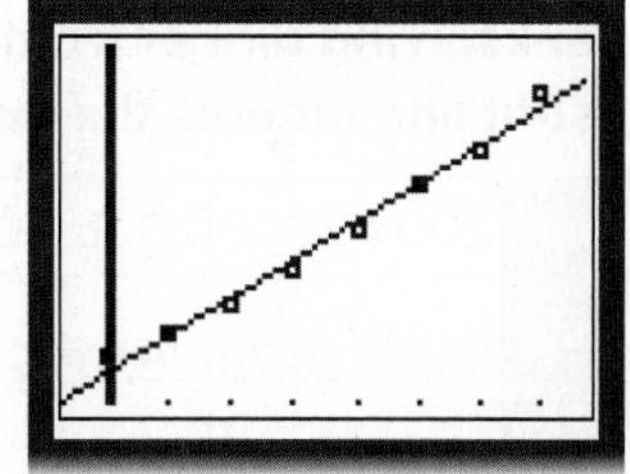

[−0.7, 7.7] scl: 1 by [583.72, 2974.28] scl: 1

Step 3 Graph the best-fit line.

Graph the best-fit line. Press **y =** **vars** and choose **Statistics**.

From the **EQ** menu, choose **RegEQ**. Press **graph**.

Step 4 Extrapolate.

Use the graph to predict the 2025 Cyber Monday sales. Change the viewing window to include the *x*-value to be evaluated, 16. Also increase **Ymax** to accommodate the increasing *y*-values. Press **2nd** CALC **enter** 16 **enter** to find that when $x = 16, y \approx 4851$.

[−0.7, 17] scl: 1 by [583.72, 5500] scl: 1

We can estimate that in 2025, Cyber Monday sales will be about $4,851,000,000.

Go Online You can complete an Extra Example online.

Think About It!

Why is it helpful to define *x* as *years since 2009* instead of *years*?

Study Tip

Assumptions Using a best-fit line to make predictions requires you to assume that the trend continues at a constant rate and that more people choose to shop on Cyber Monday each year.

Check

SOCIAL MEDIA The table shows the number of daily users on a social media site in various years.

Year	2011	2012	2013	2014	2015
Daily Users (millions)	372	526	665	802	936

Use linear regression to estimate the number of daily users in millions on the site in 2030.

A. 3187.4 users **B.** 4591.4 users

C. 285,391.4 users **D.** 3047 users

Talk About It!

Why would a residual plot where the residuals are almost on the line $y = 0$ indicate a very good fit? Explain your reasoning.

Learn Residuals

When finding a best-fit line, not all data will lie on the line. The difference between an observed y-value and its predicted y-value on a regression line is called a **residual**. When residuals are plotted on a scatter plot, they can help assess how well the best-fit line describes the data. If there is no pattern in the residual plot, then the best-fit line is a good fit.

Example 3 Graph and Analyze a Residual Plot

THANKSGIVING **The table shows the average price of a 10-person Thanksgiving dinner from 2004 to 2014. Determine whether the best-fit line models the data well by graphing a residual plot.**

Year	2004	2005	2006	2007	2008	2009	2010	2011	2012	2013	2014
Price ($)	35.68	36.78	38.10	42.26	44.61	42.91	43.47	49.20	49.48	49.04	49.41

Step 1 Find the best-fit line.

Enter the data from the table into the lists. Let 2004 be represented by 0. Then the years since 2004 are the x-values. Let the prices be the y-values. Perform the linear regression using the data in the lists.

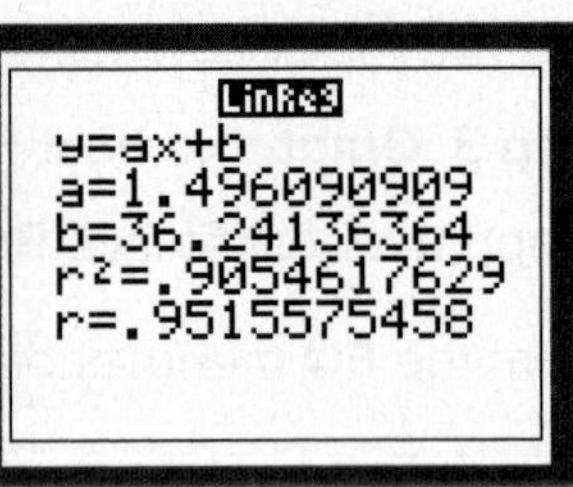

Step 2 Graph the residual plot.

Turn on **PLOT2** under the **STAT PLOT** menu and choose [scatter plot icon]. Use **L1** for the **Xlist** and **RESID** for the **Ylist**. You can obtain **RESID** by pressing **2nd** [LIST] and selecting **RESID** from the list of names. Graph the scatter plot of the residuals by pressing **zoom** and choosing **ZoomStat**. The residuals appear to be randomly scattered and centered about the line $y = 0$. Thus, the best-fit line seems to model the data well.

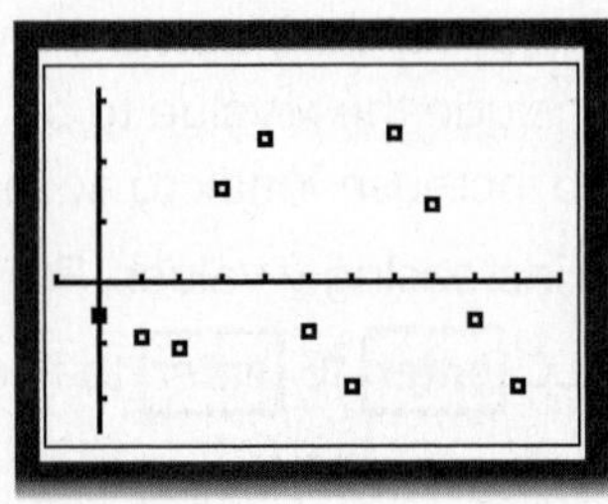

[−1, 1] scl: 1 by [−2.52, 3.21] scl: 1

Think About It!

Use a calculator to find the correlation coefficient of the best-fit line. Does the correlation coefficient also suggest a good-fit? Justify your argument.

Go Online to see how to use a graphing calculator with this example.

Go Online You can complete an Extra Example online.

Practice

Go Online You can complete your homework online.

Example 1

1. **SOCCER** The table shows the number of goals a soccer team scored each season since 2010. Let x be the number of years since 2010.

Year	2010	2011	2012	2013	2014	2015
Goals Scored	48	52	50	46	48	42

   a. Write the equation for the best-fit line for the data.

   b. Find and interpret the correlation coefficient.

2. **REVENUE** The table shows the estimated revenue earned for ringtone and ringback purchases, in millions of dollars, each year since 2010.

Year	2010	2011	2012	2013	2014	2015	2016
Revenue ($)	448	276.2	166.9	97.9	66.3	54.5	40.1

   a. Write the equation for the best-fit line for the data.

   b. Find and interpret the correlation coefficient.

3. **SALES** The table shows the sales of a health and beauty supply company, in millions of dollars, for several years. Let x be the number of years since 2010.

Year	2011	2012	2013	2014	2015
Sales	12.2	19.1	29.4	37.3	45.7

   a. Write the equation for the best-fit line for the data.

   b. Find and interpret the correlation coefficient.

Example 2

4. **PURCHASING** A supermarket chain closely monitors how many bottles of sunscreen it sells each year so that it can reasonably predict how many bottles to stock in the following year. Let x be the number of years since 2010.

Year	2013	2014	2015	2016	2017
Bottles of Sunscreen	60,200	65,000	66,300	65,200	70,600

   a. Find the equation for the best-fit line for the data.

   b. How many bottles of sunscreen should the supermarket expect to sell in 2025?

5. GOLD Ounces of gold are traded by large investment banks in commodity exchanges much the same way that shares of stock are traded. The table below shows the cost of a single ounce of gold on the last day of trading in given years. Let x be the number of years since 2000.

Year	2002	2004	2006	2008	2010	2012	2014
Price	$342.75	$435.60	$635.70	$869.75	$1420.25	$1664.00	$1199.25

a. Find the equation for the best-fit line for the data.

b. According to the equation, what would be the price of an ounce of gold on the last day of trading in 2030?

6. GOLF SCORES Emmanuel is practicing golf as part of his school's golf team. Each week he plays a full round of golf and records his total score. His scores for the first five weeks are shown.

Week	1	2	3	4	5
Golf Score	112	107	108	104	98

a. Find the equation for the best-fit line for the data.

b. What score can Emmanuel expect to get after 10 weeks?

Example 3

7. MODELING For a science project, Noah measured the effect of light on plant growth. At the end of 3 weeks, he recorded the height of each plant and how many hours of light it received each day.

Hours of Sunlight Per Day (x)	0	3	6	10	4	9	7	8	12	11	5
Height in Inches (y)	1	3	4	8	4	6	7	8	9	6	5

a. Find the equation for the best-fit line for the data.

b. Graph and analyze the residual plot.

8. STRUCTURE For his project, Darius measured the effect of fertilizer on plant growth. At the end of 3 weeks, he recorded the height of each plant and how many drops of fertilizer it received each day.

Drops of Fertilizer (x)	5	15	20	25	18	22	21	30	10	13	16
Height in Inches (y)	5	8	9	0	8	0	9	0	7	6	9

a. Find the equation of the best-fit line for the data.

b. Graph and analyze the residual plot.

Mixed Exercises

9. PHYSICAL FITNESS The table shows the percentage of students in public school who have met all six of California's physical fitness standards each year since the 2011–2012 school year.

Year	2011–2012	2012–2013	2013–2014	2014–2015
Percentage	20.5%	22.1%	22.6%	21.2%

a. Write the equation for the best-fit line for the data.

b. Find and interpret the correlation coefficient.

c. What constraints are there in the situation? Explain.

10. FARMING Some crops, such as barley, are very sensitive to how acidic the soil is. To determine the ideal level of acidity, a farmer measures how many bushels of barley he harvests in different fields with varying acidity levels.

Soil Acidity (pH)	5.7	6.2	6.6	6.8	7.1
Bushels Harvested	3	20	48	61	73

a. Find the equation for the best-fit line for the data and the correlation coefficient.

b. Use the equation of the best-fit line to estimate how many bushels the farmer would harvest if the soil had a pH of 10.

c. Could the equation of the best-fit line be used to extrapolate the data for extremely high levels of soil acidity? Explain.

11. FOOTBALL A college running back ran for 1732 total yards in the regular season. The table shows his cumulative total number of yards gained after select games.

Game Number	1	3	6	9	12
Cumulative Yards	184	431	818	1257	1732

a. Find the equation for the best-fit line for the total yards y gained after x games.

b. Find and interpret the correlation coefficient.

c. Use the trend of the data and the table to estimate when the running back will have run for 950 yards. Explain your reasoning.

d. During which game would you expect the running back to reach a total of 1000 yards?

12. REGULARITY Consider the linear regression equation that models a set of data very well to be $y = 1.43x - 4.2$. Would there be any restrictions on what the correlation coefficient value could be? Justify your reasoning.

13. STRUCTURE The table shows the number of student athletes participating in college athletics since the 2010-2011 school year.

Year	2010-2011	2011-2012	2012-2013	2013-2014	2014-2015
Student Athletes	444,077	453,347	463,202	472,625	482,533

a. Find the equation for the best-fit line for the data.

b. Find and interpret the correlation coefficient.

c. Graph and analyze the residual plot. Does this support your conclusion from **part b?**

d. Predict the number of college athletes in 2035.

Higher-Order Thinking Skills

14. WRITE How are lines of fit and linear regression similar? different?

15. CREATE For a class project, the scores that 10 randomly selected students earned on the first 8 tests of the school year are given. Explain how to find a line of best fit. Could it be used to predict the scores of the other students? Explain your reasoning.

16. ANALYZE Determine whether the following statement is *sometimes, always,* or *never* true: *If the correlation coefficient in a given situation is 0.946, the change in the independent variable* ***causes*** *change in the dependent variable.* Justify your argument.

17. PERSEVERE The table shows the number of participants in high school athletics.

Years Since 1980	0	10	20	25	30
Number of Athletes	5,356,913	5,298,671	6,705,223	7,159,904	7,667,955

a. Find an equation for the regression line.

b. According to the equation, how many participated in 2008?

Inverses of Linear Functions

Today's Goals
- Construct the inverses of relations.
- Find inverses of linear functions.

Today's Vocabulary
inverse relations
inverse functions

Learn Inverses of Relations

Two relations are **inverse relations** if and only if one relation contains points of the form (a, b) when the other relation contains points of the form (b, a). So, the x-coordinates are exchanged with the y-coordinates for each ordered pair in the relation.

Key Concept • Inverse Relations	
Words	If one relation contains the element (a, b), then the inverse relation will contain the element (b, a).
Symbols	$(a, b) \rightarrow (b, a)$
Example	A and B are inverse relations. A B $(-8, 12) \rightarrow (12, -8)$ $(-2, -5) \rightarrow (-5, -2)$ $(0, 4) \rightarrow (4, 0)$ $(7, 16) \rightarrow (16, 7)$
Graph	The graph of an inverse is the graph of the original relation reflected over the line $y = x$. For every point (a, b) on the graph of the original relation, the graph of the inverse will include (b, a).

Example 1 Inverse Relations

Determine the inverse of $\{(-8, 3), (0, 14), (11, 52), (12, -6)\}$.

Write the coordinates in the ordered pairs to complete the inverse relation.

$(-8, 3) \rightarrow (3, -8)$ $(11, 52) \rightarrow (52, 11)$

$(0, 14) \rightarrow (14, 0)$ $(12, -6) \rightarrow (-6, 12)$

The inverse relation is $\{(3, -8), (14, 0), (52, 11), (-6, 12)\}$.

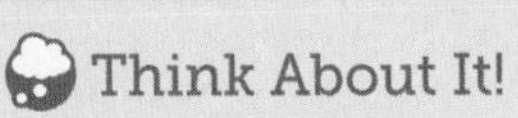

Describe the relationship between the domains and ranges of inverse relations.

Check

Determine the inverse of the relation.

$\{(-1.4, 5), (1.3, 6.5), (3, -8), (3.05, 9)\}$

Go Online You can complete an Extra Example online.

Example 2 Find Inverse Relations from a Table

Find the inverse of the relation shown in the table.

x	−11	0.3	−3	3.5
y	9	−2	−8	2

Write the coordinates in the ordered pairs to complete the inverse relation.

(−11, 9) → (9, −11) (−3, −8) → (−8, −3)

(0.3, −2) → (−2, 0.3) (3.5, 2) → (2, 3.5)

The inverse relation is {(9, −11), (−2, 0.3), (−8, −3), (2, 3.5)}.

Example 3 Graph Inverse Relations

Graph the inverse of the relation.

The graph of the relation passes through the points (−1, −5), (0, −2), (1, 1), and (2, 4).

Exchange the x-coordinates and y-coordinates to find points of the inverse relation.

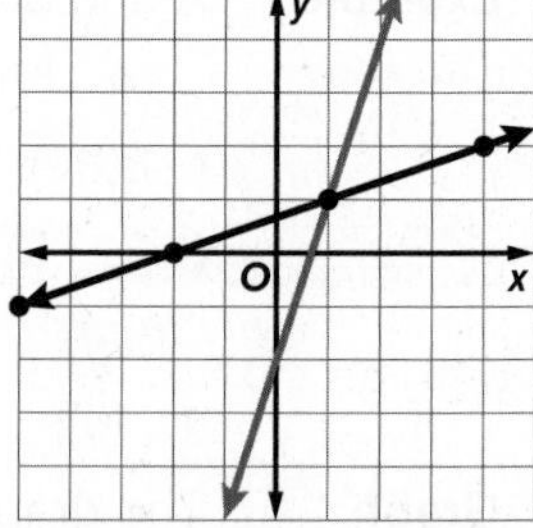

(−1, −5) → (−5, −1) (1, 1) → (1, 1)

(0, −2) → (−2, 0) (2, 4) → (4, 2)

Plot the points of the inverse relation and draw a line passing through them.

Study Tip

Plotting Points While it only takes two points to graph a line, use several points when graphing inverse relations to create more accurate graphs.

Think About It!

Describe the graph of the inverse of the horizontal line $y = 3$.

Explore Comparing a Function and Its Inverse

Online Activity Use graphing technology to complete the Explore.

INQUIRY How can you graph the inverse of a function?

Learn Inverses of Linear Functions

A linear relation that is described by a function may have an **inverse function** that can generate ordered pairs of the inverse relation. The inverse of the linear function $f(x)$ is written as $f^{-1}(x)$ and is read *f inverse of x* or *the inverse of f of x*.

Go Online You can complete an Extra Example online.

Key Concept • Finding Inverse Functions	
To find the inverse function $f^{-1}(x)$ of the linear function $f(x)$, complete the following steps.	
Step 1	Replace $f(x)$ with y in the equation for $f(x)$.
Step 2	Interchange y and x in the equation.
Step 3	Solve the equation for y.
Step 4	Replace y with $f^{-1}(x)$ in the new equation.

Watch Out!

In $f^{-1}(x)$, the -1 is not an exponent. It is a way to indicate that $f^{-1}(x)$ is an inverse of another function called $f(x)$.

Example 4 Find an Inverse Linear Function

Find the inverse of $f(x) = 5x + 10$.

Step 1 $f(x) = 5x + 10$ Original equation

$y = 5x + 10$ Replace $f(x)$ with y.

Step 2 $x = 5y + 10$ Interchange y and x.

Step 3 $x - 10 = 5y$ Subtract 10 from each side.

$\frac{x - 10}{5} = y$ Divide each side by 5.

Step 4 $\frac{x - 10}{5} = f^{-1}(x)$ Replace y with $f^{-1}(x)$.

The inverse of $f(x) = 5x + 10$ is $f^{-1}(x) = \frac{x - 10}{5}$ or $f^{-1}(x) = \frac{1}{5}x - 2$.

Example 5 Find Inverses of Linear Functions

Find the inverse of $f(x) = -\frac{2}{3}x - 8$.

Step 1 $f(x) = -\frac{2}{3}x - 8$ Original equation

$y = -\frac{2}{3}x - 8$ Replace $f(x)$ with y.

Step 2 $x = -\frac{2}{3}y - 8$ Interchange x and y.

Step 3 $x + 8 = -\frac{2}{3}y$ Add 8 to each side.

$-\frac{3}{2}(x + 8) = y$ Multiply each side by $-\frac{3}{2}$.

$-\frac{3}{2}x - 12 = y$ Simplify.

Step 4 $-\frac{3}{2}x - 12 = f^{-1}(x)$ Replace y with $f^{-1}(x)$.

The inverse of $f(x) = -\frac{2}{3}x - 8$ is $f^{-1}(x) = -\frac{3}{2}x - 12$.

Talk About It!

What is the inverse of $f(x) = -x$? How could you check your solution?

Go Online to see Example 5.

Example 6 Apply Inverse Linear Functions

BOATING **Skyler and Carmen rent a paddle boat at a state park for \$15 plus \$4 for each hour it is used. The function $C(x) = 4x + 15$ represents the total cost $C(x)$ for x hours.**

Part A Determine the inverse function.

Step 1	$C(x) = 4x + 15$	Original equation
	$y = 4x + 15$	Replace $C(x)$ with y.
Step 2	$x = 4y + 15$	Interchange y and x.
Step 3	$x - 15 = 4y$	Subtract 15 from each side.
	$\frac{x - 15}{4} = y$	Divide each side by 4.
Step 4	$\frac{x - 15}{4} = C^{-1}(x)$	Replace y with $C^{-1}(x)$.

Study Tip

Function Notation Function notation is a way to give an equation a name, such as $f(x)$, $g(x)$, or $C(x)$. In Step 1 of finding the inverse function, replace the function notion with y regardless of the name of the function.

Part B Interpret the inverse function.

x is the total cost of renting the paddle boat, and $C^{-1}(x)$ is the number of hours that Skyler and Carmen use the paddle boat.

Part C Evaluate using the inverse function.

Skyler and Carmen have \$35 to rent the paddle boat. How long can they rent it?

To find the length of time that they can rent the boat, find $C^{-1}(35)$.

$C^{-1}(x) = \frac{x - 15}{4}$	Original equation
$C^{-1}(35) = \frac{35 - 15}{4}$	Substitute 35 for x.
$= \frac{20}{4}$ or 5	Simplify.

Check

CANDLES Javi is making candles to sell at an upcoming festival. He has already made 38 candles, and he makes 24 candles each day. The function $C(x) = 24x + 38$ represents the total number of candles $C(x)$ he has in inventory, where x is the number of days since he began making more candles.

Part A Select the inverse of the function $C(x)$.

A. $C^{-1}(x) = \frac{1}{24}x - \frac{19}{12}$

B. $C^{-1}(x) = \frac{1}{24}x - 38$

C. $C^{-1}(x) = \frac{1}{38}x - \frac{19}{12}$

D. $C^{-1}(x) = \frac{12}{19}x - 24$

Part B Estimate the amount of time it would take Javi to make 350 candles. It would take Javi between ___?___ and ___?___ days to make 350 candles.

Go Online You can complete an Extra Example online.

Practice

Go Online You can complete your homework online.

Examples 1 and 2

Find the inverse of each relation.

1.

x	y
−9	−1
−7	−4
−5	−7
−3	−10
−1	−13

2.

x	y
1	8
2	6
3	4
4	2
5	0

3.

x	y
−4	−2
−2	−1
0	1
2	0
4	2

4. {(−3, 2), (−1, 8), (1, 14), (3, 20)}

5. {(5, −3), (2, −9), (−1, −15), (−4, −21)}

6. {(4, 6), (3, 1), (2, −4), (1, −9)}

7. {(−1, 16), (−2, 12), (−3, 8), (−4, 4)}

8. {(−5, 13), (6, 10.8), (3, 11.4), (−10, 14)}

9. {(−4, −49), (8, 35), (−1, −28), (4, 7)}

Example 3

Graph the inverse of each function.

10.

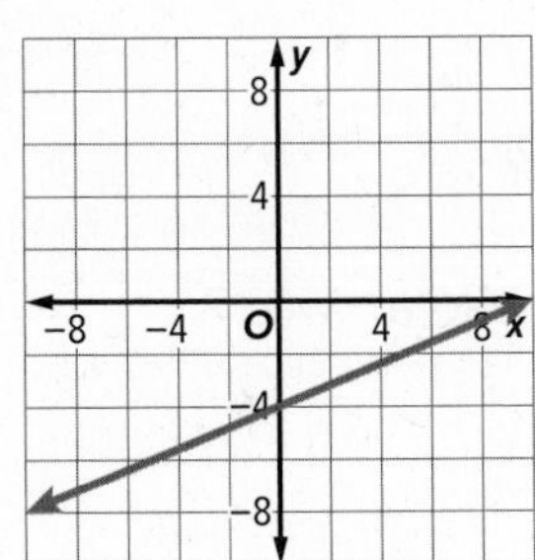

11.

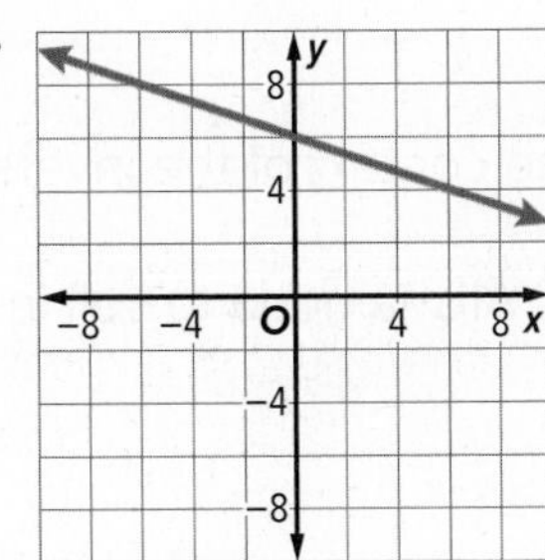

12.

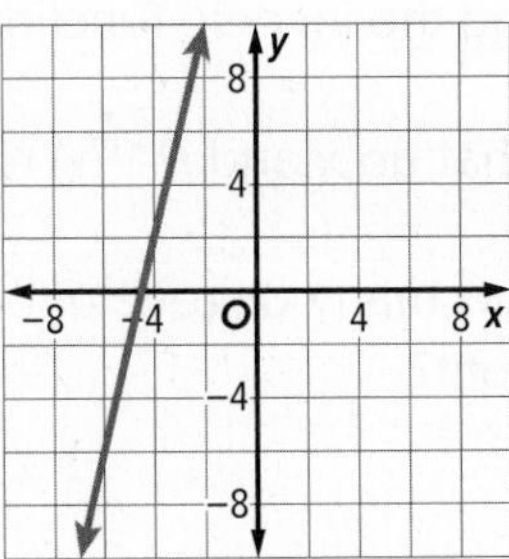

13.

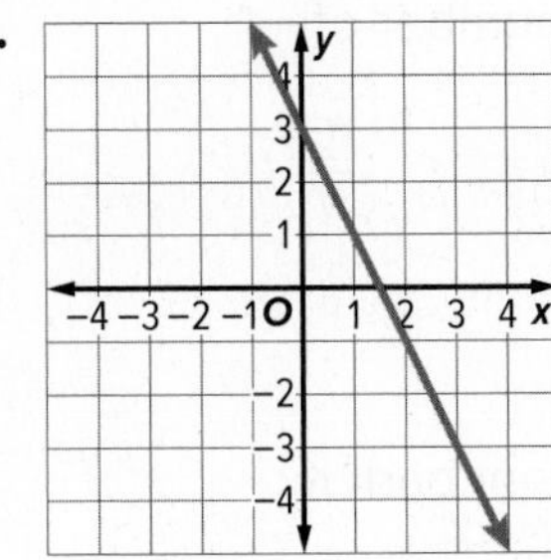

14.

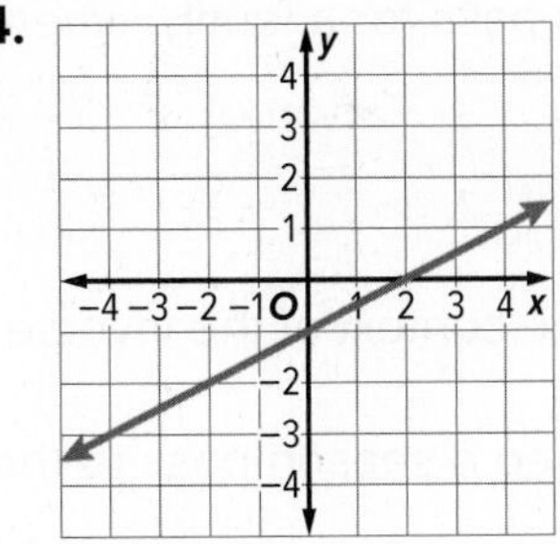

15.

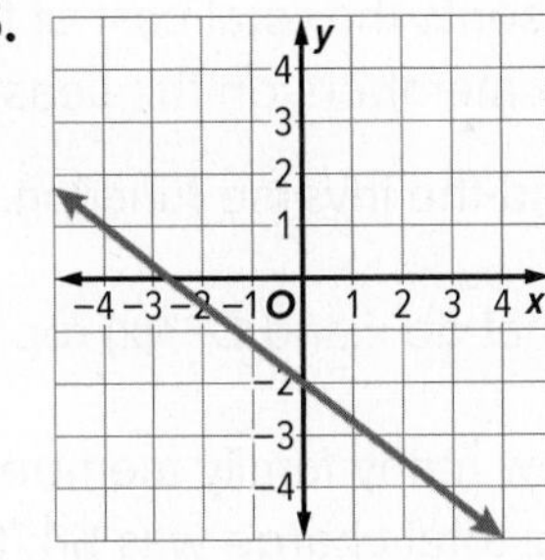

Examples 4 and 5

Find the inverse of each function.

16. $f(x) = 8x - 5$

17. $f(x) = 6(x + 7)$

18. $f(x) = \frac{3}{4}x + 9$

19. $f(x) = -16 + \frac{2}{5}x$

20. $f(x) = \frac{3x + 5}{4}$

21. $f(x) = \frac{-4x + 1}{5}$

Example 6

22. LEMONADE Bernardo spent \$15 on supplies for his lemonade stand. He charges \$1.25 per glass. The function $P(x) = 1.25x - 15$ represents his profit, where x is the number of glasses of lemonade sold.

a. Find the inverse function, $P^{-1}(x)$.

b. What do x and $P^{-1}(x)$ represent in the context of the inverse function?

c. How many glasses must Bernardo sell in order to make \$10 in profit?

23. BUSINESS Alisha started a baking business. She spent \$36 initially on supplies and can make 5 dozen brownies for \$12. She charges her customers \$10 per dozen brownies. The function $P(x) = 7.6x - 36$ represents her profit, where x is the number of dozens of brownies sold.

a. Find the inverse function, $P^{-1}(x)$.

b. What do x and $P^{-1}(x)$ represent in the context of the inverse function?

c. How many dozens of brownies does Alisha need to sell in order to make a profit?

24. SEASON PASS A season pass to an amusement park costs \$70 per family member plus an additional \$50 fee for parking. The function $C(x) = 70x + 50$ represents the total cost of the season pass for a family, where x is the number of family members on the season pass.

a. Find the inverse function, $C^{-1}(x)$.

b. What do x and $C^{-1}(x)$ represent in the context of the inverse function?

c. How many family members purchased a season pass to the amusement park if the total charge was \$470?

25. GARDENING Kara is building raised garden beds for her backyard. The total cost $C(x)$ in dollars is given by $C(x) = 125 + 16x$, where x is the number of pieces of wood required for the boxes.

a. Find the inverse function $C^{-1}(x)$.

b. If the total cost was $269 and each piece of wood was 12 feet long, how many total feet of wood were used?

26. GEOMETRY The area of the base of a cylindrical water tank is 12π square feet. The volume of water in the tank is dependent on the height of the water h and is represented by the function $V(h) = 12\pi h$.

a. Find $V^{-1}(h)$.

b. What will the height of the water be when the volume reaches 420π cubic feet?

Mixed Exercises

Write the inverse of each function in $f^{-1}(x)$ notation.

27. $3y - 12x = -72$

28. $x + 3y = 10$

29. $-42 + 6y = x$

30. $3y + 24 = 2x$

31. $-7y + 2x = -28$

32. $12y - x = 7$

Write an equation for the inverse function $f^{-1}(x)$ that satisfies the given conditions.

33. slope of $f(x)$ is 7; graph of $f^{-1}(x)$ contains the point (13, 1)

34. graph of $f(x)$ contains the points (−3, 6) and (6, 12)

35. graph of $f(x)$ contains the point (10, 16); graph of $f^{-1}(x)$ contains the point (3, −16)

36. slope of $f(x)$ is 4; $f^{-1}(5) = 2$

Match each function with the graph of its inverse.

37. $f(x) = \frac{1}{2}x + 2$

38. $f(x) = \frac{1}{2}x - 2$

39. $f(x) = x + 2$

A.
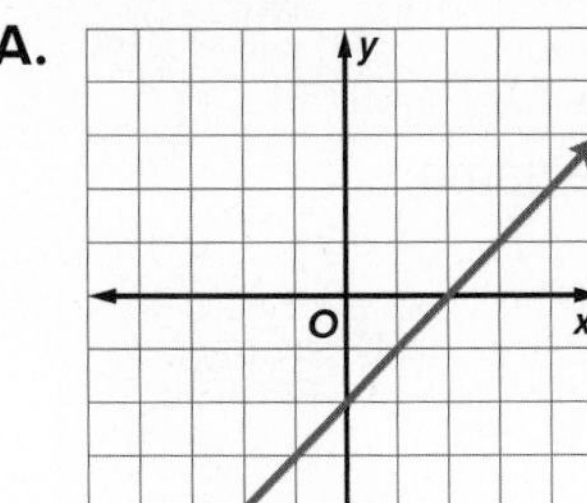

B.
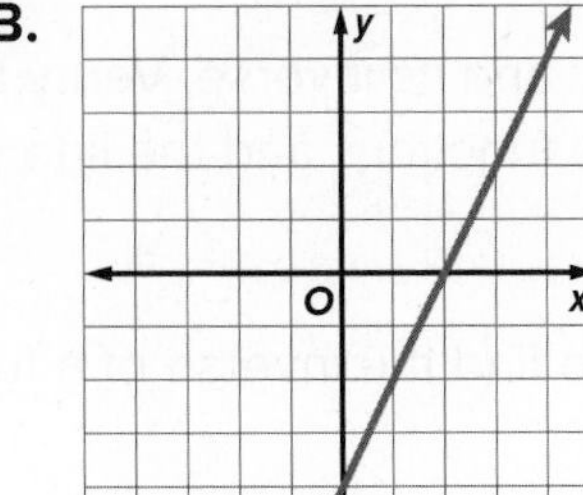

C.
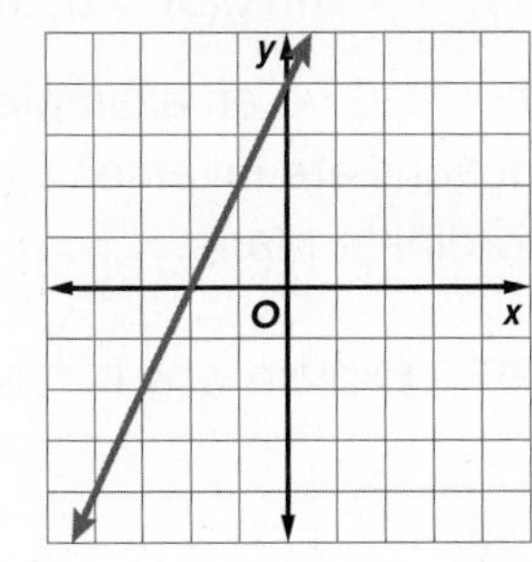

40. STRUCTURE Write the inverse of each function.

a. $f(x) = \frac{x - 10}{3}$

b. $g(x) = \frac{3}{4}x + 6$

c. $h(x) = -5x - 7$

d. Graph $h(x)$ and $h^{-1}(x)$ on the same coordinate plane to check your answer.

41. REASONING Suppose the inverse of a relation is $\{(b, -k), (-g, p), (-w, -m), (r, q)\}$. What is the relation?

Higher-Order Thinking Skills

42. ANALYZE How can you use ordered pairs to check if the inverse of a function is correct?

43. WRITE What is the relationship between the slopes of two lines that are inverse functions of one another? Give an example.

44. WRITE What is the relationship between the x- and y-intercepts of two lines that are inverse functions of one another? Give an example.

45. FIND THE ERROR A student claims that there is a simple method to find the inverse of the function $f(x)$. To find the inverse, $f^{-1}(x)$, we need only remember that raising something to the power of -1 is the same as taking its reciprocal. Is this claim correct? Include an example or counterexample.

46. PERSEVERE If $f(x) = 5x + a$ and $f^{-1}(10) = -1$, find a.

47. PERSEVERE If $f(x) = \frac{1}{a}x + 7$ and $f^{-1}(x) = 2x - b$, find a and b.

ANALYZE **Determine whether the following statements are *sometimes, always,* or *never* true. Explain your reasoning.**

48. If $f(x)$ and $g(x)$ are inverse functions, then $f(a) = b$ and $g(b) = a$.

49. If $f(a) = b$ and $g(b) = a$, then $f(x)$ and $g(x)$ are inverse functions.

50. CREATE Give an example of a function and its inverse. Verify that the two functions are inverses by graphing the functions and the line $y = x$ on the same coordinate plane.

51. WRITE Explain why it may be helpful to find the inverse of a function.

Review

Essential Question

What can a function tell you about the relationship that it represents?

Functions can tell you whether the value of the dependent variable increases or decreases as the independent variable changes. They describe trends in data and can be used to make predictions.

Module Summary

Lessons 5-1 and 5-2

Writing Equations

- Slope-intercept form is $y = mx + b$, where m is the slope of the line and b is the y-intercept.
- Point-slope form is $y - y_1 = m(x - x_1)$, where (x_1, y_1) is a given point on a nonvertical line and m is the slope of the line.
- Standard form is $Ax + By = C$, where A, B, and C are integers, $A > 0$, A and B are both not equal to 0, and the *GCF* of A, B, and C is 1.
- To write a linear equation given two points on a line, first find the slope. Then use either point to write the equation in point-slope form or find the y-intercept to write the equation in slope-intercept form.

Lessons 5-3 through 5-5

Scatter Plots

- A scatter plot shows the relationship between a set of bivariate data, graphed as ordered pairs on a coordinate plane.
- A positive correlation exists when, as x increases, y increases. A negative correlation exists when, as x increases, y decreases. No correlation exists when x and y are not related.
- A line of fit is used to describe the trend of the data in a scatter plot.
- To determine causation, determine whether one variable influences the other variable.
- The correlation coefficient tells you how well the equation for the best-fit line models the data.
- A correlation coefficient close to 1 has a strong positive correlation. A correlation coefficient close to −1 has a strong negative correlation.
- Residuals measure how much the data deviate from the regression line.

Lesson 5-6

Inverses of Linear Functions

- Two relations are inverse relations if and only if one relation contains the element (a, b) when the other relation contains the element (b, a).
- In inverse relations, the x-coordinates are exchanged with the y-coordinates for each ordered pair in the relation.
- To find the inverse of $f(x)$, replace $f(x)$ with y in the equation for $f(x)$. Interchange y and x in the equation. Solve the equation for y. Replace y with $f^{-1}(x)$ in the new equation.

Study Organizer

Foldables

Use your Foldable to review the module. Working with a partner can be helpful. Ask for clarification of concepts as needed .

Creating Linear Equations

Test Practice

1. **MULTIPLE CHOICE** What is the equation of the line that passes through the points (−2, 1) and (6, 3)? (Lesson 5-1)

   A. $y = \frac{1}{4}x + \frac{3}{2}$

   B. $y = \frac{3}{2}x + \frac{1}{4}$

   C. $y = \frac{2}{3}x + \frac{1}{4}$

   D. $y = \frac{3}{2}x + \frac{1}{4}$

2. **MULTIPLE CHOICE** Select the equation of a line with a slope of 5 that passes through the point (2, −3). (Lesson 5-1)

   A. $y = 5x + 2$

   B. $y = 5x - 3$

   C. $y = 5x + 7$

   D. $y = 5x - 13$

Use the table for exercises 3 and 4. A movie streaming service charges a set fee for membership each month, plus an additional fee for the number of movies streamed each month. This table shows the total charge for different numbers of movies.

Number of movies streamed (x)	2	4	6
Total cost (y)	\$14	\$17	\$20

3. **OPEN RESPONSE** Write the slope-intercept form of the equation that models the linear relationship in the table. (Lesson 5-1)

4. **OPEN RESPONSE** Explain the meaning of the slope and y-intercept in the context of the situation. (Lesson 5-1)

5. **MULTIPLE CHOICE** Which equation represents a line that passes through the point (3, −4) with a slope of 7? (Lesson 5-2)

   A. $y + 4 = 7(x - 3)$

   B. $y + 4 = 7(x + 3)$

   C. $y - 4 = 7(x - 3)$

   D. $y - 4 = 7(x + 3)$

6. **MULTI-SELECT** Select all of the equations that represent the line. (Lesson 5-2)

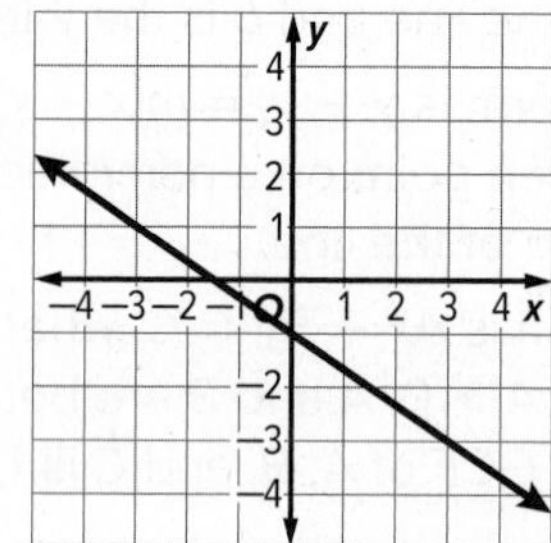

   A. $2x - 3y = -1$

   B. $2x + 3y = -3$

   C. $3x - 2y = 2$

   D. $y + 3 = -\frac{2}{3}(x - 3)$

   E. $y - 1 = -\frac{2}{3}(x + 3)$

   F. $y + 1 = -\frac{3}{2}(x + 3)$

7. **OPEN RESPONSE** A city parking garage charges \$4 to park for up to two hours. After that, an additional charge of \$2.50 per hour applies. Write an equation in point-slope form that models the total cost for parking x hours, where $x > 2$. (Lesson 5-2)

8. MULTIPLE CHOICE Which scatter plot shows the best line of fit? (Lesson 5-3)

A.

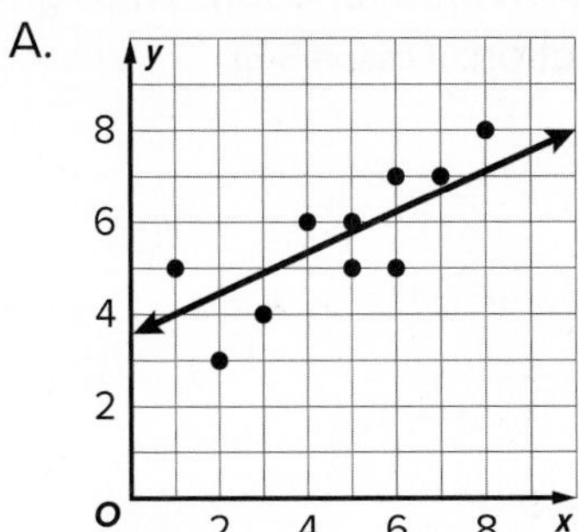

B.

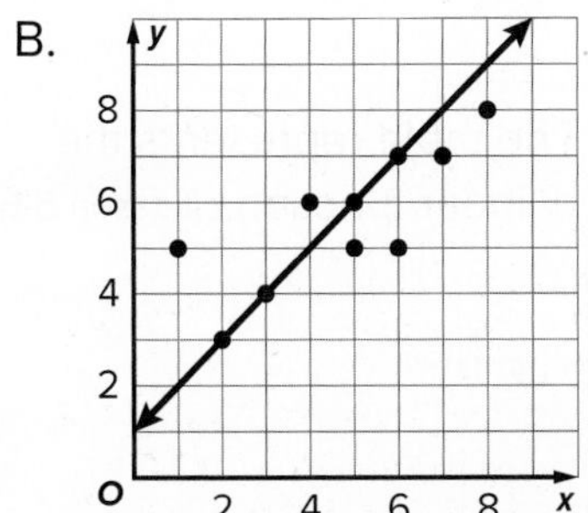

C.

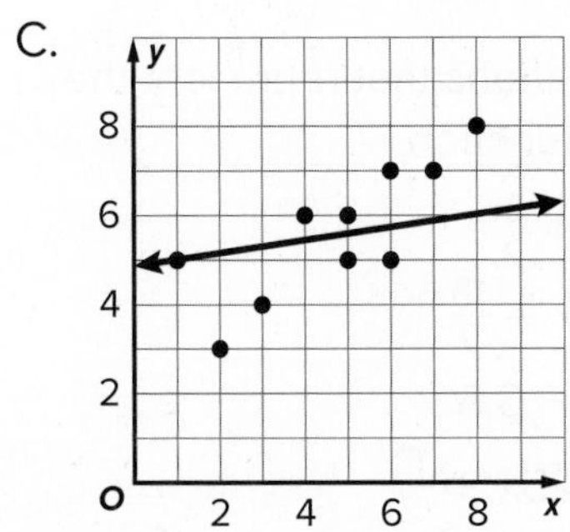

D.

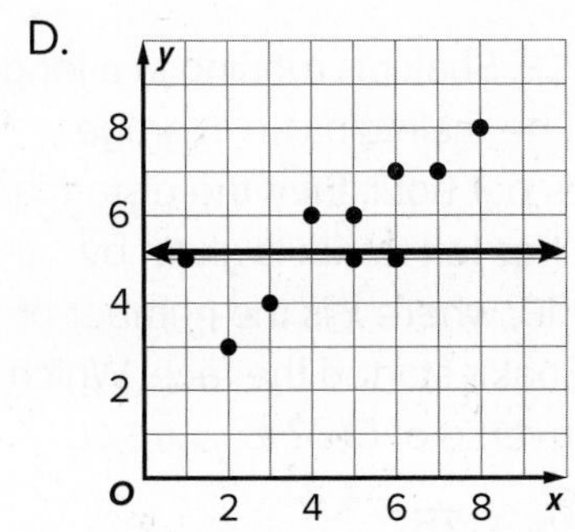

9. MULTIPLE CHOICE Which equation represents the best line of fit for the scatter plot? (Lesson 5-3)

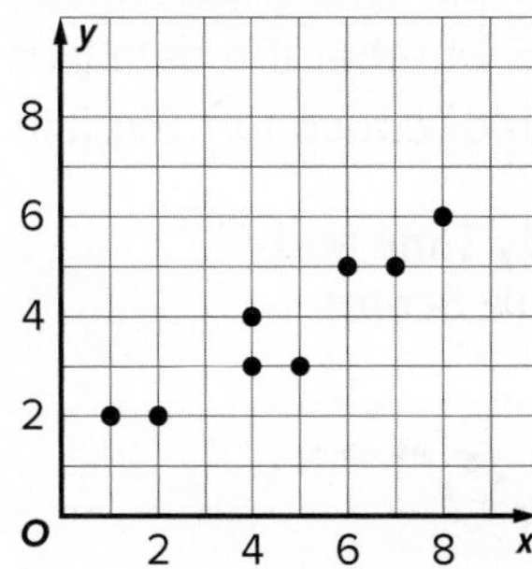

A. $y = 0.6x + 1$

B. $y = 0.5x + 2$

C. $y = x - 2$

D. $y = 0.75x$

10. OPEN RESPONSE Adriana keeps the statistics for her favorite basketball team and creates the scatter plot shown. (Lesson 5-3)

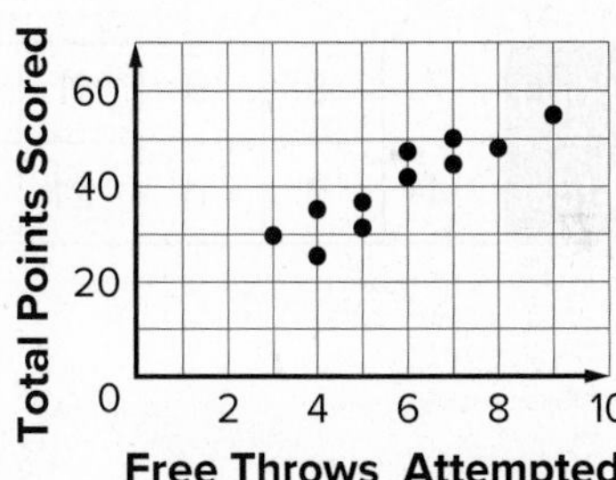

Adriana then draws a line of fit for her scatter plot. What does the slope of the line represent?

11. OPEN RESPONSE A researcher found that students who spent more time exercising each week also had higher average test scores. Describe the correlation, if any, between time spent exercising and test scores. (Lesson 5-4)

12. OPEN RESPONSE Lalita tracked the amount of time she studied each week and her score on a weekly chemistry quiz for eight weeks. She made this scatter plot from the data. Determine whether the data illustrate a *correlation* or *causation*. Explain. (Lesson 5-4)

Study Time and Quiz Scores

Quiz Scores: 100, 60, 20
20 60 100
Time (minutes)

13. MULTIPLE CHOICE Use linear regression to estimate the weight, in ounces, of a bluegill that has a length of 9.5 inches. Round your answer to the nearest tenth of an ounce. (Lesson 5-5)

Length (in.)	7	8	9	11	12	13
Weight (oz)	4	7	11	21	26	32

A. 12.9

B. 13.4

C. 14.5

D. 15.1

14. OPEN RESPONSE Use a graphing calculator and linear regression to write the equation of a best-fit line for the data in slope-intercept form. Round to the nearest tenth. (Lesson 5-5)

x	2.4	2.8	3.4	4.3	5.1	7.6	8.4	9.1
y	6.2	9.6	8.4	6.5	7.2	2.5	1.8	4.2

15. OPEN RESPONSE The graph of a function passes through (−3, 2), (−1, 1), (1, 0), and (3, −1). Find the inverse function. Then graph the inverse function. (Lesson 5-6)

16. MULTI-SELECT The table represents the coordinates of a linear function. (Lesson 5-6)

x	y
−6	5
4	1
2	−3

Select the equations that represent the inverse of the function.

A. $f^{-1}(x) = \frac{5}{2}x + \frac{5}{2}$

B. $f^{-1}(x) = -\frac{5}{2}x + \frac{13}{2}$

C. $f(x) = 2.5x - 6.5$

D. $f^{-1}(x) = -1.25x + \frac{13}{2}$

E. $f^{-1}(x) = -2.5x + 6.5$

17. MULTIPLE CHOICE Shakir is running in a long-distance race. If he maintains an average speed of 8 miles per hour, then the distance in miles that he has left to run is given by $D(x) = -\frac{2}{15}x + 10$, where x is the number of minutes since Shakir started the race. Which function is the inverse of $D(x)$? (Lesson 5-6)

A. $D^{-1}(x) = -\frac{15}{2}x + 75$

B. $D^{-1}(x) = \frac{2}{15}x - 10$

C. $D^{-1}(x) = -\frac{15}{2}x + \frac{1}{10}$

D. $D^{-1}(x) = -\frac{2}{15}x - \frac{4}{3}$

Module 6

Linear Inequalities

Essential Question

How can writing and solving inequalities help you solve problems in the real world?

What Will You Learn?

How much do you already know about each topic **before** starting this module?

KEY

— I don't know. — I've heard of it. — I know it!

	Before			After		
	I don't know.	I've heard of it.	I know it!	I don't know.	I've heard of it.	I know it!
graph linear inequalities						
solve one-step linear inequalities using addition and subtraction						
solve one-step linear inequalities using multiplication and division						
solve multi-step linear inequalities						
solve compound linear inequalities						
solve absolute value linear inequalities						
graph inequalities in two-variables						

Foldables Make this Foldable to help you organize your notes about linear inequalities. Begin with one sheet of 11″ × 17″ paper.

1. **Fold** each side so the edges meet in the center.
2. **Fold** in half.
3. **Unfold** and cut from each end until you reach the vertical line.
4. **Label** the front of each flap.

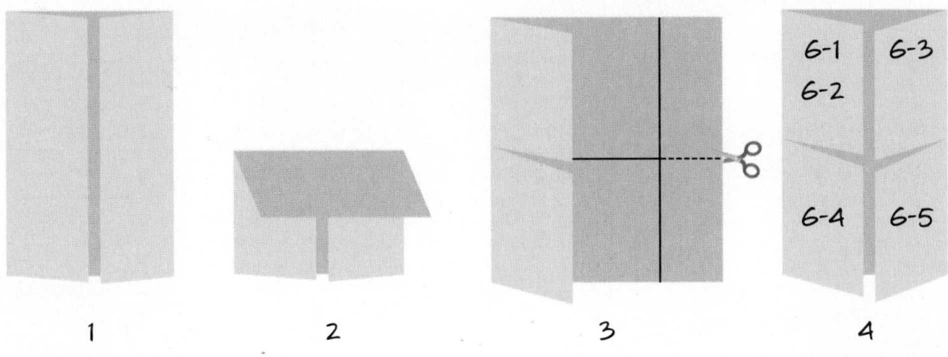

What Vocabulary Will You Learn?

- boundary
- closed half-plane
- compound inequality
- half-plane
- inequality
- intersection
- open half-plane
- set-builder notation
- union

Are you ready?

Complete the Quick Review to see if you are ready to start this module. Then complete the Quick Check.

Quick Review

Example 1

Solve $-2(x - 4) = 7x - 19$.

$-2(x - 4) = 7x - 19$	Original equation
$-2x + 8 = 7x - 19$	Distributive Property
$-2x + 8 + 2x = 7x - 19 + 2x$	Add $2x$ to each side.
$8 = 9x - 19$	Simplify.
$8 + 19 = 9x - 19 + 19$	Add 19 to each side.
$27 = 9x$	Simplify.
$3 = x$	Divide each side by 3.

Example 2

Solve $|x - 4| = 9$.

if $|x - 4| = 9$, then $x - 4 = 9$ or $x - 4 = -9$.

$x - 4 = 9$	or	$x - 4 = -9$
$x - 4 + 4 = 9 + 4$		$x - 4 + 4 = -9 + 4$
$x = 13$		$x = -5$

So, the solution set is $\{-5, 13\}$.

Quick Check

Solve each equation.

1. $2x + 1 = 9$

2. $4x - 5 = 15$

3. $9x + 2 = 3x - 10$

4. $3(x - 2) = -2(x + 13)$

Solve each equation.

5. $|x + 11| = 18$

6. $|3x - 2| = 16$

7. $|x - 7| = 8$

8. $|2x| = -9$

How did you do?

Which exercises did you answer correctly in the Quick Check?

Solving One-Step Inequalities

Explore Graphing Inequalities

Online Activity Use graphing technology to complete the Explore.

INQUIRY How can you graph the solution set of an inequality of the form $x < a$ or $x > a$ for some number a?

Today's Goals
- Graph the solutions of an inequality.
- Solve linear inequalities by using addition.
- Solve linear inequalities by using subtraction.

Today's Vocabulary
inequality
set-builder notation

Learn Graphing Inequalities

An **inequality** is a mathematical sentence that contains the symbol $<$, $>$, $\leq$, $\geq$, or $\neq$. An inequality compares the value of two numbers or expressions using these symbols.

Think About It!
Compare and contrast the graphs of $x \geq 6$ and $x > 6$.

Example 1 Graph Inequalities

Graph the solution set of $y \leq 4$.

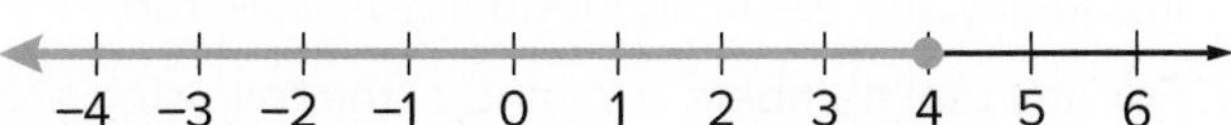

The dot at 4 shows that 4 is a solution. The heavy arrow pointing to the left shows that the solution includes all numbers less than 4.

Check

Graph the solution set of $y > \frac{1}{2}$.

Example 2 Write Inequalities from a Graph

Write an inequality that represents the graph.

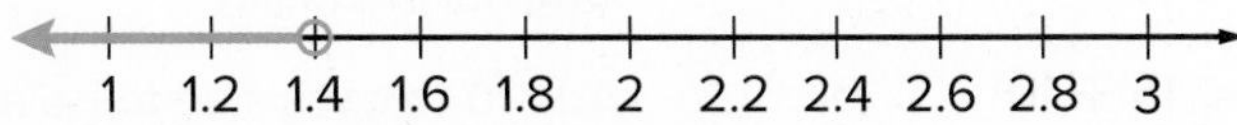

The endpoint is shown with a circle at 1.4, so 1.4 is not included in the solution. The inequality must be $<$ or $>$.

The arrow points to values less than 1.4.
The graph represents the solution of $a < 1.4$.

Think About It!
How do you know which inequality symbol to use by looking at a graph?

Check

Write an inequality that represents the graph.

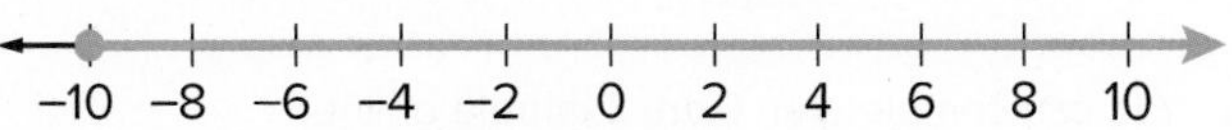

Go Online You can complete an Extra Example online.

Explore Properties of Inequalities

Online Activity Use graphing technology to complete the Explore.

> **INQUIRY** Do the properties of equality hold true for inequalities?

Learn Solving Inequalities by Using Addition and Subtraction

Addition and subtraction can be used to solve inequalities.

Key Concept • Addition Property of Inequalities	
Words	If the same number is added to each side of a true inequality, the resulting inequality is also true.
Symbols	For any real numbers a, b, and **c**, the following are true. If $a > b$, then $a + c > b + c$. If $a < b$ then $a + c < b + c$.

Key Concept • Subtraction Property of Inequalities	
Words	If the same number is subtracted from each side of a true inequality, the resulting inequality is also true.
Symbols	For any real numbers a, b, and c, the following are true. If $a > b$, then $a - c > b - c$. If a $< b$ then $a - c < b - c$.

When solving inequalities, you can write the solution set in a more concise way using **set-builder notation**. For example, $\{x | x \geq -4\}$ represents the set of all numbers x such that x is greater than or equal to -4.

Talk About It!
How many solutions of the inequality are there? Justify your argument.

Example 3 Solve Inequalities by Adding

Solve $x - 10 < 15$.

$x - 10 < 15$	Original inequality
$x - 10 + 10 < 15 + 10$	Add 10 to each side to isolate x.
$x < 25$	Simplify.

The solution set is $\{x | x < 25\}$.

Study Tip
Set-Builder Notation
$\{x | x < 25\}$ is read *the set of all numbers x such that x is less than 25.*

Check

Solve $-9 + b \leq 16$.

 **Go Online** You can complete an Extra Example online.

Example 4 Solve Inequalities by Subtracting

Solve $x + 24 \geq 61$.

$x + 24 \geq 61$	Original inequality
$x + 24 - 24 \geq 61 - 24$	Subtract 24 from each side.
$x \geq 37$	Simplify.

The solution set is $\{x \mid x \geq 37\}$.

Check

Select the solution set for $88 < x + 13$.

Example 5 Add or Subtract to Solve Inequalities with Variables on Each Side

Solve $9y + 3 \geq 10y$.

$9y + 3 \geq 10y$	Original inequality
$9y - 9y + 3 \geq 10y - 9y$	Subtract $9y$ from each side.
$3 \geq y$	Simplify.

Since $3 \geq y$ is the same as $y \leq 3$, $\{y \mid y \leq 3\}$.

Check

Write the solution set for $7x + 6 < 8x$.

 Example 6 Use an Inequality to Solve a Problem

DATA USAGE **Hassan's wireless contract allows him to use at most 5 gigabytes (GB) of data per month. At this point, Hassan has used 3.7 GB of data. How many gigabytes of data can Hassan use during the rest of the month without exceeding the maximum allowance?**

Complete the table to write an inequality to represent how many gigabytes of data Hassan can use. Then solve the inequality.

Words	Hassan can use	at most	5 GB of data.
Variables	Let g = the number of gigabytes that Hassan has left to use.		
Inequality	$3.7 + g$	$\leq$	5

(continued on the next page)

Think About It!

How can you check the solution of the inequality?

Study Tip

Writing Inequalities Simplifying the inequality so that the variable is on the left side, as in $y \leq 3$, prepares you to write the solution set in set-builder notation and graph the inequality on a number line.

Use a Source

Research data plans for wireless carriers in your area. Write and solve your own inequality to represent the amount of data remaining if you have already used 5.2 GB.

Study Tip

Inequalities Verbal problems containing phrases like *greater than* and *less than* can be solved by using inequalities. Some other phrases that include inequalities are:

$<$ less than; fewer than
$>$ greater than; more than
$\leq$ less than or equal to; at most; no more than
$\geq$ greater than or equal to; at least; no less than

Think About It!

If *a*, *b*, and *c* are positive real numbers, what must be true if *ac* is greater than or equal to *bc*? What must happen to an inequality symbol when you divide each side by a negative number if the inequality is to remain true?

$3.7 + g \leq 5$	Original inequality
$3.7 - 3.7 + g \leq 5 - 3.7$	Subtract 3.7 from each side.
$g \leq 1.3$	Simplify.

The solution set is $\{g \mid g \leq 1.3\}$.

Hassan can use up to 1.3 GB of data without exceeding his maximum allowance. Notice that negative numbers are solutions to the inequality, but they are not viable solutions to the problem because Hassan cannot use a negative amount of data.

Learn Solving Inequalities by Using Multiplication and Division

If you multiply or divide each side of an inequality by a positive number, then the inequality remains true.

If you multiply or divide each side of an inequality by a negative number, the inequality symbol changes direction.

Key Concepts • Multiplication Property of Inequalities		
Words	If each side of a true inequality is multiplied by a positive number, the resulting inequality is also true.	If each side of a true inequality is multiplied by a negative number, the direction of the inequality sign must be reversed to make the resulting inequality also true.
Symbols	For any real numbers *a* and *b* and any positive real number *c*: If $a > b$, then $ac > bc$. If $a < b$, then $ac < bc$.	For any real numbers *a* and *b* and any negative real number *c*: If $a > b$, then $ac < bc$. If $a < b$, then $ac > bc$.

Key Concepts • Division Property of Inequalities		
Words	If each side of a true inequality is divided by a positive number, the resulting inequality is also true.	If each side of a true inequality is divided by a negative number, the direction of the inequality sign must be reversed to make the resulting inequality also true.
Symbols	For any real numbers *a* and *b* and any positive real number *c*: If $a > b$, then $\frac{a}{c} > \frac{b}{c}$. If $a < b$, then $\frac{a}{c} < \frac{b}{c}$.	For any real numbers *a* and *b* and any negative real number *c*: If $a > b$, then $\frac{a}{c} < \frac{b}{c}$. If $a < b$, then $\frac{a}{c} > \frac{b}{c}$.

These properties also hold true for inequalities involving $\leq$ and $\geq$.

Apply Example 7 Write and Solve an Inequality

BOOKS **Alisa has read approximately $\frac{1}{4}$ of a novel. If she has read at least 112 pages, how many pages are there in the novel?**

1. What is the task?

Describe the task in your own words. Then list any questions that you may have. How can you find answers to your questions?

Sample answer: I know the number of pages read and the fraction of the novel read. I need to find out how many pages are in the novel.

2. How will you approach the task? What have you learned that you can use to help you complete the task?

Sample answer: I will use estimation first. Then I will write an inequality to represent the situation and solve it.

3. What is your solution?

Estimate the number of pages in the novel. 400

Write an inequality to represent this situation. Let n = the number of pages in the novel.

$$\frac{1}{4} \cdot n \geq 112$$
$$4\left(\frac{1}{4}\right)n \geq 4(112)$$
$$n \geq 448$$

There are at least 448 pages in the novel.

4. How can you know that your solution is reasonable?

Write About It! Write an argument that can be used to defend your solution.

Sample answer: Use multiplication; $448\left(\frac{1}{4}\right) = 112$, so 448 is reasonable. Also, $448 > 400$ which makes sense with our estimate of more than 400 pages.

Check

ELECTRIC CAR For every hour x that Eva's electric car charges, she can drive the car 7.5 miles. Eva needs to drive at least 60 miles tomorrow.

Part A What inequality represents the situation in terms of x hours?

Part B What is the least amount of time that Eva will need to charge her car? ___?___ hours

Math History Minute

German mathematician **Emmy Noether (1882–1935)** has been described as one of the greatest mathematicians of the twentieth century. She devised theorems for several concepts later found in Einstein's theory of relativity and was one of the founders of abstract algebra. One person wrote, "The development of abstract algebra, which is one of the most distinctive innovations of twentieth century mathematics, is largely due to her."

Go Online You can complete an Extra Example online.

Study Tip

Multiplicative Inverses The multiplicative inverse, or reciprocal, of a number can be used to undo multiplication. Multiplying $-\frac{2}{5}x$ by the reciprocal $-\frac{5}{2}$ in the example at the right is the same as dividing by $-\frac{2}{5}$, but is easier to compute mentally.

Example 8 Solve an Inequality by Multiplying

Solve $-\frac{2}{5}x \leq 11$. Graph the solution set on a number line.

$-\frac{2}{5}x \leq 11$ — Original inequality

$\left(-\frac{5}{2}\right) -\frac{2}{5}x \leq 11\left(-\frac{5}{2}\right)$ — Multiply each side by $-\frac{2}{5}$. Reverse the inequality symbol.

$x \geq -27.5$ — Simplify.

The solution set is $\{x \mid x \geq -27.5\}$.

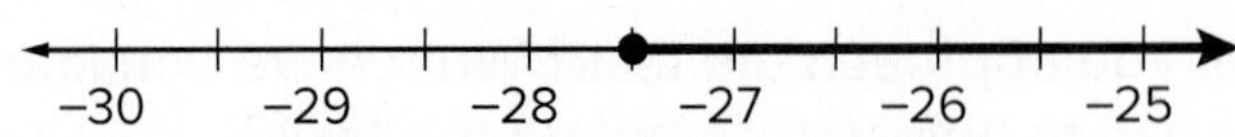

 **Think About It!**

Why was the inequality symbol reversed?

Example 9 Solve an Inequality by Dividing

Solve $20x < 4$. Graph the solution set on a number line.

$20x < 4$ — Original inequality

$\frac{20x}{20} < \frac{4}{20}$ — Divide each side by 20.

$x < \frac{1}{5}$ — Simplify.

The solution set is $\{x \mid x < \frac{1}{5}\}$.

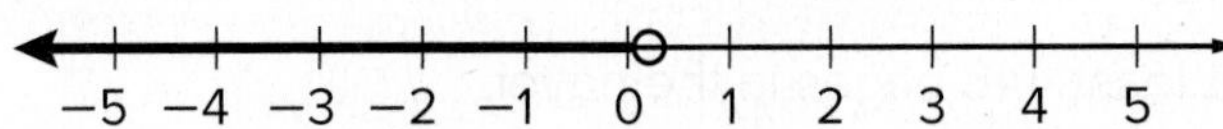

Check

Solve $7x > -161$.

 **Think About It!**

Why is the solution of $-13z \geq 117$ shaded to the left when the original inequality symbol is greater than or equal to?

Example 10 Solve an Inequality with a Negative Coefficient

Solve $-13z \geq 117$. Graph the solution set on a number line.

$-13z \geq 117$ — Original inequality

$\frac{-13z}{-13} \geq \frac{117}{-13}$ — Divide each side by −13.

$z \leq -9$ — Simplify.

The solution set is $\{z \mid z \leq -9\}$.

−16 −14 −12 −10 −8 −6 −4 −2 0 2 4

Check

Select the solution set for $-13x > -169$.

A. $\{x \mid x > 13\}$ **B.** $\{x \mid x < 13\}$ **C.** $\{x \mid x > -13\}$ **D.** $\{x \mid x < -13\}$

 **Go Online** You can complete an Extra Example online.

Study Tip

Negatives A negative sign in an inequality does not necessarily mean that the direction of the inequality symbol should change. For example, when solving $\frac{x}{3} \geq -9$, do not change the direction of the inequality symbol.

Practice

Go Online You can complete your homework online.

Example 1

Graph the solution set of each inequality.

1. $x \leq -5$

2. $y \geq -2$

3. $g > 5$

4. $h < -6$

5. $a < 7$

6. $b \leq 6$

Example 2

Write an inequality that represents each graph.

7.

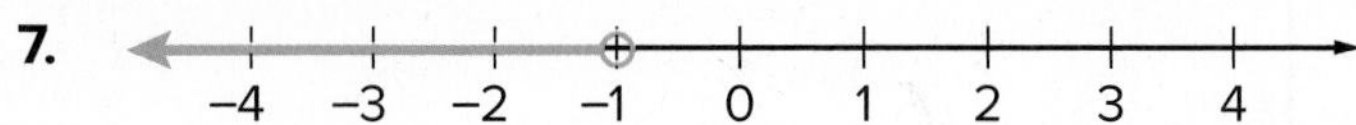

8. −8 −7 −6 −5 −4 −3 −2 −1 0

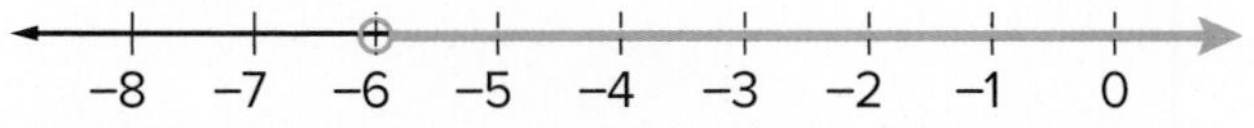

9. 0 1 2 3 4 5 6 7 8

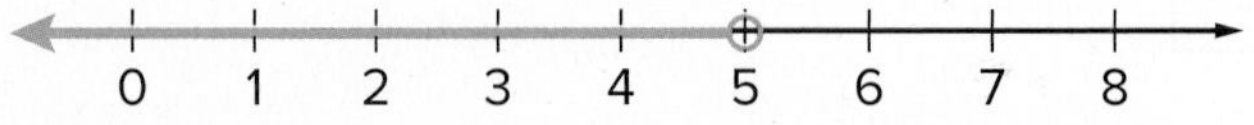

10. −4 −3 −2 −1 0 1 2 3 4

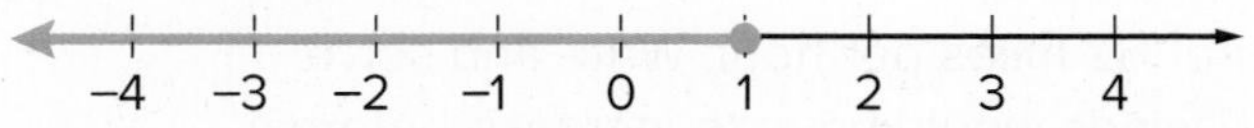

11. −8 −7 −6 −5 −4 −3 −2 −1 0

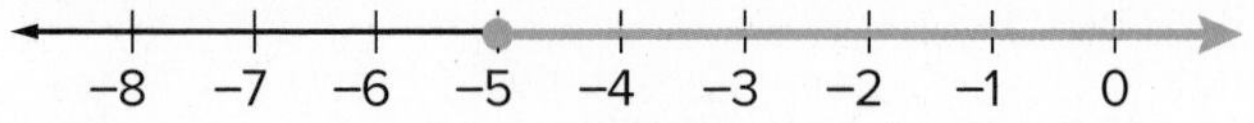

12. −8 −7 −6 −5 −4 −3 −2 −1 0

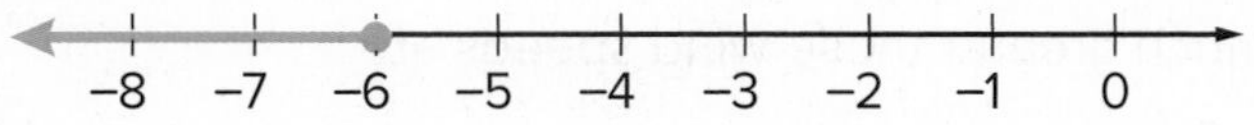

Examples 3–5

Solve each inequality.

13. $m - 4 < 3$

14. $p - 6 \geq 3$

15. $r - 8 \leq 7$

16. $t - 3 > -8$

17. $b + 2 \geq 4$

18. $13 > 18 + r$

19. $5 + c \leq 1$

20. $-23 \geq q - 30$

21. $11 + m \geq 15$

22. $h - 26 < 4$

23. $8 \leq r - 14$

24. $-7 > 20 + c$

25. $2a \leq -4 + a$

26. $z + 4 \geq 2z$

27. $w - 5 \leq 2w$

28. $3y \leq 2y - 6$

29. $6x + 5 \geq 7x$

30. $-9 + 2a < 3a$

Example 6

31. PIZZA Tara and friends order a pizza. Tara eats 3 of the 10 slices and pays $4.50 for her share. Assuming that Tara has paid at least her fair share, write and solve an inequality to represent the cost of the pizza.

32. WEATHER Theodore Fujita of the University of Chicago developed a classification of tornadoes according to wind speed and damage. The table shows the classification system.

Level	Name	Wind Speed Range (mph)
F0	Gale	40–72
F1	Moderate	73–112
F2	Significant	113–157
F3	Severe	158–206
F4	Devastating	207–260
F5	Incredible	261–318
F6	Inconceivable	319–379

Source: National Weather Service

a. Suppose an F3 tornado has winds that are 162 miles per hour. Write and solve an inequality to determine how much the winds would have to increase before the F3 tornado becomes an F4 tornado.

b. A tornado has wind speeds that are at least 158 miles per hour. Write and solve an inequality that describes how much greater these wind speeds are than the slowest tornado.

Example 7

33. GARBAGE The amount of garbage that the average American adds to a landfill each day is 4.6 pounds. If at least 2.5 pounds of a person's daily garbage could be recycled, how much would still go into a landfill?

34. SUPREME COURT The first Chief Justice of the U.S. Supreme Court, John Jay, served 2079 days as Chief Justice. He served 10,463 days fewer than John Marshall, who served as Supreme Court Chief Justice for the longest period of time. How many days must the current Supreme Court Chief Justice John Roberts serve to surpass John Marshall's record of service?

35. AIRLINES On average, at least 25,000 pieces of luggage are lost or misdirected each day by United States airlines. Of these, 98% are located by the airlines within 5 days. From a given day's lost luggage, at least how many pieces of luggage are still lost after 5 days?

36. SCHOOL Gilberto earned these scores on the first three tests in biology this term: 86, 88, and 78. What is the lowest score that Gilberto can earn on the fourth and final test of the term if he wants to have an average of at least 83?

Examples 8–10

Solve each inequality. Graph the solution on a number line.

37. $\frac{1}{4}m \leq -17$

38. $\frac{1}{2}a < 20$

39. $-11 > -\frac{c}{11}$

40. $-2 \geq -\frac{d}{34}$

41. $-10 \leq \frac{x}{-2}$

42. $-72 < \frac{f}{-6}$

43. $\frac{2}{3}h > 14$

44. $-\frac{3}{4}j \geq 12$

45. $-\frac{1}{6}n \leq -18$

46. $6p \leq 96$

47. $4r < 64$

48. $32 > -2y$

49. $-26 < 26t$

50. $-6v > -72$

51. $-33 \geq -3z$

52. $4b \leq -3$

53. $-2d < 5$

54. $-7f > 5$

Mixed Exercises

Match each inequality with its corresponding statement.

55. $3n < 9$

56. $\frac{1}{3}n \geq 9$

57. $3n \leq 9$

58. $-3n > 9$

59. $\frac{1}{3}n \leq 9$

60. $-3n \geq 9$

a. Three times a number is at most nine.

b. One third of a number is no more than nine.

c. Negative three times a number is more than nine.

d. Three times a number is less than nine.

e. Negative three times a number is at least nine.

f. One third of a number is greater than or equal to nine.

Define a variable, write an inequality, and solve each problem. Check your solution.

61. Seven more than a number is less than or equal to –18.

62. Twenty less than a number is at least 15.

63. A number plus 2 is at most 1.

64. One eighth of a number is less than or equal to 3.

65. Negative twelve times a number is no more than 84.

66. Eight times a number is at least 16.

STRUCTURE Solve each inequality. Check your solution, and then graph it on a number line.

67. $14c > 56$

68. $20b \geq -120$

69. $\frac{x}{4} < 9$

70. $\frac{x}{2.5} \leq 8$

71. $m + 3.7 < 9.1$

72. $n - \frac{1}{5} > \frac{4}{5}$

73. $c + (-1.4) \geq 2.3$

74. $k + \frac{3}{4} > \frac{1}{3}$

75. **EVENT PLANNING** The Community Center does not charge a rental fee as long as a rentee orders a minimum of $5000 worth of food. Antonio is planning a banquet. If he is expecting 225 people to attend, what is the minimum he will have to spend on food per person to avoid paying a rental fee?

76. **VITAMINS** The minimum daily requirement of vitamin C for 14-year-olds is at least 50 milligrams per day. An average-sized apple contains 6 milligrams of vitamin C. How many apples would a person have to eat each day to satisfy this requirement? Define a variable and write and solve an inequality to represent this situation.

77. **USE A SOURCE** The loudest insect is the African cicada. It produces sounds as loud as 105 decibels. The blue whale is the loudest mammal. The call of the blue whale can reach levels up to 83 decibels louder than the African cicada. Write and solve an inequality to represent the situation. How loud are the calls of the blue whale? Use a source to verify your answer.

78. **USE A MODEL** In a mathematics exam with a maximum score of 100, Machelle loses less than 27 points. The table shows the grade that matches the exam score. Compare points to grades and identify which grade Machelle can get.

Grade	Points
A	92–100
B	83–91
C	74–82
D	65–73
F	64 and below

a. Define a variable. Then write and solve an inequality to represent the number of points Machelle received on her exam.

b. Interpret the solution to your inequality. What do you know about Machelle's grade on the exam?

Higher Order Thinking Skills

79. **WHICH ONE DOESN'T BELONG** Which inequality does *not* have the solution $\{x \mid x < -2\}$?

A $-3x > 6$ B $-\frac{x}{2} < 1$ C $7x < -14$ D $\frac{4}{3}x < -\frac{8}{3}$

80. **FIND THE ERROR** Marty and Heath solved the same exercise in different ways. Is either correct? Explain your reasoning.

Marty	Heath
$3m \geq -21$	$3m \geq -21$
$\frac{3m}{3} \geq \frac{-21}{3}$	$\frac{3m}{3} \geq -\frac{21}{3}$
$m \leq -7$	$m \geq -7$

81. Solve each inequality in terms of x. Assume that a does not equal 0.

a. $ax < 7$

b. $ax \geq 12$

c. $ax > 3a$

d. $ax \geq \frac{a}{4}$

82. **ANALYZE** Determine whether the statement is *sometimes*, *always*, or *never* true. If $a > b$, then $\frac{1}{a} > \frac{1}{b}$. Justify your argument.

Solving Multi-Step Inequalities

Explore Modeling Multi-Step Inequalities

Online Activity Use algebra tiles to complete the Explore.

INQUIRY How can you model and solve a multi-step inequality?

Today's Goals

- Solve multi-step linear inequalities.

Learn Solving Inequalities Involving More Than One Step

Step 1 Isolate the variable terms on one side of the inequality using addition or subtraction.

Step 2 Multiply or divide to isolate the variable.

Study Tip

Negative Numbers When multiplying or dividing by a negative number, the direction of the inequality symbol changes. This holds true for multi-step as well as one-step inequalities.

Example 1 Apply Multi-Step Inequalities

PUBLISHING **Suzy wants to self-publish her comic book. One printing company offers to publish the book for a $220 flat rate plus $3 per copy of the book. Her maximum budget is $400.**

Part A Write an inequality.

Words	$220 flat rate	Plus	$3 per copy	is at most	$400
Inequality	220	+	$3x$	$\leq$	400

Part B Solve the inequality.

$220 + 3x \leq 400$ Original inequality

$3x \leq 180$ Subtract 220 from each side.

$x \leq 60$ Divide each side by 3.

Suzy can have up to 60 books printed while not exceeding her budget.

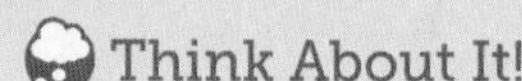

Think About It!

Can x be any real number less than or equal to 60? Explain your reasoning.

Check

TICKETS Jamal has $40 to buy tickets to a performance for himself and his friends. If he buys a $10 membership, he can buy tickets for $5 each. How many tickets can he buy while remaining within his budget?

If x represents the number of tickets Jamal purchases, write an inequality that represents the situation.

Solve the inequality.

Go Online You can complete an Extra Example online.

Example 2 Write and Solve a Multi-Step Inequality

Consider the inequality *The opposite of a number divided by two minus seventeen is less than seven.*

Translate the sentence into an inequality.

$-\frac{x}{2} - 17 < 7$

Solve the inequality.

$-\frac{x}{2} - 17 < 7$	Original inequality
$-\frac{x}{2} < 24$	Add 17 to each side.
$-x < 48$	Multiply each side by 2.
$x > -48$	Divide each side by −1, reversing the inequality symbol.

Graph the solution on a number line.

−50 −48 −46 −44 −42 −40 −38 −36 −34 −32 −30

Study Tip

Empty Set or All Real Numbers If an inequality simplifies to a false statement, then there is no solution to the inequality. The solution set is the empty set, Ø. However, if all values of a variable make the inequality true, then the solution set is all real numbers.

Example 3 Solve an Inequality with the Distributive Property

Solve the inequality $4(2x - 11) \leq -12 + 2(x - 4)$. Then graph the solution on a number line.

$4(2x - 11) \leq -12 + 2(x - 4)$	Original inequality
$8x - 44 \leq -12 + 2x - 8$	Distributive Property
$8x - 44 \leq -20 + 2x$	Simplify.
$8x \leq 24 + 2x$	Add 44 to each side.
$6x \leq 24$	Subtract $2x$ from each side.
$x \leq 4$	Divide each side by 6.

Graph $x \leq 4$ on a number line.

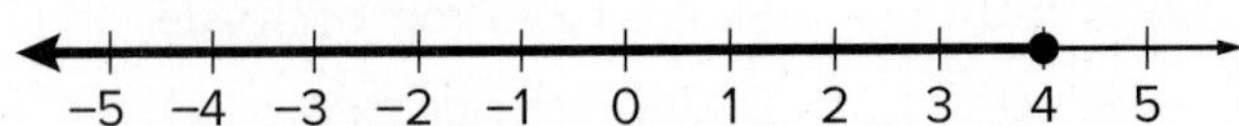

Check

Solve $88 \geq -33 + 11(x + 8)$. Then graph the inequality.

Go Online You can complete an Extra Example online.

Practice

Go Online You can complete your homework online.

Example 1

1. **BEACHCOMBING** Jay wants to rent a metal detector. A rental company charges a one-time rental fee of \$15 plus \$2 per hour to rent a metal detector. Jay has only \$35 to spend.
   a. Write an inequality to represent this situation, where h is the number of hours Jay will rent a metal detector.
   b. Solve the inequality. What is the maximum amount of time he can rent the metal detector?

2. **AGES** Pedro, Sebastian, and Manuel Martinez are each one year apart in age. The sum of their ages is greater than the age of their father, who is 60.
   a. Write an inequality to represent this situation, where x is the age of the youngest brother.
   b. Solve the inequality.
   c. How old can the oldest brother be? Explain your reasoning.

3. **RIDE SHARE** Demetri lives in the city and sometimes uses a ride share service. A ride costs \$1.50 for the first $\frac{1}{5}$ mile and \$0.25 for each additional $\frac{1}{5}$ mile. Demetri does not want to spend more than \$3.75 on a ride.
   a. Write an inequality to represent this situation, where x is the number of miles.
   b. Solve the inequality. What is the maximum distance he can travel if he does not tip the driver?
   c. Generalize a method for writing an inequality for this situation if the service charges \$1.50 for the first $\frac{1}{a}$ mile and \$0.25 for each additional $\frac{1}{a}$ mile.

4. **POST OFFICE** Keshila goes to the post office to mail a package and a few letters. Stamps cost 49 cents each. It will cost \$7.65 to mail the package. Keshila has \$10.00.
   a. Write an inequality to represent this situation, where x is the number of stamps.
   b. Solve the inequality. What is the maximum number of stamps Keshila can purchase to mail letters?

5. **BANQUET** A charity is hosting a benefit dinner. They are asking \$100 per table plus \$40 per person. Nathaniel is purchasing tickets for his friends and does not want to spend more than \$250.
   a. Write an inequality to represent this situation, where x is the number of people.
   b. Solve the inequality. What is the maximum number of people Nathaniel can invite to the dinner?

Example 2

Translate each sentence into an inequality. Then solve the inequality and graph the solution on a number line.

6. Five times a number minus one is greater than or equal to negative eleven.
7. Twenty-one is greater than the sum of fifteen and two times a number.
8. Negative nine is greater than or equal to the sum of two-fifths times a number and seven.
9. A number divided by eight minus thirteen is greater than negative six.
10. The sum of the opposite of a number and six is less than or equal to five.
11. Thirty-seven is less than the difference of seven and ten times a number.
12. Eight minus a number divided by three is greater than or equal to eleven.
13. Negative five-fourths times a number plus six is less than twelve.
14. The difference of three times a number and six is greater than or equal to the sum of fifteen and twenty-four times a number.
15. The sum of fifteen times a number and thirty is less than the difference of ten times a number and forty-five.

Example 3

Solve each inequality. Then graph the solution on a number line.

16. $-3(7n + 3) < 6n$
17. $21 \geq 3(a - 7) + 9$
18. $2y + 4 > 2(3 + y)$
19. $3(2 - b) < 10 - 3(b - 6)$
20. $7 + t \leq 2(t + 3) + 2$
21. $8a + 2(1 - 5a) \leq 20$

Mixed Exercises

Solve each inequality. Check your solution.

22. $2(x - 4) \leq 2 + 3(x - 6)$
23. $\frac{2x - 4}{6} \geq -5x + 2$
24. $5.6z + 1.5 < 2.5z - 4.7$
25. $0.7(2m - 5) \geq 21.7$
26. $2(-3m - 5) \geq -28$
27. $-6(w + 1) < 2(w + 5)$

USE TOOLS **Use a graphing calculator to solve each inequality.**

28. $3x + 7 > 4x + 9$
29. $13x - 11 \leq 7x + 37$
30. $2(x - 3) < 3(2x + 2)$
31. $\frac{1}{2}x - 9 < 2x$
32. $2x - \frac{2}{3} \geq x - 22$
33. $\frac{1}{3}(4x + 3) \geq \frac{2}{3}x + 2$

STRUCTURE Solve each inequality. Then graph it on a number line.

34. $9.1g + 4.5 < 10.1g$

35. $\frac{3}{2}p - \frac{2}{3} \leq \frac{4}{9} + \frac{1}{2}p$

36. $3.3r - 8.3 \geq 5.3r - 12.9$

37. TREEHOUSE DESIGN Devontae is building a treehouse in his backyard. He researches city restrictions on building codes. The height of the treehouse cannot exceed 13 feet. He wants to build a tree house with 2 levels of equal height that is 4 feet off the ground.

a. Write and solve an inequality.

b. What is the maximum height of one level?

c. Devontae decides to build one level higher off the ground. If the level is 8 feet tall, how high can the tree house be off the ground?

38. MEDICINE Clark's Rule is a formula used to determine pediatric dosages of over-the-counter medicines: $\frac{\text{weight of child (lb)}}{150} \times \text{adult dose} = \text{child dose}.$

a. If an adult dose of acetaminophen is 1000 milligrams and a child weighs no more than 90 pounds, what is the recommended child's dose?

b. The label below appears on a child's cold medicine. What is the adult minimum dosage in milliliters?

Weight (lb)	Age (yr)	Dose
under 48	under 6	call a doctor
48-95	6-11	2 tsp or 10 mL

c. What is the maximum adult dosage in milliliters?

39. CONSTRUCT ARGUMENTS Eric says that 15 more than 6 times the number of pencils he has is less than 20. What can you conclude about the number of pencils Eric has? Justify your argument.

40. REASONING The perimeter of a rectangular playground can be no greater than 120 meters. The width of the playground cannot exceed 22 meters. What are the possible lengths of the playground?

41. STRUCTURE Solve $10n - 7(n + 2) > 5n - 12$. Explain each step in your solution.

Higher Order Thinking Skills

42. WRITE What is the solution set of the inequality $2(2x + 4) < 4(x + 1)$? Why? How is the solution set related to the solution set of $2(2x + 4) \geq 4(x + 1)$? Explain.

43. PERSEVERE Mei got scores of 76, 80, and 78 on her last three history exams. Write and solve an inequality to determine the score she needs on the next exam so that her average is at least 82.

44. ANALYZE A triangular carpet has sides of length a feet, b feet, and c feet. The maximum perimeter is 20 feet.

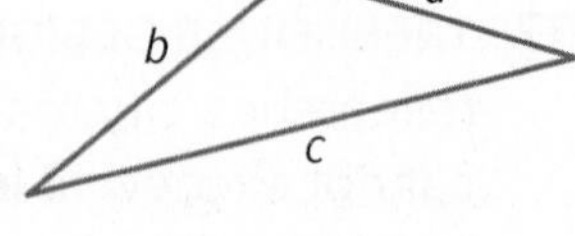

a. Side b is 2 feet longer than a and c is 2 feet longer than b. Which side is the shortest? Explain.

b. What are the possible lengths of the shortest side of the carpet? Explain.

45. CREATE Write an inequality that has the solution set graphed at the right. Solving the inequality should require the Distributive Property, Addition Property of Inequalities, and the Division Property of Inequalities.

−5 −4 −3 −2 −1 0 1 2 3 4 5

46. PERSEVERE Let $b > 2$. Describe how you would determine if $ab > 2a$.

47. CREATE Four times the number of baseball cards in Ted's collection is more than five times that number minus 15. Define a variable and write an inequality to represent Ted's baseball cards. Solve the inequality and interpret the results.

48. WRITE Explain how you could solve $-3p + 7 \geq -2$ without multiplying or dividing each side by a negative number.

49. PERSEVERE If $ax + b < ax + c$ is true for all real values x, what will be the solution of $ax + b > ax + c$? Explain your reasoning.

50. WHICH ONE DOESN'T BELONG? Name the inequality that does not belong. Justify your conclusion.

$4y + 9 > -3$	$3y - 4 > 5$	$-2y + 1 < -5$	$-5y + 2 < -13$

51. WRITE Explain when the solution set of an inequality will be the empty set or the set of all real numbers. Show an example of each.

52. PERSEVERE Solve each inequality in terms of a. Assume that a does not equal 0.

a. $ax + 5 < 11$

b. $ax - 4 \geq 12$

c. $ax - 5 > 3a$

d. $ax + 1 \geq \frac{a}{2}$

Solving Compound Inequalities

Explore Guess the Range

Online Activity Use a real-world situation to complete the Explore.

INQUIRY How can you tell if a value will satisfy a compound inequality that includes the word *and*?

Today's Goals

- Solve and graph linear inequalities containing the word *and*.
- Solve and graph linear inequalities containing the word *or*.

Today's Vocabulary

compound inequality

intersection

union

Learn Solving Compound Inequalities Using the Word *and*

A **compound inequality** is two or more inequalities that are connected by the word *and* or *or*. A compound inequality containing the word *and* is only true when both of the inequalities are true. So, its graph is where the graphs of the inequalities overlap. This overlapping section that represents the compound inequality is called the **intersection**. To determine where the graphs intersect, graph each inequality and identify where they overlap.

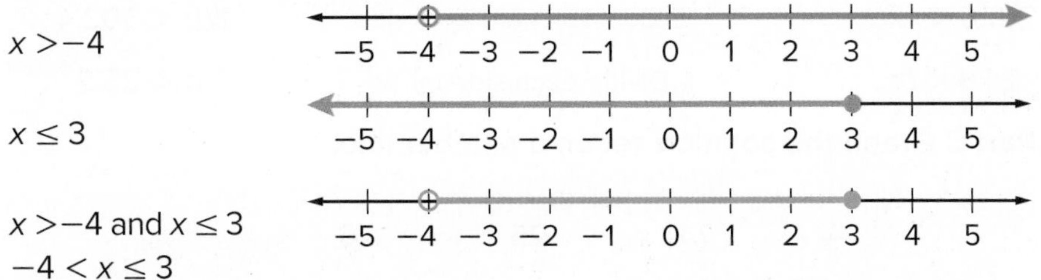

The compound inequality $-4 < x \leq 3$ can be read in two ways. It can be read as *x is greater than −4 and less than or equal to 3* or *x is between −4 and 3 including 3.*

Study Tip

Inequality Solutions For inequalities using the word *and*, a number has to be true for both inequalities in order to be a solution for the compound inequality.

Example 1 Solve and Graph an Intersection

Solve $-8 \leq h - 2 < 1$. Then graph the solution set.

Express the compound inequality as two inequalities joined by the word *and*.

$-8 \leq h - 2$	and Write the inequality using *and*.	$h - 2 < 1$
$-8 + 2 \leq h - 2 + 2$	Add 2 to each side.	$h - 2 + 2 < 1 + 2$
$-6 \leq h$	Simplify.	$h < 3$

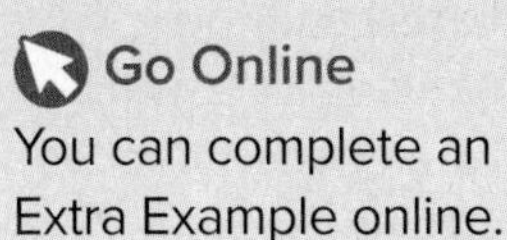

Go Online You can complete an Extra Example online.

The solution set is $\{h \mid -6 \leq h < 3\}$.

Graph the solution set on a number line.

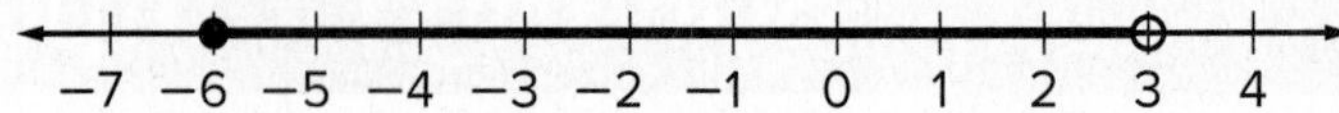

Check

Solve $-7 \leq 3x + 2 \leq 5$. Then graph the solution set.

Think About It!

What are 3 acceptable weights for a cereal box?

Example 2 Apply Compound Inequalities

MANUFACTURING **A cereal manufacturer distributes cases of cereal to grocery stores. Each case contains 12 boxes of cereal. In order to pass the manufacturer's quality assurance test, the case must weigh between 336.4 ounces and 331.6 ounces, which includes 34 ounces for the weight of the case's cardboard box.**

Part A

Write an inequality that describes the weight of a box of cereal.

$331.6 < 12b + 34 < 336.4$, where b is the weight of each box.

Part B Solve the inequality.

$331.6 < 12b + 34$	and	$12b + 34 < 336.4$
$297.6 < 12b$	Subtract 34 from each side.	$12b < 302.4$
$24.8 < b$	Divide each side by 12.	$b < 25.2$

Part C Graph the solution set on a number line.

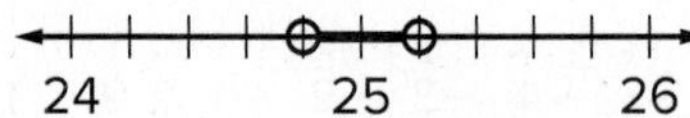

A cereal box must weigh between 24.8 and 25.2 ounces in order to pass the manufacturer's quality assurance test. The compound inequality is $\{b \mid 24.8 < b < 25.2\}$

Study Tip

Assumptions In this example, we assume that all of the cereal boxes weigh approximately the same amount and that weighing a case of boxes is an accurate test of quality control for the manufacturer. However, it is possible that a case may contain cereal boxes that are heavier or lighter than the desired weight.

Study Tip

Inclusivity and Exclusivity In application problems, *within* implies inclusivity, which means the numbers mentioned will be included and the symbols $\leq$ or $\geq$ will be used. *Between* implies exclusivity and the symbols $<$ or $>$ will be used.

Check

CARS Keshawn has been saving to buy his first car. He wants the total cost of the car and fees to be more than \$5000 but at most \$7000. The fees for buying a used car, such as title, registration, and dealership fees, will be \$700.

Graph the list price of the cars Keshawn could buy.

Go Online You can complete an Extra Example online.

Learn Solving Compound Inequalities Using the Word *or*

A compound inequality containing the word *or* is true if at least one of the inequalities is true. A **union** is the graph of a compound inequality containing *or*; the solution is a solution of either inequality, not necessarily both.

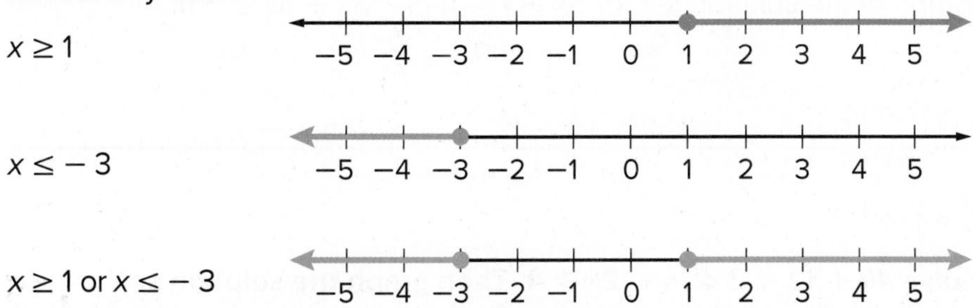

> **Go Online** You can watch a video to see how to solve inequalities involving *and* and *or*.

> **Study Tip**
> **Inequality Solutions** For inequalities using the word or, a number has to be true for at least one of the inequalities in order for it to be a solution for the compound inequality. The solution must work for the first inequality or the second inequality.

Example 3 Solve and Graph a Union

Solve $4n + 8 \leq 16$ or $-3n + 7 < -11$. Then graph the solution set.

Express the compound inequality as two inequalities joined by the word *or*.

$4n + 8 \leq 16$	or	$-3n + 7 < -11$
$4n + 8 - 8 \leq 16 - 8$	Subtract.	$-3n + 7 - 7 < -11 - 7$
$4n \leq 8$	Simplify.	$-3n < -18$
$\frac{4n}{4} \leq \frac{8}{4}$	Divide.	$\frac{-3n}{-3} > \frac{-18}{-3}$
$n \leq 2$	Simplify.	$n > 6$

Graph the solution set on a number line.

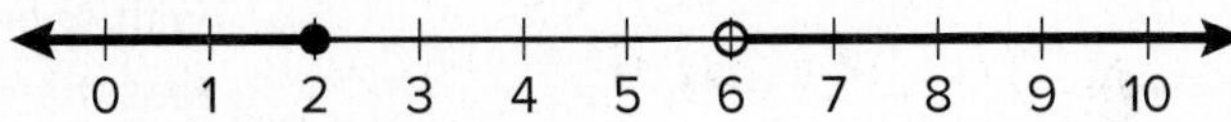

The union contains all points with coordinates less than or equal to 2 and all points with coordinates greater than 6. So, the solution set is $\{n \mid n \leq 2 \text{ or } n > 6\}$.

Go Online You can complete an Extra Example online.

Check

Solve $5x + 1 < 11$ or $-3x + 10 \leq -11$. Then graph the solution set.

Part A

Write the solution set for $5x + 1 < 11$ or $-3x + 10 \leq -11$.

Part B

Graph of the solution set for $5x + 1 < 11$ or $-3x + 10 \leq -11$.

Talk About It!

Why does the union $4k + 12 < 2$ or $4 - 2k > 4$ include $-2\frac{1}{2}$ even though one of the solutions, $k < -2\frac{1}{2}$, does not?

Example 4 Overlapping Intervals

Solve $4k + 12 < 2$ or $4 - 2k > 4$. Then graph the solution set.

Express the compound inequality as two inequalities joined by the word *or*.

$4k + 12 < 2$	or	$4 - 2k > 4$
$4k + 12 - 12 < 2 - 12$	Subtract.	$4 - 2k - 4 > 4 - 4$
$4k < -10$	Simplify.	$-2k > 0$
$\frac{4k}{4} < \frac{-10}{4}$	Divide.	$\frac{-2k}{-2} < \frac{0}{-2}$
$k < \frac{-21}{2}$	Simplify.	$k < 0$

Graph the solution set on a number line.

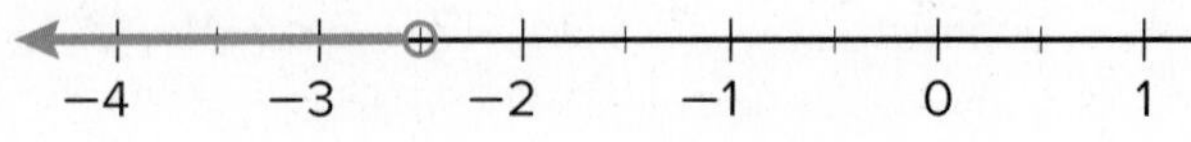

The graph of $k < -2\frac{1}{2}$ contains all points with coordinates less than $-2\frac{1}{2}$.

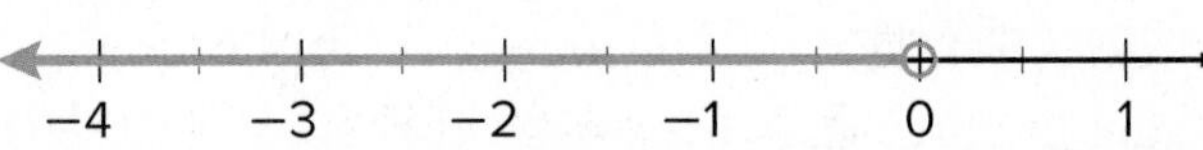

The graph of $k < 0$ contains all points with coordinates less than 0.

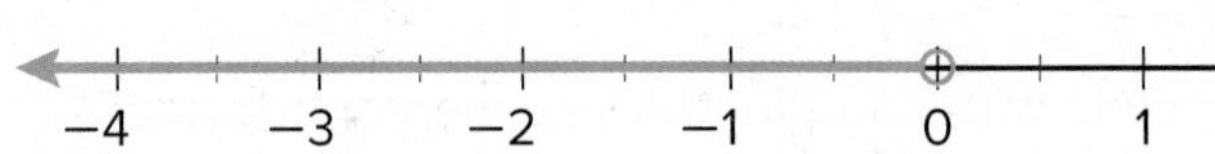

The union contains all points with coordinates less than 0.

Because $k < -2\frac{1}{2}$ is contained within $k < 0$, the solution set is $\{k \mid k < 0\}$.

Go Online You can complete an Extra Example online

Check

Solve $4m + 7 \geq 19$ or $-m + 5 \leq 0$. Then graph the solution set.

Part A

Select the solution set for $4m + 7 \geq 19$ or $-m + 5 \leq 0$.

Part B

Graph of the solution set for $4m + 7 \geq 19$ or $-m + 5 \leq 0$.

Example 5 Write a Compound Inequality for an Intersection

Write a compound inequality that describes the graph.

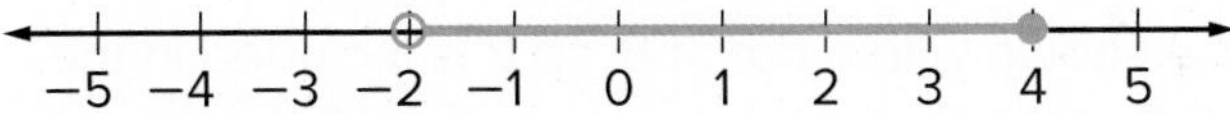

The graph shows an interval between two numbers. Because a compound inequality with the word *and* represents the intersection of two inequalities, its graph shows the overlap as an interval.

Step 1

Analyze the leftmost endpoint of the interval. The endpoint is shown with a circle at -2, so -2 is not included in the solution. Points to the right of the endpoint are shaded, so the graph represents solutions of $x > -2$.

Step 2

Analyze the rightmost endpoint of the interval. The endpoint is shown with a dot at 4, so 4 is included in the solution. Points to the left of the endpoint are shaded, so the graph represents solutions of $x \leq 4$.

Step 3

The shaded interval represents the intersection of the solutions of $x > -2$ and $x \leq 4$, so the compound inequality $-2 < x \leq 4$ describes the graph.

Go Online You can complete an Extra Example online.

Check

Write a compound inequality that describes the graph.

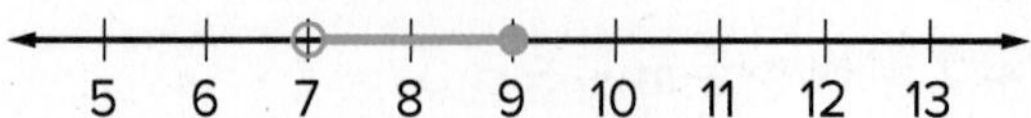

Example 6 Write a Compound Inequality for a Union

Write a compound inequality that describes the graph.

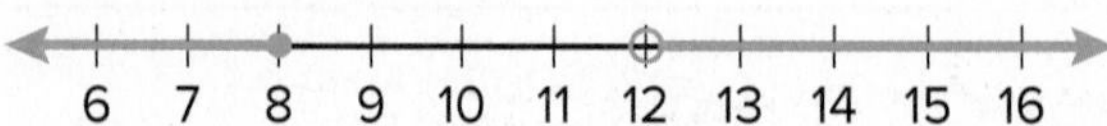

The graph shows the union of two inequalities. Because a compound inequality with the word *or* represents the union of two inequalities, its graph includes the graphs of both inequalities.

The leftmost endpoint is shown with a dot at 8, so 8 is included in the solution. Points to the left of the endpoint are shaded, so the graph represents solutions of $x \leq 8$.

The rightmost endpoint is shown with a circle at 12, so 12 is not included in the solution. Points to the right of the endpoint are shaded, so the graph represents solutions of $x > 12$.

The solutions represented on the graph represent the union of the solutions of $x \leq 8$ and $x > 12$, so the compound inequality $x \leq 8$ or $x > 12$ describes the graph.

Check

Write a compound inequality that describes the graph.

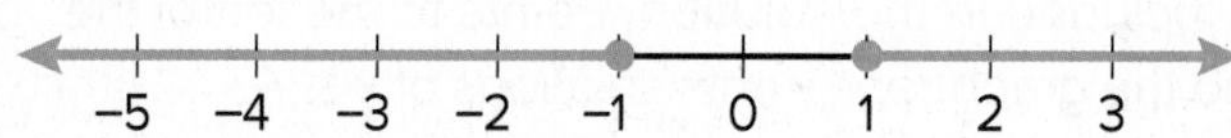

Go Online to practice what you've learned about solving linear inequalities in the Put It All Together over Lessons 6-1 through 6-3.

Go Online You can complete an Extra Example online

Practice

Go Online You can complete your homework online.

Examples 1, 3, and 4

Solve each compound inequality. Then graph the solution set.

1. $f - 6 < 5$ and $f - 4 \geq 2$

2. $n + 2 \leq -5$ and $n + 6 \geq -6$

3. $y - 1 \geq 7$ or $y + 3 < -1$

4. $t + 14 \geq 15$ or $t - 9 < -10$

5. $-5 < 3p + 7 \leq 22$

6. $-3 \leq 7c + 4 < 18$

7. $5h - 4 \geq 6$ and $7h + 11 < 32$

8. $22 \geq 4m - 2$ or $5 - 3m \leq -13$

9. $-y + 5 \geq 9$ or $3y + 4 < -5$

10. $-4a + 13 \geq 29$ and $10 < 6a - 14$

11. $3b + 2 < 5b - 6 \leq 2b + 9$

12. $-2a + 3 \geq 6a - 1 > 3a - 10$

13. $10m - 7 < 17m$ or $-6m > 36$

14. $5n - 1 < -16$ or $-3n - 1 < 8$

15. $m + 3 \geq 5$ and $m + 3 < 7$

16. $y - 5 < -4$ or $y - 5 \geq 1$

Example 2

17. **STORE SIGNS** In Randy's town, all stand-alone signs must be exactly 8 feet high. When mounted atop a pole, the combined height of the sign and pole must be less than 20 feet or greater than 35 feet so that they do not interfere with the power and phone lines.

a. Write a compound inequality to represent the possible above-ground height of the poles, x.

b. Solve the inequality. Explain any restrictions.

c. Graph the inequality.

18. **HEALTH** The human heart circulates from 770,000 to 1,600,000 gallons of blood through a person's body every year.

a. Write a compound inequality to represent the number of gallons of blood that the heart circulates through the body in one day, x.

b. Solve the inequality. Round to the nearest whole gallon.

c. Graph the inequality.

Examples 5 and 6

Write a compound inequality that describes each graph.

19. −4 −3 −2 −1 0 1 2 3 4

20. −2 −1 0 1 2 3 4 5 6

21. −4 −3 −2 −1 0 1 2 3 4

22. −4 −3 −2 −1 0 1 2 3 4

23. −4 −3 −2 −1 0 1 2 3 4

24. −4 −3 −2 −1 0 1 2 3 4

25. −4 −3 −2 −1 0 1 2 3 4

26. −3 −2 −1 0 1 2 3 4 5

Mixed Exercises

Solve each compound inequality. Then graph the solution set.

27. $4 < f + 6$ and $f + 6 < 5$

28. $w + 3 \leq 0$ or $w + 7 \geq 9$

29. $-6 < b - 4 < 2$

30. $p - 2 \leq -2$ or $p - 2 > 1$

31. $-5 \leq 2a - 1 < 9$

32. $-1 < 2x - 1 \leq 5$

Define a variable, write an inequality, and solve each problem. Check your solution.

33. A number decreased by two is at most four or at least nine.

34. The sum of a number and three is no more than eight or is more than twelve.

35. WEATHER Kenya saw this graph in the local weather forecast. It shows the predicted temperature range for the following day. Write an inequality to represent the number line.

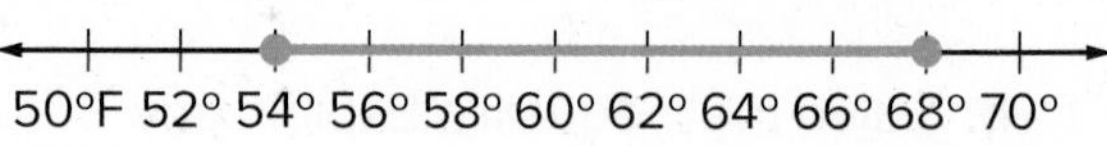

36. REASONING The pH of a person's eyes is 7.2. Therefore, the ideal pH for the water in a swimming pool is between 7.0 and 7.6. Write a compound inequality to represent pH levels that could cause physical discomfort to a person's eyes.

37. FIELD TRIP It costs $1000 to rent a bus that holds 100 students. A school is planning to rent one of these buses for a field trip to an aquarium. The trip will also have a cost of $15 per student for the tickets to the aquarium. Given that the total expense for the trip must be between $2000 and $3000, find the minimum and maximum number of students who can go on the trip. Explain.

38. **HEALTH** Body mass index (BMI) is a measure of weight status. The BMI of a person over 20 years old is calculated using the following formula.

$$\text{BMI} = 703 \times \frac{\text{weight in pounds}}{(\text{height in inches})^2}$$

The table shows the meaning of different BMI measures. Round to the nearest tenth if necessary.

BMI	Weight Status
less than 18.5	underweight
18.5 – 24.9	normal
25 – 29.9	overweight
more than 30	obese

Source: Centers for Disease Control

a. Write a compound inequality to represent the normal BMI range.

b. Write a compound inequality to represent the weight of an adult who is 6 feet tall that is within the normal BMI range.

39. **STATE YOUR ASSUMPTION** The Triangle Inequality states that in any triangle, the sum of the lengths of any two sides is greater than the length of the third side. In the figure, this means $a + b > c$, $a + c > b$, and $b + c > a$.

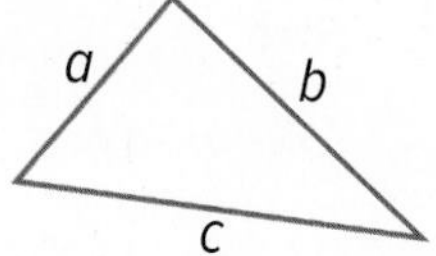

a. Suppose a triangle has a side that is 5 meters long and a perimeter of 14 meters. Let one of the unknown sides be x. Write a compound inequality that you can use to determine the value of x. Explain.

b. What assumption about the two unknown side lengths of the triangle can you make? Explain.

40. **CONSTRUCT ARGUMENTS** Bianca said that if k is a real number, then the solution set of the compound inequality $x < k$ or $x > k$ is all real numbers. Do you agree? Justify your argument.

41. **STRUCTURE** Write the solution set of the following compound inequality. Then graph the solution set: $-x + 1 < 8$ and $-x + 1 < 3$ and $-x + 1 > -4$?

42. **PHYSICS** The density, in grams per milliliter, of a substance determines whether it will float or sink in a liquid. Any object with a greater density will sink and any object with a lesser density will float. Density is given by the formula $d = \frac{m}{v}$, where m is mass and v is volume. Here is a table of common chemical solutions and their densities. Plastics vary in density when they are manufactured; therefore, their volumes are variable for a given mass. A tablet of polystyrene (a manufactured plastic) sinks in 70% isopropyl alcohol solution and floats in calcium chloride solution. The tablet has a mass of 0.4 gram. Write an inequality to represent the range of values for v, the volume of the tablet.

Solution	Density (g/mL)
concentrated calcium chloride	1.40
70% isopropyl alcohol	0.92

Source: American Chemistry Council

43. **COMPUTER SALE** Marietta is shopping during a computer store's sale. She is considering buying computers that range in cost from \$500 to \$1000.

20% OFF

a. How much are the computers after the 20% discount?

b. If sales tax is 7%, how much should Marietta expect to pay?

Higher-Order Thinking Skills

44. CREATE The figure shows the solution set of a compound inequality. Write a compound inequality that has the given solution set. Solving the inequality should require the Distributive Property, Addition Property of Inequalities, and the Division Property of Inequalities.

45. ANALYZE Which value of x is not a solution to $3x - 1 < 5$ or $7 - x \leq 3$?

A 0 **B** 2 **C** 4 **D** 5

46. FIND THE ERROR Sierra solved the compound inequality $-3x + 7x - 1 < -5$ or $-3x + 7x - 1 > 11$ as shown. What error did she make in solving the inequality?

Step	Property	Step	Property
$-3x + 7x - 1 < -5$	Original inequality	$-3x + 7x - 1 > 11$	Original inequality
$4x - 1 < -5$	Combine like terms.	$-4x - 1 > 11$	Combine like terms.
$4x < -4$	Addition Property	$-4x > 12$	Addition Property
$x < -1$	Division Property	$x < -3$	Division Property

Refer to the graphs for Exercises 47–49.

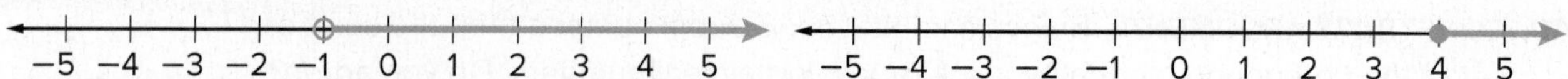

47. WRITE Write a compound inequality whose solution is the union of the two graphs. Then explain how the compound inequality can be expressed as a single inequality.

48. WRITE Write a compound inequality whose solution is the intersection of the two graphs. Then explain how the compound inequality can be expressed as a single inequality.

49. ANALYZE How are the graphs of the solution sets for the inequalities in Exercise 47 and Exercise 48 related to the given graphs?

50. PERSEVERE Jocelyn is planning to place a fence around the triangular flower bed shown. The fence costs \$1.50 per foot. Assuming that Jocelyn spends between \$60 and \$75 for the fence, what is the shortest possible length for a side of the flower bed? Use a compound inequality to explain your answer.

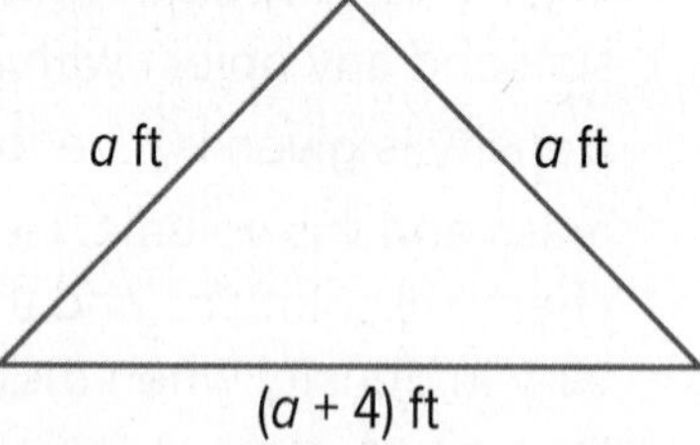

51. PERSEVERE Solve each inequality for x. Assume a is constant and $a > 0$

a. $-3 < ax + 1 \leq 5$ **b.** $-\frac{1}{a}x + 6 < 1$ or $2 - ax > 8$

52. CREATE Create an example of a compound inequality containing *or* that has infinitely many solutions.

53. ANALYZE Determine whether the following statement is *always*, *sometimes*, or *never* true. Justify your argument. *The graph of a compound inequality that involves an* or *statement is bounded on the left and right by two values of x.*

54. WRITE Give an example of a compound inequality you might encounter at an amusement park. Does the example represent an intersection or a union?

Solving Absolute Value Inequalities

Explore Solving Absolute Value Inequalities

Online Activity Use a graph to complete the Explore.

INQUIRY How is solving an absolute value inequality similar to solving an absolute value equation?

Today's Goals
- Solve absolute value inequalities (<).
- Solve absolute value inequalities (>).

Learn Solving Inequalities Involving < and Absolute Value

For a real number a, the inequality $|x| < a$ means that the distance between x and 0 is less than a.

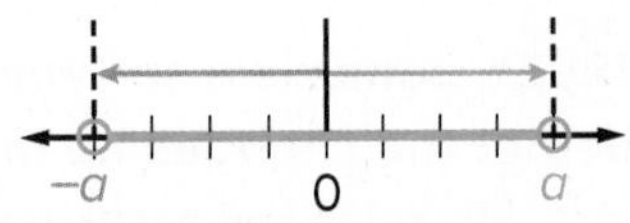

When solving absolute value inequalities, there are two cases to consider.

Case 1 The expression inside the absolute value symbols is nonnegative. If x is nonnegative, then $|x| = x$.

Case 2 The expression inside the absolute value symbols is negative. If x is negative, then $|x| = -x$.

Study Tip

Absolute Value Inequalities The inequality $|x|$ can be rewritten as $|x - 0| < a$, which is why it is read as *the distance between x and 0 is less than a.*

Example 1 Solve Absolute Value Inequalities (<)

Solve $|m + 5| < 3$. Then graph the solution set.

Part A Rewrite $|m + 5| < 3$ for Case 1 and Case 2.

Case 1 If $m + 5$ is nonnegative, $|m + 5| = m + 5$.

$m + 5 < 3$ Case 1

$m < -2$ Subtract 5 from each side.

Case 2 If $m + 5$ is negative, $|m + 5| = -(m + 5)$.

$-(m + 5) < 3$ Case 2

$-m - 5 < 3$ Distributive Property

$-m < 8$ Add 5 to each side.

$m > -8$ Divide each side by −1. Reverse the inequality symbol.

So, $m < -2$ and $m > -8$. The solution set is $\{m \mid -8 < m < -2\}$.

(continued on the next page)

Watch Out

Absolute Value Cases Assigning the correct inequality symbol in Case 2 of an absolute value inequality can be confusing. Think of an inequality like $|x - 72| < 1.8$ as *the distance from x to 72 is less than 1.8 units.* Visualizing the graph of the inequality as an interval of 1.8 units on each side of the graph of 72 can help you ensure you have the correct symbol.

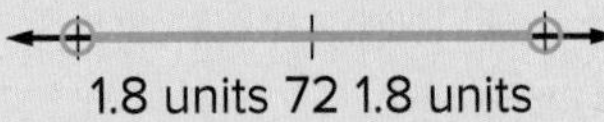

1.8 units 72 1.8 units

Part B Graph the solution set on a number line.

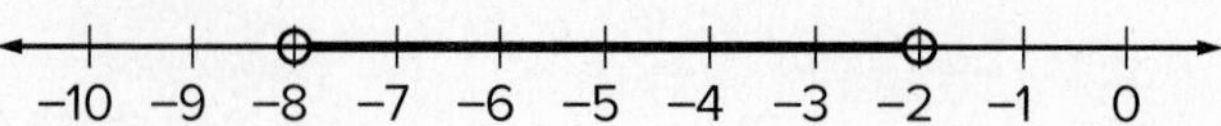

Check

Solve $|6m + 12| < 12$.
Graph the solution set.

Example 2 Absolute Value Inequalities (<) with No Solutions

Solve $|n - 1| < -5$. Then graph the solution set.

Because $|n - 1|$ is an absolute value expression, it cannot be negative. So it is not possible for $|n - 1|$ to be less than -5. Therefore, there is no solution, and the solution set is the empty set, $\varnothing$.

Example 3 Use Absolute Value Inequalities

SURVEY **Jonas is a software developer who wants to determine whether the changes he made to his program are popular with users. He releases a survey to get some feedback and finds that 72% of users like the changes. The margin of error is within 1.8%, which means that with a reasonable level of certainty, the actual percentage can be said to fall within 1.8% of 72%.**

Part A

Complete the table to write an inequality that represents the percent of users who like the changes.

Words	The difference between the actual percent and 72%	is less than or equal to	1.8%
Variable	Let x be the actual percent of users who like the changes.		
Inequality	$\lvert x - 72\rvert$	$\leq$	1.8

Part B

Case 1 $x - 72$ is nonnegative.

$x - 72 \leq 1.8$ Case 1

$x \leq 73.8$ Add 72 to each side.

Go Online You can complete an Extra Example online.

Go Online An alternate method is available for this example.

Study Tip

Absolute Value as Distance The solution set for $|n - 1| < -5$ might be easier to understand if you think of the inequality in terms of distance. The inequality can be read as *the distance between a number and 1 is less than negative five*. However, that would make the distance a negative number. Because distance cannot be negative, the solution set is $\varnothing$.

Talk About It!

Is the solution set of $|n - 1| - 6 < -5$ also the empty set? Explain your reasoning.

Case 2 $x - 72$ is negative.

$-(x - 72) \leq 1.8$	Case 2
$-x + 72 \leq 1.8$	Distributive Property
$-x \leq -70.2$	Subtract 72 from each side.
$x \geq 70.2$	Divide each side by −1. Reverse the inequality symbol.

The percent of users who favor the changes Jonas made to his software is between 70.2% and 73.8%, so the solution set is $\{x \mid 70.2 \leq x \leq 73.8\}$. This solution set is a small interval of possible values close to the percent that Jonas found, so the solution set seems reasonable for the situation.

Learn Solving Inequalities Involving > and Absolute Value

For a real number a, the inequality $|x| > a$ means that the distance between x and 0 is greater than a.

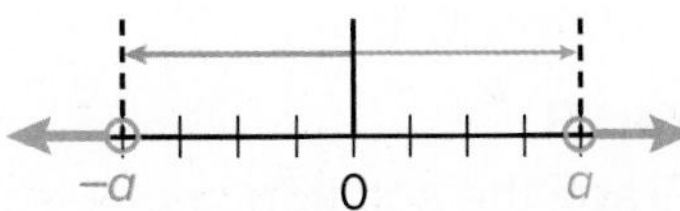

When solving absolute value inequalities, there are two cases to consider.

Case 1 The expression inside the absolute value symbols is nonnegative. If x is nonnegative, $|x| = x$.

$x > a$	Case 1

Case 2 The expression inside the absolute value symbols is negative. If x is negative, $|x| = -x$.

$-x > a$	Case 2
$\frac{-x}{-1} < \frac{a}{-1}$	Divide each side by −1. Reverse the inequality.
$x < -a$	Simplify.

The solution set is the union of the solutions to these two cases. So, $x > a$ or $x < -a$. The solution set is $\{x \mid x < -a \text{ or } x > a\}$.

> **Study Tip**
>
> **> and <** If an absolute value inequality involves > or ≥, the solution set uses the word *or*. If an absolute value inequality involves < or ≤, the solution set uses the word *and*.

Example 4 Solve Absolute Value Inequalities (>)

Solve $|2m - 9| > 13$. Then graph the solution set.

Part A Rewrite $|2m - 9| > 13$ for Case 1 and Case 2.

Case 1 If $2m - 9$ is nonnegative, $|2m - 9| = 2m - 9$.

$2m - 9 > 13$	Case 1
$2m > 22$	Add 9 to each side.
$m > 11$	Divide each side by 2.

(continued on the next page)

Go Online You can complete an Extra Example online.

Case 2 If $2m - 9$ is negative, $|2m - 9| = -(2m - 9)$.

$-(2m - 9) > 13$	Case 2
$-2m > 4$	Add 9 to each side.
$m < -2$	Divide each side by 2.

So, $m > 11$ or $m < -2$. The solution set is $\{m \mid m > 11 \text{ or } m < -2\}$.

Part B Graph the solution set on a number line.

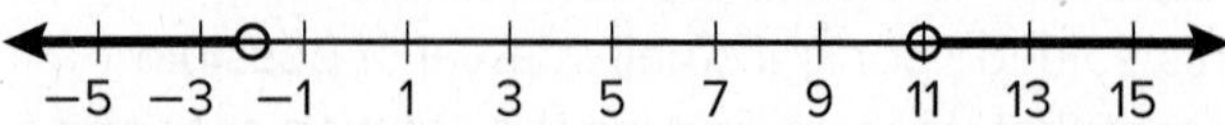

Check

Part A
Solve $|4m - 20| \geq 12$.

Part B
Graph the solution set.

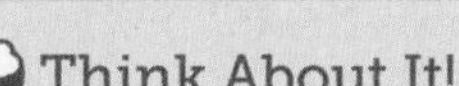

Study Tip

Overlapping Case Solutions Because the values of n for $n \geq 1$ and $n \leq 11$ overlap and extend infinitely in both directions, the solution set is all real numbers.

Think About It!

Why is the empty set the solution of $|n - 6| \leq -5$, but not the solution of $|n - 6| \geq -5$?

Example 5 Absolute Value Inequalities (>) with Overlapping Case Solutions

Solve $|n - 6| \geq -5$. Then graph the solution set.

Part A Rewrite $|n - 6| \geq -5$ for Case 1 and Case 2.

Case 1 $n - 6$ is nonnegative.

$n - 6 \geq -5$	Case 1
$n \geq 1$	Add 6 to each side.

Case 2 $n - 6$ is negative.

$-(n - 6) \geq -5$	Case 2
$-n \geq -11$	Add 9 to each side.
$n \leq 11$	Divide each side by −1. Reverse the inequality.

So, $n \geq 1$ or $n \leq 11$. The solution set is $\{n \mid n \geq 1 \text{ or } n \leq 11\}$ which is equivalent to $\{n \mid n \text{ is a real number}\}$.

Part B Graph the solution set on a number line.

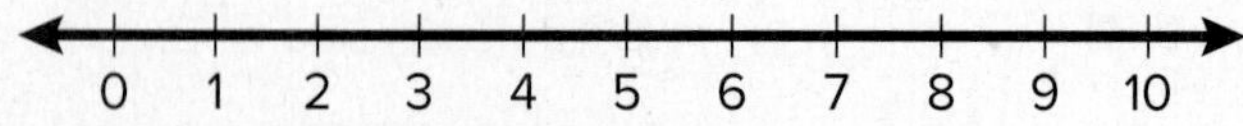

Go Online You can complete an Extra Example online.

Practice

Go Online You can complete your homework online.

Examples 1, 2, 4, 5

Solve each inequality. Then graph the solution set.

1. $|x + 8| < 16$
2. $|r + 1| \leq 2$
3. $|2c - 1| \leq 7$
4. $|3h - 3| < 12$
5. $|m + 4| < -2$
6. $|w + 5| < -8$
7. $|r + 2| > 6$
8. $|k - 4| > 3$
9. $|2h - 3| \geq 9$
10. $|4p + 2| \geq 10$
11. $|5v + 3| > -9$
12. $|-2c - 3| > -4$
13. $|4n + 3| \geq 18$
14. $|5t - 2| \leq 6$
15. $\left|\frac{3h + 1}{2}\right| < 8$
16. $\left|\frac{2p - 8}{4}\right| \geq 9$
17. $\left|\frac{7c + 3}{2}\right| \leq -5$
18. $\left|\frac{2g + 3}{2}\right| > -7$
19. $|-6r - 4| < 8$
20. $|-3p - 7| > 5$

Example 3

21. **SPEEDOMETERS** The government requires speedometers on cars sold in the United States to be accurate within ±2.5% of the actual speed of the car. If your speedometer meets this requirement, find the range of possible actual speeds at which your car could be traveling when your speedometer reads 60 miles per hour.

22. **BAKING** Pablo is making muffins for a bake sale. Before he starts baking, he goes online to research different muffin recipes. The recipes that he finds all specify baking temperatures between 350°F and 400°F, inclusive. Write an absolute value inequality to represent the possible temperatures t called for in the muffin recipes Pablo is researching.

23. **PAINT** A manufacturer claims that their cans of paint contain exactly 130 fluid ounces of paint. The amount of paint in each can of paint must be accurate within ±3.05 fluid ounces of the actual amount of paint.

    a. Write an absolute value inequality to represent the possible amount of paint, in fluid ounces, p for which the manufacturer's claim is correct.

    b. Graph the solution set of the inequality you wrote in part **a**.

24. **CATS** During a recent visit to the veterinarian's office, Mrs. Vasquez was informed that a healthy weight for her cat is approximately 10 pounds, plus or minus one pound. Write an absolute value inequality that represents unhealthy weights w for her cat.

25. **STATISTICS** In a recent year, the mean score on the mathematics section of the SAT test was 515 and the standard deviation was 114. This means that people within one deviation of the mean have SAT math scores that are no more than 114 points higher or 114 points lower than the mean.

a. Write an absolute value inequality to find the range of SAT mathematics test scores within one standard deviation of the mean.

b. What is the range of SAT mathematics test scores ±2 standard deviation from the mean?

Mixed Exercises

REGULARITY **Write an open sentence involving absolute value for each graph.**

26. 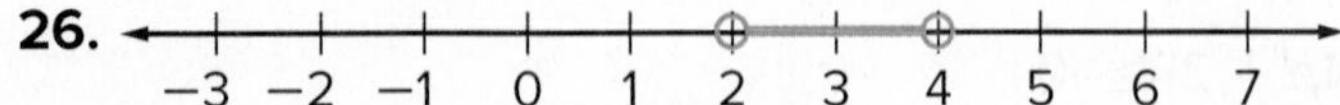

−3 −2 −1 0 1 2 3 4 5 6 7

27. −6 −5 −4 −3 −2 −1 0 1 2 3 4

28. −10 −9 −8 −7 −6 −5 −4 −3 −2 −1 0

29. −4 −3 −2 −1 0 1 2 3 4 5 6

30. 0 1 2 3 4 5 6 7 8 9 10

31. −4 −3 −2 −1 0 1 2 3 4

REASONING **Match each open sentence with the graph of its solution set.**

32. $|x| > 2$

a. −5 −4 −3 −2 −1 0 1 2 3 4 5

33. $|x - 2| \leq 3$

b. −5 −4 −3 −2 −1 0 1 2 3 4 5

34. $|x + 1| < 4$

c. −8 −7 −6 −5 −4 −3 −2 −1 0 1 2 3 4 5 6

35. $|-x + 1.5| < 3$

d. −4 −3 −2 −1 0 1 2 3 4 5 6

USE A MODEL **Express each statement using an inequality involving absolute value. Then solve and graph the absolute value inequality.**

36. The meteorologist predicted that the temperature would be within 3° of 52°F.

37. Serena will make the B team if she scores within 8 points of the team average of 92.

38. The dance committee expects attendance to number within 25 of last year's 87 students.

Solve each inequality. Then graph the solution set.

39. $\left|\frac{x-1}{2}\right| \leq 1$

40. $|2x - 1| \geq 3$

41. $\left|\frac{x+3}{3}\right| \leq 2$

42. $|x + 7| \geq 4.5$

43. $\left|\frac{2x-1}{7}\right| > 5$

44. $|-4x - 2| < 10$

45. **CONSTRUCT ARGUMENTS** Is the solution to this inequality $|x - 2| > -1$ all real numbers? Justify your argument.

46. **USE TOOLS** Forensic scientists use the equation $h = 2.4f + 46.2$ to estimate the height h of a woman given the length in centimeters f of her femur bone. Suppose the equation has a margin of error of 3 centimeters. Could a female femur bone measuring 47 centimeters be that of a woman who was 170 centimeters tall?

47. **REGULARITY** A box of cereal should weigh 516 grams. The quality control inspector randomly selects boxes to weigh. The inspector sends back any box that is not within 4 grams of the ideal weight.

a. Explain how to write an absolute value inequality to represent this situation.

b. Explain the steps to solve this inequality. What do the solutions represent?

48. **REASONING** Write a compound inequality in which the solution set is the given set.

a. $\{x \mid 4 \leq x\}$

b. $\{4\}$

49. **ANALYZE** Determine if the open sentence $|x - 2| > 4$ and the compound inequality $-2x < 4$ or $x > 6$ have the same solution set.

50. **ARCHITECTURE** An architect is designing a house for the Frazier family. In the design, she must consider the desires of the family and the local building codes. The rectangular lot on which the house will be built is 158 feet long and 90 feet wide.

a. The building codes state that one can build no closer than 20 feet to the lot line. Write an inequality to represent the possible widths of the house along the 90-foot dimension. Solve the inequality.

b. The Fraziers requested that the rectangular house contain no less than 2800 square feet and no more than 3200 square feet of floor space. If the house has only one floor, use the maximum value for the width of the house from part a, and explain how to use an inequality to find the possible lengths.

c. The Fraziers have asked that the cost of the house be about \$175,000 and are willing to deviate from this price no more than \$20,000. Write an open sentence involving an absolute value and solve. Explain the meaning of the answer.

Higher-Order Thinking Skills

51. FIND THE ERROR Jordan and Chloe are solving $|x + 3| > 10$.

Jordan	
$\lvert x + 3\rvert > 10$	
$x + 3 > 10$	$-(x + 3) > 10$
$x > 7$	$-x - 3 > 10$
	$-x > 13$
	$x < -13$

Chloe	
$\lvert x + 3\rvert > 10$	
$x + 3 > 10$	$-(x + 3) > 10$
$x > 7$	$-x + 3 > 10$
	$-x > 7$
	$x < -7$

Is either correct? Explain your reasoning.

52. CREATE Write an absolute value inequality using the numbers 3, 2, and –7. Then solve the inequality.

53. PERSEVERE Solve $2 < |n + 1| \leq 7$. Explain your reasoning and graph the solution set.

54. ANALYZE Which of the following inequalities could be represented by the graph?

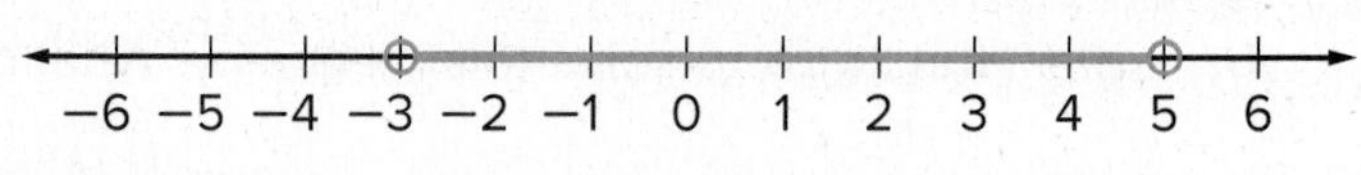

I. $|m - 1| < 4$ **II.** $3x < 15$ or $-x < 1$ **III.** $1 < 2k + 7 < 17$

A. I only **B.** I and II **C.** I and III **D.** II and III

55. FIND THE ERROR Lucita sketched a graph of her solution to $|2a - 3| > 1$. Is she correct? Explain your reasoning.

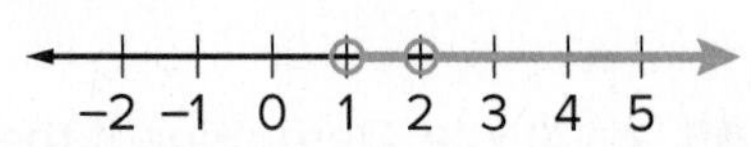

56. ANALYZE The graph of an absolute value inequality is *sometimes, always,* or *never* the union of two graphs. Explain.

57. ANALYZE Determine why the solution of $|t| > 0$ is not all real numbers. Explain your reasoning.

58. WRITE How are symbols used to represent mathematical ideas? Use an example to justify your reasoning.

59. WRITE Explain how to determine whether an absolute value inequality uses a compound inequality with *and* or a compound equality with *or.* Then summarize how to solve absolute value inequalities.

60. WHICH ONE DOESN'T BELONG? Which inequality does not belong? Justify your conclusion.

$\lvert x + 4\rvert - 7 \geq 3$	$\lvert -6x - 1\rvert \leq \frac{1}{2}$	$-2\lvert 10x + 4\rvert < 6$	$\lvert 3x + 5\rvert < -\frac{3}{5}$

Graphing Inequalities in Two Variables

Explore Graphing Linear Inequalities on the Coordinate Plane

Online Activity Use graphing technology to complete the Explore.

INQUIRY How is graphing a linear inequality on the coordinate plane similar to and different from graphing on the number line?

Today's Goals

- Graph the solutions of linear inequalities in two variables.

Today's Vocabulary

boundary

half-plane

closed half-plane

open half-plane

Learn Graphing Linear Inequalities in Two Variables

The graph of a linear inequality represents the set of all points that are solutions of the inequality.

The edge of the graph is a **boundary**. Depending on the inequality, the boundary will or will not be included in the solution set.

The boundary divides the coordinate plane into regions called **half-planes**.

When the boundary is included, the solution of the linear inequality is a **closed half-plane**.

When the boundary is not included, it is an **open half-plane**.

Go Online
You can watch a video to see how to graph inequalities in two variables.

Key Concept • Graphing Linear Inequalities	
Step 1	Graph the boundary. Use a solid boundary when the inequality contains $\leq$ or $\geq$. Use a dashed boundary when the inequality contains $<$ or $>$.
Step 2	Use a test point to determine which half-plane should be shaded.
Step 3	Shade the half-plane that contains the solution.

Example 1 Graph an Inequality with an Open Half-Plane

Graph $3x - 2y < 8$.

Step 1 Graph the boundary.

$3x - 2y < 8$	Original inequality
$-2y < -3x + 8$	Subtract $3x$ from each side.
$y > \frac{3}{2}x - 4$	Divide each side by -2.

(*continued on the next page*)

Watch Out!

Selecting a Test Point When selecting a test point to use, make sure that the point does not lie on the boundary. While using the point (0, 0) will make calculations easier, you cannot use that point if the boundary passes through the origin. Instead, try using (1, 1) or (0, 1).

Think About It!

What does the shaded area of this graph represent?

Step 2 Use a test point. Select (0, 0) as a test point.

$3x - 2y < 8$ Original inequality

$3(0) - 2(0) < 8$ $x = 0$ and $y = 0$

$0 < 8$ Simplify.

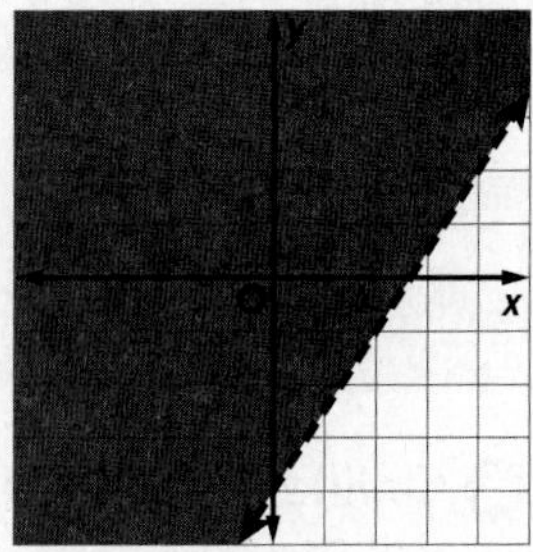

Step 3 Shade the half-plane.
Because the test point is a solution of the inequality, shade the half-plane containing the test point.

Example 2 Graph an Inequality with a Closed Half-Plane

Graph $3x + 4y \leq 0$.

Step 1 Graph the boundary.

$3x + 4y \leq 0$ Original inequality

$4y \leq -3x$ Subtract $3x$ from each side.

$y \leq \frac{-3}{4}x$ Divide each side by 4.

Step 2 Use a test point. Use (1, 1) as a test point.

$3x + 4y \leq 0$ Original inequality

$3(1) + 4(1) \leq 0$ $x = 1$ and $y = 1$

7 is not less than or equal to 0 Simplify.

Step 3 Shade the half-plane.
Because the test point *is not* a solution of the inequality, shade the half-plane that does not contain the test point.

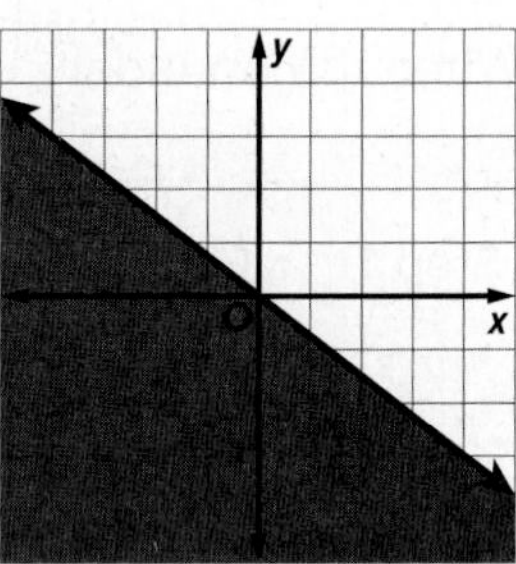

Go Online You can watch a video to to see how to use a graphing calculator with this example.

Check

Graph the inequalities.

a. $2x + y < -4$

b. $x - 2y > -4$

Go Online You can complete an Extra Example online.

Example 3 Apply Graphing Inequalities in Two Variables

REFRESHMENTS **Dominique can spend up to $20 to provide the dance squad with drinks after their practice. A bottle of water costs $0.80, and a sports drink costs $1.25. How many bottles of water and sports drinks can Dominique buy for the dance squad?**

Step 1 Write an inequality.

Words	$0.80 times the number of bottles of water	plus	$1.25 times the number of sports drinks	is less than or equal to	$20
Variables	Let x = the number of bottles of water and y = the number of sports drinks that Dominique can buy.				
Inequality	$0.8 \cdot x$	$+$	$1.25 \cdot y$	$\leq$	$20

Step 2 Solve the inequality for y.

$$0.8x + 1.25y \leq 20 \quad \text{Original inequality}$$

$$1.25y \leq -0.8x + 20 \quad \text{Subtract } 0.8x \text{ from each side.}$$

$$y \leq -0.64x + 16 \quad \text{Divide each side by 1.25.}$$

Step 3 Graph the inequality.
Because Dominique cannot buy a negative number of drinks, negative values of x and y are nonviable options. So the domain and range must be nonnegative numbers. Graph the boundary.

x	y
0	16
10	9.6
15	6.4
25	0

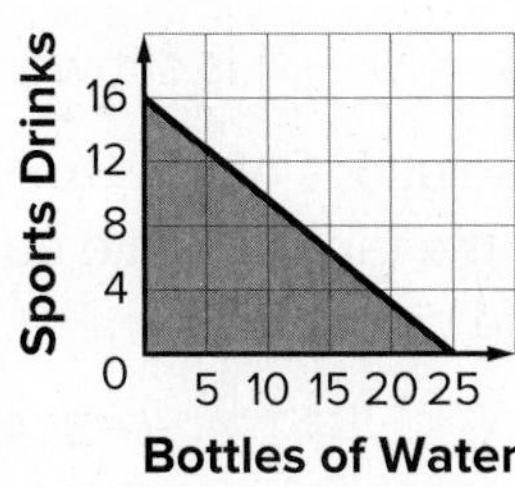

The test point (0, 0) is a solution of the inequality. Shade the closed half-plane that includes (0, 0).

Step 4 Interpret the solution in the context of the situation.
Notice that there are infinitely many solutions of the inequality. Because buying fractional bottles of water or sports drinks is not reasonable, only the solutions in which both x and y are whole numbers are viable. One viable solution is 10 bottles of water and 8 sports drinks.

Problem-Solving Tip

Use a Graph You can use a graph to visualize data, analyze trends, and make predictions.

Study Tip

Specifying Units Because we assigned the variables for the different types of drinks, it is critical to label the axes.

Talk About It!

Are there any viable solutions in which Dominique spends a total of $20? If so, where do those solutions appear on the graph?

Example 4 Solve Linear Inequalities

Graph $-4x + 7 \geq 11$.

Step 1 Graph the boundary.

$-4x + 7 \geq 11$	Original inequality
$-4x \geq 4$	Subtract 7 from each side.
$x \leq -1$	Divide each side by −4. Reverse the inequality.

Step 2 Use a test point. Use (0, 0) as a test point.

$-4x + 7 \geq 11$	Original inequality
$-4(0) + 7 \geq 11$	$x = 0$ and $y = 0$
$7 \not\geq 11$	Simplify.

Step 3 Shade the half-plane.

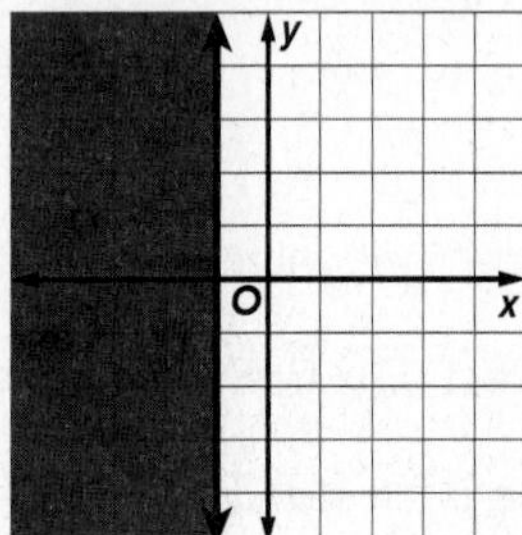

Since the test point *is not* a solution of the inequality, shade the half-plane that does not contain the test point.

Check

The graph of $y = 3x - 4$ is shown.

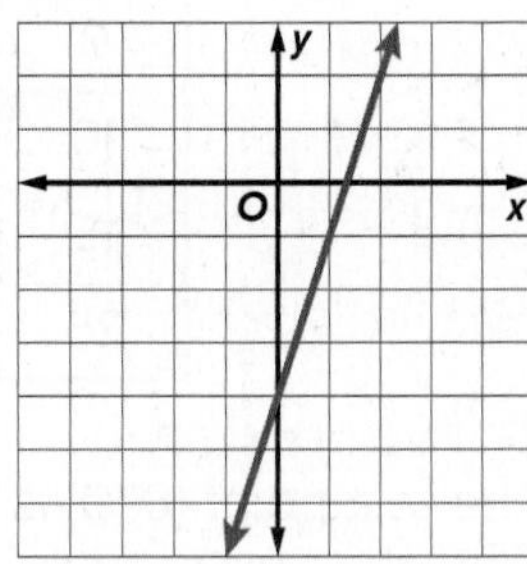

Consider the solutions of $y > 3x - 4$. Copy and complete the table to write each point in the appropriate column.

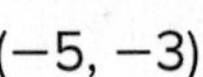

(−5, −3)	(−3, 4)
(0, −4)	(0, 0)
(1, −7)	(1, 1)
(2, 2)	(4, 2)

In Solution	Not in Solution

Go Online You can complete an Extra Example online.

Practice

Go Online You can complete your homework online.

Examples 1 and 2

Graph each inequality.

1. $y < x - 3$
2. $y > x + 12$
3. $y \geq 3x - 1$
4. $y \leq -4x + 12$
5. $6x + 3y > 12$
6. $2x + 2y < 18$
7. $5x + y > 10$
8. $2x + y < -3$
9. $-2x + y \geq -4$
10. $8x + y \leq 6$
11. $10x + 2y \leq 14$
12. $-24x + 8y \geq -48$

Example 3

13. **INCOME** In 2006 the median yearly family income was about \$48,200 per year. Suppose the average annual rate of change since then is \$1240 per year.

    a. Write and graph an inequality for the annual family incomes y that are less than the median for x years after 2006.

    b. Determine whether each of the following points is part of the solution set.

    (2, 51,000) (8, 69,200) (5, 50,000) (10, 61,000)

14. **FUNDRAISING** Troop 200 sold cider and donuts to raise money for charity. They sold small boxes of donut holes for \$1.25 and cider for \$2.50 a gallon. In order to cover their expenses, they needed to raise at least \$100. Write and graph an inequality that represents this situation.

Example 4

Graph each inequality.

15. $2y + 6 \geq 0$
16. $\frac{1}{2}x + 1 < 3$
17. $\frac{2}{3}x - \frac{10}{3} > -4$

Mixed Exercises

Graph each inequality.

18. $y < -1$
19. $y \geq x - 5$
20. $y > 3x$
21. $y \leq 2x + 4$
22. $y + x > 3$
23. $y - x \geq 1$

24. Kumiko has a \$50 gift card for a Web site that sells apps and games. Games cost \$2.50 each, and apps cost \$1.25 each.

    a. Write an inequality that represents the number of apps a and the number of games g that Kumiko can buy and describe any constraints.

    b. Graph the solution of the inequality on a coordinate plane.

    c. Use your graph to find three different combinations of apps and games that Kumiko can buy.

    d. Kumiko decides to buy the same number of apps and games, and she decides to spend as much of the \$50 as possible. How many apps and games does she buy? How much money does she have left on her gift card?

25. USE A MODEL A café sells peach smoothies and berry smoothies. The café makes a profit of \$2.25 for each peach smoothie that is sold and a profit of \$2 for each berry smoothie that is sold. The owner of the café wants to make a total profit of more than \$90 per day from the sales of smoothies.

a. Write an inequality that represents the number of peach smoothies p and berry smoothies b that the café needs to sell. Describe the constraints on the variables.

b. Graph the solution of the inequality on a coordinate plane.

c. On Monday, the café sold 20 peach smoothies and made the daily profit goal. What can you say about the number of berry smoothies that were sold on Monday?

d. On Tuesday, the café made the daily profit goal by selling the minimum number of smoothies. How many smoothies did they sell? Explain.

26. REASONING Oleg is training for a triathlon. One day, he jogged for 2 hours at x miles per hour. Then he bicycled for 2 hours at y miles per hour. Finally, he swam a distance of 2 miles. The total number of miles did not exceed 30 miles.

a. Write an inequality to represent the distance that he traveled that day. Describe the constraints on the variables.

b. Graph the solution of the inequality on the coordinate plane shown. Label the axes with a description of the quantity that each axis represents. Include the unit of measure.

c. What is the greatest possible speed that Oleg could have bicycled that day? How do you know?

27. CONSTRUCT ARGUMENTS The solution of the inequality $ax + by < c$ is a half-plane that includes the point (0, 0). What conclusion can you make about the value of c? Justify your argument.

Higher-Order Thinking Skills

28. FIND THE ERROR Reiko and Kristin are solving $4y \leq \frac{8}{3}x$ by graphing. Is either of them correct? Explain your reasoning.

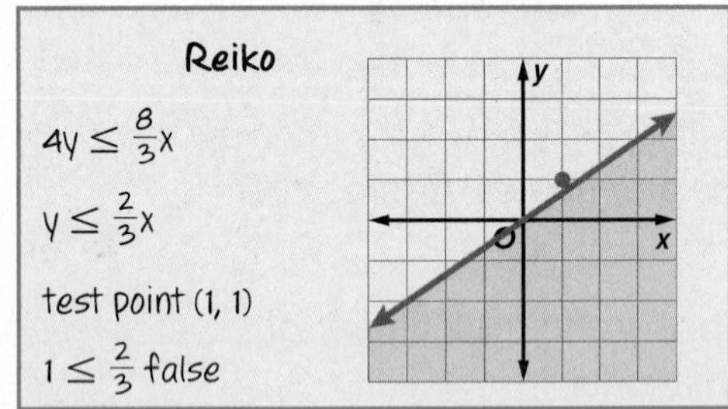

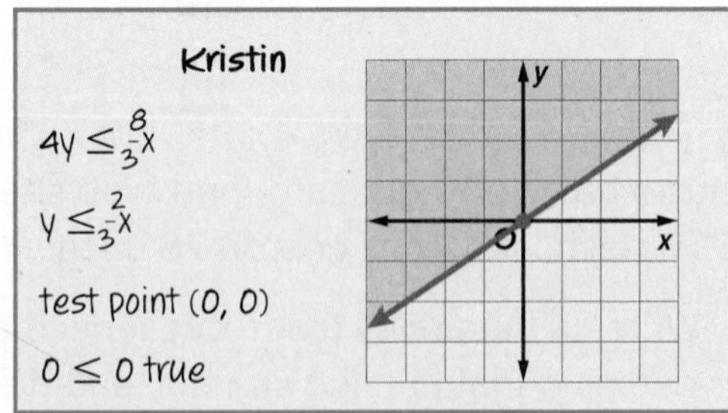

29. CREATE Write a linear inequality for which (–1, 2), (0, 1), and (3, –2) are solutions but (1, 1) is not.

30. ANALYZE Explain why a point on the boundary should not be used as a test point.

31. CREATE Write a two-variable inequality with a restricted domain and range to represent a real-world situation. Give the domain and range, and explain why they are restricted.

32. WRITE Summarize the steps to graph an inequality in two variables.

Review

 Essential Question

How can writing and solving inequalities help you solve problems in the real world?

Writing and solving inequalities can help me determine the solution sets of problems in the real world.

Module Summary

Lessons 6-1, 6-2

Solving One-Step and Multi-Step Inequalities

- A solution set can be graphed on a number line. If the endpoint is not included in the solution, use a circle; if the endpoint is included, use a dot.
- If a number is added to or subtracted from each side of a true inequality, the resulting inequality is also true.
- If each side of a true inequality is multiplied or divided by the same positive number, the resulting inequality is also true.
- If each side of a true inequality is multiplied or divided by the same negative number, the direction of the inequality symbol must be changed to make the resulting inequality true.
- Multi-step inequalities can be solved by undoing the operations in the same way you would solve a multi-step equation.

Lesson 6-3

Solving Compound Inequalities

- To determine the solution set of a compound inequality, graph each inequality and identify where they overlap.
- If a compound inequality contains *and*, the overlapping section that represents the compound inequality is an intersection.
- If a compound inequality contains *or*, its graph is a union; the solution is a solution of either inequality, not necessarily both.

Lesson 6-4

Solving Absolute Value Inequalities

- For a real number a, the inequality $|x| < a$ means the distance between x and 0 is less than a. The inequality $|x| > a$ means the distance between x and 0 is greater than a.
- When solving absolute value inequalities, there are two cases to consider. The first case is when the expression inside the absolute value symbols is nonnegative. The second case is when the expression inside the absolute value symbols is negative. The solution set is the intersection of the solutions of their union.

Lesson 6-5

Graphing Linear Inequalities in Two Variables

- To graph a linear Inequality, graph the boundary. Use a solid boundary when the inequality contains $\leq$ or $\geq$. Use a dashed boundary when the inequality contains $<$ or $>$. Then use a test point to determine which half-plane should be shaded. Finally, shade the half-plane that contains the solution.

Study Organizer

Foldables

Use your Foldable to review the module. Working with a partner can be helpful. Ask for clarification of concepts as needed.

6-1 6-2 | 6-3 | 6-4 | 6-5

Test Practice

1. **MULTIPLE CHOICE** Select the graph that shows the solution set of $7 \leq n + 5$. (Lesson 6-1)

A. 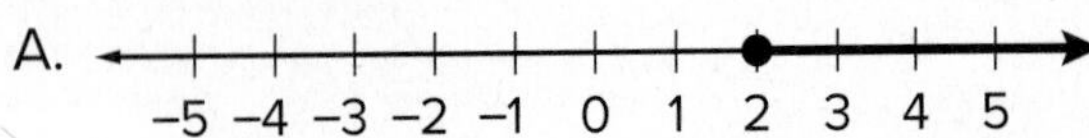

B. 

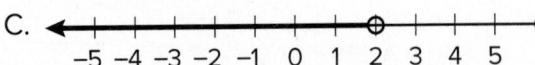

C. −5 −4 −3 −2 −1 0 1 2 3 4 5

D. −5 −4 −3 −2 −1 0 1 2 3 4 5

2. **OPEN RESPONSE** Eduardo is writing a historical novel. He wrote 16 pages today, bringing his total number of pages written to more than 50. How many pages p did Eduardo write before today? Complete the inequality that represents this situation. Then solve the inequality. (Lesson 6-1)

Inequality: ___?___ $+ p >$ ___?___

Solution: $p >$ ___?___

3. **OPEN RESPONSE** A farmer said that for every row of seeds he plants, he can harvest 6.5 bushels of tomatoes. The farmer needs to harvest at least 52 bushels of tomatoes. What is the least number of rows that the farmer will need to plant? (Lesson 6-1)

4. **OPEN RESPONSE** Find the solution set of $3d - 8 < 4d + 2$. (Lesson 6-1)

5. **MULTIPLE CHOICE** Which inequality has solutions represented by the graph? (Lesson 6-1)

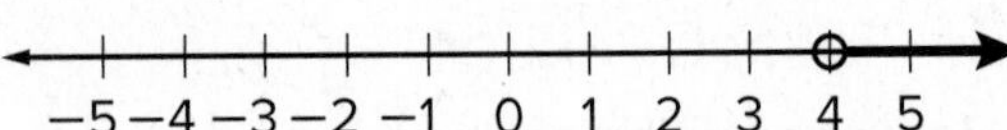

A. $5 - x > 4$

B. $2x + 1 \geq 9$

C. $2x - 5 > 3$

D. $6x - 7 > 5$

6. **MULTI-SELECT** Consider the inequality *Five plus two times a number n is less than or equal to eleven*. Select all of the representations that are solutions. (Lesson 6-2)

A. $n \leq 3$

B. $3 \leq n$

C. $n \leq 8$

D. $3 \geq n$

7. **OPEN RESPONSE** Solve $8(t + 2) + 7(t + 2) - 3(t - 2) < 0$. Write the solution using set-builder notation. (Lesson 6-2)

8. **MULTIPLE CHOICE** Solve $-\frac{4}{5}x - 3 \leq 17$. (Lesson 6-2)

A. $\{x | x \geq -25\}$

B. $\{x | x \geq -16\}$

C. $\{x | x \leq -25\}$

D. $\{x | x \leq -16\}$

9. MULTI-SELECT The science club is planning a car wash fundraiser. Halona writes and graphs an inequality to represent the number of cars c the science club needs to wash in order for the profits to cover their expenses. (Lesson 6-2)

Which inequality could Halona have graphed for this situation? Select all that apply.

A. $2(c + 1) - 50 \geq 352$

B. $310 + c \leq 3(c - 30)$

C. $20(c - 1) \geq 4000$

D. $5(c + 1) - 200 \geq 250$

E. $50(c - 5) \geq 200$

10. OPEN RESPONSE Micaela wants to plant a square garden and enclose it with a fence. She has 120 feet of fencing available and she wants the sides of her garden to be at least 6 feet long. Complete the inequality to represent the possible side lengths. (Lesson 6-3)

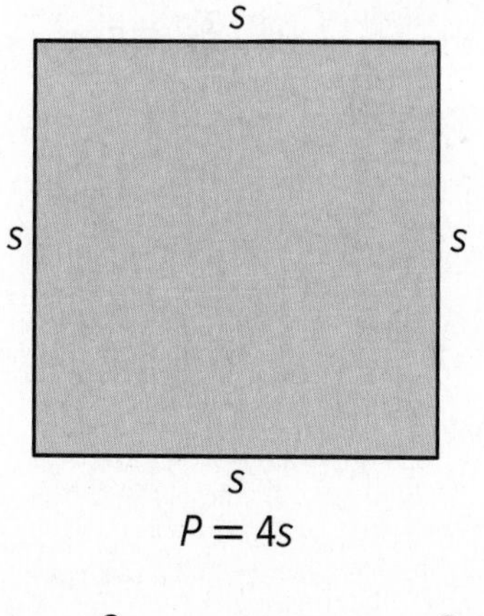

___?___ $\leq s \leq$ ___?___

11. OPEN RESPONSE Solve $4 + g \geq 3$ or $6g \geq -30$. Write the solution using set-builder notation. (Lesson 6-3)

12. MULTIPLE CHOICE Which graph shows the solution of $3n - 1 < 5$ and $-2n + 3 < 11$? (Lesson 6-3)

A. 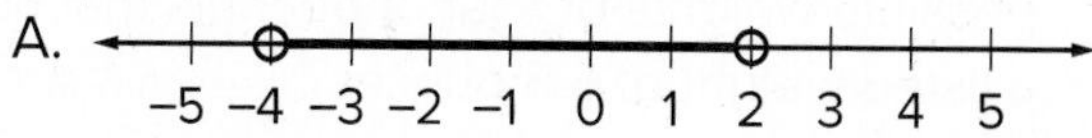

B. 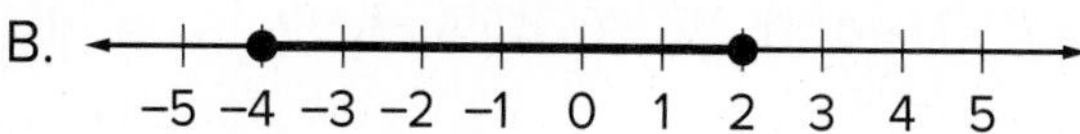

C. 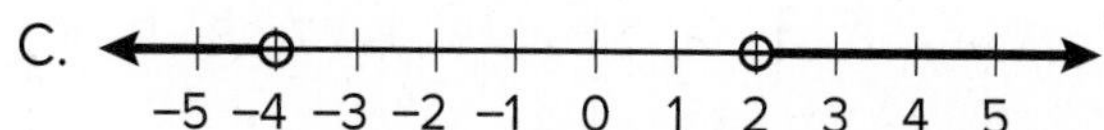

D. 

13. OPEN RESPONSE/GRAPH (Lesson 6-4)

Part A Solve $|h + 3| < 5$. Write the solution set using set-builder notation.

Part B Graph the solution set on a number line.

14. MULTIPLE CHOICE Which is NOT a true statement? (Lesson 6-4)

A. The empty set is the solution of $|k - 2| + 1 \leq -2$.

B. The solution of $|k - 2| \geq 2$ is $\{k | k \geq 4 \text{ or } k \leq 0\}$.

C. The solution of $|3k + 3| \leq -9$ is $\{k | -4 \geq k \geq 2\}$.

D. $|k - 7| < -6$ has no solution.

15. MULTIPLE CHOICE The actual weight of a jar of peanuts tends to be within 0.5 ounce of its listed weight. Which graph shows the possible weights of a jar of peanuts that has a listed weight of 6.5 ounces? (Lesson 6-4)

A.
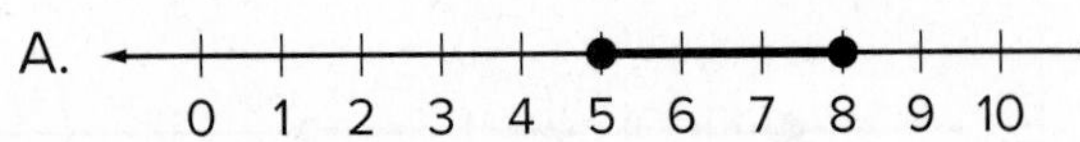

B.
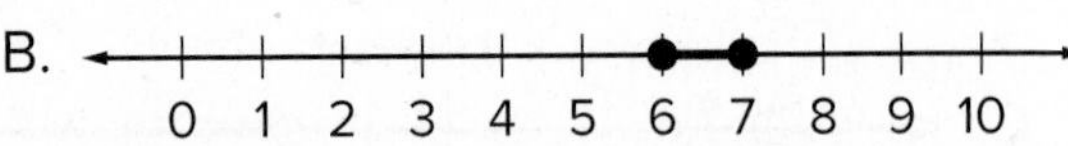

C.
0 1 2 3 4 5 6 7 8 9 10

D.
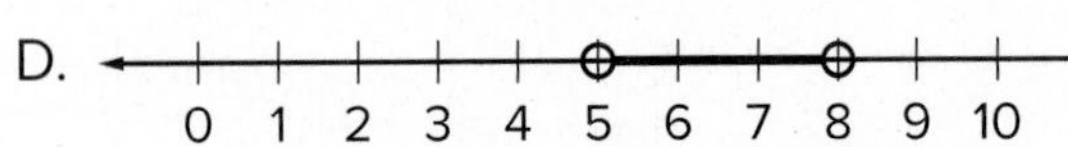

16. OPEN RESPONSE The equation for the boundary of the inequality graphed is $y = 3x - 1$. (Lesson 6-5)

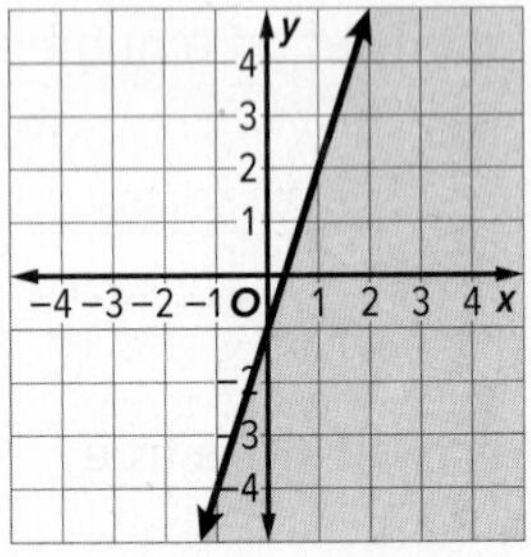

Write the inequality that represents the graph.

17. OPEN RESPONSE Consider the graphs. Write the letter of the graph that represents the solution to each of the inequalities. (Lesson 6-5)

Graphs:

A.

B.
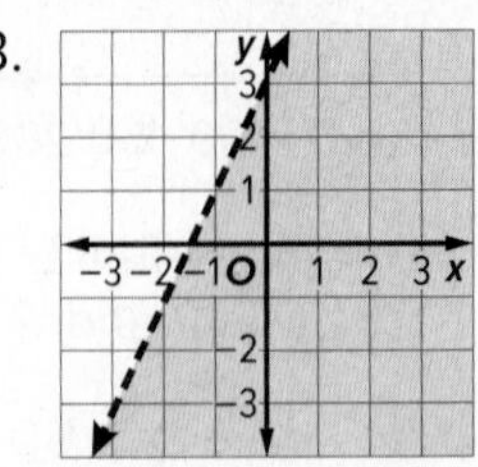

C.

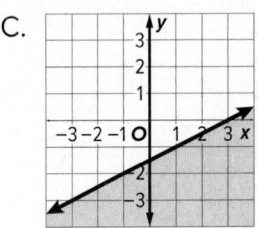

D.
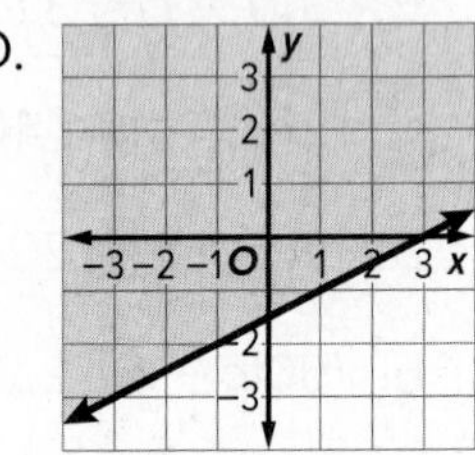

$2y - x \leq -3$

$2y - x \geq -3$

$2x - y > -3$

$2x - y < -3$

18. GRAPH Graph the inequality $\frac{2}{3}x + 5 - y < 6$. (Lesson 6-5)

Module 7

Systems of Linear Equations and Inequalities

Essential Question

How are systems of equations useful in the real world?

What Will You Learn?

How much do you already know about each topic **before** starting this module?

KEY

— I don't know. — I've heard of it. — I know it!

	Before			After		
	I don't know.	I've heard of it.	I know it!	I don't know.	I've heard of it.	I know it!
solve systems of equations by graphing						
solve systems of equations by substitution						
solve systems of equations by elimination with addition						
solve systems of equations by elimination with subtraction						
solve systems of equations by elimination with multiplication						
solve systems of inequalities by graphing						

Foldables Make this Foldable to help you organize your notes about expressions. Begin with one sheet of paper.

1. **Fold** lengthwise to the holes.
2. **Cut** 6 tabs.
3. **Label** the tabs using the lesson titles.

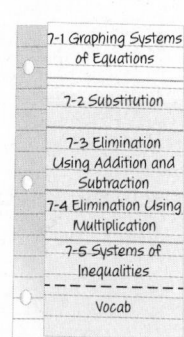

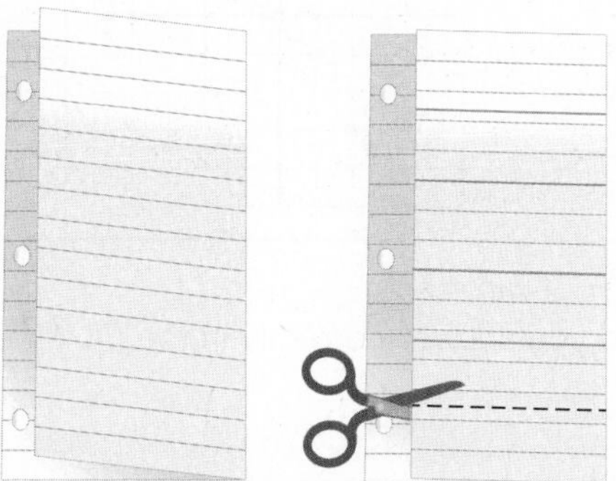

What Vocabulary Will You Learn?

- consistent
- dependent
- elimination
- inconsistent
- independent
- substitution
- system of equations
- system of inequalities

Are You Ready?

Complete the Quick Review to see if you are ready to start this module. Then complete the Quick Check.

Quick Review

Example 1

Name the ordered pair for Q on the coordinate plane.

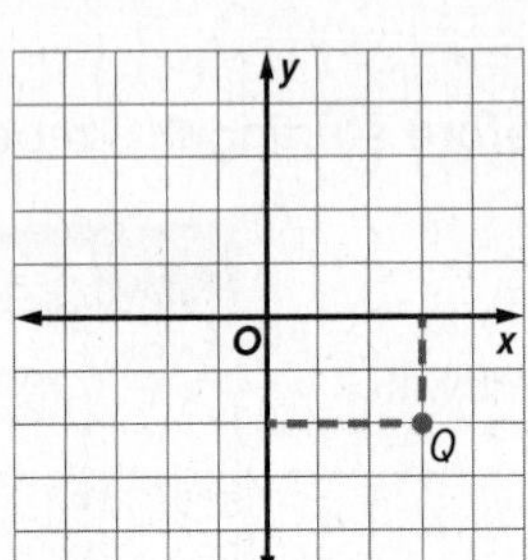

Follow a vertical line from the point to the x-axis. This gives the x-coordinate, 3.

Follow a horizontal line from the point to the y-axis. This gives the y-coordinate, −2.

The ordered pair is (3, −2).

Example 2

Solve $12x + 3y = 36$ for y.

$12x + 3y = 36$	Original equation
$12x + 3y - 12x = 36 - 12x$	Subtract $12x$ from each side.
$3y = 36 - 12x$	Simplify.
$\frac{3y}{3} = \frac{36 - 12x}{3}$	Divide each side by 3.
$\frac{3y}{3} = \frac{36}{3} - \frac{12x}{3}$	Express as a difference.
$y = 12 - 4x$	Simplify.

Quick Check

Name the ordered pair for each point.

1. A
2. B
3. C
4. D

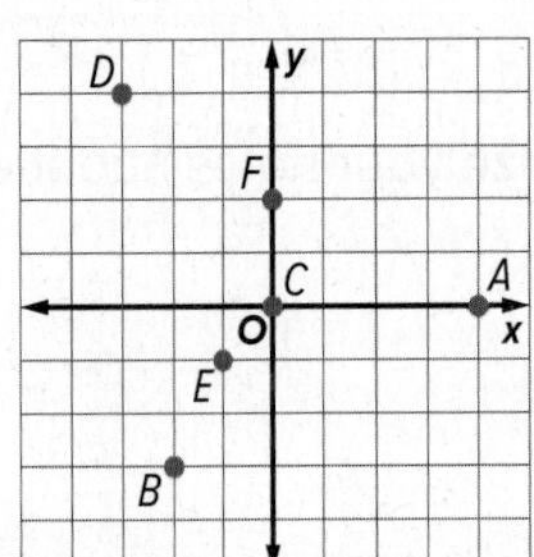

Solve each equation for the variable specified.

5. $2x + 4y = 12$, for x
6. $x = 3y - 9$, for y
7. $m - 2n = 6$, for m
8. $y = mx + b$, for x

How Did You Do?

Which exercises did you answer correctly in the Quick Check?

Graphing Systems of Equations

Today's Goals

- Determine the number of solutions of a system of linear equations.
- Solve systems of equations by graphing.
- Solve linear equations by graphing systems of equations.
- Use graphing calculators to solve systems of equations.

Today's Vocabulary

system of equations

consistent

independent

dependent

inconsistent

Go Online You may want to complete the Concept Check to check your understanding.

Explore Intersections of Graphs

Online Activity Use graphing technology to complete the Explore.

INQUIRY How can you solve a linear equation by graphing?

Learn Graphs of Systems of Equations

A set of two or more equations with the same variables is called a **system of equations**. An ordered pair that is a solution of both equations is a solution of the system. A system of two linear equations can have one solution, an infinite number of solutions, or no solution.

- A system of equations is **consistent** if it has at least one ordered pair that satisfies both equations.
- If a consistent system of equations has exactly one solution, it is said to be **independent**. The graphs intersect at one point.
- If a consistent system of equations has an infinite number of solutions, it is **dependent**. The graphs are the same line. This means that there are unlimited solutions that satisfy both equations.
- A system of equations is **inconsistent** if it has no ordered pair that satisfies both equations. The graphs are parallel.

Example 1 Consistent Systems

Use the graph to determine the number of solutions the system has. Then state whether the system of equations is *consistent* or *inconsistent* and if it is *independent* or *dependent*.

$y = -3x + 1$

$y = x - 3$

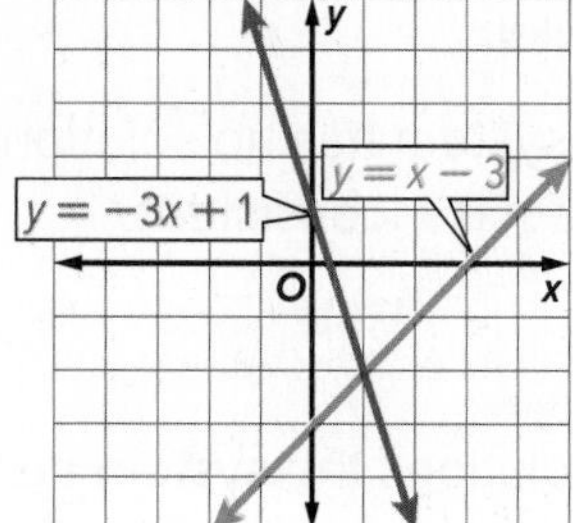

Since the graphs of these two lines intersect at one point, there is exactly one solution. Therefore, the system is consistent and independent.

Go Online You can complete an Extra Example online.

Example 2 Inconsistent Systems

Use the graph to determine the number of solutions the system has. Then state whether the system of equations is *consistent* or *inconsistent* and if it is *independent* or *dependent*.

$y = \frac{1}{2}x + 2$

$y = \frac{1}{2}x - 1$

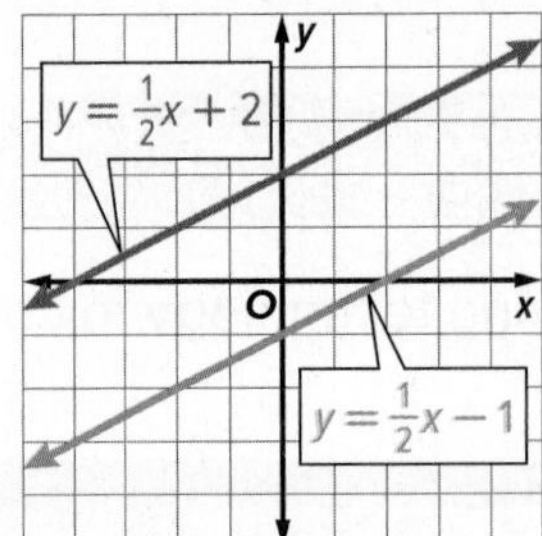

Since the graphs of these two lines are parallel, there is no solution of the system. Therefore, the system is inconsistent.

Think About It!

How could you change the equations so they form a consistent and dependent system?

Check

Determine whether each graph shows a system that is *consistent* or *inconsistent* and if it is *independent* or *dependent*.

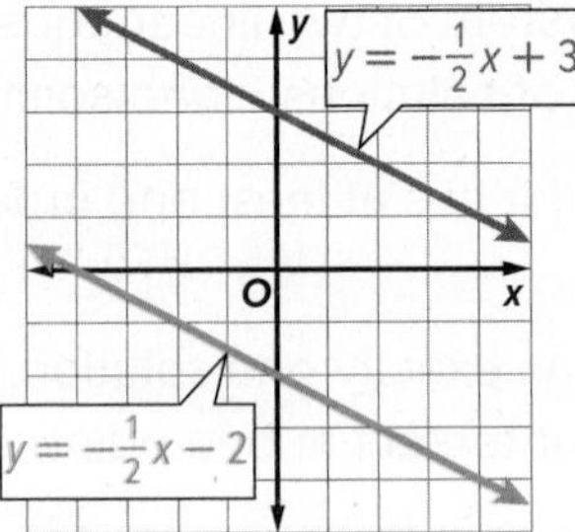

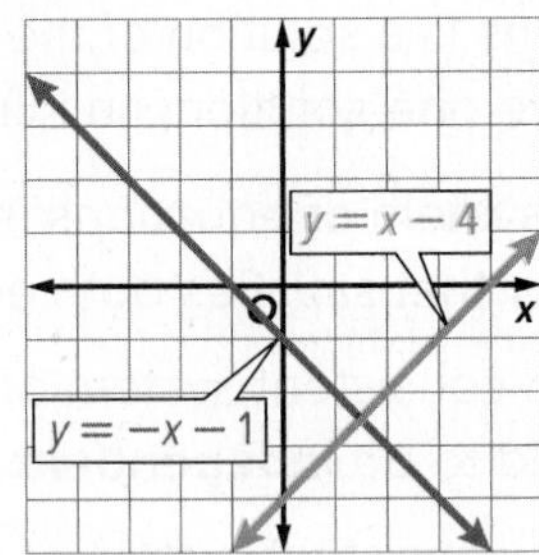

Example 3 Number of Solutions, Equations in Slope-Intercept Form

Determine the number of solutions the system has. Then state whether the system of equations is *consistent* or *inconsistent* and if it is *independent* or *dependent*.

$y = 6x + 10$	Because the slopes are the same and the *y*-intercepts are different, the lines are parallel.
$y = 6x + 4$	The system has no solution. Therefore, the system is inconsistent.

Study Tip

Parallel Lines Two lines that have the same slope are parallel and never intersect. A system of parallel lines has no solution.

Check

Determine the number of solutions the system has.

$y = \frac{4}{5}x - 2$

$4y - 5x = 9$

Go Online You can complete an Extra Example online.

Example 4 Number of Solutions, Equations in Standard Form

Determine the number of solutions the system has. Then state whether the system of equations is *consistent* or *inconsistent* and if it is *independent* or *dependent*.

$\mathbf{4y - 6x = 16}$
$\mathbf{9x - 6y = -24}$

Write both equations in slope-intercept form.

$4y - 6x = 16$	Original equation	$9x - 6y = -24$
$4y - 6x + 6x = +6x + 16$	Isolate the y-term	$9x - 6y - 9x = -9x - 24$
$4y = 6x + 16$	Simplify.	$-6y = -9x - 24$
$\frac{4y}{4} = \frac{6x}{4} + \frac{16}{4}$	Divide by coefficient of y.	$\frac{-6y}{-6} = \frac{-9x}{-6} + \frac{-24}{-6}$
$y = \frac{3}{2}x + 4$	Simplify.	$y = \frac{3}{2}x + 4$

Because the slopes are the same and the y-intercepts are the same, this is the same line.

Since the graphs of these two lines are the same, there are infinitely many solutions. Therefore, the system is consistent and dependent.

Check

Determine the number of solutions the system has.

$4x - 8y = 16$

$6x - 12y = 5$

Learn Solving Systems of Equations by Graphing

You can solve a system of equations by graphing each equation carefully on the same coordinate plane. Every point that lies on the line of one equation represents a solution of that equation. Similarly, every point on the line of the second equation in a system represents a solution of that equation. Therefore, the solution of a system of equations is the point at which the graphs intersect.

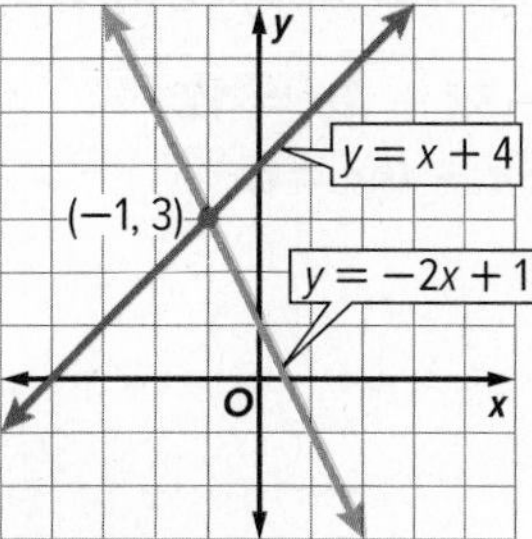

For example, the solution of this system is (−1, 3). That is the point at which the graphs intersect. Since the point of intersection lies on both lines, the ordered pair satisfies each equation in the system.

Think About It!

How many solutions will a system have if the slopes are different?

Example 5 Solve a System by Graphing

Graph the system and determine the number of solutions that it has. If it has one solution, determine its coordinates.

$y = -2x + 14$
$y = \frac{3}{5}x + 1$

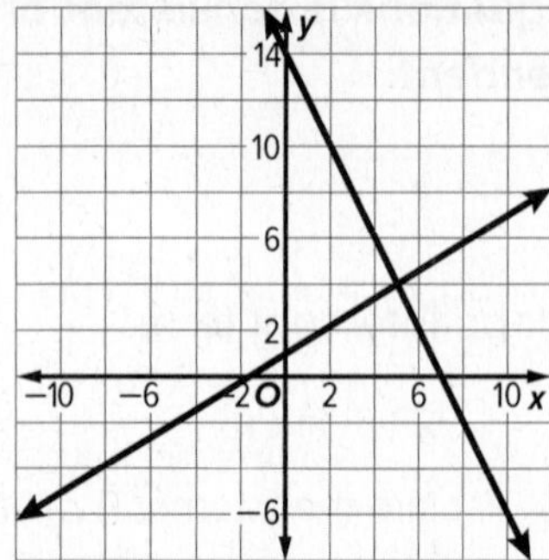

The graphs of the lines appear to intersect at the point (5, 4). If you substitute 5 for x and 4 for y into the equations, both are true. Therefore, (5, 4) is the solution of the system.

Think About It!

Why is it necessary to substitute the values of x and y into both equations to check your solution?

Check

Graph the system of equations.

$3x + 5y = 10$
$x - 5y = -10$

What is the solution of the system?

(?)

Go Online You can watch a video to see how to use a graphing calculator with this example.

Example 6 Graph and Solve a System of Equations

Graph the system and determine the number of solutions that it has. If it has one solution, determine its coordinates.

$-3x + 2y = 12$
$6x - 4y = 8$

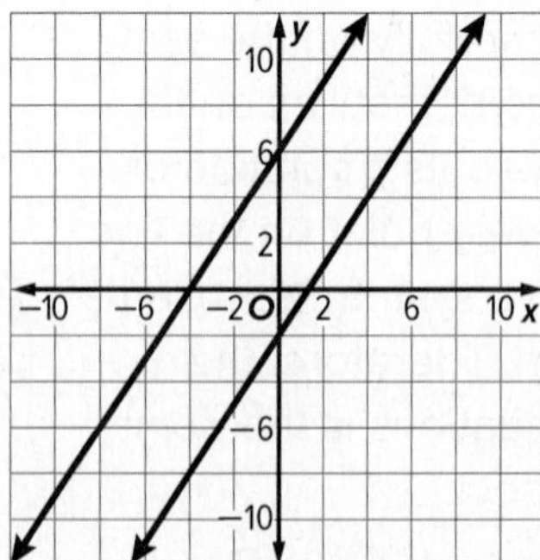

The lines have the same slope but different y-intercepts, so the lines are parallel. Since they do not intersect, this system has no solution.

Problem-Solving Tip

Tools When graphing by hand, using graph paper and a straightedge can help you make your graphs more accurate.

Go Online You can complete an Extra Example online.

Apply Example 7 Write a System of Equations

POPULATION **China and India are the two most populous countries in the world. The populations of these countries have increased steadily in recent years. In 2010, China and India had populations of about 1.34 billion and 1.19 billion, respectively. By 2016, the populations had grown to about 1.38 billion in China and 1.29 billion in India. Predict the approximate year when the populations of the two countries will be the same.**

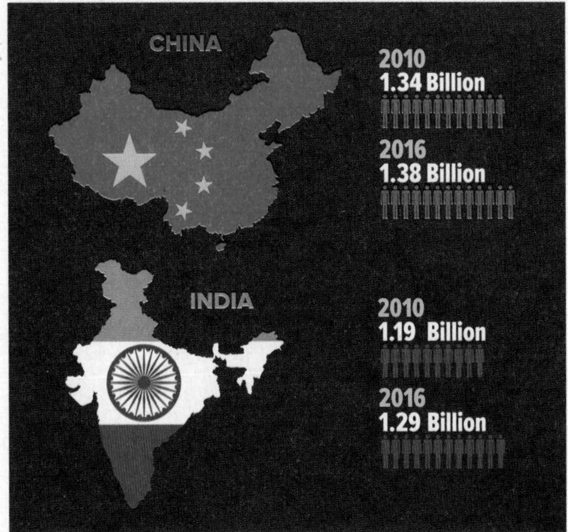

Source: IMF, CEIC

1. What is the task?

Describe the task in your own words. Then list any questions that you may have. How can you find answers to your questions?

Sample answer: I need to graph a system of equations that represents this situation and find the intersection. How can I write the system of equations that I need to graph? I can review finding the average rate of change and writing equations in slope-intercept form.

2. How will you approach the task? What have you learned that you can use to help you complete the task?

Sample answer: I will find the average rate of change for both countries. Then I will write a system of equations to represent the situation. I will graph the system and find the intersection. I will use what I have learned about graphing equations to help me graph the system.

3. What is your solution?

Use your strategy to solve the problem.
Find the average rate of change for the populations of China and India.

China: $\frac{1}{150}$

India: $\frac{1}{60}$

Write a system of equations to represent the situation. Let $x =$ the number of years after 2016. Let $y =$ population.

$y = \frac{1}{150}x + 1.38$

$y = \frac{1}{60}x + 1.29$

(continued on the next page)

Study Tip

Assumptions Populations do not typically increase at a steady rate. However, assuming that the rate of increase is constant allows you to estimate future data.

Study Tip

Labeling Axes It is important to clearly define and label axes in a real-world situation. The intersection does not mean that the population will be the same in year 9. Since the x-axis represents the number of years after 2016, you must add the x-coordinate of the intersection to 2016 to find the year.

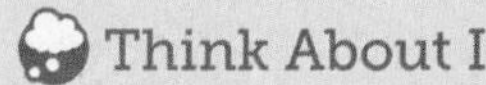

Can a value of x that is not a whole number be a viable solution? Justify your argument.

Graph the system of equations.

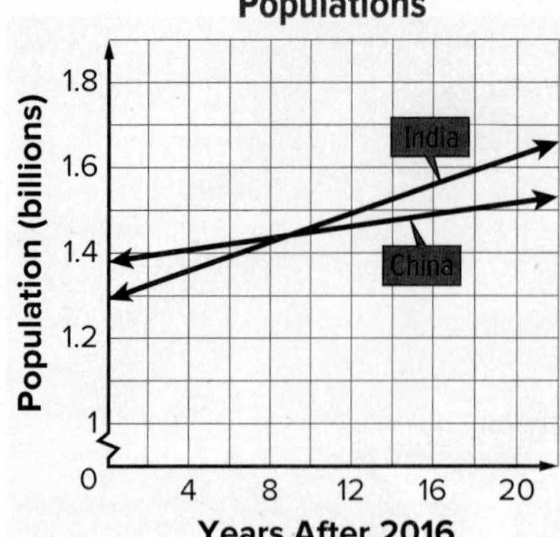

When will China and India have the same population?

2025

4. How can you know that your solution is reasonable?

Write About It! Write an argument that can be used to defend your solution.

Sample answer: From the graph, I estimated that the solution was 9 years after 2016, or in 2025. My answer is reasonable because when I substitute 9 in for x in each equation, the value of y is 1.44.

Check

OLYMPICS The number of men and women participating in the Winter Olympic Games has been steadily increasing in recent years. In the 19th Winter Olympics, 1389 men and 787 women participated. 1660 men and 1121 women participated in the 22nd Winter Olympics.

Part A

Write and graph a system to describe the number of men and women participating if x represents the number of Winter Olympics after the 22nd Winter Olympics.

Part B

Use the graph to predict the Winter Olympics when the number of men and women participating will be the same.

Go Online You can complete an Extra Example online.

Learn Using Systems to Solve Linear Equations

Key Concept • Using Systems to Solve Linear Equations	
Step 1	Write a system by setting each expression equal to y.
Step 2	Graph the system.
Step 3	Find the intersection.

Talk About It!

How do you know that the point of intersection satisfies both equations?

Example 8 Use a System to Solve a Linear Equation

Use a system of equations to solve $-6x + 8 = -4$.

Step 1 Write a system.

Write a system of equations. Set each side of $-6x + 8 = -4$ equal to y.

$y = -6x + 8$

$y = -4$

Step 2 Graph the system.

Enter the equations and graph.

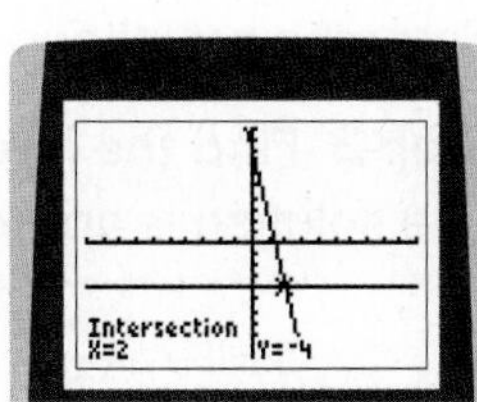

Step 3 Find the intersection.

The solution is the x-coordinate of the intersection, 2.

Step 4 Check your solution.

$-6x + 8 = -4$	Original equation
$-6(2) + 8 \stackrel{?}{=} -4$	Substitution
$-12 + 8 \stackrel{?}{=} -4$	Multiply.
$-4 = -4$	Add.

Go Online to see how to use a graphing calculator with this example.

Check

Use a system of equations and your graphing calculator to solve $-3.2x - 5.8 = 2.8x + 7$. Round to the nearest hundredth, if necessary.

$x =$ __?__

 **Go Online** You can complete an Extra Example online.

Learn Solving Systems of Equations by Using Graphing Technology

You can use a graphing calculator to graph and solve a system of equations by following these steps.

Step 1 Isolate y in each equation.

Step 2 Graph the system.

Step 3 Find the intersection.

Go Online You can watch a video to see how to graph systems of equations on a graphing calculator.

Example 9 Solve a System of Equations

Solve the system of equations.
$-1.38x - y = 5.13$
$0.62x + 2y = 1.60$

Step 1 Isolate y.

Solve each equation for y.

$-1.38x - y = 5.13$	First equation
$-y = 5.13 + 1.38x$	Add $1.38x$ to each side.
$y = -5.13 - 1.38x$	Multiply each side by -1.
$0.62x + 2y = 1.60$	Second equation
$2y = 1.60 - 0.62x$	Subtract $0.62x$ from each side.
$y = 0.80 - 0.31x$	Divide each side by 2.

Step 2 Graph the system.
Enter the equations and graph.

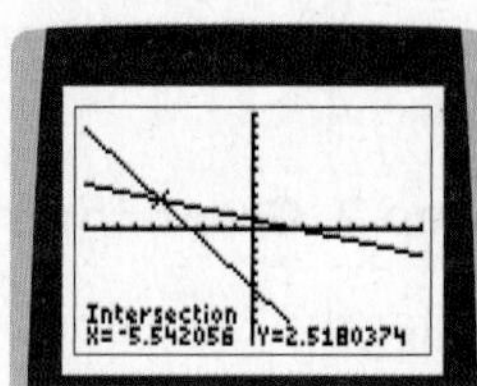

Go Online to see how to use a graphing calculator with this example.

Step 3 Find the intersection.
The solution is approximately $(-5.54, 2.52)$.

Check

What is the solution to the system of equations?

$2.29x - 4.41y = 6.52$
$4.16x + 1.11y = 4.72$
(__?__, __?__)

Example 10 Write and Solve a System of Equations

BUSINESS **Denzel is starting a food truck business to sell gourmet grilled cheese sandwiches. He has spent $34,000 on the truck, equipment, permits, and other start-up costs. Each sandwich costs about $1.32 to make, and he sells them for $7.**

How many sandwiches does Denzel need to sell to start earning a profit?

Let x = the number of grilled cheese sandwiches sold.
Let y = total cost or revenue.

Total Cost: $y = 1.32x + 34{,}000$
Total Revenue: $y = 7x$

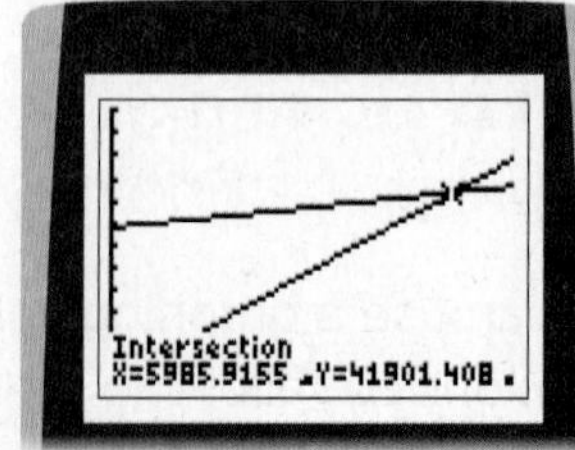

Go Online to see how to use a graphing calculator with this example.

Step 1 Graph the system.
Enter the equations and graph.

Step 2 Find the intersection.
The solution is approximately (5985.92, 41,901.41).
This means that after Denzel has sold 5986 sandwiches, he will begin to earn a profit.

Go Online You can complete an Extra Example online.

Practice

Go Online You can complete your homework online.

Examples 1 and 2

Use the graph to determine the number of solutions the system has. Then state whether the system of equations is *consistent* or *inconsistent* and if it is *independent* or *dependent*.

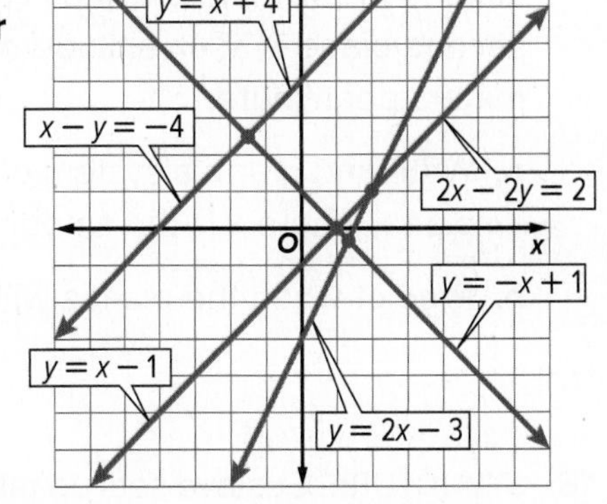

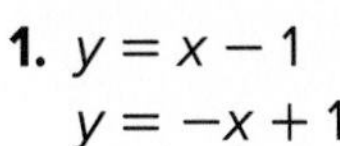

1. $y = x - 1$
   $y = -x + 1$

2. $x - y = -4$
   $y = x + 4$

3. $y = x + 4$
   $2x - 2y = 2$

4. $y = 2x - 3$
   $2x - 2y = 2$

Examples 3 and 4

Determine the number of solutions the system has. Then state whether the system of equations is *consistent* or *inconsistent* and if it is *independent* or *dependent*.

5. $y = \frac{1}{2}x$
   $y = x + 2$

6. $4x - 6y = 12$
   $-2x + 3y = -6$

7. $8x - 4y = 16$
   $-5x - 5y = 5$

8. $2x + 3y = 10$
   $4x + 6y = 12$

9. $y = -\frac{3}{2}x + 5$
   $y = -\frac{2}{3}x + 5$

10. $y = x - 3$
    $y = -4x + 3$

Examples 5 and 6

Graph each system and determine the number of solutions it has. If it has one solution, determine its coordinates.

11. $y = -3$
    $y = x - 3$

12. $y = 4x + 2$
    $y = -2x - 4$

13. $y = x - 6$
    $y = x + 2$

14. $x + y = 4$
    $3x + 3y = 12$

15. $x - y = -2$
    $-x + y = 2$

16. $2x + 3y = 12$
    $2x - y = 4$

Example 7

17. **AVIATION** An air traffic controller manages the flow of aircraft in and out of airport airspace by guiding pilots during takeoff and landing. An air traffic controller is monitoring two planes that are in flight near a local airport. The first plane is at an altitude of 1000 meters and is ascending at a rate of 400 meters per minute. The second plane is at an altitude of 5900 meters and is descending at a rate of 300 meters per minute.

    a. Write and graph a system of equations that represents the altitude of each plane, where x is the amount of time, in minutes, and y is the altitude, in meters.

    b. Predict when the planes will be at the same altitude.

18. **STRUCTURE** Gustavo sets up tables for a caterer on weekends. Each round table can seat 8 people. Each rectangular table can seat 10 people. One weekend, his boss asked him to set up tables for 124 people. He uses 2 more round tables than rectangular tables. Define variables and write a system of equations to find the number of round tables and rectangular tables Gustavo used. Then solve the system graphically.

Example 8

Write and graph a system of equations to solve each linear equation.

19. $3x + 6 = 6$

20. $2x - 17 = x - 10$

21. $-12x + 90 = 30$

22. $13x - 28 = 24$

23. $2x + 5 = 2x + 5$

24. $x + 1 = x + 3$

Example 9

Solve eCH system of equations. State the decimal solution to the nearest hundredth.

25. $2.5x + 3.75y = 10.5$
    $1.25x - 8.5y = -5.25$

26. $2.2x + 1.8y = -3.6$
    $-4.8x + 12.4y = 10.6$

27. $1.12x - 2.24y = 4.96$
    $-3.56x - 2.48y = -7.32$

Example 10

28. **USE TOOLS** An office building has two elevators. One elevator starts out on the 4th floor, 35 feet above the ground, and is descending at a rate of 2.2 feet per second. The other elevator starts out at ground level and is rising at a rate of 1.7 feet per second. Write and solve a system of equations to determine when both elevators will be at the same height. Interpret the solution.

29. **USE TOOLS** A bookstore makes a profit of \$2.50 on each book they sell, and \$0.75 on each magazine they sell. One week, the store sold x books and y magazines, for a weekly profit of \$450. The total number of publications sold that week was 260. Write and solve a system of equations to determine the number of books and magazines that the bookstore sold that week. Interpret the solution.

Mixed Exercises

Graph each system and determine the number of solutions it has. If it has one solution, determine its coordinates. Then state whether the system is *consistent* or *inconsistent* and if it is *independent* or *dependent*.

30. $2x - y = -3$
$2x + y = -1$

31. $2x + y = 4$
$y = -2x - 2$

32. $2y = 5 + x$
$3x - 6y = -15$

33. $2x - y = 5$
$x + y = -2$

34. $2y = 1.2x - 10$
$4y = 2.4x$

35. $x = 6 - \frac{3}{8}y$
$4 = \frac{2}{3}x + \frac{1}{4}y$

36. $x - y = 3$
$x - 2y = 3$

37. $x + 2y = 4$
$y = -\frac{1}{2}x + 2$

38. $y = 2x + 3$
$3y = 6x - 6$

39. $y - x = -1$
$x + y = 3$

40. **BUSINESS** The number of items sold at Store 1 can be represented by $y = 200x + 300$, where x represents the number of days and y represents the number of items sold. The number of items sold at Store 2 can be represented by $y = 200x + 100$, where x represents the number of days and y represents the number of items sold. Look at the graph of the system of equations and determine whether it has *no* solution, *one* solution, or *infinitely many* solutions.

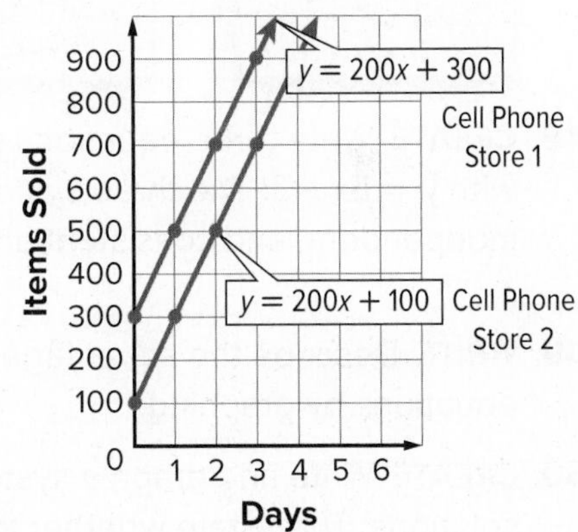

41. **USE A MODEL** Olivia and her brother William had a bicycle race. Olivia rode at a speed of 20 feet per second, while William rode at a speed of 15 feet per second. Olivia gave William a 150-foot head start, and the race ended in a tie.

a. Define variables and write a system of equations to represent this situation.

b. Graph the system of equations.

c. How far away was the finish line from where Olivia started?

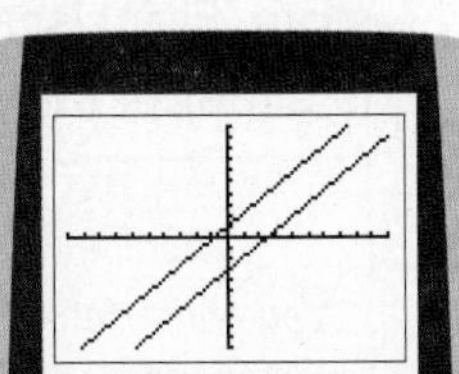

42. **CONSTRUCT ARGUMENTS** Abhijit uses a calculator to graph the system $y = 1.21x - 3$ and $5y - 5 = 6x$. He concludes that the system has no solution. Do you agree or disagree? Explain your reasoning.

43. USE A MODEL Some days, Cayley walks to school. Other days, she rides her bicycle. When she walks to school, she needs to leave home 15 minutes earlier than when she rides her bicycle. Cayley walked 3 days last week and rode her bike on 2 days. Cayley spent a total of 1 hour 10 minutes going to school last week.

a. Define variables and write a system of equations to represent this situation.

b. Graph the system of equations.

c. How long does it take Cayley to walk to school?

44. REGULARITY Maureen says that if a system of linear equations has three equations, then there could be exactly two solutions to the system. Is Maureen correct? Use a graph to justify your reasoning.

Higher-Order Thinking Skills

45. PERSEVERE Use graphing to find the solution of the system of equations $2x + 3y = 5$, $3x + 4y = 6$, and $4x + 5y = 7$.

46. ANALYZE Determine whether a system of two linear equations with (0, 0) and (2, 2) as solutions *sometimes*, *always*, or *never* has other solutions. Justify your argument.

47. WHICH ONE DOESN'T BELONG? Which one of the following systems of equations doesn't belong with the other three? Justify your conclusion.

$4x - y = 5$ $-2x + y = -1$	$-x + 4y = 8$ $3x - 6y = 6$	$4x + 2y = 14$ $12x + 6y = 18$	$3x - 2y = 1$ $2x + 3y = 18$

48. CREATE Write three equations such that they form three systems of equations with $y = 5x - 3$. The three systems should be inconsistent, consistent and independent, and consistent and dependent, respectively.

49. WRITE Describe the advantages and disadvantages of solving systems of equations by graphing.

50. CREATE Write and graph a system of equations that has the following number of solutions. Then state whether the system of equations is *consistent* or *inconsistent* and if it is *independent* or *dependent*.

a. one solution **b.** infinite solutions **c.** no solution

51. FIND THE ERROR Store A is offering a 10% discount on the purchase of all electronics in their store. Store B is offering $10 off all the electronics in their store. Francisca and Alan are deciding which offer will save them more money. Is either of them correct? Explain your reasoning.

Francisca	Alan
You can't determine which store has the better offer unless you know the price of the items you want to buy.	Store A has the better offer because 10% of the sale price is a greater discount than $10.

Substitution

Today's Goal
- Solve systems of equations by using the substitution method.

Today's Vocabulary
substitution

Explore Using Substitution

 **Online Activity** Use a system of equations to complete the Explore.

> **INQUIRY** How can you rewrite a system of equations as a single equation with only one variable?

Learn Solving Systems of Equations by Substitution

Exact solutions result when algebraic methods are used to solve systems of equations. One algebraic method is called **substitution**.

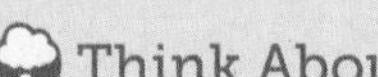

Think About It!
In algebra, what does it mean to *substitute*?

Key Concept • Substitution Method	
Step 1	When necessary, solve at least one equation for one variable.
Step 2	Substitute the resulting expression from Step 1 into the other equation to replace the variable. Then solve the equation.
Step 3	Substitute the value from Step 2 into either equation, and solve for the other variable. Write the solution as an ordered pair.

Example 1 Solve a System by Substitution

Use substitution to solve the system of equations.

$3x - y = -7$; $y = 4x + 11$

Step 1 The second equation is already solved for y.

Step 2 Substitute $4x + 11$ for y in the first equation.

$3x - (4x + 11) = -7$	Substitute $4x + 11$ for y.
$3x - 4x - 11 = -7$	Distributive Property
$-x - 11 = -7$	Combine like terms.
$-x = 4$	Add 11 to each side.
$x = -4$	Multiply each side by -1.

Step 3 Substitute -4 for x in either equation to find y.

$y = 4x + 11$	Second equation.
$y = 4(-4) + 11$	Substitution.
$y = -5$	Simplify.

The solution is $(-4, -5)$.

Go Online You can complete an Extra Example online.

Think About It!
How would the substitution process differ if the second equation were $4x - y = -11$?

 **Go Online**
You can watch a video to see how to use algebra tiles with this example.

Check

Refer to the system of equations.

$3x - 2y = -17$
$y = 2x + 2$

Part A Which expression could be substituted for y in the first equation to find the value of x?

A. $2x - 2$ **B.** $-\frac{3}{2}x + \frac{17}{2}$ **C.** $2x + 2$ **D.** $3x - 2y$

Part B What is the solution of the system?

A. $(-13, -24)$ **B.** $(-13, 24)$ **C.** $(13, -28)$ **D.** $(13, 28)$

Example 2 Solve and Then Substitute

Talk About It!
Would the solution of the system be the same if the first step used was to solve the second equation for y? Explain.

Use substitution to solve the system of equations.

$\mathbf{5x - 3y = -25}$
$\mathbf{x + 4y = 18}$

Step 1 Solve the second equation for x since the coefficient is 1.

$x + 4y = 18$	Second equation
$x = 18 - 4y$	Subtract $4y$ from each side.

Step 2 Substitute $18 - 4y$ for x in the first equation.

$5x - 3y = -25$	First equation
$5(18 - 4y) - 3y = -25$	Substitute $18 - 4y$ for x.
$90 - 20y - 3y = -25$	Distributive Property
$90 - 23y = -25$	Combine like terms.
$-23y = -115$	Subtract 90 from each side.
$y = 5$	Divide each side by -23.

Study Tip
Slope-Intercept Form If both equations are in the form $y = mx + b$, the expressions can simply be set equal to each other and then solved for x. For example, if $y = 2x - 5$ and $y = -4x + 1$, then $2x - 5 = -4x + 1$. The solution for x can then be used to find the value of y.

Step 3 Substitute 5 for y in either equation to find x.

$x + 4y = 18$	Second equation
$x + 4(5) = 18$	Substitute 5 for y.
$x + 20 = 18$	Simplify.
$x = -2$	Subtract 20 from each side.

The solution is $(-2, 5)$.

Check

Use substitution to solve the system of equations.

$5x + 3y = 5$
$x + 2y = -13$

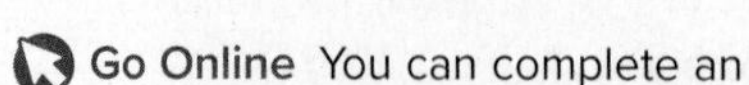
Go Online You can complete an Extra Example online.

Example 3 Use Substitution When There are No or Many Solutions

Use substitution to solve the system of equations.

$4x + 2y = -8$

$y = -2x - 4$

Substitute $-2x - 4$ for y in the first equation.

$4x + 2y = -8$	First equation
$4x + 2(-2x - 4) = -8$	Substitute $-2x - 4$ for y.
$4x - 4x - 8 = -8$	Distributive Property
$-8 = -8$	Simplify.

The equation $-8 = -8$ is an identity. Thus, there are an infinite number of solutions.

When graphed, the equations are the same line.

Check

Select the correct statement about the system of equations.

$-x + 2y = 2$

$y = \frac{1}{2}x + 1$

A. This system has no solution.

B. This system has one solution at $\left(\frac{2}{3}, \frac{4}{3}\right)$.

C. This system has one solution at $\left(\frac{4}{3}, \frac{2}{3}\right)$.

D. This system has infinitely many solutions.

Study Tip

Dependent Systems There are infinitely many solutions of the system because the equations in slope-intercept form are equivalent, and they have the same graph.

Think About It!

What would a solution like $-8 = 0$ mean? What would it look like on a graph?

Example 4 Write and Solve a System of Equations

TREE PRESERVATION **A town ordinance defines an adult tree as having a diameter greater than 10 inches and a sapling as having a diameter less than 10 inches. The ordinance requires that on a new building project, two new trees are planted for each adult tree felled and six new trees are planted for each sapling felled. Last year, there were 167 trees felled, and the community planted 742 replacement trees. How many of each type of tree were felled?**

(continued on the next page)

Study Tip

Assumptions Using systems of equations to solve real-world problems generally requires assuming that the problem actually has a solution. It is a good idea to graph the lines first, to see if a solution exists, before going through the steps to solve for each variable.

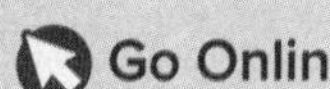

An alternate method is available for this example.

Use a Source

Research replacing trees cut for construction in an area near you. How could you find the number of trees to plant for a project similar to the one in this community?

Let a = the number of adult trees felled, and let t = the number of sapling trees felled.

$a + t = 167$	This equation represents the total number of adult trees a and sapling trees t felled with a sum of 167.
$2a + 6t = 742$	This equation represents the combinations of adult trees a and sapling trees t replaced with a total of 742.

The solution of the system of equations represents the option that meets both of the constraints.

The first equation can easily be solved for a or t.

Solve for *a*.

Step 1 Solve the first equation for a.

$a + t = 167$ First equation

$a + t - t = 167 - t$ Subtract t from each side.

$a = 167 - t$ Simplify.

Step 2 Substitute $167 - t$ for a in the second equation.

$2a + 6t = 742$ Second equation

$2(167 - t) + 6t = 742$ Substitute $167 - t$ for a.

$334 - 2t + 6t = 742$ Distributive Property

$4t + 334 = 742$ Combine like terms.

$4t = 408$ Subtract 334 from each side.

$t = 102$ Divide each side by 4.

Step 3 Substitute 102 for t in either equation to find the value of a.

$a + t = 167$ First equation

$a + 102 = 167$ Substitute 102 for t.

$a = 65$ Subtract 102 from each side.

The solution is (65 , 102).

There were 65 adult trees and 102 saplings felled. Because only whole numbers of trees can be felled, this is a viable solution.

Check

GEOMETRY Kymani has two equal-sized large pitchers and two equal-sized small pitchers. All of the pitchers together hold 40 cups of water. The capacity of one large pitcher minus the capacity of one small pitcher is 12 cups. How many cups can each type of pitcher hold?

Small pitcher = ___?___ cups

Large pitcher = ___?___ cups

Go Online You can complete an Extra Example online.

Practice

Go Online You can complete your homework online.

Examples 1–3

Use substitution to solve each system of equations.

1. $y = 5x + 1$
$4x + y = 10$

2. $y = 4x + 5$
$2x + y = 17$

3. $y = 3x - 34$
$y = 2x - 5$

4. $y = 3x - 2$
$y = 2x - 5$

5. $2x + y = 3$
$4x + 4y = 8$

6. $3x + 4y = -3$
$x + 2y = -1$

7. $y = -3x + 4$
$-6x - 2y = -8$

8. $-1 = 2x - y$
$8x - 4y = -4$

9. $x = y - 1$
$-x + y = -1$

10. $y = -4x + 11$
$3x + y = 9$

11. $y = -3x + 1$
$2x + y = 1$

12. $3x + y = -5$
$6x + 2y = 10$

13. $5x - y = 5$
$-x + 3y = 13$

14. $2x + y = 4$
$-2x + y = -4$

15. $-5x + 4y = 20$
$10x - 8y = -40$

Example 4

16. **MONEY** Harvey has some \$1 bills and some \$5 bills. In all, he has 6 bills worth \$22. Let x be the number of \$1 bills, and let y be the number of \$5 bills. Write a system of equations to represent the information, and use substitution to determine how many bills of each denomination Harvey has.

17. **REASONING** Shelby and Calvin are conducting an experiment in chemistry class. They need 5 milliliters of a solution that is 65% acid and 35% distilled water. There is no undiluted acid in the chemistry lab, but they do have two beakers of diluted acid. Beaker A contains 70% acid and 30% distilled water. Beaker B contains 20% acid and 80% distilled water.

a. Write a system of equations that Shelby and Calvin could use to determine how many milliliters they need to pour from each beaker to make their solution.

b. Solve your system of equations. How many milliliters from each beaker do Shelby and Calvin need?

Mixed Exercises

Use substitution to solve each system of equations.

18. $y = 3.2x + 1.9$
$2.3x + 2y = 17.72$

19. $y = \frac{1}{4}x - \frac{1}{2}$
$8x + 12y = -\frac{1}{2}$

20. $y = -10x - 6.8$
$-50x - 10.5y = 60.4$

21. **USE A SOURCE** Research population trends in South America. Write and solve a system of equations to predict when the population of two countries will be equal.

22. **REGULARITY** Angle A and angle B are complementary, and their measures have a difference of 20°. What are the measures of the angles? Generalize your method.

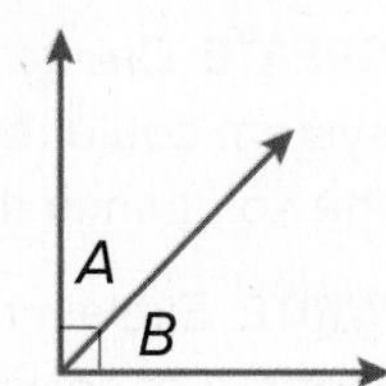

23. **STRUCTURE** A two-digit number is reduced by 45 when the digits are interchanged. The digit in the tens place of the original number is 1 more than 3 times the digit in the units place. Define variables and write a system of equations to find the original number.

24. **USE A MODEL** A zoo keeps track of the number of visitors to each exhibit. The table shows the number of visitors for two exhibits on one day.

	Big Cats	Petting Zoo
Adults	x	21
Children	1024	y

a. Three times the total number of adults was 17 less than the number of children who visited the petting zoo. Write an equation to model this relationship.

b. The total number of children who visited the zoo that day was 681 less than 10 times the number of adults who visited the big cats. Write an equation to model this relationship.

c. Solve the system of equations to find the total number of visitors to those two exhibits on that day.

d. Tickets to the zoo cost $25 for adults and $10 for children. The total ticket sales one day were $47,750. The number of children who visited the zoo was 270 more than 6 times the number of adults who visited. Write and solve a system of equations to find the total number of visitors to the zoo that day.

Higher-Order Thinking Skills

25. **FIND THE ERROR** In the system $a + b = 7$ and $1.29a + 0.49b = 6.63$, a represents pounds of apples and b represents pounds of bananas a person bought. Guillermo and Cara are finding and interpreting the solution. Is either correct? Explain your reasoning.

Guillermo

$1.29a + 0.49b = 6.63$

$1.29a + 0.49(a + 7) = 6.63$

$1.29a + 0.49a + 3.43 = 6.63$

$1.78a = 3.2$

$a = 1.8$

$a + b = 7$, so $b = 5.2$. The solution (1.8, 5.2) means that 1.8 pounds of apples and 5.2 pounds of bananas were bought.

Cara

$1.29a + 0.49b = 6.63$

$1.29(7 - b) + 0.49b = 6.63$

$9.03 - 1.29b + 0.49b = 6.63$

$-0.8b = -2.4$

$b = 3$

The solution $b = 3$ means that 3 pounds of apples and 3 pounds of bananas were bought.

26. **PERSEVERE** A local charity has 60 volunteers. The ratio of boys to girls is 7:5. Find the number of volunteers who are boys and the number who are girls.

27. **ANALYZE** Compare and contrast the solution of a system found by graphing and the solution of the same system found by substitution.

28. **CREATE** Create a system of equations that has one solution. Illustrate how the system could represent a real-world situation and describe the significance of the solution in the context of the situation.

29. **WRITE** Explain how to determine what to substitute when using the substitution method of solving systems of equations.

Lesson 7-3

Elimination Using Addition and Subtraction

Learn Solving Systems of Equations by Elimination with Addition

The **elimination** method involves eliminating a variable by combining the individual equations within a system of equations. One way to combine equations is by using addition.

Key Concept • Elimination Method Using Addition	
Step 1	Write the system so like terms with opposite coefficients are aligned.
Step 2	Add the equations, eliminating one variable. Then solve the equation.
Step 3	Substitute the value from Step 2 into one of the equations and solve for the other variable. Write the solution as an ordered pair.

Today's Goals

- Solve systems of equations by eliminating a variable using addition.
- Solve systems of equations by eliminating a variable using subtraction.

Today's Vocabulary

elimination

Example 1 Elimination Using Addition

Use elimination to solve the system of equations.

$$3x + 5y = 11$$
$$5x - 5y = 5$$

Step 1 Align terms with opposite coefficients.

Since $5y$ and $-5y$ have opposite coefficients, add the equations to eliminate the variable y.

Step 2 Add the equations.

$$3x + 5y = 11$$
$$(+)\quad 5x - 5y = 5$$

$8x = 16$ — The variable y is eliminated.

$\frac{8x}{8} = \frac{16}{8}$ — Divide each side by 8.

$x = 2$ — Simplify.

Step 3 Solve for the other variable.

$3x + 5y = 11$ — First equation

$3(2) + 5y = 11$ — Replace x with 2.

$6 + 5y = 11$ — Multiply.

$6 - 6 + 5y = 11 - 6$ — Subtract 6 from each side.

$5y = 5$ — Simplify.

$\frac{5y}{5} = \frac{5}{5}$ — Divide each side by 5.

$y = 1$ — Simplify.

The solution is (2, 1).

Go Online You can complete an Extra Example online.

Study Tip

Answer Check You can check your answer by substituting the solution into the equation you did not use in Step 3. If the equality is valid, your solution is correct.

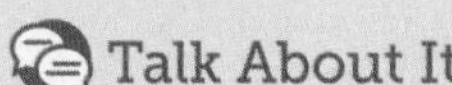

Talk About It!

When graphed, where would these lines intersect? Explain your reasoning.

Check

Use elimination to solve the system of equations.

$5x + 13y = 20$

$-5x - 3y = 30$

Example 2 Write and Solve a System of Equations Using Addition

Seven times a number minus four times another number is thirteen. Negative seven times a number plus seven times another number is fourteen. Find the numbers.

Seven times a number minus four times another number is thirteen.	Negative seven times a number plus seven times another number is fourteen.
$7x - 4y = 13$	$-7x + 7y = 14$

Steps 1 and 2 Write the equations vertically and add.

$$\begin{array}{rl} 7x - 4y = 13 & \\ (+)\ -7x + 7y = 14 & \\ \hline 3y = 27 & \text{The variable } x \text{ is eliminated.} \\ \frac{3y}{3} = \frac{27}{3} & \text{Divide each side by 3.} \\ y = 9 & \text{Simplify.} \end{array}$$

Step 3 Substitute 9 for *y* in either equation to find the value of *x*.

$$\begin{array}{rl} -7x + 7y = 14 & \text{Second equation} \\ -7x + 7(9) = 14 & \text{Replace } y \text{ with 9.} \\ -7x + 63 = 14 & \text{Multiply.} \\ -7x + 63 - 63 = 14 - 63 & \text{Subtract 63 from each side.} \\ -7x = -49 & \text{Simplify.} \\ \frac{-7x}{-7} = \frac{-49}{-7} & \text{Divide each side by } -7. \\ x = 7 & \text{Simplify.} \end{array}$$

The solution is (7, 9). So, $x = 7$ and $y = 9$.

Check

Two times a number minus six times another number is negative six. Negative two times a number plus five times another number is eighteen.

Write the system of equations.

Solve the system of equations.

Go Online You can complete an Extra Example online.

Learn Solving Systems of Equations by Elimination with Subtraction

When the coefficients of a variable are the same in two equations, you can eliminate the variable by subtracting one equation from the other.

Key Concept • Elimination Method Using Subtraction	
Step 1	Write the system so like terms with the same coefficients are aligned.
Step 2	Subtract one equation from the other, eliminating one variable. Then solve the equation.
Step 3	Substitute the value from Step 2 into one of the equations and solve for the other variable. Write the solution as an ordered pair.

Example 3 Elimination Using Subtraction

Use elimination to solve the system of equations.

$\mathbf{3x + 6y = 30}$
$\mathbf{5x + 6y = 6}$

Step 1 Align terms with the same coefficients.

Since $6y$ and $6y$ have the same coefficients, you can subtract the equations to eliminate the variable y.

Step 2 Subtract the equations.

$$\begin{aligned} 3x + 6y &= 30 \\ (-)5x + 6y &= 6 \\ \hline -2x &= 24 & &\text{The variable } y \text{ is eliminated.} \\ \frac{-2x}{-2} &= \frac{24}{-2} & &\text{Divide each side by } -2. \\ x &= -12 & &\text{Simplify.} \end{aligned}$$

Step 3 Substitute −12 for x in either equation to find the value of y.

$$\begin{aligned} 3x + 6y &= 30 & &\text{First equation} \\ 3(-12) + 6y &= 30 & &\text{Replace } x \text{ with } -12. \\ -36 + 6y &= 30 & &\text{Multiply.} \\ -36 + 36 + 6y &= 30 + 36 & &\text{Add 36 to each side.} \\ 6y &= 66 & &\text{Simplify.} \\ \frac{6y}{6} &= \frac{66}{6} & &\text{Divide each side by 6.} \\ y &= 11 & &\text{Simplify.} \end{aligned}$$

The solution is (−12, 11).

Check

Use elimination to solve the system of equations.

$-2x + 3y = 48$
$7x + 3y = 21$

Go Online You can complete an Extra Example online.

Study Tip

Adding and Subtracting Equations When the variable you want to eliminate has the same coefficient in the two equations, subtract. When the variable you want to eliminate has opposite coefficients, add.

Watch Out!

Subtracting an Equation When subtracting one equation from another in order to eliminate a variable, do not forget to distribute the negative sign to each term of the expressions on both sides of the equals sign.

Example 4 Write and Solve a System of Equations Using Subtraction

COMPUTERS **Mei and Kara build computers from parts and sell them to make a profit. Mei can build a computer in 0.9 hour, and Kara can build one in 1.2 hours. During a typical week, Mei and Kara spend a total of 15 hours building computers. One week, Mei builds twice as many computers, and the two spend a total of 24 hours on their project. How many computers do Mei and Kara each make during a typical week?**

Think About It!

What assumption was made about the rates at which Mei and Kara build computers? Why is that assumption made?

Words	Mei's time spent	plus Kara's time spent	is 15 hours.
Variables	Let m = the number of computers that Mei built and k = the number of computers that Kara built.		
Equations	$0.9m$	$+ 1.2k$	$= 15$

Words	Double Mei's time spent	plus Kara's time spent	is 24 hours.
Variables	Let m = the number of computers that Mei built and k = the number of computers that Kara built.		
Equations	$2(0.9m)$	$+ 1.2k$	$= 24$

Steps 1 and 2 Write the equations vertically and subtract.

$$0.9m + 1.2k = 15$$

$$(-)\ 2(0.9m) + 1.2k = 24$$

$$-0.9m = -9 \qquad \text{The variable } k \text{ is eliminated.}$$

$$\frac{-0.9m}{-0.9} = \frac{-9}{-0.9} \qquad \text{Divide each side by } -0.9.$$

$$m = 10 \qquad \text{Simplify.}$$

Step 3 Substitute 10 for *m* in either equation to find the value of *k*.

$$0.9m + 1.2k = 15 \qquad \text{First equation}$$

$$0.9(10) + 1.2k = 15 \qquad \text{Replace } m \text{ with 10.}$$

$$9 + 1.2k = 15 \qquad \text{Multiply.}$$

$$9 - 9 + 1.2k = 15 - 9 \qquad \text{Subtract 9 from each side.}$$

$$1.2k = 6 \qquad \text{Simplify.}$$

$$\frac{1.2k}{1.2} = \frac{6}{1.2} \qquad \text{Divide each side by 1.2.}$$

$$k = 5 \qquad \text{Simplify.}$$

During a typical week, Mei builds 10 computers and Kara builds 5. Since they cannot sell part of a computer, it makes sense that they would build a whole number of computers in a week. Therefore, 10 computers and 5 computers are viable solutions.

Go Online You can complete an Extra Example online.

Practice

Go Online You can complete your homework online.

Examples 1, 3

Use elimination to solve each system of equations.

1. $-v + w = 7$
$v + w = 1$

2. $y + z = 4$
$y - z = 8$

3. $-4x + 5y = 17$
$4x + 6y = -6$

4. $5m - 2p = 24$
$3m + 2p = 24$

5. $a + 4b = -4$
$a + 10b = -16$

6. $6r - 6t = 6$
$3r - 6t = 15$

7. $6c - 9d = 111$
$5c - 9d = 103$

8. $11f + 14g = 13$
$11f + 10g = 25$

9. $9x + 6y = 78$
$3x - 6y = -30$

10. $3j + 4k = 23.5$
$8j - 4k = 4$

11. $-3x - 8y = -24$
$3x - 5y = 4.5$

12. $6x - 2y = 1$
$10x - 2y = 5$

13. $x - y = 1$
$x + y = 3$

14. $-x + y = 1$
$x + y = 11$

15. $x + 4y = 11$
$x - 6y = 11$

16. $-x + 3y = 6$
$x + 3y = 18$

17. $3x + 4y = 19$
$3x + 6y = 33$

18. $x + 4y = -8$
$x - 4y = -8$

19. $3x + 4y = 2$
$4x - 4y = 12$

20. $3x - y = -1$
$-3x - y = 5$

21. $2x - 3y = 9$
$-5x - 3y = 30$

22. $x - y = 4$
$2x + y = -4$

23. $3x - y = 26$
$-2x - y = -24$

24. $5x - y = -6$
$-x + y = 2$

25. $6x - 2y = 32$
$4x - 2y = 18$

26. $3x + 2y = -19$
$-3x - 5y = 25$

27. $7x + 4y = 2$
$7x + 2y = 8$

Example 2

28. Twice a number added to another number is 15. The sum of the two numbers is 11. Find the numbers.

29. Twice a number added to another number is –8. The difference of the two numbers is 2. Find the numbers.

30. The difference of two numbers is 2. The sum of the same two numbers is 6. Find the numbers.

Example 4

31. GOVERNMENT The Texas State Legislature is comprised of state senators and state representatives. There is a greater number of representatives than senators. The sum of the number of representatives and the number of senators is 181. The difference of the number of representatives and number of senators is 119.

a. Write a system of equations to find the number of state representatives, r, and senators, s.

b. How many senators and how many representatives make up the Texas State Legislature?

32. SPORTS As of 2019, the New York Yankees had won the World Series more than any other team in baseball. The difference of the number of World Series championships won by the Yankees and 2 times the number of World Series championships won by the second-most-winning team, the St. Louis Cardinals, is 5. The sum of the two teams' World Series championships is 38.

a. Write a system of equations to find the number of World Series championships won by the Yankees, y, and the number of World Series championships won by the Cardinals, x.

b. How many times has each team won the World Series?

Mixed Exercises

Use elimination to solve each system of equations.

33. $4(x + 2y) = 8$
$4x + 4y = 12$

34. $3x - 5y = 11$
$5(x + y) = 5$

35. $4x + 3y = 6$
$3(x + y) = 7$

36. $0.3x - 2y = -28$
$0.8x + 2y = 28$

37. $\frac{1}{2}q - 4r = -2$
$\frac{1}{6}q - 4r = 10$

38. $\frac{1}{2}x + \frac{1}{3}y = -1$
$-\frac{1}{2}x + \frac{2}{3}y = 10$

39. **REASONING** At the end of a recent WNBA regular season, the difference of the number of wins and losses by the Phoenix Mercury was 12. The difference of the number of wins and two times the number of losses was 1. How many regular season games did the Phoenix Mercury play during that season?

40. **USE A MODEL** Marisol works for a florist that sells two types of bouquets, as shown at the right. On Monday, Marisol used 96 tulips to make the bouquets. On Tuesday, she used 192 tulips to make the same number of Spring Mix bouquets as Monday, but 3 times as many Garden Delight bouquets.

Seasonal Bouquets	
Spring Mix	12 tulips
Garden Delight	16 tulips

a. Write a system of equations that you can use to find how many bouquets of each type Marisol made. Describe what each variable represents.

b. Find the total number of tulips Marisol used to make Garden Delight bouquets on Monday and Tuesday. Explain your answer.

41. **USE A MODEL** Jeremy and Kendrick each bought snacks for their friends at a skating rink. The table shows the number of bags of popcorn and the number of plates of nachos each person bought, as well as the total cost of the snacks.

Name	Bags of Popcorn	Plates of Nachos	Total Cost
Jeremy	4	2	\$18.50
Kendrick	7	2	\$26.75

a. Write a system of equations that you can use to find the prices of the popcorn and the nachos. Describe what each variable represents.

b. Solve the system of equations and explain what your solution represents.

42. **USE A MODEL** The table shows the time Erin spent jogging and walking this weekend and the total distance she covered each day. Erin always jogs at the same rate and always walks at the same rate.

Day	Time Jogging	Time Walking	Total Distance
Saturday	15 min	30 min	3.5 mi
Sunday	1 h	30 min	8 mi

a. Write a system of equations that you can use to represent this situation. Describe what each variable represents.

b. Solve the system by elimination. Show your work. Then interpret the solution.

43. STRUCTURE Consider the system of equations $0.4x - 2y = 6$ and $0.8x + 2y = 0$.

a. What is the first step to solving the system of equations by elimination? Explain your reasoning.

b. What is the solution, as an ordered pair?

c. Graph the equations on a coordinate plane. Explain how you can use the graph to check your solution.

d. How would the solution change if the second equation was $x - 5y = 15$? Explain.

e. How would the solution change if the second equation was $0.4x - 2y = 0$? Explain.

Higher-Order Thinking Skills

44. ANALYZE Mikasi says that if you solve a system of equations using elimination by addition and the result is $0 = 0$, then the solution of the system is (0, 0). Provide a counterexample to show that his statement is false. Justify your argument.

45. ANALYZE Reece says that if you solve a system of equations using elimination by addition and the result is $0 = 2$, then the solution of the system is (0, 2). Provide a counterexample to show that her statement is false. Justify your argument.

46. CREATE Create a system of equations that can be solved by using addition to eliminate one variable. Formulate a general rule for creating such systems.

47. CREATE The solution of a system of equations is (−3, 2). One equation in the system is $x + 4y = 5$. Find a second equation for the system such that the system can be solved using elimination by addition. Explain how you derived this equation.

48. PERSEVERE The sum of the digits of a two-digit number is 8. The result of subtracting the units digit from the tens digit is −4. Define variables and write the system of equations that you would use to find the number. Then solve the system and find the number.

49. WRITE Describe when it would be most beneficial to use elimination to solve a system of equations.

Elimination Using Multiplication

Explore Graphing and Elimination Using Multiplication

Online Activity Use an interactive tool to complete the Explore.

INQUIRY How can you produce a new system of equations with the same solution as the given system?

Learn Solving Systems of Equations by Elimination with Multiplication

Key Concept • Elimination Method Using Multiplication	
Step 1	Multiply at least one equation by a constant to get two equations that contain opposite terms.
Step 2	Add the equations, eliminating one variable. Then solve the equation.
Step 3	Substitute the value from Step 2 into one of the equations and solve for the other variable. Write the solution as an ordered pair.

Example 1 Elimination Using Multiplication

Use elimination to solve the system of equations.

$$10x + 5y = 30$$
$$5x - 3y = -7$$

Step 1 Multiply an equation by a constant.

The coefficients of x will be opposites if the second equation is multiplied by -2.

$5x - 3y = -7$ Second equation.

$(-2)(5x - 3y) = (-2)(-7)$ Multiply each side by -2.

$-10 + 6y = 14$ Simplify.

Step 2 Add the equations.

$$10x + 5y = 30$$
$$\underline{(+) - 10x + 6y = 14}$$

$11y = 44$ The variable x is eliminated.

$\frac{11y}{11} = \frac{44}{11}$ Divide each side by 11.

$y = 4$ Simplify.

(continued on the next page)

Today's Goal

- Solve systems of equations by eliminating a variable using multiplication and addition.

Think About It!

How does the process of solving a system of equations by elimination using multiplication differ from elimination using just addition?

Study Tip

Common Factors If the coefficients of a variable are not the same, or are opposites, and they share a greatest common factor greater than 1, then that variable is the easiest to eliminate using multiplication. For example, the system in this example has two variables, x and y. The coefficients of the y-variable expressions are 5 and –3, which share no common factor greater than 1. However, the coefficients of the x-variable expressions are 10 and 5, which share a common factor of 5. Thus, the x-variable requires fewer steps to eliminate.

Step 3 Substitute 4 for y in either equation to find the value of x.

$10x + 5y = 30$	First equation
$10x + 5(4) = 30$	Replace y with 4.
$10x + 20 = 30$	Multiply.
$10x + 20 - 20 = 30 - 20$	Subtract 20 from each side.
$\frac{10x}{10} = \frac{10}{10}$	Divide each side by 10.
$x = 1$	Simplify.

The solution is (1, 4).

Talk About It!

Would you get the solution (1, 4) if you eliminated the y-variable instead of the x-variable? If no, explain your reasoning. If yes, explain why the y-variable was selected for elimination instead of the x-variable.

Check

Use elimination to solve the system of equations.

$13x + 14y = 59$

$4x + 7y = 37$

Example 2 Multiply Both Equations to Eliminate a Variable

Use elimination to solve the system of equations.

$3x + 4y = -22$
$-2x + 3y = -8$

Step 1 Multiply both equations by a constant.

$3x + 4y = -22$	Original equation	$-2x + 3y = -8$
$2(3x + 4y) = 2(-22)$	Multiply by a constant.	$3(-2x + 3y) = 3(-8)$
$2(3x) + 2(4y) = 2(-22)$	Distributive Property	$3(-2x) + 3(3y) = 3(-8)$
$6x + 8y = -44$	Simplify	$-6x + 9y = -24$

Step 2 Add the equations.

$6x + 8y = -44$

$\underline{(+) -6x + 9y = -24}$

$17y = -68$	The variable x is eliminated.
$\frac{17y}{17} = \frac{-68}{17}$	Divide each side by 17.
$y = -4$	Simplify.

Go Online You can complete an Extra Example online.

Step 3 Use substitution to find the value of x.

$3x + 4y = -22$	First equation
$3x + 4(-4) = -22$	Replace y with -4.
$3x - 16 = -22$	Multiply.
$3x - 16 + 16 = -22 + 16$	Add 16 to each side.
$\frac{3x}{3} = -\frac{6}{3}$	Divide each side by 3.
$x = -2$	Simplify.

The solution is $(-2, -4)$.

Check

Use elimination to solve the system of equations.

$11x - 6y = 25$

$3x + 9y = 60$

Math History Minute

German mathematician **Carl Friedrich Gauss (1777-1855)** contributed significantly to many fields, including number theory, algebra, and statistics. The elimination method is related to the Gaussian elimination method, an algorithm for solving systems of linear equations that was known to Chinese mathematicians as early as 179 B.C.

Example 3 Write and Solve a System Using Multiplication

COMICS **Jorge's comic book collection consists of single issues that cost \$4 each and paperback collections that cost \$12 each. He has 100 books in all. His collection cost him \$616. Write and solve a system of equations to determine how many single issues and paperbacks Jorge has in his collection.**

Complete the table to write the system of equations. Let c = the number of single issue comics and p = the number of paperback collections.

The number of single issue comics	plus	the number of paperback collections	is	100 books
c	+	p	=	100
\$4 per single issue comic	**plus**	**\$12 per paperback collection**	**is**	**\$616**
$4c$	+	$12p$	=	616

(continued on the next page)

Step 1 Multiply an equation by a constant.

$c + p = 100$	First equation
$-4(c + p) = -4(100)$	Multiply each side by –4.
$-4c - 4p = -4(100)$	Distributive Property
$-4c - 4p = -400$	Simplify.

Step 2 Add the equations.

$$\begin{array}{r} -4c - 4p = -400 \\ (+)4c + 12p = 616 \\ \hline \frac{8p}{8} = \frac{216}{8} \\ p = 27 \end{array}$$

The variable c is eliminated.

Divide each side by 8.

Simplify.

Step 3 Use substitution to find the value of c.

Substitute 27 for p in either equation to find the value of c.

$c + p = 100$	First equation.
$c + 27 = 100$	Replace p with 27.
$c + 27 - 27 = 100 - 27$	Subtract 27 from each side.
$c = 73$	Simplify.

Jorge has 73 single issue comics and 27 paperback collections.

Think About It!

Is (73, 27) a viable solution in the context of the situation? Explain your reasoning.

Check

SOFTWARE A software company releases two products: a home version of their photo editor, which costs $20, and a professional version, which costs $45. The company sells 1000 copies of the photo editing software, earning a total revenue of $38,075. Write and solve a system of equations to determine how many home versions and professional versions of the software the company sold.

Let h = the number of home versions sold and p = the number of professional versions sold.

Part A Write the system of equations.

Part B Solve the system of equations.

The software company sold __?__ home versions and __?__ professional versions of their photo editor.

Go Online to practice what you've learned about choosing a method to solve a system in the Put It All Together over Lessons 7-1 through 7-4.

Go Online You can complete an Extra Example online.

Practice

Go Online You can complete your homework online.

Examples 1 and 2

Use elimination to solve each system of equations.

1. $x + y = 2$
   $-3x + 4y = 15$

2. $x - y = -8$
   $7x + 5y = 16$

3. $x + 5y = 17$
   $-4x + 3y = 24$

4. $6x + y = -39$
   $3x + 2y = -15$

5. $2x + 5y = 11$
   $4x + 3y = 1$

6. $3x - 3y = -6$
   $-5x + 6y = 12$

7. $3x + 4y = 29$
   $6x + 5y = 43$

8. $8x + 3y = 4$
   $-7x + 5y = -34$

9. $8x + 3y = -7$
   $7x + 2y = -3$

10. $4x + 7y = -80$
    $3x + 5y = -58$

11. $12x - 3y = -3$
    $6x + y = 1$

12. $-4x + 2y = 0$
    $10x + 3y = 8$

Example 3

13. **SPORTS** The Fan Cost Index (FCI) tracks the average costs for attending sporting events, including tickets, drinks, food, parking, programs, and souvenirs. According to the FCI, a family of four would spend a total of $592.30 to attend two Major League Baseball (MLB) games and one National Basketball Association (NBA) game. The family would spend $691.31 to attend one MLB and two NBA games.

    a. Write a system of equations to find the family's costs for each kind of game according to the FCI.

    b. Solve the system of equations to find the cost for a family of four to attend each kind of game according to the FCI.

14. **ART** Mr. Santos, the curator of the children's museum, recently made two purchases of firing clay and polymer clay for a visiting artist to sculpt. Use the table to find the cost of each product per kilogram.

Firing Clay (kg)	Polymer Clay (kg)	Total Cost
5	24	$64.05
25	8	$51.45

    a. Write a system of equations to find the cost of each product per kilogram.

    b. Solve the system of equations to find the cost of each product per kilogram.

Mixed Exercises

15. Two times a number plus three times another number equals 13. The sum of the two numbers is 7. What are the numbers?

16. Four times a number minus twice another number is −16. The sum of the two numbers is −1. Find the numbers.

17. **FUNDRAISING** Trisha and Byron are washing and vacuuming cars to raise money for a class trip. Trisha raised \$38 by washing 5 cars and vacuuming 4 cars. Byron raised \$28 by washing 4 cars and vacuuming 2 cars. Find the amount they charged to wash a car and vacuum a car.

18. **STRUCTURE** Consider the system of equations $-2x + 3y = -5$ and $3x - 4y = 6$.
    a. Describe two different ways to solve the system by elimination.
    b. Explain why multiplying the first equation by 6 and the second equation by 5 and then adding is not useful for solving the system.
    c. Solve the system by elimination.

19. **REASONING** The owner of a juice stand wants to make a new juice drink. He would like to mix Tropical Breeze, t, and Kona Cooler, k, to make 10 quarts of a new drink that is 40% pineapple juice.

Juice Drinks	
Tropical Breeze	20% pineapple juice
Kona Cooler	50% pineapple juice

    a. Make a table to help write a system of equations that the owner of the juice stand can solve to determine the amount of each drink he should use to make the new drink.
    b. Solve the system and explain what your solution represents.
    c. Explain how you know your answer is correct.

20. **USE A MODEL** Marlene works as a cashier at a grocery store. At the end of the day, she has a total of 125 five-dollar bills and ten-dollar bills. The total value of these bills is \$990.
    a. Write a system of equations that you can use to find the number of five-dollar bills and the number of ten-dollars bills. Describe what each variable represents.
    b. Solve the system and explain what your solution represents.

21. **FIND THE ERROR** Jason and Daniela are solving a system of equations. Is either of them correct? Explain your reasoning.

Jason

$2r + 7t = 11$
$r - 9t = -7$
$2r + 7t = 11$
$(-)\ 2r = 18t = -14$
$25t = 25$
$t = 1$
$2r + 7t = 11$
$2r + 7(1) = 11$
$2r + 7 = 11$
$2r = 4$
$\frac{2r}{2} = \frac{4}{2}$
$r = 2$
The solution is (2, 1).

Daniela

$2r + 7t = 11$
$(-)\ r - 9t = -7$
$r = 18$
$2r + 7t = 11$
$2(18) + 7t = 11$
$36 + 7t = 11$
$7t = -25$
$\frac{7t}{7} = \frac{25}{7}$
$t = -3.6$
The solution is (18, -3.6).

22. **ANALYZE** Determine whether the following statement is *true* or *false*: *A system of linear equations will only have infinitely many solutions if the equations have the same coefficients.* Justify your argument.

23. **CREATE** Write a system of equations that can be solved by multiplying one equation by –3 and then adding the two equations together.

24. **PERSEVERE** The solution of the system $4x + 5y = 2$ and $6x - 2y = b$ is $(3, a)$. Find the values of a and b. Discuss the steps you used.

25. **WRITE** Why is substitution sometimes more helpful than elimination, and vice versa?

Systems of Inequalities

Today's Goal
- Solve systems of linear inequalities by graphing.

Today's Vocabulary
system of inequalities

Explore Solutions of Systems of Inequalities

Online Activity Use an interactive tool to complete the Explore.

INQUIRY How are the solutions of a system of inequalities represented on a graph?

Go Online You can watch a video to see how to solve a system of linear inequalities.

Learn Solving Systems of Inequalities by Graphing

A set of two or more inequalities with the same variables is a **system of inequalities**. The solution of a system of inequalities with two variables is the set of ordered pairs that satisfy all of the inequalities in the system. The solution is represented by the overlap, or intersection, of the graphs of the inequalities.

Example 1 Solve by Graphing

Solve the system of inequalities by graphing.

$x - 2y > -6$

$y < 3x$

Step 1 Graph one inequality of the system.
The boundary of $x - 2y > -6$ is dashed and is not included in the solution. The half-plane is shaded below the boundary, in yellow, to indicate solutions of $x - 2y > -6$.

Step 2 Graph the second inequality of the system.
The boundary of $y < 3x$ is dashed and is not included in the solution. The half-plane is shaded to the right of the boundary, in blue, to indicate solutions of $y < 3x$.

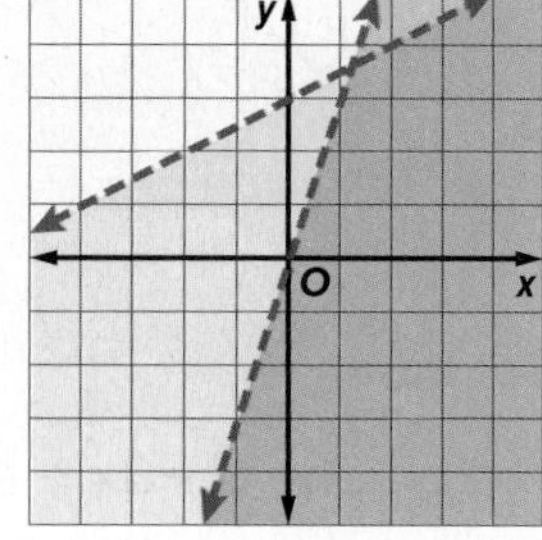

Solution
The solution of the system is the set of ordered pairs in the intersection of the graphs of $x - 2y > -6$ and $y < 3x$. The region is shaded green.

Think About It!
The boundaries for the system $y > 3$, $y \leq -2x + 1$ intersect at $(-1, 3)$. Is $(-1, 3)$ included in the solution? Explain.

Go Online You can watch a video to see how to use a graphing calculator with this example.

Go Online You can complete an Extra Example online.

Check

Graph the system of inequalities.

$\frac{1}{3}x + 2 < y$

$x \geq -3$

Example 2 Solve by Graphing, No Solution

Solve the system of inequalities by graphing.

$\mathbf{-3x + 4y > 0}$

$\mathbf{3x - 4y \geq 8}$

Step 1 Graph one inequality of the system.

The boundary of $-3x + 4y > 0$ is dashed and is not included in the solution. The half-plane is shaded above the boundary, in yellow, to indicate solutions of $-3x > 4y > 0$.

Step 2 Graph the second inequality of the system.

The boundary of $3x - 4y \geq 8$ is solid and is included in the solution. The half-plane is shaded below the boundary, in blue, to indicate solutions of $3x - 4y \geq 8$.

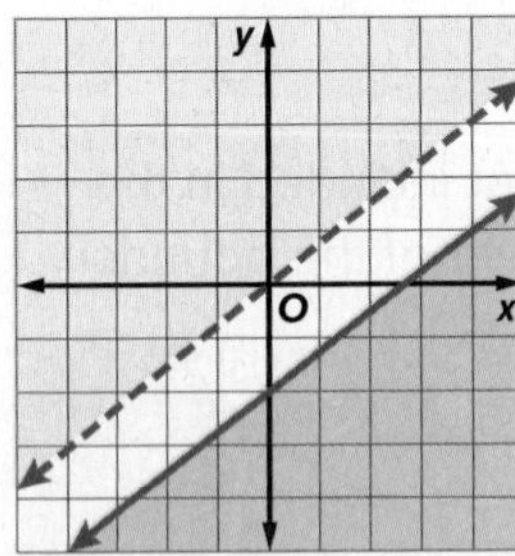

Solution

The graphs of $-3x + 4y = 0$ and $3x - 4y = 8$ are parallel lines. The regions do not intersect at any point, so the system has no solution.

Go Online You can complete an Extra Example online.

Talk About It!

Is it possible for a system of inequalities that has boundaries with different slopes to have no solution? Justify your argument.

Study Tip

Shaded Regions When graphing more than one region, it is helpful to use a different color of pencil or a different pattern for each region. This will make it easier to see where the regions intersect and find possible solutions.

Check

Graph the system of inequalities.

$x + 3 < y$

$3x - 3y \geq 12$

Example 3 Apply Systems of Inequalities

SEWING **A family and consumer sciences class is making pillows and blankets to donate to a local shelter. The class has 40 yards of fabric to use. Pillows require 1.25 yards of fabric and take 1 hour to make. Blankets use 4 yards of fabric and take 2.5 hours to make. The class has 28 hours of class time left for the semester. Determine the number of pillows and blankets the class can make for the shelter.**

Part A Define the variables, and write a system of inequalities to represent the situation.

Let p represent the number of pillows the class can make.
Let b represent the number of blankets the class can make.

Fabric and time are two constraints on the numbers of pillows and blankets the class can make.

Because pillows use 1.25 yards of fabric and blankets use 4 yards of fabric, the inequality that represents the fabric constraint is $1.25p + 4b \leq 40$.

Making a pillow takes 1 hour and making a blanket takes 2.5 hours. Because the class has only 28 hours left, the inequality that represents the time constraint is $p + 2.5b \leq 28$.

Part B Graph the system.

Part C Find a viable solution.

Only whole-number solutions make sense in this situation. One possible solution is (4, 15); 4 blankets and 15 pillows can be made.

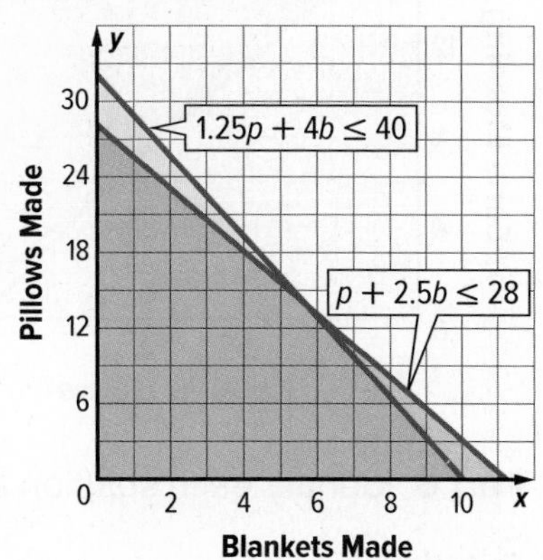

Think About It!

Can the class make 2 blankets and 24 pillows? Explain.

Think About It!

Could the graph be represented with "Pillows Made" as the *x*-axis and "Blankets Made" as the *y*-axis? Explain.

Go Online You can complete an Extra Example online.

Check

BAKERY Aisha can work up to 20 hours per week. Working at a bakery, she earns \$7 per hour most of the time and \$8.50 per hour during the early morning shift. Aisha needs to earn at least \$150 this week to pay for a trip with her friends. Determine the number of regular and early morning hours that Aisha could work.

Part A Select the correct system and graph. Let r = regular hours and m = early morning hours.

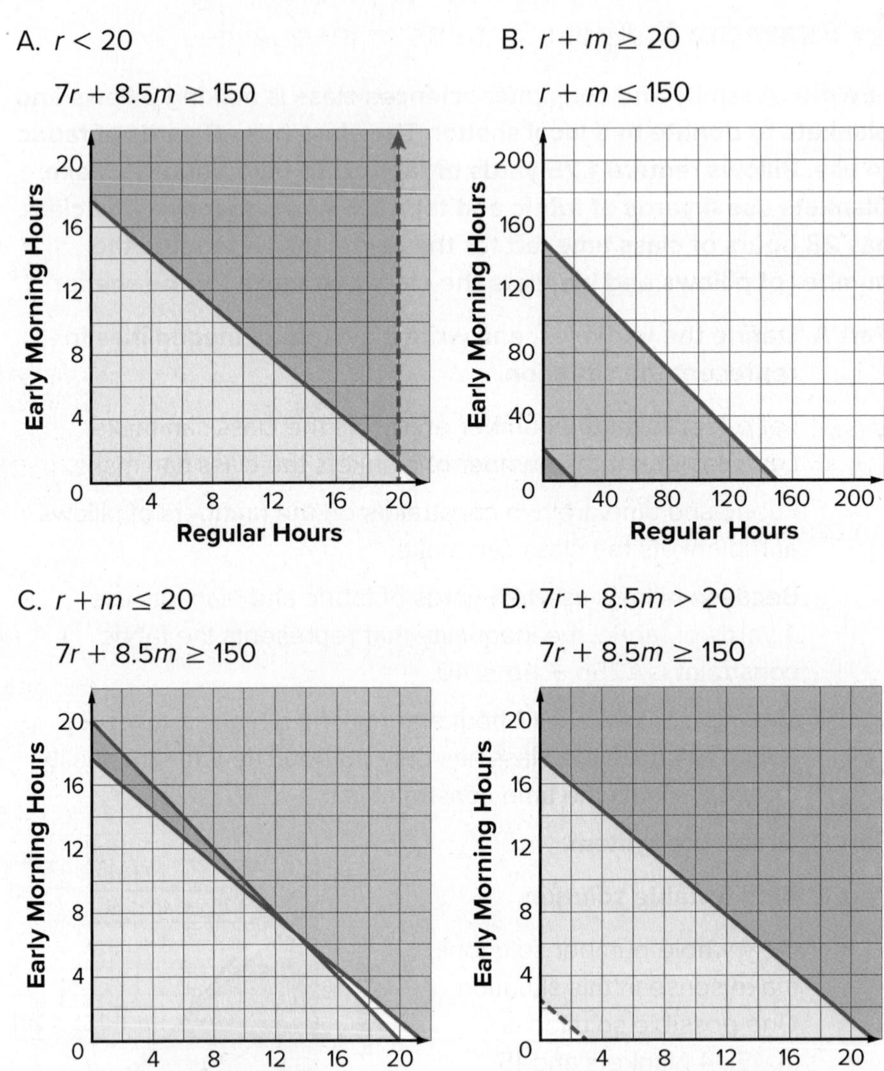

Part B Identify each solution as *viable* or *nonviable*.

Solutions

(2, 17)

(4, 7)

(5, 15)

(8, 13)

(10, 10)

(18, 6)

(21, 6)

Practice

Go Online You can complete your homework online.

Examples 1 and 2

Solve each system of inequalities by graphing.

1. $y < 6$
   $y > x + 3$

2. $y \geq 0$
   $y \leq x - 5$

3. $y \leq x + 10$
   $y > 6x + 2$

4. $y \geq x + 10$
   $y \leq x - 3$

5. $y < 5x - 5$
   $y > 5x + 9$

6. $y \geq 3x - 5$
   $3x - y > -4$

7. $x > -1$
   $y \leq -3$

8. $y > 2$
   $x < -2$

9. $y > x + 3$
   $y \leq -1$

10. $x < 2$
    $y - x \leq 2$

11. $x + y \leq -1$
    $x + y \geq 3$

12. $y - x > 4$
    $x + y > 2$

Example 3

13. **FITNESS** Diego started an exercise program in which each week he walks from 9 to 12 miles and works out at the gym from 4.5 to 6 hours.

    a. Write a system of inequalities to represent this situation. Define your variables.

    b. Graph the system.

    c. List three viable solutions.

14. **SOUVENIRS** Emiliana wants to buy turquoise stones on her trip to New Mexico to give to at least 4 of her friends. The gift shop sells stones for either \$4 or \$6 per stone. Emiliana has no more than \$30 to spend.

    a. Write a system of inequalities to represent this situation. Define your variables.

    b. Graph the system.

    c. List three viable solutions.

Mixed Exercises

Write a system of inequalities for each graph.

15. 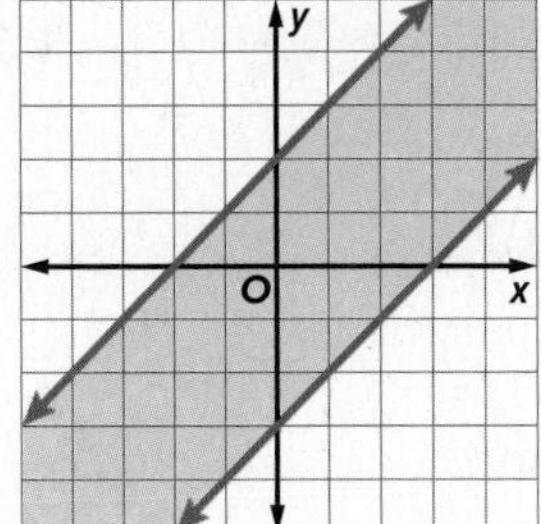

16. 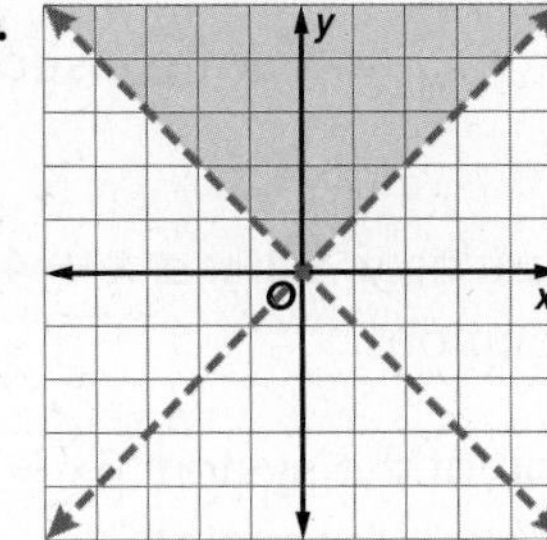

17. 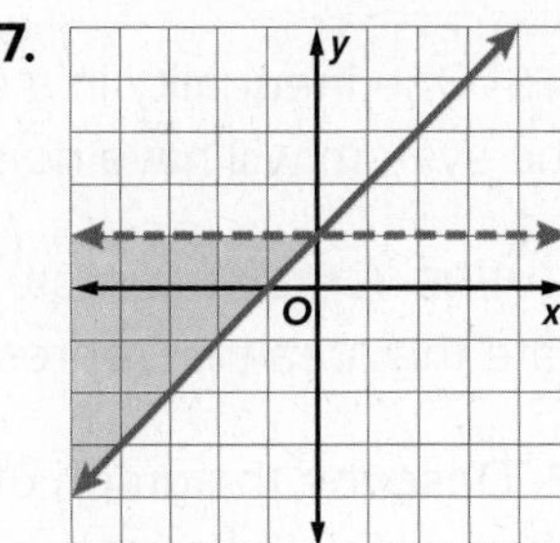

18. PRECISION Write a system of inequalities to represent the graph shown at the right.

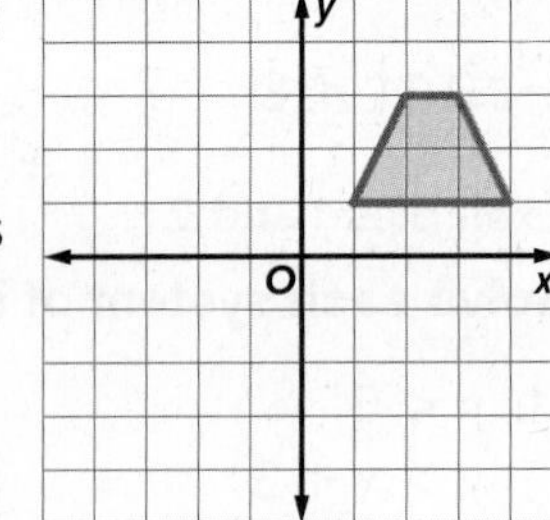

19. CONSTRUCT ARGUMENTS Is (2.5, 1) a solution of the system of inequalities $2x + 3y > 8$ and $4x - 5y \geq 2$? Justify your argument. Then explain how you can tell if the point is a solution without graphing the inequality.

20. PETS Priya's Pet Store never has more than a combined total of 20 cats and dogs and never more than 8 cats. This is represented by the inequalities $x + y \leq 20$ and $x \leq 8$. Represent the number of cats and dogs that can be at the store on a graph. Solve the system of inequalities by graphing.

21. FUNDRAISING The baseball team plans to sell tins of popcorn and peanuts as a fundraiser. The players have \$900 to spend on products and can order up to 200 tins. They want to order at least as many tins of popcorn as tins of peanuts. A tin of popcorn costs \$3, and a tin of peanuts costs \$4. Define the variables and write a system of inequalities to represent this situation. Then list any constraints for the variables.

22. BUSINESS For maximum efficiency, a factory must have at least 100 workers, but no more than 200 workers on a shift. The factory also must manufacture at least 30 units per worker.

    a. Let x be the number of workers and let y be the number of units. Write a system of inequalities expressing the conditions in the problem.

    b. Graph the systems of inequalities.

    c. Find three possible solutions.

23. DESIGN LaShawn designs Web sites for local businesses. He charges \$25 an hour to build a Web site and charges \$15 an hour to update Web sites once he builds them. He wants to earn at least \$100 every week, but he does not want to work more than 6 hours each week. What is a possible number of hours LaShawn can spend each week building Web sites x and updating Web sites y that will allow him to attain his goals? Write your answer as an ordered pair.

Higher-Order Thinking Skills

24. PERSEVERE Create a system of inequalities equivalent to $|x| \leq 4$.

25. ANALYZE State whether the following statement is *sometimes, always,* or *never* true. Justify your argument.

    Systems of inequalities with parallel boundaries have no solutions.

26. CREATE One inequality in a system is $3x - y > 4$. Write a second inequality so that the system will have no solution.

27. PERSEVERE Graph the system of inequalities $y \geq 1$, $y \leq x + 4$, and $y \leq -x + 4$. Estimate the area that represents the solution.

28. WRITE Describe the graph of the solution of the system $6x - 3y \leq -5$ and $6x - 3y \geq -5$ without graphing. Explain your reasoning.

Review

Essential Question

How are systems of equations useful in the real world?

Writing and solving systems of equations can help you find unknown values in real-world situations.

Module Summary

Lesson 7-1

Graphing Systems of Equations

- When you solve a system of equations with $y = f(x)$ and $y = g(x)$, the solution is an ordered pair that satisfies both equations. Thus, the x-coordinate of the intersection of $y = f(x)$ and $y = g(x)$ is the value of x where $f(x) = g(x)$.
- A system of equations is consistent if it has at least one ordered pair that satisfies both equations.
- A system of equations is independent if it has exactly one solution.
- A system of equations is dependent if it has an infinite number of solutions.
- A system of equations is inconsistent if it has no ordered pair that satisfies both equations.

Lessons 7-2 through 7-4

Solving Systems of Equations Algebraically

- To use the substitution method, solve at least one equation for one variable. Substitute the resulting expression into the other equation to replace the variable. Then solve the equation. Substitute this value into either equation, and solve for the other variable. Write the solution as an ordered pair.
- To use the elimination method, write the system so like terms with opposite coefficients are aligned. Add or subtract the equations, eliminating one variable. Then solve the equation. Substitute this value into one of the equations and solve for the other variable. Write the solution as an ordered pair. You may need to multiply at least one equation by a constant to get two equations that contain opposite terms.

Lesson 7-5

Systems of Inequalities

- A set of two or more inequalities with the same variables is called a system of inequalities.
- The solution of a system of inequalities with two variables is the set of ordered pairs that satisfy all of the inequalities in the system. The solution is represented by the overlap, or intersection, of the graphs of the inequalities.

Study Organizer

Foldables

Use your Foldable to review this module. Working with a partner can be helpful. Ask for clarification of concepts as needed.

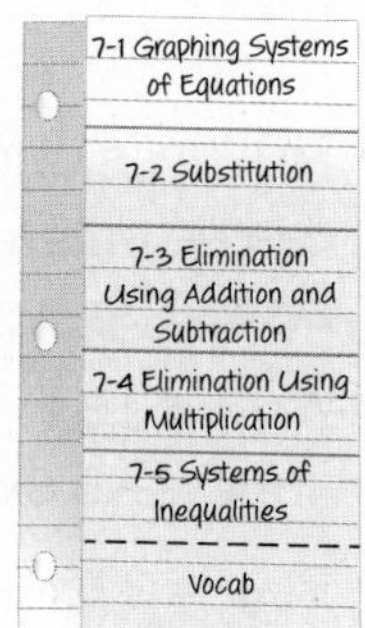

Test Practice

1. MULTI-SELECT Use the graph. Which systems of equations are consistent and independent? (Lesson 7-1)

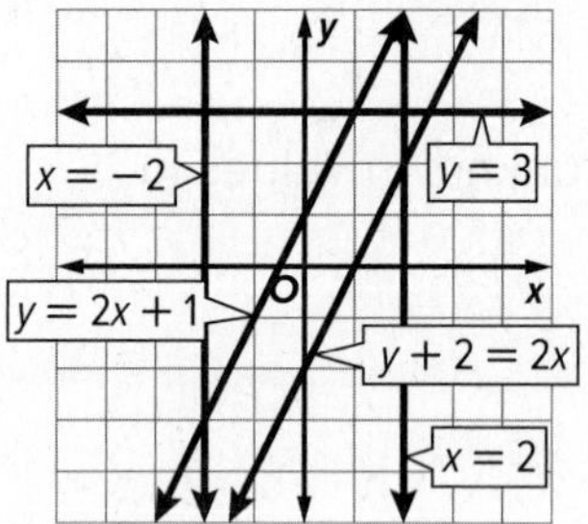

A. $y = 2x + 1$
   $y + 2 = 2x$

B. $y = 2x + 1$
   $y = 3$

C. $y = 3$
   $x = -2$

D. $x = 2$
   $y + 2 = 2x$

E. $x = -2$
   $x = 2$

2. OPEN RESPONSE Consider the system of equations. (Lesson 7-1)

$8x + 2y = 8$
$y = -4x + 4$

How many solutions are there for the system? Is the system dependent or independent?

3. MULTIPLE CHOICE Which system of equations can be entered into a graphing calculator to solve $3.5x + 18 = -5.8x + 30$? (Lesson 7-1)

A. $y = 3.5x$
   $y = -5.8x$

B. $y = 3.5x + 18$
   $y = -5.8x + 30$

C. $0 = 3.5x + 18$
   $0 = -5.8x + 30$

D. $y = 9.3x - 12$

4. MULTIPLE CHOICE Use a system of equations and a graphing calculator to solve $6.9x + 4.3 = 4.7x + 8$. Round your answer to the nearest hundredth, if necessary. (Lesson 7-1)

A. 1.06

B. 1.68

C. 2.14

D. 5.59

5. OPEN RESPONSE Taylan is selling plastic and wooden frames. He sold 7 total frames. The number of plastic frames Taylan sold was 5 less than twice the number of wooden frames. How many of each type of frame did Taylan sell? (Lesson 7-2)

6. MULTIPLE CHOICE Consider the system of equations.

$3x - 2y = 0$
$x + y = 10$

What is the solution of the system? (Lesson 7-2)

A. The solution to the system is (20, −10).

B. The solution to the system is (3, 7).

C. The solution to the system is (4, 6).

D. The solution to the system is (6, 4).

7. OPEN RESPONSE Determine whether the system has *no solution, one solution,* or *infinitely many solutions*. If the system has one solution, name it. (Lesson 7-2)

$x + y = 5$
$3x + 2y = 8$

8. OPEN RESPONSE The sum of the measures of two complementary angles is 90 degrees. Angles P and Q are complementary, and the measure of angle P is 6 degrees more than twice the measure of angle Q.

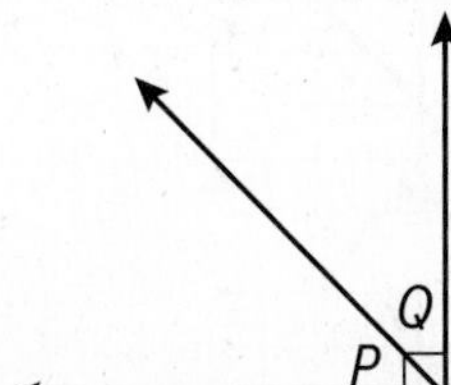

Write a system of equations and use substitution to find the measure of angles P and Q. (Lesson 7-2)

9. OPEN RESPONSE Solve the system of equations. (Lesson 7-3)

$3x + y = 34$

$0.5x - y = 1$

10. MULTIPLE CHOICE A rectangle is x inches wide and $3y$ inches long. The sum of the length and width is 36 inches and the difference between the length and twice the width is 12 inches. Find the length and width. (Lesson 7-3)

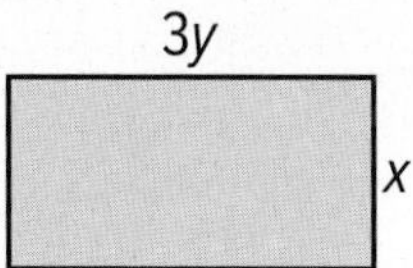

A. width: 8 inches; length: 28 inches

B. width: 8 inches; length: 9.3 inches

C. width: 12 inches; length: 24 inches

D. width: 15 inches; length: 21 inches

11. MULTI-SELECT Select all of the ways the system can be solved. (Lesson 7-4)

$9x - 2y = 4$
$3x + 8y = -12$

A. Multiply the first equation by 4, then add the equations.

B. Multiply the second equation by 3, then subtract the equations.

C. Multiply the first equation by 3, then add the equations.

D. Multiply the second equation by 3, then add the equations.

E. Multiply the first equation by 3, then subtract the equations.

12. OPEN RESPONSE Solve the system of equations. (Lesson 7-4)

$2x + 5y = 5$
$3x + 4y = -3$

13. OPEN RESPONSE Solve the system of equations. (Lesson 7-4)

$2r - t = 7$
$r - t = 1$

14. OPEN RESPONSE It takes 3 hours to paddle a kayak 12 miles downstream and 4 hours for the return trip upstream. Find the rate of the kayak in still water.

Let k = the rate of the kayak in still water and c = the rate of the current. (Lesson 7-4)

	r	t	d	$rt = d$
Downstream	$k + c$	3	12	$3(k + c) = 12$
Upstream	$k - c$	4	12	$4(k - c) = 12$

15. MULTIPLE CHOICE The graph shows the solution to the given system of inequalities. (Lesson 7-5)

$-x + 2y \leq 1$
$-3x + 2y \geq 2$

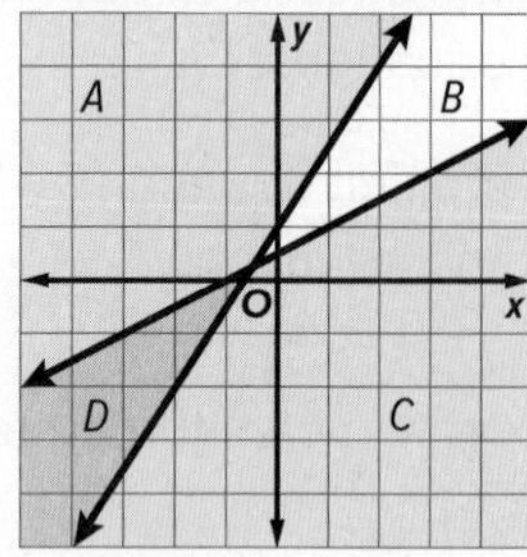

In what region is the solution set?

A. A

B. B

C. C

D. D

16. MULTIPLE CHOICE Which graph represents the solution of the system of inequalities? (Lesson 7-5)

$x - y \geq 2$
$2x + y > -3$

A. 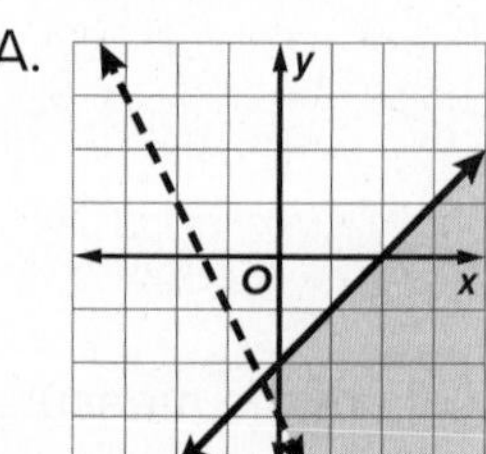

B. 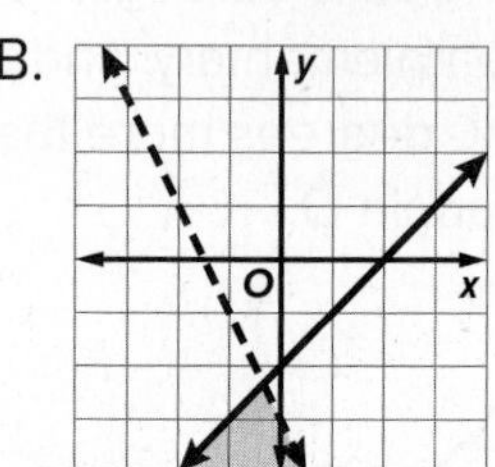

C. 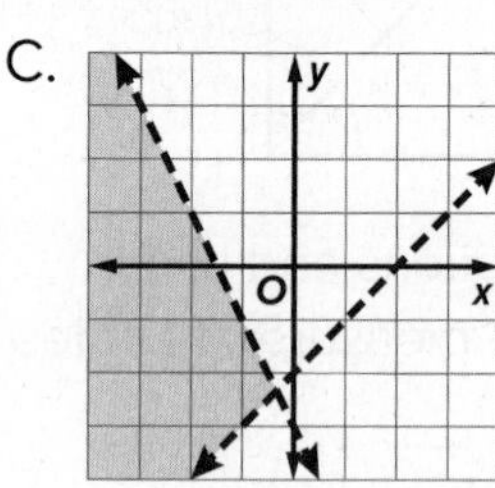

D. 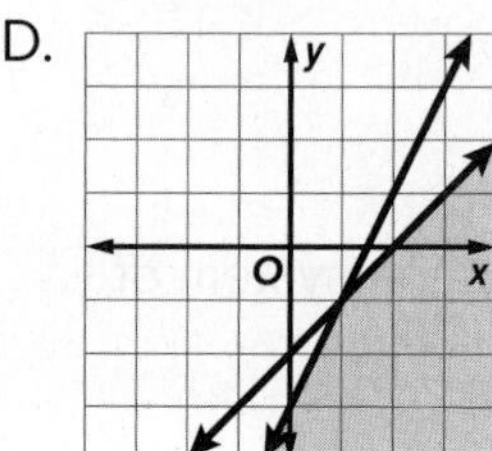

17. MULTIPLE CHOICE Diana wants to build a rectangular pen for her goats. The length of the pen should be at least 50 feet, and the perimeter of the pen should be no more than 190 feet. What is a viable solution for the dimensions of the pen? (Lesson 7-5)

A. 29 feet by 40 feet

B. 29 feet by 76 feet

C. 29 feet by 65 feet

D. 29 feet by 29 feet

Module 8

Exponential Functions

Essential Question

When and how can exponential functions represent real-world situations?

What Will You Learn?

How much do you already know about each topic **before** starting this module?

KEY

— I don't know. — I've heard of it. — I know it!

	Before			After		
	I don't know.	I've heard of it.	I know it!	I don't know.	I've heard of it.	I know it!
graph exponential growth functions						
graph exponential decay functions						
translate exponential functions						
dilate exponential functions						
reflect exponential functions						
solve problems involving exponential growth and decay						
transform exponential expressions						
generate geometric sequences						
write recursive formulas						
translate between recursive and explicit formulas						

Foldables Make this Foldable to help you organize your notes about exponential functions. Begin with a sheet of 11″ × 17″ paper and six index cards.

1. **Fold** lengthwise about 3″ from the bottom.
2. **Fold** the paper in thirds.
3. **Open** and staple the edges on either side to form three pockets.
4. **Label** the pockets as shown. Place two index cards in each pocket.

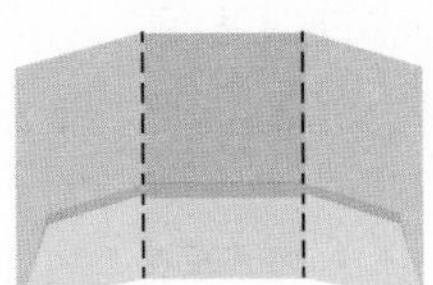

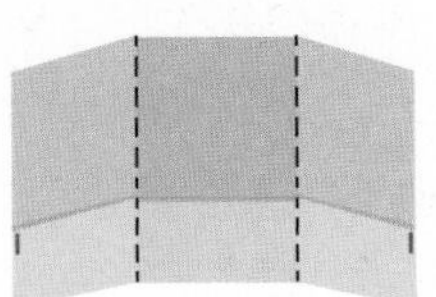

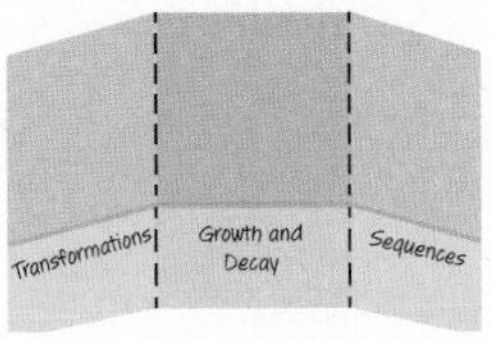

What Vocabulary Will You Learn?

- asymptote
- common ratio
- compound interest
- explicit formula
- exponential decay functions
- exponential function
- exponential growth function
- geometric sequence
- recursive formula

Are You Ready?

Complete the Quick Review to see if you are ready to start this module. Then complete the Quick Check.

Quick Review

Example 1

Evaluate $2x^3$ for $x = 5$.

$2x^3$	Original expression
$= 2(5)^3$	Substitute 5 for x.
$= 2(125)$	Evaluate the exponent.
$= 250$	Multiply.

Example 2

Divide $\frac{5}{6} \div \frac{1}{3}$.

$\frac{5}{6} \div \frac{1}{3} = \frac{5}{6} \cdot \frac{3}{1}$	Multiply by the reciprocal.
$= \frac{15}{6}$	Multiply the numerators and multiply the denominators.
$= \frac{5}{2}$	Find the simplest form.

Quick Check

Evaluate each expression for the given value.

1. $-4x^2$ for $x = 7$

2. $3x^2$ for $x = 3$

3. $0.25x^4$ for $x = 1$

4. x^5 for $x = 3$

Divide.

5. $128 \div 4$

6. $\frac{1}{3} \div 2$

7. $-9 \div 3$

8. $\frac{1}{8} \div \frac{1}{2}$

How did you do?

Which exercises did you answer correctly in the Quick Check?

Exponential Functions

Explore Exponential Behavior

Online Activity Use a table to complete the Explore.

INQUIRY How does exponential behavior differ from linear behavior?

Today's Goals

- Recognize situations modeled by linear or exponential functions.
- Graph exponential functions, showing intercepts and end behavior.

Today's Vocabulary

exponential function

exponential growth function

exponential decay function

asymptote

Learn Identifying Exponential Behavior

An **exponential function** is a function of the form $y = ab^x$, where $a \neq 0$, $b > 0$, and $b \neq 1$. Some examples of exponential functions are $y = 2(3)^x$, $y = 4^x$, and $y = \left(\frac{1}{2}\right)^x$.

Linear

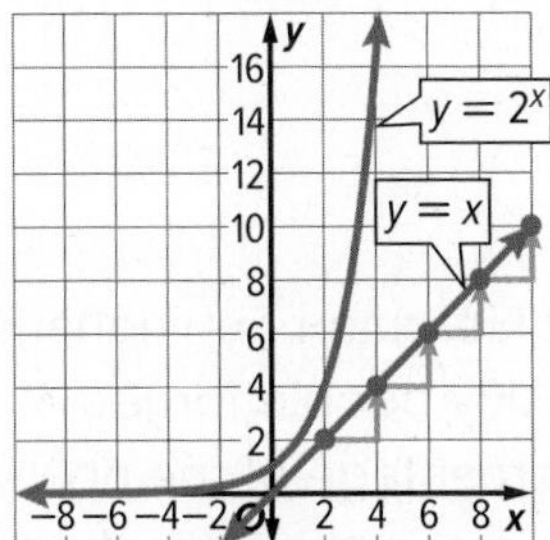

the rate of change remains constant

Exponential

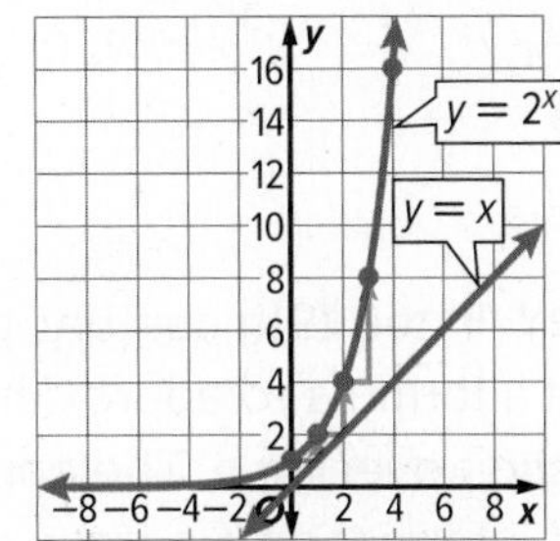

the rate of change increases by the same factor

Note that in the linear function, the rate of change remains constant. In the exponential function, the rate of change increases by the same factor.

Example 1 Identify Exponential Behavior

Think About It!
Is a function that relates an independent variable to a change in order of magnitude always exponential? Explain your reasoning.

EARTHQUAKES **The Richter Scale measures the energy that an earthquake releases and assigns a magnitude to it. These orders of magnitude can be approximated by comparing them to the explosive power of TNT. Determine whether the set of data displays exponential behavior.**

Magnitude	TNT (tons)
1	0.6
2	6
3	60
4	600
5	6000
6	60,000
7	600,000
8	6,000,000

Magnitudes 1 and 2

As the order of magnitude increases from 1 to 2, the amount of TNT that is approximately equal in magnitude increases from 0.6 tons to 6 tons. That is an increase by a factor of 10.

Magnitudes 2 and 3

As the order of magnitude increases from 2 to 3, the amount of TNT that is approximately equal in magnitude increases from 6 tons to 60 tons. That is an increase by a factor of 10.

Think About It!
Can you confirm that the entire data set displays exponential behavior by looking at the first two intervals?

Since the change in the amount of TNT increases by the same factor given an equal change in magnitude, the data set displays exponential behavior.

Check

MEMORY In the 19th century, psychologist Hermann Ebbinghaus created a formula to approximate how quickly people forget information over time. The approximate percentage of the newly learned information a person retains over time is shown in the table. Determine whether the data displays exponential behavior.

Time (days)	% Retained
0	100
1	80
2	64
3	51.2
4	40.96
5	32.768

Source: Indiana University

Go Online
An alternate method is available for this example.

The data set ___?___ display exponential behavior.

Explore Restrictions on Exponential Functions

Online Activity Use graphing technology to complete the Explore.

 **INQUIRY** Why are exponential functions defined such that $a \neq 0$, $b > 0$, and $b \neq 1$?

Learn Graphing Exponential Functions

Functions of the form $f(x) = ab^x$, where $a > 0$ and $b > 1$, are called **exponential growth functions**. Functions of the form $f(x) = ab^x$, where $a > 0$ and $0 < b < 1$, are called **exponential decay functions**.

The graphs of exponential functions have an **asymptote**. An asymptote is a line that a graph approaches.

Go Online You can watch a video to see how to graph exponential functions.

Key Concept • Types of Exponential Functions

Exponential Growth Functions	Exponential Decay Functions
Equation	
$f(x) = ab^x$, $a > 0$, $b > 1$	$f(x) = ab^x$, $a > 0$, $0 < b < 1$
Domain, Range	
D = all real numbers; R = $\{y > 0\}$	D = all real numbers; R = $\{y > 0\}$
Intercepts	
one y-intercept, no x-intercepts	one y-intercept, no x-intercepts
End Behavior	
as x increases, $f(x)$ increases; as x decreases, $f(x)$ approaches 0	as x increases, $f(x)$ approaches 0; as x decreases, $f(x)$ increases
Graph	
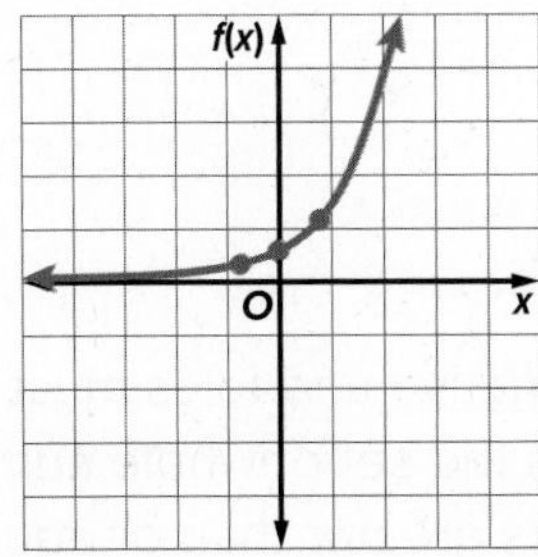	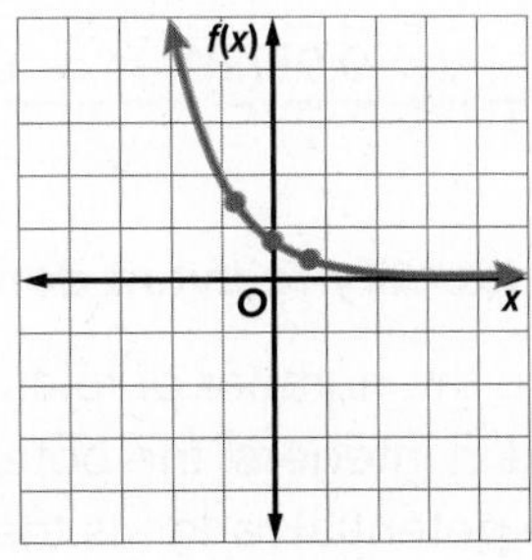

Talk About It!

How can you determine the y-intercept without substituting into the original equation? (Hint: Consider what $x = 0$ and the y-intercept mean in the context of the situation.)

Problem-Solving Tip

Make an Organized List Making an organized list of x-values and corresponding y-values is helpful in graphing the function. It can also help you identify patterns in the data.

Go Online

You can watch a video to see how to use a graphing calculator with this example.

Example 2 Exponential Growth Function

FOLDING **Each time you fold a piece of paper in half, it doubles in thickness. If a piece of paper is 0.05 millimeter thick, then you can determine the thickness y of a piece of paper given the number of folds x with the function $y = 0.05(2)^x$. Identify the key features of the function, graph it, and then identify the relevant domain and range in the context of the situation.**

Part A Identify key features.

Because $a > 0$ and $b > 1$, $y = 0.05(2)^x$ is an exponential growth function. The domain is all real numbers and the range is $y > 0$.

The y-intercept is the value of y when $x = 0$.

$$y = 0.05(2)^0$$
$$y = 0.05(1)$$
$$= 0.05$$

The y-intercept is 0.05.

Because $y = 0.05(2)^x$ is an exponential growth function, as x increases, y increases, and as x decreases, x approaches 0.

Part B Graph the function.

Make a table of values. Round to the nearest unit. Then, plot the points and draw a curve to approximate it.

x	$y = 0.05(2)^x$	y
−2	$y = 0.05(2)^{-2}$	0.0125
−1	$y = 0.05(2)^{-1}$	0.025
0	$y = 0.05(2)^0$	0.05
1	$y = 0.05(2)^1$	0.1
2	$y = 0.05(2)^2$	0.2
3	$y = 0.05(2)^3$	0.4
4	$y = 0.05(2)^4$	0.8

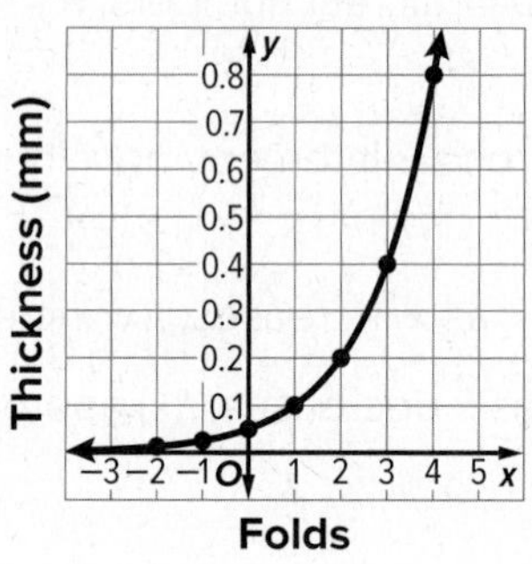

Part C Identify relevant domain and range.

Because the number of folds cannot be negative and folds must be counted in integers, the potential domain is the set of whole numbers and the potential range is the set of integers greater than or equal to 0.05. However, because the paper cannot be folded indefinitely, the thickness of the paper cannot continue to grow to infinity. So the domain will be restricted to the greatest possible number of folds, and the range will be restricted to the greatest thickness of the paper.

Go Online You can complete an Extra Example online.

Check

Consider $y = 3^x$.

Part A

List the key features that apply to $y = 3^x$. Include the domain, range, y-intercept, and end behavior of the function.

FILM The function $y = 3^x$ can be used to model a real-world situation. Sarah wants to crowdfund a film project. To spread the word, she shares the page with 3 friends, and requests that each friend share it with 3 more friends. The function that models the number of people with a link to the crowdfunding page y given the number of cycles x is defined by the function $y = 3^x$.

Part B

Select the correct graph of $y = 3^x$.

A.

B.

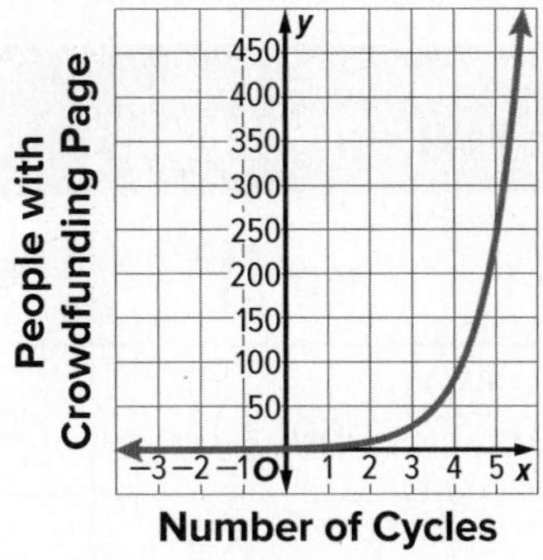

C.

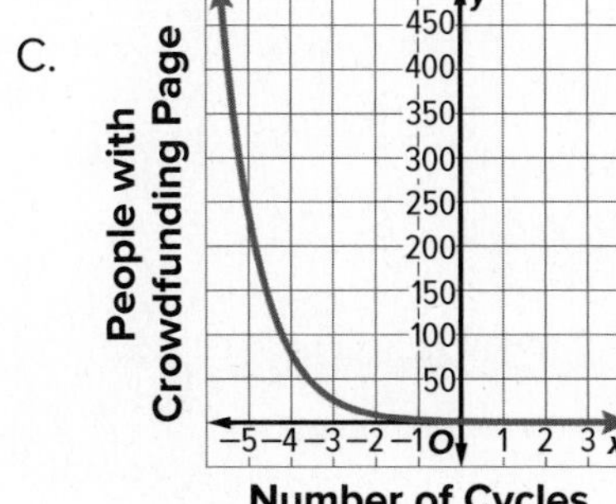

D.

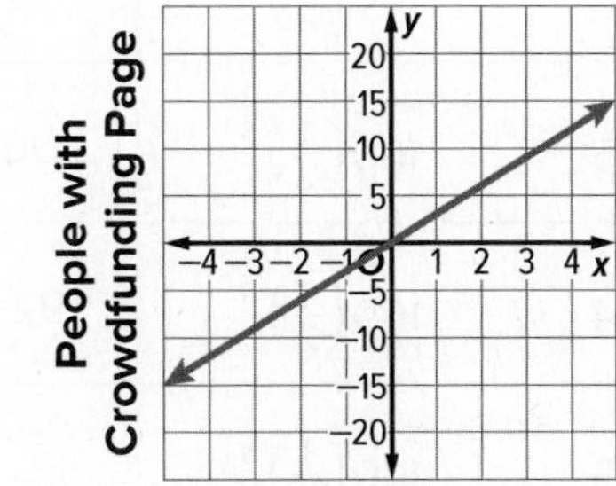

Part C

Describe the relevant range in the context of the situation.

Math History Minute

While a junior in high school, **Britney Gallivan (1985-)** showed that a single length of toilet paper 4000 feet long can be folded in half a maximum of twelve times. Britney's proof involved the exponential equation $L = \frac{\pi t}{6}(2^n + 4)(2^n - 1)$, where L is the length of the paper, t is the thickness of the material to be folded, and n represents the number of folds desired.

Use a Source

Research the amount of caffeine in another caffeinated drink. Write a function that models the amount of caffeine left in your system after x hours, and identify the key features of the function.

Example 3 Exponential Decay Function

CAFFEINE **The half-life of a substance describes how long it takes for the substance to deplete by half. The half-life of caffeine in the body of a healthy adult is approximately 5 hours, meaning that it takes 5 hours for the body to break down half of the caffeine. Suppose an energy drink contains 160 milligrams of caffeine. The amount of caffeine y left in your system after x hours is modeled by the function $y = 160\left(\frac{1}{2}\right)^{\frac{x}{5}}$. Identify the key features of the function, graph it, and then identify the relevant domain and range in the context of the situation.**

Part A Identify key features.

Because $a > 0$ and $0 < b < 1$, $y = 160\left(\frac{1}{2}\right)^{\frac{x}{5}}$ is an exponential decay function. The domain is all real numbers and the range is $y > 0$.

The y-intercept is the value of y when $x = 0$.

$$y = 160\left(\frac{1}{2}\right)^{\frac{0}{5}}$$

$$= 160(1) = 160$$

The y-intercept is 160.

Part B Graph the function.

x	$160\left(\frac{1}{2}\right)^{\frac{x}{5}}$	y
−1	$160\left(\frac{1}{2}\right)^{\frac{-1}{5}}$	184
0	$160\left(\frac{1}{2}\right)^{\frac{0}{5}}$	160
1	$160\left(\frac{1}{2}\right)^{\frac{1}{5}}$	139
2	$160\left(\frac{1}{2}\right)^{\frac{2}{5}}$	121
3	$160\left(\frac{1}{2}\right)^{\frac{3}{5}}$	106
4	$160\left(\frac{1}{2}\right)^{\frac{4}{5}}$	92
5	$160\left(\frac{1}{2}\right)^{\frac{5}{5}}$	80

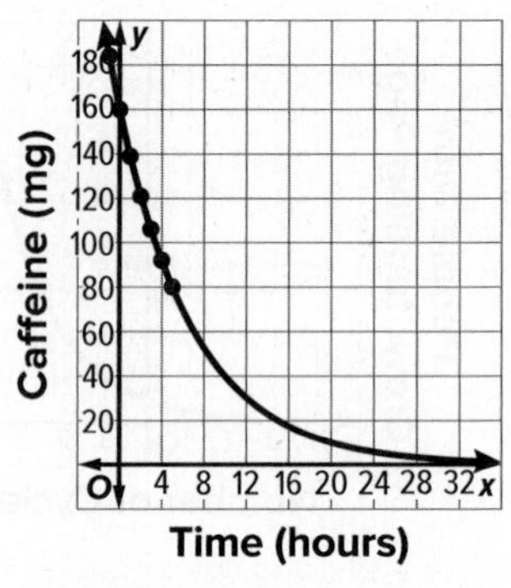

Part C Identify relevant domain and range.

Because time cannot be negative, the relevant domain is $\{x \geq 0\}$. Because the amount of caffeine cannot be negative and the amount of caffeine when $x = 0$ is 160 mg, the relevant range is $\{0 < y \leq 160\}$.

Go Online You can complete an Extra Example online.

Practice

Go Online You can complete your homework online.

Example 1

Determine whether the set of data displays exponential behavior. Write *yes* or *no*. Explain why or why not.

1.

x	−3	−2	−1	0
y	9	12	15	18

2.

x	0	5	10	15
y	20	10	5	2.5

3.

x	4	8	12	16
y	20	40	80	160

4.

x	50	30	10	−10
y	90	70	50	30

5. PICTURE FRAMES Since a picture frame includes a border, the picture must be smaller in area than the entire frame. The table shows the relationship between picture area and frame length for a particular line of frames. Is this an exponential relationship? Explain.

Frame Length (in.)	Picture Area (in^2)
5	6
6	12
7	20
8	30
9	42

Examples 2 and 3

6. WASTE Suppose the waste generated by nonrecycled paper and cardboard products in tons y after x days can be approximated by the function $y = 1000(2)^{0.3x}$.

a. Identify key features.

b. Graph the function.

c. Identify relevant domain and range.

7. IODINE Iodine 131 is a radioisotope that is related to nuclear energy, medical diagnostic and treatment procedures, and natural gas production. A scientist is testing 50 milligrams of Iodine 131. The scientist knows that the half-life of Iodine 131 is about 8.02 days. The function $y = 50\left(\frac{1}{2}\right)^{\frac{x}{8.02}}$ represents the amount of Iodine 131 remaining in milligrams y after x days.

a. Identify key features.

b. Graph the function.

c. Identify relevant domain and range.

8. DEPRECIATION Suppose a company's computer equipment is decreasing in value according to the function $y = 4000(0.87)^x$. In the equation, x represents the number of years that have elapsed since the equipment was purchased and y represents the value in dollars. What was the value 5 years after the computer equipment was purchased? Round your answer to the nearest dollar.

Mixed Exercises

MODELING Graph each function. Find the *y*-intercept and state the domain, range, and the equation of the asymptote. See margin for *y*-intercept, domain, range, and asymptote.

9. $y = 2\left(\frac{1}{6}\right)^x$

10. $y = \left(\frac{1}{12}\right)^x$

11. $y = -3(9^x)$

12. $y = -4(10^x)$

13. $y = 3(11^x)$

14. $y = 4^x + 3$

15. **METEOROLOGY** The atmospheric pressure in millibars at altitude x meters above sea level can be approximated by the function $f(x) = 1038(1.000134)^{-x}$ when x is between 0 and 10,000.

a. What is the atmospheric pressure at sea level?

b. The McDonald Observatory in Texas is at an altitude of 2000 meters. What is the approximate atmospheric pressure there?

c. As altitude increases, what happens to atmospheric pressure?

Higher-Order Thinking Skills

16. **PERSEVERE** Use tables and graphs to compare and contrast an exponential function $f(x) = ab^x + c$, where $a \neq 0$, $b > 0$, and $b \neq 1$, and a linear function $g(x) = ax + c$. Include intercepts, symmetry, end behavior, extrema, and intervals where the functions are increasing, decreasing, positive, or negative.

17. **CREATE** Write an exponential function that passes through (0, 3) and (1, 6).

18. **ANALYZE** Determine whether the graph of $y = ab^x$, where $a \neq 0$, $b > 0$, and $b \neq 1$, *sometimes, always,* or *never* has an x-intercept. Justify your argument.

19. **WRITE** Find an exponential function that represents a real-world situation, and graph the function. Analyze the graph, and explain why the situation is modeled by an exponential function rather than a linear function.

Transformations of Exponential Functions

Explore Translating Exponential Functions

 **Online Activity** Use graphing technology to complete the Explore.

INQUIRY What effect does adding to or subtracting from a function before or after it has been evaluated have on the function?

Learn Translations of Exponential Functions

Key Concept • Vertical Translations of Exponential Functions

- The graph of $g(x) = b^x + k$ is the graph of $f(x) = b^x$ translated vertically.
- If $k > 0$, the graph of $f(x)$ is translated k units up.
- If $k < 0$, the graph of $f(x)$ is translated $|k|$ units down.

Key Concept • Horizontal Translations of Exponential Functions

- The graph of $g(x) = b^{x-h}$ is the graph of $f(x) = b^x$ translated horizontally.
- If $h > 0$, the graph of $f(x)$ is translated h units right.
- If $h < 0$, the graph of $f(x)$ is translated $|h|$ units left.

Example 1 Vertical Translations of Exponential Functions

Describe the translation in $g(x) = 2^x + 3$ as it relates to the graph of the parent function $f(x) = 2^x$.

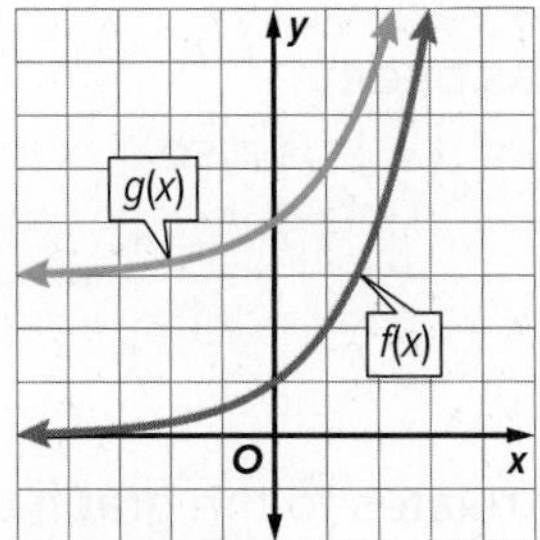

The constant k is added to the function after it has been evaluated, so k affects the output values.

The value of k is greater than 0, so the graph of $f(x) = 2^x$ is translated 3 units up.

Today's Goals
- Apply translations to exponential functions.
- Apply dilations to exponential functions.
- Apply reflections to exponential functions.
- Use transformations to identify exponential functions from graphs and write equations of exponential functions.

Go Online
You can watch a video to see how to describe translations of functions.

Think About It!
For $h < 0$, why must you move $|h|$ units left instead of h units?

Think About It!
What do you notice about the asymptote of a vertically translated exponential function compared to the asymptote of the parent function?

Example 2 Horizontal Translations of Exponential Functions

Think About It!

What do you notice about the asymptote of a horizontally translated exponential function compared to the asymptote of the parent function?

Describe the translation in $g(x) = 3^{x+1}$ as it relates to the graph of the parent function $f(x) = 3^x$.

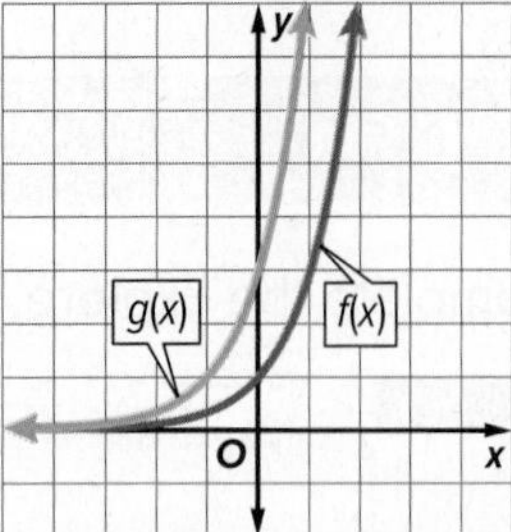

The constant h is subtracted from x before the function is performed, so h affects the input values.

The value of h is less than 0, so the graph of $f(x) = 3^x$ is translated 1 unit left.

Example 3 Multiple Translations of Exponential Functions

Think About It!

How can the placement of the constant tell you if the resulting transformation will be a vertical or horizontal translation?

Describe the translation in $g(x) = \left(\frac{1}{2}\right)^{x-2} - 4$ as it relates to the graph of the parent function $f(x) = \left(\frac{1}{2}\right)^x$.

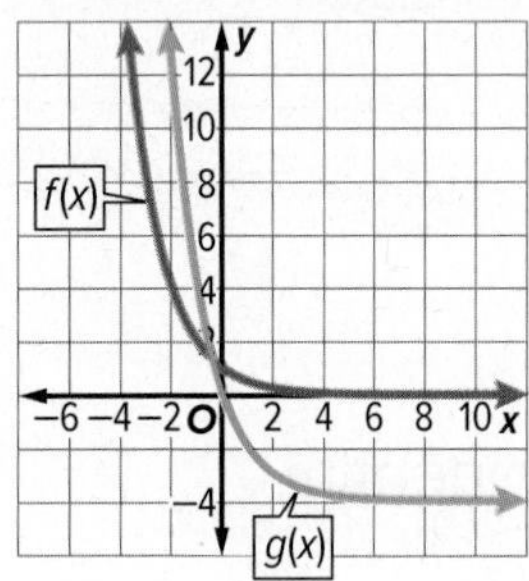

The value of h is subtracted from x before the function is performed and is greater than 0, so the graph of $f(x) = \left(\frac{1}{2}\right)^x$ is translated 2 units right.

The value of k is added to the function after it has been evaluated and is less than 0, so the graph of $f(x) = \left(\frac{1}{2}\right)^x$ is also translated 4 units down.

Check

Describe the translation in $g(x) = 2^{x+1} - 8$ as it relates to the graph of the parent function $f(x) = 2^x$.

The graph of $g(x) = 2^{x+1} - 8$ is the translation of the graph of the parent function 1 unit ___?___ and 8 units ___?___.

Go Online You can complete an Extra Example online.

Example 4 Identify Exponential Functions from Graphs (Vertical Translations)

The given graph is a translation of the parent function $f(x) = \left(\frac{1}{4}\right)^x$.

Use the graph of the function to write its equation.

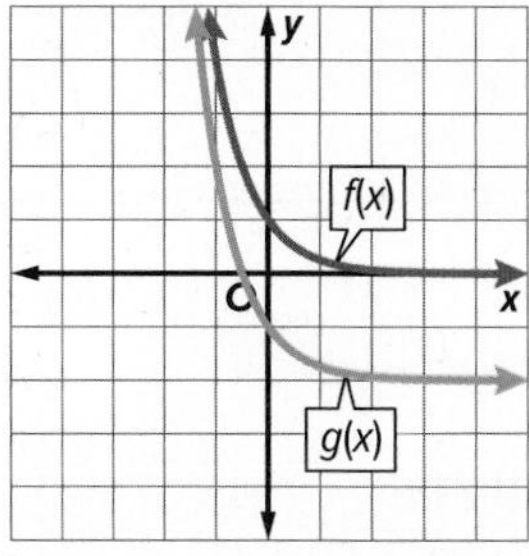

$g(x) = \left(\frac{1}{4}\right)^x - 2$

The horizontal asymptote of $g(x)$ is different from the horizontal asymptote of $f(x)$, implying a vertical translation of the form $g(x) = \left(\frac{1}{4}\right)^x + k$. The parent graph has a y-intercept at (0, 1). The translated graph has a y-intercept at (0, −1). The y-intercept is shifted 2 units down, so $k = -2$.

Study Tip

Vertical Translations Any exponential parent function has a y-intercept at (0, 1) and an asymptote at $y = 0$. By examining how far these features are shifted up or down, you can easily determine the value of k when identifying exponential functions.

Example 5 Identify Exponential Functions from Graphs (Horizontal Translations)

The given graph is a translation of the parent function $f(x) = 2^x$. Use the graph of the function to write its equation.

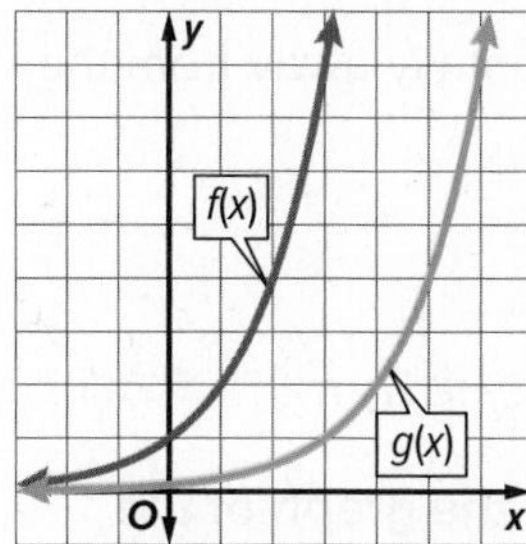

$g(x) = 2^{x-3}$

The horizontal asymptote of $g(x)$ is the same as the horizontal asymptote of $f(x)$, implying a horizontal translation of the form $g(x) = 2^{x-h}$. The parent graph passes through (0, 1). The translated graph has a y-value of 1 at (3, 1). The graph is shifted 3 units right, so $h = 3$.

Study Tip

Horizontal Translations Any exponential parent function has a y-intercept at (0, 1). By examining how far this point is shifted right or left, you can easily determine the value of h when identifying exponential functions.

Check

The given graph is a translation of $f(x) = 5^x$. Which is the equation for the function shown in the graph?

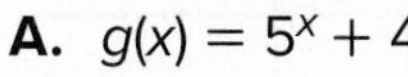

A. $g(x) = 5^x + 4$

B. $g(x) = 5^{x-4}$

C. $g(x) = 5^x - 4$

D. $g(x) = 5^{x+4}$

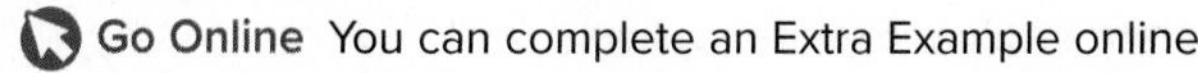

Go Online You can complete an Extra Example online.

Explore Dilating Exponential Functions

Online Activity Use graphing technology to complete the Explore.

INQUIRY What effect does multiplying a function by a value before or after it has been evaluated have on the function?

Learn Dilations of Exponential Functions

Key Concept • Vertical Dilations of Exponential Functions

- The graph of $g(x) = ab^x$ is the graph of $f(x) = b^x$ stretched or compressed vertically by a factor of $|a|$.
- If $|a| > 1$, the graph of $f(x)$ is stretched vertically away from the x-axis.
- If $0 < |a| < 1$, the graph of $f(x)$ is compressed vertically toward the x-axis.

Go Online
You can watch a video to see how to describe dilations of functions.

Key Concept • Horizontal Dilations of Exponential Functions

- The graph of $g(x) = b^{ax}$ is the graph of $f(x) = b^x$ stretched or compressed horizontally by a factor of $\frac{1}{|a|}$.
- If $|a| > 1$, the graph of $f(x)$ is compressed horizontally toward the y-axis.
- If $0 < |a| < 1$, the graph of $f(x)$ is stretched horizontally away from the y-axis.

Study Tip

Vertical Dilations For a vertical dilation, if you multiply each y-coordinate of the function $f(x)$ by a, you'll get the corresponding y-coordinate of the function $g(x)$. For the function in the example, the point $(1, 3)$ on $f(x)$ corresponds to the point $\left(1, \frac{3}{4}\right)$ on $g(x)$. The y-coordinate of $f(x)$, 3, is multiplied by a.

Example 6 Vertical Dilations of Exponential Functions

Describe the dilation in $g(x) = \frac{1}{4}(3)^x$ as it relates to the graph of the parent function $f(x) = 3^x$.

Since $f(x) = 3x$, $g(x) = a \cdot f(x)$, where $a = \frac{1}{4}$.

$g(x) = \frac{1}{4}(3)^x \rightarrow g(x) = \frac{1}{4}f(x)$

The function is multiplied by the positive constant a after it has been evaluated, and $|a|$ is between 0 and 1, so the graph of $f(x) = 3^x$ is compressed vertically by a factor of $|a|$, or $\frac{1}{4}$.

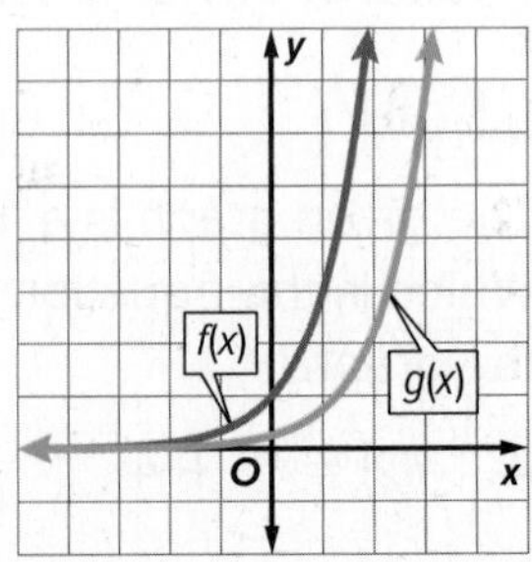

 **Go Online** You can complete an Extra Example online.

Example 7 Horizontal Dilations of Exponential Functions

Describe the dilation in $g(x) = \left(\frac{5}{3}\right)^{2x}$ as it relates to the graph of the parent function $f(x) = \left(\frac{5}{3}\right)^{x}$.

x is multiplied by the positive constant a before it has been evaluated, and $|a|$ is greater than 1, so the graph of $f(x) = \left(\frac{5}{3}\right)^{x}$ is compressed horizontally by a factor of $\frac{1}{|a|}$, or $\frac{1}{2}$.

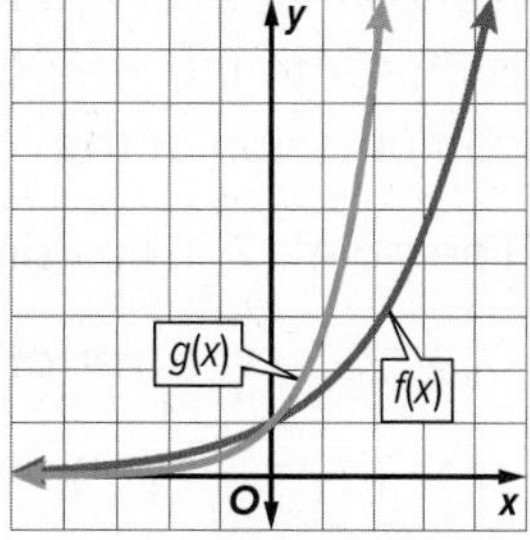

Check

Identify the dilation in each function as it relates to the parent function $f(x) = 4^x$ by copying and completing the table and writing the type of dilation and dilation factor next to each equation.

$g(x) = 4^{\frac{2}{3}x}$	$\leftrightarrow$	
$h(x) = 3(4)^x$	$\leftrightarrow$	
$k(x) = \frac{6}{7}(4)^x$	$\leftrightarrow$	
$g(x) = 4^{\frac{5}{2}x}$	$\leftrightarrow$	

Example 8 Describe Dilations of Exponential Functions

ENERGY **Since 2000, solar PV capacity in the world has been growing exponentially. It can be approximated by the function $c(x) = 0.897(1.46)^x$, where $c(x)$ is the solar PV capacity in gigawatts, x is the number of years since 2000, and 0.897 is the initial capacity. Describe the dilation in $c(x) = 0.897(1.46)^x$ as it is related to the parent function $f(x) = (1.46)^x$.**

The parent function is $f(x) = (1.46)^x$.

Then $c(x) = af(x)$, where $a = 0.897$.

$c(x) = 0.897(1.46)^x \rightarrow c(x) = 0.897f(x)$

The function is multiplied by the positive constant a after it has been evaluated and $|a|$ is between 0 and 1, so the graph of $f(x) = (1.46)^x$ is compressed vertically by a factor of $|a|$, or 0.897.

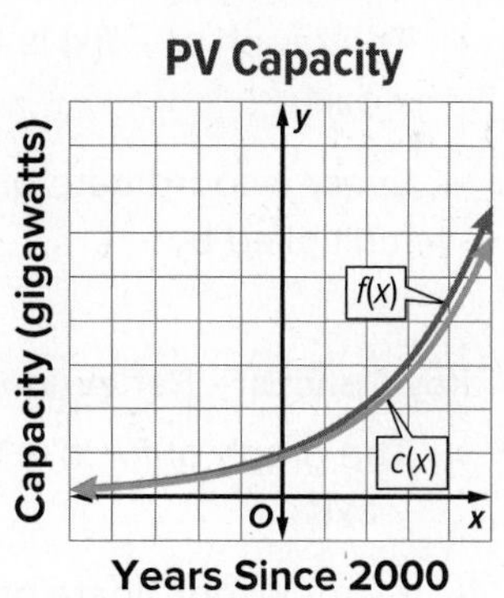

Think About It!

How can you easily tell if an exponential function is going to be horizontally dilated?

Think About It!

Why does a horizontal dilation not change the y-intercept of an exponential function? Justify your argument.

Study Tip

Assumptions Assuming that the rate at which PV capacity increases remains the same allows us to represent the situation with an exponential function.

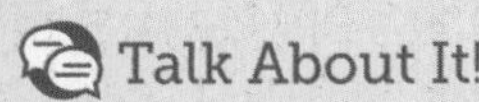

Talk About It!

Make a conjecture about the y-intercept of $g(x)$ and the value of a. Will this hold true for any vertical dilation of an exponential function? Explain your reasoning.

Example 9 Identify Exponential Functions from Graphs (Dilations)

The given graph is a dilation of the parent function $f(x) = 2^x$. Use the graph of the function to write its equation.

Notice that every point on the graph of $g(x)$ is closer to the x-axis, implying a vertical compression of the form $g(x) = a(2)^x$.

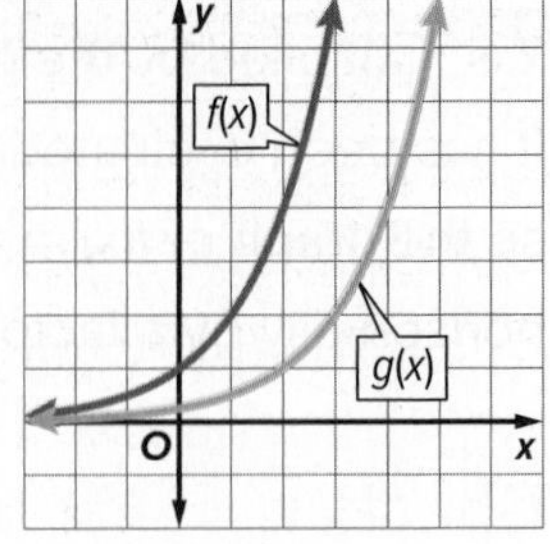

The point (2, 1) lies on the graph. Solve for a.

$$1 = a(2)^2$$

$$1 = 4a$$

$$\frac{1}{4} = a$$

$$g(x) = \frac{1}{4}(2)^x$$

Check

The given graph is a dilation of $f(x) = 3^x$. Which is the equation for the function shown in the graph?

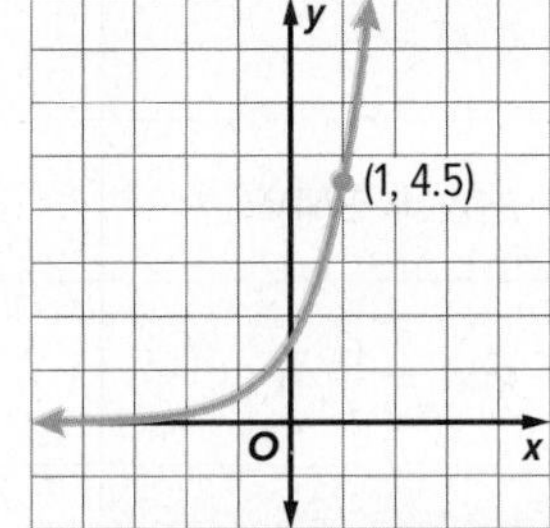

A. $g(x) = 0.007(3)^x$

B. $g(x) = \frac{2}{3}(3)^x$

C. $g(x) = 3^x + \frac{3}{2}$

D. $g(x) = \frac{3}{2}(3)^x$

Think About It!

What does the general form of an exponential function look like once it has been reflected across the x-axis?

Explore Reflecting Exponential Functions

Online Activity Use graphing technology to complete the Explore.

INQUIRY What effect does multiplying a function by −1 before or after it has been evaluated have on the function?

Learn Reflections of Exponential Functions

Go Online

You can watch a video to see how to describe reflections of functions.

Key Concept • Reflections of Exponential Functions Across the x-axis

- The graph of $-f(x)$ is the reflection of the graph of $f(x) = b^x$ across the x-axis.
- Every y-coordinate of $-f(x)$ is the corresponding y-coordinate of $f(x)$ multiplied by −1.

Key Concept • Reflections of Exponential Functions Across the y-axis

- The graph of $f(-x)$ is the reflection of the graph of $f(x) = b^x$ across the y-axis.
- Every x-coordinate of $f(-x)$ is the corresponding x-coordinate of $f(x)$ multiplied by −1.

Example 10 Vertical Reflections of Exponential Functions

Describe how the graph of $g(x) = -3(2)^x$ is related to the graph of the parent function $f(x) = 2^x$.

The function is multiplied by −1 and the positive constant a after it has been evaluated and $|a|$ is greater than 1, so the graph of $f(x) = 2^x$ is stretched vertically and reflected across the x-axis.

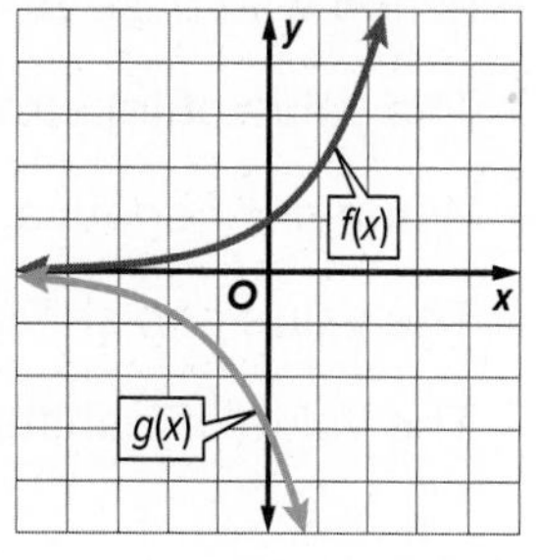

Example 11 Horizontal Reflections of Exponential Functions

Describe how the graph of $g(x) = (3)^{-2x}$ is related to the graph of the parent function $f(x) = 3^x$.

The function is multiplied by −1 and the constant a before it is evaluated and $|a|$ is greater than 1, so the graph of $f(x) = 3^x$ is compressed horizontally and reflected across the y-axis.

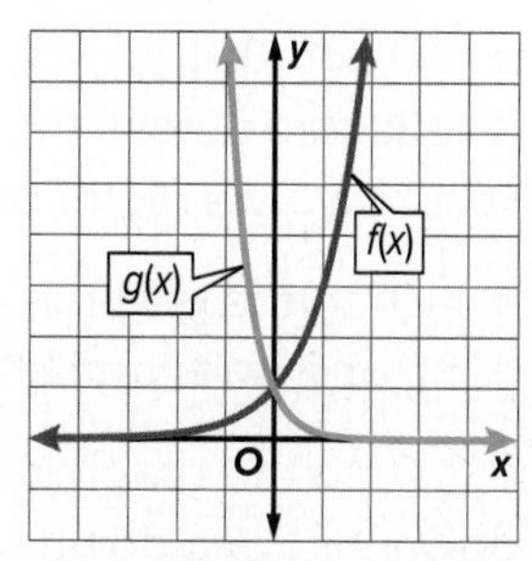

Check

Match each function with its graph.

1. $g(x) = -3^x$
2. $h(x) = \left(\frac{1}{3}\right)^{-x}$
3. $k(x) = -\left(\frac{1}{3}\right)^x$
4. $g(x) = 3^{-x}$

A.

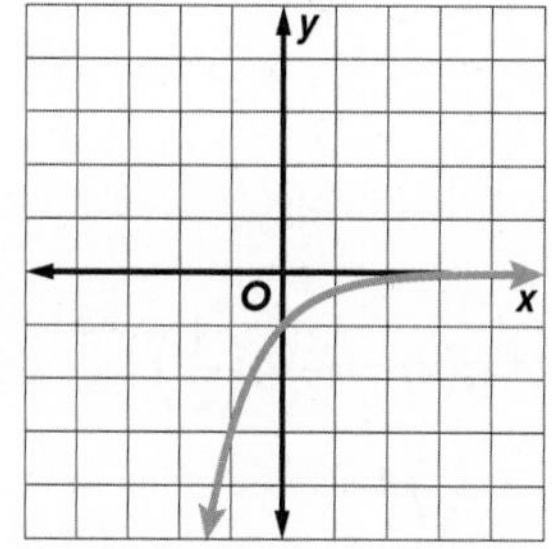

B.

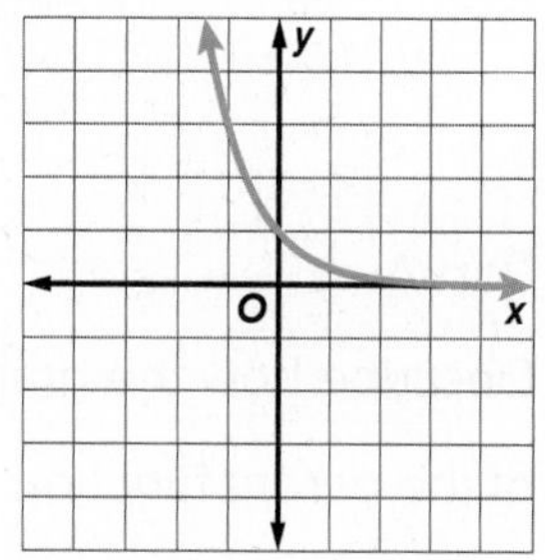

C.

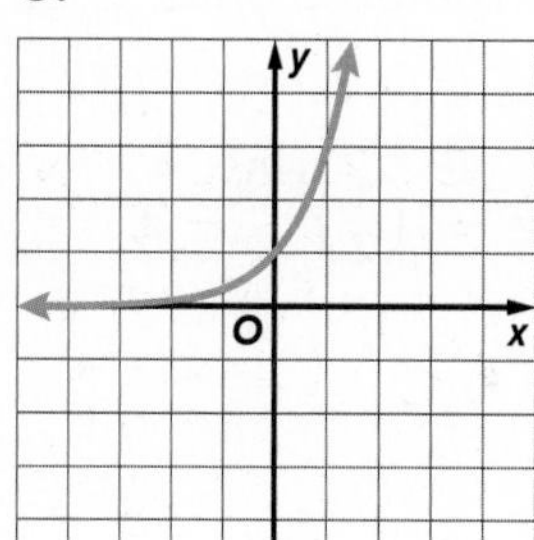

D.

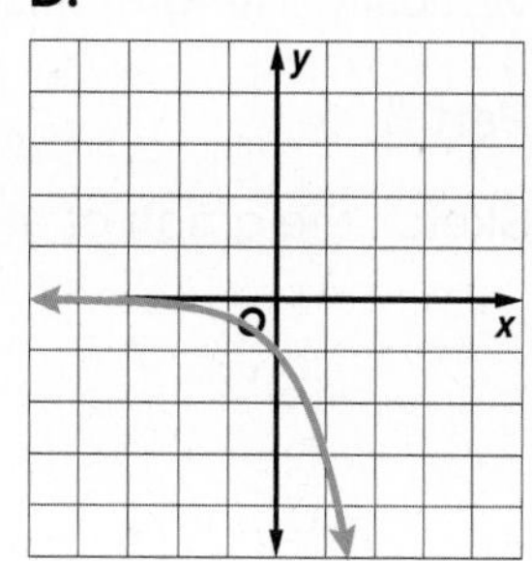

Think About It!

The example shows the reflection of an exponential function of the form $f(x) = ab^x$ over the y-axis for the case where $b > 1$. Examine the following cases and describe the effect a reflection across the y-axis would have on the end behavior of the parent function $f(x) = ab^x$.

Case 1: $g(x) = ab^{-x}$ where $b > 1$

Case 2: $g(x) = ab^{-x}$ where $0 < b < 1$

Go Online You can complete an Extra Example online.

Go Online
You can watch a video to see how to graph transformations of exponential functions using a graphing calculator.

Think About It!
Write an exponential function that is the parent function $f(x) = 2^x$ stretched vertically and translated 3 units left and 6 units up.

Learn Transformations of Exponential Functions

The general form of an exponential function is $f(x) = ab^{x-h} + k$, where a, h, and k are parameters that dilate, reflect, or translate a parent function with base b.

- The value of $|a|$ stretches or compresses (dilates) the parent graph.
- When a is negative, the graph is reflected across the x-axis.
- The value of h shifts (translates) the parent graph right or left.
- The value of k shifts (translates) the parent graph up or down.

Example 12 Multiple Transformations of Exponential Functions

Describe how the graph of $g(x) = -\frac{1}{2}(3)^{x-2} - 1$ is related to the graph of the parent function $f(x) = 3^x$.

$a < 0$ and $0 < |a| < 1$, so the graph of $f(x) = 3^x$ is reflected across the x-axis and compressed vertically by a factor of $|a|$, or $\frac{1}{2}$.

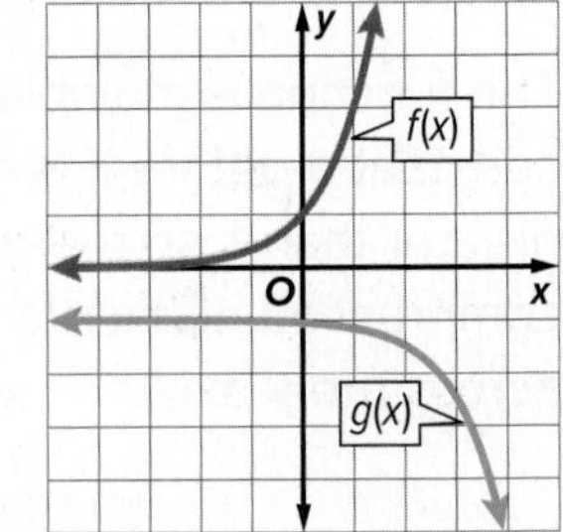

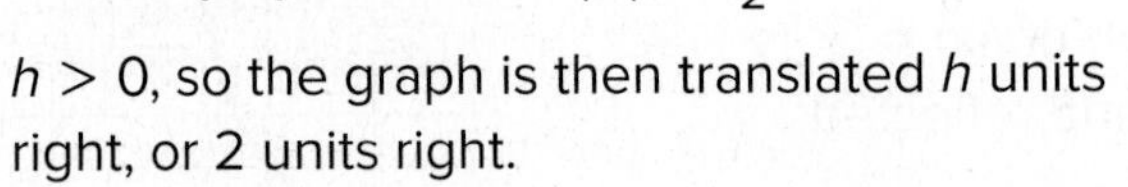

$h > 0$, so the graph is then translated h units right, or 2 units right.

$k < 0$, so the graph is then translated $|k|$ units down, or 1 unit down.

$g(x) = -\frac{1}{2}(3)^{x-2} - 1$ is the graph of the parent function compressed vertically, reflected across the x-axis, and translated 2 units right and 1 unit down.

Check

Part A

Describe how the graph of $g(x) = -4\left(\frac{1}{2}\right)^{x+5} - 2$ is related to the graph of the parent function $f(x) = \left(\frac{1}{2}\right)^x$.

The graph of $f(x) = \left(\frac{1}{2}\right)^x$ is reflected across the __?__ and __?__ vertically. The graph is translated 5 units __?__ and 2 units __?__.

Part B

Sketch the graph of $g(x) = -4\left(\frac{1}{2}\right)^{x+5} - 2$.

Go Online You can complete an Extra Example online.

Practice

Go Online You can complete your homework online.

Examples 1–3, 6–7, 10–12

Describe the transformation of $g(x)$ as it relates to the parent function $f(x)$.

1. $f(x) = 6^x$; $g(x) = 6^x + 8$
2. $f(x) = 5^x$; $g(x) = -5^x$
3. $f(x) = 3^x + 1$; $g(x) = 3^{2x} + 1$
4. $f(x) = 4^x - 3$; $g(x) = 4^{0.5x} - 3$
5. $f(x) = 2.3^x$; $g(x) = -2.3^{x-1}$
6. $f(x) = 2^x$; $g(x) = 2^{-x} + 1$
7. $f(x) = 5^x + 2$; $g(x) = 5^{-x} + 6$
8. $f(x) = 1.4^x - 1$; $g(x) = -1.4^x + 6$
9. $f(x) = 3^x + 1$; $g(x) = 2(3^x + 1)$
10. $f(x) = -4^x$; $g(x) = \frac{1}{3}(-4^x)$
11. $f(x) = 4^x$; $g(x) = 4^{x-3}$
12. $f(x) = \left(\frac{1}{2}\right)^x + 5$; $g(x) = \left(\frac{1}{2}\right)^x$

Examples 4–5, 9

Each graph is a transformation of the parent function $y = 2^x$. Use the graph of the function to write its equation.

13.

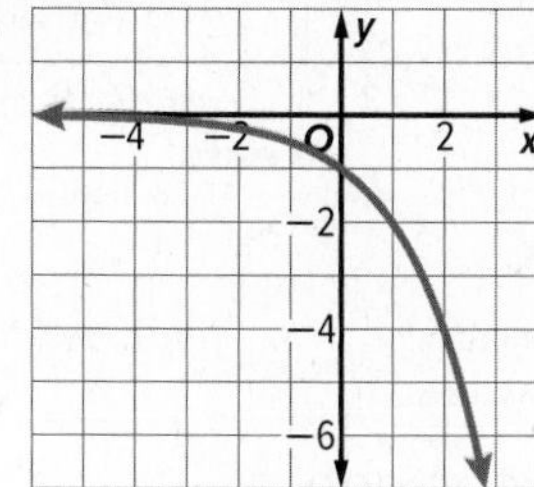

14.

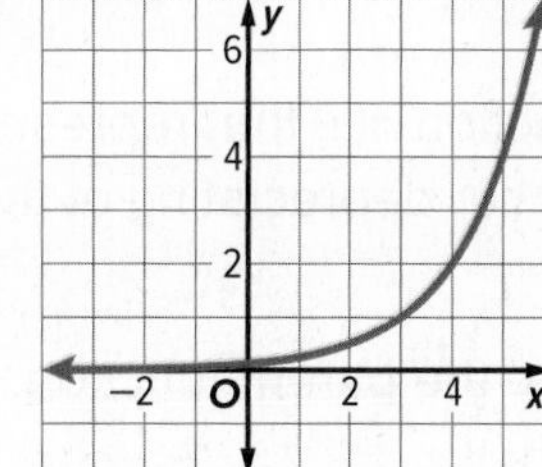

15.

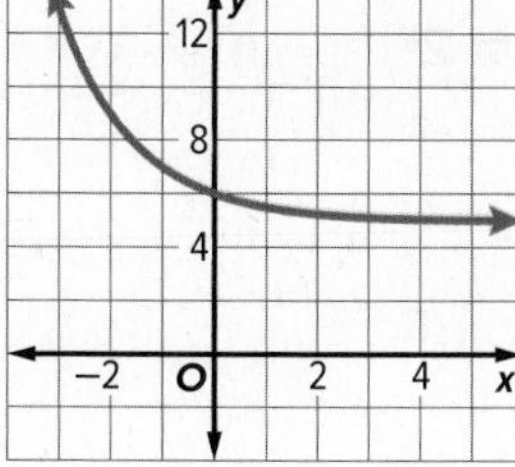

16.

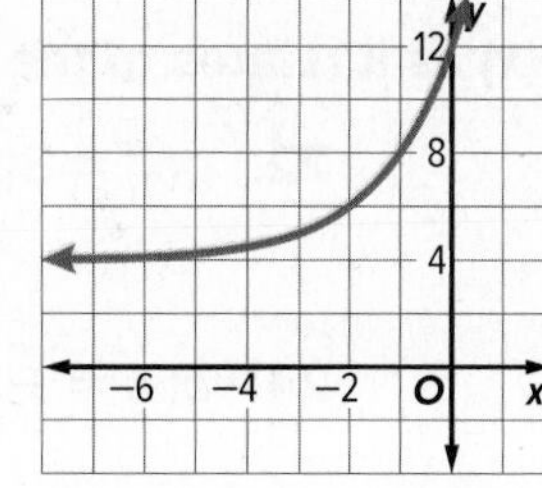

Example 8

17. SAVING Celia invests \$2000 in a savings account that earns 1.25% interest per year compounded annually. The amount of money in her bank account after x years can be modeled by $g(x) = 2000(1.0125)^x$. Describe the dilation in $g(x)$ as it relates to the parent function $f(x) = 1.0125^x$.

18. CAFFEINE Suppose an 8-ounce cup of coffee contains 100 milligrams of caffeine. The rate at which caffeine is eliminated from an adult's body is 11% per hour. The function $f(x) = 100(0.89)^x$ can be used to model the amount of caffeine left in a person's bloodstream after x hours of consuming the cup of coffee. Suppose the function $g(x) = 25(0.89)^x$ represents the amount of caffeine left in a person's bloodstream after x hours of consuming an 8-ounce cup of green tea. Describe $g(x)$ as a transformation of $f(x)$.

19. VISITORS The number of visitors to a new skateboarding park can be modeled by the exponential function $g(x) = 20(2^x)$, where x represents the number of months since the park's grand opening. Explain how the number of visitors during that first month is a dilation of the parent function $f(x) = 2^x$.

20. DEPRECIATION Depreciation is the decrease in the value of an item resulting from its age or wear. When an item loses about the same percent of its value each year, an exponential function can be used to model its decreasing value over time. The function $g(x) = 12{,}000(0.85)^x$ can be used to model the value of a \$12,000 car as it depreciates at an annual rate of 15% over x years. $g(x)$ is a dilation of the parent function $f(x) = 0.85^x$. The graph shows the function $g(x)$.

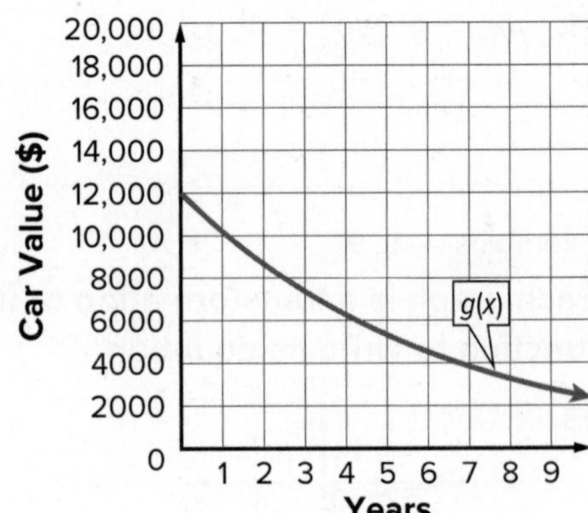

a. Write the equation of a function $h(x)$ that represents the depreciation of a \$20,000 car depreciating at the same rate over x years.

b. Describe $h(x)$ as it relates to the parent function.

c. What is the difference between the values of the \$12,000 car and the \$20,000 car after 5 years?

Mixed Exercises

Describe the transformation of $g(x)$ as it relates to the parent function $f(x) = 2^x$.

21. $g(x) = 2^x + 6$

22. $g(x) = 3(2)^x$

23. $g(x) = -\frac{1}{4}(2)^x$

24. $g(x) = -3 + 2^x$

25. $g(x) = 2^{-x}$

26. $g(x) = -5(2)^x$

Write a function $g(x)$ to represent the transformation of the parent function $f(x)$.

27. $f(x) = 2^x$ translated 3 units up

28. $f(x) = 8^x$ translated 1 unit down

29. $f(x) = 5^x$ translated 2 units right

30. $f(x) = 3^x$ translated 4 units left

31. $f(x) = 6^x + 7$ translated 2 units down

32. $f(x) = -2^x + 3$ translated 5 units right

33. $f(x) = 4^x$ is compressed vertically by a factor of $\frac{1}{2}$

34. $f(x) = 3^x$ is stretched vertically by a factor of 5

35. $f(x) = 2^x$ is compressed horizontally by a factor of $\frac{1}{3}$

36. $f(x) = 5^x$ is stretched horizontally by a factor of 4

Each graph is a transformation of the parent function $f(x) = 5^x$. Use the graph of the function to write its equation.

37.

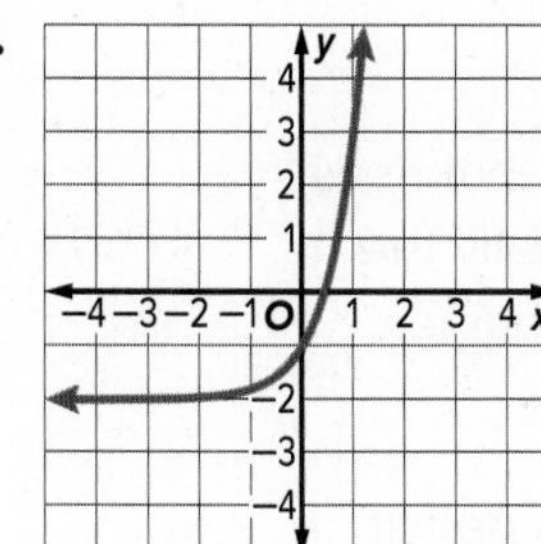

38.

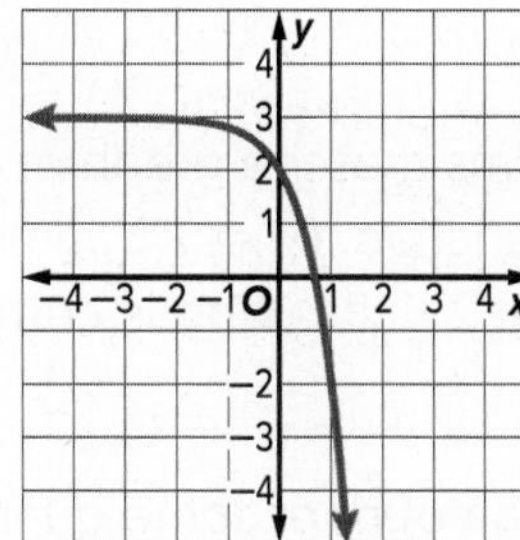

39.

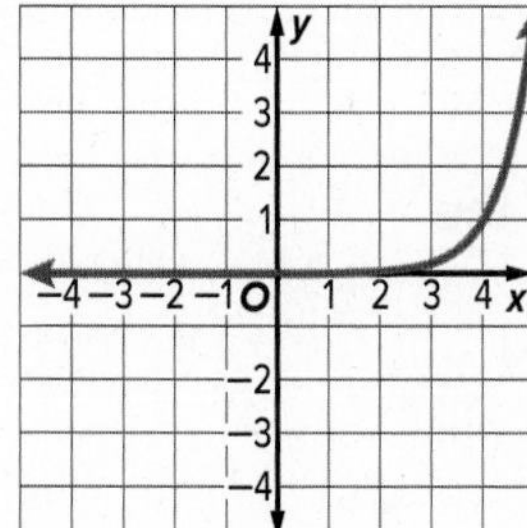

40.

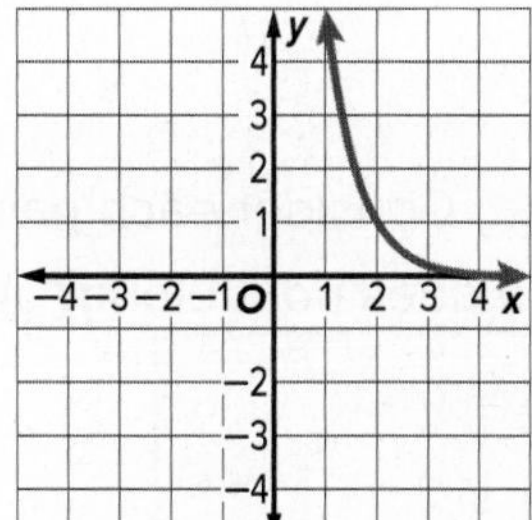

41. ASSETS Thomas and Rebecca each put $1000 into a bank account that earns 1.5% interest per year compounded annually. Thomas also has an antique toy automobile. The graph shows the amount of their assets over time.

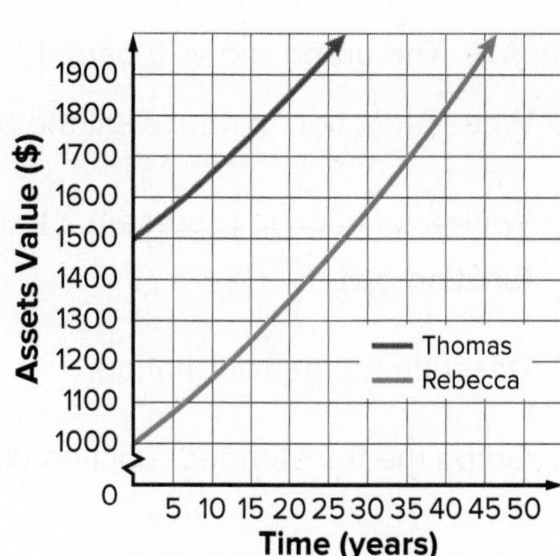

a. Describe the graph of Thomas' assets as a transformation of Rebecca's assets.

b. Use the graph to extrapolate the value of Thomas' antique toy automobile.

The graph of $g(x)$ is a transformation of the parent function $f(x)$. Graph $g(x)$ and describe the transformation in each function as it relates to the parent function.

42. $f(x) = 3^x$
$g(x) = 5 \cdot 3^{x+2} - 4$

43. $f(x) = 2^x$
$g(x) = -3 \cdot 2^{-x} + 1$

44. **CONSTRUCT ARGUMENTS** Name the coordinates of the point at which the graphs of $g(x) = 2^x + 3$ and $h(x) = 5^x + 3$ intersect. Explain your reasoning.

45. **STRUCTURE** Describe the similarities between the graph of $f(x) = 4^{x+2}$ and the graph of $g(x) = 16 \cdot 4^x$. Use the properties of exponents to justify your answer.

Higher-Order Thinking Skills

46. **ANALYZE** What would happen to the shape of the graph of an exponential function if the function is multiplied by a number between 0 and −1? What would happen to its shape if the exponent is multiplied by a number between 0 and −1? Justify your argument.

47. **FIND THE ERROR** Jennifer claims that the graph of $g(x) = 2(2^x)$ is a graph that rises more rapidly than its parent function $f(x) = 2^x$. James claims that it is actually the parent graph shifted to the left 2 units. Who is correct? Explain your reasoning.

48. **WRITE** A deficit is a negative amount of some quantity, such as money. A deficit that is growing exponentially can be modeled by $y = ab^{c(x-h)} + k$. Describe the constraints on a, b, and c.

49. **WHICH ONE DOESN'T BELONG?** Consider each pair of transformations of the function $f(x)$ to $g(x)$. Which one does not belong? Justify your conclusion.

$f(x) = 3^{x+2}$ $g(x) = 3^{x-1} + 2$	$f(x) = 2^x$ $g(x) = 2^{x+3} + 2$	$f(x) = 4^{x+1}$ $g(x) = 4^{x+4} + 2$

50. **CREATE** The graph shows a parent function $f(x)$.

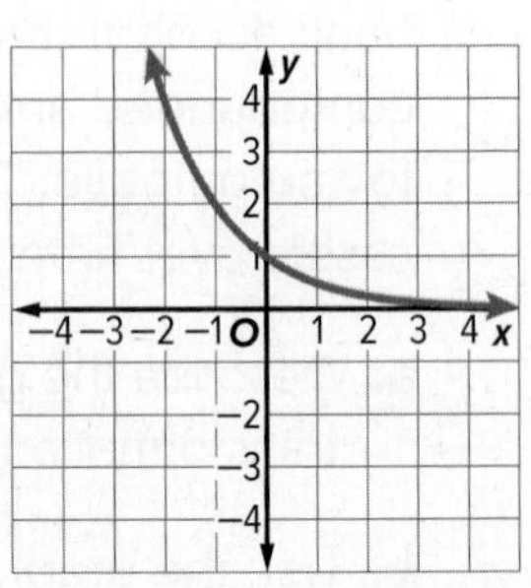

a. Write a function to represent the parent function $f(x)$.

b. Write a function to represent a transformation of the parent function $g(x)$.

c. Describe the transformation.

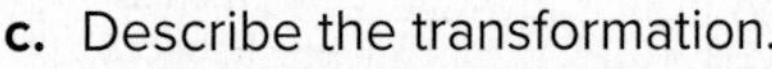

d. Graph the transformed function $g(x)$.

Writing Exponential Functions

Explore Writing an Exponential Function to Model Population Growth

 **Online Activity** Use a real-world situation to complete the Explore.

INQUIRY How can you find an equation that models the population growth of a colony of organisms that grows exponentially?

Today's Goals

- Construct exponential functions by using a graph, a description, or two points.
- Create equations and solve problems involving exponential growth.
- Create equations and solve problems involving exponential decay.

Today's Vocabulary

compound interest

Learn Constructing Exponential Functions

If you are given two points, a graph, or a description of an exponential function, you can write an exponential function to model the data.

Given Two Points

Substitute the values of x and y into the equation $y = ab^x$. The result will be a system of equations in two variables, each with a and b as unknowns. You can then solve the system using substitution.

Given a Graph

Use any two points to write an equation in the form $y = ab^x$. Substitute the values of x and y into the equation, and solve the resulting system of equations for a and b using substitution.

Given a Description

Use the information to create a table or a graph. Then look for a pattern between the input and output values. Keep an eye out for words such as *exponential*, *linear*, *multiple*, *constant*, and *factor*, which can help you determine whether the function is exponential.

Study Tip

Graphs If you write an exponential equation based on a graph that includes the y-intercept, you can substitute the y-value of the intercept for a in $y = ab^x$. Since $b^0 = 1$, $y = a$. You can then use another point to find the common ratio b.

Example 1 Write an Exponential Function Given Two Points

Write an exponential function for the graph that passes through (1, 6) and (3, 24).

Substitute x and y into $y = ab^x$ to get a system of two equations.

$y = ab^x$	General form	$y = ab^x$	General form
$6 = ab^1$	$y = 6$ and $x = 1$	$24 = ab^3$	$y = 24$ and $x = 3$

(continued on the next page)

Solve the system of equations using substitution.

Solve the first equation for a.

$6 = ab^1$	First equation
$\frac{6}{b} = a$	Division Property of Equality

Substitute $\frac{6}{b}$ for a in the second equation to find b.

$24 = ab^3$	Second equation
$24 = \frac{6}{b}b^3$	Substitute $\frac{6}{b}$ for a.
$24 = \frac{6b^3}{b}$	Multiply.
$24 = 6b^2$	Quotient of Powers
$4 = b^2$	Division Property of Equality; then simplify
$2 = b$	Definition of exponent

Substitute 2 for b in either equation to find a.

$6 = ab^1$	Second equation
$6 = a(2)^1$	Substitute 2 for b.
$3 = a$	Simplify.

Write the equation.

$$y = 3 \times 2^x$$

Study Tip

Assumptions The graphs of multiple functions pass through the points (1, 6) and (3, 24). For example, the linear equation $y = 9x - 3$ also passes through these points. However, because you are told to find the exponential function that passes through those points, you can assume that the exponential relationship is the correct one.

Check

Write an exponential function that passes through (−1, 20) and (1, 5).

$y =$ __?__ $\times$ __?__

Example 2 Write an Exponential Function Given a Graph

Write an exponential function for the graph.

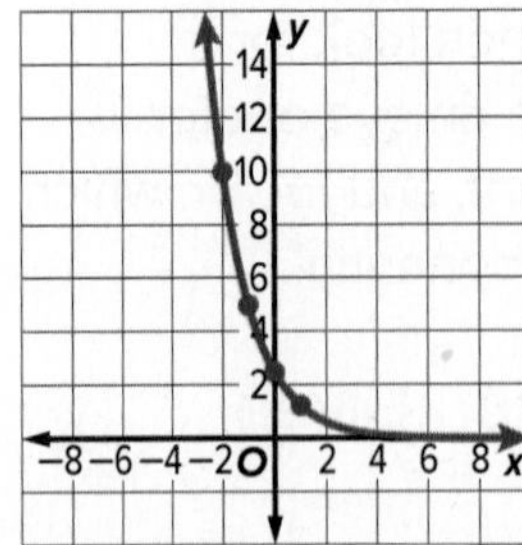

x	y
−2	10
−1	5
0	2.5
1	1.25

You can use any two points to write a system of two equations using $y = ab^x$.

$2.5 = ab^0$

$1.25 = ab^1$

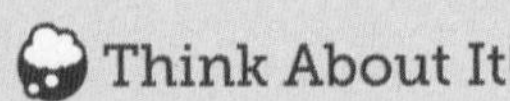

Think About It!

Use the graph to estimate the value of y when $x = 2$. Then use the function to find y when $x = 2$.

Since there is a zero exponent in the first equation, solve it for a.

$a = 2.5$

Then substitute this value into the second equation to find b.

$b = 0.5$

The graph can be modeled by the function $y = 2.5(0.5)^x$.

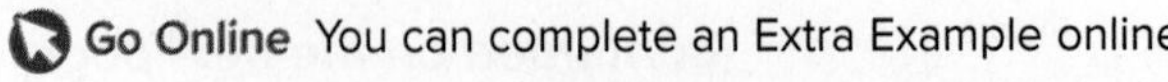

Go Online You can complete an Extra Example online.

Check

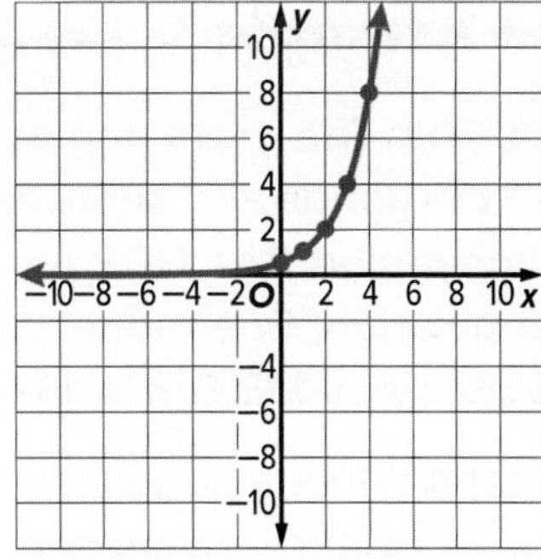

Part A Use the graph to estimate the value of y when $x = 4$.

Part B Write an exponential function that models the graph.

Part C Use the function in Part B to find y when $x = 4$.

Example 3 Write an Exponential Function Given a Description

CONTEST **A radio station is giving away \$1000 to the first listener who answers a question correctly. If the question goes unanswered for one hour, the prize increases by 10% until it is answered correctly. Write a function to describe this situation.**

Hour (h)	Prize Total (P)
0	1000
1	1100
2	1210
3	1331

This is an example of exponential growth, so $b > 1$. Divide each value of P by the preceding term to find a common ratio of 1.1.

The value of a can be found by identifying the y-intercept.

$$a = 1000$$

Write the function that best models the situation.

$$P = 1000(1.1)^h$$

Talk About It!

Why is the common ratio 1.1 and not 0.1? Explain.

Learn Solving Problems Involving Exponential Growth

Key Concept • Equation for Exponential Growth

$$y = a(1 + r)^t$$

y is the final amount.

t is time.

a is the initial amount.

r is the rate of growth expressed as a decimal, $r > 0$.

Think About It!

Why is the constant 1 in the exponential growth formula? What does it represent?

Key Concept • Equation for Compound Interest

$$A = P\left(1 + \frac{r}{n}\right)^{nt}$$

A is the current amount.

r is the annual interest rate expressed as a decimal, $r > 0$.

n is the number of times the interest is compounded each year.

t is time in years.

P is the principal, or the initial amount.

Think About It!

Why do you not substitute 2016 for t to determine the GDP per capita in 2016?

Study Tip

Assumptions The actual rates of change for the GDP are calculated annually and have varied from −11% to 18%, depending on the current economy. For the purposes of this lesson, we will assume a constant rate of change is present.

Example 4 Exponential Growth

GOODS The gross domestic product (GDP) is the monetary value of all the goods and services produced within a country in a specific time period. The GDP per capita is this value divided by the population. One model says that the GDP per capita in the United States was $13,513 in 1946, and it has increased by 2% every year.

Part A Write an equation to represent the GDP per capita after t years, where a represents the initial GDP in 1946, and r represents the rate of growth each year.

If the initial year was 1946, the initial GDP was $13,513. So $a = 13{,}513$. The GDP grew 2% each year, so $r = 2\%$ or 0.02.

$y = a(1 + r)^t$	Equation for exponential growth
$y = 13{,}513(1 + 0.02)^t$	$a = 13{,}513$ and $r = 2\%$ or 0.02
$y = 13{,}513(1.02)^t$	Simplify.

In this equation, y is the GDP per capita, and t is the time in years since 1946.

Part B If this trend continued, calculate the GDP per capita in 2016.

$t = 70$

Using $y = 13{,}513(1.02)^t$, the GDP per capita in 2016 was 54,046.

Check

POPULATION From 2013 to 2014, the city of Austin, Texas, saw one of the highest population growth rates in the country at 2.9%. The population of Austin in 2014 was estimated to be about 912,000.

Part A If the trend were to continue, which equation represents the estimated population t years after 2014?

A. $y = 912{,}000(0.029)^t$

B. $y = 912{,}000(3.9)^t$

C. $y = 1.029(912{,}000)^t$

D. $y = 912{,}000(1.029)^t$

Part B To the nearest person, predict the population of Austin in 5 years.

___?___ people

Go Online You can complete an Extra Example online.

Apply Example 5 Compound Interest

COLLEGE PLANNING **Maria invests $5500 into a college savings account that pays 3.25% compounded quarterly. How much money will there be in the account after 5 years?**

1 What is the task?

Describe the task in your own words. Then list any questions that you may have. How can you find answers to your questions?

Sample answer: I need to determine how much money will be in Maria's account after 5 years. How can I write an equation to represent this situation? I can apply what I have learned about different types of functions.

2 How will you approach the task? What have you learned that you can use to help you complete the task?

Sample answer: I will substitute the information I know into the compound interest equation and simplify. Then I will find the amount of money in Maria's account after 5 years. I will use the properties of exponents to simplify my equation.

3 What is your solution?

Use your strategy to solve the problem.

Write an equation to represent the amount of money in Maria's account after t years.

$A = 5500(1.008125)^{4t}$

How much money will be in Maria's account after 5 years?

$6466.22

4 How can you know that your solution is reasonable?

Write About It! Write an argument that can be used to defend your solution.
Sample answer: My solution is reasonable because if I find the amount of money in the account after each quarter, I get the same answer as I do when I use the compound interest equation.

Check

BANKING Twin brothers Amare and Jermaine each received $1000 for graduation. Amare invests his money in an account that pays 2.25% compounded daily. Jermaine invests his money in an account that pays 2.25% compounded annually.

Part A Which brother will have more money at the end of 10 years?

A. Amare

B. Jermaine

C. The accounts will be equal.

Part B To the nearest cent, how much more money? $ ___?___

Go Online You can complete an Extra Example online.

Think About It!

What major difference do you notice between the equation for exponential growth and the equation for exponential decay?

Think About It!

Why is the rate subtracted from 1 in the equation for exponential decay, but added to 1 in the equation for exponential growth?

Go Online

to practice what you've learned about exponential growth and decay in the Put It All Together over Lessons 8-1 through 8-3.

Learn Solving Problems Involving Exponential Decay

Key Concept • Equation for Exponential Decay

$$y = a(1 - r)^t$$

y is the final amount.

t is time.

a is the initial amount.

r is the rate of decay expressed as a decimal, $0 < r < 1$.

Example 6 Exponential Decay

BANKS **In banking, a dormant account is one that has not been used in over a year. A bank charges a monthly fee on dormant accounts of 0.8% of the account balance. One dormant account initially had a balance of $1609.**

Part A Write an equation to represent the balance in the account after t months.

$y = a(1 - r)^t$	Equation for exponential decay
$y = 1609(1 - 0.008)^t$	$a = 1609$ and $r = 0.8\%$ or 0.008
$y = 1609(0.992)^t$	Simplify.

The equation is $y = 1609(0.992)^t$, where y is the balance in the account after t months.

Part B Estimate the balance in the account after a year.

$t = 12$

$y = 1609(0.992)^t = 1461$

Check

CITY PLANNING A city has been experiencing a slight population loss over the last few years. In 2014, the population was 1.8503 million, representing a 0.18% decrease from the previous year.

Part A If the trend were to continue, which equation represents the estimated population in millions after t years?

A. $y = 1.8503(0.18)^t$

B. $y = 1.8503(0.9982)^t$

C. $y = 1.8503(1.0018)^t$

D. $y = 1.8503(1.9982)^t$

Part B To the nearest ten thousandth, predict the population in 15 years. __?__ million

Go Online You can complete an Extra Example online.

Practice

Go Online You can complete your homework online.

Example 1

Write an exponential function for a graph that passes through the points.

1. (2, 16) and (3, 32)

2. (1, 1) and (3, 0.25)

3. (2, 90) and (4, 810)

4. (−2, 4) and (1, 0.5)

5. (1, 12) and (3, 192)

6. (1, 18) and (3, 72)

Example 2

Write an exponential function for the graph.

7.

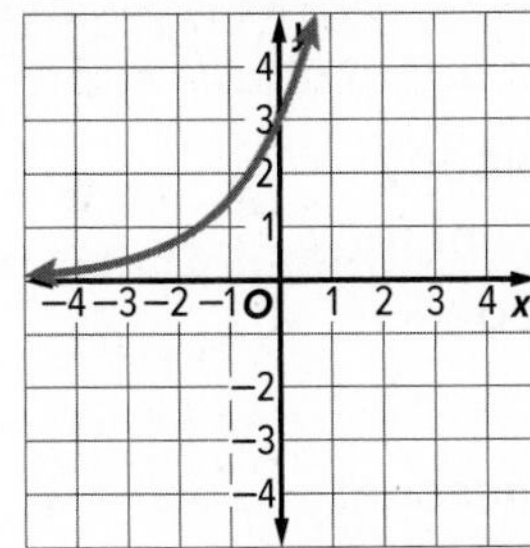

8.

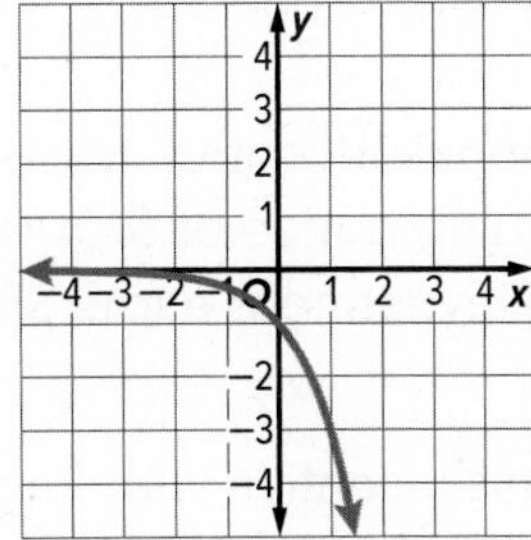

9.

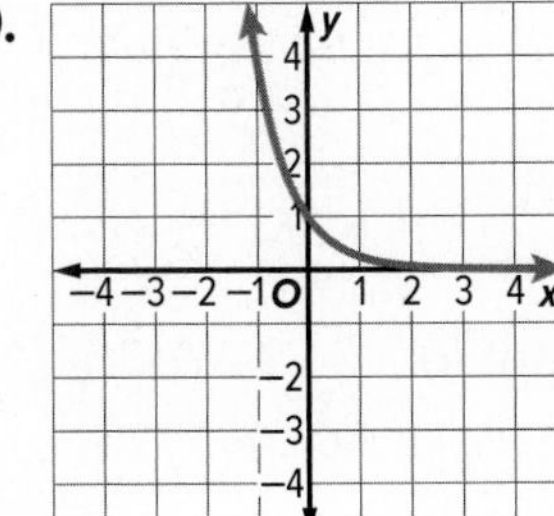

10.

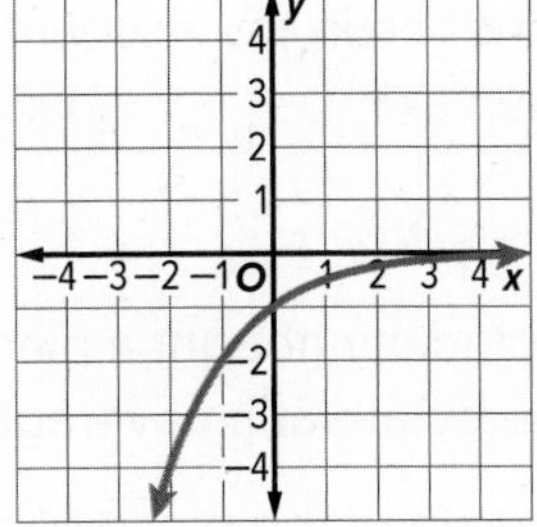

Example 3

11. BIOLOGY A certain species of bacteria in a laboratory culture begins with 50 cells and doubles in number every 30 minutes. Write a function to model the situation.

12. DEPRECIATION Amrita bought a new delivery van for $32,500. The value of this van depreciates at a rate of 12% each year. Write a function to model the value of the van after x years of ownership.

13. COMMUNICATION Cell phone usage grew about 23% each year from 2010 to 2016. If cell phone usage in 2010 was 43 million, write a function to model U.S. cell phone usage over that time period.

Examples 4–6

14. **INVESTING** Robyn invests \$1500 at 4.85% compounded quarterly. Write an equation to represent the amount of money she will have in t years.

15. **POPULATION** The population of New York City increased from 8,192,426 in 2010 to 8,550,405 in 2015. The annual rate of population increase for the period was about 0.9%.

    a. Write an equation for the population, P, t years after 2010.

    b. Use the equation to predict the population of New York City in 2025.

16. **SAVINGS** A company has a bonus incentive for its employees. The company pays employees an initial signing bonus of \$1000 and invests that amount for the employees. Suppose the investment earns 8% interest compounded quarterly.

    a. If an employee receiving this incentive withdraws the balance of the account after 5 years, how much will be in the account?

    b. If an employee receiving this incentive withdraws the balance of the account after 35 years, how much will be in the account?

17. **MANUFACTURING** A textile company bought a piece of weaving equipment for \$60,000. It is expected to depreciate at an average rate of 10% per year.

    a. Write an equation for the value of the piece of equipment Z after t years.

    b. Find the value of the piece of equipment after 6 years.

18. **HIGHER EDUCATION** The table lists the average annual costs of attending a four-year college in the United States during a recent year.

College Sector	Tuition and Fees	Room and Board
Four-year Public	\$9,410	\$10,138
Four-year Private	\$32,410	\$11,516

Source: College Board

Rayelle's parents plan to invest \$15,000 in a mutual fund earning an average of 4.5 percent interest, compounded monthly. After 15 years, for how many years will this investment be able to cover the tuition, fees, room, and board for Rayelle at a public college if costs stay the same? Round your answer to the nearest month.

19. DEPRECIATION The value of a home theater system depreciates by about 7% each year. Aeryn purchases a home theater system for \$3000. What is its value 4 years after purchase? Round your answer to the nearest hundred.

20. MONEY Hans opens a savings account by depositing \$1200. The account earns 0.2 percent interest compounded weekly. How much will be in the account in 10 years if he makes no more deposits? Assume that there are exactly 52 weeks in a year, and round your answer to the nearest cent.

21. POPULATION In 2016 the U.S. Census Bureau estimated the population of the United States at 322 million. If the annual rate of growth was about 0.81%, find the expected population at the time of the 2030 census. Round your answer to the nearest ten million.

Mixed Exercises

Write an exponential function for a graph that passes through the points.

22. (2, 1.4) and (4, 5.6) **23.** (1, 10.4) and (4, 665.6) **24.** (1, 42) and (3, 2688)

25. POPULATION The population of Camden, New Jersey, has been decreasing by 0.12% a year on average. If this trend continues, and the population was 79,318 in 2006, estimate Camden's population in 2025.

26. MEDICINE When doctors prescribe medication, they have to consider the rate at which the body filters a drug from the bloodstream. Suppose it takes the human body 6 days to filter out half of a certain vaccine. The amount of the vaccine remaining in the bloodstream x days after an injection is given by the equation $y = y_0(0.5)^{\frac{x}{6}}$, where y_0 is the initial amount. Suppose a doctor injects a patient with 20 μg (micrograms) of the vaccine.

   a. How much of the vaccine will remain after 1 day? Round your answer to the nearest tenth, if necessary.

   b. How much of the vaccine will remain after 12 days? Round your answer to the nearest tenth, if necessary.

   c. After how many days will the amount of vaccine be less than 1 μg?

27. USE TOOLS Graham invested money to save for a car. After x years, the value of Graham's investment can be modeled by the equation $y = 2400(0.95)^x$. How much did Graham originally invest? Is the value of his investment increasing or decreasing? Explain your reasoning. Use technology to find when the investment will be worth half of its starting value.

28. USE A MODEL There is a leak in a container that holds a certain nontoxic gas. Each hour, it loses 10% of its volume.

a. Write an equation that models the amount of gas left in the container after x hours, assuming there were 300 cubic centimeters in the container before the leak. Then use your equation to determine the amount of gas left in the container after 11 hours. Round your answer to the nearest tenth.

b. Dewanda believes a graph of this function should be a scatter plot instead of a continuous curve. Do you agree? Explain how this relates to the domain of the function.

29. STRUCTURE A wildlife researcher is studying the population of deer in a forest.

Years of Study	0	1	2	3
Population	128	160	200	250

a. The table shows the estimated number of deer in the forest over a 3-year period. Write an exponential function that fits this data and can be used to predict the deer population in future years.

b. The average rate of change is the change in the value of the dependent variable divided by the change in the value of the independent variable. What was the average rate of change in population during those three years?

c. If the population growth follows the model from **part a**, do you expect the deer population to continue to increase by the value you came up with in **part b**? Explain.

d. Use the values in the table to show how you know the function is exponential, not linear.

Higher-Order Thinking Skills

30. ANALYZE Determine the growth rate (as a percent) of a population that quadruples every year. Justify your argument.

31. PERSEVERE Santos invested $1200 into an account with an interest rate of 8% compounded monthly. Use a calculator to approximate how long it will take for Santos's investment to reach $2500.

32. ANALYZE The amount of water in a container doubles every minute. After 8 minutes, the container is full. After how many minutes was the container half-full? Justify your argument.

33. WRITE What should you consider when using exponential models to make decisions?

34. WRITE Compare and contrast the exponential growth formula and the exponential decay formula.

35. CREATE Honovi purchased a new car for $25,000 and has $5000 left to invest.

a. Choose an interest rate between 4% and 7% for Honovi's investment, and find the length of time it would take for the investment to double.

b. Choose an annual depreciation rate from 8% to 10% for the new car that Honovi purchased, and find the length of time it would take for the car's value to be equal to one-half of the purchase price.

c. Using the rates from part **a** and part **b**, find the length of time it would take for the investment to be equal to the value of the car. What is the value at that time?

Transforming Exponential Expressions

Example 1 Write Equivalent Exponential Expressions

BANKING **Savewell Bank offers a savings account with 0.15% interest compounded monthly, and Second Local Bank offers a savings account with 2% interest compounded annually.**

To compare the two accounts, we need to compare rates with the same compounding frequency. One way to do this is to compare the approximate monthly interest rates offered by each bank, which is also called the *effective* monthly interest rate.

Part A Compare monthly rates.

Write a function to represent the amount A that would be earned after t years with Second Local Bank. Then write an equivalent function that represents monthly compounding.

For convenience, let the initial amount of the investment be $1.

$y = a(1 + r)^t$	Equation for exponential growth
$A(t) = 1(1 + 0.02)^t$	$y = A(t)$, $a = 1$, $r = 2\%$ or 0.02
$A(t) = 1.02^t$	Simplify.

Then write a function that represents 12 compoundings per year, a power of $12t$, instead of 1 compounding per year, a power of $1t$.

$A(t) = 1.02^{1t}$	Original function
$A(t) = 1.02^{\left(\frac{1 \cdot 12}{12}\right)t}$	$1 \text{ year} = \frac{1 \text{ year}}{12 \text{ months}} \cdot 12 \text{ months}$
$A(t) = \left(1.02^{\frac{1}{12}}\right)^{12t}$	Power of a Power
$A(t) \approx 1.00165^{12t}$	$(1.02)^{\frac{1}{12}} = \sqrt[12]{1.02}$ or about 1.00165

The effective monthly interest rate offered by Second Local Bank is about 0.00165 or about 0.165% per month. It is slightly more than the 0.15% offered by Savewell Bank. So, Second Local Bank is a better choice.

Part B Compare annual rates.

Write a function to represent the amount A earned after t months by Savewell Bank. Then write an equivalent function that represents annual compounding.

$y = a(1 + r)^t$	Equation for exponential growth
$A(t) = 1(1 + 0.0015)^t$	$y = A(t)$, $a = 1$, $r = 0.15\%$ or 0.0015
$A(t) = 1.0015^t$	Simplify.

(continued on the next page)

Today's Goal
- Use the properties of exponents to transform expressions for exponential functions.

Go Online You can watch a video to see how to compare savings accounts.

$A(t) = 1.0015^t$ represents the amount earned with a savings account at Savewell Bank after t months.

Talk About It

Does the result in **Part B** make sense compared to the result of **Part A**? Explain.

Write an equivalent function that represents 1 compounding per year. Since there are 12 months in a year, the exponent should be $\frac{1}{12}t$.

$A(t) = 1.0015^{1t}$	Original function
$A(t) = 1.0015^{(12 \cdot \frac{1}{12})t}$	1 year = 12 months $\cdot \frac{\text{1 year}}{\text{12 months}}$
$A(t) = (1.0015^{12})^{\frac{1}{12}t}$	Power of a Power
$A(t) = (1.0181)^{\frac{1}{12}t}$	$1.0015^{12} \approx 1.0181$

From this expression, we can determine that the effective annual interest rate of Savewell Bank is about 0.0181, or about 1.81%, which is less than the 2% interest rate offered by Second Local Bank.

Check

SAVINGS Tareq is planning to invest money into a savings account. Oak Hills Financial offers 3.1% interest compounded annually. First City Bank has savings accounts with a quarterly compounded interest rate of 0.7%.

Part A Write the expression $A(t)$ to represent the amount that Tareq earns after t quarters through Oak Hills Financial.

$A(t) \approx$ __?__

What is the effective quarterly interest rate of Oak Hills Financial, rounded to the nearest hundredth?

Part B Write the expression $A(t)$ to represent the amount that Tareq earns after t years through First City Bank.

$A(t) \approx$ __?__

What is the effective annual rate of First City Bank, rounded to the nearest hundredth?

__?__ is the better bank for Tareq's savings account.

Go Online You can complete an Extra Example online.

Practice

Go Online You can complete your homework online.

Example 1

1. **INVESTING** Kimiyo is planning to invest money in a savings account. She is comparing the interest rates of savings accounts at two banks. Bank A offers a savings account with 2.1% interest compounded annually. Bank B offers a savings account with a quarterly compounded interest rate of 0.8%.

   a. Write a function to represent the amount A that Kimiyo would earn after t years through Bank A, assuming an initial investment of $1. Then write an equivalent function that represents quarterly compounding.

   b. Which is the better plan? Explain.

   c. What is the approximate effective annual interest rate at Bank B? How does your result relate to your answer to part **b**?

2. **COLLECTIONS** Keandra is comparing the growth rates in the value of two items in a collection. The value of a necklace increases by 3.2% per year. The value of a ring increases by 0.33% per month.

   a. Write a function to represent the value A of the necklace after t years, assuming an initial value of $1. Then write an equivalent function that represents monthly compounding.

   b. Which item is increasing in value at a faster rate? Explain.

   c. What is the approximate annual rate of growth of the ring? How does your result relate to your answer to part **b**?

3. **SAVINGS** Amir is trying to decide between two savings account plans at two different banks. He finds that Bank A offers a quarterly compounded interest rate of 0.95%, while Bank B offers 3.75% interest compounded annually. Which is the better plan? Explain.

4. **BACTERIA** The scientist found that Bacteria A has a growth rate of 0.99% per minute, while Bacteria B has a growth rate of 0.018% per second. Determine which bacterium has a faster growth rate. Explain.

5. **POPULATION** The population of Species A is decreasing at a rate of about 0.25% per quarter. The population of Species B is decreasing at a rate of about 1.34% per year. Determine which species has a population that is decreasing at a faster rate. Explain.

Mixed Exercises

6. **POPULATION** The table shows the population of two small towns that experience increases in population.

Year	Population Town A	Population Town B
2012	8,000	9,500
2013	8,480	9,975
2014	8,989	10,474
2015	9,528	10,997
2016	10,100	11,547

   a. Write a function that can be used to estimate the population $P(t)$ of Town A t years after 2012.

   b. Write a function that can be used to estimate the population $P(t)$ of Town B t years after 2012.

   c. Use your equations and properties of exponents to find the approximate effective monthly increase in the populations of Town A and Town B.

7. **ACCOUNTS** Dominic is trying to decide between two checking account plans. Plan A offers a monthly compounded interest rate of 0.05%, while Plan B offers 0.5% interest compounded annually. Which is the better plan?

8. **CAR DEPRECIATION** Juana is deciding between two cars to purchase. Car A depreciates annually at a rate of 3.5%, while Car B depreciates monthly at a rate of 0.32%. Which car has a better effective rate of depreciation?

9. **INVESTMENT** As a wedding gift, Dotty and Brad received $10,000 cash from Dotty's grandparents. The couple is trying to decide where to invest the money. Account A offers 2.3% interest compounded semi-annually. Account B offers 4.2% interest compounded annually. Which account has the better rate? Explain.

10. **SAVINGS** Hernando is deciding between two certificate of deposit accounts. Account Y offers 4.5% interest compounded annually. Account Z offers 1.13% interest compounded quarterly. Which is the better deal? Explain.

11. **FINANCE** Gita is deciding between two retirement accounts. Account A offers 0.5% interest compounded monthly. Account B offers 2.5% interest compounded annually. Which is the better deal? Explain.

12. **WILDLIFE** The table shows that the population of hawks in two different nature preserves has been decreasing.

Year	Hawk Population (Nature Preserve A)	Hawk Population (Nature Preserve B)
2013	114	120
2014	111	115
2015	108	110
2016	105	106

   a. Write a function that can be used to estimate the population $P(t)$ of the hawks in Nature Preserve A t years after 2013.

   b. Write a function that can be used to estimate the population $P(t)$ of the hawks in Nature Preserve B t years after 2013.

   c. Use your equations and properties of exponents to find the approximate effective quarterly decrease in population of hawks in Nature Preserve A and Nature Preserve B.

Higher-Order Thinking Skills

13. **PERSEVERE** The rate at which an object cools is related to the temperature of the surrounding environment. At the time of an experiment, Mrs. Haubner's lab temperature was 72°F. The approximate temperature of the water at time t in minutes in Mrs. Haubner's lab is predicted by the function $T(t) = 72 + (212 - 72)2.72^{-0.4t}$, where −0.4° per minute is defined as the rate of cooling. Rewrite this function so that the coefficient of t in the exponent is 1.

14. **WRITE** Explain why it is important for a consumer to compare rates in the same unit before making a purchase.

15. **CREATE** Write a scenario that compares two accounts with interest rates compounded at different rate units. Then determine which account has the better rate.

16. **FIND THE ERROR** Marsha is opening a savings account. Eagle Savings Bank is offering her an account with a 0.13% monthly interest rate, while Admiral Savings Bank is offering an account with a 1% annual interest rate. Marsha believes the account at Admiral Savings bank is better because 1% is a greater interest rate than 0.13%. Why is Marsha incorrect? Explain your reasoning.

Geometric Sequences

Explore Modeling Geometric Sequences

Online Activity Use a real-world situation to complete the Explore.

INQUIRY How can you create a formula to predict how a ball bounces?

Learn Geometric Sequences

A **geometric sequence** is a pattern of numbers that begins with a nonzero term and each term after is found by multiplying the previous term by a nonzero constant *r*. This constant is called the **common ratio**. Dividing a term by the previous term results in the common ratio.

To find the common ratio, divide each term by the previous term. Then, write the ratio in simplest form.

Example 1 Geometric Sequences

Determine whether the sequence −432, 144, −48, 16, ... is geometric. Explain.

$\frac{144}{-432} = -\frac{1}{3}$ $\quad \frac{-48}{144} = -\frac{1}{3}$ $\quad \frac{16}{-48} = -\frac{1}{3}$

Since the ratio is the same for all of the terms, $-\frac{1}{3}$, the sequence is geometric.

Example 2 Identify Geometric Sequences

Determine whether the sequence 16, 12, 8, 4, ... is geometric. Explain.

$\frac{12}{16} = \frac{3}{4}$ $\quad \frac{8}{12} = \frac{2}{3}$ $\quad \frac{4}{8} = \frac{1}{2}$

The ratios are not the same, so the sequence is not geometric.

Go Online You can complete an Extra Example online.

Today's Goals
- Identify and generate geometric sequences.
- Construct and use exponential functions for geometric sequences.

Today's Vocabulary
geometric sequence
common ratio

Talk About It!
Why must neither the first term nor the common ratio of a geometric sequence be zero?

Watch Out!
Find the Common Ratio Be sure to write the term as the numerator and the previous term as the denominator when finding the common ratio. Otherwise, you will be calculating the reciprocal of the common ratio.

Check

Determine whether each sequence is geometric. If so, determine its common ratio.

a.

n	1	2	3	4	...
a_n	8	20	50	125	...

This sequence __?__ a geometric sequence. It has a common ratio of __?__.

b. −0.7, 0.07, −0.007, 0.0007, ...

This sequence __?__ a geometric sequence. It has a common ratio of __?__.

Think About It!

How could you determine a term prior to any given term of a geometric sequence?

Example 3 Find Terms of Geometric Sequences

Find the next three terms in each geometric sequence.

a. 64, 16, 4, 1, ...

Step 1 Find the common ratio.

$\frac{16}{64} = \frac{1}{4}$ $\quad$ $\frac{4}{16} = \frac{1}{4}$ $\quad$ $\frac{1}{4} = \frac{1}{4}$

The common ratio is $\frac{1}{4}$.

Step 2 Multiply by the common ratio.

$1 \times \frac{1}{4} = \frac{1}{4}$ $\quad$ $\frac{1}{4} \times \frac{1}{4} = \frac{1}{16}$ $\quad$ $\frac{1}{16} \times \frac{1}{4} = \frac{1}{64}$

The next three terms are $\frac{1}{4}$, $\frac{1}{16}$, and $\frac{1}{64}$.

b.

n	**1**	**2**	**3**	**4**	**...**
a_n	**8**	**12**	**18**	**27**	**...**

Step 1 Find the common ratio.

$\frac{12}{8} = 1.5$ $\quad$ $\frac{18}{12} = 1.5$ $\quad$ $\frac{27}{18} = 1.5$

The common ratio is 1.5.

Step 2 Multiply by the common ratio.

$27 \times 1.5 = 40.5$ $\quad$ $40.5 \times 1.5 = 60.75$ $\quad$ $60.75 \times 1.5 = 91.125$

The next three terms are 40.5, 60.75, and 91.125.

Check

Find the next three terms in each geometric sequence.

a. 729, 243, 81, __?__, __?__, __?__, ...

b. −4, −44, −484, ____?____, ____?____, ____?____, ...

Go Online You can complete an Extra Example online.

Learn Geometric Sequences as Exponential Functions

Key Concept • *n*th Term of a Geometric Sequence

The *n*th term a_n of a geometric sequence with first term a_1 and common ratio r is given by the following formula, where n is any positive integer and $a_1, r \neq 0$.

$$a_n = a_1 r^{n-1}$$

Think About It!

When finding the *n*th term of a geometric sequence, why is r raised to the $n - 1$ power instead of to the *n*th power?

Example 4 Find the *n*th Term of a Geometric Sequence

Use an explicit formula to find the 11th term of each geometric sequence.

512, 256, 128, 64, ...

The first term of the sequence is 512. So, $a_1 = 512$.

Find the common ratio.

$\frac{256}{512} = \frac{1}{2}$ $\frac{128}{256} = \frac{1}{2}$ $\frac{64}{128} = \frac{1}{2}$

The common ratio is $\frac{1}{2}$.

Use the common ratio to find the 11th term of the sequence.

$a_n = a_1 r^{n-1}$	Formula for the *n*th term
$a_n = 512\left(\frac{1}{2}\right)^{n-1}$	$a_1 = 512$ and $r = \frac{1}{2}$
$a_{11} = 512\left(\frac{1}{2}\right)^{11-1}$	To find the eleventh term, $n = 11$.
$= 512\left(\frac{1}{2}\right)^{10}$	Simplify.
$= 512\left(\frac{1}{1024}\right)$	$\left(\frac{1}{2}\right)^{10} = \left(\frac{1}{1024}\right)$
$= \frac{1}{2}$	Simplify.

Watch Out!

Exponents Remember that the base, which is the common ratio, is raised to $n - 1$ instead of n.

Check

Write the equation for the *n*th term of the geometric sequence.

n	1	2	3	4	...
a_n	729	243	81	27	...

$a_n =$ ___?___

Find the 8th term of the sequence.

Go Online You can complete an Extra Example online.

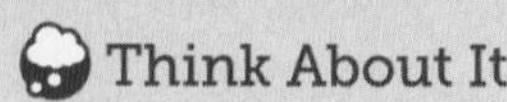

What assumption is made when calculating the population of North Dakota in 2030?

Example 5 Use a Geometric Sequence

POPULATION North Dakota's population is increasing more quickly than any other state's population. In 2011, the population was 685,242, and it has been increasing by an average of 2.5% each year. If this trend continues, determine the estimated population in 2030.

Since the population is growing exponentially, we can apply the equation for exponential growth, $y = a(1 + r)^t$ to determine the common ratio. An increase of 2.5% means that the population is being multiplied by $1 + 0.025$, or 1.025, each year. So, $r = 1.025$. Since $a_1 = 685{,}242$ in 2011, the population in 2030 is represented by the twentieth term, a_{20}.

$a_n = a_1 r^{n-1}$	Formula for the *n*th term
$a_{20} = 685{,}242(1.025)^{20-1}$	$a_1 = 685{,}242$, $r = 1.025$, $n = 20$
$a_{20} = 685{,}242(1.025)^{19}$	Simplify.
$a_{20} = 1{,}095{,}462$	Use a calculator.

In 2030, the estimated population of North Dakota will be 1,095,462.

Check

BOUNCES A rubber bouncy ball is dropped from a height of 5 feet. Each time the ball bounces back to 85% of the height from which it fell. Determine the height of the ball after 6 bounces and after 10 bounces. Round to the nearest hundredth.

$a_6 =$ ___?___ ft

$a_{10} =$ ___?___ ft

Dmitry Kaminsky/Shutterstock

Go Online You can complete an Extra Example online.

Practice

Go Online You can complete your homework online.

Examples 1 and 2

Determine whether each sequence is geometric. Explain.

1. 4, 1, 2, ...

2. 10, 20, 30, 40, ...

3. 4, 20, 100, ...

4. 212, 106, 53, ...

5. −10, −8, −6, −4, ...

6. 5, −10, 20, 40, ...

7. −96, −48, −24, −12, ...

8. 7, 13, 19, 25, ...

9. 3, 9, 81, 6561, ...

10. 108, 66, 141, 99, ...

11. $\frac{3}{8}, -\frac{1}{8}, -\frac{5}{8}, -\frac{9}{8}, \ldots$

12. $\frac{7}{3}$, 14, 84, 504, ...

Example 3

Find the next three terms in each geometric sequence.

13. 2, −10, 50, ...

14. 36, 12, 4, ...

15. 4, 12, 36, ...

16. 400, 100, 25, ...

17. −6, −42, −294, ...

18. 1024, −128, 16, ...

19. 2, 6, 18, ...

20. 2500, 500, 100, ...

21. $\frac{4}{5}, \frac{2}{5}, \frac{1}{5}, \ldots$

22. −4, 24, −144, ...

23. 72, 12, 2, ...

24. −3, −12, −48, ...

Example 4

Use an explicit formula to find the 10th term of each geometric sequence.

25. 1, 9, 81, 729, ...

26. 2, 8, 32, 128, ...

27. −9, 27, −81, 243, ...

28. 6, −24, 96, −384, ...

Example 5

29. MUSEUMS The table shows the annual visitors to a museum in millions. Write an equation for the projected number of visitors after n years.

Year	Visitors (millions)
1	4
2	6
3	9
4	$13\frac{1}{2}$
n	?

30. WORLD POPULATION The CIA estimates that the world population is growing at a rate of 1.167% each year. The world population in 2015 was about 7.3 billion.

a. Write an equation for the world population after n years.

b. Find the estimated world population in 2025.

31. DEPRECIATION Te'Andra has a computer system that she bought for $5000. Each year, the computer system loses one-fifth of its then-current value. How much money will the computer system be worth after 6 years?

Mixed Exercises

32. POPULATION The table shows the projected population of the United States through 2060. Does this table show an *arithmetic sequence*, a *geometric sequence,* or neither? Explain.

Year	Population
2020	334,503,000
2030	359,402,000
2040	380,219,000
2050	398,328,000
2060	416,795,000

Source: U.S. Census Bureau

33. SAVINGS ACCOUNTS A bank offers a savings account that earns 0.5% interest each month.

a. Write an equation for the balance of the savings account after n months.

b. Given an initial deposit of $500, what will the account balance be after 15 months?

34. Write an equation for the nth term of the geometric sequence 3, −24, 192, ... Then find the 9th term of this sequence.

35. Write an equation for the nth term of the geometric sequence $\frac{9}{16}, \frac{3}{8}, \frac{1}{4}, \ldots$. Then find the 7th term of this sequence.

36. Write an equation for the nth term of the geometric sequence 1000, 200, 40, ...Then find the 5th term of this sequence.

37. Write an equation for the *n*th term of the geometric sequence $-8, -2, -\frac{1}{2}, \ldots$.

Find the 8th term of this sequence.

38. Write an equation for the *n*th term of the geometric sequence 32, 48, 72, ...

Find the 6th term of this sequence.

39. **USE A SOURCE** Research the average annual salary for a 25-year-old and the average rate of increase in salary per year. Then write an equation for the *n*th year of employment. Find the 20th term of this sequence, and explain what it means.

40. **STRUCTURE** For each of the geometric sequences below, fill in the missing terms, write the corresponding exponential equation, and use the exponential equation to determine the 10th term in the sequence.

a. 0.5, 6, __________; $f(x) =$ ________; 10th term: __________

b. __, 10, __, 40, __; $g(x) =$ ______; 10th term: ____

41. **REASONING** Find the previous three terms of the geometric sequence −192, −768, −3072, ...

42. **STATE YOUR ASSUMPTION** Consider two different geometric sequences. Each starts with the same constant. The common ratio producing subsequent terms in the first is positive and is the reciprocal of the common ratio producing subsequent terms in the second. How would the graphs of the two sequences compare? Think about intercepts, asymptotes, and end behavior. Then graph an example of the situation.

43. **REASONING** You have just been offered a part-time job. The employer offers two different methods of payment. They are shown in the table.

Month	Method 1 Payment	Method 2 Payment
1	$100.00	$0.01
2	$108.00	$0.02
3	$116.00	$0.04
4	$124.00	$0.08

a. Describe the two different methods of payment being offered.

b. What kind of mathematical equations can you use to model each situation? How do you know? Write each equation.

c. You are planning to work at this job for two years. Your manager promises to raise your salary the way it is described in the table, as long as you meet the minimum performance rating each month. Which payment plan would you choose? Explain your reasoning.

44. **CONSTRUCT ARGUMENTS** The terms of a geometric sequence are defined by the equation $a_n = 512(0.5)^x$. A second sequence contains the terms $b_3 = 7168$ and $b_7 = 28$.

a. Determine which sequence has the greater common ratio.

b. What is the initial term of each sequence? Explain your reasoning.

45. REGULARITY The sum of the interior angles of a triangle is 180°. The interior angles of a pentagon add to 540°. Is the relationship between the number of sides in a polygon and the sum of interior angles a geometric sequence? Use the sum of the measures of the interior angles of a square to justify your answer.

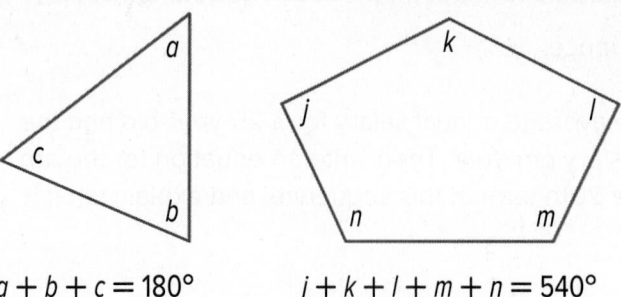

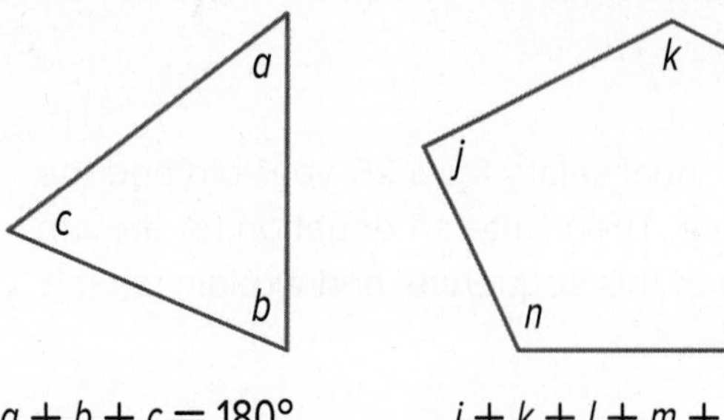

$a + b + c = 180°$ $\qquad$ $j + k + l + m + n = 540°$

Higher-Order Thinking Skills

46. PERSEVERE Write a sequence that is both geometric and arithmetic. Explain your answer.

47. FIND THE ERROR Haro and Matthew are finding the ninth term of the geometric sequence −5, 10, −20, … . Is either of them correct? Explain your reasoning.

Haro	Matthew
$r = \frac{10}{-5}$ or -2	$r = \frac{10}{-5}$ or -2
$a_9 = -5(-2)^{9-1}$	$a_9 = -5 \cdot (-2)^{9-1}$
$= -5(512)$	$= -5 \cdot -256$
$= -2560$	$= 1280$

48. ANALYZE Write a sequence of numbers that form a pattern but are neither arithmetic nor geometric. Justify your argument.

49. WRITE How are graphs of geometric sequences and exponential functions similar? How are they different?

50. WRITE Summarize how to find a specific term of a geometric sequence.

51. CREATE Give a counterexample for the following statement:
As n increases in a geometric sequence, the value of a_n will move farther away from zero.

52. CREATE Write a geometric sequence. Then explain why your sequence is geometric.

Recursive Formulas

Learn Using Recursive Formulas

An **explicit formula** allows you to find any term of a sequence by using a formula written in terms of n. A **recursive formula** allows you to find the nth term of a sequence by performing operations to one or more of the preceding terms.

Example 1 Recursive Formula for an Arithmetic Sequence

Find the first five terms of the sequence $a_1 = 7$ and $a_n = a_{n-1} - 9$ if $n \geq 2$.

Use $a_1 = 7$ and the recursive formula to find the next four terms.

$a_2 = a_{2-1} - 9$	$n = 2$	$a_3 = a_{3-1} - 9$	$n = 3$
$= a_1 - 9$	Simplify.	$= a_2 - 9$	Simplify.
$= 7 - 9$	$a_1 = 7$	$= -2 - 9$	$a_2 = -2$
$= -2$	Simplify.	$= -11$	Simplify.
$a_4 = a_{4-1} - 9$	$n = 4$	$a_5 = a_{5-1} - 9$	$n = 5$
$= a_3 - 9$	Simplify.	$= a_4 - 9$	Simplify.
$= -11 - 9$	$a_3 = -11$	$= -20 - 9$	$a_4 = -20$
$= -20$	Simplify.	$= -29$	Simplify.

The first five terms of the sequence are 7, −2, −11, −20, and −29.

Example 2 Recursive Formula for a Geometric Sequence

Find the first five terms of the sequence $a_1 = 5$ and $a_n = 3a_{n-1}$ if $n \geq 2$.

n	$a_n = 3a_{n-1}$	a_n
1	—	5
2	$a_n = 3(5)$	15
3	$a_n = 3(15)$	45
4	$a_n = 3a_{4-1}$	135
5	$a_n = 3a_{5-1}$	405

The first five terms of the sequence are 5, 15, 45, 135, and 405.

Go Online You can complete an Extra Example online.

Today's Goals
- Calculate terms in sequences by using recursive formulas.
- Write arithmetic and geometric sequences recursively and use them to model situations.

Today's Vocabulary
explicit formula
recursive formula

Study Tip

Recursive and Explicit Formulas Recursive formulas are used for generating sequences of numbers. They are not as useful for finding, for example, the fiftieth term of a sequence since you would first have to find terms one through forty-nine. For this type of calculation, it is better to use an explicit formula.

Explore Writing Recursive Formulas from Sequences

Online Activity Use an interactive tool to complete the Explore.

> **INQUIRY** How can you write a formula that relates the numbers in a geometric sequence?

Talk About It!

When writing a recursive formula for an arithmetic or geometric sequence, how do you know which formula to use?

Learn Writing Recursive Formulas

Key Concept • Writing Recursive Formulas	
Step 1	Determine whether the sequence is arithmetic or geometric by finding a common difference or a common ratio.
Step 2	Write a recursive formula. Arithmetic Sequence $a_n = a_{n-1} + d$, where d is the common difference Geometric Sequence $a_n = r \cdot a_{n-1}$, where r is the common ratio
Step 3	State the first term and domain for n.

Example 3 Write a Recursive Formula Using a List

Write a recursive formula for 16, 48, 144, 432, ...

Step 1 Determine whether a common difference or ratio exists.

Subtract each term from the term that follows it to check for a common difference.

$48 - 16 = 32 \qquad 144 - 48 = 96 \qquad 432 - 144 = 288$

There is no common difference.

Check for a common ratio by dividing each term by the term that precedes it.

$\frac{48}{16} = 3 \qquad \frac{144}{48} = 3 \qquad \frac{432}{144} = 3$

The common ratio is 3. The sequence is geometric.

Step 2 Write a recursive formula.

$a_n = r \cdot a_{n-1}$ Recursive formula for geometric sequence

$a_n = 3 \cdot a_{n-1}$ $r = 3$

Step 3 State the first term and domain for n.

The first term a_1 is 16, and the domain of the function is $n \geq 1$.

A recursive formula for the sequence is $a_1 = 16$, $a_n = 3\, a_{n-1}$, $n \geq 2$.

Notice that n must be greater than or equal to 2 in the recursive formula.

Study Tip

***n*th Term** For the nth term of a sequence, the value of n must be a positive integer. Although we must still state the domain of n from this point forward, we will assume that n is an integer.

Go Online

You may want to complete the Concept Check to check your understanding.

Check

SOCIAL MEDIA The table shows the total number of views at the end of each day for a video.

Day	Views
1	100
2	9000
3	17,900
4	26,800

Write a recursive formula for the sequence.

$a_1 =$ ___?___

$a_n = a_{n-1}$ ___?___

Example 4 Write a Recursive Formula Using a Graph

Write a recursive formula for the graph.

Step 1 Find a common difference or common ratio, or determine that neither exists.

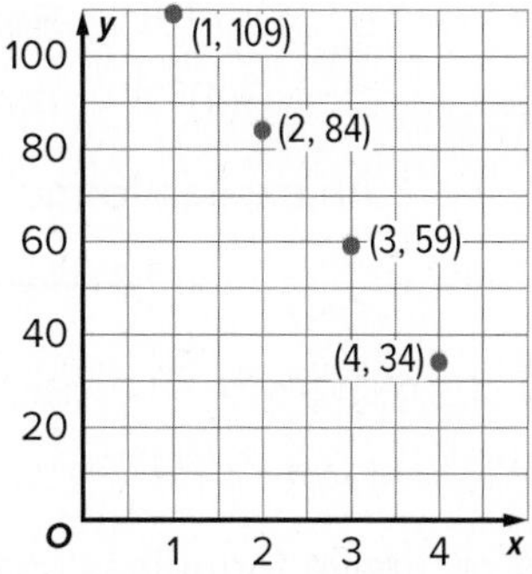

$84 - 109 = -25$

$59 - 84 = -25$

$34 - 59 = -25$

The common difference is -25. The sequence is arithmetic.

Step 2 Write a recursive formula.

$a_n = a_{n-1} + d$ Recursive formula for arithmetic sequence

$a_n = a_{n-1} + (-25)$ $d = -25$

Step 3 State the first term and domain for n.

The first term a_1 is 109, and $n \geq 1$.

A recursive formula for the sequence is $a_1 = 109$, $a_n = a_{n-1} - 25, n \geq 2$.

Example 5 Write Recursive and Explicit Formulas

MOVIES The premise of a movie is that a new virus is spreading, turning infected persons into zombie-like creatures. The table outlines the total number of infected persons at the end of each day.

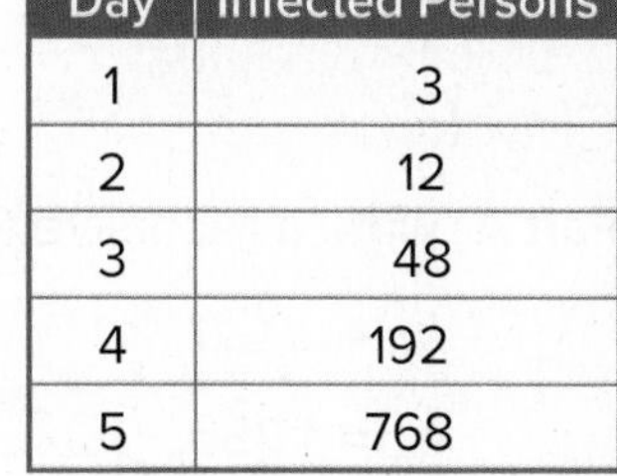

Day	Infected Persons
1	3
2	12
3	48
4	192
5	768

a. Write a recursive formula for the sequence.

Step 1 Find a common difference or common ratio.

$12 - 3 = 9$ $48 - 12 = 36$ $192 - 48 = 144$

There is no common difference. Check for a common ratio by dividing each term by the term that precedes it.

$\frac{12}{3} = 4$ $\frac{48}{12} = 4$ $\frac{192}{48} = 4$ $\frac{768}{192} = 4$

There is a common ratio of 4. The sequence is geometric.

(continued on the next page)

Think About It!

How can you make sure that your recursive formula is correct?

Math History Minute

Hungarian mathematician **Rózsa Péter** (1905–1977) was the first Hungarian female mathematician to become an Academic Doctor of Mathematics. She helped to establish the modern field of recursive function theory, and she was the author of *Playing with Infinity: Mathematical Explorations and Excursions.*

Step 2 Write a recursive formula.

$a_n = r \cdot a_{n-1}$ Recursive formula for geometric sequence

$a_n = 4a_{n-1}$ $r = 4$

Step 3 State the first term and domain for n.

The first term a_1 is 3, and $n \geq 1$. A recursive formula for the sequence is $a_1 = 3$, $a_n = 4a_{n-1}$, $n \geq 2$.

b. Write an explicit formula for the sequence.

Steps 1 and 2 The common ratio is 4.

Step 3 Use the formula for the nth term of a geometric sequence.

$a_n = a_1 r^{n-1}$ Formula for the nth term

$= 3(4)^{n-1}$ $a_1 = 3$ and $r = 4$

An explicit formula for the sequence is $a_n = 3(4)^{n-1}$.

Think About It!

For the sequence in part **b**, find a_2.

Example 6 Translate Between Recursive and Explicit Formulas

A recursive formula is useful when finding a number of successive terms in a sequence. An explicit formula is useful when finding the nth term of a sequence. Therefore, it may be necessary to translate between the two forms.

a. Write a recursive formula for $a_n = 0.5n + 2$.

$a_n = 0.5n + 2$ is an explicit formula for an arithmetic sequence with $d = 0.5$ and $a_1 = 0.5(1) + 2$ or 2.5.

Therefore, a recursive formula for a_n is $a_1 = 2.5$, $a_n = a_{n-1} + 0.5$, $n \geq 2$.

Study Tip

Geometric Sequences Recall that the formula for the nth term of a geometric sequence is $a_n = a_1 r^{n-1}$.

b. Write an explicit formula for $a_1 = 1011$, $a_n = 1.25a_{n-1}$, $n \geq 2$.

$a_n = 1.25a_{n-1}$ is a recursive formula for a geometric sequence with $a_1 = 1011$ and $r = 1.25$.

Therefore, an explicit formula for a_n is $a_n = 1011 \cdot (1.25)^{n-1}$.

Check

Part A Write a recursive formula for $a_n = \frac{1}{2} + (n - 1)10$.

$a_1 =$ __?__

$a_n = 1.25a_{n-1}$ __?__

Part B Write an explicit formula for $a_1 = -60$, $a_n = 1.5a_{n-1}$, $n \geq 2$.

$a_n =$ __?__ (__?__)$^{n-1}$

Go Online You can complete an Extra Example online.

Practice

Go Online You can complete your homework online.

Examples 1 and 2

Find the first five terms of each sequence.

1. $a_1 = 23, a_n = a_{n-1} + 7, n \geq 2$

2. $a_1 = 48, a_n = -0.5a_{n-1} + 8, n \geq 2$

3. $a_1 = 8, a_n = 2.5a_{n-1}, n \geq 2$

4. $a_1 = 12, a_n = 3a_{n-1} - 21, n \geq 2$

5. $a_1 = 13, a_n = -2a_{n-1} - 3, n \geq 2$

6. $a_1 = \frac{1}{2}, a_n = a_{n-1} + \frac{3}{2}, n \geq 2$

Example 3

Write a recursive formula for each sequence.

7. 12, −1, −14, −27, ...

8. 27, 41, 55, 69, ...

9. 2, 11, 20, 29, ...

10. 100, 80, 64, 51.2, ...

11. 40, −60, 90, −135, ...

12. 81, 27, 9, 3, ...

Example 4

Write a recursive formula for each graph.

13.

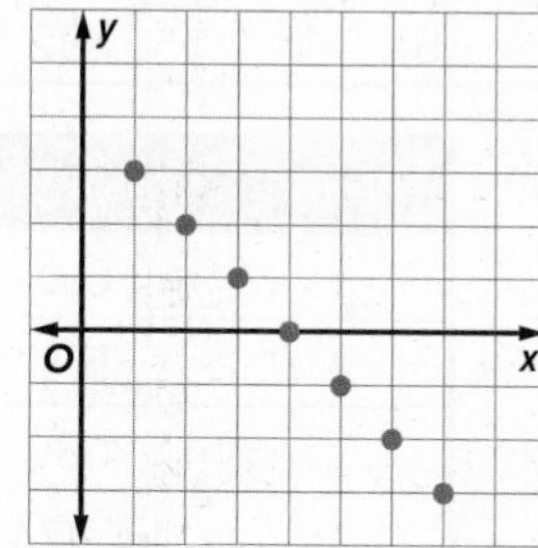

14.

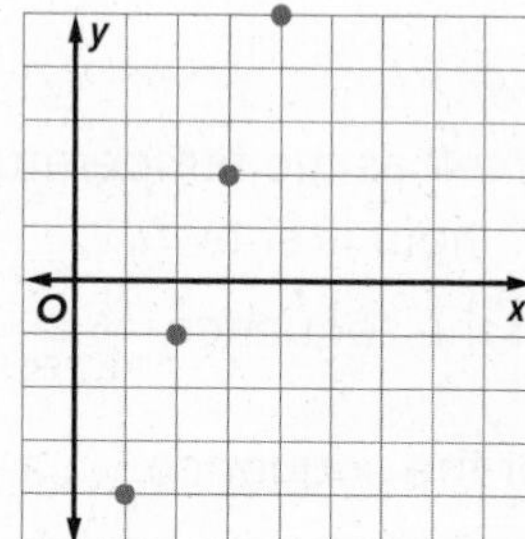

15.

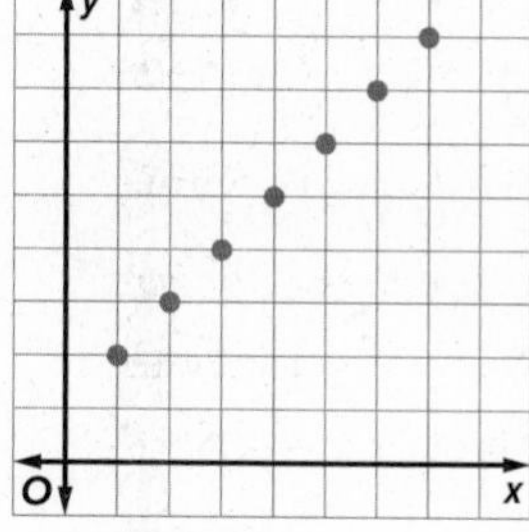

16.

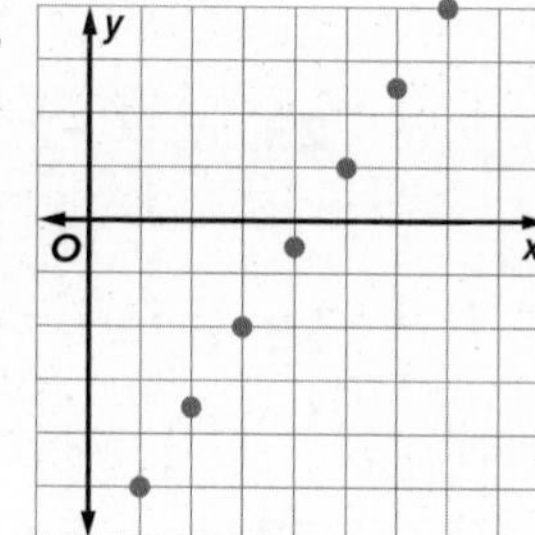

Write a recursive formula for each graph.

17.

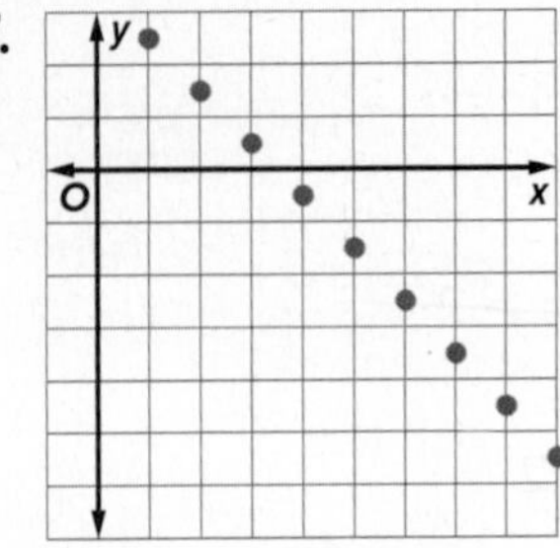

18.

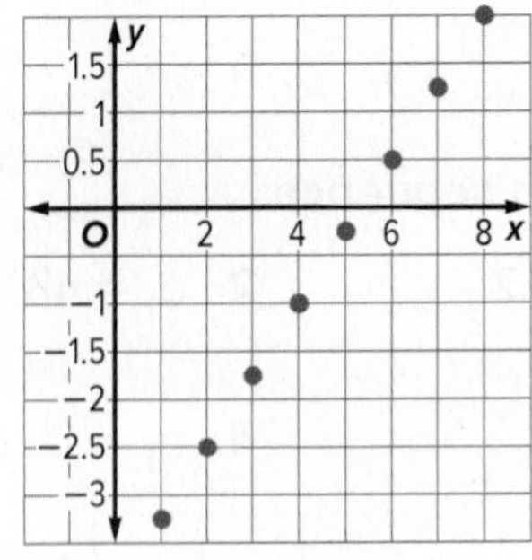

Example 5

19. VIRAL VIDEOS A viral video got 175 views in one hour, 350 views in two hours, 525 views in three hours, 700 views in four hours, and so on.

a. Find the next 5 terms in the sequence.

b. Write a recursive formula for the sequence.

c. Write an explicit formula for the sequence.

20. PAPER A piece of paper is folded several times. The number of sections into which the piece of paper is divided after each fold is shown.

a. Write a recursive formula for the sequence.

b. Write an explicit formula for the sequence.

Number of Folds	Sections
1	2
2	4
3	8
4	16
5	32

21. SNOW A snowman begins to melt as the temperature rises. The height of the snowman in feet after each hour is shown.

a. Write a recursive formula for the sequence.

b. Write an explicit formula for the sequence.

Hour	Height (ft)
1	6.0
2	5.4
3	4.86
4	4.374

Example 6

For each recursive formula, write an explicit formula. For each explicit formula, write a recursive formula.

22. $a_n = 3(4)^{n-1}$

23. $a_1 = -2$, $a_n = a_{n-1} - 12$, $n \geq 2$

24. $a_1 = 38$, $a_n = \frac{1}{2}a_{n-1}$, $n \geq 2$

25. $a_n = -7n + 52$

26. $a_1 = 38$, $a_n = a_{n-1} - 17$, $n \geq 2$

27. $a_n = 5n - 16$

28. $a_n = 50(0.75)^{n-1}$

29. $a_1 = 16$, $a_n = 4a_{n-1}$, $n \geq 2$

Mixed Exercises

30. CLEANING An equation for the cost a_n in dollars that a carpet cleaning company charges for cleaning n rooms is $a_n = 50 + 25(n - 1)$. Write a recursive formula to represent the cost a_n.

31. SAVINGS A recursive formula for the balance of a savings account a_n in dollars at the beginning of year n is $a_1 = 500$, $a_n = 1.05a_{n-1}$, $n \geq 2$. Write an explicit formula to represent the balance of the savings account a_n.

32. USE TOOLS In 2010, County A had a population of 1.3 million people. The largest factory in the area produced 1700 million widgets per year. The population of County A is projected to grow at 1.2% per year, and the number of widgets produced is expected to grow by 10 million per year.

a. Develop explicit formulas for the population and annual widget production, in millions, as functions of the number of years n after 2010.

b. The graph of $y = 1700 + \frac{10x}{1.3}(1.012)^x$ represents the annual widget production per person for County A from 2010 to 2020, where x is the number of years after 2010. The next-highest widget-producing county produces widgets at a constant rate of 1200 widgets per person. Use a graphing calculator to extend the graph and find the year when County A will no longer be the leader in widget production. Explain your results.

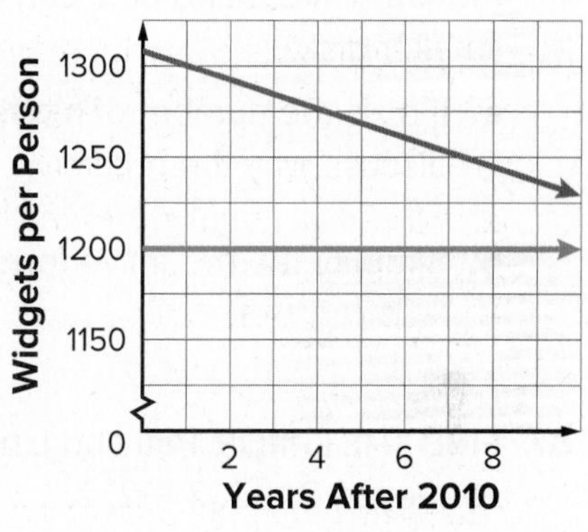

33. USE A MODEL Ramon has been tracing his family tree with his parents. He claims that he has over 250 great- great- great- great- great- great-grandparents. Is this possible? Write both an explicit and recursive formula for this situation.

34. REASONING Carl Friedrich Gauss, a German mathematician of the 1700s, was asked as a young boy for the sum of the integers from 1 to 100, and he unhesitatingly replied with the correct answer.

a. Identify the type of the sequence 1, 2, 3, ... 100, and explore a way to find its sum based on grouping pairs of numbers from each end of the sequence. Then explain how Guass was able to find the sum so quickly.

b. Find an explicit formula for the sum S of n terms of an arithmetic sequence whose first term is a_1 and whose nth, or last, term is a_n.

35. REGULARITY The first ten numbers in the Fibonacci sequence can be defined by $a_{n+1} = a_n + a_{n-1}$, and each ratio $\frac{a_n}{a_{n-1}}$ can be computed using a spreadsheet (see column C).

a. Which spreadsheet formulas could have been used to calculate the entries in cells B3 and C2?

b. Compute the ratio $\frac{a_n}{a_{n-1}}$ up to $n = 50$. What do you observe?

Fibonacci sequence

	A	B	C
1	1	1	
2	2	1	1
3	3	2	2
4	4	3	1.5
5	5	5	1.666667
6	6	8	1.6
7	7	13	1.625
8	8	21	1.615385
9	9	34	1.619048
10	10	55	1.617647

Sheet 1 / Sheet 2 / Sheet 3

36. STRUCTURE There is a famous puzzle called the "Tower of Hanoi." There are three pegs, and a certain number of disks of varying sizes can be set on each peg. The puzzle starts with the disks in a stack on the left-most peg, with the largest disk on the bottom and the disks getting smaller as they are stacked. The goal is to move the disks from the left-most peg to the right-most peg while obeying three rules. First, only one disk can be moved at a time. Second, only the top disk on any peg can be moved. Third, at no time can a larger disk be placed on a smaller disk.

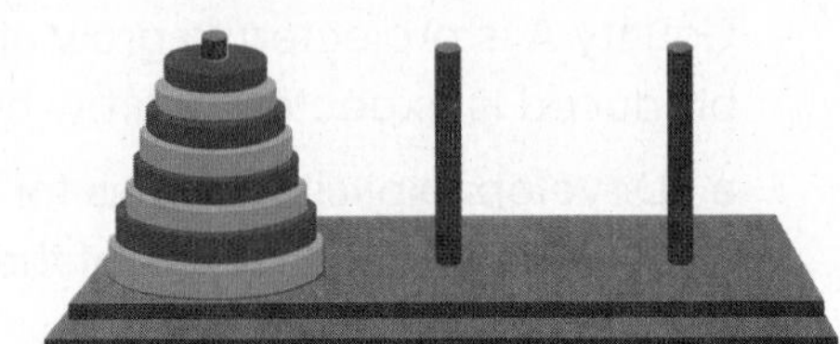

a. If a_n is the number of moves it takes to solve a puzzle consisting of n disks, discuss why the recursive formula $a_n = a_{n-1} + 1 + a_{n-1}$ makes sense.

b. Simplify the recursive formula. What is a_1? Why?

37. FIND THE ERROR Pati and Linda are working on a math problem that involves the sequence 2, −2, 2, −2, 2, … . Pati thinks that the sequence can be written as a recursive formula. Linda believes that the sequence can be written as an explicit formula. Is either of them correct? Explain your reasoning.

38. PERSEVERE Find a_1 for the sequence in which $a_4 = 1104$ and $a_n = 4a_{n-1} + 16$.

39. ANALYZE Determine whether the following statement is *true* or *false*. Justify your argument. *There is only one recursive formula for every sequence.*

40. PERSEVERE Find a recursive formula for 4, 9, 19, 39, 79, … .

41. WRITE Explain the difference between an explicit formula and a recursive formula.

42. CREATE Give a counterexample for the following statement: In a recursive sequence, if $a_1 = a_2$, then $a_2 = a_3$, and so on.

Review

Essential Question

When and how can exponential functions represent real-world situations?

Exponential functions can be used in real life to represent situations that grow or decay. One example is representing compound interest.

Module Summary

Lesson 8-1

Exponential Functions

- Functions of the form $y = ab^x$, where $a \neq 0$ and $b > 1$, are exponential growth functions.
- Functions of the form $y = ab^x$, where $a \neq 0$ and $0 < b < 1$, are exponential decay functions.
- The graphs of exponential functions have an asymptote.

Lessons 8-2 through 8-4

Transforming and Writing Exponential Functions

- The graph $f(x) = b^x$ is a parent graph of an exponential function.
- The graph of $g(x) = b^x + k$ is the graph of $f(x) = b^x$ translated vertically.
- The graph of $g(x) = b^{x-h}$ is the graph of $f(x) = b^x$ translated horizontally.
- The graph $g(x) = ab^x$ is the graph of $f(x) = b^x$ stretched or compressed vertically by a factor of $|a|$.
- The graph $g(x) = b^{ax}$ is the graph of $f(x) = b^x$ stretched or compressed horizontally by a factor of $\frac{1}{|a|}$.
- When an exponential function $f(x)$ is multiplied by -1, the result is a reflection across the x- or y-axis.
- In the equation $y = a(1 + r)^t$, y is the final amount, a is the initial amount, r is the rate of change expressed as a decimal, and t is time.

Lesson 8-5

Geometric Sequences

- A geometric sequence is a pattern of numbers that begins with a nonzero term and each term after is found by multiplying the previous term by a nonzero constant r.
- The nth term a_n of a geometric sequence with first term a_1 and common ratio r is given by the formula $a_n = a_1r^{n-1}$, where n is any positive integer, $a_1 \neq 0$, and $r \neq 0$.

Lesson 8-6

Recursive Functions

- An explicit formula allows you to find any term a_n of a sequence by using a formula written in terms of n.
- To write a recursive formula for an arithmetic or geometric sequence, determine whether the sequence is arithmetic or geometric by finding a common difference or a common ratio.

Study Organizer

Foldables

Use your Foldable to review this module. Working with a partner can be helpful. Ask for clarification of concepts as needed.

Transformations | Growth and Decay | Sequences

Test Practice

1. GRAPH The table shows the function $y = 2^x - 1$. (Lesson 8-1)

x	y
0	0
1	1
2	3
3	7

Graph the function.

2. MULTIPLE CHOICE The table shows the number of text messages Ernesto sent each month. (Lesson 8-1)

Month	Text Messages
April	2
May	6
June	18
July	54

What type of behavior is shown in the table?

A. linear

B. piece-wise

C. exponential

D. none of the above

3. OPEN RESPONSE Describe the end behavior of the graph of the exponential function shown on the graph. (Lesson 8-1)

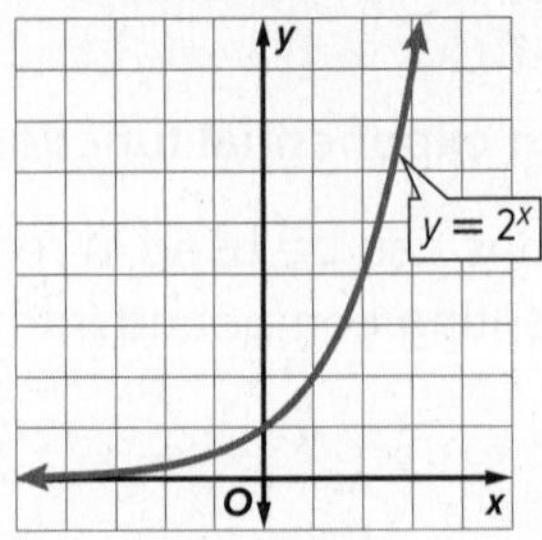

4. MULTIPLE CHOICE Consider the graph. Which function represents the reflection of the parent function $f(x) = 3^x$ across the y-axis? (Lesson 8-2)

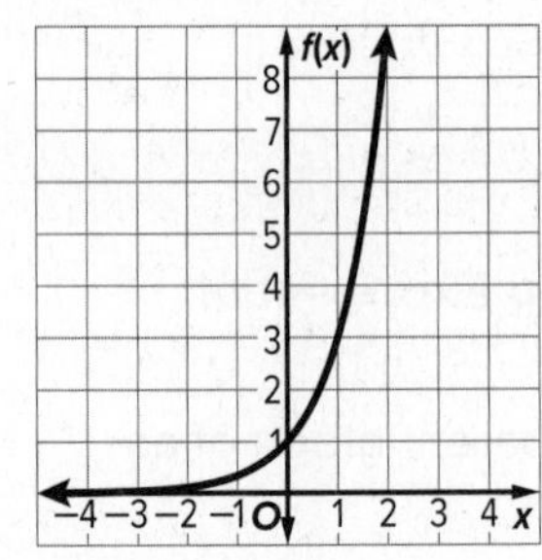

A. $f(x) = -3^x$

B. $f(x) = 3^{-x}$

C. $f(x) = 3^{-2x}$

D. $f(x) = -2(3)^x$

5. **MULTIPLE CHOICE** Describe the translation in $h(x) = 2^x + 5$ as it relates to the parent function $h(x) = 2^x$. (Lesson 8-2)

A. Up 5 units

B. Down 5 units

C. Right 5 units

D. Left 5 units

6. **OPEN RESPONSE** Horticulturists can estimate the number of hybrid plants of a certain type they will sell based on the parent function $b(x) = 2.5^x$. Suppose a new facility starts with 4 of these plants to hybridize, which can be modeled with the function $b(x) = 4(2.5)^x$. Describe the effect on the graph as it relates to the parent function. (Lesson 8-2)

7. **MULTIPLE CHOICE** Which exponential function models the graph? (Lesson 8-3)

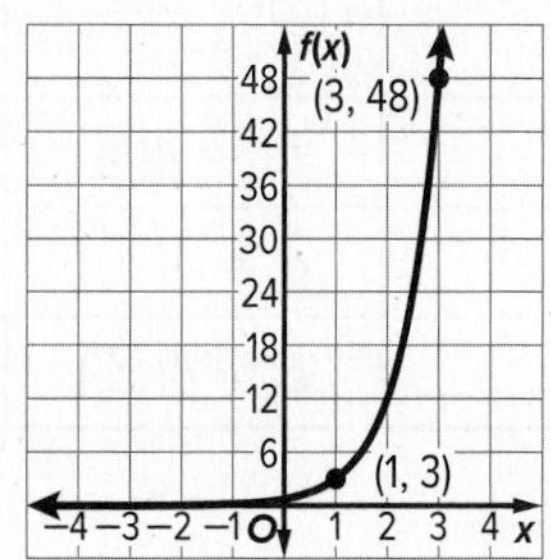

A. $y = \frac{3}{4}(4)^x$

B. $y = 4\left(\frac{3}{4}\right)^x$

C. $y = \frac{3}{4}(x)^4$

D. $y = 4(x)^x$

8. **OPEN RESPONSE** A population, $f(x)$, after x years may be modeled with $f(x) = 2(3)^x$. What is the initial amount, growth rate, domain and range? (Lesson 8-3)

Use the table below for Exercises 9–11.
Joey wants to invest money in a savings account. The table compares two banks he is considering. Joey needs to decide which is the better deal for investing his money.

	Interest Rate	Compound Frequency
First & Loan	0.6%	monthly
Local Credit Union	9%	annually

9. **MULTIPLE CHOICE** What is the effective monthly interest rate offered by Local Credit Union? (Lesson 8-4)

A. 5.5%

B. 2.2%

C. 0.75%

D. 0.72%

10. **MULTIPLE CHOICE** What is the effective annual interest rate offered by First & Loan? (Lesson 8-4)

A. 7.4%

B. 7.2%

C. 1.006%

D. 0.6%

11. **OPEN RESPONSE** Which bank gives Joey the better savings plan? Justify your answer. (Lesson 8-4)

12. MULTIPLE CHOICE Whitney invests $3000 in an account earning 4.5% interest that is compounded annually. How much money will be in Whitney's account after 10 years? (Lesson 8-3)

A. $1893.02

B. $4658.91

C. $4700.98

D. $123,254.07

13. OPEN RESPONSE Attendance for local baseball games has been increasing by an average of 10% per year for the last few years. In 2018, the average attendance was 100 people.

Predict the average number of people attending local baseball games in 2022 if this trend continues. Round to the nearest whole number. (Lesson 8-5)

14. MULTIPLE CHOICE What equation can be written for the nth term of this geometric sequence? (Lesson 8-5)

n	1	2	3	4
a_n	100	−50	25	−12.5

A. $a_n = 100(2)^{n-1}$

B. $a_n = 100(-2)^{n-1}$

C. $a_n = 100\left(-\frac{1}{2}\right)^{n-1}$

D. $a_n = 100\left(\frac{1}{2}\right)^{n-1}$

15. OPEN RESPONSE The table shows the number of pages Aaron read in his book each day. Write a recursive formula for the sequence. (Lesson 8-6)

Day	1	2	3	4
Pages Read	20	35	50	65

16. OPEN RESPONSE What are the first five terms of the sequence for $a_1 = -2$ and $a_n = 2a_{n-1} + 5$ if $n \geq 2$. (Lesson 8-6)

17. OPEN RESPONSE Copy and complete the table for the geometric sequence. (Lesson 8-6) $a_1 = 3$ and $a_n = 4a_{n-1}$, if $n \geq 2$

n	formula	a_n
1	—	3
2	$a_n = 4(3)$	?
3	$a_n = 4(\underline{?})$	?
4	$a_n = 4(\underline{?})$	?

Module 9

Statistics

Essential Question

How do you summarize and interpret data?

What will you learn?

How much do you already know about each topic **before** starting this module?

KEY

👎 — I don't know. 👉 — I've heard of it. 👍 — I know it!

	Before			After		
	👎	👉	👍	👎	👉	👍
find measures of center in a data set						
calculate percentiles						
represent data in dot plots, bar graphs, and histograms						
collect data and analyze bias						
represent data in box plots						
calculate standard deviation						
analyze data distributions						
transform linear data						
compare two data sets						
represent data in two-way frequency tables						
find frequencies, including marginal and conditional relative frequencies						

Foldables Make this Foldable to help you organize your notes about statistics. Begin with 8 sheets of $8\frac{1}{2}''$ by 11″ paper.

1. **Fold** each sheet of paper in half. Cut 1 inch from the end to the fold. Then cut 1 inch along the fold.
2. **Write** the lesson number and title on each page.
3. **Label** the inside of each sheet with *Definitions* and *Examples*.
4. **Stack** the sheets. Staple along the left side. Write *Statistics* on the first page.

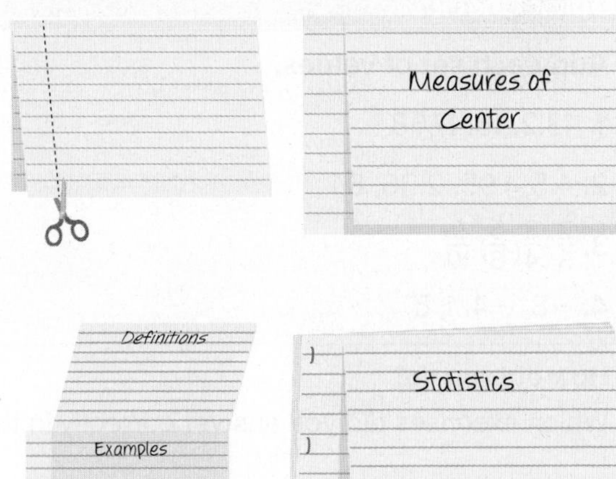

What Vocabulary Will You Learn?

- bar graph
- bias
- box plot
- categorical data
- conditional relative frequency
- distribution
- dot plot
- extreme values
- five-number summary
- histogram
- interquartile range
- joint frequencies
- linear transformation
- lower quartile
- marginal frequencies
- measurement data
- measures of center
- measures of spread
- median
- negatively skewed distribution
- outlier
- percentile
- population
- positively skewed distribution
- quartile
- range
- relative frequency
- sample
- standard deviation
- statistic
- symmetric distribution
- two-way frequency table
- two-way relative frequency table
- univariate data
- upper quartile
- variable
- variance

Are You Ready?

Complete the Quick Review to see if you are ready to start this module.
Then complete the Quick Check.

Quick Review

Example 1

Add the set of values.

12.5, 3.4, 1.75, 9

$$\begin{array}{r} 12.5 \\ 3.4 \\ 1.75 \\ +\ 9 \\ \hline 26.65 \end{array}$$

Align the numbers at the decimal.

The sum is 26.65.

Example 2

Write the fraction $\frac{33}{80}$ as a percent. Round to the nearest tenth.

$\frac{33}{80} \approx 0.413$	Simplify and round.
$0.413 \cdot 100 = 41.3$	Multiply the decimal by 100.
$\frac{33}{80} \approx 41.3\%$	Write as a percent.

Quick Check

Add each set of values.

1. 13.2, 15, 17.68

2. 4.5, 1.95, 2.36, 8.1

3. $\frac{2}{3}, \frac{3}{4}, \frac{5}{6}, \frac{9}{10}$

4. −8, −4, 1, 5

Write each fraction as a percent. Round to the nearest tenth.

5. $\frac{14}{17}$

6. $\frac{7}{8}$

7. $\frac{107}{125}$

8. $\frac{625}{1024}$

How did you do?

Which exercises did you answer correctly in the Quick Check?

Measures of Center

Learn Mean, Median, and Mode

- A **variable** is any characteristic, number, or quantity that can be counted or measured. A variable is an item of data.
- Data that have units and can be measured are called **measurement data** or quantitative data.
- Data that can be organized into different categories are called **categorical data** or qualitative data.
- Measurement data in one variable, called **univariate data**, are often summarized using a single number to represent what is average, or typical.
- Measures of what is average are called **measures of center** or central tendency. The most common measures of center are mean, median, and mode.

  Mode: the value of the elements that appear most often in a set of data

  Mean: the sum of the elements of a data set divided by the total number of elements in the set

  Median: the middle element, or the mean of the two middle elements, in a set of data when the data are arranged in numerical order

Today's Goals

- Represent sets of data by using measures of center.
- Represent sets of data by using percentiles.

Today's Vocabulary

variable
measurement data
categorical data
univariate data
measures of center
percentile

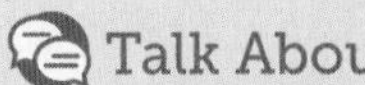

Talk About It

A set of data can have only one value for the mean and median. How many values can a set of data have for the mode? Explain your reasoning.

Example 1 Measures of Center

BASKETBALL **The table shows the total number of points scored in several NCAA Championship Basketball Games. Find the mean, median, and mode of the data.**

Year	Score	Year	Score
2016	150	2008	143
2015	131	2007	159
2014	114	2006	130
2013	158	2005	145
2012	126	2004	155
2011	94	2003	159
2010	120	2002	116
2009	161		

(continued on the next page)

Study Tip

Mean When calculating the mean, your answer will always be between the least and greatest values of the data set. It can never be less than the least value or greater than the greatest value.

Watch Out!

Median If there is an even number of values in the set of data, you will have to find the average of the two middle values to find the median. To do this, divide the sum of the two middle values by 2.

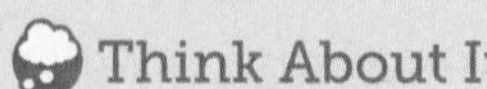

Think About It!

Carlos says that a set of data cannot have the same mean and mode. Do you agree or disagree? Explain your reasoning or provide a counterexample.

Study Tip

Tools To quickly calculate the mean $\overline{x}$ and the median **Med** of a data set, enter the data as **L1** in a graphing calculator, and then use the **1-VAR Stats** feature from the **CALC** menu.

Mean

To find the mean, find the sum of all the points and divide by the number of years in the data set.

$$= \frac{150 + 131 + 114 + 158 + 126 + 94 + 120 + 161 + 143 + 159 + 130 + 145 + 155 + 159 + 116}{15}$$

$$= \frac{2061}{15} \text{ or about } 137.4.$$

The mean is about 137 points.

Median and Mode

To find the median, order the points from least to greatest and find the middle value.

94, 114, 116, 120, 126, 130, 131, 143, 145, 150, 155, 158, 159, 159, 161

median (143) mode (159, 159)

The median is 143.

From the arrangement of data values, we can see that 159 is the only value that appears more than once. So, the mode is 159.

The mean and median are close together, so they both represent the average of the scores well. Notice that the median is greater than the mean. This indicates that the scores less than the median are more spread out than the scores greater than the median. The mode is greater than most of the scores.

Check

FOOTBALL The data show the number of interceptions thrown during one regular season for each team in the NFC. Find the mean, median, and mode. Round to the nearest whole number, if necessary.

13	17	10	12	22	14	8	11
9	12	14	18	12	8	15	11

Mean: ___?___

Median: ___?___

Mode: ___?___

Go Online You can complete an Extra Example online.

Explore Finding Percentiles

Online Activity Use a real-world situation to complete the Explore.

> INQUIRY How can you describe a data value based on its position in the data set?

Go Online
You can watch a video to see how to find percentile rank.

Learn Percentiles

A **percentile** is a measure that is often used to report test data, such as standardized test scores. It tells us what percent of the total scores were below a given score.

- Percentiles measure rank from the bottom.
- There is no 0 percentile rank. The lowest score is at the 1st percentile.
- There is no 100th percentile rank. The highest score is at the 99th percentile.

Key Concept • Finding Percentiles

To find the percentile rank of an element of a data set, use these steps.

Step 1 Order the data values from greatest to least.

Step 2 Find the number of data values less than the chosen element. Divide that number by the total number of values in the data set.

Step 3 Multiply the value from Step 2 by 100.

Study Tip

Percent vs Percentile
Percent and *percentile* mean two different things. For example, a score at the 40th percentile means that 40% of the scores are either the same as the score at the 40th percentile or less than the score at that rank. It does not mean that the person scored 40% of the possible points.

Example 2 Find Percentiles

FIGURE SKATING **The table shows the total points scored by each country in the team figure skating event in the 2014 Olympic Winter Games. Find the United States' percentile rank.**

Country	Score
Canada	65
China	20
France	22
Germany	17
Great Britain	8
Italy	52
Japan	51
Russia	75
Ukraine	10
United States	60

(continued on the next page)

Go Online You can complete an Extra Example online.

Step 1 Order the data.

Order the data values from greatest to least.

Country	Score
Russia	75
Canada	65
United States	60
Italy	52
Japan	51
France	22
China	20
Germany	17
Ukraine	10
Great Britain	8

Study Tip

Percentiles The team with the highest score is at the 99th percentile rank, and the team with the lowest score is at the 1st percentile rank.

Step 2 Divide.

Divide the number of teams with scores lower than the United States by the total number of teams.

$$\frac{\text{number of teams below the United States}}{\text{Total number of teams}} = \frac{7}{10}$$

Step 3 Multiply by 100.

$\frac{7}{10} \cdot 100$ or 70

The United States figure skating team scored at the 70th percentile in the 2014 Olympics.

Think About It!

Which team scored at the 40th percentile?

Check

DRUM CORPS The table shows the scores of the corps that competed in the Drum Corps International World Championship World Class Finals in 2015.

Corps	Score
Bluecoats	96.925
Blue Devils	97.650
Blue Knights	91.850
Blue Stars	85.150
Boston Crusaders	86.800
Carolina Crown	97.075
Crossmen	85.025
Madison Scouts	88.750
Phantom Regiment	90.325
Santa Clara Vanguard	93.850
The Cadets	95.900
The Cavaliers	88.325

The Cavaliers scored at the ___?___ th percentile. The ______?______ scored at the 75th percentile.

Go Online You can complete an Extra Example online.

Practice

Go Online You can complete your homework online.

Example 1

Find the mean, median, and mode for each data set.

1. {17, 11, 8, 15, 28, 20, 10, 16}

2. {2.5, 6.4, 7.0, 5.3, 1.1, 6.4, 3.5, 6.2, 3.9, 4.0}

3.

2	1	1	5	7
3	2	4	6	2

4.

50	30	40	10
20	80	60	90
10	30	110	70

5. number of students helping at a booth each hour: 3, 5, 8, 1, 4, 11, 3

6. weight in pounds of boxes loaded onto a semi-truck: 201, 201, 200, 199, 199

7. car speeds in miles per hour observed by a highway patrol officer: 60, 53, 53, 52, 53, 55, 55, 57

8. number of songs downloaded by students last week in Ms. Turner's class: 3, 7, 21, 23, 63, 27, 29, 95, 23

9. ratings of an online video: 2, 5, 3.5, 4, 4.5, 1, 1, 4, 2, 1.5, 2.5, 2, 3, 3.5

Example 2

MARCHING BAND **A competition was recently held for 12 high school marching bands. Each band received a score from 0 through 100, with 100 being the highest.**

Band	Score	Band	Score
Freeport	78	Madison	69
Ross	85	Monmouth	67
Hamilton	88	Carlisle	65
Groveport	94	Dupont	48
Lakehurst	56	Cave City	90
Benton	77	Monroe	80

10. Find Hamilton High School's percentile rank.

11. Find Monmouth High School's percentile rank.

12. Find Freeport High School's percentile rank.

Mixed Exercises

13. REASONING The mean number of people at the movies on Saturday nights throughout the year is 425, and the median is 412. Explain why the mean could be slightly higher.

14. REASONING The mode length of time it takes to fly from New York City to Chicago is 2 hours 35 minutes, and the mean is 3 hours 15 minutes. Explain why the mode could be slightly lower.

15. **FOOTBALL** Find the mean, median, and mode for the data set. The weights in pounds of 5 offensive linemen of a football team: 217, 212, 285, 245, 301.

16. **WEB SITES** The ratings for a new recipe Web site varied from very low, 1 point, to very high, 10 points, with half of the scores receiving a rating of 7. If a new rating of 7 were added to the data set, how would the mode be affected? Explain.

17. Find a mean of {16, 19, 22, 27, 33, 19, 25}.

18. **SINGING** In a singing competition that involved 50 contestants, Reina's score ranked higher than 40 of the contestants. In what percentile did Reina score?

19. **SPORTS** The table shows the number of points scored by a basketball team during their first several games. Find the mean, median, and mode of the number of points scored.

Game	1	2	3	4	5	6	7	8
Points	43	50	52	47	55	61	48	56

20. **PERFORMANCE** At a bodybuilding competition, Shawnte earned a score of 42 points. There were 19 competitors who received a lower score than Shawnte and 5 competitors who earned a higher score. What was Shawnte's percentile rank in the bodybuilding competition?

21. **GRADES** On her first four quizzes, Rachael has earned scores of 21, 24, 23, and 17 points. What score must Rachael earn on her fifth and final quiz so that both the mean and median of her quiz scores is 21?

22. **QUIZZES** Sequon scored 95, 86, 81, 83, and 95 on his math quizzes this quarter. Find the mode of his quiz scores.

23. **BAND** Out of the 30 bands at the competition, Coastal High School's band scored higher than 27 others. Find the percentile rank for Coastal High School's band.

24. SHOPPING The table shows the prices of comparable laptop computers at different retailers.

Retailer	Price ($)
A	389
B	425
C	350
D	499
E	475
F	360
G	319
H	425
I	299
J	379

a. Find the mean, median, and mode of the prices.

b. Why are the mean and median much lower than the mode?

c. After deliberation, Nikki is interested in buying a laptop from either retailer C, F, or J. What are the percentile ranks for the laptop at each of these retailers?

25. NOVELS The table shows the lengths (in words) of the seven novels on the required reading list of Miguel's language arts class.

Novel	Number of Words
Old Yeller	35,968
Lord of the Flies	59,900
Moby Dick	206,052
Jane Eyre	183,858
Great Expectations	183,349
Call of the Wild	31,750
The Color Purple	66,556

a. Find the mean, median, and mode for the data set.

b. Predict which novels will be lower than the 50th percentile in length. Verify your prediction.

c. If *Lord of the Flies* comes with an additional 10,020-word online reading assignment, then how will the median length be affected? How will the mean be affected?

26. VOLUNTEERING The table shows the number of hours different students spent volunteering as part of a community outreach program. Find the mean, median, and mode of the data set.

Volunteer Hours				
25	30	35	40	35
25	50	45	25	90

27. USE A SOURCE Research the total medal counts for Canada, France, Japan, Russia, Brazil, and Great Britain at the 2016 Rio de Janeiro Olympics. Make a table of the data you collect. Then find the percentile rank of each country.

28. CONSTRUCT ARGUMENTS The table shows the number of pet adoptions each week for a shelter over a two-month period. If there are 65 pets adopted next week, how are the mean, median, and mode affected?

Pet Adoptions			
7	19	26	20
23	21	24	20

29. STATE YOUR ASSUMPTION A data set has a mean of 37, a median of 36.5, and a mode of 37. What assumption(s) can you make about the dataset?

30. BOWLING The table shows Lucinda's score for each of her last ten bowling games.

a. Find the mean, median, and mode of the scores. Round to the nearest whole number.

b. Why is the mean slightly higher than the median?

Game	Score
1	220
2	235
3	255
4	210
5	240
6	220
7	225
8	220
9	250
10	210

31. DANCE COMPETITION At a dance competition, Pascal earned a score of 73 points. There were 12 competitors who received a lower score than Pascal and 3 competitors who earned a higher score. What was Pascal's percentile rank in the dance competition?

32. CREATE Create a data set that has a mean of 11, a median of 10, and a mode of 8.

33. WRITE Describe how an outlier value that is greater than the numbers in the data set affects each measure of center.

34. ANALYZE Determine whether the statement is *true* or *false*. If it is false, explain how to make the statement true.

To find percentile rank, divide the selected value by the total of all the values.

35. PERSEVERE Describe the effect on the mean, median, and mode of a set when all the items in the set are multiplied by the same number.

36. WHICH ONE DOESN'T BELONG? Analyze each situation. Which situation is NOT best described by the median of the data? Explain.

An art gallery has many items for sale that are reasonably priced, but it also carries luxury priced paintings.	Most of the students volunteered 2 hours each week, but James volunteered 8 hours per week.	The amusement park had about the same number of attendees each day. On the annual bring-a-friend-for-free day, the number of attendees tripled.

37. FIND THE ERROR Julio is studying botany and has been tracking the growth of 10 tomato plants each week. The first week, the plants measured the following growth: 1 in., 1.5 in., 2.2 in., 0.5 in., 1 in., 1.25 in., 1.4 in., 2 in., 2.1 in., 1.9 in. In his research paper, Julio includes the median growth value for the week. Has Julio chosen the best measure of center to describe the plant growth? Explain.

38. STRUCTURE Explain how you determine that a data set is best described by the mean.

39. WRITE Explain in your own words the process for finding a percentile rank.

Representing Data

Learn Dot Plots

One way to represent data is by using a **dot plot**, which is a diagram that shows the frequency of data on a number line.

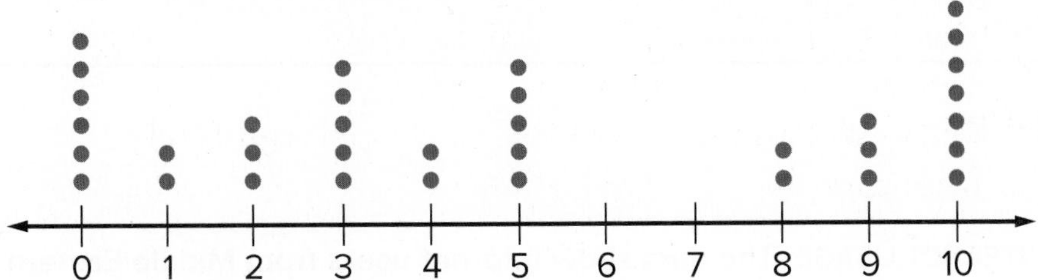

Key Concept • Making Dot Plots

Step 1 Write the data points in order from least to greatest.

Step 2 Make a number line that starts at the least data point and ends at the greatest data point. Choose an appropriate scale.

Step 3 Plot the dots on the number line. Stack the points when there is more than one data point with the same number.

Step 4 If appropriate, include a label for the number line and title for the dot plot.

Example 1 Make a Dot Plot

Represent the data as a dot plot.

11, 12, 14, 15, 12, 13, 15, 13, 9, 15, 12, 13, 15, 15, 11

Step 1 Write the data points in order from least to greatest.

9, 11, 11, 12, 12, 12, 13, 13, 13, 14, 15, 15, 15, 15, 15

Step 2 Make a number line.

The data are whole numbers ranging from 9 to 15. So, make a number line starting at 9 with intervals of 1.

Step 3 Plot the dots on the number line.

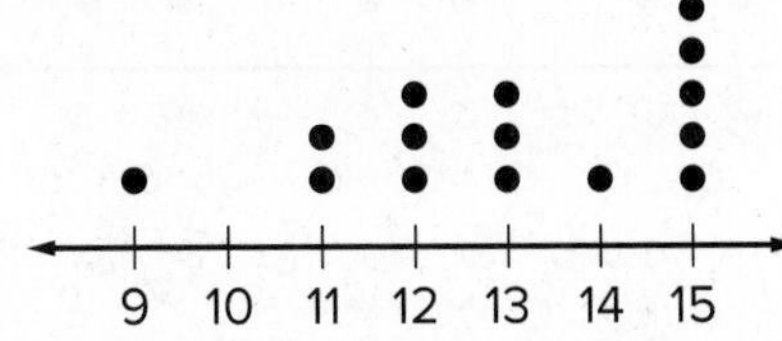

Step 4 If appropriate, include a label for the number line and title for the dot plot.

Because no information is given regarding what these data represent, no title is needed for this dot plot.

Go Online You can complete an Extra Example online.

Today's Goals

- Represent sets of data by using dot plots.
- Determine whether discrete or continuous graphical representations are appropriate, and represent sets of data by using bar graphs or histograms.

Today's Vocabulary

dot plot

bar graph

histogram

Think About It!

What would be the benefit of representing data in a dot plot?

Study Tip

Accuracy Count the listed data and check that the number of data points matches the sum of the frequency table. Missing just one or two pieces of data can change the values of statistics.

Check

Represent the data as a dot plot.

8, 6, 0, 2, 7, 1, 8, 1, 4, 8, 0, 1, 2, 8, 4, 7, 1, 5, 9, 1

Example 2 Make a Dot Plot by Using a Scaled Number Line

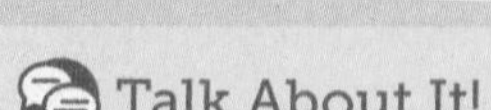

Talk About It!

How would these data appear if the number line were scaled by 1 instead of 10? Do you think a scale of 1 would be a good way to represent the data? Explain.

INTERNET USAGE **The data show Internet users from Middle Eastern countries as a percentage of their total population. Represent the data as a dot plot.**

96.4	**57.2**	**33.0**	**74.7**	**86.1**	**78.7**	**80.4**
78.6	**64.6**	**91.9**	**65.9**	**28.1**	**93.2**	**22.6**

Step 1 Write the data points in order from least to greatest.

22.6, 28.1, 33.0, 57.2, 64.6, 65.9, 74.7, 78.6, 78.7, 80.4, 86.1, 91.9, 93.2, 96.4

Step 2 Make a number line.

The data range from 22.6 to 96.4. Since these data represent a broad range with specific values, it is unlikely that any data point is represented more than once. To represent the data in a meaningful way, scale the number line.

Step 3 Plot the dots on the number line.

Internet Usage from Middle Eastern Countries

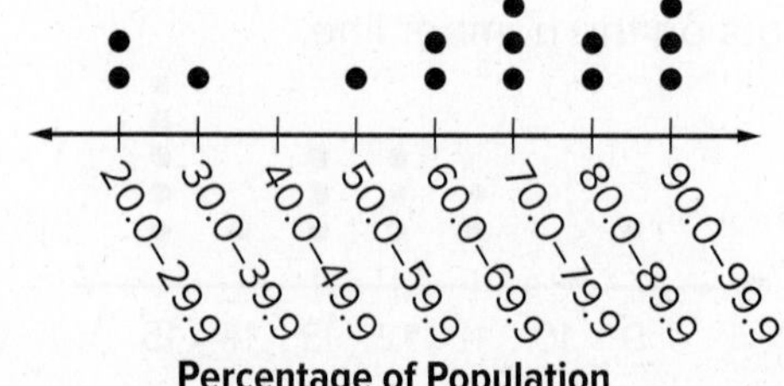

Percentage of Population

Step 4 If appropriate, include a label for the number line and title for the dot plot.

Go Online You can complete an Extra Example online.

Check

MOUNTAINS The data give the elevation of the highest mountain peaks in the United States. Create a dot plot that best represents the data.

20,308	18,009	17,402	16,421	16,391
16,237	15,325	14,951	14,829	14,573

Learn Bar Graphs and Histograms

A **bar graph** is a graphical display that compares categories of data using bars of different heights. Bar graphs are used when the data are discrete. To indicate this, there is a space between each of the bars.

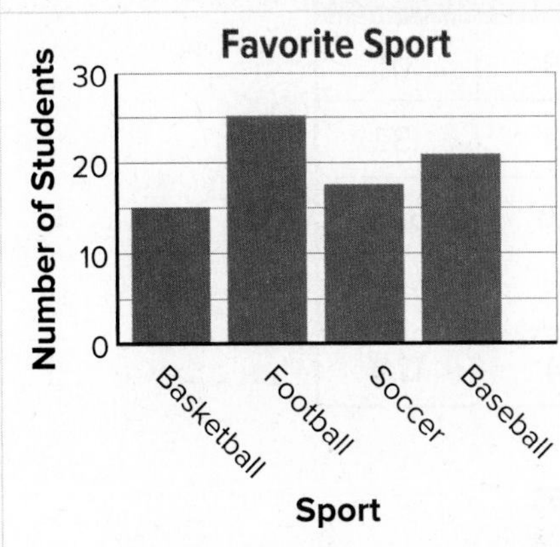

A **histogram** is a graphical display that uses bars to display numerical data that have been organized in equal intervals. A histogram represents continuous data, so the bins have no spaces between them.

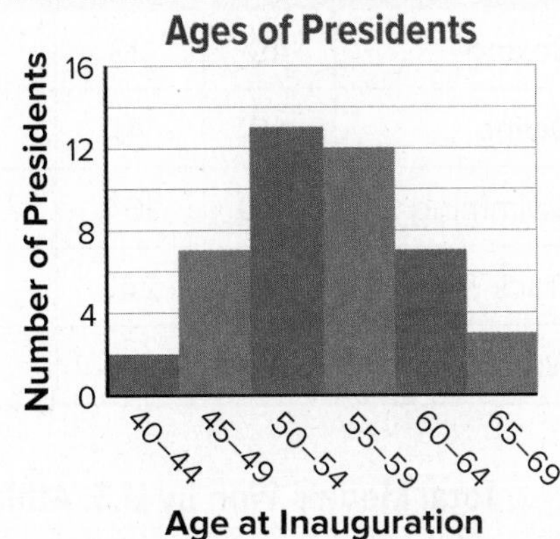

Key Concept • Making a Bar Graph or Histogram

Step 1 Determine whether the data should be represented as a bar graph or histogram.

Step 2 Determine appropriate categories or bins, and tally the data, if necessary.

Step 3 Draw bars to represent each category or bin.

Step 4 Label the axes. If appropriate, include a title for the graph.

Apply Example 3 Determine an Appropriate Graph for Discrete Data

OLYMPICS **The table shows the total number of Olympic medals won by U.S. athletes competing in selected events from the first Summer Olympics in 1896 through 2012. Make a graph of the data to show the total medals won for each sport.**

1 What is the task?

Describe the task in your own words. Then list any questions that you may have. How can you find answers to your questions?

Determine whether the data should be represented as a bar graph or histogram. These data represent discrete, categorical data, so use a bar graph.

2 How will you approach the task? What have you learned that you can use to help you complete the task?

I'll tally the number of medals won for each sport and then I'll create the bar graph. Each event will be represented by one bar. I have learned how to make a bar graph.

3 What is your solution?

Use your strategy to solve the problem.

Event	Gold	Silver	Bronze	Total
Boxing	49	23	39	111
Diving	48	41	43	132
Swimming	230	164	126	520
Track & Field	319	247	193	759
Wrestling	52	43	34	129

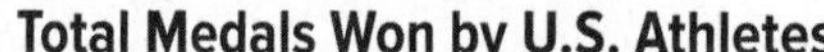

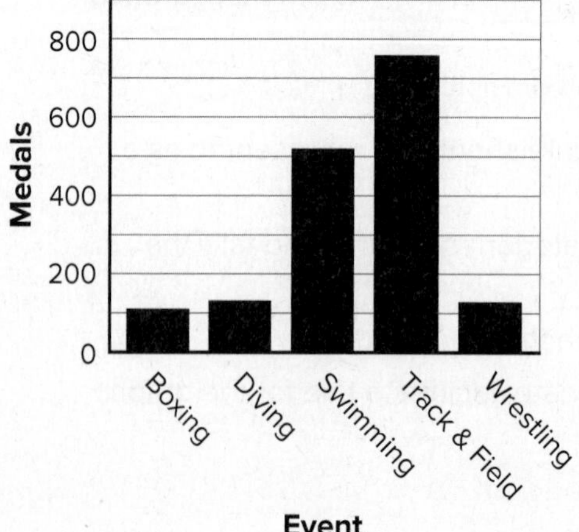

4 How can you know that your solution is reasonable?

Write About It! Write an argument that can be used to defend your solution.

Sample answer: Because the graph is supposed to represent the total number of medals in each event, it makes sense for the data to be organized in categories. Categorical data should be represented by a bar graph.

Check

VIDEO GAMES The table shows the number of active video game players in each country. Make a graph that best displays the data.

Country	Australia	Brazil	France	Germany	Italy	Poland	Spain	Turkey	UK	US
Players (millions)	9.5	40.2	25.3	38.5	18.6	11.8	17	21.8	33.6	157

Example 4 Determine an Appropriate Graph for Continuous Data

MARATHON **The results of the top finishers of the 2015 New York City Marathon, wheelchair division, are given below. Determine whether the data are *discrete* or *continuous*. Then make a graph.**

1:30:54 1:30:55 1:34:05 1:35:19 1:35:21 1:35:37 1:35:38 1:36:45
1:36:59 1:38:39 1:39:22 1:39:22 1:39:27 1:39:27 1:40:36 1:43:04

Step 1 Because racers can finish with any time, the data are continuous and you can use a histogram.

(*continued on the next page*)

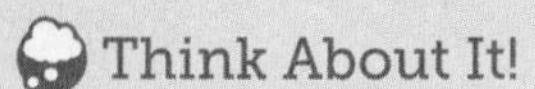

Think About It!

Describe the histogram. What does it show you about the racers' times? What do the gaps in the graph represent?

Step 2 Because the data are spread over several minutes, group the data by the minute. Then, tally each interval.

Time (h:m:s)	Frequency
1:30:00–1:30:59	2
1:31:00–1:31:59	0
1:32:00–1:32:59	0
1:33:00–1:33:59	0
1:34:00–1:34:59	1
1:35:00–1:35:59	4
1:36:00–1:36:59	2

Time (h:m:s)	Frequency
1:37:00–1:37:59	0
1:38:00–1:38:59	1
1:39:00–1:39:59	3
1:40:00–1:40:59	1
1:41:00–1:41:59	0
1:42:00–1:42:59	0
1:43:00–1:43:59	1

Steps 3 and 4 Draw a bar to represent each bin. Label the axes. Include a title for the graph.

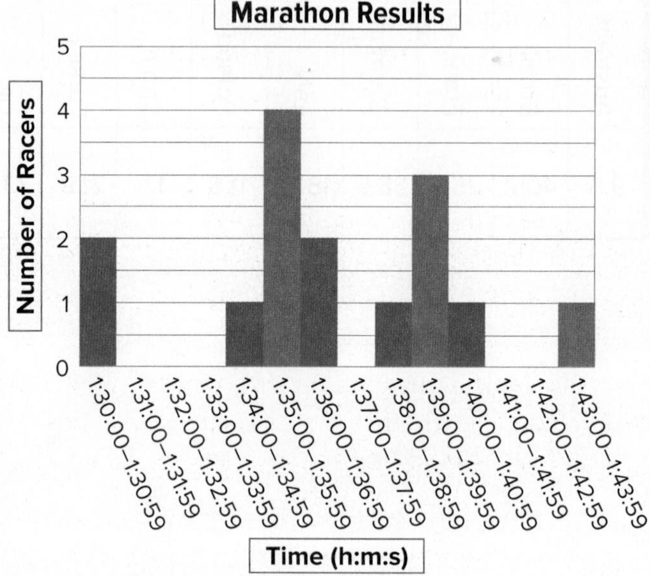

Check

PHOTO SHARING The table shows the users of a photo sharing app by age group. Make a graph that best displays the data.

Age	18–24	25–34	35–44	45–54	55–64	65+
Users (%)	45	26	13	10	6	1

Go Online You can complete an Extra Example online.

Practice

Go Online You can complete your homework online.

Examples 1 and 2

1. **READING** The table shows the number of books read by students in a summer reading program. Make a dot plot of the data.

Number of Books Read				
3	8	5	6	5
4	5	5	4	5
5	6	8	8	4

2. **QUIZ SCORES** Represent the quiz scores as a dot plot. Scale the number line as needed.

   50, 45, 24, 28, 27, 38, 21, 22, 23, 42, 41, 35, 37, 25, 43

Examples 3 and 4

3. **SURVEY** A survey was conducted among students in Mr. Dalton's science class to determine a field trip destination. The results are shown in the table at the right. Make a graph to display the data.

Destination	Number of Votes
zoo	6
museum	4
observatory	11
state park	7

4. **MOVIES** In a survey, students were asked to name their favorite type of movie. Of those surveyed, 8 chose action movies, 6 chose comedies, 5 chose horror movies, 3 chose dramas, and 7 chose science fiction movies (sci-fi). Determine whether the data are discrete or continuous. Then make a graph.

5. **CONCERT** The table shows the number of attendees by age at a concert. Determine whether the data should be shown in a *bar graph* or *histogram*. Then make an appropriate graph for the data.

Concert Attendees	
0–9	400
10–19	1440
20–29	2400
30–39	2000
40–49	960
50–59	560
60–69	240

Mixed Exercises

6. **PRIZES** The table shows the number of prizes won by customers at a carnival game each of the past several days. Determine whether the data are discrete or continuous. Then make an appropriate graph for the data.

Prizes Won				
37	29	53	32	42
21	41	45	17	27
44	34	24	34	31
19	51	48	35	54
46	38	39	49	25

7. **JOGGING** The number of miles Lisa jogged each of the last 10 days are 3, 4, 6, 2, 5, 8, 7, 6, 4, and 5.
   a. Choose the most appropriate type of data display and graph the data.
   b. How many days did Lisa jog at least 4 miles?
   c. What was the greatest number of miles she jogged in a day?

8. **MOVIES** The number of movies that are released theatrically each year are shown in the table.
   a. Select an appropriate display for the data. Explain your reasoning.

Year	2010	2011	2012	2013	2014	2015	2016
Number of Movies Released	563	609	678	661	709	708	718

Source: comScore, MPAA

   b. Make a graph of the data.

9. **RUNNING** The ages of the participants in a 10K race at Masonville are 65, 47, 23, 70, 41, 55, 32, 29, 56, 39, 12, 57, 25, 33, 15, 18, 35, 22, 63, 49, 23, 30, 37, 40, and 50.

   a. Construct an appropriate data display for the data.

   b. How many participants are less than 30 years old?

   c. In what interval is the most frequent age?

10. **ORCHESTRA** The ages of the members of an orchestra are 39, 43, 31, 53, 41, 25, 35, 46, 27, 34, 37, 26, 51, 29, 36, 40, 33, 28, 48, 26, 42, and 38 years. Make a graph of the data.

11. **PRECISION** A scientific research study tracks the growth of an insect in millimeters. The growth data for each insect in the study during week 1 are 1.1, 1.25, 1.3, 1.67, 1.9, 2.35, 2.1, 2.3, 1.5, 1.7, 2.25, 2.1, 2.45, 1.37, 1.83. The scientist is preparing a histogram to show the distribution of growth across the population. How should the scientist break down his data into categories?

12. **PETS** The pets owned by Liza's classmates are rabbit: 2, dog: 6, cat: 3, horse: 2, bird: 5, mouse: 1, fish: 3, and other: 1.

   a. Make a dot plot of the data.

   b. How many types of pets are represented by the dot plot?

   c. Which pet is the most popular?

13. **ANALYZE** Make two conclusions about a product that received the ratings shown in the dot plot. Justify your conclusions.

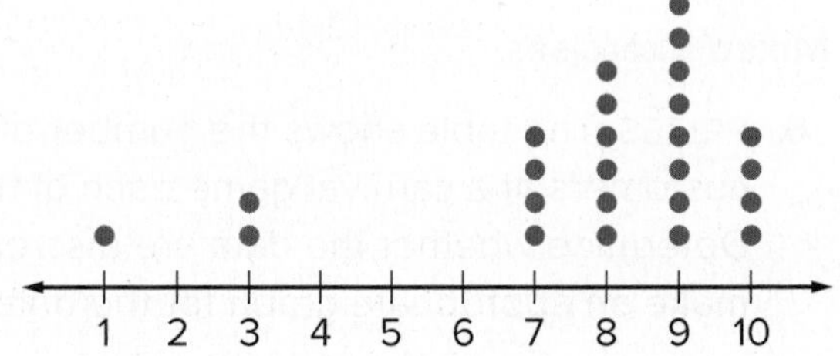

14. **REGULARITY** Explain when a histogram is the best model for data, and describe the process of creating a histogram.

Higher-Order Thinking Skills

15. **WRITE** Explain why it may be necessary to scale the number line of a dot plot.

16. **PERSEVERE** Using the data provided in the double bar graph about peanut butter, what are two conclusions the grocery store could infer?

17. **STRUCTURE** How is a bar graph similar to a histogram? How is it different?

Peanut Butter Sales (in 1000's of dollars)

	Peanut Butter 2016	Peanut Butter 2017
Creamy	200	250
Crunchy	300	400
With Jelly	50	35
Natural	60	120
Sugar Free	75	100

Using Data

Explore Phrasing Questions

Online Activity Use a real-world situation to complete the Explore.

INQUIRY How can the way you collect data affect the results?

Today's Goals

- Identify potential bias in sampling methods and questions.
- Identify potential bias in statistics and representations of data.

Today's Vocabulary

population

sample

bias

statistic

Learn Collecting Data

A **population** consists of all the members of a group of interest about which data will be collected. Since it may be impractical to examine every member of a population, a subset of the group, called a **sample**, is sometimes selected to represent the population. The sample can then be analyzed to draw conclusions about the entire population.

Sample data are often used to estimate a characteristic of a population. Therefore, a sample should be selected so that it closely represents the entire population. Also, the larger the sample size, or the more samples taken, the better it represents the population.

A **bias** is an error that results in a misrepresentation of a population. If a sample favors one conclusion over another, the sample is biased and the data are invalid.

Talk About It!

Some polls use both landlines and cell phones. How might this alleviate the issue of bias with the landline-only sampling method?

Example 1 Sample Bias

POLLS **Before the 2010 elections for members of the U.S. House of Representatives, pollsters called American households on their landline phones to see how they planned to vote. What kind of sample bias might have affected the poll?**

Step 1 Identify the intended population.

The population is all likely voters.

Step 2 Identify the sample method.

The data for this poll were collected over landline phones, so the sample consists of likely voters who have a landline.

Step 3 Determine potential bias.

Because not all likely voters have landline phones, the results could be skewed because not all likely voters are available for this sample.

Go Online You can complete an Extra Example online.

Check

SOCIAL MEDIA Shia wants to determine the age of the average internet user. He posts a poll to his friends on a social media site, asking their age. Is this a good sample? If not, what kind of sample bias might have affected the poll?

A. This is a good sample.

B. This is not a good sample. He asked only people on the Internet.

C. This is not a good sample. He asked only about users' ages.

D. This is not a good sample. He asked only his friends on a specific Web site.

Think About It!
Rewrite the question to try to remove the bias.

Example 2 Question Bias

SOFT DRINKS **A survey organization wants to see what percent of New York City citizens support a ban on soft drinks. The question posed is, "Do you support a ban on soft drinks, which contribute to heart disease and tooth decay?"**

Part A Identify bias.

Step 1 Identify the purpose of the question: To find the percent of New York citizens who support a ban on soft drinks.

Step 2 Identify potential bias in the question: The question lists some of the health risks of soft drinks. This might make respondents more likely to respond that they do support a ban.

Part B Identify interests.

The bias in the question might make respondents more likely to support a ban. This bias could serve the interests of health groups who want to ban soft drinks or companies who sell competing drinks, like juices.

Study Tip
Assumptions When you identify and try to remove bias, you will make assumptions about what information is important for the question and what might create undue bias. For example, you need to consider whether details of the plan are relevant and whether the question includes persuasive language.

Check

FILM One of your friends wants to determine whether people in your class prefer to watch movies or television. She asks, "Do you prefer to watch movies or television?"

Part A Does this question potentially bias the results?

A. No; the question is as neutral as possible.

B. Yes; your friend asked only people in your class.

C. Yes; your friend provided only two options.

D. Yes; the framing of the question influences the respondent to choose *television*.

Part B Whose interests might be served by asking the question in this way?

Go Online You can complete an Extra Example online.

Learn Using Statistics and Representations

A **statistic** is a measure that describes a characteristic of a sample. Like data, statistics and representations of data are nonneutral. When the average of a set of data is discussed, it uses a measure of center: mean, median, or mode. However, depending on the data set and what information is being conveyed, one measure of center might not give the whole picture of the data. Even if the data is being discussed in whole, it can also be misrepresented. For example, a person might manipulate the scales of the axes of a graph or how the data are represented graphically to misrepresent the data.

Math History Minute

With M. A. Girschick, **David Blackwell** (1919–2010) authored the classic book *Theory of Games and Statistical Decisions*. In 1965, he became the first African American president of the American Statistical Society.

Example 3 Data Summaries

TEACHING **A teacher wants to tell his students how the average student did on an exam, so he looks at the scores in his gradebook. Two students scored a 0 because they stopped showing up for class in the last month and did not take the exam. He uses the mean, 71, as the measure of center. Does the mean accurately represent these data?**

0, 0, 82, 83, 85, 87, 88, 88, 91, 91, 91

Step 1 Identify the other measures of center. Round your answer to the nearest unit.

median = 87

mode = 91

Step 2 Analyze the measures of center and how they align with the information the teacher wants to convey.

Mean: The mean, 71, is affected by the two 0 scores. However, no one who showed up for the exam scored below an 82, so the mean does not do a good job of indicating the performance of the students who took the exam.

Median: The median, 87, is not affected by the extreme values. It provides a more accurate average for how students performed on the test because it includes the scores of the two students who did not take the exam at all.

Mode: The mode, 91, is both the score most students received and the highest score received on the exam, but it does not accurately portray how students performed on average.

Because the teacher wants to discuss the performance of students who took the exam, the mean is not the best measure of center. It indicates that all students who took the exam performed worse than their actual scores.

Go Online You can complete an Extra Example online.

Check

READING Karen writes down the number of books she has read for each of her classes so far: 5, 6, 4, 5, 5, 6, 5, 4, 5. Using the mean, 5, she says that she reads around 5 books on average for an English course. Does the mean accurately represent the data for the situation? Explain.

A. No; the mean is overly influenced by the low numbers.

B. No; the mean is overly influenced by the high numbers.

C. No; the mean doesn't tell how many pages are in the average book.

D. Yes; the mean accurately represents the data for the situation.

Example 4 Data Representation

SOCCER **A group compares two soccer players who play the same position for different teams. They make a graph of the number of goals scored throughout the season for each player. Do the graphs misrepresent the data? Whose interests might be served by the representation?**

Think About It!

If the two graphs had the same scales but were comparing a goalkeeper and a forward, how might the data be misleading when comparing the skill of the two players?

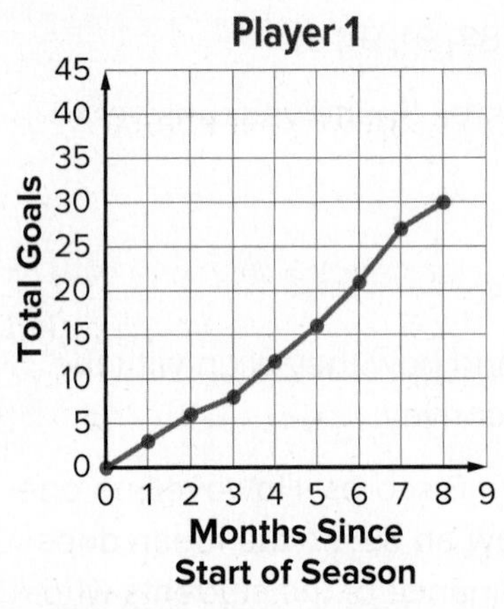

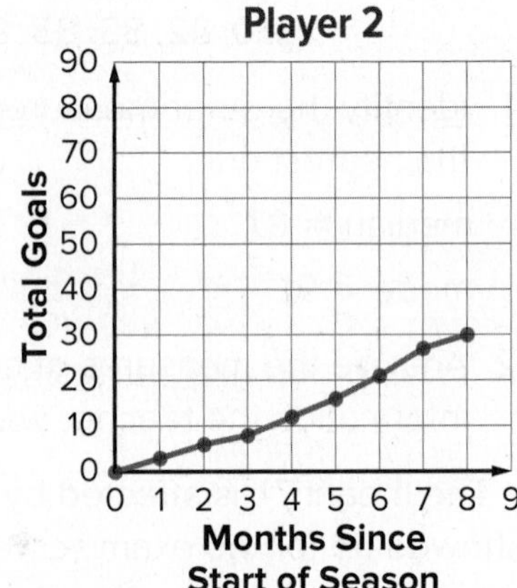

Part A Identify misleading representations.

Step 1 Identify the purpose of the graphs. The purpose of the graphs is to compare the number of goals each soccer player scored.

Step 2 Identify differences in the graphs. The graphs appear to be the same in terms of the data being represented, but the *y*-axis for the second player goes up to 90 in increments of 10, whereas the first goes up to 45 in increments of 5.

Step 3 Identify how this affects the representation of the data. Although the numbers are the same, the scale for Player 1 makes it look like he is scoring more goals than Player 2.

Part B Identify interests.

The misleading representation of the data makes it appear that Player 1 scores more goals than Player 2. This bias could serve the interests of the team, sponsors of Player 1's team, or Player 1's agent.

Use a Source

Find a graph or set of statistics online. Ask yourself, are the data accurately represented? Whose interests are served by the graph or statistics?

Go Online You can complete an Extra Example online.

Practice

Example 1

1. **SPORTS** Awan wants to know what the favorite sport is among students. To find out, he asks everyone he sees leaving school after basketball practice. Identify the intended population and determine the potential sample bias.

2. **STORES** Raya wants to conduct a survey at a nearby mall to determine which are the mall's most popular stores. How could she choose a sample that is unbiased?

Example 2

3. **MUSIC** Shea is shopping online, and a survey question pops up that says, "Music education enriches student learning. Do you support music education in schools?"

   a. Identify potential bias in the question.

   b. Identify whose interests may be served by the question.

4. **CANDIDATES** There are three candidates for mayor. To investigate how the townspeople feel about the candidates, a newspaper posts a poll that lists the three candidates and asks which candidate people support. The poll appears on the same page as an opinion piece in support of one of the candidates.

   a. Identify potential bias in the question.

   b. Identify whose interests may be served by the question.

Example 3

5. **BUTTERFLIES** Tania recorded the number of butterflies she saw on her daily runs each day for a week. The numbers are: 1, 8, 2, 2, 5, 6, and 4. Find the mean, median, and mode of the data. Which measure(s) are appropriate to accurately summarize the data?

6. **OUTLIERS** In a data set with an outlier, which measure of center, mean or median, is the better measure to use to describe the center of the data? Explain your reasoning.

Example 4

7. **SALES** The graphs show the number of T-shirts sold at a baseball tournament for two years by two different vendors. The tournament director wants to compare the vendors. Do the graphs misrepresent the data? How does that difference affect the interpretation?

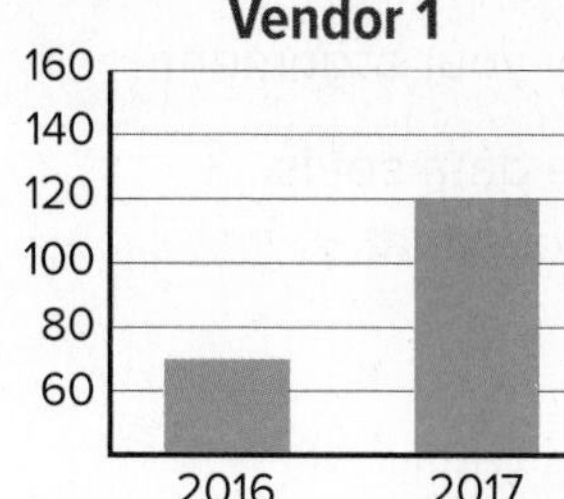

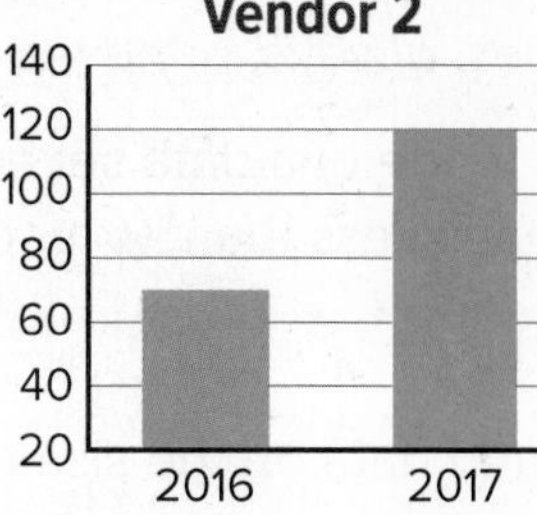

8. **SCALE** If the same set of data is graphed with a scale of 0 to 10 on the y-axis and then with a scale of 0–100 on the y-axis, what effect does that have on the representation of the data?

Mixed Exercises

9. **SCIENCE** A school wants to know which area of science; physics, biology, or chemistry, is most interesting to its students. Would it be better to survey students in a class that is an elective or required to get a sample with the least bias? Explain your reasoning.

10. **TAX** Before surveying people about whether they favor or oppose a proposed tax, the surveyors want to present information about the tax. Suppose the surveyors give facts about the tax without giving opinions. How could the facts given by the surveyors introduce bias?

11. **CONSTRUCT ARGUMENTS** The weights, in pounds, of several dolphins at a sea animal care facility are 185, 222, 755, 801, 835, 990, and 1104. Which measure of center best represents the data? Justify your conclusion.

12. **FOOD DRIVE** The chart shows the number of canned goods collected by Valley High School in 2012 and 2017. Is the graph misleading? Explain.

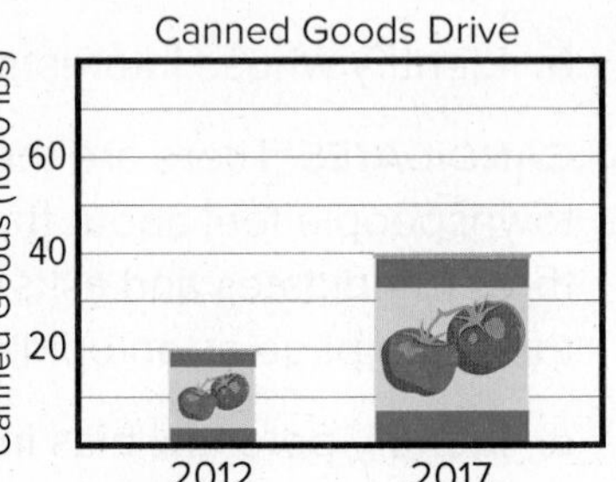

13. **REASONING** The number of participants at reading club for six weeks are 11, 12, 10, 13, 10, and 10. Without calculating the measures of center, how would adding an outlier of 24 participants affect which measure of center most appropriately represents the data?

14. **PRECISION** A community garden has 8 tomato plants with heights ranging from 0.4 to 0.9 meters. Regina found the median to be 0.7 meters, which she rounded and reported as 1 meter. Is Regina's report of the median accurate? Explain your reasoning.

15. **REGULARITY** Describe a general method for assessing a sample for bias.

16. **STRUCTURE** How are the median and mean scores affected if all data values in a set are increased by a specific value, such as 10?

17. **CREATE** Create two sets of data and display them in a graph or chart that shows bias toward one of the sets of data.

18. **WRITE** Write two scenarios that have different examples of sample bias. Have a classmate rewrite your statements without bias.

19. **CREATE** Think of a topic about which you can survey the teachers at your school. Conduct the survey. Explain whether your survey question(s) introduce bias.

20. **ANALYZE** Is a biased sample *sometimes*, *always*, or *never* valid? Justify your argument.

21. **PERSEVERE** If the mean, median, and mode of a data set are equal, the data set is symmetric. If a data set has a mean that is less than its median, what does that tell you about the data set?

22. **FIND THE ERROR** Two students collected data on the sizes of box turtle shells. Olivia measured 8 turtles from a pond near her school. Caleb measured 2 turtles from each of 4 ponds around town. Which is more likely to be free of sample bias? Explain your reasoning.

Measures of Spread

Explore Using Measures of Spread to Describe Data

 Online Activity Use a real-world situation to complete the Explore.

INQUIRY Why might you describe a data set with more than the mean?

Today's Goals

- Determine measures of spread, including the range and interquartile range, of a set of data.
- Determine the standard deviation of a data set.

Today's Vocabulary

measures of spread
range
quartiles
five-number summary
median
lower quartile
upper quartile
box plot
interquartile range
standard deviation

Learn Range and Interquartile Range

Statisticians use **measures of spread** or variation to describe how widely data values vary. One such measure is the **range**, which is the difference between the greatest and least values in a set of data.

Quartiles divide a data set arranged in ascending order into four groups, each containing about one-fourth, or 25%, of the data. A **five-number summary** contains the minimum, quartiles, and maximum of a data set.

The **median** marks the second quartile, Q_2, and separates the data into upper and lower halves.

The **lower quartile**, Q_1, is the median of the lower half.

The **upper quartile**, Q_3, is the median of the upper half.

A **box plot**, or box-and-whisker plot, is a graphical representation of the five-number summary of a data set. A box is drawn from Q_1 to Q_3 with a vertical line at the median. This box represents the **interquartile range**, or IQR, which is the difference between the upper and lower quartiles. The whiskers of the box plot are drawn from Q_1 to the minimum and from Q_3 to the maximum.

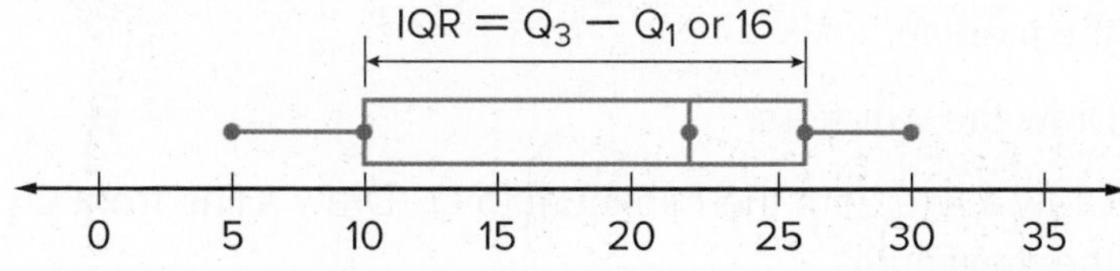

Study Tip

Ordering Because the range involves only the greatest and least values in a data set, it can be determined without ordering the numbers. However, it is often useful to order the data to avoid missing a number.

Example 1 Range

GRADES **What is the range of the scores?**

79, 83, 88, 62, 91, 99, 70

Step 1 Arrange the data in ascending order. 62, 70, 79, 83, 88, 91, 99

Step 2 Determine the range.

$$\text{range} = \text{greatest value} - \text{least value}$$
$$= 99 - 62 \text{ or } 37$$

Go Online You can complete an Extra Example online.

Example 2 Make a Box Plot

BOX OFFICE **A financial analyst for a movie studio wants to determine how much most of the top-earning movies have grossed to compare his studio's recent grosses. The worldwide grosses, in millions of dollars, for the top 10 highest-grossing films of all time are given. Determine the five-number summary and draw a box plot of the data to see the spread of the data.**

2788	2187	2060	1670	1520
1516	1405	1342	1277	1215

Part A Determine the five-number summary.

Step 1 Arrange the data in ascending order.

1215, 1277, 1342, 1405, 1516, 1520, 1670, 2060, 2187, 2788

Step 2 Determine the five-number summary of the data.

1215, 1277, 1342, 1405, 1516, 1520, 1670, 2060, 2187, 2788

min $\quad Q_1 = 1342 \quad Q_2 = \frac{1516 + 1520}{2}$ or 1518 $\quad Q_2 = 2060 \quad$ max

Part B Construct a box plot.

Step 1 Construct a number line.

Because the minimum is 1215 and the maximum is 2788, your number line must include those values.

Step 2 Draw the box.

Draw and label a box from Q_1 to Q_3, with a vertical line at the median.

Step 3 Draw the whiskers.

Draw a line from the minimum to Q_1. Draw a line from Q_3 to the maximum.

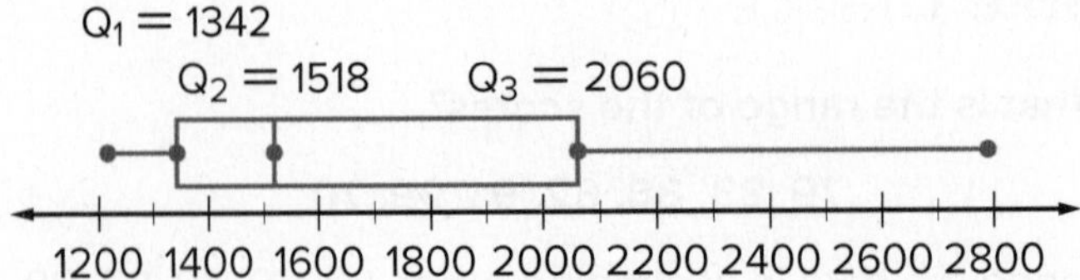

Go Online You can complete an Extra Example online.

Think About It!

If Q_1 were located between two numbers, how would you determine its value?

Check

MARRIAGE The average age at which women first get married differs by country. The ages for eight countries are shown. Select the box plot for the data.

25, 26, 21, 27, 31, 31, 29, 20

A. **Average Age of Women at First Marriage**

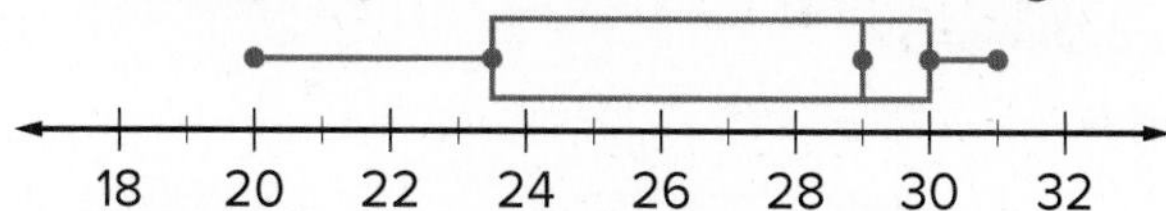

B. **Average Age of Women at First Marriage**

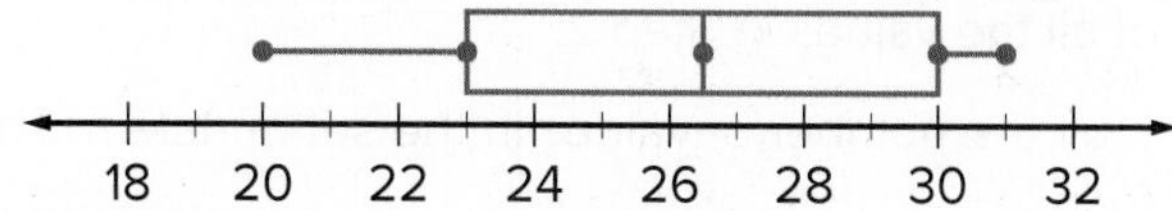

C. **Average Age of Women at First Marriage**

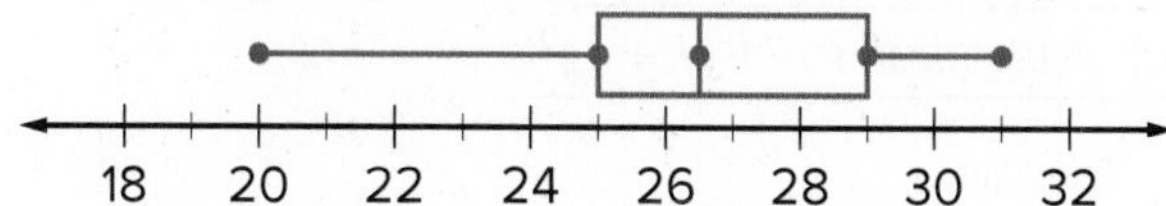

D. **Average Age of Women at First Marriage**

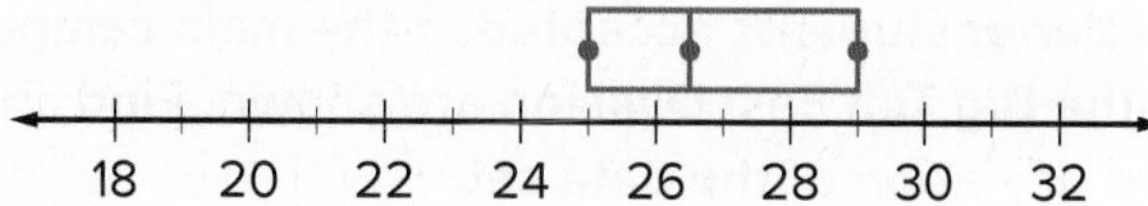

Example 3 Interquartile Range

AUDIO **Sarah wants to upload a song that she recorded and share it with her friends. She wants to know whether her song, which is currently −12 decibels on average, is the right volume compared to other songs online so listeners do not have to adjust their volume. She writes down the average volume of seven songs. Find the interquartile range of these average volumes.**

−18, −20, −8, −13, −14, −15, −18

Step 1 Order the data: −20, −18, −18, −15, −14, −13, −8

Step 2 Determine Q_1 and Q_3.

Step 3 Determine the *IQR*.

$IQR = Q_3 - Q_1 = -13 - (-18)$ or 5

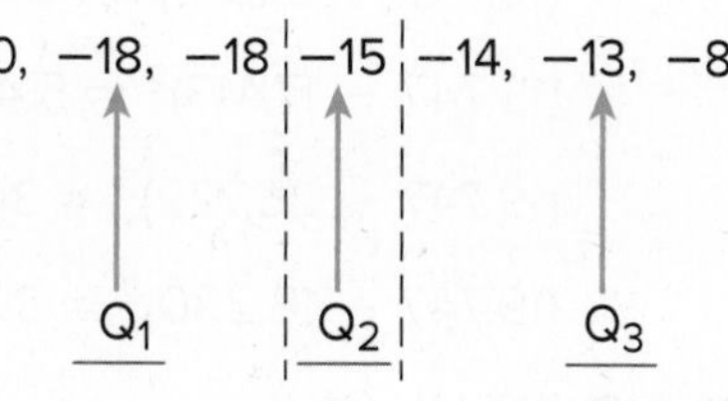

Check

WRITING Cora is writing a novel and tracks the number of pages she writes each day for a week. The number of pages she wrote each day for a week is shown. Find the interquartile range of the data set.

6, 5, 0, 3, 8, 1, 4

Go Online You can complete an Extra Example online.

Think About It!

If all the data in a set were the same value, then what would the standard deviation be?

Learn Standard Deviation

In a data set, the **standard deviation** shows how the data deviate from the mean. A majority of the data in a set, approximately two-thirds, is contained within 1 standard deviation below and above the mean. So, if two data sets have the same mean, but one has a greater standard deviation, then the data in that set is more spread out from the mean.

Key Concept • Standard Deviation

Step 1 Find the mean μ.

Step 2 Find the square of the difference between each data value x_n and the mean, $(\mu - x_n)^2$.

Step 3 Find the sum of all the values in Step 2.

Step 4 Divide the sum by the number of values in the set of data n. This value is the variance.

Step 5 Take the square root of the variance.

Formula $\sigma = \sqrt{\frac{(\mu - x_1)^2 + (\mu - x_2)^2 + \cdots + (\mu - x_n)^2}{n}}$

Example 4 Calculate Standard Deviation

UNIVERSITY **The number of students accepted at the main campus of each university in the Big Ten East Division are shown. Find and interpret the standard deviation of the data set.**

27,300	**12,333**	**15,570**	**21,610**
17,413	**25,772**	**18,230**	

Step 1 Find the mean μ.

$$\mu = \frac{27{,}300 + 12{,}333 + 15{,}570 + 21{,}610 + 17{,}413 + 25{,}772 + 18{,}230}{7} \text{ or } 19{,}747$$

Step 2 Find the square of the differences, $(\mu - x_n)^2$.

$(19{,}747 - 27{,}300)^2 = 57{,}047{,}809$

$(19{,}747 - 12{,}333)^2 = 54{,}967{,}396$

$(19{,}747 - 15{,}570)^2 = 17{,}447{,}329$

$(19{,}747 - 21{,}610)^2 = 3{,}470{,}769$

$(19{,}747 - 17{,}413)^2 = 5{,}447{,}556$

$(19{,}747 - 25{,}772)^2 = 36{,}300{,}625$

$(19{,}747 - 18{,}230)^2 = 2{,}301{,}289$

Step 3 Find the sum.

$57{,}047{,}809 + 54{,}967{,}396 + \ldots + 2{,}301{,}289 = 176{,}982{,}773$

Step 4 Divide by the number of values. This value is the variance.

$$\frac{176{,}982{,}773}{7} \approx 25{,}283{,}253$$

Step 5 Find the standard deviation.

$$\sqrt{25{,}283{,}253} \approx 5028$$

Think About It!

If another data point less than 14,719 or greater than 24,775 is added to the data set, how would that change the standard deviation?

Go Online

You can complete an Extra Example online.

Practice

Go Online You can complete your homework online.

Example 1

Find the range of each data set.

1. 12, 27, 43, 52, 43, 18, 45, 53, 26
2. 132, 127, 129, 130, 141, 125, 138, 129
3. 56, 101, 78, 49, 55, 108, 111, 64
4. 5.9, 6.2, 3.9, 3.7, 8.5, 6.2, 9.0, 8.7, 4.5, 9.3
5. **EXERCISE** Kent tracked his daily number of minutes of exercise. Find the range of the data set.

Number of Minutes of Exercise					
30	35	25	28	40	38
36	29	34	45	42	39

Example 2

Determine the five-number summary and draw a box plot of the data.

6. prices in dollars of smartphones: 311, 309, 312, 314, 399, 312
7. attendance at an event for the last nine years: 68, 99, 73, 65, 67, 62, 80, 81, 83
8. books a student checks out of the library: 17, 9, 10, 17, 18, 5, 2
9. ounces of soda dispensed into 36-ounce cups: 36.1, 35.8, 35.2, 36.5, 36.0, 36.2, 35.7, 35.8, 35.9, 36.4, 35.6
10. ages of riders on a roller coaster: 45, 17, 16, 22, 25, 19, 20, 21, 32, 37, 19, 21, 24, 20, 18, 22, 23, 19

Example 3

Find the interquartile range of each data set.

11. 43, 36, 51, 68, 50, 27, 38, 81, 33
12. 201, 225, 217, 240, 232, 252, 228, 231
13. 94, 87, 105, 99, 118, 97, 102, 85
14. 8.4, 7.1, 6.3, 6.8, 9.2, 7.3, 8.8, 7.9, 5.3, 8.2
15. **HEART RATE** A nurse tracked the heart rates of several patients. Find the interquartile range (IQR) of the data set: 108, 88, 119, 75, 96, 88, 100, 99, 125, 81.

Example 4

Find the standard deviation.

16. {10, 9, 11, 6, 9}
17. {6, 8, 2, 3, 2, 9}
18. {23, 18, 28, 36, 15}
19. {44, 35, 40, 37, 43, 38, 40}
20. **PARKING** A city councilor wants to know how much revenue the city would earn by installing parking meters on Main Street. He counts the number of cars parked on Main Street each weekday: {64, 79, 81, 53, 63}. Find the standard deviation.

Mixed Exercises

21. **REASONING** A hockey team keeps track of how many goals it scores each game: {2, 4, 0, 3, 7, 2}. Find and interpret the standard deviation of the data.

22. **FOOTBALL** The table shows information about the number of carries a running back had over a number of years. Find and interpret the standard deviation of the number of carries.

Year	Number of Carries
2006	31
2007	90
2008	105
2009	115
2010	162

23. **MOVIES** The manager at a movie theater kept track of the age of each person in a matinee movie: 67, 62, 65, 38, 69, 67, 59, 41, 43, 36, 45, 22, 69, 68, 18, 15, 9, 60, 64.

   a. Determine the five-number summary for the data set.

   b. Draw a box plot of the data.

24. **GAS PRICES** Renee is planning a road trip to her aunt's house. To estimate how much the trip will cost, she goes online and finds the price of a gallon of gasoline for 5 randomly selected gas stations along the route: \$2.09, \$2.19, \$3.99, \$2.39, \$2.29.

   a. Determine the five-number summary for the data set.

   b. Draw a box plot of the data.

Find the range, five-number summary, interquartile range, and standard deviation for each data set. Then draw a box plot of the data.

25. **SEASHELLS** Jorja collected the following number of seashells for the last nine trips to the beach: 5, 11, 7, 12, 13, 17, 3, 15, 14.

26. **SHOE SIZE** The following shoe sizes of students at a high school were randomly recorded for one hour: 6, 8, 8.5, 10, 12, 6.5, 7, 8, 8.5, 7.5, 9, 11.5, 10, 13, 5.5, 6.5, 5, 9.5.

Higher-Order Thinking Skills

27. **FIND THE ERROR** Jennifer and Megan are determining one way to decrease the size of the standard deviation of a set of data. Is either correct? Explain your reasoning.

Jennifer	Megan
Remove the outliers from the data set.	Add data values to the data set that are equal to the mean.

28. **ANALYZE** Determine whether the statement *Two random samples taken from the same population will have the same mean and standard deviation* is *sometimes, always,* or *never* true. Justify your argument.

29. **CREATE** Write your own survey question and collect data about your question from 8 classmates. Use that data to find the range, five-number summary, interquartile range, and standard deviation for the data set. Then draw a box plot of the data.

30. **WRITE** What does the interquartile range tell you about how data clusters around the median of the data?

Distributions of Data

Learn Shapes of Distributions

Analyzing the shape of a **distribution** can help you learn a lot about the data it represents. When data are graphed, the shape of the distribution can be seen.

Key Concept • Symmetric and Skewed Distributions

Histograms	Box Plots	Dot Plots
In a **symmetric distribution**, the mean and median are approximately equal.		
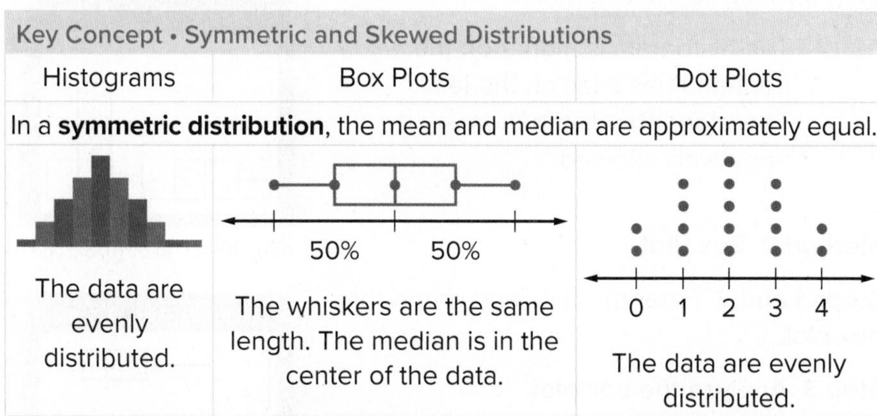		
The data are evenly distributed.	The whiskers are the same length. The median is in the center of the data.	The data are evenly distributed.
A **negatively skewed distribution** typically has a median greater than the mean.		
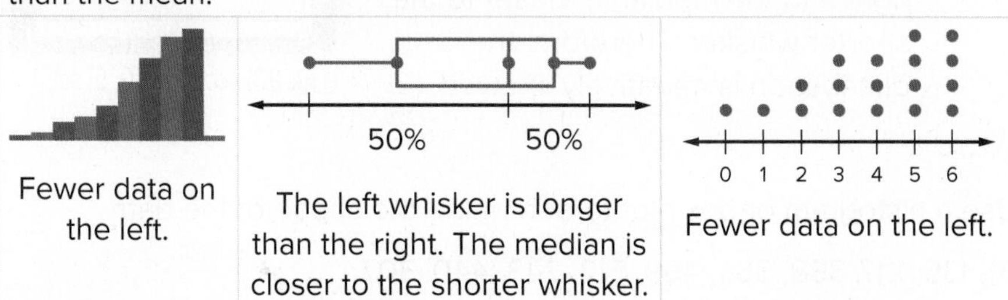		
Fewer data on the left.	The left whisker is longer than the right. The median is closer to the shorter whisker.	Fewer data on the left.
A **positively skewed distribution** typically has a mean greater than the median.		
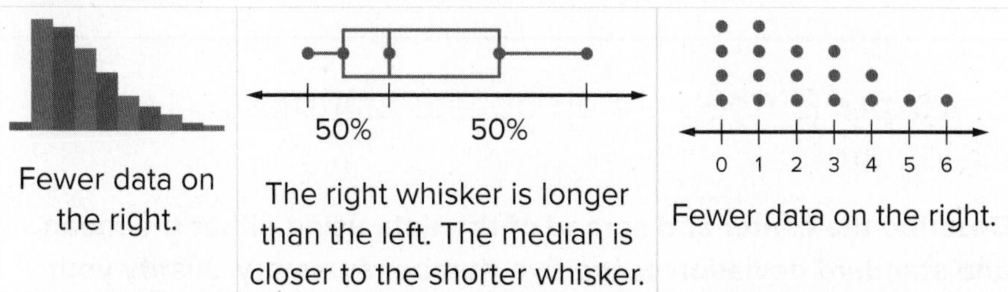		
Fewer data on the right.	The right whisker is longer than the left. The median is closer to the shorter whisker.	Fewer data on the right.

Analyzing Distribution

Negatively Skewed	Symmetric	Positively Skewed
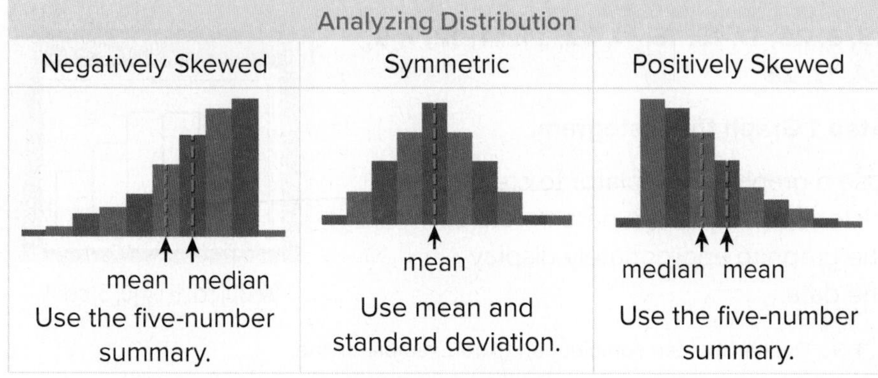		
mean median	mean	median mean
Use the five-number summary.	Use mean and standard deviation.	Use the five-number summary.

Today's Goals

- Interpret differences in the shapes of distributions.
- Account for the possible effects of extreme data points.

Today's Vocabulary

distribution
symmetric distribution
negatively skewed distribution
positively skewed distribution
outlier

Go Online

You may want to complete the Concept Check to check your understanding.

Study Tip

Distribution Shape For a histogram, drawing a curve over the data bars may help you see the distribution.

Example 1 Analyze Distribution by Using Technology

Use a graphing calculator to construct a histogram and box plot for the data. Then describe the shape of the distribution.

78, 53, 24, 75, 76, 83, 78, 60, 64, 53, 36, 47, 32, 75, 54, 68, 68, 74, 85, 42

Method 1 Histogram

Steps 1 and 2 Enter the data and then graph the histogram.

Step 3 Analyze the histogram.

The histogram is higher on the right and has a tail on the left. Therefore, the distribution is negatively skewed.

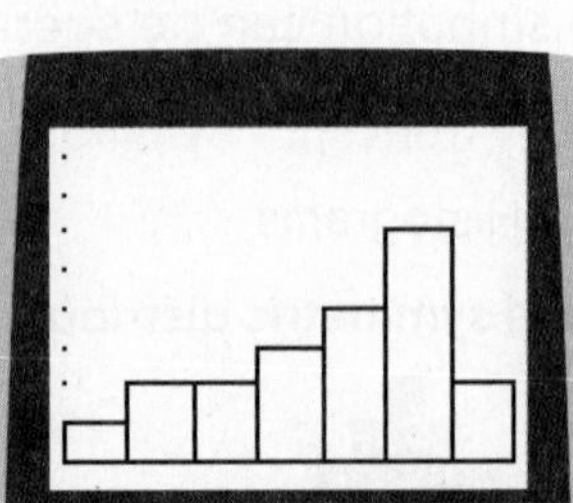

[20, 90] scl: 10 by [0, 8] scl: 1

Method 2 Box Plot

Steps 1 and 2 Enter the data and graph the box plot.

Step 3 Analyze the box plot.

The left whisker is longer than the right and the median is closer to the shorter whisker. Therefore, the distribution is negatively skewed.

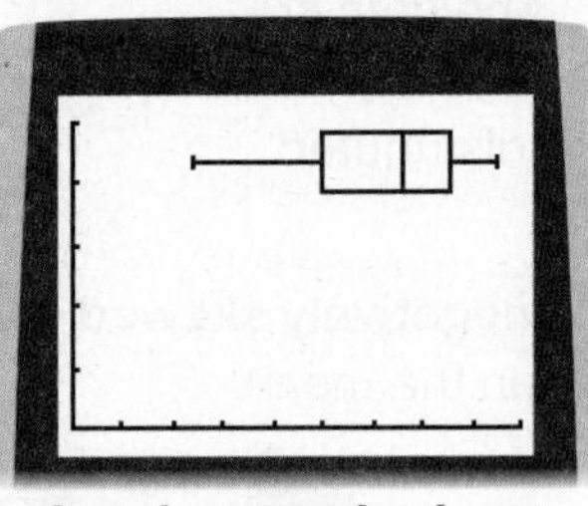

[0, 90] scl: 10 by [0, 5] scl: 1

Check

Use a histogram or box plot to determine the shape of the data.

61, 135, 217, 388, 354, 459, 512, 243, 440, 307

The shape of the distribution is ___?___.

Example 2 Choose Appropriate Statistics by Using a Histogram

Describe the center and spread of the data using either the mean and standard deviation or the five-number summary. Justify your choice by constructing a histogram for the data.

18, 3, 28, 17, 13, 18, 11, 22, 21, 14, 12, 7, 9, 24, 17, 28

Step 1 Graph the histogram.

Use a graphing calculator to create a histogram. Adjust the parameters of the graph to appropriately display the data.

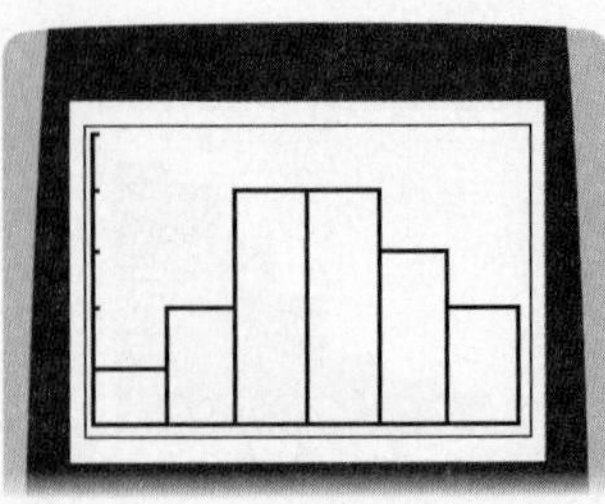

[0, 30] scl: 5 by [0, 5] scl: 1

 **Go Online** You can complete an Extra Example online.

Talk About It!

How could different bin widths of a histogram affect the shape of the distribution? Explain your reasoning.

Study Tip:

Window Settings On a TI-84, use **ZoomStat** from the **Zoom** menu to get a basic fitting view window. Then, adjust the window parameters and bin width.

Go Online

to see how to use a graphing calculator with these examples.

This graph shows that the frequency of the data in the middle is high while frequency of data to the left and right are low. Therefore, the distribution is symmetric.

Step 2 Calculate statistics.

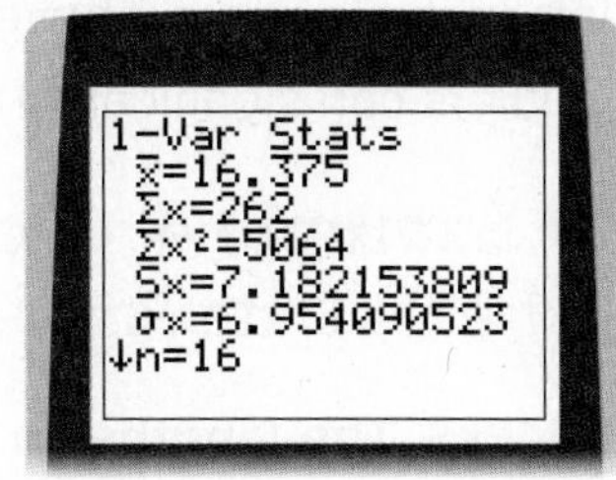

The distribution is symmetric, so the mean and standard deviation are good statistics to represent the data.

To display the statistics, press **STAT**, access the **CALC** menu, select **1-VAR Stats**, and press **ENTER**.

The mean $\overline{x}$ is about 16.4 with a standard deviation σ of about 7.0.

Think About It!

Why is mean an appropriate statistic to represent the center for these data?

Example 3 Choose Appropriate Statistics by Using a Box Plot

Describe the center and spread of the data using either the mean and standard deviation or the five-number summary. Justify your choice by constructing a box plot for the data.

202, 148, 21, 60, 74, 140, 462, 157, 225, 23, 88, 241, 59, 139, 351

Step 1 Graph the box plot.

Use a graphing calculator to create a box plot. Adjust the parameters of the graph to appropriately display the data.

[0, 500] scl: 100 by [0, 5] scl: 1

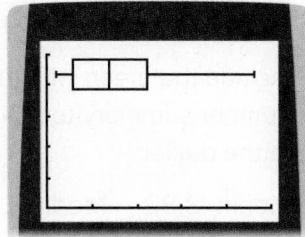

The right whisker is longer than the left and the median is slightly closer to the right whisker. So, this distribution is positively skewed.

Step 2 Calculate statistics.

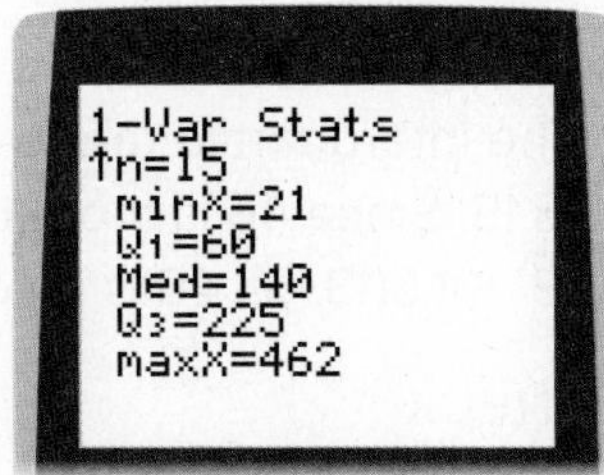

The distribution is positively skewed, so use the five-number summary.

To display the statistics, press **stat**, access the **CALC** menu, select **1-VAR Stats**, and press **enter**. Use the down arrow key to display more statistics.

Maximum: 462

Minimum: 21

Median: 140

Lower Quartile: 60

Upper Quartile: 225

 **Go Online** You can complete an Extra Example online.

Learn Extreme Data Points

The least and greatest values in a set of data are called **extreme values**. An **outlier** is a value that is more than 1.5 times the interquartile range above the third quartile or below the first quartile. Outliers can significantly skew the mean and standard deviation.

Example 4 Choose Appropriate Statistics with Extreme Data Points

SHARKS **The lengths, in feet, of adult sharks of various species are shown. Describe the center and spread of the data using appropriate statistics, and identify the effect of extreme data points.**

33	5.2	12.5	11.5	12	0.6	11	20	23	10
6.5	18	13	5	12	4	1	20	12	46

Think About It!

Suppose a new species of shark is discovered that has an average length of 50 feet. How would two extreme data points affect the measures of center?

Step 1 Make a box plot.

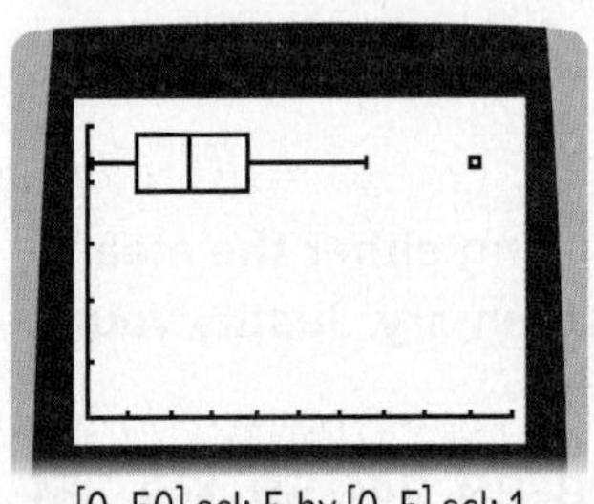

[0, 50] scl: 5 by [0, 5] scl: 1

Step 3 Calculate statistics.

Include the mean with the five-number summary to see the effect of the outlier.

Mean: 13.82 Median: 12

Max: 46 Min: 0.6

Lower Quartile: 5.85

Upper Quartile: 19

The interquartile range is 13.15. Since 46 is more than 19 + 1.5(13.15), 46 is an outlier.

Step 2 Analyze the graph.

Notice that the right whisker is longer than the left. So, this distribution is positively skewed. The plot also shows that there is an outlier.

Step 4 Describe the effect of the outlier.

Since 46 is an outlier, it has affected the mean. To see how much, remove 46 from the set of data and display the statistics again.

Mean: 12.12 Median: 12

Notice that the median did not change when the extreme data point was removed, but the mean did. Without the outlier, the mean and median are closer to the same value.

Go Online to see how to use a graphing calculator with this example.

Check

PRECIPITATION The table shows the annual rainfall in Death Valley, CA.

Year	2006	2007	2008	2009	2010	2011	2012	2013
Rainfall (in.)	0.85	0.18	1.04	0.26	2.41	0.98	0.40	0.51

Part A What year(s) represent outlier(s)?

Part B Because of the outlier, the mean is ___?___.

Go Online You can complete an Extra Example online.

Practice

Go Online You can complete your homework online.

Example 1

Use a graphing calculator to construct a histogram and a box plot for the data. Then describe the shape of the distribution.

1. 55, 65, 70, 73, 25, 36, 33, 47, 52, 54, 55, 60, 45, 39, 48, 55, 46, 38
   50, 54, 63, 31, 49, 54, 68, 35, 27, 45, 53, 62, 47, 41, 50, 76, 67, 49

2. 42, 48, 51, 39, 47, 50, 48, 51, 54, 46, 49, 36, 50, 55, 51, 43, 46, 37
   50, 52, 43, 40, 33, 51, 45, 53, 44, 40, 52, 54, 48, 51, 47, 43, 50, 46

Example 2

Describe the center and spread of the data using either the mean and standard deviation or the five-number summary. Justify your choice by constructing a histogram for the data.

3. 32, 44, 50, 49, 21, 12, 27, 41, 48, 30, 50, 23, 37, 16, 49, 53, 33, 25
   35, 40, 48, 39, 50, 24, 15, 29, 37, 50, 36, 43, 49, 44, 46, 27, 42, 47

4. 82, 86, 74, 90, 70, 81, 89, 88, 75, 72, 69, 91, 96, 82, 80, 78, 74, 94
   85, 77, 80, 67, 76, 84, 80, 83, 88, 92, 87, 79, 84, 96, 85, 73, 82, 83

Example 3

Describe the center and spread of the data using either the mean and standard deviation or the five-number summary. Justify your choice by constructing a box plot for the data.

5. 47, 16, 70, 80, 28, 33, 91, 55, 60, 45, 86, 54, 30, 98, 34, 87, 44, 35
   64, 58, 27, 67, 72, 68, 31, 95, 37, 41, 97, 56, 49, 71, 84, 66, 45, 93

6. 64, 36, 32, 65, 41, 38, 50, 44, 39, 34, 47, 35, 46, 36, 53, 35, 68, 40
   36, 62, 34, 38, 59, 46, 63, 38, 67, 39, 59, 43, 39, 36, 60, 47, 52, 45

Example 4

7. FLYING The various prices of a flight from Los Angeles to New York are shown.
   $182, $234, $264, $271, $277, $314, $317, $455
   a. Make a box plot of the data.
   b. Calculate the statistics that best represent the data.
   c. Describe the effect of the outlier.

8. EXERCISE Yoshiko tracked her minutes of exercise each day for 10 days as shown. 57, 60, 53, 59, 57, 61, 61, 54, 62, 10
   a. Make a box plot of the data.
   b. Calculate the statistics that best represent the data.
   c. Describe the effect of the outlier.

Mixed Exercises

USE TOOLS **Use a graphing calculator to construct a histogram and a box plot for the data. Then describe the shape of the distribution.**

9. 14, 71, 63, 42, 24, 76, 34, 77, 37, 69, 54, 64, 47, 74, 59, 43, 76, 56
   78, 52, 18, 54, 39, 28, 56, 74, 68, 36, 20, 49, 67, 47, 69, 68, 72, 69

10. 53, 34, 36, 38, 43, 49, 52, 36, 39, 37, 58, 45, 37, 38, 46, 52, 45, 39
55, 39, 40, 55, 38, 40, 42, 38, 45, 36, 46, 39, 35, 41, 49, 43, 52, 34

11. 51, 19, 46, 64, 29, 51, 58, 30, 55, 31, 34, 31, 50, 37, 40, 39, 40, 41
42, 32, 24, 48, 43, 45, 38, 43, 58, 47, 34, 36, 50, 54, 46, 28, 60, 22

12. **TRACK** Daryn recorded the number of laps he walked around the track each week. Use a graphing calculator to construct a histogram for the data, and describe the shape of the distribution.

17, 21, 23, 26, 27, 28, 28, 27, 33, 34, 33, 27, 29, 22, 19, 28, 35

13. **GOLF** Mr. Swatsky's geometry class's miniature golf scores are shown below. Use a graphing calculator to construct a box plot for the data, and describe the shape of the distribution.

Scores
36, 38, 38, 39, 40, 42, 44, 46, 46, 47, 48, 48, 50,
52, 52, 53, 54, 55, 56, 56, 56, 60, 57, 58, 63

14. **HAIR LENGTH** Ruth recorded the lengths, in centimeters, of hair of students in her school. Describe the center and spread of the data using either the mean and standard deviation or the five-number summary. Justify your choice by creating a box plot for the data.

40, 39, 37, 26, 25, 40, 35, 34, 26, 39, 42, 33, 26, 25, 34, 38, 41, 34
37, 39, 32, 30, 22, 38, 36, 28, 27, 39, 34, 26, 36, 38, 25, 39, 23, 8

15. **PRESIDENTS** The ages of the presidents of the United States at the time of their inaugurations are shown. Describe the center and spread of the data using either the mean and standard deviation or the five-number summary. Justify your choice by creating a box plot for the data.

Ages of Presidents
57, 61, 57, 57, 58, 57, 61, 54, 68, 51, 49, 64, 50, 48, 65,
52, 56, 46, 54, 49, 51, 47, 55, 55, 54, 42, 51, 56, 55, 51,
54, 51, 60, 62, 43, 55, 56, 61, 52, 69, 64, 46, 54, 47

16. **AUTOMOTIVE** A service station tracks the number of cars they service per day.

Cars Serviced
40, 47, 37, 42, 46, 31, 50, 41, 17, 43, 36, 45, 21, 43, 45, 23, 49, 50,
48, 26, 42, 46, 35, 52, 27, 51, 31, 44, 35, 27, 46, 39, 33, 50, 45, 50

a. Use a graphing calculator to construct a histogram for the data, and describe the shape of the distribution.
b. Describe the center and spread of the data using either the mean and standard deviation or the five-number summary. Justify your choice.

17. COMMUTE The number of miles that Armando drove each week during a 15-week period is shown.

Distance (miles)
62, 110, 92, 430, 73, 84, 525, 123, 86, 290, 114, 98, 103, 312, 71

a. Use a graphing calculator to construct a box plot. Describe the center and spread of the data.

b. Armando visited four colleges during this period, and these visits account for the four highest weekly totals. Remove these four values from the data set. Use a graphing calculator to construct a box plot that reflects this change. Then describe the center and spread of the new data set.

c. Calculate and compare the mean and median for the original data set to the mean and median for the data set from **part b**.

18. ELEVATION The table contains data about 10 elevations in the United States.

a. Use a graphing calculator to construct a box plot for the data, and describe the shape of the distribution.

b. Describe the center and spread of the data using either the mean and standard deviation or the five-number summary. Justify your choice.

c. If there is an outlier, describe its effect on the statistics.

Elevations in the US	
Mt McKinley, AK	20,237
Mt Whitney, CA	14,494
Mt Elbert, CO	14,433
Mt Rainier, WA	14,410
Gannett Peak, WY	13,804
Mauna Kea, HI	13,796
Kings Peak, UT	13,528
Wheeler Peak NM	13,161
Boundary Peak, NV	13,140
Granite Peak, MT	12,799

19. USE A MODEL The histograms show the weight of sample boxes of two brands of pasta.

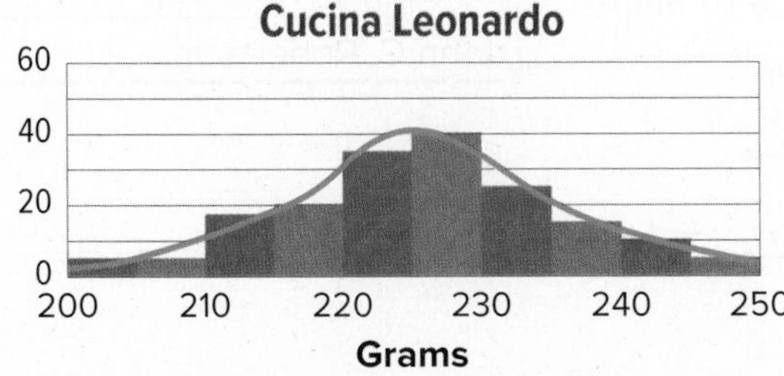

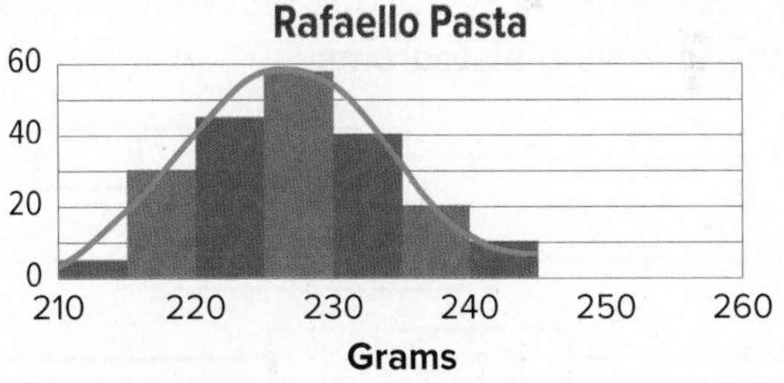

a. Do the two packages of pasta likely have the same advertised weight? Which manufacturer's quantity control appears better? Explain your answers based on the distributions.

b. Infer the two population distribution shapes by analyzing the smooth curves across the tops of the histograms. Describe the shapes you observe.

20. STRUCTURE The United States has been sending astronauts up in the Space Shuttle since 1981. The table provides data regarding the duration of Space Shuttle flights from 1981 to 1985, and then from 2005 to 2011.

Length of Flights from 1981–1985 (days)	Length of Flights from 2005–2011 (days)
Days: 2, 2, 8, 7, 5, 5, 6, 6, 10, 8, 7, 6, 8, 8, 3, 7, 7, 7, 8, 7, 4, 7, 7	Days: 14, 13, 12, 13, 14, 13, 15, 13, 16, 14, 15, 13, 13, 16, 14, 11, 14, 15, 12, 13, 16, 13

Choose and calculate the statistics appropriate for the distribution of the data sets. Use the statistics to compare the two sets.

21. REASONING Gerardo live streams with 15 of his friends. Most of his streams have lasted 10–15 days so far, however he has two streams that have lasted 93 days. Describe what Gerardo's data distribution would look like currently and how it would be affected if he lost his longest streams.

22. CONSTRUCT ARGUMENTS Examine the two box plots shown. Without knowing the data points but assuming the same scale, what conclusion can be made? Justify your argument.

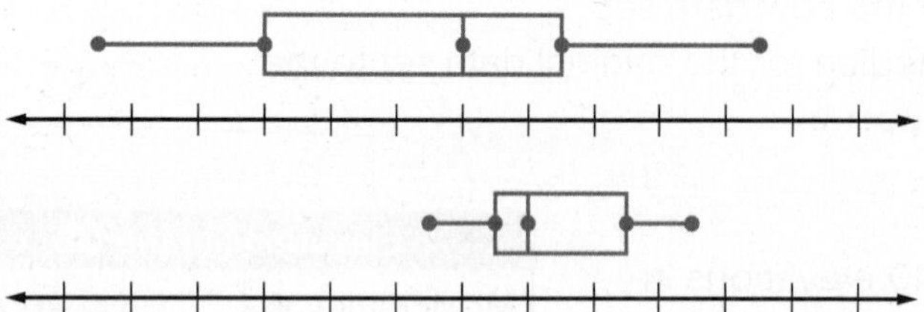

23. SUPREME COURT The table gives the ages of the Supreme Court Justices in 2017.

a. Use a graphing calculator to construct a histogram for the data, and describe the shape of the distribution.

b. Describe the center and spread of the data using appropriate statistics. Justify your choice.

c. If there is an outlier, describe its effect on the statistics.

Supreme Court Justices	
Neil Gorsuch	49
Elena Kagan	57
Sonia Sotomayor	62
Samuel Anthony Alito	67
Stephen G. Breyer	78
Ruth Bader Ginsburg	84
Clarence Thomas	68
Anthony M Kennedy	80
John G. Roberts Jr.	62

Higher-Order Thinking Skills

24. PERSEVERE Identify the box plot that corresponds to each of the following histograms.

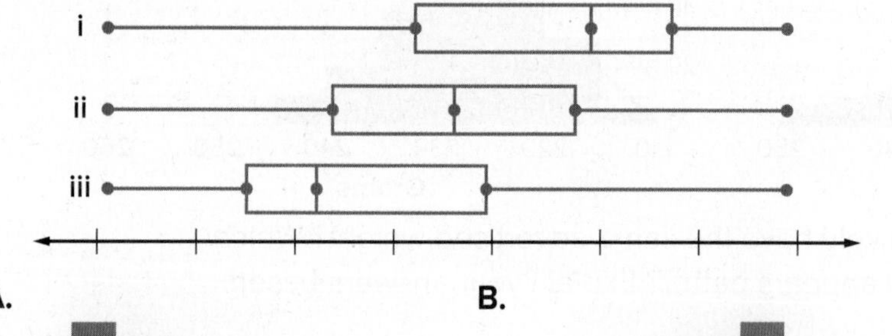

A.

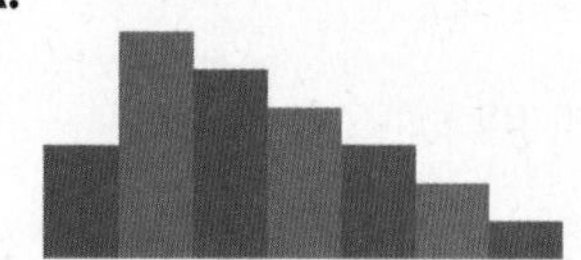

B.

C.

25. ANALYZE Research and write a definition for a *bimodal distribution*. How can the measures of center and spread of a bimodal distribution be described?

26. CREATE Give an example of a set of real-world data with a distribution that is symmetric and one with a distribution that is not symmetric.

27. WRITE Explain why the mean and standard deviation are used to describe the center and spread of a symmetrical distribution and the five-number summary is used to describe the center and spread of a skewed distribution.

Comparing Sets of Data

Explore Transforming Sets of Data by Using Addition

Online Activity Use graphing technology to complete the Explore.

INQUIRY How can you find the measures of center and spread of a set of data that has been transformed using addition?

Explore Transforming Sets of Data by Using Multiplication

Online Activity Use graphing technology to complete the Explore.

INQUIRY How can you find the measures of center and spread of a set of data that has been transformed using multiplication?

Learn Linear Transformations of Data

A **linear transformation** is one or more operations performed on a set of data that can be written as a linear function. Common linear transformations are adding a constant to or multiplying a constant by every value in the set of data.

Key Concept • Linear Transformations of Data

Transformations Using Addition	Transformations Using Multiplication
A real number k is added to every value in a set of data, $k \neq 0$.	Every value in a set of data is multiplied by a constant k, $k > 0$.
Measures of Center	
The mean, median, and mode of the new set of data can be found by adding k to the mean, median, and mode of the original set of data.	The mean, median, and mode of the new set of data can be found by multiplying each original statistic by k.
Measures of Spread	
The range and standard deviation of the new set of data will be unchanged.	The range and standard deviation of the new set of data can be found by multiplying each original statistic by k.

Today's Goal

- Describe the effects that linear transformations have on measures of center and spread.

Today's Vocabulary

linear transformation

Example 1 Transformations Using Addition

Find the mean, median, mode, range, and standard deviation of the data set obtained after adding 6 to each value.

8, 11, 3, 6, 15, 3, 5, 7, 14, 3, 5, 4

Method 1 Add k to the measures of center and spread of the original set of data.

Find the mean, median, mode, range, and standard deviation of the original data set.

Mean 7 | Mode 3 | Standard Deviation 4

Median 5.5 | Range 12

Add 6 to the mean, median, and mode. The range and standard deviation are unchanged.

Mean 13 | Mode 9 | Standard Deviation 4

Median 11.5 | Range 12

Method 2 Add k to each data value of the original set of data. Add 6 to each data value.

14, 17, 9, 12, 21, 9, 11, 13, 20, 9, 11, 10

Mean 13 | Mode 9 | Standard Deviation 4

Median 11.5 | Range 12

Study Tip

1-Var Stats To quickly calculate the mean $\bar{x}$, median **Med**, standard deviation σ, and range of a data set, enter the data as **L1** in a graphing calculator, and then use the **1-VAR Stats** feature from the **CALC** menu. Subtract **minX** from **maxX** to find the range.

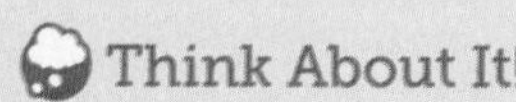

Think About It!

Compare and contrast the two methods.

Example 2 Transformations Using Multiplication

Find the mean, median, mode, range, and standard deviation of the data set obtained after multiplying each value by 4.

12, 18, 20, 12, 14, 18, 11, 21, 13, 18, 11, 24

Find the measures of center and spread for the original data set.

Mean	Median	Mode	Range	Standard Deviation
16	16	18	13	4.2

Multiply the measures of center and spread by 4.

Mean	Median	Mode	Range	Standard Deviation
64	64	72	52	16.8

Check

Find the mean, median, mode, range, and standard deviation of the data set obtained after multiplying each value by 0.6. Round to the nearest tenth, if necessary.

45, 33, 43, 51, 39, 48, 34, 39, 30, 39, 47, 44

Go Online You can complete an Extra Example online.

Think About It!

Use a calculator and the data set to examine the effect of multiplying by a constant k, $k<0$. How can the mean, median, mode, range, and standard deviation of the new data set be found?

Example 3 Compare Symmetric Distributions of Data

RESTAURANTS **The numbers of customers eating at a restaurant during breakfast, lunch, and dinner each day are shown below.**

Breakfast: 71, 58, 65, 48, 44, 56, 68, 64, 51, 67, 74, 62, 59, 53, 62, 73, 54, 49, 63, 55

Lunch: 115, 105, 87, 108, 117, 110, 92, 101, 114, 91, 109, 96, 100, 98, 103, 111, 95, 94, 102, 106

Dinner: 76, 62, 91, 76, 79, 68, 65, 89, 81, 76, 90, 82, 79, 74, 71, 73, 84, 87, 81, 64

Part A Construct a histogram or box plot for each set of data. Then describe the shape of each distribution.

Method 1 Histogram

Enter the data in **L1**, **L2**, and **L3**. From the **STAT PLOT** menu, enter **L1** as the **Xlist** for Plot 1, **L2** for Plot 2, and **L3** for Plot 3. Select [histogram icon] as the plot type for each Plot. View each histogram by turning on Plot 1, Plot 2, and then Plot 3. Use the same window dimensions and bin width for each graph.

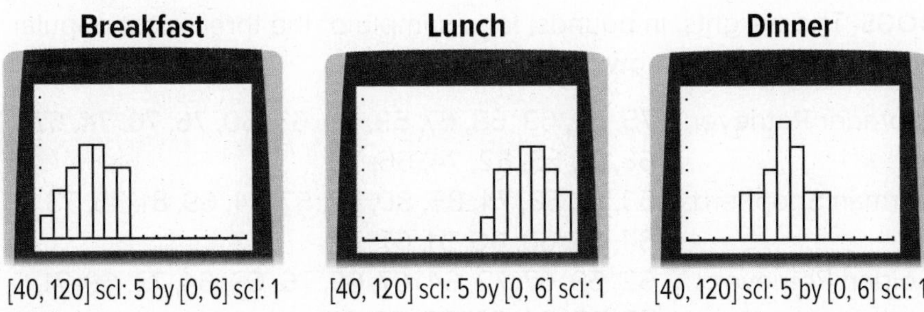

[40, 120] scl: 5 by [0, 6] scl: 1 [40, 120] scl: 5 by [0, 6] scl: 1 [40, 120] scl: 5 by [0, 6] scl: 1

For each time of day, the distribution is high in the middle and low on the left and right. Therefore, all of the distributions are symmetric.

Method 2 Box Plot

Enter the data using the same process. Select [box plot icon] as the plot type for each set of data. To view all of the box plots at once, turn on Plot 1, Plot 2, and Plot 3 and graph.

For each time of day, the lengths of the whiskers are approximately equal, and the median is in the middle of the data. The left and right sides are approximately mirror images of one another. Therefore, all of the distributions are symmetric.

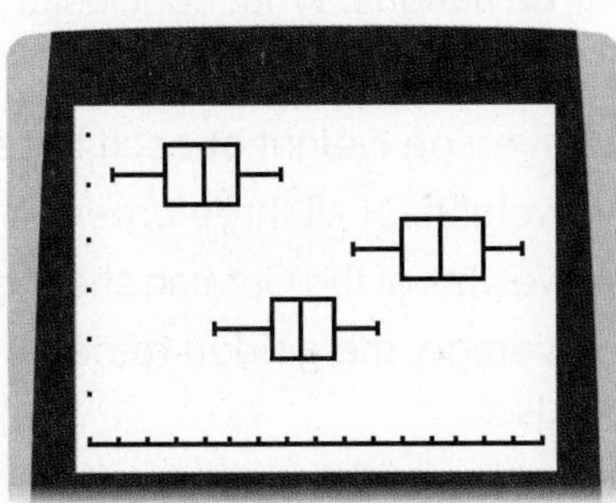

[40, 120] scl: 5 by [0, 6] scl: 1

(continued on the next page)

Go Online You can complete an Extra Example online.

Study Tip

Window and Bin Settings When setting the window dimensions for multiple sets of data, try setting the minimum and maximum as the least and greatest values of all the sets. When selecting a bin width, consider the context of the situation. For example, if the data does not include fractional numbers, as would be the case with number of people, use a whole number as the bin width.

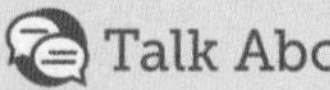

Talk About It!

Describe how this data could be used to make decisions about the restaurant.

Part B Compare the data sets using the means and standard deviations.

All of the distributions are symmetric, so use the means and standard deviation to describe the centers and spreads.

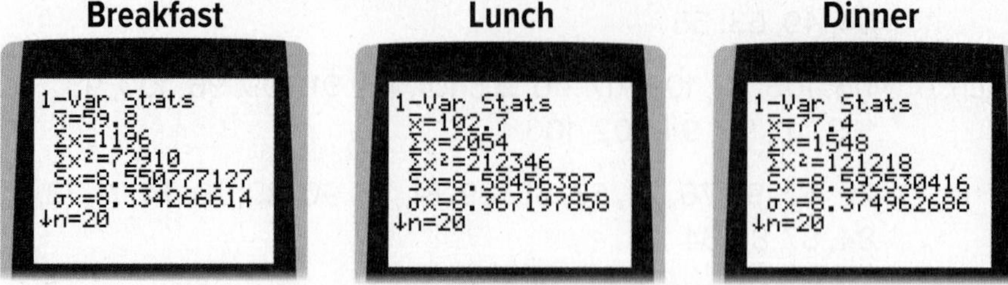

The means vary, with breakfast having the lowest average number of customers and lunch having the highest average number of customers. However, the standard deviations are approximately equal. This means that, while the average number of customers for each time of day is very different, the number of customers for each time of day generally varies by the same amount from day to day.

Go Online to see how to use a graphing calculator with this example.

Check

DOGS The weights, in pounds, for a sample of the three most popular breeds of dogs are shown below.

Labrador Retriever: 75, 59, 63, 68, 67, 59, 69, 63, 60, 76, 70, 74, 67, 68, 71, 65, 62, 74, 66, 78

German Shepherd: 53, 61, 58, 74, 85, 80, 72, 57, 64, 69, 81, 75, 73, 64, 76, 68, 66, 51, 67, 73

Golden Retriever: 62, 59, 67, 72, 64, 67, 69, 76, 63, 64, 73, 69, 71, 75, 59, 64, 69, 59, 74, 68

Part A Use a graphing calculator to construct a histogram or box plot for each set of data. Then complete the statement about the shape of each distribution.

All of the distributions are ___?___.

Part B Compare the data sets using the means and standard deviations. What conclusion(s) can you make about the sets of data? Select all that apply.

A. The average weight of each breed is about the same.

B. The weights of all three breeds are very close to their means.

C. The weights of the German shepherds vary more than the other breeds.

D. On average, the golden retrievers weigh much more than the other breeds.

E. The means of the weights differ by less than 1.5 pounds.

F. The weights of the Labrador retrievers and golden retrievers are generally closer to their means than the German shepherds' weights are to their mean.

Example 4 Compare Skewed Distributions of Data

SPORTS **The numbers of high school boys and girls, in hundred thousands, participating in tennis from 2001–2015 are shown below.**

Boys (hundred thousands)	Girls (hundred thousands)
144, 139, 145, 153, 149, 153, 157, 156, 157, 163, 161, 160, 157, 161, 157	164, 160, 163, 168, 169, 174, 177, 172, 178, 182, 182, 181, 181, 184, 183

Part A Construct a histogram or box plot for each set of data. Then describe the shape of each distribution.

Method 1 Histogram

Enter the data in **L1** and **L2**. From the **STATPLOT** menu, enter **L1** as the **Xlist** for Plot 1 and **L2** for Plot 2. Select [histogram icon] as the plot type for each Plot. View each histogram by turning on Plot 1, and then Plot 2. Use the same window dimensions and bin width for each graph.

Boys

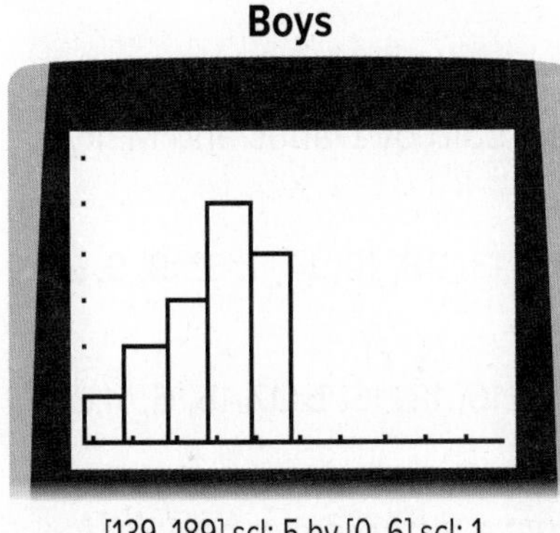

[139, 189] scl: 5 by [0, 6] scl: 1

Girls

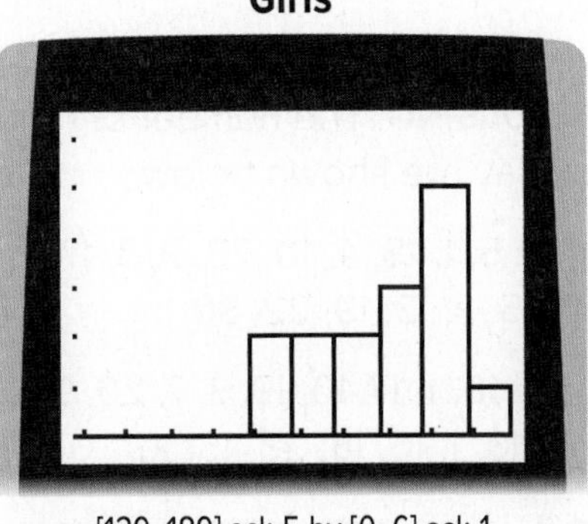

[139, 189] scl: 5 by [0, 6] scl: 1

Both distributions are high on the right and have tails on the left. Therefore, both distributions are negatively skewed.

Method 2 Box Plot

Enter the data using the same process. Select [box plot icon] as the plot type for each set of data. To view both box plots at once, turn on Plot 1 and Plot 2 and graph.

For each distribution, the left whisker is longer than the right, and the median is closer to the right whisker. Therefore, both distributions are negatively skewed.

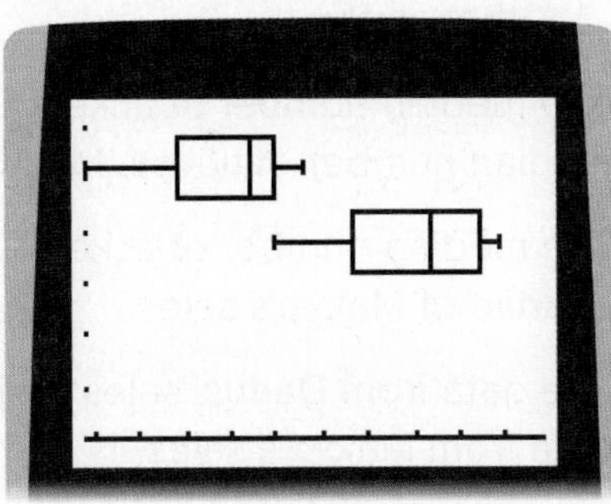

[139, 189] scl: 5 by [0, 6] scl: 1

Part B Compare the data sets using the five-number summaries.

Both distributions are skewed, so use the five-number summary to compare the data.

(continued on the next page)

Go Online You can complete an Extra Example online.

Go Online to see how to use a graphing calculator with this example.

Think About It!
Compare two other statistics from the five-number summaries. What does this tell you about the number of girls and boys that participated in tennis?

The upper quartile for the number of boys that participated in tennis is 160, while the minimum number of girls that participated is 160. This means there were only 160,000 or more boys participating in tennis for 25% of the years, while at least 160,000 girls participated every year.

Boys

```
1-Var Stats
↑n=15
 minX=139
 Q1=149
 Med=157
 Q3=160
 maxX=163
```

We can conclude that many more girls participated in tennis from 2001 to 2015 than boys.

Girls

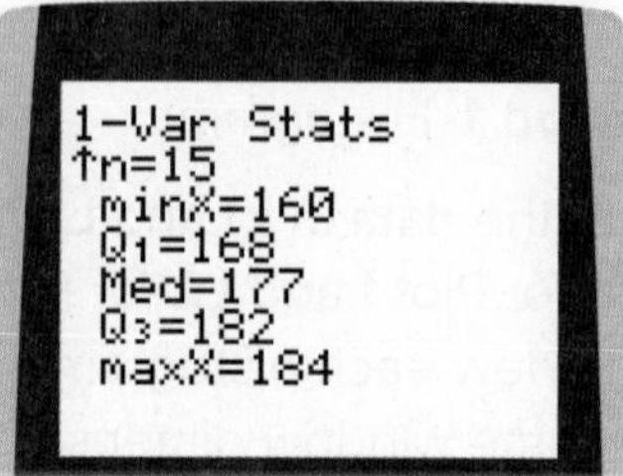

Check

FUNDRAISING The number of raffle tickets sold by Darius and Makya each day are shown below.

Darius: 5, 1, 15, 4, 10, 23, 9, 3, 17, 2, 6, 21, 5, 13, 28, 10, 14, 7, 5, 19, 9, 22, 10, 8, 15, 9, 13, 19, 22, 30

Makya: 18, 1, 17, 10, 19, 3, 7, 20, 9, 22, 12, 13, 16, 18, 16, 5, 17, 15, 6, 11, 18, 14, 16, 18, 1, 16, 18, 23, 15, 10

Part A Use a graphing calculator to construct a histogram or box plot for each set of data. Then complete the statement about the shape of each distribution.

The distribution of Darius' raffle ticket sales is ___?___.

The distribution of Makya's raffle ticket sales is ___?___.

Part B Compare the data sets using the five-number summaries. What conclusion(s) can you make about the sets of data? Select all that apply.

A. The median number of tickets Darius sold is much higher than the median number of tickets Makya sold.

B. The median number of tickets Darius sold is the same as the lower quartile of Makya's sales.

C. The data from Darius' sales is spread over a wider range than the data from Makya's sales.

D. The median number of tickets each student sold was the same.

E. The fewest number of tickets each student sold in a day was 1.

F. The upper 50% of Darius' data spans from 10 to 30, while the upper 75% of Makya's data spans from 10 to 23.

Go Online You can complete an Extra Example online.

Practice

Go Online You can complete your homework online.

Example 1

Find the mean, median, mode, range, and standard deviation of each data set that is obtained after adding the given constant to each value.

1. 52, 53, 49, 61, 57, 52, 48, 60, 50, 47; +8
2. 101, 99, 97, 88, 92, 100, 97, 89, 94, 90; +(−13)
3. 27, 21, 34, 42, 20, 19, 18, 26, 25, 33; +(−4)
4. 72, 56, 71, 63, 68, 59, 77, 74, 76, 66; +16

Example 2

Find the mean, median, mode, range, and standard deviation of each data set that is obtained after multiplying each value by the given constant.

5. 11, 7, 3, 13, 16, 8, 3, 11, 17, 3; ×4
6. 64, 42, 58, 40, 61, 67, 58, 52, 51, 49; ×0.2
7. 33, 37, 38, 29, 35, 37, 27, 40, 28, 31; ×0.8
8. 1, 5, 4, 2, 1, 3, 6, 2, 5, 1; ×6.5

Examples 3 and 4

9. BASEBALL The total wins per season for the first 17 seasons of the Marlins are shown. The total wins over the same time period for the Cubs are also shown.

Marlins
64, 51, 67, 80, 92, 54, 64, 79, 76, 79, 91,
83, 83, 78, 71, 84, 87

Cubs
84, 49, 73, 76, 68, 90, 67, 65, 88, 67, 88,
89, 79, 66, 85, 97, 83

   a. Use a graphing calculator to construct a box plot for each set of data. Then describe the shape of each distribution.

   b. Compare the data sets using either the means and standard deviations or the five-number summaries. Justify your choice.

10. HEALTH CLUBS To plan their future equipment purchases, the Northville Health Club randomly chooses 8 patrons and tracks how many minutes they spend on the treadmill.

   a. Use a graphing calculator to construct a histogram for each set of data. Then describe the shape of each distribution.

   b. Compare the data sets using either the means and standard deviations or the five-number summaries. Justify your choice.

Minutes on Treadmill Last Week	Minutes on Treadmill This Week
30	20
30	30
45	45
20	45
60	30
30	60
30	50
45	45

Mixed Exercises

Find the mean, median, mode, range, and standard deviation of each data set that is obtained after adding or multiplying each value by the given constant(s).

11. 98, 95, 97, 89, 88, 95, 90, 81, 87, 95; +2

12. 32, 30, 27, 29, 25, 33, 38, 26, 23, 31; ×1.6

13. 14, 17, 13, 9, 15, 7, 12, 16, 8, 9; ×5

14. 5, 12, 7, 3, 8, 5, 7, 1, 4, 7, 3, 9; +22

15. 12, 15, 16, 12, 12, 15, 17, 19, 22, 27, 42, 42; +5

16. 49, 43, 26, 39, 40, 30, 33, 64, 26, 45, 23, 26; ×3, +(−8)

17. 71, 72, 68, 70, 72, 67, 68, 72, 65, 70; ×0.2

18. 112, 91, 108, 129, 80, 99, 78, 80; +(−15)

19. 57, 38, 42, 51, 39, 44, 33, 55; +(−7), ×2

20. 55, 50, 58, 52, 56, 57, 50, 55, 50; ×2, +5

21. **BOWLING** The scores of 15 bowlers are shown in the table.

Score
211, 123, 183, 176, 224, 115, 109, 136, 152, 177, 127, 196, 143, 166, 170

a. Find the mean, median, mode, range, and standard deviation of the scores.

b. The handicap of the bowling team will add 56 points to each score. Find the statistics of the scores while including the handicap.

22. **COMPETITION** The distances that 18 participants threw a football are shown in the table.

Distance (feet)
96, 94, 114, 85, 96, 109, 90, 109, 67, 82, 98, 79, 69, 70, 106, 96, 112, 84

a. Find the mean, median, mode, range, and standard deviation of the participants' distances.

b. Find the statistics of the participants' distances in yards.

23. **TEMPERATURE** The monthly average high temperatures for Lexington, Kentucky, are shown in the table.

Temperature (°F)
40, 45, 55, 65, 74, 82, 86, 85, 78, 67, 55, 44

a. Find the mean, median, mode, range, and standard deviation of the temperatures.

b. Find the statistics of the temperatures in degrees Celsius. Recall that $C = \frac{5}{9}(F - 32)$.

24. FANTASY SPORTS The weekly total points of Scott's and Azumi's fantasy baseball teams are shown in the tables.

Scott's Team
109, 99, 121, 137, 131, 141, 77,
83, 139, 92, 42, 133, 98, 153, 124,
102, 113, 117, 112, 128, 107, 147

Azumi's Team
113, 121, 98, 104, 106, 123, 175,
141, 109, 129, 49, 110, 112, 144,
106, 119, 127, 88, 132, 93, 137, 123

a. Use a graphing calculator to construct a box plot for each set of data. Then describe the shape of each distribution.

b. Compare the data sets using either the means and standard deviations or the five-number summaries. Justify your choice.

c. How does eliminating the outliers of each data set affect the statistics and comparison from **part b**?

25. BUSINESS Saeed owns an electronics store. He is revising his pricing for phone accessories. His current prices for an assortment of accessories are listed at the right. He has also determined that the mean price for the same assortment of accessories at a rival store is $10.99.

Saeed's Price Data ($)		
14.99	4.49	9.99
18.49	12.99	6.99
8.49	21.99	13.49
13.99	9.99	10.99
12.49	4.49	12.99

a. Saeed wants to match his rival's prices. Make a table to list the new prices. Explain.

b. Compare the mean and standard deviation of the current prices to the new prices.

26. REASONING Two different samples on the shell diameter of a species of snail are shown.

Sample A (mm)		
45	35	37
40	42	40
28	38	31

Sample B (mm)		
26	44	40
27	35	28
26	39	31

a. Use the median and interquartile range to compare the samples.

b. Based on your findings and on the data points in each sample, which sample appears to be more representative? Explain your reasoning.

27. STRUCTURE Height data samples of 17-year-old male and female students are shown. Use the mean and standard deviation to compare the samples.

Heights of Male Students (inches)		
71	69	67
68	69	70
72	74	68
71	69	72

Heights of Female Students (inches)		
67	62	69
65	71	66
63	65	68
66	63	70

28. CONSTRUCT ARGUMENTS Francisca is planning a two-week vacation to one of two cities and wants to base her decision on the weather history for the same dates as her vacation. She has collected the number of days that it has rained during this two-week period for each city over the past 10 years. The results are shown.

City A	
5	0
7	6
5	6
6	6
3	2

City B	
4	4
6	5
3	7
4	3
5	7

a. Determine the shape of each distribution, and use the appropriate statistics to find the center and spread for each set of data.

b. Which city do you think Francisca should visit on her vacation? Justify your argument.

29. WRITE Compare and contrast the benefits of displaying data using histograms and box plots.

30. ANALYZE If every value in a set of data is multiplied by a constant k, $k < 0$, then how can the mean, median, mode, range, and standard deviation of the new data set be found?

31. PERSEVERE A salesperson has 15 SUVs priced between \$33,000 and \$37,000 and 5 luxury cars priced between \$44,000 and \$48,000. The average price for all of the vehicles is \$39,250. The salesperson decides to reduce the prices of the SUVs by \$2000 per vehicle. What is the new average price for all of the vehicles?

32. ANALYZE If k is added to every value in a set of data, and then each resulting value is multiplied by a constant m, $m > 0$, how can the mean, median, mode, range, and standard deviation of the new data set be found? Justify your argument.

33. WRITE Explain why the mean and standard deviation are used to compare the center and spread of two symmetrical distributions, and the five-number summary is used to compare the center and spread of two skewed distributions or a symmetric distribution and a skewed distribution.

Summarizing Categorical Data

Explore Categorical Data

Online Activity Use a real-world situation to complete the Explore.

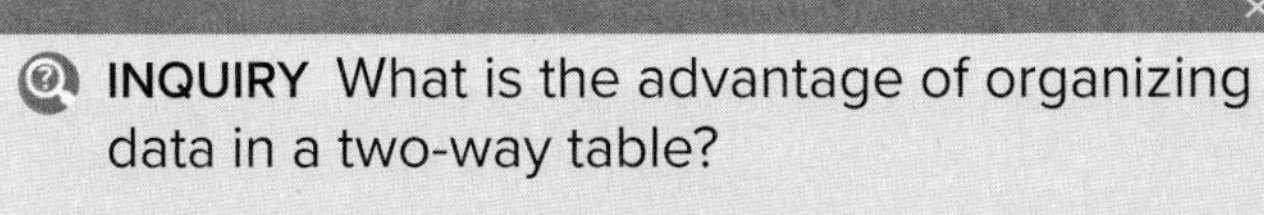
INQUIRY What is the advantage of organizing data in a two-way table?

Today's Goals

- Organize categorical data in a two-way frequency table.
- Determine and interpret the values in a two-way relative frequency table.

Today's Vocabulary

two-way frequency table

joint frequencies

marginal frequencies

relative frequency

two-way relative frequency table

conditional relative frequency

Learn Two-Way Frequency Tables

A **two-way frequency table** or *contingency table* is used to show the frequencies of data from a survey or experiment classified according to two categories, with the rows indicating one category and the columns indicating the other.

Suppose you are constructing a two-way frequency table based on two categories, grade level and employment. The table is constructed below for sample values.

Grade	Employed	Unemployed	Totals
Junior	8	12	20
Senior	15	10	25
Totals	23	22	45

Subcategories: The subcategories are the column and row headers that represent the two different types of categories. In this case, Employed, Unemployed, Junior, and Senior are the subcategories.

Joint frequencies: **Joint frequencies** are the values for every combination of subcategories. So, 8 is a joint frequency that represents the number of students who are employed and juniors.

Marginal frequencies: **Marginal frequencies** are the totals of each subcategory. So, 20 is a marginal frequency that represents the total number of juniors.

Example 1 Use a Two-Way Frequency Table

NAMES **Unisex names are names often used for both males and females. At one point, the most common unisex names in the U.S. were Casey and Riley, with 176,544 Caseys and 154,861 Rileys. During that time, there were 104,161 males with the name Casey and 75,882 females with the name Riley. Organize the data in a two-way frequency table.**

Steps 1 and 2 Enter the given data in a table. Then use the information given to fill in the rest of the cells.

Top Unisex Names in the U.S.			
	Casey	Riley	Totals
Male	104,161	78,979	183,140
Female	72,383	75,882	148,265
Totals	176,544	154,861	331,405

Male Rileys: 154,861 − 75,882 = 78,979

Total Males: 104,161 + 78,979 = 183,140

Female Caseys: 176,544 − 104,161 = 72,383

Total Females: 72,383 + 75,882 = 148,265

Totals: 176,544 + 154,861 = 331,405

Check

TECHNOLOGY Pew Research Center released a survey that asked whether participants thought technological advancements in the future will make people's lives better or worse. Of the people interviewed, 423 earned less than $50,000 per year and 328 earned $50,000 or more. Of those earning less than $50,000 per year, 262 thought that people's lives would get better, and 240 of those who earned $50,000 or more thought the same. Copy and complete the two-way frequency table.

Will technological advancements in the future make people's lives better or worse?			
	Better	Worse	Totals
$<$			
$\geq$			
Totals			

Go Online You can complete an Extra Example online.

Study Tip

Check For each cell, you can see if your calculations are correct by calculating the value for that cell using different data from the table. For example, you could calculate the total number of female participants in the study either by adding the number of women named Casey or Riley, or by subtracting the number of men from the total number of people named Casey or Riley. Either way, you should get the same number.

Use a Source

Create your own two-way frequency table. Find data online that divides a group of subjects into two categories, with each subject fitting into one subcategory of each. For example, in the data shown the categories are whether each person is male or female and whether each person's name is Casey or Riley. Determine the subcategories, enter the given data, and fill in any cells for which values are not provided.

Learn Two-Way Relative Frequency Tables

A **relative frequency** is the ratio of the number of observations in a category to the total number of observations. A **two-way relative frequency table** can help you see patterns of association in the data. To create a two-way relative frequency table, divide each of the values by the total number of observations and replace them with their corresponding decimals or percents.

A **conditional relative frequency** is the ratio of the joint frequency to the marginal frequency. Because each two-way frequency table has two categories, each two-way relative frequency table can provide two different conditional relative frequency tables.

Example 2 Use a Two-Way Relative Frequency Table

PARENTING **Many parents monitor their teenagers' Internet usage. The Pew Research Center conducted a survey of whether parents do or do not check what sites their teens had visited and whether they are the parent of a teen between the ages of 13 and 14 or between the ages of 15 and 17. The results of the survey are shown. Organize the data in a relative frequency table by age group, and interpret the data.**

How Parents Monitor Teenagers' Internet Usage			
Teen's Age	Does Check	Does Not Check	Totals
13 to 14	299	140	439
15 to 17	348	273	621
Totals	647	413	1060

Part A Organize the data in a relative frequency table.

How Parents Monitor Teenagers' Internet Usage			
Teen's Age	Does Check	Does Not Check	Totals
13 to 14	$\frac{299}{1060} \approx 28.2\%$	$\frac{140}{1060} \approx 13.2\%$	$\frac{439}{1060} \approx 41.4\%$
15 to 17	$\frac{348}{1060} \approx 32.8\%$	$\frac{273}{1060} = 25.8\%$	$\frac{621}{1060} \approx 58.6\%$
Totals	$\frac{647}{1060} \approx 61.0\%$	$\frac{413}{1060} \approx 39.0\%$	$\frac{1060}{1060} \approx 100\%$

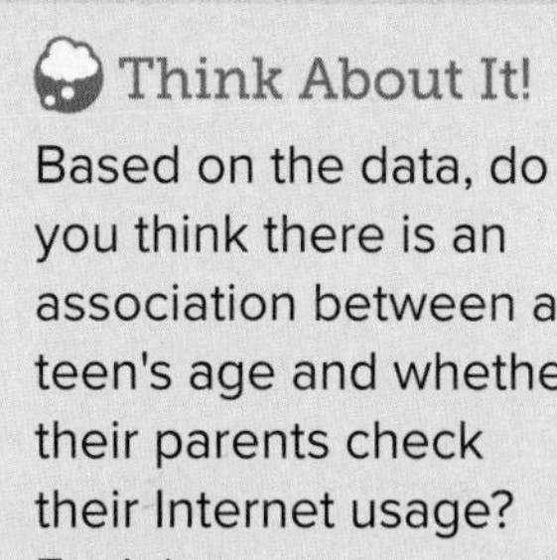
Think About It!
Based on the data, do you think there is an association between a teen's age and whether their parents check their Internet usage? Explain.

Part B Interpret the data.

Do more parents check what sites their teens have visited, or do more parents not check?

61% of parents do check the sites their teens have visited compared to 39% who do not.

Think About It!

If the condition for the relative frequency table were whether a person had voted or not rather than age group, the joint frequency for people between the ages of 18 and 24 who voted would be 52.5%. If someone claims that this indicates 52.5% of all people who voted in the 2012 election were between the ages of 18 and 24, would they be correct? Justify your argument.

Example 3 Use a Two-Way Conditional Relative Frequency Table

VOTING **According to the U.S. Census Bureau, voter turnout describes how many eligible voters show up to vote in an election. The table shows the number of eligible voters who did and did not vote in 2012 for the oldest and youngest eligible age groups. Organize the data in a conditional relative frequency table by age group, and interpret the data.**

Voter Turnout			
Age Group	**Voted**	**Did Not Vote**	**Totals**
18 to 24	12,515	13,275	**25,790**
75 and over	11, 344	5380	**6,724**
Totals	**23,859**	**18,655**	**42,514**

Part A Organize the data in a conditional relative frequency table by age group.

Step 1 Determine which marginal frequencies to use.

The conditional relative frequency relates the number of voters or nonvoters to the age group, so the relevant marginal frequencies are the total numbers of voters for each age group.

Step 2 Determine the ratios of the joint frequencies to the marginal frequencies.

Voter Turnout			
Age Group	**Voted**	**Did Not Vote**	**Totals**
18 to 24	$\frac{12,515}{25,790} \approx 48.5\%$	$\frac{13,275}{25,790} \approx 51.5\%$	$\frac{25,790}{25,790} = 100\%$
75 and over	$\frac{11,344}{16,724} \approx 67.8\%$	$\frac{5380}{16,724} \approx 32.2\%$	$\frac{16,724}{16,724} = 100\%$

Part B Interpret the data.

Which age group has the higher voter turnout?

The percent of eligible voters aged 18 to 24 that voted is 48.5%, and the percent for those aged 75 and over is 67.8%. Based on the data, there is an association between age and whether a person voted. People aged 18 to 24 were more likely to not have voted than people aged 75 and over.

Go Online You can complete an Extra Example online.

Practice

Go Online You can complete your homework online.

Example 1

TREATS **The owner of a snow cone stand keeps track of the sizes and flavors sold one afternoon. He sold 125 snow cones in all. Of these, 40% were large snow cones, 32% were grape, and 12% were small watermelon snow cones. The stand sold 15 more cherry snow cones than grape. The most popular snow cone of the day was small cherry, with a total of 35 sales.**

1. Construct a two-way frequency table to organize the data.
2. How many large grape snow cones were sold?
3. How many watermelon snow cones were sold in all?
4. How many more small snow cones were sold than large snow cones?

Example 2

FOREIGN LANGUAGE **Christy surveyed several students at her school and asked each person what foreign language he or she is studying. The results are shown in the table.**

	Male	Female	Total
Spanish	18	20	38
French	16	12	28
German	6	8	14
Total	40	40	80

5. Construct a relative frequency table by converting the data in the table to percentages. Round to the nearest tenth, if necessary.
6. Find the joint relative frequency of a female student who is studying French.
7. Interpret the data.

Example 3

CLASS PRESIDENT **In a poll for senior class president, 68 of the 145 male students said they planned to vote for Santiago. Out of 139 female students, 89 planned to vote for his opponent, Measha.**

8. Construct a conditional relative frequency table based on voter preference. Show your calculations.
9. What does each conditional relative frequency represent?
10. What is the probability that a vote for Measha will come from a female student? How is this different from the probability that a female student intends to vote for Measha?

Mixed Exercises

VETERINARIAN **The two-way frequency table shows the number of dogs and cats that were seen at a veterinarian's office and the primary purpose of their visit.**

	Dog	Cat	Total
Exam	12	5	17
Shots	6	3	9
Grooming	7	2	9
Total	25	10	35

11. How many dogs were seen for an exam today?

12. How many more dogs than cats were seen at the veterinarian's office?

BIRD WATCHING **A group of bird-watchers has been tracking the number of tree swallows, cardinals, and goldfinches in a region. Over the weekend, a total of 40 birds were observed. Of those, 45% were male, 37.5% were cardinals, and 12.5% were male tree swallows. Twice as many female cardinals were observed as male cardinals. There were 5 female goldfinches spotted.**

13. Construct a two-way frequency table to organize the data.

14. How many more female tree swallows were seen than male cardinals?

15. How many male goldfinches and female cardinals were seen?

16. How many more female birds were seen than male birds?

SCHOOL ACTIVITIES **The two-way frequency table shows the number of students who participate in school sports or clubs at Monroe High School.**

	Sports or Clubs	No Sports or Clubs	Total
Freshmen	48	60	108
Sophomores	60	72	132
Juniors	51	69	120
Seniors	57	63	120
Total	216	264	480

17. Construct a relative frequency table by converting the data in the table to percentages. Round to the nearest tenth, if necessary.

18. Find the joint relative frequency of a sophomore who participates in school sports or clubs.

19. What percentage of freshmen do not participate in school sports or clubs? Round to the nearest tenth percent, if necessary.

20. What percentage of seniors participate in school sports or clubs? Round to the nearest tenth percent, if necessary.

SCHOOL MASCOT **The freshmen and sophomores at Lakeview High School are tasked with adopting a new school mascot next school year. The district asked a representative group of students to vote for one of the three mascot finalists and to indicate to which grade they belong. The results are shown in the table.**

School Mascot Vote Results			
	Freshmen	Sophomores	Total
Panthers	30	36	66
Hornets	17	33	50
Lions	28	31	59
Total	75	100	175

21. How many students voted for Panthers?

22. How many students voted for Lions?

23. How many sophomores were in the representative group?

24. Of the students who voted for Hornets, how many of them are freshmen?

25. Of the students who voted for Lions, how many of them are sophomores?

26. To the nearest whole, what percent of all the students voted for Lions?

27. To the nearest whole, what percent of all the students voted for Panthers?

28. To the nearest whole, what percent of all the students who voted were freshmen?

THANKSGIVING PIE **An online poll collected a sample of Thanksgiving pie preferences for different U.S. regions.**

Region	Apple	Sweet Potato	Pumpkin	Totals
West	77	4	13	
Midwest	32		54	
South		63	24	
Northeast	92	2		
Total	213	75	117	

29. PRECISION Copy and complete the table. Then find each relative frequency to the nearest tenth of a percent.

30. USE A MODEL Assuming the poll is representative of the whole population, what is a reasonable estimate of the probability that a family will be from the northeast and will be eating pumpkin pie on Thanksgiving?

31. STRUCTURE Construct a table of conditional relative frequencies based on pie preference. Round each percent to the nearest tenth. Interpret the meaning of the probabilities in the context of the problem.

32. REGULARITY If we had found the conditional relative frequencies by dividing by the total replies from each region, what would be the meaning of the probability in each cell?

VEHICLES The table shows the relative frequencies of drive systems for different vehicle types in a school parking lot. There are 215 vehicles in the lot.

Vehicle Type	2WD	AWD	Totals
Hatchbacks	42%	4%	
Sedans	28%	6%	
SUVs	1%	19%	
Total			215

33. USE TOOLS Construct a table to show the joint and marginal frequencies.

34. REASONING Without calculating individual frequencies, how many times greater will the conditional relative frequencies based on drive systems for AWD be than the relative frequencies for AWD, and why?

Higher-Order Thinking Skills

35. PERSEVERE Len conducted a survey among a random group of 1000 families in his home state of California. He wanted to determine whether there is an association between gasoline prices and distances traveled on family vacations. He collected the following information. According to Len's two-way frequency table, does there appear to be an association between gasoline prices and vacation distances traveled? Explain.

	$1.75–$3.24 per gallon	$3.25-$4.74 per gallon	Total
Less than 250 miles	109	255	364
More than 250 miles	329	86	415
No vacation travel	34	187	221
Total	472	528	1000

36. CREATE Select your own data for a two-way frequency table, write a question related to the data in the table, and provide the solution.

37. WRITE Compare two-way relative frequency tables and two-way conditional relative frequency tables.

38. FIND THE ERROR Magdalena took a survey of students in her school to find out what snack was most popular.

Favorite Snack Vote Results			
	Freshmen	Sophomores	Total
Fruit Snack	65	61	126
Granola	27	21	48
Yogurt	21	18	39
Total	113	100	213

a. Interpret the data based on the conditional relative frequency related to age groups.

b. Magdalena claims that fruit snack is the most popular snack for freshmen and sophomores, and Ben claims that a higher percentage of sophomores prefer fruit snack than do freshmen. Is either correct? Explain your reasoning.

Normal Distributions

Learn Probability Distributions

A **random variable** is a variable with possible values that are the outcomes of a random event. A **probability distribution** is a mapping of those outcomes to their probabilities of occurrence. It is usually shown as a histogram or bar graph. For example, if the random variable X represents the outcomes of 20 coin flips, a bar graph representing the results 8 heads and 12 tails might have bars showing $\frac{8}{20}$, or 0.4 for heads and $\frac{12}{20}$, or 0.6 for tails.

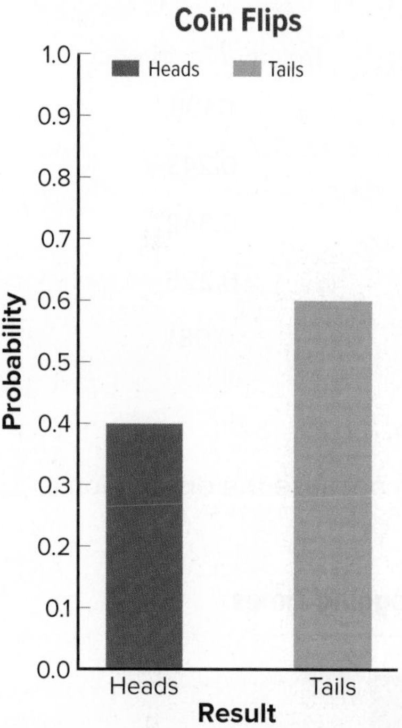

Today's Vocabulary
random variable
probability distribution
discrete random variable
continuous random variable
normal distribution

Key Concept • Conditions for Probability Distributions

The probability distribution of a random variable X must satisfy these conditions.

- The probability of each value of the random variable X must be between 0 and 1.
- The sum of the probabilities of all of the values of X must equal 1.

Think About It!
Why must the probability of each value of a random variable be between 0 and 1?

There are two types of random variables and distributions. A **discrete random variable** is finite and can be counted. The outcomes for flipping a coin are discrete because there are only two—heads or tails. A discrete distribution can be represented by a bar graph.

A **continuous random variable** can take on any value. The outcomes for teens' ages are continuous. A continuous distribution can be represented by a histogram.

Study Tip

Discrete vs. Continuous Variables representing height, weight, and capacity will always be continuous variables because they can take on any positive value.

Example 1 Analyze a Probability Distribution

GROCERIES **A grocery store chain measures the lengths of time the employees take to scan and bag customers' items. The frequency of each time interval is given. Construct a probability distribution to represent the data.**

Time (X)	Frequency
0–0:59	12
1–1:59	27
2–2:59	38
3–3:59	25
4–4:59	9

Step 1 Construct a relative frequency table.

The relative frequency table converts the frequencies to probabilities. Divide each frequency by the total number of measurements, 111. Round your answers to the nearest thousandth.

Time (X)	Frequency	Relative Frequency
0–0:59	12	0.108
1–1:59	27	0.243
2–2:59	38	0.342
3–3:59	25	0.225
4–4:59	9	0.081

Step 2 Graph the probability distribution.

The bars are not separated on the graph because the distribution is continuous.

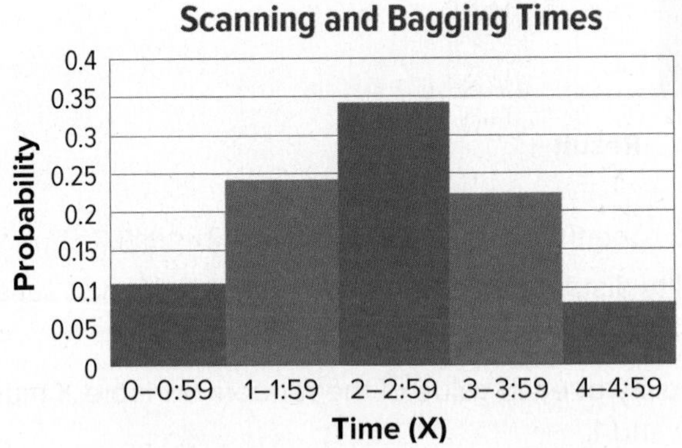

Talk About It!

If a histogram that measures frequency is symmetric about the mean, will its probability distribution necessarily be symmetric? Explain your reasoning.

Learn The Normal Distribution

The **normal distribution** is the most common continuous probability distribution. It is bell-shaped and symmetric about the mean.

Go Online You can complete an Extra Example online.

Key Concept • The Normal Distribution

- The graph of a normal distribution is continuous, bell-shaped, and symmetric with respect to the mean.
- The mean, median, and mode are equal and located at the center.
- The curve approaches, but never touches, the x-axis.
- The total area under the curve is equal to 1, or 100%.

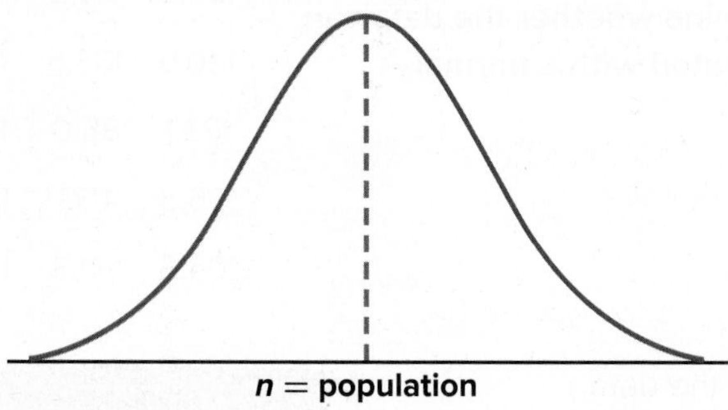

The area under the normal curve is 1 because the probability of a data point falling between the lowest and highest possible values is 1. Thus, the area under the curve between two values for X represents the probability that a data point will fall in that interval.

Learn The Empirical Rule

When a set of data is normally distributed, or approximately normal, the Empirical Rule can be used to determine the area under the normal curve at specific intervals.

Key Concept • The Empirical Rule

In a normal distribution with mean μ and standard deviation σ,

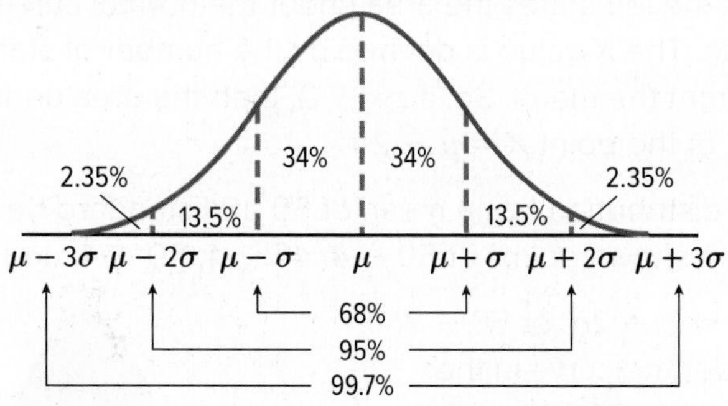

- approximately 68% of the data fall within 1σ of the mean,
- approximately 95% of the data fall within 2σ of the mean, and
- approximately 99.7% of the data fall within 3σ of the mean.

When a set of data is *not* approximately normal, it cannot be represented by the Empirical Rule. Skewed data like the graph at the right is one example of a set of data that is not approximately normal.

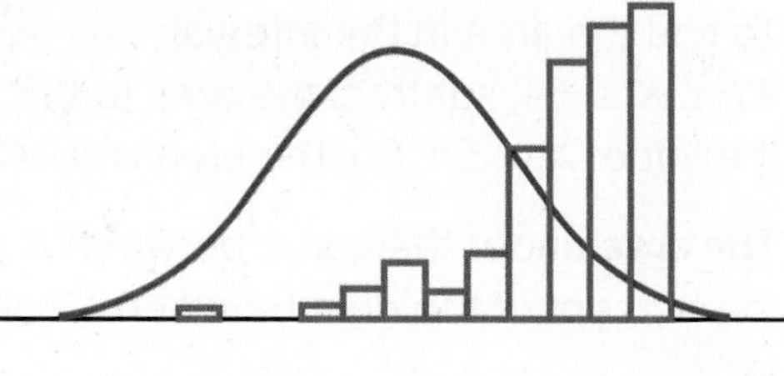

Study Tip

Normal Distributions In order for a distribution to be approximately normal, there must be a large number of values in the data set.

Example 2 Approximate Data by Using a Normal Distribution

HOUSING **The values of several houses on a street are given in thousands of dollars. Create a histogram of the set of data. Determine whether the data can be approximated with a normal distribution.**

Values of Houses (Thousands of Dollars)			
138.8	127.2	101.3	134.9
120.5	133.5	128.7	118.7
123.1	85.0	136.7	119.4
135.5	117.1	124.0	99.4
104.6	131.3	128.6	132.4

Step 1 Enter the data.

Step 2 Graph the histogram.

Step 3 Analyze the histogram.

The data are positively skewed. Thus, the data cannot be approximated with the normal distribution.

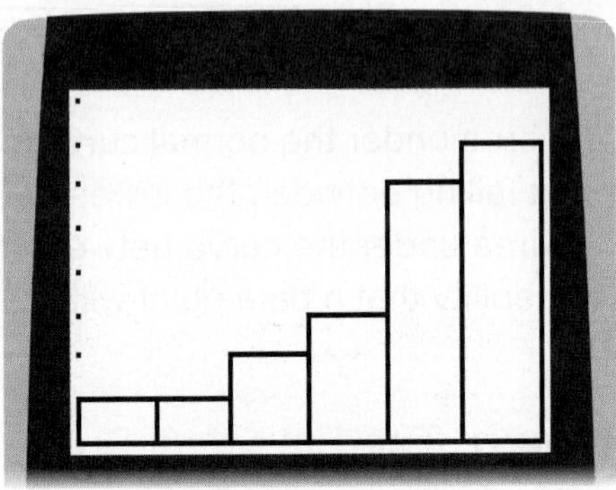

[80, 140] scl: 10 by [0, 8] scl: 1

Example 3 Use the Empirical Rule to Analyze Data

A normal distribution has a mean of 50 and a standard deviation of 4. Find the percent of the data between 42 and 54.

Standard Deviations from the Mean (z)	Area Under Curve for $X \leq z$
−3	0.0013
−2	0.0228
−1	0.1587
0	0.5000
1	0.8413
2	0.9772
3	0.9987

The table at the left states the area under the normal curve to the left of an X-value. The X-value is defined by the number of standard deviations from the mean. So, if $z = -2$, then the area under the curve is to the left of the point $X = \mu - 2\sigma$.

The normal distribution has a mean of 50 and standard deviation of 4. So, the graph shows points at $50 - 4$, $46 - 4$, $50 + 4$, and so on.

Because $42 = \mu - 2\sigma$, or $50 - 2(4)$, the area under the curve in the interval $X \geq 42$ is 0.0228.

Because $54 = \mu + \sigma$, the area under the curve in the interval $X \leq 54$ is 0.8413.

38 42 46 50 54 58 62

To find the area in the interval $42 \leq X \leq 54$, subtract the area to the left of $X = 42$ from the area to the left of $X = 54$. So, the area is $0.8413 - 0.0228$, or 0.8185.

The area under the curve between $X = 42$ and $X = 54$ is 0.8185. Thus, the percent of the data between 42 and 54 is approximately 81.85%.

Go Online An alternate method is available for this example.

Go Online You can complete an Extra Example online.

Practice

Go Online You can complete your homework online.

Example 1

Construct a probability distribution to represent each set of data.

1. FOOD SERVICE The table shows the length of time each customer spent in the drive-thru line one day at a fast-food restaurant.

Time (X)	Frequency
0–0:29	4
0:30–0:59	22
1:00–1:29	131
1:30–1:59	49
2:00–2:29	18

2. FOOD The table shows the numbers of packages of the five most popular flavors of bagels sold in the U.S. in a recent year.

Flavor (X)	Packages (millions)
plain	136
cinnamon raisin	56
everything	40
blueberry	38
100% whole wheat	21

3. SOCIAL MEDIA The table shows the responses teens had to the question "How many new friends have you met online?"

Time (X)	Frequency
0	456
1	22
2–5	233
6 or more	307

4. GOVERNMENT The table shows the ages of the U.S. presidents at their inauguration.

Age (X)	Presidents
41–45	2
46–50	8
51–55	16
56–60	9
61–65	7
66–70	3

Example 2

Determine whether each set of data can be approximated with a normal distribution. Explain your reasoning.

5.

Values of Used Cars (thousands of dollars)				
4.8	7.2	10.1	4.9	13.9
12.7	13.1	4.2	11.8	12.6
13.1	5.6	11.7	13.4	11.3
9.5	7.7	12.1	9.4	5.6

6.

Speeds of Cars on I–71 (mph)				
65	66	61	69	68
68	71	62	66	65
67	60	72	67	65
68	62	66	67	68
60	66	69	71	66

7.

Men's Shot Put Distances (m)				
21.30	19.49	18.58	20.08	19.70
18.91	18.21	18.97	19.26	18.49
18.31	18.73	19.53	18.81	19.63
18.94	17.57	17.09	20.38	18.89
18.60	17.19	18.63	18.52	18.67

8.

Women's 400 m Relay Times (s)				
42.91	44.41	44.58	43.34	43.45
44.73	43.08	45.09	44.71	44.63
44.44	44.27	43.85	43.76	44.65
44.85	43.73	45.12	44.47	44.92
44.54	44.51	44.68	44.61	44.71

Example 3

A normal distribution has a mean of 455 and a standard deviation of 24.

9. Find the percent of the data between 407 and 455.

10. What percent of the data are greater than 479?

11. Find the percent of the data that are less than 407.

12. What percent of the data are between 431 and 503?

Mixed Exercises

13. **BIRTHS** The table shows the numbers of births in the United States in a recent year.

a. Complete the relative frequency column.

b. Construct a probability distribution to represent the data.

Births	States	Relative Frequency
0–19,999	12	
20,000–39,999	9	
40,000–59,999	6	
60,000–79,999	7	
80,000–99,999	4	
100,000–119,999	3	
120,000–139,999	3	
140,000–159,999	2	
160,000 or more	4	

Source: Annie E. Casey Foundation

14. **REACTION TIME** In a test of 1200 teenagers, the reaction times to a visual cue were normally distributed with a mean of 0.25 second and a standard deviation of 0.05 second.

a. About how many teenagers had reaction times between 0.15 and 0.35 second?

b. What is the probability that a teenager selected at random had a reaction time greater than 0.3 second?

Higher-Order Thinking Skills

15. **ANALYZE** The graphing calculator screen shows the graph of a normal distribution for a large set of data that has a mean of 50 and a standard deviation of 10. If every data point in the set is increased by 5 points, describe how the mean, standard deviation, and graph of the data changes.

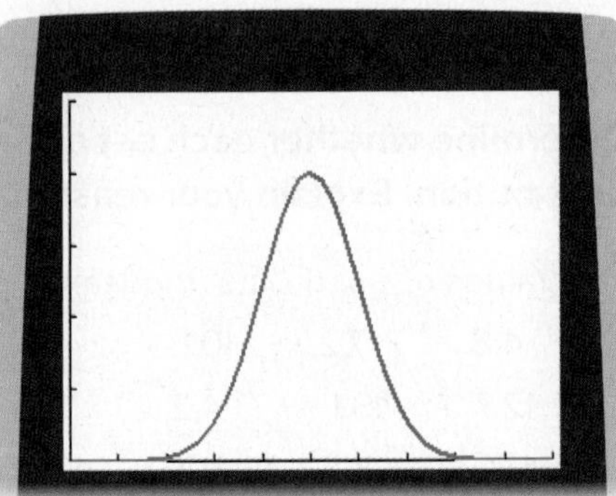

[0, 80] scl: 10 by [0, 0.05] scl: 0.01

16. **FIND THE ERROR** Courtney says that the graphs all represent normal distributions with the same mean but different standard deviations. Michael says that only the middle graph represents a normal distribution. Is either correct? Explain.

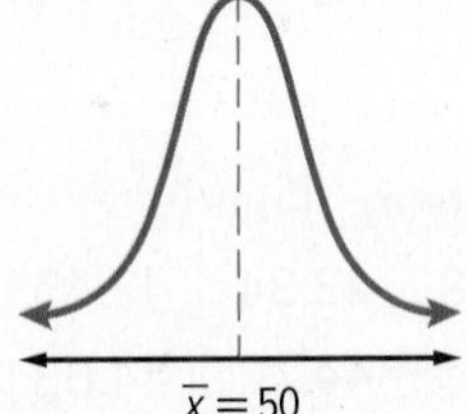

$\bar{x} = 50$

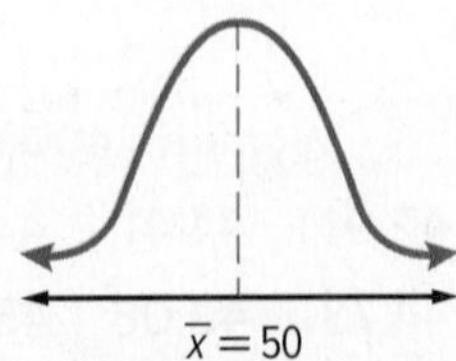

$\bar{x} = 50$

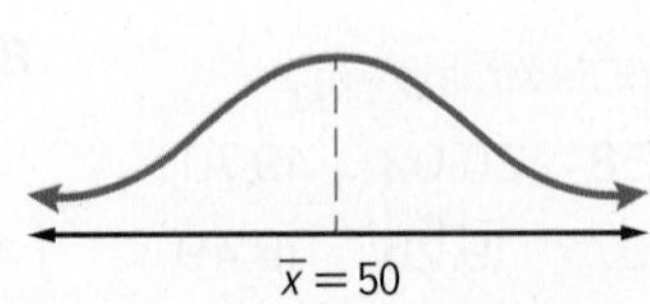

$\bar{x} = 50$

17. **CREATE** Create a probability distribution in which one possible value of the random variable is twice as likely to occur as one other possible value of the random variable.

18. **PERSEVERE** The boxes of cereal in a shipment are normally distributed with a mean weight of 17.1 ounces and a standard deviation of 0.2 ounce. Nine of the boxes weigh more than 17.5 ounces. How many boxes are in the shipment?

Review

Essential Question

How do you summarize and interpret data?

By using statistics, you can analyze data to find meaningful results. Calculating measures of center and spread and making a dot plot, bar graph, or histogram can help you interpret the data.

Module Summary

Lessons 9-1 and 9-4

Measures of Center and Spread

- The mean of a data set is the sum of the elements of the data set divided by the total number of elements in the set.
- The median of a data set is the middle element or the mean of the two middle elements in the set of data when the data are arranged in numerical order.
- The mode of a data set is the value of the elements that appear most often in the set of data.
- The formula for standard deviation, with mean $\overline{x}$ and n terms is

$$\sigma = \sqrt{\frac{(\overline{x} - x_1)^2 + (\overline{x} - x_2)^2 + \ldots + (\overline{x} - x_n)^2}{n}}.$$

Lessons 9-2 and 9-3

Representing and Using Data

- Dot plots, bar graphs, and histograms are commonly used to represent data.
- Bar graphs are used with discrete data, and histograms are used with continuous data.
- A population is all members of a group of interest about which data will be collected. A sample is a subset of the population.
- A bias is an error that results in a misrepresentation of a population.

Lesson 9-5

Distributions of Data

- In a symmetric distribution, the mean and median are approximately equal.
- A negatively skewed distribution typically has a median greater than the mean. A positively skewed distribution typically has a mean greater than the median.
- An outlier is a value that is more than 1.5 times the interquartile range above the third quartile or below the first quartile.

Lesson 9-6

Comparing Sets of Data

- A linear transformation is one or more operations performed on a set of data that can be written as a linear function.
- Common linear transformations are adding a constant to or multiplying a constant by every value in the set of data.

Lesson 9-7

Two-Way Frequency Tables

- A two-way frequency table shows the frequencies of data classified according to two or more categories.

Study Organizer

Foldables

Use your Foldable to review this module. Working with a partner can be helpful. Ask for clarification of concepts as needed.

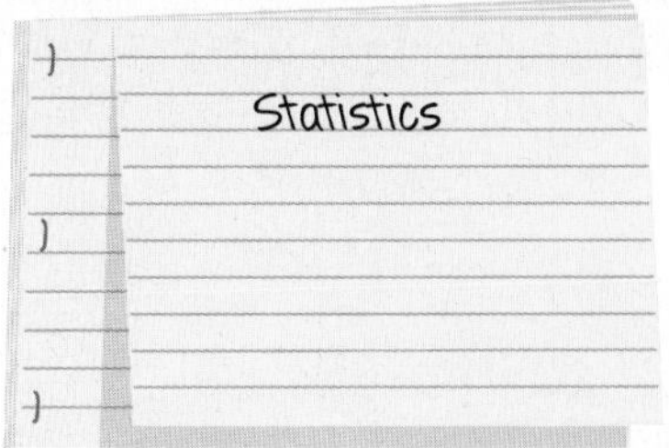

Test Practice

1. GRAPH Make a dot plot of the quiz scores of Ms. Perez's third period class.

Quiz Scores			
85	88	75	100
90	90	88	72
72	79	88	85

(Lesson 9-2)

2. OPEN RESPONSE When is it a good idea to scale the number line when making a dot plot?

(Lesson 9-2)

3. MULTI-SELECT Which of the statements are true regarding dot plots, bar graphs, and histograms? Select all that apply.

Lesson 9-2

A. Dot plots use a number line and dots to represent very large amounts of data.

B. Bar graphs are used to represent data that is continuous.

C. Histograms are used to represent data that is continuous.

D. Bar graphs are used to represent data that is discrete.

E. Histograms are used to represent data that is discrete.

4. MULTIPLE CHOICE Which dot plot correctly models these data values?

36, 38, 42, 36, 36, 40, 42, 38, 38, 39, 40, 38, 38, 38, 40 (Lesson 9-2)

A.

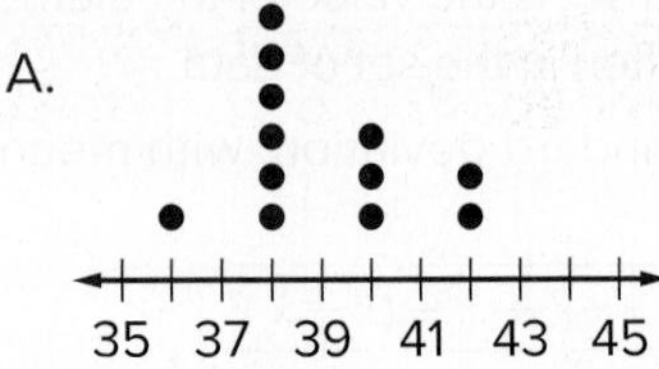

B.

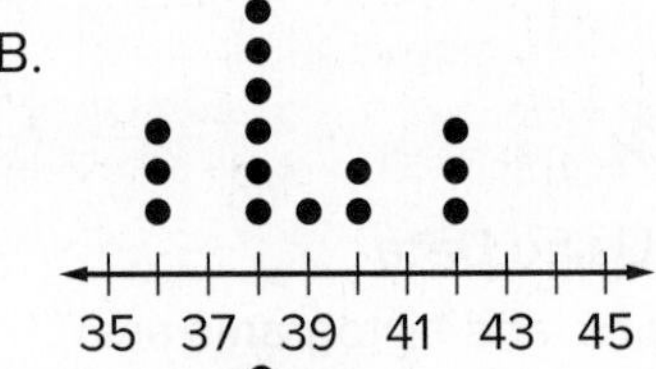

C.

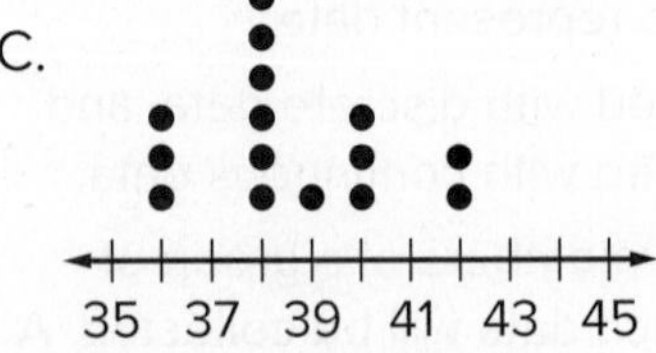

D.

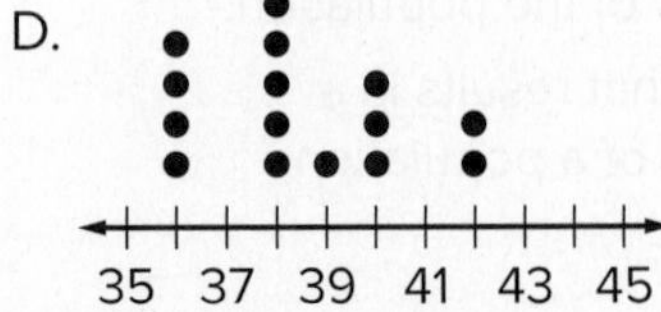

5. **OPEN RESPONSE** Given the set of data in the table, describe what size intervals could be used when making a histogram. (Lesson 9-2)

Ages of Guests at a Picnic
74, 26, 32, 4, 61, 56, 16, 15, 17, 28, 39, 42, 47, 72, 66, 12, 16, 38, 35, 8, 16, 11, 10, 41, 47, 5, 13, 77, 24, 30, 9, 62

6. **GRAPH** A survey was conducted among students in Mr. Sadiq's history class to determine their favorite major topic covered in class this semester. The results are shown in the table. Make a bar graph to display the data. (Lesson 9-2)

Topic	Number of Votes
Civil War	12
Revolutionary War	5
The Industrial Revolution	8
Westward Expansion	10

7. **OPEN RESPONSE** Akeem wants to determine how long it took students in his class to complete a 1-mile run.

Running Time (min)
18.5, 8.4, 10.2, 27.1, 9.5, 10.9, 17.0, 5.3, 6.1, 8.4, 8.4, 9.9, 10.0, 7.4, 8.4

State two types of displays Akeem could use to appropriately display his data. (Lesson 9-2)

8. **MULTIPLE CHOICE** Which box plot correctly models these data values? 95, 72, 84, 98, 87, 75, 100, 86, 90, 81, 93, 90 . (Lesson 9-4)

A. 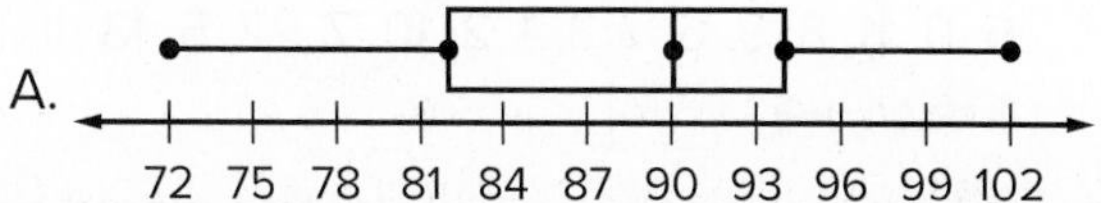

B. 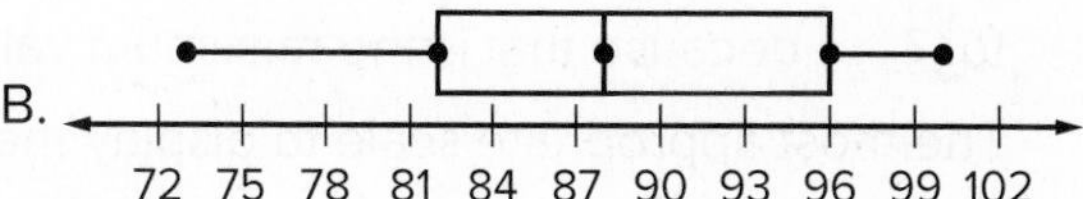

C. 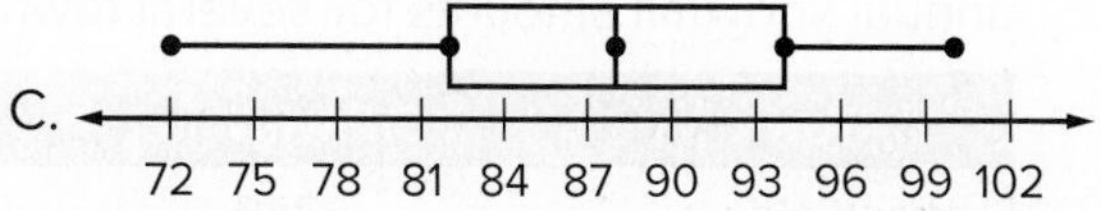

D. 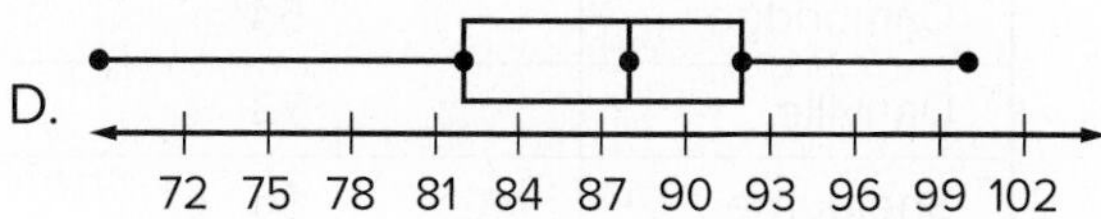

9. **GRAPH** The table shows the number of pages each student read in one night.

Pages Read
13, 15, 8, 22, 11, 17, 15, 9, 14, 16, 13

Create a box plot to represent the set of data. (Lesson 9-4)

10. **OPEN RESPONSE** Fifteen people in their fifties were surveyed about the number of apps they have on their cell phone. (There was an assumption that all 15 of them owned a cell phone). The results are listed, below. 6, 0, 11, 8, 9, 6, 7, 3, 1, 2, 10, 7, 22, 5, 13 (Lesson 9-4)

A box plot to represent this data would have to begin at __?__ because that is the minimum value, and would have to extend to __?__ because that is the maximum value.

The most appropriate scale to display the data in the box plot should be __?__ or 2.

11. **MULTIPLE CHOICE** The table shows the annual snowfall amounts for several towns.

Town	Snowfall (in.)
Westfield	246
Brattleboro	73
Cambridge	54
Danville	73
Shelburne	86
Lowell	67

Which measure(s) of center and measure(s) of spread best describe the set of data? (Lesson 9-5)

A. mean

B. median

C. standard deviation

D. five-number summary

12. **OPEN RESPONSE** *True* or *false*: Histogram B has more variability than Histogram A. (Lesson 9-6)

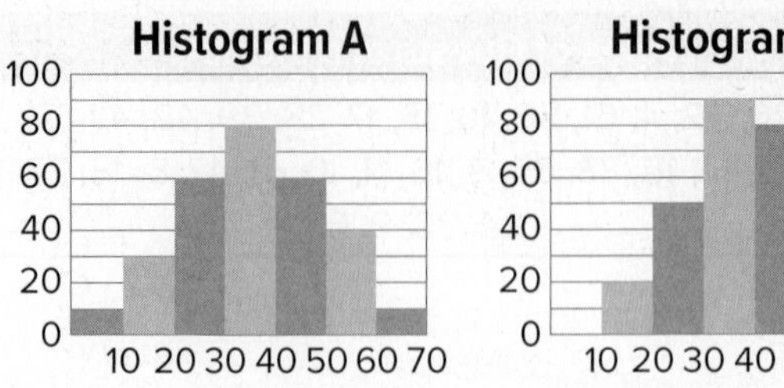

13. **MULTIPLE CHOICE** The junior varsity dance team is selecting the color of their new uniforms. The team consists of 28 freshmen and sophomores. Of the 16 freshmen, 7 want red uniforms and 9 want black uniforms. Only 4 of the sophomores want black uniforms. How many total team members want red uniforms? (Lesson 9-7)

A. 7

B. 13

C. 15

D. 21

14. **OPEN RESPONSE** The table shows the frequencies of positions for different offensive players on a school football team. There are 38 offensive players on the team.

Position	Senior	Junior	Sophomore
Quarterback	1	1	0
Running Back	2	1	1
Receiver	3	2	2
Lineman	13	6	6

Suppose a junior player was picked at random. What is the probability that player is a receiver? (Lesson 9-7)

15. **OPEN RESPONSE** A normal distribution has a mean of 347.2 and a standard deviation of 13.9. (Lesson 9-8)

The data that is less than 319.4 represents __?__% of the data.

The data that is greater than 361.1 represents __?__% of the data.

Module 10
Tools of Geometry

Essential Question

How are points, lines, and segments used to model the real world?

What Will You Learn?

How much do you already know about each topic **before** starting this module?

KEY

👎 — I don't know. 👉 — I've heard of it. 👍 — I know it!

	Before			After		
	👎	👉	👍	👎	👉	👍
analyze axiomatic systems and identify types of geometry						
analyze figures to identify points, lines, planes, and intersections of lines and planes						
find measures of line segments						
apply the Distance Formula to find lengths of line segments						
find points that partition directed line segments on number lines						
find points that partition directed line segments on the coordinate plane						
find midpoints and bisect line segments						

Foldables Make this Foldable to help you organize your notes about geometric concepts. Begin with four sheets of 11″ × 17″ paper.

1. **Fold** the four sheets of paper in half.
2. **Cut** along the top fold of the papers. Staple along the side to form a book.
3. **Cut** the right sides of each paper to create a tab for each lesson.
4. **Label** each tab with a lesson number.

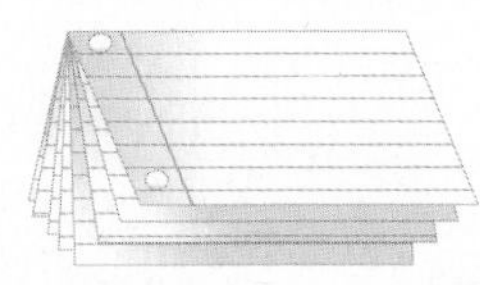

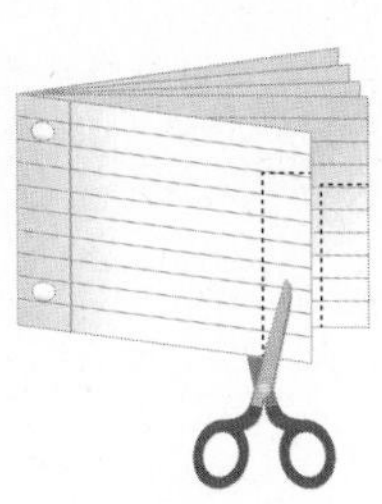

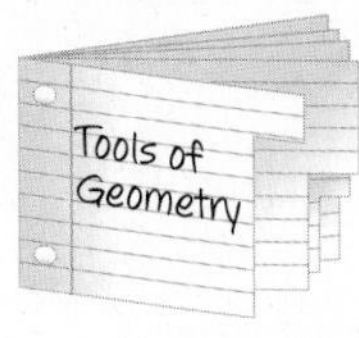

What Vocabulary Will You Learn?

- analytic geometry
- axiom
- axiomatic system
- betweenness of points
- bisect
- collinear
- congruent
- congruent segments
- coplanar
- defined term
- definition
- directed line segment
- distance
- equidistant
- fractional distance
- intersection
- line
- line segment
- midpoint
- plane
- point
- postulate
- segment bisector
- space
- synthetic geometry
- theorem
- undefined terms

Are You Ready?

Complete the Quick Review to see if you are ready to start this module. Then complete the Quick Check.

Quick Review

Example 1

Graph and label the point Q(−3, 4) in the coordinate plane.

Start at the origin. Because the *x*-coordinate is negative, move 3 units to the left. Then move 4 units up because the *y*-coordinate is positive. Draw a dot and label it *Q*.

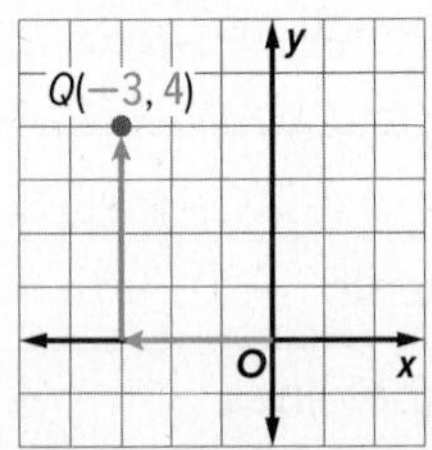

Example 2

Evaluate the expression $[-2-(-7)]^2+(1-8)^2$.

Follow the order of operations.

$[-2-(-7)]^2+(1-8)^2$

$= 5^2+(-7)^2$ Subtract in parentheses.

$= 25+49$ Evaluate exponents.

$= 74$ Add.

Quick Check

Graph and label each point on the coordinate plane.

1. $W(-5, 2)$

2. $X(0, 4)$

3. $Y(-3, -1)$

4. $Z(4, -2)$

Evaluate each expression.

5. $(4-2)^2+(7-3)^2$

6. $(-5-3)^2+(3-4)^2$

7. $[-1-(-9)]^2+(5-3)^2$

8. $[-3-(-4)]^2+[-1-(-6)]^2$

How did you do?

Which exercises did you answer correctly in the Quick Check?

The Geometric System

Explore Using a Game to Explore Axiomatic Systems

Online Activity Use a real-world situation to complete the Explore.

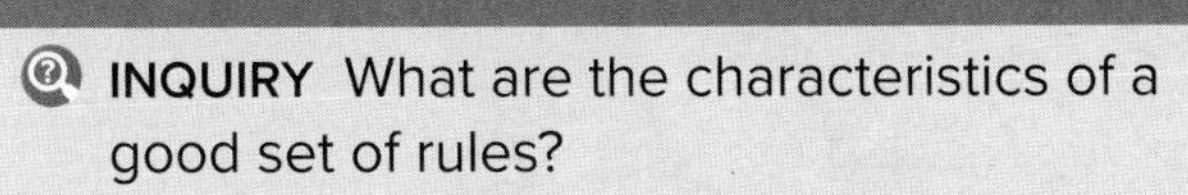
INQUIRY What are the characteristics of a good set of rules?

Today's Goals
- Apply axioms to draw conclusions.
- Identify examples of synthetic and analytic geometry.

Today's Vocabulary
axiomatic system
undefined terms
definition
defined term
axiom
postulate
theorem
synthetic geometry
analytic geometry

Learn The Axiomatic System of Geometry

Geometry is an axiomatic system based on logical reasoning and axioms.

The Axiomatic System of Geometry	
An **axiomatic system** has a set of axioms from which theorems can be derived.	
undefined terms	words, usually readily understood, that are not formally explained by means of more basic words and concepts
definition	assigns properties to a mathematical object
defined term	a term that has a definition and can be explained using undefined terms and/or defined terms
axiom or **postulate**	statement that is accepted as true without proof
theorem	statement or conjecture that can be proven true using undefined terms, definitions, and axioms

Undefined terms are used to write definitions. Undefined terms and definitions are used to create axioms. Undefined terms, definitions, and axioms are used to prove theorems.

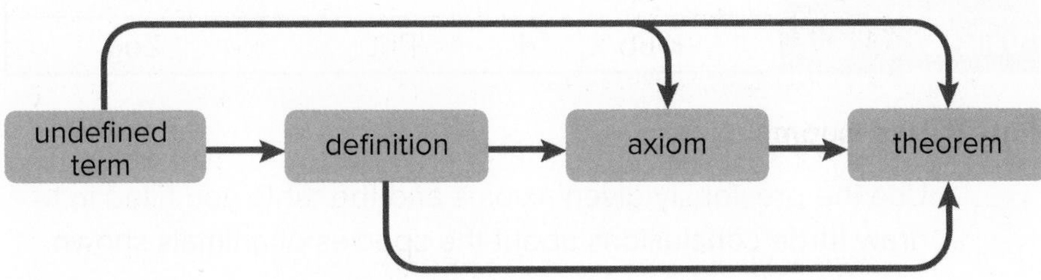

One real-world axiomatic system that is probably familiar to you is the set of rules to a game. The rules are the axioms, and they are used to evaluate the legality of each play.

Math History Minute

Thales (c. 624–546 B.C.) was a Greek mathematician, philosopher, and astronomer, and is the first known individual attributed with a mathematical discovery. He inspired Euclid, Plato, and Aristotle, who considered him to be the first philosopher in the Greek tradition.

Example 1 Apply an Axiomatic System

ANIMALS In the fictional country of Rythoth, blue animals are from the mountains, and red animals are from the valleys. These animals are categorized into three distinct classes: mammals, birds, and reptiles. Mammals are covered by hair or fur, birds are covered by feathers, and reptiles are covered by scales.

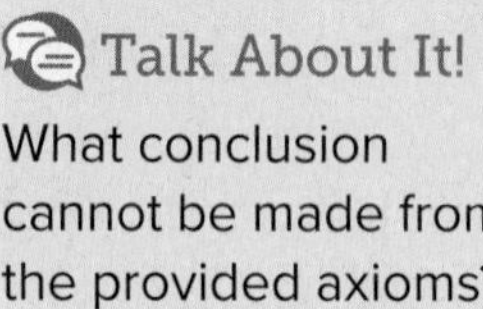

Talk About It!

What conclusion cannot be made from the provided axioms?

Part A Categorize the animals.

Write the name of each animal in the corresponding categories in the table.

Birthplace	Mammal	Bird	Reptile
Mountains	Rorx	Pax	Awub
Valleys	Klub	Prit	Zog

Part B Use axioms.

Use the previously given axioms and the table you filled in to draw three conclusions about the species of animals shown.

- The Rorx is a mammal from the mountains of Rythoth.
- The Zog is a reptile from the valleys of Rythoth.
- The Prit is a bird from the valleys of Rythoth.

Study Tip

Theorems Theorems, or conclusions, made from a set of axioms must be true in every situation. It takes only one example that contradicts the conjecture to show that a theorem or conclusion is not true.

Check

PLANETS The fictional galaxy of Yogul contains at least 20 planets including Mothera, Sothera, and Kothera. An animal can live on any planet in the Yogul galaxy that contains its biome. Lizards and scorpions live in the desert. Frogs and monkeys live in tropical forests. Bears and foxes can be found in the tundra. The biomes of each planet are permanent and will not change over time.

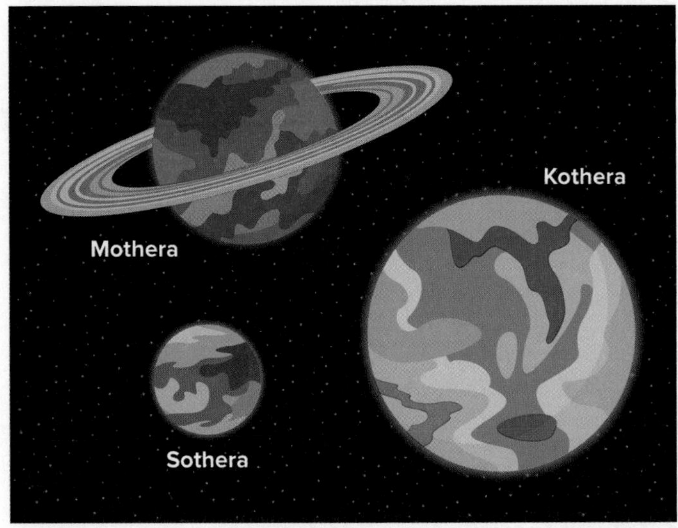

Color	Biome
	desert
	tropical forest
	tundra

Use the axioms given to determine what conclusions can be made about the planets of Yogul. Select all that apply.

A. Bears and foxes can live on Sothera.

B. Lizards and scorpions can only live on Mothera.

C. Only frogs and monkeys can survive on Kothera.

D. Bears and foxes can survive on Sothera at temperatures as low as −20°F.

E. All animals can live on Kothera.

F. Scorpions and lizards can live on Mothera.

Learn Types of Geometry

There are several types of geometry that are built upon different sets of postulates including synthetic geometry and analytic geometry.

Synthetic geometry is the study of geometric figures without the use of coordinates. Synthetic geometry is sometimes called *pure geometry* or *Euclidean geometry*.	**Analytic geometry** is the study of geometry using a coordinate system. Analytic geometry is sometimes called *coordinate geometry* or *Cartesian geometry*.

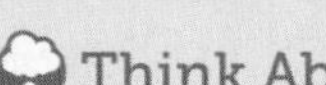

Think About It!
What is an advantage of using analytic geometry instead of synthetic geometry?

Go Online You can complete an Extra Example online.

Example 2 Identify Types of Geometry

Classify each figure as illustrating *synthetic geometry* or *analytic geometry*.

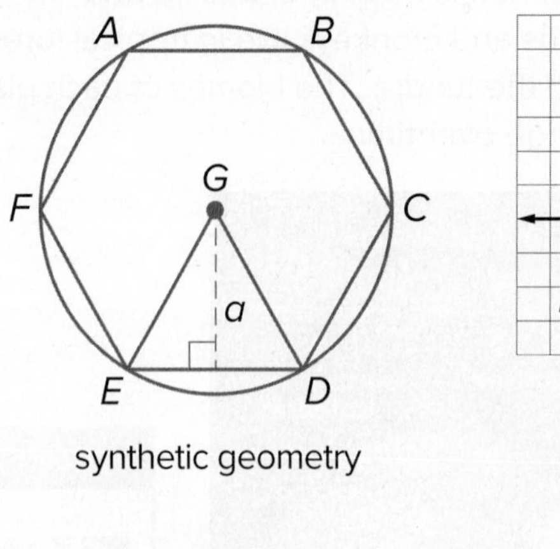

synthetic geometry

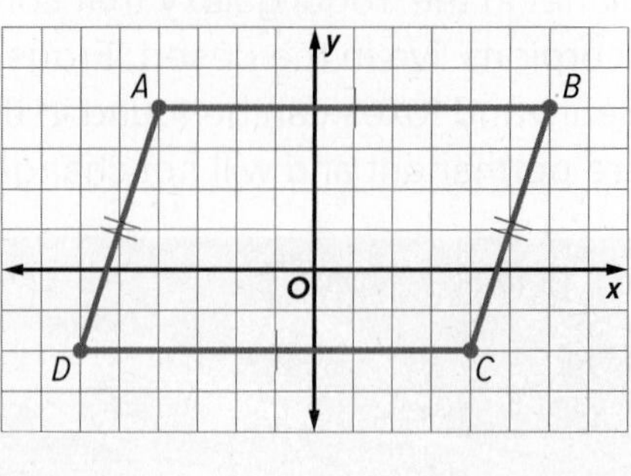

analytic geometry

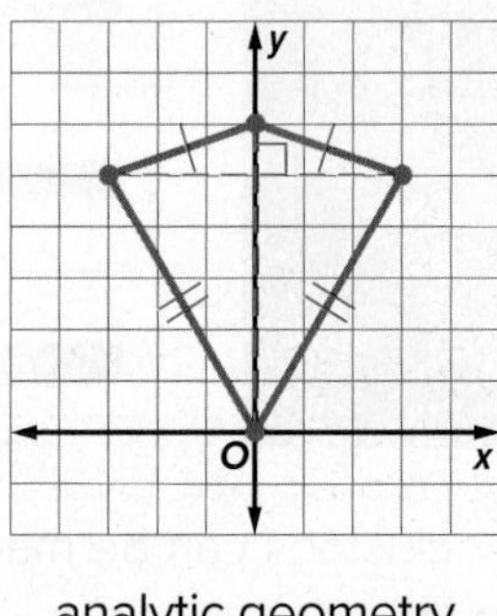

analytic geometry

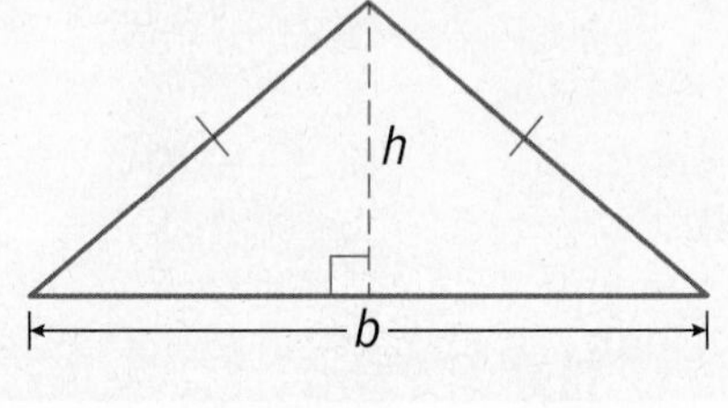

synthetic geometry

Check

Classify each figure as illustrating *synthetic geometry* or *analytic geometry*.

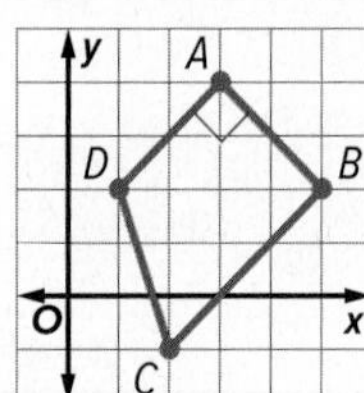

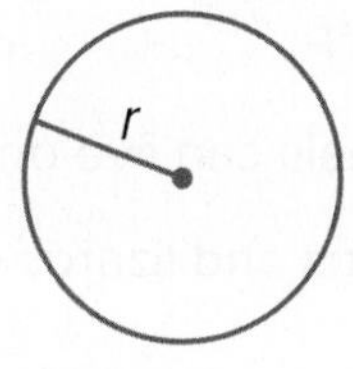

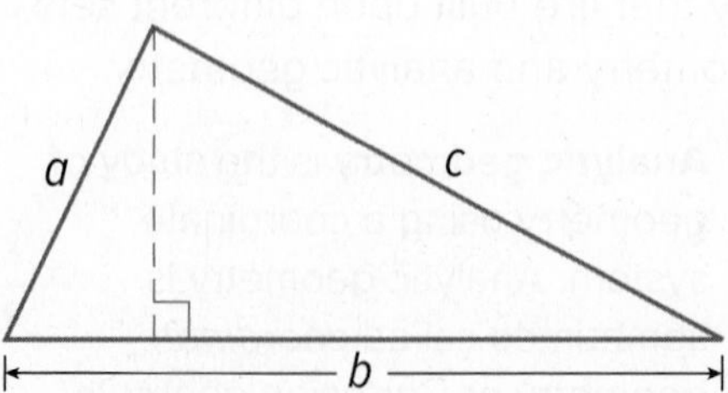

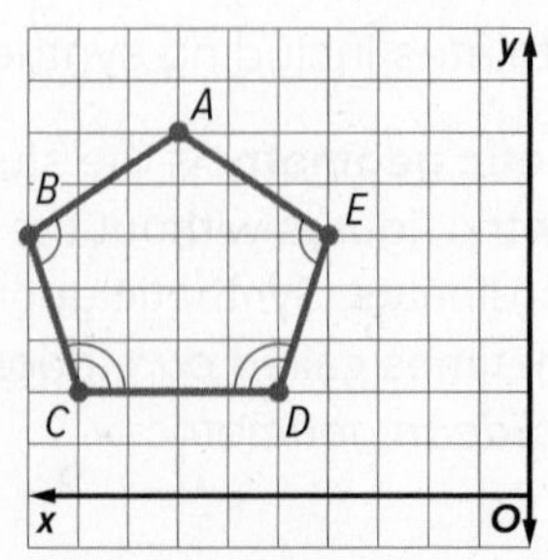

Practice

Go Online You can complete your homework online.

Example 1

1. **BASKETBALL** The Badgers' basketball team has 10 players. During practice, half of the players wear red jerseys numbered 1–5, and the other half wear yellow jerseys numbered 6–10. The yellow team wins the practice game 32-26.
   - Kylie wears number 5 and scores 9 points.
   - Kelsey's team wins the game.
   - Marie and Kylie are on opposing teams.

   Use the axioms to make three conclusions about the game played.

2. **PRINTING** Rico's T-shirt Company sells customized short sleeve T-shirts, long sleeve T-shirts, and sweatshirts. Each type of shirt sells in multiples of 5. It costs \$25.00 for 5 short sleeve T-shirts, \$30.00 for 5 long sleeve T-shirts, and \$40.00 for 5 sweatshirts. Short sleeve and long sleeve T-shirts can be made in any color except navy or black. Sweatshirts are only made in navy and black.
   - Mercedes bought green shirts for \$55.00.
   - Quinn bought 10 navy sweatshirts.
   - Rachel paid \$30.00 for several red shirts.
   - Hector bought black and yellow shirts for \$65.00.

   Use the axioms to make four conclusions about the shirts sold.

3. **LANDSCAPING** Tom owns a landscaping business. He charges \$40 for a yard cleanup, \$50 to mow a lawn, and \$75 to mulch a yard. On average, it takes Tom 25 minutes for a yard cleanup, 40 minutes to mow a lawn, and 2 hours to mulch a yard. Tom's clients are Mr. Hansen, Ms. Martinez, and Mrs. Johnson.
   - Mr. Hansen paid \$125 for lawn services this week.
   - Tom spent more than an hour at Ms. Martinez' house this week.
   - Mrs. Johnson wrote Tom a check for \$165 for the week.
   - Tom made \$405 from his three clients this week.

   Use the axioms to make four conclusions about the landscaping that Tom did.

4. **CUPCAKES** Olivia's Cupcake Shoppe sells small and large cupcakes in three flavors.
   - Niamh paid \$3 for a cupcake with buttercream icing.
   - Bethany bought a small vanilla cupcake.
   - Mateo paid \$3.50 for a cupcake with strawberry icing and a chocolate cupcake.

   Use the axioms to make two conclusions about the cupcakes that were purchased.

Olivia's
cupcake shoppe
Flavors
Chocolate with vanilla icing
Vanilla with strawberry icing
Strawberry with buttercream icing
Sizes
Small........\$1.75
Large........\$3.00

Example 2

Classify each figure as illustrating *synthetic geometry* or *analytic geometry*.

5.

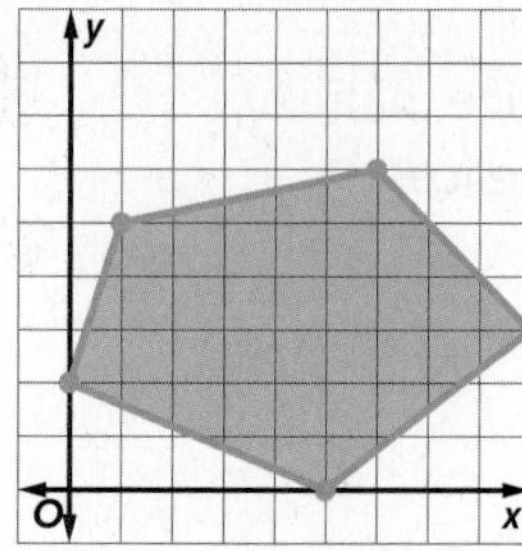

6.

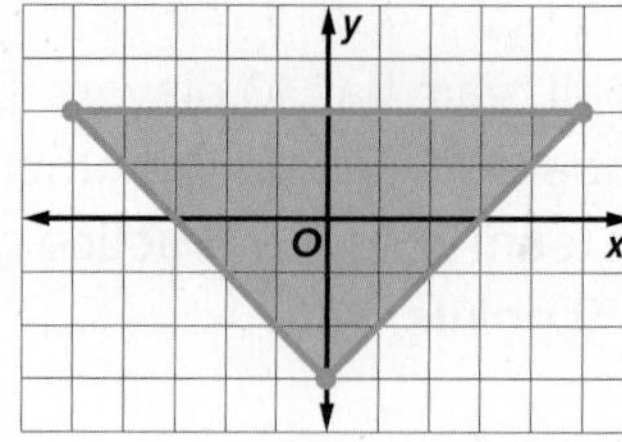

7.

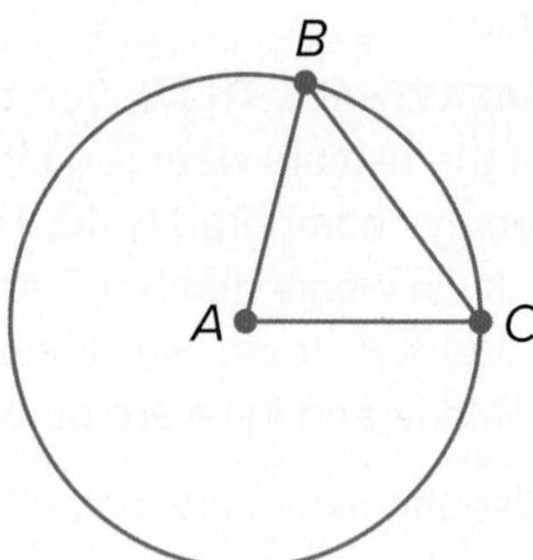

8.

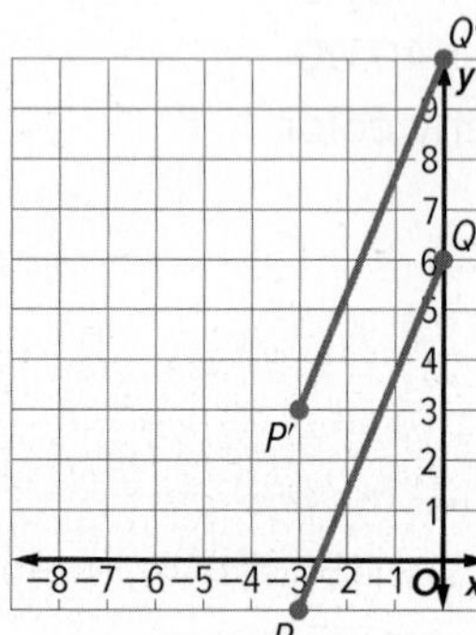

9.

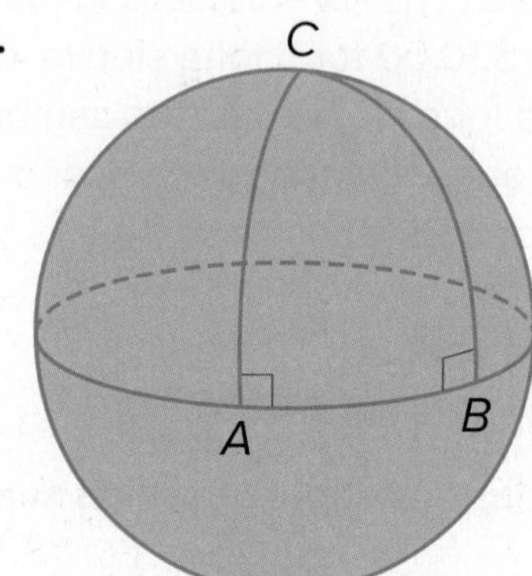

10.

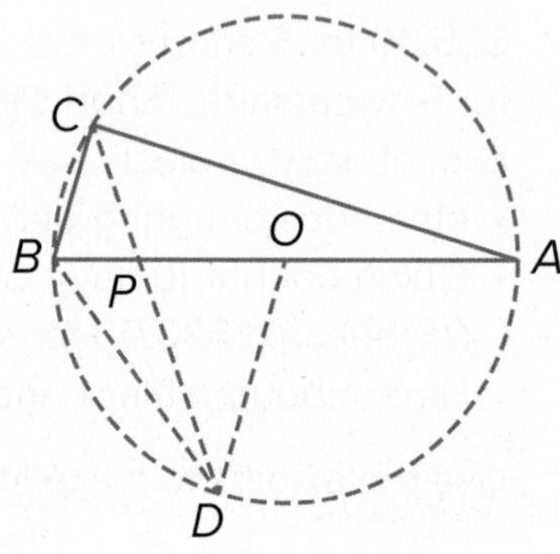

Mixed Exercises

11. RESTAURANT Damon sells three types of salads at his restaurant: cobb, wedge, and spinach. Each salad is served with 2 dinner rolls. The price of the cobb salad is \$7.99, the price of the wedge salad is \$8.99, and the price of the spinach salad is \$5.99. Grilled chicken can be added to any salad for an additional \$2.00.

- Malik spent \$7.99 on a salad.
- Pedro and Deandra each spent \$8.99 on their salads.
- Rafael ate a wedge salad.
- Drake did not add chicken to his salad.

Use the axioms to make a conclusion about the salads that are eaten.

12. CLASSROOM Mrs. Fields teaches high school geometry. Her classroom tools include a compass, straightedge, pencil, and protractor. Does Mrs. Fields likely teach *analytic geometry* or *synthetic geometry*? Explain your reasoning.

13. REASONING Theo is stuck on a problem on a test. The problem is asking him to use a given formula to find the distance between two points on a graph. Is Theo using analytic *geometry* or *synthetic geometry*? Explain your reasoning.

14. USE A SOURCE Survey a group of students in your classroom about favorite colors. Write three axioms about the data you collected. Then use your axioms to write a conclusion. Explain your reasoning.

15. **STATE YOUR ASSUMPTION** Sydney is an engineer. She is using a blueprint for a project that is drawn on a grid, as shown. Is Sydney likely using *analytic geometry* or *synthetic geometry*? Explain any assumptions that you make.

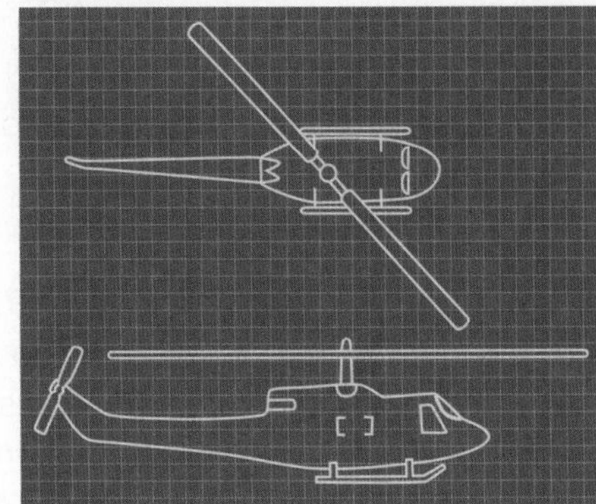

16. Mr. Sail assigns a project where students identify shapes that represent real-world objects. Is this an example of *analytic geometry* or *synthetic geometry?* Explain your reasoning.

Atomic Imagery/Getty Images, C Squared Studios/Photodisc/Getty Images, Tewin Kijthamrongworakul/Alamy Stock Photo

17. **CONSTRUCT ARGUMENTS** Consider the following axiomatic system for bus routes.
    - Each bus route lists the stops in the order at which they are visited by the bus.
    - Each route visits at least four distinct stops.
    - No route visits the same stop twice, except for the first stop, which is always the same as the last stop.
    - There is a stop called Downtown, which is visited by each route.
    - Every stop other than Downtown is visited by at most two routes.

    The city has stops at Downtown, King St, Maxwell Ave, Stadium District, State St, Grace Blvd, and Charlotte Ave. Are the following three routes a model for the axiomatic system? Justify your argument.
    ROUTE 1: Downtown, King St, Stadium District, State St, Downtown
    ROUTE 2: Stadium District, State St, Grace Blvd, Maxwell Ave, Downtown, Stadium District
    ROUTE 3: King St, Stadium District, Downtown, Maxwell Ave, Stadium District, King St

18. **SHOPPING** The Clothing Shop is having a sale. All clothes are 20% off, and all accessories are 30% off.
    - Jaisa bought two necklaces.
    - Sheree bought a shirt and a purse.

    Use the axioms to make one conclusion about Jaisa or Sheree's purchases.

19. **WRITE** Write a comparison of the rules and plays of a game and the elements of an axiomatic system. Then choose a game or sport for which you know the rules. Explain a rule from the game or sport and a play from the game. Does the play violate or fall within the rule? Explain.

20. **CREATE** Given the following list of axioms, draw a model to properly represent the information.
- There exist five points.
- Each line contains only these five points.
- There exist two lines.
- Each line contains at least two points.

21. **WHICH ONE DOESN'T BELONG?** Three-point geometry is a finite subset of geometry with the following four axioms:
- There exists exactly three distinct points.
- Each pair of distinct points are on exactly one line.
- Not all the points are on the same line.
- Each pair of distinct lines intersect in at least one point.

Which of the following does not satisfy all the axioms of three-point geometry? Justify your conclusion.

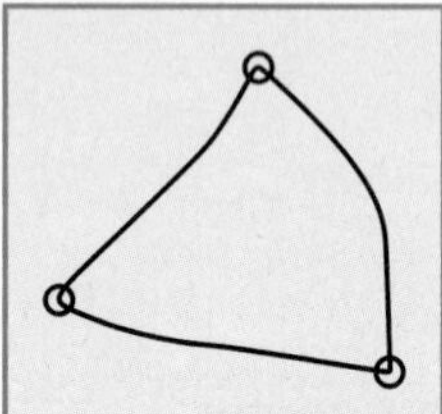
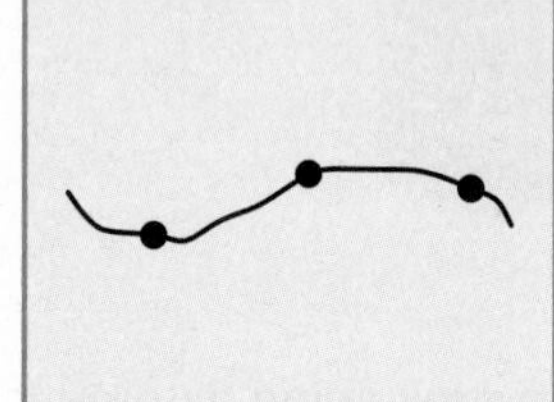
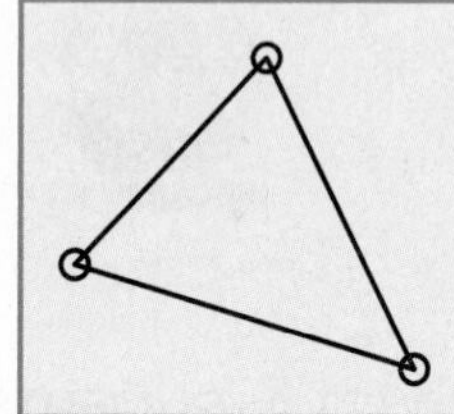

22. **FIND THE ERROR** Grant read the following axioms for a video game he is playing.
- There are four keys hidden on each level.
- Each level ends when the player collects the third key.
- The game has 10 levels.

From these axioms, Grant concluded:
- to complete the game, he will need to find 30 keys.
- there are 40 keys in the game.
- he can collect all 40 keys in the game.

Are Grant's conclusions correct? Explain your reasoning.

23. **WHICH ONE DOESN'T BELONG?** Using your understanding of analytic and synthetic geometry, which of the following figures does not belong? Justify your conclusion.

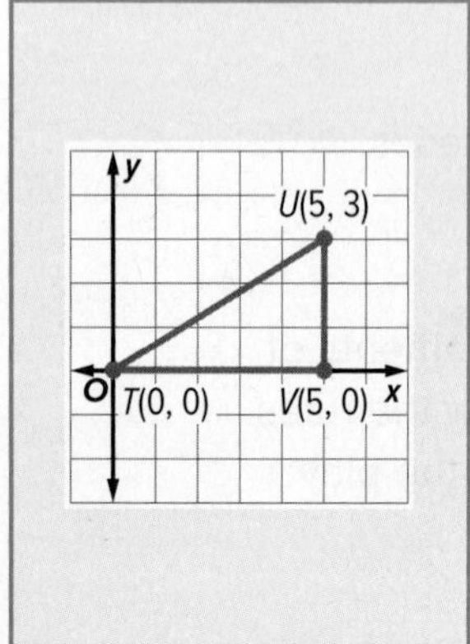

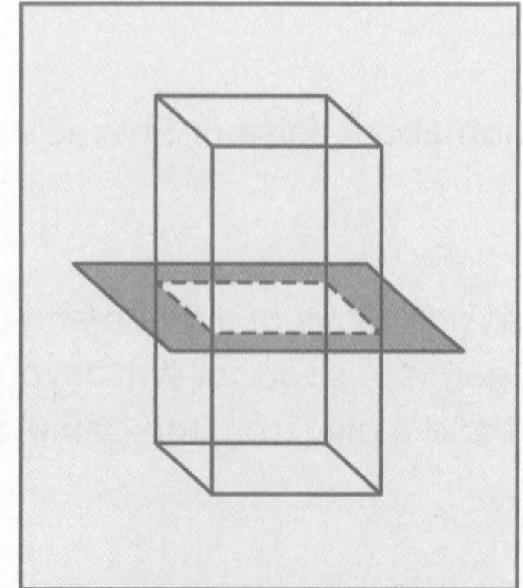
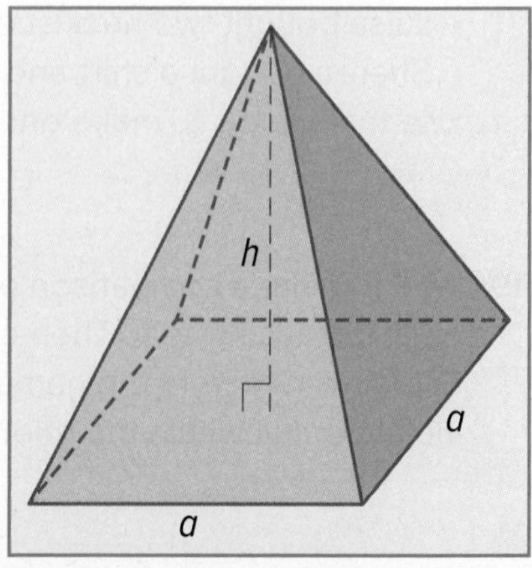

Points, Lines, and Planes

Learn Points, Lines, and Planes

In geometry, *point*, *line*, and *plane* are considered undefined terms because they are usually readily understood and are not formally explained by means of more basic words and concepts.

You are already familiar with the terms point, line, and plane from algebra. You graphed on a coordinate *plane* and found ordered pairs that represented *points* on *lines*. In geometry, these terms have a similar meaning.

Undefined Terms	
A **point** is a location. It has neither shape nor size. Named by a capital letter Example point *A*	A
A **line** is made up of points and has no thickness or width. There is exactly one line through any two points. Named by the letters representing two points on the line or a lowercase script letter Example line m, line *PQ* or $\overleftrightarrow{PQ}$, line *QP* or $\overleftrightarrow{QP}$	P Q m
A **plane** is a flat surface made up of points that extends infinitely in all directions. There is exactly one plane through any three points not on the same line. Named by a capital script letter or by the letters naming three points that are not all on the same line Example plane $\mathcal{K}$, plane *BCD*, plane *CDB*, plane *DCB*, plane *DBC*, plane *CBD*, plane *BDC*	D B C K

Space is defined as a boundless three-dimensional set of all points. Space can contain lines and planes.

Collinear points are points that lie on the same line. *Noncollinear* points do not lie on the same line.

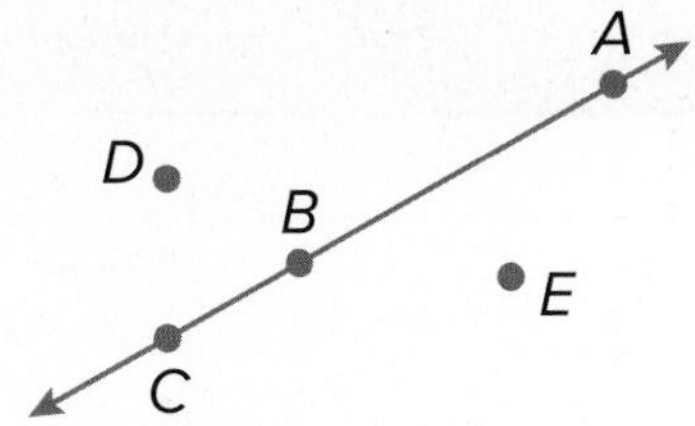

Points *A*, *B*, and *C* are collinear.

Coplanar points are points that lie in the same plane. *Noncoplanar* points do not lie in the same plane.

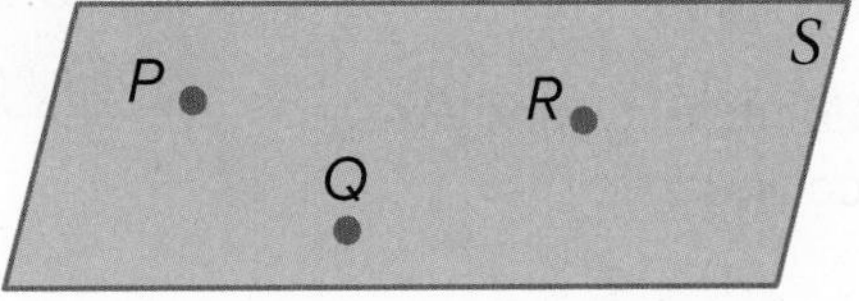

Points *P*, *Q*, and *R* are coplanar.

Today's Goals
- Identify points, lines, and planes.
- Identify intersections of lines and planes.

Today's Vocabulary
point
line
plane
space
collinear
coplanar
intersection

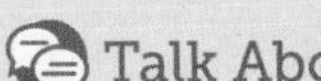

Talk About It!
Can three points be both noncollinear and noncoplanar? Justify your argument.

Example 1 Name Lines and Planes

Use the figure to name each of the following.

a. a line containing point *Q*

The line can be named as line c, or any two of the three points on the line can be used to name the line.

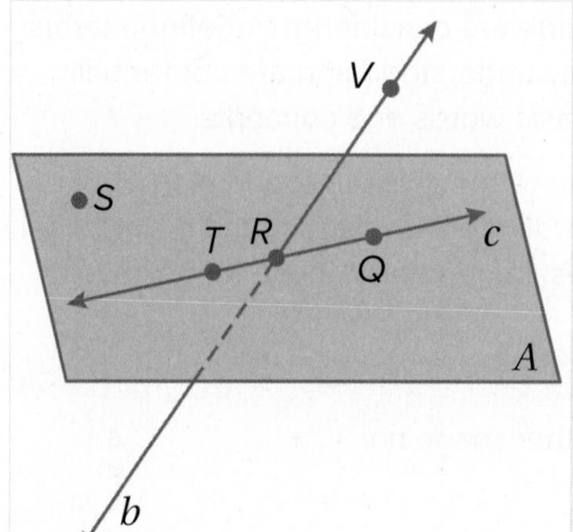

Write the additional names for line c below.

$\overleftrightarrow{TR}$ $\overleftrightarrow{RT}$ $\overleftrightarrow{TQ}$ $\overleftrightarrow{QT}$ $\overleftrightarrow{RQ}$ $\overleftrightarrow{QR}$

b. a plane containing point *S* and point *T*

One plane that can be named is plane A. You can also use the letters of any three *noncollinear* points to name this plane. Plane *TRS* and plane *TQS* can be used to name this plane.

Circle another correct name for plane A.

(plane *QST*) plane *STV* plane *QVS* plane *VST*

> **Study Tip**
>
> **Additional Planes** Although not drawn, there is another plane that contains point *S* and point *T*. Because points *S*, *T*, and *V* are noncollinear, points *S* and *T* are in plane *STV*.

Example 2 Model Points, Lines, and Planes

STUDENT DESK **Name the geometric terms modeled by the objects in the picture.**

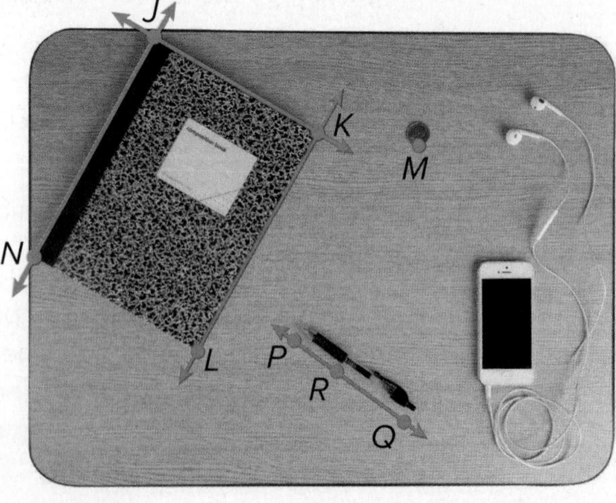

The notebook models plane *JKL* or *NJK*.

The edges of the notebook model lines *JK*, *KL*, and *JN*.

The quarter models point *M* in space.

Points *N*, *L*, and *K* are coplanar.

Points *P*, *Q*, and *R* are collinear.

Go Online You can complete an Extra Example online.

Explore Intersections of Three Planes

 **Online Activity** Use a concrete model to complete the Explore.

> **INQUIRY** What figures can be formed by the intersection of three planes?

Learn Intersections of Lines and Planes

The **intersection** of two or more geometric figures is the set of points they have in common. Two lines intersect in a point. Lines can intersect planes, and planes can intersect each other.

Example 3 Draw Geometric Figures

Draw and label a figure to represent the relationship.

$\overleftrightarrow{QR}$ and $\overleftrightarrow{ST}$ intersect at U for $Q(-3, -2)$, $R(4, 1)$, $S(2, 3)$, and $T(-1, -5)$ on the coordinate plane. Point V is coplanar with these points but not collinear with $\overleftrightarrow{QR}$ and $\overleftrightarrow{ST}$.

Graph each point and draw $\overleftrightarrow{QR}$ and $\overleftrightarrow{ST}$.

Label the intersection point as U.

An infinite number of points are coplanar with Q, R, S, T, and U but are not collinear with $\overleftrightarrow{QR}$ and $\overleftrightarrow{TS}$. In the graph, one such point is $V(-2, 3)$.

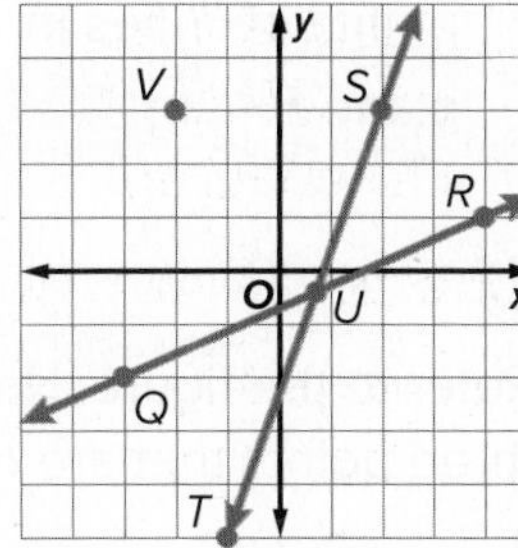

Check

Draw and label a figure to represent the relationship.

$\overleftrightarrow{JK}$ and $\overleftrightarrow{LM}$ intersect at P for $J(-4, 3)$, $K(6, -3)$, $L(-4, -5)$, and $M(3, 3)$ on the coordinate plane. Point Q is coplanar with these points, but not collinear with $\overleftrightarrow{JK}$ and $\overleftrightarrow{LM}$.

Go Online You can complete an Extra Example online.

Study Tip

Dimensions A point has no dimension. A line exists in one dimension. However, a circle is two-dimensional, and a pyramid is three-dimensional.

Study Tip

Three-Dimensional Drawings Because it is impossible to show an entire plane in a figure, edged shapes with different shades of color are used to represent planes.

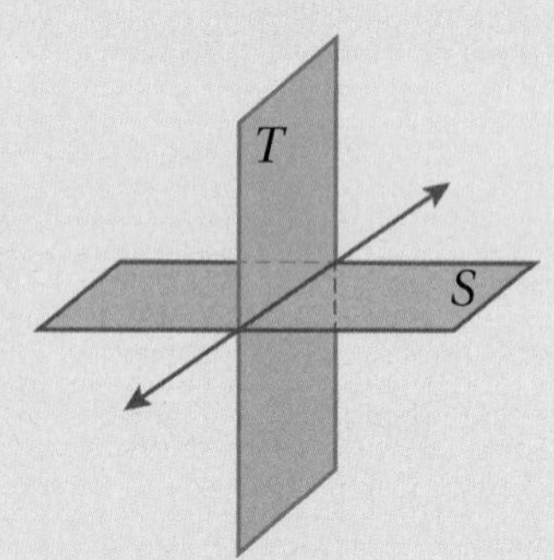

Example 4 Interpret Drawings

Refer to the figure.

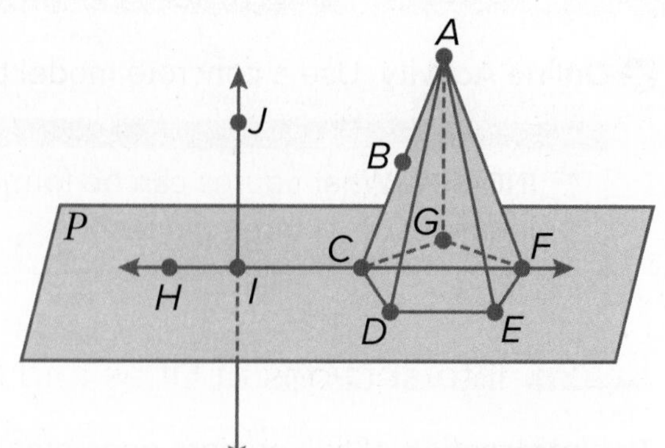

a. How many planes appear in this figure?

six: plane P, plane CAG, plane GFA, plane EFA, plane DEA, and plane DCA

b. Name four points that are collinear.

Points H, I, C, and F are collinear.

c. Name the intersection of plane GAC and plane P.

Plane GAC intersects plane P in $\overleftrightarrow{GC}$.

d. At what point do $\overleftrightarrow{JI}$ and $\overleftrightarrow{DC}$ intersect? Explain.

It does not appear that these lines intersect. $\overleftrightarrow{DC}$ lies in plane P, but only point I of $\overleftrightarrow{JI}$ lies in plane P.

Check

Refer to the figure. Name three points that are collinear.

Points 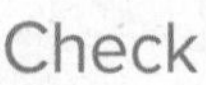 ___?___, ___?___, and ___?___ are collinear.

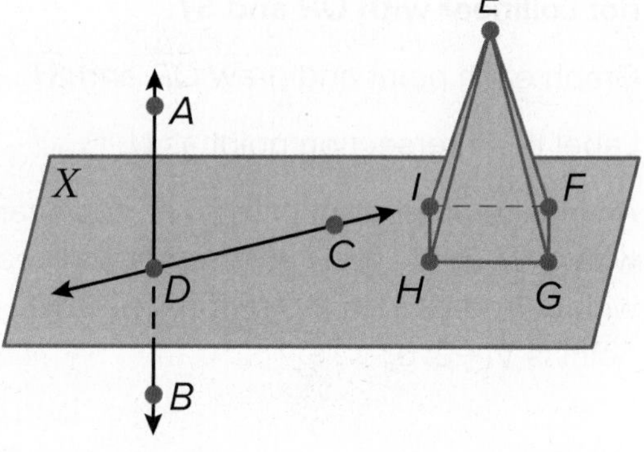

Example 5 Model Intersections

AVIATION A biplane has two main wings that are stacked one above the other. Struts connect the wings and are used for support. Flying wires run diagonally from the main body of the plane to the wings and between the stacked wings.

Complete the statements regarding the geometric terms modeled by the biplane.

Each wing models a plane.

The intersection of a strut and a wing models a point.

The crossing of two flying wires models a point.

Go Online You can complete an Extra Example online.

Practice

Go Online You can complete your homework online.

Example 1

Refer to the figure for Exercises 1–7.

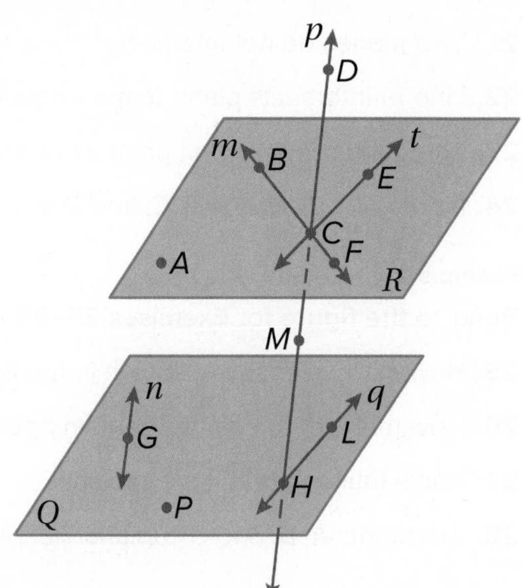

1. Name the lines that are only in plane Q.
2. How many planes are labeled in the figure?
3. Name the plane containing the lines m and t.
4. Name the intersection of lines m and t.
5. Name a point that is *not* coplanar with points *A*, *B*, and *C*.
6. Are points *F*, *M*, *G*, and *P* coplanar? Explain.
7. Does line n intersect line q? Explain.

Example 2

Name the geometric terms modeled by each object or phrase.

8. roof of a house
9. a tabletop
10. bridge support beam

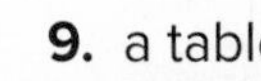

11. a chessboard
12.

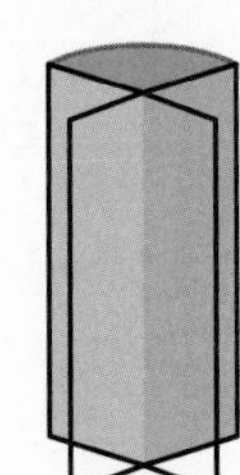

13.

14. a wall and the floor
15. the edge of a table
16. two connected walls

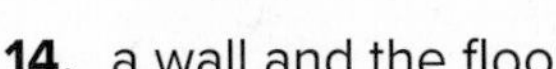

17. a blanket
18. a telephone pole
19. a tablet computer

Example 3

USE TOOLS **Draw and label a figure for each relationship.**

20. Points *X* and *Y* lie on $\overleftrightarrow{CD}$.

21. Two planes do not intersect.

22. Line m intersects plane R at a single point.

23. Three lines intersect at point *J* but do not all lie in the same plane.

24. Points *A*(2, 3), *B*(2, −3), *C*, and *D* are collinear, but *A*, *B*, *C*, *D*, and *F* are not.

Example 4

Refer to the figure for Exercises 25–28.

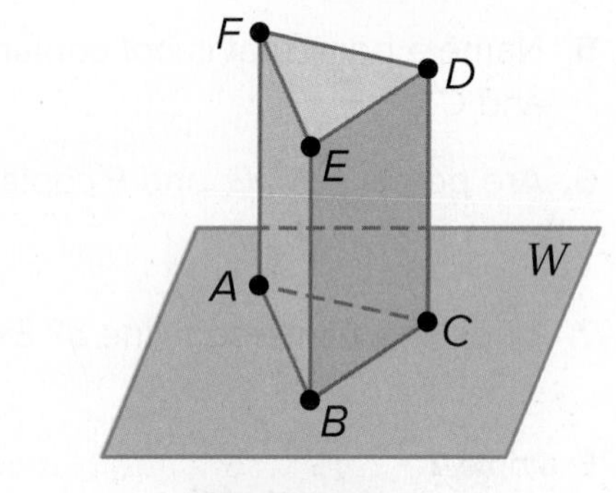

25. How many planes are shown in the figure?

26. How many of the planes contain points *F* and *E*?

27. Name four points that are coplanar.

28. Are points *A*, *B*, and *C* coplanar? Explain.

Example 5

29. BUILDING The roof and exterior walls of a house represent intersecting planes. Using the image, name all the lines that are formed by the intersecting planes.

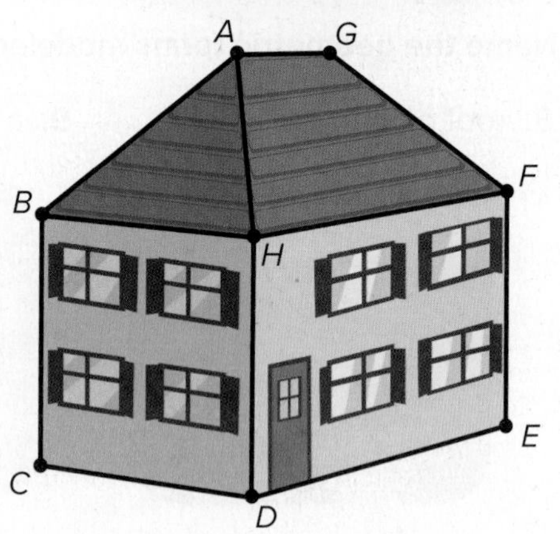

30. If the surface of a lake represents a plane, what geometric term is represented by the intersection of a fishing line and the lake's surface?

31. ART Perspective drawing is a method that artists use to create paintings and drawings of three-dimensional objects. The artist first draws the horizon line and two vanishing points along the horizon. Buildings or other objects are created by drawing receding lines and vertical lines.

a. Where do the receding lines and horizon lines intersect?

b. Identify examples of planes within this picture.

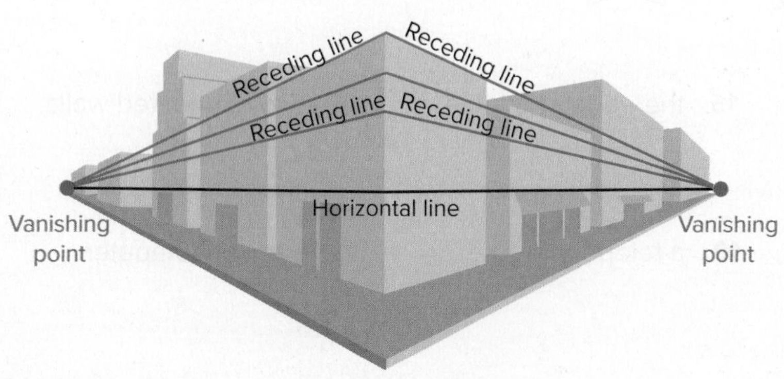

Mixed Exercises

USE TOOLS **Draw and label a figure for each relationship.**

32. $\overleftrightarrow{LM}$ and $\overleftrightarrow{NP}$ are coplanar but do not intersect.

33. $\overleftrightarrow{FG}$ and $\overleftrightarrow{JK}$ intersect at $P(4, 3)$, where point F is at $(-2, 5)$ and point J is at $(7, 9)$.

34. Lines s and t intersect, and line v does not intersect either one.

Refer to the figure for Exercises 35-38.

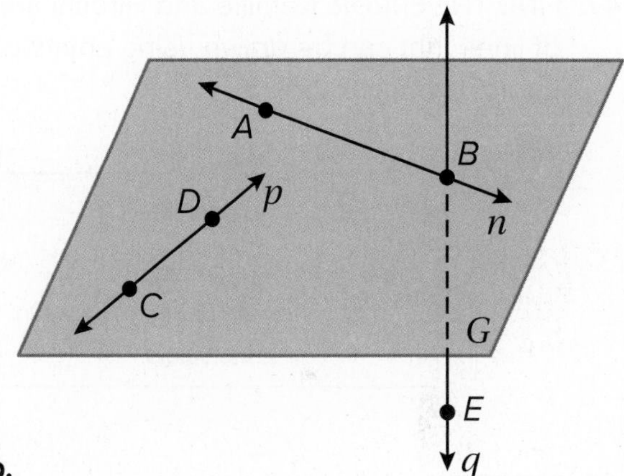

35. Name a line that contains point E.

36. Name a point contained in line n.

37. What is another name for line p?

38. Name the plane containing lines n and p.

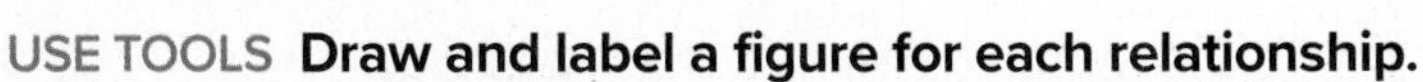

USE TOOLS **Draw and label a figure for each relationship.**

39. Point K lies on $\overleftrightarrow{RT}$.

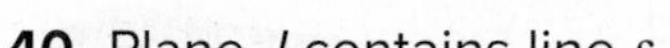

40. Plane J contains line s.

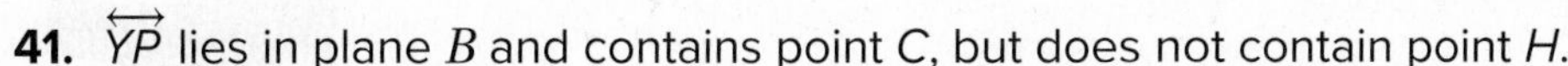

41. $\overleftrightarrow{YP}$ lies in plane B and contains point C, but does not contain point H.

42. Lines q and f intersect at point Z in plane U.

43. Name the geometric term modeled by the object.

44. Name the geometric term modeled by a partially-opened folder.

Higher-Order Thinking Skills

45. **CREATE** Sketch three planes that intersect in a point.

46. **ANALYZE** Is it possible for two points on the surface of a prism to be neither collinear nor coplanar? Justify your argument.

47. **FIND THE ERROR** Camille and Hiroshi are trying to determine the greatest number of lines that can be drawn using any two of four random points. Is either correct? Explain your reasoning.

Camille	Hiroshi
Because there are four points, $4 \cdot 3$ or 12 lines can be drawn between the points.	You can draw $3 + 2 + 1$ or 6 lines between the points.

48. **PERSEVERE** What is the greatest number of planes determined using any three of the points *A*, *B*, *C*, and *D* if no three points are collinear?

49. **WRITE** A *finite plane* is a plane that has boundaries or does not extend indefinitely. The sides of the cereal box shown are finite planes. Give a real-life example of a finite plane. Is it possible to have a real-life object that is an infinite plane? Explain your reasoning.

50. **CREATE** Sketch three planes that intersect in a line.

Line Segments

Today's Goals
- Calculate measures of line segments.
- Apply the definition of congruent line segments to find missing values.

Today's Vocabulary
line segment
betweenness of points
congruent
congruent segments

Explore Using Tools to Determine Betweenness of Points

Online Activity Use a pencil and straightedge to complete the Explore.

INQUIRY How can a line segment be divided into any number of line segments?

Learn Betweenness of Points

A **line segment** is a measurable part of a line that consists of two points, called endpoints, and all of the points between them. The two endpoints are used to name the segments.

You know that for any two real numbers a and b, there is a real number n between a and b such that $a < n < b$. This relationship also applies to points on a line and is called **betweenness of points.**

Key Concept • Betweenness of Points
Point C is between A and B if and only if A, B, and C are collinear and $AC + CB = AB$.

Talk About It!

What is an example of how the betweenness of points can be applied to the real world?

Example 1 Find Measurements by Adding

Find the measure of $\overline{XZ}$.

X
11.3 cm
Y
Z
3.8 cm

XZ is the measure of $\overline{XZ}$. Point Y is between X and Z. Find XZ by adding XY and YZ.

$XY + YZ = XZ$	Betweenness of points
$11.3 + 3.8 = XZ$	Substitution
$15.1 \text{ cm} = XZ$	Add.

Check

Find the measure of $\overline{DF}$.

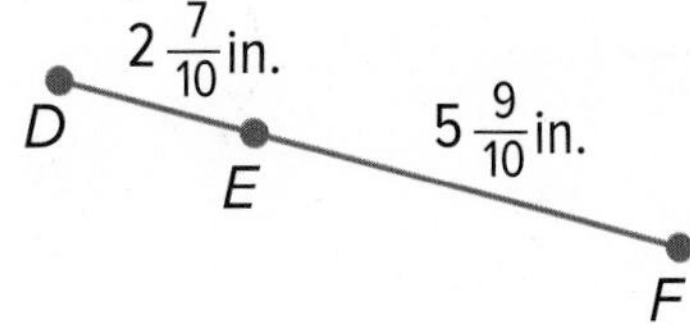

Go Online You can complete an Extra Example online.

Example 2 Find Measurements by Subtracting

Find the measure of $\overline{QR}$.

Point Q is between points P and R.

$PQ + QR = PR$	Betweenness of points
$6\frac{5}{8} + QR = 13\frac{3}{4}$	Substitution
$QR = 7\frac{1}{8}$ ft	Subtract $6\frac{5}{8}$ from each side and simplify.

Problem-Solving Tip

Draw a Diagram Draw a diagram to help you see and correctly interpret a situation that has been described in words.

Check

Find the measure of $\overline{PQ}$. Round your answer to the nearest tenth, if necessary.

___?___ cm

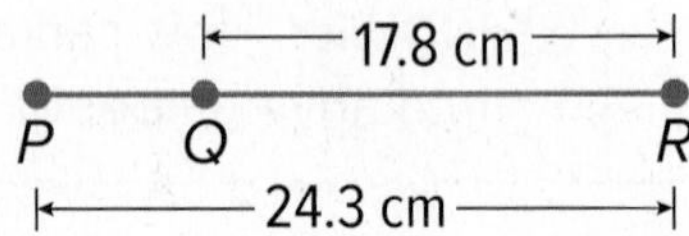

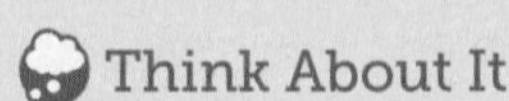

Think About It!

How can you check your solution for x?

Example 3 Write and Solve Equations to Find Measurements

Find the value of x and BC if B is between A and C, $AC = 4x - 12$, $AB = x$, and $BC = 2x + 3$.

Step 1 Plot two points and label them A and C. Connect the points.

Step 2 Plot point B between points A and C.

Step 3 Label segments AB, BC, and AC with their given measures.

Step 4 Use betweenness of points to write an equation and solve for x.

$AC = AB + BC$	Betweenness of points
$4x - 12 = x + 2x + 3$	Substitution
$4x - 12 = 3x + 3$	Combine like terms.
$x - 12 = 3$	Subtract $3x$ from each side. Simplify.
$x = 15$	Add 12 to each side. Simplify.

Now find BC.

$BC = 2x + 3$	Given
$= 2(15) + 3$	$x = 15$
$= 33$	

Think About It!

Once you find BC, how could you find AC without evaluating $AC = 4x - 12$?

Go Online You can complete an Extra Example online.

Apply Example 4 Use Betweenness of Points

SPACE NEEDLE **Darrell is visiting the Space Needle in Seattle, Washington. He knows that the total height of the Space Needle is 605 feet. The distance from the ground to the observation deck is 10 feet more than six times the distance from the observation deck to the top of the Space Needle. Help Darrell find the distance from the ground to the observation deck.**

1 What is the task?

Describe the task in your own words. Then list any questions that you may have. How can you find answers to your questions?

Sample answer: I need to find the distance from the ground to the observation deck. How does the distance from the ground to the observation deck compare to the total height of the Space Needle? I can express the information that I am given in the exercise as an equation, solve for any missing information, and then use that information to find the answer.

2 How will you approach the task? What have you learned that you can use to help you complete the task?

Sample answer: I will express the information that I am given into an equation that represents the total height of the Space Needle. I have learned how to convert written information into expressions, and I have learned how to solve equations.

3 What is your solution?

Use your strategy to solve the problem.

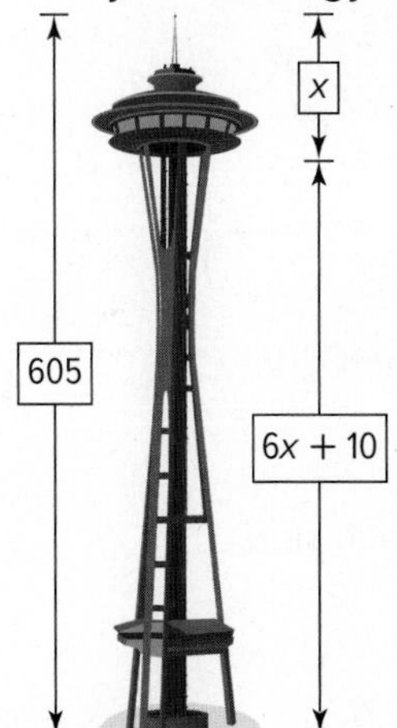

x

$6x + 10$

What equation represents the distance from the ground to the top of the Space Needle?

$x + 6x + 10 = 605$

What is the distance from the ground to the observation deck?

520 ft

4 How can you know that your solution is reasonable?

Write About It! Write an argument that can be used to defend your solution.

Sample answer: 520 feet seems reasonable for the distance from the ground to the observation deck. The distance from the observation deck to the top of the Space Needle is 85 feet. The combined heights are realistic compared to the total height.

Study Tip

Congruent Segments Use a consecutive number of tick marks for each new pair of congruent segments in a figure. The segments with two tick marks are congruent, and the segments with three tick marks are congruent.

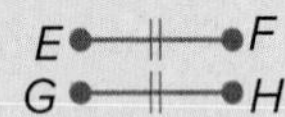

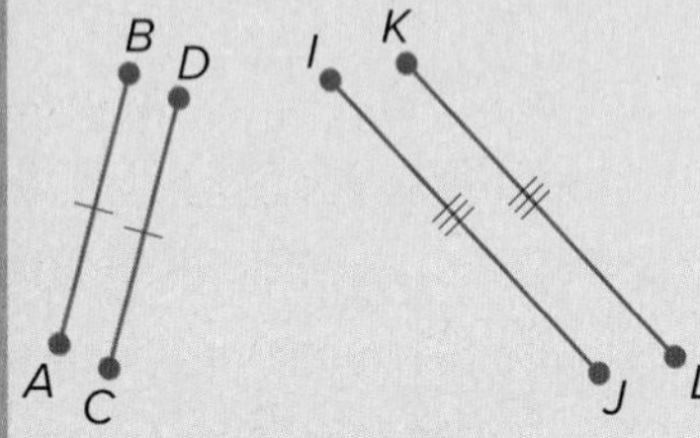

Learn Line Segment Congruence

If two geometric figures have exactly the same shape and size, then they are **congruent**. Two segments that have the same measure are **congruent segments**.

Key Concept • Congruent Segments

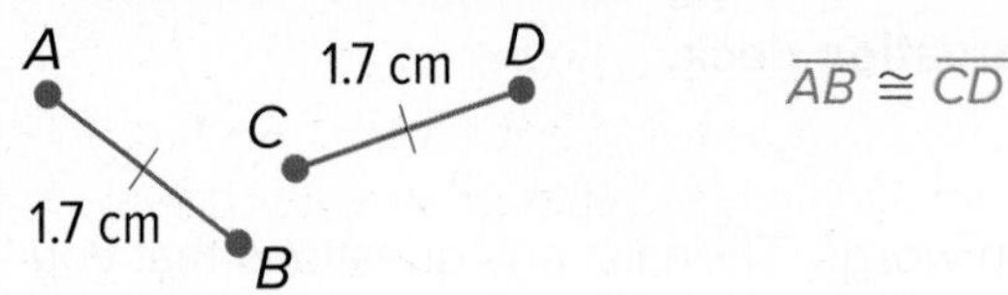

$\overline{AB} \cong \overline{CD}$

$\cong$ is read *is congruent to*. Red slashes on the figure also indicate congruence.

Segment *AB* is congruent to segment *CD*.

Congruent segments have the same measure.

Example 5 Write and Solve Equations by Using Congruence

Find the value of *x* if *Q* is between *P* and *R*, $PQ = 6x + 20$, $QR = 2(x + 6)$, and $\overline{PQ} \cong \overline{QR}$.

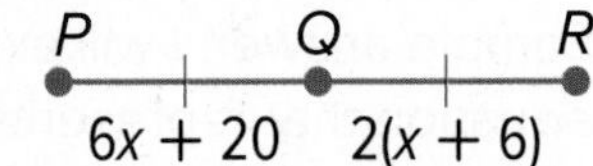

Write the justifications in the correct order. You may use a justification more than once.

Definition of congruence Distributive Property
Divide each side by 4. Simplify. Substitution
Subtract 2*x* from each side. Subtract 20 from each side.

$PQ = QR$	Definition of congruence
$6x + 20 = 2(x + 6)$	Substitution
$6x + 20 = 2x + 12$	Distributive Property
$6x + 20 - 2x = 2x + 12 - 2x$	Subtract 2*x* from each side.
$4x + 20 = 12$	Simplify.
$4x + 20 - 20 = 12 - 20$	Subtract 20 from each side.
$4x = -8$	Simplify.
$\frac{4x}{4} = \frac{-8}{4}$	Divide each side by 4.
$x = -2$	Simplify.

Study Tip

Equal vs. Congruent Lengths are *equal*, and segments are *congruent*. It is correct to say that $AB = CD$ and $\overline{AB} \cong \overline{CD}$. However, it is not correct to say that $\overline{AB} = \overline{CD}$ or that $AB \cong CD$.

Watch Out!

Check Your Answer Sometimes solutions will result in negative segment lengths. If this occurs, review your work carefully. Either an error was made, or there is no solution.

Check

Find the value of *x* if *U* is between *T* and *V*, $TU = 7x + 35$, $UV = 4(x + 7)$, and $\overline{TU} \cong \overline{UV}$.

$x =$ __?__

Go Online You may want to complete the construction activities for this lesson.

Practice

Go Online You can complete your homework online.

Examples 1 and 2

Find the measure of each segment.

1. $\overline{PR}$

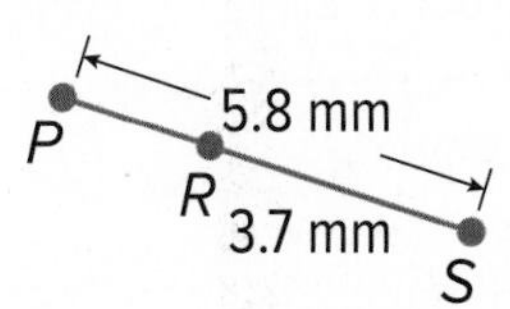

2. $\overline{EF}$

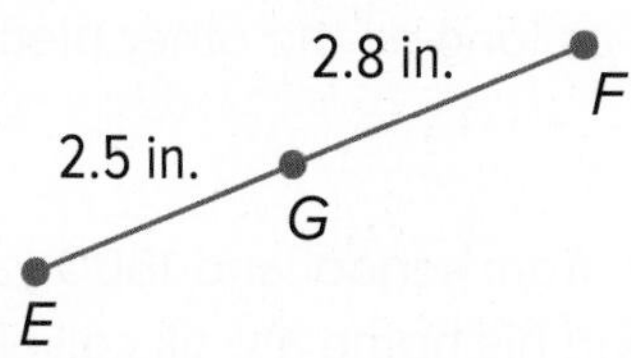

3. $\overline{JL}$

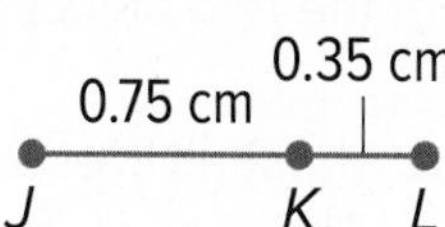

4. $\overline{HJ}$

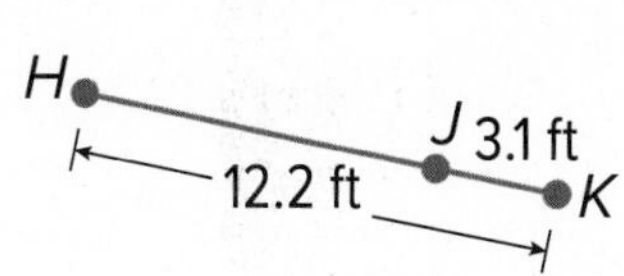

5. $\overline{AC}$

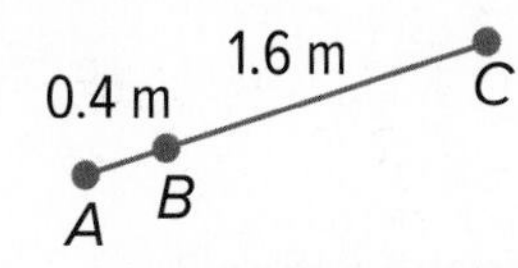

6. $\overline{SV}$

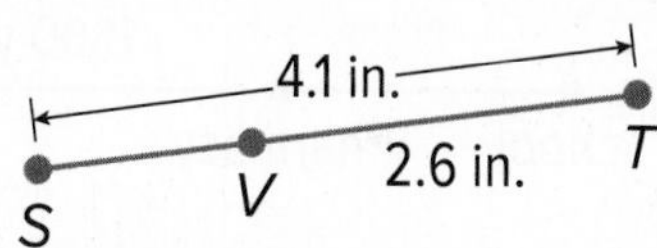

7. $\overline{NQ}$

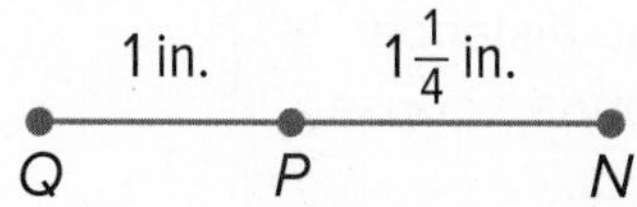

8. $\overline{AC}$

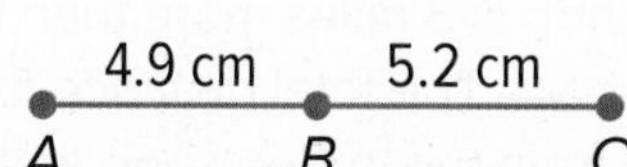

9. $\overline{GH}$

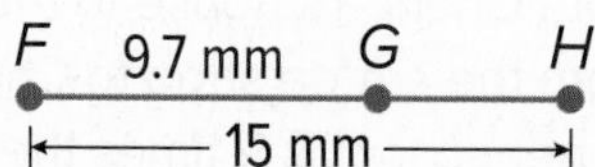

Example 3

Find the value of the variable and YZ if Y is between X and Z.

10. $XY = 11$, $YZ = 4c$, $XZ = 83$

11. $XY = 6b$, $YZ = 8b$, $XZ = 175$

12. $XY = 7a$, $YZ = 5a$, $XZ = 6a + 24$

13. $XY = 5.5$, $YZ = 2c$, $XZ = 8.9$

14. $XY = 5n$, $YZ = 2n$, $XZ = 91$

15. $XY = 4w$, $YZ = 6w$, $XZ = 12w - 8$

16. $XY = 11d$, $YZ = 9d - 2$, $XZ = 5d + 28$

17. $XY = 4n + 3$, $YZ = 2n - 7$, $XZ = 20$

18. $XY = 3a - 4$, $YZ = 6a + 2$, $XZ = 5a + 22$

19. $XY = 3k - 2$, $YZ = 7k + 4$, $XZ = 4k + 38$

20. $XY = 4x$, $YZ = x$, and $XZ = 25$

21. $XY = 4x$, $YZ = 3x$, and $XZ = 42$

22. $XY = 12$, $YZ = 2x$, and $XZ = 28$

23. $XY = 2x + 1$, $YZ = 6x$, and $XZ = 81$

Example 4

24. RAILROADS A straight railroad track is being built to connect two cities. The measured distance of the track between the two cities is 160.5 miles. A mail stop is 28.5 miles from the first city. How far is the mail stop from the second city?

25. CARPENTRY A carpenter has a piece of wood that is 78 inches long. He wants to cut it so that one piece is five times as long as the other piece. What are the lengths of the two pieces?

26. WALKING Marshall lives 2300 yards from school and 1500 yards from the pharmacy. The school, pharmacy, and his home are all collinear, as shown in the figure.

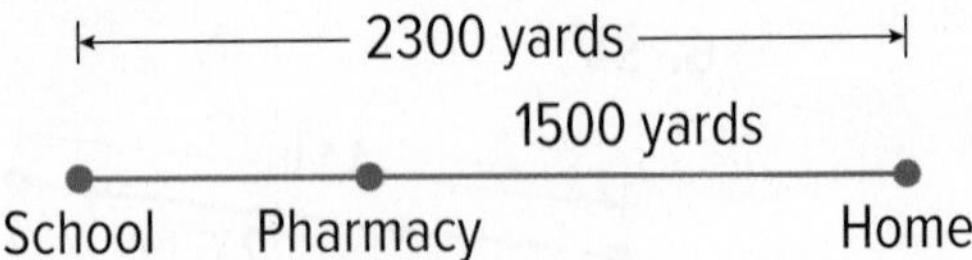

What is the distance from the pharmacy to the school?

27. COFFEE SHOP Chenoa wants to stop for coffee on her way to school. The distance from Chenoa's house to the coffee shop is 3 miles more than twice the distance from the coffee shop to Chenoa's school. The total distance from Chenoa's house to her school is 5 times the distance from the coffee shop to her school.

a. What is the distance from Chenoa's house to the coffee shop? Write your answer as a decimal, if necessary.

b. What assumptions did you make when solving this problem?

Example 5

Find the measure of each segment.

28. $\overline{MO}$

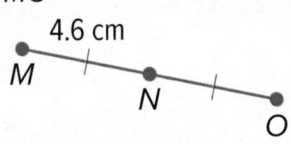

4.6 cm
M N O

29. $\overline{WY}$

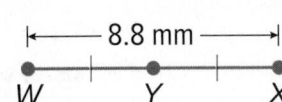

8.8 mm
W Y X

30. $\overline{FG}$

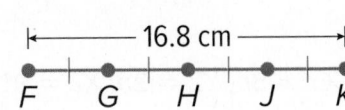

16.8 cm
F G H J K

31. $\overline{QT}$

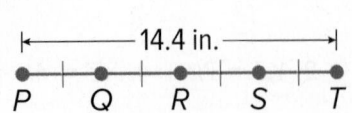

14.4 in.
P Q R S T

32. $\overline{DE}$

$2x + 7$ $4(x - 3)$
C D E

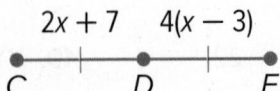

33. $\overline{UX}$

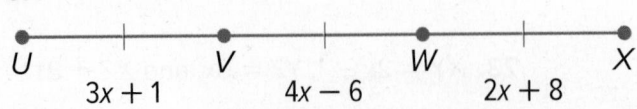

U V W X
$3x + 1$ $4x - 6$ $2x + 8$

Mixed Exercises

34. Find the length of $\overline{UW}$ if W is between U and V, $UV = 16.8$ centimeters, and $VW = 7.9$ centimeters.

35. Find the value of x if $RS = 24$ centimeters.

$6x - 4$ 10 cm
R T S

36. Find the length of $\overline{LO}$ if M is between L and O, $LM = 7x - 9$, $MO = 14$ inches, and $LO = 10x - 7$.

37. Find the value of x if $\overline{PQ} \cong \overline{RS}$, $PQ = 9x - 7$, and $RS = 29$.

38. Find the measure of $\overline{NL}$.

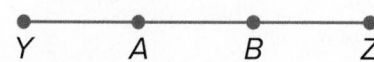

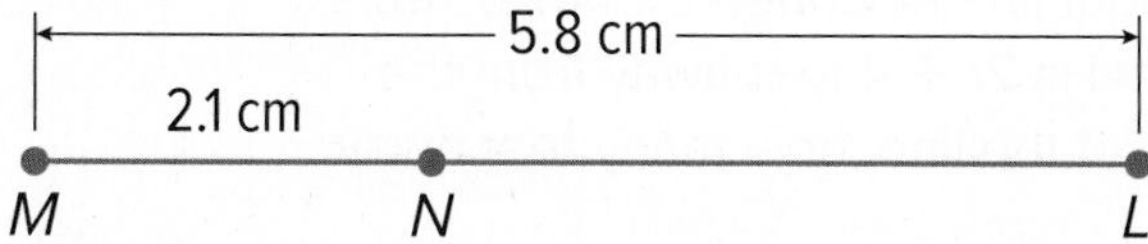

39. **PRECISION** If point P is between A and M, write a true statement.

40. **HIKING** A hiking trail is 20 kilometers long. Park organizers want to build 5 rest stops for hikers with one on each end of the trail and the other 3 spaced evenly between. How much distance will separate successive rest stops?

41. **RACE** The map shows the route of a race. You are at Y, 6000 feet from the first checkpoint A. The second checkpoint B is located at the midpoint between A and the end of the race Z. The total race is 3.1 miles. How far apart are the two checkpoints?

Y A B Z

42. **FIELD TRIP** The marching band at Jefferson High School is taking a field trip from Lansing, Michigan, to Detroit, Michigan. The bus driver was told to stop 53 miles into the trip. If the rest of the trip is 41 miles and the entire journey can be represented by the expression $3x + 16$, find the value of x.

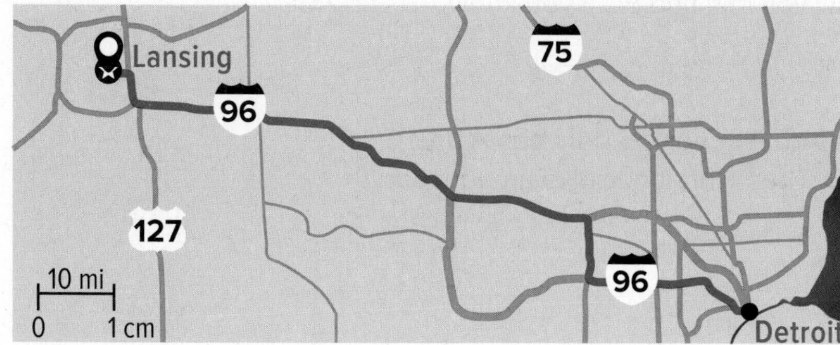

43. DISTANCE Madison lives between Anoa and Jamie as depicted on the line segment. The distance between Anoa's house and Madison's house is represented by $3x + 2$ miles, the distance between Madison's house and Jamie's house is represented by $3x + 4$ miles, and the distance between Anoa's house and Jamie's house is represented by $9x - 3$ miles. Find the value of x. Then find the distance between Madison's house and Jamie's house.

44. FIREFIGHTING A firefighter training course is taking place in a high-rise building. The high-rise building where they practice is 48 stories high. If the emergency happens on the top floor and the firefighters have already gone 29 stories, how many stories do they still need to go?

45. CAFE You are waiting at the end of a long straight line at Coffee Express. Your friend Denzel is $r + 12$ feet in front of you. Denzel is $2r + 4$ feet away from the front of the line. If Denzel is in the exact middle of the line, how many feet away are you from the front of the line?

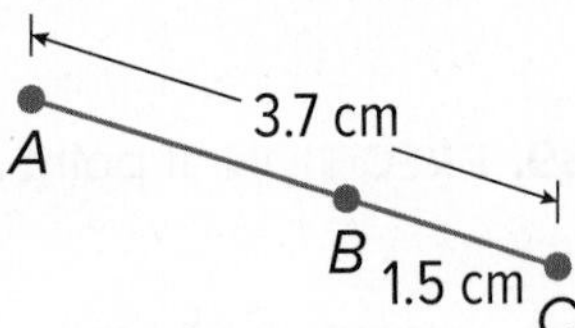

46. REASONING For $\overline{AC}$, write and solve an equation to find AB.

Higher-Order Thinking Skills

47. PERSEVERE Point K is between points J and L. If $JK = x^2 - 4x$, $KL = 3x - 2$, and $JL = 28$, find JK and KL.

48. ANALYZE Determine whether the statement *If point M is between points C and D, then CD is greater than either CM or MD* is *sometimes, always,* or *never* true. Justify your argument.

49. PERSEVERE Point C is located between points B and D. Also, $BC = 5x + 7$, $CD = 3y + 4$, $BD = 38$, and $BD = 2x + 8y$. Find the values of x and y.

50. WRITE If point B is between points A and C, explain how you can find AC if you know AB and BC. Explain how you can find BC if you know AB and AC.

51. CREATE Sketch line segment AC. Plot point B between A and C. Use a ruler to find AC and AB. Then write and solve an equation to find BC.

Distance

Learn Distance on a Number Line

The **distance** between two points is the length of the segment between the points. The coordinates of the points can be used to find the length of the segment.

Key Concept • Distance Formula on Number Line

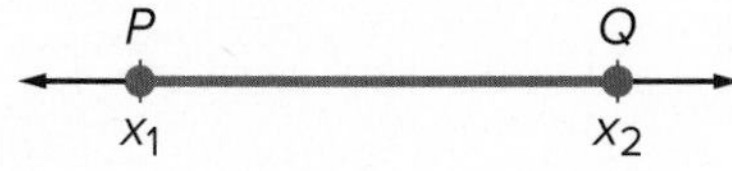

If P has coordinate x_1 and Q has coordinate x_2, then $PQ = |x_2 - x_1|$ or $|x_1 - x_2|$.

Because $\overline{PQ}$ is the same as $\overline{QP}$, the order in which you name the endpoints is not important when calculating distance.

Example 1 Find Distance on a Number Line

Use the number line.

A B C D E F

−5 −4 −3 −2 −1 0 1 2 3 4 5

Find CF.

$CF = \|x_2 - x_1\|$	Distance Formula
$= \|5 - (-1)\|$	$x_1 = -1$ and $x_2 = 5$
$= 6$	Simplify.

Check

Use the number line.

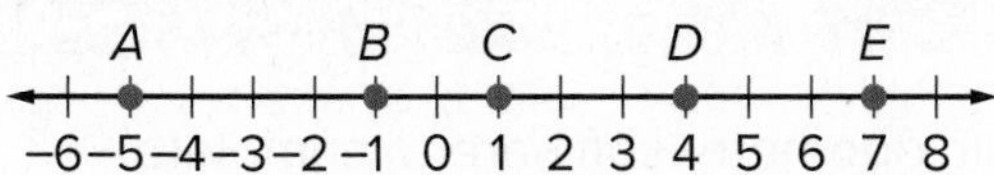

Find AE.

A. −12 B. 2 C. 12 D. 13

Go Online You can complete an Extra Example online.

Today's Goals

- Find the length of a line segment on a number line.
- Find the distance between two points on the coordinate plane.

Today's Vocabulary

distance

Think About It!
Why do you think the Distance Formula uses absolute value?

Think About It!
Compare and contrast the length of $\overline{CF}$ and the length of $\overline{FC}$.

Example 2 Determine Segment Congruence

Determine whether $\overline{CB}$ and $\overline{DF}$ are congruent.

A B C D E F

−5 −4 −3 −2 −1 0 1 2 3 4 5

The coordinates of C and B are −1 and −3. The coordinates of D and F are 2 and 5. Find the length of each segment.

$CB = \lvert x_2 - x_1 \rvert$	Distance Formula
$= \lvert -3 - (-1) \rvert$	Substitute.
$= \lvert -2 \rvert$	Subtract.
$= 2$	Simplify.

The length of $\overline{CB}$ is 2 units.

$DF = \lvert x_2 - x_1 \rvert$	Distance Formula
$= \lvert 5 - 2 \rvert$	Substitute.
$= \lvert 3 \rvert$	Subtract.
$= 3$	Simplify.

The length of $\overline{DF}$ is 3 units.

Because $CB \neq DF$, the segments are not congruent.

Check

Determine whether $\overline{AC}$ and $\overline{BD}$ are congruent.

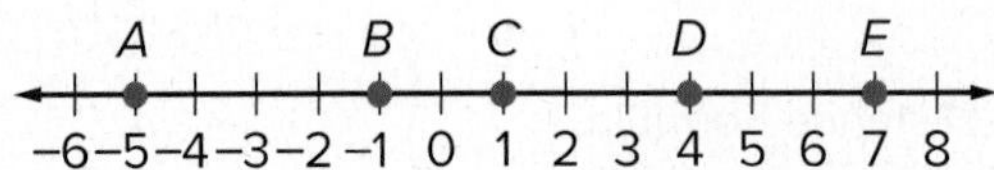

The segments ___?___ congruent.

Watch Out!

Subtraction with Negatives Remember that subtracting a negative number is like adding a positive number.

Explore Use the Pythagorean Theorem to Find Distances

Online Activity Use dynamic geometry software to complete the Explore.

INQUIRY How can you find the distance between two points on the coordinate plane?

Go Online A derivation of the Distance Formula is available.

Learn Distance on the Coordinate Plane

The endpoints of a segment on the coordinate plane can be used to find the length of that segment by using the Distance Formula.

Key Concept • Distance Formula on the Coordinate Plane	
If P has coordinates (x_1, y_1) and Q has coordinates (x_2, y_2), then $PQ = \sqrt{(x_2 - x_1)^2 + (y_2 - y_1)^2}$.	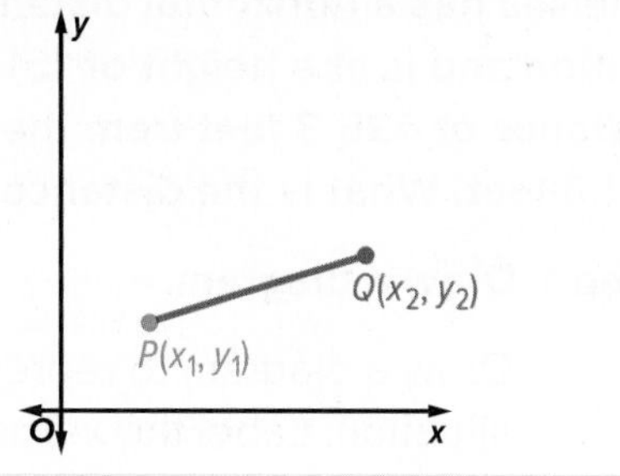

Example 3 Find Distance on the Coordinate Plane

Find the distance between $J(4, 3)$ and $K(-3, -7)$.

Let $J(4, 3)$ be (x_1, y_1) and $K(-3, -7)$ be (x_2, y_2).

$JK = \sqrt{(x_2 - x_1)^2 + (y_2 - y_1)^2}$	Distance Formula
$= \sqrt{(-3 - x_1)^2 + (-7 - y_1)^2}$	Substitute x_2 and y_2.
$= \sqrt{(-3 - 4)^2 + (-7 - 3)^2}$	Substitute x_1 and y_1.
$= \sqrt{(-7)^2 + (-10)^2}$	Subtract.
$= \sqrt{49 + 100}$	Simplify.
$= \sqrt{149}$	Simplify.

The distance between J and K is $\sqrt{149}$ or approximately 12.2 units.

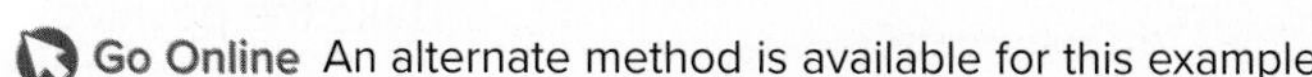
Go Online An alternate method is available for this example.

Check

Find the distance between A and B.

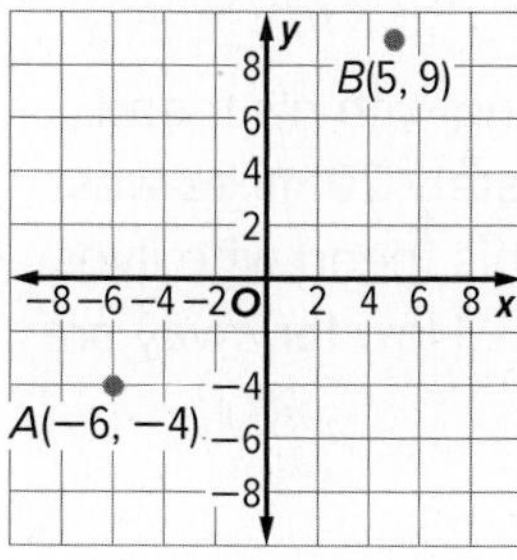

Go Online You can complete an Extra Example online.

Think About It!
Compare and contrast the Distance Formula on a number line with the Distance Formula on the coordinate plane.

Watch Out!
Simplify Radicals Do not forget to leave your answer in simplest radical form when using the Distance Formula or the Pythagorean Theorem.

Example 4 Calculate Distance

INCLINE **Chelsea and Amie are sitting in separate cars on the Monongahela Incline. Chelsea is traveling up Mount Washington and Amie is traveling down. When the two girls notice each other, Chelsea has a horizontal distance of 212.0 feet from the lower station and is at a height of 151.6 feet. Amie has a horizontal distance of 435.3 feet from the lower station and is at a height of 311.3 feet. What is the distance between the two girls?**

Step 1 Draw a diagram.

Draw a diagram to represent the situation. Label the x-axis as the "Horizontal Distance from Lower Station (in feet)." Label the y-axis as the "Height (in feet)." Use a scale of 50 on the x-axis and the y-axis.

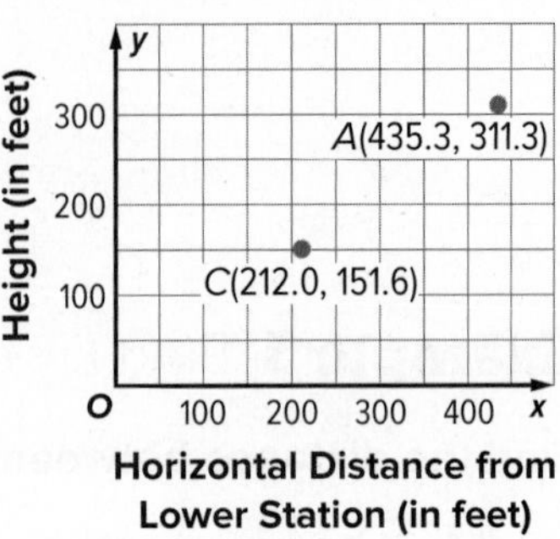

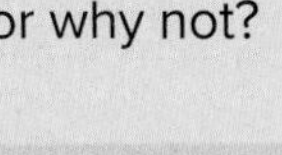

Think About It!

Does your answer seem reasonable? Why or why not?

Step 2 Use the Distance Formula.

$(x_1, y_1) = (212.0, 151.6)$ and $(x_2, y_2) = (435.3, 311.3)$

$D = \sqrt{(x_2 - x_1)^2 + (y_2 - y_1)^2}$ Distance Formula

$= \sqrt{(435.3 - 212.0)^2 + (311.3 - 151.6)^2}$ Substitute.

$= \sqrt{223.3^2 + 159.7^2}$ Subtract.

$= \sqrt{49{,}862.89 + 25{,}504.09}$ Square each term.

$= \sqrt{75{,}366.98}$ Add.

≈ 274.5 Take the positive square root.

Chelsea and Amie are approximately 274.5 feet apart.

Check

SNOWBOARDING Manuel wants to go snowboarding with his friend. The closest ski and snowboard resort is approximately 20 miles west and 50 miles north of his house. Manuel picks up his friend who lives 15 miles south and 10 miles east of Manuel's house. How far away are the two boys from the resort?

___?___ mi

Go Online You can complete an Extra Example online.

Practice

Go Online You can complete your homework online.

Example 1

Use the number line to find each measure.

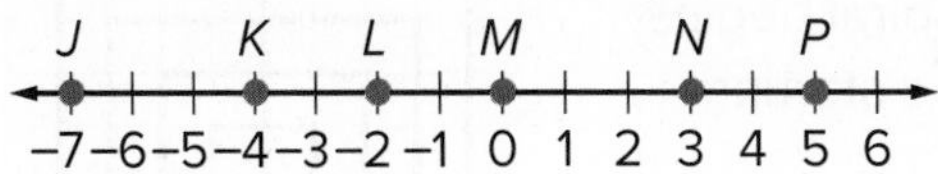

1. JL
2. JK
3. KP
4. NP
5. JP
6. LN

Use the number line to find each measure.

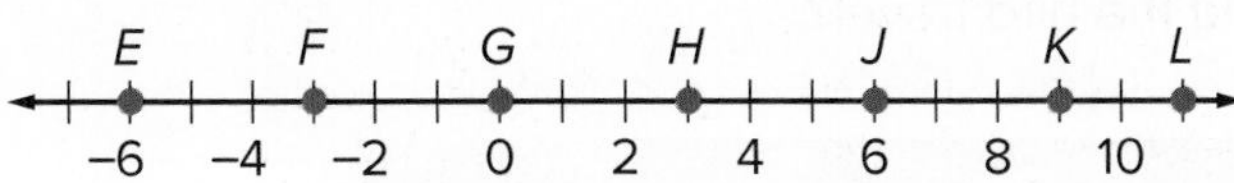

7. JK
8. LK
9. FG
10. JG
11. EH
12. LF

Use the number line to find each measure.

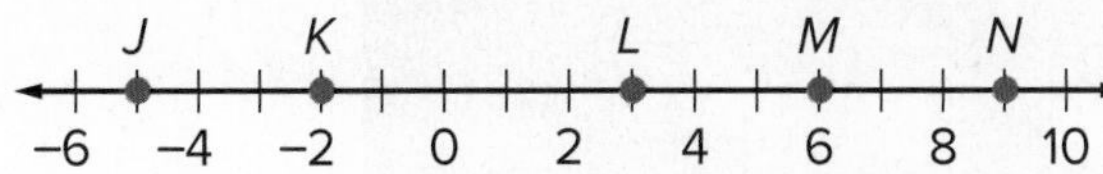

13. LN
14. JL

Example 2

Determine whether the given segments are congruent. Write *yes* or *no*.

15. $\overline{AB}$ and $\overline{EF}$
16. $\overline{BD}$ and $\overline{DF}$
17. $\overline{AC}$ and $\overline{CD}$
18. $\overline{AC}$ and $\overline{DE}$
19. $\overline{BE}$ and $\overline{CF}$
20. $\overline{CD}$ and $\overline{DF}$

Example 3

Find the distance between each pair of points.

21.

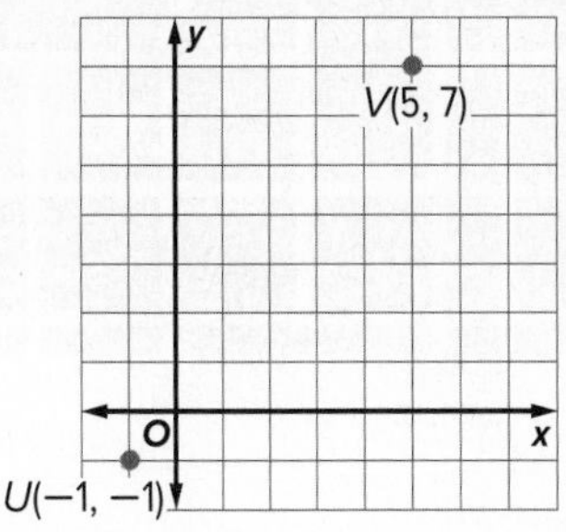

22.

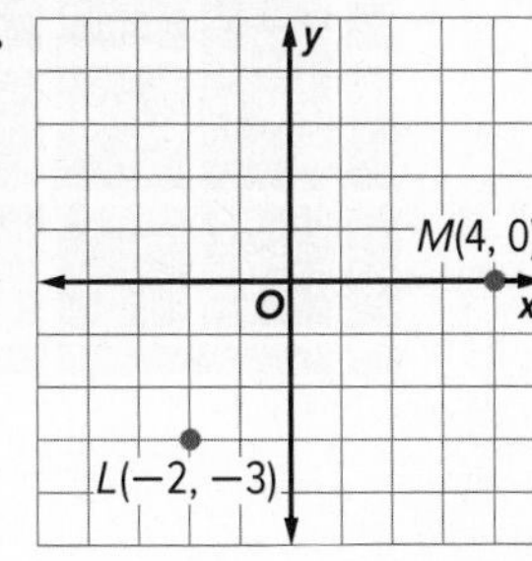

23.

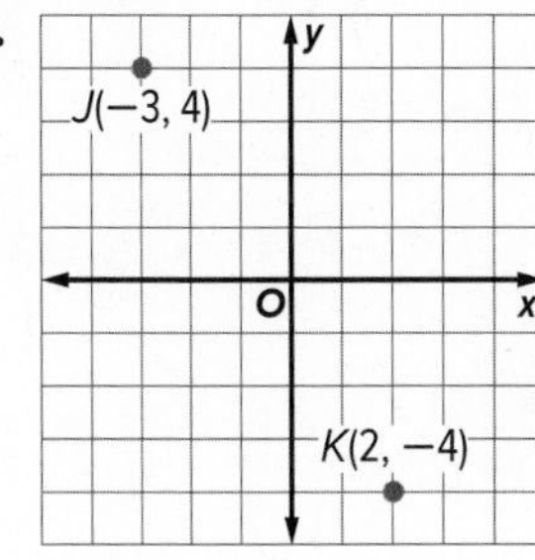

24. $A(2, 6), N(5, 10)$

25. $R(3, 4), T(7, 2)$

26. $X(-3, 8), Z(-5, 1)$

Example 4

27. **SPIRALS** Denise traces the spiral shown in the figure. The spiral begins at the origin. What is the shortest distance between Denise's starting point and her ending point?

28. **ZOOLOGY** A tiny songbird called the blackpoll warbler migrates each fall from North America. A tracking study showed one bird flew from Vermont at map coordinates (63, 45) to Venezuela at map coordinates (67, 10) in three days. If each map coordinate represents 75 kilometers, how far did the bird travel?

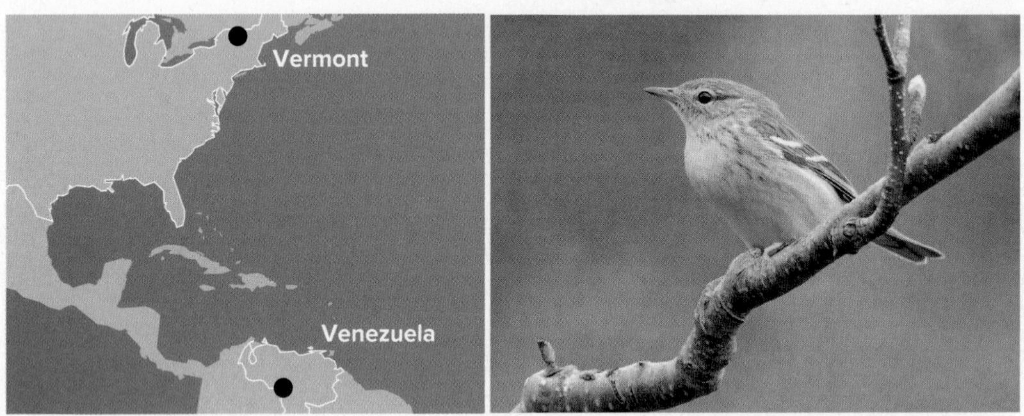

29. **CONSTRUCT ARGUMENTS** Mariah is training for a sprint-distance triathlon. She plans on cycling from her house to the library, shown on the grid with a scale in miles. If the cycling portion of the triathlon is 12 miles, will Mariah have cycled at least $\frac{2}{3}$ of that distance during her bike ride? Justify your argument.

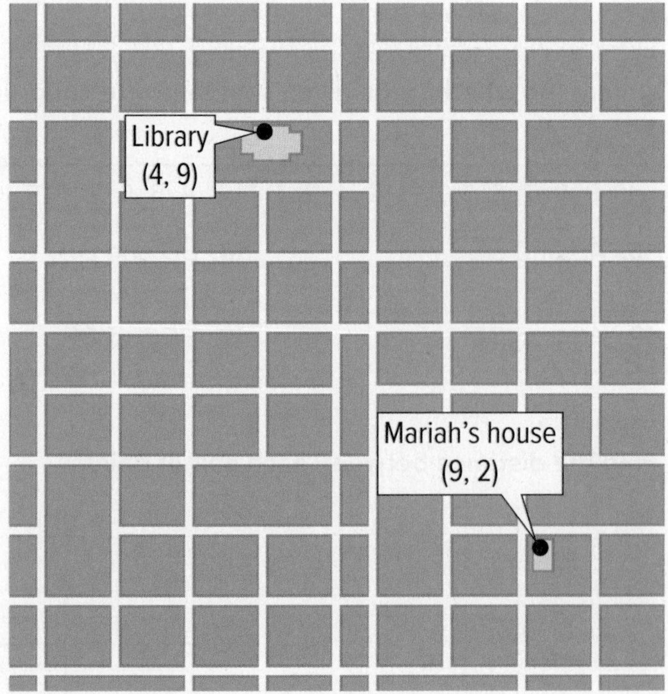

30. SPORTS The distance between each base on a baseball infield is 90 feet. The third baseman throws a ball from third base to point P. To the nearest foot, how far did the player throw the ball?

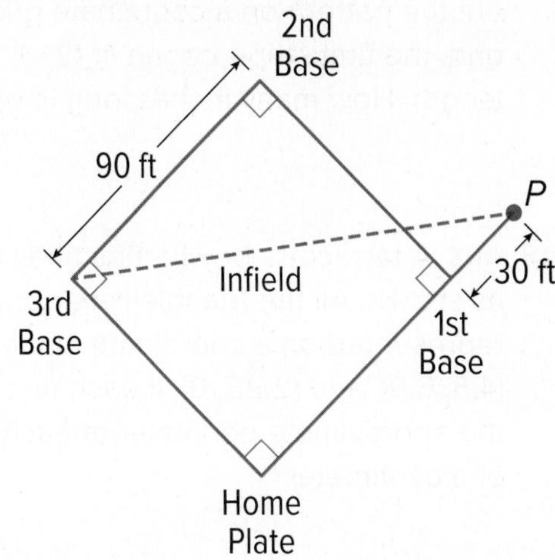

Mixed Exercises

Find the distance between each pair of points. Round to the nearest tenth, if necessary.

31. $M(-4, 9)$, $N(-5, 3)$

32. $C(2, 4)$, $D(5, 7)$

33. $A(5, 1)$, $B(3, 6)$

34. $V(4, 4)$, $X(5, 8)$

35. $S(6, 4)$, $T(3, 2)$

36. $M(-1, 8)$, $N(-3, 3)$

37. $W(-8, 1)$, $Y(0, 6)$

38. $B(3, -4)$, $C(5, -5)$

39. $R(6, 11)$, $T(3, -7)$

40. $A(-3, 8)$ and $B(-1, 4)$

41. $M(4, -3)$ and $N(-2, 1)$

42. $X(-3, 5)$ and $Y(4, 2)$

43. Use the number line to determine whether $\overline{SV}$ and $\overline{UX}$ are congruent. Write *yes* or *no*.

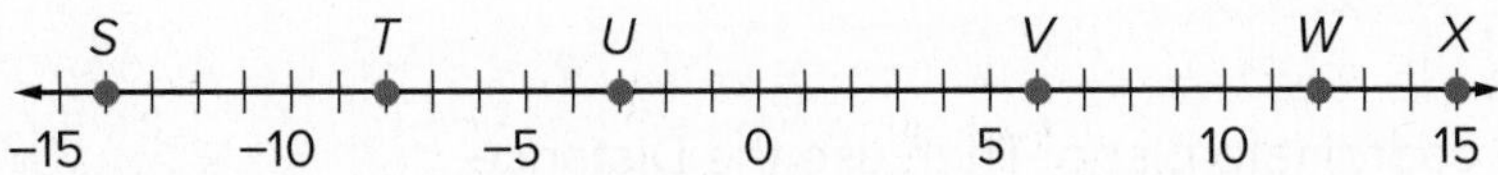

Name the point(s) that satisfy the given condition.

44. two points on the x-axis that are 10 units from (1, 8)

45. two points on the y-axis that are 25 units from (–24, 3)

46. Refer to the figure. Are $\overline{VT}$ and $\overline{SU}$ congruent?

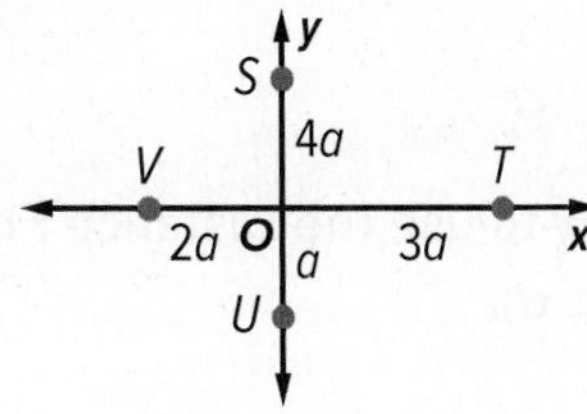

47. KNITTING Mei is knitting a scarf with diagonal stripes. Before she began, she laid out the pattern on a coordinate grid where each unit represented 2 inches. On the grid, the first stripe began at (2, 0) and ended at (5, 4). All the stripes are the same length. How many inches long is each stripe on the scarf?

48. ART A terracotta bowl artifact has a triangular pattern around the top, as shown. All the triangles are about the same size and can be represented on a coordinate plane with vertices at points (0, 6.8), (4.5, 6.8), and (2.25, 0). If each unit represents 1 centimeter, what is the approximate perimeter of each triangle, to the nearest tenth of a centimeter?

Higher-Order Thinking Skills

49. ANALYZE Consider rectangle $QRST$ with $QR = ST = 4$ centimeters and $RS = QT = 2$ centimeters. If point U is on $\overline{QR}$ such that $QU = UR$ and point V is on $\overline{RS}$ such that $RV = VS$, then is $\overline{QU}$ congruent to $\overline{RV}$? Justify your argument.

50. WRITE Explain how the Pythagorean Theorem and the Distance Formula are related.

51. PERSEVERE Point P is located on the segment between point $A(1, 4)$ and point $D(7, 13)$. The distance from A to P is twice the distance from P to D. What are the coordinates of point P?

52. CREATE Plot points Y and Z on a coordinate plane. Then use the Distance Formula to find YZ.

53. PERSEVERE Suppose point A is located at (1, 3) on a coordinate plane. If AB is 10 and the x-coordinate of point B is 9, explain how to use the Distance Formula to find the y-coordinate of point B.

54. WRITE Explain how to use the Distance Formula to find the distance between points (a, b) and (c, d).

Locating Points on a Number Line

Today's Goals
- Find a point on a directed line segment on a number line that is a given fractional distance from the initial point.
- Find a point that partitions a directed line segment on a number line in a given ratio.

Today's Vocabulary
directed line segment
fractional distance

Explore Locating Points on a Number Line with Fractional Distance

Online Activity Use dynamic geometry software to complete the Explore.

INQUIRY What general method can you use to locate a point some fraction of the distance from one point to another point on a number line?

Learn Locating Points on a Number Line with Fractional Distance

While a line segment has two endpoints, a **directed line segment** has an initial endpoint and a terminal endpoint.

Using a directed line segment enables you to calculate the coordinate of an intermediary point some fraction of the length of the segment, or **fractional distance,** from the initial endpoint.

Key Concept • Locating a Point at Fractional Distances on a Number Line

Find the coordinate of a point that is $\frac{a}{b}$ of the distance from point C to point D.

Step 1 Calculate the difference of the coordinates of point C and point D.	C, D, x_1, x_2 $(x_2 - x_1)$
Step 2 Multiply the difference by the given fraction. The fractional distance is given by $\frac{a}{b}(x_2 - x_1)$.	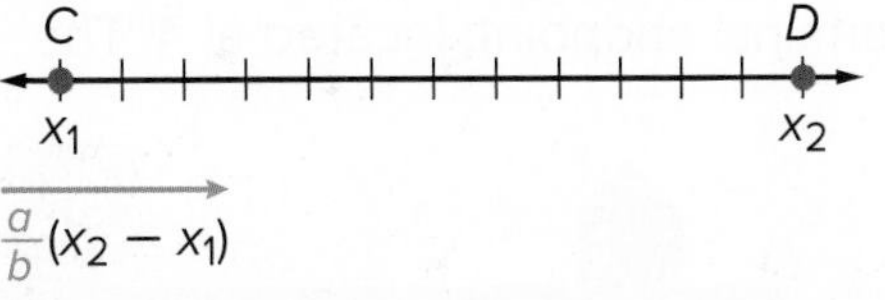
Step 3 Add the fractional distance to the coordinate of the initial point x_1. The coordinate of point P is given by $x_1 + \frac{a}{b}(x_2 - x_1)$.	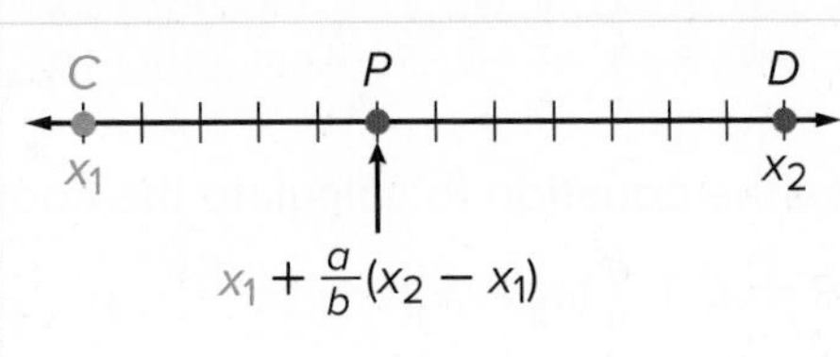

The coordinate of a point on a line segment with endpoints x_1 and x_2 is given by $x_1 + \frac{a}{b}(x_2 - x_1)$, where $\frac{a}{b}$ is the fraction of the distance.

Watch Out!

Don't Use Absolute Value When finding the distance from an initial endpoint to a terminal endpoint on a directed line segment, don't use absolute value. The difference created by $(x_2 - x_1)$ can be positive or negative. The sign of the difference will indicate the direction of the directed line segment.

Talk About It!

In the Key Concept, what phrase helped you identify the initial endpoint? What phrase helped you identify the terminal endpoint?

Example 1 Locate a Point at a Fractional Distance

Find B on $\overline{AC}$ that is $\frac{1}{4}$ of the distance from A to C.

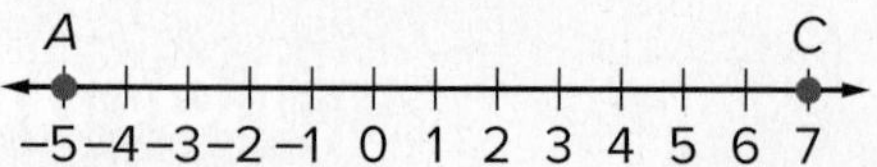

Point A is the initial endpoint, and point C is the terminal endpoint.

A C

−5 −4 −3 −2 −1 0 1 2 3 4 5 6 7

Use the equation to calculate the coordinate of point B.

$B = x_1 + \frac{a}{b}(x_2 - x_1)$	Coordinate equation
$= -5 + \frac{1}{4}(7 - (-5))$	$x_1 = -5$, $x_2 = 7$, and $\frac{a}{b} = \frac{1}{4}$
$= -2$	Simplify.

> **Think About It!**
> How would you check your solution?

Point B is located at −2 on the number line.

A B C

−5 −4 −3 −2 −1 0 1 2 3 4 5 6 7

Check

Find X on $\overline{BE}$ that is $\frac{3}{5}$ of the distance from B to E.

A. 2 **B.** 3 **C.** 5 **D.** 6

Example 2 Locate a Point at a Fractional Distance in the Real World

BIKING **Julio is biking from his house to the library. His house is 8 blocks west of the school, and the library is 4 blocks east of the school. If he stops to rest $\frac{1}{3}$ of the distance from his house to the library, at what point does he stop?**

Julio's house is the initial endpoint, located at −8, and the library is the terminal endpoint, located at 4. The school is at 0.

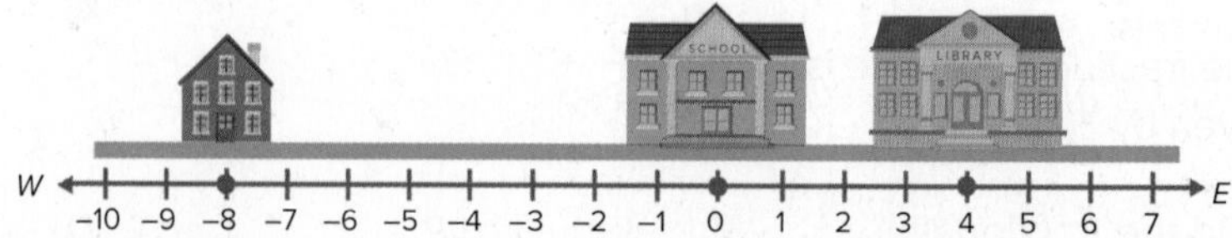

> **Think About It!**
> What would the coordinate be if Julio wanted to rest $\frac{1}{3}$ of the distance if he is going from the library to his house?

Use the equation to calculate the coordinate of Julio's resting point.

$B = x_1 + \frac{a}{b}(x_2 - x_1)$	Coordinate equation
$= -8 + \frac{1}{3}[4 - (-8)]$	$x_1 = -8$, $x_2 = 4$, and $\frac{a}{b} = \frac{1}{3}$
$= -4$	Simplify.

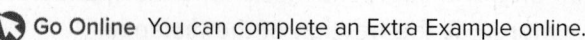

Go Online You can complete an Extra Example online.

Check

DECORATING Taji is hanging a picture $\frac{5}{8}$ of the distance from the floor to the ceiling. If the distance between the floor and the ceiling is 12 feet, how high should he hang the picture?

Learn Locating Points on a Number Line with a Given Ratio

You can calculate the coordinate of an intermediary point that partitions the directed line segment into a given ratio.

Key Concept • Section Formula on a Number Line

If C has coordinate x_1 and D has coordinate x_2, then a point P that partitions the line segment in a ratio of $m:n$ is located at coordinate $\frac{nx_1 + mx_2}{m + n}$, where $m \neq -n$.

$\frac{nx_1 + mx_2}{m + n}$

C P D

x_1 $m : n$ x_2

Example 3 Locate a Point on a Number Line When Given a Ratio

Find B on $\overline{AC}$ such that the ratio of AB to BC is 3:4.

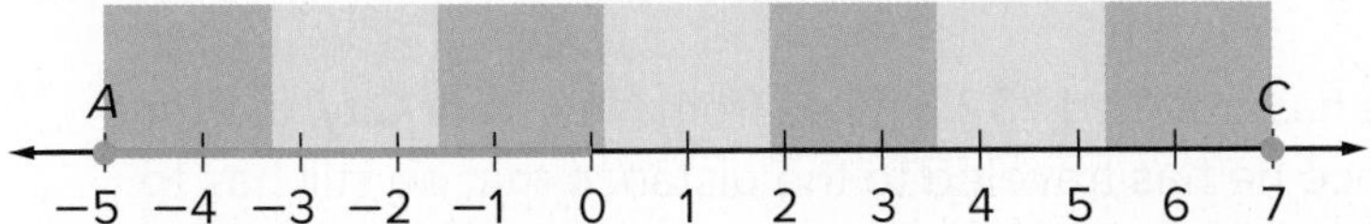

Use the Section Formula to determine the coordinate of point B.

$B = \frac{nx_1 + mx_2}{m + n}$ Section Formula

$= \frac{4(-5) + 3(7)}{3 + 4} = \frac{1}{7}$ $m = 3, n = 4, x_1 = -5$, and $x_2 = 7$

A B C

−5 −4 −3 −2 −1 0 1 2 3 4 5 6 7

So, B is located at $\frac{1}{7}$ on the number line.

(continued on the next page)

Go Online You may want to complete the Concept Check to check your understanding.

Study Tip

Checking Solutions When using the Section Formula, you can check your solution by converting the given ratio into a fraction. Use this fraction and the coordinate equation to find the fractional distance from your initial endpoint to your terminal endpoint. If you don't calculate the same coordinate, you have made an error.

Check

Find P on $\overline{AF}$ such that the ratio of AP to PF is 1:3.

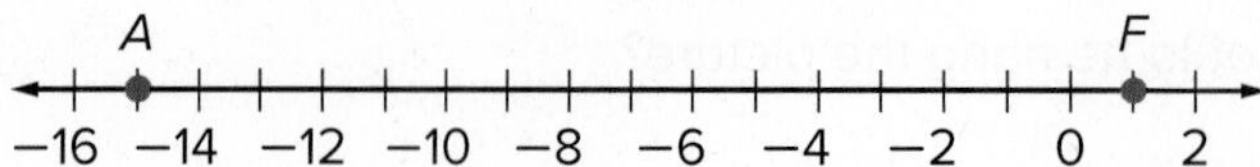

P is located at ___?___ on the number line.

Example 4 Partition a Directed Line Segment

Think About It!

How can you use estimation to check your answer?

ROAD TRIP **Jorge is traveling 2563 miles from New York City to San Francisco by car. He plans on stopping for gas when the ratio of the distance he has already traveled to the distance he still has to travel is 2:5. How far has Jorge traveled when he stops for gas?**

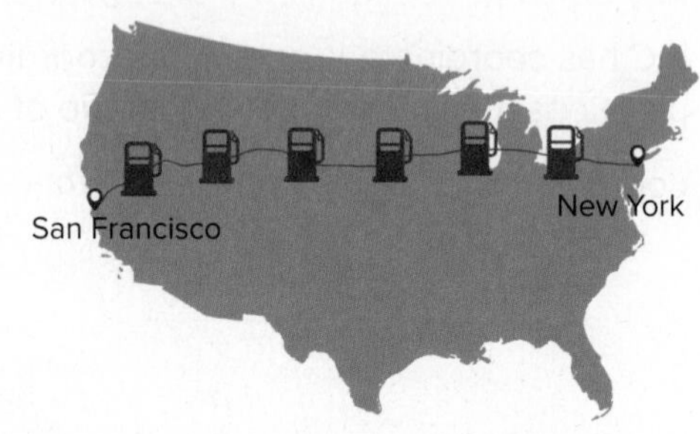

Use the Section Formula to determine how far Jorge will travel before he stops for gas.

$B = \dfrac{nx_1 + mx_2}{m + n}$ — Section Formula

$= \dfrac{5(0) + 2(2563)}{2 + 5} = 732.3$ — $m = 2, n = 5, x_1 = 0,$ and $x_2 = 2563$

When Jorge has traveled 732.3 miles from New York City, the ratio of the distance he has traveled to the distance that he still has to travel is 2:5.

Check

ERRANDS Eduardo travels 30 miles from his house to the bike shop. When Eduardo goes to the bike shop, he always stops at a local pizza place that is along the way. The ratio of the distance Eduardo travels from his house to the pizza place to the distance he travels from the pizza place to the bike shop is 2:3.

How far is the pizza place from Eduardo's house?

___?___ mi

Go Online You can complete an Extra Example online.

Practice

Go Online You can complete your homework online.

Examples 1 and 3

Refer to the number line.

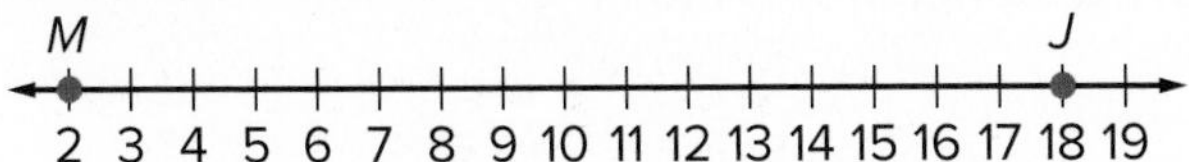

1. Find the coordinate of point B that is $\frac{1}{4}$ of the distance from M to J.

2. Find the coordinate of point C that is $\frac{7}{8}$ of the distance from M to J.

3. Find the coordinate of point D that is $\frac{7}{16}$ of the distance from M to J.

4. Find the coordinate of point X such that the ratio of MX to XJ is 3:1.

5. Find the coordinate of point X such that the ratio of MX to XJ is 2:3.

6. Find the coordinate of point X such that the ratio of MX to XJ is 1:1.

Refer to the number line.

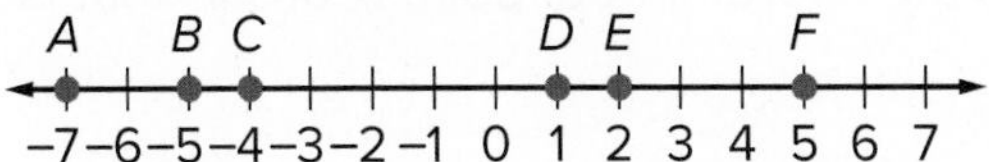

7. Find the coordinate of point G that is $\frac{2}{3}$ of the distance from B to D.

8. Find the coordinate of point H that is $\frac{1}{5}$ of the distance from C to F.

9. Find the coordinate of point J that is $\frac{1}{6}$ of the distance from A to E.

10. Find the coordinate of point K that is $\frac{4}{5}$ of the distance from A to F.

11. Find the coordinate of point X such that the ratio of AX to XF is 1:3.

12. Find the coordinate of point X such that the ratio of BX to XF is 3:2.

13. Find the coordinate of point X such that the ratio of CX to XE is 1:1.

14. Find the coordinate of point X such that the ratio of FX to XD is 5:3.

Refer to the number line.

A B C D F

−5 −4 −3 −2 −1 0 1 2 3 4 5

15. Find the coordinate of point X on $\overline{AF}$ that is $\frac{1}{3}$ of the distance from A to F.

16. Find the coordinate of point Y on $\overline{AC}$ that is $\frac{1}{4}$ of the distance from A to C.

Refer to the number line.

A W X Y Z E

−10 −9 −8 −7 −6 −5 −4 −3 −2 −1 0 1 2 3 4 5 6 7 8 9 10

17. Which point on $\overline{AE}$ is $\frac{2}{3}$ of the distance from A to E?

18. Point X is what fractional distance from E to A?

19. Find the coordinate of point M on $\overline{AE}$ that is $\frac{1}{5}$ of the distance from A to E.

Refer to the number line.

F G H J K L

−15 −10 −5 0 5

20. The ratio of FX to XK is 1:1. Which point is located at X?

21. Find the coordinate of Q on $\overline{FL}$ such that the ratio of FQ to QL is 12:7.

Examples 2 and 4

22. **TRAVEL** Caroline is taking a road trip on I-70 in Kansas. She stops for gas at mile marker 36. Her destination is at mile marker 353 in Topeka, but she decides to stop at an attraction $\frac{3}{4}$ of the way after stopping for gas. At about which mile marker did Caroline stop to visit the attraction?

23. HIKING A hiking trail is 24 miles from start to finish. There are two rest areas located along the trail.

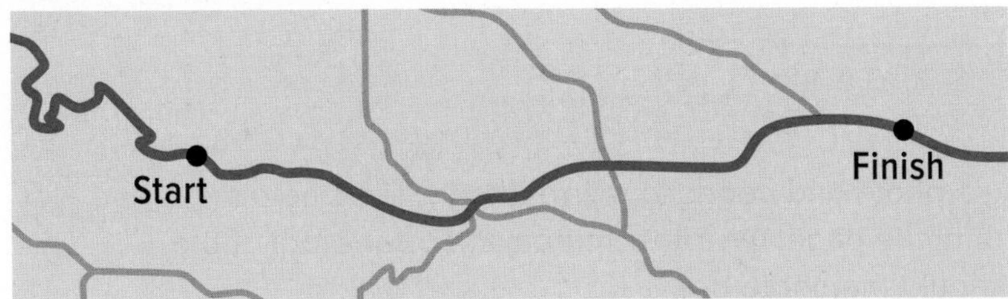

a. The first rest area is located such that the ratio of the distance from the start of the trail to the rest area and the distance from the rest area to the end of the trail is 2:9. To the nearest hundredth of a mile, how far is the first rest area from the starting point of the trail?

b. Kadisha claims that the distance she has walked and that the distance she has left to walk has a ratio of 5:7. How many miles has Kadisha walked?

24. Melany wants to hang a canvas, which is 8 feet wide, on his wall. Where on the canvas should Melany mark the location of the hangers if the canvas requires a hanger every $\frac{1}{5}$ of its length, excluding the edges? Justify your answer.

25. MIGRATION Many American White Pelicans migrate each year, with hundreds of them stopping to rest in various locations along the way. The ratio of the distance some flocks travel from their summer home to one stopover to the distance from the stopover to the winter home is 3:4. If the total distance that the pelicans migrate is 1680 miles, how long is the distance from the summer home to the stopover?

Mixed Exercises

26. Write an equation that can be used to find the coordinate of point K that is $\frac{2}{5}$ of the distance from Q to R.

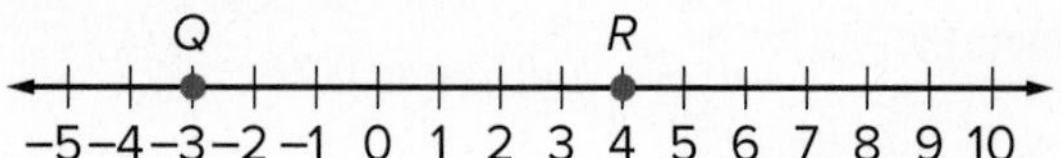

27. **SOCIAL MEDIA** Tito is posting a photo and needs to resize it to fit. The photo's width should fill $\frac{4}{5}$ of the width of the page. On Tito's screen, the total width of the page is 3 inches. How wide should the photo be?

28. **NEONATAL** At birth, the ratio of a baby's head length to the length of the rest of its body is 1:3. If a baby's total body length is 22 inches, how long is the baby's head?

29. **CREATE** Draw a segment and label it $\overline{AB}$. Using only a compass and a straightedge, construct a segment $\overline{CD}$ such that $CD = 5\frac{1}{4}AB$. Explain and then justify your construction.

30. **WRITE** Naoki wants to center a canvas, which is 8 feet wide, on his bedroom wall, which is 17 feet wide. Where on the wall should Naoki mark the location of the nails, if the canvas requires nails every $\frac{1}{5}$ of its length, excluding the edges? Explain your solution process.

31. **ANALYZE** Determine whether the following statement is *sometimes, always,* or *never* true. Justify your argument.

If $\overline{XY}$ is on a number line and point W is $\frac{2}{5}$ of the distance from X to Y, then the coordinate of point W is greater than the coordinate of point X.

32. **PERSEVERE** On a number line, point A is at 5, and point B is at −10. Point C is on $\overline{AB}$ such that the ratio of AC to CB is 1:3. Find D on $\overline{BC}$ that is $\frac{3}{8}$ of the distance from B to C.

Locating Points on a Coordinate Plane

Explore Applying Fractional Distance

Online Activity Use a real-world situation to complete the Explore.

INQUIRY How do we use fractional distances in the real world?

Learn Locating Points on the Coordinate Plane with Fractional Distance

You can find a point on a directed line segment that is a fractional distance from an endpoint on the coordinate plane.

Key Concept • Locating a Point at a Fractional Distance on the Coordinate Plane

The coordinates of a point on a line segment that is $\frac{a}{b}$ of the distance from initial endpoint $A(x_1, y_1)$ to terminal endpoint $C(x_2, y_2)$ are given by $\left(x_1 + \frac{a}{b}(x_2 - x_1), y_1 + \frac{a}{b}(y_2 - y_1)\right)$, where $\frac{a}{b}$ is the fraction of the distance if $b \neq 0$.

Example 1 Fractional Distances on the Coordinate Plane

Find C on $\overline{AB}$ that is $\frac{3}{4}$ of the distance from A to B.

Step 1 Identify the endpoints.

Identify the initial and terminal endpoints.

$(x_1, y_1) = (-7, -5)$ and $(x_2, y_2) = (6, 8)$

Step 2 Find the x- and y-coordinates.

Find the coordinates of C using the formula for fractional distance.

$\left(x_1 + \frac{a}{b}(x_2 - x_1), y_1 + \frac{a}{b}(y_2 - y_1)\right)$ — Fractional Distance Formula

$\left(-7 + \frac{3}{4}[6 - (-7)], -5 + \frac{3}{4}[8 - (-5)]\right)$ — Substitution

Point C is located at (2.75, 4.75).

Go Online You can complete an Extra Example online.

Today's Goals

- Find a point on a directed line segment on the coordinate plane that is a given fractional distance from the initial point.
- Find a point that partitions a directed line segment on the coordinate plane in a given ratio.

Watch Out!

Determine the Initial Endpoint Direction is important when determining a point that is a fractional distance on a directed line segment. Identify the initial endpoint you move from and the terminal endpoint you move toward.

Study Tip

Checking Coordinates You can check that you have computed the coordinates of C correctly by finding the lengths of $\overline{AC}$ and $\overline{AB}$. If $\frac{AC}{AB}$ is not equal to $\frac{3}{4}$, then you have made an error.

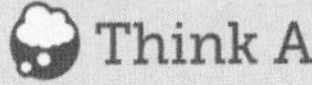

Think About It!

What are the coordinates of a point that is $\frac{3}{4}$ of the distance from B to A?

Check

Find P on $\overline{QR}$ that is $\frac{1}{6}$ of the distance from Q to R.

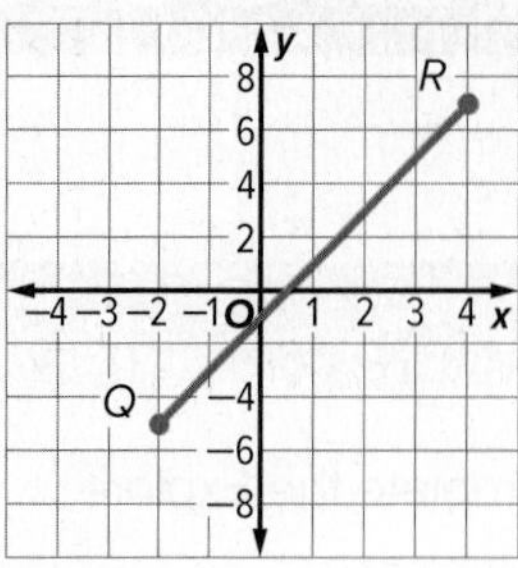

Coordinates of point P ___?___

Learn Locating Points on the Coordinate Plane with a Given Ratio

The Section Formula can be used to locate a point that partitions a directed line segment on the coordinate plane.

Key Concept • Section Formula on the Coordinate Plane

If A has coordinates (x_1, y_1) and C has coordinates (x_2, y_2), then a point B that partitions the line segment in a ratio of $m{:}n$ has coordinates $B\left(\frac{nx_1 + mx_2}{m+n}, \frac{ny_1 + my_2}{m+n}\right)$, where $m \neq n$.

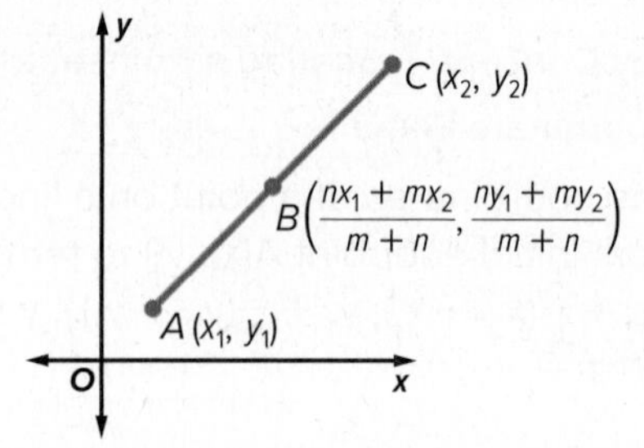

Example 2 Locate a Point on the Coordinate Plane When Given a Ratio

Find C on $\overline{AB}$ such that the ratio of AC to CB is 1:2.

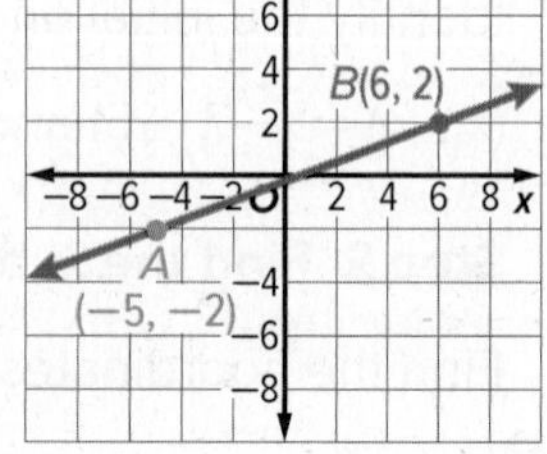

Use the Section Formula to determine the coordinates of point C.

$\left(\frac{nx_1 + mx_2}{m+n}, \frac{ny_1 + my_2}{m+n}\right)$ Section Formula

$= \left(\frac{2(-5) + 1(6)}{1+2}, \frac{2(-2) + 1(2)}{1+2}\right)$ Substitute.

$= \left(-\frac{4}{3}, -\frac{2}{3}\right)$ Simplify.

Point C is located at $\left(-\frac{4}{3}, -\frac{2}{3}\right)$.

Talk About It!

How could you check the coordinates of point C?

Check

Find S on $\overline{QR}$ such that the ratio of QS to SR is 2:1.

A. (4, 8)

B. (2, 3)

C. (1, 1)

D. (0, –1)

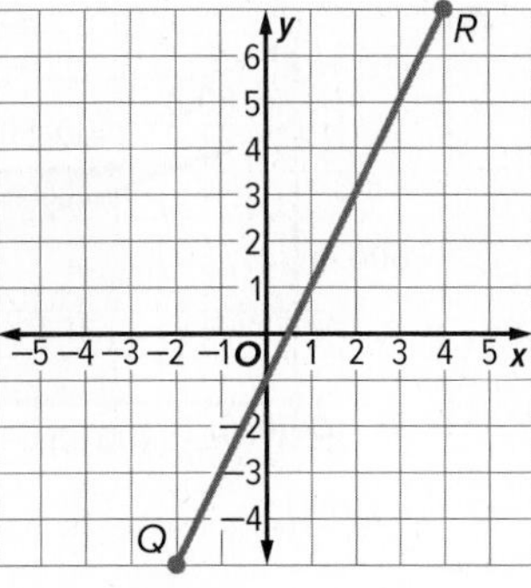

Example 3 Partition a Directed Line Segment on the Coordinate Plane

ZIP LINES **Kendrick is riding a zip line. The zip line is 1800 meters long and starts at a platform 600 meters above the ground. After he jumps, someone takes a picture of his descent. When the picture is taken, the ratio of the distance Kendrick has traveled to the distance he has remaining is 1:2. The picture will show the horizontal distance from 400 meters to 1200 meters from the base of the platform and the vertical distance from ground level to a height of 500 meters. Will Kendrick be in the frame of the picture?**

To determine whether Kendrick is in the frame of the picture, first, determine the horizontal distance x of the zip line. Then, use this information to determine Kendrick's location using the Section Formula.

Step 1 Determine the horizontal distance x of the zip line.

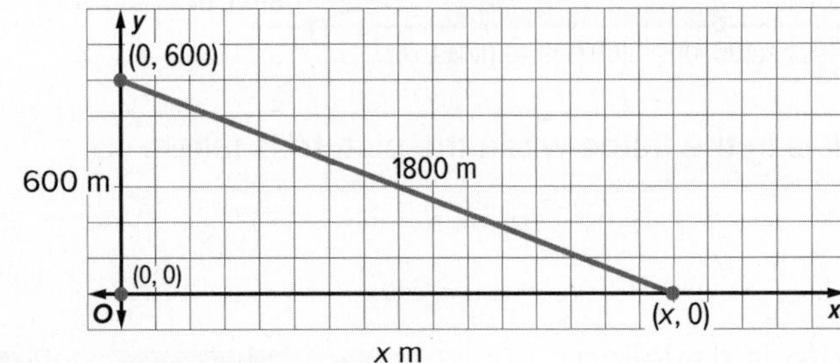

$a^2 + b^2 = c^2$	Pythagorean Theorem
$600^2 + x^2 = 1800^2$	Substitute.
$x \approx 1697.1$	Solve.

The horizontal distance of the zip line is about 1697.1 meters.

(continued on the next page)

Go Online You can complete an Extra Example online.

Step 2 Model the area captured by the photograph.

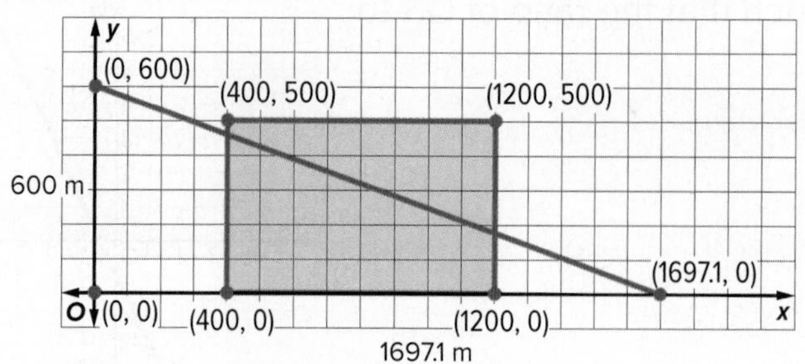

Step 3 Determine Kendrick's location on the zip line.

Use the Section Formula to calculate Kendrick's coordinates.

$\left(\frac{nx_1 + mx_2}{m + n}, \frac{ny_1 + my_2}{m + n}\right)$ Section Formula

$= \left(\frac{2(0) + 1(1697.1)}{1 + 2}, \frac{2(600) + 1(0)}{1 + 2}\right)$ Substitute.

$= (565.7, 400)$ Simplify.

Kendrick is at (565.7, 400) when the picture is taken.

Step 4 Graph Kendrick's location to determine whether he is in the frame.

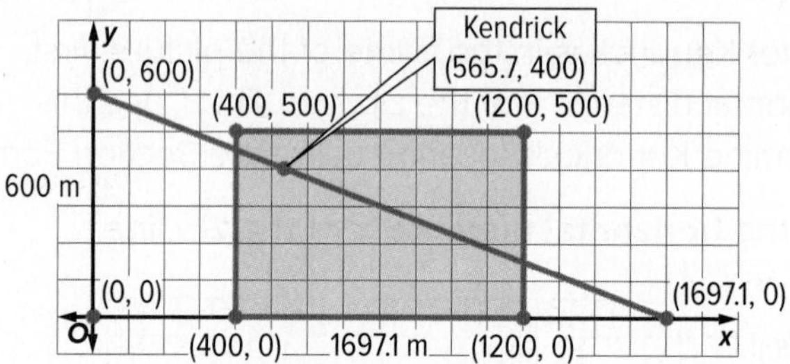

Yes. Kendrick is in the frame when the picture is taken.

Check

TRAVEL Andre is traveling from Jeffersonville to Springfield. He plans to stop for a break when the distance he has traveled and the distance he has left to travel have a ratio of 3:7. Where should Andre stop for his break?

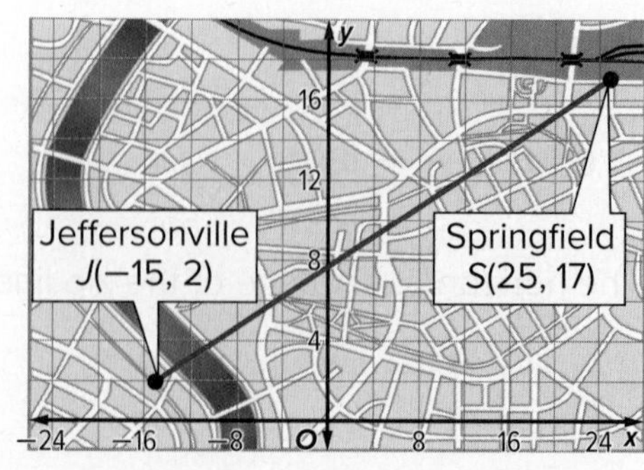

A. (13, 12.5) B. (22, 12.5) C. (−3, 6.5) D. (−12, 6.5)

Go Online You can complete an Extra Example online.

Practice

Go Online You can complete your homework online.

Example 1

Find the coordinates of point *X* on the coordinate plane for each situation.

1. Point *X* on $\overline{AB}$ is $\frac{1}{5}$ of the distance from *A* to *B*.

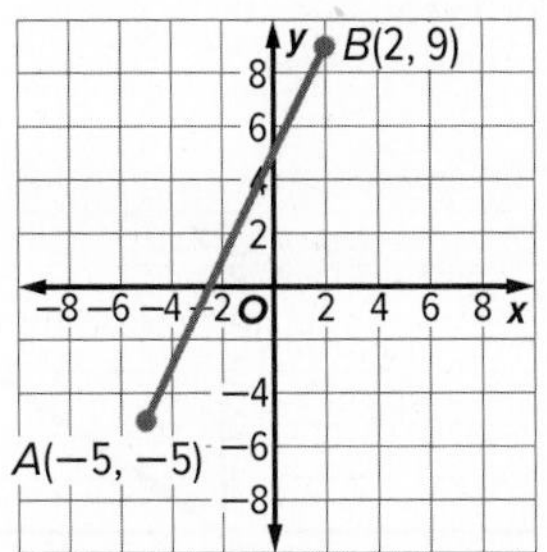

2. Point *X* on $\overline{RS}$ is $\frac{1}{6}$ of the distance from *R* to *S*.

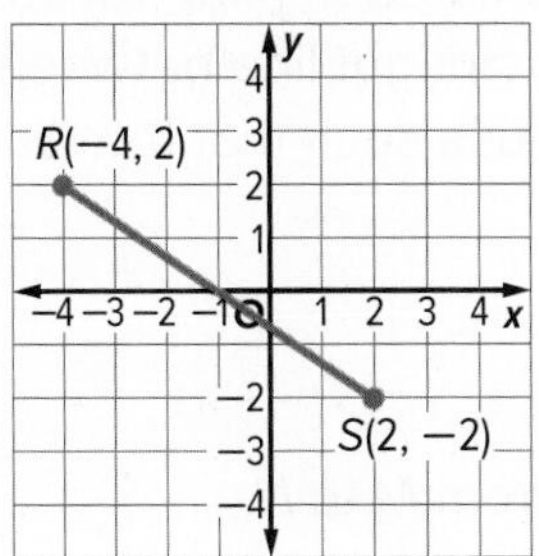

3. Point *X* on $\overline{JK}$ is $\frac{1}{3}$ of the distance from *J* to *K*.

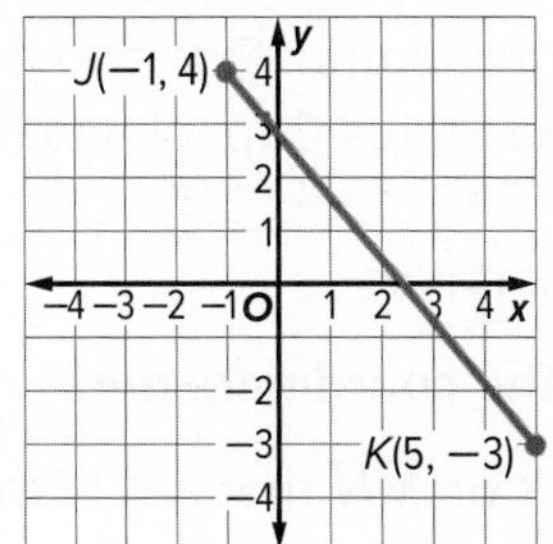

Example 2

Refer to the coordinate grid.

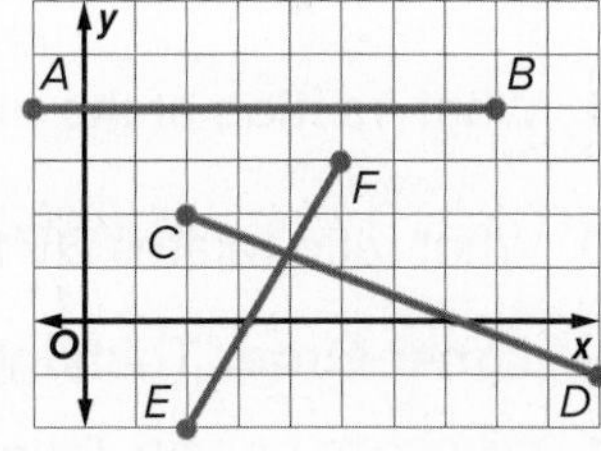

4. Find point *X* on $\overline{AB}$ such that the ratio of *AX* to *XB* is 1:3.

5. Find point *Y* on $\overline{CD}$ such that the ratio of *DY* to *YC* is 2:1.

6. Find point *Z* on $\overline{EF}$ such that the ratio of *EZ* to *ZF* is 2:3.

Examples 1 and 2

Refer to the coordinate grid.

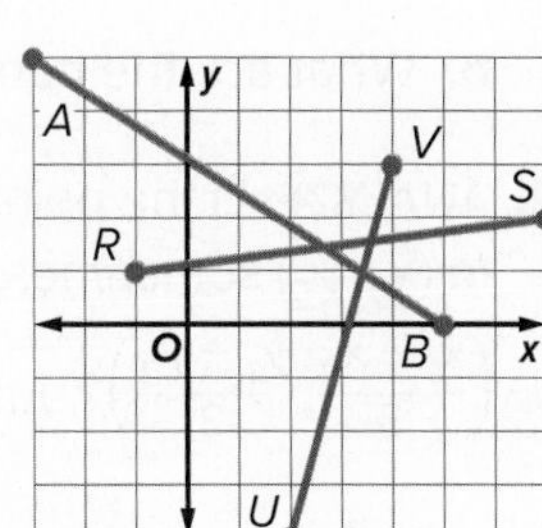

7. Find point *C* on $\overline{AB}$ that is $\frac{1}{5}$ of the distance from *A* to *B*.

8. Find point *Q* on $\overline{RS}$ that is $\frac{5}{8}$ of the distance from *R* to *S*.

9. Find point *W* on $\overline{UV}$ that is $\frac{1}{7}$ of the distance from *U* to *V*.

10. Find point *D* on $\overline{AB}$ that is $\frac{3}{4}$ of the distance from *A* to *B*.

11. Find point *Z* on $\overline{RS}$ such that the ratio of *RZ* to *ZS* is 1:3.

12. Find point *G* on $\overline{AB}$ such that the ratio of *AG* to *GB* is 3:2.

13. Find point *E* on $\overline{UV}$ such that the ratio of *UE* to *EV* is 3:4.

Example 3

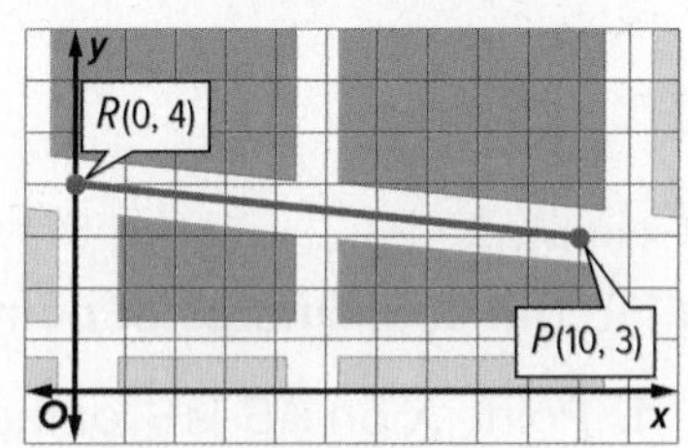

14. MAPS Leila is walking from the park at point P to a restaurant at point R. She wants to stop for a break when the distance she has traveled and the distance she has left to travel has a ratio of 3:5. At which point should Leila stop for her break?

15. CITY PLANNING The United States Capitol is located at (2, −4) on a coordinate grid. The White House is located at (−10, 16) on the same coordinate grid. Find two points on the straight line between the United States Capitol and the White House such that the ratio is 1:3.

Mixed Exercises

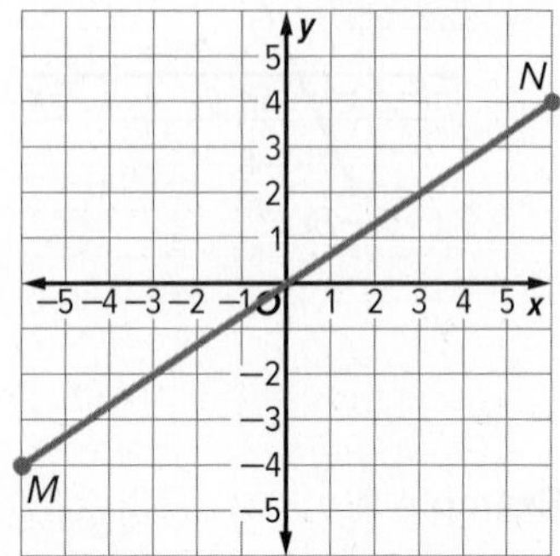

Refer to the coordinate grid.

16. Find X on $\overline{MN}$ that is $\frac{3}{4}$ of the distance from M to N.

17. Find Y on $\overline{MN}$ such that the ratio of MY to YN is 1:3.

Point D is located on $\overline{MV}$. The coordinates of D are $\left(0, -\frac{3}{4}\right)$.

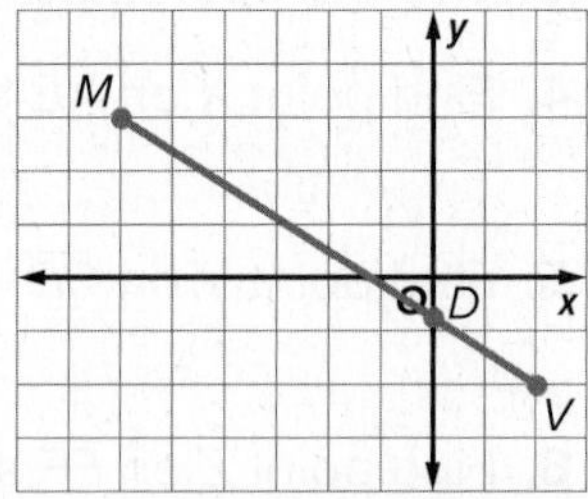

18. What ratio relates MD to DV?

19. What fraction of the distance from M to V is MD?

20. What ratio relates DV to MD?

Higher-Order Thinking Skills

21. FIND THE ERROR Point W is located at (0, 7), and point X is located at (4, 0). Julianne wants to find point F on $\overline{WX}$ such that the ratio of WF to FX is 2:3.

a. What error did Julianne make when solving this problem?

b. What are the correct coordinates of point F?

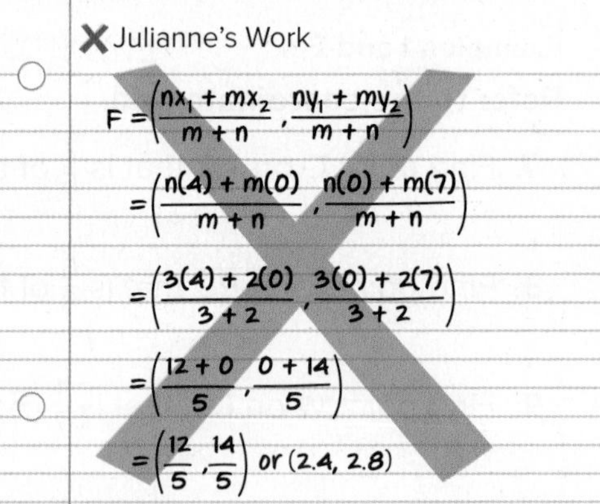

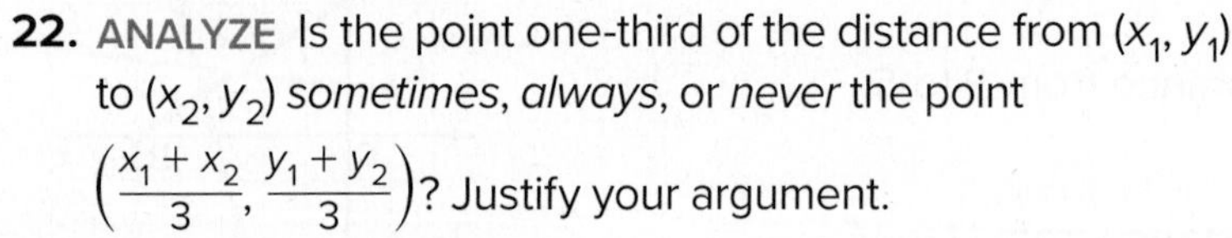

22. ANALYZE Is the point one-third of the distance from (x_1, y_1) to (x_2, y_2) *sometimes*, *always*, or *never* the point $\left(\frac{x_1 + x_2}{3}, \frac{y_1 + y_2}{3}\right)$? Justify your argument.

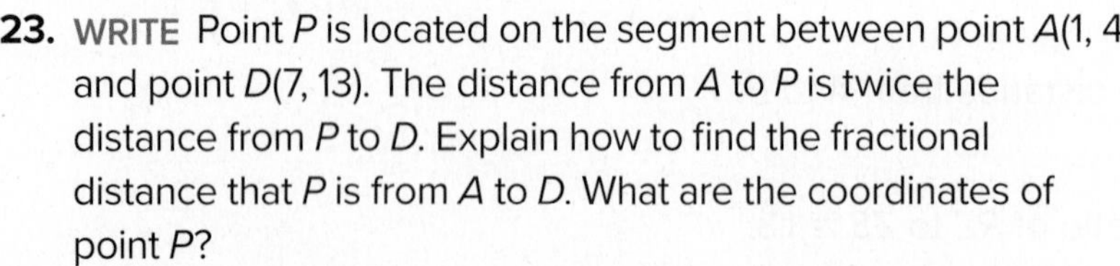

23. WRITE Point P is located on the segment between point $A(1, 4)$ and point $D(7, 13)$. The distance from A to P is twice the distance from P to D. Explain how to find the fractional distance that P is from A to D. What are the coordinates of point P?

24. PERSEVERE Point $C(6, 9)$ is located on the segment between point $A(4, 8)$ and point B. Point C is $\frac{1}{4}$ of the distance from A to B. What are the coordinates of point B?

25. CREATE Draw a line on a coordinate plane. Label two points on the line F and G. Locate a third point on the line between points F and G and label this point H. The point H on $\overline{FG}$ is what fractional distance from F to G?

Midpoints and Bisectors

Explore Midpoints

Online Activity Use paper folding to complete the Explore.

> INQUIRY What general formula can you use to find the midpoint of a line segment?

Learn Midpoints on a Number Line

The **midpoint** of a segment is the point halfway between the endpoints of the segment. A point is **equidistant** from other points if it is the same distance from them. The midpoint separates the segment into two segments with a ratio of 1:1. So, you can use the Section Formula to derive the Midpoint Formula.

Key Concept • Midpoint on a Number Line

If $\overline{AB}$ has endpoints at x_1 and x_2 on a number line, then the midpoint M of $\overline{AB}$ has coordinate $M = \frac{x_1 + x_2}{2}$.

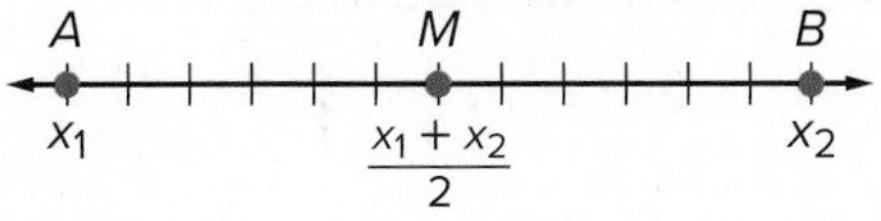

Example 1 Find the Midpoint on a Number Line

What is the midpoint of $\overline{XZ}$?

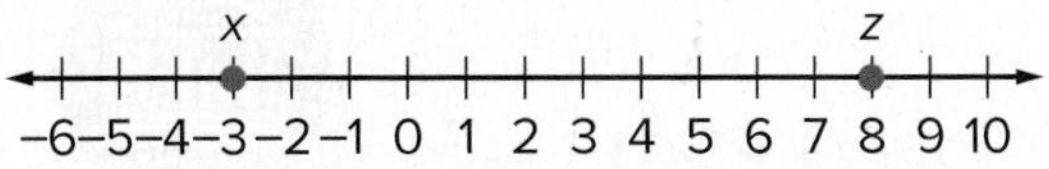

$M = \frac{x_1 + x_2}{2}$ Midpoint Formula

$= \frac{8 + (-3)}{2}$ Substitution

$= \frac{5}{2}$ or 2.5 Simplify.

The midpoint of $\overline{XZ}$ is 2.5.

Go Online You can complete an Extra Example online.

Today's Goals
- Find the coordinate of a midpoint on a number line.
- Find the coordinates of the midpoint or endpoint of a line segment on the coordinate plane.
- Find missing values using the definition of a segment bisector.

Today's Vocabulary
midpoint
equidistant
bisect
segment bisector

Watch Out!

Ratios Remember that 1:1 refers to the ratio of the distances, not to the measures of the segments.

Think About It!

Would your answer be different if you reversed the order of x_1 and x_2?

Check

What is the midpoint of $\overline{AF}$?

Example 2 Midpoints in the Real World

Think About It!

How else could Aponi have located the midpoint?

SIGNS **Aponi works at a vintage clothing store. She wants to hang a new sign so it is centered above the dressing-room doors. Given that the dressing-room doors have the same width, find the point along the wall that Aponi should hang the new sign.**

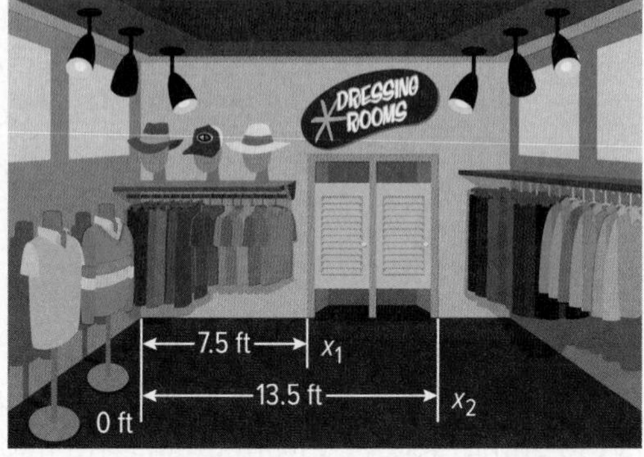

$M = \frac{x_1 + x_2}{2}$ Midpoint Formula

$= \frac{7.5 + (13.5)}{2}$ Substitution

$= \frac{21}{2}$ or 10.5 Simplify.

Aponi should hang the sign 10.5 feet from the left side of the wall.

Check

DISTANCE Jorge travels from his school on 38th Street to the library on 62nd Street. He stops halfway there to take a break. Where does Jorge stop to rest?

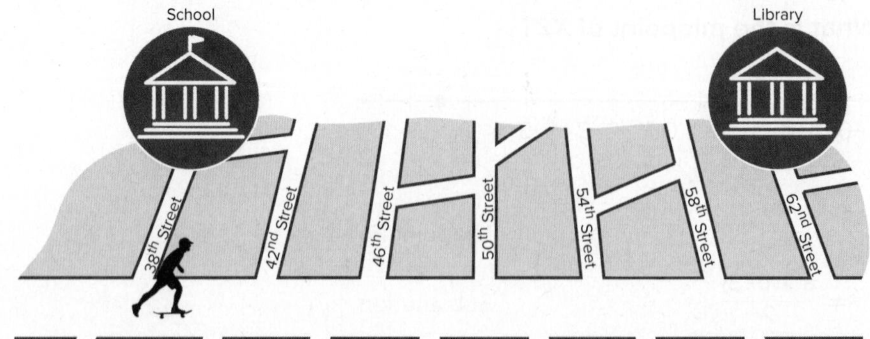

Jorge stops at ____?____.

Go Online You can complete an Extra Example online.

Learn Midpoints on the Coordinate Plane

The Section Formula can be used to derive the Midpoint Formula for a segment on the coordinate plane.

Because the midpoint separates the line segment into a ratio of 1:1, substitute 1 for m and n into the formula.

$M = \left(\frac{nx_1 + mx_2}{m+n}, \frac{ny_1 + my_2}{m+n}\right)$ Section Formula

$= \left(\frac{(1)x_1 + (1)x_2}{1+1}, \frac{(1)y_1 + (1)y_2}{1+1}\right)$ Substitution

$= \left(\frac{x_1 + x_2}{2}, \frac{y_1 + y_2}{2}\right)$ Midpoint Formula

Key Concept • Midpoint Formula on the Coordinate Plane

If $\overline{PQ}$ has endpoints at $P(x_1, y_1)$ and $Q(x_2, y_2)$ on the coordinate plane, then the midpoint M of $\overline{PQ}$ has coordinates $M\left(\frac{x_1 + x_2}{2}, \frac{y_1 + y_2}{2}\right)$.

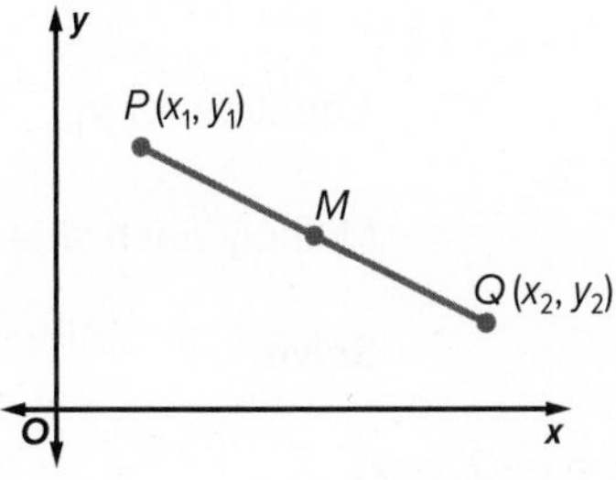

Example 3 Find the Midpoint on the Coordinate Plane

Find the coordinates of M, the midpoint of $\overline{AB}$, for $A(-2, 1)$ and $B(8, 3)$.

$M = \left(\frac{x_1 + x_2}{2}, \frac{y_1 + y_2}{2}\right)$ Midpoint Formula

$= \left(\frac{-2 + 8}{2}, \frac{1 + 3}{2}\right)$ Substitution

$= \left(\frac{6}{2}, \frac{4}{2}\right)$ or (3, 2) Simplify.

Check

Find the coordinates of B, the midpoint of $\overline{AC}$, for $A(-3, -2)$ and $C(5, 10)$.

(__?__, __?__)

Talk About It!

Would the coordinates of the midpoint be different if you use point A as (x_2, y_2) and point B as (x_1, y_1)? Explain.

Go Online You can complete an Extra Example online.

Watch Out!

Midpoint Formula The Midpoint Formula only uses addition and division. Think of the midpoint as the average of the *x*- and *y*-coordinates of the given endpoints.

Study Tip

Check for Reasonableness Always graph the given information and the calculated coordinates of the midpoint to check the reasonableness of your answer.

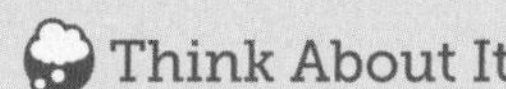

How can you use the graph to determine whether your answer is reasonable?

Example 4 Find Missing Coordinates

Find the coordinates of *A* if $P\left(3, \frac{1}{2}\right)$ is the midpoint of $\overline{AB}$ and *B* has coordinates (8, 3).

First, substitute the known information into the Midpoint Formula. Let *A* be (x_1, y_1) and *B* be (x_2, y_2).

$M = \left(\frac{x_1 + x_2}{2}, \frac{y_1 + y_2}{2}\right)$ Midpoint Formula

$\left(3, \frac{1}{2}\right) = \left(\frac{x_1 + 8}{2}, \frac{y_1 + 3}{2}\right)$ Substitution

Next, write two equations to solve for x_1 and y_1.

$3 = \frac{x_1 + 8}{2}$ Equation for x_1

$6 = x_1 + 8$ Multiply each side by 2.

$-2 = x_1$ Solve.

$\frac{1}{2} = \frac{y_1 + 3}{2}$ Equation for y_1

$1 = y_1 + 3$ Multiply each side by 2.

$-2 = y_1$ Solve.

The coordinates of *A* are (−2, −2).

Plot the points on a coordinate plane to check your answer for reasonableness.

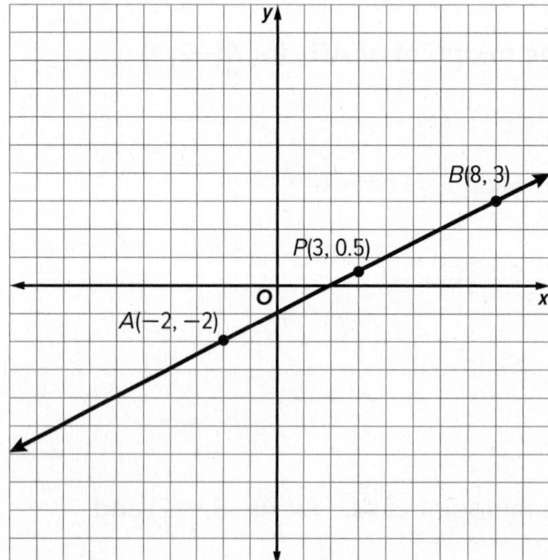

Check

Find the coordinates of *Q* if *R*(6, −1) is the midpoint of $\overline{QS}$ and *S* has coordinates (12, 4).

Go Online You can complete an Extra Example online.

Learn Bisectors

Because the midpoint separates the segment into two congruent segments, we can say that the midpoint **bisects** the segment. Any segment, line, plane, or point that bisects a segment is called a **segment bisector**.

Example 5 Find Missing Measures

Find the measure of $\overline{RT}$ if T is the midpoint of $\overline{RQ}$.

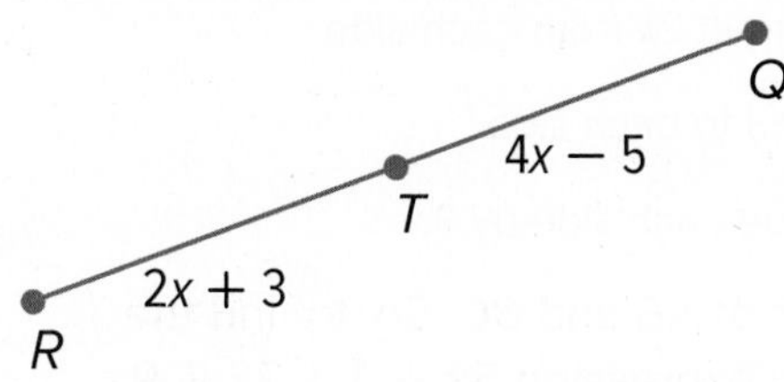

Because T is the midpoint, $RT = TQ$. Use this equation to solve for x.

$RT = TQ$	Definition of midpoint
$2x + 3 = 4x - 5$	Substitution
$3 = 2x - 5$	Subtract $2x$ from each side.
$8 = 2x$	Add 5 to each side.
$4 = x$	Divide each side by 2.

Substitute 4 for x in the equation for RT.

$RT = 2x + 3$	Equation for RT
$= 2(4) + 3$	Substitution
$= 11$	Simplify.

Check

Find the measure of $\overline{RS}$ if S is the midpoint of $\overline{RT}$.

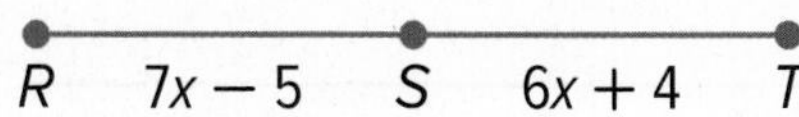

A. 56

B. 58

C. 112

D. 116

Go Online You can complete an Extra Example online.

Think About It!

Is there a way to find the length of $\overline{TQ}$ without calculating when you know the length of $\overline{RT}$? Why or why not?

Example 6 Find the Total Length

Find the measure of $\overline{AC}$ if B is the midpoint of $\overline{AC}$.

A $5x - 3$ B $2x + 9$ C

Because B is the midpoint, $AB = BC$. Use this equation to solve for x.

$AB = BC$	Definition of midpoint
$5x - 3 = 2x + 9$	Substitution
$3x - 3 = 9$	Subtract $2x$ from each side.
$3x = 12$	Add 3 to each side.
$x = 4$	Divide each side by 3.

> **Think About It!**
> What concept are we using when we say that $AC = AB + BC$?

The length of $\overline{AC}$ is equal to the sum of AB and BC. So, to find the length of $\overline{AC}$, substitute 4 for x in the expression $5x - 3 + 2x + 9$.

$AC = 5x - 3 + 2x + 9$	Length of $\overline{AC}$
$= 5(4) - 3 + 2(4) + 9$	$x = 4$
$= 20 - 3 + 8 + 9$	Multiply.
$= 34$	Simplify.

The measure of $\overline{AC}$ is 34.

Check

Find the measure of $\overline{AC}$ if B is the midpoint of $\overline{AC}$. Round your answer to the nearest tenth, if necessary.

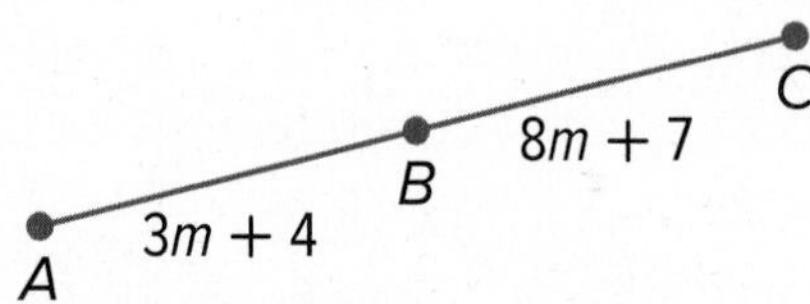

Pause and Reflect

Did you struggle with anything in this lesson? If so, how did you deal with it?

> **Go Online**
> You may want to complete the construction activities for this lesson.

Go Online You can complete an Extra Example online.

Practice

Go Online You can complete your homework online.

Example 1

Use the number line to find the coordinate of the midpoint of each segment.

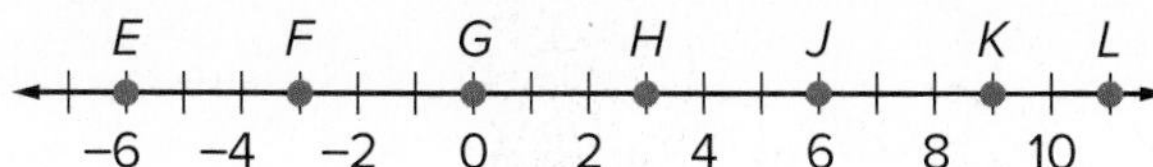

1. $\overline{KM}$
2. $\overline{JP}$
3. $\overline{LN}$
4. $\overline{MP}$
5. $\overline{LP}$
6. $\overline{JN}$

Use the number line to find the coordinate of the midpoint of each segment.

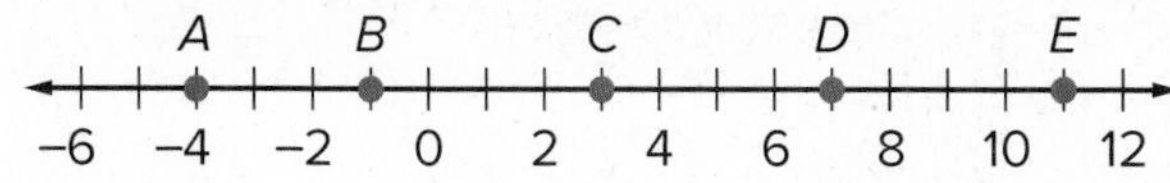

7. $\overline{FK}$
8. $\overline{HK}$
9. $\overline{EF}$
10. $\overline{FG}$
11. $\overline{JL}$
12. $\overline{EL}$

USE TOOLS **Use the number line to find the coordinate of the midpoint of each segment.**

A B C D E
−6 −4 −2 0 2 4 6 8 10 12

13. $\overline{DE}$
14. $\overline{BC}$
15. $\overline{BD}$
16. $\overline{AD}$

Example 2

17. HOME IMPROVEMENT Callie wants to build a fence halfway between her house and her neighbor's house. How far away from Callie's house should the fence be built?

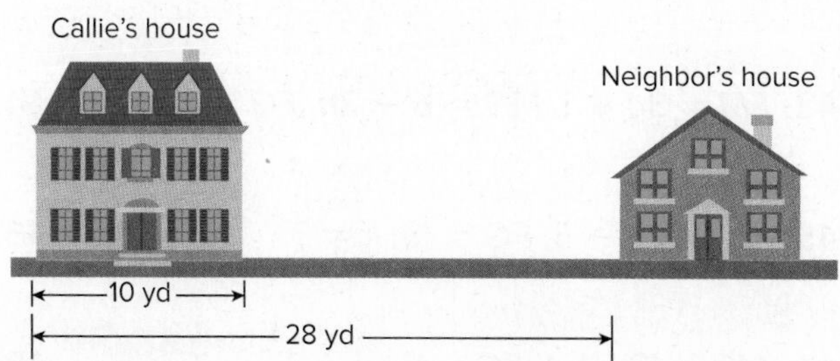

18. DINING Calvino's home is located at the midpoint between Fast Pizza and Pizza Now. Fast Pizza is a quarter mile away from Calvino's home. How far away is Pizza Now from Calvino's home? How far apart are the two pizzerias?

Example 3

Find the coordinates of the midpoint of a segment with the given endpoints.

19. (5, 11), (3, 1)

20. (7, −5), (3, 3)

21. (−8, −11), (2, 5)

22. (7, 0), (2, 4)

23. (−5, 1), (2, 6)

24. (−4, −7), (12, −6)

25. (2, 8), (8, 0)

26. (9, −3), (5, 1)

27. (22, 4), (15, 7)

28. (12, 2), (7, 9)

29. (−15, 4), (2, −10)

30. (−2, 5), (3, −17)

31. (2.4, 14), (6, 6.8)

32. (−11.2, −3.4), (−5.6, −7.8)

Example 4

Find the coordinates of the missing endpoint if *B* is the midpoint of $\overline{AC}$.

33. $C(-5, 4)$, $B(-2, 5)$

34. $A(1, 7)$, $B(-3, 1)$

35. $A(-4, 2)$, $B(6, -1)$

36. $C(-6, -2)$, $B(-3, -5)$

37. $A(4, -0.25)$, $B(-4, 6.5)$

38. $C\left(\frac{5}{3}, -6\right)$, $B\left(\frac{8}{3}, 4\right)$

Examples 5 and 6

Suppose *M* is the midpoint of $\overline{FG}$. Find each missing measure.

39. $FM = 5y + 13$, $MG = 5 - 3y$, $FG = ?$

40. $FM = 3x - 4$, $MG = 5x - 26$, $FG = ?$

41. $FM = 8a + 1$, $FG = 42$, $a = ?$

42. $MG = 7x - 15$, $FG = 33$, $x = ?$

43. $FM = 3n + 1$, $MG = 6 - 2n$, $FG = ?$

44. $FM = 12x - 4$, $MG = 5x + 10$, $FG = ?$

45. $FM = 2k - 5$, $FG = 18$, $k = ?$

46. $FG = 14a + 1$, $FM = 14.5$, $a = ?$

47. $MG = 13x + 1$, $FG = 15$, $x = ?$

48. $FG = 11x - 15.6$, $MG = 10.9$, $x = ?$

Mixed Exercises

Find the coordinates of the missing endpoint if *P* is the midpoint of $\overline{NQ}$.

49. $N(2, 0)$, $P(5, 2)$

50. $N(5, 4)$, $P(6, 3)$

51. $Q(3, 9)$, $P(-1, 5)$

52. Find the value of y if M is the midpoint of $\overline{LN}$.

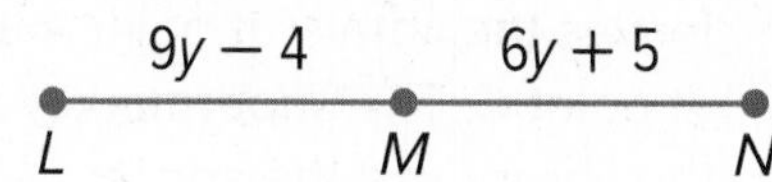

53. **CAMPING** Troop 175 is designing a new campground by first mapping everything on a coordinate grid. They found locations for the mess hall and their cabins. They want the bathrooms to be halfway between these two places. What are the coordinates of the location of the bathrooms?

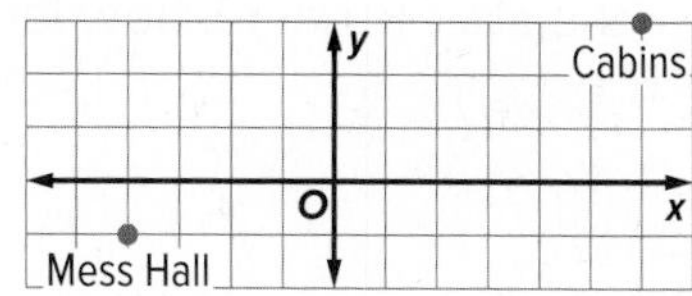

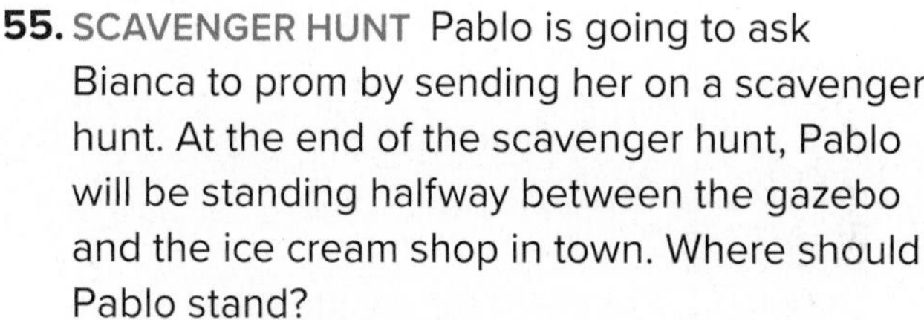

54. **GAME DESIGN** A computer software designer is creating a new video game. The designer wants to create a secret passage that is halfway between the castle and the bridge. Where should the secret passage be located?

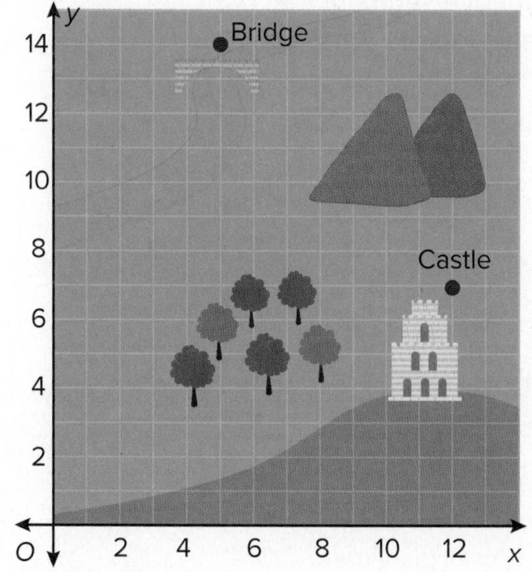

55. **SCAVENGER HUNT** Pablo is going to ask Bianca to prom by sending her on a scavenger hunt. At the end of the scavenger hunt, Pablo will be standing halfway between the gazebo and the ice cream shop in town. Where should Pablo stand?

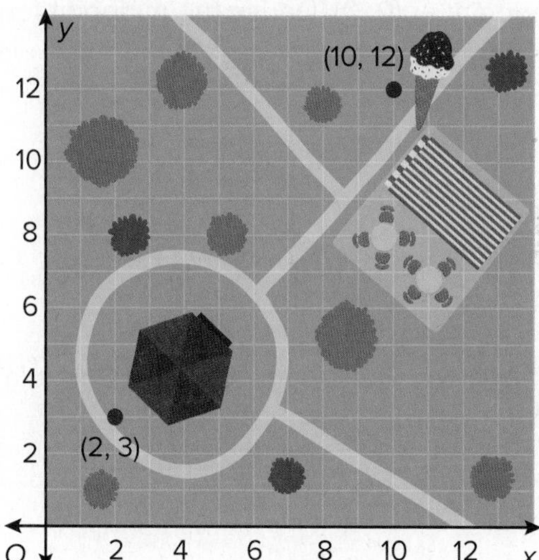

56. **WALKING** Javier walks from his home at point K to the Internet café at point O. If the school at point W is exactly halfway between Javier's house and the Internet café, how far does Javier walk?

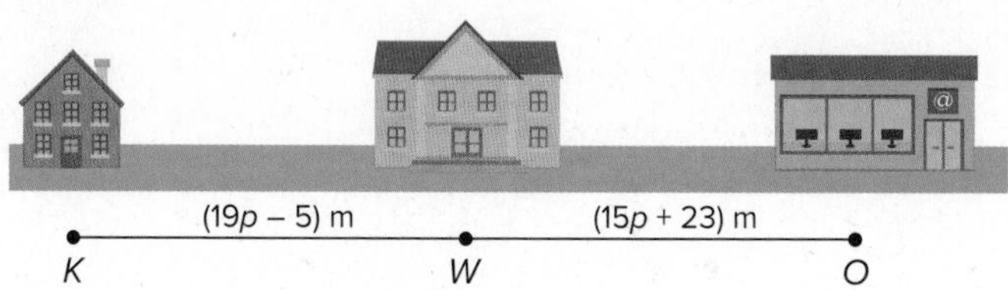

57. SCHOOL LIFE Bryan is at the library doing a research paper. He leaves the library at point A and walks to the soccer field for a game at point C. The supermarket at point B is exactly halfway between the library and the soccer field. After Bryan's first soccer game, he walks to the supermarket to buy a snack, and then he walks back to the soccer field for his second game. Not including the time spent at the soccer game, how far does Bryan walk?

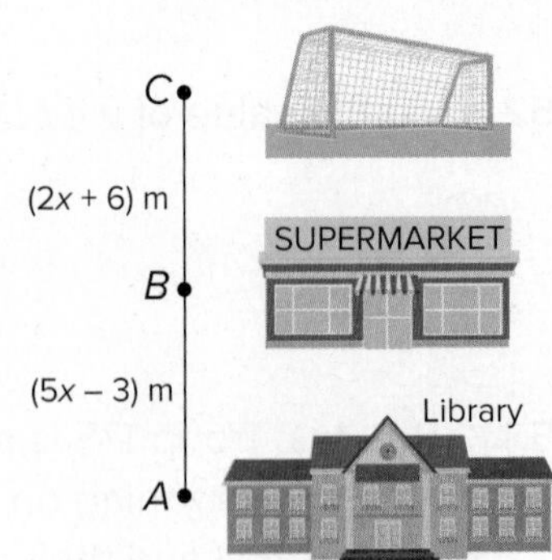

58. REASONING A drone flying over a field of corn identifies a dry area. The coordinates of the vertices of the area are shown. To what coordinates should the portable irrigation system be sent to water the dry area? Explain your reasoning.

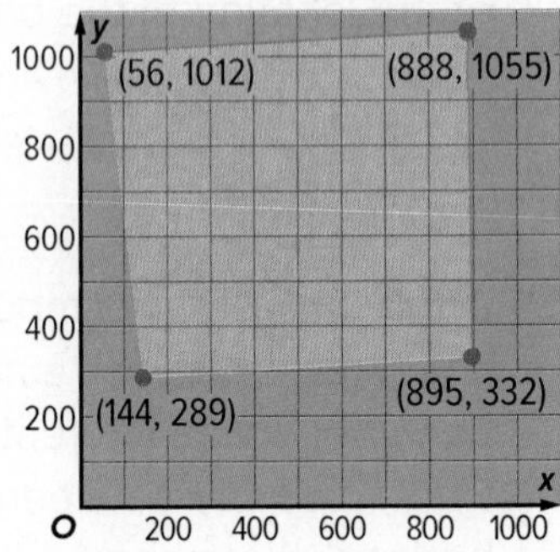

59. PERSEVERE Describe a method of finding the midpoint of a segment that has one endpoint at (0, 0). Derive the midpoint formula, give an example using your method, and explain why your method works.

60. WRITE Explain how the Midpoint Formula is a special case of the Section Formula.

61. CREATE Construct $\overline{AC}$ given $\overline{AB}$ if B is the midpoint of $\overline{AC}$.

Review

Essential Question

How are points, lines, and segments used to model the real world?

Points, lines, and segments allow something that is abstract to be seen as a drawing. It in turn allows for certain calculations to be made to solve for missing measures.

Module Summary

Lesson 10-1

The Geometric System

- An axiomatic system has a set of axioms from which theorems can be derived.
- Synthetic geometry is the study of geometric figures without the use of coordinates.
- Analytic geometry is the study of geometry using a coordinate system.

Lessons 10-2 through 10-4

Points, Lines, Line Segments, and Planes

- The terms *point*, *line*, and *plane* are undefined terms because they are readily understood and are not formally explained by means of more basic words and concepts.
- Collinear points are points that lie on the same line. Coplanar points are points that lie in the same plane.
- The intersection of two or more geometric figures is the set of points they have in common.
- Point C is between A and B if and only if A, B, and C are collinear and $AC + CB = AB$.
- Two segments that have the same measure are congruent segments.
- The distance between two points on a number line is the absolute value of their difference.
- The distance between two points on a coordinate plane, (x_1, y_1) and (x_2, y_2), is $\sqrt{(x_2 - x_1)^2 + (y_2 - y_1)^2}$.

Lessons 10-5 and 10-6

Locating Points

- If C has coordinate x_1 and D has coordinate x_2, then a point P that partitions the line segment in a ratio of $m{:}n$ is located at coordinate $\frac{nx_1 + mx_2}{m + n}$.
- The coordinates of point B that is $\frac{a}{b}$ of the distance from point $A(x_1, y_1)$ to point $C(x_2, y_2)$ are $\left(x_1 + \frac{a}{b}(x_2 - x_1), y_1 + \frac{a}{b}(y_2 - y_1)\right)$.

Lesson 10-7

Midpoints and Bisectors

- If $\overline{AB}$ has endpoints at x_1 and x_2 on a number line, then the midpoint M of $\overline{AB}$ has coordinate $M = \frac{x_1 + x_2}{2}$
- A midpoint separates a segment into two congruent parts, so it bisects the segment.

Study Organizer

Foldables

Use your Foldable to review this module. Working with a partner can be helpful. Ask for clarification of concepts as needed.

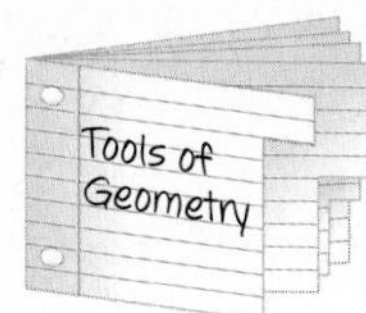

Test Practice

1. **MULTI-SELECT** Select all real-world objects that model a line. (Lesson 10-2)

   A. electric tablet

   B. pool stick

   C. scoop of ice cream

   D. light pole

   E. emoji

2. **MULTI-SELECT** Use the figure to name all planes containing point *W*. (Lesson 10-2)

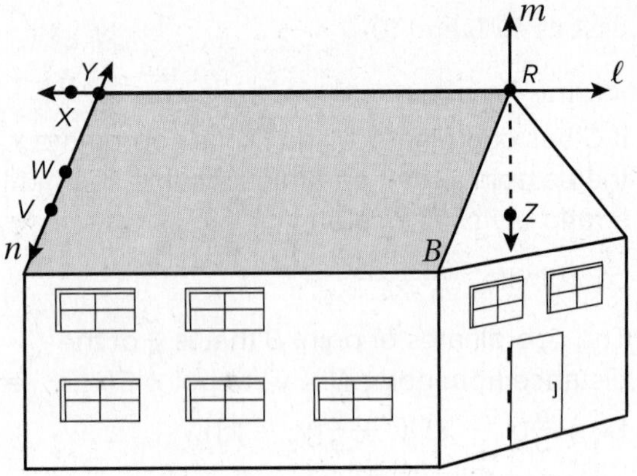

   A. plane *VWY*

   B. plane *VWX*

   C. plane *RYV*

   D. plane *VWZ*

   E. plane *RYX*

3. **OPEN RESPONSE** What geometric figures do the pages of the book represent? (Lesson 10-2)

4. **MULTIPLE CHOICE** Which sequence identifies the correct order for completing the construction to copy a line segment using a compass and straightedge? (Lesson 10-3)

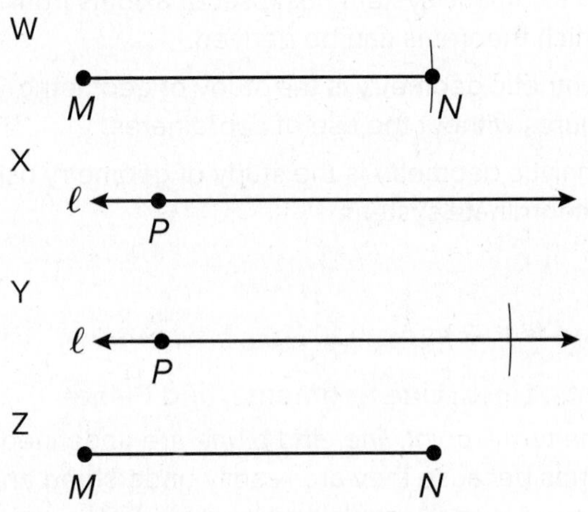

   A. X, Y, Z, W

   B. W, Z, X, Y

   C. W, Y, X, Z

   D. Z, X, W, Y

5. OPEN RESPONSE Find the value of x if Q is between P and R, $PQ = 5x - 10$, $QR = 3(x + 4)$, and $\overline{PQ} \cong \overline{QR}$. (Lesson 10-3)

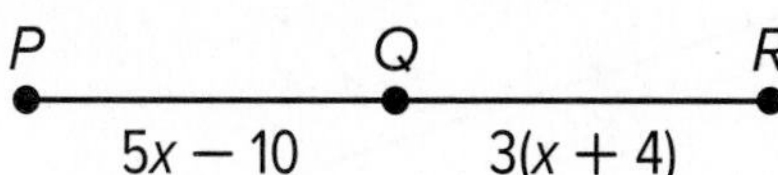

6. OPEN RESPONSE On a straight highway, the distance from Loretta's house to a park is 43 miles. Her friend Jamal lives along this same highway between Loretta's house and the park. The distance from Loretta's house to Jamal's house is 31 miles. How many miles is it from Jamal's house to the park? (Lesson 10-3)

7. MULTIPLE CHOICE Find the distance between the two points on a coordinate plane. (Lesson 10-4)

$A(5, 1)$ and $B(-3, -3)$

A. $4\sqrt{5}$

B. $4\sqrt{3}$

C. $2\sqrt{2}$

D. $2\sqrt{3}$

8. OPEN RESPONSE True or false: $\overline{XY} \cong \overline{WZ}$ (Lesson 10-4)

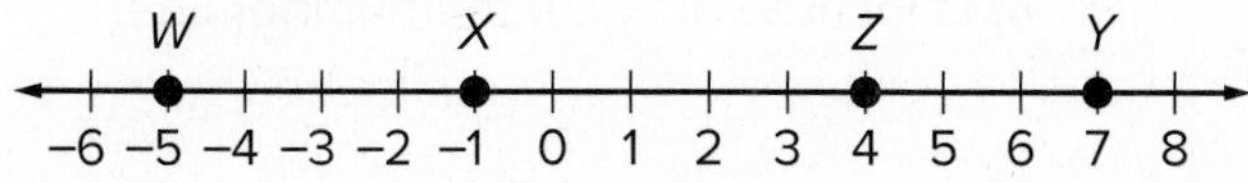

9. MULTIPLE CHOICE The coordinates of A and B on a number line are -7 and 9. The coordinates of C and D on a number line are -4 and 12. Are $\overline{AB}$ and $\overline{CD}$ congruent? If yes, what is the length of each segment? (Lesson 10-4)

A. no

B. yes; 16

C. yes; -16

D. yes; 8

10. OPEN RESPONSE The coordinate of point X on $\overline{PQ}$ that is $\frac{3}{4}$ of the distance from P to Q is ___. (Lesson 10-5)

11. MULTIPLE CHOICE On a number line, point S is located at -3 and point T is located at 9. Where is point R located on $\overline{ST}$ if the ratio of SR to RT is 3:4? (Lesson 10-5)

A. $\frac{27}{7}$

B. $2\frac{1}{4}$

C. $1\frac{1}{4}$

D. $\frac{15}{7}$

12. MULTIPLE CHOICE Find point R on $\overline{ST}$ such that the ratio of SR to RT is 1:2. (Lesson 10-6)

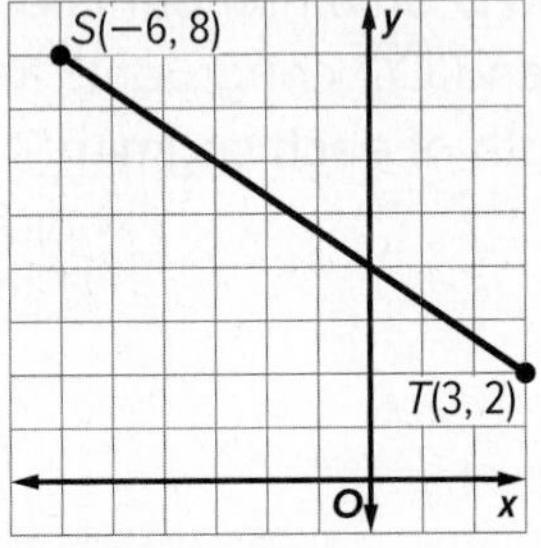

A. $R(-5, 6)$

B. $R(-3, 6)$

C. $R(-1.5, 5)$

D. $R(0, 4)$

13. OPEN RESPONSE Alonso plans to go to the animal shelter to adopt a dog and then take the dog to Precious Pup Grooming Services. The shelter is located at $(-1, 9)$ on the coordinate plane, while Precious Pup Grooming Services is located at $(11, 0)$ on the coordinate plane. Find the location of Alonso's home if it is $\frac{1}{3}$ of the distance from the shelter to Precious Pup Grooming Services. (Lesson 10-6)

14. OPEN RESPONSE Find the coordinates of A if $M(6, -1)$ is the midpoint of $\overline{AB}$, and B has the coordinates $(8, -7)$. (Lesson 10-7)

15. MULTIPLE CHOICE Find the measure of $\overline{YZ}$ if Y is the midpoint of $\overline{XZ}$. (Lesson 10-7)

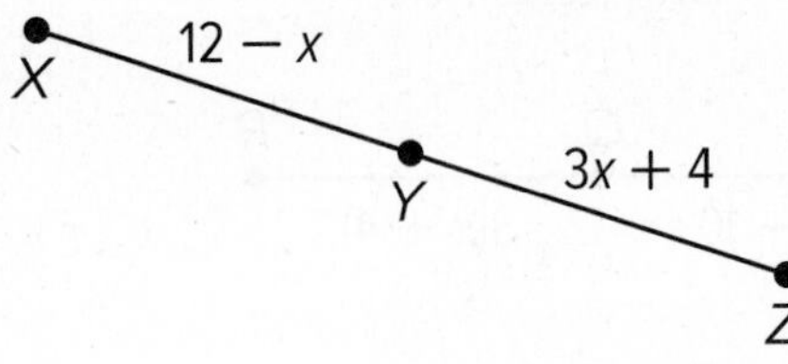

A. 2

B. 10

C. 16

D. 20

16. MULTIPLE CHOICE Find the y-coordinate of the point M, the midpoint of $\overline{AB}$, for $A(-3, 3)$ and $B(5, 7)$. (Lesson 10-7)

A. −1

B. 1

C. 2

D. 5

17. MULTIPLE CHOICE Points A and B are plotted on a number line. What is the location of M, the midpoint of $\overline{AB}$, for A at −9 and B at 28? (Lesson 10-7)

A. M is located at 18.5 on the number line.

B. M is located at 14 on the number line.

C. M is located at 9.5 on the number line.

D. M is located at $\frac{10}{3}$ on the number line.

Module 11

Angles and Geometric Figures

Essential Question

How are angles and two-dimensional figures used to model the real world?

What Will You Learn?

How much do you already know about each topic **before** starting this module?

KEY

— I don't know. — I've heard of it. — I know it!

	Before			After		
	I don't know.	I've heard of it.	I know it!	I don't know.	I've heard of it.	I know it!
apply the definitions of angles, parts of angles, congruent angles, and angle bisectors to calculate angle measures						
apply the characteristics of complementary and supplementary angles and parallel and perpendicular lines to calculate angle measures						
apply the characteristics of perpendicular lines to calculate angle measures						
find perimeters, circumferences, and areas of two-dimensional geometric shapes						
reflect, translate, and rotate figures						
solve for unknown measures of three-dimensional figures by calculating surface areas and volumes						
model three-dimensional geometric figures with orthographic drawings						
determine levels of precision and accuracy						
determine the correct numbers of significant figures in recorded measurements						

Foldables Make this Foldable to help you organize your notes about angles and geometric figures. Begin with two sheets of grid paper.

1. **Fold** in half along the width.
2. **On** the first sheet, cut 5 centimeters along the fold at the ends.
3. **On** the second sheet, cut in the center, stopping 5 centimeters at the ends.
4. **Insert** the first sheet through the second sheet and align the folds. Label with lesson numbers.

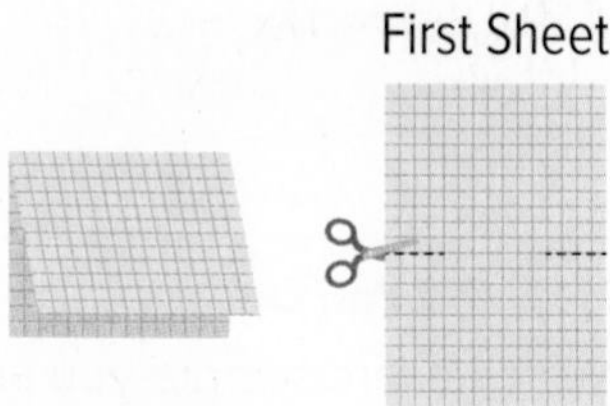

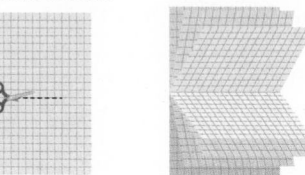

What Vocabulary Will You Learn?

- accuracy
- adjacent angles
- angle
- angle bisector
- angle of rotation
- approximate error
- area
- base of a pyramid or cone
- bases of a prism or cylinder
- center of rotation
- circumference
- complementary angles
- component form
- concave
- cone
- congruent angles
- convex
- cylinder
- edge of a polyhedron
- equiangular polygon
- equilateral polygon
- exterior
- face of a polyhedron
- geometric model
- image
- interior
- line of reflection
- linear pair
- net
- opposite rays
- orthographic drawing
- perimeter
- perpendicular
- Platonic solid
- polygon
- Polyhedron
- precision
- Preimage
- prism
- pyramid
- ray
- reflection
- regular polygon
- regular polygon
- regular polyhedron
- rigid motion
- rotation
- sides
- significant figures
- sphere
- straight angle
- supplementary angles
- surface area
- transformation
- translation
- translation vector
- vertex
- vertex of a polyhedron
- vertical angles
- volume

Are You Ready?

Complete the Quick Review to see if you are ready to start this module.
Then complete the Quick Check.

Quick Review

Example 1		Example 2	
Solve $5x + 2 = 90$.		**Evaluate $2(3)(4) + 2(3)(5) + 2(4)(5)$.**	
$5x + 2 = 90$	Original equation.	$2(3)(4) + 2(3)(5) + 2(4)(5)$	Original expression
$5x = 88$	Subtract 2 from each side.	$= 24 + 30 + 40$	Multiply.
$x = 17.6$	Divide each side by 5.	$= 94$	Add.

Quick Check

Solve each equation.	Evaluate each expression.
1. $3x - 9 = 180$	**5.** $6(15)(22)$
2. $2x + 10x - 9 = 90$	**6.** $0.5(8)(9)$
3. $15x + 42 = 12x + 51$	**7.** $2(6)(7) + 2(6)(10) + 2(7)(10)$
4. $9x + 1 = 17x - 31$	**8.** $0.5(5)(12) + 0.5(5)(12) + 5(14) + 12(14) + 13(14)$

How Did You Do?

Which exercises did you answer correctly in the Quick Check?

Angles and Congruence

Explore Angles Formed by Intersecting Lines

Online Activity Use dynamic geometry software to complete the Explore.

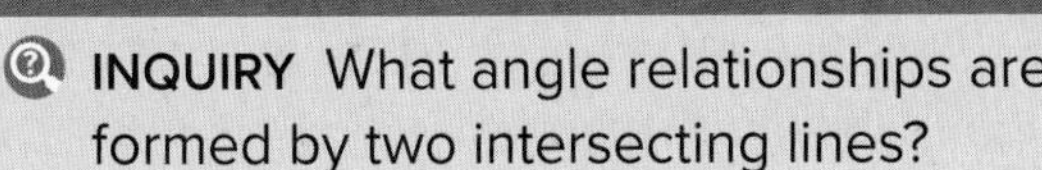

INQUIRY What angle relationships are formed by two intersecting lines?

Learn Angles

Lines and portions of lines intersect to form angles.

A **ray** is the part of a line consisting of a point on the line, called the *endpoint of the ray,* together with all of the collinear points on one side of the endpoint.

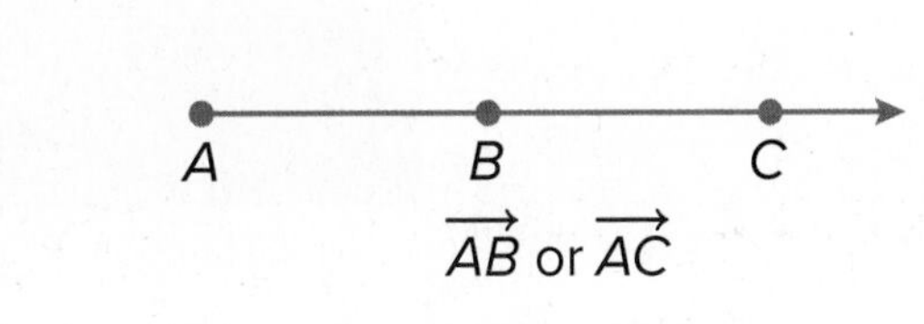

Two collinear rays with a common endpoint are **opposite rays**. Opposite rays form a **straight angle**, which has a measure of 180°.

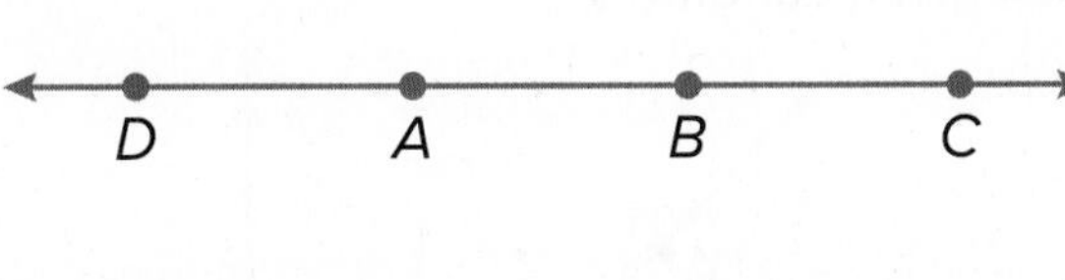

An **angle** is a pair of rays that have a common endpoint.

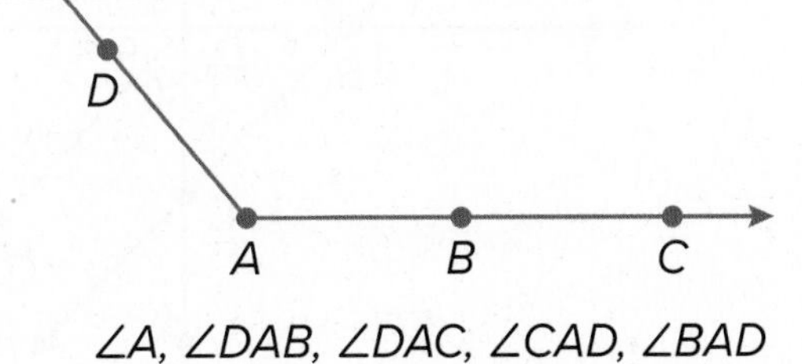

The rays are called **sides** of the angle. The common endpoint is the **vertex**.

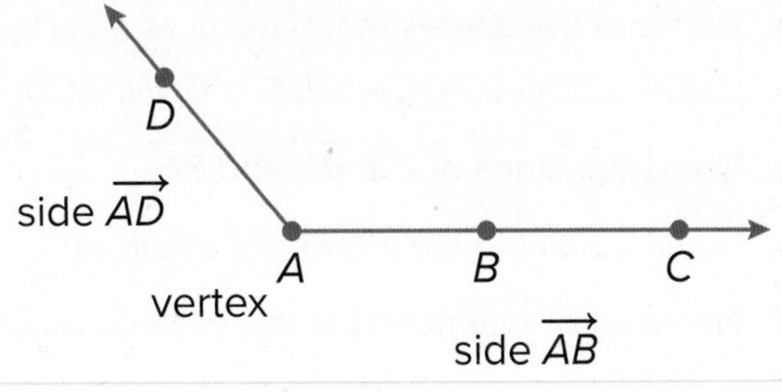

(continued on the next page)

Today's Goals

- Analyze figures using the definitions of angles and parts of angles.
- Calculate angle measures using the definitions of congruent angles and angle bisectors.
- Analyze figures using the characteristics of adjacent angles, linear pairs of angles, and vertical angles.

Today's Vocabulary

ray
opposite rays
straight angle
angle
sides
vertex
interior
exterior
congruent angles
angle bisector
adjacent angles
linear pair
vertical angles

Study Tip

Naming Angles When naming an angle using three letters, the first letter represents a point on one side of the angle, the second letter must always represent the vertex, and the third letter represents a point on the other side of the angle. Name an angle using a single letter only when there is exactly one angle located at that vertex.

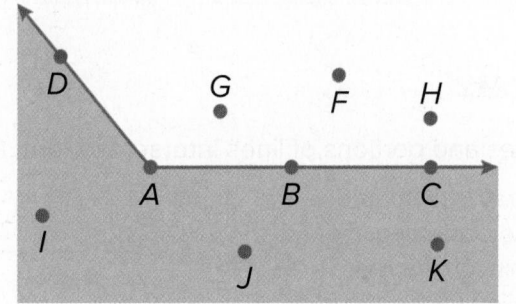

Go Online You can watch a video to see how to use a protractor to measure and draw angles.

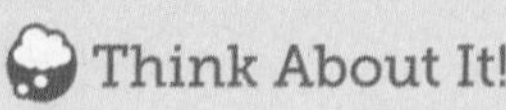

Think About It!

Can a point be in the interior of one angle and the exterior of another angle? If so, give an example.

An angle divides a plane into three distinct parts.

Points *D*, *A*, *B*, and *C* lie on the angle.	
Points *G*, *F*, and *H* lie in the **interior** of the angle.	
Points *I*, *J*, and *K* lie in the **exterior** of the angle.	

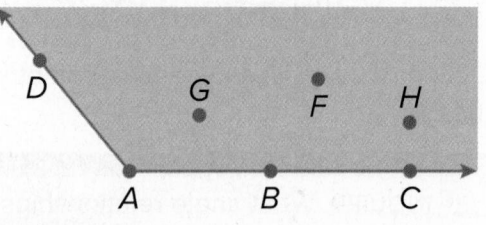

Example 1 Identify Angles

Use the figure to identify the angles or parts of angles that satisfy each given condition.

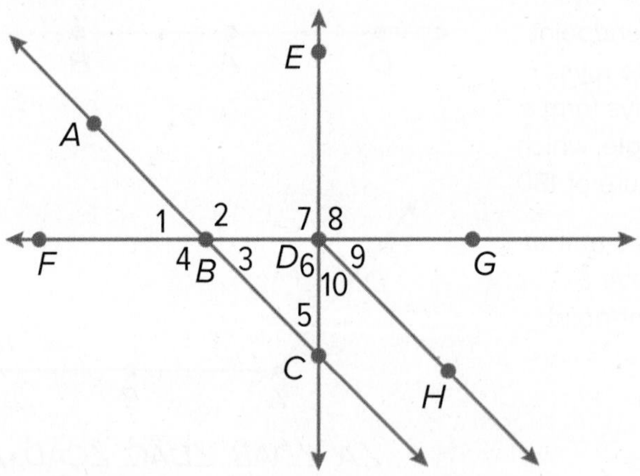

a. Name all the angles that have *D* as a vertex.
$\angle EDF$, $\angle EDG$, $\angle FDC$, $\angle GDC$, $\angle GDH$, $\angle CDH$, $\angle FDH$

b. Name the sides of $\angle 2$. $\overrightarrow{BA}$ and $\overrightarrow{BG}$

c. Name a point in the interior of $\angle FDE$. *A*

d. Name a point or points in the exterior of $\angle FDE$. *C*, *H*, and *G*

Go Online You can complete an Extra Example online.

Check

Use the figure to identify the angles or parts of angles that satisfy the given condition. Which angle has sides $\overrightarrow{DB}$ and $\overrightarrow{DC}$? Select all that apply.

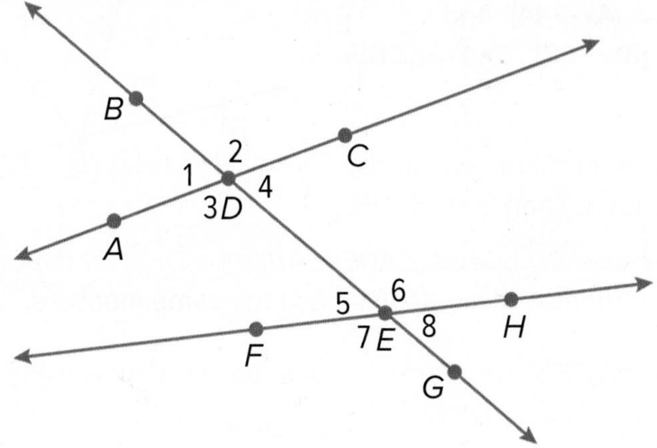

A. ∠2

B. ∠3

C. ∠ADB

D. ∠BDC

E. ∠CDB

F. ∠EDC

Learn Congruent Angles

The measure of an angle is the measure in degrees of the space between the sides of an angle. Angles that have the same measure are **congruent angles**. Congruent angles are indicated on the figure by matching numbers of arcs.

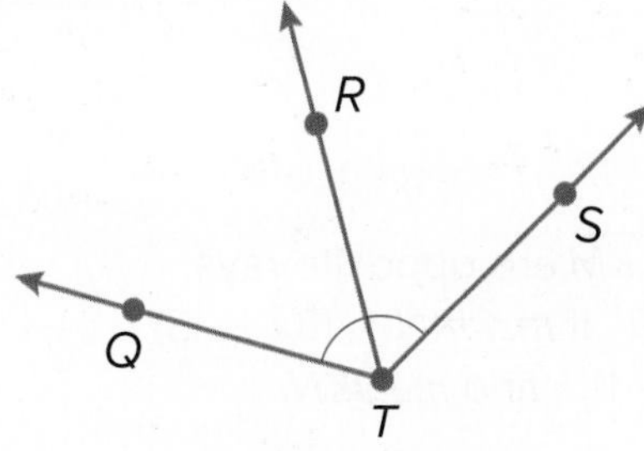

$\angle QTS \cong \angle STR$

A ray or segment that divides an angle into two congruent parts is an **angle bisector**. In the figure, $\overrightarrow{TR}$ bisects $\angle QTS$.

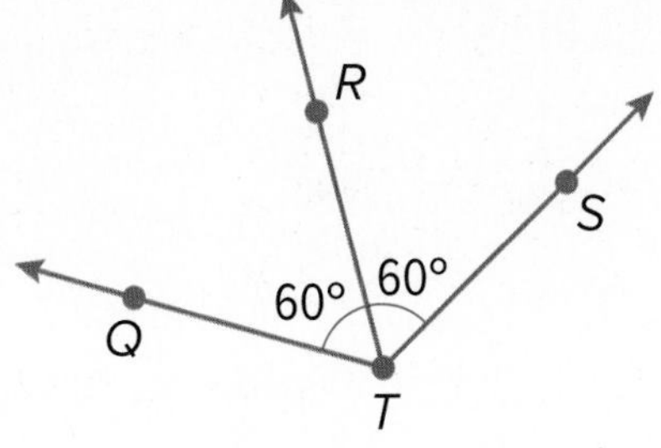

$m\angle QTR \cong m\angle STR$

Example 2 Congruent Angles and Angle Bisectors

In the figure, $\overrightarrow{BA}$ and $\overrightarrow{BC}$ are opposite rays and $\overrightarrow{BD}$ bisects $\angle ABE$. If $m\angle ABD = (4x + 14)°$ and $m\angle DBE = (8x - 32)°$, find $m\angle DBE$.

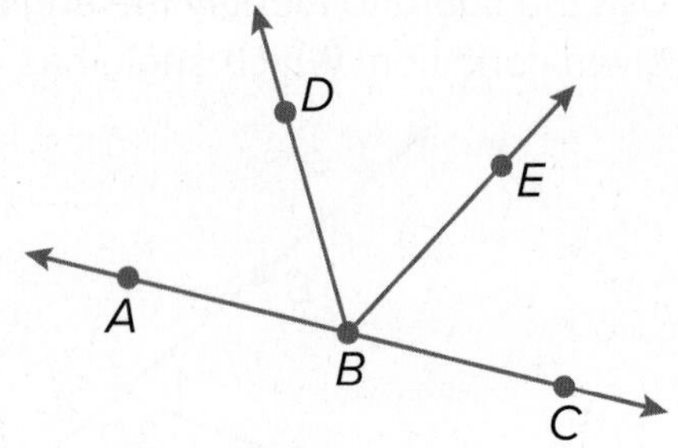

We can solve for this in two steps. First, solve for x. Then find $m\angle DBE$.

Step 1: Because $\overrightarrow{BD}$ bisects $\angle ABE$, $\angle ABD \cong \angle DBE$. By the definition of congruence, these angles have the same measure.

$m\angle ABD = m\angle DBE$	Definition of congruent angles
$4x + 14 = 8x - 32$	Substitution
$14 = 4x - 32$	Subtract $4x$ from each side.
$46 = 4x$	Add 32 to each side.
$11.5 = x$	Divide each side by 4.

Talk About It!

Suppose $\overrightarrow{BE}$ is an angle bisector of $\angle DBC$. What is $m\angle EBC$? Explain your solution process.

Step 2: Because we are asked to find $m\angle DBE$, we substitute 11.5 for x in the expression.

$m\angle DBE = 8x - 32$	Given
$= 8(11.5) - 32$	Substitute.
$= 92 - 32$	Multiply.
$= 60$	Subtract.

$m\angle DBE = 60°$

Check

In the figure, $\overrightarrow{KJ}$ and $\overrightarrow{KM}$ are opposite rays, and $\overrightarrow{KN}$ bisects $\angle JKL$. If $m\angle JKN = (8x - 13)°$ and $m\angle NKL = (6x + 11)°$, find $m\angle JKN$.

$m\angle JKN =$ ___?___ °

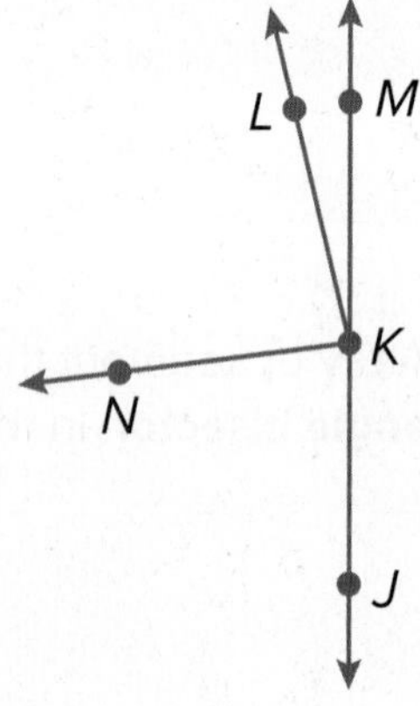

Go Online You can complete an Extra Example online.

Learn Special Angle Pairs

There are three special angle pairs.

Key Concept • Special Angle Pairs

Special Angle Pair Definition	Examples	Nonexamples
Adjacent angles are two angles that lie in the same plane, have a common vertex and a common side, but have no common interior points.	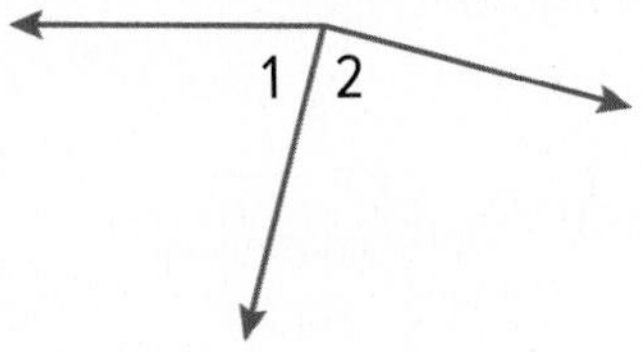 ∠1 and ∠2 are adjacent angles.	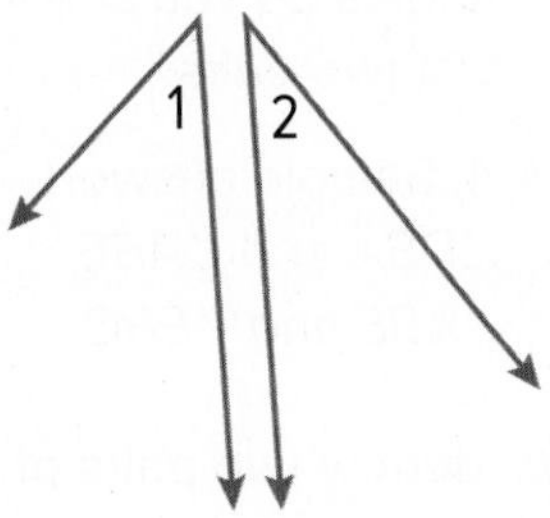
A **linear pair** is a pair of adjacent angles with noncommon sides that are opposite rays. The sum of the angle measures is 180°.	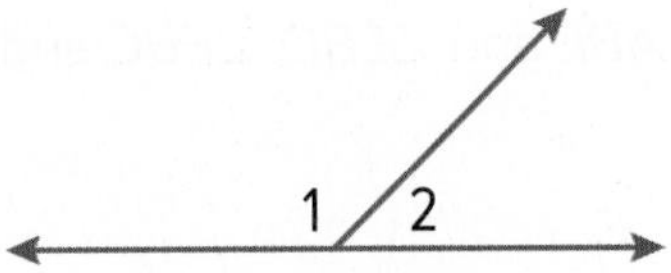 ∠1 and ∠2 are a linear pair. 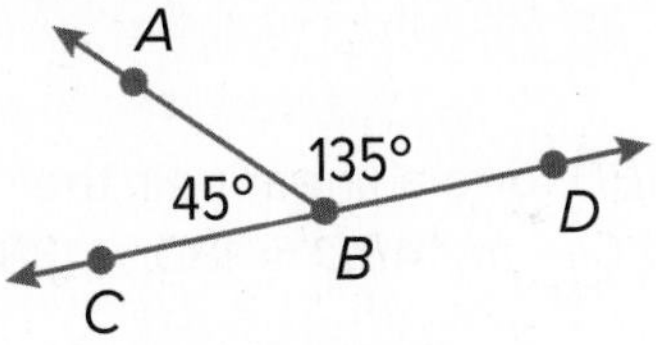 A, 45°, 135°, B, C, D	
Vertical angles are the two nonadjacent angles formed by two intersecting lines. Vertical angles are congruent.	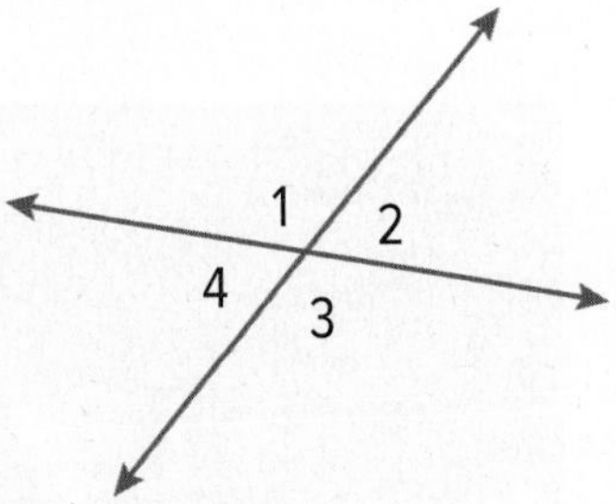 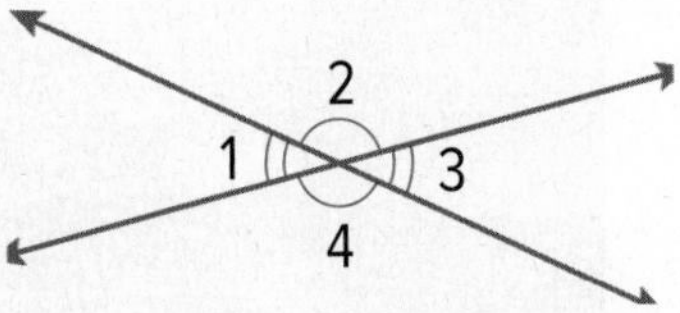 ∠1 and ∠3 and ∠2 and ∠4 are vertical angles.	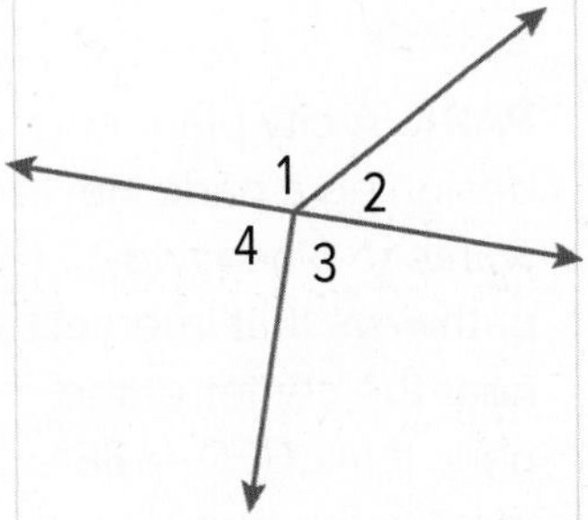

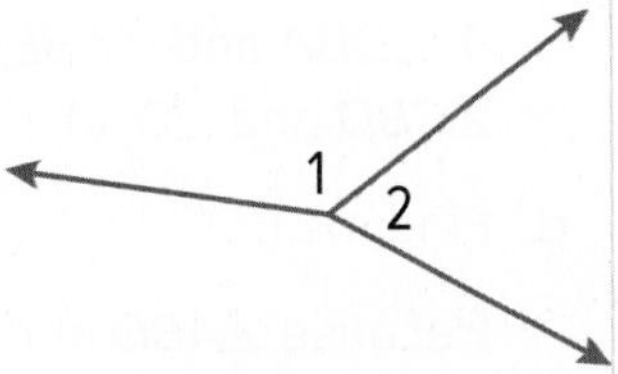

Example 3 Vertical Angles and Angle Pairs

HOME DECOR **The office lamp is made using two intersecting metal bars.**

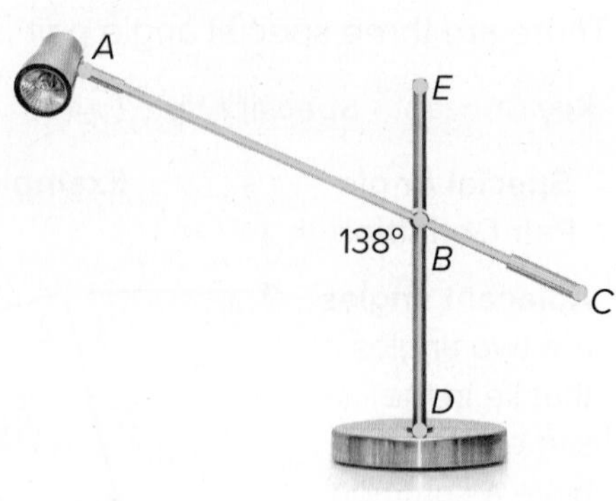

a. How many pairs of adjacent angles do you see in the figure? List two pairs.

4; Sample answer: $\angle DBA$ and $\angle ABE$, $\angle ABE$ and $\angle EBC$

b. Identify two pairs of vertical angles in the figure.

$\angle DBA$ and $\angle EBC$, $\angle ABE$ and $\angle CBD$

c. How many linear pairs do you see in the figure? List each pair.

4; $\angle DBA$ and $\angle ABE$, $\angle ABE$ and $\angle EBC$, $\angle EBC$ and $\angle CBD$, $\angle CBD$ and $\angle DBA$

d. Find $m\angle EBC$.

Because $\angle ABD$ and $\angle EBC$ are formed by intersecting line segments, they are vertical angles. Because vertical angles are congruent, $m\angle EBC$ is the same as $m\angle ABD$, 138°.

e. Find $m\angle ABE$.

Because $\angle ABE$ and $\angle ABD$ form a linear pair, their measures add to 180°. Thus, $m\angle ABE = 180 - m\angle ABD = 180 - 138 = 42°$.

Think About It!
Can vertical angles also be adjacent angles? Explain.

Check

PARK A city planner is designing a park. He wants to place two pathways that intersect near the center of the park. If $m\angle GED = 88°$, identify the true statement(s).

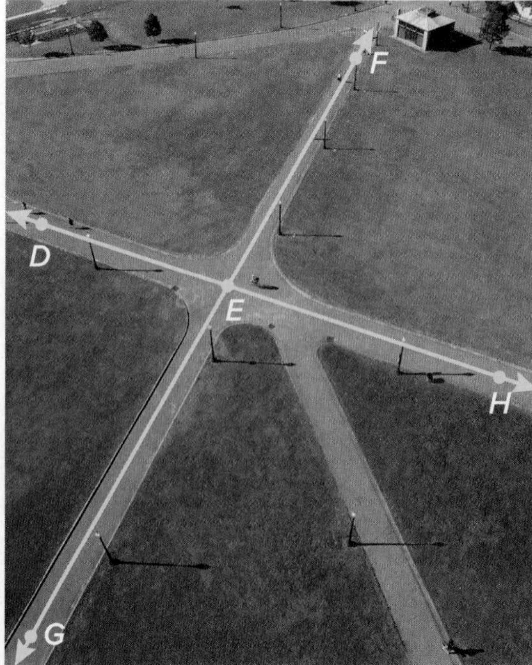

firina/123RF, Glowimages/Getty Images

A. $m\angle DEF = 92°$

B. $m\angle DEG = 92°$

C. $m\angle FEH = 88°$

D. $m\angle DEH = 92°$

E. $m\angle GEH = 88°$

Go Online You may want to complete the construction activities for this lesson.

Go Online You can complete an Extra Example online.

Practice

Go Online You can complete your homework online.

Example 1

Use the figure to identify angles and parts of angles that satisfy each given condition.

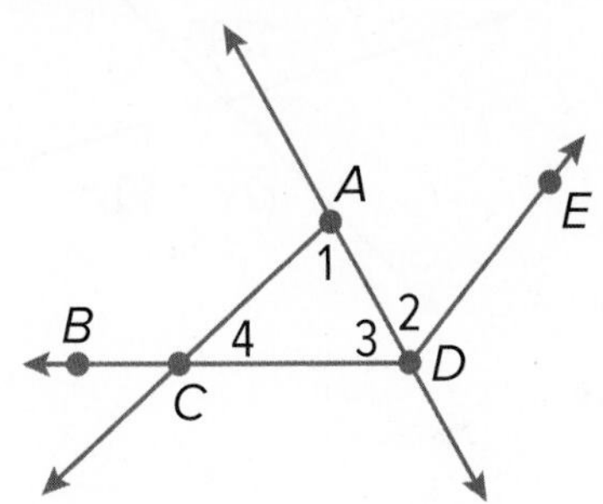

1. Name the vertex of $\angle 1$.

2. Name the sides of $\angle 4$.

3. What is another name for $\angle 3$?

4. What is another name for $\angle CAD$?

Example 2

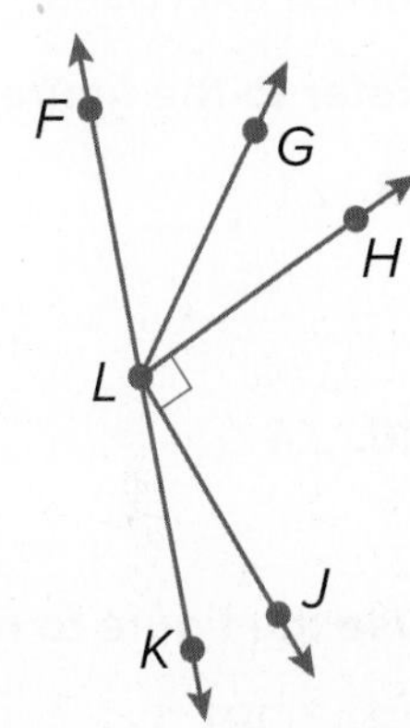

5. In the figure, $\overrightarrow{LF}$ and $\overrightarrow{LK}$ are opposite rays. $\overrightarrow{LG}$ bisects $\angle FLH$. If $m\angle FLG = 14x + 5$ and $m\angle HLG = 17x - 1$, find $m\angle FLH$.

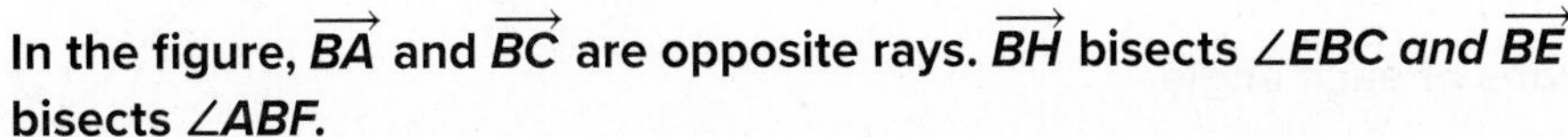
In the figure, $\overrightarrow{BA}$ and $\overrightarrow{BC}$ are opposite rays. $\overrightarrow{BH}$ bisects $\angle EBC$ and $\overrightarrow{BE}$ bisects $\angle ABF$.

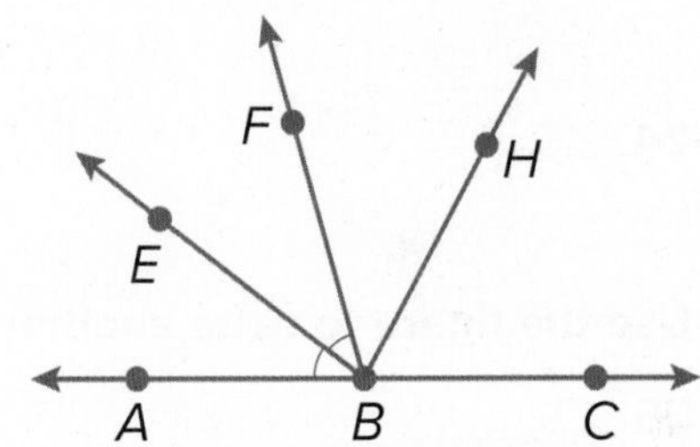

6. If $m\angle ABE = 2n + 7$ and $m\angle EBF = 4n - 13$, find $m\angle ABE$.

7. If $m\angle EBH = 6x + 12$ and $m\angle HBC = 8x - 10$, find $m\angle EBH$.

8. If $m\angle ABF = 7b - 24$ and $m\angle ABE = 2b$, find $m\angle EBF$.

9. If $m\angle EBC = 31a - 2$ and $m\angle EBH = 4a + 45$, find $m\angle HBC$.

10. If $m\angle ABF = 8w - 6$ and $m\angle ABE = 2(w + 11)$, find $m\angle EBF$.

11. If $m\angle EBC = 3r + 10$ and $m\angle ABE = 2r - 20$, find $m\angle EBF$.

Example 3

Refer to the figure.

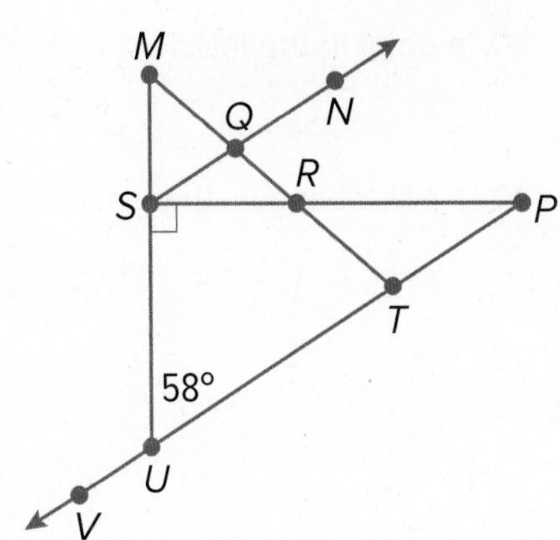

12. Name two adjacent angles.

13. Name two vertical angles.

14. Find $m\angle SUV$.

Find the value of each variable.

15.

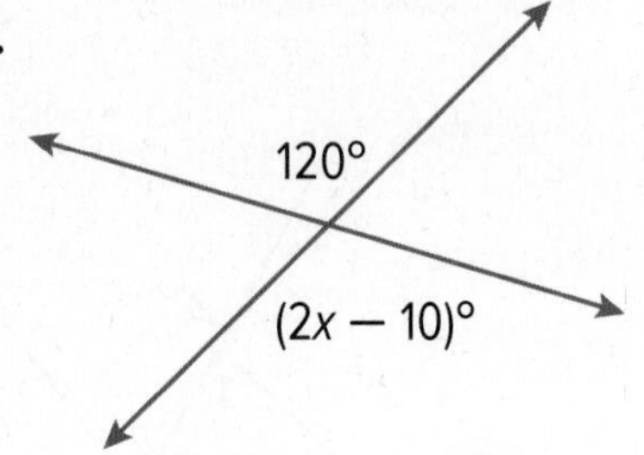

16.

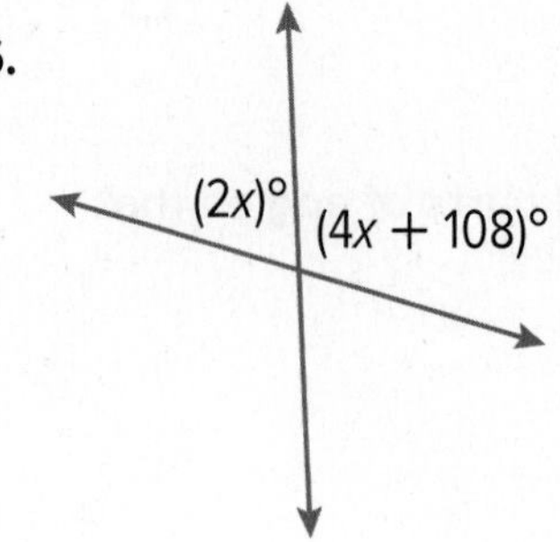

17.

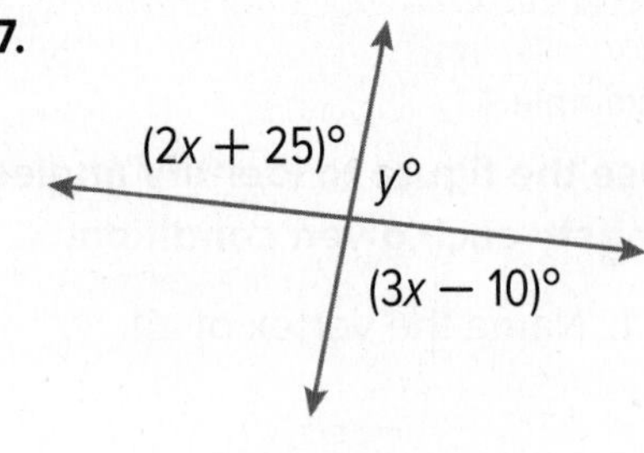

Mixed Exercises

Refer to the figure to name the vertex of each angle.

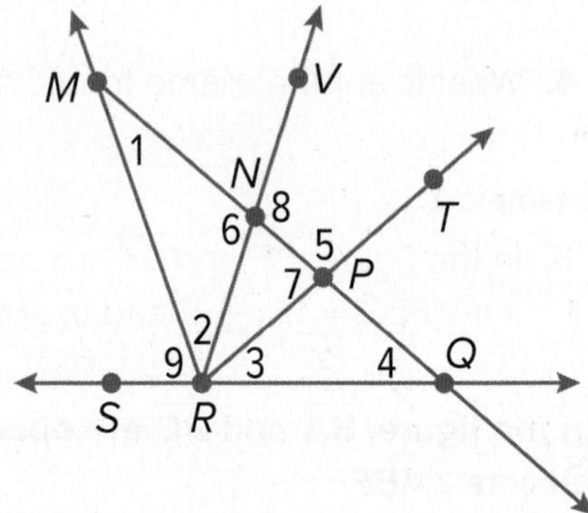

18. $\angle 1$

19. $\angle 2$

20. $\angle 4$

21. $\angle 7$

Use the figure to name the sides of each angle.

22. $\angle QPT$

23. $\angle MNV$

24. $\angle 6$

25. $\angle 3$

Use the figure to write another name for each angle.

26. $\angle 9$

27. $\angle QPT$

28. $\angle MQS$

29. $\angle 5$

Use the figure above to name each angle, point, or pair of angles.

30. a point in the interior of $\angle VRQ$

31. a point in the exterior of $\angle MRT$

32. a pair of angles that share exactly one point

33. a pair of angles that share more than one point

Find the value of each variable.

34.

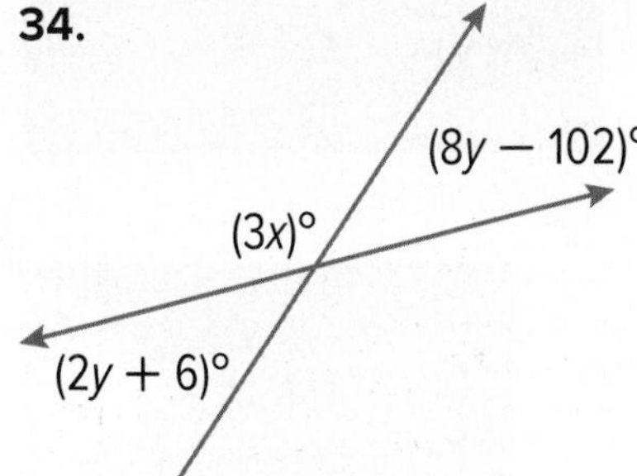

35.

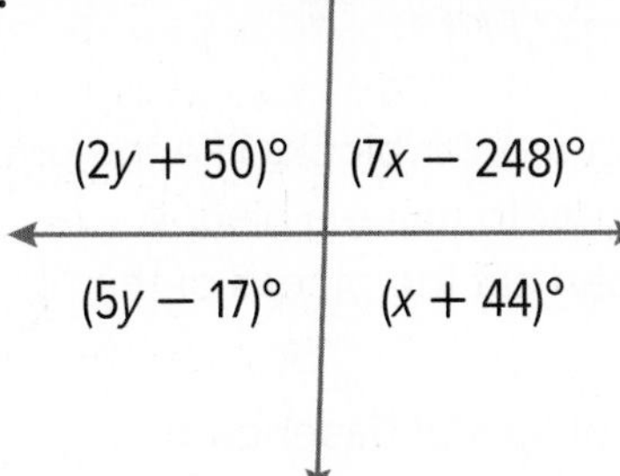

36.

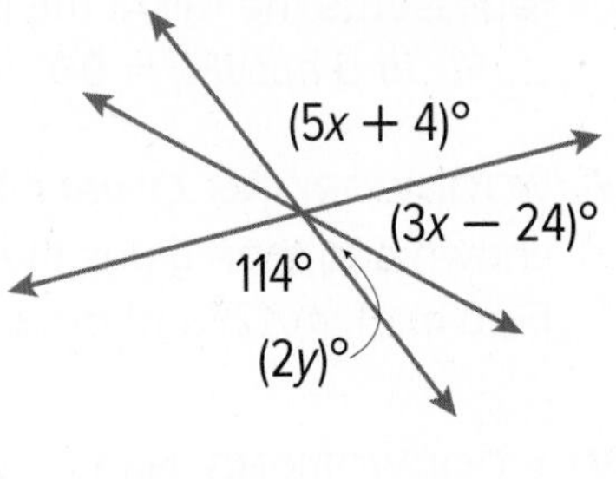

Name an angle or angle pair that satisfies each condition.

37. two adjacent angles

38. two vertical angles

39. a linear pair that has vertex F

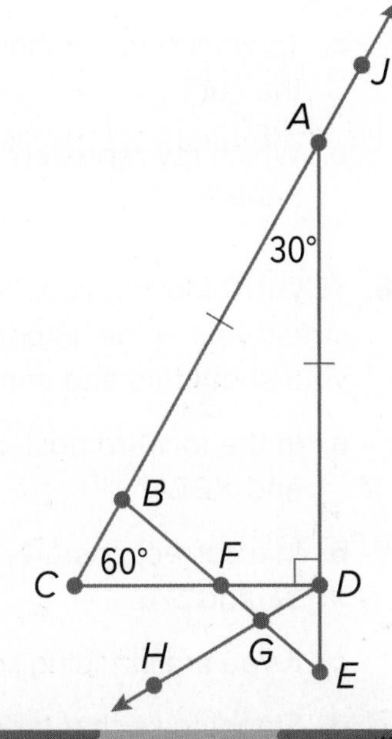

Use the picture at the right.

40. Name four rays.

41. Name three angles.

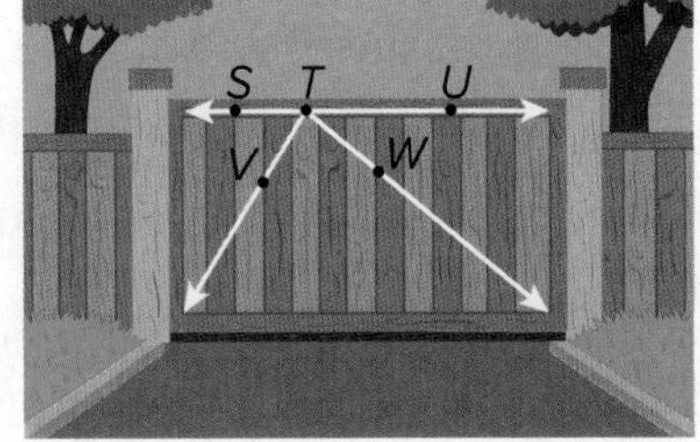

TRAFFIC In the traffic circle around the Arc de Triomphe in Paris, France, there are eight lanes of traffic. Tell whether each angle pair satisfies the given condition.

42. vertical angles

a. $\angle ZCY$ and $\angle TCU$ **b.** $\angle XCW$ and $\angle SCT$

c. $\angle QCR$ and $\angle WCV$ **d.** $\angle TCU$ and $\angle UCT$

43. linear pair

a. $\angle RCU$ and $\angle WCU$ **b.** $\angle QCR$ and $\angle SCR$

c. $\angle VCX$ and $\angle WCY$ **d.** $\angle ZCR$ and $\angle UCW$

44. adjacent angles

a. $\angle WCU$ and $\angle RCU$ **b.** $\angle QCS$ and $\angle SCR$

c. $\angle VCW$ and $\angle QCR$ **d.** $\angle VCX$ and $\angle VCU$

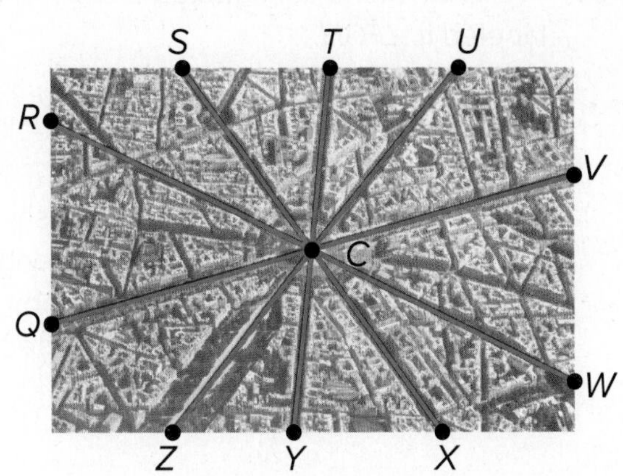

45. **POOL** Felipe uses a computer program to model the paths of pool balls. $\angle GFH$ is a straight angle that represents the rail of the pool table. If $\overrightarrow{FK}$ bisects $\angle JFL$, and $m\angle JFL = 90°$, what is $m\angle LFK$?

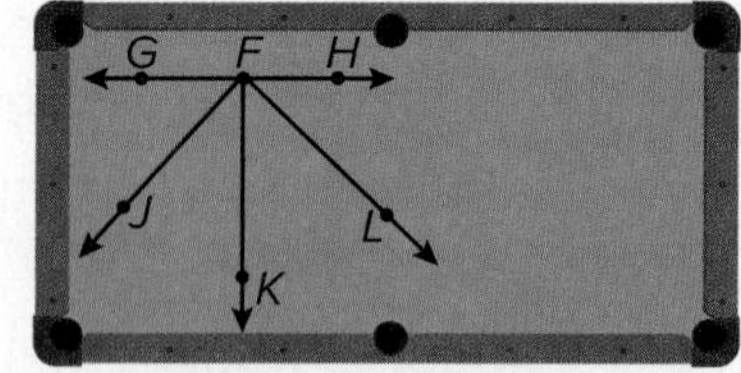

46. **WOODWORKING** Oliver makes rectangular blocks like the one shown and then glues them together to make a plaque. Find $m\angle 1$, $m\angle 2$, and $m\angle 3$, so he can cut the pieces of the plaque.

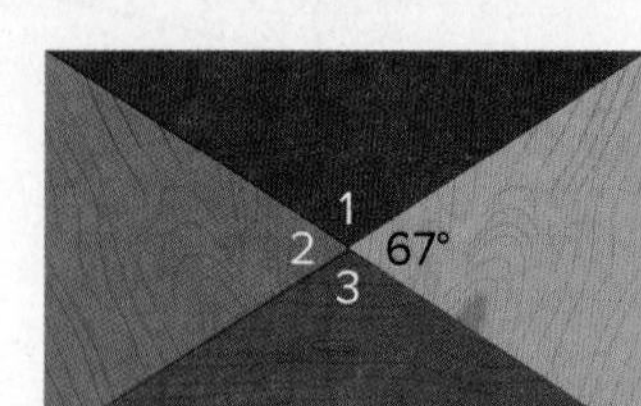

47. **WOODWORKING** Naomi cuts two pieces of baseboard molding to meet in a corner at a 90° angle.

a. To what degree should she set her table saw for the cut?

b. Which ray represents the angle bisector of the molding angle?

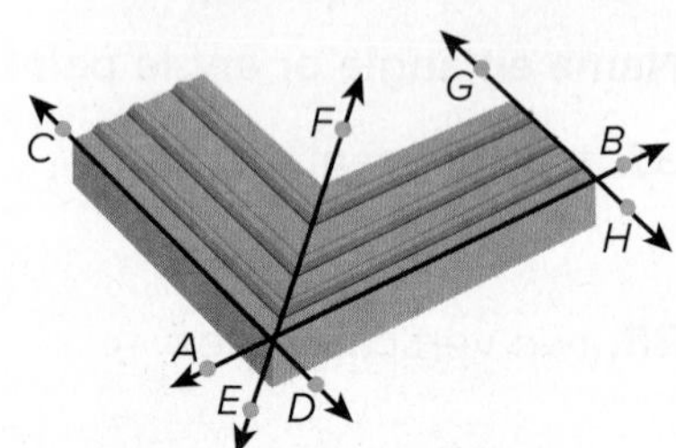

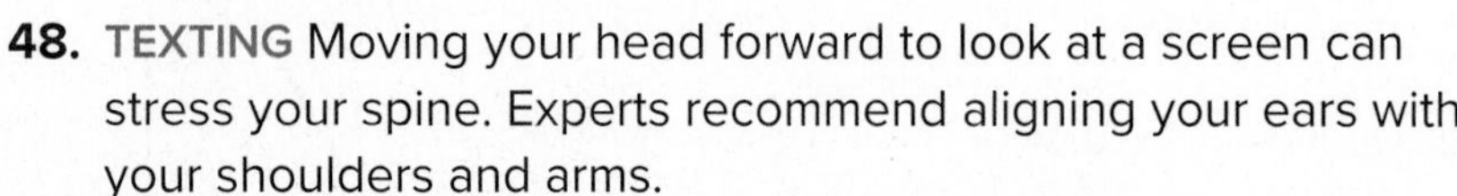

48. **TEXTING** Moving your head forward to look at a screen can stress your spine. Experts recommend aligning your ears with your shoulders and arms.

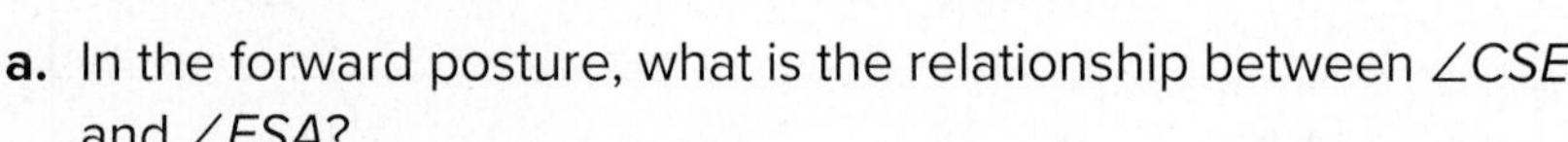

a. In the forward posture, what is the relationship between $\angle CSE$ and $\angle ESA$?

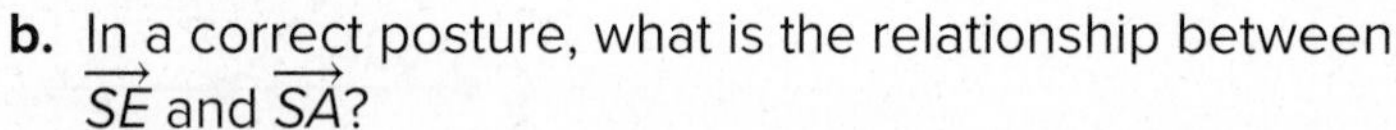

b. In a correct posture, what is the relationship between $\overrightarrow{SE}$ and $\overrightarrow{SA}$?

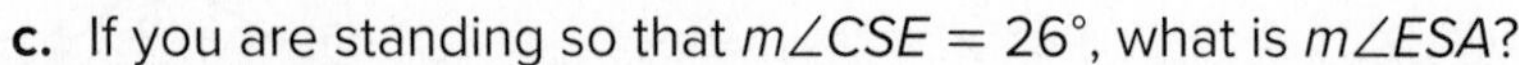

c. If you are standing so that $m\angle CSE = 26°$, what is $m\angle ESA$?

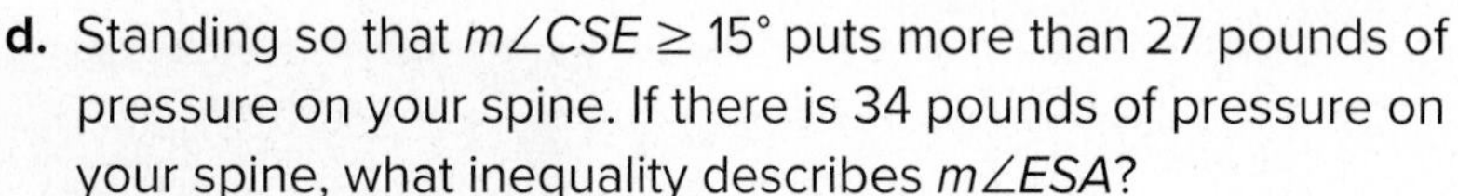

d. Standing so that $m\angle CSE \geq 15°$ puts more than 27 pounds of pressure on your spine. If there is 34 pounds of pressure on your spine, what inequality describes $m\angle ESA$?

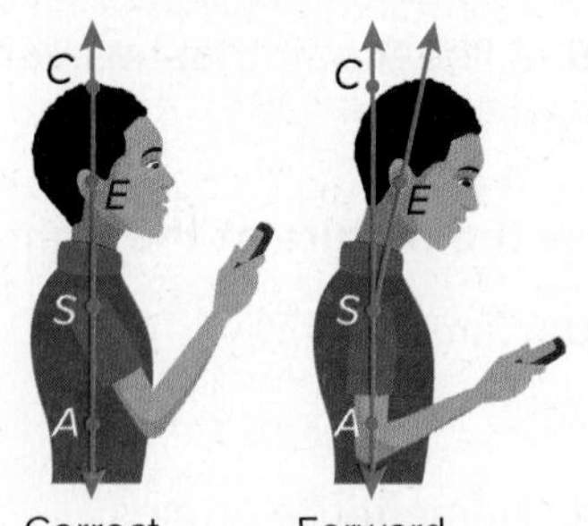

Correct posture   Forward posture

Higher-Order Thinking Skills

49. **PERSEVERE** $\overrightarrow{MP}$ bisects $\angle LMN$, $\overrightarrow{MQ}$ bisects $\angle LMP$, and $\overrightarrow{MR}$ bisects $\angle QMP$. If $m\angle RMP = 21°$, find $m\angle LMN$. Explain your reasoning.

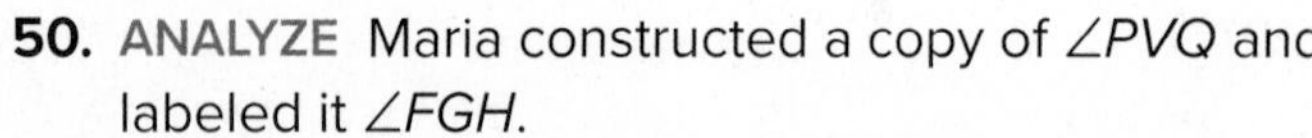

50. **ANALYZE** Maria constructed a copy of $\angle PVQ$ and labeled it $\angle FGH$.

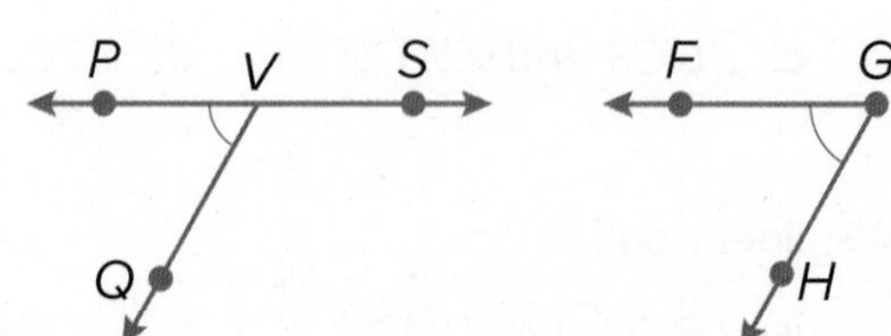

a. Are $\angle FGH$ and $\angle QVS$ a linear pair? Explain.

b. Maria must also copy $\angle QVS$. Sal says she can create a copy of $\angle QVS$ if she extends $\overrightarrow{GH}$ past G. Mona says Maria can create a copy of $\angle QVS$ by extending $\overrightarrow{GF}$ past G. Who is correct? Justify your argument.

Angle Relationships

Explore Complementary and Supplementary Angles

Online Activity Use dynamic geometry software to complete the Explore.

INQUIRY How do complementary angles compare to supplementary angles?

Learn Complementary and Supplementary Angles

Complementary and Supplementary Angles

Complementary Angles	**Supplementary Angles**
Definition	
two angles with measures that have a sum of 90°	two angles with measures that have a sum of 180°
Examples	
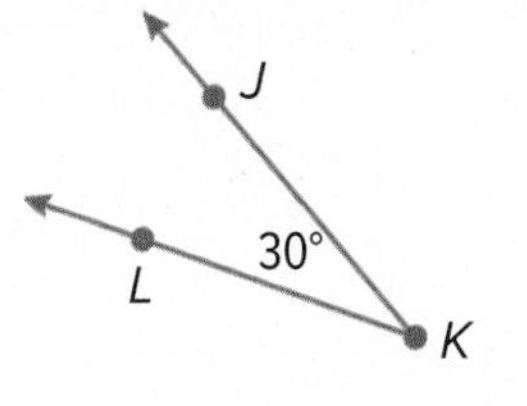 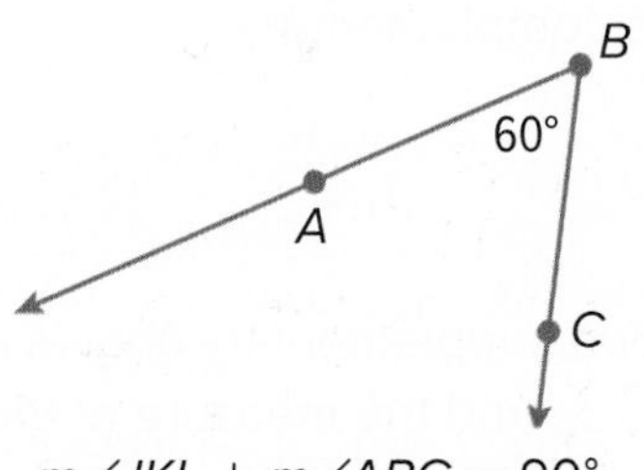 $m\angle JKL + m\angle ABC = 90°$ $30° + 60° = 90°$	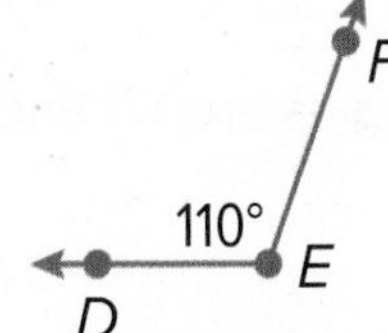 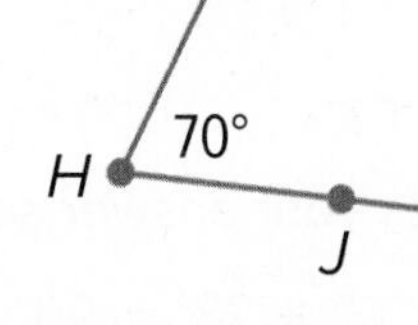 $m\angle DEF + m\angle GHJ = 180°$ $110° + 70° = 180°$
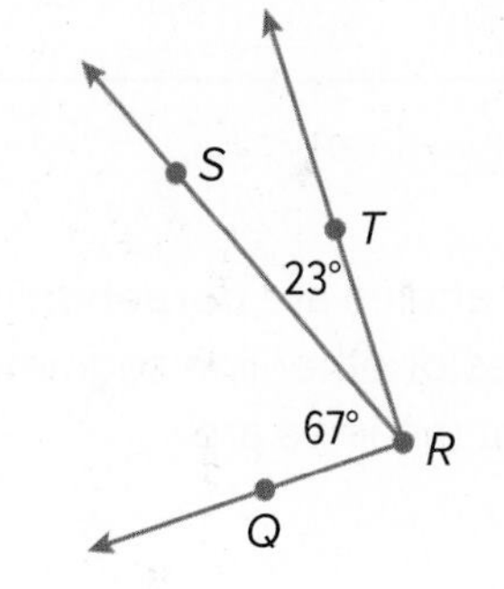 $m\angle QRS + m\angle SRT = 90°$ $67° + 23° = 90°$	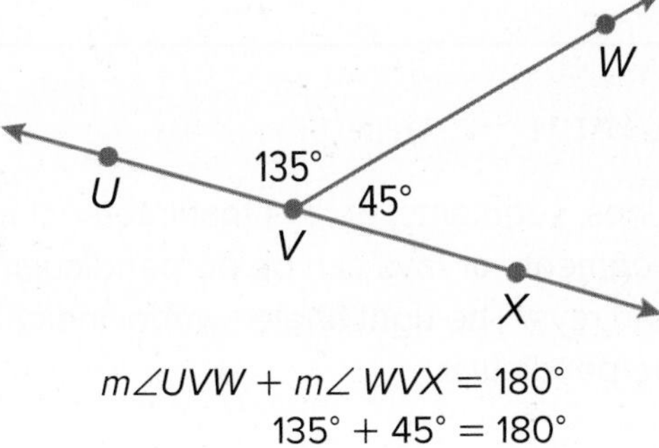 $m\angle UVW + m\angle WVX = 180°$ $135° + 45° = 180°$

Today's Goals

- Calculate angle measures using the characteristics of complementary and supplementary angles.
- Calculate angle measures using the characteristics of perpendicular lines.
- Demonstrate understanding of what can and cannot be assumed from a diagram.

Today's Vocabulary

complementary angles

supplementary angles

perpendicular

Think About It!

A linear pair is 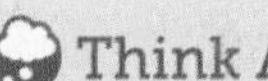 __?__, supplementary while two supplementary angles are __?__ a linear pair.

Study Tip

Complementary and Supplementary Angles Pairs of angles that are complementary or supplementary do not have to be adjacent angles.

Example 1 Complementary and Supplementary Angles

Find the measures of two complementary angles if the measure of the larger angle is five more than four times the measure of the smaller angle.

If two angles are complementary, then the sum of the angle measures is 90°. To find the measures of each angle, first write an equation. Let x = the measure of the smaller angle. Then the measure of the larger angle is $4x + 5$.

Step 1

First, solve for x.

$x + 4x + 5 = 90$	Complementary angle measures add to 90°.
$5x + 5 = 90$	Combine like terms.
$5x = 85$	Subtract 5 from each side.
$x = 17$	Divide each side by 5.

So, the measure of the smaller angle is 17°.

Step 2

Next, find the measure of the larger angle.

$4x + 5 = 4(17) + 5$	Substitute 17 for x.
$= 68 + 5$	Multiply.
$= 73$	Solve.

The measures of the angles are 17° and 73°.

CHECK

Does your answer seem reasonable?

Yes; 17° + 73° = 90°, so the two angles are complementary.

Talk About It!

Adrian claims that if two complementary angles are both acute, then a pair of supplementary angles must both be obtuse. Do you agree? Explain why or why not.

Check

The difference between the measures of two supplementary angles is 18°. The measure of the smaller angle is ___?°___, and the measure of the larger angle is ___?°___.

Learn Perpendicularity

Lines, segments, or rays that intersect at right angles are **perpendicular**. Segments or rays can be perpendicular to lines or other line segments and rays. The right angle symbol indicates that the lines are perpendicular.

Go Online You can complete an Extra Example online.

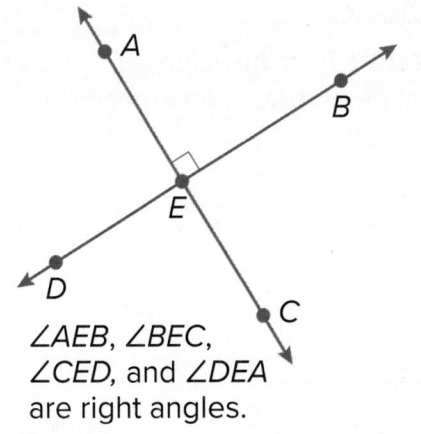

Perpendicular lines intersect to form four right angles.	A, B, E, D, C $\angle AEB$, $\angle BEC$, $\angle CED$, and $\angle DEA$ are right angles.
Perpendicular lines intersect to form congruent adjacent angles.	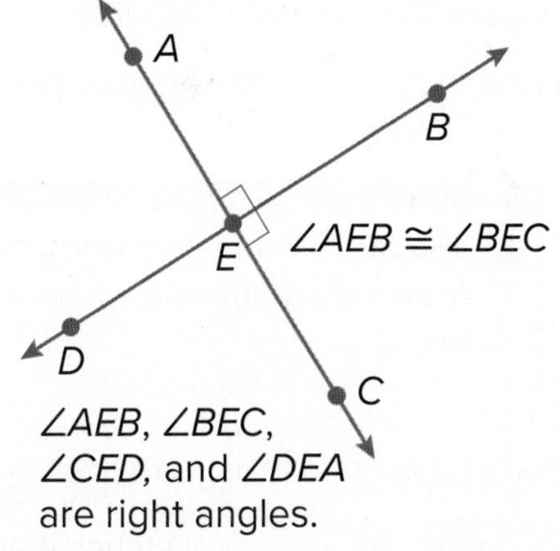 $\angle AEB$, $\angle BEC$, $\angle CED$, and $\angle DEA$ are right angles.

Study Tip

Symbols $\perp$ is read is *perpendicular to.* Example: $\overline{AC} \perp \overline{DB}$

Example 2 Perpendicular Lines

TANGRAMS **The tangram is a puzzle consisting of seven flat shapes called *tans* which are put together to form shapes. Find the values of *x* and *y* such that $\overleftrightarrow{AD}$ and $\overleftrightarrow{EC}$ in the tangram are perpendicular.**

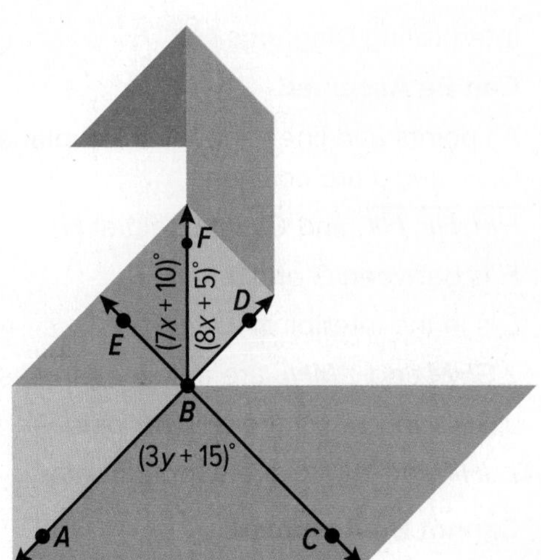

If $\overleftrightarrow{AD}$ and $\overleftrightarrow{EC}$ are perpendicular, then $m\angle ABC = 90°$ and $m\angle EBD = 90°$.

Step 1 Solve for *y*.

$3y + 15 = 90$	$m\angle ABC = 90°$
$y = 25$	Solve for *y*.

Step 2 Solve for *x*.

$m\angle EBF + m\angle FBD = m\angle EBD$	sum of parts = whole
$7x + 10 + 8x + 5 = 90$	Substitution
$x = 5$	Solve for *x*.

Think About It!

Besides right angles, how else can you describe $\angle ABC$ and $\angle EBD$?

Check

DESIGN Find the values of x and y such that $\overleftrightarrow{PR}$ and $\overleftrightarrow{QS}$ are perpendicular.

$x =$ ___?___

$y =$ ___?___

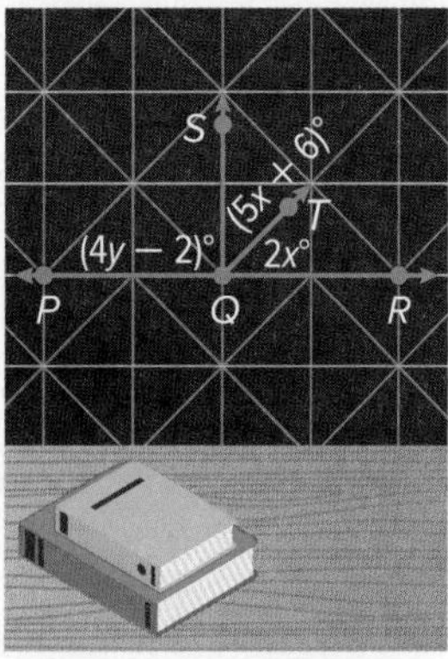

Explore Interpreting Diagrams

 **Online Activity** Use dynamic geometry software to complete the Explore.

INQUIRY What information can be assumed from a diagram, and what information cannot be assumed?

Learn Interpreting Diagrams

In geometry, figures are sketches that are used to depict a situation. They are not drawn to reflect total accuracy. Certain relationships can be assumed from a figure, but most cannot.

Interpreting Diagrams	
Can Be Assumed All points and lines shown are coplanar. G, H, and J are collinear. $\overrightarrow{HM}$, $\overrightarrow{HL}$, $\overrightarrow{HK}$, and $\overleftrightarrow{GJ}$ intersect at H. H is between G and J. L is in the interior of $\angle MHK$. $\angle GHM$ and $\angle MHL$ are adjacent angles. $\angle GHL$ and $\angle LHJ$ are a linear pair. $\angle JHK$ and $\angle KHG$ are supplementary.	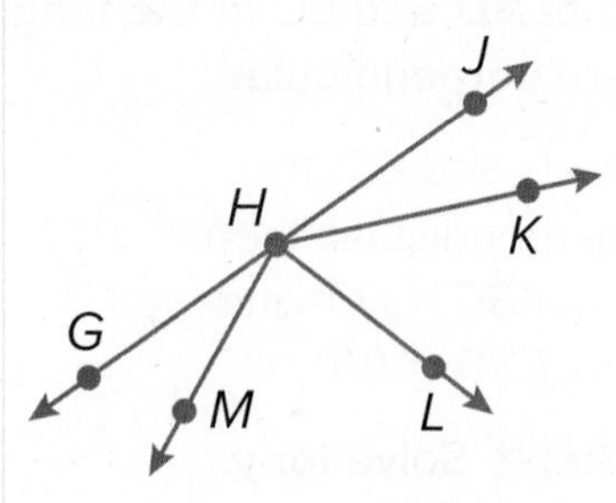
Cannot Be Assumed Lines that appear perpendicular may not be perpendicular. Angles that appear congruent may not be congruent. Segments that appear congruent may not be congruent.	

The list of statements that can be assumed is not a complete list. There are more special pairs of angles than those listed.

 **Go Online** You can complete an Extra Example online.

Because points of intersection can be assumed, you can identify vertical angles from the figure.

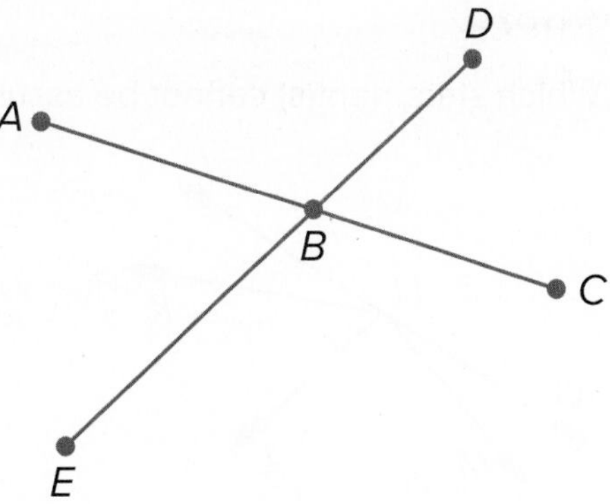

Because linear pairs can be assumed from the figure, you can apply known characteristics of a linear pair, such as supplementary angles.

$\angle ABD$ and $\angle CBE$ are vertical angles.

$\angle EBA$ and $\angle ABD$ form a linear pair, so $m\angle EBA + m\angle ABD = 180°$.

Example 3 Interpreting Diagrams

Determine whether each statement can be assumed from the figure. Explain.

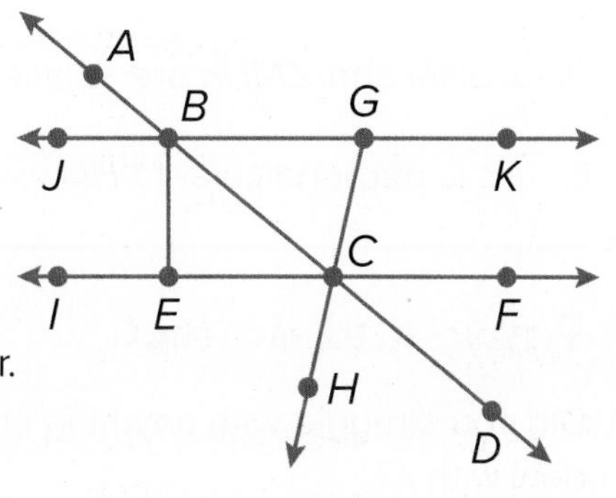

a. $\overrightarrow{CE}$ and $\overrightarrow{CF}$ are opposite rays.

Yes; C is a common endpoint.

b. $\angle BGC$ and $\angle KGC$ form a linear pair.

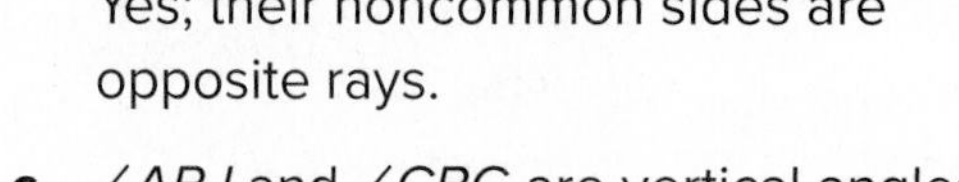

Yes; their noncommon sides are opposite rays.

c. $\angle ABJ$ and $\angle CBG$ are vertical angles.

Yes; these angles are nonadjacent and are formed by two intersecting lines.

d. $\angle BCG$ and $\angle DCF$ are congruent.

No; these angles are not vertical angles. There isn't enough information given to determine this.

e. $\overline{BE}$ and $\overleftrightarrow{IF}$ are perpendicular.

No; there isn't enough information given to determine this.

f. $\angle EBC$ and $\angle GBC$ are complementary angles.

No; there isn't any information about perpendicularity or angle measure so this cannot be determined.

g. $\angle ICH$ and $\angle HCD$ are adjacent angles.

Yes; these angles share a common side.

h. $\overline{BC}$ is an angle bisector of $\angle ECG$.

No; there isn't any information about congruent angles so this cannot be determined.

Think About It!

If you are given that $\overline{BE} \perp \overline{IC}$, can you determine whether $\angle BEI \cong \angle BEC$? Explain your solution process.

Watch Out!

Congruence and Perpendicularity
Remember that congruent angles or segments and perpendicular or parallel lines cannot be assumed from a figure.

Check

Which statement(s) cannot be assumed from the figure?

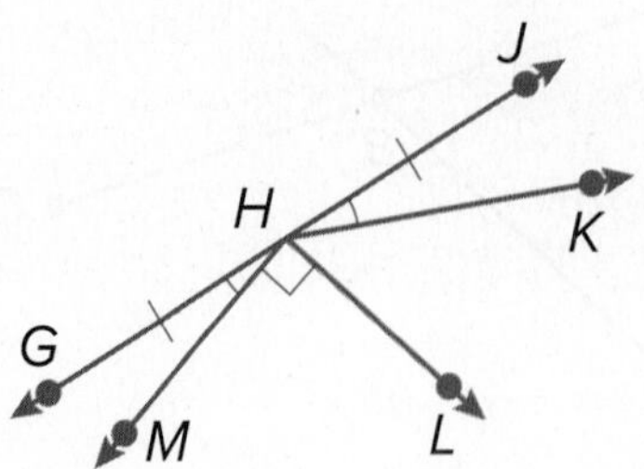

A. $\angle KHJ$ and $\angle GHM$ are complementary.

B. $\angle GHK$ and $\angle JHK$ are a linear pair.

C. $\overrightarrow{HL}$ is perpendicular to $\overrightarrow{HJ}$.

D. $\angle GHM$ and $\angle MHK$ are adjacent angles.

E. HL is perpendicular to HM.

Pause and Reflect

Did you struggle with anything in this lesson? If so, how did you deal with it?

Go Online You may want to complete the construction activities for this lesson.

Go Online You can complete an Extra Example online.

Practice

Go Online You can complete your homework online.

Example 1

1. Find the measures of two supplementary angles if the difference between the measures of the two angles is 35°.

2. $\angle E$ and $\angle F$ are complementary. The measure of $\angle E$ is 54° more than the measure of $\angle F$. Find the measure of each angle.

3. The measure of an angle's supplement is 76° less than the measure of the angle. Find the measures of the angle and its supplement.

4. $\angle Q$ and $\angle R$ are complementary. The measure of $\angle Q$ is 26° less than the measure of $\angle R$. Find the measure of each angle.

5. The measure of the supplement of an angle is three times the measure of the angle. Find the measures of the angle and its supplement.

6. The bascule bridge shown is opening from its horizontal position to its fully vertical position. So far, the bridge has lifted 35° in 21 seconds. At this rate, how much longer will it take for the bridge to reach its vertical position?

Nick Tzolov/E+/Getty Images

Example 2

7. Rays BA and BC are perpendicular. Point D lies in the interior of $\angle ABC$. If $m\angle ABD = (3r + 5)°$ and $m\angle DBC = (5r - 27)°$, find $m\angle ABD$ and $m\angle DBC$.

8. $\overleftrightarrow{WX}$ and $\overleftrightarrow{YZ}$ intersect at point V. If $m\angle WVY = (4a + 58)°$ and $m\angle XVY = (2b - 18)°$, find the values of a and b such that $\overleftrightarrow{WX}$ is perpendicular to $\overleftrightarrow{YZ}$.

9. Refer to the figure at the right. If $m\angle 2 = (a + 15)°$ and $m\angle 3 = (a + 35)°$, find the value of a such that $\overrightarrow{HL} \perp \overrightarrow{HJ}$.

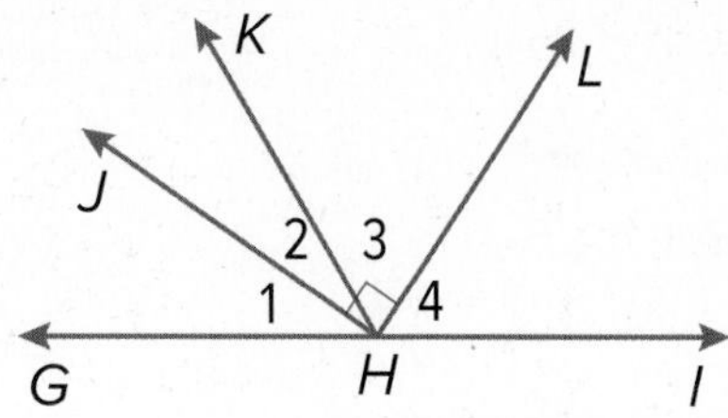

10. Rays DA and DC are perpendicular. Point B lies in the interior of $\angle ADC$. If $m\angle ADB = (3a + 10)°$ and $m\angle BDC = 13a°$, find a, $m\angle ADB$, and $m\angle BDC$.

Example 3

Determine whether each statement can be assumed from the given figure. Explain.

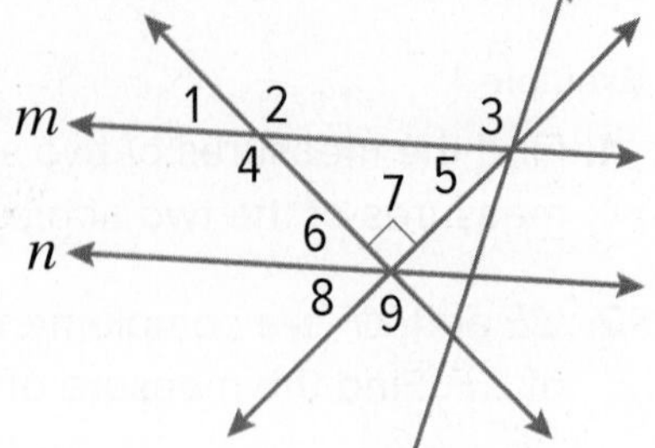

11. $\angle 6$ and $\angle 8$ are complementary.

12. $\angle 7$ and $\angle 8$ form a linear pair.

13. $\angle 2$ and $\angle 4$ are vertical angles.

14. $m\angle 9 = m\angle 6 + m\angle 8$

Mixed Exercises

15. The measure of the supplement of an angle is 60° less than four times the measure of the complement of the angle. Find the measure of the angle.

16. $\angle 6$ and $\angle 7$ form a linear pair. Twice the measure of $\angle 6$ is twelve more than four times the measure of $\angle 7$. Find the measure of each angle.

Refer to the figure at the right.

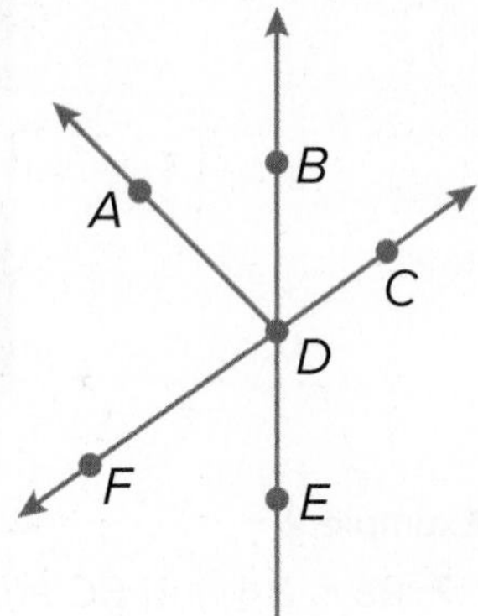

17. If $m\angle ADB = (6x - 4)°$ and $m\angle BDC = (4x + 24)°$, find the value of x such that $\angle ADC$ is a right angle.

18. If $m\angle FDE = (3x - 15)°$ and $m\angle FDB = (5x + 59)°$, find the value of x such that $\angle FDE$ and $\angle FDB$ are supplementary.

19. If $m\angle BDC = (8x + 12)°$ and $m\angle FDB = (12x - 32)°$, find $m\angle FDE$.

Determine whether each statement can be assumed from the given figure. Explain.

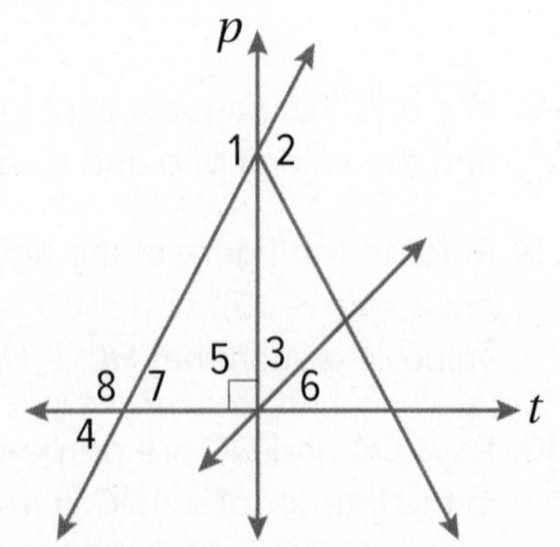

20. $\angle 4$ and $\angle 7$ are vertical angles.

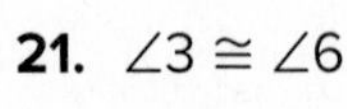

21. $\angle 3 \cong \angle 6$

22. $m\angle 5 = m\angle 3 + m\angle 6$

23. $\angle 5$ and $\angle 7$ form a linear pair.

For Exercises 24 and 25, lines p and q intersect to form adjacent angles 1 and 2.

24. If $m\angle 1 = (7x + 6)^\circ$ and $m\angle 2 = (8x - 6)^\circ$, find the value of x such that p is perpendicular to q.

25. If $m\angle 1 = (4x - 3)^\circ$ and $m\angle 2 = (3x + 8)^\circ$, find the value of x such that $\angle 1$ is supplementary to $\angle 2$.

26. COLOR GUARD Shannon is designing a new rectangular flag for the school's color guard and is determining the angles at which to cut the fabric. She wants the measure of $\angle 2$ to be three times as great as the measure of $\angle 1$. She thinks the measures of $\angle 3$ and $\angle 4$ should be equal. Finally, she wants the measure of $\angle 6$ to be half that of $\angle 5$. Determine the measures of the angles.

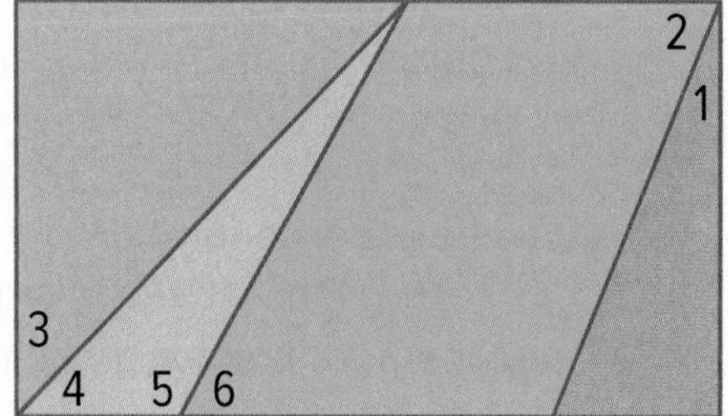

27. STRING ART String art is created by wrapping string around nails or wires to form patterns. Use the string art pattern to find the values of x, y, and z.

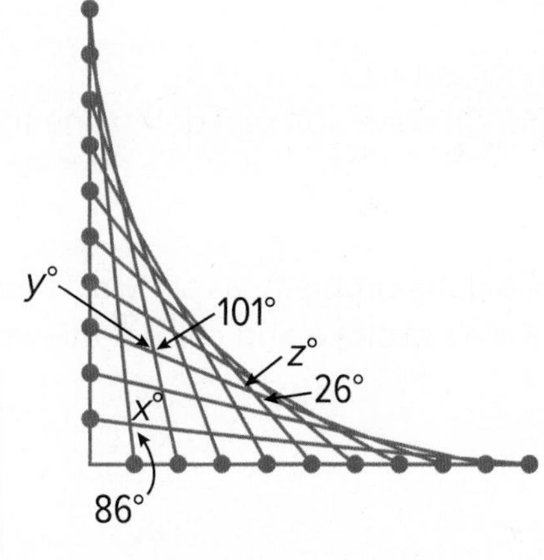

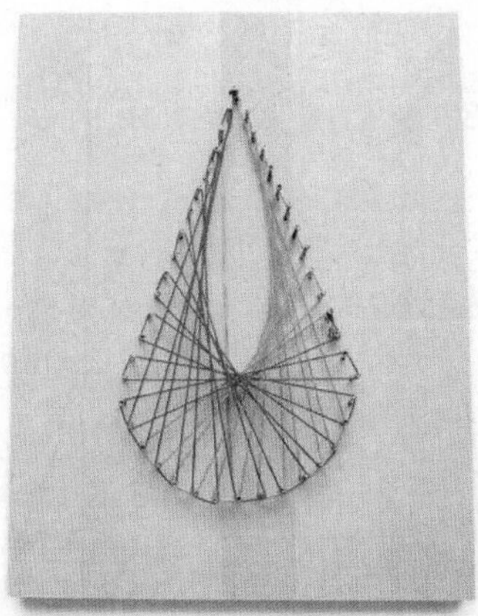

McGraw-Hill Education

28. USE TOOLS Draw an acute angle, $\angle ABC$. Let $m\angle ABC = (6x - 1)^\circ$.

a. Use a protractor to determine the measure of $\angle ABC$. Use this measure to determine the value of x.

b. Explain how you would determine the measure of an angle that is complementary to $\angle ABC$.

c. Explain how you would determine the measure of an angle that is supplementary to $\angle ABC$.

Higher-Order Thinking Skills

29. ANALYZE Are there angles that do not have a complement? Justify your argument.

30. PERSEVERE If a line, line segment, or ray is perpendicular to a plane, then it is perpendicular to every line, line segment, or ray in the plane that intersects it.

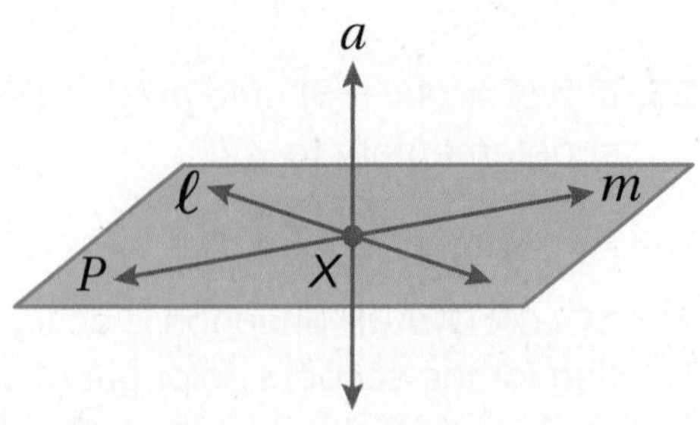

a. If a line is perpendicular to each of two intersecting lines at their point of intersection, then the line is perpendicular to the plane determined by them. If line a is perpendicular to line ℓ and line m at point X, what must also be true?

b. If a line is perpendicular to a plane, then any line perpendicular to the given line at the point of intersection with the given plane is in the given plane. If line a is perpendicular to plane P and line m at point X, what must also be true?

c. If a line is perpendicular to a plane, then every plane containing the line is perpendicular to the given plane. If line a is perpendicular to plane P, what must also be true?

31. WRITE Describe three different ways you can determine that an angle is a right angle.

32. FIND THE ERROR Kaila solved the problem, as shown. Is her solution correct? If it is, explain your reasoning. If not, explain Kaila's mistake and correct the work.

If $m\angle F = (6x - 9)^\circ$ and $m\angle G = (2x + 13)^\circ$, find the value of x such that $\angle F$ and $\angle G$ are supplementary.

$$(6x - 9)^\circ + (2x + 13)^\circ = 90^\circ$$
$$8x^\circ - 4^\circ = 90^\circ$$
$$8x^\circ = 86^\circ$$
$$x = 10.75$$

33. CREATE Create $\angle 1$ along with its complement and supplement by drawing only a line and two rays.

34. WHICH ONE DOESN'T BELONG Three students used the figure to write a statement. Is each statement correct? Justify your conclusion.

Samar: $\angle WZU$ is a right angle.

Jana: $\angle YZU$ and $\angle UZV$ are supplementary.

Antonio: $\angle VZU$ is adjacent to $\angle YZX$.

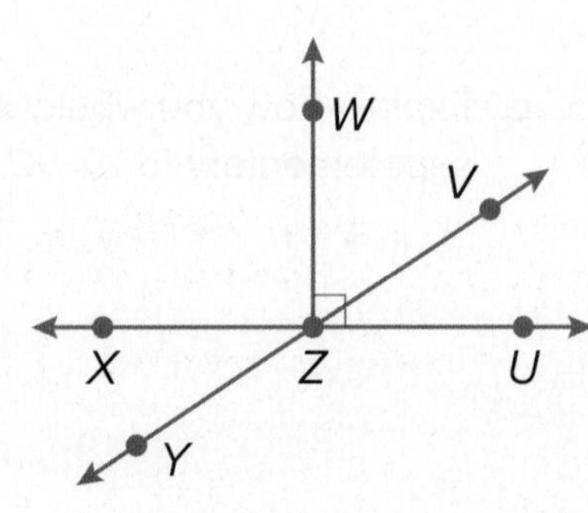

35. ANALYZE Do all angles have a supplement? Explain.

Two-Dimensional Figures

Learn Perimeter, Circumference, and Area

A **polygon** is a closed plane figure with at least three straight sides.

The **perimeter** of a polygon is the sum of the lengths of the sides of the polygon. Some shapes have special formulas for perimeter, but all are derived from the basic definition of perimeter.

The **circumference** of a circle is the distance around the circle.

Area is the number of square units needed to cover a surface.

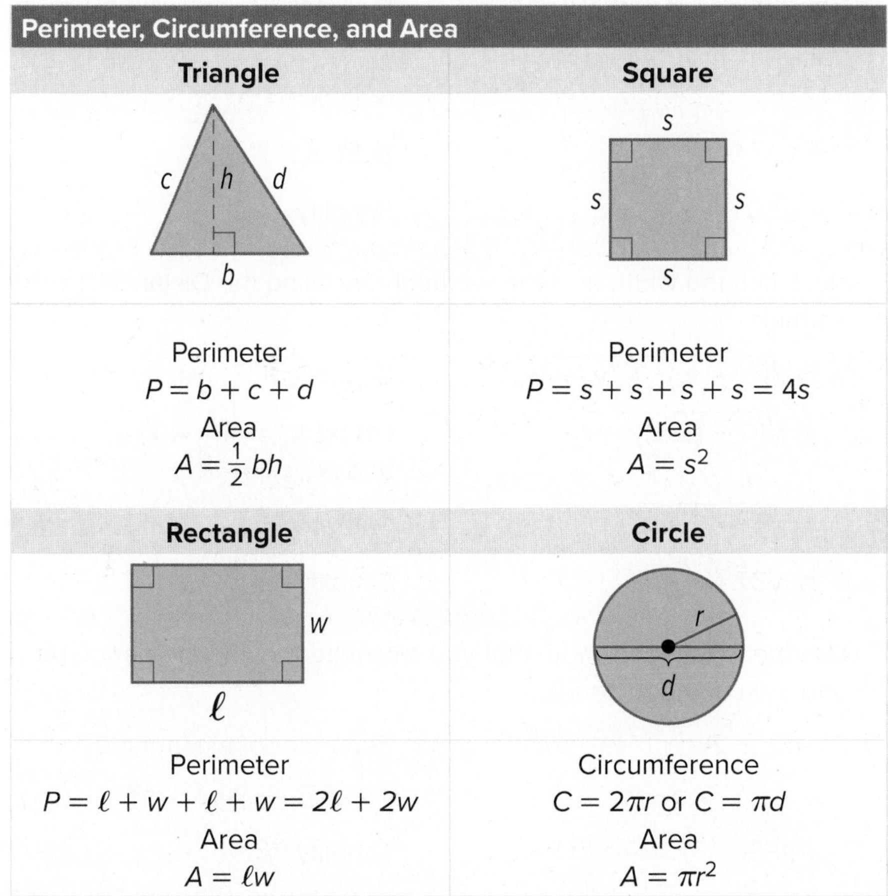

Perimeter, Circumference, and Area

Triangle	Square
Perimeter $P = b + c + d$ Area $A = \frac{1}{2}bh$	Perimeter $P = s + s + s + s = 4s$ Area $A = s^2$

Rectangle	Circle
Perimeter $P = \ell + w + \ell + w = 2\ell + 2w$ Area $A = \ell w$	Circumference $C = 2\pi r$ or $C = \pi d$ Area $A = \pi r^2$

You can use the Distance Formula to find the perimeter and area of a polygon graphed on a coordinate plane. You can also use the Distance Formula to calculate the radius of a circle and then use the appropriate equations for circumference and area.

An **equilateral polygon** has all sides congruent. An **equiangular polygon** has all angles congruent. A **regular polygon** is a convex polygon that is both equilateral and equiangular.

Today's Goals

- Find perimeters, circumferences, and areas of two-dimensional geometric shapes.
- Calculate the measures of real-world objects.

Today's Vocabulary

polygon
perimeter
circumference
area
equilateral polygon
equiangular polygon
regular polygon
concave
convex
geometric model

Go Online You can watch a video to see how to find the perimeter and area of a figure on the coordinate plane.

Example 1 Find Perimeter, Circumference, and Area

Find the perimeter or circumference and area of each figure.

a. Rectangle *ABCD*

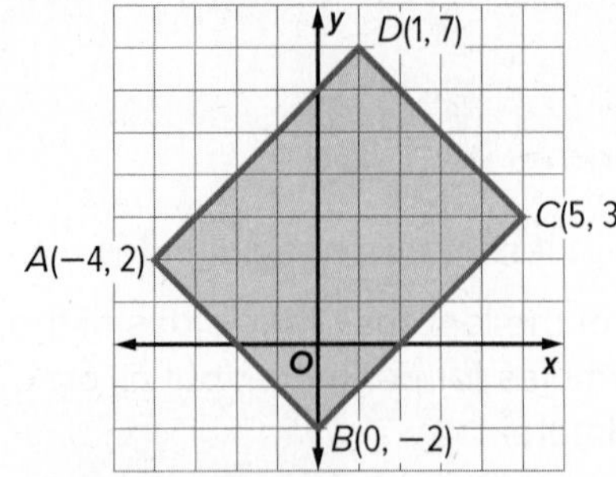

First, find the length ℓ of the rectangle by using the Distance Formula.

$\ell = \sqrt{(x_2 - x_1)^2 + (y_2 - y_1)^2}$	Distance Formula
$= \sqrt{[1 - (-4)]^2 + (7 - 2)^2}$	Let $(x_1, y_1) = A(-4, 2)$ and $(x_2, y_2) = D(1, 7)$.
$= \sqrt{5^2 + 5^2}$	Subtract.
$= \sqrt{50}$	Simplify.

Next, find the width w of the rectangle by using the Distance Formula.

$w = \sqrt{(x_2 - x_1)^2 + (y_2 - y_1)^2}$	Distance Formula
$= \sqrt{[0 - (-4)]^2 + [(-2) - 2]^2}$	Let $(x_1, y_1) = A(-4, 2)$ and $(x_2, y_2) = B(0, -2)$.
$= \sqrt{4^2 + (-4)^2}$	Subtract.
$= \sqrt{32}$	Simplify.

Use the length and width that you calculated to find the perimeter and area of the rectangle.

$P = 2\ell + 2w$	Perimeter of a rectangle
$= 2\sqrt{50} + 2\sqrt{32}$	$\ell = \sqrt{50}$ and $w = \sqrt{32}$
≈ 25.5	Simplify.

The perimeter is about 25.5 units.

$A = \ell w$	Area of a rectangle
$= \sqrt{50} \times \sqrt{32}$	$\ell = \sqrt{50}$ and $w = \sqrt{32}$
$= 40$	Simplify.

The area is 40 square units.

Study Tip

Concave and Convex

Polygons can be **concave** or **convex**. Suppose the line containing each side is drawn. If any of the lines contain any point in the interior of the polygon, then it is concave. Otherwise, it is convex.

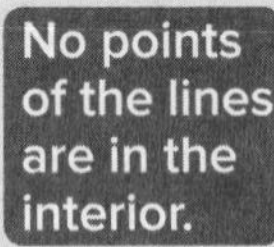

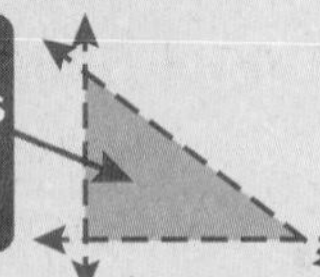

convex polygon

Some of the lines pass through the interior.

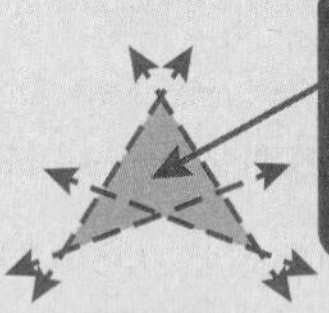

concave polygon

Study Tip

Perimeter vs. Area

Because calculating the area of a figure involves multiplying two dimensions, *square units* are used. There is only one dimension used when finding the perimeter, thus, it is given simply in *units*.

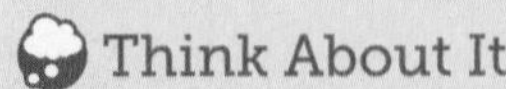

Think About It!

How can you use the coordinate grid to check your answer?

b. Circle C

C(4, 8)
D(6, 3)

Use the Distance Formula to calculate the length of the radius of the circle.

$r = \sqrt{(x_2 - x_1)^2 + (y_2 - y_1)^2}$	Distance Formula
$= \sqrt{(6 - 4)^2 + (3 - 8)^2}$	$C(4, 8)$ and $D(6, 3)$
$= \sqrt{2^2 + (-5)^2}$	Subtract.
$= \sqrt{29}$	Simplify.

Use the value of *r* to find the circumference and area of the circle.

$C = 2\pi r$	Circumference
$= 2\pi\sqrt{29}$ or about 33.8	$r = \sqrt{29}$

The circumference of the circle is about 33.8 units.

$A = \pi r^2$	Area of a circle
$= \pi(\sqrt{29})^2$	$r = \sqrt{29}$
$= 29\pi$ or about 91.1	Simplify.

The area of the circle is about 91.1 square units.

Check

Find the circumference and area of the circle. Round to the nearest tenth if necessary.

$C \approx$ ___?___ units

$A \approx$ ___?___ units^2

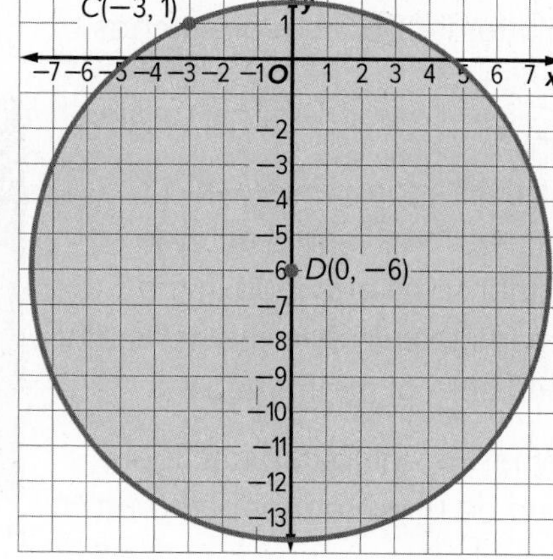

Explore Modeling Objects by Using Two-Dimensional Figures

 **Online Activity** Use real-world objects to complete the Explore.

INQUIRY How can you apply the properties of two-dimensional figures to solve real-world problems?

Learn Modeling with Two-Dimensional Figures

A **geometric model** is a geometric figure that represents a real-world object. A good model shows all the important characteristics of the object it represents, although some of the detail may be lost.

Drafters use two-dimensional geometric models to create technical drawings that communicate an object's function or construction. Scientists may use two-dimensional models to record an object's general shape or mechanics in a field notebook. You can use two-dimensional models to estimate the perimeter, circumference, and area of objects.

Study Tip

Approximations Because the corners of the serving platter are rounded, we can only approximate the perimeter and area of the serving platter using the formulas for a square.

Example 2 Modeling with Two-Dimensional Figures

Use an appropriate two-dimensional model and the dimensions provided in the image to calculate the perimeter and area of the plate.

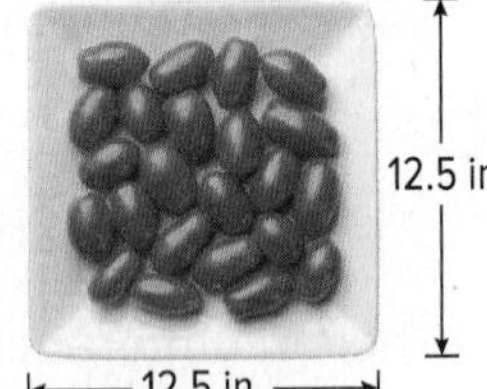

What two-dimensional figure can be used to model the serving platter? square

What are the perimeter and area of the serving platter? Round to the nearest tenth, if necessary.

Perimeter $= 4s = 4(12.5) = 50$ in.

Area $= s^2 = (12.5)^2 = 156.3$ in^2

Because the platter is a square, the perimeter of the platter is 4 multiplied by the length of the side. The area is the length of the side squared. The perimeter of the platter is 50 inches, and the area of the platter is 156.3 square inches.

Check

Use an appropriate two-dimensional model and the dimensions provided in the image to calculate the perimeter and area of the framed art.

What two-dimensional figure can be used to model the art?

$P =$ __?__ cm; $A =$ __?__ cm^2

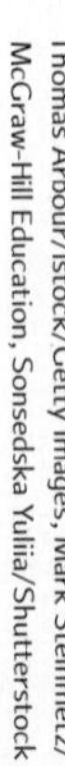

Example 3 Using a Two-Dimensional Model

BUSINESS Isaiah owns a small café.

Part A A new fire code states that there must be 15 square feet of free space for every customer in the café. How many people can be in the café?

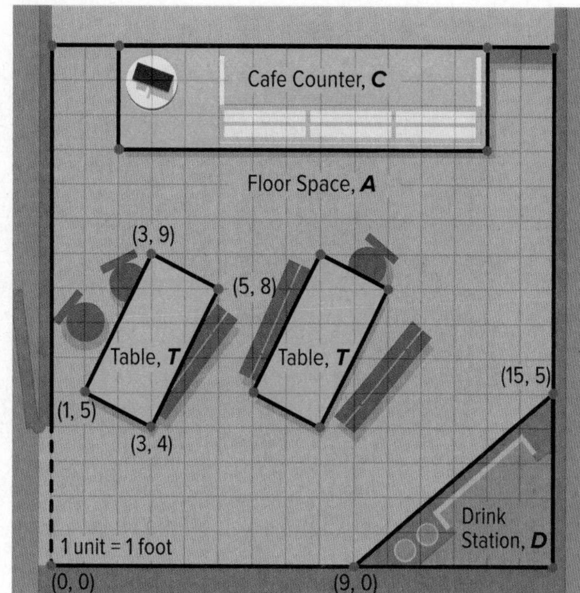

Step 1 Find the amount of free space available.

Find the total area of the café.

Area of the café =
15 × 15 or 225 ft^2

Then, find the area of the counter and the drink station.

$C = 3 \times 11$ or 33 ft^2

$D = \frac{1}{2}(5 \cdot 6)$ or 15 ft^2

Find the areas of the tables by using the Distance Formula.

$\ell = \sqrt{(3-1)^2 + (9-5)^2}$ or $\sqrt{20}$ and $w = \sqrt{(3-1)^2 + (4-5)^2}$ or $\sqrt{5}$

$T = \ell \cdot w = \sqrt{20} \cdot \sqrt{5}$ or 10 ft^2

Find the amount of free space available for Isaiah's customers.

A = area of the café $- C - D - 2T$

$= 225 - 33 - 15 - (2 \times 10)$ or 157 ft^2

Step 2 Find the number of people that can be in the café.

$157 \text{ ft}^2 \cdot \frac{1 \text{ person}}{15 \text{ ft}^2} \approx 10.5$ or 10 people

The café can hold 10 people.

Part B Isaiah wants to hang garland around the tables and the drink station. How much garland does Isaiah need?

Find the sum of the perimeters of the tables and drink station.

length of garland = 2 • perimeter of table + perimeter of drink station

$= 2(2\sqrt{20} + 2\sqrt{5}) + (6 + 5 + \sqrt{(15-9)^2 + (5-0)^2})$

≈ 45.6 feet

Isaiah would need at least 45.6 feet of garland.

Problem-Solving Tip

Evaluate Your Answer It can be tempting to complete the final calculation in a multi-step exercise and conclude that you have arrived at the answer. However, always remember to define appropriate quantities when solving a real-world problem. In this example, it does not make sense to have 10.5 people. You can determine that a correct answer for this exercise must be a whole number.

Study Tip

Radical Form Leave answers in radical form until the last calculation. This will prevent compounding errors caused by rounding throughout steps within a problem.

Check

LANDSCAPING Monica is redesigning her backyard. She has created the following blueprint to model her design.

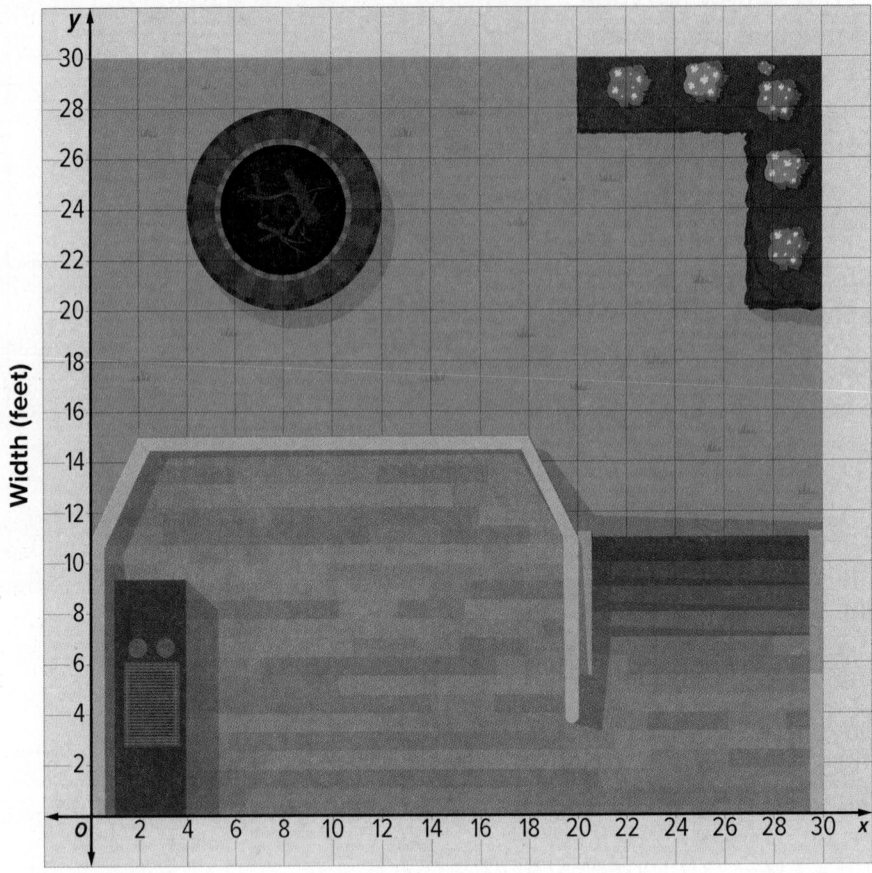

Part A

Monica wants to have at least 300 square feet of grass available in the backyard for her dog. Is there enough space for her dog? If there is, then how much area is available?

A. no

B. yes; 387.7 ft^2

C. yes; 396.7 ft^2

D. yes; 472.7 ft^2

Part B

Monica wants to build a fence in the backyard. She does not want to enclose the edge of the deck that extends from (0, 0) to (30, 0). If Monica wants to enclose the rest of the backyard, including the side edges of the deck and the side edge of the stairs, then how many feet of material are needed to complete the project?

_?_ feet

Go Online You can complete an Extra Example online.

Practice

Go Online You can complete your homework online.

Example 1

Find the perimeter or circumference and area of each figure if each unit on the graph measures 1 centimeter. Round answers to the nearest tenth, if necessary.

1.

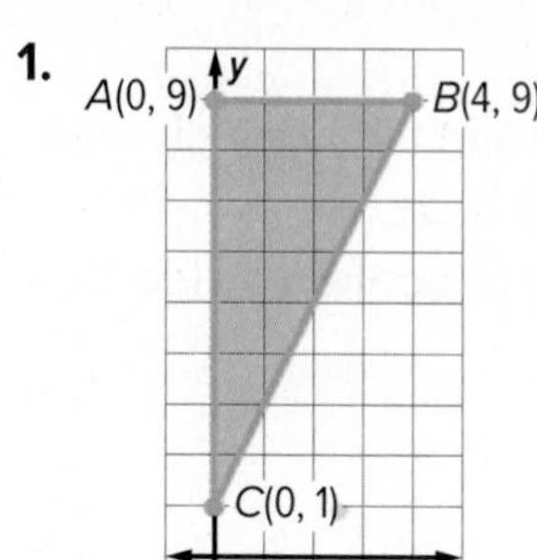

2.

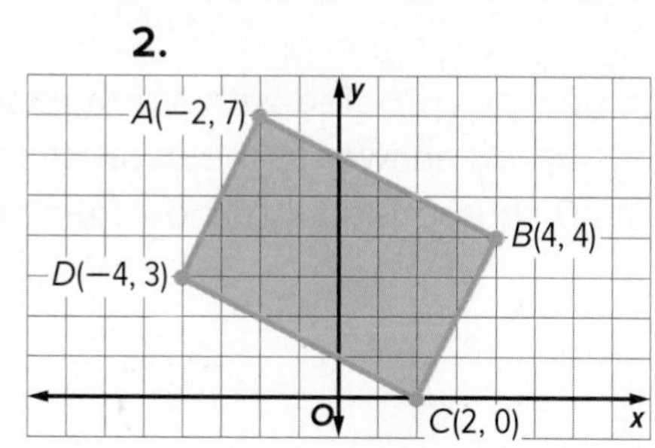

3.

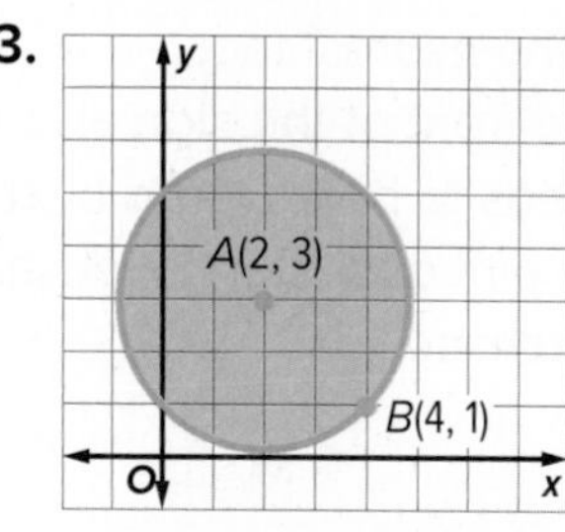

4.

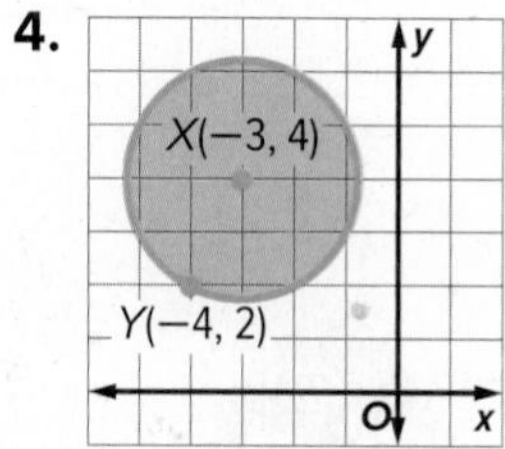

5.

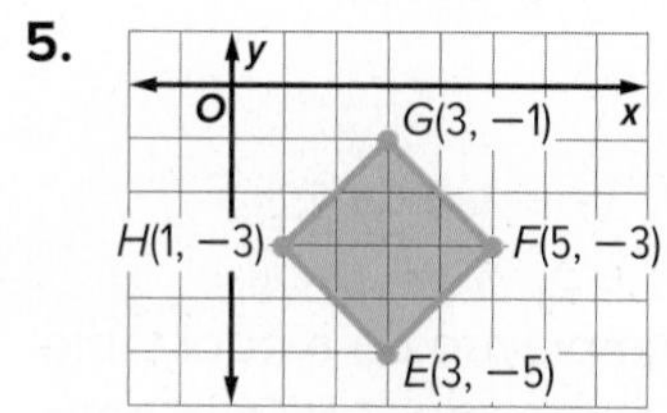

6.

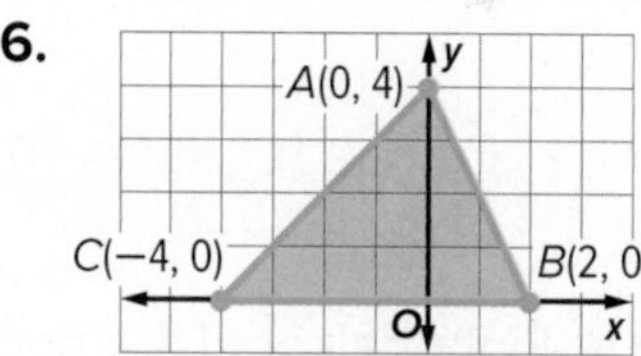

Magnetix/Shutterstock.com

Example 2

Use a two-dimensional model and the dimensions provided to calculate the perimeter or circumference and area of each object. Round to the nearest tenth, if necessary.

7.

8.

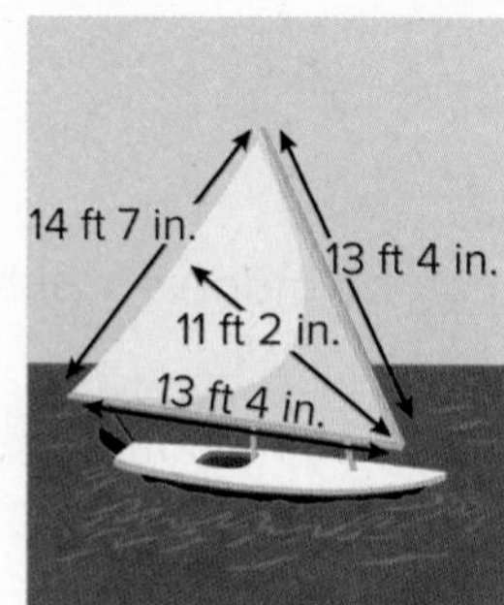

9.

Example 3

10. **DESIGN** Dev is designing a new sign for his art studio. However, he needs to make several improvements to the sign before it is ready to be hung.

a. Dev wants to add a metal trim around the perimeter of the sign. How much trim should Dev purchase? Round answer to the nearest foot.

b. The front of the sign also needs to be waterproofed with a protective sealer. How much area needs to be covered by the sealer? Round answer to the nearest square foot.

c. If a pint of sealer covers an area of 20 square feet, then how many pints of sealer should Dev purchase?

11. **WORLD RECORD** The world's largest ice cream cake was created on May 10, 2011, in Toronto, Canada. The cake was 4.45 meters long, 4.06 meters wide, and 1 meter tall. All surfaces of the cake except the bottom were covered with a cookie crumble topping. Use an appropriate two-dimensional model to approximate the area covered by the cookie crumble topping. Round the answer to the nearest tenth of a square meter.

12. **POOL** Eight-ball pool is a popular game played on a pool table that has six pockets. In eight-ball pool, there are 7 striped balls, 7 solid-colored balls, and a black eight ball. At the beginning of each game, players position the 15 balls in a rack in preparation for the first shot.

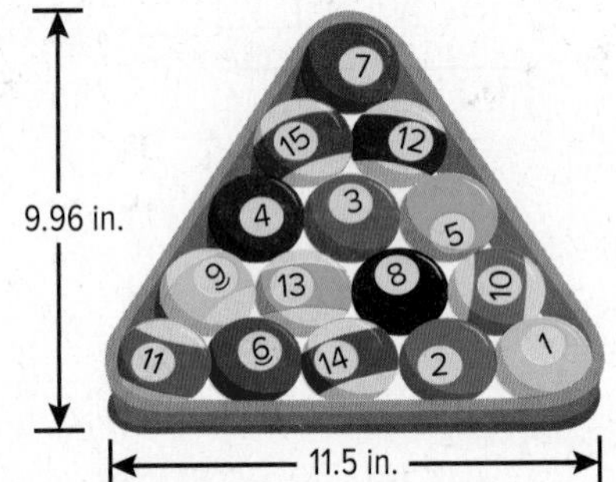

a. Find the area contained by the rack using an appropriate two-dimensional model. Round the answer to the nearest tenth of a square inch.

b. Approximate the area covered by a single ball to the nearest tenth of a square inch.

13. **TRACK** A 400-meter Olympic-size track can be modeled with a rectangle and two semicircles.

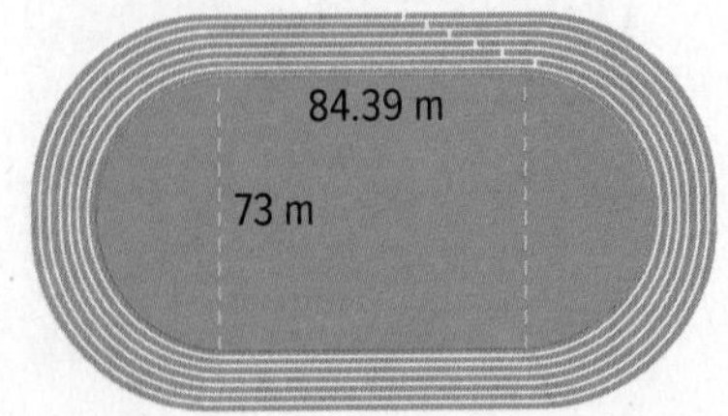

a. If an athlete runs around the track once, then how far has the athlete traveled to the nearest meter?

b. What assumption can be used to explain the difference between your answer in **part a** and the actual length around the track?

c. Each lane is 1.22 meters wide. If the athlete runs in the center of the inside lane, then how far has she traveled after a single lap to the nearest meter?

d. How far inside the track should the athlete be positioned to run exactly 400 meters? Round the answer to the nearest centimeter.

Mixed Exercises

Identify the figure with the given vertices. Find the perimeter and area of the figure.

14. $A(3, 5)$, $B(3, 1)$, $C(0, 1)$

15. $Q(-3, 2)$, $R(1, 2)$, $S(1, -4)$, $T(-3, -4)$

16. $G(-4, 1)$, $H(4, 1)$, $I(0, -2)$

17. $K(-1, 1)$, $L(3, 4)$, $M(6, 0)$, $N(2, -3)$

18. Rectangle *WXYZ* has a length that is 5 more than three times its width.

a. Draw and label a figure for rectangle *WXYZ*.

b. Write an algebraic expression for the perimeter of the rectangle.

c. Find the width if the perimeter is 58 millimeters. Explain how you can check that your answer is correct.

d. Use a ruler to draw and label $\overline{PQ}$, which is congruent to the segment representing the length of rectangle *WXYZ*. What is the measure of $\overline{PQ}$?

19. FENCING The figure shows Derek's house and his backyard on a coordinate grid. Derek is planning to fence in the play area in his backyard. Part of the play area is enclosed by the house and does not need to be fenced. Each unit on the coordinate grid represents 5 feet. The cost for the fencing materials and installation is \$10 per foot. How much will it cost Derek to install the fence? Explain.

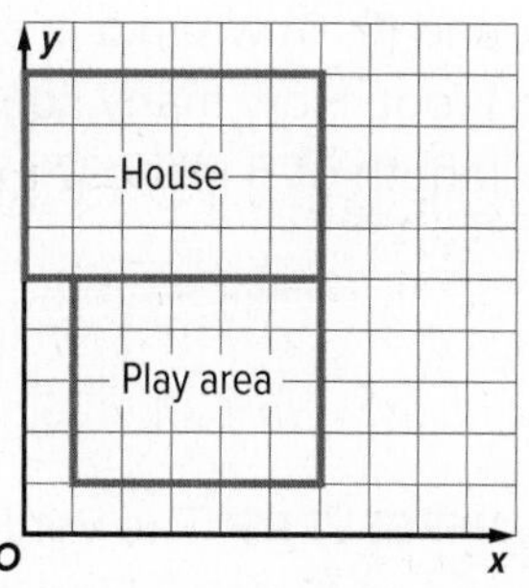

20. Explain a method to find the area of $\triangle QRS$ given that $\overline{RT} \perp \overline{QS}$. Then find the area. Show your work.

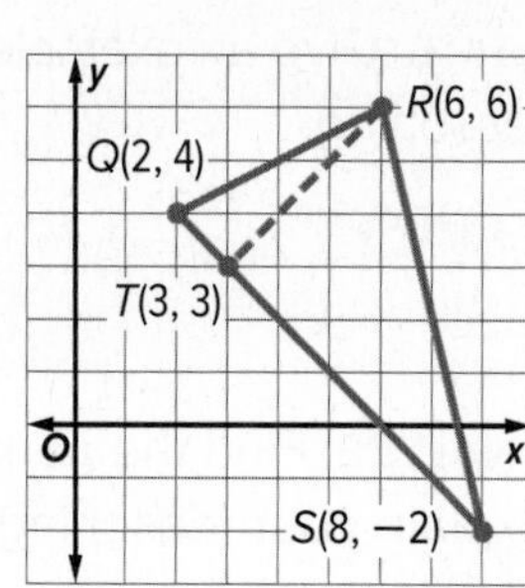

21. **SONAR** Sonar is used by oceanographers to locate marine animals and to map the contours of the ocean floor. Sonar sends out sound pulses, called pings, and receives the returning sound echo. Sonar uses the returning sound echo to detect the location of animals or the distance from a rock formation. If each unit on the coordinate grid measures 1 mile, then what area does the sonar system cover? Round to the nearest tenth.

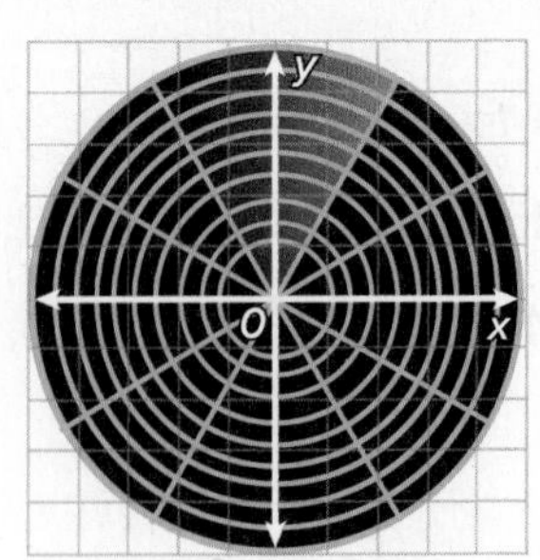

22. Two vertices of square *ABCD* are *C*(5, 8) and *D*(2, 4).

    a. Do you need to find the coordinates for the other two vertices to find the perimeter and area of the square? Justify your argument.

    b. Find the perimeter and area of square *ABCD*. Show your work.

23. The coordinate grid shows an equilateral triangle that fits inside a square.

    a. Find the area of the square. Show your work.

    b. Find the area of the triangle. Show your work.

    c. Find the area of the square that is not covered by the triangle. Write an exact value and then round to the nearest tenth. Justify your reasoning.

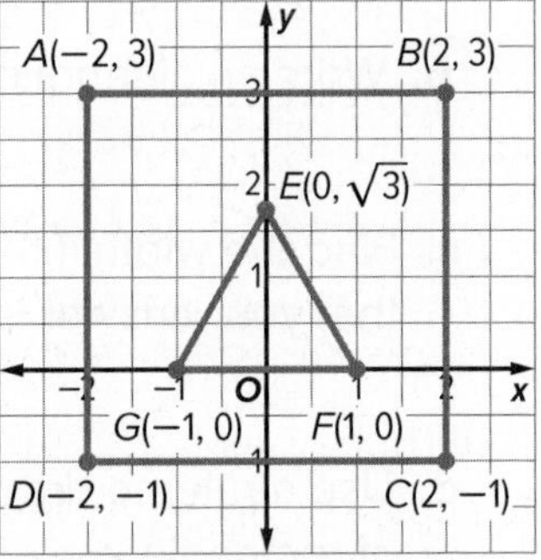

Higher-Order Thinking Skills

24. **PERSEVERE** The floor plan of a rectangular room has the coordinates (0, 12.5), (20, 12.5), (20, 0), and (0, 0) when it is placed on the coordinate plane. Each unit on the coordinate plane measures 1 foot. How many square tiles will it take to cover the floor of the room if the tiles have a side length of 5 inches? Explain.

25. **PERSEVERE** The vertices of a rectangle with side lengths of 10 and 24 units are on a circle of radius 13 units. Find the area between the figures.

26. **WRITE** Give an example of a polygon that is equiangular but not a regular polygon. Explain your reasoning.

27. **ANALYZE** Find the perimeter of equilateral triangle *KLM* given the vertices *K*(−2, 1) and *M*(10, 6). Explain your reasoning.

Transformations in the Plane

Today's Goals

- Analyze figures to identify the types of rigid motions represented.
- Calculate the coordinates of the vertices of images given the coordinates of the preimages.

Today's Vocabulary

transformation
preimage
image
rigid motion
reflection
translation
rotation
line of reflection
center of rotation
translation vector
component form
angle of rotation

Explore Introducing Transformations

 **Online Activity** Use graphing technology to complete the Explore.

INQUIRY How are reflections, translations, and rotations similar?

Learn Identifying Transformations

A **transformation** is a function that takes points in the plane as inputs and gives other points as outputs. In a transformation, the **preimage** is mapped onto the **image**. A **rigid motion**, also called a *congruence transformation* or an *isometry*, is a transformation that preserves distance and angle measure.

The three main types of rigid motions are shown below. The preimage is shown in blue, and the image is shown in green. Prime notation is used to indicate transformations. If A is the preimage, then A' is the image after one transformation.

Study Tip

Rigid Motion A rigid motion is also called a *rigid transformation*. The two terms can be used interchangeably.

Key Concept • Reflections, Translations, and Rotations

A **reflection** or *flip* is a transformation in a line called the **line of reflection**. Each point of the preimage and its image are the same distance from the line of reflection.	A **translation** or *slide* is a transformation that moves all points of the original figure the same distance in the same direction.	A **rotation** or *turn* is a transformation about a fixed point (called the **center of rotation**), through a specific angle, and in a specific direction. Each point of the original figure and its image are the same distance from the center.
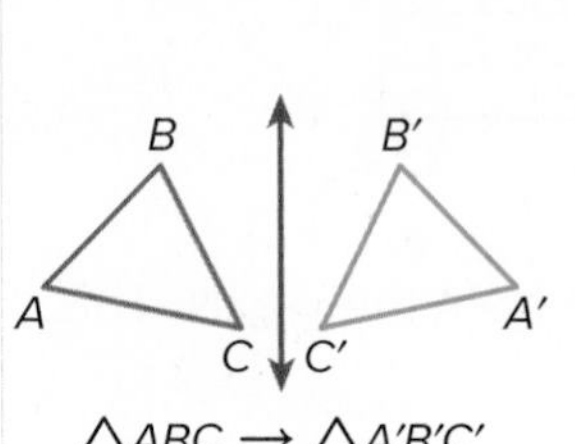 $\triangle ABC \to \triangle A'B'C'$	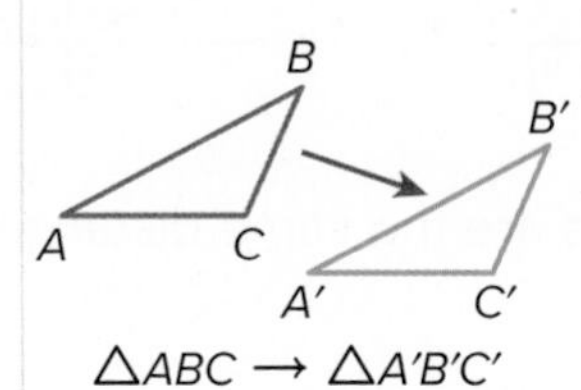 $\triangle ABC \to \triangle A'B'C'$	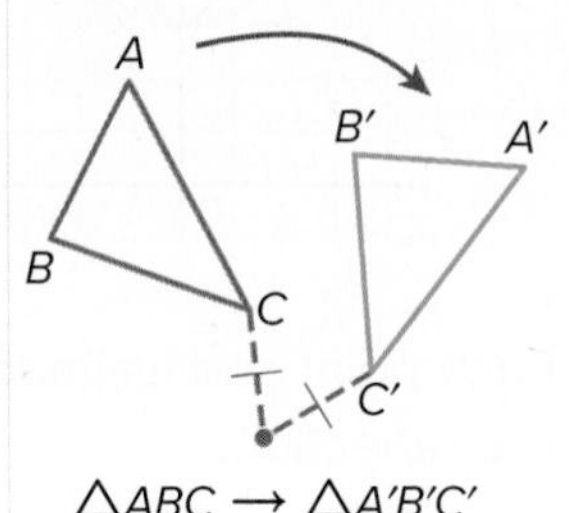 $\triangle ABC \to \triangle A'B'C'$

Study Tip

Identifying Transformations Look for lines of reflection or centers of rotation when identifying transformations. In a reflection, each point of the preimage and its corresponding point of the image are the same distance from the line of reflection. In a rotation, each point of the preimage and its corresponding point of the image are the same distance from the center of rotation.

Example 1 Identify Transformations in the Real World

HOBBIES **Identify the type of rigid motion shown in the puzzle as a *reflection, translation,* or *rotation.***

The landscape is mirrored in the water. This is an example of a reflection.

Check

CHECKERS In the game of checkers, players move their pieces on the diagonal. Identify the type of rigid motion shown as a *reflection*, *translation*, or *rotation*.

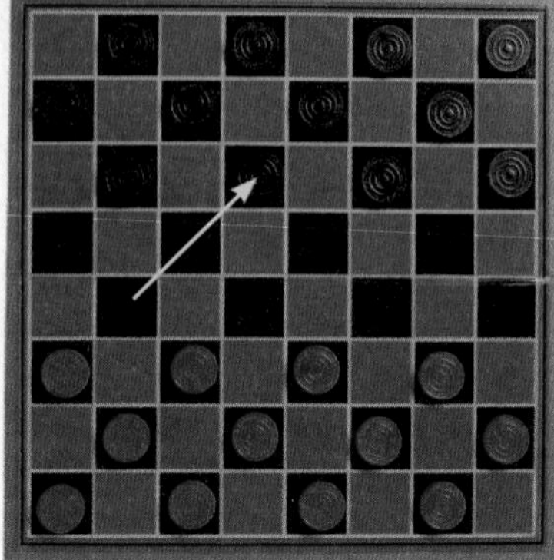

The type of rigid motion is a ___?___.

Example 2 Identify Transformations on the Coordinate Plane

Identify the type of rigid motion shown as a *reflection, translation,* or *rotation.*

a.

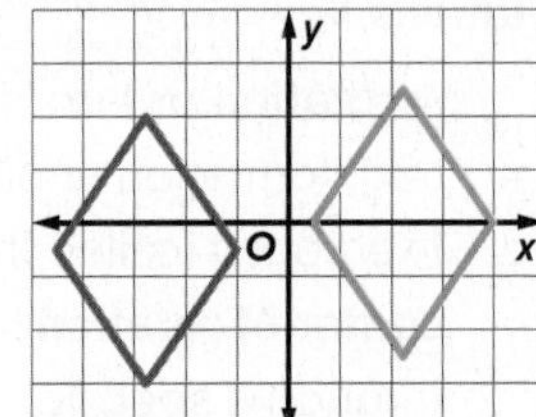

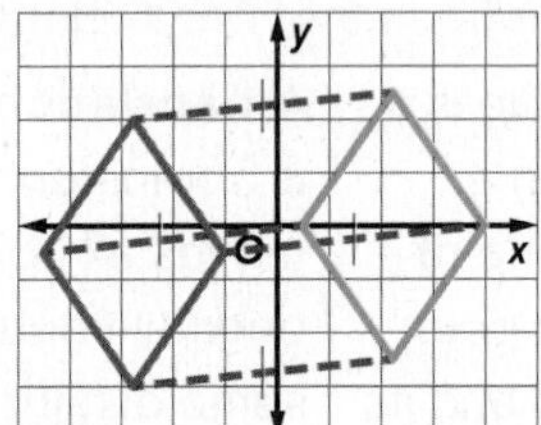

Each vertex and its image can be connected by lines with the same length and slope. This is a translation.

b.

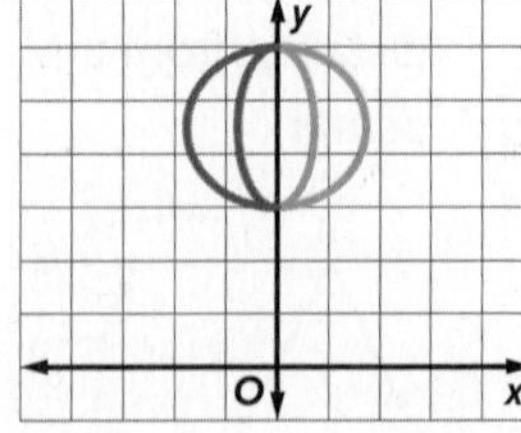

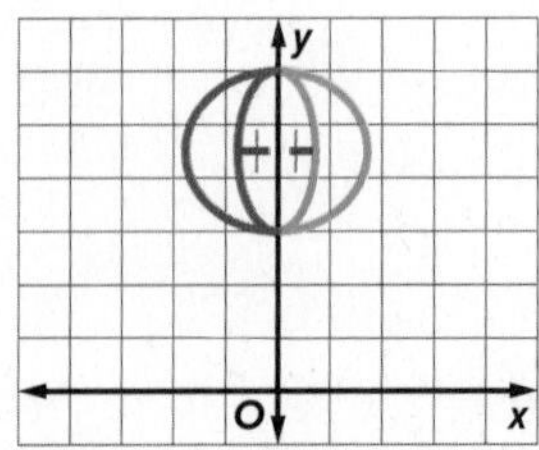

Each point and its image are the same distance from the y-axis. This is a reflection.

Go Online You can complete an Extra Example online.

c.

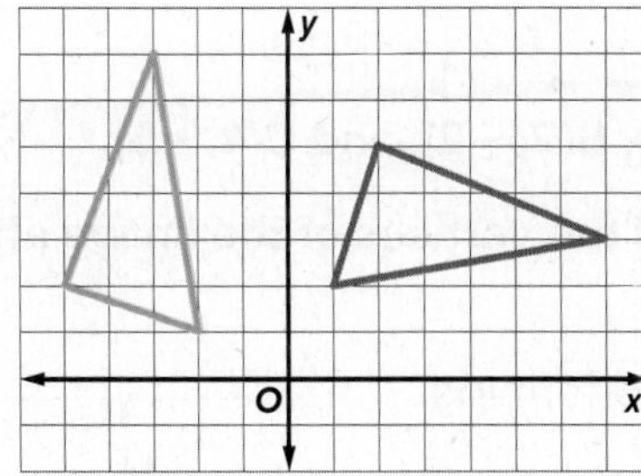

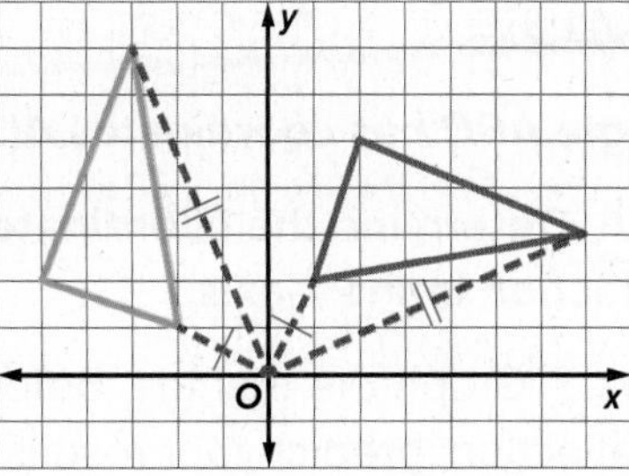

Each vertex and its image are the same distance from the origin. The angles formed by each pair of corresponding points and the origin are congruent. This is a rotation.

Check

The type of rigid motion shown is a ____?____.

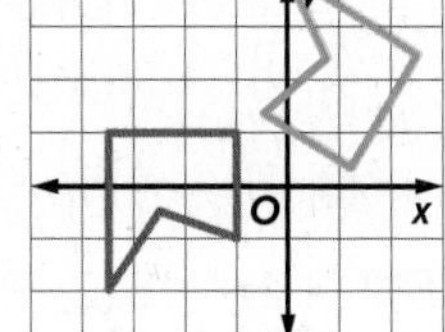

Learn Representing Reflections

In a reflection, each point of the preimage and its corresponding point on the image are the same distance from the line of reflection.

A reflection can be described as a function in which the preimage is reflected in the line of reflection. The points of the preimage are the input, and the corresponding points on the image are the output.

Key Concept • Reflections in the *x*- or *y*-axis

	Reflections in the *x*-axis	Reflections in the *y*-axis
Words	To reflect a point in the *x*-axis, multiply its *y*-coordinate by −1.	To reflect a point in the *y*-axis, multiply its *x*-coordinate by −1.
Symbols	$(x, y) \rightarrow (x, -y)$	$(x, y) \rightarrow (-x, y)$
Example	y; O; x; $A(4, 1)$; $B(7, 3)$; $A'(4, -1)$; $B'(7, -3)$	y; O; x; $A'(-2, 3)$; $A(2, 3)$; $B'(-6, -4)$; $B(6, -4)$

Study Tip

What Is Preserved? Because it is a rigid motion, all lengths and angle measures are preserved in a reflection.

Example 3 Reflection in the x- or y-Axis

Triangle *ABC* has coordinates *A*(3, 2), *B*(2, −2), and *C*(4, −5).

Part A Determine the coordinates of the vertices of the image after a reflection in the *x*-axis.

> **Think About It!**
> Suppose the coordinates of *A* are (5, −2) and the coordinates of *A′* are (5, 2). Describe the transformation of *A*.

PREDICT Graph the triangle. Before performing the reflection, predict your results.
The image of a reflection in the x-axis will be a triangle in the first and fourth quadrants.
Multiply the y-coordinate of each vertex by −1.

Find the coordinates of the vertices of the image.

$(x, y) \rightarrow (x, -y)$

$A(3, 2) \rightarrow A'(3, -2)$

$B(2, -2) \rightarrow B'(2, 2)$

$C(4, -5) \rightarrow C'(4, 5)$

CHECK The image matches the prediction.

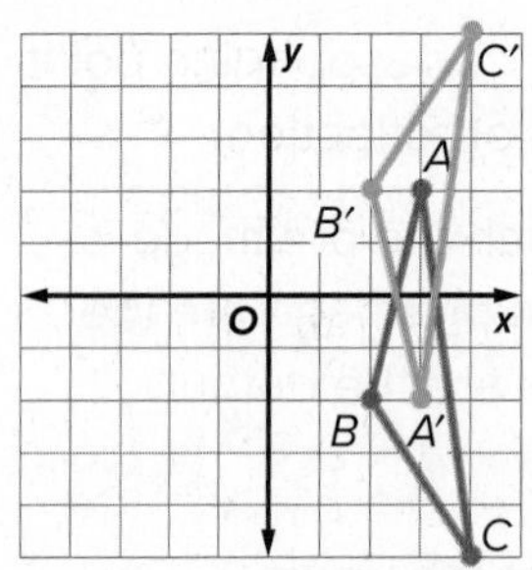

Part B Reflect △*ABC* in the *y*-axis. Determine the coordinates of the image.

PREDICT Before performing the reflection, predict your results.
The image of a reflection in the y-axis will be a triangle in the second and third quadrants.
Multiply the x-coordinate of each vertex by −1.

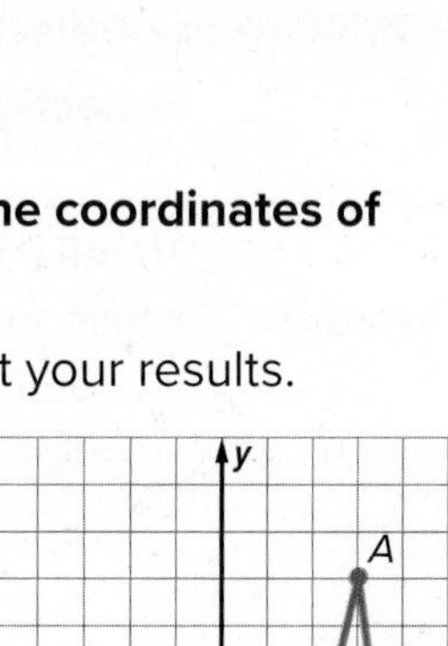

Find the coordinates of the vertices of the image.

$(x, y) \rightarrow (-x, y)$

$A(3, 2) \rightarrow A'(-3, 2)$

$B(2, -2) \rightarrow B'(-2, -2)$

$C(4, -5) \rightarrow C'(-4, -5)$

Go Online You can complete an Extra Example online.

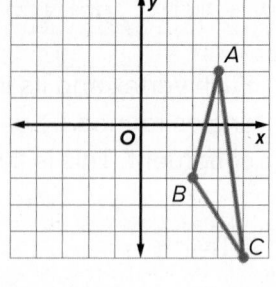

CHECK The image matches the prediction.

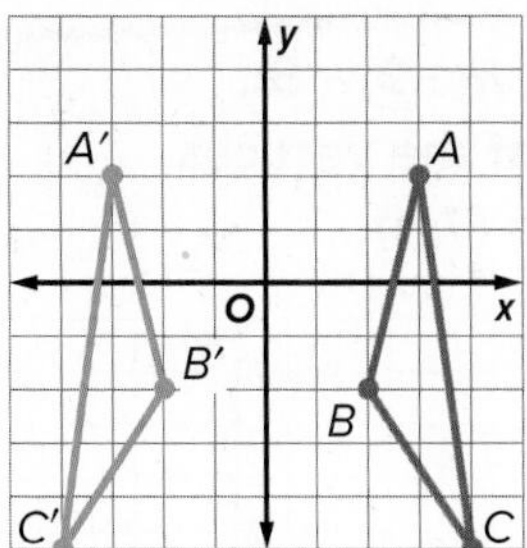

Check

Triangle *JKL* has coordinates *J*(2, −8), *K*(6, −7), and *L*(4, −2). Determine the coordinates of the vertices of the image after a reflection in the *x*-axis.

A. *J*′(2, 8), *K*′(6, 7), *L*′(4, 2)

B. *J*′(−2, −8), *K*′(−6, −7), *L*′(−4, −2)

C. *J*′(−2, 8), *K*′(−6, 7), *L*′(−4, 2)

D. *J*′(2, −8), *K*′(6, −7), *L*′(4, −2)

Learn Representing Translations

A translation is a function in which all of the points of a figure move the same distance in the same direction.

A preimage is translated along a **translation vector.** The translation vector describes the magnitude and direction of the slide if the magnitude is the length of the vector from its initial point to its terminal point.

To describe a translation in the coordinate plane, it is helpful to write the vector in component form. A vector in **component form** is written as $\langle x, y\rangle$, which describes the vector in terms of its horizontal component *x* and vertical component *y*.

Key Concept • Translations	
Words	To translate a point along vector $\langle a, b\rangle$, add *a* to the *x*-coordinate and add *b* to the *y*-coordinate.
Symbols	$(x, y) \rightarrow (x + a, y + b)$
Example	*P*(−2, 3) translated along vector $\langle 7, 4\rangle$ is *P*′(−2 + 7, 3 + 4) or *P*′(5, 7).

Study Tip

What Is Preserved? Because it is a rigid motion, all lengths and angle measures are preserved in a translation.

Example 4 Translations

For quadrilateral *QRST* with vertices $Q(-8, -2)$, $R(-9, -5)$, $S(-4, -7)$, and $T(-4, -2)$, find the coordinates of the vertices of the image after a translation along the vector $\langle 7, 1\rangle$.

PREDICT Graph the quadrilateral. Before performing the translation, predict your results.

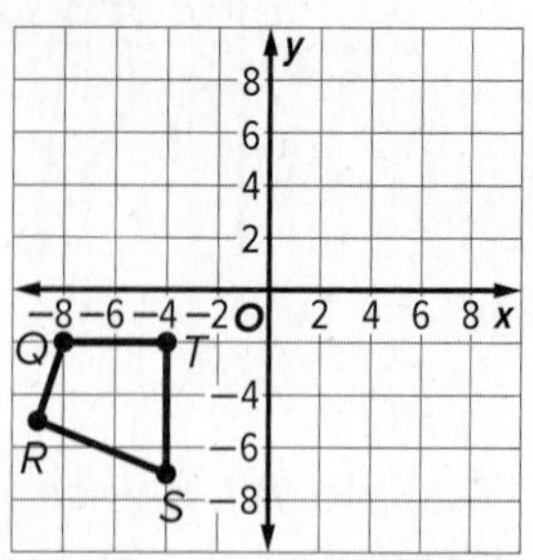

The image of a translation along vector $\langle 7, 1\rangle$ will be a quadrilateral in the third and fourth quadrants.

A translation along $\langle 7, 1\rangle$ will move the figure 7 units to the right and 1 unit up.

Find the coordinates of the vertices of the image.

$(x, y) \rightarrow (x + 7, y + 1)$

$Q(-8, -2) \rightarrow Q'(-8 + 7, -2 + 1)$ or $Q'(-1, -1)$

$R(-9, -5) \rightarrow R'(-9 + 7, -5 + 1)$ or $R'(-2, -4)$

$S(-4, -7) \rightarrow S'(-4 + 7, -7 + 1)$ or $S'(3, -6)$

$T(-4, -2) \rightarrow T'(-4 + 7, -2 + 1)$ or $T'(3, -1)$

CHECK The image matches the prediction.

Check

Quadrilateral *ABCD* has vertices $A(-3, 1)$, $B(-5, 3)$, $C(-2, 5)$, and $D(-1, 3)$. What are the coordinates of the vertices of the image after a translation along vector $\langle 5, -3\rangle$?

A. $A'(2, -2)$, $B'(0, 0)$, $C'(3, 2)$, and $D'(4, 0)$

B. $A'(-8, -2)$, $B'(-10, 0)$, $C'(-7, 2)$, and $D'(-6, 0)$

C. $A'(2, 4)$, $B'(0, 6)$, $C'(3, 8)$, and $D'(4, 6)$

D. $A'(-8, 4)$, $B'(-10, 6)$, $C'(-7, 8)$, and $D'(-6, 6)$

Pause and Reflect

Did you struggle with anything in this lesson? If so, how did you deal with it?

Go Online You can complete an Extra Example online.

Learn Representing Rotations

A rotation is a function that moves every point of a preimage through a specified angle and direction about a fixed point, called the center of rotation. Under a rotation, each point and its image are at the same distance from the center of rotation. In this lesson, you can assume that the origin is the center of rotation. The specified angle is called the **angle of rotation**.

The direction of a rotation can be clockwise or counterclockwise. In this course, you can assume that all rotations are counterclockwise unless stated otherwise.

clockwise

counterclockwise

When a point is rotated 90°, 180°, or 270° counterclockwise about the origin, you can use the following rules. A rotation of 360° will map the image onto the preimage.

Key Concept • Rotations in the Coordinate Plane

90° Rotation To rotate a point 90° counterclockwise about the origin, multiply the y-coordinate by −1 and then interchange the x- and y-coordinates. **Symbols** $(x, y) \rightarrow (-y, x)$	**Example** 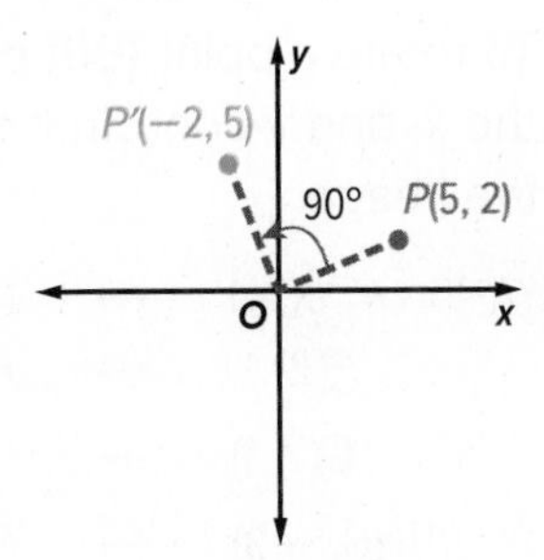
180° Rotation To rotate a point 180° counterclockwise about the origin, multiply the x- and y-coordinates by −1. **Symbols** $(x, y) \rightarrow (-x, -y)$	**Example** 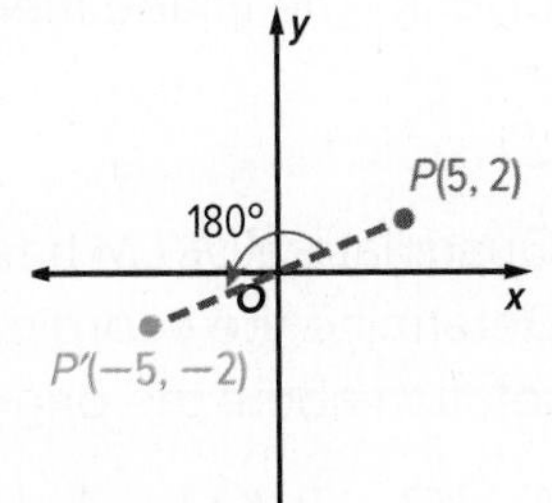
270° Rotation To rotate a point 270° counterclockwise about the origin, multiply the x-coordinate by −1 and then interchange the x- and y-coordinates. **Symbols** $(x, y) \rightarrow (y, -x)$	**Example** 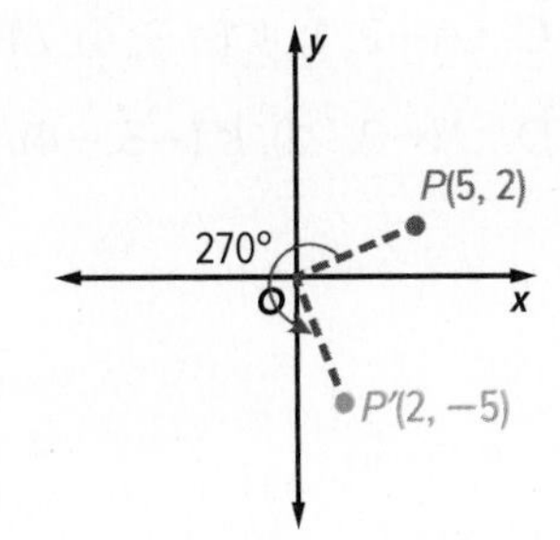

Study Tip

What Is Preserved? Because it is a rigid motion, all lengths and angle measures are preserved in a rotation.

Talk About It!

Would two successive 90° rotations counterclockwise about the origin result in the same image as a 180° rotation clockwise about the origin? Explain.

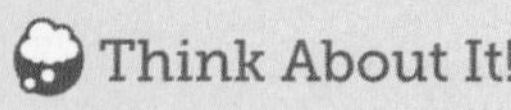

Think About It!

Describe the figure after the rotation.

Example 5 Rotations

Parallelogram *FGHJ* has vertices *F*(2, 1), *G*(7, 1), *H*(6, −3), and *J*(1, −3). What are the coordinates of the vertices of its image after a rotation of 180° about the origin?

PREDICT Graph parallelogram *FGHJ*.

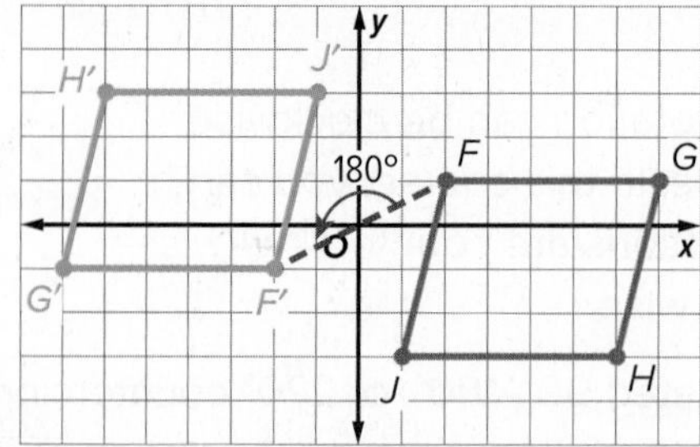

Before performing the rotation, predict your results.

The image of the parallelogram rotated 180° will be a parallelogram in the second and third quadrants.

To rotate a point 180° counterclockwise about the origin, multiply the *x*- and *y*-coordinates by −1. Find the coordinates of the vertices of the image.

$(x, y) \rightarrow (-x, -y)$

$F(2, 1) \rightarrow F'(-2, -1)$

$G(7, 1) \rightarrow G'(-7, -1)$

$H(6, -3) \rightarrow H'(-6, 3)$

$J(1, -3) \rightarrow J'(-1, 3)$

CHECK The image meets the prediction.

Check

Quadrilateral *JKLM* has coordinates *J*(1, 2), *K*(4, 3), *L*(6, 1), and *M*(3, 1). Determine the coordinates of the vertices of the image after a 270° rotation about the origin.

A. *J*′(2, −1), *K*′(3, −4), *L*′(1, −6), and *M*′(1, −3)

B. *J*′(2, 1), *K*′(3, 4), *L*′(1, 6), and *M*′(1, 3)

C. *J*′(−2, 1), *K*′(−3, 4), *L*′(−1, 6), and *M*′(−1, 3)

D. *J*′(−2,−1), *K*′(−3,−4), *L*′(−1,−6), and *M*′(−1,−3)

Study Tip

Rotations
Verify the coordinates of the image by graphing the preimage and image on the coordinate grid.

Go Online You can complete an Extra Example online.

Practice

Go Online You can complete your homework online.

Examples 1 and 2

Identify the type of rigid motion shown as a *reflection*, *translation*, or *rotation*.

1. 

2. 

3. 

4. 

5. 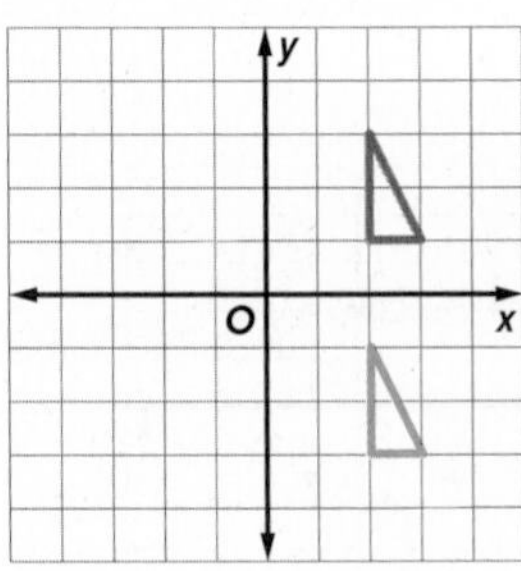

6. 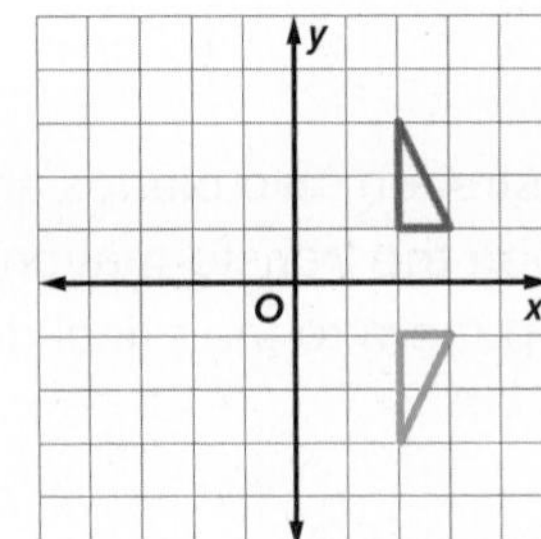

7. 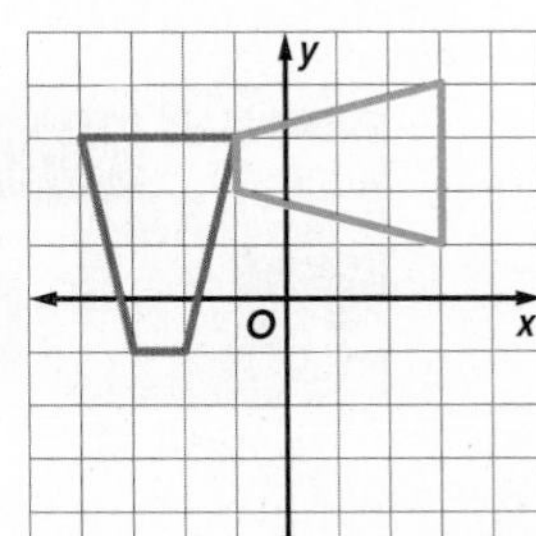

8. 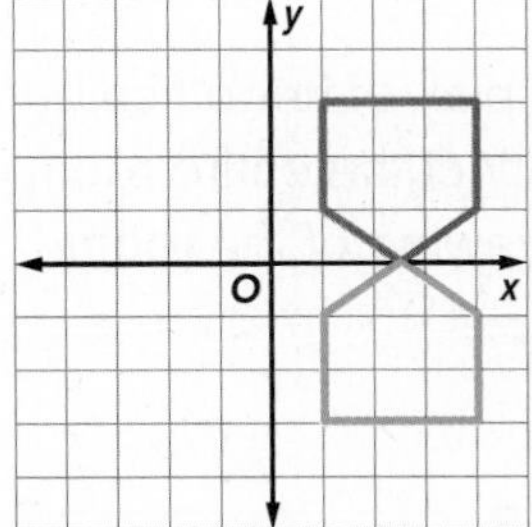

Examples 3–5

Triangle *ABC* has coordinates *A*(2, 0), *B*(−1, 5), and *C*(4, 3). Determine the coordinates of the vertices of the image after each transformation.

9. reflection in *x*-axis

10. reflection in *y*-axis

11. translation along the vector $\langle 0, 2 \rangle$

12. translation along the vector $\langle 3, -4 \rangle$

13. rotation 180° about the origin

14. rotation 90° counterclockwise about the origin

Triangle *DEF* has coordinates *D*(4, −1), *E*(5, 2), and *F*(1, 2). Determine the coordinates of the vertices of the image after each transformation.

15. reflection in *x*-axis

16. reflection in *y*-axis

17. translation along the vector $\langle 1, 0 \rangle$

18. translation along the vector $\langle -3, 1 \rangle$

19. rotation 180° about the origin

20. rotation 270° counterclockwise about the origin

21. AIR SHOW At a flight demonstration, two planes are flying in a synchronized pattern. Describe the transformation that represents the planes' flight pattern to their final destinations at (−30, 20) and (0, 20).

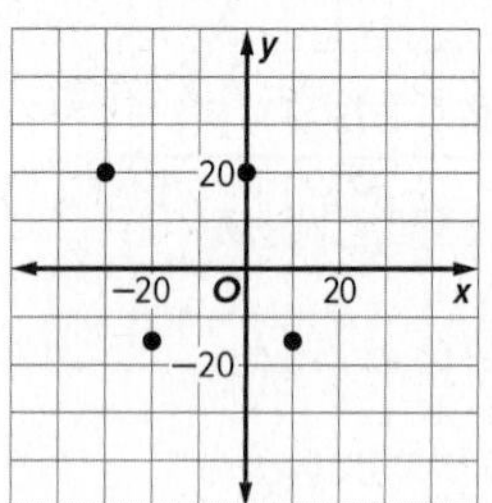

Mixed Exercises

22. OFFICE Francesca draws a plan of her office before she rearranges the furniture. She decides to reflect the entire room over a vertical line through the center of the drawing of the room.

Which is the reflected plan?

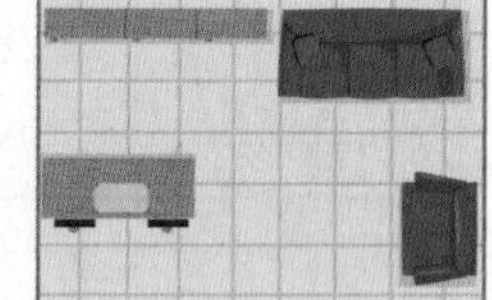

A. 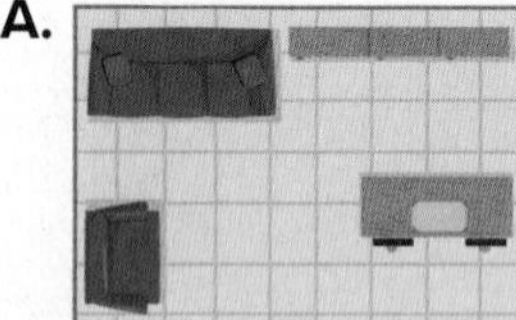

B. 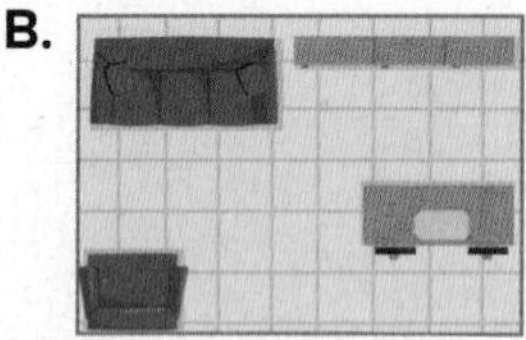

C. 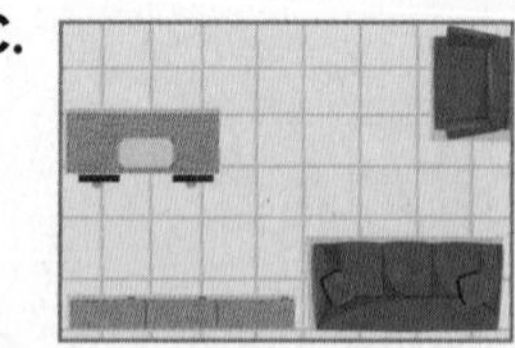

23. BASKETBALL James spins a basketball on his finger and then passes the ball to his friend.

a. What type of transformation is used when James spins the basketball on his finger?

b. What type of transformation is used to pass the basketball?

24. COMBINATION LOCKS Benicio locks his safe by setting each of the three dials to 8. To unlock the safe, he turns the left dial 90° counterclockwise, the middle dial 270° clockwise, and the right dial 180° counterclockwise. Which three numbers, in order, unlock the safe?

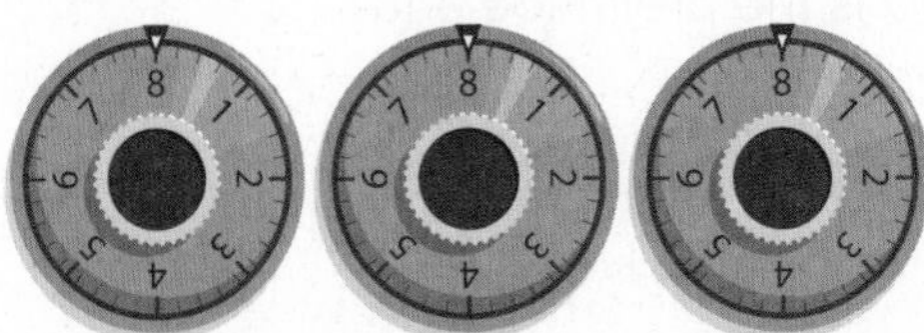

25. BEEKEEPING A beekeeper uses a frame of partial honeycomb cells that bees fill with honey and complete with wax. When the honey is ready for harvest, the beekeeper turns the tap allowing the honey to flow out of the hive without disturbing the bees. By what transformation are the sides of the partial honeycomb cells related when the tap is closed? when the tap is open?

Tap Closed **Tap Open**

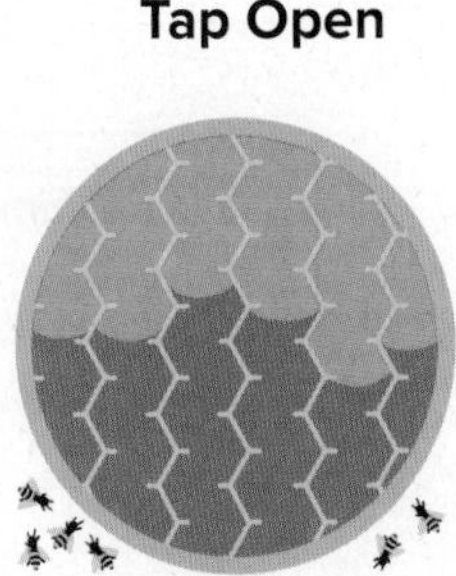

Find the coordinates of the figure with the given coordinates after the transformation on the plane. Then graph the preimage and image.

26. preimage: $J(-3, 0)$, $K(-2, 4)$, $L(-1, 0)$, image: triangle QRS, translation of JKL along vector $\langle 5, -4 \rangle$

27. preimage: $A(1, 3)$, $B(1, 1)$, $C(4, 1)$, image: triangle DEF, rotation of ABC 270° counterclockwise about the origin

28. FIND THE ERROR Saurabh and Elena visit a craft fair and notice a quilt with a pattern. Saurabh claims the pattern is made using translations. Elena believes that the pattern is made using rotations. Who is correct? Justify your argument.

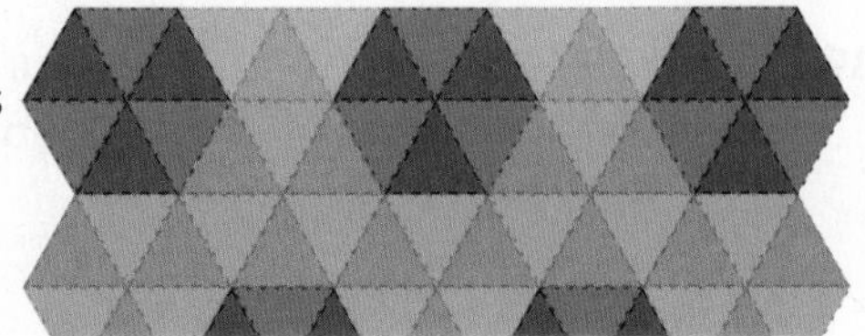

29. The vertices of $\triangle ABC$ are $A(-1, 1)$, $B(4, 2)$, and $C(1, 5)$. The vertices of $\triangle DEF$ are $D(-1, -1)$, $E(4, -2)$, and $F(1, -5)$ such that $\triangle ABC \cong \triangle DEF$. Identify the congruence transformation.

30. STRUCTURE $\overline{XY}$ has endpoints $X(-5, 6)$ and $Y(0, 4)$, the image of $\overline{XY}$ has the endpoints $X'(6, 5)$ and $Y'(4, 0)$, and $\overline{XY} \cong \overline{X'Y'}$. Identify the transformation.

31. STRUCTURE The vertices of quadrilateral $FGHJ$ are $F(2, -3)$, $G(-2, -5)$, $H(-3, 6)$, and $J(3, 5)$. The vertices of quadrilateral $KLMN$ are $K(5, -5)$, $L(1, -7)$, $M(0, 4)$, and $N(6, 3)$ such that $FGHJ \cong KLMN$. If quadrilateral $FGHJ$ is the preimage and quadrilateral $KLMN$ is the image, identify the transformation.

Higher-Order Thinking Skills

32. ANALYZE The image of $\triangle ABC$ reflected in the y-axis is $\triangle A'B'C'$.

a. Describe the result of reflecting $\triangle A'B'C'$ in the y-axis. Explain.

b. Describe the result of reflecting $\triangle A'B'C'$ in the x-axis. Explain.

33. FIND THE ERROR Antwan and Diamond are finding the coordinates of the image of $P(2, 3)$ after a reflection in the x-axis. Is either of them correct? Explain your reasoning.

Antwan	Diamond
$P'(2, -3)$	$P'(2, -3)$

34. WRITE In the diagram, $\triangle DEF$ is called a *glide reflection* of $\triangle ABC$. Based on the diagram, define a glide reflection. Explain your reasoning.

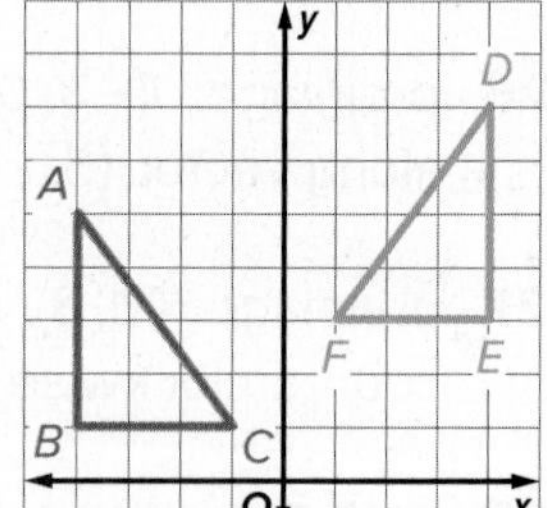

35. CREATE Draw a polygon on the coordinate plane that when reflected in the y-axis looks exactly like the original figure.

36. ANALYZE Is the reflection of a figure in the x-axis equivalent to the rotation of that same figure 180° about the origin? Explain.

Three-Dimensional Figures

Learn Identifying Three-Dimensional Figures

A **polyhedron** is a closed three-dimensional figure made up of flat polygonal regions. A **face of a polyhedron** is a flat surface on the polyhedron. An **edge of a polyhedron** is a line segment where the faces of the polyhedron intersect. The **vertex of a polyhedron** is the intersection of three edges of the polyhedron. The **bases of a prism or cylinder** are the two parallel congruent faces of the solid. The **base of a pyramid or cone** is the face of the solid opposite the vertex of the solid.

Types of Solids

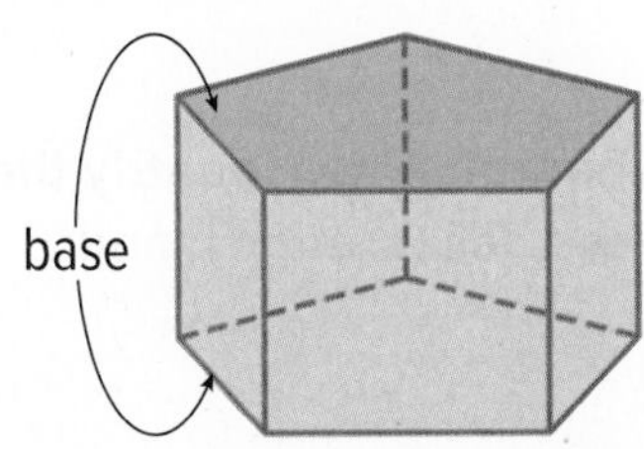

A **prism** is a polyhedron that has two parallel congruent bases, connected by parallelogram faces.

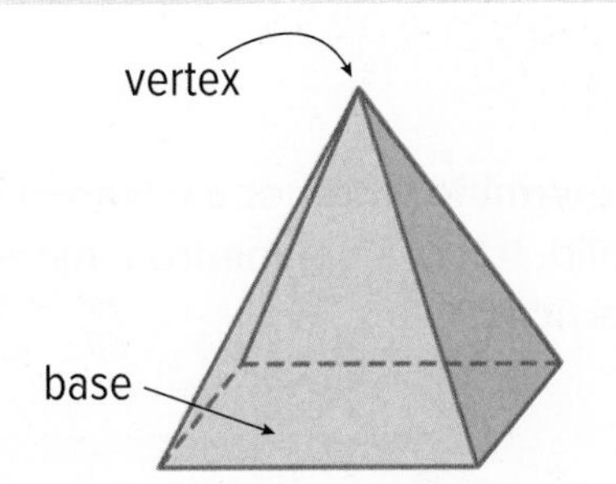

A **pyramid** is a polyhedron that has a polygonal base and three or more triangular faces that meet at a common vertex.

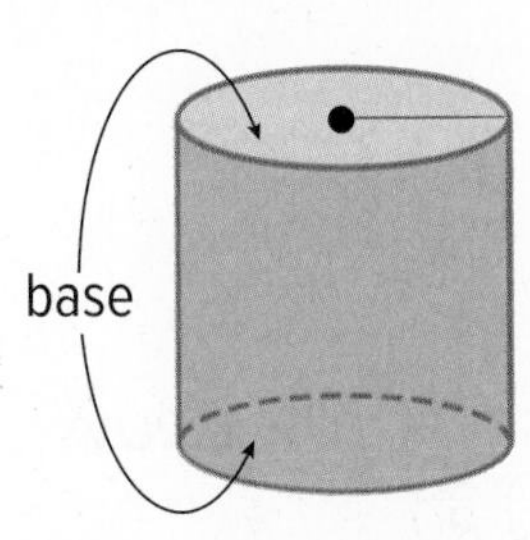

A **cylinder** is a solid figure that has two congruent and parallel circular bases connected by a curved surface.

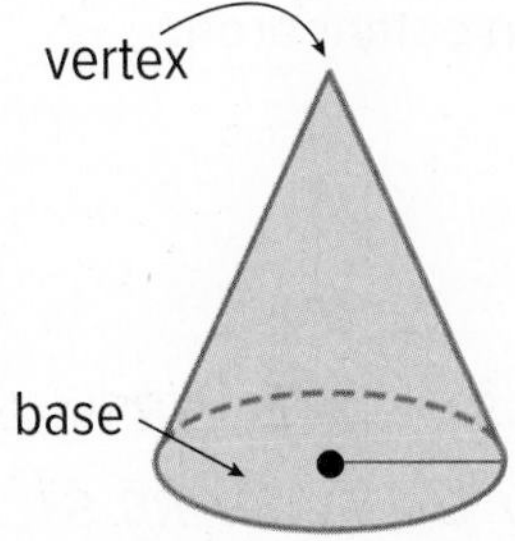

A **cone** is a solid figure that has a circular base connected by a curved surface to a single vertex.

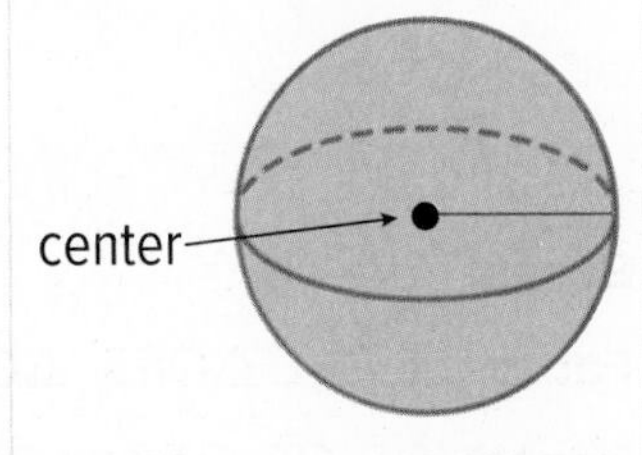

A **sphere** is a set of all points in space equidistant from a given point called the center of the sphere. A sphere has no faces, edges, or vertices.

Polyhedra, or *polyhedrons,* are named by the shapes of their bases.

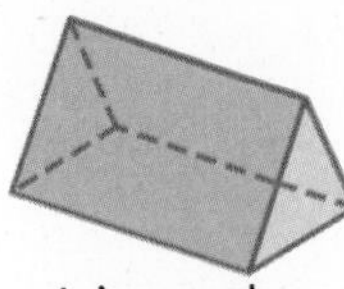
triangular prism

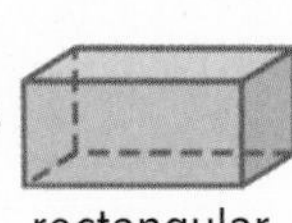
rectangular prism

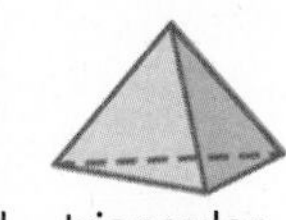
pentagonal prism

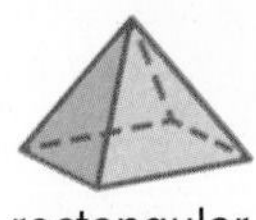
triangular pyramid

rectangular pyramid

pentagonal pyramid

Today's Goals

- Identify and determine characteristics of three-dimensional figures.
- Calculate surface areas and volumes.

Today's Vocabulary

polyhedron
face of a polyhedron
edge of a polyhedron
vertex of a polyhedron
bases of a prism or cylinder
base of a pyramid or cone
prism
pyramid
cylinder
cone
sphere
regular polyhedron
Platonic solids
surface area
volume

Study Tip

Right vs. Oblique In right prisms, the bases are connected to each other by rectangular faces. However, in oblique prisms, at least one face is not a rectangle.

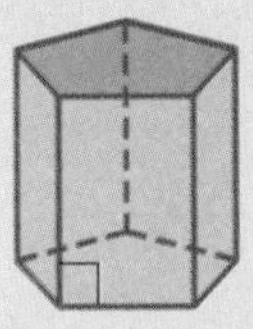
right prism

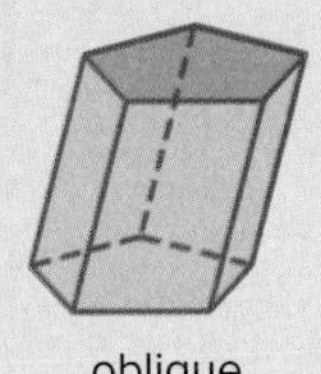
oblique prism

A polyhedron is a **regular polyhedron** if all of its faces are regular congruent polygons and all of the edges are congruent. There are exactly five types of regular polyhedra, called **Platonic solids** because Plato used them extensively.

Platonic Solids				
Tetrahedron	Hexahedron or Cube	Octahedron	Dodecahedron	Icosahedron
4 equilateral triangular faces	6 square faces	8 equilateral triangular faces	12 regular pentagonal faces	20 equilateral triangular faces

Math History Minute

The Platonic solids are named for the Greek philosopher **Plato (c. 428 BCE–c. 348 BCE)**, who theorized in his dialogue, the *Timaeus*, that the classical elements were made of them. Of all the Platonic solids, the Greeks believed that the dodecahedron represented the entire universe.

Example 1 Identify Properties of Three-Dimensional Figures

Determine whether each solid is a polyhedron. Then identify the solid. If it is a polyhedron, name the bases, faces, edges, and vertices.

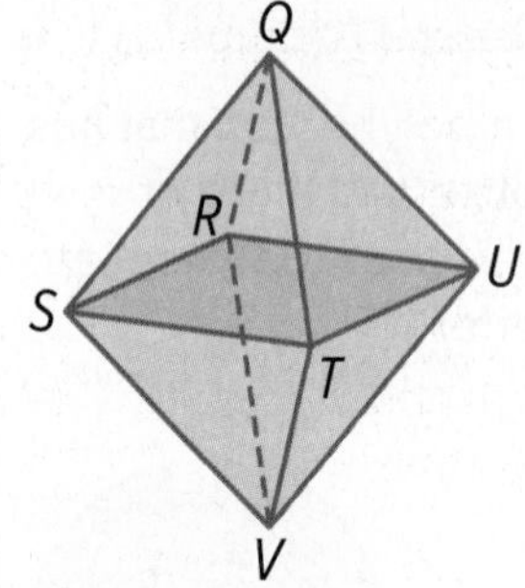

The solid is formed by polygonal faces, so it is a polyhedron. There is no base, but the solid has 8 equilateral triangular faces, so it is an octahedron.

Bases: none

Faces: $\triangle QRU$, $\triangle QRS$, $\triangle QST$, $\triangle QTU$, $\triangle RSV$, $\triangle STV$, $\triangle TUV$, $\triangle RUV$

Edges: $\overline{QR}$, $\overline{QS}$, $\overline{QT}$, $\overline{QU}$, $\overline{RV}$, $\overline{UV}$, $\overline{TV}$, $\overline{SV}$, $\overline{RS}$, $\overline{ST}$, $\overline{TU}$, $\overline{RU}$

Vertices: Q, R, S, T, U, V

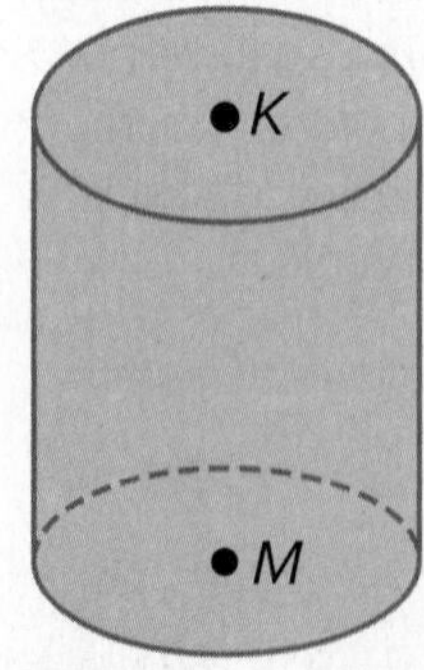

The solid has a curved surface, so it is not a polyhedron. It has two congruent and parallel circular bases and parallel sides, so it is a cylinder.

 **Go Online** You can complete an Extra Example online.

Example 2 Model Three-Dimensional Figures

Identify the three-dimensional figure that can model the beverage container. State whether the model is a polyhedron.

The beverage container can be modeled by a cylinder. Because the model has a curved surface it is not a polyhedron.

Check

Identify the three-dimensional figure that can model the top of the camping lodge. State whether the model is a polyhedron.

The top of the camping lodge can be modeled by a ___?___. The model ___?___ a polyhedron.

Study Tip

Approximations When modeling a real-world object, often the object cannot be perfectly modeled by a three-dimensional figure. Thus, three-dimensional figures provide only approximate measures for an object.

Explore Measuring Real-World Objects

Online Activity Use dynamic geometry software to complete the Explore.

INQUIRY How can you apply the properties of three-dimensional figures to solve real-world problems?

Talk About It!

What is the relationship between the volume of a prism and the volume of a regular pyramid that have the same base and height? How does this compare to the relationship between the volume of a cylinder and the volume of a cone that have the same height and congruent bases?

Learn Measuring Three-Dimensional Figures

Often a geometric figure is used to model a real-world object to estimate a measurement. **Surface area** is the sum of the areas of all faces and side surfaces of a three-dimensional figure. **Volume** is the measure of the amount of space enclosed by a three-dimensional figure.

Prism	Right Pyramid	Cylinder	Cone	Sphere
P, h	ℓ	r, h	r, h, ℓ	r
$S = Ph + 2B$	$S = \frac{1}{2}P\ell + B$	$S = 2\pi rh + 2\pi r^2$	$S = \pi r\ell + \pi r^2$	$S = 4\pi r^2$
$V = Bh$	$V = \frac{1}{3}Bh$	$V = \pi r^2 h$	$V = \frac{1}{3}\pi r^2 h$	$V = \frac{4}{3}\pi r^3$

S = total surface area V = volume h = height of a solid
P = perimeter of the base B = area of base ℓ = slant height, r = radius

Watch Out!

Height vs. Slant Height The *height* of a pyramid or cone is not the same as its *slant height*. The height of a pyramid or cone is the perpendicular distance from the center of the base to the vertex. The slant height is the height of the face or curved surface of a pyramid or cone, respectively, that joins the base to the vertex.

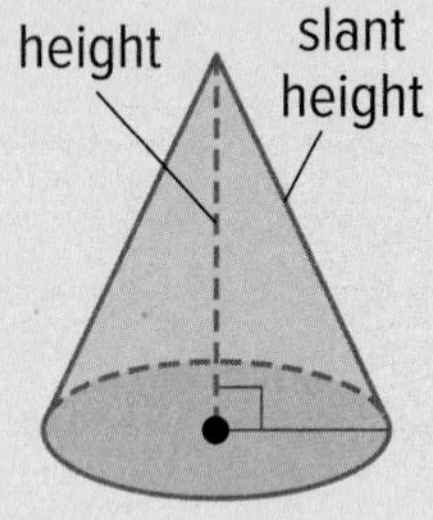

Watch Out!

Square and Cubic Units Remember to use square units when measuring surface area and cubic units when measuring volume.

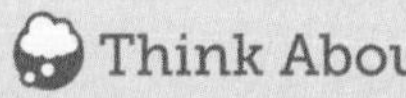

Think About It!

How could you write the surface area of the cone in terms of pi?

Example 3 Find Measurements of Three-Dimensional Figures

Find the surface area and volume of the cone. Round each measure to the nearest tenth, if necessary.

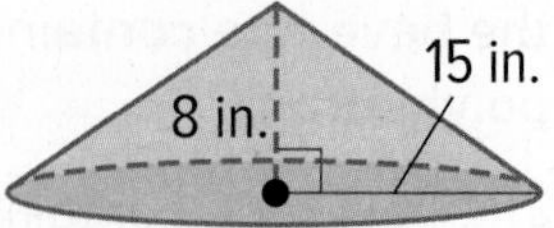

Surface Area

Because the radius of the base is 15 inches and the height of the cone is 8 inches, you can use the Pythagorean Theorem to find the slant height.

$c^2 = a^2 + b^2$	Pythagorean Theorem
$c^2 = 8^2 + 15^2$	$a = 8$ and $b = 15$
$c^2 = 289$	Simplify.
$c = \sqrt{289}$ or 17	Simplify.

The slant height is 17 inches, and the radius is 15 inches. Use the formula for the surface area of a cone.

$S = \pi r\ell + \pi r^2$	Surface area of cone
$= \pi (15)(17) + \pi (15)^2$	$r = 15$ in. and $\ell = 17$ in.
≈ 1508.0	Use a calculator.

The surface area of the cone is about 1508.0 square inches.

Volume

Use the formula for the volume of a cone.

$V = \frac{1}{3}\pi r^2 h$	Volume of cone
$= \frac{1}{3}\pi (15)^2 (8)$ or about 1885.0	$r = 15$ in. and $h = 8$ in.

The volume of the cone is about 1885.0 cubic inches.

Check

Find the surface area and volume of the rectangular prism. Round each measure to the nearest tenth, if necessary.

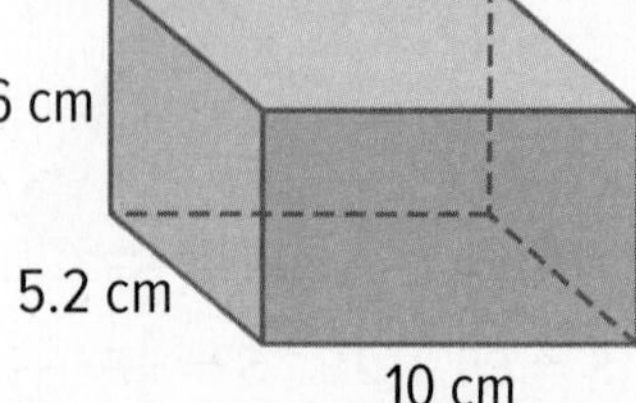

$S =$ __?__ cm^2

$V =$ __?__ cm^3

Example 4 Calculate Measurements by Using Three-Dimensional Models

NEW YEAR'S EVE **The New Year's Eve ball is a geodesic sphere that is 12 feet in diameter. It weighs 11,875 pounds, is lit by 32,256 LED lights, and is covered with 2688 crystal triangles.**

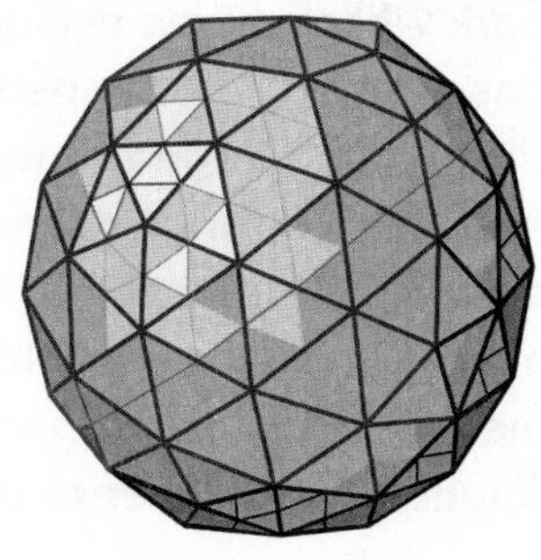

Part A How many lights are contained on the ball's surface within an area of 4 square feet?

Step 1 Find the surface area of the ball.

Because the diameter is 12 feet, the radius of the sphere is 6 feet.

$S = 4\pi r^2$	Surface area of a sphere
$= 4\pi(6)^2$	$r = 6$
$= 144\pi$ or about 452.4	Use a calculator.

The surface area of the ball is about 452.4 square feet.

Step 2 Determine the number of lights within an area of 4 square feet.

$$4\ \cancel{ft^2} \times \frac{32{,}256 \text{ lights}}{452.4\ \cancel{ft^2}} = 285.2 \text{ or } 285 \text{ lights}$$

There are 285 lights within an area of 4 square feet.

Part B Tony is repairing a section of the ball that has a volume of 8 cubic feet. How much does the section weigh?

Step 1 Find the volume of the ball.

$V = \frac{4}{3}\pi r^3$	Volume of a sphere
$= \frac{4}{3}\pi(6)^3$	$r = 6$
$= 288\pi$ or about 904.8	Use a calculator.

Step 2 Determine the weight of the section.

$$8\ \cancel{ft^3} \times \frac{11{,}875 \text{ lb}}{904.8\ \cancel{ft^3}} = 105.0$$

The section of the ball weighs about 105.0 pounds.

Check

POOLS Mateo is building a new pool. A cross section of the pool is shown.

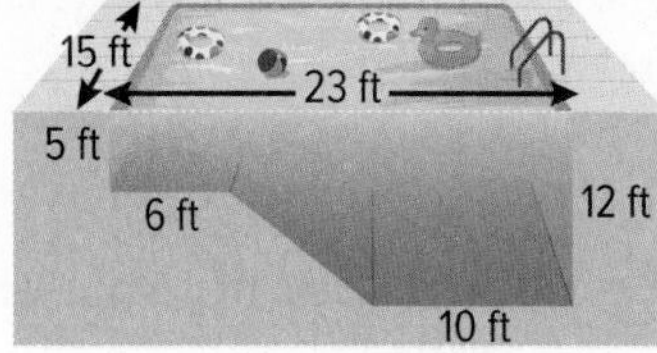

Part A What is the volume of the pool to the nearest tenth?

$V =$ ___?___ ft^3

Part B Mateo needs to install a protective liner to cover the walls and flat base of the deep end of the pool. How much liner is required to cover the deep end of the pool in square feet?

A. 570 ft^2 B. 750 ft^2 C. 900 ft^2 D. 1800 ft^2

Think About It!

What assumption did you make about the New Year's Eve ball to solve the problem?

Apply Example 5 Solve for Unknown Values

Study Tip

Assumptions When modeling, you assume that a three-dimensional object is perfectly modeled by a geometric solid. Because this is not always true, you should assume that your calculations for surface area and volume will be approximations.

WATER PARK **Destiny visits a water park where a new cyclone water ride has opened. On the new ride, a tunnel flows into a large funnel. The length from the entrance of the funnel to its edge is 29 meters, and the surface area of the funnel is 910 square meters. When the funnel is at its widest, what is the diameter? Round your answer to the nearest tenth.**

1 What is the task?

Describe the task in your own words. Then list any questions that you may have. How can you find answers to your questions?

Sample answer: I need to find the diameter of the funnel at its widest location. What three-dimensional solid can I use to model the funnel? Are there any special considerations that I need to make when modeling this real-world object? I can review the three-dimensional solids for which I know the formulas for surface area. Then, I can compare these solids to the funnel to see how the two objects are different.

2 How will you approach the task? What have you learned that you can use to help you complete the task?

Sample answer: I will draw a diagram that represents the funnel and all of the information that I know about the funnel. I will then identify the equation that I will use to find the diameter of the funnel, I will substitute any known information into the equation, and then I will solve for the length of the diameter.

3 What is your solution?

Use your strategy to solve the problem.

What three-dimensional solid can you use to model the funnel?

Sample answer: I can model the funnel using a cone. Because the mouth of the funnel is open, the cone that models the funnel would not have a base.

What equation will you use?

$S = \pi r \ell$

What is the diameter of the funnel?

20.0 meters

4 How can you know that your solution is reasonable?

Write About It! Write an argument that can be used to defend your solution.

Sample answer: A diameter of 20 meters seems reasonable when compared to the slant height of the funnel, which is 29 meters.

Go Online You can complete an Extra Example online.

Practice

Go Online You can complete your homework online.

Example 1

Determine whether each solid is a polyhedron. Then identify the solid. If it is a polyhedron, name the bases, faces, edges, and vertices.

1.

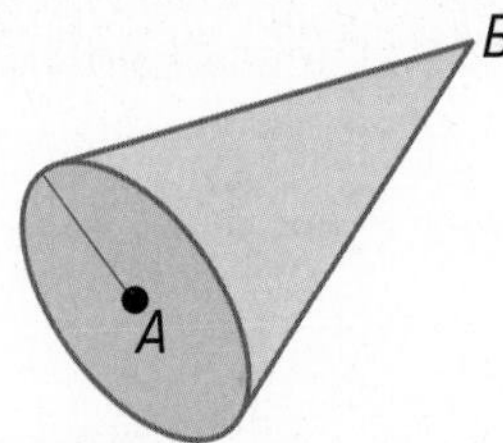

2.

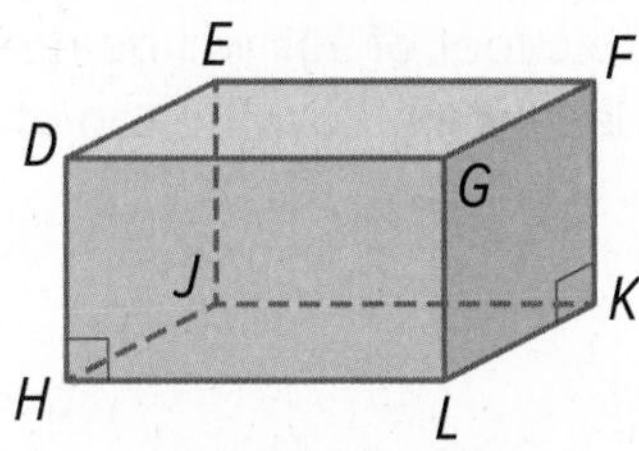

3.

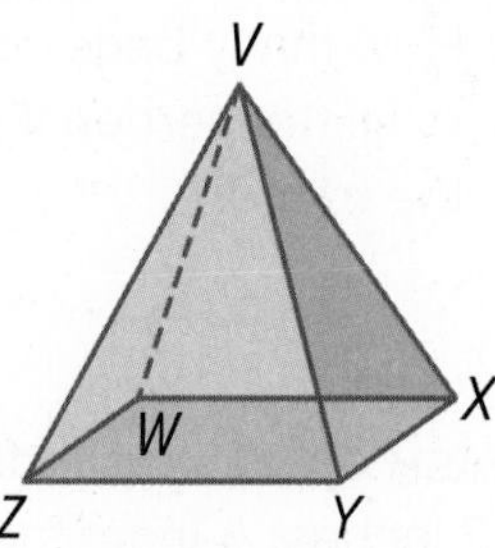

Example 2

Identify the three-dimensional figure that can model each object. State whether the model is or is not a polyhedron.

4.

5.

6.

Example 3

Find the surface area and volume of each solid. Round each measure to the nearest tenth, if necessary.

7.

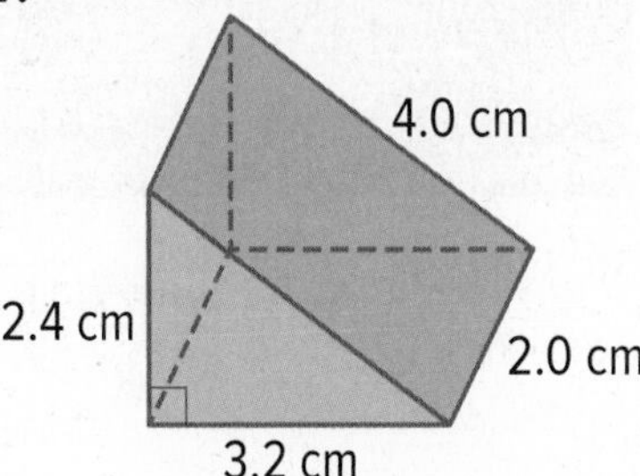

8.

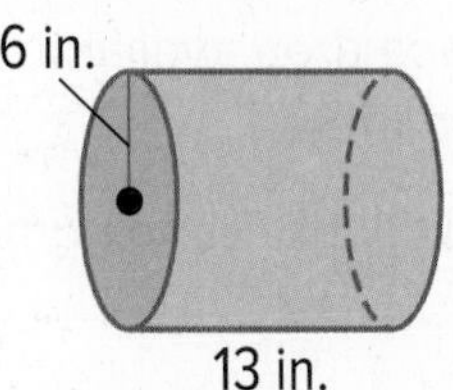

9.

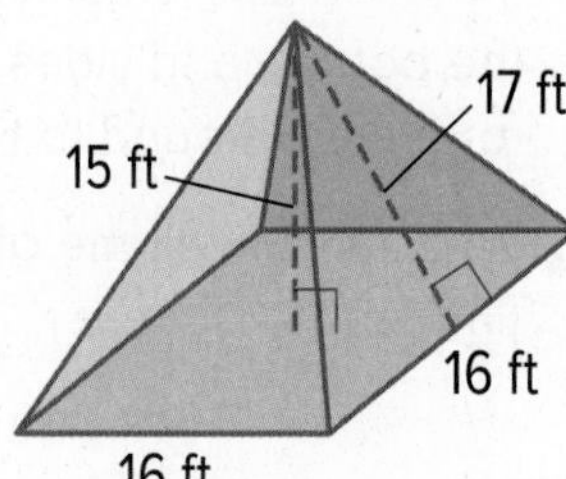

10.

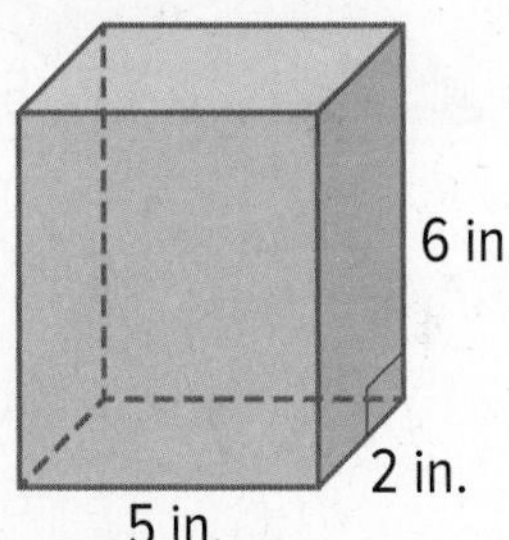

11.

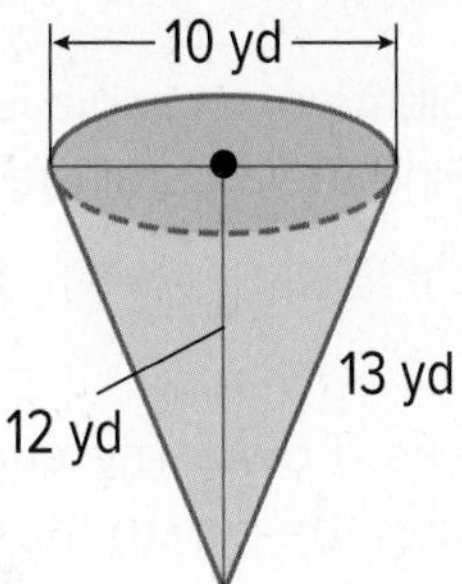

12.

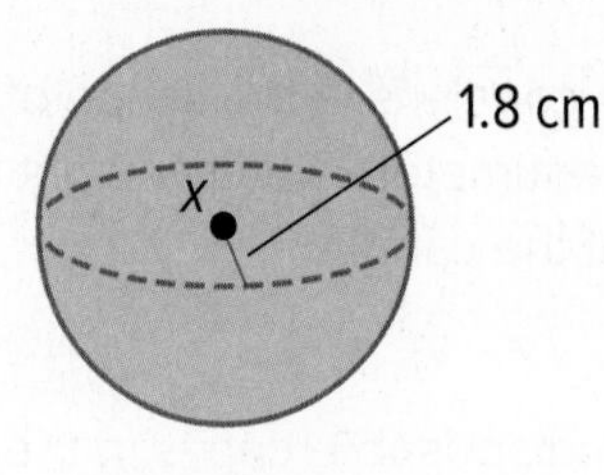

Example 4

13. **GARDENING** The plans for constructing a raised vegetable garden use corrugated metal in a wooden frame. The finished garden is 4 feet long, 30 inches wide, and 32 inches tall.

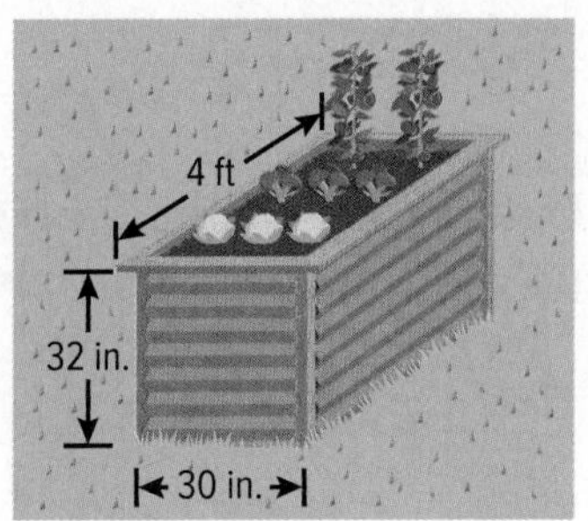

a. The metal is only used on the lateral faces, so how many square feet of metal should be purchased? Round to the nearest square foot.

b. How many bags containing 2 cubic feet of soil will be needed to fill the garden if the soil level is 1 inch below the top of the frame?

14. **TRASH CANS** A cylindrical trash can is 30 inches high and has a base radius of 7 inches. A manufacturer wants to know the surface area of this trash can, including the top of the lid. What is the surface area? Round to the nearest square inch.

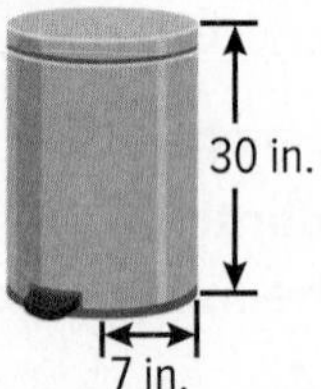

15. **ALGAE** A scientist has a fish tank in the shape of a rectangular prism. The tank is 18 inches high, 14 inches wide, and 30 inches long. After one month, the scientist found that the sides and bottom of his fish tank were covered with algae. The scientist wants to run tests on the algae to help determine why it started to grow. How much algae is there for the scientist to test?

16. **GEOLOGY** A *tiankeng* is a sinkhole with nearly vertical walls. The Tianpingmiao tiankeng is approximately cylindrical with a diameter of 180 meters and a depth of 420 meters.

a. If the top of the tiankeng is open and plants can grow on the bottom and sides, what is the surface area available for plants? Round to the nearest square meter.

b. What is the volume of water that could fill the Tianpingmiao tiankeng?

Example 5

17. The model of a roof is in the shape of a square pyramid, as shown. If the surface area of the model is 64 cm^2, what is the slant height?

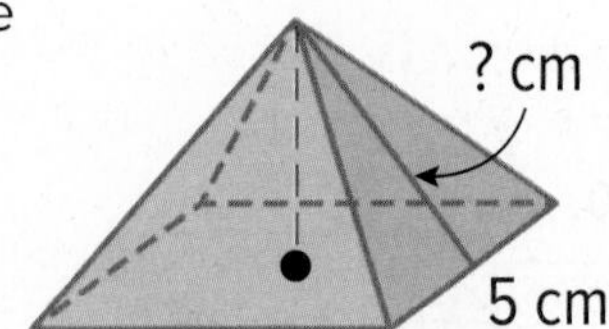

18. A candle is in the shape of a pyramid. The volume of a candle is 27 cubic centimeters and its height is 6 centimeters. Find the area of the base of the candle.

19. A disposable cup is in the shape of a cone, as shown. The cup has a volume of about 48.8 in^3. What is the radius of the cup to the nearest inch?

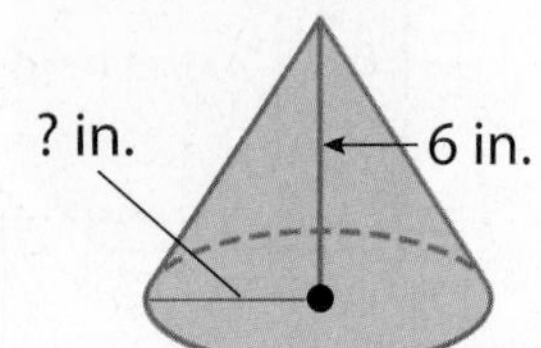

Mixed Exercises

20. PLANETS For a time, Johannes Kepler thought that the Platonic solids were related to the orbits of the planets. He made models of each of the Platonic solids. He made a frame of each of the Platonic solids by fashioning together wooden edges. How many edges did Kepler have to make for the cube?

21. SILO A silo used for storing grain is shaped like a cylinder with a cone on top. The radius of the base of the cylinder and cone is 8 feet. The height of the cylindrical part is 25 feet, and the height of the cone is 6 feet.

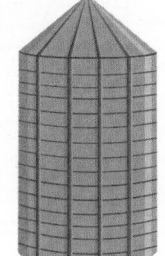

a. What is the volume of the cylindrical part of the silo? Round to the nearest cubic foot.

b. What is the volume of the conical part of the silo? Round to the nearest cubic foot.

c. What is the volume of the entire silo? Round to the nearest cubic foot.

22. USE A SOURCE Find a real object that can be modeled with one or more three-dimensional figures. Identify the best three-dimensional model and calculate the surface area and volume of the object.

23. A garden shop sells pyramid-shaped lawn ornaments that each have a base area of 900 square centimeters and a height of 40 centimeters. The lawn ornaments are made of concrete, granite, or marble.

Material	Density (kg/m^3)
Concrete	2371
Granite	2691
Marble	2711

a. What is the volume of one lawn ornament in cubic meters? Explain.

b. Find the weights of three of these ornaments that are each made from a different material. Round to the nearest tenth of a kilogram.

c. What generalization can you make about the relationships among the volume of an ornament, the weight of the lawn ornament, and the density of the material used to make it?

24. REASONING The volume of a new extra large toy tennis ball for pets is about 221 cubic centimeters. If 3 extra large toy tennis balls are packaged and sold in cylindrical package as shown, what is the approximate volume of the cylindrical package? Explain.

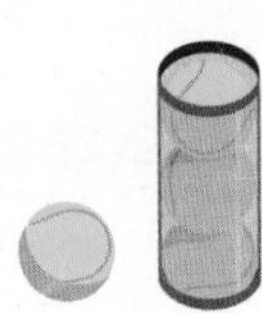

Higher-Order Thinking Skills

25. FIND THE ERROR Alex and Sia are calculating the surface area of the rectangular prism shown. Is either of them correct? Explain your reasoning.

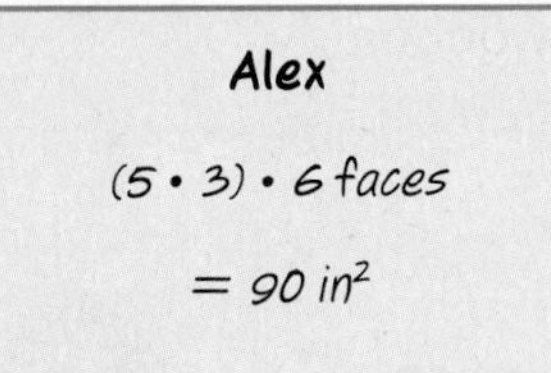

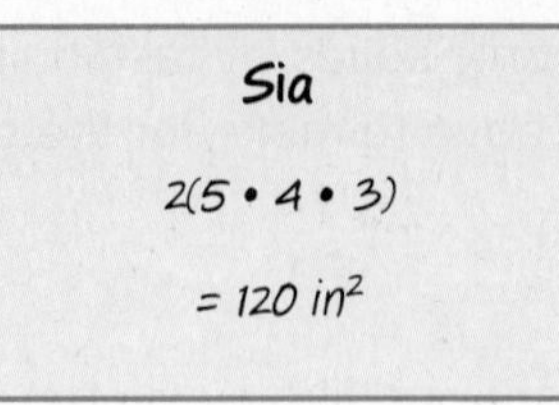

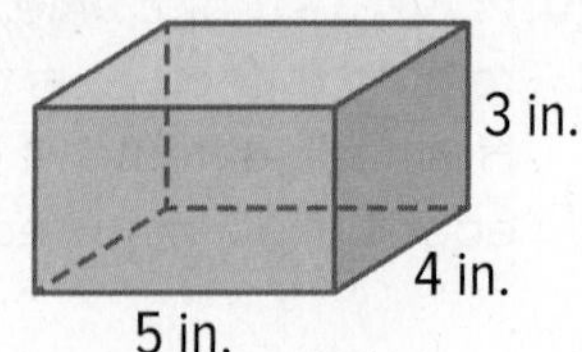

26. ANALYZE When a polygon is *inscribed* in a circle, all of the vertices of the polygon lie on the circle. Consider a pyramid and a prism that have bases that are regular polygons inscribed in a circle. What solid results if the number of sides of the bases is increased infinitely?

27. WRITE Which solid has a greater volume: cone with a base radius of 7 centimeters and a height of 28 centimeters or a pyramid with base area of 154 square centimeters and height of 28 centimeters? Explain your reasoning.

28. CREATE Draw an irregular 14-sided polyhedron that has two congruent bases.

29. PERSEVERE Find the volume of a cube that has a total surface area of 54 square millimeters.

30. ANALYZE Is a cube a regular polyhedron? Justify your argument.

Lesson 11-6

Two-Dimensional Representations of Three-Dimensional Figures

Today's Goals
- Identify the orthographic drawings that best model selected three-dimensional figures.
- Calculate surface areas of three-dimensional figures represented by nets, and determine the correct nets for three-dimensional geometric figures.

Today's Vocabulary
orthographic drawing
net

Explore Representing Three-Dimensional Figures

 **Online Activity** Use two-dimensional drawings to complete the Explore.

INQUIRY How can you accurately represent a three-dimensional figure with two-dimensional drawings?

Learn Representing Three-Dimensional Figures with Orthographic Drawings

The two-dimensional views of the top, left, front, and right sides of an object are called an **orthographic drawing**.

Example 1 Make a Model from an Orthographic Drawing

Make a model of a figure from the orthographic drawing shown.

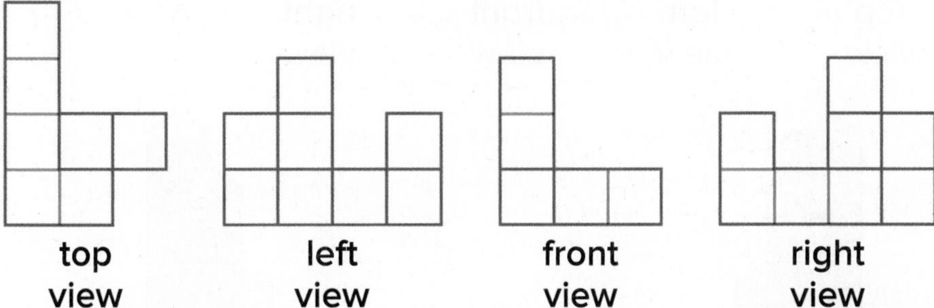

Step 1 Create the base of the model. Start with a base that matches the top view.

front right

Step 2 Use the front view.
- The front left side is 3 blocks high.
- The front middle and right sides are 1 block high.
- Highlighted segments indicate breaks where columns or rows of blocks appear at different depths.
- The highest block in the front left column is farther back than the 2 blocks below it.
- The third block on the bottom row is farther back than the first 2 blocks in the row.

front right

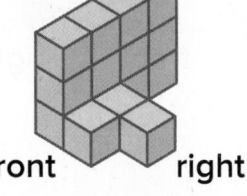

(*continued on the next page*)

Step 3 Use the left view. Use the left view to find where the breaks in the front view occur.

- The first column is 2 blocks high.
- The second column is 1 block high.
- The third column is 3 blocks high.
- The fourth column is 2 blocks high.
- Remove any unnecessary blocks.

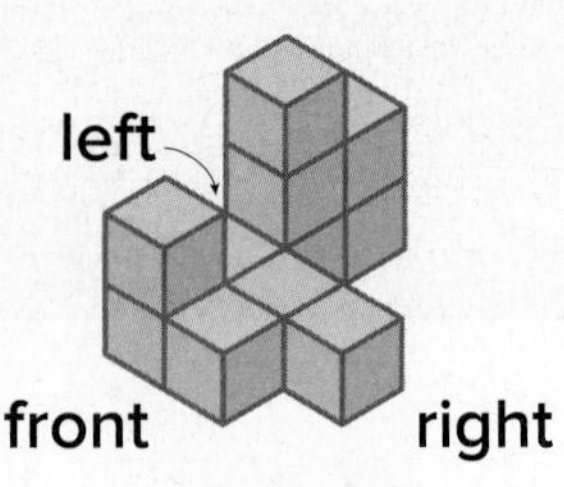

Step 4 Check your model. Use the right view to confirm that you have made the correct model.

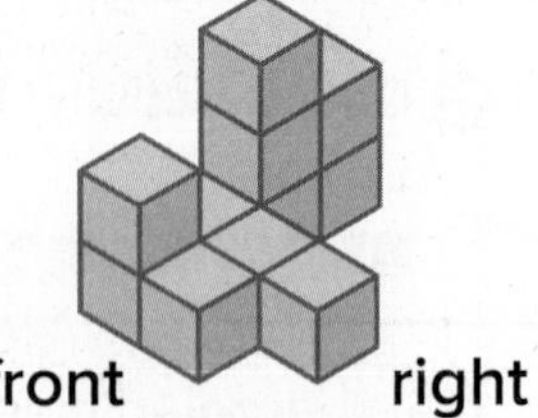

Study Tip

Left View It is often difficult to accurately interpret the left view of an orthographic drawing. Imagine holding a model and rotating it counterclockwise to see the left side. The outline after this rotation will be the left view of the model.

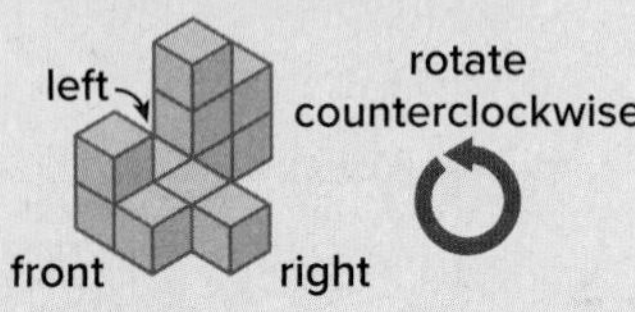

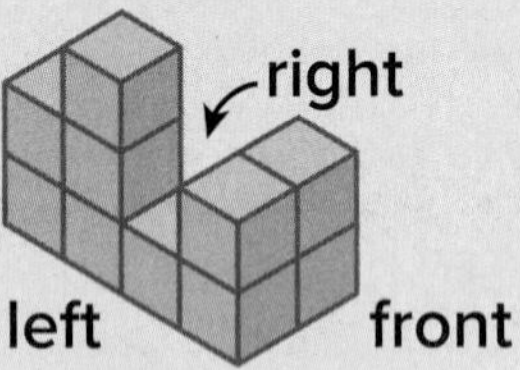

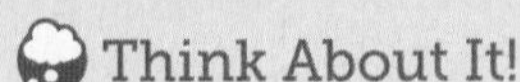

Think About It!

Why is a back view not used in an orthographic drawing?

Check

Which model corresponds to the orthographic drawing?

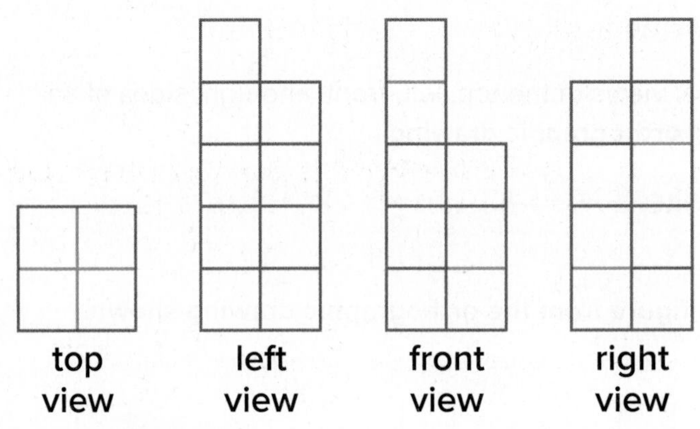

A.

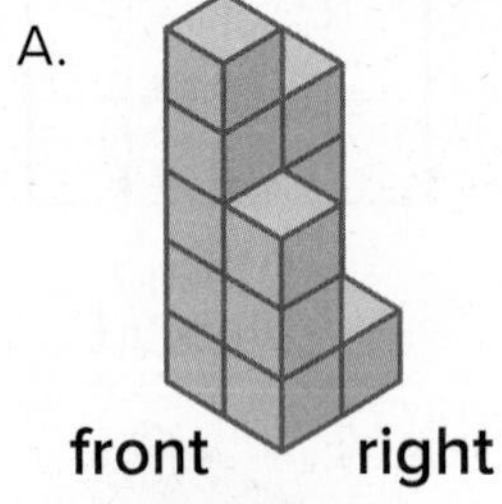

B.

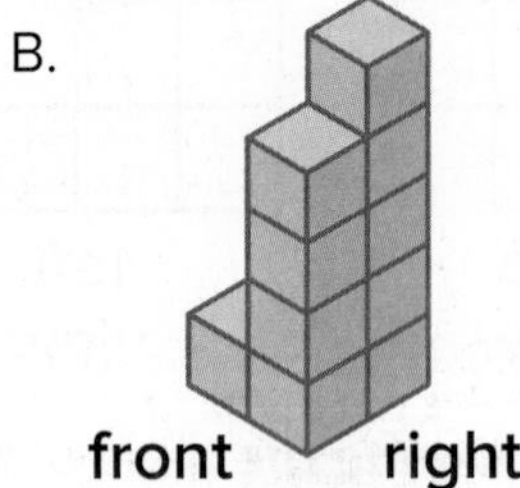

C.

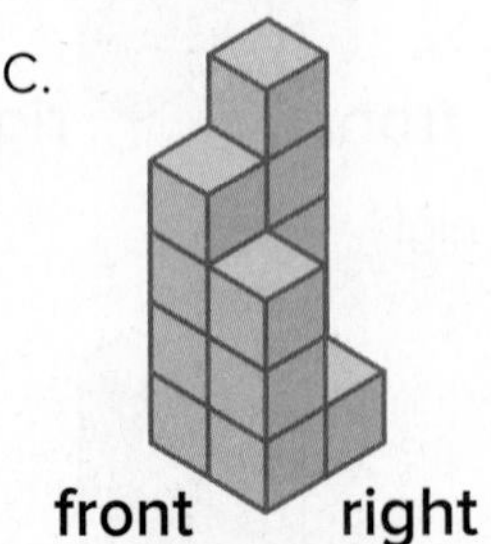

D.

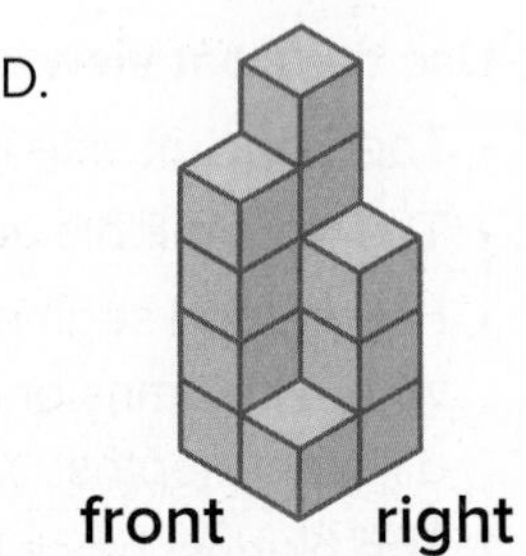

 **Go Online** You can complete an Extra Example online.

Example 2 Make an Orthographic Drawing

Make an orthographic drawing of the figure shown.

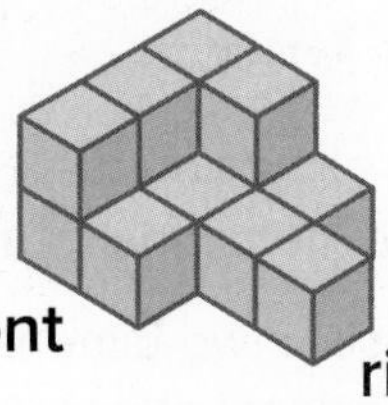

front

right

Step 1 Draw the visible features of each view.

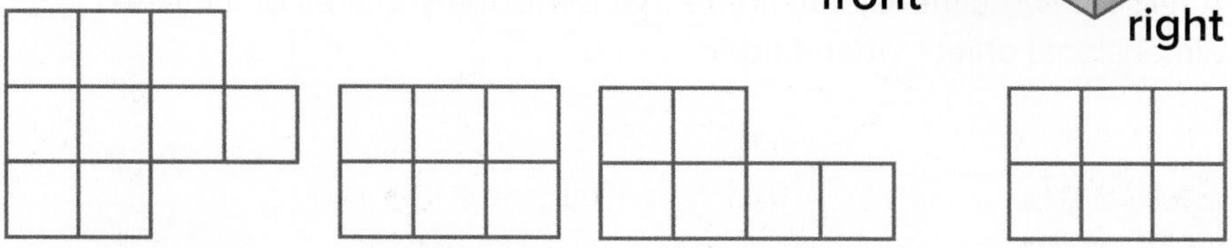

top view left view front view right view

Step 2 Mark each segment where a break occurs.

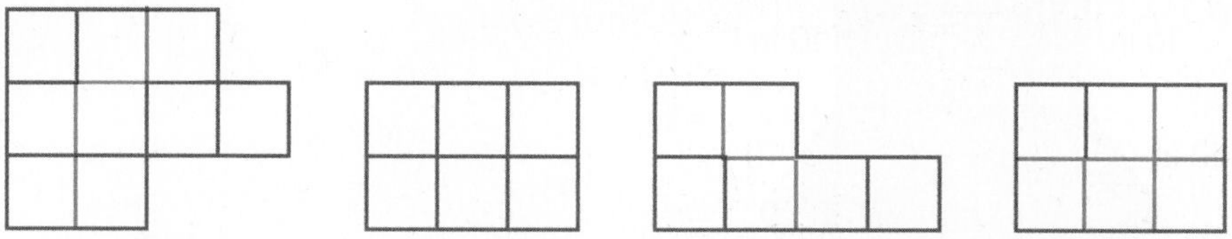

top view left view front view right view

Check

Make an orthographic drawing of the figure shown. Write the letter of the drawing that represents the correct view.

front

right

? top view	? left view	? front view	? right view

A 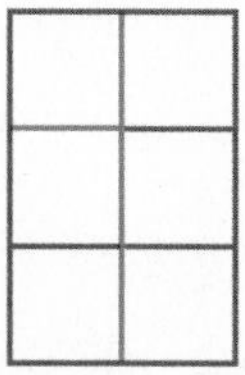 B 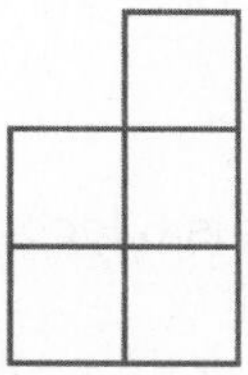 C 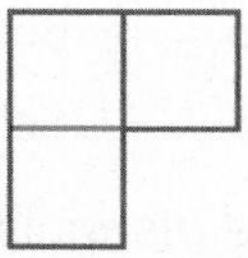 D

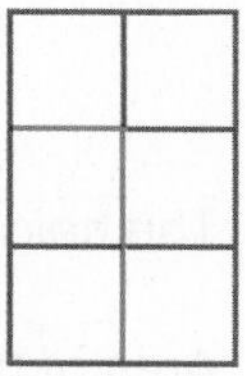

E 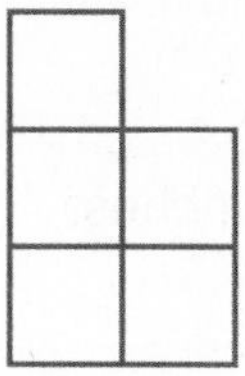 F 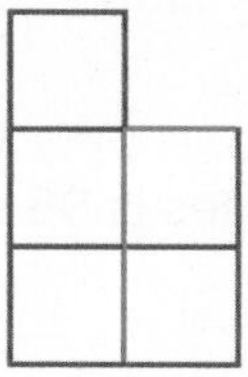 G 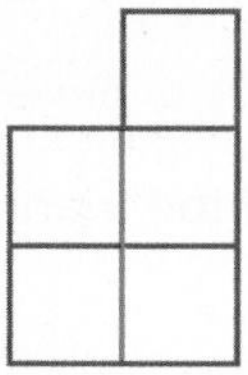 H

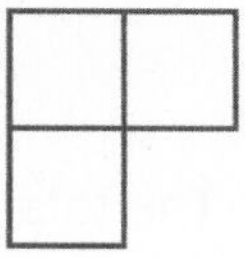

Go Online You can complete an Extra Example online.

Talk About It!

What profession do you think utilizes orthographic drawings? Explain.

Learn Representing Three-Dimensional Figures with Nets

Nets allow you to see all the surfaces of a three-dimensional figure in a two-dimensional drawing.

A **net** is a two-dimensional figure that forms the surfaces of a three-dimensional object when folded.

Example 3 Use a Net to Find Surface Area

Identify the solid that is represented by the net. Then find its surface area.

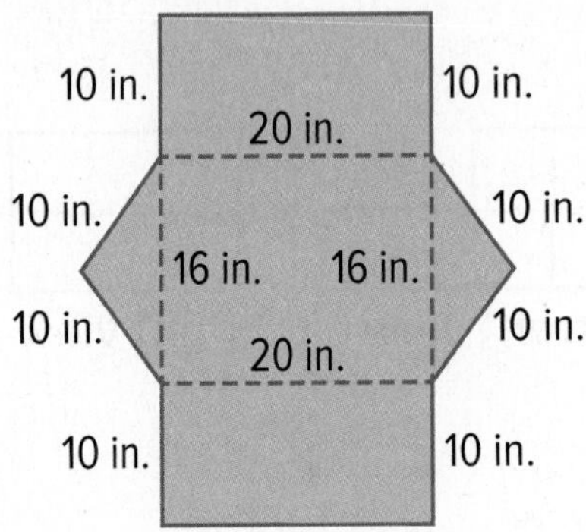

Because this net has two congruent triangular bases, when it is folded, it will form a triangular prism.

Use the net to find the surface area of the solid.

Step 1 Find the area of the triangular bases.

Use the Pythagorean Theorem to find the height of the congruent triangles.

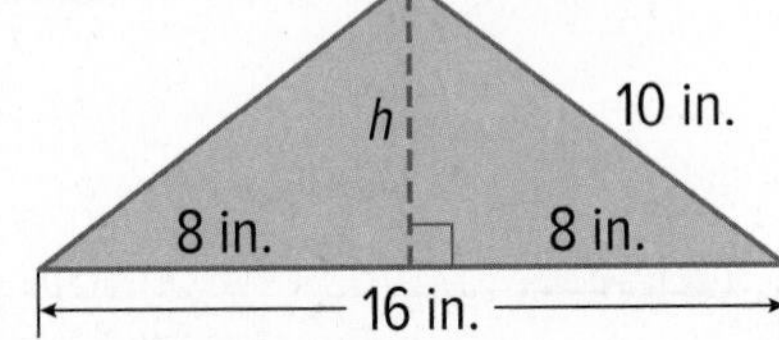

$a^2 + b^2 = c^2$	Pythagorean Theorem
$h^2 + 8^2 = 10^2$	Substitute.
$h^2 + 64 = 100$	Simplify.
$h^2 = 36$	Subtract.
$h = 6$	Solve.

The height of the triangular bases is 6 inches.

Area of triangular bases $= 2 \times \frac{1}{2}bh$	Area of $2 \cong \triangle$s
$= 2 \times \frac{1}{2}(16)(6)$	$b = 16$ and $h = 6$
$= 96$	Simplify.

The total area of the triangular bases is 96 square inches.

Step 2 Find the total surface area of the triangular prism.

$S = 96 + 2(10)(20) + 16(20)$ Area of triangular bases plus area of three rectangles

$= 96 + 400 + 320$ or 816 in^2 Simplify.

The surface area of the triangular prism is 816 square inches.

> **Think About It!**
> If a solid can be represented by more than one net, will the surface area of the solid change? Explain.

Check

Identify the solid that is represented by the net. Then find its surface area.

A. square pyramid; 104 cm^2

B. tetrahedron; $64 + 64\sqrt{21} \text{ cm}^2$

C. tetrahedron; 88 cm^2

D. square pyramid; $64 + 32\sqrt{21} \text{ cm}^2$

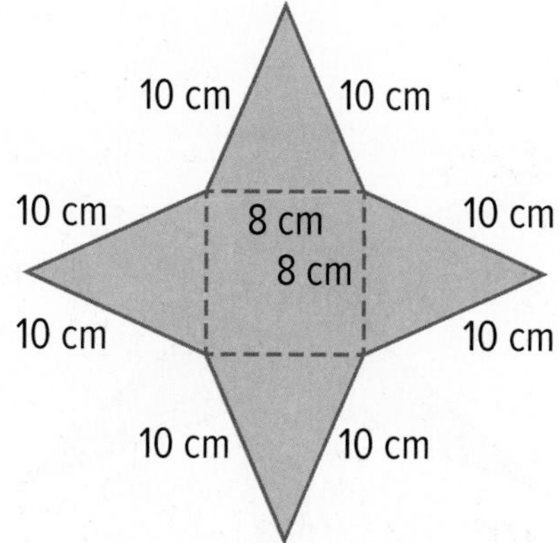

Example 4 Identify Platonic Solids

Identify the Platonic solid that is represented by the net.

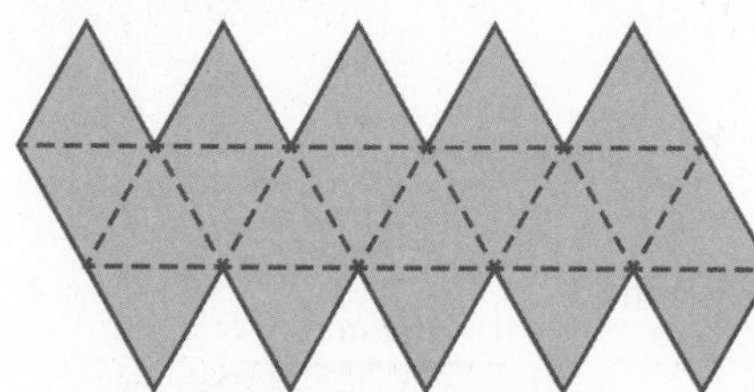

Because this net has 20 equilateral triangles, it represents a net of a(n) icosahedron.

Check

Identify the Platonic solid that is represented by the net.

A. decahedron

B. pentagonal prism

C. dodecahedron

D. icosahedron

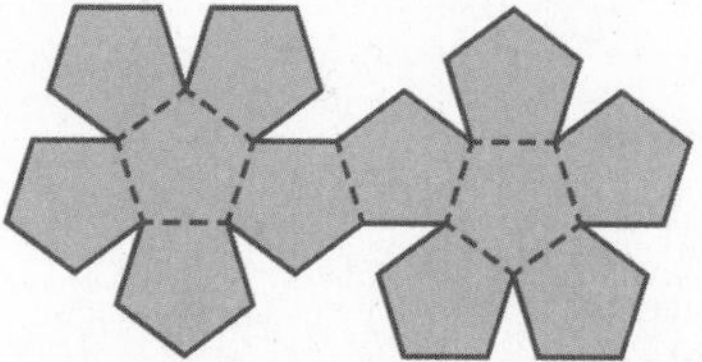

 **Go Online** You can complete an Extra Example online.

Example 5 Draw Nets for Three-Dimensional Figures

Draw a net for the hexagonal pyramid.

To draw the net of a three-dimensional solid, visualize cutting the solid along one or more of its edges, opening up the solid, and flattening it completely.

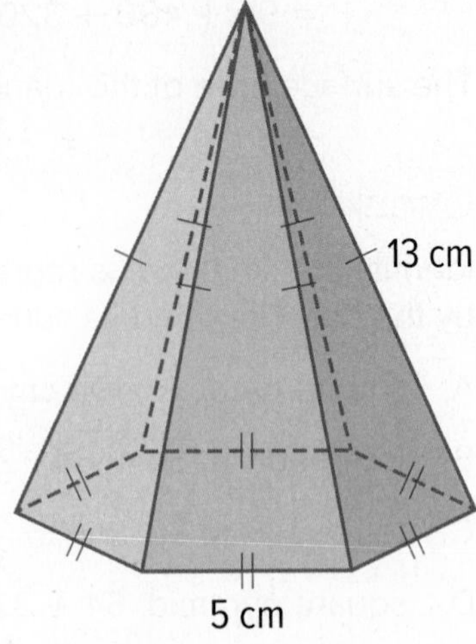

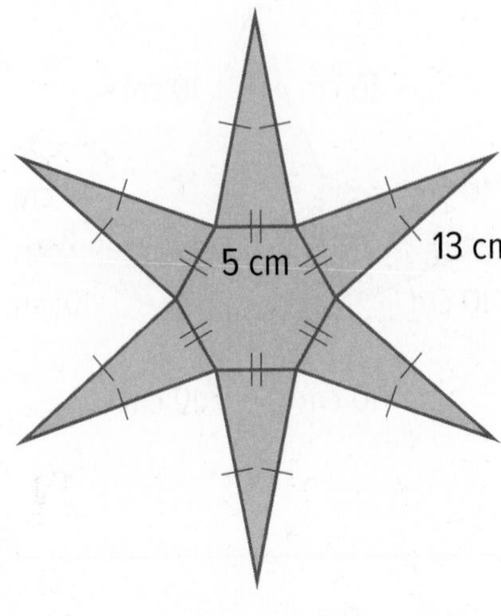

Study Tip

Approximations A net for a sphere can be created using several adjoining pointed ellipses or by creating a polyhedron with a large number of sides. However, because paper can only curve in one direction, it is impossible to make a perfect sphere. Thus, the spheres made from nets will be approximations.

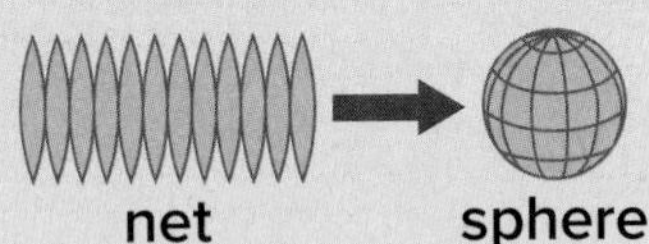

Check

Draw a net for the regular pentagonal prism.

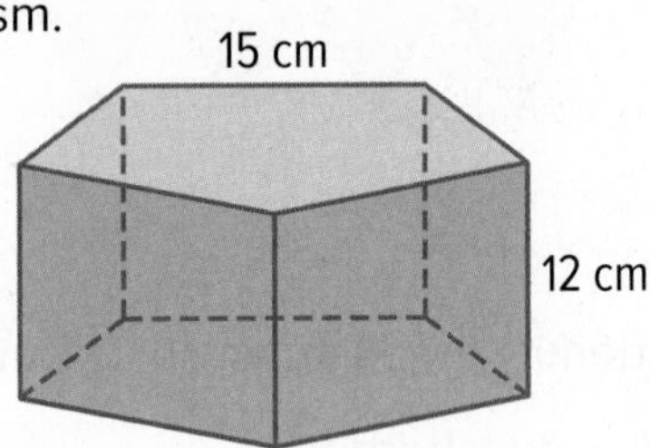

Go Online You can complete an Extra Example online.

Example 6 Represent a Real-World Object with a Net

TENTS **Draw a net to represent the three-dimensional figure that can be used to model the tent.**

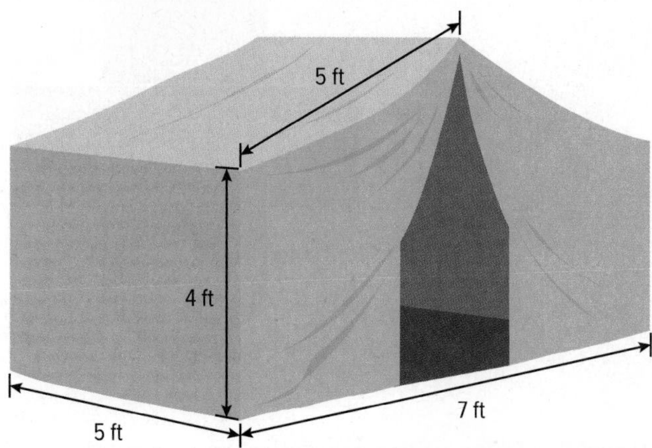

The tent can be modeled by a(n) pentagonal prism.

Step 1 Start by drawing the bottom of the tent.

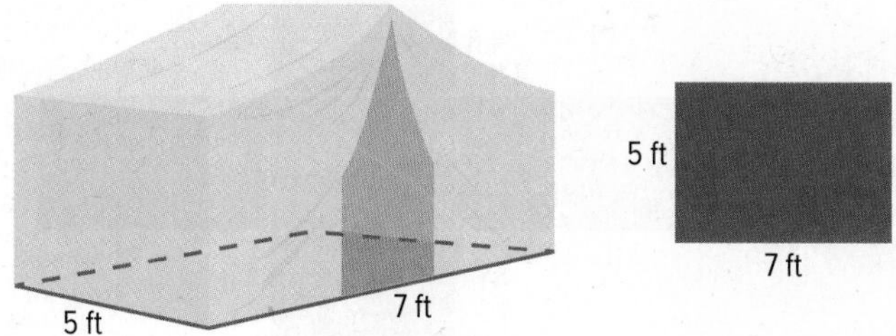

Step 2 Next, draw the pentagonal bases of the prism in the net. The pentagonal faces will attach to the rectangle at the 7-foot edges.

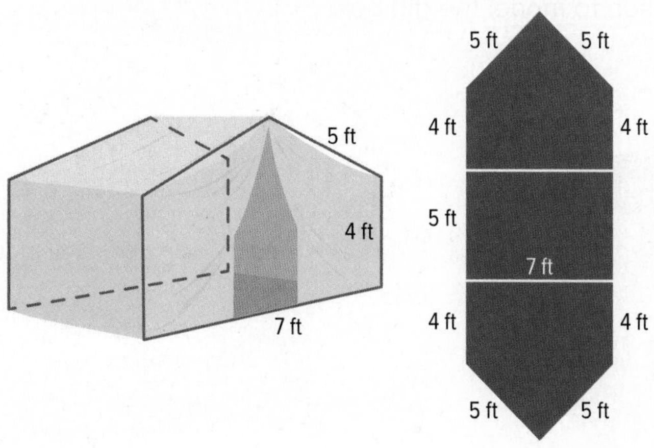

(*continued on the next page*)

Step 3 Draw the rectangular faces of the prism that represent the sides of the tent.

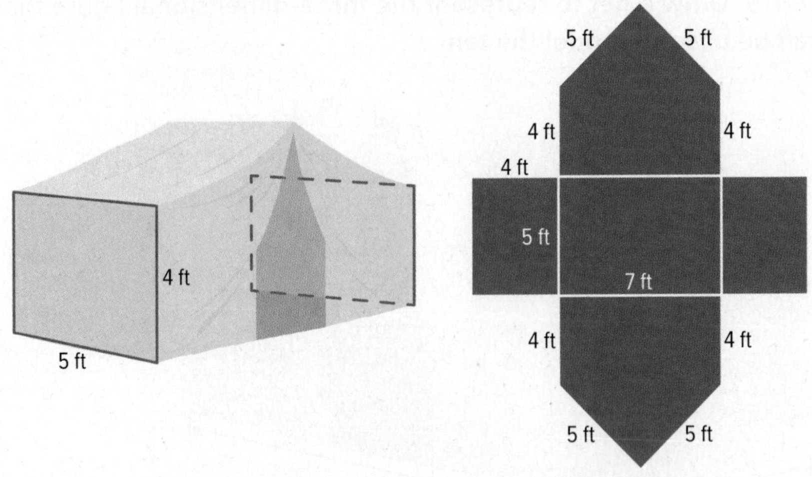

Think About It!

Is there more than one way to correctly draw a net? Explain using the net from Example 6.

Step 4 Draw the rectangular faces of the prism that represent the roof of the tent. Compare your net to the original figure to ensure that the dimensions are correct.

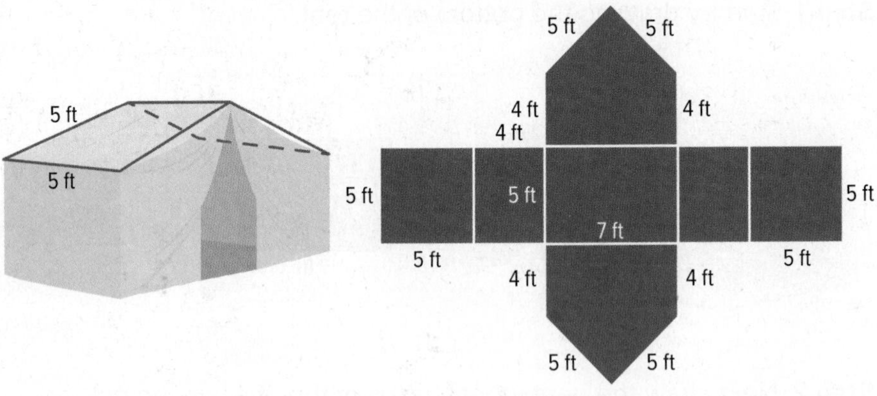

Check

GIFT WRAPPING Draw a net to represent the three-dimensional figure that can be used to model the gift box.

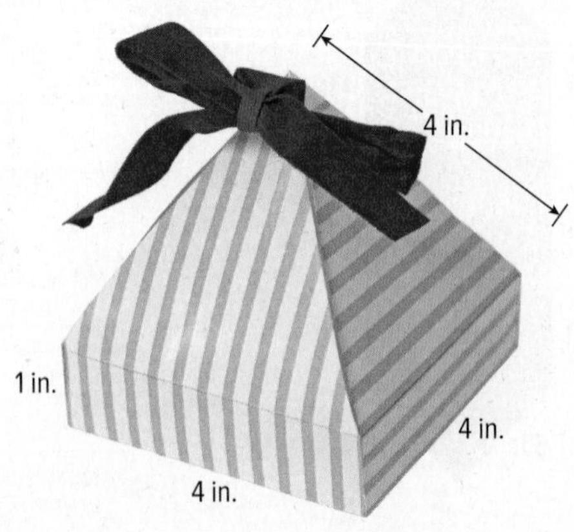

 **Go Online** You can complete an Extra Example online.

Practice

Go Online You can complete your homework online.

Example 1

Make a model of a figure for each orthographic drawing.

1.

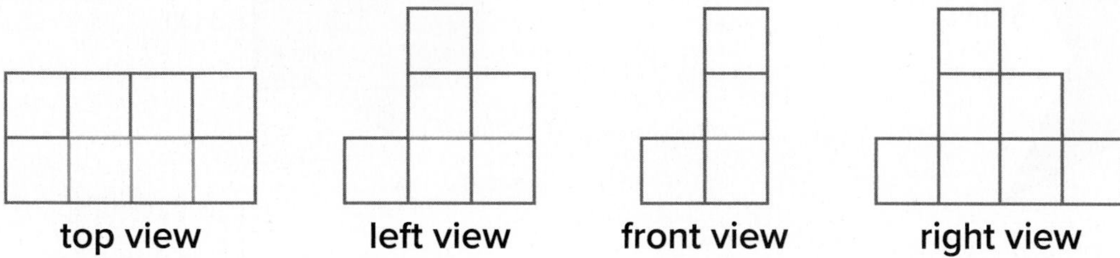

2.

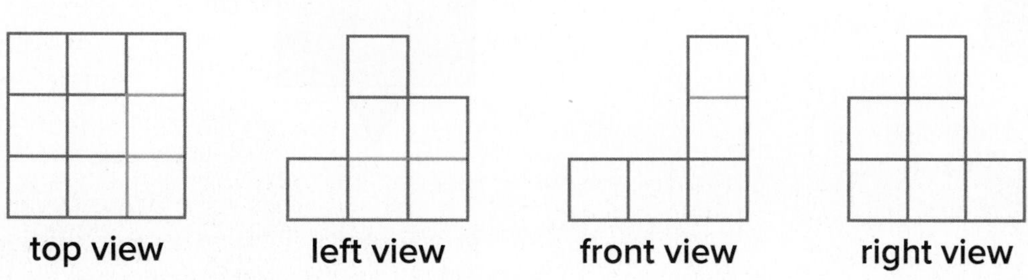

Example 2

Make an orthographic drawing of each figure.

3.

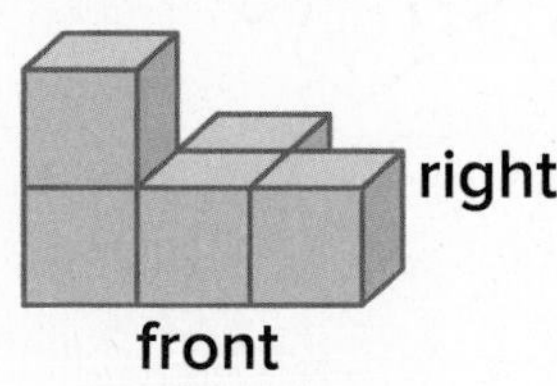

4.

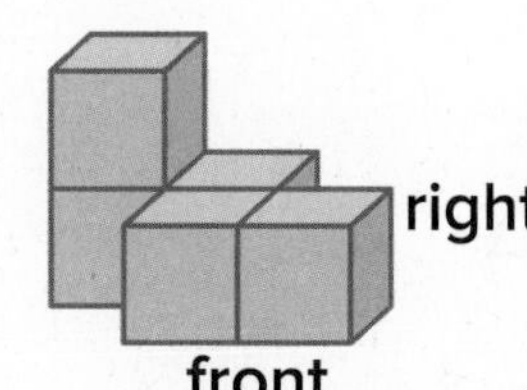

5.

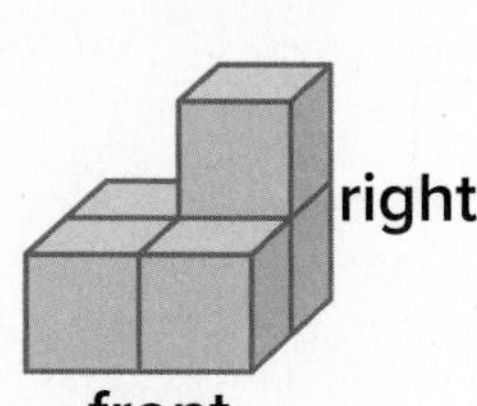

6.

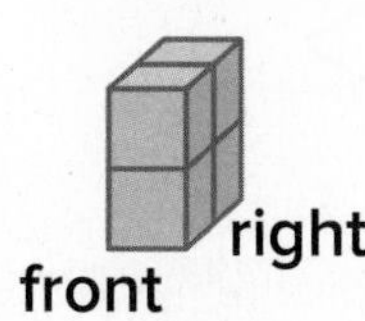

7.

8.

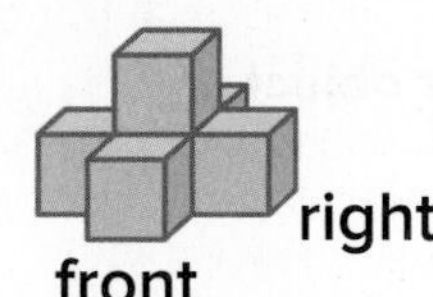

Example 3

Make a model of the solid that is represented by each net. Then identify the solid and find its surface area.

9.

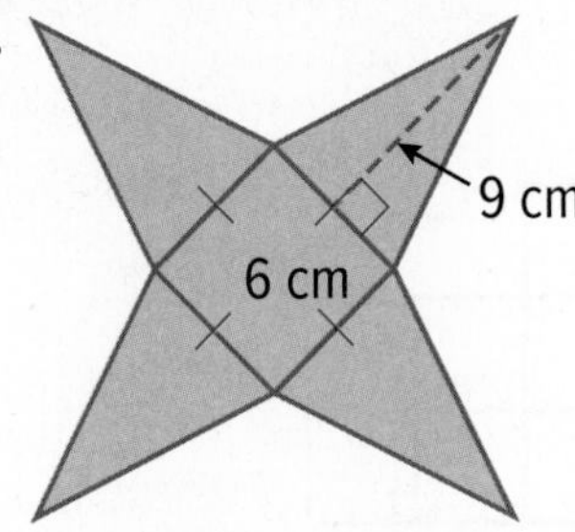

10.

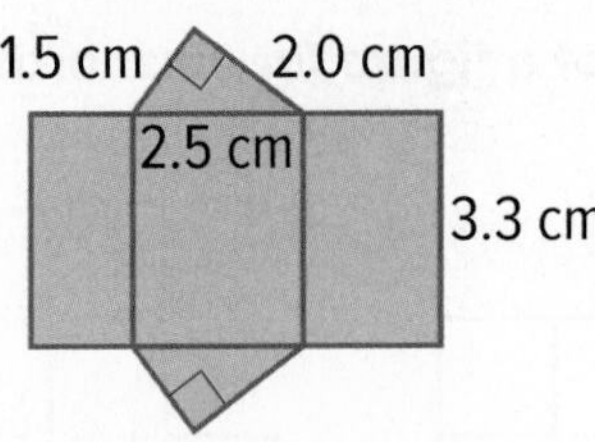

11.

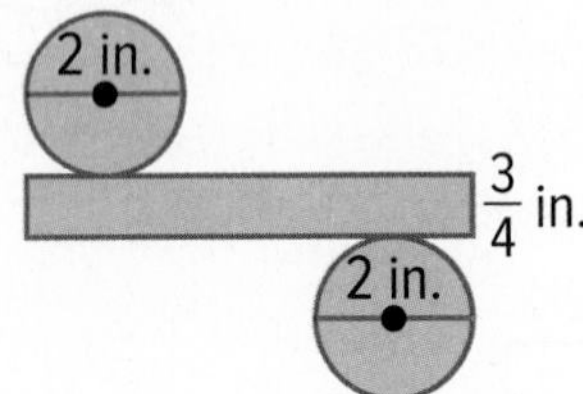

12.

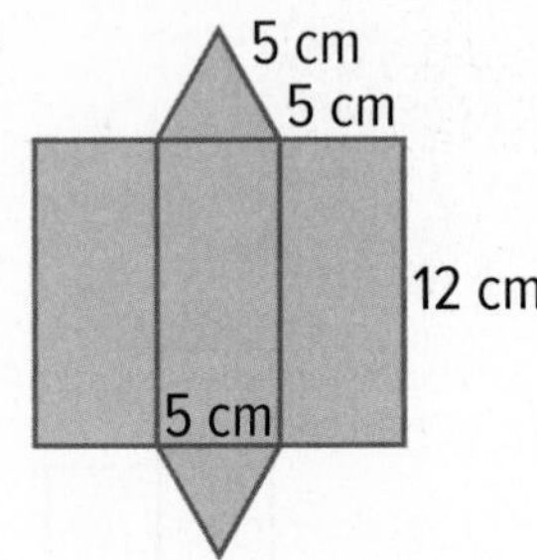

Example 4

Identify the Platonic solid that is represented by the net.

13.

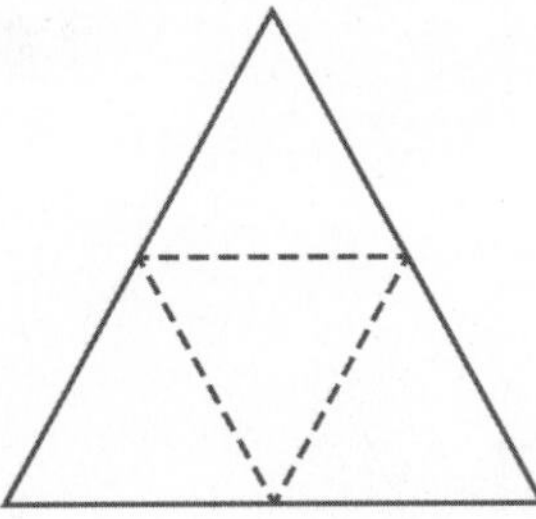

14.

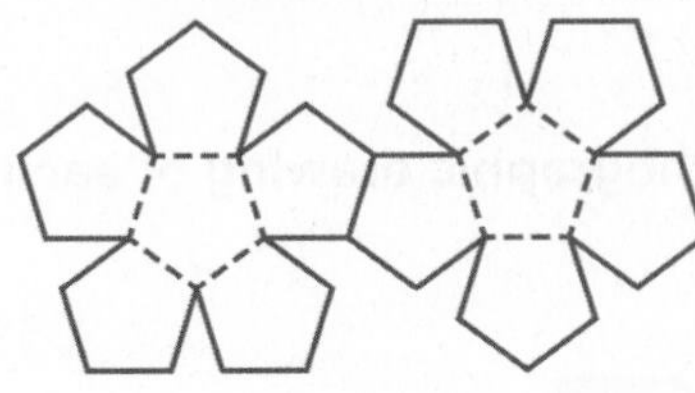

15.

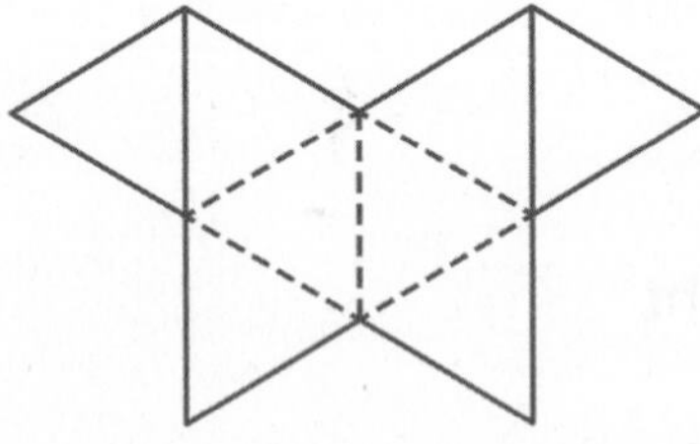

16.

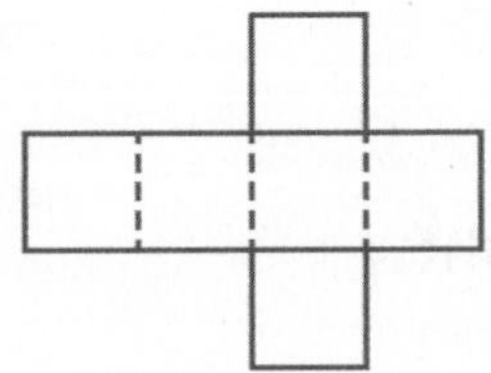

Examples 5 and 6

Draw a net for each solid or object.

17.

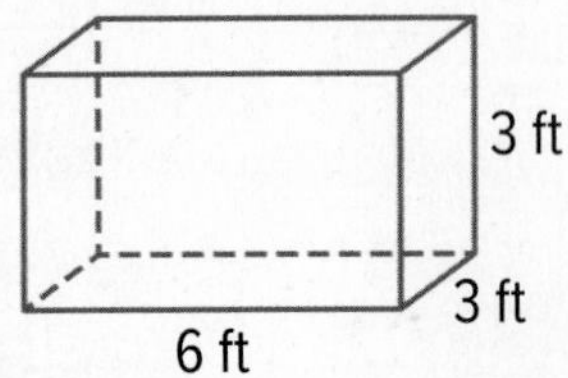

18.

Draw a net for each solid or object.

19.

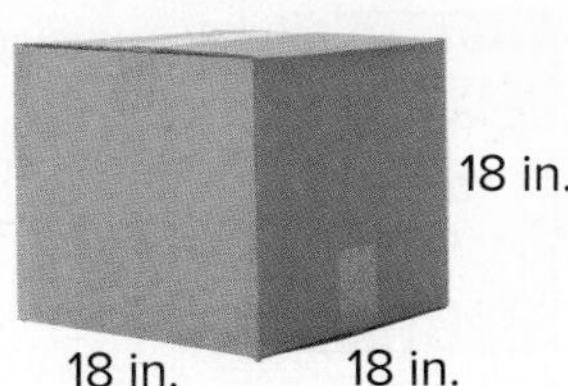

20.

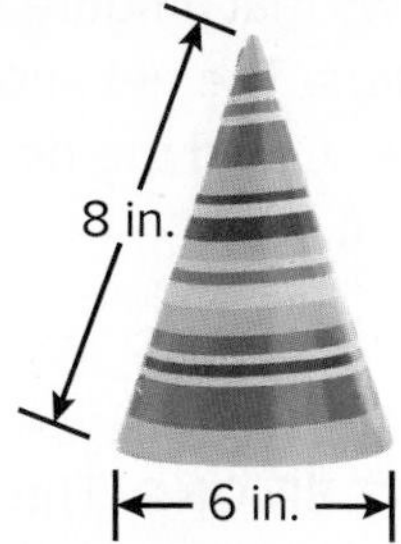

21.

22.

Mixed Exercises

23. GAMING Candela is playing a game that has game pieces. Use the orthographic drawing to make a model of the game piece.

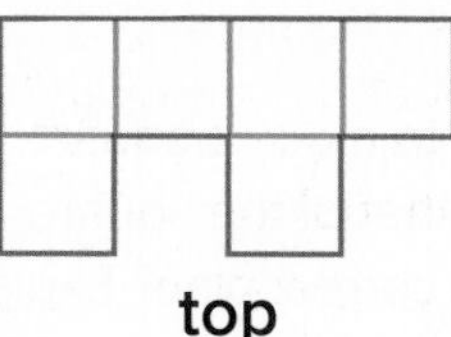
top

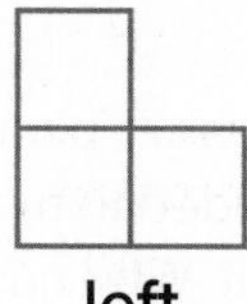
left

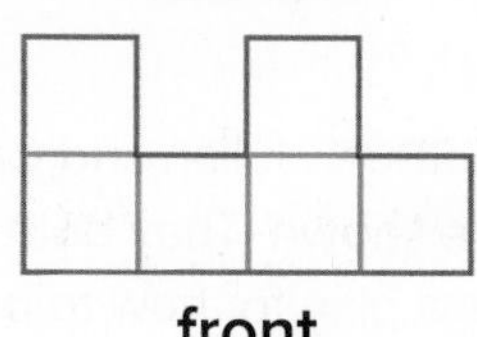
front

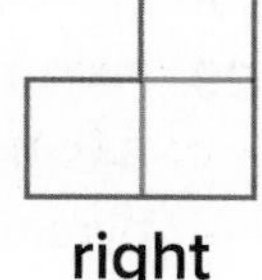
right

24. FURNITURE Make an orthographic drawing to show the top, front, left, and right views of the storage cabinet.

25. MODELING Identify a real-world object that can be represented by the net shown.

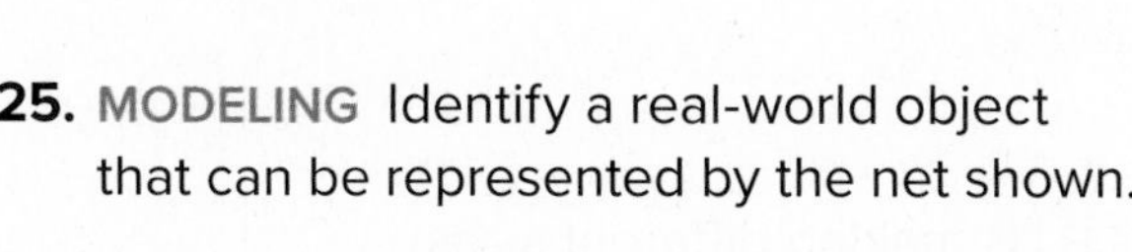

26. GIFT WRAP Olive is wrapping a gift for her sister in a box that has the net shown here. How many square inches of wrapping paper will it take to cover the box?

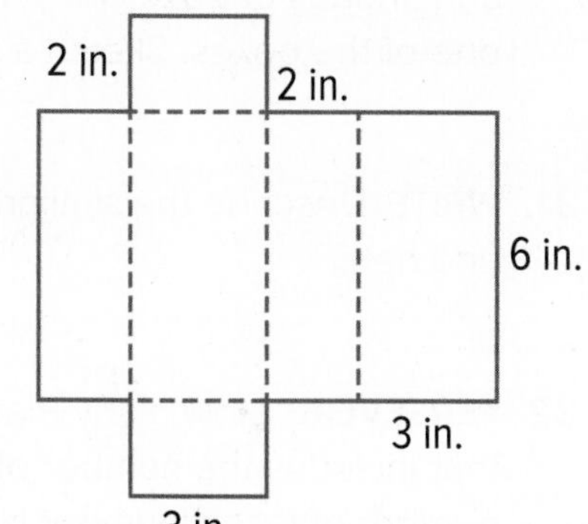

Higher-Order Thinking Skills

27. **ANALYZE** Julia knows that a figure has a surface area of 40 square centimeters. The net shown has 5-centimeters and 2-centimeters edges. Could the net represent the figure? Justify your argument.

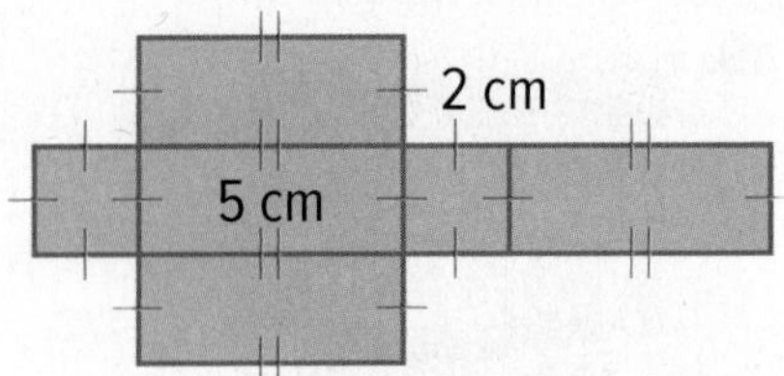

28. **WHICH ONE DOESN'T BELONG** The model represents a building. Which orthographic drawing does not belong? Justify your conclusion.

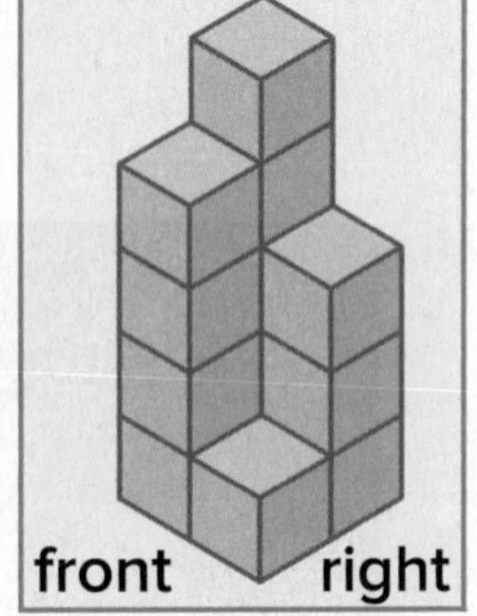

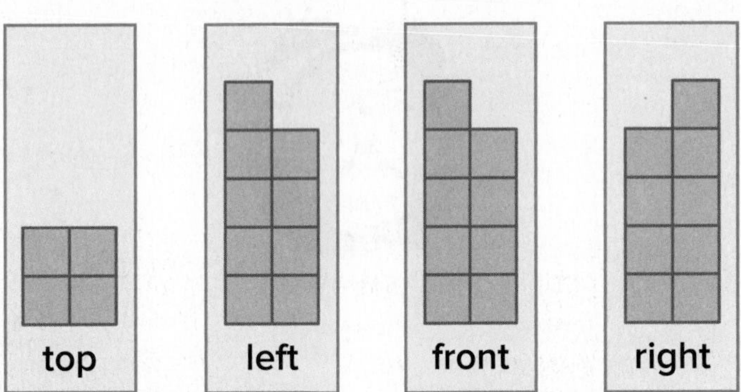

29. **FIND THE ERROR** Julian and Caleb were planning to make a square pyramid like the one shown. They both decided to make a net of the square pyramid as a plan for how to build it. Who has the correct plan? Explain your reasoning.

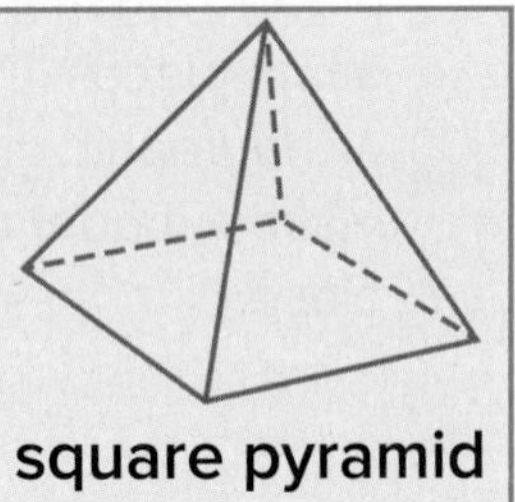

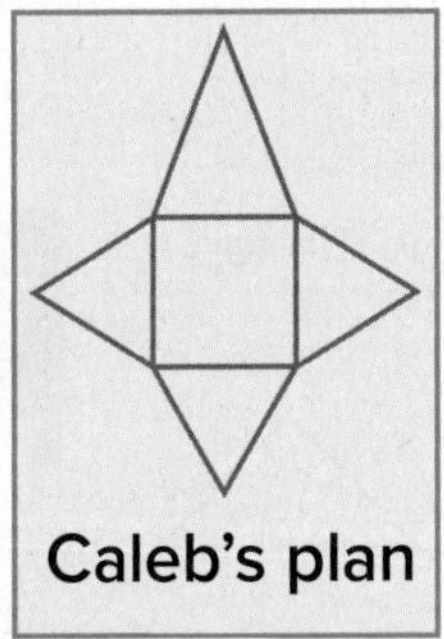

30. **CREATE** Adriana works for a company called Boxes R Us making different sizes and shapes of boxes for packages. Adriana's boss wants her to sketch nets of one of the boxes. Sketch a possible net that Adriana could have drawn.

31. **WRITE** Describe the similarities and differences in orthographic drawings and nets.

32. **PERSEVERE** How many Platonic solids are there? Give a description of each solid that includes the number of two-dimensional shapes that meet at each vertex, number of faces, number of vertices, and number of edges.

Precision and Accuracy

Explore Precision and Accuracy in Basketball

Online Activity Use a real-world situation to complete the Explore.

INQUIRY How are the concepts of precision and accuracy similar, and how are they different?

Today's Goals

- Determine the levels of precision and accuracy in real-world scenarios.
- Calculate the approximate error of measurements.
- Choose the appropriate level of accuracy of measurements when reporting quantities.

Today's Vocabulary

precision

accuracy

approximate error

Learn Precision and Accuracy

Precision is the repeatability, or reproducibility, of a measurement. It depends only on the smallest unit of measure available on a measuring tool. Suppose you are told that a segment measures 8 centimeters. The length, to the nearest centimeter, of each segment shown below is 8 centimeters.

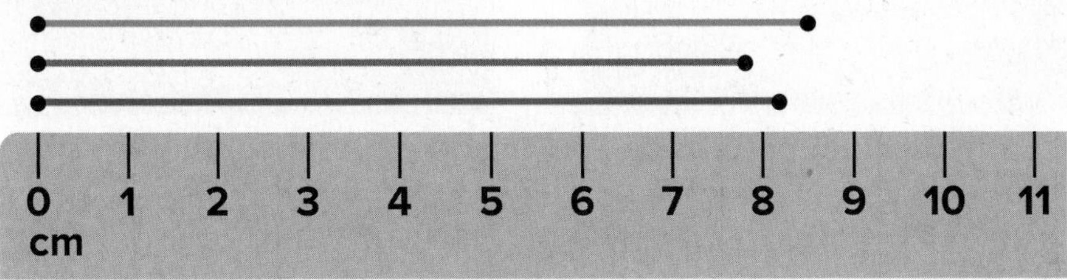

Accuracy is the nearness of a measurement to the true value of the measure. Consider the target practice results shown below.

The targets demonstrate the various levels of precision and accuracy.

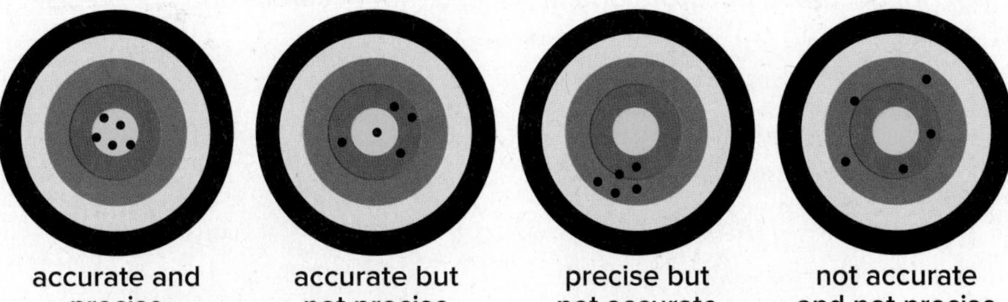

accurate and precise | accurate but not precise | precise but not accurate | not accurate and not precise

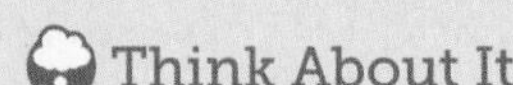

Think About It!

How do you determine how to round when measuring a line segment?

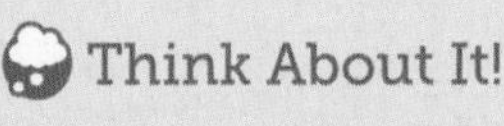

Think About It!

Why are only the first and second targets accurate?

Study Tip

Accuracy vs. Precision Remember that precision refers to the clustering of a group of measurements and is dependent on reproducing a certain measurement. Accuracy only refers to how close a measurement is to the true value of the measure. A measurement can be accurate without being precise or vice versa.

Example 1 Identify Precision and Accuracy

HORSESHOES **Ally and Isha are playing horseshoes. In this game, each player has two horseshoes that are thrown as close as possible to the stake in the middle of the pit. The results for four innings are shown below.**

Label each horseshoe pit as *accurate but not precise, precise but not accurate, accurate and precise,* or *not accurate and not precise.*

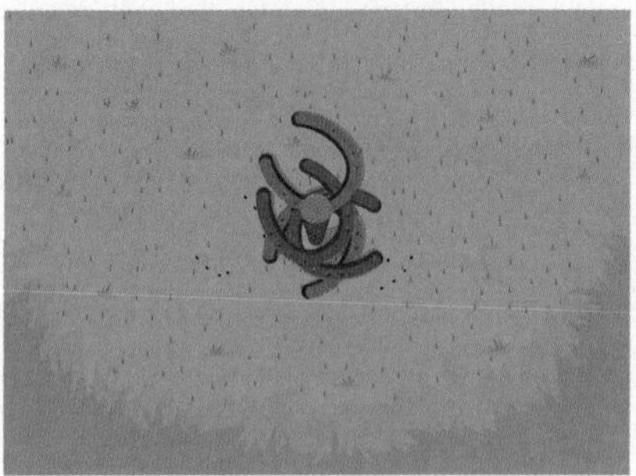

accurate and precise

precise but not accurate

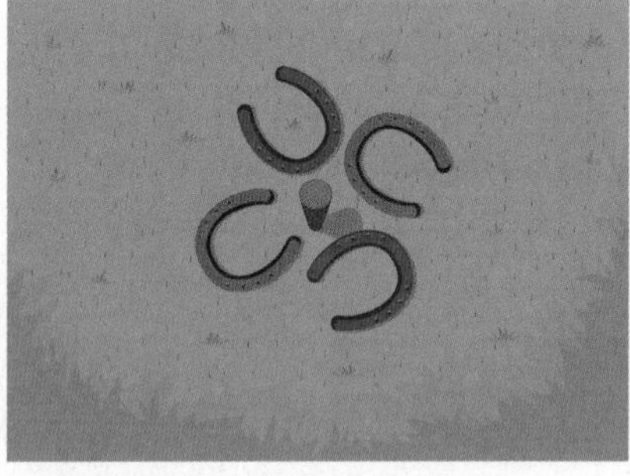

accurate but not precise

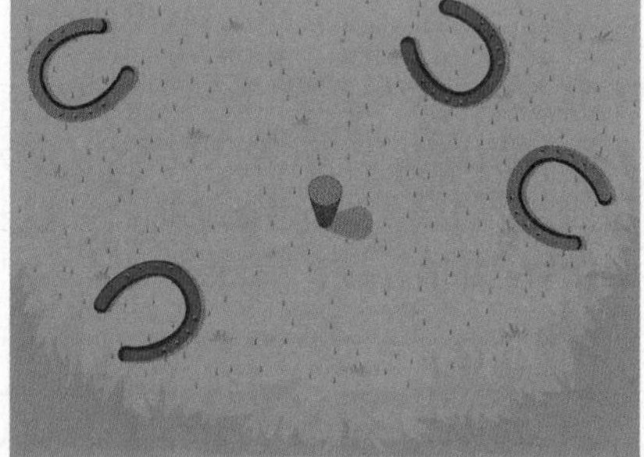

not accurate and not precise

Check

LAWN GAMES Kate is playing bean bag toss with her friends. Teams of two take turns tossing bean bags at a raised platform with a hole at the far end. When a team throws, or knocks, their own bag into the hole, they receive 3 points. Label each board as *accurate but not precise, precise but not accurate, accurate and precise,* or *not accurate and not precise.*

Go Online You can complete an Extra Example online.

Learn Approximate Error

In the physical world, measurements are always approximate. The **approximate error** of a measurement can help you determine how accurate your calculations can be using the measurement.

Key Concept • Approximate Error

The positive difference between an actual measurement and an approximate or estimated measurement is its approximate error E_a.

$E_a = |\text{actual measurement} - \text{estimated measurement}|$

Think About It!

In what real-world situation would it be helpful to find an approximate error?

Example 2 Find Approximate Error

A student weighs a 10-gram precision mass on three different scales. Find the approximate error for each measurement.

a. spring scale: 9.86 grams

$E_a = |\text{actual measurement} - \text{estimated measurement}|$
$= |10 - 9.86|$ or 0.14 g

b. lab scale: 9.92 grams

$E_a = |\text{actual measurement} - \text{estimated measurement}|$
$= |10 - 9.92|$ or 0.08 g

c. food scale: 10.3 grams

$E_a = |\text{actual measurement} - \text{estimated measurement}|$
$= |10 - 10.3|$ or 0.3 g

Think About It!

Why is it important to calculate approximate errors when using scales?

Check

The temperature in Portland, Oregon, is 35° F. Declan measures the temperature outside his house. The thermometer measures 34.2° F. What is the approximate error of the temperature?

__?__ °F

Learn Calculating with Rounded Measurements

When rounding to a place value, look at the value immediately to the right of that position. If the value is 5 or greater, then round up.

42.64 rounds to 42.6. Because $4 < 5$, do not round to the next tenth.

42.57 rounds to 42.6. Because $7 \geq 5$, round to the next tenth.

Given a measurement of 42.6 centimeters rounded to the nearest tenth, the actual measurement could be any value in a range of values that round to 42.6.

$42.55 \leq \text{actual measurement} < 42.65$

Example 3 Calculate with Rounded Measurements

CARPETING Alejandro wants to carpet his bedroom. He measures the dimensions of his bedroom and rounds to the nearest foot. The carpet he chose costs $2.63 per square foot.

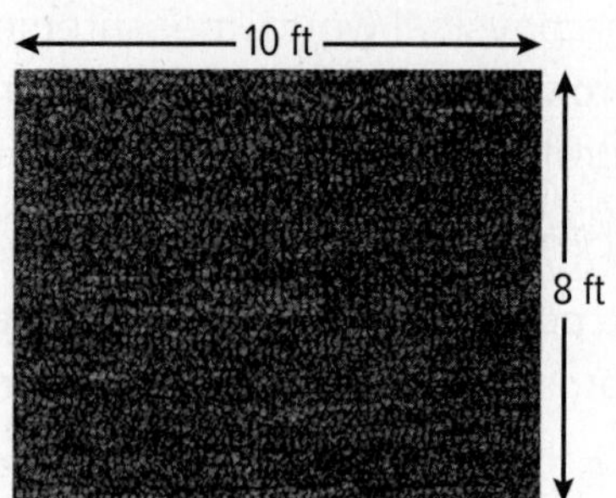

Part A What is the possible range for how much it will cost to carpet Alejandro's bedroom?

Step 1 **Find the possible range for the area of the room.**

9.5 feet $\leq$ actual length $<$ 10.5 feet

7.5 feet $\leq$ actual width $<$ 8.5 feet

least possible area $= 9.5 \cdot 7.5$ or 71.25 ft^2
greatest possible area $= 10.5 \cdot 8.5$ or 89.25 ft^2

The area is at least 71.25 square feet but less than 89.25 square feet.

Step 2 **Find the cost to buy carpet for the room.**

cost for least possible area: $71.25 \cdot \$2.63 = \$187.39$
cost for greatest possible area: $89.25 \cdot \$2.63 = \$234.73$

The cost would be at least $187.39 but less than $234.73.

Part B Alejandro checks the dimensions of the room, and measures it to be 9.8 feet by 8.2 feet to the nearest tenth of a foot. How does this change the range for the cost of the carpeting?

Step 1 **Find the possible range for the area of the room.**

9.75 feet $\leq$ actual length $<$ 9.85 feet

8.15 feet $\leq$ actual width $<$ 8.25 feet

least possible area $= 9.75 \cdot 8.15$ or 79.4625 ft^2
greatest possible area $= 9.85 \cdot 8.25$ or 81.2625 ft^2

Step 2 **Find the cost to buy carpet for the room.**

cost for least possible area: $79.4625 \cdot \$2.63 = \$208.99$
cost for greatest possible area: $81.2625 \cdot \$2.63 = \$213.72$

The cost would be at least $208.99 but less than $213.72.

When the measurements were rounded to the nearest foot, the range of costs was more than $45. With the measurements rounded to the nearest tenth, the range of costs is about $4.75. Rounding to the nearest tenth creates a more accurate range for the cost of the carpeting.

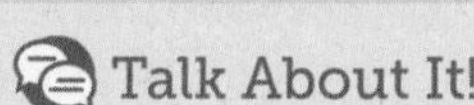

Talk About It!

How would a recorded time be affected in a stopwatch rounded to the nearest second versus to the nearest millisecond?

Practice

Go Online You can complete your homework online.

Example 1

1. PRECISION A manufacturer claims that its rice cakes are packaged with 20 in each package. A sample of 12 packages is counted for accuracy. The sample yields a count of {18, 17, 17, 17, 18, 18, 18, 17, 18, 17, 18, 17} rice cakes. How accurate and precise is the manufacturer's claim? Explain your reasoning.

Example 2

2. SCALES A 10-pound weight is weighed on two different scales. Find the approximate error of each weight.

   a. digital bathroom scale: 9.59 pounds

   b. food scale: 10.09 pounds

3. PHYSICS A circuit has amperage of 0.01 milliamp. A multimeter measures the amperage of the circuit at 0.06 milliamp. What is the approximate error?

4. CARPENTRY A door frame is 2.13 meters high. Chandra measures the height of the door frame with a carpenter's rule. She measures 2.22 meters. What is the approximate error of the height?

5. COOKING Water boils at 212.0°F. Jeremiah uses a kitchen thermometer to measure the temperature of a pot of boiling water. The thermometer measures 213.1°F. What is the approximate error of the temperature?

Example 3

6. CONSTRUCTION The public works department is repaving some of the roads in the city. The materials needed to repave this section of the road cost $2.50 per square foot.

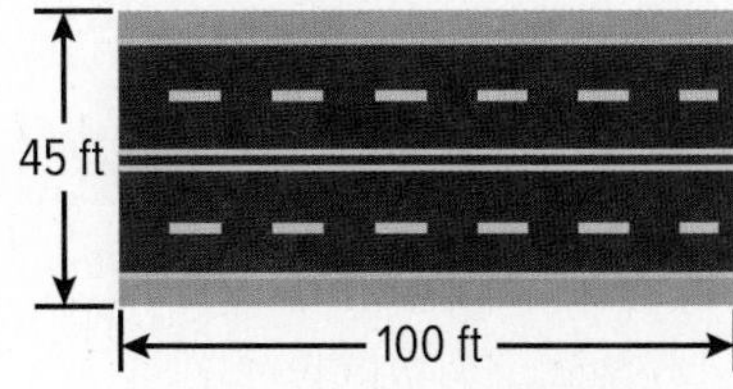

   a. What is the possible range for the area of the road?

   b. What is the possible range for the cost, c, of the materials needed to repave this section of the road?

7. GRASS SEED George is buying grass seed for his lawn. Grass seed is sold at $0.40 per square yard.

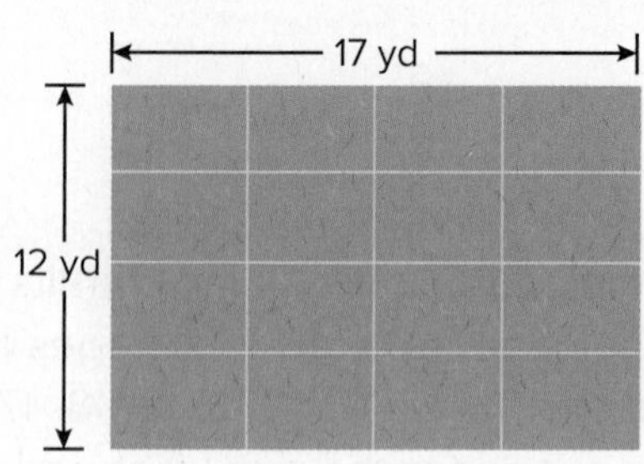

a. What is the least value for the length of the lawn?

b. What is the greatest value for the width of the lawn?

c. What is the possible range for the area of the lawn?

d. What is the possible range for the cost of the grass seed?

Mixed Exercises

8. THERMOMETER The thermostat on a heated pool is set at 76.5°F. A thermometer in the pool is shown. What is the approximate error of the temperature?

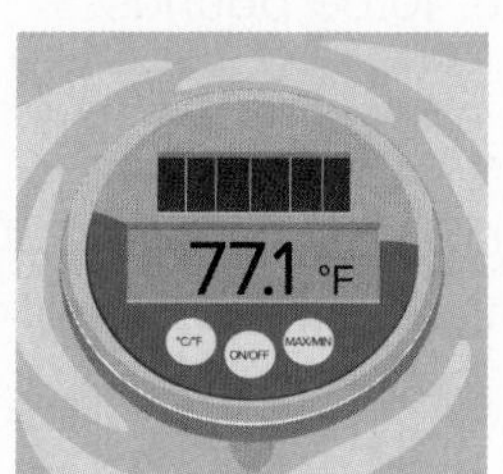

9. SANDWICHES A sandwich shops claims to sell foot-long sandwiches. Xiao uses a ruler to measure her sandwich. The ruler measures 11 inches. What is the approximate error of the length?

10. SPEED A police officer uses a radar detector to measure the speed of Roya's car. Roya's speedometer reads 55 miles per hour. The radar detector measures her speed at 56.71 miles per hour. What is the approximate error of the speed?

11. BICYCLES An assembly line supervisor weighs three 25-pound bicycle frames on a scale. Find the approximate error of each weight.

a. Bicycle A: 25.11 pounds

b. Bicycle B: 24.99 pounds

c. Bicycle C: 24.36 pounds

12. DELI Josephina works at a deli. She is testing the scales at the deli to make sure they are accurate. She uses a weight that is exactly 1 pound and gets the following results shown in the table. Which scale is the most accurate?

Scale	Weight (lb)
1	1.013
2	1.01
3	0.97

13. **GARDEN** Mr. Granger wants to spread fertilizer on his vegetable garden that has dimensions 41.5 feet by 30.8 feet. The fertilizer he chose costs $0.75 per square foot for adequate coverage. What is the possible range for the cost of the fertilizer that is needed to cover the vegetable garden?

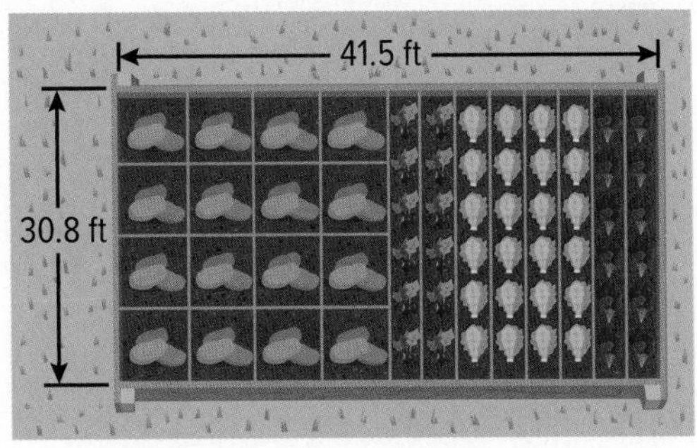

14. **HEIGHT** Lucas was proud of how much he had grown over the last six months since his grandma had seen him last. He told her that he was 6 feet 3 inches. His grandma didn't believe him, so she measured him again, and he was 6 feet 1 inch. What is the approximate error of Lucas' height?

15. **PAINT** You measure a wall of your room as 8 feet high and 12 feet wide. You want to apply wallpaper to only this wall. The wallpaper is expensive and will cost $1.25 per square foot. What is the possible range for the cost of the wallpaper?

16. Four measurements were taken three different times. The correct measurement is 52.4 cm. Determine whether the set of measurements is *accurate, precise, both,* or *neither*. Explain your reasoning.

    a. 56.1 cm, 48.9 cm, 24.2 cm, 5 cm

    b. 73.1 cm, 74.0 cm, 73.5 cm, 73.7 cm

    c. 52.6 cm, 52.5 cm, 52.2 cm, 52.3 cm

Higher-Order Thinking Skills

17. **WRITE** Many people confuse the definitions of accuracy and precision. What is the difference between accuracy and precision? Give an example of a set of four numbers that represents accurate and precise measurements for a cut of meat at a steakhouse that advertises a 16-ounce ribeye steak special on Tuesday nights.

18. **WRITE** Isabel says that if a set of measurements is accurate, then it is also precise. If you agree, explain your reasoning. If you disagree, provide a counterexample.

19. **PERSEVERE** Jayden measures and labels the dimensions of a box.

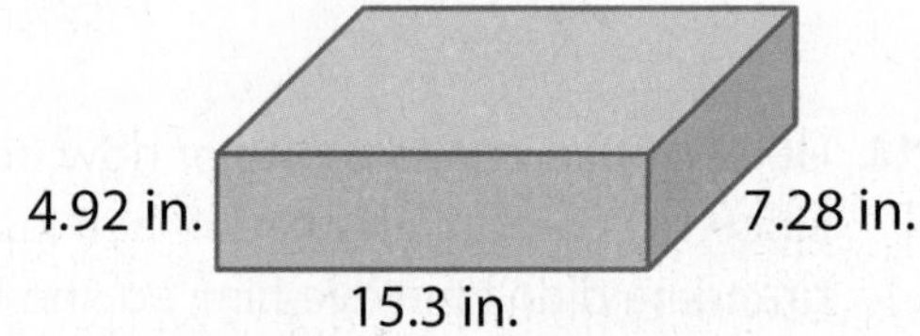

a. Calculate the areas of the faces of the box.

b. Determine the surface area of the box.

c. Determine the range of values that should contain the actual (true) measure of the surface area of the box. Explain your reasoning.

d. Suppose that Jayden had incorrectly measured the first dimension as 15.1 inches. Find the surface area of the box using this measure.

20. **CREATE** A manufacturer claims that its bags of sweetener contain 9.7 ounces in each bag. Create a sample of weights of 10 bags of sweetener such that the sample is precise and accurate. Explain your reasoning.

Representing Measurements

Explore Significant Figures

Online Activity Use the guiding exercises to complete the Explore.

INQUIRY How can you determine the number of significant figures in a measurement?

Today's Goals
- Determine the correct number of significant figures in recorded measurements.
- Round measurements to the correct number of significant figures.

Today's Vocabulary
significant figures

Learn Determining Significant Figures

Using significant figures allows you to maintain the correct level of precision when you are working with measurements. The **significant figures**, or *significant digits*, of a number are the digits that are used to express a measure to the appropriate degree of accuracy.

Key Concept • Significant Figures

Rules	Examples
Nonzero digits are always significant.	2.14 **3 significant figures: 2, 1, and 4**
In whole numbers, zeros are significant if they fall between nonzero digits.	5078 **4 significant figures: 5, 0, 7, and 8**
In decimal numbers greater than or equal to 1, every digit is significant.	7.60 **3 significant figures: 7, 6, and 0**
In decimal numbers less than 1, the first nonzero digit and every digit to the right are significant.	0.029 **2 significant figures: 2 and 9**

Watch Out!

Look for the Decimal
If a number has no decimal place, then zeros are only significant if they fall between nonzero digits. For example, 165,000 has only 3 significant figures because all three zeros fall after the 5. If a number does have a decimal place, then follow the rules listed.

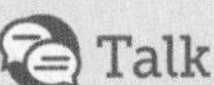

Talk About It!

What is the purpose of using significant figures?

Example 1 Determine Significant Figures

Determine the number of significant figures in each measurement.

0.0320 inches
This is a decimal number less than 1. The first nonzero digit is 3, and there are two digits to the right of 3; 2 and 0. So, this measurement has 3 significant figures.

107,000 centimeters
Because this is a whole number, zeros are only significant if they fall between nonzero digits. There is one zero that falls between 1 and 7. So, this measurement has 3 significant figures.

Think About It!

Create three different measurements that each have four significant figures.

Check

Determine the number of significant figures in each measurement.

a. 0.03927 milliliter has ___?___ significant figures

b. 5,134,180 pounds has ___?___ significant figures

Example 2 Find Significant Figures by Using Tools

Find the possible range for the length of the segment using the correct number of significant figures.

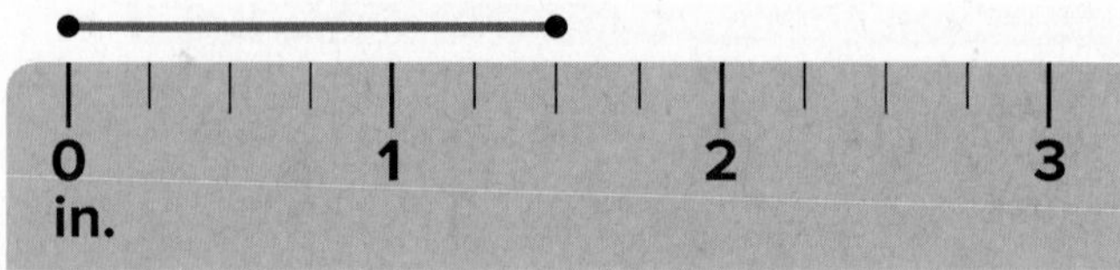

The length of the segment is approximately $1\frac{1}{2}$ inches.

This measurement was given to the nearest $\frac{1}{4}$ inch, so the possible range of this measurement is within $\frac{1}{2}\left(\frac{1}{4}\right)$ or $\frac{1}{8}$ inch of the measured length.

The exact measurement is between $1\frac{3}{8}$ and $1\frac{5}{8}$ inches or 1.375 and 1.625 inches.

Due to the precision of the ruler, the length of the segment has 4 significant figures.

Check

Find the possible range for the length of the segment.

A. 2.0 cm to 2.2 cm

B. 2.00 cm to 3.00 cm

C. 2.15 cm to 2.25 cm

D. 2.08 cm to 2.12 cm

Go Online You can complete an Extra Example online.

Learn Calculating with Significant Figures

When you are calculating with significant figures, the accuracy of the result is limited by the least accurate measurement.

Key Concept • Calculations with Significant Figures	
Addition and Subtraction	**Multiplication and Division**
When using addition and subtraction, a calculation cannot have more digits to the right of the decimal point than either of the original numbers.	When using multiplication and division, the number of significant figures in the final product or quotient is determined by the original number that has the *fewest* number of figures.

Numbers that are not measured are not considered when determining significant figures. For example, if you have 5 cereal boxes that weigh 14 ounces each, then the significant figures used in a calculation would be determined from the measurement, 14 ounces, not the quantity.

Significant figures are also not affected by conversion factors. For example, when using the conversion 12 inches = 1 foot, the significant figures are determined by the original measurement being converted.

Talk About It!

Why is it important to have a standard method for calculations with significant figures?

Example 3 Calculate with Significant Figures

Find each measurement rounded to the correct number of significant figures.

a. volume of an 837.24-mL sample after 276.516 mL is removed

837.24 has 2 digits after the decimal, and 276.516 has 3. So the result should have 2 digits after the decimal.

Find the difference. Then round to the hundredths place.

$837.24 - 276.516 = 560.724$ or 560.72 mL.

b. area of the rectangle

$A = (4.25)(2.5)$

$= 10.625$

Using significant figures, the area is 11 square inches.

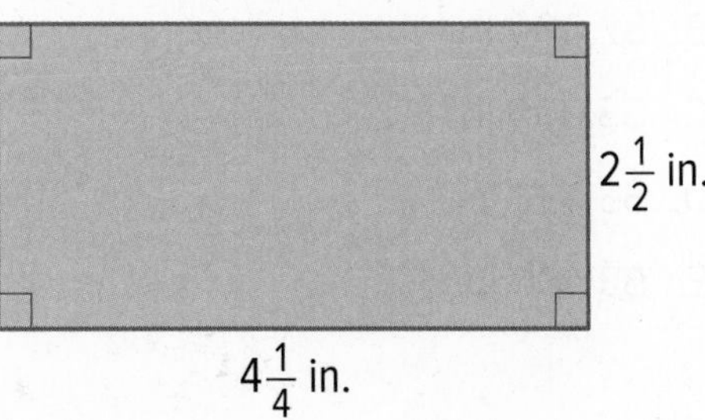

Check

A mixing bowl contains 8.5 fluid ounces of water. If 4.25 fluid ounces are removed from the bowl, how many fluid ounces of water remain? Round to the correct number of significant figures.

___?___ fl oz

 **Go Online** You can complete an Extra Example online.

Example 4 Use Significant Figures in the Real World

Think About It!

What assumption did you make while solving this problem?

FLOWER GARDEN The world's second largest flower garden is Keukenhof Park in Lisse, the Netherlands, which covers 79 acres and contains 7 million flower bulbs. About 30 gardeners work together each year to design the flower formations. How many bulbs do they use in each square yard of the garden? Round to the correct number of significant figures.

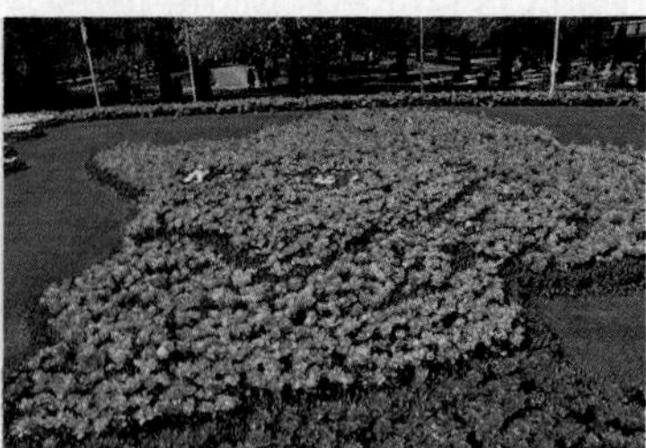

$$\frac{\text{total bulbs}}{\text{total area}} = \frac{7{,}000{,}000 \text{ bulbs}}{79 \text{ acres}}$$

$$= \frac{7{,}000{,}000 \text{ bulbs}}{79 \text{ acres}} \times \frac{1 \text{ acre}}{4840 \text{ yd}^2} \quad 1 \text{ acre} = 4840\text{yd}^2$$

$$= 18.30735433 \text{ bulbs/yd}^2 \quad \text{Simplify.}$$

Because the number of bulbs is a quantity, not a measurement, the number of significant figures is determined by the given measurement, which is 79 acres. Because 79 has 2 significant figures, the final product should also have 2 significant figures. Thus, they use 18 bulbs in each square yard of the garden.

Check

SNOW REMOVAL Mark's snow plow truck can clear 1600 tons of snow in an hour. How many pounds can Mark's truck clear in a minute? Round to the correct number of significant figures. (Hint: 1 T = 2000 lb)

A. 20,000 lb

B. 32,000 lb

C. 53,000 lb

D. 53,330 lb

E. 53,400 lb

Go Online You can complete an Extra Example online.

siraanamwong/iStock/Getty Images Plus/Getty Images

Example 5 Use Tools to Calculate Measurements

The radius of a circle has the measurement shown. What is the possible range for the area of the circle? Round to the correct number of significant figures.

Step 1 Find the possible range for the length of the radius.

The approximate length of the segment is 7.6 centimeters. This measurement is given to the nearest 0.1 centimeter, so the approximate error is $\frac{1}{2}(0.1)$ or 0.05 centimeter. Therefore, the exact length is between 7.55 and 7.65 centimeters.

Step 2 Determine the number of significant figures.

Because the range of the length is between 7.55 and 7.65 centimeters, the length has 3 significant figures.

Step 3 Calculate the area of the circle.

The area of a circle is equal to πr^2, where r is the length of the radius. Complete the expressions to calculate the least and greatest possible areas of the circle.

least possible area: $\pi(7.55)^2 \approx 179.0786352 \text{ cm}^2$

greatest possible area: $\pi(7.65)^2 \approx 183.8538561 \text{ cm}^2$

Using significant figures, the area of the circle is between 179 and 184 square centimeters.

Check

The radius of a circle has the measurement shown. What is the possible range for the area of the circle? Round to the correct number of significant figures.

The area of the circle is between __?__ and __?__ square inches.

Pause and Reflect

Did you struggle with anything in this lesson? If so, how did you deal with it?

Practice

Example 1

Determine the number of significant digits in each measurement.

1. 54.023 **2.** 0.923 **3.** 0.30

4. 100.58 **5.** 0.0002 **6.** 101.01

7. ACADEMICS The students in Miss Li's class are measuring the height of a chalkboard. Miss Li asked the students to write the measurement with 4 significant digits. Which student correctly followed her instructions?

Student	Measure
Sasha	48.5 centimeters
Michelle	48.53 centimeters
Alwan	49 centimeters
Remmie	48.530 centimeters

Example 2

Find the possible range for each length of the segment using the correct number of significant figures.

8.

9.

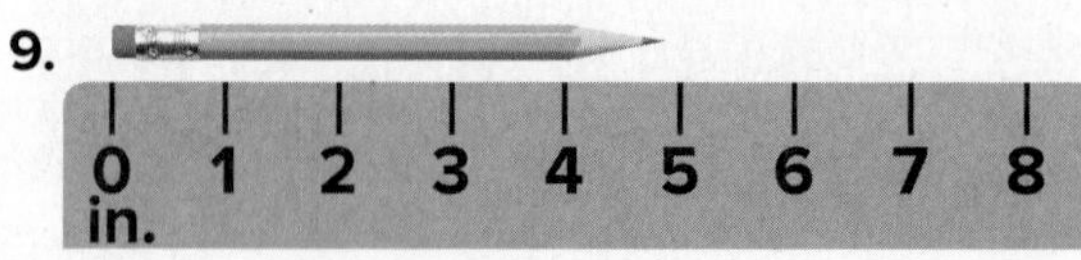

10.

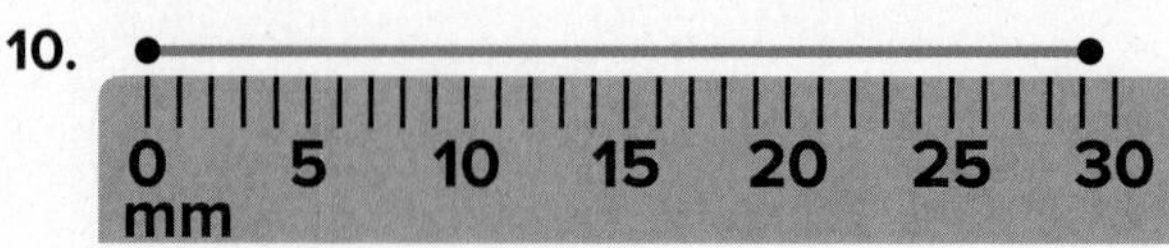

11.

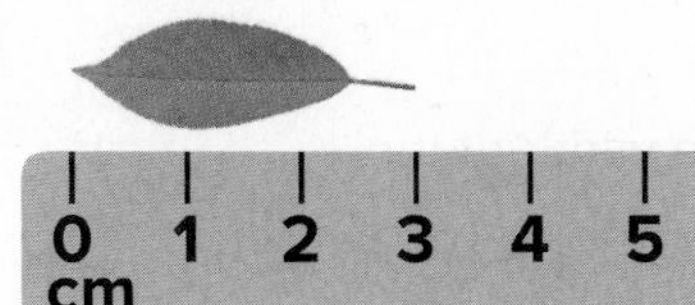

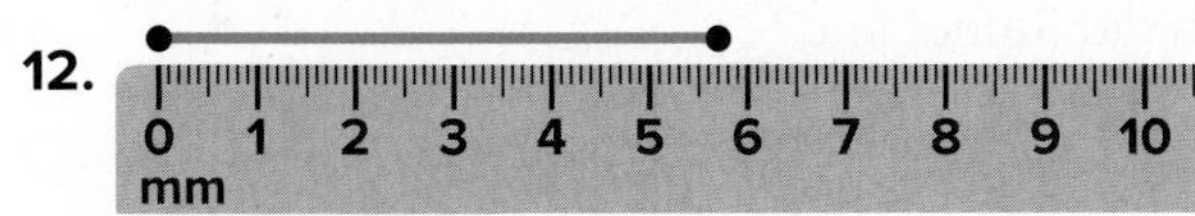

12.

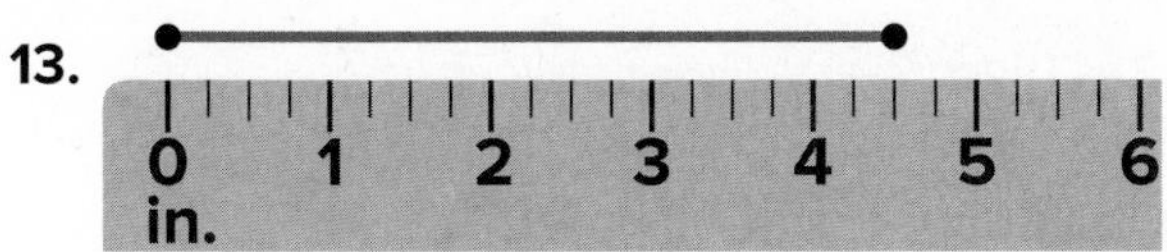

13.

Example 3

The base of a triangle is fixed at 2.218 millimeters. Determine the number of significant figures of the area of the triangle with each given height.

14. 1.86 mm

15. 0.099 mm

16. 0.1279 mm

17. 2.109 mm

18. 11.0 mm

19. 1.7 mm

20. Using significant figures, which of the following students wrote a calculation that could have a sum or difference of 51.9?

Student	Calculation
Juliana	48.222 + 3.769
Lori	48.22 + 3.76
Nobu	48.222 + 3.7
Jerome	48.2 + 3.769

Example 4

21. CHEMISTRY Angel has 8.341 mL of saline. She pours 1.1 mL of saline into another solution. How much saline does Angel have left? Round your measurement to the correct number of significant figures.

22. A parallelogram with base b and height h has area, A, given by the formula $A = bh$. Find the area of the given parallelogram. Round your measurement to the correct number of significant figures.

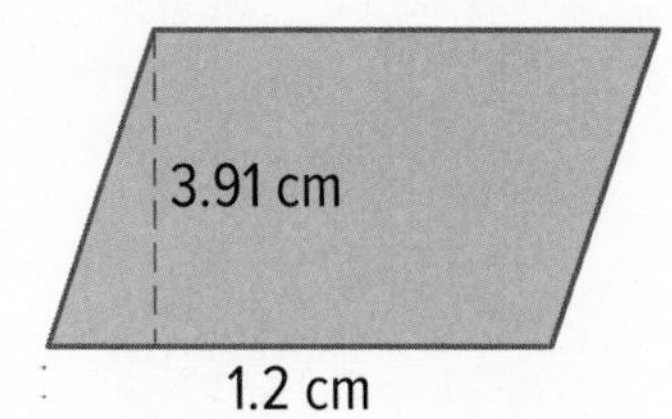

23. AREA Find the area of a triangle with a height of 4.90 centimeters and a base length of 6.174 centimeters. Round your measurement to the correct number of significant figures.

24. DIMENSIONS Rafael is building a horseshoe pit in his backyard. The width of the pit is 29.71 inches, and the length is 30.1 inches.

Part A Estimate the area of Rafael's horseshoe pit.

Part B Rafael finds the exact area of the horseshoe pit and rounds his answer to the correct number of significant figures. What area did Rafael find?

Example 5

25. AREA The radius of a circle has the measurement shown. What is the possible range for the area of the circle? Round to the correct number of significant figures.

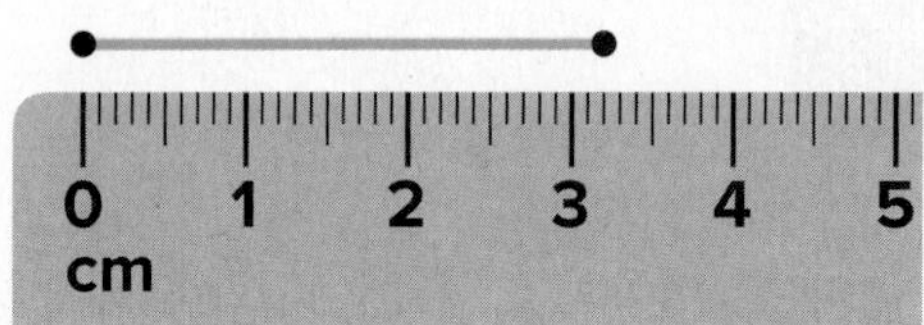

26. CIRCUMFERENCE The radius of a circle has the measurement shown. What is the possible range for the circumference of the circle? Round to the correct number of significant figures.

Mixed Exercises

Determine the number of significant digits in each measurement.

27. 53.74

28. 0.03298

29. 10.500

30. 6,102.0

31. 7,109,100

32. 0.110

33. CHEMISTRY A beaker contains a sample of NaCl weighing 49.8767 grams. If the empty beaker weighs 49.214 grams, what is the weight of the NaCl? Round to the correct number of significant figures.

34. COOKING Jordan makes a sandwich on a paper plate weighing 32.47 grams. The bread weighs 60.13 grams. Jordan adds 12.3 grams of turkey, 2.4 grams of mayonnaise, and 3.0 grams of lettuce. What is the final weight of the plate and sandwich?

35. **CHEMISTRY** Three chemists weigh an item using different scales. The values they report are shown on the scales. How many significant figures should be used for each measurement?

36. **SWIMMING POOL** A rectangular swimming pool measures 24.2 feet by 76 feet.

a. Find the perimeter of the pool. Round to the correct number of significant figures.

b. Find the area of the pool. Round to the correct number of significant figures.

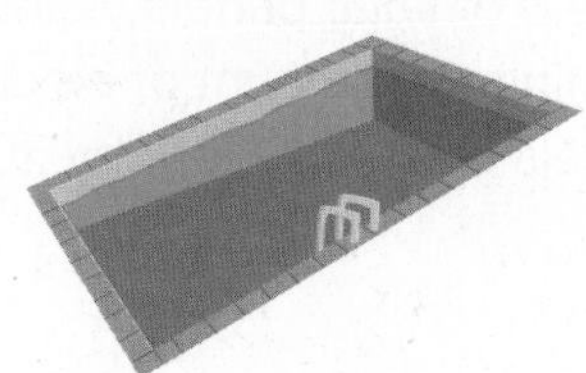

Andrew Cribb/iStockphoto/Getty Images

37. **AREA** Find the area of the given triangle. Round your measure to the correct number of significant figures.

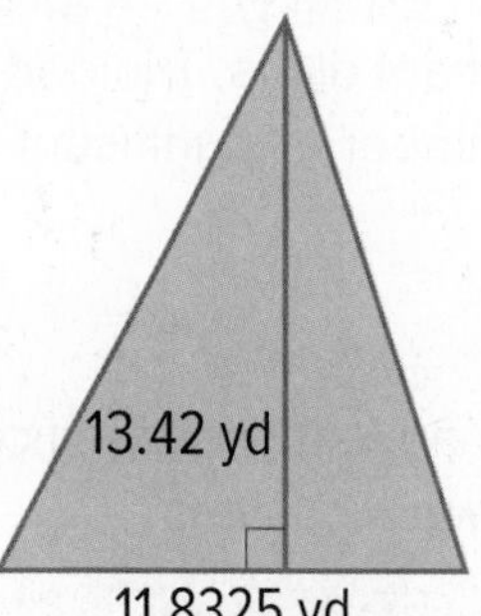

38. **MURAL** Krista is painting a rectangular wall that has an area of 247 square feet. If she can paint 5.25 square feet in an hour, about how long will it take for Krista to finish the mural on her own? Round to the correct number of significant figures.

39. **MASS** Suppose that you measured the volume of a rock to be 2.3 cm^3 and you know the density to be 3.6 g/cm^3. What is the mass of the rock? Round your measure to the correct number of significant figures.

40. **DRIVING** Keandra is taking a trip to visit her extended family and makes a stop somewhere during her trip. The distance between Cincinnati, where Keandra started, and Dayton, where Keandra made the stop is 54 miles. The distance between Dayton and Toledo, where Keandra's family lives is 150.2 miles. How far did Keandra travel on her trip? Round to the correct number of significant figures.

41. **VOLUME** A rectangular box has a length of 10.876 inches, a width of 4.34 inches, and a height of 13.22 inches. What is the volume of the rectangular prism? Round to the correct number of significant figures.

42. **TRAVEL** You estimate that your car gets 28 miles per gallon. The cost of gas per gallon is shown. How much does it cost you to travel 455 miles? Round to the correct number of significant figures.

Higher-Order Thinking Skills

43. FIND THE ERROR A student found that the dimensions of a rectangle were 1.40 meters and 1.60 meters. She was asked to report the area using the correct number of significant figures. She reported the area as 2.2 square meters. What error did the student make? Explain your reasoning.

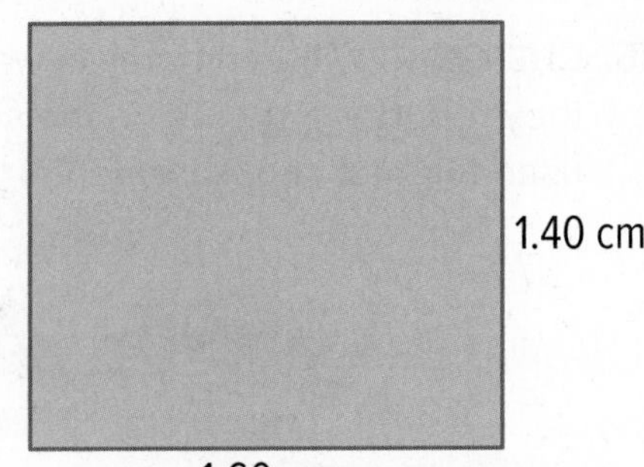

44. PERSEVERE The Sun is an excellent source of electrical energy. A field of solar panels yields 19.23 Watts per square foot. Determine the amount of electricity produced by a field of solar panels that is 410 feet by $201\frac{1}{3}$ yards.

45. WRITE When explaining the process of finding the perimeter of a triangle using significant digits, Trinidad claimed that 0.045 inch and 0.0045 inch have the same number of significant figures. Is she correct? Explain your answer.

46. WRITE How do you use significant figures to determine how to report a sum or product of two measures?

47. ANALYZE Determine whether the following statement is *sometimes, always,* or *never* true. Justify your argument.
Zeros are significant figures.

48. CREATE The swim team measures time to the hundredth of a second. Amanda's time was slower than Jocelyn's time in the 100-meter freestyle. What are possible times for Amanda and Jocelyn if each has times with 4 significant digits?

Review

Essential Question

How are angles and two-dimensional figures used to model the real world?

Two-dimensional figures can be drawn to represent real-world objects. Two-dimensional figures can model three-dimensional figures in the form of nets so that computations about those objects can be made more easily. Angles and sides should be labeled in these representations to help when computations are made.

Module Summary

Lessons 11-1 and 11-2

Angles

- Angles that have the same measure are congruent angles.
- A ray or segment that divides an angle into two congruent parts is an angle bisector.
- Relationships between special angle pairs can be used to find missing measures.
- Complementary angles are two angles with measures that have a sum of 90°. Supplementary angles are two angles that have measures that have a sum of 180°.
- Certain relationships can be assumed from a figure, but most cannot.

Lessons 11-3 and 11-4

Two-Dimensional Figures

- The perimeter of a polygon is the sum of the lengths of the sides of the polygon.
- The circumference of a circle is the distance around the circle.
- Area is the number of square units needed to cover a surface.
- A transformation is a function that takes points in the plane as inputs and gives other points as outputs.
- A rigid motion is a transformation that preserves distance and angle measure.
- The three main types of rigid motions are reflection, translation, and rotation.

Lessons 11-5 and 11-6

Three-Dimensional Figures

- Surface area is the sum of the areas of all faces and side surfaces of a three-dimensional figure.
- Volume is the measure of the amount of space enclosed by a three-dimensional figure.
- A three-dimensional figure can be modeled by an orthographic drawing, which shows its top, left, front, and right views.
- A net is a two-dimensional figure that forms the surfaces of a three-dimensional object when folded.

Lessons 11-7 and 11-8

Measurements

- Precision is the repeatability, or reproducibility, of a measurement.
- Accuracy is the nearness of a measurement to the true value of the measure.
- The significant figures, or significant digits, of a number are the digits that contribute to its precision in a measurement.

Study Organizer

 **Foldables**

Use your Foldable to review this module. Working with a partner can be helpful. Ask for clarification of concepts as needed.

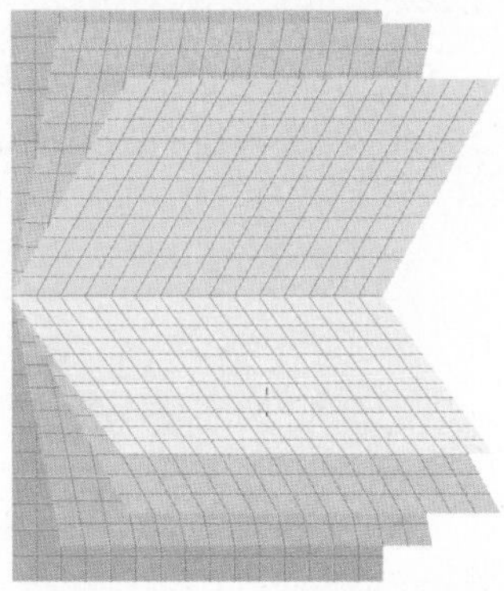

Test Practice

1. MULTI-SELECT Select all the angles for which $\overrightarrow{HA}$ and $\overrightarrow{HE}$ are the sides. (Lesson 11-1)

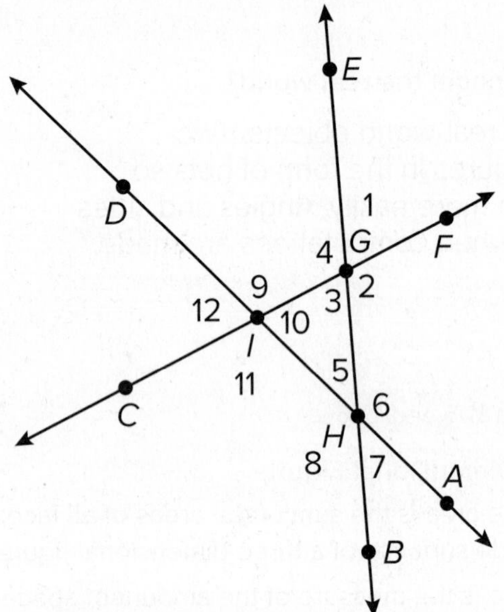

A. $\angle AHE$

B. $\angle AGE$

C. $\angle EHA$

D. $\angle EGA$

E. $\angle AHB$

2. OPEN RESPONSE In the figure, $\overrightarrow{CD}$ and $\overrightarrow{CB}$ are opposite rays, and $\overrightarrow{CA}$ bisects $\angle BCE$.

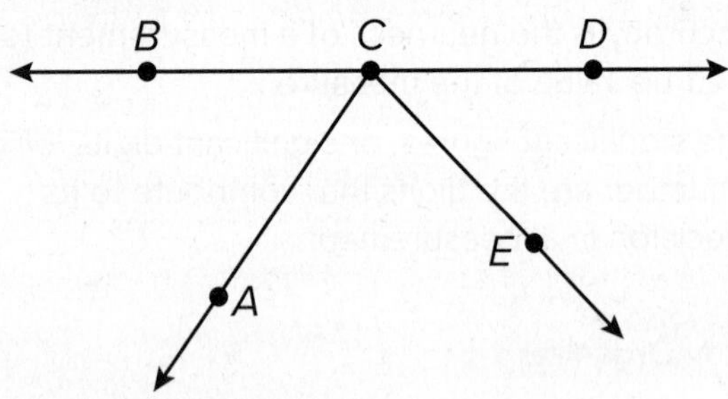

Suppose $m\angle ECA = 14x - 2$ and $m\angle ACB = 12x + 8$. What is $m\angle ECA$? (Lesson 11-1)

3. OPEN RESPONSE Describe how you would construct an angle bisector using paper-folding. (Lesson 11-1)

4. MULTIPLE CHOICE Two angles are supplementary. The measure of the larger angle is 12 less than 3 times the measure of the smaller angle. Find the measure of the larger angle. (Lesson 11-2)

A. 25.5°

B. 48°

C. 64.5°

D. 132°

5. MULTIPLE CHOICE Which value of x will make $\overleftrightarrow{AB}$ perpendicular to $\overleftrightarrow{CD}$? (Lesson 11-2)

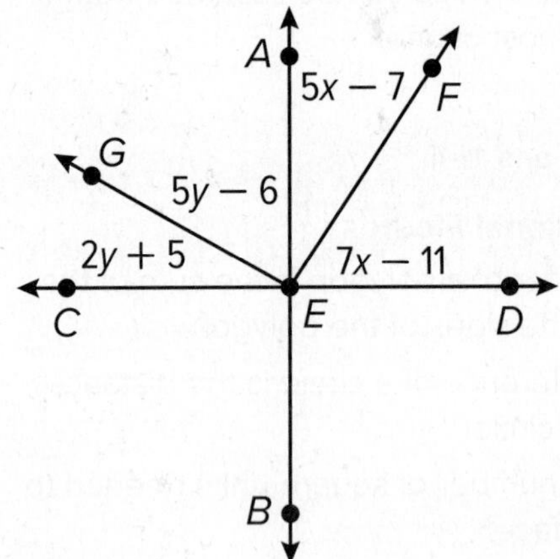

A. 6

B. 9

C. 11

D. 13

6. **MULTIPLE CHOICE** What is the **best** estimate for the area of the triangle, in square units? (Lesson 11-3)

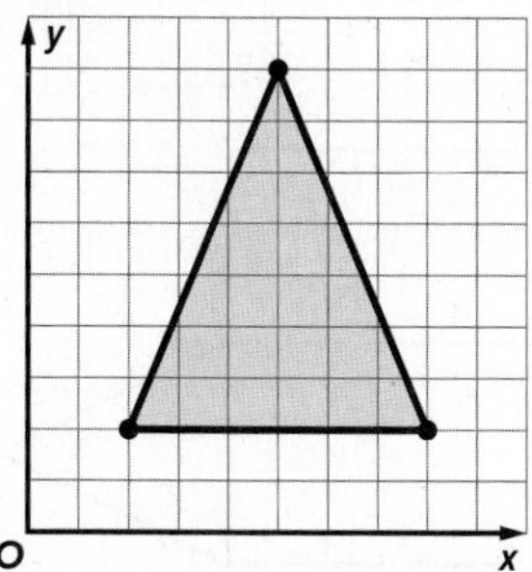

A. 5.25 square units

B. 10.5 square units

C. 21 square units

D. 42 square units

7. **OPEN RESPONSE** Find the perimeter of the rectangle. Then, find the area of the rectangle. (Lesson 11-3)

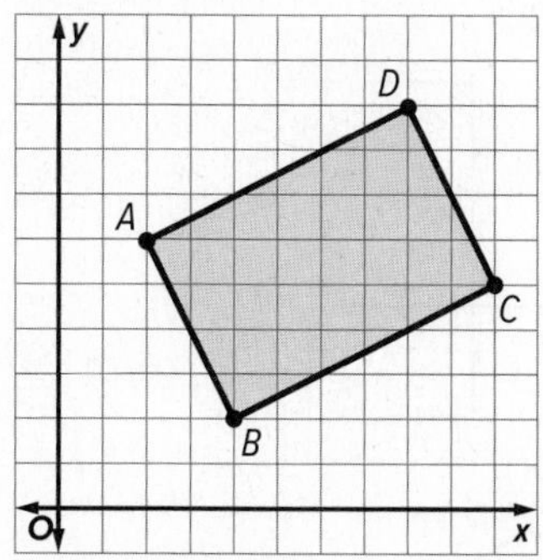

8. **OPEN RESPONSE** Find the perimeter of the triangle. Round your answer to the nearest hundredth. (Lesson 11-3)

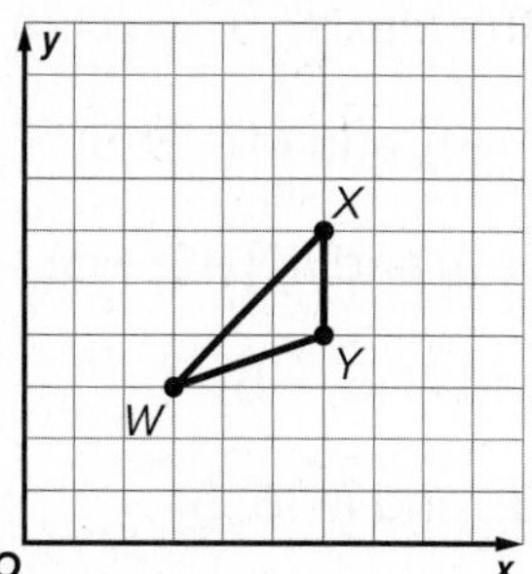

9. **MULTIPLE CHOICE** What are the coordinates of the image of $\triangle JKL$ after a 180° clockwise rotation about the origin? (Lesson 11-4)

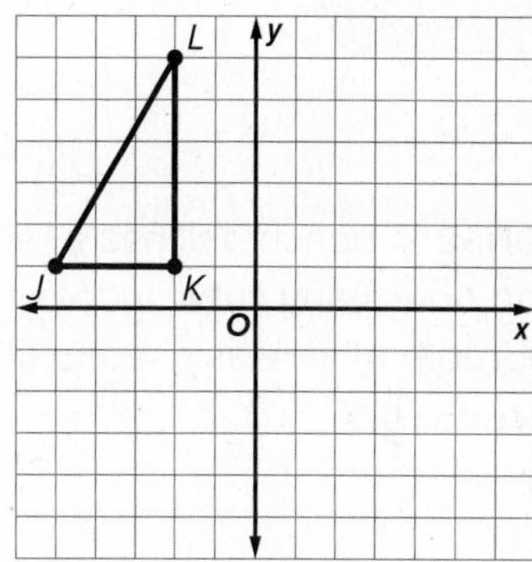

A. $J'(-5, 1)$, $K'(-2, 1)$, and $L'(-2, 6)$

B. $J'(5, -1)$, $K'(2, -1)$, and $L'(2, -6)$

C. $J'(-1, -5)$, $K'(-1, -2)$, and $L'(-6, -2)$

D. $J'(1, 5)$, $K'(1, 2)$, and $L'(6, 2)$

10. **OPEN RESPONSE** Kyle is creating a video game. Every time the main character jumps in the game, the image follows a translation using the function mapping $(x, y) \rightarrow (x + 4, y + 9)$. If the main character is located at $(-5, 0)$, what will the new location be after a jump? (Lesson 11-4)

11. MULTIPLE CHOICE $\triangle KLM$ has coordinates $K(4, -2)$, $L(6, -1)$, and $M(5, 5)$. What would be the coordinates of the vertices of the image after a reflection in the x-axis? (Lesson 11-4)

A. $K'(-4, -2)$, $L'(-6, -1)$, and $M'(-5, 5)$

B. $K'(-4, 2)$, $L'(-6, -1)$, and $M'(-5, -5)$

C. $K'(4, 2)$, $L'(6, 1)$, and $M'(5, -5)$

D. $K'(-2, 4)$, $L'(-1, 6)$, and $M'(5, 5)$

12. OPEN RESPONSE What is the volume, in cubic centimeters, of the cylinder? (Lesson 11-5)

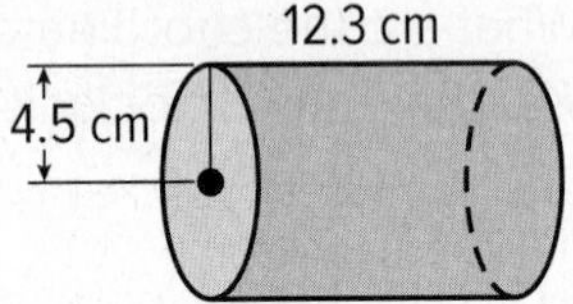

13. OPEN RESPONSE A beach ball has a radius of 8 inches. Find how many cubic inches to the nearest hundredth of air was used to fill the beach ball. (Lesson 11-5)

14. OPEN RESPONSE Identify two three-dimensional shapes that are represented by the grain silo. (Lesson 11-6)

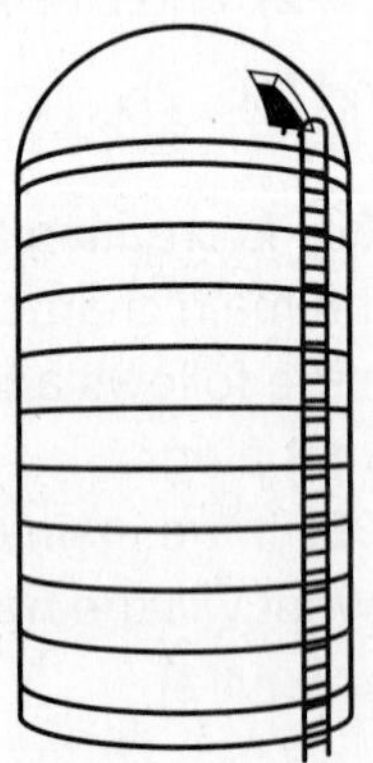

15. MULTIPLE CHOICE Which net could be used to represent the storage shed? (Lesson 11-6)

A.

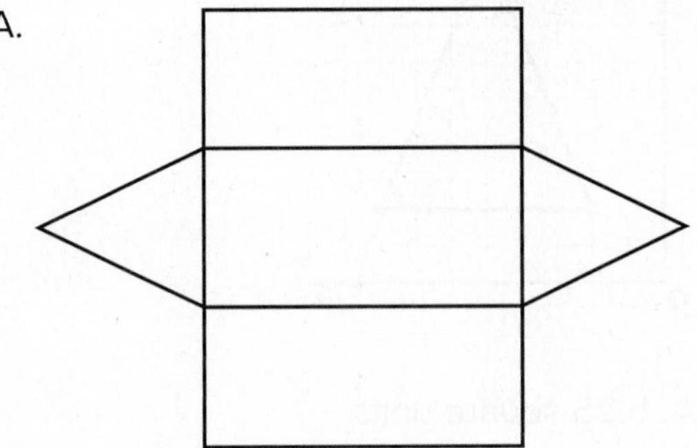

B.

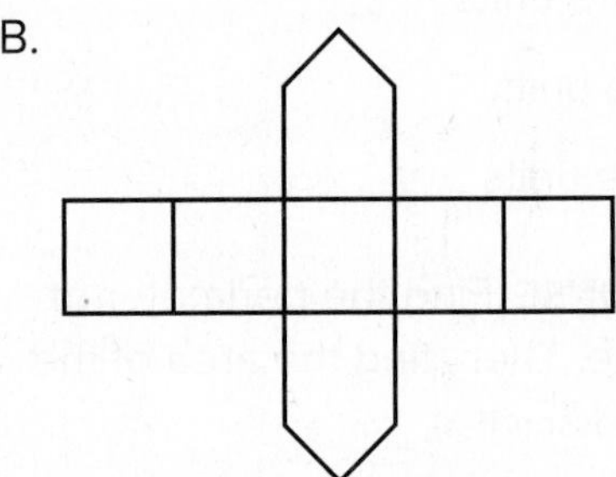

C.

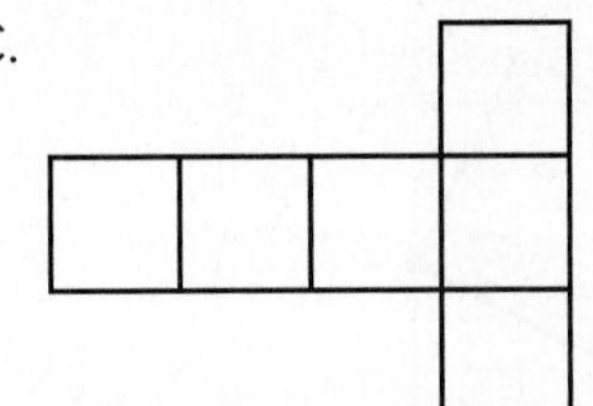

D.

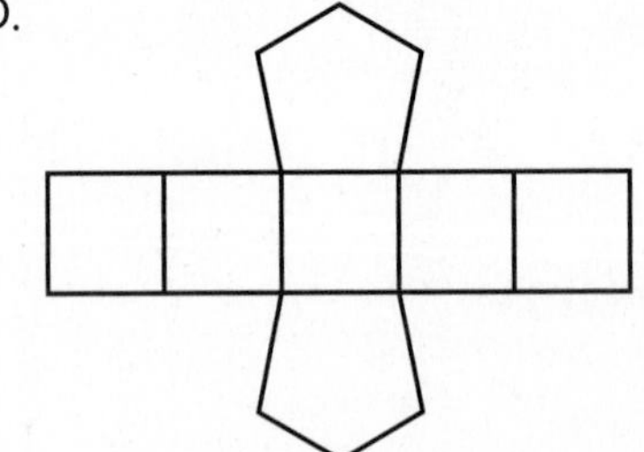

Module 12

Logical Arguments and Line Relationships

Essential Question

What makes a logical argument, and how are logical arguments used in geometry?

What Will You Learn?

How much do you already know about each topic **before** starting this module?

KEY

— I don't know. — I've heard of it. — I know it!

	Before			After		
make and analyze conjectures based on inductive reasoning						
disprove conjectures by using counterexamples						
determine truth values of statements, negations, conjunctions, and disjunctions						
write and analyze conditionals and biconditionals using logic						
distinguish correct logic or reasoning from that which is flawed using the Laws of Detachment and Syllogism						
construct viable arguments by writing paragraph proofs						
construct viable arguments by writing flow proofs						
prove statements about segments and angles by writing two-column proofs						
identify and use relationships between pairs of angles						
identify and use parallel and perpendicular lines using the slope criteria						
solve problems using distances and parallel and perpendicular lines						

Foldables Make this Foldable to help you organize your notes about logic, reasoning, and proof. Begin with one sheet of notebook paper.

1. **Fold** lengthwise to the holes.
2. **Cut** five tabs in the top sheet.
3. **Label** the tabs as shown.

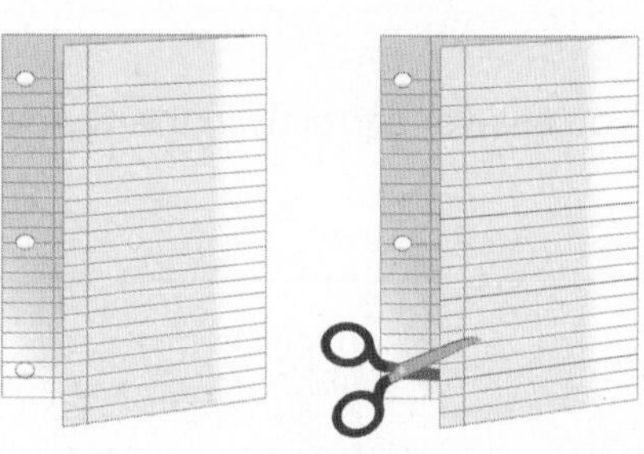

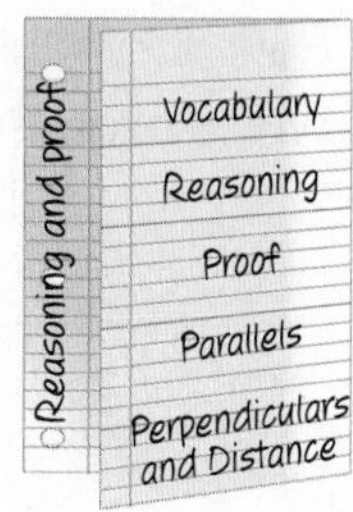

What Vocabulary Will You Learn?

- alternate exterior angles
- alternate interior angles
- biconditional statement
- compound statement
- conclusion
- conditional statement
- conjecture
- conjunction
- consecutive interior angles
- contrapositive
- converse
- corresponding angles
- counterexample
- deductive argument
- deductive reasoning
- disjunction
- equidistant
- exterior angles
- flow proof
- hypothesis
- if-then statement
- inductive reasoning
- interior angles
- inverse
- logically equivalent
- negation
- paragraph proof
- parallel lines
- parrallel planes
- proof
- skew lines
- slope
- slope criteria
- statement
- transversal
- truth value
- two-column proof
- valid argument

Are You Ready?

Complete the Quick Review to see if you are ready to start this module. Then complete the Quick Check.

Quick Review

Example 1

Solve $36x - 14 = 16x + 58$.

$36x - 14 = 16x + 58$	Original equation
$20x - 14 = 58$	Subtract $16x$ from each side.
$20x = 72$	Add 14 to each side.
$x = 3.6$	Divide each side by 20.

Example 2

If $m\angle BXA = 3x + 5$ and $m\angle DXE = 56$, find x.

$m\angle BXA = m\angle DXE$

Vertical $\angle$s are $\cong$.

$3x + 5 = 56$	Substitution.
$3x = 51$	Subtract 5 from each side.
$x = 17$	Divide each side by 3.

Quick Check

Solve each equation.

1. $8x - 10 = 6x$

2. $18 + 7x = 10x + 39$

3. $3(11x - 7) = 13x + 25$

4. $3x + 8 = 0.5x + 35$

Refer to the figure above.

5. Identify a pair of vertical angles that appear to be obtuse.

6. If $m\angle DXB = 116$ and $m\angle EXA = 3x + 2$, find x.

7. If $m\angle BXC = 90$, $m\angle CXD = 6x - 13$, and $m\angle DXE = 10x + 7$, find x.

How did you do?

Which exercises did you answer correctly in the Quick Check?

Conjectures and Counterexamples

Explore Using Inductive Reasoning to Make Conjectures

 **Online Activity** Use dynamic geometry software to complete the Explore.

INQUIRY How can you use observations and patterns to make predictions?

Today's Goals
- Write and analyze conjectures by using inductive reasoning.
- Disprove conjectures by using counterexamples.

Today's Vocabulary
inductive reasoning
conjecture
counterexample

Learn Inductive Reasoning and Conjecture

Inductive reasoning is the process of reaching a conclusion based on a pattern of examples. When you assume that an observed pattern will continue, you are applying inductive reasoning. You can use inductive reasoning to make an educated guess based on known information and specific examples. This educated guess is also known as a **conjecture**.

Example 1 Patterns and Conjectures

Write a conjecture that describes the pattern in the sequence. Then use your conjecture to find the next term in the sequence.

Appointment times: 8:30 A.M., 9:15 A.M., 10:00 A.M., 10:45 A.M., . . .

Step 1 Look for a pattern.

8:30 A.M. → 9:15 A.M. → 10:00 A.M. → 10:45 A.M.
+45 min, +45 min, +45 min

Step 2 Make a conjecture.

Each appointment time is 45 minutes after the previous appointment time. The next appointment time will be 10:45 A.M. + 0:45 or 11:30 A.M.

Check

Write a conjecture that describes the pattern in the sequence. Then use your conjecture to find the next term in the sequence.

$\frac{1}{2}$, 1, 2, 4, . . .

The next number in the sequence is __?__ the preceding number.

The next number in the sequence is __?__.

Go Online You can complete an Extra Example online.

Example 2 Algebraic Conjectures

Make a conjecture about the sum of the squares of two consecutive natural numbers. List or draw some examples that support your conjecture.

Step 1 List examples.

$1^2 + 2^2 = 5$ $6^2 + 7^2 = 85$

$2^2 + 3^2 = 13$ $10^2 + 11^2 = 221$

Step 2 Look for a pattern.

Notice that all the sums are odd numbers.

Step 3 Make a conjecture.

The sum of the squares of two consecutive natural numbers is an odd number.

Think About It!

It is possible for more than one conjecture to correctly describe a pattern. Each number in the sequence 2, 4, 8, 16, 32, . . . is 2^n where $n \geq 1$. What other conjecture can be made about the values?

Check

Make a conjecture about the sum of two odd numbers.

The sum of two odd numbers is always a(n) ___?___ number.

Example 3 Geometric Conjectures

Make a conjecture about the relationship between the segments joining opposite vertices of isosceles trapezoids.

Step 1 Draw several examples.

An isosceles trapezoid is a trapezoid with two opposite congruent legs.

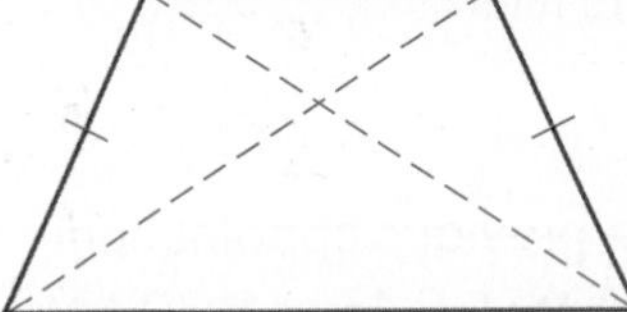

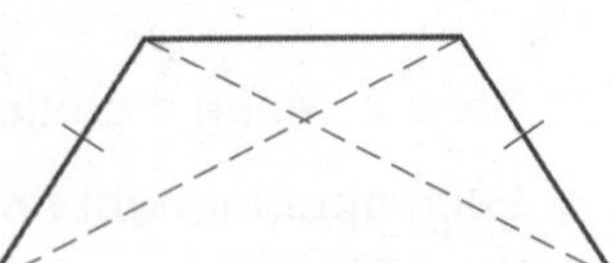

Step 2 Look for a pattern.

Notice that the segments joining opposite vertices of each isosceles trapezoid appear to have the same measure. Use a ruler or compass to confirm this.

Step 3 Make a conjecture.

The segments joining opposite vertices of an isosceles trapezoid are congruent.

 **Go Online** You can complete an Extra Example online.

Check

Make a conjecture about the relationships between AD and AB, if C is the midpoint of $\overline{AB}$ and D is the midpoint of $\overline{AC}$.

AD is ___?___ of AB.

Example 4 Make Conjectures from Data

GAS PRICES **The table shows the average price of gasoline in the United States for the years 2010 through 2018. Make a conjecture about the price of gas in 2019. Explain how this conjecture is supported by the data given.**

Year	Price (dollars per gallon)
2010	2.84
2011	3.58
2012	3.68
2013	3.58
2014	3.44
2015	2.43
2016	2.14
2017	2.42
2018	2.84

Look for patterns in the data.

The price of gasoline increased from 2010 to 2012. From 2012 to 2016, the price of gas decreased, at first at a steady rate, and then more dramatically. Beginning in 2017, the price of gas began to increase at a steady rate.

The data shows that the price of gas follows an oscillating pattern, increasing in price for several years before decreasing in price for several years.

Conjecture: In 2019, the price of gas will continue to increase.

Check

HEARING LOSS Almost 50% of young adults between the ages of 12 and 35 years old are exposed to damaging levels of sound from the use of personal electronic devices. The intensity of a sound and the time spent listening to a sound highly affects the amount of damage that can be done to someone's hearing. The intensity of a sound to the human ear is measured in A-weighted decibels, or dBA. For every 3 decibels over 85 decibels, the exposure time it takes to cause hearing damage is cut in half. How long does it take to cause hearing damage at 106 decibels? Write your answer as a decimal.

Decibel Level (dBA)	Exposure Time (hours)
85	8
88	4
91	2
94	1
97	$\frac{1}{2}$
100	$\frac{1}{4}$

___?___ minutes

Go Online You can complete an Extra Example online.

Think About It!

Could the pattern of the data change over time? Explain your reasoning.

Use a Source

Find data about the digital music revenue in the United States in recent years. Make a conjecture about the future trends in digital music revenue.

Learn Counterexamples

To show that a conjecture is true for all cases, you must prove it. It only takes one example that contradicts the conjecture, however, to show that a conjecture is not always true. This example is called a **counterexample**, and it can be a number, a drawing, or a statement.

Think About It!

What is a counterexample for the conjecture *All silvery coins are nickels*?

Example 5 Find Counterexamples

Find a counterexample to show that each conjecture is false.

a. If n is a real number, then $-n$ is a negative number.

When n is -4, $-n$ is $-(-4)$ or 4, which is a positive number. Because $-n$ is not negative, this is a counterexample.

What does the prefix *counter-* mean? How does this meaning relate to a counterexample?

b. If $\angle ABC \cong \angle DBE$, then $\angle ABC$ and $\angle DBE$ are vertical angles.

When points A, B, and D are noncollinear and points E, B, and C are noncollinear, the conjecture is false.

In the figure, $\angle ABC \cong \angle DBE$, but $\angle ABC$ and $\angle DBE$ are not vertical angles.

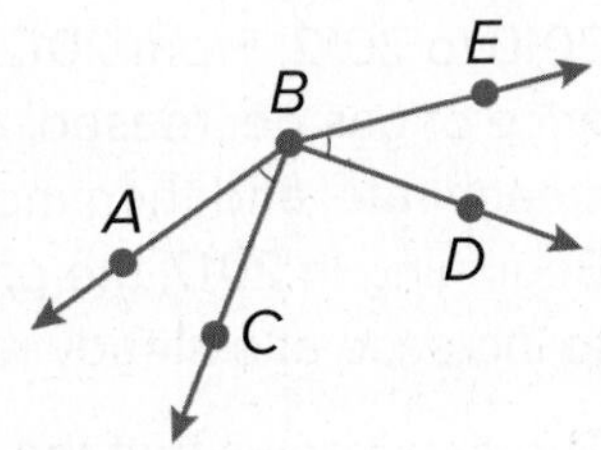

Check

Find a counterexample to show that each conjecture is false.

a. If n is a real number, then $\frac{1}{n} < n$. Select all that apply.

A. $n = -3$ B. $n = \frac{1}{4}$ C. $n = 1$ D. $n = 5$ E. $n = 100$

b. If a line intersects a segment at its midpoint, then the line is perpendicular to the segment. Draw a diagram to represent the counterexample.

Problem-Solving Tip

Draw a Diagram
Remember that a counterexample can be a number, a drawing, or a statement that proves a conjecture to be false. If you are struggling to find a counterexample, try drawing a diagram. This will allow you to analyze the situation and determine the validity of the conjecture.

Go Online You can complete an Extra Example online.

Practice

Go Online You can complete your homework online.

Example 1

Write a conjecture that describes the pattern in each sequence. Then use your conjecture to find the next term in the sequence.

1. 4, 8, 12, 16, 20

2. 2, 22, 222, 2222

3. $1, \frac{1}{2}, \frac{1}{4}, \frac{1}{8}$

4. $6, \frac{11}{2}, 5, \frac{9}{2}, 4$

5. Arrival times: 3:00 P.M., 12:30 P.M., 10:00 A.M., . . .

6. Percent humidity: 100%, 93%, 86%, . . .

7.

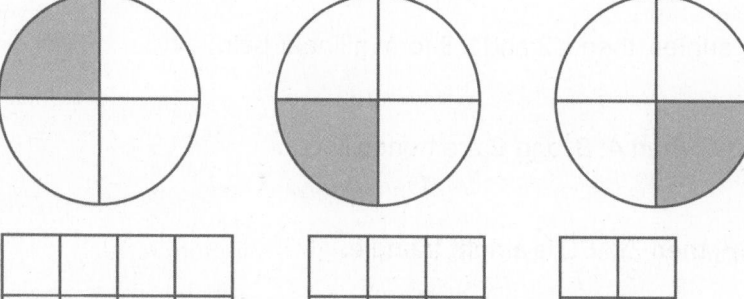

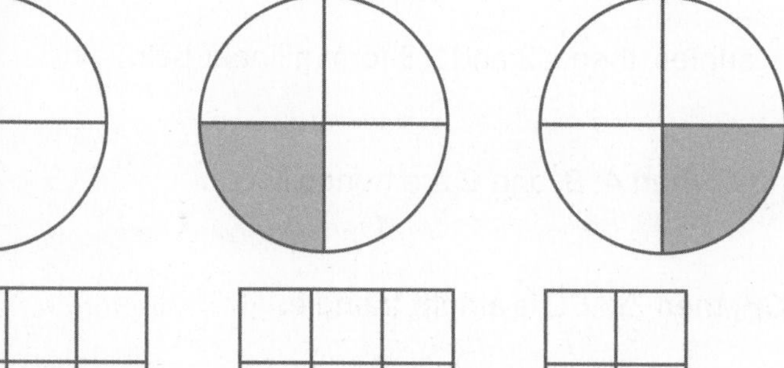

8.

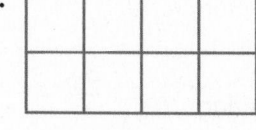

Examples 2 and 3

Make a conjecture about each value or geometric relationship.

9. the product of two odd numbers

10. the product of two and a number, plus one

11. the relationship between a and c if $ab = bc, b \neq 0$

12. the relationship between a and b if $ab = 1$

13. the relationship between two intersecting lines that form four congruent angles

14. the relationship between the angles of a triangle with all sides congruent

15. the relationship between NP and PQ if point P is the midpoint of $\overline{NQ}$

16. the relationship between the volume of a prism and a pyramid with the same base and equal heights

Example 4

17. RAMPS Xio is rolling marbles down a ramp. Every second that passes, she measures how far the marbles travel. She records the information in the table shown below.

Second	1st	2nd	3rd	4th
Distance (cm)	20	60	100	140

Make a conjecture about how far the marble will roll in the fifth second.

Example 5

Determine whether each conjecture is *true* or *false*. Find a counterexample for any false conjecture.

18. If n is a prime number, then $n + 1$ is not prime.

19. If x is an integer, then $-x$ is positive.

20. If $\angle 2$ and $\angle 3$ are supplementary angles, then $\angle 2$ and $\angle 3$ form a linear pair.

21. If you have three points A, B, and C, then A, B, and C are noncollinear.

22. If in $\triangle ABC$, $(AB)^2 + (BC)^2 = (AC)^2$, then $\triangle ABC$ is a right triangle.

23. If the area of a rectangle is 20 square meters, then the length is 10 meters and the width is 2 meters.

Mixed Exercises

24. REASONING Given: $2a^2 = 72$. Conjecture: $a = 6$. Write a counterexample.

25. CONSTRUCT ARGUMENTS Barbara is in charge of the award medals for a sporting event. She has 31 medals to present to various individuals on 6 competing teams. She asserts that at least one team will end up with more than 5 medals. Do you believe her assertion? Justify your argument.

26. USE TOOLS Miranda is developing a chart that shows her ancestry. She makes the three sketches shown. The first dot represents herself. The second sketch represents herself and her parents. The third sketch represents herself, her parents, and her grandparents. Sketch what you think would be the next figure in the sequence.

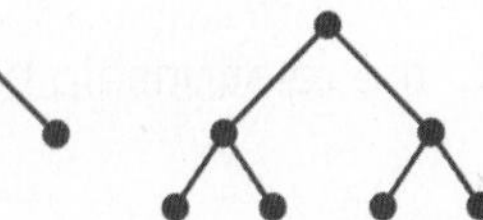

27. **REGULARITY** The figure shows a sequence of squares each made out of identical square tiles.

a. Starting from zero tiles, how many tiles do you need to make the first square? How many tiles do you have to add to the first square to get the second square? How many tiles do you have to add to the second square to get the third square?

b. Make a conjecture about the list of numbers that you started writing in your answer to **part a**.

c. Make a conjecture about the sum of the first n odd numbers.

28. **STRUCTURE** Adric made the following pattern by connecting points with line segments.

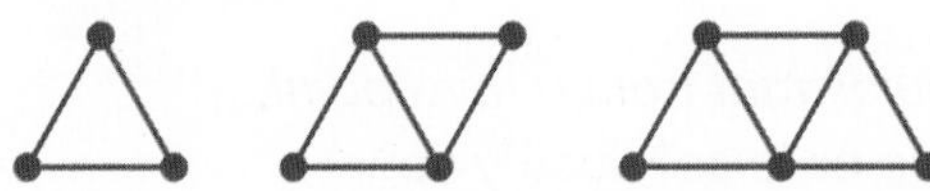

a. Suppose Adric continues the pattern. How many line segments will he need to make 4 triangles? 5 triangles?

b. Suppose Adric makes n triangles. Make a conjecture about the number of line segments he will need to make the triangles.

c. Compare the number of line segments to the number of points in each step of the pattern. How many more line segments than points will there be if Adric continues the pattern to 4 triangles? 5 triangles? Extend the pattern to make a conjecture stating how many more line segments than points are needed to draw n triangles.

29. A prime number is a number, other than 1, that is divisible by only itself and 1. Lucille read that prime numbers are very important in cryptography, so she decided to find a systematic way of producing prime numbers. After some experimenting, she conjectured that $2^n - 1$ is a prime for all whole numbers $n > 1$. Find a counterexample to this conjecture.

30. A line segment of length 1 is repeatedly shortened by removing one third of its remaining length, as shown.

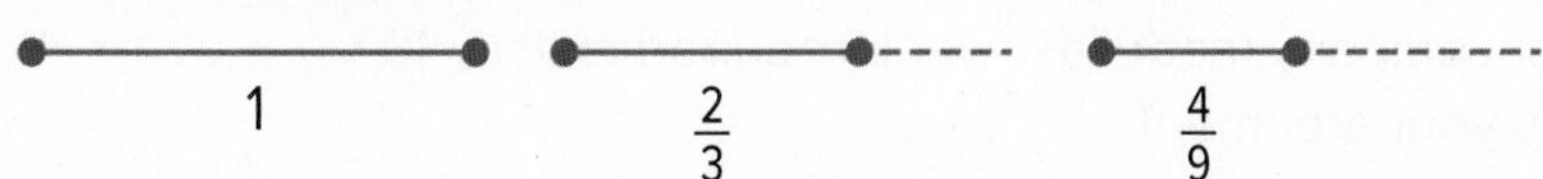

Find and use a pattern to make a conjecture about the length of the line segment after being shortened n times.

Higher-Order Thinking Skills

31. **PERSEVERE** If you draw points on a circle and connect every pair of points, then the circle is divided into regions. For example, two points form two regions, three points form four regions, and four points form eight regions.

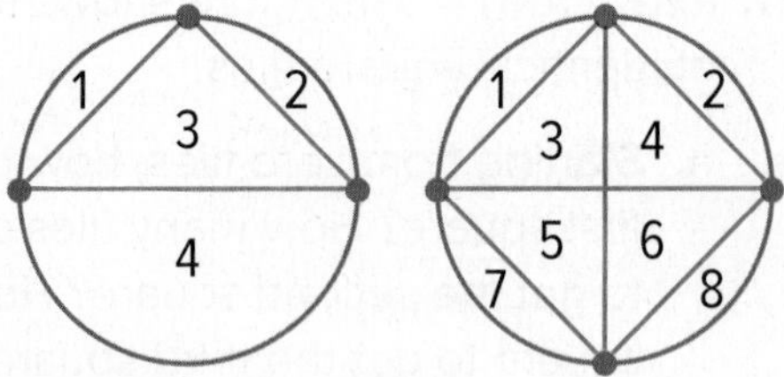

   a. Make a conjecture about the relationship between the number of points on a circle and the number of regions formed in the circle.

   b. Does your conjecture hold true when there are six points? Support your answer with a diagram.

32. **CREATE** Write a number sequence that can be generated by two different patterns. Explain your patterns.

33. **ANALYZE** Consider the conjecture *If two points are equidistant from a third point, then the three points are collinear.* Is the conjecture *true* or *false*? Justify your argument. If false, give a counterexample.

In Exercises 34 and 35, use undefined terms, definitions, and/or postulates to explain why each conjecture is true.

34. **WRITE** Wei drew the figure at the right. Then she stated the following conjecture: A plane contains at least two lines.

35. **WRITE** Andre drew the figure at the right. Then he stated the following conjecture: Every line contains at least one line segment.

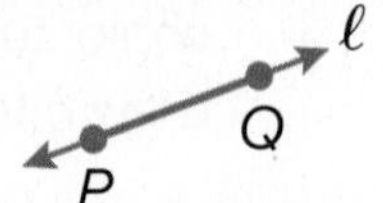

36. **ANALYZE** Kayla owns a company that makes patios and garden paths out of square tiles. The figures show the patterns used to make paths of different lengths.

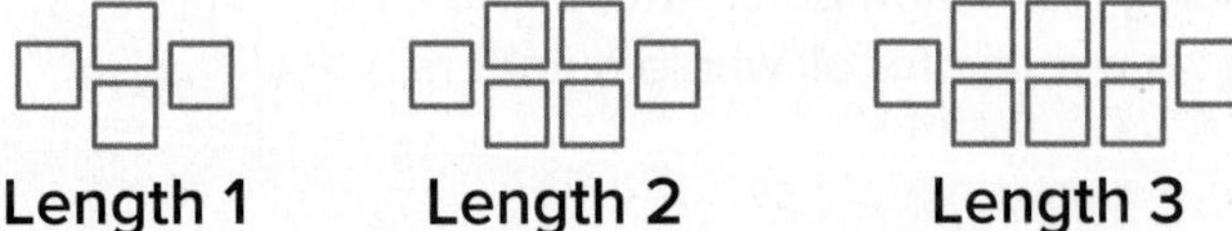

Length 1 Length 2 Length 3

   a. Kayla would like to find the number of tiles needed to make a path of any length. Look for a pattern and make a conjecture about the number of tiles needed to make a path of length n.

   b. Kayla said that one path her company made last week required exactly 103 tiles. Is this possible? Justify your argument.

Statements, Conditionals, and Biconditionals

Explore Truth Values

Online Activity Use the video to complete the Explore.

INQUIRY How can you determine the truth value of a statement?

Learn Using Logic

A **statement** is any sentence that is either true T or false F, but not both. **Truth value** is the truth or falsity of a statement. Statements are often represented using a letter such as p or q.

If a statement is represented by p, then *not p* is the **negation** of the statement. The negation of a statement has the opposite meaning, as well as the opposite truth value, of the original statement. The negation of a statement p is *not p* or $\sim p$.

Two or more statements joined by the word *and* or *or* form a **compound statement**. A compound statement using the word *and* is called a **conjunction**. A conjunction is true only when both statements that form it are true. A conjunction is written as p and q or $p \wedge q$.

A compound statement using the word *or* is called a **disjunction**. A disjunction is true if at least one of the statements is true. A disjunction is written as *p or q* or $p \vee q$.

Example 1 Truth Values of Conjunctions

Use the statements to write a compound statement for each conjunction. Then find the truth values. Explain your reasoning.

p: The figure is a trapezoid.

q: The figure has four congruent sides.

r: The figure has four right angles.

a. *p* and *r*

p and r: The figure is a trapezoid, and the figure has four right angles. Although r is true, p is false. So, *p and r* is false.

b. $\sim p \wedge q$

$\sim p \wedge q$: The figure is not a trapezoid, and the figure has four congruent sides. Both $\sim p$ and q are true, so $\sim p \wedge q$ is true.

Go Online You can complete an Extra Example online.

Today's Goals

- Write compound statements for conjunctions and disjunctions and determine truth values of statements.
- Identify hypotheses and conclusions of conditional statements and write related conditionals.
- Write and analyze biconditional statements and determine truth values of biconditional statements.

Today's Vocabulary

statement
truth value
negation
compound statement
conjunction
disjunction
conditional statement
if-then statement
hypothesis
conclusion
converse
inverse
contrapositive
logically equivalent
biconditional statement

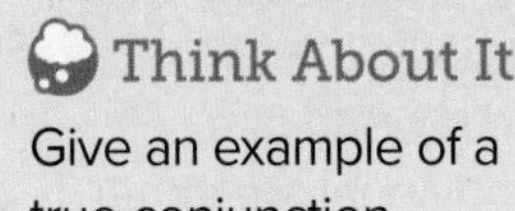

Think About It!

Give an example of a true conjunction.

Watch Out!

Negation Just as the opposite of an integer is not always negative, the negation of a statement is not always false. The negation of a statement has the opposite truth value of the original statement.

Study Tip

If and Then The word *if* is not part of the hypothesis, and the word *then* is not part of the conclusion. However, these words can indicate where the hypothesis and conclusion begin. Consider the conditional below.

If Felipe has band practice, then he will come home after dinner.

Felipe has band practice is the hypothesis, and *Felipe will come home after dinner* is the conclusion.

Study Tip

Logically Equivalent A conditional and its contrapositive are either both true or both false. Similarly, the converse and inverse of a conditional are either both true or both false. Statements with the same truth value are said to be **logically equivalent**.

Example 2 Truth Values of Disjunctions

Use the statements to write a compound statement for the disjunction $p \vee \sim r$. Then find its truth value. Explain your reasoning.

***p*: ∠*ABC* and ∠*CBD* are complementary.**

***q*: ∠*ABC* and ∠*CBD* are vertical angles.**

***r*: $\overline{AB} \cong \overline{BD}$**

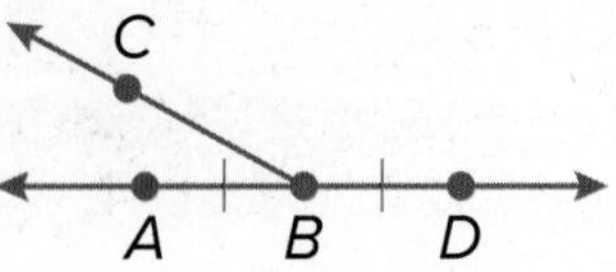

$p \vee \sim r$: ∠*ABC* and ∠*CBD* are complementary, or $\overline{AB}$ and $\overline{BD}$ are not congruent.

$p \vee \sim r$ is false, because p is false and $\sim r$ is false.

Learn Conditionals

A **conditional statement** is a compound statement that consists of of a premise, or *hypothesis*, and a *conclusion*, which is false only when its premise is true and its conclusion is false.

Conditional Statements and Related Conditionals

Words	Examples
An **if-then statement** is a compound statement of the form "if p, then q," where p and q are statements. Symbols: $p \rightarrow q$; read *if* p, *then* q, or p *implies* q	If it rains, then the parade will be canceled.
The **hypothesis** of a conditional statement is the phrase immediately following the word *if*. Symbols: $p \rightarrow q$; read *if* p, *then* q, or p *implies* q	
The **conclusion** of a conditional statement is the phrase immediately following the word *then*. Symbols: $p \rightarrow q$; read *if* p, *then* q, or p *implies* q	
The **converse** is formed by exchanging the hypothesis and conclusion of the conditional. Symbols: $q \rightarrow p$, read *if* q, then p, or q *implies* p	If the parade is canceled, then it has rained.
The **inverse** is formed by negating both the hypothesis and conclusion of the conditional. Symbols: $\sim p \rightarrow \sim q$, read *if not* p, *then not* q	If it does not rain, then the parade will not be canceled.
The **contrapositive** is formed by negating both the hypothesis and the conclusion of the converse of the conditional. Symbols: $\sim q \rightarrow \sim p$, read *if not* q, *then not* p	If the parade is not canceled, then it does not rain.

Go Online You can complete an Extra Example online.

Example 3 Identify the Hypothesis and Conclusion

Identify the hypothesis and conclusion of each conditional statement.

a. If a polygon has six sides, then it is a hexagon.

Hypothesis: A polygon has six sides.

Conclusion: The polygon is a hexagon.

b. Another performance will be scheduled if the first one is sold out.

Notice that the word *if* appears in the second portion of the sentence.

Hypothesis: The first performance is sold out.

Conclusion: Another performance will be scheduled.

Think About It!

If a conditional is true, are the converse and inverse *sometimes*, *always*, or *never* true? Support your answer with an example.

Check

Identify the hypothesis and conclusion of each conditional statement.

a. If the forecast is rain, then I will take an umbrella.

Hypothesis: ?____________

Conclusion: ?____________

b. A number is divisible by 10 if its last digit is a 0.

Hypothesis: ?____________

Conclusion: ?____________

Example 4 Write a Conditional in If-Then Form

Identify the hypothesis and conclusion for each conditional statement. Then write the statement in if-then form.

a. Four quarters can be exchanged for a $1 bill.

Hypothesis: You have four quarters.

Conclusion: You can exchange them for a $1 bill.

If-then: If you have four quarters, then you can exchange them for a $1 bill.

b. The sum of the measures of two supplementary angles is 180°

Hypothesis: Two angles are supplementary.

Conclusion: The sum of their measures is 180°.

If-then: If two angles are supplementary, then the sum of their measures is 180°.

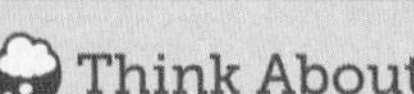

Think About It!

How do you identify the hypothesis and conclusion of a conditional statement when the statement is not in if-then form?

Go Online You can complete an Extra Example online.

Check

Identify the hypothesis and conclusion of the conditional statement *A polygon with two sets of parallel sides is a parallelogram.* Then write the statement in if-then form.

Hypothesis: ? ______

Conclusion: ? ______

? ______

______

Example 5 Related Conditionals

NATURE The tang is a saltwater fish that inhabits shallow coral reefs in tropical areas. Tangs are a part of the Acanthuridae family along with surgeonfish and unicornfish. All members of the Acanthuridae family are saltwater fish. Write the converse, inverse, and contrapositive of the true conditional statement *Tangs are fish that live in salt water.* Determine whether each related conditional is *true* or *false*. If a statement is false, then find a counterexample.

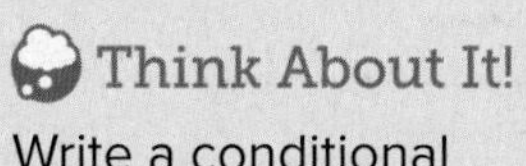

Think About It!

Write a conditional statement in which the converse, inverse, and contrapositive are true.

Conditional: *If a fish is a tang, then it lives in salt water.*

Converse: *If a fish lives in salt water, then it is a tang.*

Counterexample: A surgeonfish lives in salt water, but it is not a tang.

Therefore, the converse is false.

Inverse: *If a fish is not a tang, then it does not live in salt water.*

Counterexample: A surgeonfish is not a tang, but it does live in salt water. Therefore, the inverse is false.

Contrapositive: *If a fish does not live in salt water, then it is not a tang.*

Based on the information above, this statement is true.

Check

MUSIC Symphony orchestras contain instruments from 4 musical families: strings, woodwinds, brass, and percussion. However, string orchestras only contain string instruments. String instruments include the violin, viola, cello, bass, and harp. Write the converse, inverse, and contrapositive of the true conditional statement *If an orchestra is a string orchestra, then it contains string instruments*. Determine whether each related conditional is *true* or *false*. If the statement is false, find a counterexample.

Converse: ? ______

Inverse: ? ______

Contrapositive: ? ______

Go Online You can complete an Extra Example online.

Learn Biconditionals

You can use logic and *biconditional statements* to indicate exclusivity in situations. For example, Aarón is applying for admission into culinary school. He must earn a 3.5 GPA or higher this semester to be accepted. You can express this as two if-then statements.

- If he earns a 3.5 GPA or higher this semester, then he will be accepted.
- If Aarón is accepted into culinary school, then he has earned a 3.5 GPA or higher for the semester.

Biconditional Statement	
Words	A **biconditional statement** is the conjunction of a conditional and its converse.
Symbols	$(p \rightarrow q) \wedge (q \rightarrow p) \rightarrow (p \leftrightarrow q)$, read *p if and only if q*

So, the biconditional statement for the example above is *Aarón will be accepted into culinary school if and only if he earns a 3.5 GPA or higher this semester.*

Study Tip

If and Only If The phrase *if and only if* can be abbreviated with *iff*.

Think About It!

Compare the mathematical meanings of the symbols $\rightarrow$ and $\leftrightarrow$ in $p \rightarrow q$ and $p \leftrightarrow q$.

Example 6 Write Biconditionals

Write the conditional and converse for each statement. Determine the truth values of the conditionals and converses. If false, find a counterexample. Write a biconditional statement if possible.

a. Rasha listens to music when she is in study hall.

Conditional: If Rasha is in study hall, then she is listening to music.

Is the conditional statement *true* or *false*? If false, provide a counterexample. true

Converse: If Rasha is listening to music, then she is in study hall.

Is the converse *true* or *false*? If false, provide a counterexample.
False; sample answer: Rasha could be listening to music in the cafeteria.

Because the converse is false, a biconditional statement cannot be written.

b. If two lines are parallel, then they have the same slope.

Conditional: If two lines are parallel, then they have the same slope.

Converse: If two lines have the same slope, then they are parallel.

The conditional and the converse are true. So, a biconditional can be written.

Biconditional: Two lines are parallel if and only if they have the same slope.

Think About It!

If a biconditional is true, what do you know about the conditional and converse? If a biconditional is false, what do you know about the conditional and converse?

Go Online You can complete an Extra Example online.

Check

Write the conditional and converse for the statement. Determine the truth values of the conditional and converse. If false, find a counterexample. Write a biconditional statement if possible.

Isosceles triangles have at least two congruent sides.

Conditional: If a triangle is isosceles, then it has at least two congruent sides.

Converse: If a triangle has at least two congruent sides, then it is isosceles.

The conditional is true, and the converse is true.

Biconditional: A triangle is isosceles iff it has at least two congruent sides.

Example 7 Determine Truth Values of Biconditionals

Write each biconditional as a conditional and its converse. Then determine whether the biconditional is *true* or *false*. If it is false, give a counterexample.

Two angles are complements if and only if their measures have a sum of 90°.

Write the biconditional statement as a conditional.

Sample answer: If two angles are complements, then their measures have a sum of 90°.

Write the converse of your conditional statement.

Sample answer: If the measures of two angles have a sum of 90°, then the two angles are complements.

The conditional and the converse are true, so the biconditional is true.

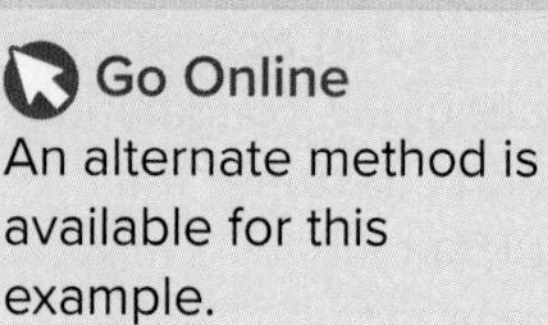

Go Online
An alternate method is available for this example.

Check

Write the biconditional as a conditional and its converse. Then, determine whether the biconditional is *true* or *false*. If false, give a counterexample.

$x > -2$, if and only if x is positive.

Conditional: ___?___

Converse: ___?___

The biconditional is ___?___, because ___?___.

Go Online You can complete an Extra Example online.

Practice

Go Online You can complete your homework online.

Examples 1 and 2

Use the statements to write a compound statement for each conjunction or disjunction. Then find the truth values. Explain your reasoning.

***p*: $-3 - 2 = -5$**

***q*: Vertical angles are congruent.**

***r*: $2 + 8 > 10$**

1. p and q

2. $p \wedge r$

3. $q \vee \sim r$

4. $r \vee q$

5. $\sim p \wedge \sim q$

6. $\sim r \vee \sim p$

Example 3

Identify the hypothesis and conclusion of each conditional statement.

7. "If there is no struggle, there is no progress." (Frederick Douglass).

8. If two angles are adjacent, then they have a common side.

9. If you lead, then I will follow.

10. If $3x - 4 = 11$, then $x = 5$.

11. If two angles are vertical, then they are congruent.

Example 4

Identify the hypothesis and conclusion for each conditional statement. Then write each statement in if-then form.

12. Get a free water bottle with a one-year membership.

13. Everybody at the party received a gift.

14. The intersection of two planes is a line.

15. The area of a circle is πr^2.

16. Collinear points lie on the same line.

17. A right angle measures 90 degrees.

Example 5

Write the converse, inverse, and contrapositive of each true conditional statement. Determine whether each related conditional is *true* or *false*. If a statement is false, then find a counterexample.

18. AIR TRAVEL Ulma is waiting to board an airplane. Over the speakers she hears a flight attendant say "If you are seated in rows 10 to 20, you may now board."

19. RAFFLE If you have five dollars, then you can buy five raffle tickets.

20. GEOMETRY If two angles are complementary, then the angles are acute.

21. MEDICATION A medicine bottle says "If you will be driving, then you should not take this medicine."

Example 6

Write the conditional and converse for each statement. Determine the truth values of the conditionals and converses. If false, find a counterexample. Write a biconditional statement if possible.

22. 89 is an even number if it is divisible by 2.

23. The game will be cancelled if it is raining.

24. Laura's soccer team plays on Saturdays.

Example 7

Write each biconditional as a conditional and its converse. Then determine whether the biconditional is *true* or *false*. If it is false, give a counterexample.

25. A polygon is a quadrilateral if and only if it has four sides.

26. An angle is acute if and only if it has a measure less than 90°.

Mixed Exercises

27. Find the truth value of $(p \wedge q) \vee r$.

p: $(-4)^2 > 0$

q: An isosceles triangle has at least two congruent sides.

r: Two angles, whose measure have a sum of 90, are supplements.

28. Suppose p and q are both false. What is the truth value of $(p \wedge {\sim}q) \vee {\sim}p$?

29. What is the truth value of $({\sim}p \vee q) \wedge r$ if p is true, q is false, and r is true?

30. What is the truth value of $({\sim}p \wedge q) \vee r$ if p is true, q is false, and r is true?

31. **CHOCOLATE** Luca has a bag of miniature chocolate bars that come in two distinct types: dark and milk. Luca picks a chocolate out of the bag. Use the following statements to determine whether the statement ${\sim}({\sim}p \vee {\sim}q)$ is true.

p: the chocolate bar is dark chocolate

q: the chocolate bar is milk chocolate

32. Clark says that a parallelogram is a quadrilateral with equal opposite angles. Write his statement in if-then form.

33. **REASONING** Kala asked Elijah whether his hockey team won the game last night and whether he scored a goal. Elijah said "yes." Kalia then asked Goldi whether she or Elijah scored a goal at the game. Goldi said "yes." What can you conclude about whether or not Goldi scored?

34. **PRECISION** If I roll two 6-sided dice and the sum of the numbers is 11, then one die must be a 5. Write the converse, inverse, and contrapositive of the true conditional statement. Determine whether each related conditional is *true* or *false*. If a statement is false, then find a counterexample.

For Exercises 35 and 36, use the following statement.

If a ray bisects an angle, then it divides the angle into two congruent angles.

35. Write the inverse of the given statement.

36. Write the contrapositive of the given statement.

37. Write the statement *All right angles are congruent* in if-then form.

38. Use the segment to write a statement that has the same truth value as $3 = 5$.

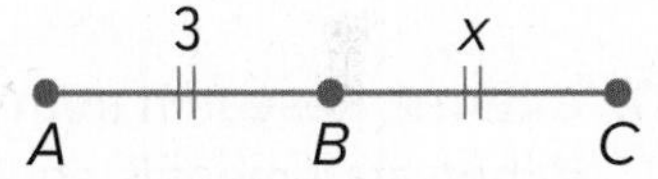

Higher-Order Thinking Skills

39. **CREATE** Consider a situation that can be represented with an if-then statement.

a. Write a true if-then statement for which the converse is false.

b. Write the converse, inverse, and contrapositive of your sentence.

c. Give the truth value of each statement you wrote for part **b.**

40. **ANALYZE** You are evaluating a conditional statement in which the hypothesis is true, but the conclusion is false. Is the inverse of the statement true or false? Justify your argument.

PERSEVERE To negate a statement containing the words *all* or *for every*, you can use the phrase *at least one* or *there exists*. To negate a statement containing the phrase *there exists*, use the phrase *for all* or *for every*.

p: All polygons are convex. $\sim p$: *At least one* polygon is *not* convex.

q: *There exists* a problem that has no solution. $\sim q$: *For every* problem, there is a solution.

Sometimes there are phrases that may be implied. For example, *The square of a real number is nonnegative* implies the following conditional and its negation.

p: *For every* real number x, $x^2 \geq 0$.

$\sim p$: *There exists* a real number x, such that $x^2 < 0$.

Use the information above to write the negation of each statement.

41. Every student at Hammond High School has a locker.

42. All squares are rectangles.

43. There exists a real number x, such that $x^2 = x$.

44. There exists a student who has at least one class in the C-Wing.

45. Every real number has a real square root.

46. There exists a segment that has no midpoint.

47. **CREATE** Research truth tables online. Then make a truth table to prove that an if-then statement is equivalent to its contrapositive and its inverse is equivalent to its converse.

48. **WRITE** Describe the relationship among a conditional, its converse, its inverse, and its contrapositive.

49. **FIND THE ERROR** Nicole and Kiri are evaluating the conditional *If 15 is prime, then 20 is divisible by 4*. Both think that the conditional is true, but their reasoning differs. Is either of them correct? Explain your reasoning.

Nicole	Kiri
The conclusion is true because 20 is divisible by 4. So, the conditional is true.	The hypothesis is false because 15 is not prime. So, the conditional is true.

50. **CREATE** Write a conditional statement for which the converse, inverse, and contrapositive are all true. Explain your reasoning.

Deductive Reasoning

Explore Applying Laws of Deductive Reasoning by Using Venn Diagrams

 **Online Activity** Use dynamic geometry software to complete the Explore.

INQUIRY How can you use Venn diagrams to determine the truth value of a statement?

Today's Goals
- Apply the Law of Detachment to determine the validity of conclusions.
- Apply the Law of Syllogism to make valid conclusions from given statements.

Today's Vocabulary
deductive reasoning
valid argument

Learn The Law of Detachment

Unlike inductive reasoning, which uses a specific pattern of examples or observations to make a general conclusion, **deductive reasoning** uses general facts, rules, definitions, or properties to reach specific *valid* conclusions from given statements. An argument is **valid** if it is impossible for all the premises, or supporting statements, of the argument to be true and for its conclusion to be false. One law related to deductive reasoning is the Law of Detachment.

Key Concept • Law of Detachment	
Words	If $p \rightarrow q$ is a true statement and p is true, then q is true.
Example	*Given*: If a car is out of gas, then it will not start. Sarah's car is out of gas. *Valid Conclusion*: Sarah's car will not start.

Example 1 Inductive and Deductive Reasoning

Determine whether each conclusion is based on *inductive* or *deductive* reasoning.

a. If a student is late returning a library book, then he or she will be charged a $2 late fee. Chang returned a library book late, so he concludes that he will be charged a $2 late fee.

Chang is basing his conclusion on the library's policies, so he is using deductive reasoning.

b. Every time Tamika has worn her favorite jersey to a football game, her school's team has won the game. Tamika is wearing her favorite jersey to the football game tonight, so she concludes that her school's team will win the game.

Tamika is basing her conclusion on a specific pattern of observations, so she is using inductive reasoning.

Go Online You can complete an Extra Example online.

Check

Determine whether each conclusion is based on *inductive* or *deductive* reasoning.

a. Newton's first law of motion states that an object at rest will remain at rest unless acted on by an unbalanced force. Elisa watches a soccer ball roll across the field. She concludes that an unbalanced force has acted upon the soccer ball ____?____.

b. Mrs. Jackson notices that her family's data usage is increasing by approximately 2500 megabytes of data every month. So, she concludes that her family's data usage next month will be 2500 megabytes greater than this month's data usage ____?____.

Study Tip

Given Information All information provided after the *Given* can be assumed to be true. The *Conclusion* cannot be assumed to be true.

Study Tip

Arguments An argument consists of reasons, proof, or evidence to support a position. A *logical argument*, such as the one shown in part **a**, is supported by the rules of logic. These rules include the Law of Detachment. This is different from a *statistical argument*, which is supported by examples or data.

Example 2 The Law of Detachment

Determine whether each conclusion is valid based on the given information. Write *valid* or *invalid*. Explain your reasoning.

a. Given: To go on the field trip, a student must turn in a permission slip. Mariana turned in her permission slip.

Conclusion: Mariana can go on the field trip.

Step 1 Identify the hypothesis and conclusion.

Because a student must turn in a permission slip to go on the field trip, the phrase *a student must turn in a permission slip* is the hypothesis of the conditional statement.

p: A student turns in a permission slip.

q: The student can go on the field trip.

Step 2 Analyze the conclusion.

The given statement *Mariana turned in her permission slip* satisfies the hypothesis, so p is true. By the Law of Detachment, *Mariana can go on the field trip,* which matches q, is a true or valid conclusion.

b. Given: If a figure is a square, then it is a polygon.

Figure *A* is a polygon.

Conclusion: Figure *A* is a square.

Step 1 Identify the hypothesis and conclusion.

p: A figure is a square.

q: It is a polygon.

Step 2 Analyze the conclusion.

The given statement *Figure A is a polygon* satisfies the conclusion q of a true conditional. However, knowing that a conditional statement and its conclusion are true does not make the hypothesis true. Figure *A* could be a triangle. The conclusion is *invalid*.

Go Online You can complete an Extra Example online.

Check

Determine whether the conclusion is valid based on the given information. Select the correct answer and justification.

a. Given: If three points are noncollinear, then they determine a plane. Points *A*, *B*, and *C* lie in plane G.

Conclusion: Points *A*, *B*, and *C* are noncollinear.

A. Valid; points *A*, *B*, and *C* determine plane G. Therefore, they are noncollinear.

B. Valid; because points *A*, *B*, and *C* are noncollinear, they determine plane G.

C. Invalid; points *A*, *B*, and *C* determine plane G. Therefore, they are noncollinear.

D. Invalid; points *A*, *B*, and *C* can be collinear and lie in plane G.

b. Given: If Dakota goes to the video game store, then he will buy a new game. Dakota went to the video game store this afternoon.

Conclusion: Dakota bought a new game.

A. Invalid; because the statement *Dakota bought a new game* does not satisfy the hypothesis of the conditional statement, the conclusion is not true.

B. Valid; because the statement *Dakota went to the video game store this afternoon* satisfies the conclusion of the conditional statement, the hypothesis of the conditional is true.

C. Valid; because the statement *Dakota went to the video game store this afternoon* satisfies the hypothesis of the conditional statement, the conclusion is true.

D. Invalid; because the statement *Dakota went to the video game store this afternoon* satisfies only the hypothesis, the conclusion is not true.

Learn The Law of Syllogism

One law that is related to deductive reasoning is the Law of Syllogism. This law allows you to draw conclusions from two true conditional statements when the conclusion of one statement is the hypothesis of the other.

Key Concept • Law of Syllogism	
Words	If $p \rightarrow q$ and $q \rightarrow r$ are true statements, then $p \rightarrow r$ is a true statement.
Example	Given: If you get a job, then you will earn money. If you earn money, then you will buy a car. Valid Conclusion: If you get a job, then you will buy a car.

Talk About It!

Do you think that the order of the given statements is important when applying the Law of Syllogism? Justify your argument.

Example 3 The Law of Syllogism

SLEEP **Scientists have found that the quality and amount of sleep greatly impact learning and memory. Lack of sleep causes students to have trouble focusing and receiving new information. Sleep deprivation also makes it difficult to retrieve previously-learned information. Draw a valid conclusion from the given statements, if possible.**

Given: If you are tired, then you will not do well on your test.
If you do not get enough sleep, then you will be tired.

Step 1 Identify the hypothesis and conclusion that are the same.

Determine whether the conclusion of one statement is the hypothesis of the other statement.

Given: If you are tired, then you will not do well on your test.

If you do not get enough sleep, then you will be tired.

Reorder the given statements so the conclusion of the first statement is the hypothesis of the second statement. This will allow you to make a valid conclusion using the Law of Syllogism.

Given: If you do not get enough sleep,	*p*: You do not get enough sleep.
then you will be tired.	*q*: You will be tired.
If you are tired, then you will not do well on your test.	*r*: You will not do well on your test.

Step 2 Represent the statements with symbols.

Let *p*, *q*, and *r* represent the parts of the given conditional statements. Analyze the logic of the given conditional statements using symbols.

Statement 1: $p \rightarrow q$

Statement 2: $q \rightarrow r$

Because both statements are true and the conclusion of the first statement is the hypothesis of the second statement, $p \rightarrow r$ by the Law of Syllogism. A valid conclusion is *If you do not get enough sleep, then you will not do well on your test.*

Check

GRAND CANYON The Grand Canyon covers an area of 1900 square miles and contains 277 miles of the Colorado River. Since the Grand Canyon became a national park in 1919, over 193 million people have visited.

Draw a valid conclusion from the given statements, if possible.

Given: If Ebony takes a vacation, then she will go to the Grand Canyon. If Ebony goes to the Grand Canyon, then she will hike to the Colorado River.

Go Online You can complete an Extra Example online.

Study Tip

True vs. Valid Conclusions A true conclusion is not the same as a valid conclusion. True conclusions reached using invalid reasoning are still invalid.

Think About It!

Can the Law of Syllogism be applied if the two given statements have the same conclusion? Justify your argument.

Go Online to practice what you've learned about deductive reasoning in the Put It All Together over Lessons 12-1 through 12-3.

Practice

Go Online You can complete your homework online.

Example 1

Determine whether each conclusion is based on *inductive* or *deductive* reasoning.

1. At Fumio's school, if a student is late five times, then the student will receive a detention. Fumio has been late to school five times. Therefore, he will receive a detention.

2. A dental assistant notices that a patient has never been on time for an appointment. She concludes that the patient will be late for her next appointment.

3. A person must have a membership to work out at a gym. Jessie is working out at that gym. Jessie has a membership to that gym.

4. If Emilio decides to go to a concert tonight, then he will miss football practice. Tonight, Emilio went to a concert. Emilio missed football practice.

5. Every Wednesday, Jacy's mother calls. Today is Wednesday, so Jacy concludes that her mother will call.

6. Whenever Juanita has attended a tutoring session, she notices that her grades have improved. Juanita attends a tutoring session, and she concludes her grades will improve.

Example 2

Determine whether each conclusion is valid based on the given information. Write *valid* or *invalid*. Explain your reasoning.

7. **Given:** Right angles are congruent. $\angle 1$ and $\angle 2$ are right angles.

   Conclusion: $\angle 1 \cong \angle 2$

8. **Given:** If a figure is a square, then it has four right angles. Figure *ABCD* has four right angles.

   Conclusion: Figure *ABCD* is a square.

9. **Given:** If you leave your lights on while your car is off, then your battery will die. Your battery is dead.

   Conclusion: You left your lights on while your car was off.

10. **Given:** If Dennis gets a part-time job, then he can afford a car payment. Dennis can afford a car payment.

    Conclusion: Dennis got a part-time job.

11. **Given:** If 75% of the prom tickets are sold, then the prom will be held at the country club. 75% of the prom tickets were sold.

    Conclusion: The prom will be held at the country club.

Example 3

Use the Law of Syllogism to draw a valid conclusion from each set of given statements, if possible. If no valid conclusion can be drawn, write *no valid conclusion* and explain your reasoning.

12. If you interview for a job, then you wear a suit.
If you interview for a job, then you will update your resume.

13. If Tina has a grade point average of 3.0 or greater, she will be on the honor roll.
If Tina is on the honor roll, then she will have her name in the school paper.

14. If two lines are perpendicular, then they intersect to form right angles.
Lines *s* and *r* form right angles.

15. If the measure of an angle is between 90° and 180°, then it is obtuse.
If an angle is obtuse, then it is not acute.

16. If two lines in a plane are not parallel, then they intersect.
If two lines intersect, then they intersect in a point.

17. If a number ends in 0, then it is divisible by 2.
If a number ends in 4, then it is divisible by 2.

Mixed Exercises

CONSTRUCT ARGUMENTS **Draw a valid conclusion from the given statements, if possible. Then state whether your conclusion was drawn using the Law of Detachment or the Law of Syllogism. If no valid conclusion can be drawn, write *no valid conclusion*. Justify your argument.**

18. Given: If a figure is a square, then all the sides are congruent. Figure *ABCD* is a square.

19. Given: If two angles are complementary, the sum of the measures of the angles is 90°. $\angle 1$ and $\angle 2$ are complementary angles.

20. Given: Ballet dancers like classical music. If you like classical music, then you enjoy the opera.

21. Given: If you are an athlete, then you enjoy sports. If you are competitive, then you enjoy sports.

22. Given: If a polygon is regular, then all of its sides are congruent. All of the sides of polygon *WXYZ* are congruent.

23. Given: If Terryl completes a course with a grade of C, then he will not receive credit. If Terryl does not receive credit, he will have to take the course again.

USE TOOLS **Determine whether each conclusion is valid based on the given information. Write *valid* or *invalid*. Explain your reasoning using a Venn diagram.**

24. Given: If the temperature drops below 32°F, it may snow. The temperature did not drop below 32°F on Monday.

Conclusion: It did not snow on Monday.

25. Given: All vegetarians do not eat meat. Theo is a vegetarian.

Conclusion: Theo does not eat meat.

26. TUTORING Marla sometimes stays after school to tutor classmates. If it is Tuesday, then Marla tutors chemistry. If Marla tutors chemistry, then she arrives home at 4 P.M. Today Marla arrived home at 4 P.M. Can it be concluded that today is Tuesday? Explain your reasoning.

27. MUSIC Composer Ludwig van Beethoven wrote 9 symphonies and 5 piano concertos. If you lived in Vienna in the early 1800s, then you could attend a concert conducted by Beethoven himself. Write a valid conclusion to the hypothesis: *If Mozart could not attend a concert conducted by Beethoven, . . .*

28. DIRECTIONS Paolo has an appointment to see a financial advisor on the fifteenth floor of an office building. When he gets to the building, the people at the front desk tell him that if he wants to go to the fifteenth floor, then he must take the red elevator. While looking for the red elevator, a guard informs him that if he wants to find the red elevator, then he must find the replica of Michelangelo's David. When he finally got to the fifteenth floor, his financial advisor greeted him asking, "What did you think of the Michelangelo?" How did Paolo's financial advisor conclude that Paolo must have seen the Michelangelo statue?

29. SIGNS Two signs are posted outside a trampoline park. Inside the trampoline park, you see a child with a parent. Write a valid conclusion based on the given information about the age of the child.

NO ONE UNDER 5 ALLOWED
NO ONE UNDER 8 ALLOWED WITHOUT A PARENT

30. LOGIC As Maite's mother left for work, she quickly gave Maite some instructions. "If you need me, call my cell phone. If I do not answer, then it means I'm in a meeting. The meeting will not last more than 30 minutes, and I will call you back when the meeting is over." Later that day, Maite tried to call her mother's cell phone, but her mother was in a meeting and could not answer the phone. Maite concludes that she will have to wait no more than 30 minutes before she gets a call back from her mother. What law of logic did Maite use to draw this conclusion?

31. ENERGY Use deductive reasoning to draw a valid conclusion from the following statements: If a heat wave occurs, then air conditioning will be used more frequently; if air conditioning is used more frequently, then energy costs will be higher; there is a heat wave in Florida. If no valid conclusion can be drawn, then write *no valid conclusion* and explain your reasoning.

Higher-Order Thinking Skills

32. WRITE Explain why the Law of Syllogism cannot be used to draw a conclusion from these conditionals.

If you wear winter gloves, then you will have warm hands.

If you do not have warm hands, then your gloves are too thin.

33. PERSEVERE Use symbols for *conjunction*, *disjunction*, and *implies* to represent the Law of Detachment and the Law of Syllogism symbolically. Let p represent the hypothesis, and let q represent the conclusion.

34. CREATE Write a pair of statements in which the Law of Syllogism can be used to reach a valid conclusion. Specify the conclusion that can be reached.

35. ANALYZE Students in Mr. Kendrick's class are divided into two groups for an activity. Students in Group A must always tell the truth. Students in Group B must always lie. Jonah and Janeka are in Mr. Kendrick's class. When asked whether he and Janeka are in group A or B, Jonah says, "We are both in Group B." To which group does each student belong? Justify your argument.

36. WRITE Compare and contrast inductive and deductive reasoning when making conclusions and proving conjectures.

37. CREATE Write three statements that illustrate the Law of Syllogism.

38. CREATE Write three statements that illustrate the Law of Detachment.

39. WHICH ONE DOESN'T BELONG? Use statements (1) and (2). Determine which statement does not belong. Justify your conclusion.

(1) *If a triangle is equilateral, then it has three congruent sides.*

(2) *If all the sides of a triangle are congruent, then each angle measures 60°.*

A. If a triangle is not equilateral, then it cannot have congruent angles.

B. A figure with three congruent sides is always an equilateral triangle.

C. If a triangle is not equilateral, then none of the angles measures 60°.

D. If a triangle is equilateral, then each of its angles measures 60°.

Writing Proofs

Explore Algebraic Proof

Online Activity Use guiding exercises to complete the Explore.

INQUIRY How can you write an algebraic proof?

Learn Postulates About Points, Lines, and Planes

Recall that a postulate or axiom is a statement accepted as true without proof. The postulates listed below about points, lines, and planes cannot be proven, but they can be used as reasons in proofs.

Postulates: Points, Lines, and Planes	
12.1	Through any two points, there is exactly one line.
12.2	Through any three noncollinear points, there is exactly one plane.
12.3	A line contains at least two points.
12.4	A plane contains at least three noncollinear points.
12.5	If two points lie in a plane, then the entire line containing those points lies in that plane.
12.6	If two lines intersect, then their intersection is exactly one point.
12.7	If two planes intersect, then their intersection is a line.

Example 1 Identify Postulates

ARCHITECTURE **Explain how the photo illustrates that each statement is true. Then state the postulate that can be used to show that the statement is true.**

a. Lines n and p intersect at point *D*.

The top edges of the building are represented by lines n and p. The lines intersect at the corner point, *D*. Postulate 12.6 states that if two lines intersect, then their intersection is exactly one point.

(*continued on the next page*)

Go Online You can complete an Extra Example online.

Today's Goals

- Analyze figures to identify and use postulates about points, lines, and planes.
- Analyze and construct viable arguments in a two-column format.
- Analyze and construct viable arguments in a flow proof format.
- Analyze and construct viable arguments in a paragraph proof format.

Today's Vocabulary

proof
two-column proof
deductive argument
flow proof
paragraph proof

b. Points *A*, *B*, and *D* determine a plane.

Points *A*, *B*, and *D* are three noncollinear points on the front face of the building. By Postulate 12.2, through any three noncollinear points, there is exactly one plane.

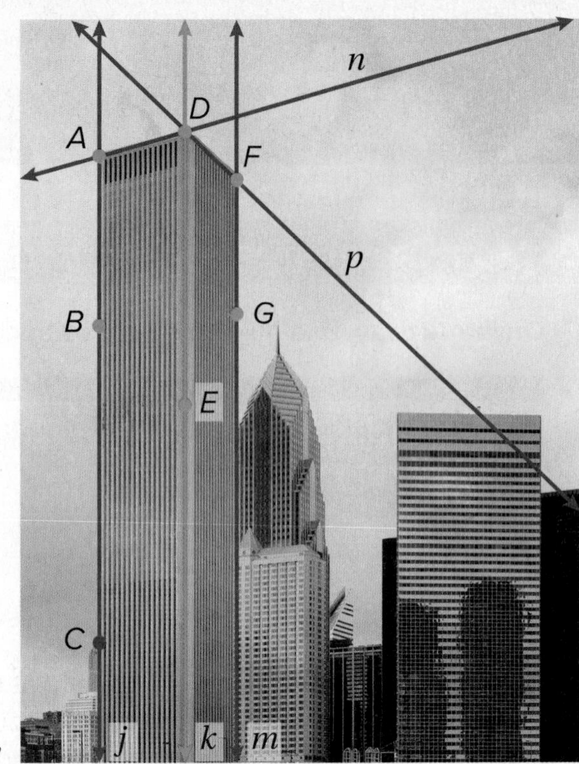

c. The plane that contains *A*, *D*, and *E* intersects the plane that contains *F*, *G*, and *E* in line k.

The front face of the building can be represented by the plane that contains *A*, *D*, and *E*. The side face of the building can be represented by the plane that contains *F*, *G*, and *E*. These planes intersect at the corner of the building represented by line k. Postulate 12.7 states that if two planes intersect, then their intersection is a line.

Check

ANCIENT MONUMENTS The image illustrates the statement $\overleftrightarrow{AB}$ *is the only line through A and B*. Which postulate proves that this statement is true?

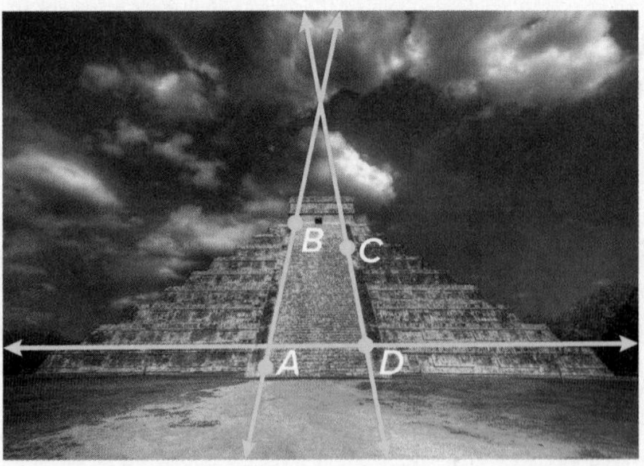

George Hammerstein/Fancy/Corbis/Glow Images, Dollia Sheombar/Moment/Getty Images

Example 2 Use Postulates

Determine whether each statement is *always, sometimes,* or *never* true. Justify your argument.

a. The intersection of three planes is a line.

Sometimes; if three planes intersect, then their intersection could be a line or a point.

b. Line *r* contains only point *P*.

Never; Postulate 12.3 states that a line contains at least two points.

c. Through points *H* and *K*, there is exactly one line.

Always; Postulate 12.1 states that through any two points, there is exactly one line.

Think About It!

Martin claims that this is a true statement. *Through any three points, there is exactly one plane*. Do you agree? Explain.

Check

Determine whether the statement is *always*, *sometimes*, or *never* true. Justify your argument.

Two intersecting lines determine a plane.

Learn Two-Column Proofs

A **proof** is a logical argument in which each statement is supported by a statement that is accepted as true. These supporting statements can include definitions, postulates, and theorems. A **two-column proof** is a proof that contains statements and reasons that are organized in a two-column format. You can develop a **deductive argument** to prove a statement by building a logical chain of statements and reasons.

Key Concept • How to Write a Proof	
Step 1	List the given information. Draw a diagram if needed.
Step 2	Create a deductive argument that links the given information to the statement that you are proving.
Step 3	Justify each statement with a reason. Reasons include definitions, postulates, theorems, and algebraic properties.
Step 4	State what it is that you have proven.

Go Online You can complete an Extra Example online.

Example 3 Two-Column Proof

Complete the two-column proof by selecting the correct statements and reasons.

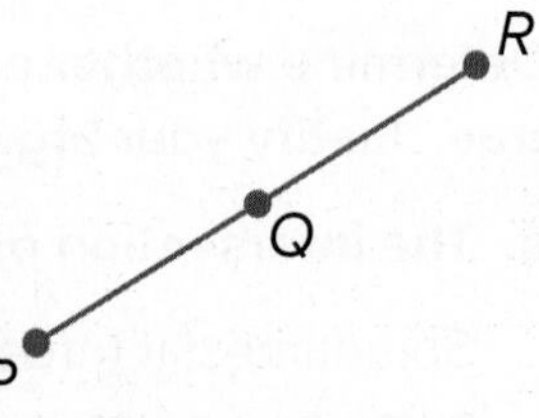

Given: Q is the midpoint of $\overline{PR}$.

Prove: $\overline{PQ} \cong \overline{QR}$

STATEMENTS/REASONS:
Definition of midpoint
Definition of congruence
$PQ = QR$
Betweenness of points
Q is between P and R.

Statements	Reasons
1. Q is the midpoint of $\overline{PR}$.	1. Given
2. $PQ = QR$	2. Definition of midpoint
3. $\overline{PQ} \cong \overline{QR}$	3. Definition of congruence

Once a conjecture has been proven true, it can be used as a reason in other proofs. The conjecture proven above is known as the Midpoint Theorem.

Theorem 12.1: Midpoint Theorem

If M is the midpoint of $\overline{AB}$, then $\overline{AM} \cong \overline{MB}$.

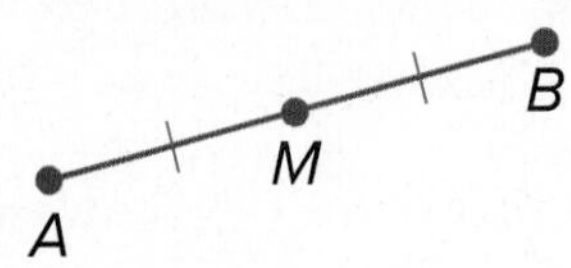

Study Tip

Midpoint Theorem and Definition The definition of *midpoint* is in terms of equality, and the Midpoint Theorem is in terms of congruence.

Check

Copy and complete the two-column proof by selecting the correct statements and reasons.

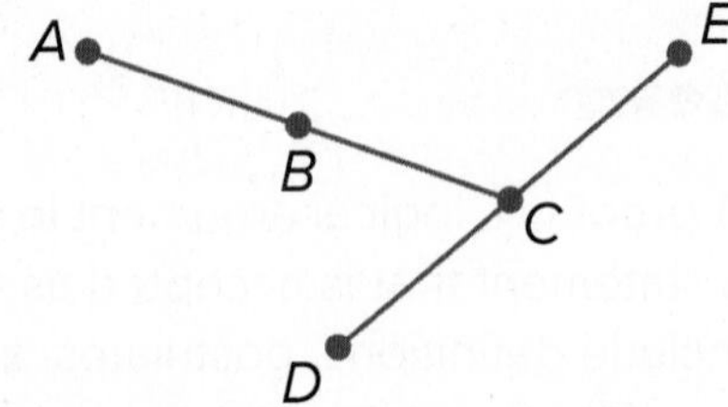

Given: B is the midpoint of $\overline{AC}$. C is the midpoint of $\overline{DE}$. $AB = CE$.

Prove: $BC = DC$

STATEMENTS/REASONS:
$AB = BC, DC = CE$
Substitution
$BC = CE$
Transitive Property
$AC = DE$
$AB = DC, BC = CE$

Statements	Reasons
1. B is the midpoint of $\overline{AC}$. C is the midpoint of $\overline{DE}$.	1. Given
2. ?	2. Definition of midpoint
3. $AB = CE$	3. Given
4. ?	4. ?
5. $BC = DC$	5. Substitution

Learn Flow Proofs

A **flow proof** uses boxes and arrows to show the logical progression of an argument. The statement is in the box, and the reason is below it. Arrows indicate the order of the steps.

Go Online You can complete an Extra Example online.

Example 4 Flow Proofs

Write each statement and reason in the correct box to complete the flow proof.

Given: P is the midpoint of $\overline{JK}$.

Prove: $x = 2$

Proof:

P is the midpoint of $\overline{JK}$.
Given

$\overline{JP} \cong \overline{PK}$
Midpoint Theorem

$JP = PK$
Definition of congruence

$x = 5x - 8$
Substitution

$x - 5x = 5x - 5x - 8$
Subtraction Property

$-4x = -8$
Substitution

$\frac{-4x}{-4} = \frac{-8}{-4}$
Division Property

$x = 2$
Substitution

STATEMENTS/REASONS:

$x = 2$

Substitution

$x = 5x - 8$

Substitution

$\overline{JP} \cong \overline{PK}$

Midpoint Theorem

Think About It!

Can you eliminate a step from the proof? Explain.

Check

Copy and complete the flow diagram by writing each statement and reason in the correct box.

Given: $FG = HK$

Prove: $x = 7$

Proof:

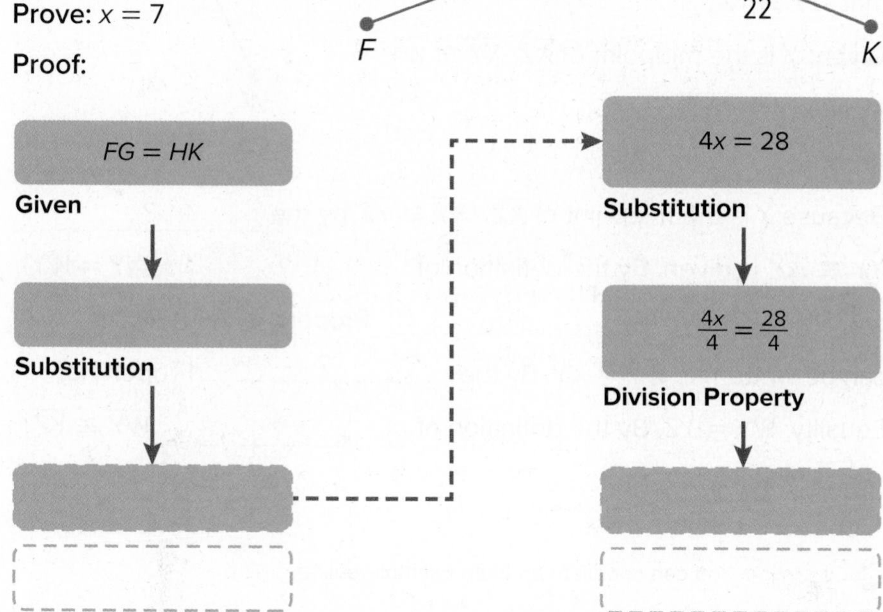

STATEMENTS/REASONS:

Addition Property

$x = 7$

Substitution

$4x - 6 = 22$

$4x - 6 + 6 = 22 + 6$

Learn Paragraph Proofs

Another way to prove a conjecture is to write a paragraph that explains why the conjecture for a given situation is true. This is called a **paragraph proof**. A paragraph proof includes the theorems, definitions, or postulates that support each statement.

Example 5 Paragraph Proof

Given that C is between A and B and $\overline{AC} \cong \overline{CB}$, write a paragraph proof to show that C is the midpoint of $\overline{AB}$.

Step 1: Write the given and prove statements.

Given: C is between A and B and $\overline{AC} \cong \overline{CB}$.

Prove: C is the midpoint of $\overline{AB}$.

Step 2: Draw a diagram and label any given information.

Example:

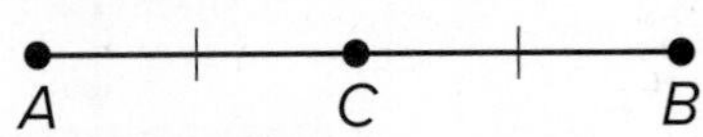

Step 3: Write the proof.

If C is between points A and B, then by the definition of betweenness, A, B, and C are collinear and $AC + CB = AB$. If $\overline{AC} \cong \overline{CB}$, then by the definition of congruence, the segments have the same measure, which means that $AC = CB$. From the definition of midpoint of a segment, if C is between points A and B and $AC = CB$, then C is the midpoint of $\overline{AB}$.

Avoid a Common Error

A common error in writing proofs is to skip steps in the logical progression of the argument.

Study Tip

Properties of Equality
The following properties are true for any real numbers a, b, and c.

Reflexive Property of Equality	$a = a$
Symmetric Property of Equality	If $a = b$, then $b = a$.
Transitive Property of Equality	If $a = b$ and $b = c$, then $a = c$.

Check

Given that Y is the midpoint of $\overline{XZ}$ and $\overline{XY} \cong \overline{WY}$, write a paragraph proof to show that $\overline{WY} \cong \overline{YZ}$.

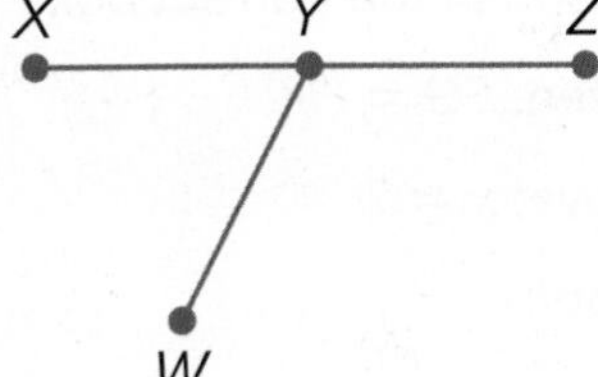

Given: Y is the midpoint of $\overline{XZ}$; $\overline{XY} \cong \overline{WY}$

Prove: $\overline{WY} \cong \overline{YZ}$

Proof:

Because Y is the midpoint of $\overline{XZ}$, $\overline{XY} \cong \overline{YZ}$ by the ___?___ $\overline{XY} \cong \overline{WY}$ is given. By the definition of ___?___, $XY = WY$ and $XY = YZ$. By the ___?___ Property of Equality, $XY = WY$ can be written as $WY = XY$. By the ___?___ Property of Equality, $WY = YZ$. By the definition of ___?___ $\overline{WY} \cong \overline{YZ}$.

Go Online You can complete an Extra Example online.

Practice

Go Online You can complete your homework online.

Example 1

MUSIC **Explain how the figure illustrates that each statement is true. Then state the postulate that can be used to show that each statement is true.**

1. Planes O and M intersect in line t.

2. Line p lies in plane N.

SIGNS **In the figure, $\overleftrightarrow{DG}$ and $\overleftrightarrow{DP}$ are in plane *J* and *H* lies on $\overleftrightarrow{DG}$. State the postulate that can be used to show that each statement is true.**

3. Points *G* and *P* are collinear.

4. Points *D*, *H*, and *P* are coplanar.

Example 2

CONSTRUCT ARGUMENTS **Determine whether each statement is *always*, *sometimes*, or *never* true. Justify your argument.**

5. There is exactly one plane that contains noncollinear points *A*, *B*, and *C*.

6. There are at least three lines through points *J* and *K*.

7. If points *M*, *N*, and *P* lie in plane *X*, then they are collinear.

8. Points *X* and *Y* are in plane *Z*. Any point collinear with points *X* and *Y* is in plane *Z*.

9. The intersection of two planes can be a point.

10. Points *A*, *B*, and *C* determine a plane.

Example 3

11. PROOF Point *Y* is the midpoint of $\overline{XZ}$. Point *W* is collinear with *X*, *Y*, and *Z*. *Z* is the midpoint of $\overline{YW}$. Write a two-column proof to prove that $\overline{XY} \cong \overline{ZW}$.

12. PROOF Write a two-column proof to prove that $w = 3.5$.

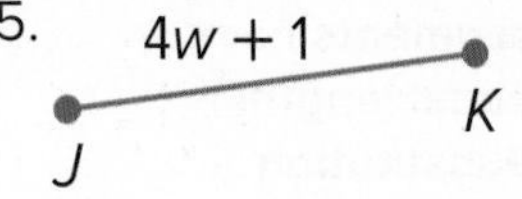

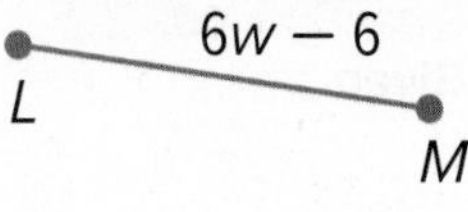

Given: $\overline{JK} \cong \overline{LM}$

Prove: $w = 3.5$

13. **PROOF** Copy and complete the two-column proof.

Given: $SR = RT$, $SR = UR$, and $RT = RV$

Prove: R is the midpoint of $\overline{ST}$. R is the midpoint of $\overline{UV}$.

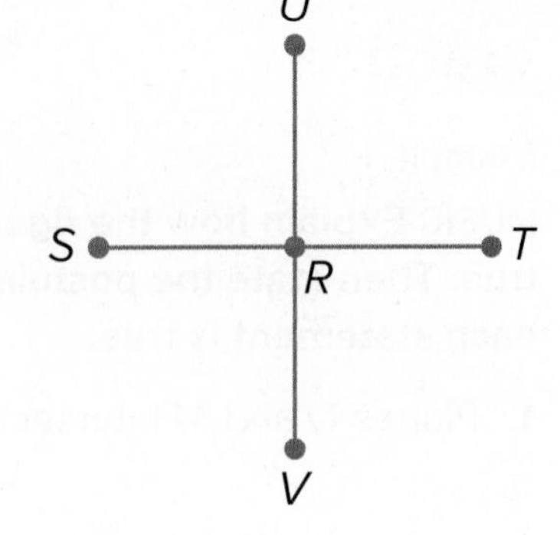

Statements	Reasons
1. $SR = RT$	1. ___?___
2. ___?___	2. Definition of midpoint
3. $SR = UR$ and $RT = RV$	3. ___?___
4. ___?___	4. Substitution
5. ___?___	5. ___?___

14. **PROOF** Copy and complete the two-column proof to prove that $x = 1.25$.

Given: H is the midpoint of $\overline{FG}$.

Prove: $x = 1.25$

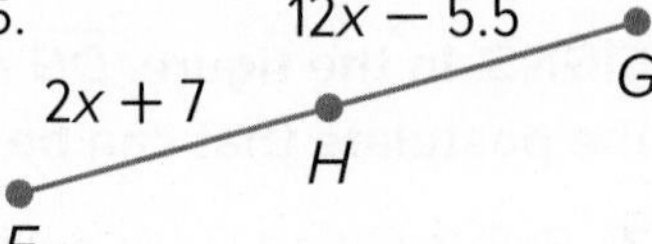

Statements	Reasons
1. H is the midpoint of $\overline{FG}$.	1. ___?___
2. ___?___	2. Midpoint Theorem
3. ___?___	3. Congruent segments have equal lengths.
4. $2x + 7 = 12x - 5.5$	4. ___?___
5. ___?___	5. Addition Property of Equality
6. ___?___	6. ___?___
7. ___?___	7. Division Property of Equality
8. $x = 1.25$	8. ___?___

Example 4

15. **PROOF** Point L is the midpoint of $\overline{JK}$. $\overline{JK}$ intersects $\overline{MK}$ at K. If $\overline{MK} \cong \overline{JL}$, write a flow proof to prove that $\overline{LK} \cong \overline{MK}$.

16. **PROOF** Copy and complete the flow proof to prove that if $\overline{MN} \cong \overline{PQ}$, $MN = 5x - 10$, and $PQ = 4x + 10$, then $MN = 90$.

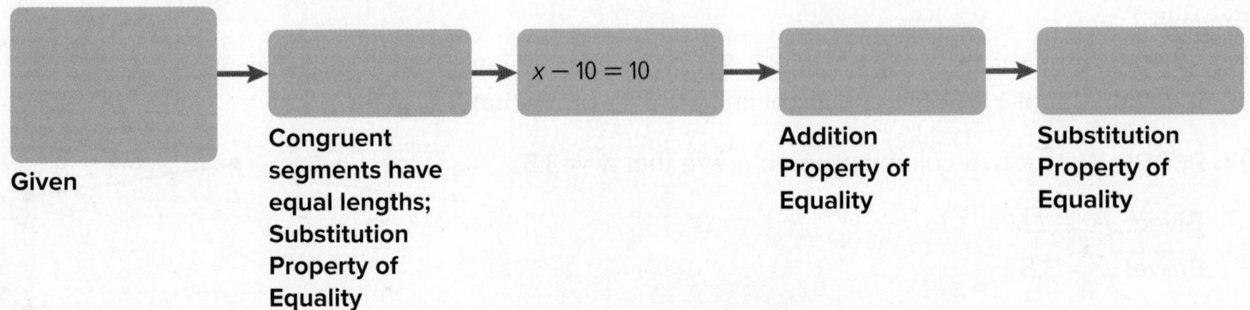

Example 5

17. PROOF In the figure at the right, point B is the midpoint of $\overline{AC}$ and point C is the midpoint of $\overline{BD}$. Write a paragraph proof to prove that $AB = CD$.

A B C D

18. PROOF Write a paragraph proof to prove that if $PQ = 4(x - 3) + 1$, $QR = x + 10$, and $x = 7$, then $\overline{PQ} \cong \overline{QR}$.

$4(x - 3) + 1$ $x + 10$

P Q R

Mixed Exercises

19. What postulate can be used to show the following statement is true?
Line m contains points A and F.

20. ROOFING Fai and Max are building a new roof. They wanted a roof with two sloping planes that intersect in a curved arch. Is this possible?

21. Carson claims that a line will always intersect a plane at only one point, and he draws this picture to show his reasoning. Iza thinks it is possible for a line to intersect a plane at more than one point. Who is correct? Explain.

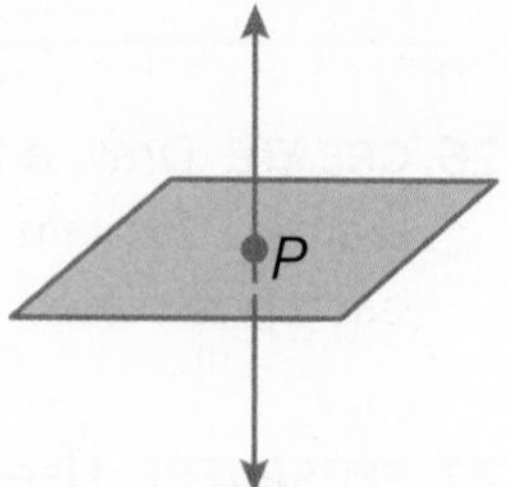

22. REASONING The figure shows a straight portion of the course for a city marathon. The water station W is located at the midpoint of $\overline{AB}$.

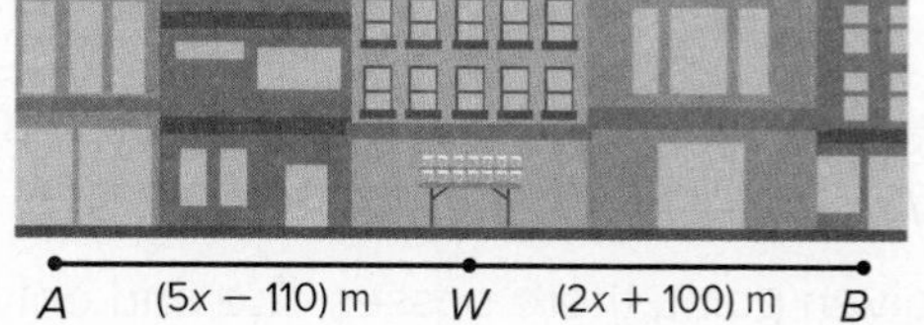

A $(5x - 110)$ m W $(2x + 100)$ m B

a. What is the length of the course from point A to point W?

b. Write a paragraph proof for your answer to part **a**.

c. Explain how you used a definition in your paragraph proof.

23. AIRLINES An airline company wants to provide service to San Francisco, Los Angeles, Chicago, Dallas, Washington D.C., and New York City. The company's president draws lines between each pair of cities in the list on a map. No three of the cities are collinear. How many lines did the president draw?

24. SMALL BUSINESSES A small company has 16 employees. The owner placed 16 points on a sheet of paper in such a way that no 3 were collinear. Each point represented a different employee. He then connected two points with a line segment if they represented coworkers in the same department.

a. What is the maximum number of line segments that can be drawn between pairs among the 16 points?

b. When the owner finished the diagram, he found that his company was split into two groups, one with 10 people and the other with 6. All the people within a group were in the same department, but nobody from one group was from the other group. How many line segments were there?

Higher-Order Thinking Skills

25. FIND THE ERROR Omair and Ana were working on a paragraph proof to prove that if $\overline{AB}$ is congruent to $\overline{BD}$ and A, B, and D are collinear, then B is the midpoint of $\overline{AD}$. Each student started his or her proof in a different way. Is either of them correct? Explain your reasoning.

Omair	Ana
If B is the midpoint of $\overline{AD}$, then B divides $\overline{AD}$ into two congruent segments.	$\overline{AB}$ is congruent to $\overline{BD}$, and A, B, and D are collinear.

26. CREATE Draw a figure that satisfies five of the seven postulates you have learned. Explain which postulates you chose and how your figure satisfies each postulate.

27. PERSEVERE Use the following true statements and the definitions and postulates you have learned to answer each question.

Two planes are perpendicular if and only if one plane contains a line perpendicular to the second plane.

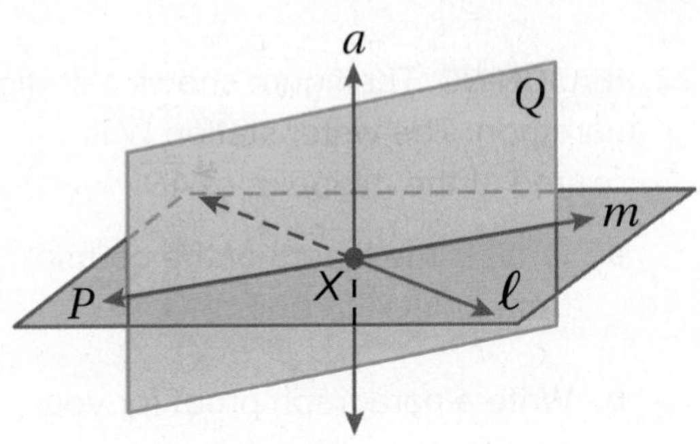

a. Through a given point, there passes one and only one plane perpendicular to a given line. If plane Q is perpendicular to line ℓ at point X and line ℓ lies in plane P, what must also be true?

b. Through a given point, there passes one and only one line perpendicular to a given plane. If plane Q is perpendicular to plane P at point X and line a lies in plane Q, what must also be true?

28. WRITE How does writing a proof require logical thinking?

ANALYZE Determine whether each statement is *sometimes*, *always*, or *never* true. Justify your argument.

29. Through any three points, there is exactly one plane.

30. A plane contains at least two distinct lines.

Proving Segment Relationships

Explore Segment Relationships

Online Activity Use dynamic geometry software to complete the Explore.

INQUIRY How can you use what you have already learned to prove segment relationships?

Today's Goals

- Prove theorems about line segments by using the Segment Addition Postulate.
- Prove theorems about line segments by using properties of segment congruence.

Learn Segment Addition

When you use a ruler to measure the length of an object, you match the mark for zero at one endpoint of the object. Then you look for the ruler mark that corresponds to the other endpoint. This illustrates the Ruler Postulate.

Postulate 12.8: Ruler Postulate	
Words	The points on any line or line segment can be put into one-to-one correspondence with real numbers.
Example	Given any two points A and B on a line, if A corresponds to zero, then B corresponds to a positive real number. 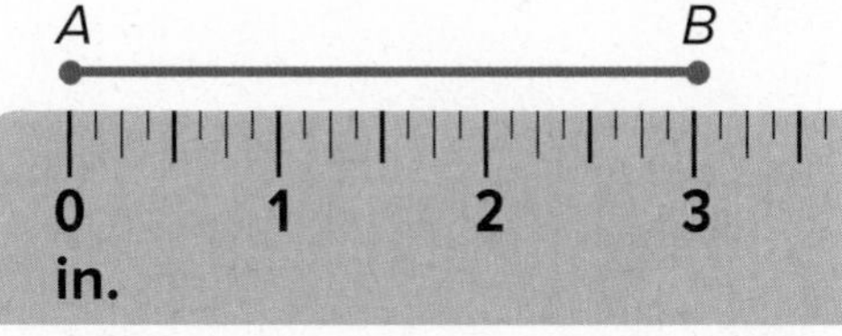

In this figure, point B is said to be between points A and C. You can also say that $AB + BC = AC$ by the Segment Addition Postulate.

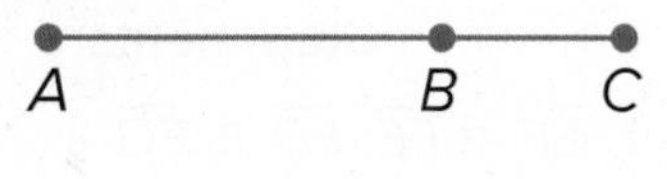

Postulate 12.9: Segment Addition Postulate	
Words	If A, B, and C are collinear, then point B is between A and C if and only if $AB + BC = AC$.
Example	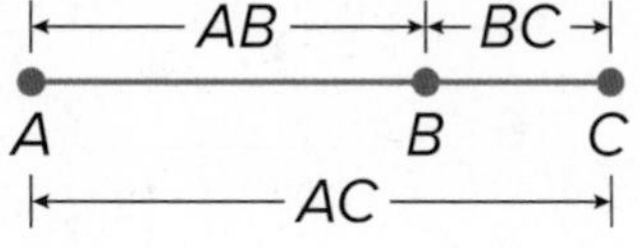

Think About It!

Determine whether the statement is true or false. If it is false, provide a counterexample.

If A, B, C, D, and E are collinear with $AC = 10$, B between A and C, C between B and D, D between C and E, and $AC = BD = CE$, then $AB = BC = DE$.

Example 1 Segment Addition Postulate

STATEMENTS/REASONS:

Segment Addition Postulate

$QT = RV$

Substitution Property of Equality

$QR + RT = RT + TV$

Write the correct statements and reasons to complete the two-column proof.

Given: $\overline{QT} \cong \overline{RV}$

Prove: $\overline{QR} \cong \overline{TV}$

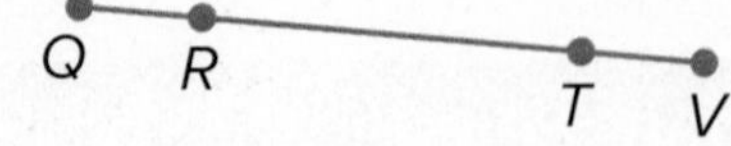

Proof:

Statements	Reasons
1. $\overline{QT} \cong \overline{RV}$	1. Given
2. $QT = RV$	2. Definition of congruence
3. $QR + RT = QT$; $RT + TV = RV$	3. Segment Addition Postulate
4. $QR + RT = RT + TV$	4. Substitution Property of Equality
5. $QR + RT - RT = RT + TV - RT$	5. Subtraction Property of Equality
6. $QR = TV$	6. Substitution Property of Equality
7. $\overline{QR} \cong \overline{TV}$	7. Definition of congruence

Check

REASONS:

Definition of congruence

Given

Segment Addition Postulate

Substitution Property of Equality

Transitive Property of Equality

Copy and complete the two-column proof by writing the correct statement and reason.

Given: $\overline{CE} \cong \overline{FE}$; $\overline{ED} \cong \overline{EG}$

Prove: $\overline{CD} \cong \overline{FG}$

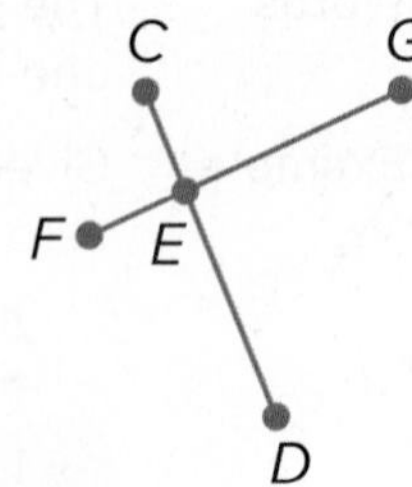

Proof:

Statements	Reasons
1. $\overline{CE} \cong \overline{FE}$; $\overline{ED} \cong \overline{EG}$	1. ___?___
2. $CE = FE$; $ED = EG$	2. ___?___
3. $CE + ED = CD$	3. ___?___
4. $FE + EG = CD$	4. ___?___
5. $FE + EG = FG$	5. ___?___
6. $CD = FG$	6. ___?___
7. $\overline{CD} \cong \overline{FG}$	7. ___?___

Go Online You can complete an Extra Example online.

Learn Segment Congruence

You learned that segment measures are reflexive, symmetric, and transitive. Because segments with the same measure are congruent, these properties apply to segment congruence.

Theorem 12.2: Properties of Segment Congruence

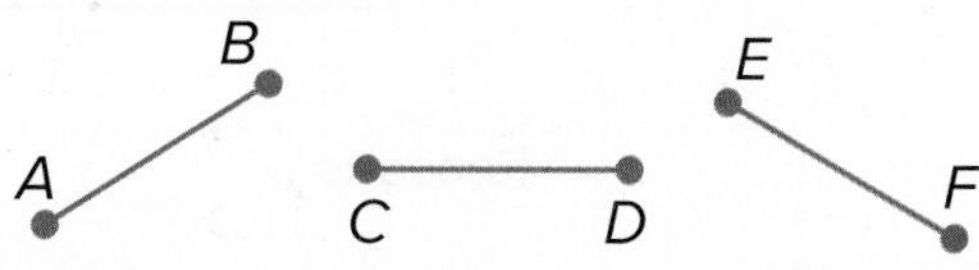

Reflexive Property of Congruence	$\overline{AB} \cong \overline{AB}$
Symmetric Property of Congruence	If $\overline{AB} \cong \overline{CD}$, then $\overline{CD} \cong \overline{AB}$.
Transitive Property of Congruence	If $\overline{AB} \cong \overline{CD}$, and $\overline{CD} \cong \overline{EF}$, then $\overline{AB} \cong \overline{EF}$.

You will prove the Reflexive and Symmetric Properties of Congruence in Exercise 12.

Go Online A proof of Theorem 12.2 is available.

Talk About It! Is there an Addition Property of Congruence? Explain.

Example 2 Prove Segment Congruence

Write the correct statement and reasons to complete the two-column proof.

Given: R is the midpoint of $\overline{QS}$.
T is the midpoint of $\overline{VS}$.
$\overline{QR} \cong \overline{VT}$

Prove: $\overline{RS} \cong \overline{TS}$

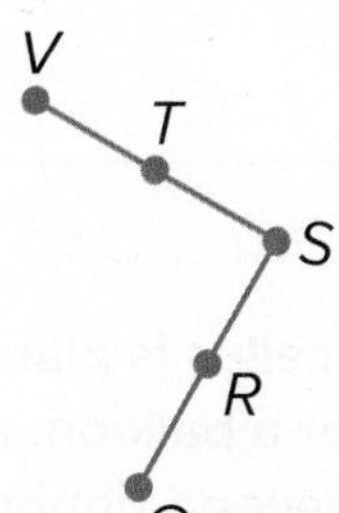

Proof:

Statements	Reasons
1. R is the midpoint of $\overline{QS}$. T is the midpoint of $\overline{VS}$.	1. Given
2. $\overline{QR} \cong \overline{RS}$; $\overline{VT} \cong \overline{TS}$	2. Midpoint Theorem
3. $\overline{QR} \cong \overline{VT}$	3. Given
4. $\overline{QR} \cong \overline{TS}$	4. Transitive Property of Congruence
5. $\overline{RS} \cong \overline{QR}$	5. Symmetric Property of Congruence
6. $\overline{RS} \cong \overline{TS}$	6. Transitive Property of Congruence

STATEMENTS/REASONS

Midpoint Theorem

$\overline{RS} \cong \overline{QR}$

Transitive Property of Congruence

 **Go Online** You can complete an Extra Example online.

REASONS:
Add.
Definition of congruence
Divide by 2.
Given
Midpoint Theorem
Segment Addition Postulate
Substitution Property of Equality

Check

Copy and complete the two-column proof by writing the correct reasons.

Given: $\overline{GJ} \cong \overline{GI}$
K is the midpoint of $\overline{GJ}$.
H is the midpoint of $\overline{GI}$.

Prove: $\overline{GK} \cong \overline{GH}$

Proof:

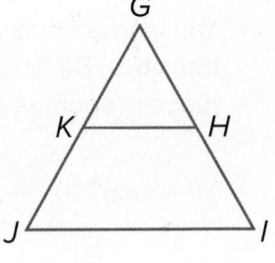

Statements	Reasons
1. K is the midpoint of $\overline{GJ}$; H is the midpoint of $\overline{GI}$	1. ?
2. $\overline{GK} \cong \overline{KJ}$; $\overline{GH} \cong \overline{HI}$	2. ?
3. $GK = KJ$; $GH = HI$	3. ?
4. $\overline{GJ} \cong \overline{GI}$	4. ?
5. $GJ = GI$	5. ?
6. $GJ = GK + KJ$; $GI = GH + HI$	6. ?
7. $GK + KJ = GH + HI$	7. ?
8. $GK + GK = GH + GH$	8. ?
9. $2GK = 2GH$	9. ?
10. $GK = GH$	10. ?
11. $\overline{GK} \cong \overline{GH}$	11. ?

Example 3 Determine Congruence

CITY PLANNING **Marcellus is planning a birthday party. He measures a length of ribbon for a balloon, and then uses this ribbon to measure and cut a second ribbon. He continues this pattern of using the last ribbon that he cut to measure the next ribbon until 10 ribbons have been cut for balloons. Is the last ribbon that he cut the same length as the first ribbon? Justify your argument.**

Yes; because the first ribbon is the same length as the second ribbon and the third ribbon is the same length as the second ribbon, the first ribbon is the same length as the third ribbon by the Transitive Property of Congruence.

This logic can be applied until the last ribbon is shown to be the same length as the first ribbon.

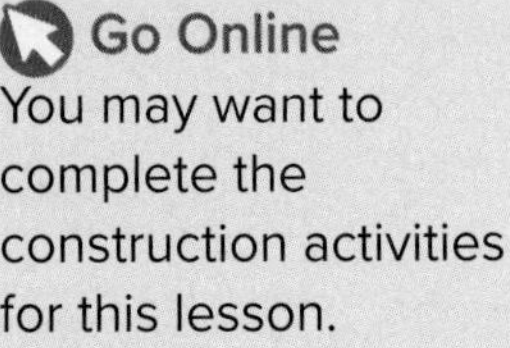

Go Online You may want to complete the construction activities for this lesson.

Check

CITY PLANNING A city council plans to convert a section of a city block into green space for the community. The north sidewalk is congruent to the south sidewalk, which is congruent to the west sidewalk. Therefore, the north sidewalk is congruent to the west sidewalk. What theorem, postulate, or property justifies this statement?

Go Online You can complete an Extra Example online.

Practice

Go Online You can complete your homework online.

Example 1

1. **PROOF** Write the correct statements and reasons to complete the two-column proof.

Given: C is the midpoint of $\overline{AE}$.
C is the midpoint of $\overline{BD}$.
$\overline{AE} \cong \overline{BD}$

Prove: $\overline{AC} \cong \overline{CD}$

Proof:

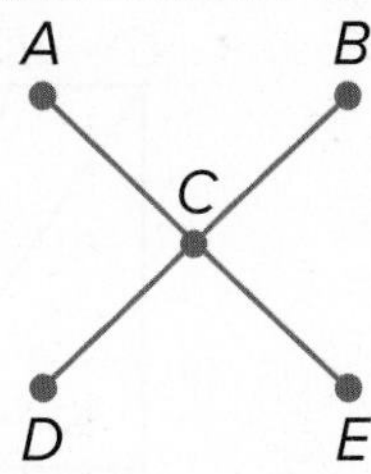

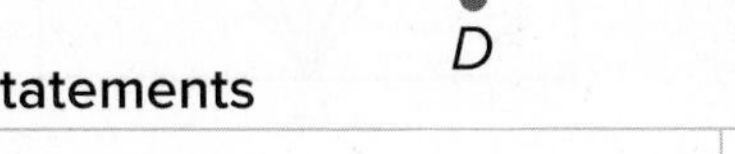

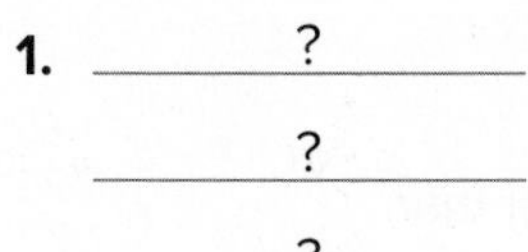

Statements	Reasons
1. ___?___ ___?___ ___?___	1. Given
2. $AC = CE$ $BC = CD$	2. ___?___
3. $AE = BD$	3. ___?___
4. ___?___ ___?___	4. Segment Addition Property
5. $AC + CE = BC + CD$	5. ___?___
6. $AC + AC = CD + CD$	6. ___?___
7. ___?___	7. Simplify.
8. ___?___	8. Division Property
9. $\overline{AC} \cong \overline{CD}$	9. ___?___

2. **PROOF** Write the correct statements and reasons to complete the two-column proof.

Given: $\overline{SU} \cong \overline{LR}$
$\overline{TU} \cong \overline{LN}$

Prove: $\overline{ST} \cong \overline{NR}$

Proof:

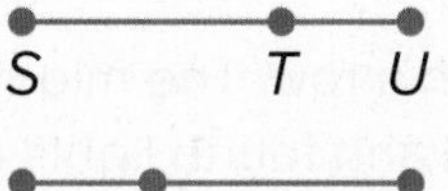

Statements	Reasons
1. $\overline{SU} \cong \overline{LR}$, $\overline{TU} \cong \overline{LN}$	1. ___?___
2. ___?___	2. Definition of $\cong$ segments
3. $SU = ST + TU$ $LR = LN + NR$	3. ___?___
4. $ST + TU = LN + NR$	4. ___?___
5. $ST + LN = LN + NR$	5. ___?___
6. $ST + LN - LN = LN + NR - LN$	6. ___?___
7. ___?___	7. Substitution Property
8. $\overline{ST} \cong \overline{NR}$	8. ___?___

Example 2

PROOF Write a two-column proof to prove each geometric relationship.

3. If $\overline{VZ} \cong \overline{VY}$ and $\overline{WY} \cong \overline{XZ}$, then $\overline{VW} \cong \overline{VX}$.

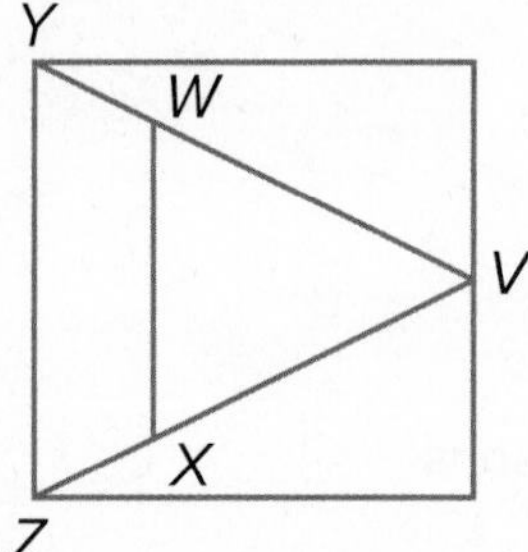

4. If E is the midpoint of $\overline{DF}$ and $\overline{CD} \cong \overline{FG}$, then $\overline{CE} \cong \overline{EG}$.

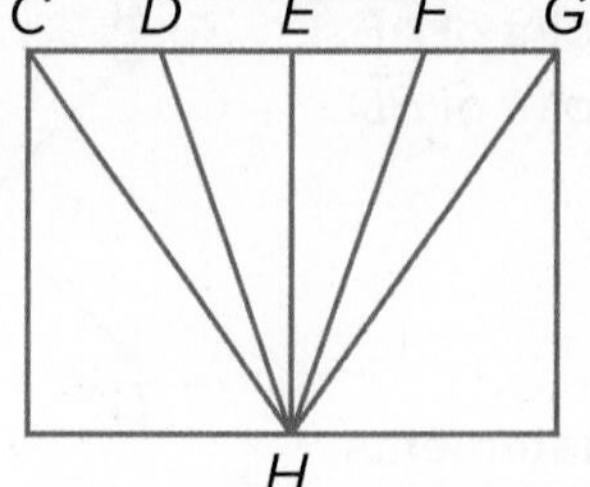

Example 3

5. **FAMILY** Maria is 11 inches shorter than her sister Clara. Luna is 11 inches shorter than her brother Chad. If Maria is shorter than Luna, how do the heights of Clara and Chad compare? What else can be concluded if Maria and Luna are the same height?

6. **LUMBER** Byron works in a lumberyard. His boss just cut a dozen planks and asked Byron to double check that they are all the same length. The planks were numbered 1 through 12. Byron took out plank number 1 and checked that the other planks are all the same length as plank 1. He concluded that they must all be the same length. Explain how you know that plank 7 and plank 10 are the same length even though they were never directly compared to each other.

7. **NEIGHBORHOODS** Karla, Lola, and Mandy live in three houses that are on the same line. Lola lives between Karla and Mandy. Karla and Mandy live a mile apart. Is it possible for Lola's house to be a mile from both Karla's and Mandy's houses?

Mixed Exercises

8. **PROOF** Five lights, A, B, C, D, and E, are aligned in a row. The middle light is the midpoint of the segment between the second and fourth lights and also the midpoint of the segment between the first and last lights.

a. Draw a figure to illustrate the situation.

b. Complete this proof.

Given: C is the midpoint of $\overline{BD}$ and $\overline{AE}$.

Prove: $AB = DE$

Statement	Reason
1. C is the midpoint of $\overline{BD}$ and $\overline{AE}$.	1. Given
2. $BC = CD$ and ___?___	2. ___?___
3. $AC = AB + BC$, $CE = CD + DE$	3. ___?___
4. $AC - BC = AB$	4. ___?___
5. ___?___	5. Substitution Property
6. $CE - CD = DE$	6. ___?___
7. $AB = CE - CD$	7. Symmetric Property of Equality
8. ___?___	8. ___?___

9. **PROOF** $\overline{AC} \cong \overline{GI}$, $\overline{FE} \cong \overline{LK}$, and $AC + CF + FE = GI + IL + LK$. Prove that $\overline{CF} \cong \overline{IL}$.

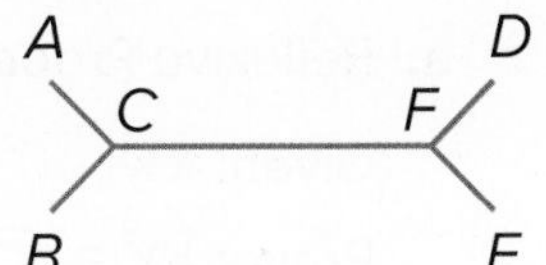

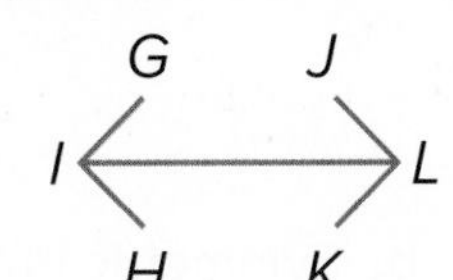

10. **PROOF** Consider $\overleftrightarrow{PS}$.

   a. Complete the two-column proof.

   Given: $\overline{PQ} \cong \overline{RS}$

   Prove: $\overline{PR} \cong \overline{QS}$

Statement	Reason
1. $\overline{PQ} \cong \overline{RS}$	1. ___?___
2. ___?___	2. Congruent segments have equal lengths.
3. $PQ + QR = PR$ and $QR + RS = QS$	3. ___?___
4. $RS + QR = PR$	4. ___?___
5. $QR + RS = PR$	5. ___?___
6. $QS = PR$	6. ___?___
7. $PR = QS$	7. Symmetric Property of Equality
8. ___?___	8. Segments with equal lengths are congruent.

   b. Can it also be proved that $\overline{PQ} \cong \overline{RS}$ if $\overline{PR} \cong \overline{QS}$? Explain.

11. **PROOF** A city planner is designing a new park. The park has two straight paths, $\overline{AB}$ and $\overline{CD}$, which are the same length. A monument, M, is located at the midpoint of both paths.

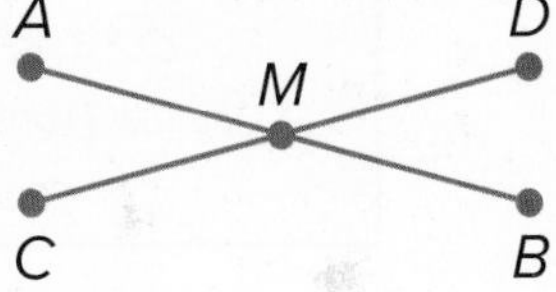

   a. The city planner thinks that the length of $\overline{AM}$ will be the same as the length of $\overline{CM}$. Explain why this makes sense.

   b. Complete the two-column proof.

   Given: $\overline{AB} \cong \overline{CD}$; M is the midpoint of $\overline{AB}$ and $\overline{CD}$.

   Prove: $\overline{AM} \cong \overline{CM}$

Statement	Reason
1. ___?___	1. Given
2. $AB = CD$	2. ___?___
3. $\overline{AM} \cong \overline{MB}$; $\overline{CM} \cong \overline{MD}$	3. ___?___
4. $AM = MB$; $CM = MD$	4. Congruent segments have equal lengths.
5. $AM + MB = AB$; $CM + MD = CD$	5. ___?___
6. $AM + MB = CM + MD$	6. ___?___
7. $AM + AM = CM + CM$	7. Substitution Property of Equality
8. $2AM = 2CM$	8. ___?___
9. ___?___	9. Division Property of Equality
10. ___?___	10. Segments with equal lengths are congruent.

12. PROOF Write a paragraph proof for each property of segment congruence.

a. Reflexive Property of Segment Congruence

Given: $\overline{XY}$

Prove: $\overline{XY} \cong \overline{XY}$

b. Symmetric Property of Segment Congruence

Given: $\overline{AB} \cong \overline{CD}$

Prove: $\overline{CD} \cong \overline{AB}$

Higher-Order Thinking Skills

13. FIND THE ERROR In the diagram, $\overline{AB} \cong \overline{CD}$ and $\overline{CD} \cong \overline{BF}$. Examine the conclusions made by Leslie and Shantice. Is either of them correct? Explain your reasoning.

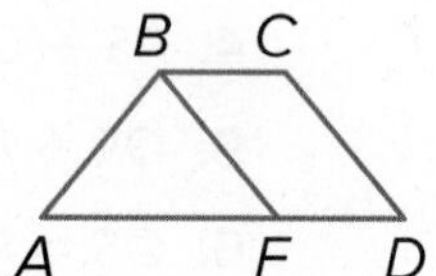

Leslie	Shantice
Because $\overline{AB} \cong \overline{CD}$ and $\overline{CD} \cong \overline{BF}$, $\overline{AB} \cong \overline{AF}$ by the Transitive Property of Congruence.	Because $\overline{AB} \cong \overline{CD}$ and $\overline{CD} \cong \overline{BF}$, $\overline{AB} \cong \overline{BF}$ by the Reflexive Property of Congruence.

14. PROOF $ABCD$ is a square. Prove that $\overline{AC} \cong \overline{BD}$.

15. CREATE Draw a representation of the Segment Addition Postulate in which the segment is two inches long, contains four collinear points, and contains no congruent segments.

16. CREATE Write an example of the Transitive Property and the Substitution Property that illustrates the difference between them.

17. FIND THE ERROR Justin knows that point R is the midpoint of $\overline{QS}$, and he knows that this means that $QR = RS$. He says that $PR = PQ + QR$ by the Segment Addition Postulate. So, $PR = PQ + RS$ by substitution. Do you agree with Justin's reasoning? Explain your reasoning.

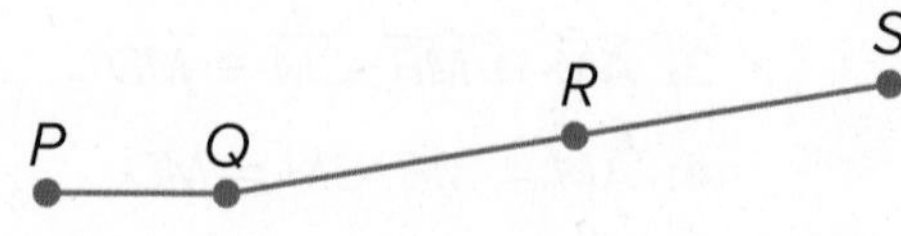

18. WRITE Compare and contrast paragraph proofs and two-column proofs.

19. PROOF Write a paragraph proof to prove that if P, Q, R, and S are collinear, $\overline{PQ} \cong \overline{RS}$, and Q is the midpoint of $\overline{PR}$, then R is the midpoint of $\overline{QS}$.

Proving Angle Relationships

Today's Goals

- Prove theorems about angles by using the Angle Addition Postulate.
- Prove theorems about angles by using properties and theorems of angle congruence.
- Prove theorems about right angles.

Explore Angle Relationships

 **Online Activity** Use dynamic geometry software to complete the Explore.

INQUIRY How is the complement of a given $\angle A$ related to an angle congruent to $\angle A$?

Learn Angle Addition

A protractor is used to measure angles. The Protractor Postulate illustrates the relationship between angle measures and real numbers. You will use these theorems and postulates to find angle measures.

Postulate 12.10: Protractor Postulate
The measure of any angle has a measure that is between 0 and 180.

Postulate 12.11: Angle Addition Postulate
D is in the interior of $\angle ABC$ if and only if $m\angle ABD + m\angle DBC = m\angle ABC$.

Theorem 12.3: Supplement Theorem
If two angles form a linear pair, then they are supplementary angles.

Theorem 12.4: Complement Theorem
If the noncommon sides of two adjacent angles form a right angle, then the angles are complementary angles.

You will prove Theorems 12.3 and 12.4 in Exercises 19–20.

Example 1 Angle Addition Postulate

What is $m\angle 3$ if $m\angle 1 = 23°$ and $m\angle ABC = 131°$?

Choose from the reasons provided to justify each step.

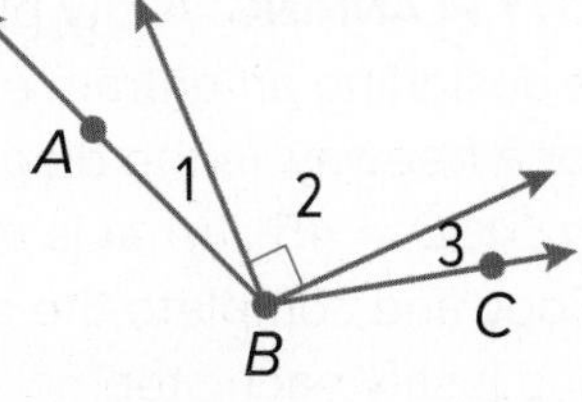

$m\angle 1 + m\angle 2 + m\angle 3 = m\angle ABC$	Angle Addition Postulate
$23° + 90° + m\angle 3 = 131°$	Substitution Property
$113° + m\angle 3 = 131°$	Substitution Property
$113° + m\angle 3 - 113° = 131° - 113°$	Subtraction Property
$m\angle 3 = 18°$	Substitution Property

Reasons:
Angle Addition Postulate
Betweenness of points
Subtraction Property
Substitution Property

Check

What is $m\angle 3$ if $m\angle 2 = 26°$ ___?___

What is $m\angle 4$ if $m\angle 5 = (2x°)$ and $m\angle 4 = (x + 9)°$ ___?___

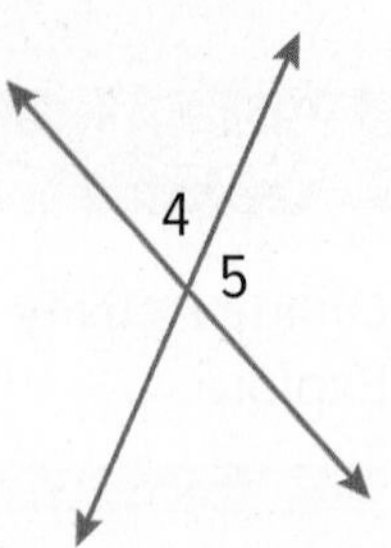

Example 2 Complement and Supplement Theorems

SHELVING **Mae Lin is installing shelves in her room. One of the brackets she chose for her shelves is shown. If $m\angle 3 = 55°$, what is $m\angle 4$?**

Choose from the reasons provided to justify each step.

REASONS:

Complement Theorem

Substitution Property

Subtraction Property

Supplement Theorem

$m\angle 3 + m\angle 4 = 180°$	Supplement Theorem
$55° + m\angle 4 = 180°$	Substitution Property
$m\angle 4 = 125°$	Subtraction Property

Check

CITY PLANNING A city planner is designing an entrance ramp for a freeway. In the diagram, $m\angle ACD = 45°$. What is $m\angle BCA$? Copy and complete the calculations and justify each step.

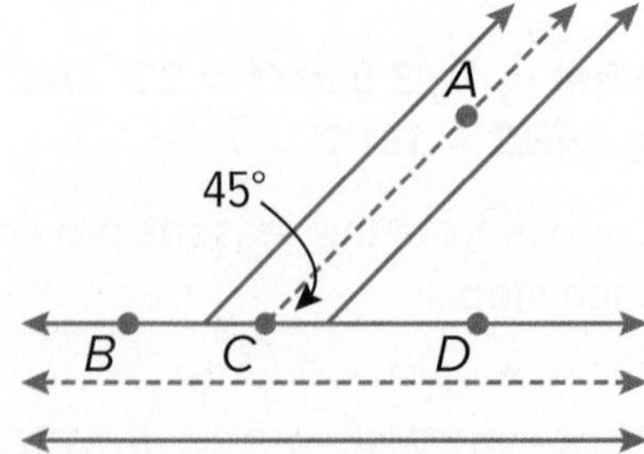

$m\angle BCA + m\angle ACD =$ ___?___

$m\angle BCA +$ ___?___ $=$ ___?___

$m\angle BCA =$ ___?___

Go Online You can complete an Extra Example online.

Learn Congruent Angles

The properties of algebra that apply to the congruence of segments and the equality of their measures also hold true for the congruence of angles and the equality of their measures.

Theorem 12.5: Properties of Angle Congruence	
Reflexive Property of Congruence $\angle 1 \cong \angle 1$	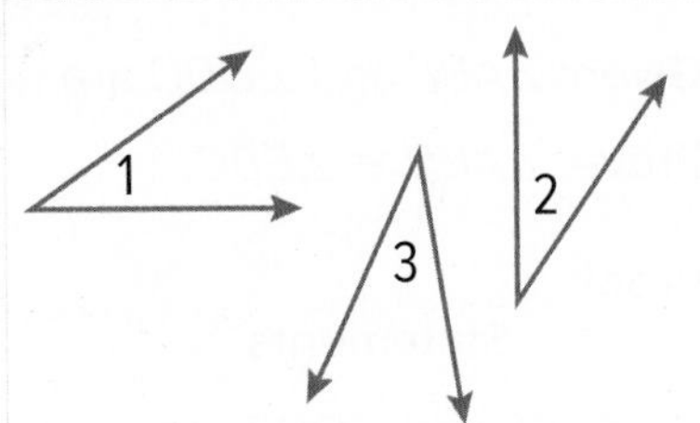
Symmetric Property of Congruence If $\angle 1 \cong \angle 2$, then $\angle 2 \cong \angle 1$.	
Transitive Property of Congruence If $\angle 1 \cong \angle 2$ and $\angle 2 \cong \angle 3$, then $\angle 1 \cong \angle 3$.	

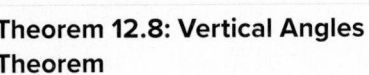

Go Online Proofs of the Reflexive Property of Congruence and the Transitive Property of Congruence are available.

Proof: Symmetric Property of Congruence

Given: $\angle J \cong \angle K$

Prove: $\angle K \cong \angle J$

Paragraph Proof:
We are given that $\angle J \cong \angle K$. By the definition of congruent angles, $m\angle J = m\angle K$. Using the Symmetric Property of Equality, $m\angle K = m\angle J$. Thus, $\angle K \cong \angle J$ by the definition of congruent angles.

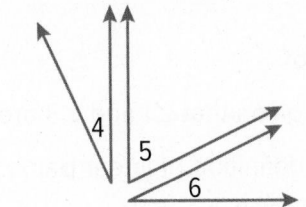

Theorems	
Theorem 12.6: Congruent Supplements Theorem Angles supplementary to the same angle or to congruent angles are congruent. **Abbreviation** $\angle$s suppl. to same $\angle$ or $\cong$ $\angle$s are $\cong$.	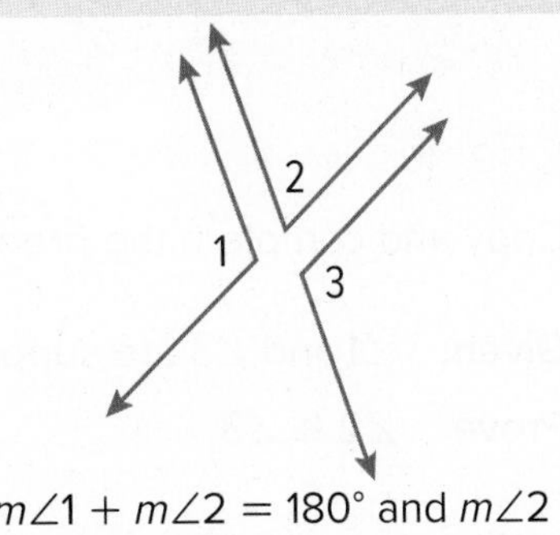 If $m\angle 1 + m\angle 2 = 180°$ and $m\angle 2 + m\angle 3 = 180°$, then $\angle 1 \cong \angle 3$.
Theorem 12.7: Congruent Complements Theorem Angles complementary to the same angle or to congruent angles are congruent. **Abbreviation** $\angle$s compl. to same $\angle$ or $\cong$ $\angle$s are $\cong$.	4 5 6 If $m\angle 4 + m\angle 5 = 90°$ and $m\angle 5 + m\angle 6 = 90°$, then $\angle 4 \cong \angle 6$.
Theorem 12.8: Vertical Angles Theorem If two angles are vertical angles, then they are congruent.	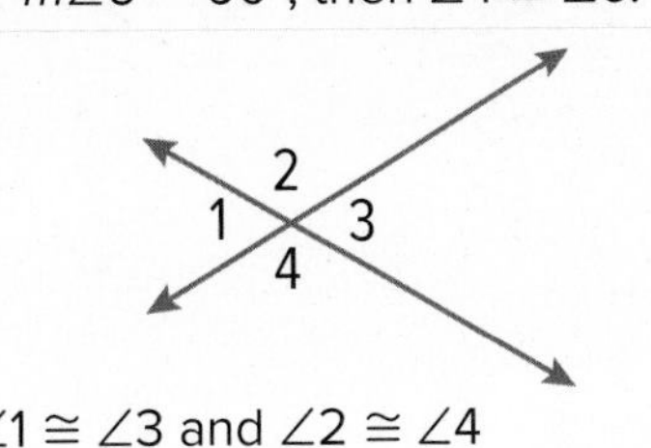 $\angle 1 \cong \angle 3$ and $\angle 2 \cong \angle 4$

Talk About It!
Explain the difference between the Complement Theorem and the Congruent Complements Theorem.

You will prove one case of Theorems 12.6 and 12.7 in Exercises 21-22. You will prove the second case of each theorem in Exercise 31.

Example 3 Congruent Supplements and Complements

In the figure, $\angle ABE$ and $\angle DBC$ are right angles.

Select from the reasons provided to complete the proof.

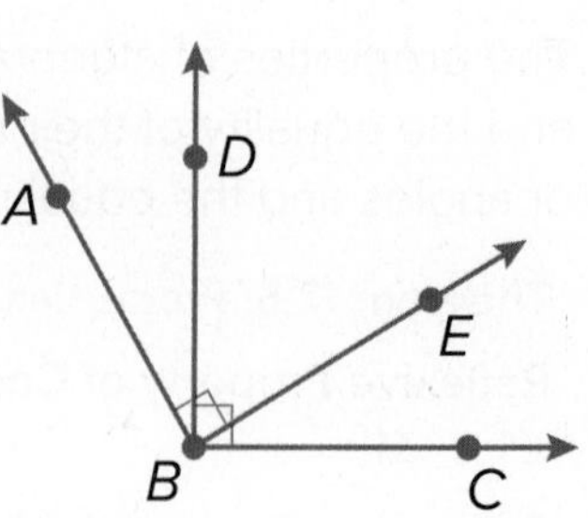

Given: $\angle ABE$ and $\angle DBC$ are right angles.

Prove: $\angle ABD \cong \angle EBC$

REASONS:
Complement Theorem
Congruent Complements Theorem
Definition of complementary angles
Definition of congruence

Proof:

Statements	Reasons
1. $\angle ABE$ and $\angle DBC$ are right angles.	1. Given
2. $m\angle ABE = 90°$; $m\angle DBC = 90°$	2. Definition of right angle
3. $\angle ABD$ and $\angle DBE$ are complementary; $\angle DBE$ and $\angle EBC$ are complementary.	3. Complement Theorem
4. $\angle ABD \cong \angle EBC$	4. Congruent Complements Theorem

Check

Copy and complete the proof.

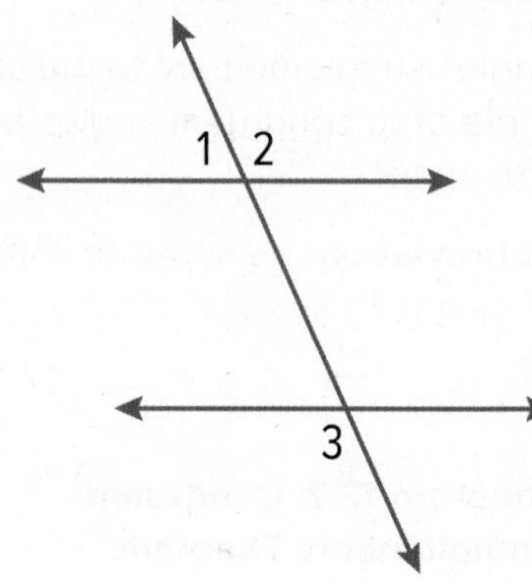

Given: $\angle 1$ and $\angle 3$ are supplementary.

Prove: $\angle 2 \cong \angle 3$

Proof:

It is given that $\angle 1$ and $\angle 3$ are ______?______. By the definition of linear pair, $\angle 1$ and $\angle 2$ are a linear pair. So, by the ______?______, $\angle 1$ and $\angle$ are supplementary. Thus, $\angle 2 \cong \angle 3$ by the ______?______.

Go Online You can complete an Extra Example online.

Example 4 Vertical Angles

Complete the proof.

Choose from the statements and reasons provided.

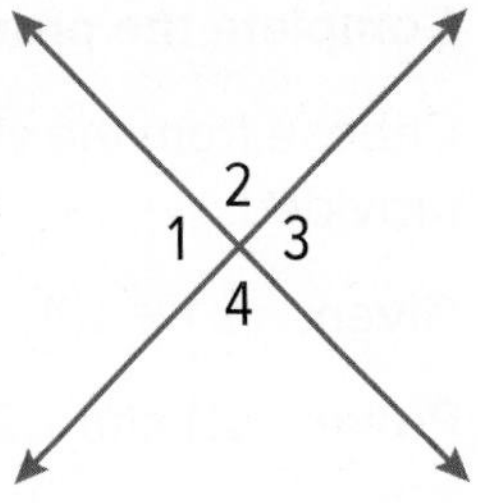

Given: $\angle 1 \cong \angle 2$

Prove: $\angle 3 \cong \angle 4$

Proof:

Statements	Reasons
1. $\angle 1 \cong \angle 2$	1. Given
2. $\angle 1 \cong \angle 3$; $\angle 2 \cong \angle 4$	2. Vertical Angles Theorem
3. $\angle 3 \cong \angle 1$	3. Symmetric Property of Congruence
4. $\angle 3 \cong \angle 2$	4. Transitive Property of Congruence
5. $\angle 3 \cong \angle 4$	5. Transitive Property of Congruence

STATEMENTS/REASONS:

Symmetric Property of Congruence

Transitive Property of Congruence

Vertical Angles Theorem

$\angle 3 \cong \angle 2$

$\angle 4 \cong \angle 2$

$\angle 4 \cong \angle 3$

Check

Copy and complete the proof. Choose from the reasons provided.

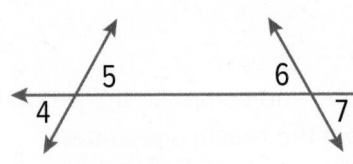

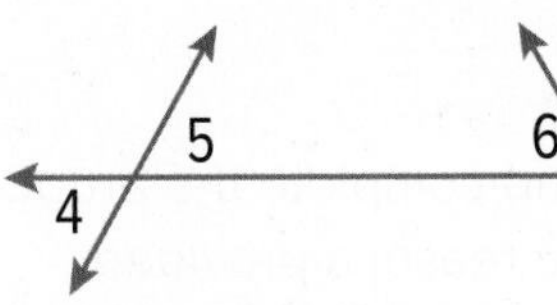

Given: $\angle 4 \cong \angle 7$

Prove: $\angle 5 \cong \angle 6$

Proof:

Statements	Reasons
1. $\angle 4 \cong \angle 7$	1. Given
2. $\angle 5 \cong \angle 4$ and $\angle 7 \cong \angle 6$	2. ___?___
3. $\angle 5 \cong \angle 7$	3. ___?___
4. $\angle 5 \cong \angle 6$	4. ___?___

REASONS:

Vertical Angles Theorem

Definition of vertical angles

Transitive Property of Congruence

Symmetric Property of Congruence

Supplement Theorem

Definition of linear pair

Go Online You can complete an Extra Example online.

Learn Right Angle Theorems

You can prove the following theorems about right angles using what you already know about angle measures.

Theorem 12.9	Perpendicular lines intersect to form four right angles.
Theorem 12.10	All right angles are congruent.
Theorem 12.11	Perpendicular lines form congruent adjacent angles.
Theorem 12.12	If two angles are congruent and supplementary, then each angle is a right angle.
Theorem 12.13	If two congruent angles form a linear pair, then they are right angles.

You will prove Theorem 12.9 and Theorems 12.11 through 12.13 in Exercises 23–26.

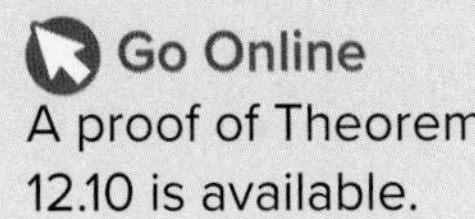

Go Online
A proof of Theorem 12.10 is available.

Example 5 Right Angle Theorems in Proofs

Complete the proof.

Choose from the statements and reasons provided.

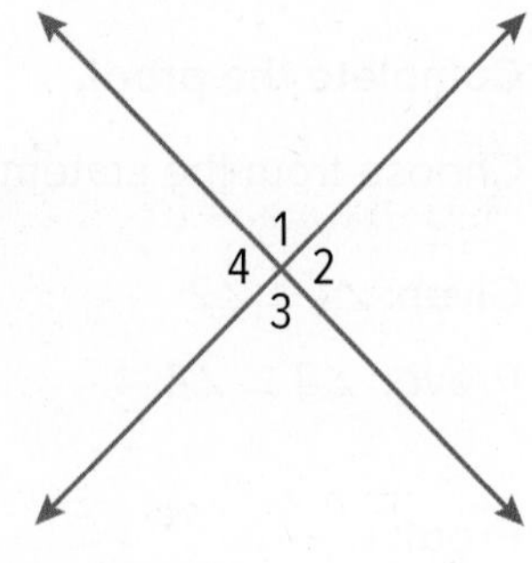

STATEMENTS/REASONS:

Reflexive Property of Congruence

Symmetric Property of Congruence

Transitive Property of Congruence

Vertical Angles Theorem

$\angle 1$ and $\angle 2$ are right angles. $\angle 1 \cong \angle 2$

Given: $\angle 1 \cong \angle 4$

Prove: $\angle 1$ and $\angle 2$ are right angles.

Proof:

Statements	Reasons
1. $\angle 1 \cong \angle 4$	1. Given
2. $\angle 2 \cong \angle 4$	2. Vertical Angles Theorem
3. $\angle 4 \cong \angle 2$	3. Symmetric Property of Congruence
4. $\angle 1 \cong \angle 2$	4. Transitive Property of Congruence
5. $\angle 1$ and $\angle 2$ are right angles.	5. If two angles are congruent and supplementary, then each angle is a right angle.

Check

Copy and complete the proof. Choose from the reasons provided.

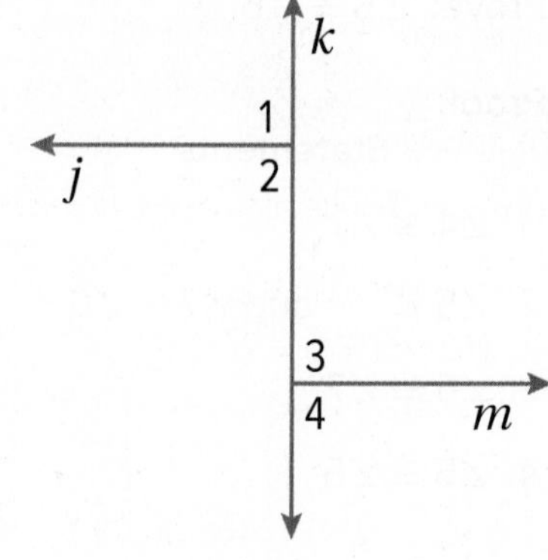

REASONS:

Perpendicular lines intersect to form four right angles.

All right angles are congruent.

Perpendicular lines form congruent adjacent angles.

If two angles are congruent and supplementary, then each angle is a right angle.

If two congruent angles form a linear pair, then they are right angles.

Given: $\angle 1 \cong \angle 4$
Lines j and k are perpendicular.

Prove: $\angle 2 \cong \angle 4$

Proof:

Statements	Reasons
1. Lines j and k are perpendicular.	1. Given
2. $\angle 2 \cong \angle 1$	2. ____?____
3. $\angle 1 \cong \angle 4$	3. Given
4. $\angle 2 \cong \angle 4$	4. Transitive Property of Congruence

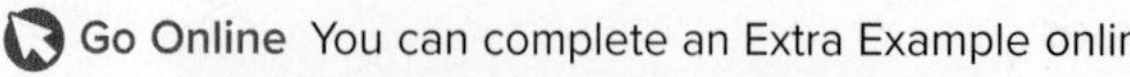

Practice

Go Online You can complete your homework online.

Example 1

Find the measure of each angle.

1. Find $m\angle ABC$ if $m\angle ABD = 70°$ and $m\angle DBC = 43°$.
2. If $m\angle EBC = 55°$ and $m\angle EBD = 20°$, find $m\angle 2$.
3. Find $m\angle ABD$ if $m\angle ABC = 110°$ and $m\angle 2 = 36°$.

Example 2

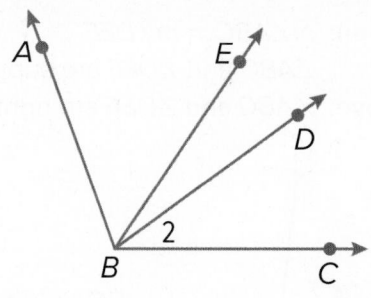

4. **FLAGS** The Alabama state flag is white and has two diagonal red stripes. If the $m\angle 1 = 112°$, what is $m\angle 2$?

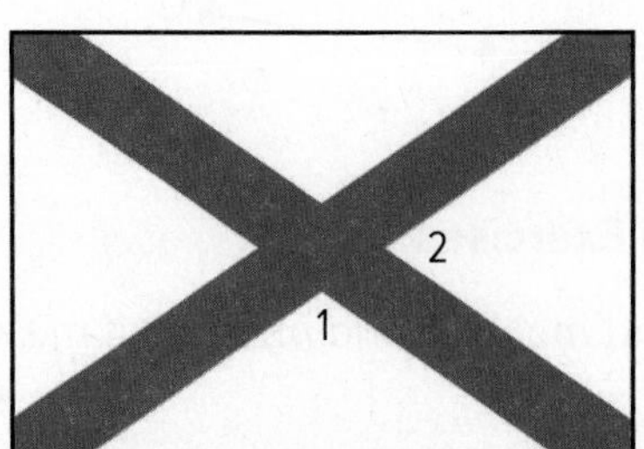

5. **CONSTRUCTION** Aaron has installed a new window above the entrance of an office building. If $m\angle 2 = 44°$, what is $m\angle 1$?

Example 3

PROOF **Write a two-column proof.**

6. **Given:** $\angle 2 \cong \angle 4$
   Prove: $\angle 1 \cong \angle 3$

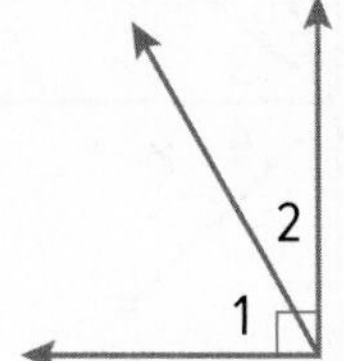

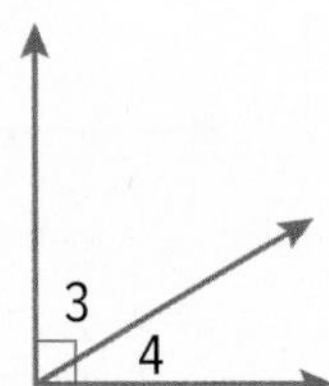

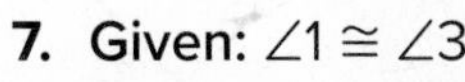

7. **Given:** $\angle 1 \cong \angle 3$
   Prove: $\angle 2 \cong \angle 4$

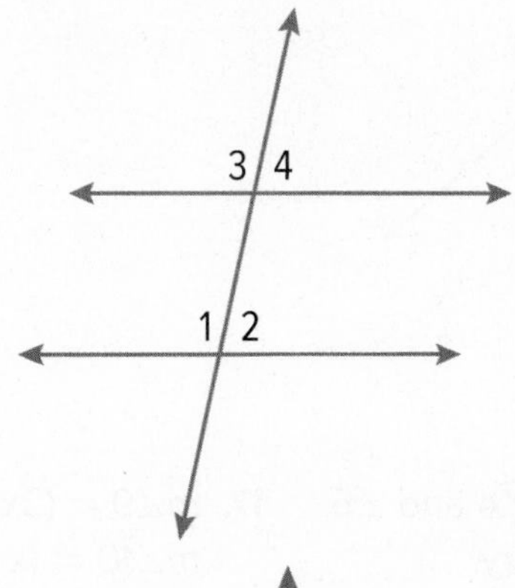

Example 4

PROOF **Write a two-column proof.**

8. **Given:** $\angle 5 \cong \angle 7$
   Prove: $\angle 5 \cong \angle 8$

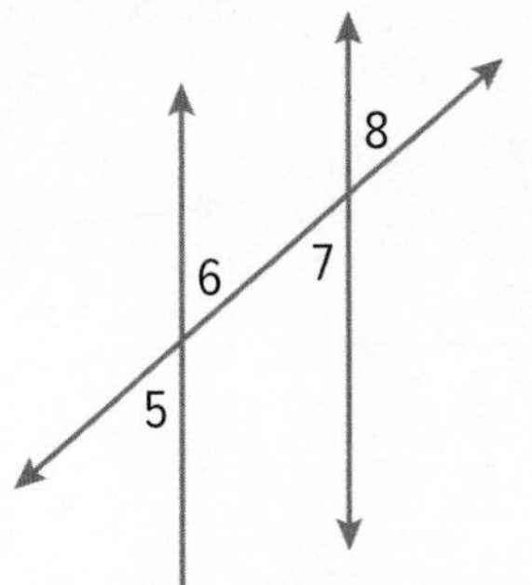

Example 5

PROOF Write a two-column proof.

9. Given: $m\angle ABC = m\angle DEF$

$\angle ABC$ and $\angle DEF$ are supplementary.

Prove: $\angle ABC$ and $\angle DEF$ are right angles.

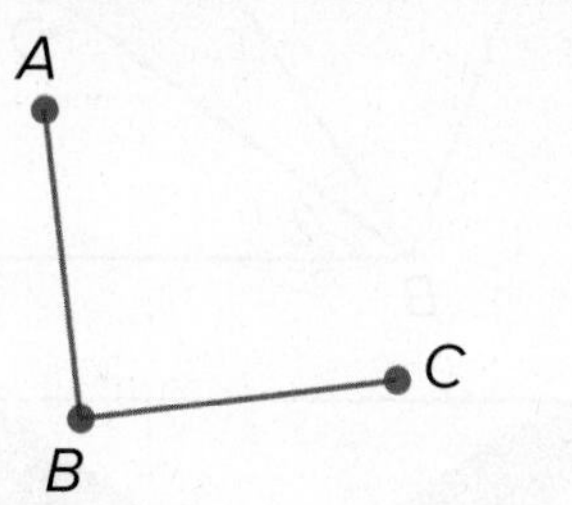

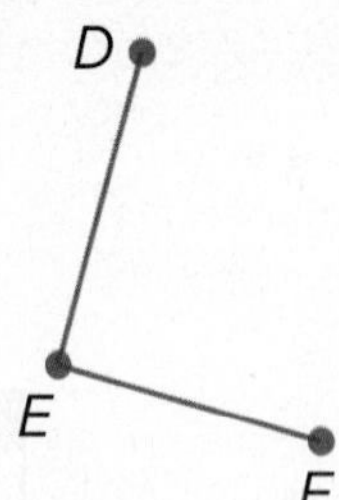

10. Given: $\angle 1 \cong \angle 2$; $m \perp p$

Prove: $\angle 2 \cong \angle 3$

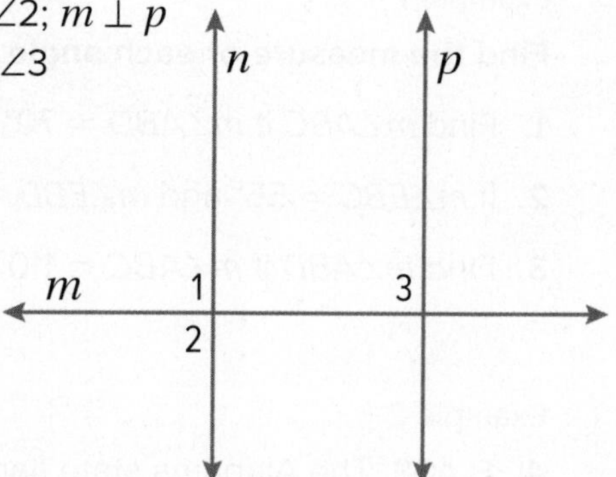

Mixed Exercises

11. Find $m\angle ABC$ and $m\angle CBD$ if $m\angle ABD = 120°$.

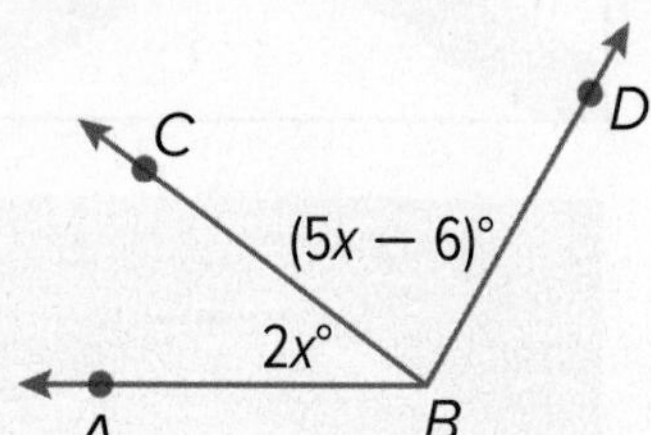

12. Find $m\angle JKL$ and $m\angle LKM$ if $m\angle JKM = 140°$.

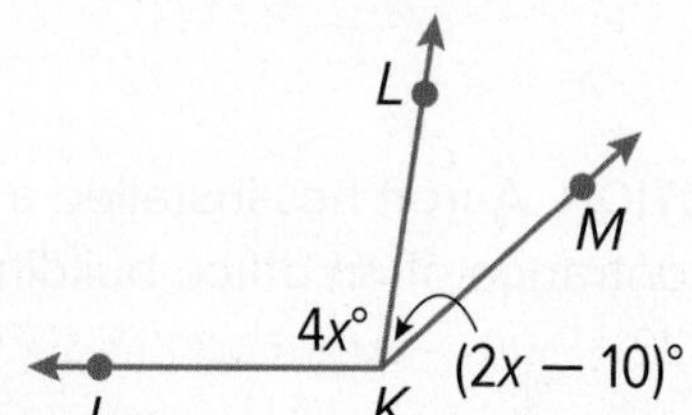

Find the measure of each numbered angle and name the theorems that you used to justify your work.

13. $m\angle 6 = (2x - 21)°$

$m\angle 7 = (3x - 34)°$

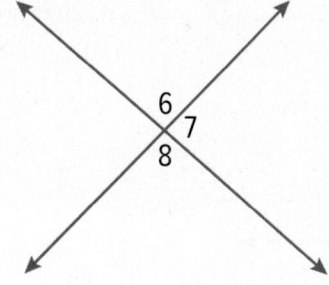

14. $m\angle 5 = m\angle 6$

15. $\angle 2$ and $\angle 3$ are complementary.

$\angle 1 \cong \angle 4$ and $m\angle 2 = 28°$.

16. $\angle 2$ and $\angle 4$ and $\angle 4$ and $\angle 5$ are supplementary.

$m\angle 4 = 105°$.

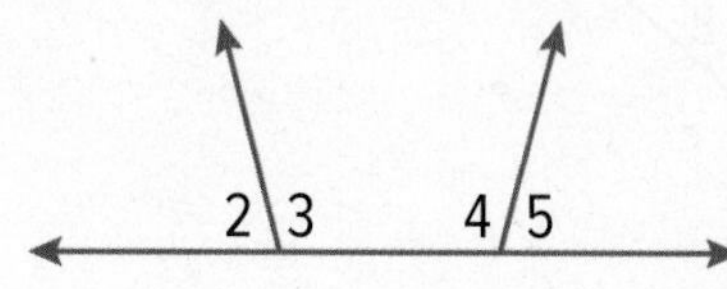

17. $m\angle 9 = (3x + 12)°$

$m\angle 10 = (x - 24)°$

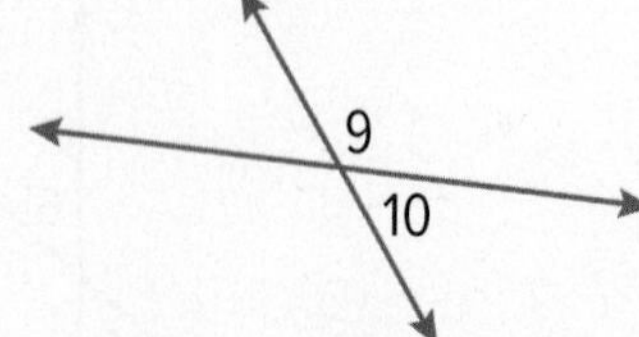

18. $m\angle 3 = (2x + 23)°$

$m\angle 4 = (5x - 112)°$

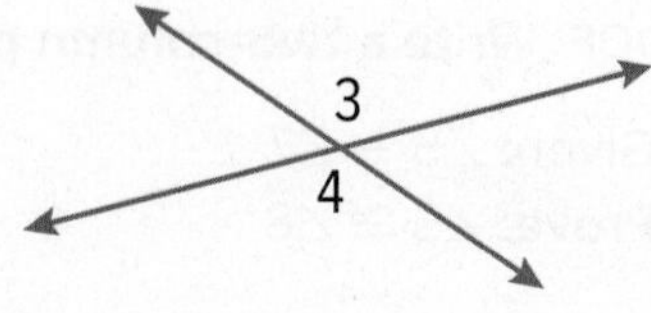

PROOF Write a two-column proof for each theorem.

19. Supplement Theorem

Given: $\angle PQT$ and $\angle TQR$ form a linear pair.

Prove: $\angle PQT$ and $\angle TQR$ are supplementary.

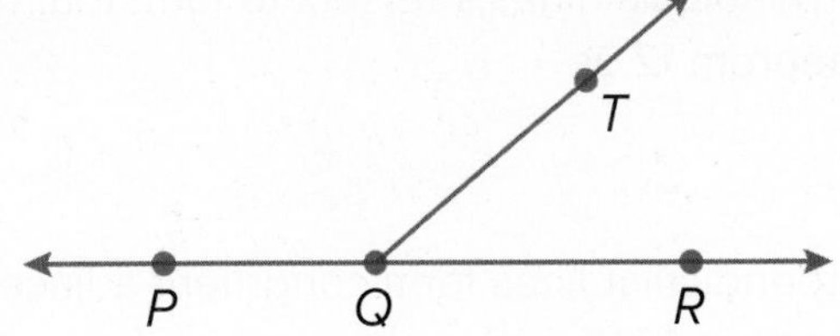

20. Complement Theorem

Given: $\angle ABC$ is a right angle.

Prove: $\angle ABD$ and $\angle CBD$ are complementary.

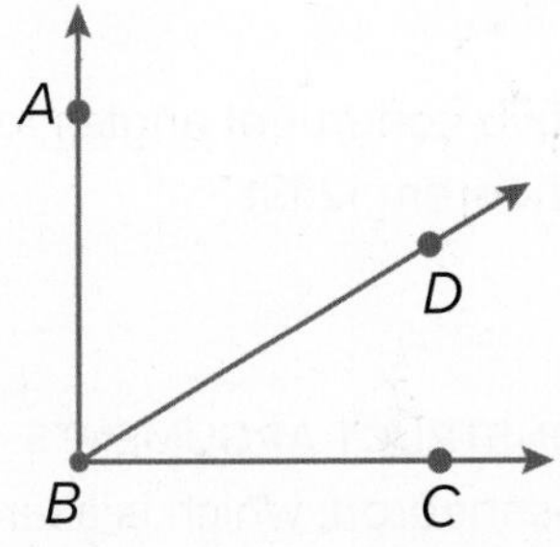

21. Congruent Supplements Theorem (Case 1)

Given: $\angle 1$ and $\angle 2$ are supplementary.
$\angle 2$ and $\angle 3$ are supplementary.

Prove: $\angle 1 \cong \angle 3$

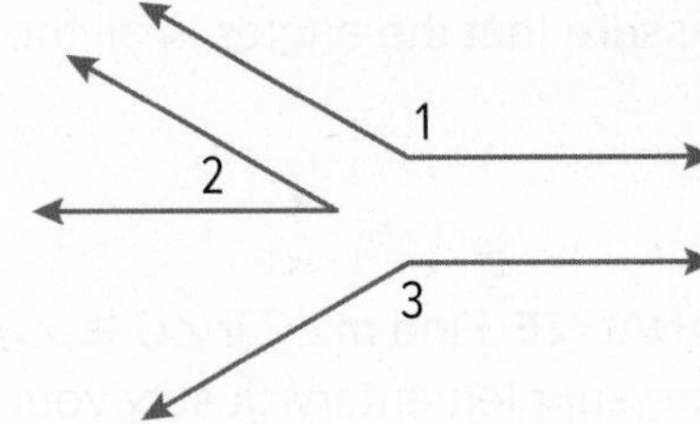

22. Congruent Complements Theorem (Case 1)

Given: $\angle 4$ and $\angle 5$ are complementary.
$\angle 5$ and $\angle 6$ are complementary.

Prove: $\angle 4 \cong \angle 6$

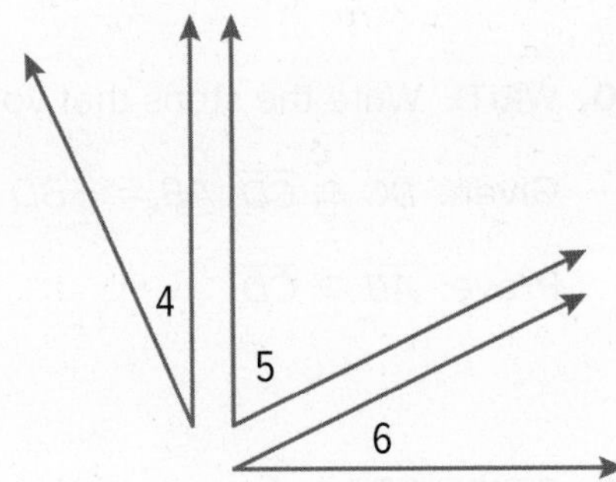

PROOF **Use the figure to write a proof of each theorem.**

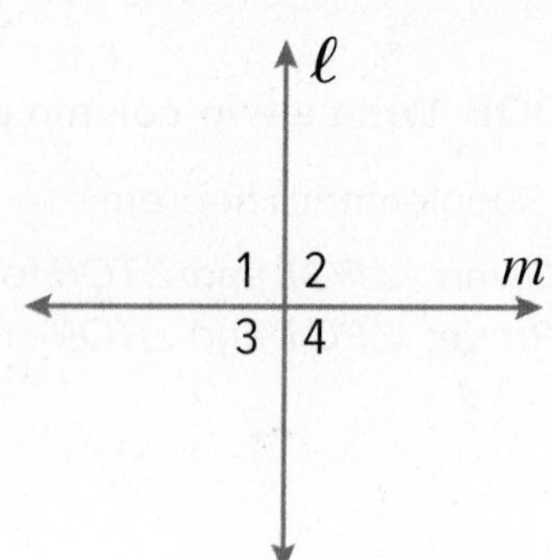

23. Perpendicular lines intersect to form four right angles. (Theorem 12.9)

24. Perpendicular lines form congruent adjacent angles. (Theorem 12.11)

25. If two angles are congruent and supplementary, then each angle is a right angle. (Theorem 12.12)

26. If two congruent angles form a linear pair, then they are right angles. (Theorem 12.13)

27. **CONSTRUCT ARGUMENTS** For a school project, students are making a giant icosahedron, which is a large solid with twenty identical triangular faces. John is in charge of quality control. He must make sure that the measures of all the angles in all the triangles are the same. He does this by using a precut template and comparing the corner angles of every triangle to the template. How does this assure that the angles in all the triangles will be congruent to each other?

28. **ANALYZE** Find $m\angle C$ if $\angle C \cong \angle A$, $m\angle A = 3x°$, $m\angle B = (x + 20)°$, and $\angle A$ and $\angle B$ are supplementary. Justify your argument.

29. **CREATE** Draw $\angle WXZ$ such that $m\angle WXZ = 45°$. Construct $\angle YXZ \cong \angle WXZ$. Make a conjecture about the measure of $\angle WXY$, and then prove your conjecture.

30. **WRITE** Write the steps that you would use to complete the proof.

Given: $\overline{BC} \cong \overline{CD}$, $AB = \frac{1}{2}BD$

A B C D

Prove: $\overline{AB} \cong \overline{CD}$

31. **PERSEVERE** In Exercises 21 and 22, you proved one case of the Congruent Supplements Theorem and one case of the Congruent Complements Theorem. Explain why there is another case for each of these theorems. Then write a proof of this second case for each theorem.

Parallel Lines and Transversals

Learn Parallel Lines and Transversals

If two lines do not intersect, then they are either parallel or skew.

Parallel and Skew

Parallel Lines

Parallel lines are coplanar lines that do not intersect.

Example $\overleftrightarrow{JK} \parallel \overleftrightarrow{LM}$

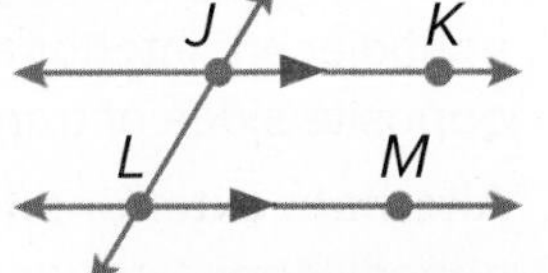

Skew Lines

Skew lines are lines that do not intersect and are not coplanar.

Example Lines ℓ and m are skew.

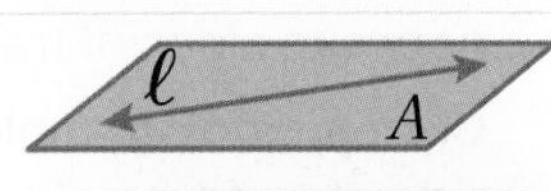

Parallel Planes

Parallel planes are planes that do not intersect.

Example Planes A and B are parallel.

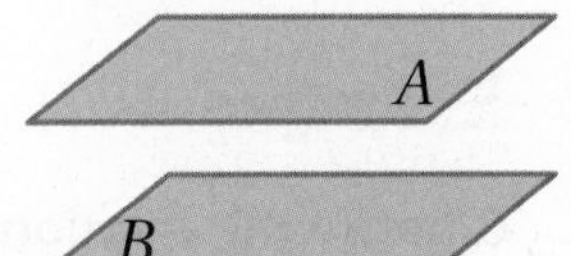

If segments or rays are contained within lines that are parallel or skew, then the segments or rays are parallel or skew.

Today's Goals
- Identify special angle pairs, parallel and skew lines, and transversals.
- Find values by applying theorems about parallel lines and transversals.

Today's Vocabulary
parallel lines
skew lines
parallel planes
transversal
interior angles
exterior angles
consecutive interior angles
alternate interior angles
alternate exterior angles
corresponding angles

Study Tip

Parallel Lines The statement $\overleftrightarrow{JK} \parallel \overleftrightarrow{LM}$ is read as *line JK is parallel to line LM*. In a figure, arrowheads are used to indicate that lines are parallel.

Example 1 Identify Parallel and Skew Relationships

Identify each of the following using the cube shown. Assume lines and planes that appear to be parallel or perpendicular are parallel or perpendicular, respectively.

a. all lines skew to $\overleftrightarrow{BC}$

$\overleftrightarrow{AF}$, $\overleftrightarrow{DE}$, $\overleftrightarrow{FG}$, and $\overleftrightarrow{HE}$

b. all lines parallel to $\overleftrightarrow{EH}$

$\overleftrightarrow{AB}$, $\overleftrightarrow{CD}$, or $\overleftrightarrow{FG}$

c. all planes parallel to plane *DCH*

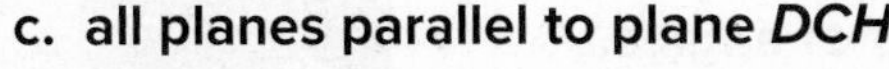

Plane *ABG* is the only plane parallel to plane *DCH*.

A line that intersects two or more lines in a plane at different points is called a **transversal**. In the diagram, line t is a transversal of lines q and r. Notice that line t forms a total of eight angles with lines q and r. These angles and specific pairings of these angles are given special names.

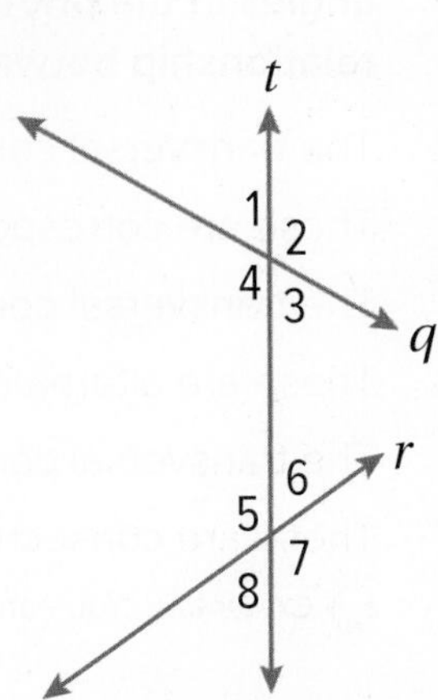

Talk About It!

Can a two-dimensional figure contain skew lines? Justify your argument.

Transversal Angle Pair Relationships	
Four **interior angles** lie in the region between lines q and r.	$\angle 3$, $\angle 4$, $\angle 5$, $\angle 6$
Four **exterior angles** lie in the two regions that are not between lines q and r.	$\angle 1$, $\angle 2$, $\angle 7$, $\angle 8$
Consecutive interior angles are interior angles that lie on the same side of transversal t.	$\angle 4$ and $\angle 5$, $\angle 3$ and $\angle 6$
Alternate interior angles are nonadjacent interior angles that lie on opposite sides of transversal t.	$\angle 3$ and $\angle 5$, $\angle 4$ and $\angle 6$
Alternate exterior angles are nonadjacent exterior angles that lie on opposite sides of transversal t.	$\angle 1$ and $\angle 7$, $\angle 2$ and $\angle 8$
Corresponding angles lie on the same side of transversal t and on the same side of lines q and r.	$\angle 1$ and $\angle 5$, $\angle 2$ and $\angle 6$, $\angle 3$ and $\angle 7$, $\angle 4$ and $\angle 8$

Study Tip

Same-Side Interior Angles Consecutive interior angles are also called *same-side interior angles*.

Example 2 Classify Angle Pair Relationships

Classify the relationship between each pair of angles as *alternate interior, alternate exterior, corresponding,* or *consecutive interior angles.*

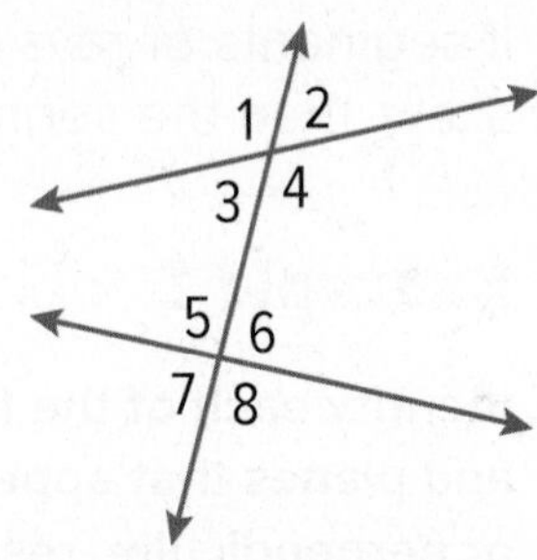

$\angle 4$ and $\angle 5$ are alternate interior angles.

$\angle 3$ and $\angle 7$ are corresponding angles.

$\angle 3$ and $\angle 5$ are consecutive interior angles.

$\angle 1$ and $\angle 8$ are alternate exterior angles.

Example 3 Identify Transversals and Classify Angle Pairs

Identify the transversal connecting each pair of angles in the photo. Then classify the relationship between each pair of angles.

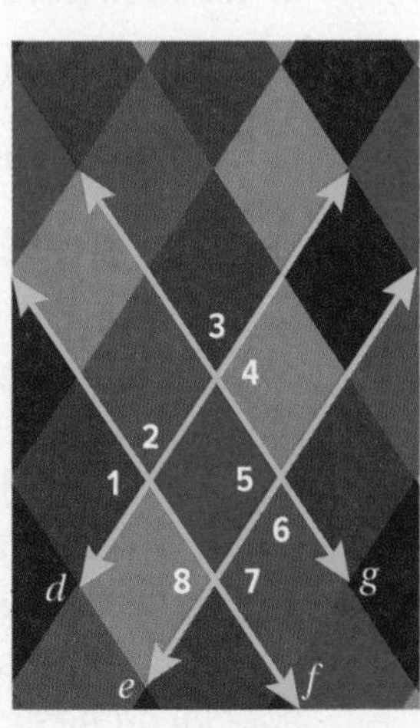

The transversal connecting $\angle 1$ and $\angle 8$ is line f.

These are corresponding angles.

The transversal connecting $\angle 3$ and $\angle 6$ is line g.

These are alternate exterior angles.

The transversal connecting $\angle 6$ and $\angle 7$ is line e.

These are consecutive interior angles.

Go Online You can complete an Extra Example online.

Explore Relationships Between Angles and Parallel Lines

 **Online Activity** Use dynamic geometry software to complete the Explore.

INQUIRY How do parallel lines affect the relationships between special angle pairs?

Learn Angles and Parallel Lines

If two lines are parallel and cut by a transversal, then there are special relationships in the angle pairs formed by the lines.

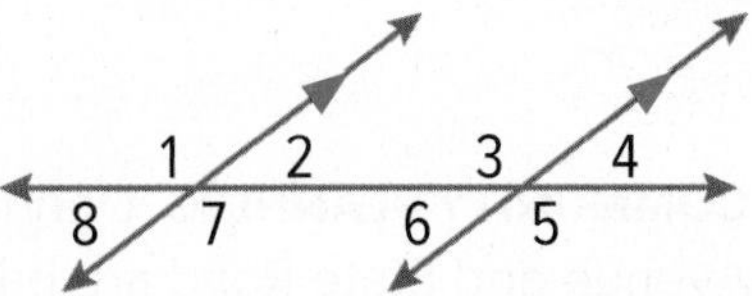

Theorem 12.14: Corresponding Angles Theorem	
If two parallel lines are cut by a transversal, then each pair of corresponding angles is congruent.	$\angle 1 \cong \angle 3$, $\angle 2 \cong \angle 4$, $\angle 5 \cong \angle 7$, $\angle 6 \cong \angle 8$
Theorem 12.15: Alternate Interior Angles Theorem	
If two parallel lines are cut by a transversal, then each pair of alternate interior angles is congruent.	$\angle 2 \cong \angle 6$, $\angle 3 \cong \angle 7$
Theorem 12.16: Consecutive Interior Angles Theorem	
If two parallel lines are cut by a transversal, then each pair of consecutive interior angles is supplementary.	$\angle 1$ and $\angle 3$, $\angle 6$ and $\angle 7$
Theorem 12.17: Alternate Exterior Angles Theorem	
If two parallel lines are cut by a transversal, then each pair of alternate exterior angles is congruent.	$\angle 1 \cong \angle 5$, $\angle 4 \cong \angle 8$

You will prove Theorems 12.16 and 12.17 in Exercises 47 and 48.

A special relationship also exists when the transversal of two parallel lines is a perpendicular line.

Theorem 12.18: Perpendicular Transversal Theorem
In a plane, if a line is perpendicular to one of two parallel lines, then it is perpendicular to the other.

You will prove Theorem 12.18 in Exercise 48.

Study Tip

Angle Relationships Theorems 12.15–12.17 generalize the relationships between specific pairs of angles. If you get confused about the relationships, you can verify them using only corresponding angles, vertical angles, and linear pairs.

Go Online Proofs of Theorems 12.14 and 12.15 are available.

 **Go Online** You can complete an Extra Example online.

Example 4 Use Theorems About Parallel Lines

RAILROADS Crossties i and k are parallel. Both crossties are intersected by crosstie h. If $m\angle 1 = 42°$, find $m\angle 7$.

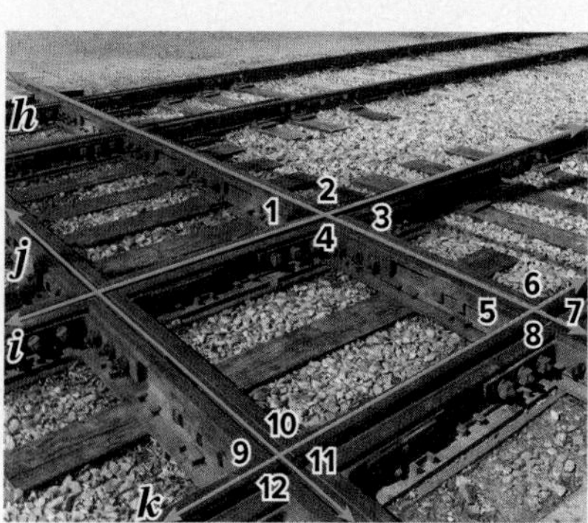

$\angle 7 \cong \angle 1$	Alternate Exterior Angles Theorem
$m\angle 7 = m\angle 1$	Definition of congruent angles
$m\angle 7 = 42°$	Substitution

The measure of $\angle 7$ is 42°.

Check

COMMUNITY PLANNING Dennis Avenue and State Road are parallel streets that intersect Newport Lane along the south side of Oak Creek Park. If $m\angle 3 = 62°$, find $m\angle 4$ ___?___

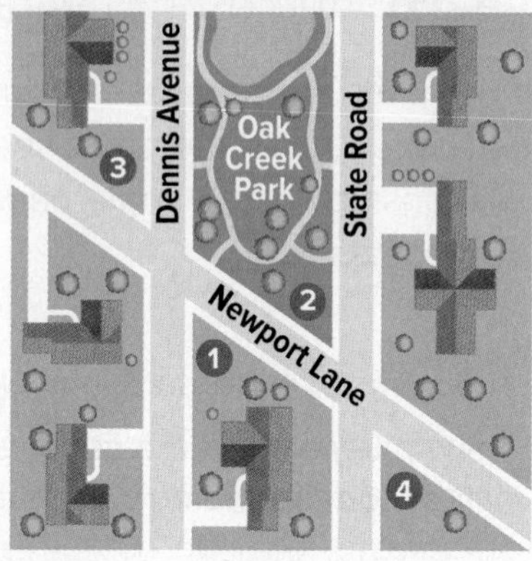

Study Tip

Precision Theorems 12.14–12.17 only apply to *parallel* lines cut by a transversal. You should assume that lines are parallel only if the information is given or the lines are marked with parallel arrows.

Go Online An alternate method is available for this example.

Example 5 Find Values of Variables

Use the figure to find the value of the indicated variable. Justify your reasoning.

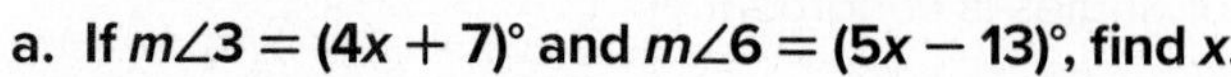

a. If $m\angle 3 = (4x + 7)°$ and $m\angle 6 = (5x - 13)°$, find x.

$\angle 3 \cong \angle 6$	Alternate Interior Angles Theorem
$m\angle 3 = m\angle 6$	Definition of congruent angles
$4x + 7 = 5x - 13$	Substitution
$x = 20$	Simplify.

b. Find y if $m\angle 8 = 68°$ and $m\angle 3 = (3y - 2)°$.

$\angle 5 \cong \angle 8$	Vertical Angles Theorem
$m\angle 5 = m\angle 8$	Definition of congruent angles
$m\angle 5 = 68°$	Substitution

Because lines j and k are parallel, $\angle 5$ and $\angle 3$ are supplementary by the Consecutive Interior Angles Theorem.

$m\angle 3 + m\angle 5 = 180°$	Definition of supplementary angles
$3y - 2 + 68 = 180$	Substitution
$3y + 66 = 180$	Simplify.
$y = 38$	Simplify.

 **Go Online** You can complete an Extra Example online.

Practice

Go Online You can complete your homework online.

Example 1

Identify each of the following using the figure shown. Assume lines and planes that appear to be parallel or perpendicular are parallel or perpendicular, respectively.

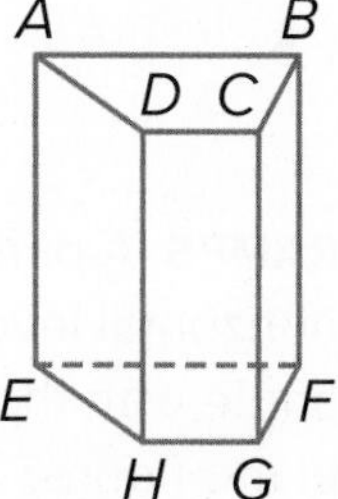

1. three segments parallel to $\overline{AE}$

2. a segment skew to $\overline{AB}$

3. a pair of parallel planes

4. a segment parallel to $\overline{AD}$

5. three segments parallel to $\overline{HG}$

6. five segments skew to $\overline{BC}$

7. How could you characterize the relationship between faces *ABCD* and *DCGH*? Explain.

Examples 2 and 3

Identify the transversal connecting each pair of angles. Then classify the relationship between each pair of angles as *alternate interior*, *alternate exterior*, *corresponding*, or *consecutive interior* angles.

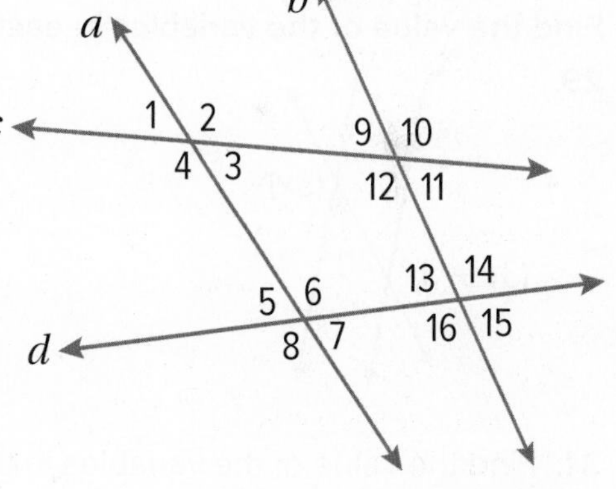

8. ∠4 and ∠5

9. ∠5 and ∠15

10. ∠12 and ∠14

11. ∠7 and ∠15

12. ∠2 and ∠12

13. ∠3 and ∠6

14. ∠1 and ∠9

15. ∠3 and ∠9

16. ∠10 and ∠16

17. ∠5 and ∠13

For Exercises 18 and 19, use the figure.

18. What type of angles are ∠3 and ∠10?

19. State the transversal that connects ∠11 and ∠13.

20. ESCALATORS An escalator at a shopping mall runs up several levels. The escalator railing can be modeled by a straight line running past horizontal lines that represent the floors. Describe the relationships of these lines.

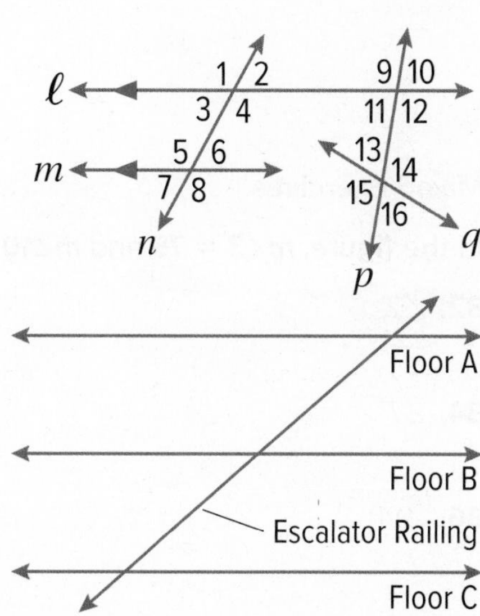

Example 4

In the figure, $m\angle 7 = 100°$. Find the measure of each angle.

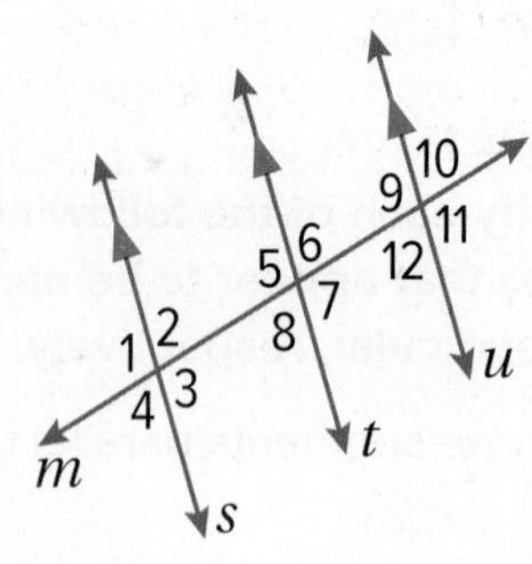

21. $\angle 9$

22. $\angle 6$

23. $\angle 8$

24. $\angle 2$

25. $\angle 5$

26. $\angle 11$

27. RAMPS A parking garage ramp rises to connect two horizontal levels of a parking lot. The ramp makes a 10° angle with the horizontal. What is the measure of angle 1 in the figure?

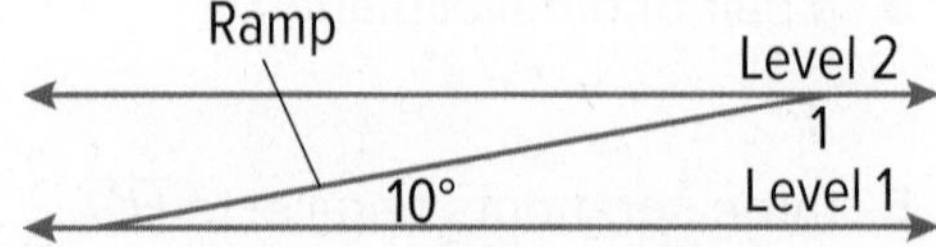

28. CITY ENGINEERING Seventh Avenue runs perpendicular to 1st and 2nd Streets, which are parallel. However, Maple Avenue makes a 115° angle with 2nd Street. What is the measure of angle 1?

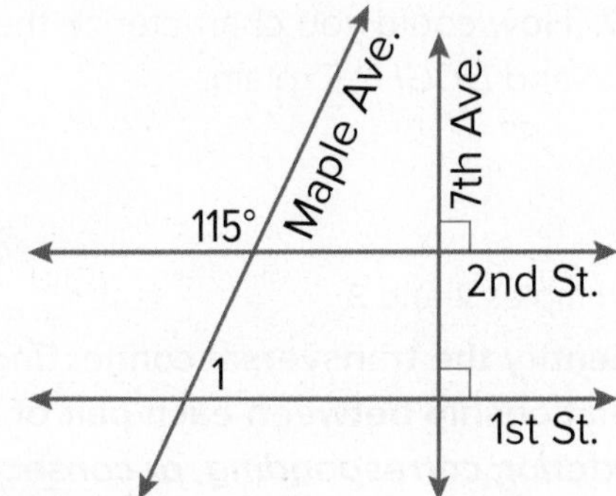

Example 5

Find the value of the variables in each figure. Explain your reasoning.

29.

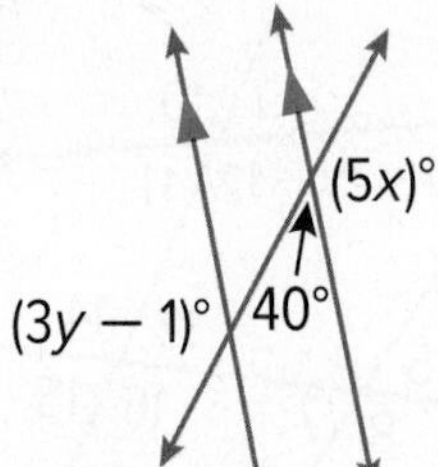

30.

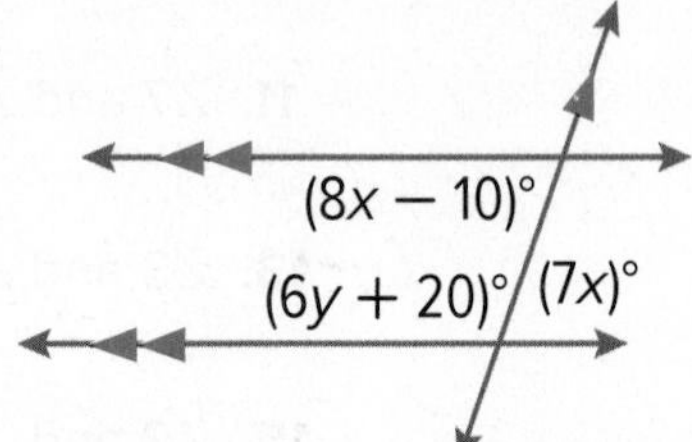

31. Find the value of the variables in the figure.

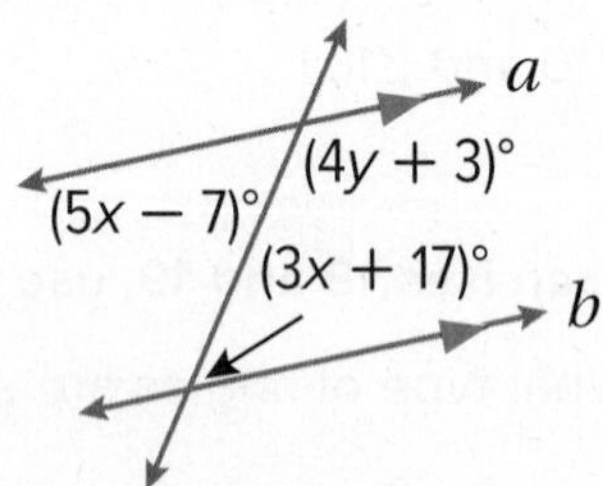

Mixed Exercises

In the figure, $m\angle 3 = 75$ and $m\angle 10 = 105°$. Find the measure of each angle.

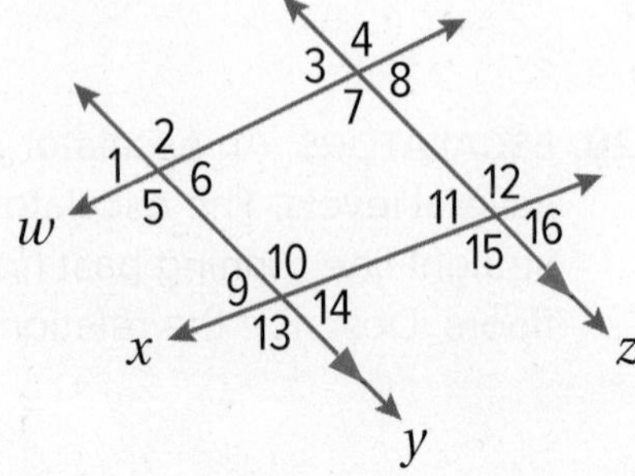

32. $\angle 2$

33. $\angle 5$

34. $\angle 7$

35. $\angle 15$

36. $\angle 14$

37. $\angle 9$

USE A MODEL **Lines a and b are parallel and are cut by transversal t to form interior angles ∠7, ∠8, ∠9, and ∠10. ∠7 and ∠8 are consecutive interior angles, and $m\angle 7 = 94°$. ∠8 and ∠10 are alternate interior angles. Find the measure of each angle.**

38. ∠10

39. ∠9

40. ∠8

41. **CARPENTRY** A carpenter is building a podium. The side panel of the podium is cut from a rectangular piece of wood. The rectangle must be sawed along the dashed line in the figure. What is the measure of ∠1? Explain your reasoning.

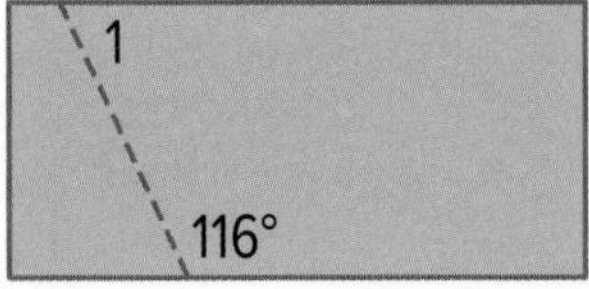

42. **MAPPING** Copy the figure.

a. Connor lives at the angle that forms an alternate interior angle with Georgia's residence. Label the location of Connor's home on the map.

b. Quincy lives at the angle that forms a consecutive interior angle with Connor's residence. Label the location of Quincy's home on the map.

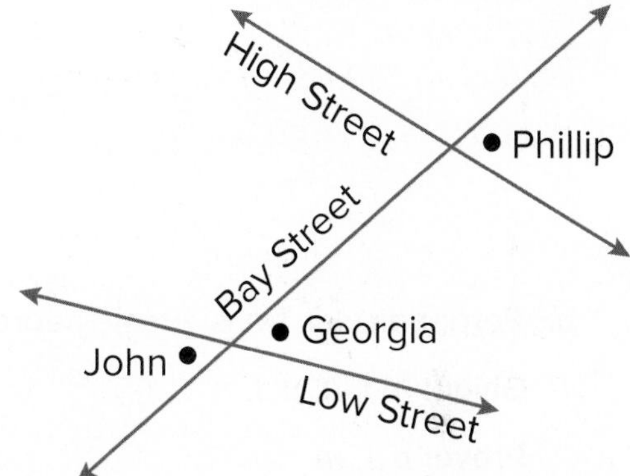

43. **USE A SOURCE** Research the flag for the Solomon Islands. Sketch the flag. Label angles formed by the yellow stripe, or transversal. Describe the relationship between the angles you labeled on the flag.

44. **PRECISION** Find the values of x and y in the trapezoid. Justify your answer.

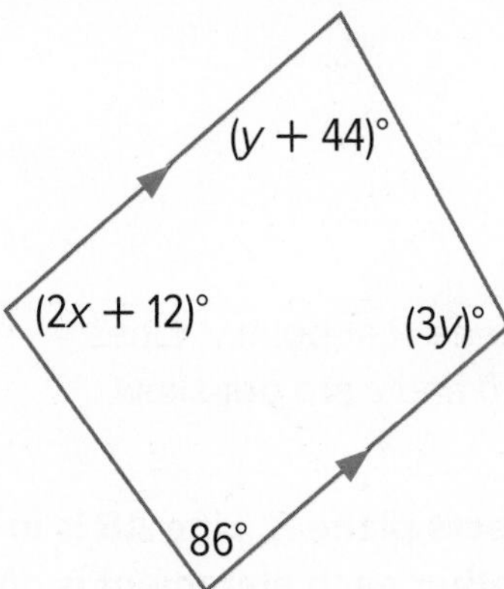

45. **PROOF** In the figure, lines m and n are parallel and lines p and q are parallel. Write a paragraph proof to prove that if $m\angle 1 - m\angle 4 = 25°$, then $m\angle 9 - m\angle 12 = 25°$.

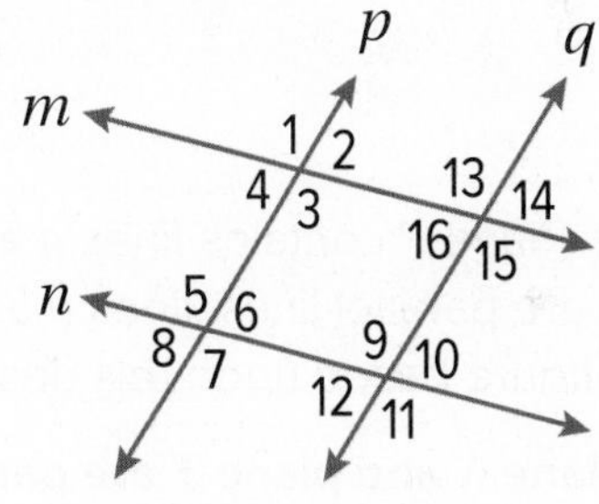

46. In the figure, $m\angle 4 = 118°$. Find each angle measure. Justify each step.

a. $m\angle 8$

b. $m\angle 7$

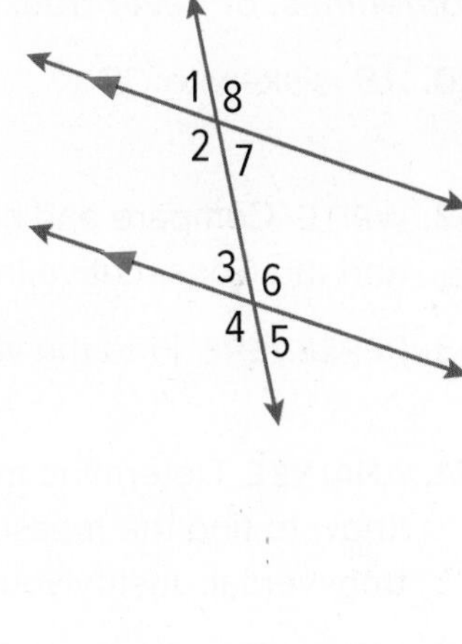

47. **PROOF** Write a paragraph proof of the Alternate Exterior Angles Theorem. Given: $q \parallel r$; Prove: $\angle 1 \cong \angle 7$.

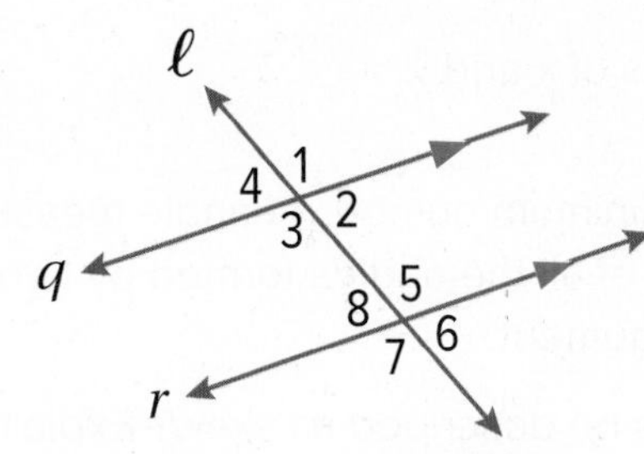

48. PROOF **Write a two-column proof to prove each theorem.**

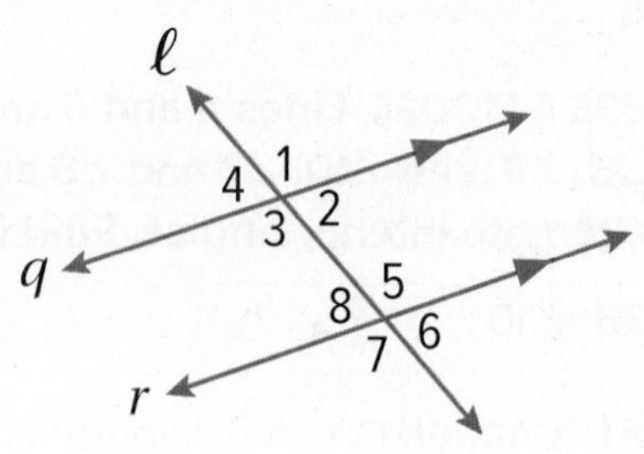

a. Consecutive Interior Angles Theorem

Given: $q \parallel r$

Prove: $\angle 2$ and $\angle 5$ are supplementary

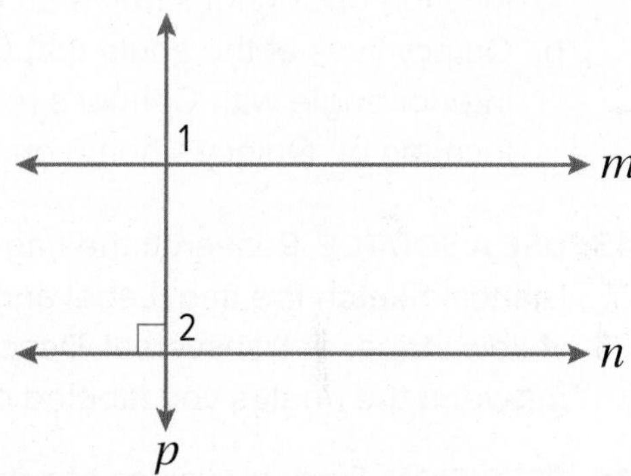

b. Perpendicular Transversal Theorem.

Given: $m \parallel n$; $p \perp n$

Prove: $p \perp m$

49. CREATE Plane P contains lines a and b. Line c intersects plane P at point J. Lines a and b are parallel lines, lines a and c are skew, and lines b and c are not skew. Draw a figure based upon this description.

ANALYZE **Plane X and plane Y are parallel and plane Z intersects plane X. Line $\overleftrightarrow{AB}$ is in plane X, line $\overleftrightarrow{CD}$ is in plane Y, and line $\overleftrightarrow{EF}$ is in plane Z. Determine whether each statement is *always*, *sometimes*, or *never* true. Justify your argument.**

50. $\overleftrightarrow{AB}$ is skew to $\overleftrightarrow{CD}$.

51. $\overleftrightarrow{AB}$ intersects $\overleftrightarrow{EF}$.

52. WRITE Compare and contrast the Alternate Interior Angles Theorem and the Consecutive Interior Angles Theorem.

53. PERSEVERE Find the values of x and y.

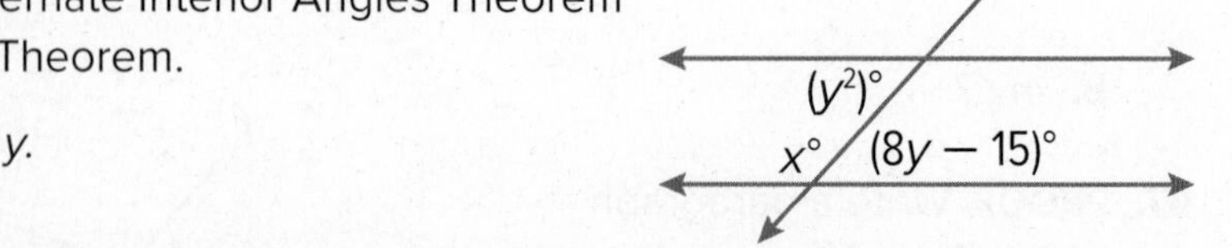

54. ANALYZE Determine the minimum number of angle measures you would have to know to find the measures of all the angles formed by two parallel lines cut by a transversal. Justify your argument.

55. WRITE Can a pair of planes be described as skew? Explain.

Slope and Equations of Lines

Learn Slope Criteria for Parallel and Perpendicular Lines

Slope is the ratio of the change in the *y*-coordinate (rise) to the corresponding change in the *x*-coordinate (run) as you move from one point to another along a line. The **slope criteria** outlines a method for proving the relationship between lines based on a comparison of the slopes of the lines. You can use the slopes of two lines to determine whether the lines are parallel, perpendicular, or neither.

Postulate 12.12: Slope Criteria for Parallel and Perpendicular Lines
Slopes of Parallel Lines Two distinct nonvertical lines have the same slope if and only if they are parallel. All vertical lines are parallel.
Slopes of Perpendicular Lines Two nonvertical lines are perpendicular if and only if the product of their slopes is −1. Vertical and horizontal lines are perpendicular.

Example 1 Determine Line Relationships When Given Points

Determine whether $\overleftrightarrow{AB}$ and $\overleftrightarrow{CD}$ are *parallel, perpendicular,* or *neither* for *A*(3, 6), *B*(−9, 2), *C*(5, 4), and *D*(2, 3). Graph each line to verify your answer.

Step 1 Find the slope of each line.

slope $= \frac{y_2 - y_1}{x_2 - x_1}$, where $x_1 \neq x_2$

slope of $\overleftrightarrow{AB} = \frac{6 - 2}{3 - (-9)} = \frac{4}{12}$ or $\frac{1}{3}$

slope of $\overleftrightarrow{CD} = \frac{4 - 3}{5 - 2}$ or $\frac{1}{3}$

Step 2 Determine the relationship.

The two lines have the same slope, so they are parallel.

Check

Determine whether $\overleftrightarrow{AB}$ and $\overleftrightarrow{CD}$ are *parallel, perpendicular,* or *neither* for *A*(14, 13), *B*(− 11, 0), *C*(−3, 7), and *D*(−4, −5). Graph each line to verify your answer.

Go Online You can complete an Extra Example online.

Today's Goals

- Classify lines as parallel, perpendicular, or neither by comparing the slopes of the lines.
- Classify lines as parallel, perpendicular, or neither by comparing the equations of the lines.

Today's Vocabulary

slope

slope criteria

Go Online

You may want to complete the Concept Check to check your understanding.

Talk About It!

Feng argues that you could have graphed the points and determined whether the lines were parallel, perpendicular, or neither just by looking at the graph. Do you agree? What useful question would you ask Feng to determine whether his argument is reasonable?

Example 2 Determine Line Relationships When Given Graphs

Determine whether each pair of lines is *parallel, perpendicular,* or *neither.*

a. $\overleftrightarrow{RS}$ and $\overleftrightarrow{TU}$

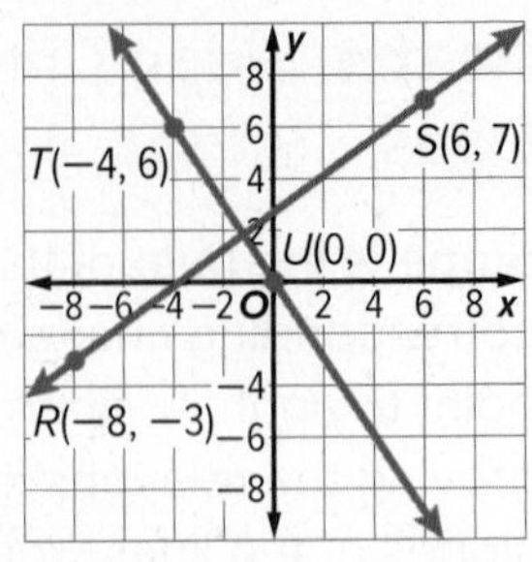

Step 1 Find the slope of each line.

$$\text{slope} = \frac{y_2 - y_1}{x_2 - x_1}, \text{ where } x_1 \neq x_2$$

$$\text{slope of } \overleftrightarrow{RS} = \frac{7 - (-3)}{6 - (-8)} = \frac{10}{14} \text{ or } \frac{5}{7}$$

$$\text{slope of } \overleftrightarrow{TU} = \frac{0 - 6}{0 - (-4)} = -\frac{6}{4} \text{ or } -\frac{3}{2}$$

Step 2 Determine the relationship, if any, between the lines.

The two line do not have the same slope, so they are not parallel. The product of the slopes of the lines is $\left(\frac{5}{7}\right)\left(-\frac{3}{2}\right)$ or $-\frac{15}{14}$. Because the product of the slopes is not −1, the two lines are not perpendicular. So, $\overleftrightarrow{RS}$ and $\overleftrightarrow{TU}$ are neither parallel nor perpendicular.

Study Tip

Slopes of Perpendicular Lines If a line ℓ has a slope of $\frac{a}{b}$, then the slope of a line perpendicular to line ℓ is the negative reciprocal, $-\frac{b}{a}$, because $\frac{a}{b}\left(-\frac{b}{a}\right) = -1$.

b. $\overleftrightarrow{EF}$ and $\overleftrightarrow{DG}$

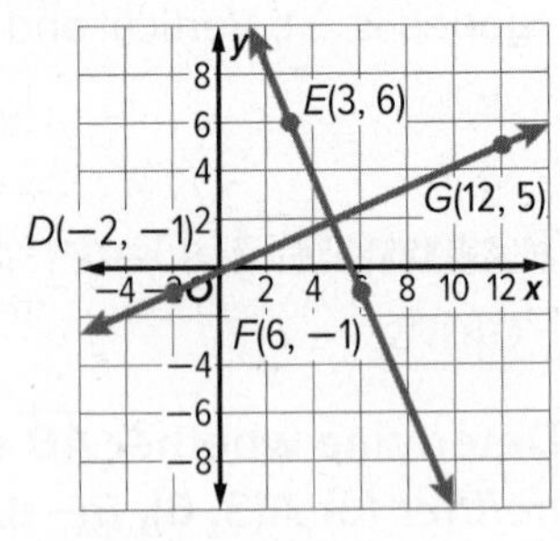

Step 1 Find the slope of each line.

$$\text{slope} = \frac{y_2 - y_1}{x_2 - x_1}, \text{ where } x_1 \neq x_2$$

$$\text{slope of } \overleftrightarrow{EF} = \frac{-1 - 6}{6 - 3} = -\frac{7}{3}$$

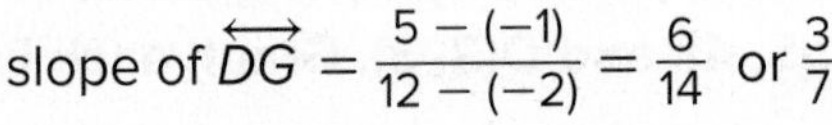

$$\text{slope of } \overleftrightarrow{DG} = \frac{5 - (-1)}{12 - (-2)} = \frac{6}{14} \text{ or } \frac{3}{7}$$

Step 2 Determine the relationship, if any, between the lines.

The two lines do not have the same slope, so they are not parallel. To determine whether the lines are perpendicular, find the product of their slopes.

$-\frac{7}{3}\left(\frac{3}{7}\right) = -1$ Product of slopes for $\overleftrightarrow{EF}$ and $\overleftrightarrow{DG}$

Because the product of their slopes is −1, the two lines are perpendicular.

Check

Determine whether the pair of lines is *parallel, perpendicular,* or *neither.*

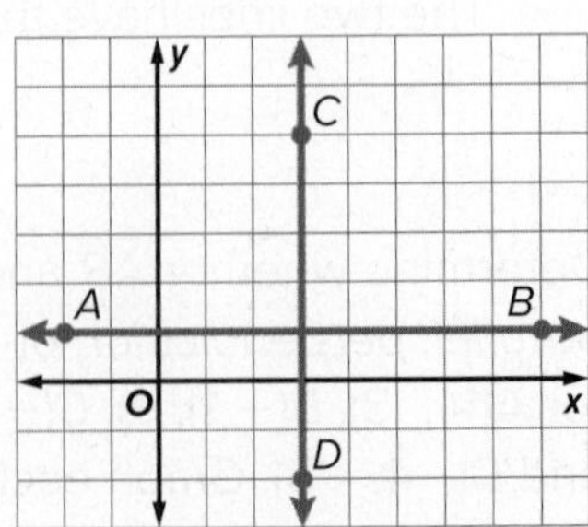

Go Online You can complete an Extra Example online.

Explore Equations of Lines

Online Activity Use dynamic geometry software to complete the Explore.

> **INQUIRY** How do the equations of parallel lines compare to the equations of perpendicular lines?

Learn Equations of Lines

An equation of a nonvertical line can be written in different but equivalent forms.

Key Concept • Nonvertical Line Equations

The slope-intercept form of a linear equation is $y = mx + b$, where m is the slope of the line and b is the y-intercept.	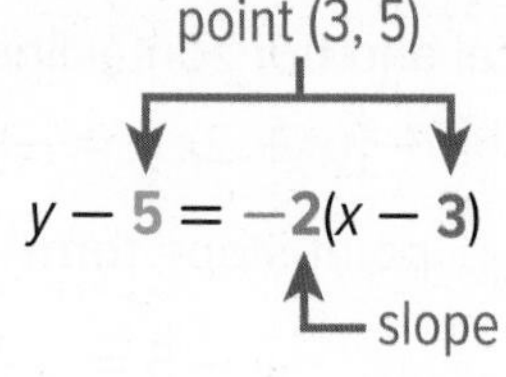
The point-slope form of a linear equation is $y - y_1 = m(x - x_1)$, where (x_1, y_1) is any point on the line and m is the slope of the line.	point (3, 5) $y - 5 = -2(x - 3)$ slope

The equations of horizontal and vertical lines involve only one variable.

Key Concept • Horizontal and Vertical Line Equations

The equation of a horizontal line is $y = b$, where b is the y-intercept of the line.	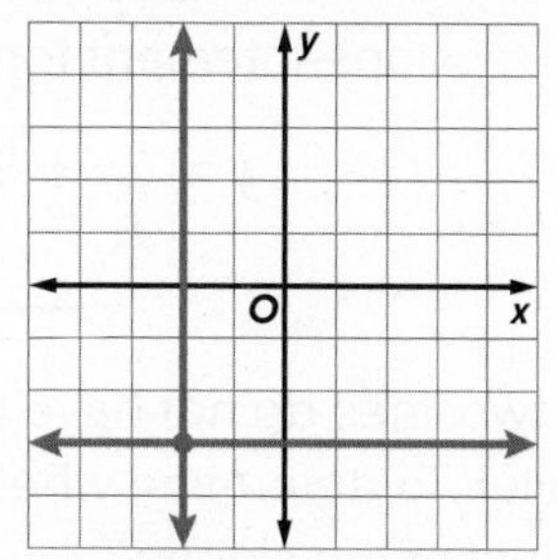
The equation of a vertical line is $x = a$, where a is the x-intercept of the line.	

When given the equations of two lines, you can compare the equations to determine the relationship between the lines.

Math History Minute

French mathematician **Gaspard Monge (1746–1818)** is known as the father of the point-slope form of the linear equation. He is also credited with first stating in print the relationship between the slopes of perpendicular lines as $aa' + 1 = 0$. For his work in mathematics, his name is one of 72 names inscribed on the base of the Eiffel Tower.

Example 3 Determine Line Relationships When Given Equations

Determine whether each pair of lines is *parallel, perpendicular,* or *neither.*

a. $y = 3x - 2;\ y - 0 = -\frac{1}{3}(x - 2)$

slope-intercept form — point-slope form

$$y = 3x - 2 \qquad y - 0 = -\frac{1}{3}(x - 2)$$

slope

The two lines do not have the same slope, so the lines are not parallel. To determine whether the lines are perpendicular, find the product of the slopes.

$3\left(-\frac{1}{3}\right) = -1$ Product of slopes

Because the product of their slopes is −1, the two lines are perpendicular.

b. $y = 3;\ x = 1$

$y = 3$	$x = 1$
horizontal line	vertical line
slope of 0	undefined slope

Vertical and horizontal lines are always perpendicular.

c. $y - 5 = -\frac{3}{4}(x + 2);\ y = -\frac{3}{4}x + 2$

point-slope form — slope-intercept form

$$y - 5 = -\frac{3}{4}(x + 2) \qquad y = -\frac{3}{4}x + 2$$

slope

Because the slopes of both lines are $-\frac{3}{4}$, the lines are parallel.

d. $y = 2x + 3;\ y - 1 = \frac{1}{2}(x + 2)$

slope-intercept form — point-slope form

$$y = 2x + 3 \qquad y - 1 = \frac{1}{2}(x + 2)$$

slope

The two lines do not have the same slope, so the lines are not parallel. To determine whether the lines are perpendicular, find the product of the slopes.

$2\left(\frac{1}{2}\right) = 1$ Product of slopes

Because the product of the slopes is not −1, the two lines are not perpendicular. So, the two lines are neither parallel nor perpendicular.

e. $x = -2;\ x = 4$

Both lines are vertical with undefined slope. Vertical lines are always parallel.

Go Online You can complete an Extra Example online.

Watch Out!

Vertical Lines If you calculate the slope of the line $x = 1$ using the slope formula, you get $m = \frac{-1-4}{1-1} = -\frac{5}{0}$ or an undefined slope. You cannot find the product of the slope of $x = 1$ and $y = 3$. However, vertical and horizontal lines are always perpendicular.

Study Tip

Zero and Undefined Slope If the change in y values is 0, then the line is horizontal. If the change in x values is 0, then the line is vertical.

Check

Determine whether each pair of lines is *parallel*, *perpendicular*, or *neither*.

a. $y = 3x - 9$; $y = -\frac{1}{3}x + 2$

b. $y = \frac{9}{7}x - \frac{19}{7}$; $y - 1 = \frac{9}{7}(x + 3)$

c. $x = -3$; $x = 4$

Example 4 Use Slope to Graph a Line

DESIGN **Valentina is designing a park using grid paper. She wants to build a sidewalk that connects with the fountain at $P(0, 1)$ and is perpendicular to the existing sidewalk that passes through points $Q(-6, -2)$ and $R(0, -6)$. Graph the line that represents the new sidewalk.**

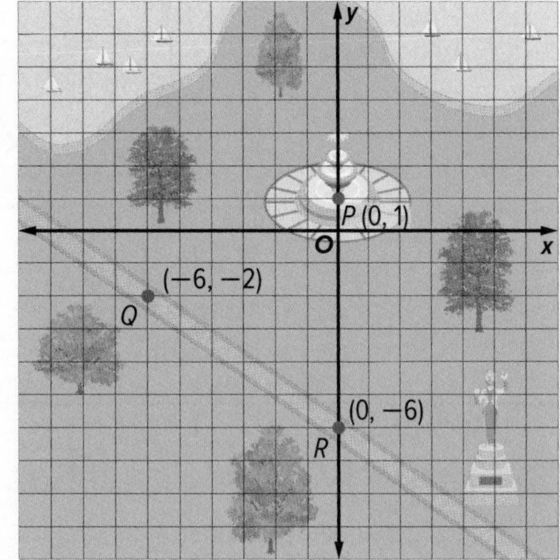

The slope of the existing sidewalk, $\overleftrightarrow{QR}$ is $\frac{-6-(-2)}{0-(-6)} = -\frac{4}{6}$ or $-\frac{2}{3}$.

Because $-\frac{2}{3}\left(\frac{3}{2}\right) = -1$, the slope of the line perpendicular to $\overleftrightarrow{QR}$ through P is $\frac{3}{2}$.

Graph the line that represents the new sidewalk.

Step 1 Plot a point at $P(0, 1)$.

Step 2 Move up 3 units and then right 2 units. Plot a second point at this location.

Step 3 Graph the line connecting these two points.

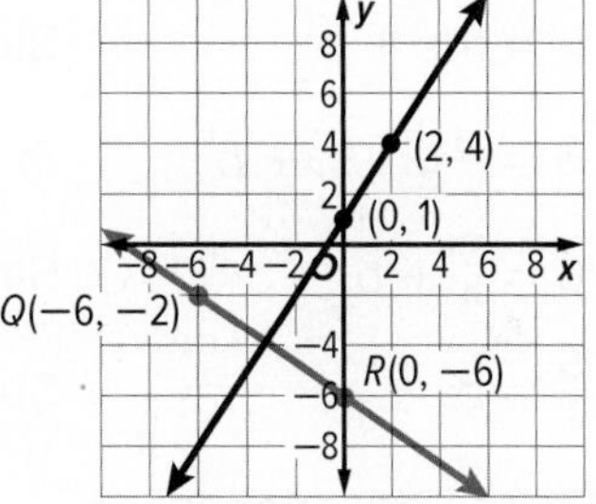

Go Online You can complete an Extra Example online.

Check

MAPS Isabella is creating a map of her town's metro lines. She knows that the A Line and the E Line are parallel. On her map, the equation that represents the A Line is $y = 8x + 11$ and the E Line passes through (9, 5). Write the equation in slope-intercept form that represents the E Line.

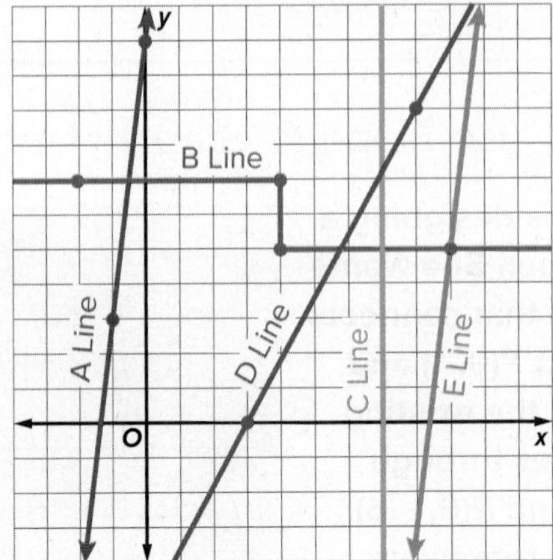

> **Think About It!**
> Kennedy suggests that there is another line parallel to $y = -\frac{3}{4}x + 3$ that contains the point $(-3, 6)$. She says that the equation of the line is $y - 6 = -\frac{3}{4}(x + 3)$. Do you agree? Explain your reasoning.

Example 5 Write Equations of Parallel and Perpendicular Lines

Write an equation in slope-intercept form for the line parallel to $y = -\frac{3}{4}x + 3$ containing $(-3, 6)$.

The slope of $y = -\frac{3}{4}x + 3$ is $-\frac{3}{4}$, so the slope of the line parallel to it is $-\frac{3}{4}$.

$y = mx + b$	Slope-intercept form
$6 = -\frac{3}{4}(-3) + b$	$m = -\frac{3}{4}$ and $(x, y) = (-3, 6)$
$6 = \frac{9}{4} + b$	Simplify.
$\frac{15}{4} = b$	Subtract $\frac{9}{4}$ from each side.

So, the equation is $y = -\frac{3}{4}x + \frac{15}{4}$.

> **Go Online**
> An alternate method is available for this example.

Check

Write an equation in slope-intercept form for the line parallel to $y = \frac{1}{2}x + \frac{5}{2}$ containing $\left(\frac{3}{2}, 1\right)$.

Go Online You can complete an Extra Example online.

Practice

Go Online You can complete your homework online.

Example 1

Determine whether $\overleftrightarrow{AB}$ and $\overleftrightarrow{CD}$ are *parallel, perpendicular,* or *neither.* Graph each line to verify your answer.

1. $A(1, 5), B(4, 4), C(9, -10), D(-6, -5)$
2. $A(-6, -9), B(8, 19), C(0, -4), D(2, 0)$
3. $A(4, 2), B(-3, 1), C(6, 0), D(-10, 8)$
4. $A(8, -2), B(4, -1), C(3, 11), D(-2, -9)$
5. $A(8, 4), B(4, 3), C(4, -9), D(2, -1)$
6. $A(4, -2), B(-2, -8), C(4, 6), D(8, 5)$

Example 2

Determine whether each pair of lines is *parallel, perpendicular,* or *neither.*

7. 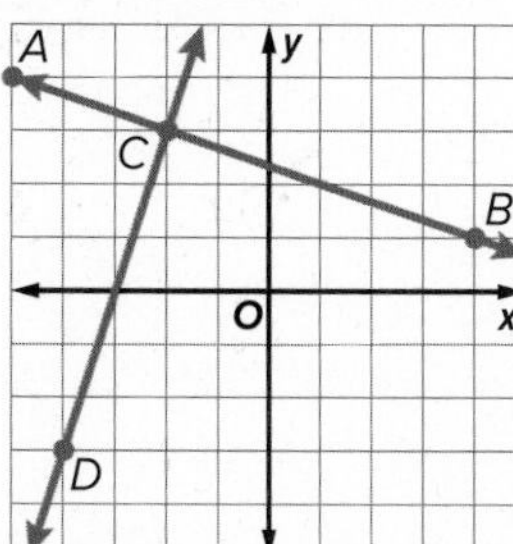

8. 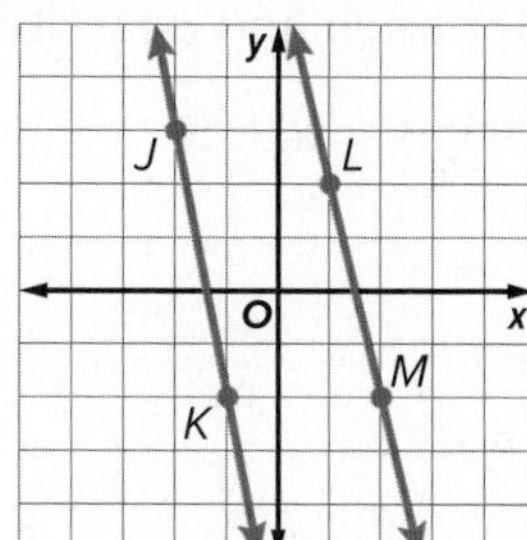

9. 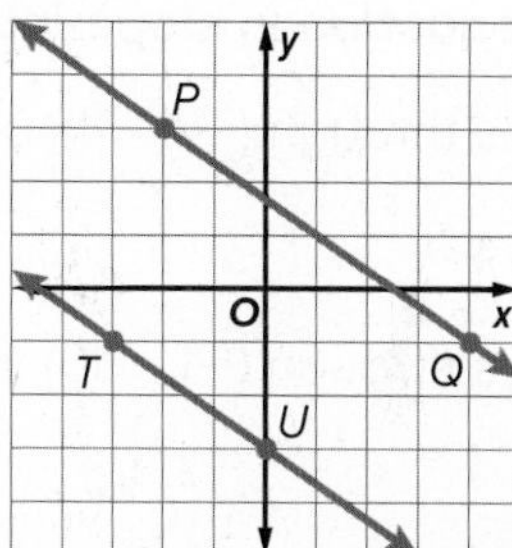

Example 3

Determine whether each pair of lines is *parallel, perpendicular,* or *neither.*

10. $y = 2x + 4, y = 2x - 10$
11. $y = -\frac{1}{2}x - 12, y - 3 = 2(x + 2)$
12. $y - 4 = 3(x + 5), y + 3 = -\frac{1}{3}(x + 1)$
13. $y - 3 = 6(x + 2), y + 3 = -\frac{1}{3}(x - 4)$
14. $x = -2, y = 10$
15. $y = 5, y = -3$

Example 4

Graph the line that satisfies each condition.

16. passes through $A(2, -5)$, parallel to $\overleftrightarrow{BC}$ with $B(1, 3)$ and $C(4, 5)$
17. passes through $X(1, -4)$, parallel to $\overleftrightarrow{YZ}$ with $Y(5, 2)$ and $Z(-3, -5)$
18. passes through $K(3, 7)$, perpendicular to $\overleftrightarrow{LM}$ with $L(-1, -2)$ and $M(-4, 8)$
19. passes through $D(-5, -6)$, perpendicular to $\overleftrightarrow{FG}$ with $F(-2, -9)$ and $G(1, -5)$
20. **SKIING** Gavin is working on an animated film about skiing. The figure shows a ski slope, represented by $\overleftrightarrow{AB}$, and one of the chairs on the chair lift, represented by point C.

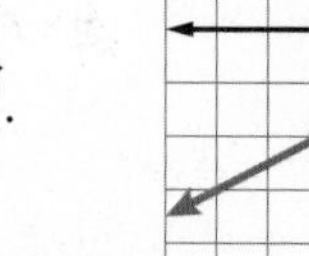

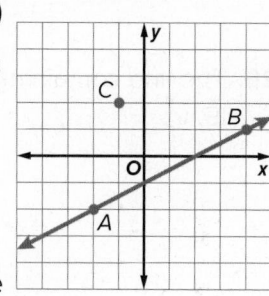

   a. The chair needs to move along a straight line that is parallel to $\overleftrightarrow{AB}$. What is the equation of this line?
   b. The top of the chair lift occurs at $y = 20$. Explain how Gavin can find the coordinates of the chair when it reaches the top of the chair lift.

21. REASONING The director of a marching band uses a coordinate plane to design the band's formations. During one formation, a drummer marches from point A to point B and then turns 90° to her right and marches until she reaches the x-axis.

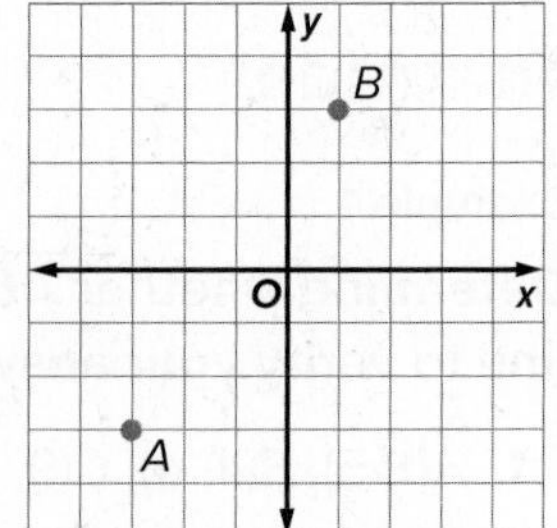

a. When the drummer marches from point B to the x-axis, what is the equation of the line that she marches along?

b. The director wants to know whether the drummer will cross the x-axis at a point where the x-coordinate is greater than or less than 5. Explain how the director can answer this question.

Example 5

Write an equation in slope-intercept form for each line described.

22. passes through (−7, −4), perpendicular to $y = \frac{1}{2}x + 9$

23. passes through (−1, −10), parallel to $y = 7$

24. passes through (6, 2), parallel to $y = -\frac{2}{3}x + 1$

25. passes through (−2, 2), perpendicular to $y = -5x - 8$

Mixed Exercises

Find the value of x or y that satisfies the given conditions. Then graph the line.

26. The line containing (4, −2) and (x, −6) is perpendicular to the line containing (−2, −9) and (3, −4).

27. The line containing (−4, 9) and (4, 3) is parallel to the line containing (−8, 1) and (4, y).

28. The line containing (8, 7) and (7, −6) is perpendicular to the line containing (2, 4) and (x, 3).

29. The line containing (1, −3) and (3, y) is parallel to the line containing (5, −6) and (9, y).

Write equations in slope-intercept form for a line that is *parallel* and a line that is *perpendicular* to the given line and that passes through the given point.

30. passes through $B(6, 3)$

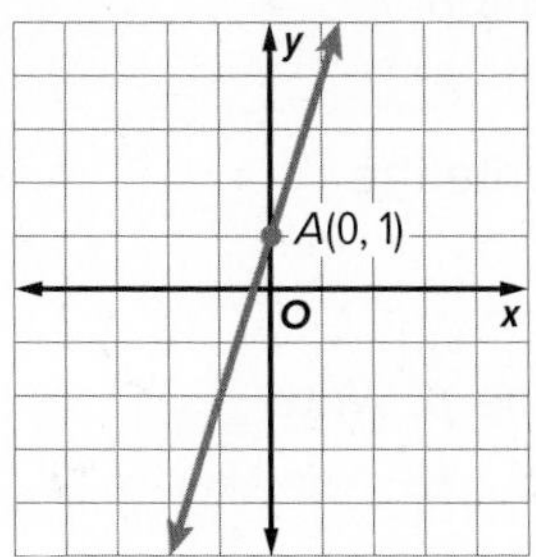

31. passes through $S(-2, -4)$

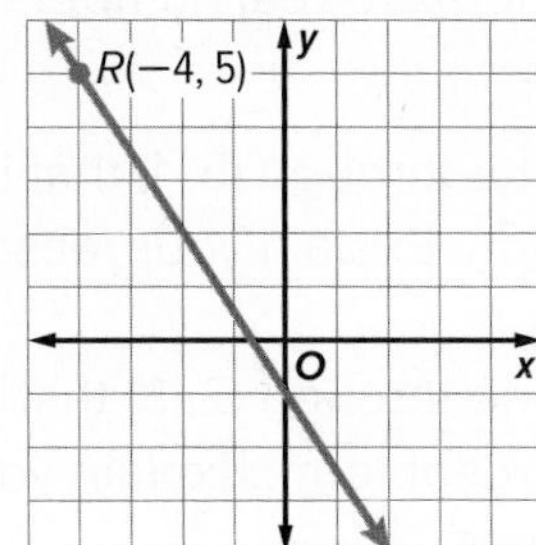

PRECISION **Determine whether any of the lines in each figure are parallel or perpendicular. Justify your answers.**

32.

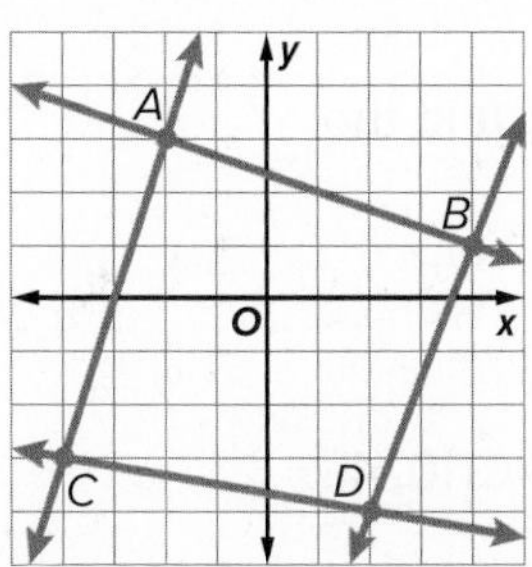

33.

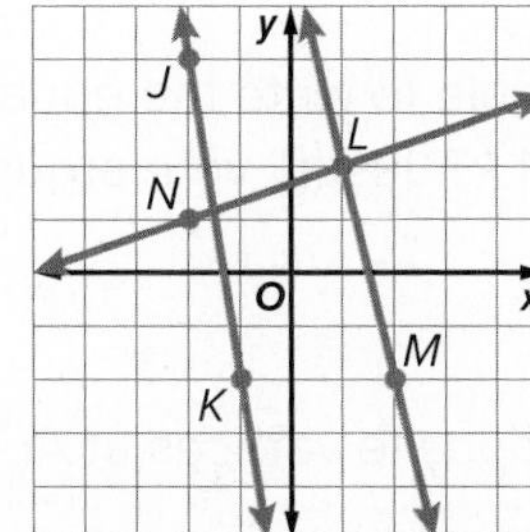

34.

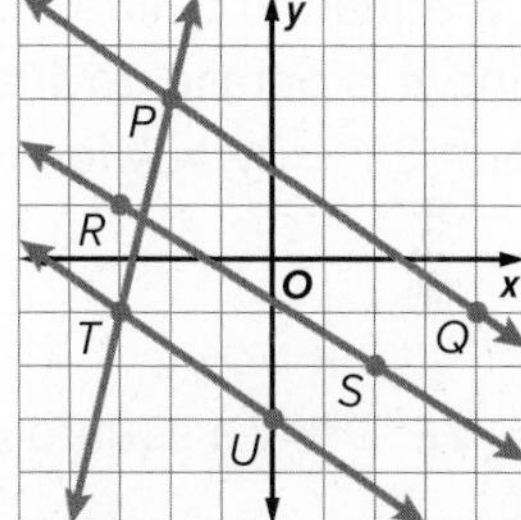

35. CITY BLOCKS The figure shows a map of part of a city consisting of two pairs of parallel roads. If a coordinate grid is applied to this map, Ford Street would have a slope of −3.

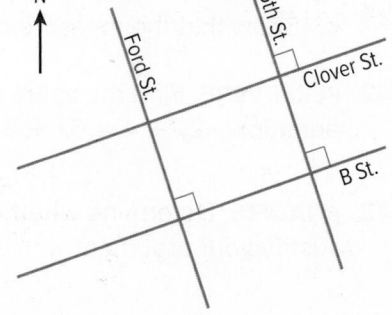

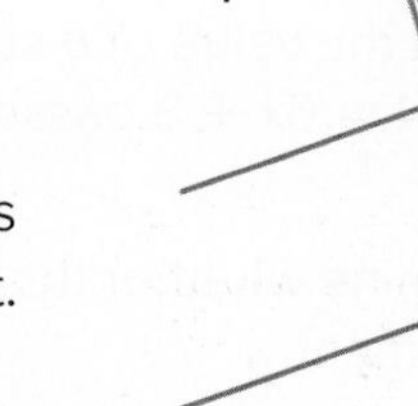

a. The intersection of B Street and Ford Street is 150 yards east of the intersection of Ford Street and Clover Street. How many yards south is it?

b. What is the slope of 6th Street? Explain.

c. What are the slopes of Clover and B Streets? Explain.

d. The intersection of B Street and 6th Street is 600 yards east of the intersection of B Street and Ford Street. How many yards north is it?

36. REASONING $\overleftrightarrow{AB}$ is parallel to $\overleftrightarrow{CD}$. The coordinates of A, B, and C are $A(-3, 1)$, $B(6, 4)$, and $C(1, -1)$. What is a possible set of coordinates for point D? Describe the reasoning you used to find the coordinates.

37. USE A MODEL A video game designer is using a coordinate plane to plan the path of a helicopter. She has already determined that the helicopter will move along straight segments from P to Q to R. The designer wants the next part of the path, $\overline{RS}$, to be perpendicular to $\overline{QR}$, and she wants point S to lie on the y-axis. What should the coordinates of point S be? Justify your answer.

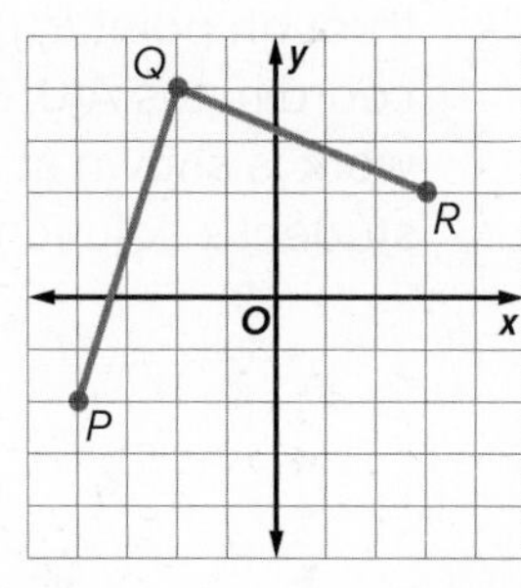

38. Line p passes through (1, 3) and (4, 7), and line q passes through (0, −2) and (a, b).

a. Find the slopes of lines p and q.

b. Find possible values of a and b if $p \parallel q$.

39. **STRUCTURE** Let a and b be nonzero real numbers. Line p has the equation $y = ax + b$.

a. Find the equation of the line through (5, 1) that is parallel to line p. Write the equation in point-slope form. Explain your reasoning.

b. Find the equation of the line through (2, 3) that is perpendicular to line p. Write the equation in slope-intercept form. Explain your reasoning.

40. **CONSTRUCT ARGUMENTS** The equation of line ℓ is $3y - 2x = 6$.

a. Line m is perpendicular to line ℓ and passes through the point $P(6, -2)$. Find the equation of line m.

b. Line n is parallel to line m. Is it possible to write the equation of line n in the form $2x + 3y = k$ for some constant k? Justify your argument.

41. **ANALYZE** Draw a square $ABCD$ with opposite vertices at $A(2, -4)$, and $C(10, 4)$.

a. Find the other two vertices of the square and label them B and D.

b. Show that $\overline{AD} \parallel \overline{BC}$ and $\overline{AB} \parallel \overline{DC}$.

c. Show that the measure of each angle inside the square is equal to 90°.

42. **PERSEVERE** Find the value of n so that the line perpendicular to the line with the equation $-2y + 4 = 6x + 8$ passes through the points $(n, -4)$ and $(2, -8)$.

43. **ANALYZE** Determine whether the points at (−2, 2), (2, 5), and (6, 8) are collinear. Justify your argument.

44. **CREATE** Write equations for a pair of perpendicular lines that intersect at the point at (−3, −7).

45. **WRITE** Write biconditionals to determine whether lines are parallel or perpendicular using slopes.

46. **FIND THE ERROR** A student was asked to find the equation of the line perpendicular to $\overleftrightarrow{AB}$ that passes through point P, given that A, B, and P have coordinates $A(0, 3)$, $B(2, 2)$, and $P(1, 4)$. The student's work is shown at the right. Do you agree with the student's solution? Explain your reasoning.

> Slope of $\overleftrightarrow{AB} = \frac{2-3}{2-0} = -\frac{1}{2}$.
>
> So, the slope of the required line is 2. The equation of this line is $y = 2x + b$. The line passes through $P(1,4)$.
>
> To find b: $1 = 2(4) + b$
>
> $1 = 8 + b$
>
> $-7 = b$
>
> So, the equation is $y = 2x - 7$.

Proving Lines Parallel

Explore Intersecting Lines

 **Online Activity** Use dynamic geometry software to complete the Explore.

> **INQUIRY** If a pair of alternate exterior or alternate interior angles is congruent, what relationship is formed?

Today's Goals

- Apply angle relationship theorems to identify parallel lines and find missing values.

Learn Identifying Parallel Lines

Corresponding angles are congruent when the lines cut by the transversal are parallel. The converse of this relationship is also true.

Theorem 12.19: Converse of Corresponding Angles Theorem

If two lines are cut by a transversal so that corresponding angles are congruent, then the lines are parallel.

Postulate 12.13: Parallel Postulate

If given a line and a point not on the line, then there exists exactly one line through the point that is parallel to the given line.

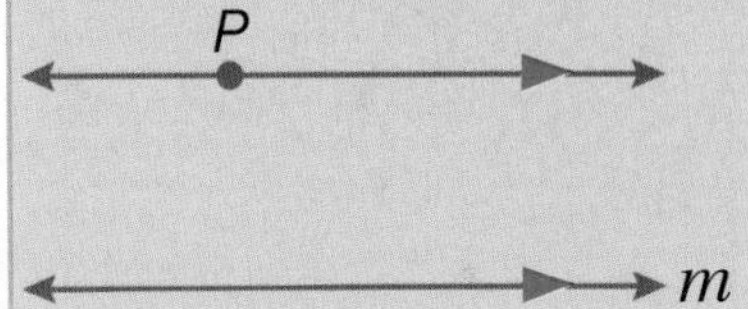

Study Tip

Euclid's Postulates The father of modern geometry, Euclid (c. 300 B.C.), realized that only a few postulates were needed to prove the theorems in his day. The Parallel Postulate is one of Euclid's five original postulates.

Parallel lines that are cut by a transversal create several pairs of congruent angles. These special angle pairs can be used to prove that a pair of lines is parallel.

Theorem 12.20: Alternate Exterior Angles Converse	
If two lines in a plane are cut by a transversal so that a pair of alternate exterior angles is congruent, then the lines are parallel.	If $\angle 1 \cong \angle 5$, then $a \parallel b$.
Theorem 12.21: Consecutive Interior Angles Converse	
If two lines in a plane are cut by a transversal so that a pair of consecutive interior angles is supplementary, then the lines are parallel.	If $m\angle 7 + m\angle 6 = 180°$, then $a \parallel b$.
Theorem 12.22: Alternate Interior Angles Converse	
If two lines in a plane are cut by a transversal so that a pair of alternate interior angles is congruent, then the lines are parallel.	If $\angle 7 \cong \angle 3$, then $a \parallel b$.
Theorem 12.23: Perpendicular Transversal Converse	
If two lines in a plane are perpendicular to the same line, then the lines are parallel.	

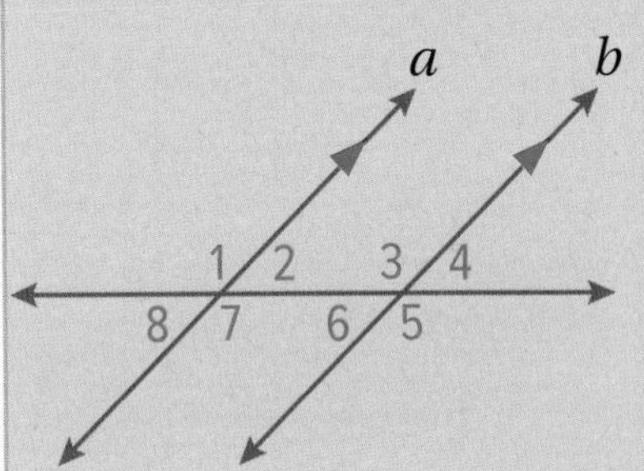

You will prove Theorems 12.20, 12.22, and 12.23 in Exercises 20, 19, and 18, respectively.

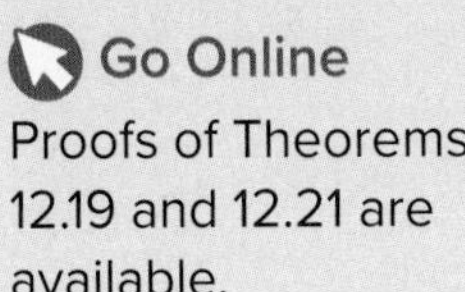

Proofs of Theorems 12.19 and 12.21 are available.

Example 1 Identify Parallel Lines

Use the given information to determine which lines, if any, are parallel. State the postulate or theorem that justifies your answer.

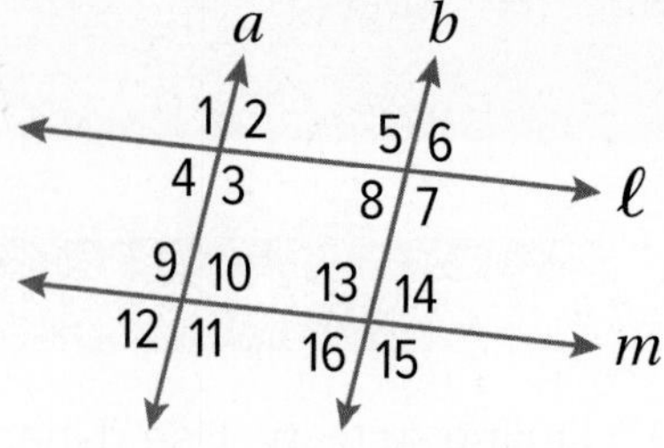

a. ∠2 ≅ ∠8

∠2 and ∠8 are alternate interior angles of lines a and b.

Because ∠2 ≅ ∠8, $a \parallel b$ by the Alternate Interior Angles Converse.

b ∠3 ≅ ∠11

∠3 and ∠11 are corresponding angles of lines ℓ and m. Because ∠3 ≅ ∠11, $\ell \parallel m$ by the Converse of the Corresponding Angles Theorem.

c. ∠12 ≅ ∠14

∠12 and ∠14 are alternate exterior angles of lines a and b. Because ∠12 ≅ ∠14, $a \parallel b$ by the Alternate Exterior Angles Converse.

Watch Out!

Error Analysis

Students should recognize that line ℓ is the transversal in part **a**, line a is the transversal in part **b**, and line m is the transversal in part **c**.

Check

Use the given information to determine which lines, if any, are parallel. State the postulate or theorem that justifies your answer.

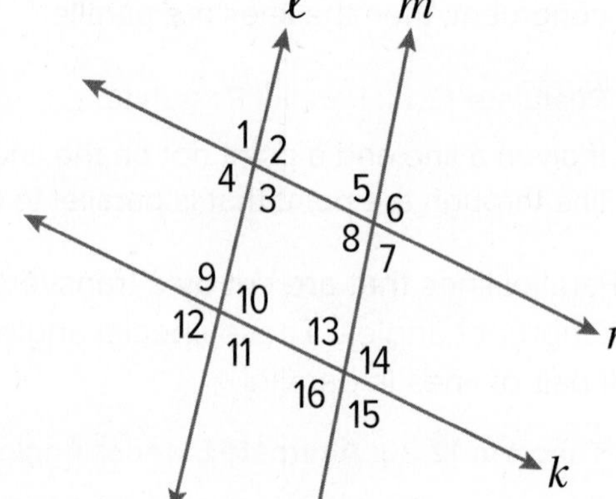

a. ∠1 ≅ ∠15

A. $\ell \parallel m$; Alternate Exterior Angles Converse

B. $n \parallel k$; Alternate Exterior Angles Converse

C. $\ell \parallel m$; Converse of Corresponding Angles Theorem

D. It is not possible to determine whether the lines are parallel.

b. $m\angle 3 + m\angle 10 = 180$

c. ∠3 ≅ ∠5

A. $\ell \parallel m$; Alternate Interior Angles Converse

B. $\ell \parallel m$; Consecutive Interior Angles Converse

C. $n \parallel k$; Alternate Interior Angles Converse

D. It is not possible to determine whether the lines are parallel.

Go Online You can complete an Extra Example online.

Example 2 Use Angle Relationships

Find the value of y so that $e \parallel f$.

From the figure, you know that line d is perpendicular to line e. For lines e and f to be parallel, line f must also be perpendicular to line d. If line f is perpendicular to line d, then $(4y + 10) = 90°$. Solve for y.

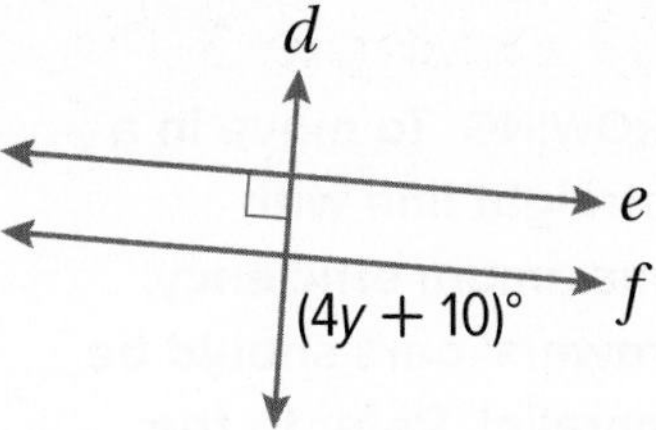

$90 = 4y + 10$	Definition of perpendicular
$80 = 4y$	Subtract 10 from each side.
$20 = y$	Divide each side by 4.

Check

a. Find the value of x so that $m \parallel n$. Identify the postulate or theorem you used.

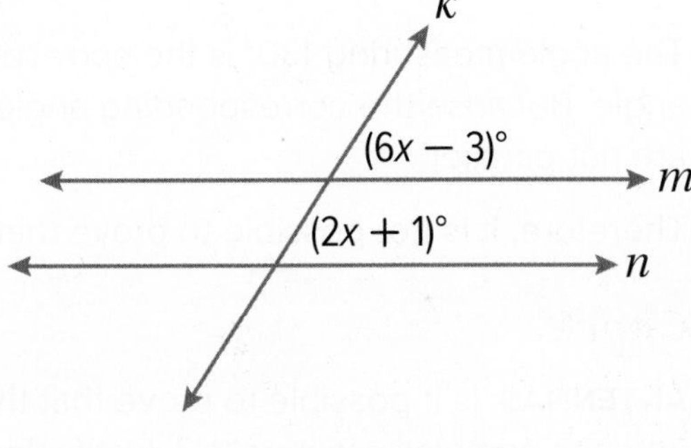

$x =$ ___?___ ; ___?___

b. Find $m\angle LMN$ so that $a \parallel b$.

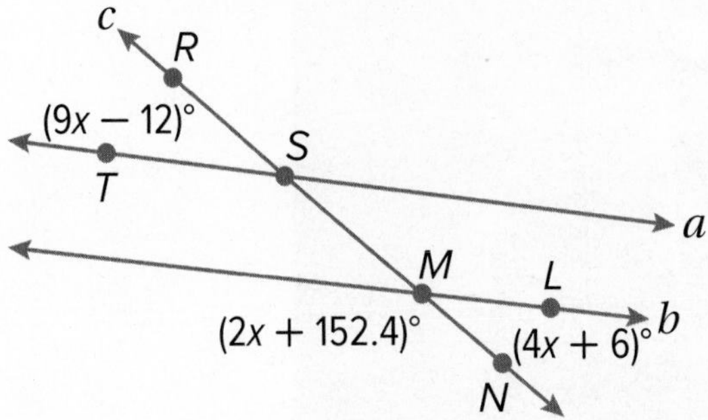

A. 3.6° B. 20.4° C. 100° D. 159.6°

Think About It!

Lakeisha argues that we do not have enough information to determine the correct value of y. She says that the two angles are not corresponding angles; therefore, it does not help to assume that the two angles are congruent. What theorems can you use to prove that line e is parallel to line f?

Go Online You can complete an Extra Example online.

The angle pair relationships formed by a transversal can be used to prove that two lines are parallel.

Study Tip

Assumptions When applying geometric theorems to real-world objects, we often make assumptions about the relationships between the objects being represented. In Example 3, the boat is the transversal and we are trying to determine whether the oars are parallel lines.

Example 3 Prove Lines Parallel

ROWING **To move in a straight line with maximum efficiency, rowers' oars should be parallel. Refer to the photo at the right. Is it possible to prove that any of the oars are parallel? Justify your answer.**

The angle that forms a linear pair with the 50° angle has a measure of 180° − 50° or 130°.

The angle measuring 130° is the corresponding angle to the 124° angle. Because the corresponding angles are not congruent, the lines are not parallel.

Therefore, it is not possible to prove that the oars are parallel.

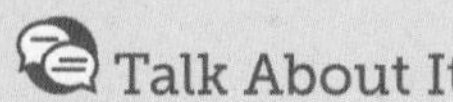

Talk About It!

What other information can be used to show that the oars are not parallel?

Check

ANTENNAS Is it possible to prove that the support poles of the antenna complex are parallel? Justify your answer.

A. No; because the consecutive interior angles are not supplementary, the support poles cannot be parallel.

B. No; because the alternate interior angles are not supplementary, the support poles cannot be parallel.

C. Yes; because the alternate interior angles are supplementary, the support poles are parallel.

D. Yes; because the consecutive interior angles are congruent, the support poles are parallel.

Go Online You can complete an Extra Example online.

Study Tip

Proving Lines Parallel When two parallel lines are cut by a transversal, the angle pairs formed are either congruent or supplementary. When a pair of lines forms angles that do not meet this criterion, the lines cannot be parallel.

Go Online

You may want to complete the construction activities for this lesson.

Ivan Smuk/Shutterstock; schankz/Shutterstock

Practice

Go Online You can complete your homework online.

Example 1

Use the given information to determine which lines, if any, are parallel. State the postulate or theorem that justifies your answer.

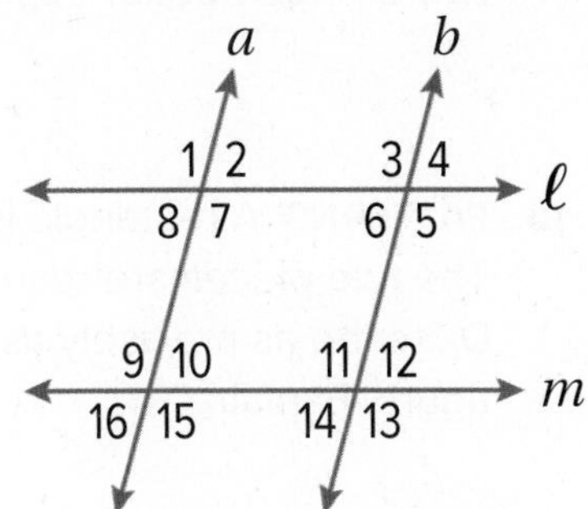

1. $\angle 3 \cong \angle 7$

2. $\angle 9 \cong \angle 11$

3. $\angle 2 \cong \angle 16$

4. $m\angle 5 + m\angle 12 = 180°$

Given the following information, determine which lines, if any, are parallel. State the theorem that justifies your answer.

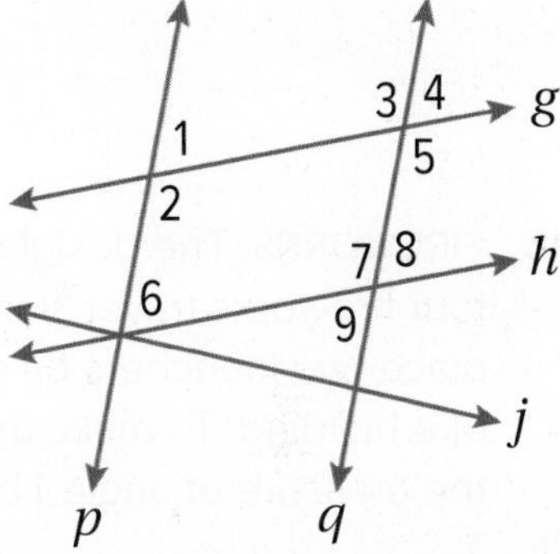

5. $\angle 1 \cong \angle 6$

6. $m\angle 7 + m\angle 6 = 180°$

Example 2

Find the value of x so that $\ell \parallel m$.

7.

8. 

9. 

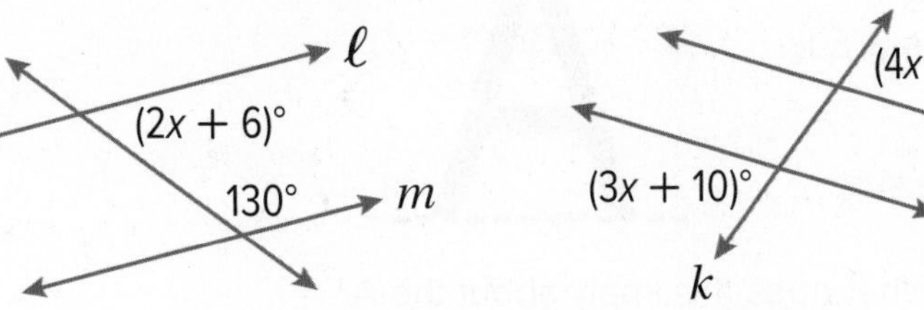

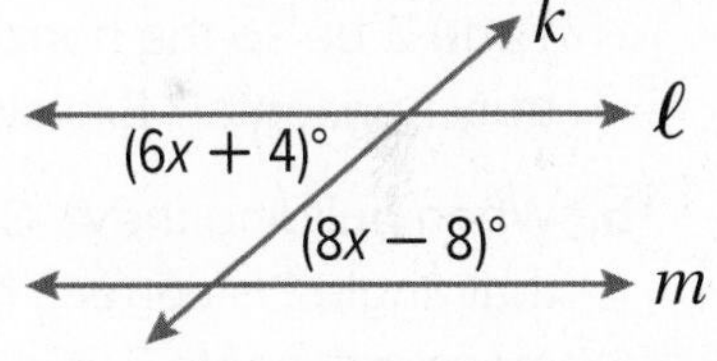

10.

11. 

12.

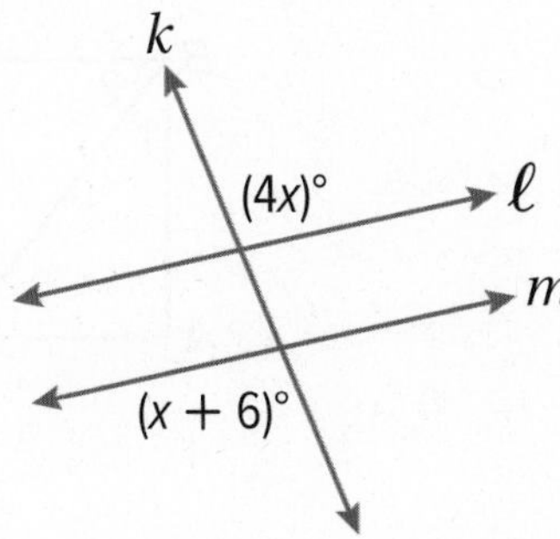

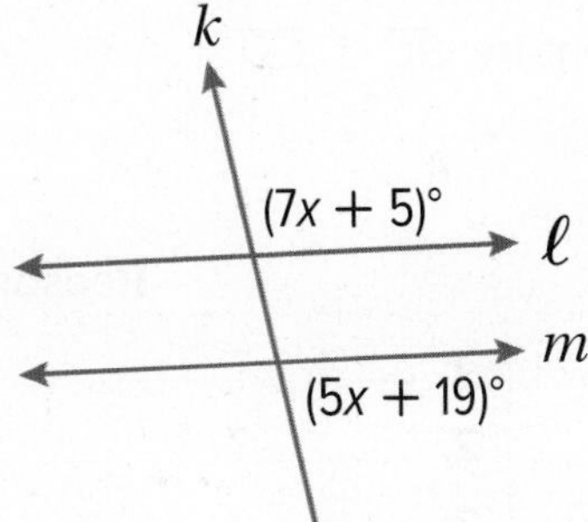

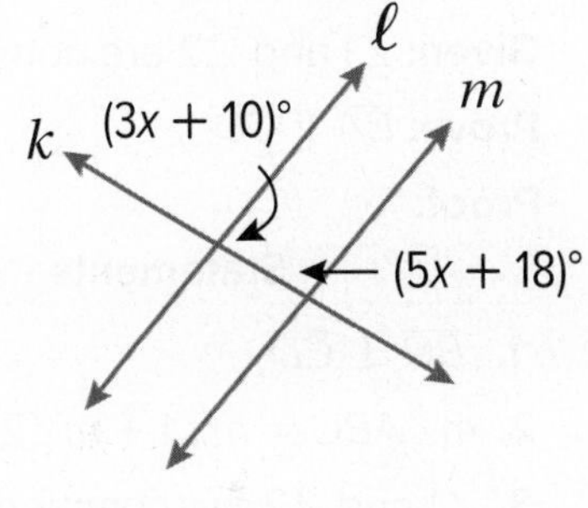

13. Find the value of x so that $\ell \parallel m$.

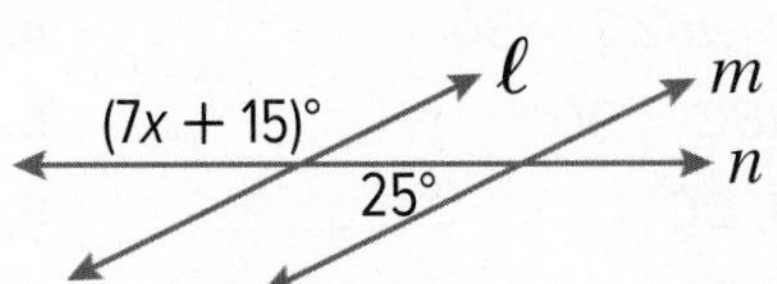

Example 3

14. **BOOKS** Each orange book on the bookshelf makes a 70° angle with the base of the shelf. What more can you say about these two orange books? Explain.

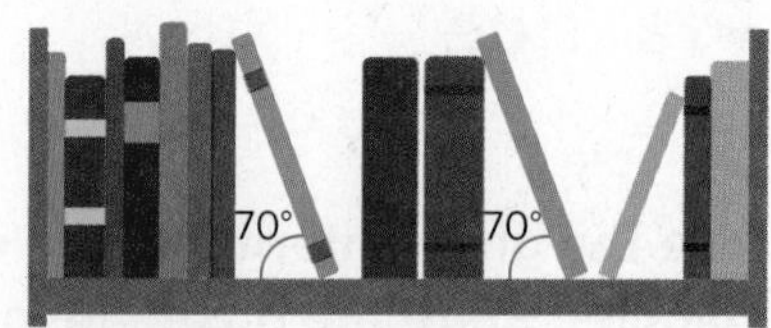

15. **PATTERNS** A rectangle is cut along the slanted, dashed line shown in the figure. The two pieces are rearranged to form another figure. Describe as precisely as you can the shape of the new figure. Explain.

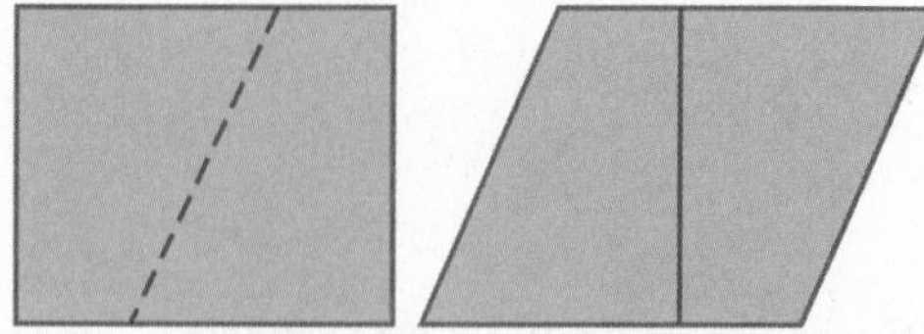

16. **FIREWORKS** The designers of a fireworks display want to have four fireworks travel along parallel trajectories. They decide to place two launchers on a dock and two launchers on the roof of a building. To make this display work correctly, what should the measure of angle 1 be? Explain.

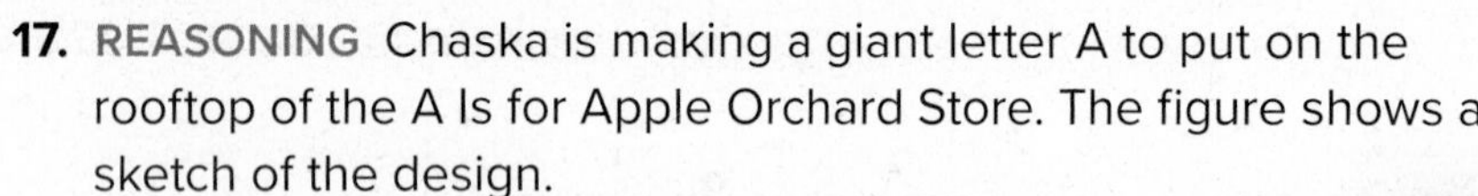

17. **REASONING** Chaska is making a giant letter A to put on the rooftop of the A Is for Apple Orchard Store. The figure shows a sketch of the design.

a. What should the measures of angles 1 and 2 be so the horizontal part of the A is truly horizontal? Explain.

b. When building the A, Chaska makes sure that angle 1 is correct, but when he measures angle 2, it is not correct. What does this imply about the A?

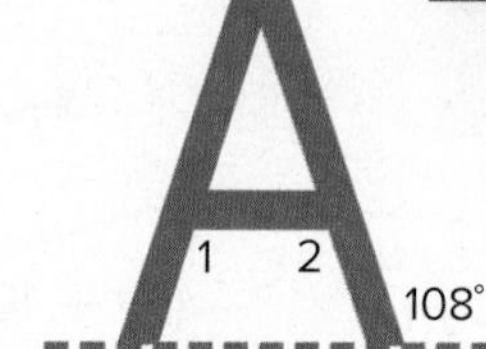

Mixed Exercises

18. **PROOF** Provide a reason for each statement in the proof of the Perpendicular Transversal Converse.

Given: $\angle 1$ and $\angle 2$ are complementary; $\overline{BC} \perp \overline{CD}$

Prove: $\overline{BA} \parallel \overline{CD}$

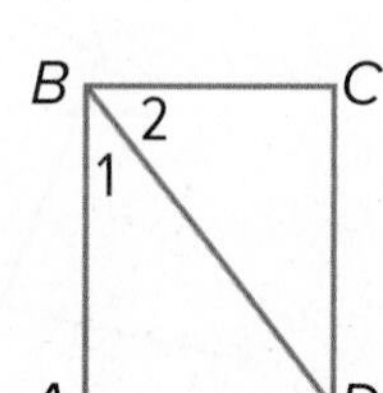

Proof:

Statements	Reasons
1. $\overline{BC} \perp \overline{CD}$	**1.** ___?___
2. $m\angle ABC = m\angle 1 + m\angle 2$	**2.** ___?___
3. $\angle 1$ and $\angle 2$ are complementary.	**3.** ___?___
4. $m\angle 1 + m\angle 2 = 90$	**4.** ___?___
5. $m\angle ABC = 90$	**5.** ___?___
6. $\overline{BA} \perp \overline{BC}$	**6.** ___?___
7. $\overline{BA} \parallel \overline{CD}$	**7.** ___?___

19. PROOF Write a paragraph proof to prove the Alternate Interior Angles Converse.

Given: $\angle 1 \cong \angle 2$

Prove: $\ell \parallel m$

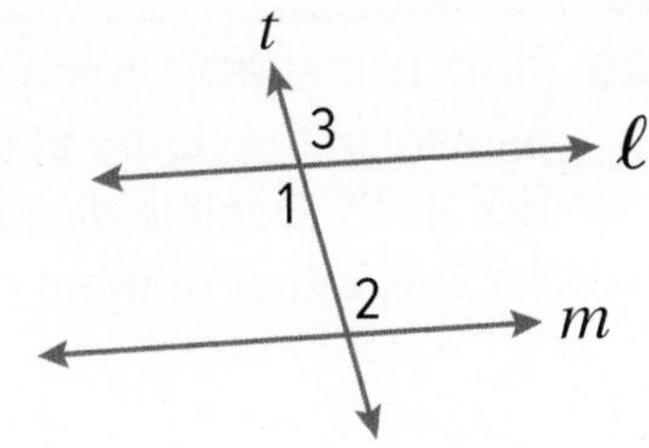

20. PROOF Write a paragraph proof to prove the Alternate Exterior Angles Converse.

Given: $\angle 1 \cong \angle 2$

Prove: $\ell \parallel m$

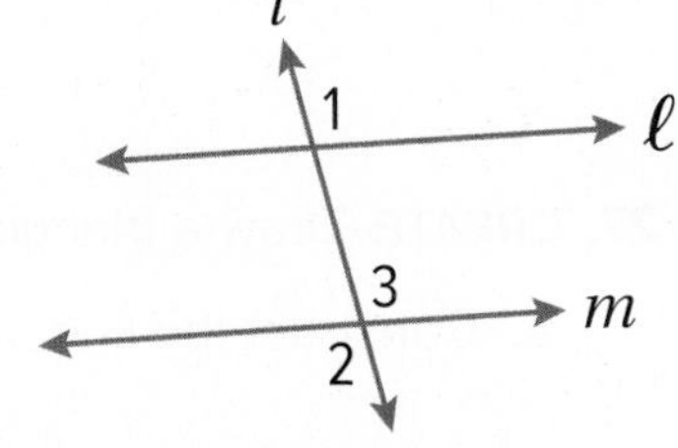

USE TOOLS **Use a compass and straightedge to construct the line through point *P* that is parallel to line *q*.**

21.

P

q

22.

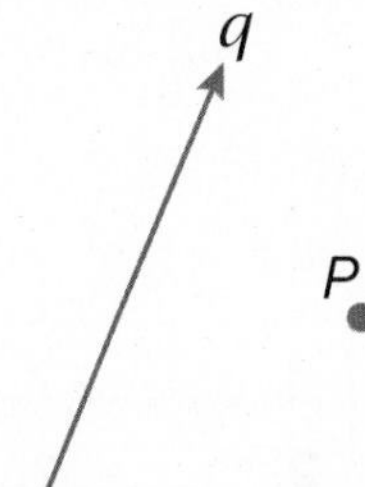

q

P

23. PICTURE FRAMES Lindy is making a wooden picture frame. She cuts the top and bottom pieces at a 45° angle. If the corners are right angles, explain how Lindy knows that each pair of opposite sides is parallel.

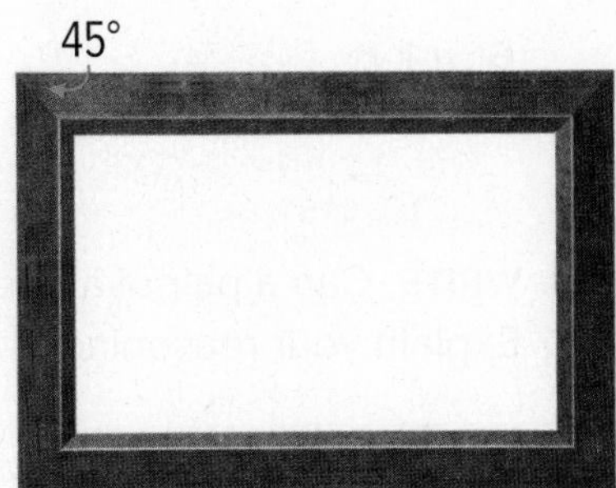

24. REASONING Jim made a frame for a painting. He wants to check to make sure that opposite sides are parallel by measuring the angles at the corners and seeing whether they are right angles. How many corners must he check to be sure that the opposite sides are parallel?

Higher Order Thinking Skills

25. FIND THE ERROR Sean and Daniela are determining which lines are parallel in the figure at the right. Sean says that because $\angle 1 \cong \angle 2$, $\overline{WY} \parallel \overline{XZ}$. Daniela disagrees and says that because $\angle 1 \cong \angle 2$, $\overline{WX} \parallel \overline{YZ}$. Is either of them correct? Explain your reasoning.

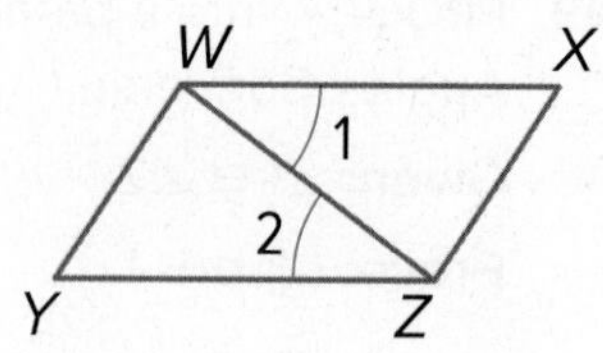

26. ANALYZE Is Theorem 3.23 still true if the two lines are not coplanar? Draw a figure to justify your argument.

27. CREATE Draw a triangle ABC.

a. Construct the line parallel to $\overline{BC}$ through point A.

b. Use measurements to justify that the line you constructed is parallel to $\overline{BC}$.

c. Justify the construction.

28. PROOF Use the figure at the right to complete the two-column proof to prove that two lines parallel to a third line are parallel to each other.

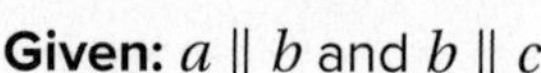

Given: $a \parallel b$ and $b \parallel c$

Prove: $a \parallel c$

Proof:

Statements	Reasons
1. $a \parallel b$ and $b \parallel c$	1. ___?___
2. $\angle 1 \cong \angle 3$	2. ___?___
3. ___?___	3. ___?___
4. ___?___	4. ___?___
5. ___?___	5. ___?___
6. $a \parallel c$	6. ___?___

29. WRITE Can a pair of angles be supplementary and congruent? Explain your reasoning.

30. PROOF Refer to the figure at the right.

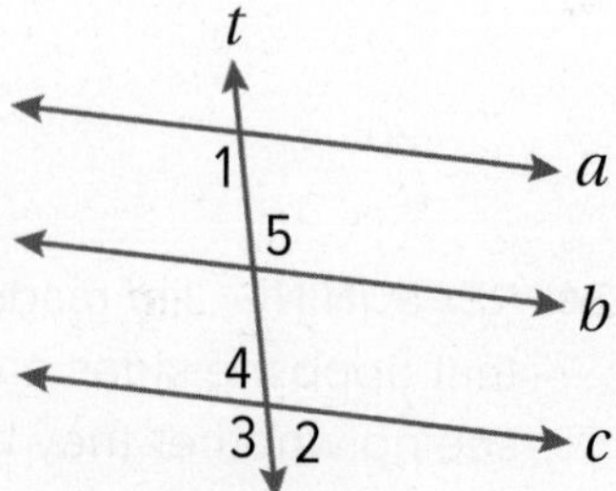

a. If $m\angle 1 + m\angle 2 = 180°$, prove that $a \parallel c$.

b. Given that $a \parallel c$, if $m\angle 1 + m\angle 3 = 180°$, prove that $t \perp c$.

Perpendiculars and Distance

Today's Goals

- Use perpendicular lines to find the distance between a point and a line.
- Find the distance between parallel lines by using perpendicular distance.

Today's Vocabulary

equidistant

Explore Distance from a Point to a Line

 **Online Activity** Use dynamic geometry software to complete the Explore.

INQUIRY How do you measure the distance between a point and a line?

Learn Distance Between a Point and a Line

There is an infinite number of lines that intersect a line and pass through a given point not on the line. However, when determining the distance between the line and the point, you must find the shortest distance between the two. This distance is the length of the segment that is perpendicular to the line through the point.

Key Concept • Distance Between a Point and a Line

The distance between a line and a point not on the line is the length of the segment perpendicular to the line from the point.

Given $\overleftrightarrow{AB}$ and point C not on the line, there are an infinite number of lines that pass through the point and intersect the line. The shortest distance between the point and the line is the length of the segment that is perpendicular to the line through the point. So, the distance between C and $\overleftrightarrow{AB}$ is CD.

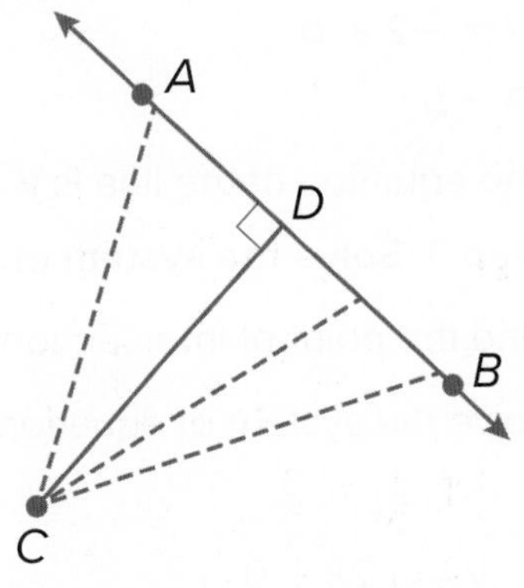

Just as there is one shortest distance from C to $\overleftrightarrow{AB}$, there is exactly one line that passes through C and is perpendicular to $\overleftrightarrow{AB}$.

Postulate 12.14: Perpendicular Postulate

If given a line and a point not on the line, then there exists exactly one line through the point that is perpendicular to the given line.

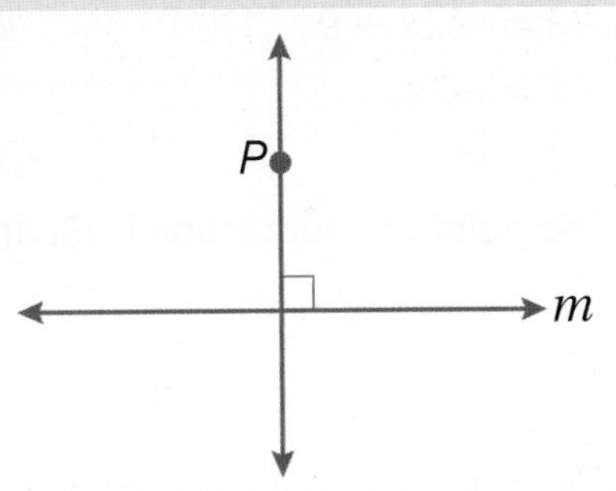

 **Go Online**

You can watch a video to see how to find the shortest distance between a point and a line on the coordinate plane.

 **Think About It!**

If line m and point P are on the coordinate plane and P is not on line m, how do you find the distance between P and line m?

Example 1 Distance from a Point to a Line on the Coordinate Plane

Line ℓ contains points (1, 2) and (5, 4). Find the distance between line ℓ and the point $P(1, 7)$.

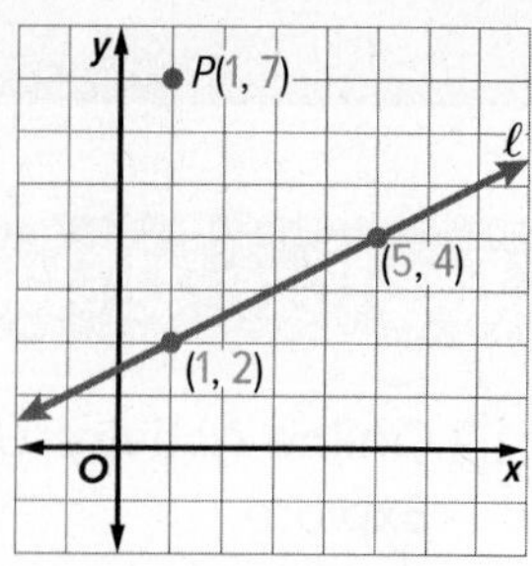

Step 1 Find the equation of line ℓ.

Begin by finding the slope of the line through points (1, 2) and (5, 4).

$m = \frac{y_2 - y_1}{x_2 - x_1} = \frac{4 - 2}{5 - 1} = \frac{2}{4}$ or $\frac{1}{2}$

Then write the equation of the line using the point (1, 2).

$y = mx + b$	Slope-intercept form
$2 = \frac{1}{2}(1) + b$	$m = \frac{1}{2}$ and $(x, y) = (1, 2)$
$2 = \frac{1}{2} + b$	Simplify.
$\frac{3}{2} = b$	Subtract $\frac{1}{2}$ from each side.

The equation of line ℓ is $y = \frac{1}{2}x + \frac{3}{2}$.

Step 2 Find the equation of the line perpendicular to line ℓ.

Write the equation of line w that is perpendicular to line ℓ and contains $P(1, 7)$. Because the slope of line ℓ is $\frac{1}{2}$, the slope of line w is -2. Write the equation of line w through $P(1, 7)$ with slope -2.

$y = mx + b$	Slope-intercept form
$7 = -2\,(1) + b$	$m = -2, (x, y) = (1, 7)$
$7 = -2 + b$	Simplify.
$9 = b$	Add 2 to each side.

The equation of the line is $y = -2x + 9$.

Step 3 Solve the system of equations.

Find the point of intersection of lines ℓ and w.

Solve the system of equations to determine the point of intersection.

$y = \frac{1}{2}x + \frac{3}{2}$	Equation of line ℓ
$y = -2x + 9$	Equation of line w
$y = 3$	Solve for y.

Solve for x.

$y = -2x + 9$	Equation of line w
$3 = -2x + 9$	Substitute 3 for y.
$-6 = -2x$	Subtract 9 from each side.
$3 = x$	Divide each side by -2.

The point of intersection is (3, 3). Let this be point Q.

Talk About It!

Why do you use the perpendicular distance to find the distance between a line and a point not on the line?

Study Tip

Solving Systems of Equations Systems of equations can be solved by graphing, substitution, or elimination. Keep this in mind when you are finding the intersection point of perpendicular lines.

Go Online You can complete an Extra Example online.

Step 4 Calculate the distance between P and Q.

Use the Distance Formula to determine the distance between $P(1, 7)$ and $Q(3, 3)$.

$d = \sqrt{(x_2 - x_1)^2 + (y_2 - y_1)^2}$ Distance Formula

$= \sqrt{(3 - 1)^2 + (3 - 7)^2}$ $x_2 = 3$, $x_1 = 1$ and $y_2 = 3$, and $y_1 = 7$

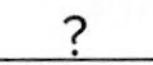

$= \sqrt{20}$ Simplify.

The distance between point P and line ℓ is $\sqrt{20}$ or about 4.47 units.

Check

Line n contains points $(-5, 3)$ and $(4, -6)$. Find the distance between line n and point $Q(2, 4)$. Round to the nearest tenth, if necessary.

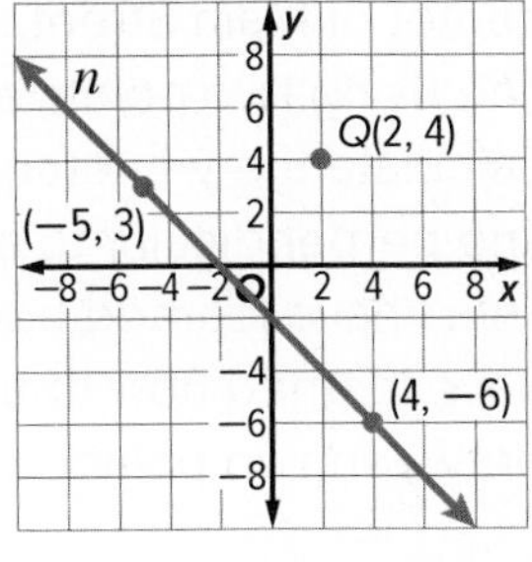

__?__

Study Tip

Units of Measure When finding the distance between a point and a line on the coordinate plane, your final measurement should be labeled with *units* unless the problem is set in a real-world context.

Apply Example 2 Solve a Design Problem by Using Distance

AMUSEMENT PARK **The developers of an amusement park want to build a new attraction. According to park regulations, the entrance to each attraction must be at least 10 yards from the center of Main Street. In the design plans, the entrance to the new attraction is located at $A(-6, -10)$, and Main Street contains the points $(-1, 3)$ and $(11, -9)$. If each unit represents 1 yard, will the new attraction comply with park regulations?**

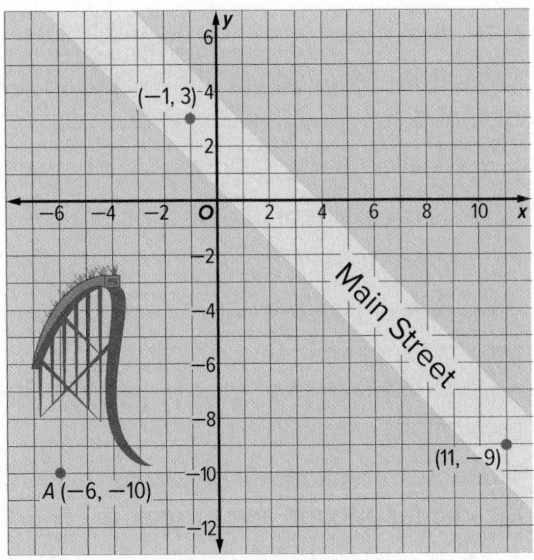

1 What is the task?

Describe the task in your own words. Then list any questions that you may have. How can you find answers to your questions?

Sample answer: I need to determine whether the entrance to the new attraction is at least 10 yards from the center of Main Street. How can I represent Main Street as a linear equation? How can I find the perpendicular distance from the entrance to the center of Main Street? I can review using points to write the equation of a line, and I can review finding the perpendicular distance between a point and a line.

(*continued on the next page*)

Study Tip

Estimation You can also use the horizontal distance between a line and a point not on the line to estimate the distance between the point and the line. When you graph the horizontal and perpendicular lines that contain the point and intersect the given line, a right triangle is created. The horizontal distance between the given point and line is the same as the length of the hypotenuse of the right triangle. So, you know the perpendicular distance, or the length of the right triangle's leg, must be less than the horizontal distance between the point and the given line.

2 How will you approach the task? What have you learned that you can use to help you complete the task?

Sample answer: I will use the given points to write an equation in slope-intercept form that can be used to represent Main Street. Then, I will find the equation of a line that is perpendicular to Main Street that also passes through point A. I will find the point of intersection of Main Street and the perpendicular line that I found. Finally, I will use the Distance Formula, the point of intersection, and point A to calculate the distance between the entrance of the new attraction and the center of Main Street. I have learned how to calculate the slope of a line using two points and how to use a slope and a given point to calculate the y-intercept. I have learned how to find the equation of a line perpendicular to a given line and through a point not on the given line. I have learned how to find the point of intersection of two lines. I have learned how to use the Distance Formula to find the distance between two points.

3 What is your solution?

Use your strategy to solve the problem.

What is the equation of the line in slope-intercept form that represents Main Street?

$y = -x + 2$

What is the equation of the line in slope-intercept form that is perpendicular to Main Street and passes through point A?

$y = x - 4$

What is the point of intersection of these two lines?

$(3, -1)$

What is the distance between the entrance to the new attraction and the center of Main Street? Will the new attraction be located far enough away from the center of Main Street to comply with park regulations?

The new attraction will be located about 12.7 yards, so it will be located far enough away from the center of Main Street.

4 How can you know that your solution is reasonable?

Write About It! Write an argument that can be used to defend your solution.

Sample answer: I can use the coordinate grid to estimate the distance between the entrance of the new attraction and the center of Main Street. After sketching a line that appears to be perpendicular to Main Street through point A, I estimate the number of units between point A and Main Street. My estimation supports my solution.

Go Online You can complete an Extra Example online.

Check

ZONING Javier wants to build a shed on his property. According to zoning laws, the shed must be at least 20 feet from his property line. Javier knows that points A and B fall on his property line. If Javier plans to build the shed behind his house at point C, will he satisfy the zoning laws? If yes, how far away will the shed be from Javier's property line? Each unit on the coordinate plane represents 1 foot.

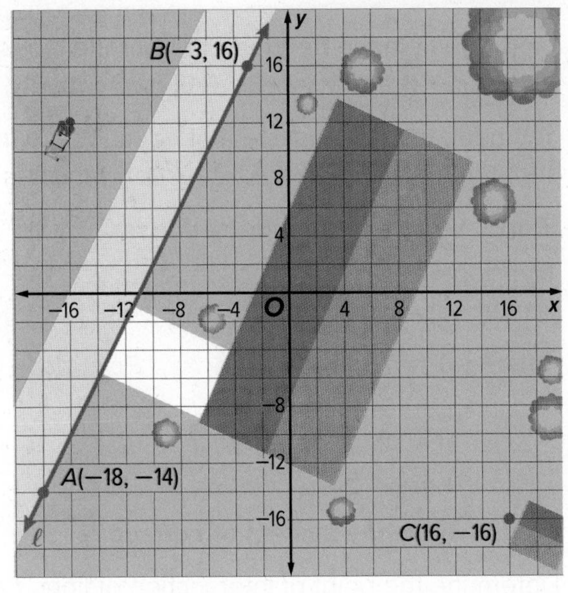

A. no B. yes; 31.3 ft C. yes; 34.1 ft D. yes; 37.2 ft

Learn Distance Between Parallel Lines

By definition, parallel lines do not intersect. An alternate definition states that two lines in a plane are parallel if they are always equidistant. Two lines are **equidistant** from each other if the distance between the two lines, measured along a perpendicular line or segment to the two lines, is always the same.

Think About It!

How would you find the distance between two parallel planes?

Key Concept • Distance Between Parallel Lines

The distance between two parallel lines is the perpendicular distance between one of the lines and any point on the other line.

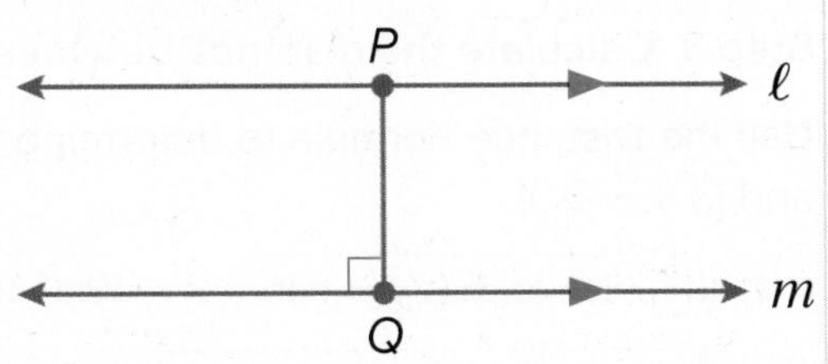

Theorem 12.24: Two Lines Equidistant from a Third

In a plane, if two lines are each equidistant from a third line, then the two lines are parallel to each other.

Example If line w and line v are equidistant from line x, then line w and line v are parallel.

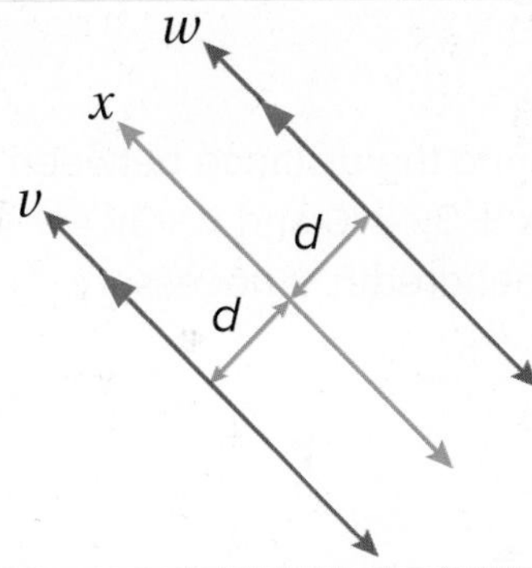

You will prove Theorem 12.24 in Exercise 28.

Example 3 Distance Between Parallel Lines

Find the distance between the parallel lines r and t with equations $y = -3x - 5$ and $y = -3x + 6$, respectively.

You need to solve a system of equations to find the endpoints of a segment perpendicular to lines r and t. Lines r and t have slope -3.

Step 1 Write an equation of line q.

The slope of q is the opposite reciprocal of -3, or $\frac{1}{3}$. Use the y-intercept of line r, $(0, -5)$, as a point through which line q will pass.

$(y - y_1) = m(x - x_1)$	Point-slope form
$[y - (-5)] = \frac{1}{3}(x - 0)$	$x_1 = 0$, $y_1 = -5$, and $m = \frac{1}{3}$
$y = \frac{1}{3}x - 5$	Solve

Step 2 Solve the system of equations.

Determine the point of intersection of lines t and q.

t: $y = -3x + 6$	q: $y = \frac{1}{3}x - 5$
$-3x + 6 = \frac{1}{3}x - 5$	Substitute.
$6 + 5 = \frac{1}{3}x + 3x$	Group like terms.
$\frac{33}{10} = x$	Solve.

Solve for y when $x = \frac{33}{10}$.

$y = \frac{1}{3}\left(\frac{33}{10}\right) - 5$	Substitute $\frac{33}{10}$ for x in the equation for q.
$y = -\frac{39}{10}$	Simplify.

The point of intersection is $\left(\frac{33}{10}, -\frac{39}{10}\right)$ or $(3.3, -3.9)$.

Step 3 Calculate the distance between lines r and t.

Use the Distance Formula to determine the distance between $(0, -5)$ and $(3.3, -3.9)$.

$d = \sqrt{(x_2 - x_1)^2 + (y_2 - y_1)^2}$	Distance Formula
$= \sqrt{(3.3 - 0)^2 + [-3.9 - (-5)]^2}$	$x_2 = 3.3$, $x_1 = 0$, $y_2 = -3.9$, and $y_1 = -5$
≈ 3.5	Use a calculator.

The distance between the lines is about 3.5 units.

Check

Find the distance between parallel lines a and b with equations $x + 3y = 6$ and $x + 3y = -14$, respectively. Round to the nearest hundredth, if necessary.

___?___

Go Online You can complete an Extra Example online.

Study Tip

Perpendicular Lines When you are finding the equation of a line perpendicular to a pair of parallel lines, using the y-intercept of one of the parallel lines as a point contained by the perpendicular line will allow for easier calculations. The y-intercept is also used because it can easily be determined when the equation of a line is given in slope-intercept form.

Think About It!

Compare and contrast the processes for finding the distance between a point and a line and for finding the distance between parallel lines.

Practice

Go Online You can complete your homework online.

Examples 1 and 2

Find the distance between point P and line ℓ.

1. Line ℓ contains points (0, −3) and (7, 4). Point P has coordinates (4, 3).

2. Line ℓ contains points (11, −1) and (−3, −11). Point P has coordinates (−1, 1).

3. Line ℓ contains points (−2, 1) and (4, 1). Point P has coordinates (5, 7).

4. Line ℓ contains points (4, −1) and (4, 9). Point P has coordinates (1, 6).

5. Line ℓ contains points (1, 5) and (4, −4). Point P has coordinates (−1, 1).

6. Line ℓ contains points (−8, 1) and (3, 1). Point P has coordinates (−2, 4).

7. DESIGN Dante is designing a poster for prom using a design program with a coordinate grid. He starts by creating a geometric border. Dante wants the text on the poster to be at least 3 inches away from the top left-hand corner of the border. The border contains the points (0, 7) and (7, 14). If Dante places the text at (7, 8), is the text at least 3 inches away from the border? If yes, how far away is the text from the border? Let every unit represent an inch. Round your answer to the nearest hundredth, if needed.

8. PHYSICS Mrs. Holmes's physics class is using 3D-printing software to create miniature bridges that can hold at least 5 pounds. Teams will print multiple parts of the bridges and then assemble the parts. One team wants there to be at least 6 inches between the upper rail and the lower rail of the bridge. The lower rail of the bridge contains the points (2, 10) and (16, 3). If the upper rail contains the point (5, 1), will the bridge meet the team's specifications? If yes, how far apart are the rails? Let every unit represent an inch. Round your answer to the nearest hundredth, if needed.

Example 3

Find the distance between each pair of parallel lines with the given equations.

9. $y = 7$
$y = -1$

10. $x = -6$
$x = 5$

11. $y = 3x$
$y = 3x + 10$

12. $y = -5x$
$y = -5x + 26$

13. $y = x + 9$
$y = x + 3$

14. $y = -2x + 5$
$y = -2x - 5$

15. $y = \frac{1}{4}x + 2$
$4y - x = -60$

16. $3x + y = 3$
$y + 17 = -3x$

17. $y = -\frac{5}{4}x + 3.5$
$4y + 10.6 = -5x$

Mixed Exercises

Find the distance from the line to the given point.

18. $y = -3$; $(5, 2)$

19. $y = \frac{1}{6}x + 6$; $(-6, 5)$

20. $x = 4$; $(-2, 5)$

21. TELEPHONE WIRES Isaiah works for a telephone company. He rewired some telephone wires on a pole. How can Isaiah use perpendicular distances to confirm that the wires are parallel?

lkpro/Shutterstock

22. **STATE YOUR ASSUMPTION** A city planner is designing a new park using a map of the city on a coordinate plane. The planner wants the entrance of the park to be at least 4 meters away from Washington Avenue. On the map, Washington Avenue contains the points (2, −4) and (11, −1).

a. If the city planner wants to build the entrance of the park at (3, 3), will the entrance be at least 4 meters away from Washington Avenue? If yes, how far away will the entrance be from the street? Let every unit represent 1 meter. Round your answer to the nearest hundredth, if needed.

b. What assumption did you make while solving this problem?

23. Construct the line through G perpendicular to $\overleftrightarrow{EF}$.

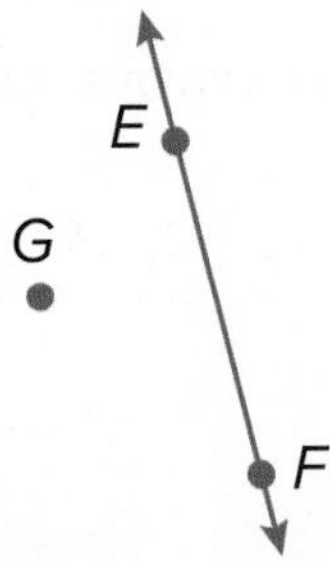

24. Construct the line through P perpendicular to m.

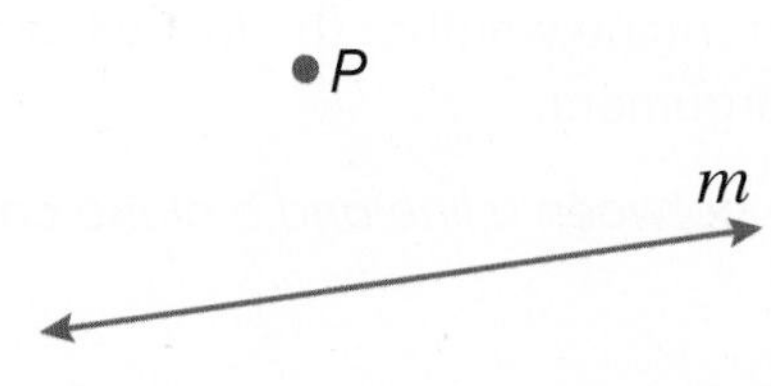

25. **REASONING** The diagram at the right shows the path that Mark walked from the tee box to where his ball landed on the green. Is the path the shortest possible one from the tee box to the golf ball? Explain why or why not.

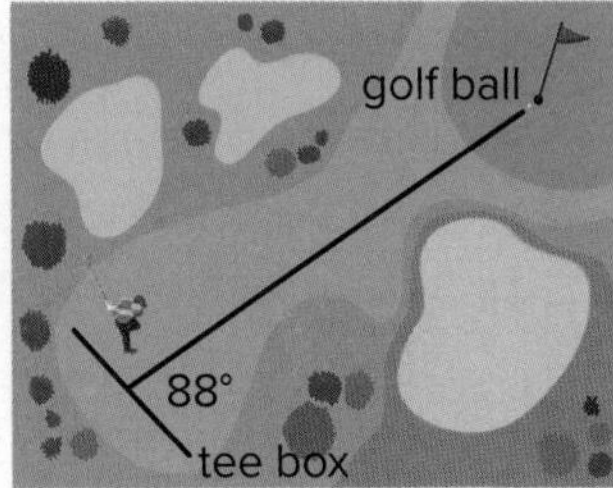

26. $\overline{AB}$ has a slope of 2 and midpoint $M(3, 2)$. A segment perpendicular to $\overline{AB}$ has midpoint $P(4, -1)$ and shares endpoint B with $\overline{AB}$.

a. Graph the segments.

b. Find the coordinates of A and B.

27. What does it mean if the distance between a point P and a line ℓ is zero? If the distance between two lines is zero?

28. **PROOF** Copy and complete the two-column proof of Theorem 12.24.

Given: ℓ is equidistant to m, and n is equidistant to m.

Prove: $\ell \parallel n$

Proof:

Statements	Reasons
1. ℓ is equidistant to m, and n is equidistant to m.	1. ___?___
2. ___?___	2. Definition of equidistant
3. ___?___	3. Definition of parallel lines
4. ___?___	4. Transititve Property of Equality
5. $\ell \parallel n$	5. ___?___

29. WRITE Summarize the steps that are necessary to find the distance between a pair of parallel lines given the equations of the two lines.

30. PERSEVERE Suppose a line perpendicular to a pair of parallel lines intersects the lines at the points $(a, 4)$ and $(0, 6)$. If the distance between the parallel lines is $\sqrt{5}$, find the value of a and the equations of the parallel lines.

31. ANALYZE Determine whether the following statement is *sometimes, always,* or *never* true. Justify your argument.

The distance between a line and a plane can be found.

32. CREATE Draw an irregular convex pentagon using a straightedge.

a. Use a compass and straightedge to construct a perpendicular line between one vertex and a side opposite the vertex.

b. Use measurement to justify that the constructed line is perpendicular to the side chosen.

c. Use mathematics to justify this conclusion.

33. WRITE Rewrite Theorem 12.24 in terms of two planes that are equidistant from a third plane. Sketch an example.

34. FIND THE ERROR Harold draws the segments AB and CD shown below using a straightedge. He claims that these two lines, if extended in both directions, will never intersect. Olga claims that the lines will eventually intersect. Who is correct? Explain your reasoning.

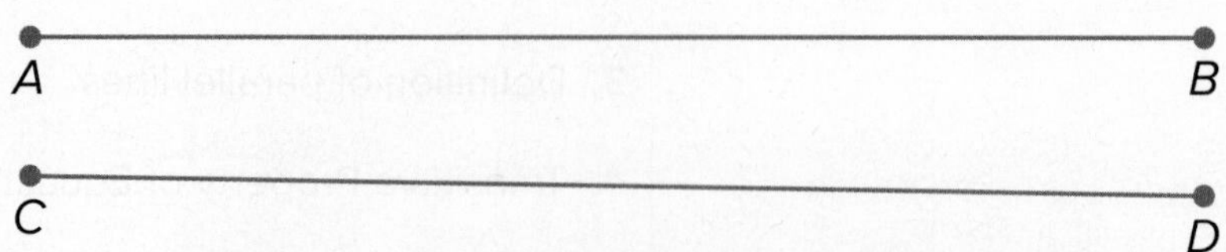

Review

 Essential Question

What makes a logical argument, and how are logical arguments used in geomtetry?

A logical argument is well organized and has statements that can be justified using postulates, theorems, and definitions.

Module Summary

Lessons 12-1 and 12-2

Conjectures and Logical Statements

- To show that a conjecture is not true for all cases, find a counterexample.
- An if-then statement is a compound statement of the form "if *p*, then *q*," where *p* and *q* are statements.
- The converse of a conditional statement is formed by exchanging the hypothesis and conclusion of the conditional statement. The inverse is formed by negating the hypothesis and the conclusion of the conditional statement. The contrapositive is formed by negating the hypothesis and the conclusion of the converse of the conditional statement.

Lessons 12-3 and 12-4

Reasoning and Proof

- If $p \rightarrow q$ is a true statement and *p* is true, then *q* is true.
- If $p \rightarrow q$ and $q \rightarrow r$ are true statements, then $p \rightarrow r$ is a true statement.
- A postulate or axiom is a statement that is accepted as true without proof.
- A proof contains statements and reasons that are organized to show progression from given information to a conclusion. Proofs can be in a two-column format, a flow format (using boxes and arrows), or in a paragraph format.

Lessons 12-5 and 12-6

Proving Segment and Angle Relationships

- The Angle Addition Postulate can be used with other angle relationships to prove theorems about supplementary and complementary angles.
- The properties of algebra that apply to the congruence of segments and the equality of their measures also hold true for the congruence of angles and the equality of their measures.

Lessons 12-7 through 12-10

Relationships Among Angles and Lines

- When two parallel lines are cut by a transversal, there are relationships between specific pairs of angles.
- If two lines are cut by a transversal so corresponding angles are congruent, then the lines are parallel.
- The distance between a line and a point not on the line is the length of the segment perpendicular to the line from the point.

Study Organizer

 Foldables

Use your Foldable to review this module. Working with a partner can be helpful. Ask for clarification of concepts as needed.

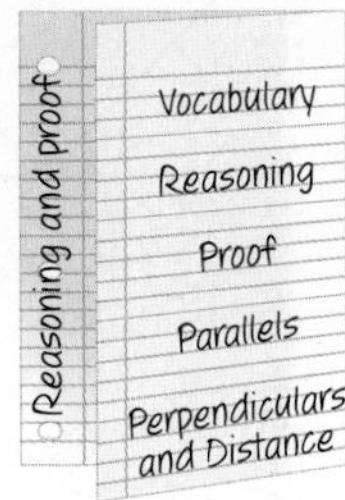

Test Practice

1. **OPEN RESPONSE** Point B is the midpoint of $\overline{AC}$, and point C is the midpoint of $\overline{AD}$. If $CD = 12$, what is AB? (Lesson 12-5)

2. **OPEN RESPONSE** Points X, Y, and Z are collinear, and Y is the midpoint of $\overline{XZ}$. Find the value of b. (Lesson 12-5)

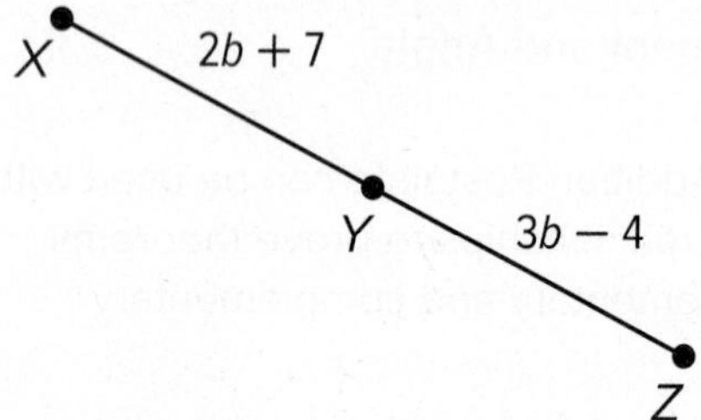

3. **MULTIPLE CHOICE** Point B is the midpoint of $\overline{AC}$. $AB = 2x + 5$ and $BC = 5x - 1$. What is the length of $\overline{AB}$? (Lesson 12-5)

   A. 2 units

   B. 9 units

   C. 18 units

   D. 21 units

4. **MULTIPLE CHOICE** If $m\angle 1 = (2x)°$ and $m\angle 3 = (3x)°$, what is $m\angle 1$ in degrees? (Lesson 12-6)

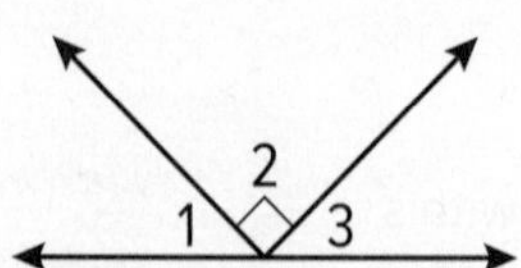

   A. 18

   B. 36

   C. 54

   D. 72

5. **OPEN RESPONSE** Use the figure below to find x, $m\angle ABD$, and $m\angle DBC$. (Lesson 12-6)

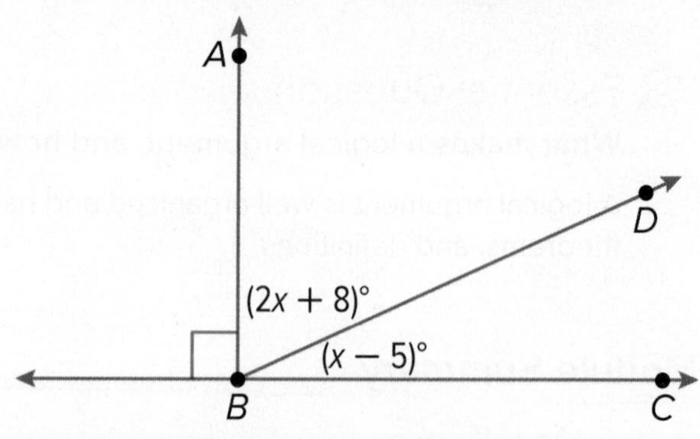

6. **OPEN RESPONSE** If Pine Street is parallel to Locust Street, find the values of a and b. (Lesson 12-7)

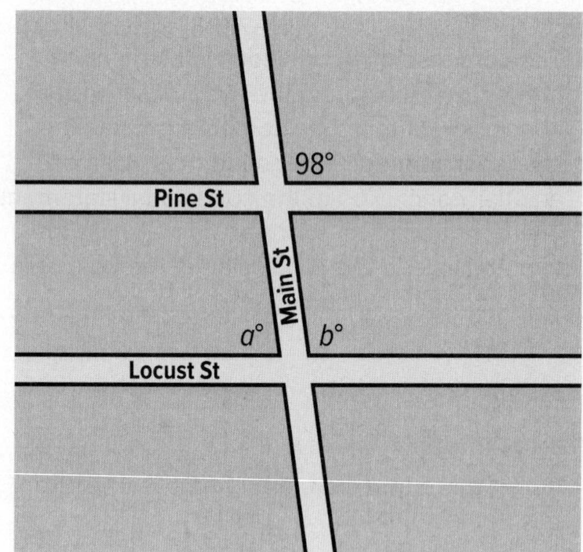

7. MULTI-SELECT Select all the statements that describe parallel lines. (Lesson 12-7)

A. If lines are parallel, then they are coplanar.

B. If lines are parallel, then they are not coplanar.

C. If lines are parallel, then they intersect.

D. If lines are parallel, then they do not intersect.

E. If lines are parallel, then they are not skew.

8. OPEN RESPONSE A line passes through points at (9, 5) and (4, 3). What is the slope of the line perpendicular to this line? (Lesson 12-8)

9. OPEN RESONSE Three lines have these equations:

Line m: $y = \frac{2}{3}x - 7$

Line n: $y = -\frac{2}{3}(x + 1)$

Line p: $y = -\frac{3}{2}x + 4$

Identify the lines that have a perpendicular relationship. (Lesson 12-8)

10. OPEN RESPONSE Write an equation in slope-intercept form for the line that passes through $(-3, 2)$, perpendicular to $y = \frac{1}{2}x + 9$. (Lesson 12-8)

11. OPEN RESPONSE If $\overleftrightarrow{AB} \parallel \overleftrightarrow{CD}$, what is $m\angle ACD$? (Lesson 12-9)

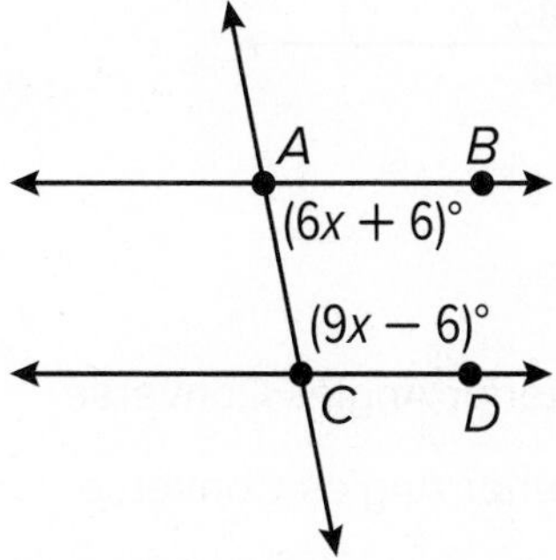

12. MULTIPLE CHOICE In the diagram, $\angle GDE$ and $\angle DEF$ are supplementary, but $\angle GDE$ is not congruent to $\angle EFG$.

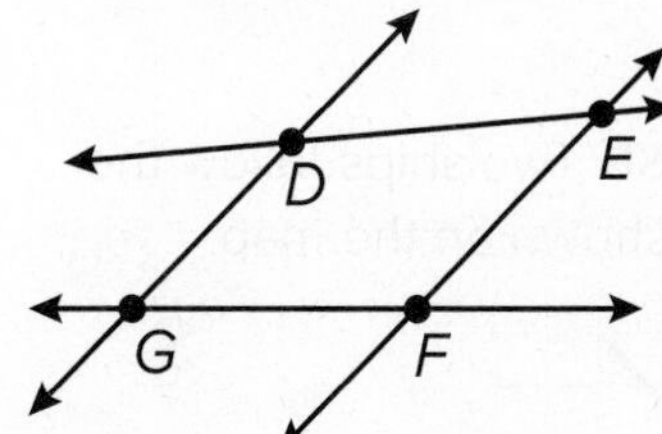

Which lines are parallel? (Lesson 12-9)

A. $\overleftrightarrow{DE} \parallel \overleftrightarrow{GF}$

B. $\overleftrightarrow{DG} \parallel \overleftrightarrow{EF}$

C. $\overleftrightarrow{DE} \parallel \overleftrightarrow{GF}$ and $\overleftrightarrow{DG} \parallel \overleftrightarrow{EF}$

D. Neither pair of lines is parallel.

13. **MULTI-SELECT** Using the given figure, which theorem(s) could be used to prove the lines are parallel? Select all that apply. (Lesson 12-9)

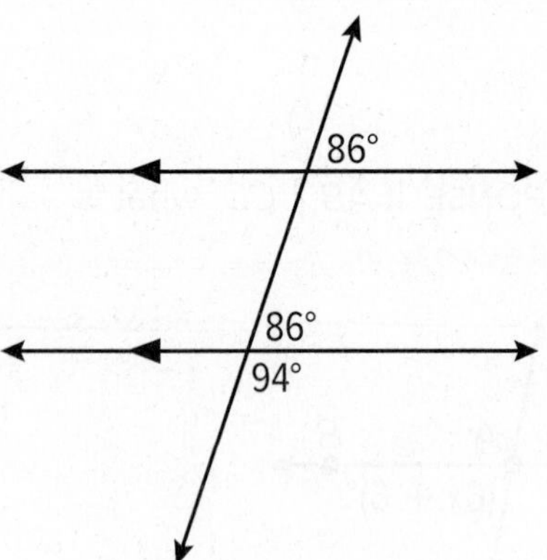

A. Alternate Exterior Angles Converse

B. Alternate Interior Angles Converse

C. Consecutive Interior Angles Converse

D. Corresponding Angles Converse

E. Perpendicular Transversal Converse

F. None of the above

14. **OPEN RESPONSE** Two ships follow the parallel paths shown on the map.

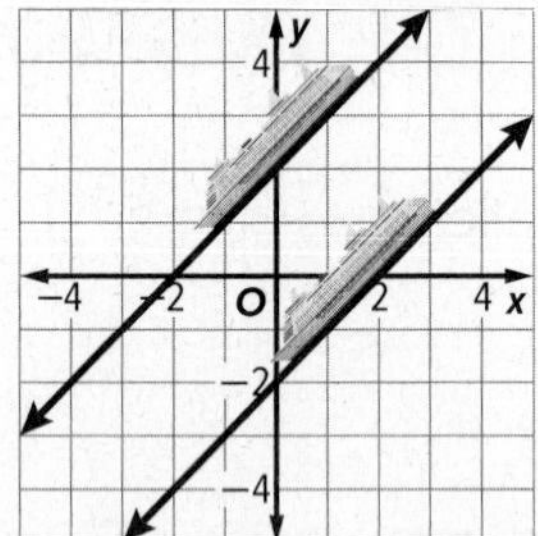

If one unit is 1 nautical mile, what is the shortest distance between the two paths? Round your answer to the nearest tenth. (Lesson 12-10)

15. **MULTIPLE CHOICE** Which indicates the correct order of steps for the construction of a perpendicular line through a point on the line using dynamic software? (Lesson 12-10)

W

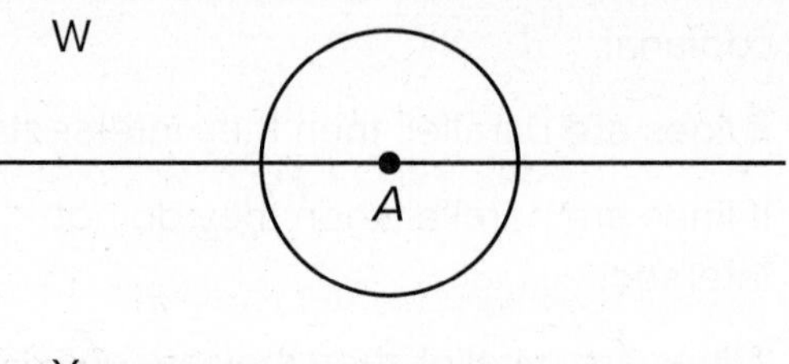

X

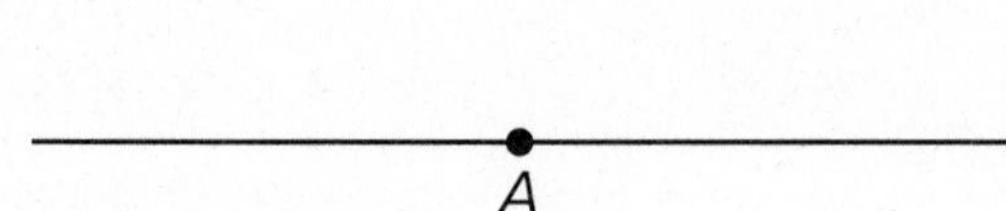

Y

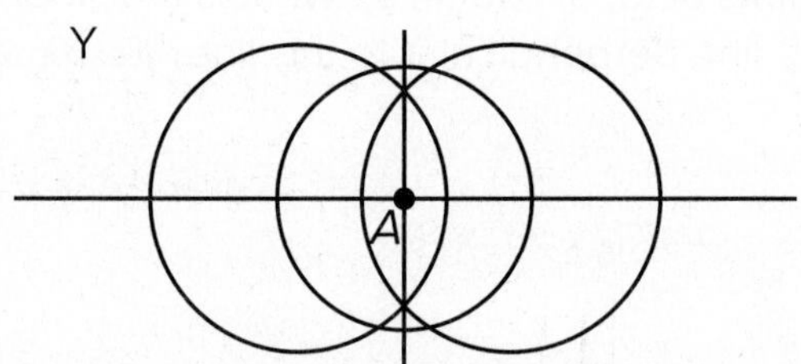

Z

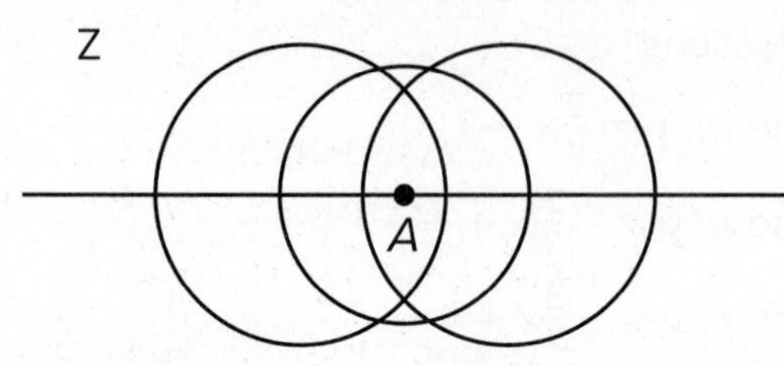

A. X, W, Y, Z

B. X, W, Z, Y

C. W, X, Y, Z

D. W, X, Z, Y

Module 13
Transformations and Symmetry

Essential Question

How are rigid motions used to show geometric relationships?

What Will You Learn?

How much do you already know about each topic **before** starting this module?

KEY

👎 — I don't know. 👉 — I've heard of it. 👍 — I know it!

	Before			After		
	👎	👉	👍	👎	👉	👍
define congruence in terms of rigid motions						
reflect figures						
draw and analyze reflected figures						
translate figures						
draw and analyze translated figures						
rotate figures						
draw and analyze rotated figures						
draw and analyze figures under multiple transformations						
identify tessellations						
identify line symmetries in two-dimensional figures						
identify rotational symmetries in two-dimensional figures						

Foldables Make this Foldable to help you organize your notes about transformations and symmetry. Begin with two sheets of paper.

1. **Fold** each sheet of paper in half.
2. **Open** the folded papers and fold each paper lengthwise two inches, to form a pocket.
3. **Glue** the sheets side-by-side to create a booklet.
4. **Label** each of the pockets as shown.

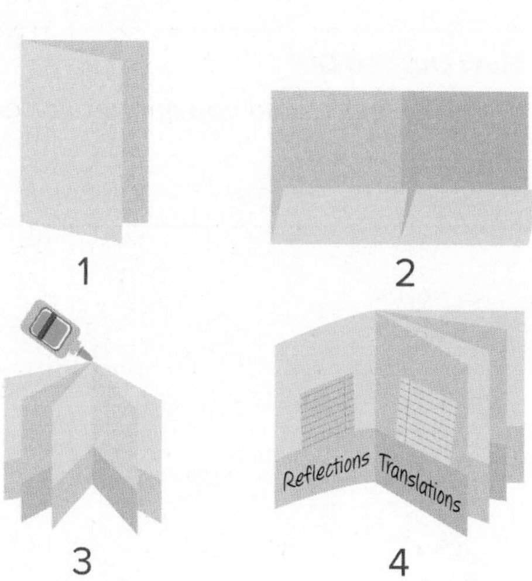

What Vocabulary Will You Learn?

- center of symmetry
- composition of transformations
- glide reflection
- line of symmetry
- line symmetry
- magnitude
- magnitude of symmetry
- order of symmetry
- point of symmetry
- point symmetry
- regular tessellation
- rotational symmetry
- semiregular tessellation
- symmetry
- tessellation
- uniform tessellation

Are You Ready?

Complete the Quick Review to see if you are ready to start this module. Then complete the Quick Check.

Quick Review

Example 1

Find the sum of $7 + (-2)$.

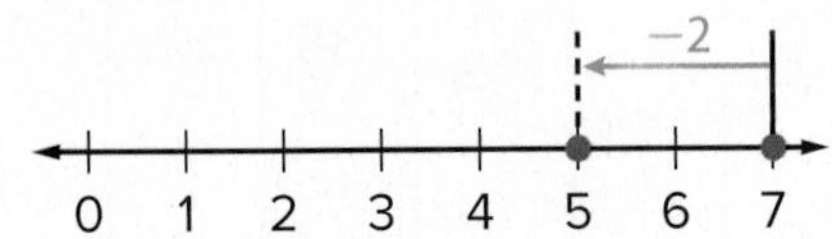

$7 + (-2) = 5$

Example 2

Identify the ordered pair for *H*.

The point is 4 units left and 3 units up.

H is located at $(-4, 3)$.

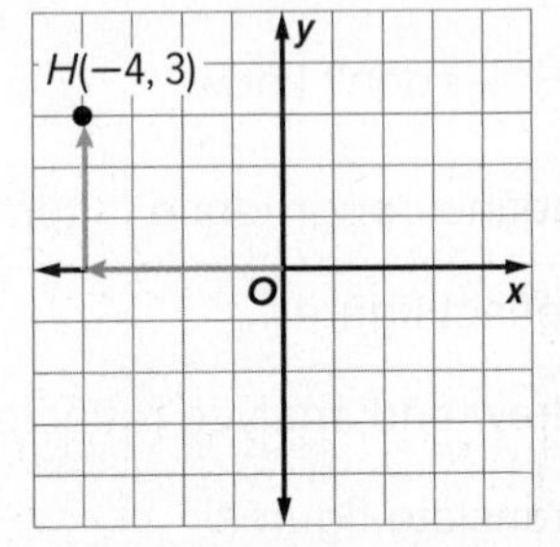

Quick Check

Find each sum.

1. $-9 + (-5)$
2. $6 + (-4)$
3. $1 + (-3)$
4. $-1 + (-7)$

Identify each ordered pair.

5. *A*
6. *B*
7. *C*
8. *D*

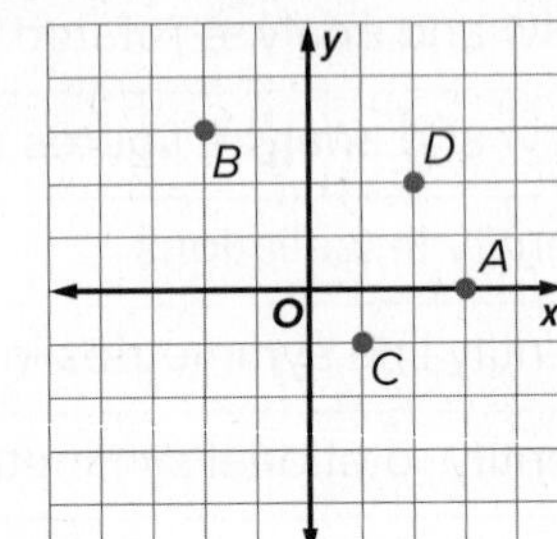

How Did You Do?

Which exercises did you answer correctly in the Quick Check?

Reflections

Explore Developing the Definition of a Reflection

Online Activity Use dynamic geometry software to complete the Explore.

INQUIRY How can you define a reflection?

Today's Goals
- Use rigid motions to reflect figures on the coordinate plane and describe the effects of the reflections.

Learn Reflections

You've learned that when a figure is reflected in a line, each point of the preimage and its corresponding point on the image are the same distance from the line of reflection.

Key Concept • Reflection
Reflection in a Vertical Line
When a figure is reflected in a vertical line that is not the *y*-axis, the *y*-coordinates of the image remain the same as the preimage. The distance from a point in the preimage to the line of reflection is the same as the distance from the corresponding point in the image to the line of reflection.
Reflection in a Horizontal Line
When a figure is reflected in a horizontal line that is not the *x*-axis, the *x*-coordinates of the image remain the same as the preimage. The distance from a point in the preimage to the line of reflection is the same as the distance from the corresponding point in the image to the line of reflection.
Reflection in $y = x$
To reflect a point in the line $y = x$ interchange the *x*- and *y*-coordinates; $(x, y) \rightarrow (y, x)$.

Example 1 Reflection in a Horizontal or Vertical Line

Consider quadrilateral *RSTV* with vertices at *R*(2, 1), *S*(2, 4), *T*(5, 4), and *V*(5, 3). Graph the image of quadrilateral *RSTV* under each reflection. Determine the coordinates of the image.

a. in the line $y = -1$

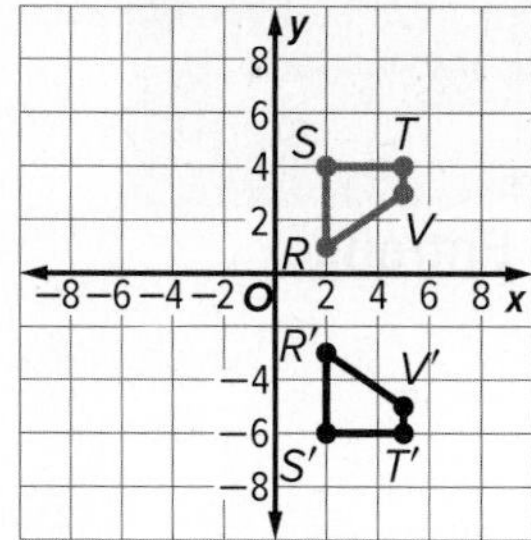

$R'(2, -3)$, $S'(2, -6)$, $T'(5, -6)$, $V'(5, -5)$

b. in the line $x = -2$

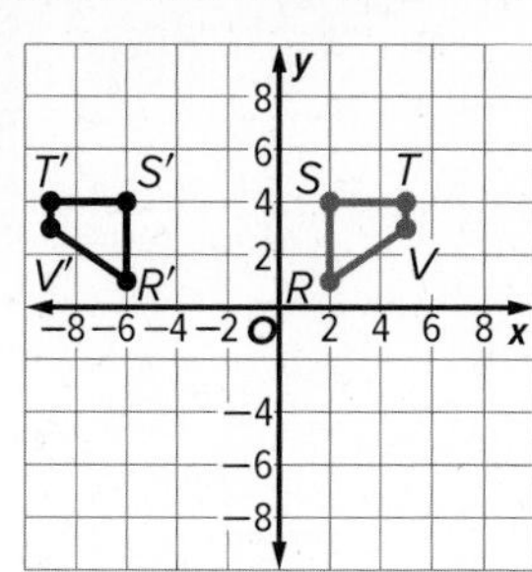

$R'(-6, 1)$, $S'(-6, 4)$, $T'(-9, 4)$, $V'(-9, 3)$

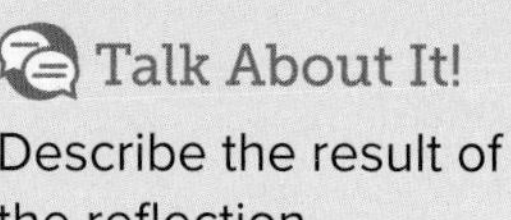

Talk About It!
Describe the result of the reflection.

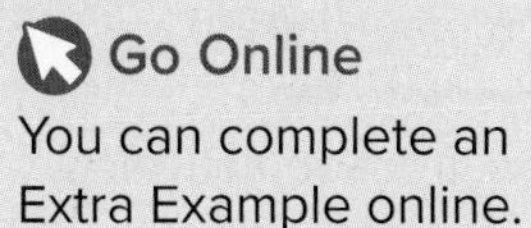

Go Online
You can complete an Extra Example online.

Check

Triangle BCD has coordinates $B(-3, 3)$, $C(1, 4)$, and $D(-2, -4)$.

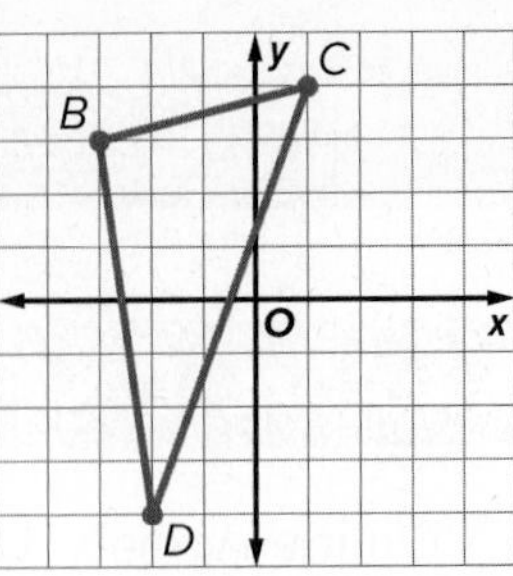

Select the coordinates of the vertices of the image after a reflection in the line $x = 3$.

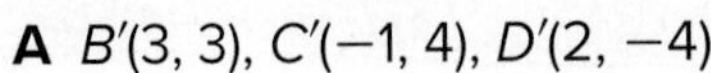

A $B'(3, 3)$, $C'(-1, 4)$, $D'(2, -4)$

B $B'(-3, -3)$, $C'(1, -4)$, $D'(-2, 4)$

C $B'(9, 3)$, $C'(5, 4)$, $D'(8, -4)$

D $B'(-3, -3)$, $C'(1, 2)$, $D'(-2, 10)$

Example 2 Reflection in the Line $y = x$

DESIGN **Winona is designing a logo for her blog header. She graphs a figure on the coordinate plane and wants to reflect it in the line $y = x$ to complete the basic shape for her logo design. What are the coordinates of the vertices of the image after the reflection?**

$(x, y) \rightarrow (y, x)$

$A(-2, 1) \rightarrow A'(1, -2)$

$B(1, 8) \rightarrow B'(8, 1)$

$C(1, 4) \rightarrow C'(4, 1)$

$D(4, 4) \rightarrow D'(4, 4)$

$E(2, 2) \rightarrow E'(2, 2)$

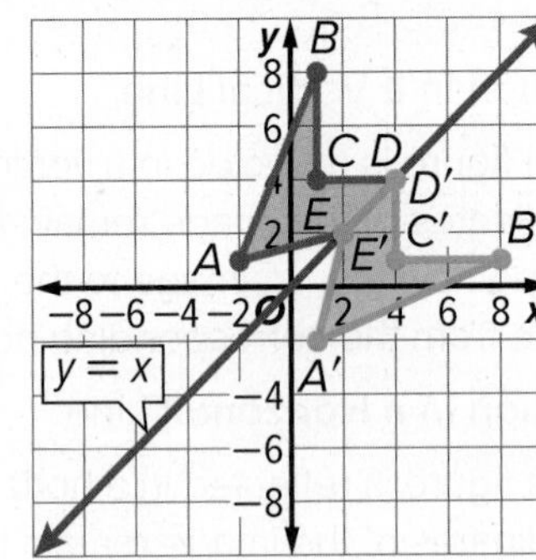

Check

LANDSCAPE Tomas is designing a sculpture garden for an art museum. There is a sidewalk connecting the center of the museum entrance to the edge of the lawn. Tomas has a set of 4 sculptures in a series that he wants to be equidistant from this sidewalk. He plotted positions for Q and X on the graph. If pieces Q and R are a pair and X and Y are a pair, where will pieces R and Y be placed in the garden?

A $R(3, 1)$, $Y(4, 2)$

B $R(3, -1)$, $Y(4, -2)$

C $R(-3, 1)$, $Y(-4, 2)$

D $R(-3, -1)$, $Y(-4, -2)$

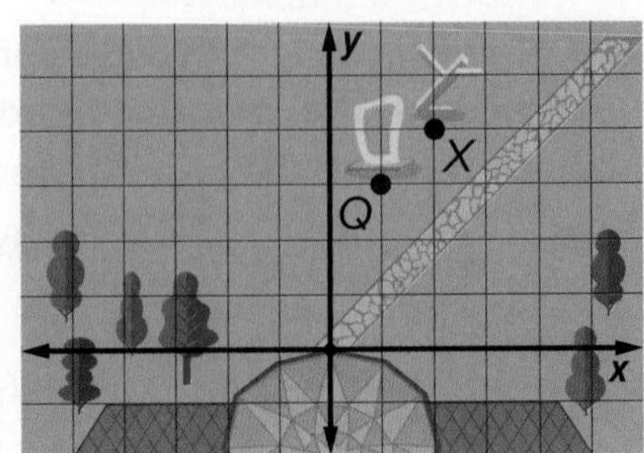

Museum Entrance

Go Online You may want to complete the construction activities for this lesson.

 **Go Online** You can complete an Extra Example online.

Practice

Examples 1 and 2

Graph the image of each figure under the given reflection. Determine the coordinates of the image.

1. $\triangle ABC$ in the line $y = x$

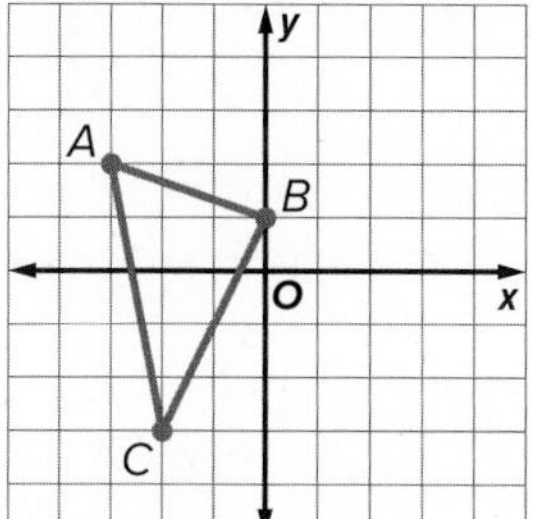

2. trapezoid $DEFG$ in the line $x = -1$

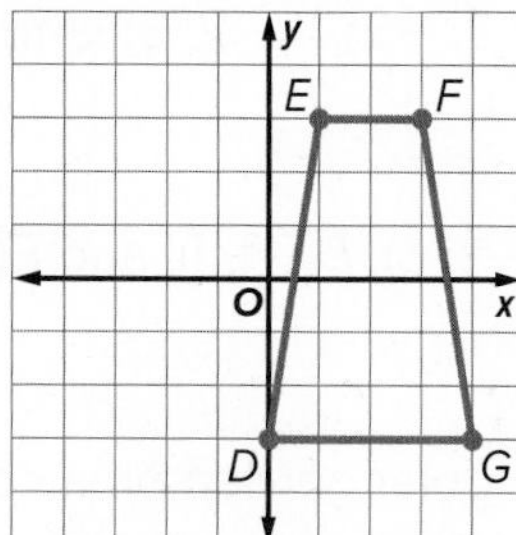

3. parallelogram $RSTU$ in the line $y = x$

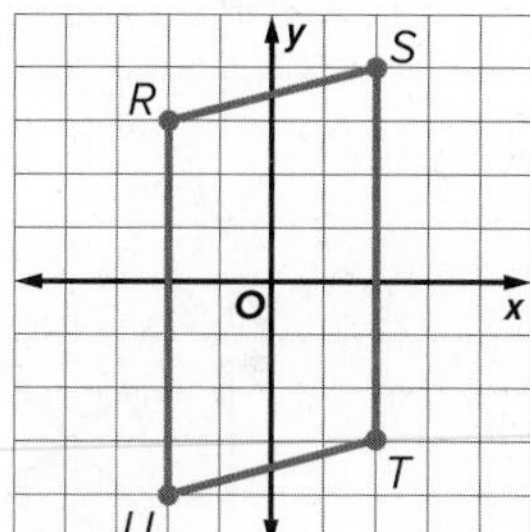

4. square $KLMN$ in the line $y = -2$

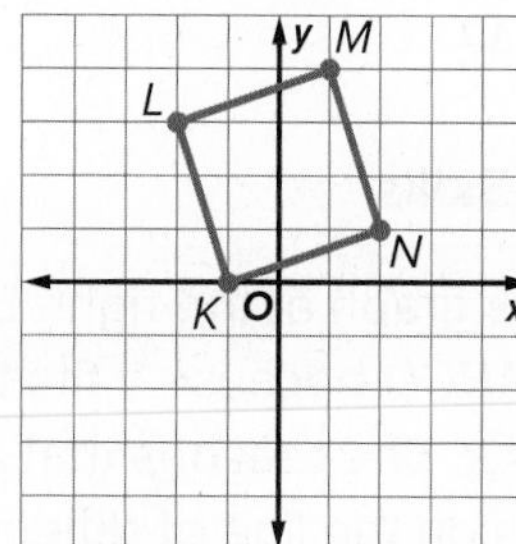

5. Determine the coordinates of $S(-7, 1)$ after a reflection in the line $y = 3$.

6. Determine the coordinates of $Q(6, -4)$ after a reflection in the line $x = 2$.

7. BANNERS Fiona is making a banner in the shape of a triangle for a school project. She graphs the banner on a coordinate plane with vertices at $P(0, 4)$, $Q(2, 8)$, and $R(-3, 6)$. She wants to reflect the banner over the line $x = 1$. Draw the image of the banner reflected in the line $x = 1$.

8. SANDBOX Aliyah is drawing the top view of a square sandbox on a coordinate plane with vertices at $D(1, 1)$, $E(1, 6)$, $F(6, 6)$, and $G(6, 1)$. She wants to change the location of the sandbox so that it is in the shade. She reflects the sandbox in the line $x = 1$. Find the coordinates of the image of the sandbox.

Mixed Exercises

9. Determine the coordinates of $W(-7, 4)$ after a reflection in the line $y = 9$.

Graph each figure and its image under the given reflection.

10. rectangle $ABCD$ with vertices $A(-5, 2)$, $B(1, 2)$, $C(1, -1)$, and $D(-5, -1)$ in the line $y = -2$

11. $\triangle FGH$ with vertices $F(-3, 2)$, $G(-4, -1)$, and $H(-6, -1)$ in the line $y = x$

12. $\triangle STU$ with vertices $S(-3, -2)$, $T(-2, 3)$, and $U(2, 2)$ in the line $y = x$

13. $\triangle CDE$ with vertices $C(-3, 6)$, $D(-1, 1)$, and $E(3, 5)$ in the line $y = 3x$

14. Naveen plotted his triangular garden on a coordinate plane. What are the vertices of the image of his garden if it is reflected in the line $y = x$?

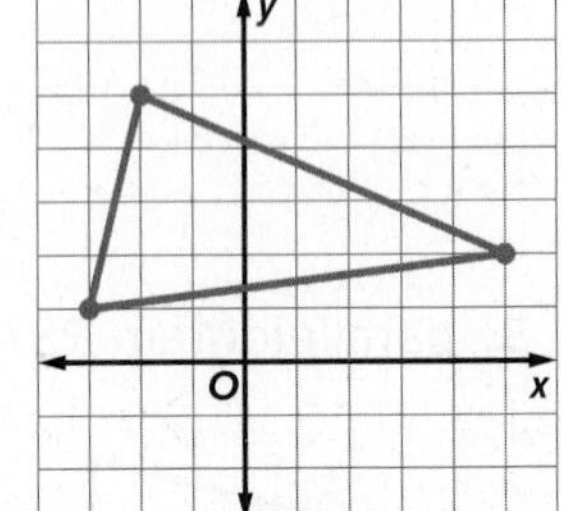

15. The image of $A(-1, 1)$ after a reflection is $A'(-1, -3)$. Which reflection produces the image of A?

Higher-Order Thinking Skills

16. **FIND THE ERROR** For the graph at the right, Evelyn maintains that $AEFG$ is a reflection of $ABCD$ because it fits the definition of a reflection in the line $y = x$. She reasons that A is the same point in each figure because it is on the line of reflection and the remaining vertices are equidistant from that line. Do you agree with Evelyn's analysis? Explain your reasoning.

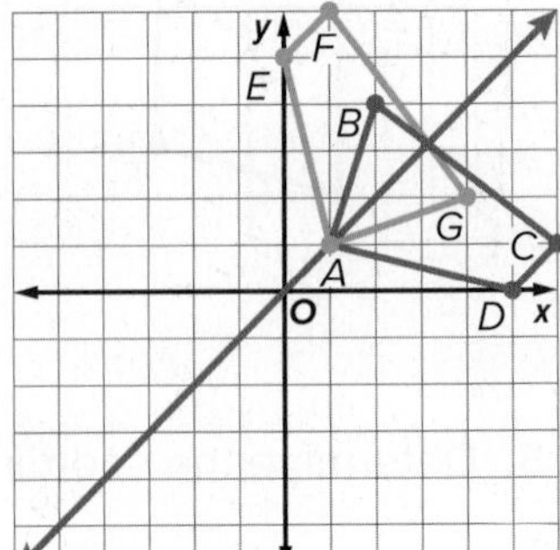

17. **CREATE** Create five points on the coordinate plane to form the letter M. Find their image under a reflection in the line $y = x$.

18. **WRITE** Describe how to reflect a figure not on the coordinate plane in a line.

19. **PERSEVERE** A point in the second quadrant with coordinates $(-a, b)$ is reflected in the line $y = -x$. What are the coordinates of the image?

20. **ANALYZE** Is the image of a point reflected in a line *sometimes, always,* or *never* located on the other side of the line of reflection? Justify your argument.

Translations

Explore Developing the Definition of a Translation

Online Activity Use dynamic geometry software to complete the Explore.

INQUIRY How can you define a translation?

Today's Goals
- Determine the translation vector.

Today's Vocabulary
magnitude

Learn Translations

You've learned that a translation is a function in which all of the points of a figure move the same distance in the same direction as described by a translation vector.

When a translation has been applied to a figure:

1. The distance between each pair of corresponding vertices is the same.
2. The segments that connect each pair of corresponding vertices are parallel.

Recall that a translation vector describes the magnitude and direction of the translation. The **magnitude** of a vector is its length from the initial point to the terminal point.

Example 1 Determine a Translation Vector

Determine whether a translation maps △*JKL* onto △*J′K′L′*. If so, find the translation vector. If not, explain why.

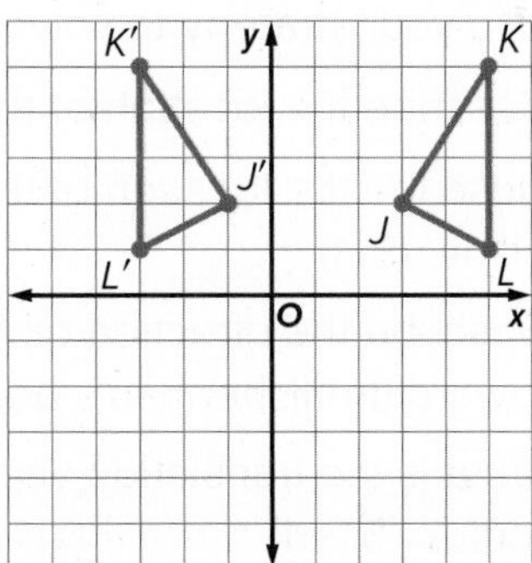

△*J′K′L′* is not a translation of △*JKL*. The distances between corresponding vertices are not equal.

Study Tip

Translations If the distance between each pair of corresponding vertices is not the same, then the figure was not translated. You do not have to check that the slopes are the same.

Check

Determine whether a translation maps △*JKL* onto △*J′K′L′*. If so, find the translation vector. If not, explain why.

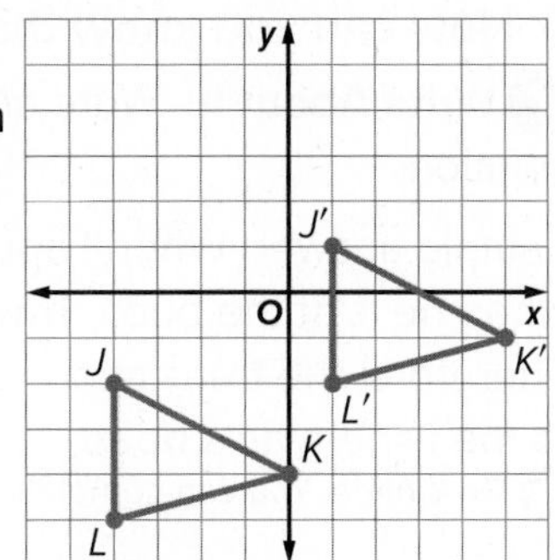

A yes; ⟨5, 3⟩
B yes; ⟨3, 5⟩
C No; this is a reflection.
D No; the triangles are not congruent.

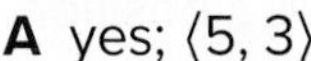

Go Online You can complete an Extra Example online.

Apply Example 2 Translations on the Coordinate Plane

Talk About It!

Is the shortest distance between the buoy and the boat the same length as the translation vector? Justify your argument.

SCAVENGER HUNT **Travis is on a scavenger hunt at the lake. He needs to swim to the nearest buoy, pick up a card, and then swim until he reaches a fisherman who will give him the next clue in exchange for the card. Describe the translation from the buoy to the fisherman's boat by using a translation vector.**

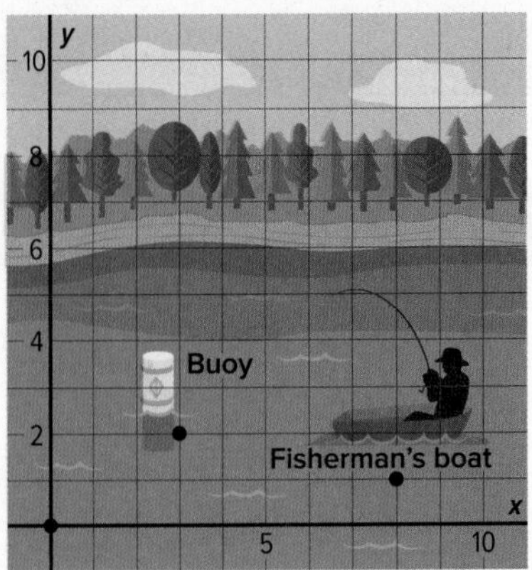

1 What is the task?

Describe the task in your own words. Then list any questions that you may have. How can you find answers to your questions?

Sample answer: I need to determine the translation vector that describes Travis's route from the buoy to the fisherman's boat. What are the locations of the buoy and the fisherman's boat? How can I determine the magnitude and direction of the translation vector? I can reread the problem and review the definition of vector to find the magnitude and direction of the translation vector.

2 How will you approach the task? What have you learned that you can use to help you complete the task?

Sample answer: I will find the coordinates of the buoy and the fisherman's boat. I will determine the length in the *x*- and *y*-directions and the general direction of Travis's route. Then, I will determine the translation vector that describes that route. I have learned how to understand points on the coordinate plane, and I have learned how to describe translations using translation vectors.

3 What is your solution?

Use your strategy to solve the problem.

Describe the location of the buoy on the coordinate plane. (3, 2)

Describe the location of the fisherman's boat on the coordinate plane. (8, 1)

Describe the direction of Travis's route. Sample answer: from the buoy toward the fisherman's boat

What is the translation vector that can be used to describe Travis's route? $\langle 5, -1 \rangle$

4 How can you know that your solution is reasonable?

Write About It! Write an argument that can be used to defend your solution.

Sample answer: When I apply the translation vector to Travis's position when he is at the buoy, Travis will arrive at the fisherman's boat. Therefore, the translation vector describes Travis's route from the buoy to the fisherman's boat.

Go Online You can complete an Extra Example online.

Practice

Examples 1 and 2

1. Determine whether a translation maps $\triangle JKL$ onto $\triangle J'K'L'$. If so, find the translation vector. If not, explain why.

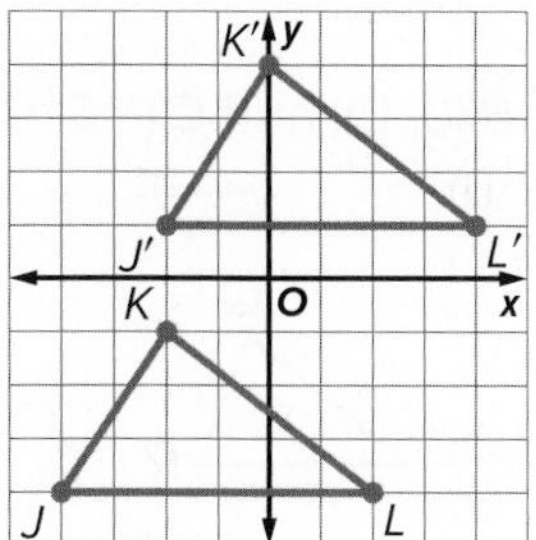

2. Determine whether a translation maps quadrilateral $LMNP$ onto quadrilateral $L'M'N'P'$. If so, find the translation vector. If not, explain why.

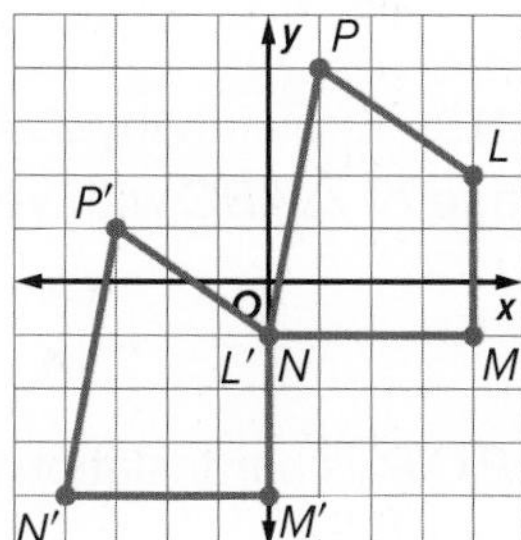

3. WALLPAPER A wallpaper design consists of repeated translations of a single isosceles triangle. The pattern is shown overlaid on a coordinate plane. The space above the triangle around the coordinate (5, 1) should be filled with a missing triangle. What are the coordinates of the vertices of the triangle that fill this space consistently with the rest of the pattern?

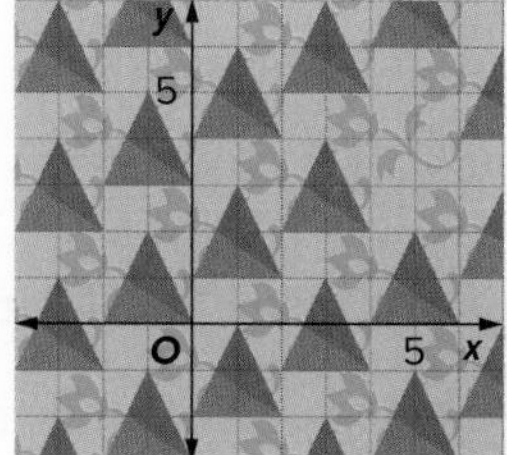

4. FURNITURE Alejandro plotted the location of a reclining chair and an end table on a coordinate plane. The end table is represented by the circle, and the chair is represented by the square with solid sides. The image of the chair along a translation is represented by the square with dashed sides.

a. Describe this translation of the chair.

b. Draw the image of the end table under the same translation that you described in part **a**.

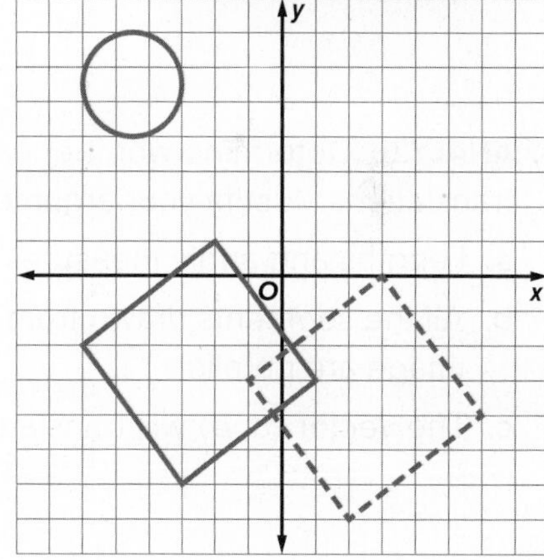

Mixed Exercises

Copy the graph. Draw and label the image of each figure after the given translation.

5. 3 units to the left

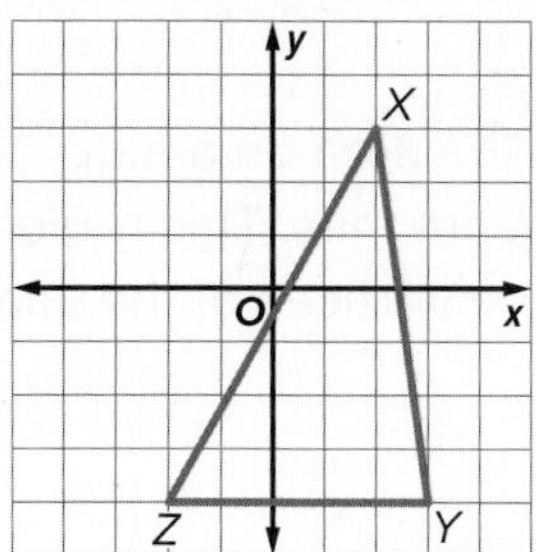

6. translation vector $\langle 1, -2.5\rangle$

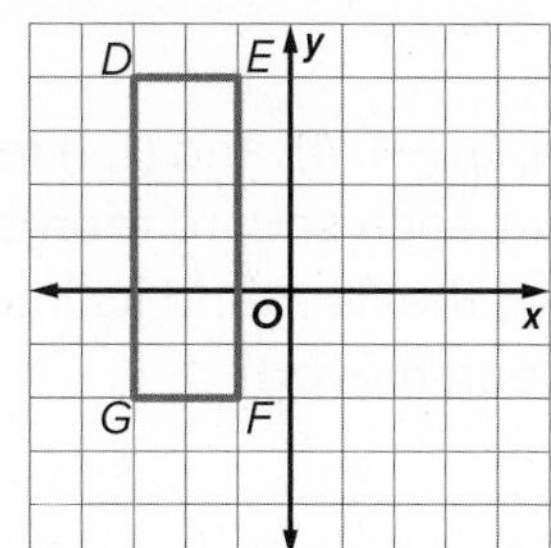

7. translation vector $\langle -5, -7\rangle$

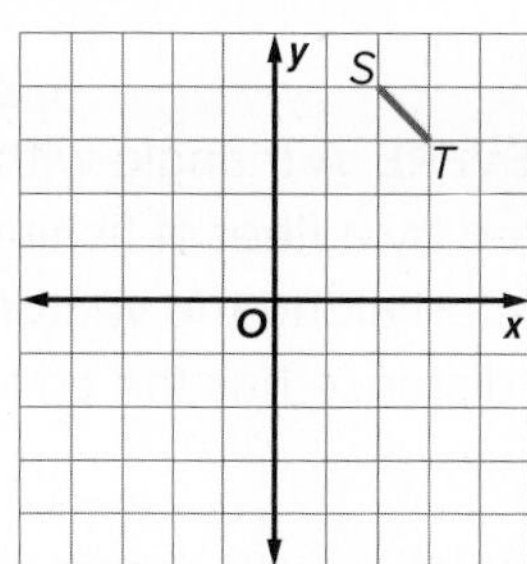

Name the image of each point after the given translation vector.

8. $F(-3, 1)$; $\langle 5, -1 \rangle$

9. $Q(4, -2)$; $\langle -2, -5 \rangle$

10. $P(9, 1.5)$; $\langle 3, -0.5 \rangle$

11. The image of $A(-3, -5)$ under a translation is $A'(6, -1)$. Find the image of $B(3, -2)$ under the same translation.

12. CONSTRUCT ARGUMENTS Explain why $\triangle A'B'C'$ with vertices $A'(-1, -2)$, $B'(0, 0)$, and $C'(-6, 0)$ is not a translation image of $\triangle ABC$ with vertices $A(1, 2)$, $B(0, 0)$, and $C(6, 0)$.

13. Determine whether $\triangle P'Q'R'$ is a translation image of $\triangle PQR$. Explain.

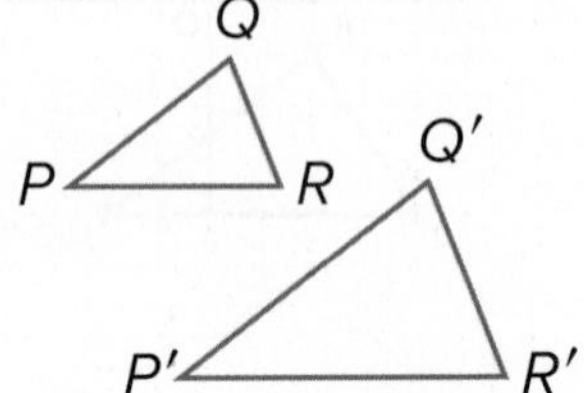

14. Determine the translation vector that moves every point of a preimage 4 units left and 6 units up.

Higher-Order Thinking Skills

15. PERSEVERE Yolanda reflects an object in the line $y = -1$. Then she reflects it in the line $y = 1$. Describe the translation.

16. ANALYZE Determine whether each statement is *always, sometimes*, or *never* true for translations. Justify your argument.

a. Lengths and angle measures of the image are preserved.

b. All the segments drawn from a vertex of the preimage to the corresponding vertex of the image are parallel.

c. The vector $\langle a, b \rangle$ will translate each coordinate of a preimage a units right and b units up.

17. WRITE A square in the coordinate plane has vertices of (2, 3), (4, 3), (2, 1), and (4, 1). It is translated such that one of the vertices is at the origin. Find the coordinates of each vertex of the image if the translation vector has the least possible length. Explain your reasoning. Draw the image and preimage on a coordinate plane.

18. PERSEVERE A triangle with vertices (−3, 1), (−1, 4), and (1, 1) represents the area on a map covered by a fleet of fishing ships, where each square represents a square mile. This region is translated along the vector $\langle 4, -5 \rangle$. Draw the fleet and its image. List the vertices of the image. What distance has the coverage area been moved?

Rotations

Today's Goals

- Use rigid motions to rotate figures about points that are not the origin and describe the effects of the rotations.

Explore Developing the Definition of a Rotation

Online Activity Use dynamic geometry software to complete the Explore.

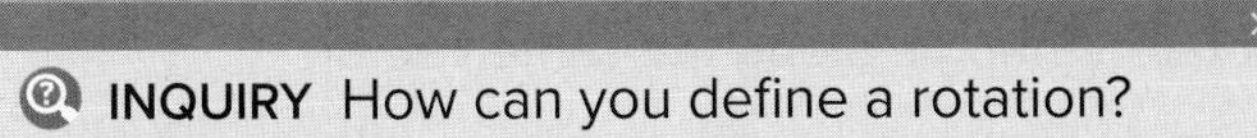

Learn Rotations About Points that Are Not the Origin

Key Concept • Rotation

A rotation about a fixed point P through an angle of $a°$ is a function that maps point M to point M' such that:

- point P does not move,
- $m\angle MPM'$ is $a°$, and
- $MP = M'P$

When a point is rotated 90°, 180°, or 270° counterclockwise about the origin, you can use the following rules to determine the coordinates of an image. A rotation of 360° will map an image onto the preimage.

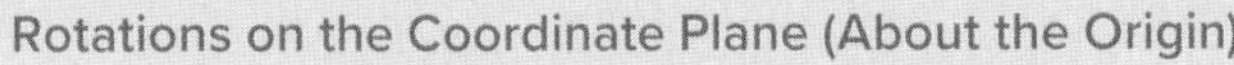

Rotations on the Coordinate Plane (About the Origin)

90° Rotation $(x, y) \rightarrow (-y, x)$

180° Rotation $(x, y) \rightarrow (-x, -y)$

270° Rotation $(x, y) \rightarrow (y, -x)$

When combined with translations, these rules can also be used to rotate figures about points that are not the origin.

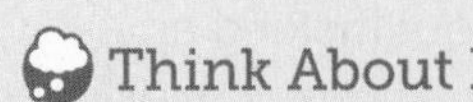

Why does MP have to be equal to $M'P$ for the rotation to occur?

Example 1 Rotation About a Point That Is Not the Origin

Triangle ABC has vertices $A(-8, 5)$, $B(-6, 9)$, and $C(-3, 6)$. Graph $\triangle ABC$ and its image after a rotation of 180° about $(-5, 3)$.

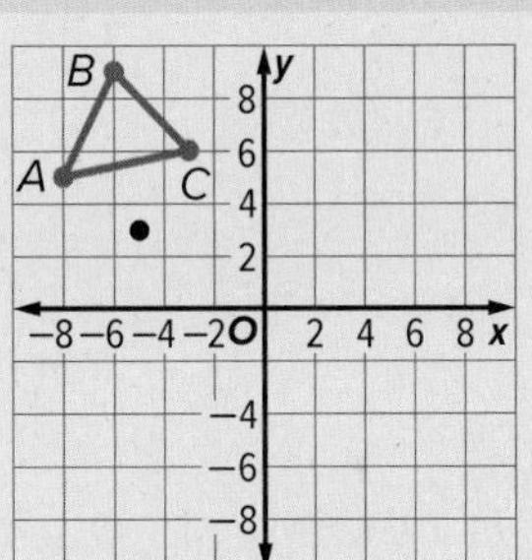

Step 1 Graph $\triangle ABC$.

Step 2 Map the center of rotation to the origin.

To map the center of rotation to the origin, translate the center of rotation along the vector $\langle 5, -3 \rangle$. Then translate the vertices of $\triangle ABC$ along the same vector.

$(x, y) \rightarrow (x + 5, y - 3)$

$A(-8, 5) \rightarrow (-3, 2)$ $B(-6, 9) \rightarrow (-1, 6)$ $C(-3, 6) \rightarrow (2, 3)$

(continued on the next page)

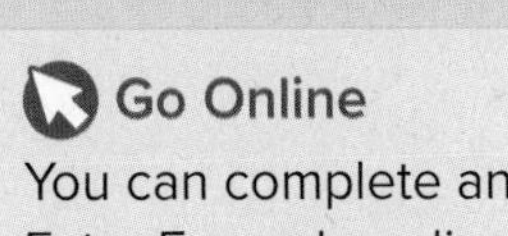

You can complete an Extra Example online.

Step 3 **Rotate 180° about the origin.**

$(x, y) \rightarrow (-x, -y)$

$A(-3, 2) \rightarrow (3, -2)$

$B(-1, 6) \rightarrow (1, -6)$

$C(2, 3) \rightarrow (-2, -3)$

Step 4 **Map the center of rotation to its original position.**

To map the center of rotation to its original position, translate the center of rotation along the vector $\langle -5, 3 \rangle$. Then translate the vertices of the rotated triangle along the same vector.

$(x, y) \rightarrow (x - 5, y + 3)$

$A(3, -2) \rightarrow A'(-2, 1)$

$B(1, -6) \rightarrow B'(-4, -3)$

$C(-2, -3) \rightarrow C'(-7, 0)$

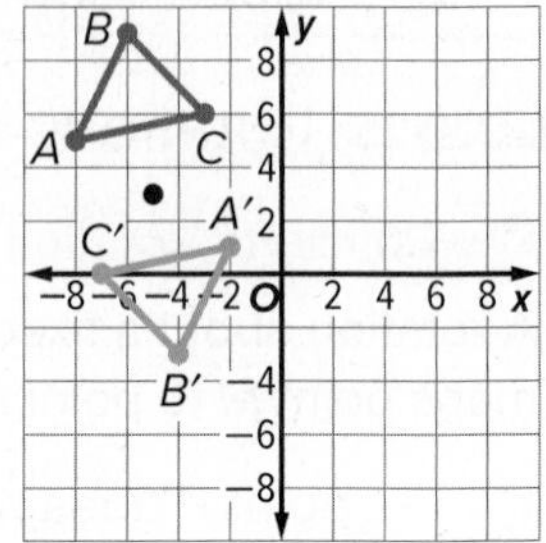

Study Tip

Approximations When you are describing the effects of a rotation, you can approximate the location of the rotated figure without making any calculations. You will know the shape and size of the image because angle measures and lengths are preserved.

Check

Triangle PQR has vertices $P(2, 1)$, $Q(2, 4)$, and $R(5, 1)$. Graph $\triangle PQR$ and its image after a rotation 270° counterclockwise about (7, 5).

Use a Source

Find an example of a flag that has elements that can be created using a rotation. Describe the center and angle of rotation.

Example 2 Describe the Effect of a Rotation

FLAGS **Kendrick is working with a team in his social studies class to create a new country and its government. Kendrick is responsible for creating the country's flag. He is using geometry software to design the flag on the coordinate plane. Describe how the two yellow stars would be affected if they were rotated 90° counterclockwise about the center of the white star.**

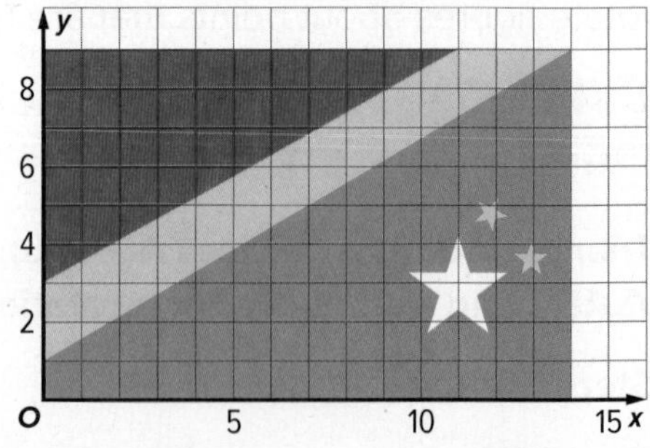

If the stars were rotated, they would curve around the top-left sides of the white star. Together the preimage and the image would create a semicircle of yellow stars above the white star.

Go Online You can complete an Extra Example online.

Practice

Go Online You can complete your homework online.

Examples 1 and 2

1. Triangle XYZ has vertices $X(0, 2)$, $Y(4, 4)$, and $Z(3, -1)$. Graph $\triangle XYZ$ and its image after a rotation of 180° about $(2, -3)$.

2. Triangle ABC has vertices $A(1, 7)$, $B(3, 2)$, and $C(-2, -2)$. Graph $\triangle ABC$ and its image after a rotation of 270° counterclockwise about $(-4, 2)$.

3. Triangle FGH has vertices $F(-3, 4)$, $G(2, 0)$, and $H(-1, -2)$. Graph $\triangle FGH$ and its image after a rotation of 180° about $(-3, -6)$.

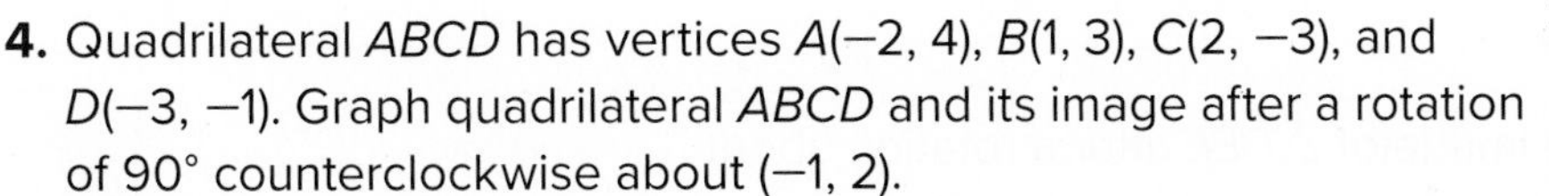

4. Quadrilateral $ABCD$ has vertices $A(-2, 4)$, $B(1, 3)$, $C(2, -3)$, and $D(-3, -1)$. Graph quadrilateral $ABCD$ and its image after a rotation of 90° counterclockwise about $(-1, 2)$.

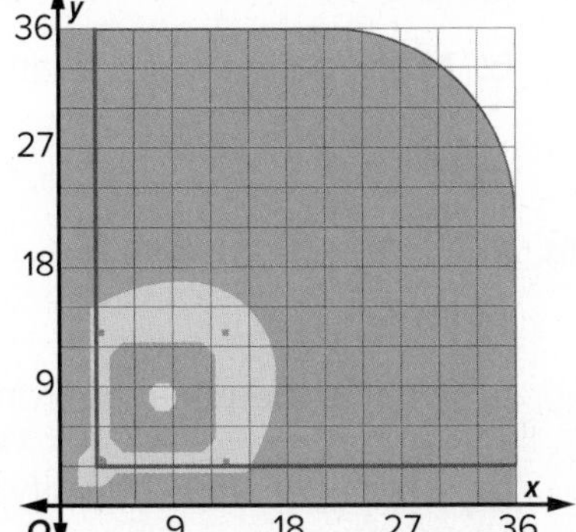

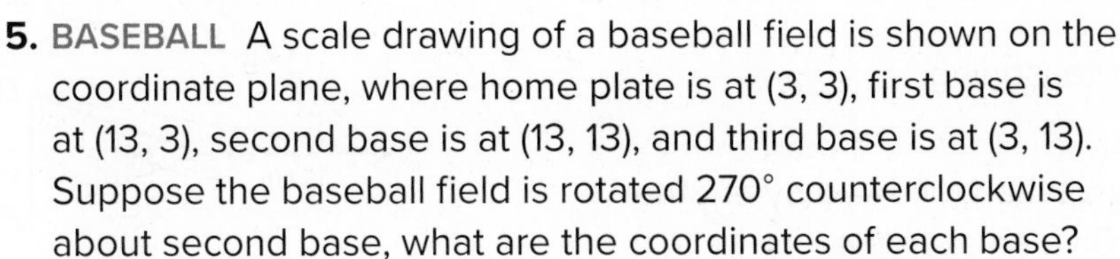

5. **BASEBALL** A scale drawing of a baseball field is shown on the coordinate plane, where home plate is at (3, 3), first base is at (13, 3), second base is at (13, 13), and third base is at (3, 13). Suppose the baseball field is rotated 270° counterclockwise about second base, what are the coordinates of each base?

Mixed Exercises

6. Point Q with coordinates $(4, -7)$ is rotated 270° clockwise about $(5, 1)$. What are the coordinates of its image?

7. Parallelogram $JKLM$ has vertices $J(2, 1)$, $K(7, 1)$, $L(6, -3)$, and $M(1, -3)$. What are the coordinates of the image of K if the parallelogram is rotated 270° counterclockwise about $(-2, -1)$?

8. **USE TOOLS** Use a protractor and ruler to draw a rotation of $\triangle PQR$ 210° about T.

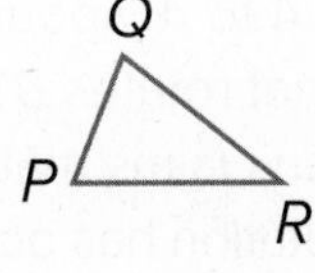

9. The line segment XY with endpoints $X(3, 1)$ and $Y(2, -2)$ is rotated 90° counterclockwise about $(-6, 4)$. What are the endpoints of $\overline{X'Y'}$?

10. **HIKING** A damaged compass points northwest instead of north. If you travel west by the compass, what is your angle of rotation to true north?

11. A circular dial with the digits 0 through 9 evenly spaced around its edge is rotated clockwise 36°. How many times would you have to perform this rotation to bring the dial back to its original position?

12. Under a rotation about the origin, the point $A(5, -1)$ is mapped to the point $A'(1, 5)$. What is the image of the point $B(-4, 6)$ under this rotation? Explain.

Higher-Order Thinking Skills

13. **CREATE** Draw a right triangle ABC and point P not on the triangle.

a. Rotate triangle ABC about point P 90° counterclockwise.

b. Name a clockwise rotation that would map triangle ABC onto triangle $A'B'C'$.

14. In the figure, $\triangle D'E'F'$ is the image of $\triangle DEF$ after a rotation about point Z.

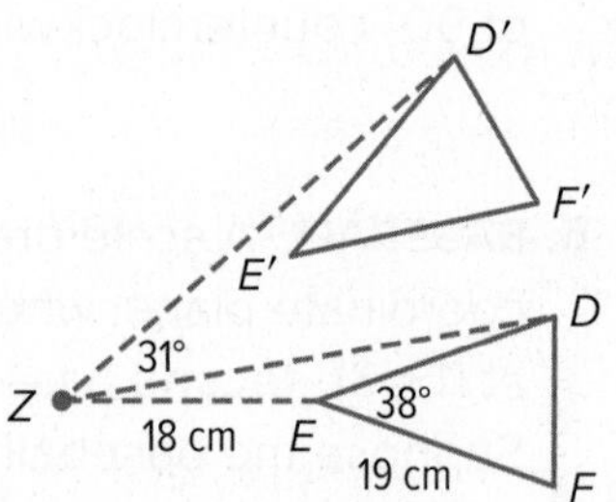

a. What is the distance from E' to Z? Justify your reasoning.

b. What is $m\angle FZF'$? Justify your reasoning.

15. **ANALYZE** What is the result of a rotation followed by another rotation about the same point? Give an example.

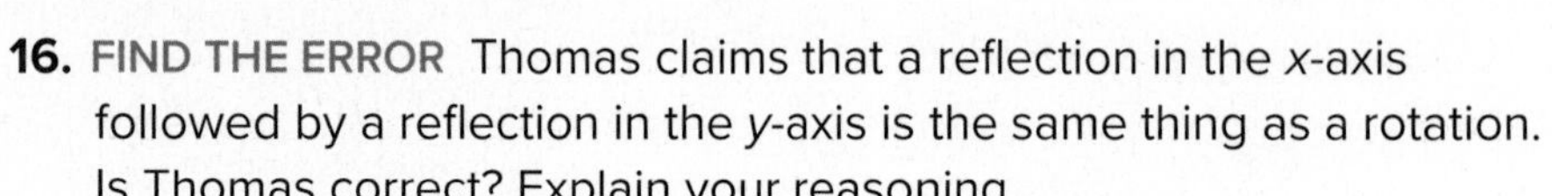

16. **FIND THE ERROR** Thomas claims that a reflection in the x-axis followed by a reflection in the y-axis is the same thing as a rotation. Is Thomas correct? Explain your reasoning.

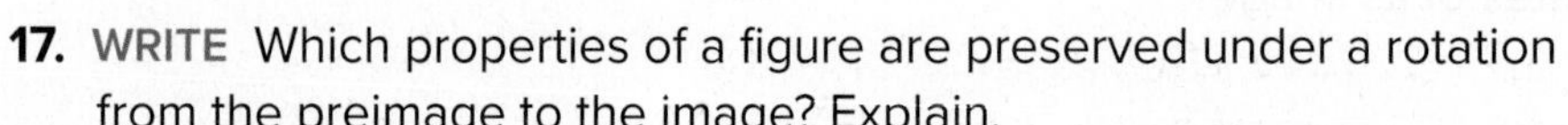

17. **WRITE** Which properties of a figure are preserved under a rotation from the preimage to the image? Explain.

18. **FIND THE ERROR** Shanice is looking at the figure shown, which shows two congruent triangles. She measures the angle that rotates A to A' about O and finds it to be 30°. She measures the angle that rotates B to B' about O and also finds it to be 30°. She then claims that because the two triangles are congruent, a 30° rotation has occurred about point O. Is Shanice correct? Explain your reasoning.

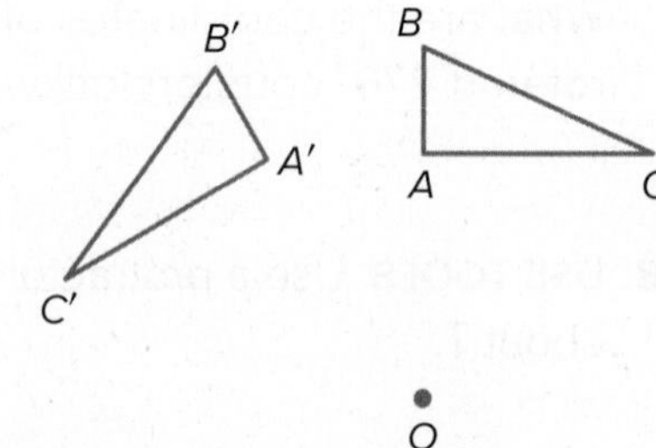

19. **WRITE** Are collinearity and betweenness of points maintained under rotations? Explain.

Compositions of Transformations

Explore Reflections in Two Lines

 **Online Activity** Use dynamic geometry software to complete the Explore.

INQUIRY How is a figure affected by reflections in two lines?

Learn Compositions of Transformations

When a transformation is applied to a figure and then another transformation is applied to its image, the result is called a **composition of transformations**. A glide reflection is one type of composition of transformations.

A **glide reflection** is the composition of a translation followed by a reflection in a line parallel to the translation vector.

Theorem 13.1: Composition of Isometries
The composition of two (or more) isometries is an isometry.

You will prove one case of Theorem 13.1 in Exercise 29.

So, the composition of two or more isometries—reflections, translations, or rotations—results in an image that is congruent to its preimage.

Example 1 Glide Reflection

Triangle *PQR* has vertices *P*(1, 1), *Q*(2, 5), and *R*(4, 2). Determine the coordinates of the vertices of the image after a translation along ⟨−4, 0⟩ and a reflection in the *x*-axis.

Step 1 Graph △*PQR*.

Step 2 Graph the image of △*PQR* after a translation along the vector ⟨−4, 0⟩.

$(x, y) \rightarrow (x - 4, y)$
$P(1, 1) \rightarrow P'(-3, 1)$
$Q(2, 5) \rightarrow Q'(-2, 5)$
$R(4, 2) \rightarrow R'(0, 2)$

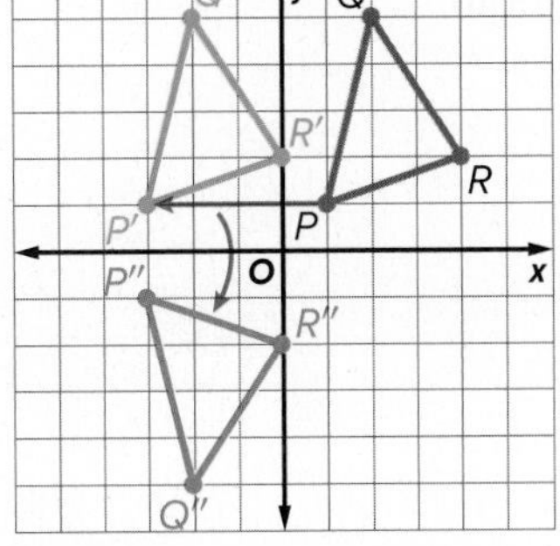

Step 3 Graph the image of △*P′Q′R′* after a reflection in the *x*-axis.

$(x, y) \rightarrow (x, -y)$
$P'(-3, 1) \rightarrow P''(-3, -1)$
$Q'(-2, 5) \rightarrow Q''(-2, -5)$
$R'(0, 2) \rightarrow R''(0, -2)$

Go Online You can complete an Extra Example online.

Today's Goals
- Determine the image of a figure after a composition of transformations.
- Describe the transformation that produces the same image as a reflection in two lines.

Today's Vocabulary
composition of transformations
glide reflection

 **Think About It!**
Compare and contrast glide reflections and compositions of transformations.

Study Tip
Compositions of Transformations Use double primes to indicate the image created by the second transformation in the composition.

Check

Triangle JKL has vertices $J(6, -1)$, $K(10, -2)$, and $L(5, -3)$. Determine the coordinates of the vertices of the image after a translation along $\langle 0, 4\rangle$ and a reflection in the y-axis.

$J''($? , ? $)$, $K''($? , ? $)$, $L''($? , ? $)$

 Talk About It!

Do any points remain invariant, or unchanged, under glide reflections? under compositions of transformations? Explain.

Example 2 Composition of Isometries

Triangle ABC has vertices $A(-6, -2)$, $B(-5, -5)$, and $C(-2, -1)$. Graph $\triangle ABC$ and its image after a rotation 180° about the origin and a translation along $\langle -2, 4\rangle$.

Step 1 Graph $\triangle ABC$.

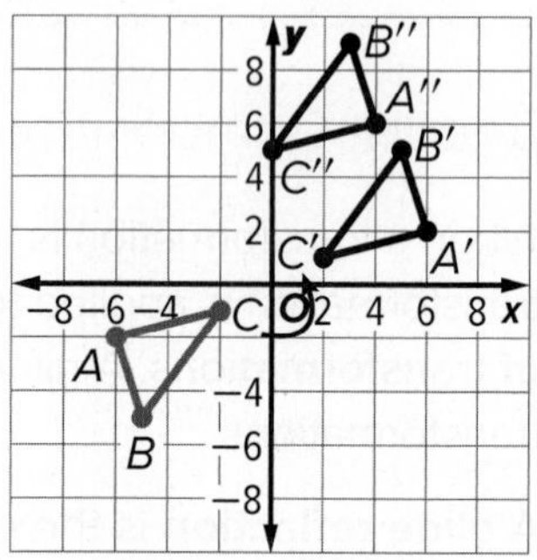

Step 2 Graph the image of $\triangle ABC$ after rotation 180° about the origin.

$(x, y) \rightarrow (-x, -y)$

$A(-6, -2) \rightarrow A'(6, 2)$

$B(-5, -5) \rightarrow B'(5, 5)$

$C(-2, -1) \rightarrow C'(2, 1)$

Step 3 Graph the image of $\triangle A'B'C'$ after a translation along $\langle -2, 4\rangle$.

$(x, y) \rightarrow (x - 2, y + 4)$

$A'(6, 2) \rightarrow A''(4, 6)$

$B'(5, 5) \rightarrow B''(3, 9)$

$C'(2, 1) \rightarrow C''(0, 5)$

Check

The endpoints of $\overline{CD}$ are $C(-7, 1)$ and $D(-3, 2)$. Graph $\overline{CD}$ and its image after a reflection in the x-axis and a rotation 90° about the origin.

Go Online You can complete an Extra Example online.

Learn Compositions of Two Reflections

The composition of two reflections can result in the same image as a translation or rotation.

Theorem 13.2: Reflections in Parallel Lines

The composition of two reflections in parallel lines can be described by a translation vector that is

- perpendicular to the two lines and
- twice the distance between the two lines.

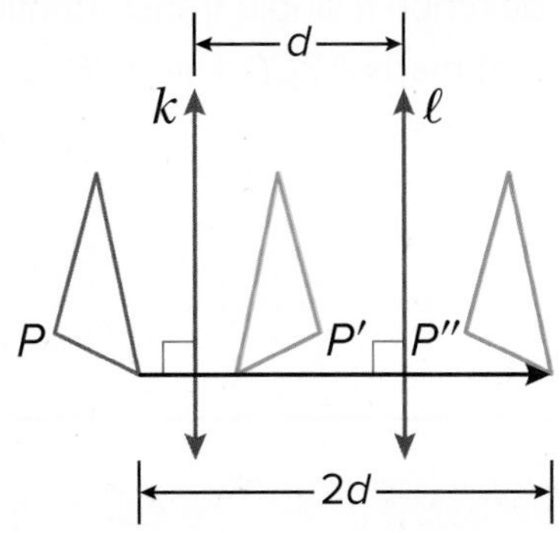

Theorem 13.3: Reflections in Intersecting Lines

The composition of two reflections in intersecting lines can be described by a rotation that is

- about the point where the lines intersect and
- through an angle that is twice the measure of the acute or right angle formed by the lines.

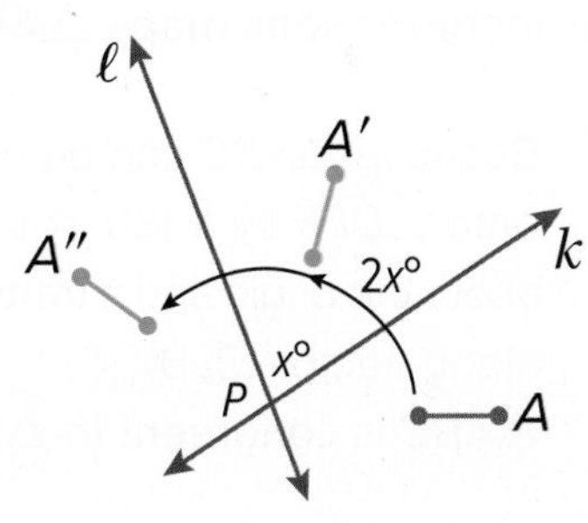

You will prove Theorem 13.2 in Exercise 30.

Think About It!

In what type of intersecting lines will a composition of two reflections form a 180° rotation?

Go Online

A proof of Theorem 13.3 is available.

Example 3 Reflect a Figure in Two Lines

Reflect each figure in line n and then line q. Then describe a single transformation that maps the preimage onto the final image.

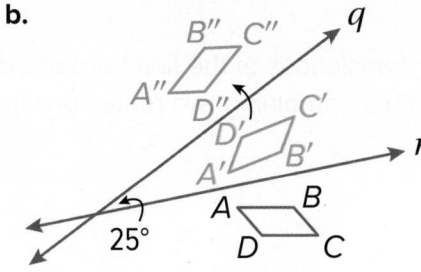

a. By Theorem 13.2, the composition of two reflections in parallel horizontal lines n and q is equivalent to a vertical translation down $2 \cdot \frac{3}{8}$ or $\frac{3}{4}$ inches.

b. By Theorem 13.3, the composition of two reflections in intersecting lines n and q is equivalent to a $2 \cdot 25°$ or 50° counterclockwise rotation about the intersection point of lines n and q.

Go Online You can complete an Extra Example online.

Check

Copy the diagram. Reflect quadrilateral *ABCD* in line m and then line n. Then describe a single transformation that maps *ABCD* onto *A″B″C″D″*.

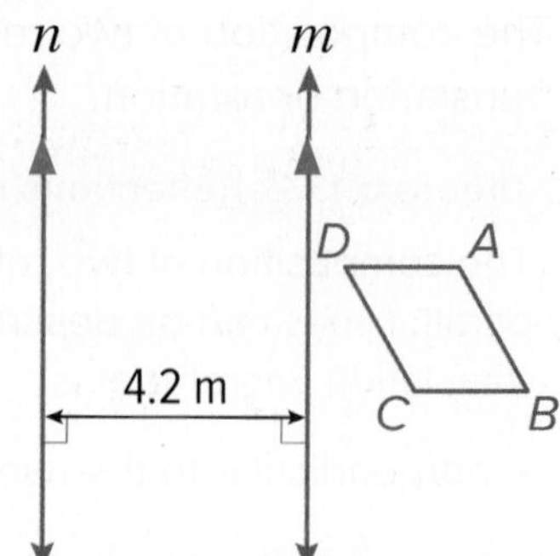

Think About It!

How can you tell that $\triangle ABC$ needs to be rotated 180° before being translated? Use the position of $\triangle DEF$ to justify your argument.

Go Online

An alternate method is available for this example.

Example 4 Determine Congruence

Are triangles *ABC* and *DEF* congruent? If so, what composition of transformations maps $\triangle ABC$ onto $\triangle DEF$?

Because $\triangle ABC$ can be mapped onto $\triangle DEF$ by a 180° rotation about the origin and a translation along vector $\langle 3, 0\rangle$, $\triangle ABC$ is congruent to $\triangle DEF$.

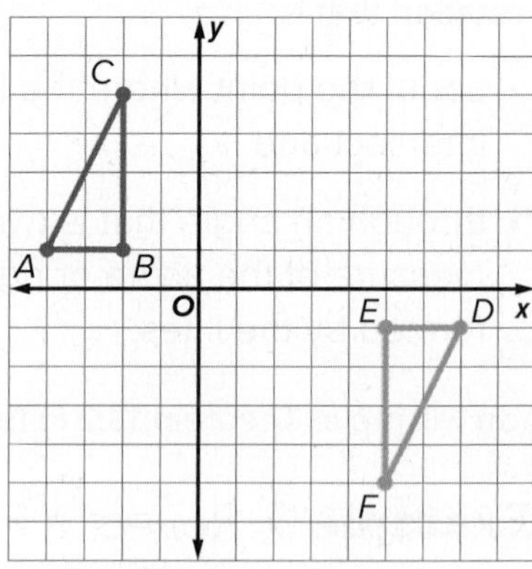

Example 5 Describe Transformations

DESIGN PATTERNS **Describe the transformations that are combined to create the pattern shown.**

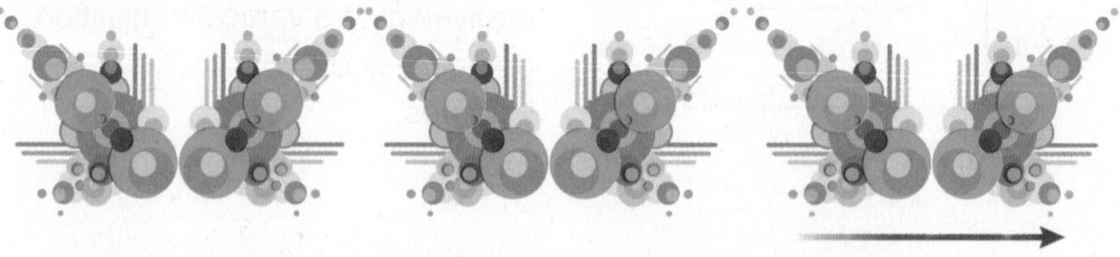

The pattern is created by successive translations of the first third of the design. So this pattern can be created by combining two reflections in a pair of parallel lines.

 **Go Online** You can complete an Extra Example online.

Practice

Go Online You can complete your homework online.

Example 1

Graph each figure with the given vertices and its image after the indicated glide reflection.

1. $\triangle RST$: $R(1, -4)$, $S(6, -4)$, $T(5, -1)$
Translation: along $\langle 2,0 \rangle$
Reflection: in x-axis

2. $\triangle JKL$: $J(1, 3)$, $K(5, 0)$, $L(7, 4)$
Translation: along $\langle -3, 0 \rangle$
Reflection: in x-axis

3. $\triangle DFG$: $D(2, 8)$, $F(1, 2)$, $G(4, 6)$
Translation: along $\langle 3, 3 \rangle$
Reflection: in $y = x$

4. $\triangle MPQ$: $M(-4, 3)$, $P(-5, 8)$, $Q(-1, 6)$
Translation: along $\langle -4, -4 \rangle$
Reflection: in $y = x$

Example 2

Graph each figure with the given vertices and its image after the indicated composition of transformations.

5. $\overline{WX}$: $W(-4, 6)$ and $X(-4, 1)$
Reflection: in x-axis
Rotation: 90° about origin

6. $\overline{AB}$: $A(-3, 2)$ and $B(3, 8)$
Rotation: 90° about origin
Translation: along $\langle 4, 4 \rangle$

7. $\overline{FG}$: $F(1, 1)$ and $G(6, 7)$
Reflection: in x-axis
Rotation: 180° about origin

8. $\overline{RS}$: $R(2, -1)$ and $S(6, -5)$
Translation: along $\langle -2, -2 \rangle$
Reflection: in y-axis

Example 3

Copy and reflect each figure in line u and then line v. Then describe a single transformation that maps the preimage onto the image.

9.

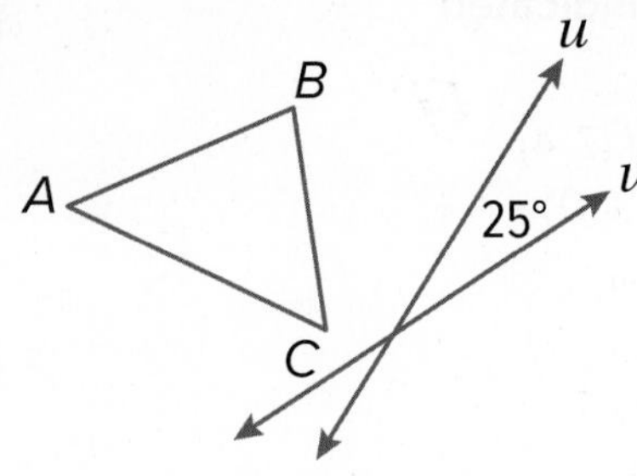

10.

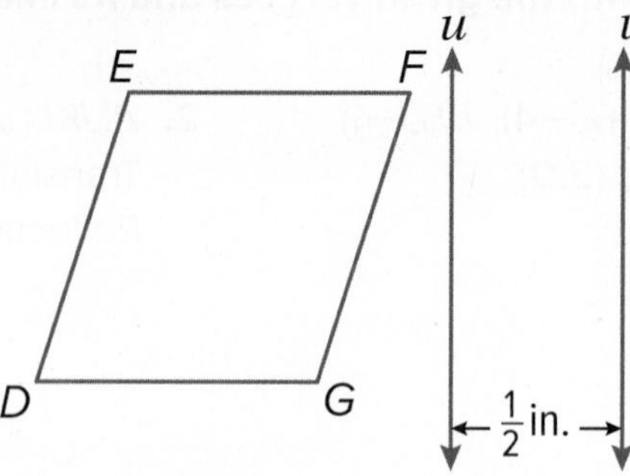

11.

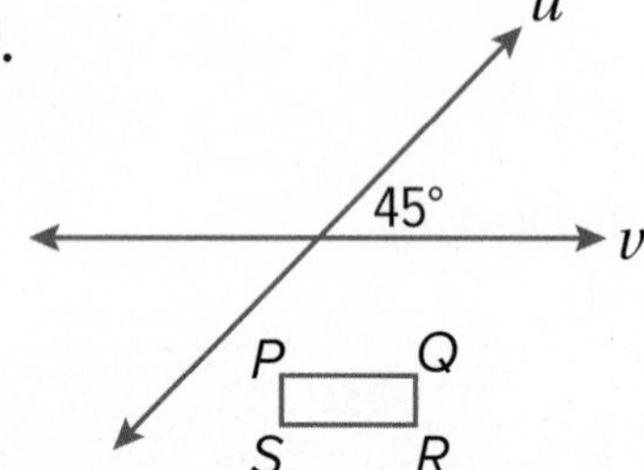

12.

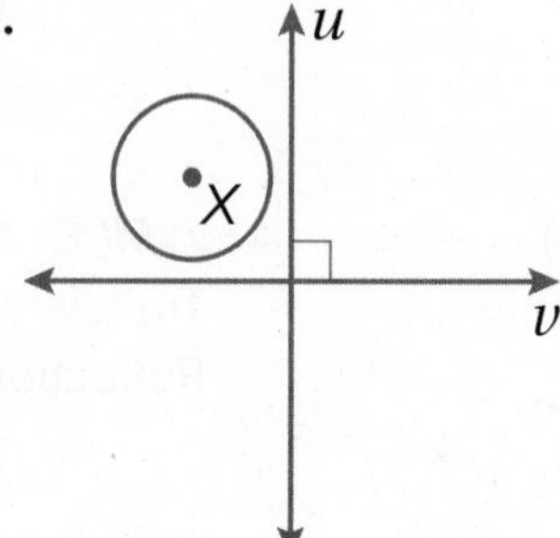

Example 4

Is △*JKL* congruent to △*MNP*? If so, what composition of transformations maps △*JKL* onto △*MNP*?

13.

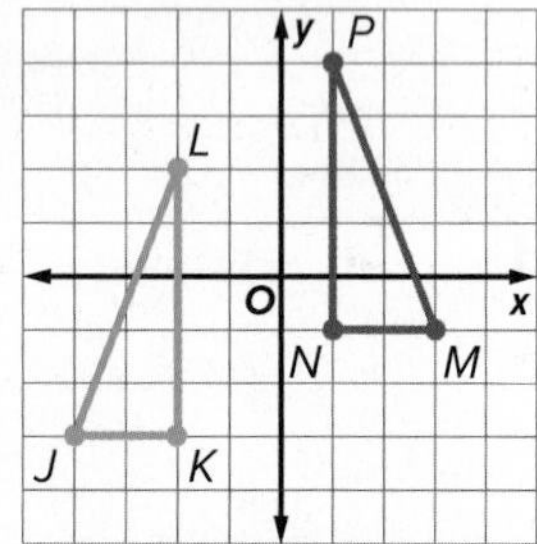

14.

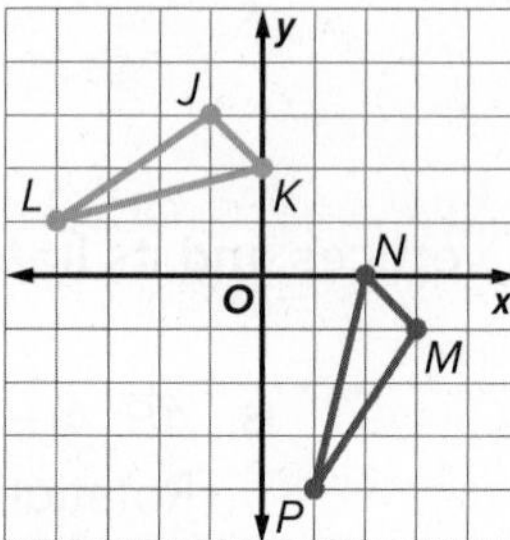

15.

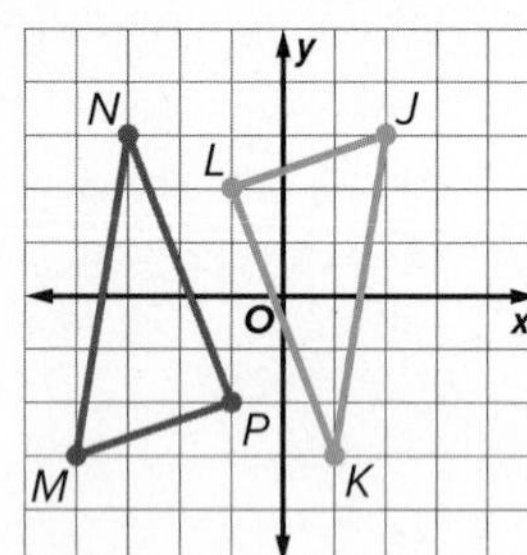

16.

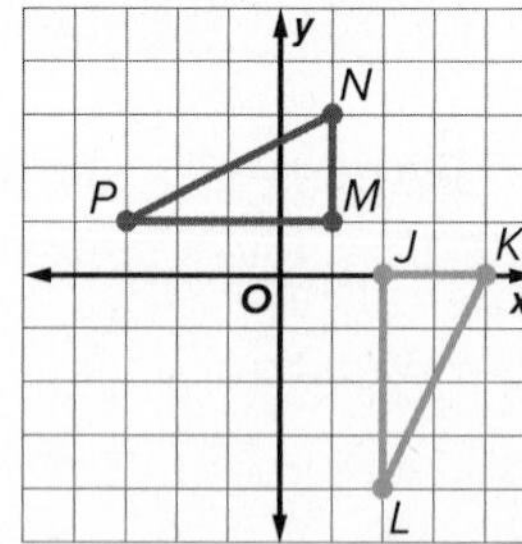

Example 5

17. Describe the transformations that are combined to create the border.

18. Describe the transformations that are combined to create the pattern.

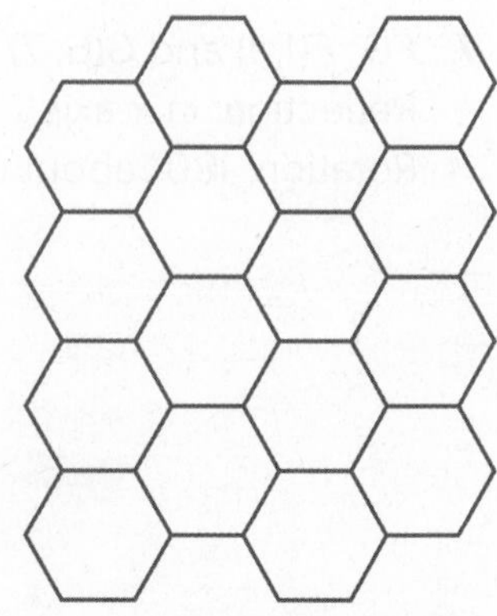

Mixed Exercises

Draw and label the image of each figure after the given composition of transformations.

19. 270° rotation about the origin followed by translation along $\langle 2,-2\rangle$

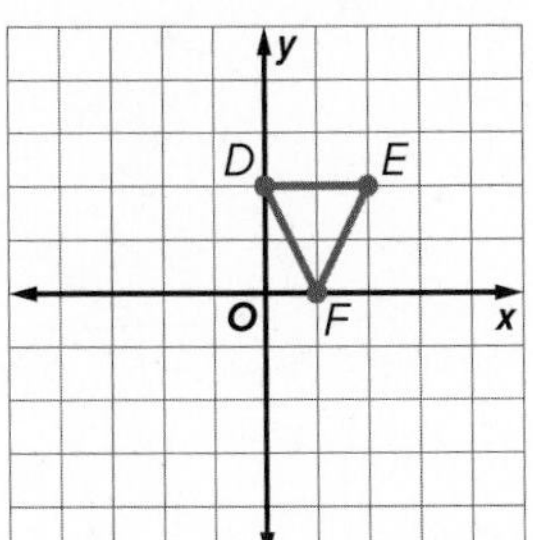

20. reflection in the y-axis followed by 180° rotation about the origin

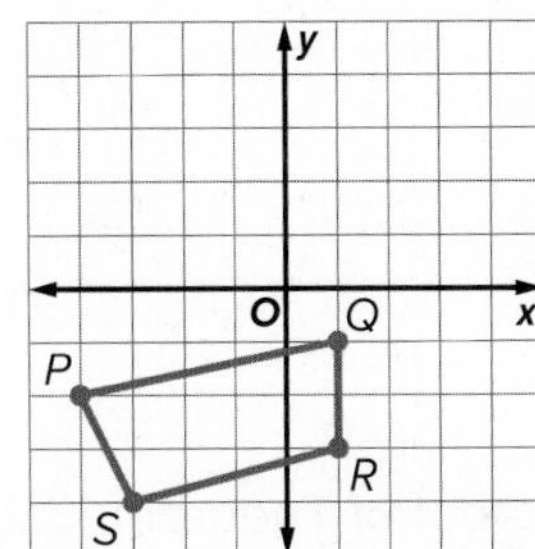

Determine the coordinates of the preimage given the image and composition of transformations.

21. reflection in the x-axis, reflection in the y-axis

22. rotation 180° about origin, translation 3 units up

23. reflection in the y-axis, translation 2 units left

24. Point K is reflected over line p and then over line d. If lines p and d are parallel and 2.8 feet apart, what single translation maps K onto K''?

Determine whether each statement is *always*, *sometimes*, or *never* true. Justify your argument.

25. A composition of two reflections is a rotation.

26. A composition of two translations is a rotation.

27. A reflection in the x-axis followed by a reflection in the y-axis leaves a point in its original location.

28. A translation along $\langle a, b\rangle$ followed by the translation along $\langle c, d\rangle$ is the translation along $\langle a + c, b + d\rangle$.

29. **PROOF** Write a paragraph proof for one case of the Composition of Isometries Theorem.

Given: A translation along $\langle a, b\rangle$ maps R to R' and S to S'. A reflection in a maps R' to R'' and S' to S''.

Prove: $\overline{RS} \cong \overline{R''S''}$

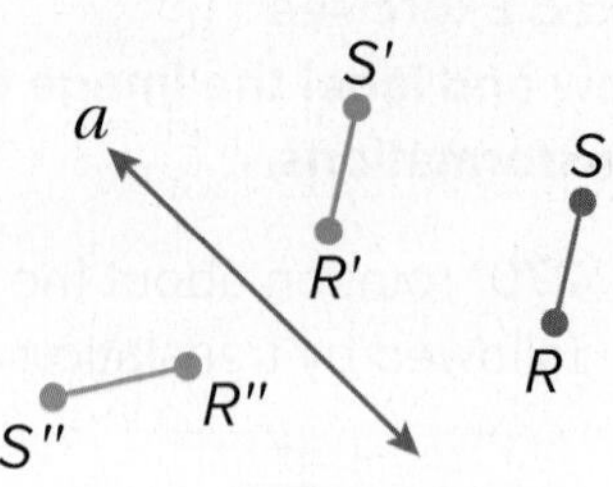

30. **PROOF** Write a two-column proof of Theorem 13.2

Given: A reflection in line p maps $\overline{BC}$ to $\overline{B'C'}$. A reflection in line q maps $\overline{B'C'}$ to $\overline{B''C''}$. $p \parallel q$, $AD = x$

Prove: $\overline{BB''} \perp p$, $\overline{BB''} \perp q$; $BB'' = 2x$

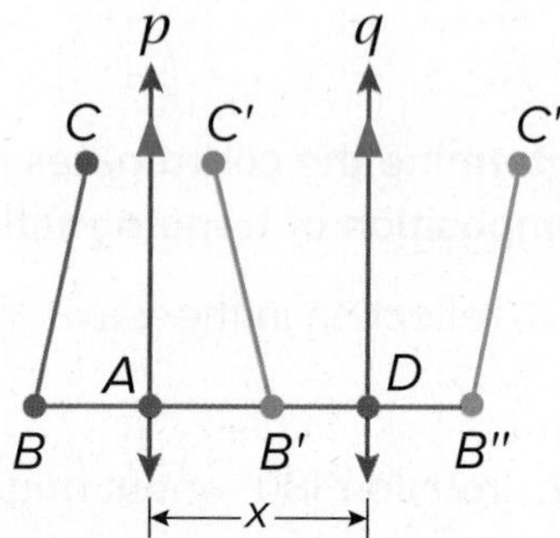

Higher-Order Thinking Skills

31. **ANALYZE** When a rotation and a reflection are performed as a composition of transformations on a figure, does the order of the transformations *sometimes*, *always*, or *never* affect the location of the final image? Justify your argument.

32. **FIND THE ERROR** Daniel and Lolita are translating $\triangle XYZ$ along $\langle 2, 2\rangle$ and reflecting it in the line $y = 2$. Daniel says that the transformation is a glide reflection. Lolita disagrees and says that the transformation is a composition of transformations. Is either of them correct? Explain your reasoning.

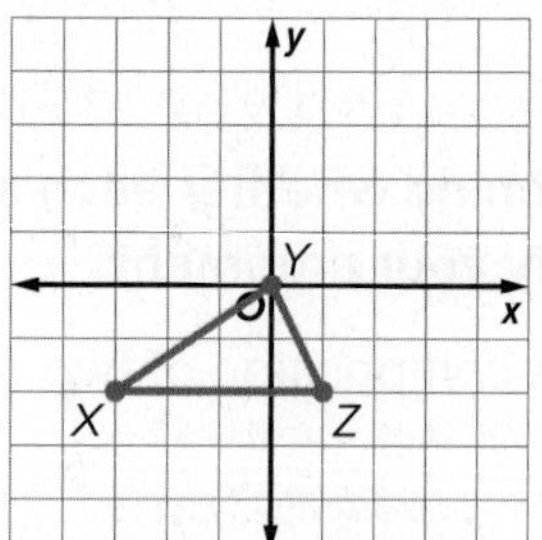

33. **PERSEVERE** If $PQRS$ is translated along $\langle 3, -2\rangle$, reflected in $y = -1$, and rotated 90° about the origin, what are the coordinates of $P'''Q'''R'''S'''$?

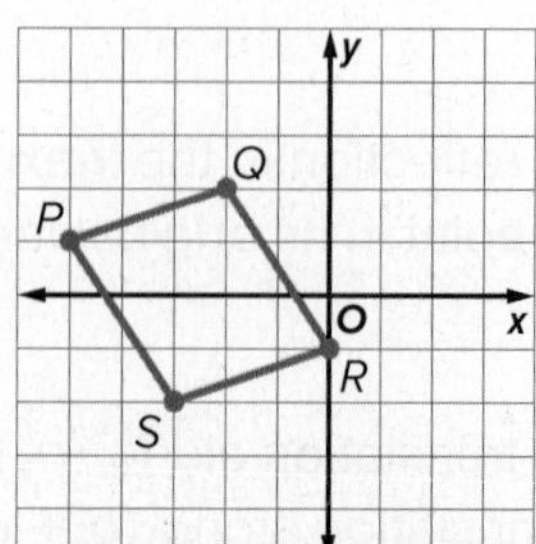

34. **ANALYZE** If an image will be reflected in the line $y = x$ and the x-axis, does the order of reflections affect the final image? Explain.

Tessellations

Explore Creating Tessellations

Online Activity Use graphing technology to complete the Explore.

INQUIRY When will a regular polygon not tessellate the plane?

Learn Types of Tessellations

Compositions of transformations can be used to create patterns from polygons. A **tessellation** is a repeating pattern of one or more figures that covers a plane with no overlapping or empty spaces. A tessellation can be created by transforming the same figure or set of figures in a plane. The sum of the measures of the angles around a vertex of a tessellation is 360°.

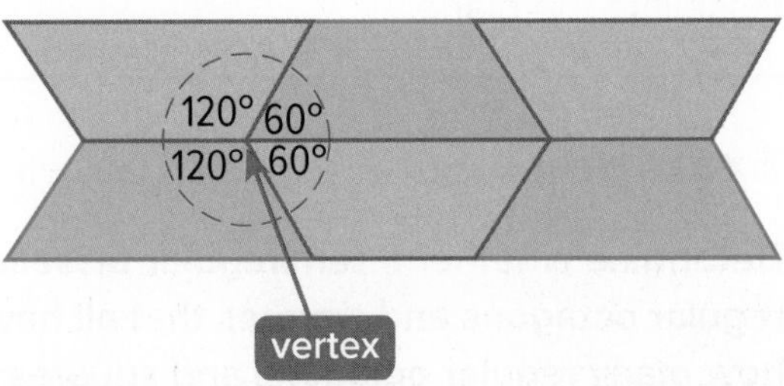
120°
60°
120°
60°
vertex

A **regular tessellation** is formed by only one type of regular polygon. A regular polygon will tessellate if it has an interior angle measure that is a factor of 360°.

A **semiregular tessellation** is formed by two or more regular polygons. The tessellation shown is made up of only equilateral triangles, so it is a regular tessellation.

A tessellation can contain any type of polygon. A tessellation is a **uniform tessellation** if it contains the same arrangement of shapes and angles at each vertex.

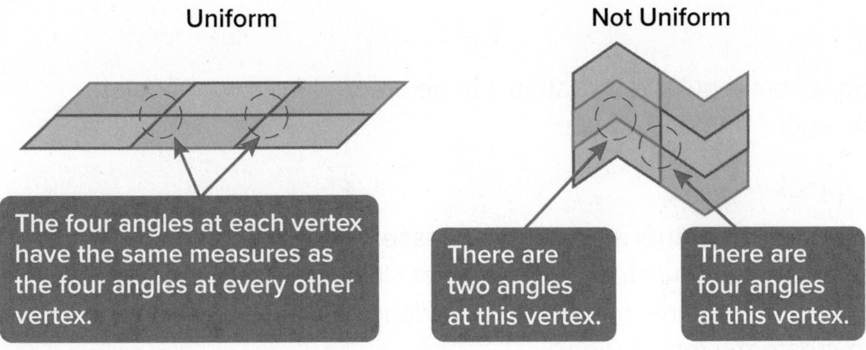
Uniform
Not Uniform
The four angles at each vertex have the same measures as the four angles at every other vertex.
There are two angles at this vertex.
There are four angles at this vertex.

Go Online You can complete an Extra Example online.

Today's Goals

- Use transformations to classify tessellations and identify figures that tessellate the plane.
- Determine whether given polygons tessellate the plane and describe transformations used to create tessellations.

Today's Vocabulary

tessellation
regular tessellation
semiregular tessellation
uniform tessellation

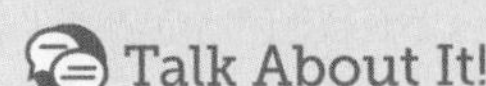

Talk About It!

Can an isosceles trapezoid be used to create a tessellation? a regular tessellation? Justify your arguments.

Example 1 Regular Tessellation

Determine whether a regular 16-gon will tessellate the plane. Explain.

Let x represent the measure of an interior angle of a regular 16-gon.

$$x = \frac{180(n-2)}{n}$$

$$= \frac{180(16-2)}{16}$$

$$= 157.5°$$

Because 157.5° is not a factor of 360°, a regular 16-gon will not tessellate the plane.

Check

Determine whether a regular decagon will tessellate the plane. Explain.

Because ___?___ a factor of 360°, a regular decagon ___?___ tessellate the plane.

Math History Minute

Although she had only a high-school education, **Marjorie Rice (1923–2017)** devoted her life to finding ways to tessellate a plane with pentagons. She eventually discovered four new types of tessellating pentagons and more than 60 distinct tessellations by pentagons.

Example 2 Semiregular Tessellation

Determine whether a semiregular tessellation can be created from regular octagons and squares that all have sides 1 unit long. If so, how many regular octagons and squares are needed at each vertex to create the tessellation.

Try to draw a pattern that has no empty spaces using only regular octagons and squares. In the pattern, the vertices are formed by two regular octagons and one square.

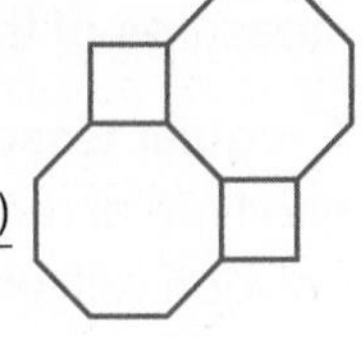

Each interior angle of a regular octagon measures $\frac{180(8-2)}{8}$ or 135°. Each interior angle of a square measures 90°.

The sum of the measures of the angles around a vertex of a tessellation is 360°. If there are x regular octagons and y squares at a vertex, then the equation $135x + 90y = 360$ can be used to verify that if there are two regular octagons at a vertex, then there is also a square at the vertex.

Let $x = 2$.	$135x + 90y = 360$	Original equation
	$135(2) + 90y = 360$	Substitution
	$270 + 90y = 360$	Simplify.
	$y = 1$	Solve for y.

So, a semiregular tessellation can be created from two regular octagons and one square.

Check

Determine whether a semiregular tessellation can be created from squares and equilateral triangles that all have sides 1 unit long. If so, how many squares and equilateral triangles are needed at each vertex to create the tessellation?

Go Online
You can complete an Extra Example online.

Example 3 Classify a Tessellation

TILES Tiles for kitchen backsplashes come in many shapes that can create unique patterns. The pattern shown is created with rectangular tiles. Determine whether the pattern is a tessellation. If so, describe it as *uniform, not uniform, regular, not regular,* or *semiregular.*

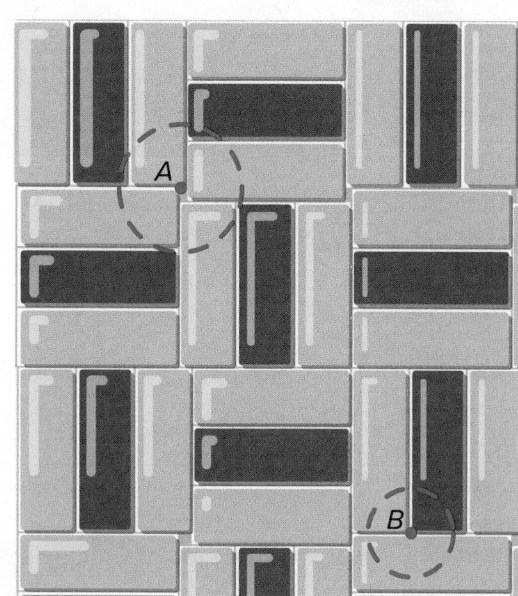

The pattern is a tessellation because there are no empty spaces and the sum of the angles at the different vertices is 360°.

The tessellation is not uniform because at vertex *A* there are four angles and at vertex *B* there are three angles.

The tessellation is not regular because a rectangular tile is used to create the pattern and a rectangle is not a regular polygon.

Check

WEAVING Basket weaving is one of the oldest art forms of human civilization, dating back to 5000 B.C. Throughout the years, different cultures have created hundreds of basket patterns. Which terms describe the pattern shown?

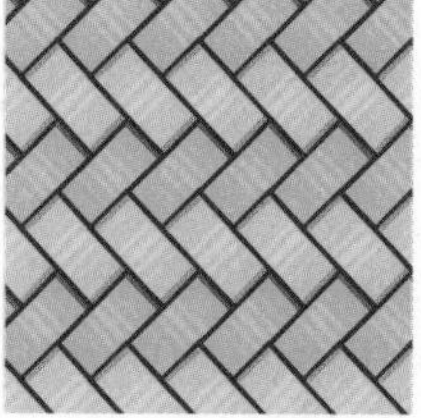

Learn Transformations in Tessellations

Not all polygons have to be regular to tessellate the plane. Any triangle is capable of tessellating the plane because the sum of the measures of its interior angles is 180°.

Any quadrilateral is capable of tessellating the plane. Because a quadrilateral can be formed by two triangles, the sum of the interior angles of a quadrilateral is 2 • 180° or 360°.

Even though all triangles and quadrilaterals can tessellate the plane, not all polygons can. Only fifteen known types of convex pentagons and three types of convex hexagons can tessellate the plane. If a convex polygon has seven or more sides, then it cannot tessellate the plane.

Go Online You may want to complete the Concept Check to check your understanding.

 **Go Online** You can complete an Extra Example online.

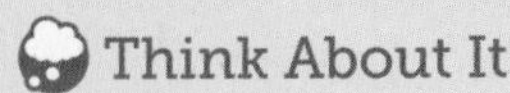

Can the same tessellation be created using only two types of transformations? If so, describe the transformations.

Example 4 Identify Transformations in a Tessellation

Will an isosceles triangle *sometimes*, *always*, or *never* tessellate the plane? Describe the transformation(s) that can be used to create the tessellation shown below.

Because all triangles tessellate the plane, an isosceles triangle will always tessellate the plane.

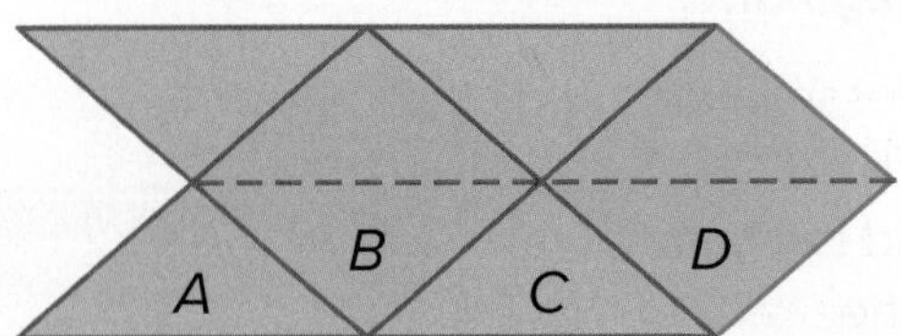

Triangle *A* can be rotated 180° about the midpoint of its right leg to create Triangle *B*.

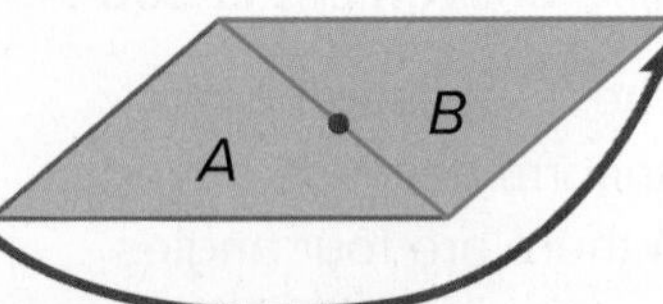

Triangles *C* and *D* can be created by translating *A* and *B* along a vector.

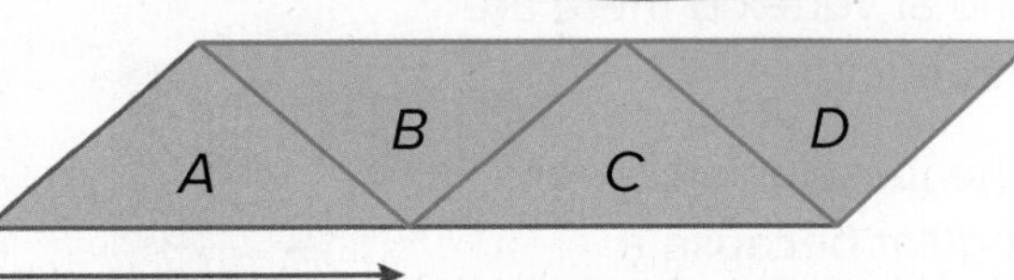

Triangles *A*, *B*, *C*, and *D* can be reflected in the line that contains the bases of Triangles *B* and *D* to create the tessellation.

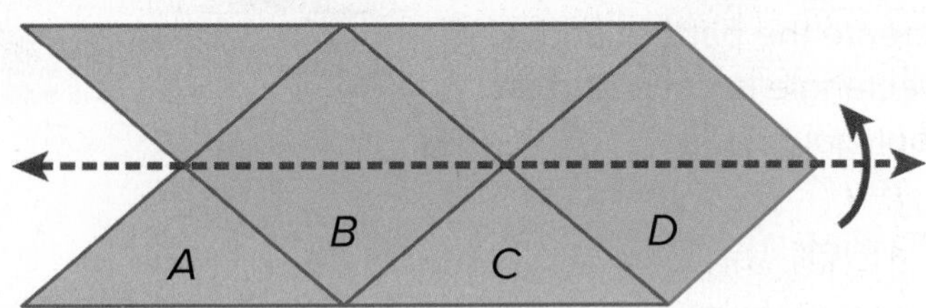

So, the tessellation can be created using rotations, translations, and reflections.

Check

Will a kite *sometimes, always,* or *never* tessellate the plane?

Describe the transformation(s) that can be used to create the tessellation shown. Select all that apply.

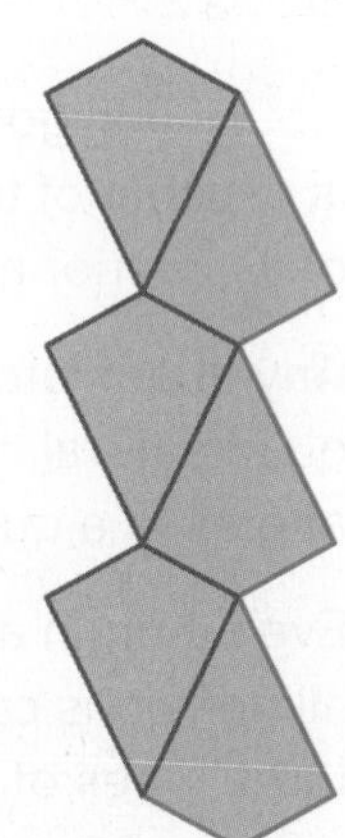

A. rotation and translation

B. rotation and reflection

C. reflection and translation

D. translation and translation

E. reflection and rotation

Go Online You can complete an Extra Example online.

Practice

Go Online You can complete your homework online.

Example 1

Determine whether each regular polygon will tessellate the plane. Explain.

1. pentagon
2. hexagon
3. 9-gon

Example 2

Determine whether a semiregular uniform tessellation can be created from the given shapes, assuming that all sides are 1 unit long. If so, determine the number of each shape needed at each vertex to create the tessellation.

4. regular pentagons and squares

5. regular hexagons and equilateral triangles

Example 3

Determine whether the pattern is a tessellation. If so, describe it as *uniform, not uniform, regular, not regular, or semiregular.*

6. 

7. 

8. 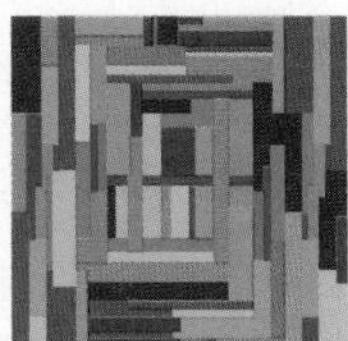

Example 4

Determine whether a tessellation can be created from each figure. If so, describe the transformation(s) that can be used to create the tessellation and draw a picture to support your reasoning.

9. scalene triangle

10. rhombus

11. Determine whether a tessellation can be created from a regular dodecagon. If so, describe the transformation(s) that can be used to create the tessellation. Will a regular dodecagon *sometimes, always,* or *never* tessellate the plane? Justify your argument.

12. Sketch a tessellation that can be created from an isosceles trapezoid. Describe the transformation(s) that can be used to create the tessellation.

13. Will a regular 15-gon *sometimes, always,* or *never* tessellate the plane? Justify your argument.

14. Determine whether a tessellation can be created from a parallelogram. If so, describe the transformation(s) that can be used to create the tessellation and draw a picture to support your reasoning.

Mixed Exercises

Determine the transformation(s) used to make each tessellation.

15. 

16. 

17. 

18. HOME IMPROVEMENT A hardware store sells various shapes of regular polygon paving stones. Kiyoko wants a simple design and only wants to buy one shape of stone. To build a solid base floor for her patio, what type of shape should Kiyoko buy?

19. GIFTS Matthew wants to surprise his girlfriend with a homemade gift. He wants to make a puzzle by tessellating one piece with a picture of a heart on it. What types of transformations can Matthew perform to create his puzzle? Explain.

20. FIND THE ERROR Heather says that if an interior angle of a regular n-gon measures 180°, then the n-gon will tessellate because 180° is a factor of 360°. Do you agree? Explain your reasoning.

21. CREATE Draw a tessellation that can be created by translations or rotations.

22. WRITE How would you accurately describe a tessellation to a person who had never heard the term before?

Symmetry

Explore Symmetry in Figures

 **Online Activity** Use dynamic geometry software to complete the Explore.

INQUIRY How can you tell when a figure can be mapped onto itself?

Today's Goal

- Use line symmetry to describe the reflections that carry a figure onto itself.
- Use rotational symmetry to describe the rotations that carry a figure onto itself.

Today's Vocabulary

symmetry
line symmetry
line of symmetry
rotational symmetry
center of symmetry
order of symmetry
magnitude of symmetry
point symmetry
point of symmetry

Learn Line Symmetry

A figure has **symmetry** if there exists a rigid motion—reflection, translation, rotation, or glide reflection—that maps the figure onto itself. Figures that have symmetry are self-congruent. One type of symmetry is *line symmetry*.

A figure in the plane has **line symmetry** (or *reflectional symmetry*) if each half of the figure matches the other side exactly. When a figure has line symmetry, the figure can be mapped onto itself by a reflection in a line, called the **line of symmetry** (or *axis of symmetry*).

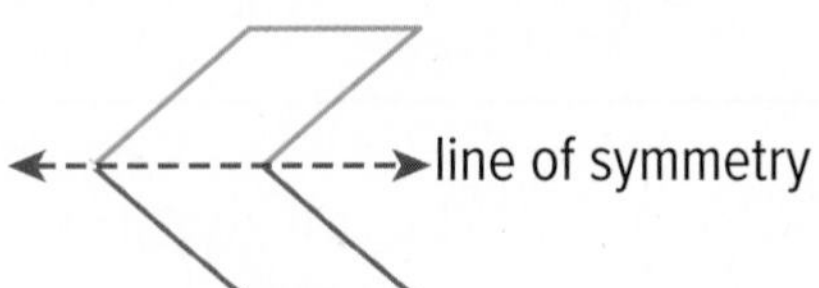

 **Talk About It!**

Do you think that a figure can have multiple lines of symmetry? Justify your argument.

Example 1 Identify Line Symmetry

Determine whether each figure has a line of symmetry. If so, draw the lines of symmetry and state how many lines of symmetry it has.

a.

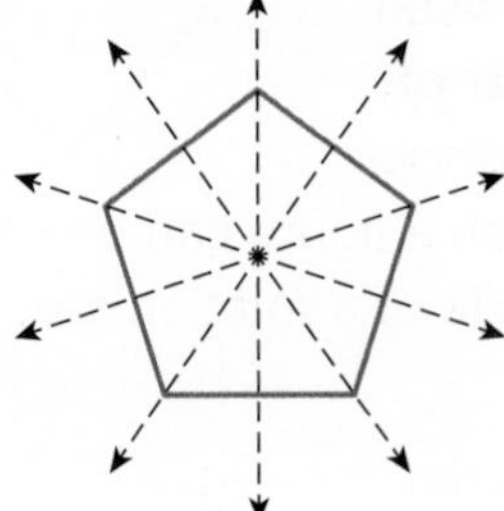

5 lines of symmetry

b.

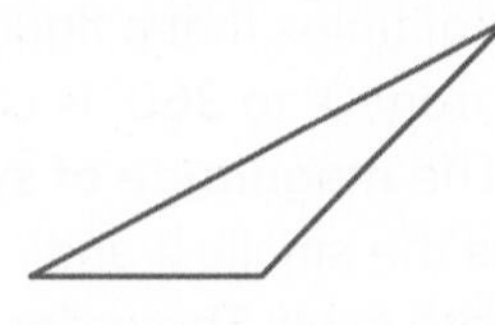

0 lines of symmetry

Go Online You can complete an Extra Example online.

Check

Determine whether each figure has a line of symmetry. If so, copy the figure and draw the lines of symmetry and state how many lines of symmetry it has.

This figure has __?__ line(s) of symmetry.

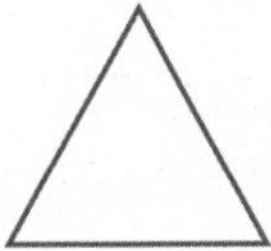

This figure has __?__ line(s) of symmetry.

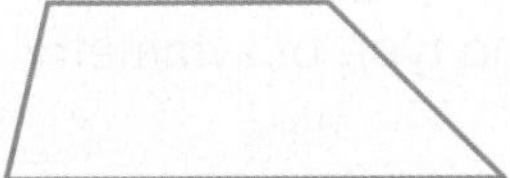

This figure has __?__ line(s) of symmetry.

Think About It!

Josefina argues that you can count the number of lines of symmetry in a circle. Do you agree or disagree? Justify your argument.

Learn Rotational Symmetry

A figure in the plane has **rotational symmetry** (or *radial symmetry*) if the figure can be mapped onto itself by being rotated less than 360° about the center of the figure so the image and the preimage are indistinguishable. The point in which a figure can be rotated onto itself is called the **center of symmetry** (or *point of symmetry*).

This figure has rotational symmetry because a rotation of 90°, 180°, or 270° maps the figure onto itself.

The number of times that a figure maps onto itself as it rotates from 0° to 360° is called the **order of symmetry.** The **magnitude of symmetry** (or *angle of rotation*) is the smallest angle through which a figure can be rotated so it maps onto itself. The order and magnitude of a rotation are related by the following equation.

$$\text{magnitude} = 360° \div \text{order}$$

This figure has order 4 and magnitude 90°.

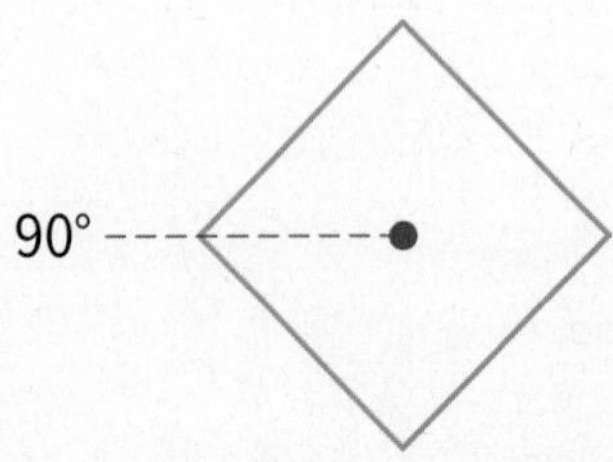

Key Concept • Point Symmetry

A figure has **point symmetry** if it can be mapped onto itself by a rotation of 180°. If a figure has point symmetry, then the center of symmetry in the figure is called the **point of symmetry**.

Example	A rhombus has point symmetry because it looks the same right-side up as upside down.

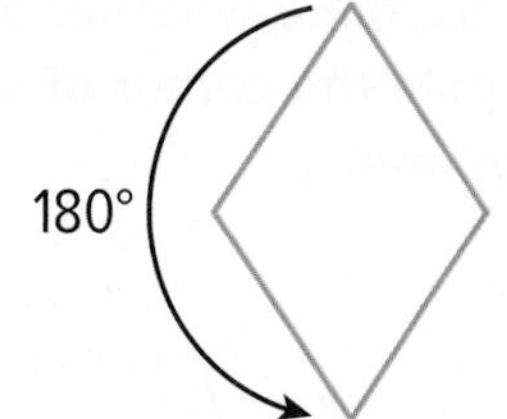

Go Online You may want to complete the Concept Check to check your understanding.

Example 2 Identify Rotational Symmetry

NATURE **Objects found in nature often have rotational symmetry.**

Determine whether each figure has rotational symmetry. Explain.

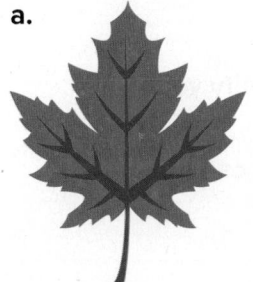

a. No; no rotation less than 360° maps the leaf onto itself.

b. Yes; the flower can map onto itself with a rotation that is less than 360°.

c. Yes; the clover can map onto itself with a rotation that is less than 360°.

Check

HOUSEHOLD Below are several objects that you might find around your house. Determine whether each figure has rotational symmetry. Explain.

Yes; the orange slice can map onto itself with a rotation that is less than 360°.

No; no rotation less than 360° maps the pair of scissors onto itself.

Yes; the tablet can map onto itself with a rotation that is less than 360°.

Example 3 Determine Order and Magnitude of Symmetry

Part A State the order and magnitude of symmetry.

Determine whether each figure has rotational symmetry. If so, locate the center of symmetry and state the order and magnitude of symmetry.

a.

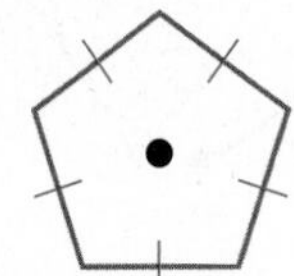

rotational symmetry: yes

order = 5
magnitude = 72°

b.

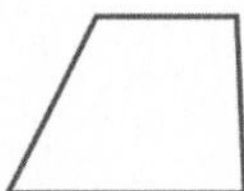

rotational symmetry: no

order = none
magnitude = none

c.

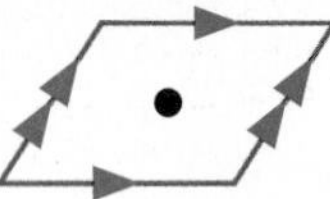

rotational symmetry: yes

order = 2
magnitude = 180°

Part B Identify point symmetry.

Which figure(s) in Part A has point symmetry? Justify your reasoning.

The parallelogram has point symmetry because it can be rotated 180° about its center so that it maps onto itself.

Check

Determine whether each figure has rotational symmetry. If so, copy the figure and locate the center of symmetry and state the order and magnitude of symmetry.

a.

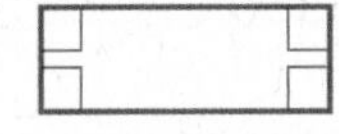

rotational symmetry: ___?___

order = ___?___
magnitude = ___?___

b.

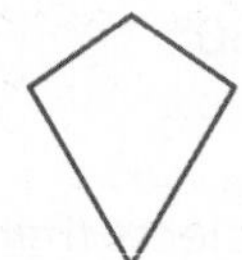

rotational symmetry: ___?___

order = ___?___
magnitude = ___?___

c.

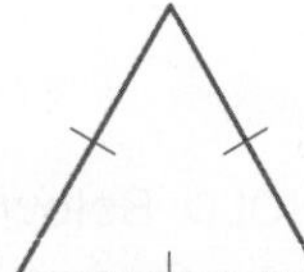

rotational symmetry: ___?___

order = ___?___
magnitude = ___?___

Which figure(s) has point symmetry? Justify your answer.

Think About It!
Is the following statement *sometimes*, *always*, or *never* true? Explain. *A polygon with order 4 rotational symmetry has point symmetry.*

Go Online to practice what you've learned about transformations and symmetry in the Put It All Together over Lessons 13-1 through 13-6.

Go Online You can complete an Extra Example online.

Practice

Go Online You can complete your homework online.

Example 1

Determine whether each figure has a line of symmetry. If so, draw the lines of symmetry and state how many lines of symmetry it has.

1.

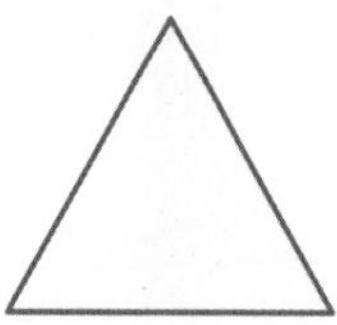

2.

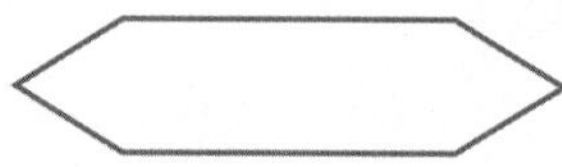

3.

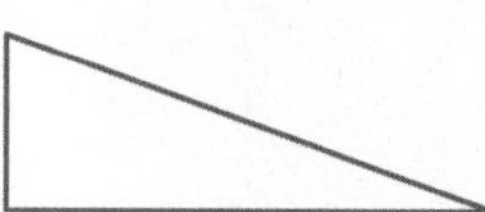

4.

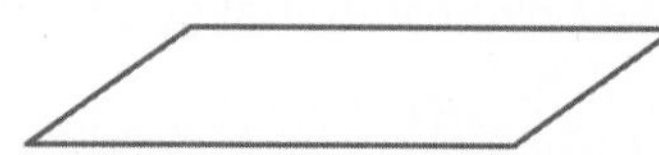

5.

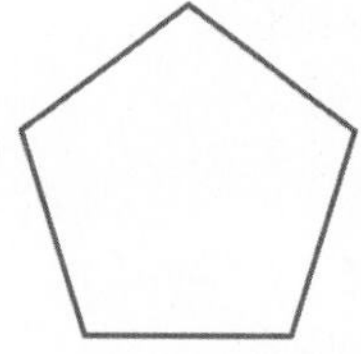

6.

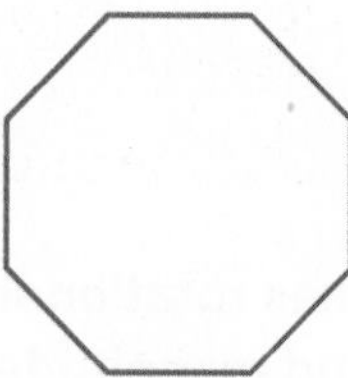

Example 2

7. **CARS** Steve found the hubcaps shown below at his local junkyard. Determine whether each hubcap has rotational symmetry. Explain.

a. 

b. 

c. 

8. **FLAGS** The figure shows the Union Jack, which is the flag of the United Kingdom. Does the flag have rotational symmetry? Explain.

Stefan Ataman/Shutterstock

9. **RECYCLING** A waste management company offers recycling programs for its clients. Recycling is denoted by the symbol shown. Does the recycling symbol have rotational symmetry? Explain.

10. **VACATION** Annabel and her family went to a beach for vacation. While she was on the beach, Annabel collected seashells. Does the seashell shown have rotational symmetry? Explain.

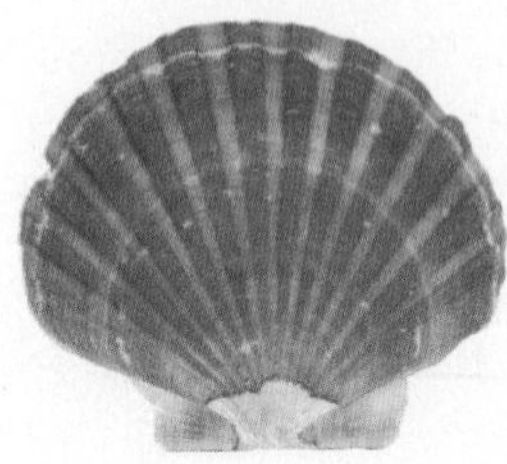

Example 3

Determine whether each figure has rotational symmetry. If so, locate the center of symmetry, and state the order and magnitude of symmetry.

11.

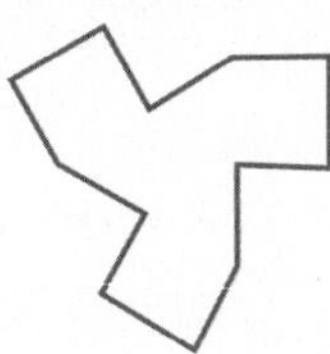

12.

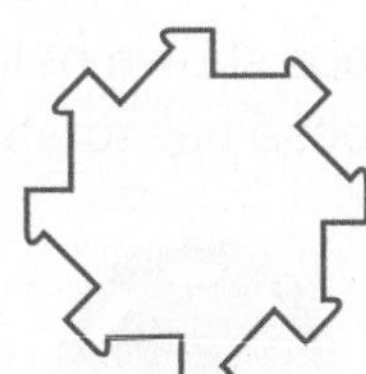

13.

14.

Frazer Cunningham/Image Source

Mixed Exercises

Refer to the figure at the right.

15. Draw the line(s) of symmetry in the figure.

16. Locate the center of symmetry for the figure.

17. What is the order and magnitude of symmetry for the figure?

18. LETTERS Examine each capital letter in the alphabet. Determine which letters have 180° rotational symmetry about a point in the center of the letter.

19. STRUCTURE A regular polygon has rotational symmetry with an order of 5 and a magnitude of 72°. What is the figure?

20. CONSTRUCT ARGUMENTS Consider the symmetry of a circle.

a. How many lines of symmetry does a circle have? Justify your argument.

b. What is the order of rotation for a circle? Justify your argument.

State whether each figure has rotational symmetry. If so, describe the rotations that map the figure onto itself by giving the order of symmetry and magnitude of symmetry.

21. equilateral triangle

22. scalene triangle

23. regular hexagon

Higher-Order Thinking Skills

24. PERSEVERE Draw a three-dimensional object that has a base with line symmetry.

25. CREATE Draw an object that has at least one line of symmetry. Describe the lines of symmetry in this object.

26. ANALYZE The figure shows the floor plan for a new gallery in an art museum. Describe every reflection or rotation that maps the gallery onto itself.

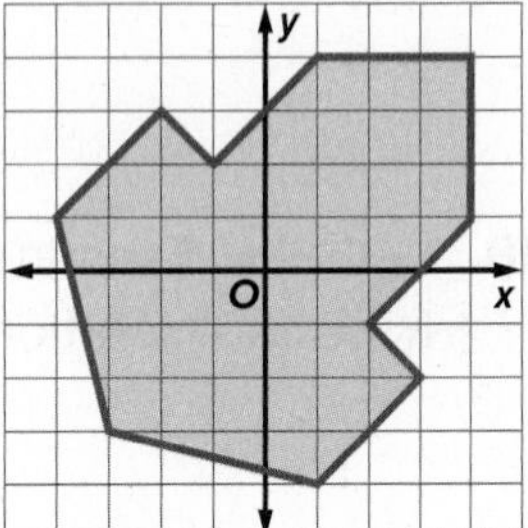

27. WRITE A regular polygon has magnitude of symmetry 15°. How many sides does the polygon have? Explain.

28. FIND THE ERROR Jaime says that Figure A has only line symmetry, and Jewel says that Figure A has only rotational symmetry. Is either of them correct? Explain your reasoning.

Figure A

29. PERSEVERE A quadrilateral in the coordinate plane has exactly two lines of symmetry, $y = x - 1$ and $y = -x + 2$. Find possible vertices for the figure. Graph the figure and the lines of symmetry.

30. CREATE Draw a figure that has line symmetry but not rotational symmetry. Explain.

31. WRITE How are line symmetry and rotational symmetry related?

Review

Essential Question

How are rigid motions used to show geometric relationships?

Rigid motions are used to show that figures are congruent. If no series of rigid motions exists from one figure to another, then the figures are not congruent.

Module Summary

Lessons 13-1 through 13-3

Reflections, Translations, and Rotations

- When a figure is reflected in a line, each point of the preimage and its corresponding point on the image are the same distance from the line of reflection.
- A translation is a function in which all the points of a figure move the same distance in the same direction as described by a translation vector.
- A translation vector describes the magnitude and direction of the translation. The magnitude of a vector is its length from the initial point to the terminal point.
- A rotation about a fixed point P through an angle of $a°$ is a function that maps point M to point M' such that point P does not move, $m\angle MPM'$ is $a°$, and $MP = M'P$.

Lesson 13-4

Compositions of Transformations

- When a transformation is applied to a figure and then another transformation is applied to its image, the result is called a composition of transformations.
- A glide reflection is the composition of a translation followed by a reflection in a line parallel to the translation vector.
- The composition of two reflections can result in the same image as a translation or rotation.

Lessons 13-5 and 13-6

Tessellations and Symmetry

- A regular polygon will tessellate if it has an interior angle measure that is a factor of 360°.
- A semiregular tessellation is formed by two or more regular polygons.
- A figure has symmetry if there exists a rigid motion—reflection, translation, rotation, or glide reflection—that maps the figure onto itself.
- A figure in the plane has line symmetry (or reflectional symmetry) if each half of the figure matches the other side exactly.
- A figure in the plane has rotational symmetry (or radial symmetry) if the figure can be mapped onto itself by being rotated less than 360° about the center of the figure so the image and the preimage are indistinguishable.

Study Organizer

 Foldables

Use your Foldable to review this module. Working with a partner can be helpful. Ask for clarification of concepts as needed.

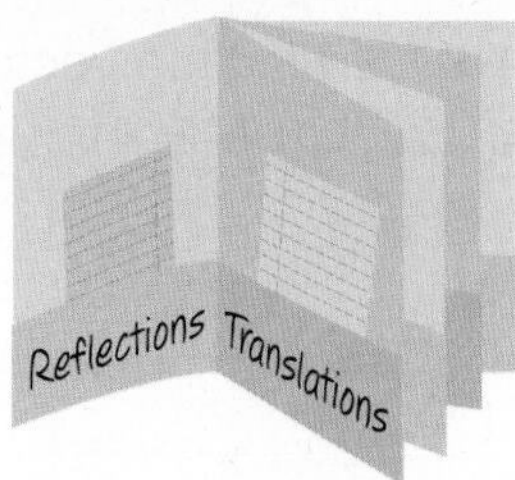

1. **GRAPH** Graph the image of $\triangle ABC$ with vertices at $A(-5, 0)$, $B(-3, 5)$, $C(-1, 2)$ after a reflection in the line $x = -2$. (Lesson 13-1)

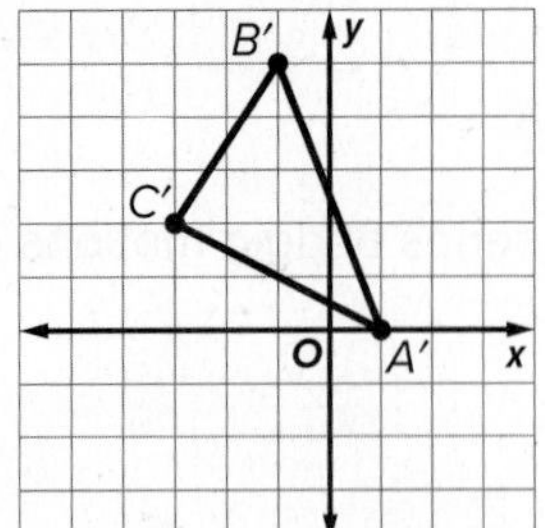

2. **MULTIPLE CHOICE** Which **best** describes a possible step that is used to determine the location of the image of point B when it is reflected in the line $y = x$? (Lesson 13-1)

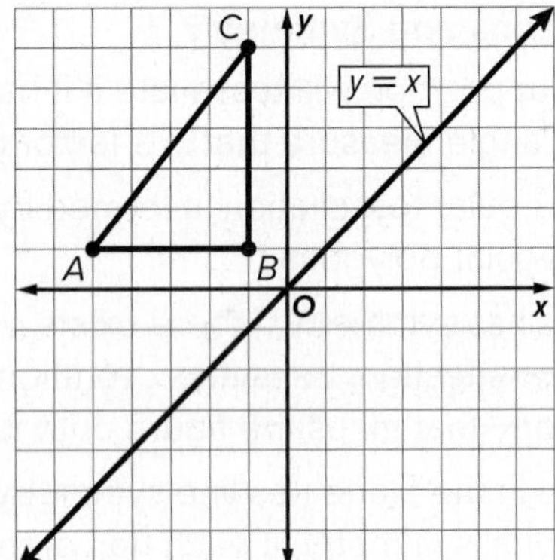

A. Move down one and right one from (0, 0).

B. Move down one and right one from (−1, 1).

C. Move right two from (1, 1).

D. Move down two from (−1, −1).

3. **OPEN RESPONSE** When point F is reflected in the line $y = x$, the image is located at $F'(6, -9)$. Find the coordinates of point F. (Lesson 13-1)

4. **MULTIPLE CHOICE** Find the vector that translates $A(-2, 7)$ to $A'(6, 4)$. (Lesson 13-2)

A. $\langle -8, 3\rangle$

B. $\langle -3, 8\rangle$

C. $\langle 3, 8\rangle$

D. $\langle 8, -3\rangle$

5. **OPEN RESPONSE** Refer to the graph.

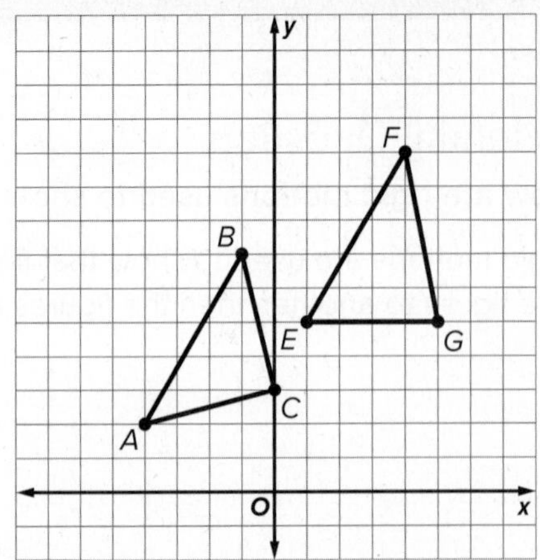

Explain why a translation does not map $\triangle ABC$ to $\triangle EFG$. (Lesson 13-2)

6. **MULTIPLE CHOICE** Which is the image of $P(-5, 11)$ along the vector $\langle 3, -8\rangle$? (Lesson 13-2)

A. $P'(-8, 19)$

B. $P'(-8, 3)$

C. $P'(-2, -3)$

D. $P'(-2, 3)$

7. **MULTIPLE CHOICE** Juan is designing a new playground for the elementary school. He needs to determine the shortest distance from the monkey bars to the slide to create a path. Which statement **best** describes the translation from the monkey bars to the slide? (Lesson 13-2)

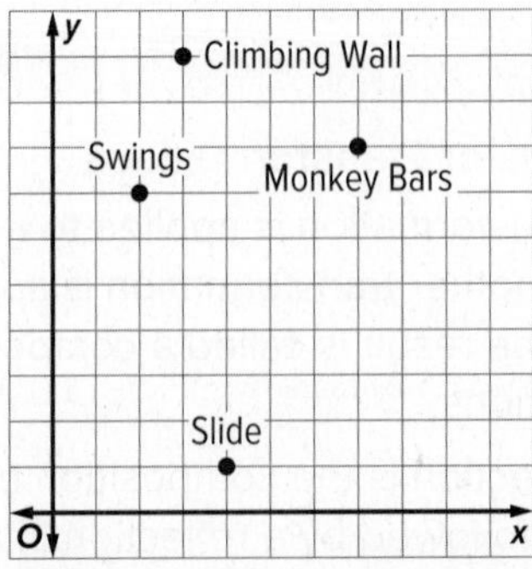

A. a translation right 11 units and up 9 units

B. a translation right 3 units and up 7 units

C. a translation left 3 units and down 7 units

D. a translation left 11 units and down 9 units

8. **MULTIPLE CHOICE** Which is the image of $P(3, 0)$ after a counterclockwise rotation of 90° about (2, 4)? (Lesson 13-3)

A. $P'(-2, 5)$

B. $P'(6, 5)$

C. $P'(6, 3)$

D. $P'(-4, 2)$

9. **OPEN RESPONSE** Refer to the graph.

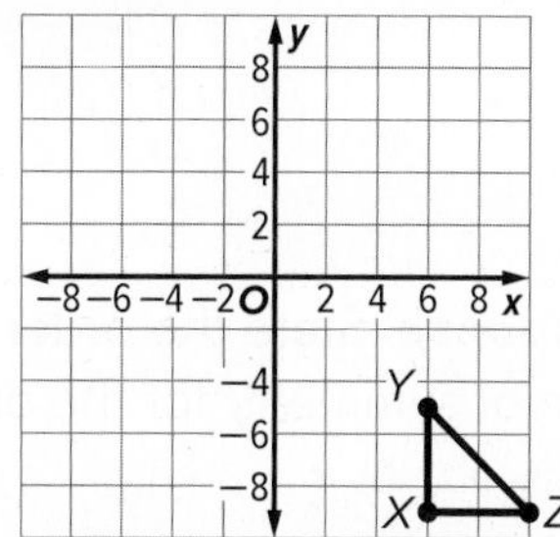

In which quadrant will the image be after a rotation of 180° about (1, −2)? (Lesson 13-3)

10. **MULTIPLE CHOICE** Which is the image of $F(-2, -7)$ after a counterclockwise rotation of 180° about (−1, 5)? (Lesson 13-3)

A. $F'(-1, 5)$

B. $F'(0, 17)$

C. $F'(1, 12)$

D. $F'(2, 7)$

11. **OPEN RESPONSE** *True or false:* Rotating $M(-5, 1)$ 180° about the origin and then translating along $\langle -3, 4 \rangle$ will give the same result as translating along $\langle -3, 4 \rangle$ and then rotating 180°. (Lesson 13-4)

12. **MULTPLE CHOICE** Triangle ABC is shown. (Lesson 13-4)

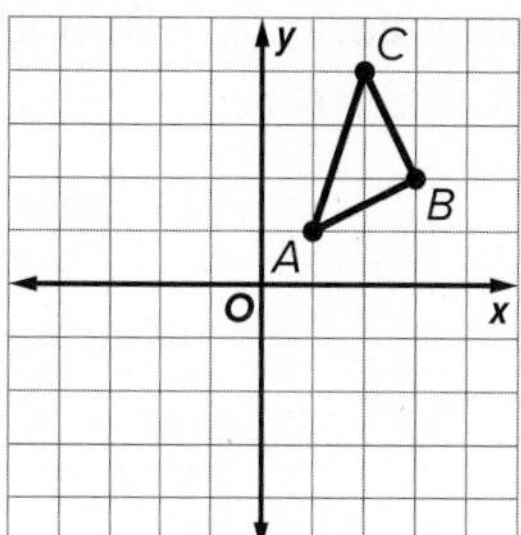

Triangle ABC is rotated 90° counterclockwise about the origin and then translated along $\langle -2, 3 \rangle$. What is the location of the image of point B?

A. (0, 0)

B. (1, 1)

C. (−4, 6)

D. (−5, 1)

13. **MULTI-SELECT** Select all transformations or compositions of transformations that would map $\triangle ABC$ to $\triangle A'B'C'$. (Lesson 13-4)

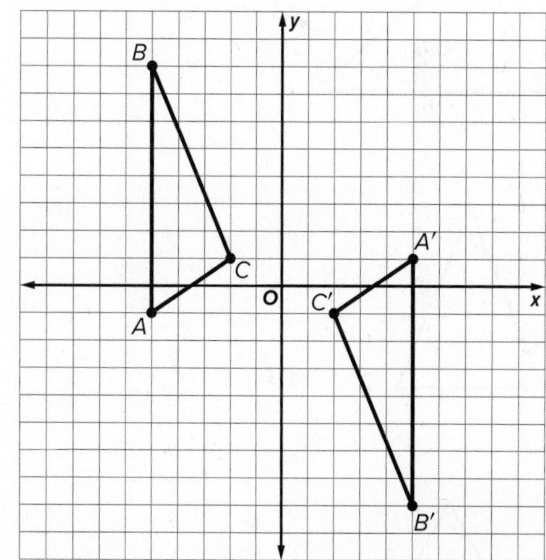

A. Reflection in the x-axis followed by a reflection in the y-axis

B. Reflection in the x-axis followed by a rotation of 90° counterclockwise about the origin

C. Reflection in $y = -x$

D. Rotation of 180° about the origin

E. Reflection in $y = x$

14. GRAPH Graph the image of $\triangle PQR$ with vertices at $P(4, 7)$, $Q(7, 3)$, and $R(2, 2)$ after a translation along $\langle 1, -9 \rangle$ and a reflection in $x = 2$. (Lesson 13-4)

15. MULTIPLE CHOICE Which figure has 3 lines of symmetry? (Lesson 13-6)

A. 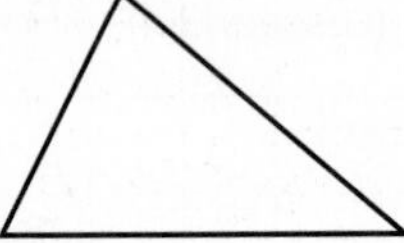

B. 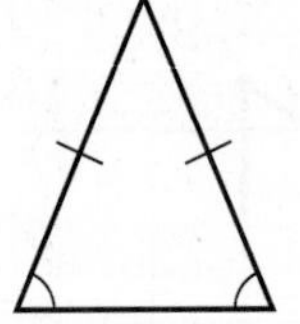

C. 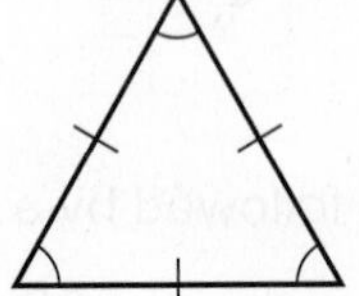

D. 

16. OPEN RESPONSE How many lines of symmetry does this figure have? Describe the reflections, if any, that map the figure onto itself. (Lesson 13-6)

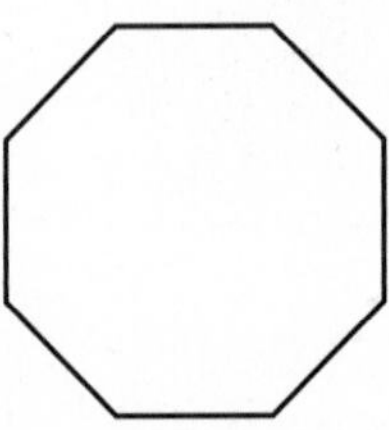

17. Open Response State the order and magnitude of symmetry for the object below. (Lesson 13-6)

Module 14

Triangles and Congruence

Essential Question

How can you prove congruence and use congruent figures in real-world situations?

What Will You Learn?

How much do you already know about each topic **before** starting this module?

KEY

👎 — I don't know. 👉 — I've heard of it. 👍 — I know it!

	Before			After		
	👎	👉	👍	👎	👉	👍
solve problems using the Triangle Angle-Sum Theorem						
solve problems using the Exterior Angle Theorem						
show that triangles are congruent						
identify corresponding parts of congruent triangles						
solve problems using the SSS Congruence Postulate						
solve problems using the SAS Congruence Postulate						
solve problems using the ASA Congruence Postulate						
solve problems using the AAS Congruence Theorem						
construct congruent triangles						
solve problems using the LL, HA, LA, and HL Theorems						
solve problems involving isosceles and equilateral triangles						
write coordinate proofs						

Foldables Make this Foldable to help you organize your notes about triangles and congruence. Begin with one sheet of paper.

1. **Fold** a sheet of paper as shown, cutting off the excess paper strip to form a taco.
2. **Open** the fold and refold the square the opposite way to form another taco and an X-fold pattern.
3. **Open** and fold the corners toward the center point of the X, forming a small square.
4. **Label** the flaps as shown.

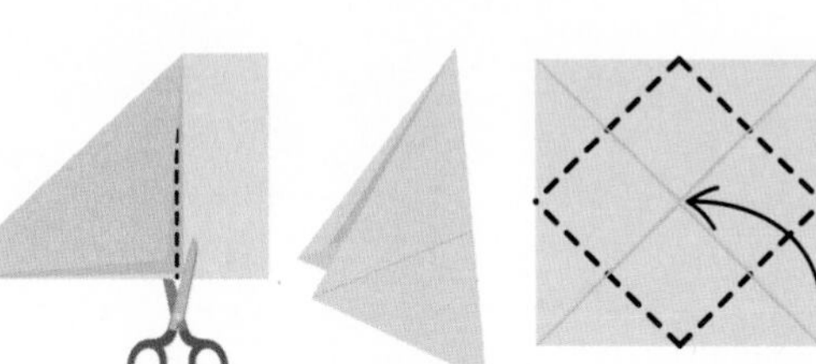

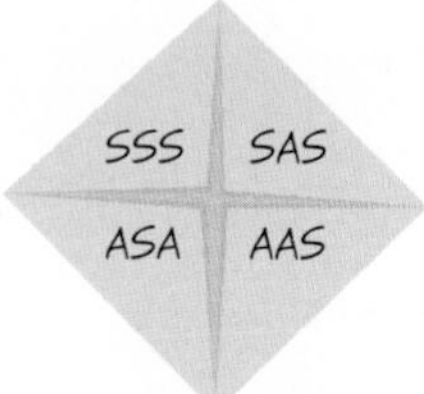

What Vocabulary Will You Learn?

- auxiliary line
- base angles of an isosceles triangle
- congruent polygons
- coordinate proofs
- corollary
- corresponding parts
- exterior angle of a triangle
- included angle
- included side
- interior angle of a triangle
- isosceles triangle
- legs of an isosceles triangle
- principle of superposition
- remote interior angles
- vertex angle of an isosceles triangle

Are You Ready?

Complete the Quick Review to see if you are ready to start this module. Then complete the Quick Check.

Quick Review

Example 1

Classify each angle as *right*, *acute*, or *obtuse*.

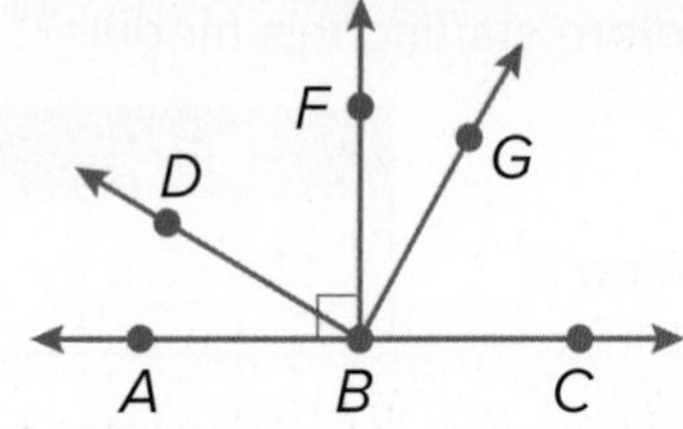

a. $\angle ABG$

Point G on $\angle ABG$ lies on the exterior of right angle $\angle ABF$, so $\angle ABG$ is an obtuse angle.

b. $\angle DBA$

Point D on $\angle DBA$ lies on the interior of right angle $\angle FBA$, so $\angle DBA$ is an acute angle.

Example 2

Find the distance between $J(5, 2)$ and $K(11, -7)$.

$JK = \sqrt{(x_2 - x_1)^2 + (y_2 - y_1)^2}$ — Distance Formula

$= \sqrt{(11 - 5)^2 + [(-7) - 2]^2}$ — Substitute.

$= \sqrt{6^2 + (-9)^2}$ — Subtract.

$= \sqrt{36 + 81}$ — Simplify.

$= \sqrt{117}$ or about 10.8 — Add.

Quick Check

Classify each angle as *right*, *acute*, or *obtuse*.

1. $\angle VQS$
2. $\angle TQV$
3. $\angle PQV$
4. $\angle SQR$

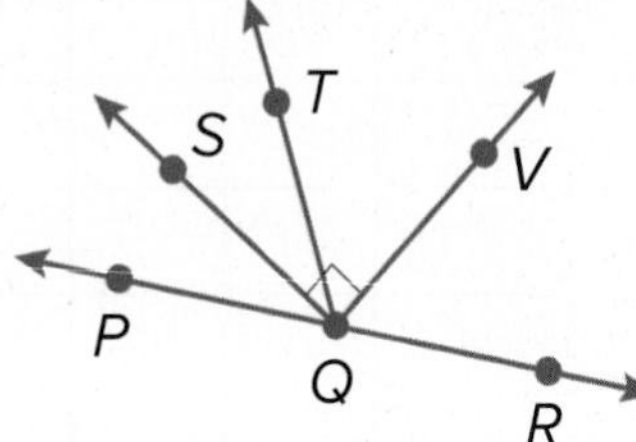

Find the distance between each pair of points. Round to the nearest tenth.

5. $F(3, 6)$ and $G(7, -4)$
6. $X(-2, 5)$ and $Y(1, 11)$
7. $R(8, 0)$ and $S(-9, 6)$
8. $A(14, -3)$ and $B(9, -9)$

How did you do?

Which exercises did you answer correctly in the Quick Check?

Angles of Triangles

Explore Triangle Angle Sums

 **Online Activity** Use dynamic geometry software to complete the Explore.

INQUIRY Is there a relationship associated with the interior angles of a triangle? If so, how do we prove that this relationship is always true?

Learn Interior Angles of Triangles

An **interior angle of a triangle** is the angle at a vertex of a triangle. Because a triangle has three vertices, it also has three interior angles. The Triangle Angle-Sum Theorem describes the relationships among the interior angle measures of any triangle.

Theorem 14.1: Triangle Angle-Sum Theorem
The sum of the measures of the interior angles of a triangle is 180°.

Go Online A proof of Theorem 14.1 is available.

Apply Example 1 Use the Triangle Angle-Sum Theorem

Find the measure of each numbered angle.

J, K, L, M, 3, 28°, 71°, 1, 57°, 2

1 What is the task?

Describe the task in your own words. Then list any questions that you may have. How can you find answers to your questions?
Sample answer: I need to find the measures of $\angle 1$, $\angle 2$, and $\angle 3$. What are the relationships between the angle measures that are given and the angle measures that I need to find? I can use the theorems and postulates that I have learned to find the information that I need.

2 How will you approach the task? What have you learned that you can use to help you complete the task?
I will use the Triangle Angle-Sum Theorem and the definition of supplementary angles to solve for the missing angle measures.

3 What is your solution?

Use your strategy to solve the problem.
$m\angle 1 = 123°$ $\quad m\angle 2 = 52°$ $\quad m\angle 3 = 29°$

4 How can you know that your solution is reasonable?

Write About It! Write an argument that can be used to defend your solution.
Sample answer: The sums of the measures of the interior angles of $\triangle JKL$ and $\triangle LKM$ should be 180°. When I add up the angle measures for each triangle, the sum equals 180°.

Go Online You can complete an Extra Example online.

Today's Goals
- Prove the Triangle Angle-Sum Theorem and apply the theorem to solve problems.
- Prove the Exterior Angle Theorem and apply the theorem to solve problems.
- Prove the corollaries to the Triangle Angle-Sum Theorem and apply the corollaries to solve problems.

Today's Vocabulary
interior angle of a triangle
exterior angle of a triangle
remote interior angles
corollary

Watch Out!
Triangle Angle-Sum Theorem When you are finding missing angle measures of a triangle, check the solution by seeing whether the sum of the measures of the angles of the triangle is 180°.

Talk About It!
Ellie believes that she can solve for $m\angle 3$ before solving for $m\angle 1$. What useful questions can you ask to understand her approach?

Check

Find the measure of each numbered angle.

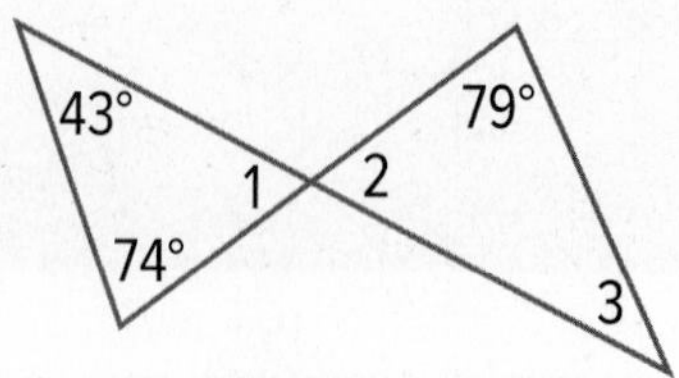

Learn Exterior Angles of Triangles

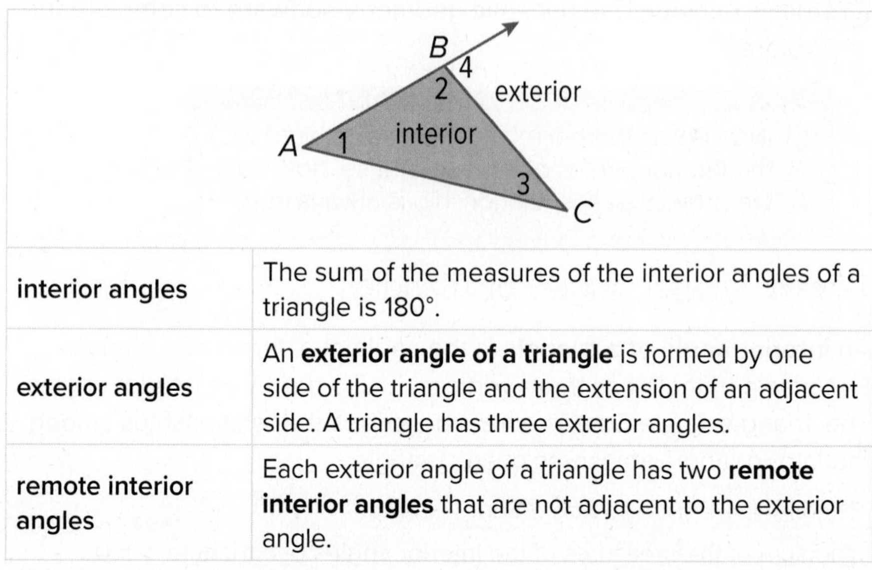

interior angles	The sum of the measures of the interior angles of a triangle is 180°.
exterior angles	An **exterior angle of a triangle** is formed by one side of the triangle and the extension of an adjacent side. A triangle has three exterior angles.
remote interior angles	Each exterior angle of a triangle has two **remote interior angles** that are not adjacent to the exterior angle.

Theorem 14.2 Exterior Angle Theorem

The measure of an exterior angle of a triangle is equal to the sum of the measures of the two remote interior angles.

Given: $\triangle ABC$

Prove: $m\angle A + m\angle B = m\angle 1$

Proof:

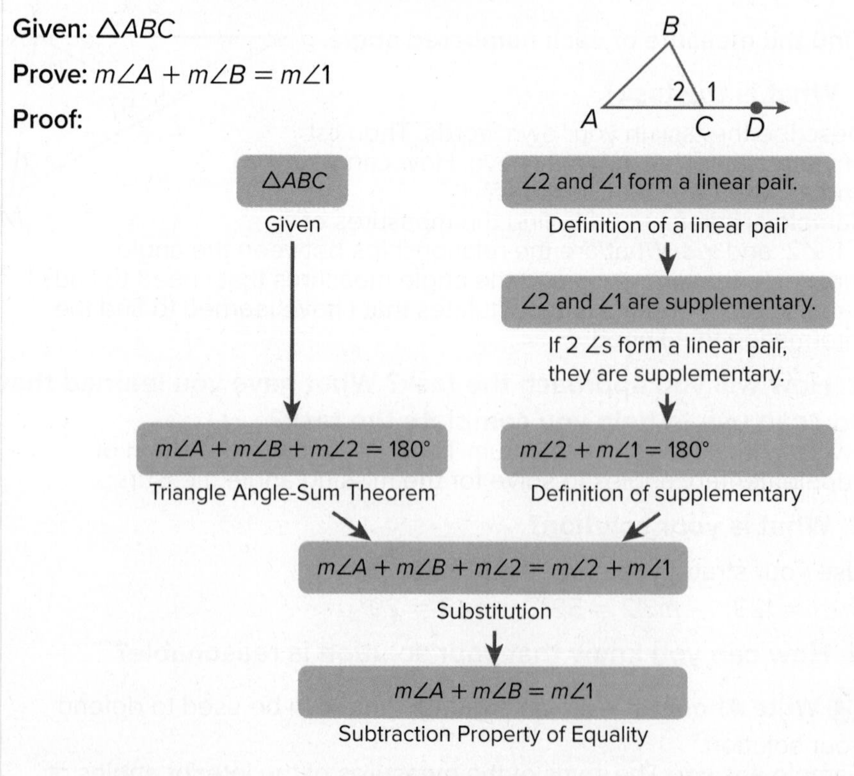

Example 2 Use the Exterior Angle Theorem

ARCHITECTURE **Find the measure of $\angle DAB$ in the front face of the building.**

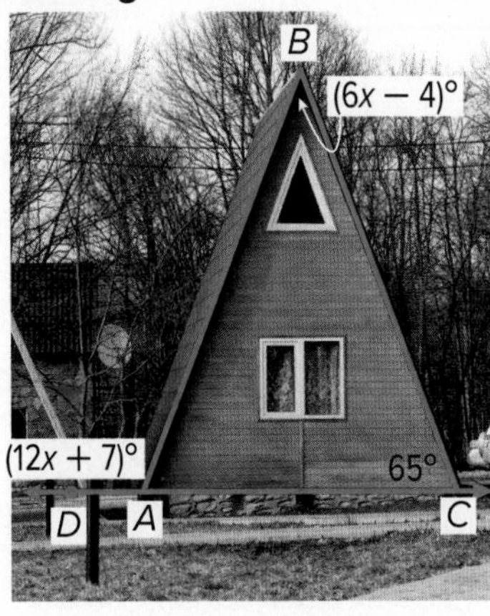

$m\angle DAB = m\angle ABC + m\angle BCA$	Exterior Angle Theorem
$12x + 7 = 6x - 4 + 65$	Substitution
$x = 9$	Solve.
$m\angle DAB = 12(9) + 7$ or $115°$	

Check

PUZZLES Find the measure of $\angle XYZ$ created by the triangle.

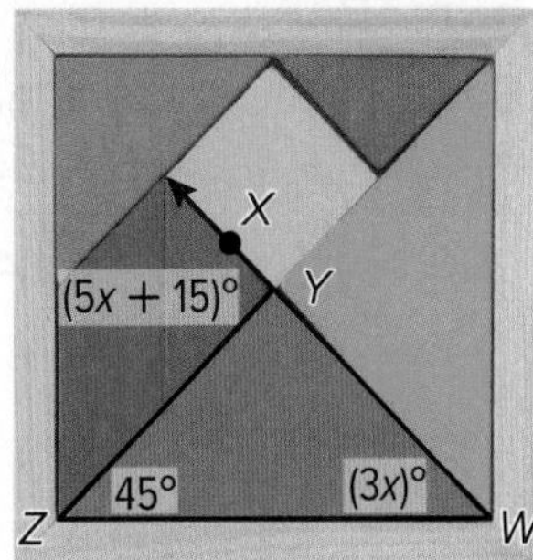

Go Online You can complete an Extra Example online.

Think About It!
What theorems and definitions can you use to check your answer for reasonableness?

What assumption did you make when you were modeling the front face of the building as a triangle?

Ruslan Kokarev/123RF, Olga Popova/Shutterstock

Learn Triangle Angle-Sum Corollaries

A **corollary** is a theorem with a proof that follows as a direct result of another theorem. As with a theorem, a corollary can be used as a reason in a proof. The corollaries below follow directly from the Triangle Angle-Sum Theorem.

Corollary 14.1
The acute angles of a right triangle are complementary.

Corollary 14.2
There can be at most one right or obtuse angle in a triangle.

You will prove Corollary 14.1 and 14.2 in Exercises 19 and 20, respectively.

Example 3 Find Angle Measures in Right Triangles

Find each measure.

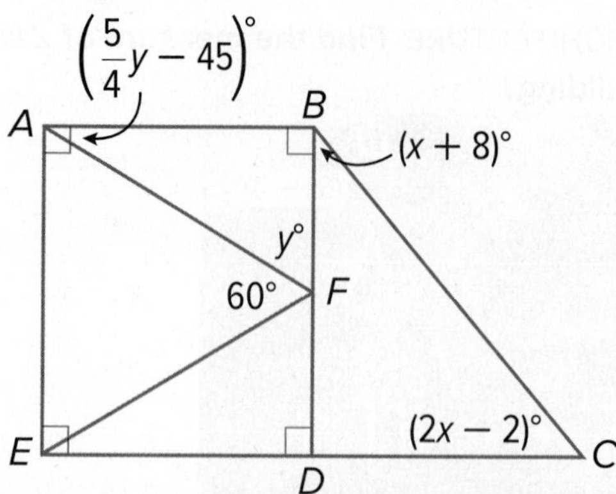

a. $m\angle BCD$

Because $\angle BDC$ and $\angle EDF$ form a linear pair and $m\angle EDF = 90°$, $m\angle BDC = 90°$ by the Supplement Theorem. Therefore, $m\angle BCD + m\angle DBC = 90°$ because the acute angles of a right triangle are complementary.

$m\angle BCD + m\angle DBC = 90°$	Corollary 14.1
$(x + 8) + (2x - 2) = 90°$	Substitution
$x = 28$	Solve.

$m\angle BCD = (2(28) - 2)°$ or $54°$

b. $m\angle BAF$

Because the acute angles of a right triangle are complementary, $m\angle BAF + m\angle AFB = 90°$.

$90° = m\angle BAF + m\angle AFB$	Corollary 14.1
$= (\frac{5}{4}y - 45) + y$	Substitution
$60 = y$	Solve.

$m\angle BAF = \frac{5}{4}(60) - 45$ or $30°$

Study Tip

Check for Reasonableness
When you are solving for the measure of one or more angles of a triangle, always check to make sure that the sum of the angle measures is 180°.

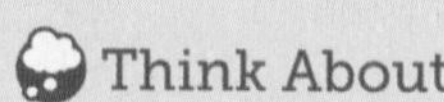

Think About It!

Do you have enough information to solve for $m\angle FED$? If so, explain your solution process.

Check

Find each measure.

$m\angle BAC =$ ___?___

$m\angle BCA =$ ___?___

$m\angle DCF =$ ___?___

$m\angle CDF =$ ___?___

$m\angle CFD =$ ___?___

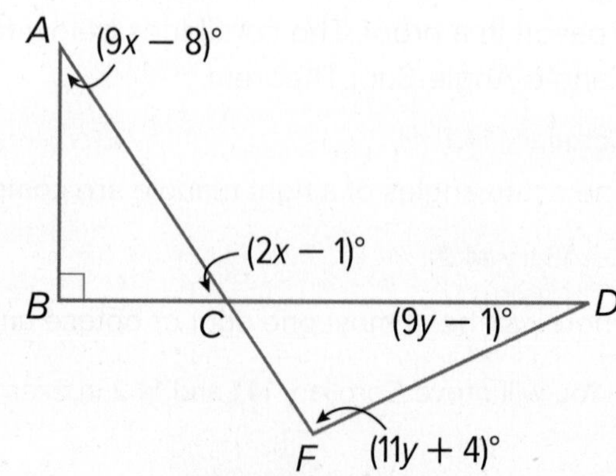

 **Go Online** You can complete an Extra Example online.

Practice

Go Online You can complete your homework online.

Example 1

Find the measure of each numbered angle.

1. 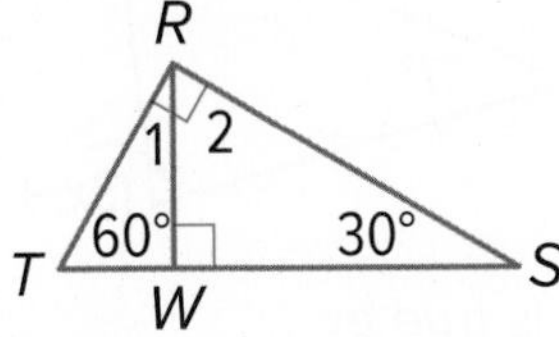

2. 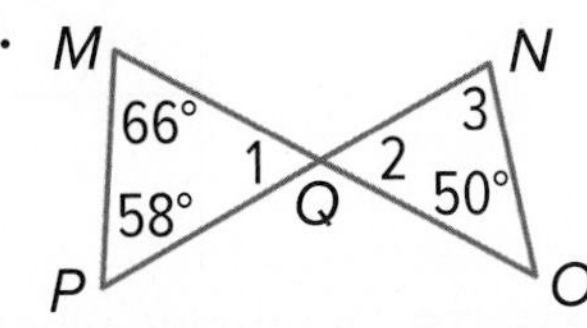

3. 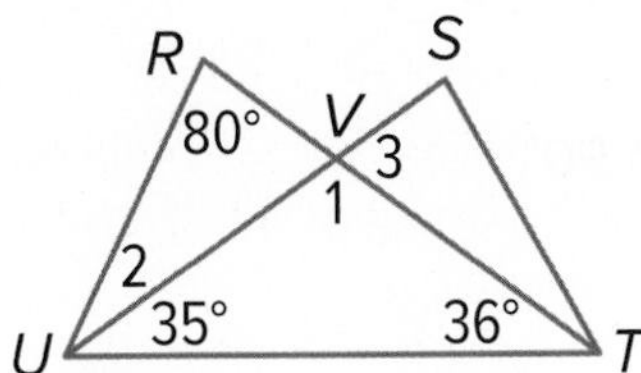

4. 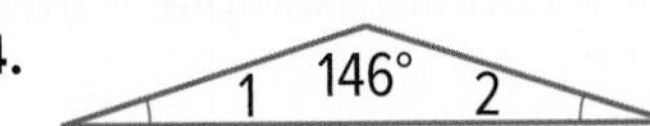

Example 2

Find each measure.

5. $m\angle ABC$

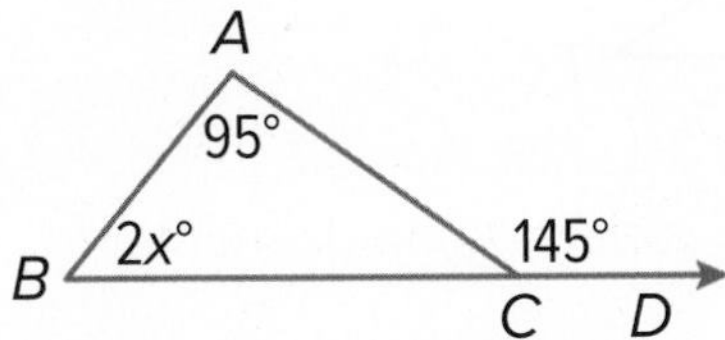

6. $m\angle F$

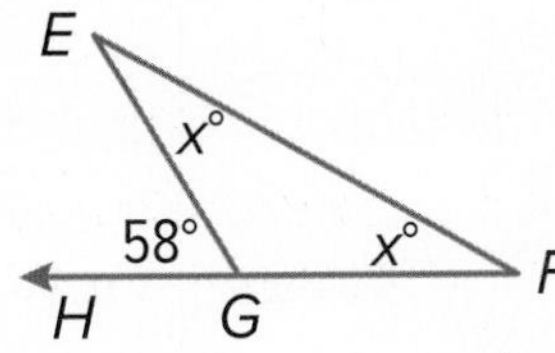

7. **TOWERS** A lookout tower sits on a network of struts and posts. Leslie measured three angles on the tower. If $m\angle 1 = (7x - 7)°$, $m\angle 2 = (4x + 2)°$, and $m\angle 3 = (2x + 6)°$, what is $m\angle 1$?

8. **GARDENING** A gardener uses a grow light to grow vegetables indoors. If $m\angle 1 = 8x°$ and $m\angle 2 = (7x - 4)°$, what is $m\angle 1$?

Example 3

Find each measure.

9. $m\angle 1$

10. $m\angle 2$

11. $m\angle 3$

12. $m\angle 4$

13. $m\angle 5$

14. $m\angle 6$

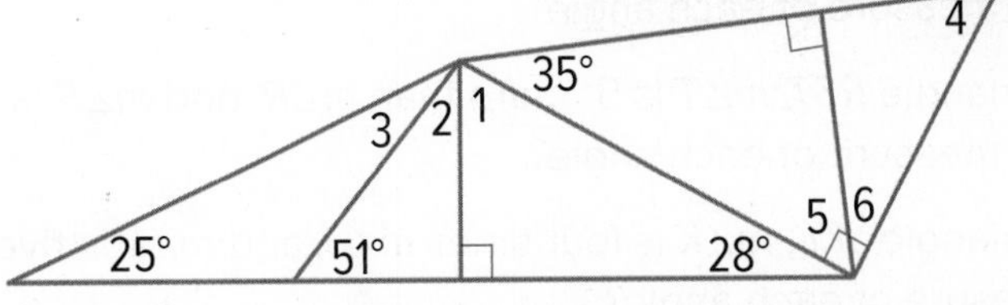

Mixed Exercises

Find the value of *x*. Then find the measure of each angle.

15.

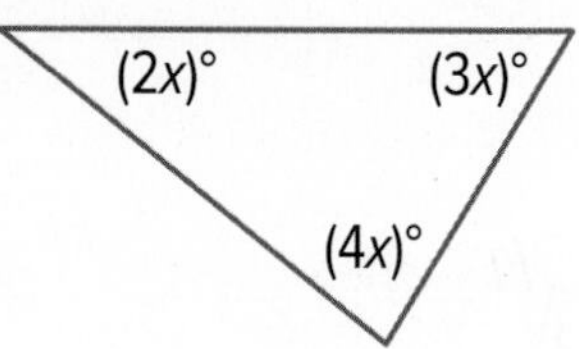

16.

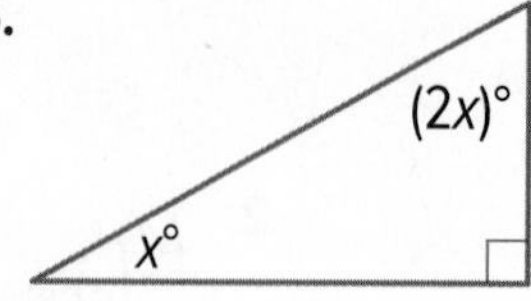

17.

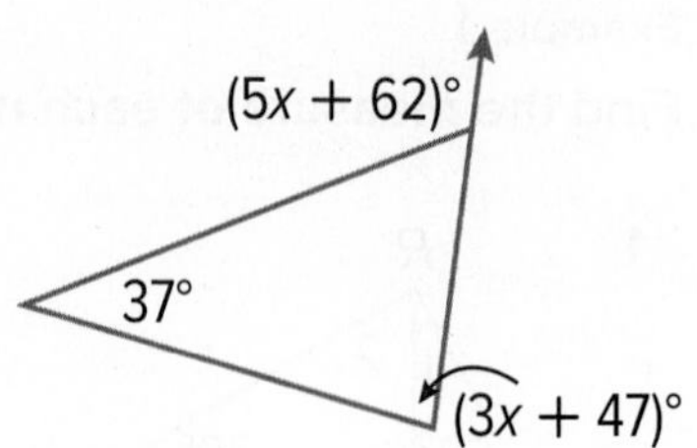

18. CONSTRUCT ARGUMENTS Determine whether the following statement is *true* or *false*. If false, give a counterexample. If true, give an argument to support your conclusion.

If the sum of two acute angles of a triangle is greater than 90°, then the triangle is acute.

PROOF **Write the specified type of proof for each corollary.**

19. flow proof of Corollary 14.1
Given: $\angle R$ is a right angle.
Prove: $\angle S$ and $\angle T$ are complementary.

S
R T

20. paragraph proof of Corollary 14.2

a. Case 1
Given: $\triangle MNO$; $\angle M$ is a right angle.
Prove: There can be at most one right angle in a triangle.

b. Case 2
Given: $\triangle PQR$; $\angle P$ is an obtuse angle.
Prove: There can be at most one obtuse angle in a triangle.

REASONING **Solve each problem.**

21. In triangle *DEF*, $m\angle E$ is three times $m\angle D$, and $m\angle F$ is 9° less than $m\angle E$. What is the measure of each angle?

22. In triangle *RST*, $m\angle T$ is 5° more than $m\angle R$, and $m\angle S$ is 10° less than $m\angle T$. What is the measure of each angle?

23. In triangle *JKL*, $m\angle K$ is four times $m\angle J$, and $m\angle L$ is five times $m\angle J$. What is the measure of each angle?

24. Classify the triangle shown by its angles. Justify your reasoning.

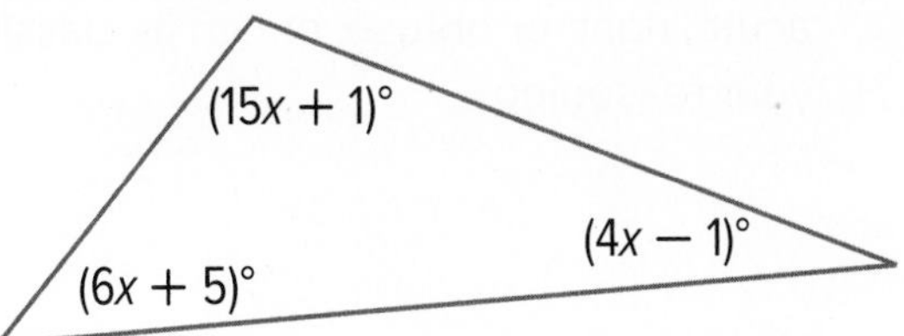

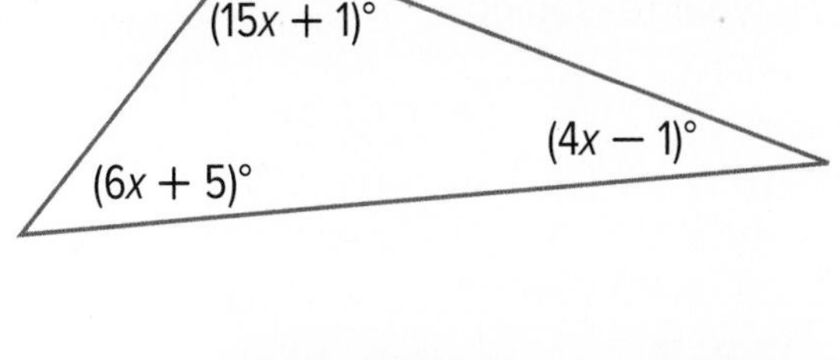

25. In $\triangle XYZ$, $m\angle X = 157°$, $m\angle Y = y°$, and $m\angle Z = z°$. Write an inequality to describe the possible measures of $\angle Z$. Justify your reasoning.

26. AUTOMOBILES Refer to the image at the right.

a. Find $m\angle 1$ and $m\angle 2$.

b. If the brown hood prop rod were shorter than the one shown, how would $m\angle 1$ change? Explain.

c. If the brown hood prop rod were shorter than the one shown, how would $m\angle 2$ change? Explain.

27. BASKETBALL Sam, Kendra, and Tony are passing a basketball. If Sam is looking at Kendra, then he needs to turn 40° to pass to Tony. If Tony is looking at Sam, then he needs to turn 50° to pass to Kendra. How many degrees would Kendra have to turn her head to look at Tony if she is looking at Sam?

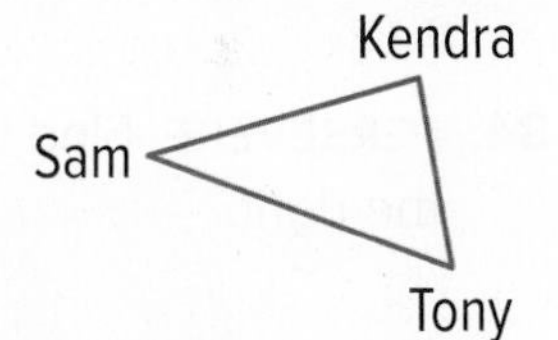

28. CONSTRUCTION The diagram shows an example of the Pratt Truss used in bridge construction. Find $m\angle 1$.

Find the measure of each numbered angle.

29.

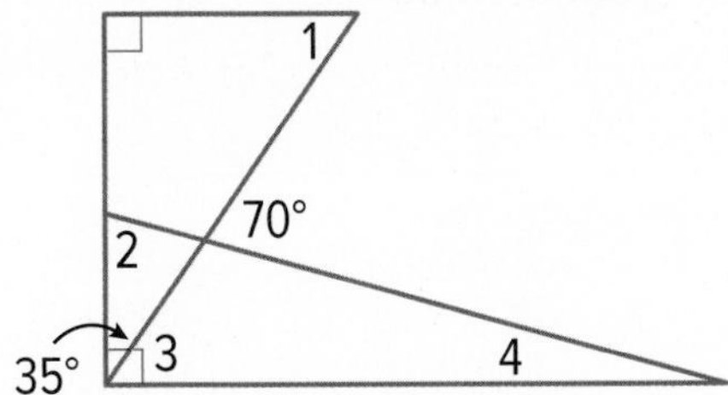

30.

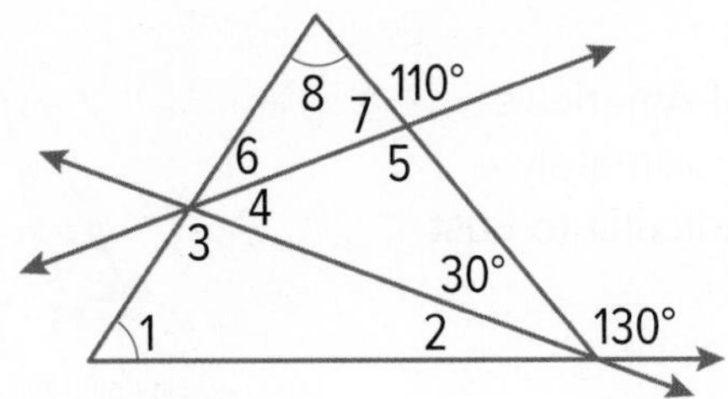

31. USE TOOLS Use tracing paper to verify the Triangle Angle-Sum Theorem. Describe your method and include a sketch.

32. ANALYZE In $\triangle ABC$, if an exterior angle adjacent to $\angle A$ is acute, is the triangle acute, right, or obtuse, or can its classification not be determined? Explain your reasoning.

33. WRITE Explain why a triangle cannot have an obtuse, acute, and a right exterior angle.

34. PERSEVERE Find the values of y and z in the figure at the right.

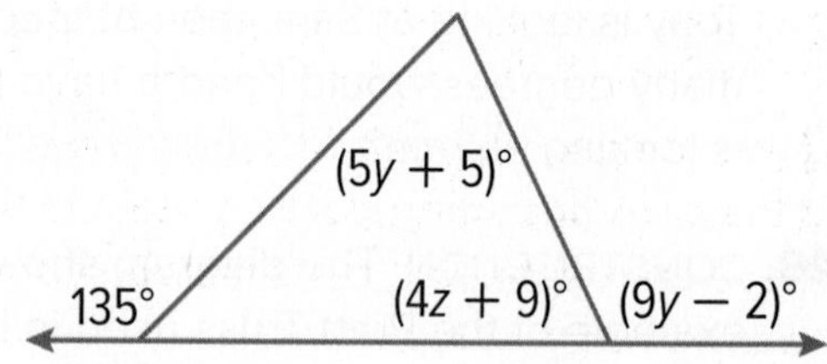

35. CREATE Construct a right triangle and measure one of the acute angles. Calculate the measure of the second acute angle and explain your method. Confirm your result using a protractor.

36. PERSEVERE The Flatiron Building in New York City is one of America's oldest skyscrapers, completed in 1902. Its floor plan is approximately a right triangle. As shown in the figure, 5th Avenue is perpendicular to East 22nd Street, and $m\angle B$ is 10 less than 3 times $m\angle C$.

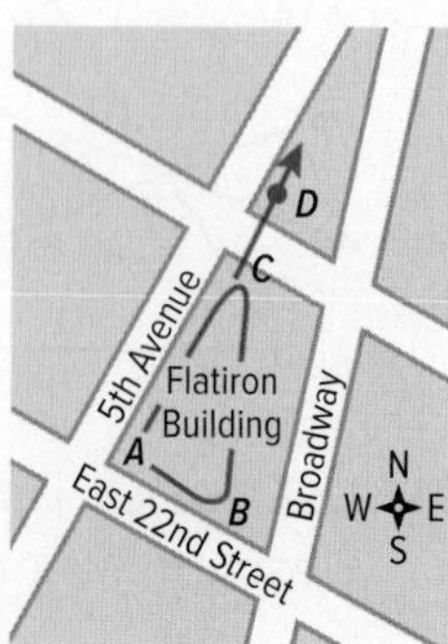

a. Find the angle measures in the floor plan.

b. Find $m\angle BCD$ in two ways. Explain each method.

Congruent Triangles

Explore Relationships in Congruent Triangles

Online Activity Use dynamic geometry software to complete the Explore.

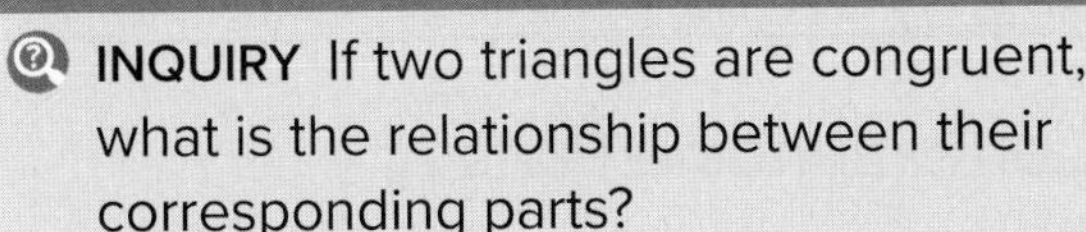

INQUIRY If two triangles are congruent, what is the relationship between their corresponding parts?

Learn Congruent Triangles

The **principle of superposition** states that two figures are congruent if and only if there is a rigid motion or series of rigid motions that maps one figure exactly onto the other. Recall that congruent figures have exactly the same shape and size.

In two **congruent polygons**, all the parts of one polygon are congruent to the **corresponding parts,** or matching parts, of the other polygon. These corresponding parts include *corresponding angles* and *corresponding sides*.

Key Concept • Congruent Triangles

Two triangles are congruent if and only if their corresponding parts are congruent.

For triangles, we say *Corresponding parts of congruent triangles are congruent*, or CPCTC.

Example 1 Identify Corresponding Congruent Parts

Show that the polygons are congruent by identifying all the congruent corresponding parts. Then write a congruence statement.

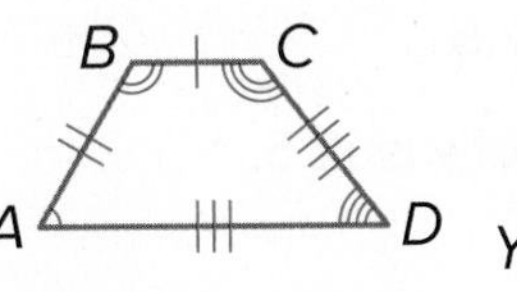

Angles: $\angle A \cong \angle W$; $\angle B \cong \angle X$; $\angle C \cong \angle Y$; $\angle D \cong \angle Z$

Sides: $\overline{BC} \cong \overline{XY}$; $\overline{AB} \cong \overline{WX}$; $\overline{DA} \cong \overline{ZW}$; $\overline{CD} \cong \overline{YZ}$

All corresponding parts of the two polygons are congruent. Therefore, polygon $ABCD \cong$ polygon $WXYZ$.

 **Go Online** You can complete an Extra Example online.

Today's Goals

- Use congruence criterion of corresponding congruent parts of triangles to solve problems.
- Use the Third Angles Theorem and the properties of triangle congruence to solve problems and to prove relationships in geometric figures.

Today's Vocabulary

principle of superposition

congruent polygons

corresponding parts

Go Online

You can watch a video to see how to use transformations to determine whether two triangles are congruent.

Problem-Solving Tip

Get a New Perspective When comparing two figures, it may be helpful to redraw the figures so they have the same orientation. This would make it easier to compare the corresponding sides and angles.

Check

Show that the polygons are congruent by identifying all the congruent corresponding parts. Then write a congruence statement.

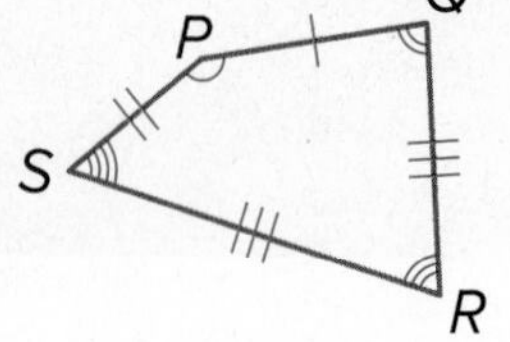

$\angle P \cong$? ; $\angle Q \cong$? ; $\angle R \cong$? ; $\angle S \cong$?

$\overline{PQ} \cong$? ; $\overline{QR} \cong$? ; $\overline{RS} \cong$? ; $\overline{SP} \cong$?

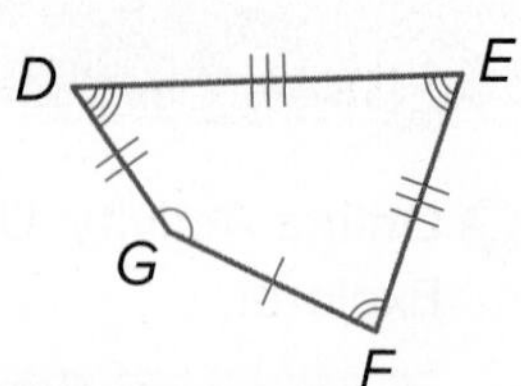

Complete the congruence statement.

Polygon $PQRS \cong$ Polygon ?

Study Tip

Congruence Statements Valid congruence statements for congruent polygons list corresponding vertices in the same order.

Example 2 Use Corresponding Parts of Congruent Triangles

In the diagram, $\triangle RSV \cong \triangle TVS$. Find the values of x and y.

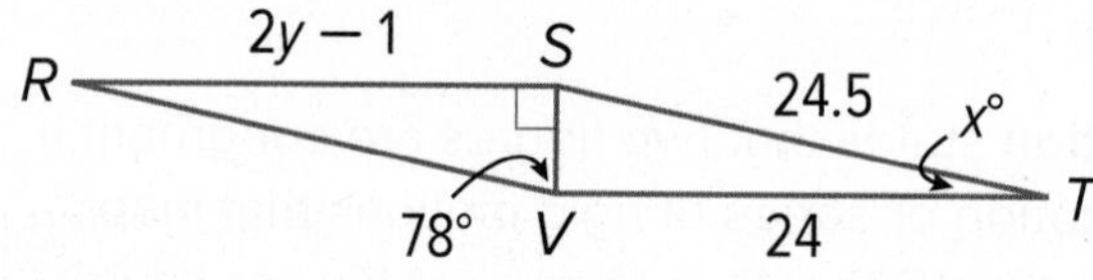

Part A Find the value of x.

$\angle T \cong \angle R$	CPCTC
$m\angle T = m\angle R$	Definition of congruence
$= 180° - 90° - 78°$	Triangle Angle-Sum Theorem
$= 12°$	Solve.

The value of x is 12.

Part B Find the value of y.

$\overline{RS} \cong \overline{TV}$	CPCTC
$RS = TV$	Definition of congruence
$2y - 1 = 24$	Substitution
$y = 12.5$	Solve.

The value of y is 12.5.

Study Tip

Use a Congruence Statement You can use a congruence statement to help you correctly identify corresponding sides.

$\triangle RSV \cong \triangle TVS$

$\overline{RS} \cong \overline{TV}$

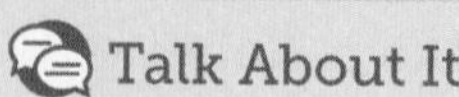

Talk About It!

Suppose the congruence statement $\triangle RSV \cong \triangle TVS$ was *not* given. Would you be able to solve this problem? Explain.

Check

In the diagram, $\triangle ABC \cong \triangle EDC$. Find the values of x and y.

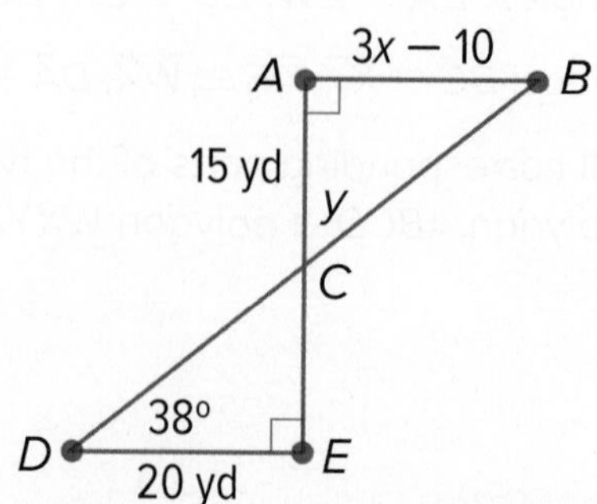

$x =$? ; $y =$?

Go Online You can complete an Extra Example online.

Learn Third Angles Theorem and Triangle Congruence

Theorem 14.3: Third Angles Theorem	
Words	If two angles of one triangle are congruent to two angles of a second triangle, then the third angles of the triangles are congruent.
Example	If $\angle C \cong \angle K$ and $\angle B \cong \angle J$, then $\angle A \cong \angle L$.

You will prove Theorem 14.3 in Exercise 25.

Like congruence of segments and angles, congruence of triangles is reflexive, symmetric, and transitive.

Theorem 14.4: Properties of Triangle Congruence
Reflexive Property of Triangle Congruence $\triangle ABC \cong \triangle ABC$
Symmetric Property of Triangle Congruence If $\triangle ABC \cong \triangle EFG$, then $\triangle EFG \cong \triangle ABC$.
Transitive Property of Triangle Congruence If $\triangle ABC \cong \triangle EFG$ and $\triangle EFG \cong \triangle JKL$, then $\triangle ABC \cong \triangle JKL$.

Go Online
Proofs of Theorem 14.4 are available.

Example 3 Use the Third Angles Theorem

ORIGAMI **Aika is folding origami dragons for a party she is hosting. If $\angle ABD \cong \angle CBD$ and $m\angle BAD = 58°$, find $m\angle CBD$.**

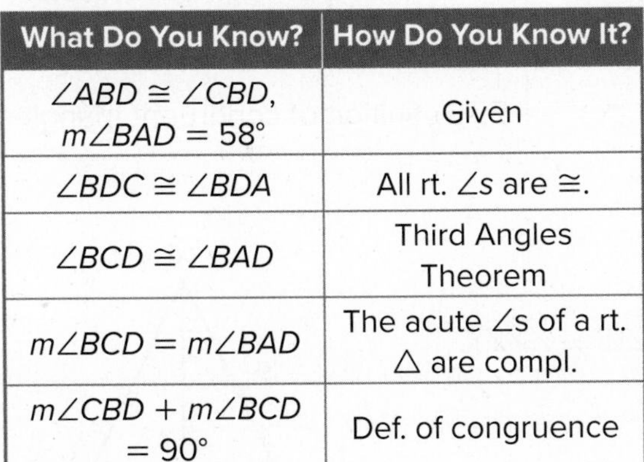

What Do You Know?	How Do You Know It?
$\angle ABD \cong \angle CBD$, $m\angle BAD = 58°$	Given
$\angle BDC \cong \angle BDA$	All rt. ∠s are ≅.
$\angle BCD \cong \angle BAD$	Third Angles Theorem
$m\angle BCD = m\angle BAD$	The acute ∠s of a rt. △ are compl.
$m\angle CBD + m\angle BCD = 90°$	Def. of congruence

$m\angle BCD = 58°$ Substitute.

$m\angle CBD + 58° = 90°$ Substitute.

$m\angle CBD = 32°$ Solve.

The measure of $\angle CBD$ is 32°.

Go Online You can complete an Extra Example online.

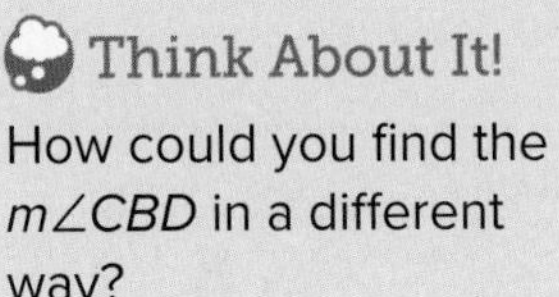

Think About It!
How could you find the $m\angle CBD$ in a different way?

Check

KITES The kite shown is made of two congruent triangles. If $m\angle BAD = m\angle BCD = 45°$, find $m\angle ABD$. $m\angle ABD =$ __?__

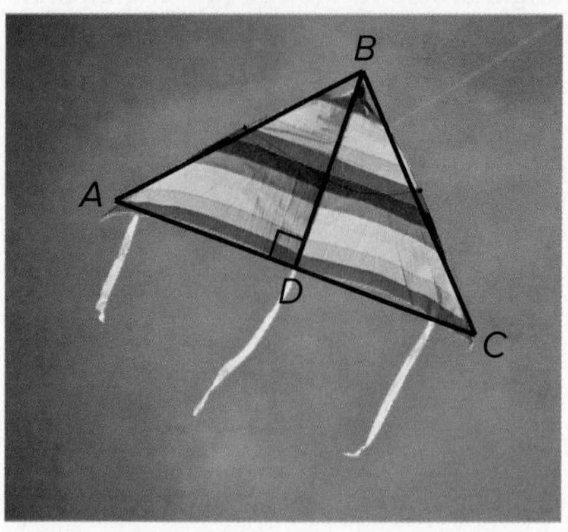

Example 4 Prove that Two Triangles Are Congruent

Study Tip

Symbols To indicate that two triangles are not congruent, write $\triangle ABC \ncong \triangle EFG$. $\triangle ABC \ncong \triangle EFG$ is read as *triangle ABC is not congruent to triangle EFG.*

Write a two-column proof.

Given: $\angle J \cong \angle P$, $\overline{JK} \cong \overline{PM}$, $\overline{JL} \cong \overline{PL}$, and L bisects $\overline{KM}$.

Prove: $\triangle JLK \cong \triangle PLM$

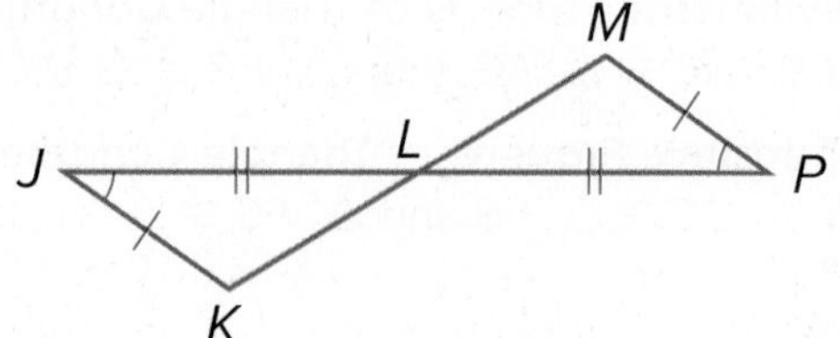

Statements	Reasons
1. $\angle J \cong \angle P$, $\overline{JK} \cong \overline{PM}$, $\overline{JL} \cong \overline{PL}$ and L bisects $\overline{KM}$.	1. Given
2. $\angle JLK \cong \angle PLM$	2. Vertical angles are congruent.
3. $\overline{LK} \cong \overline{LM}$	3. Definition of segment bisector
4. $\angle K \cong \angle M$	4. Third Angles Theorem
5. $\triangle JLK \cong \triangle PLM$	5. Definition of congruent triangles

Check

Write a paragraph proof.

Given: $\angle WXZ \cong \angle YXZ$, $\angle XZW \cong \angle XZY$, $\overline{WX} \cong \overline{YX}$, $\overline{WZ} \cong \overline{YZ}$

Prove: $\triangle WXZ \cong \triangle YXZ$

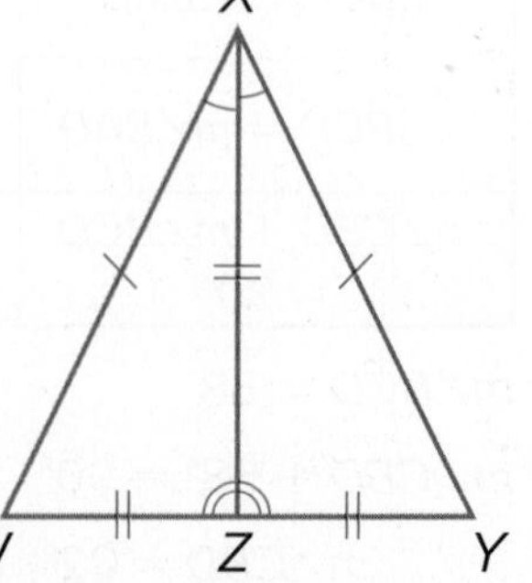

It is given that $\overline{WX} \cong \overline{YX}$ and $\overline{WZ} \cong \overline{YZ}$. By the __?__ Property, $\overline{XZ} \cong \overline{XZ}$. It is also given that $\angle WXZ \cong \angle YXZ$ and $\angle XZW \cong \angle XZY$. So, by __?__ the Theorem, $\angle W \cong \angle Y$. By the definition of congruent triangles, $\triangle WXZ \cong \triangle YXZ$.

Go Online You can complete an Extra Example online.

Clicknique/iStock/Getty Images

Practice

Go Online You can complete your homework online.

Example 1

Show that the polygons are congruent by identifying all the congruent corresponding parts. Then write a congruence statement.

1. 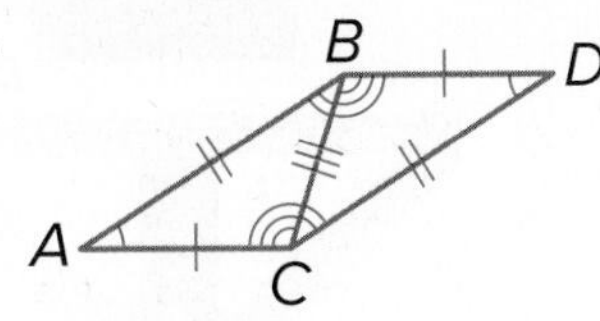

2. 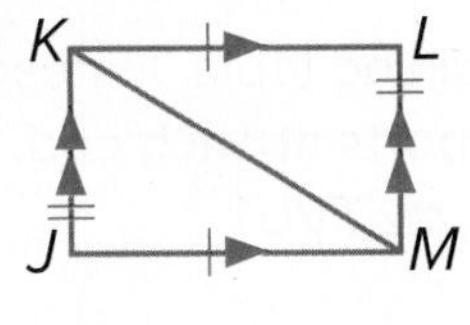

3. 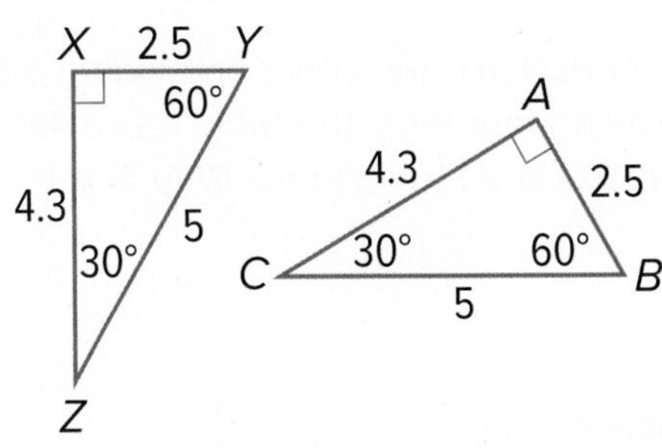

4. 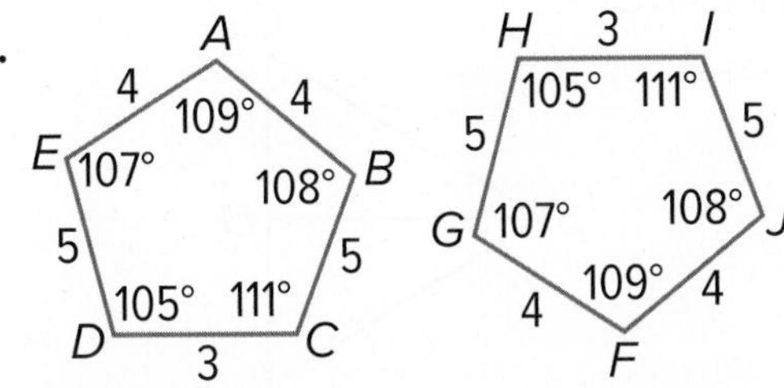

5. 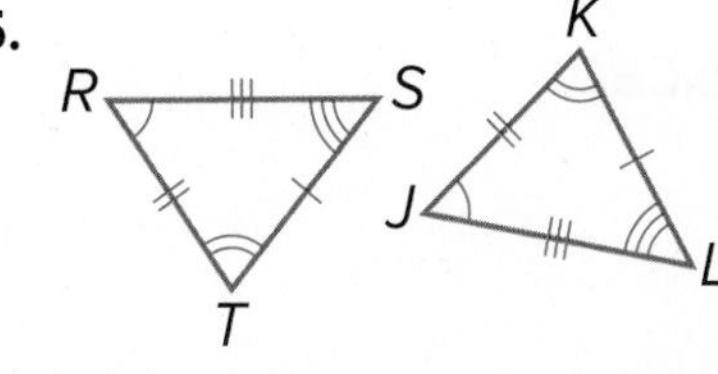

Example 2

In the diagram, $\triangle ABC \cong \triangle FDE$.

6. Find the value of x.

7. Find the value of y.

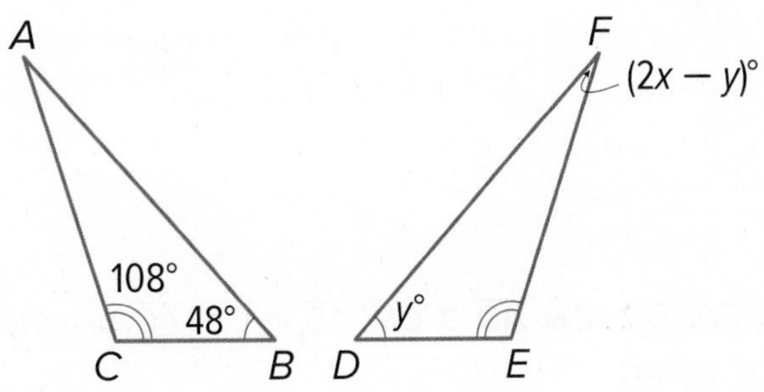

In the diagram, polygon $ABCD \cong$ polygon $PQRS$.

8. Find the value of x.

9. Find the value of y.

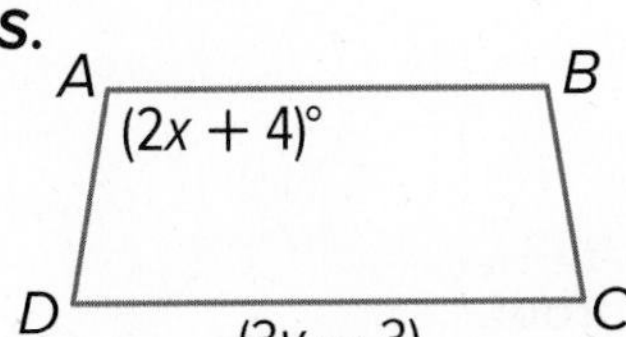

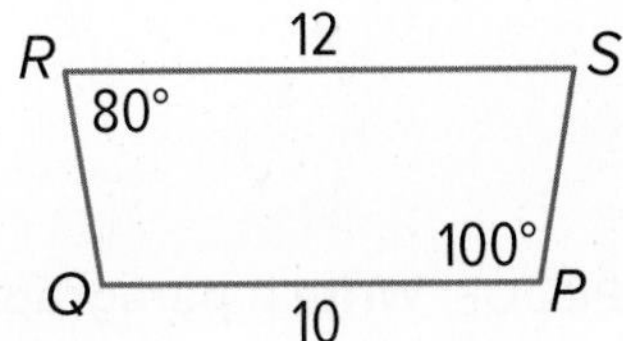

In the diagram, $\triangle ABC \cong \triangle DEF$.

10. Find the value of x.

11. Find the value of y.

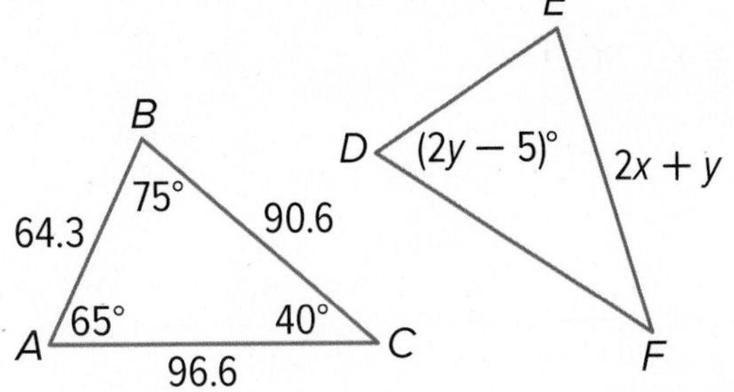

Example 3

12. **DESIGN** Camila is designing a new image for her cell phone case. If $m\angle ABC = 35°$, $m\angle BAC = 29°$, and $\angle ACB \cong \angle DEB$, what is $m\angle DEB$?

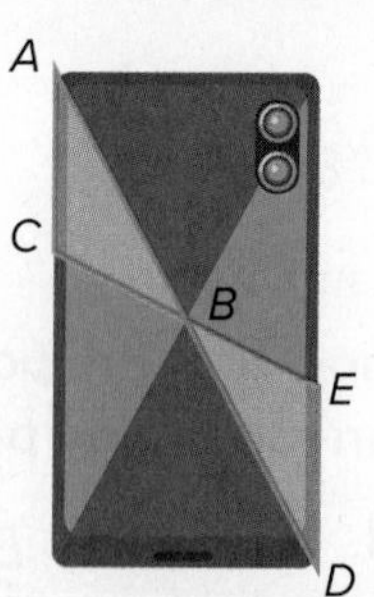

13. **CARPENTRY** Mr. Lewis is building a rustic dining table. Instead of having four legs, the table has a set of supports at each end. If $\angle PRQ \cong \angle TVU$ and $m\angle RPQ = 49°$, what is $m\angle TVU$?

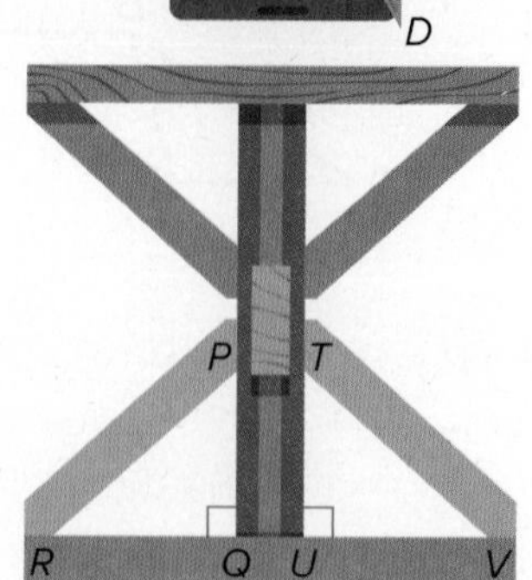

Example 4

PROOF **For 14–16, write a two-column proof.**

14. Given: $\overline{AB} \cong \overline{CB}$, $\overline{AD} \cong \overline{CD}$, $\angle BAD \cong \angle BCD$, $\overline{BD}$ bisects $\angle ABC$.

Prove: $\triangle ABD \cong \triangle CBD$

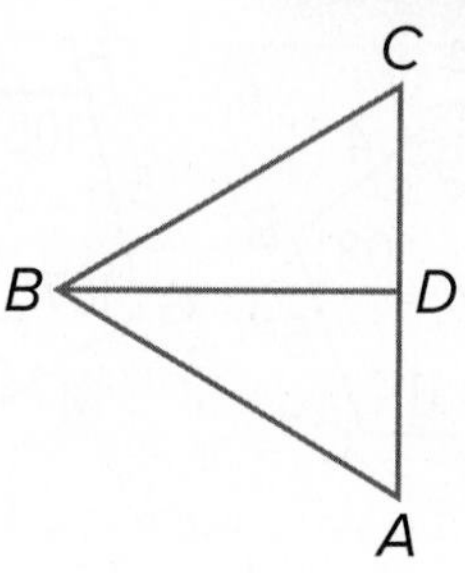

15. Given: $\overline{AB} \cong \overline{CB}$, $\overline{AD} \cong \overline{CD}$, $\angle ABD \cong \angle CBD$, $\angle ADB \cong \angle CDB$

Prove: $\triangle ABD \cong \triangle CBD$

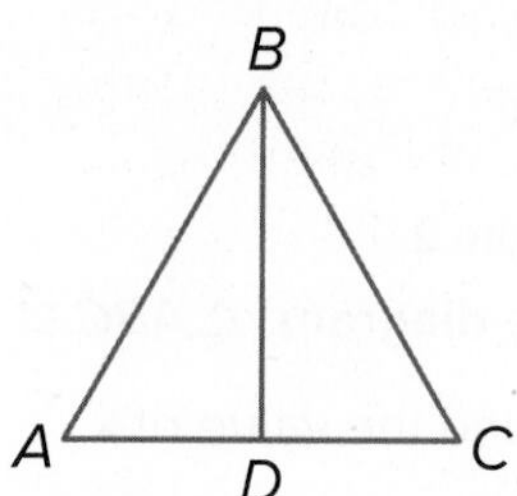

16. Given: $\angle A \cong \angle C$, $\angle D \cong \angle B$, $\overline{AD} \cong \overline{CB}$, $\overline{AE} \cong \overline{CE}$, $\overline{AC}$ bisects $\overline{BD}$.

Prove: $\triangle AED \cong \triangle CEB$

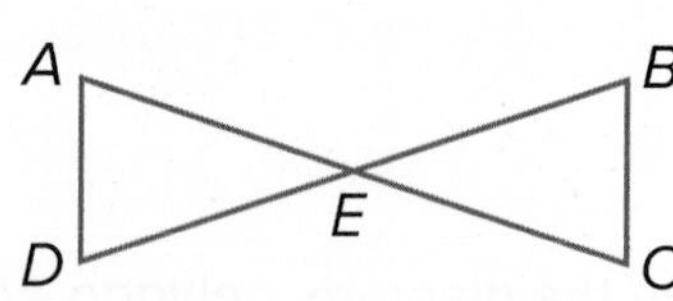

PROOF **Write a paragraph proof.**

17. Given: $\overline{BD}$ bisects $\angle ABC$ and $\angle ADC$, $\overline{AB} \cong \overline{CB}$, $\overline{AD} \cong \overline{CD}$

Prove: $\triangle ABD \cong \triangle CBD$

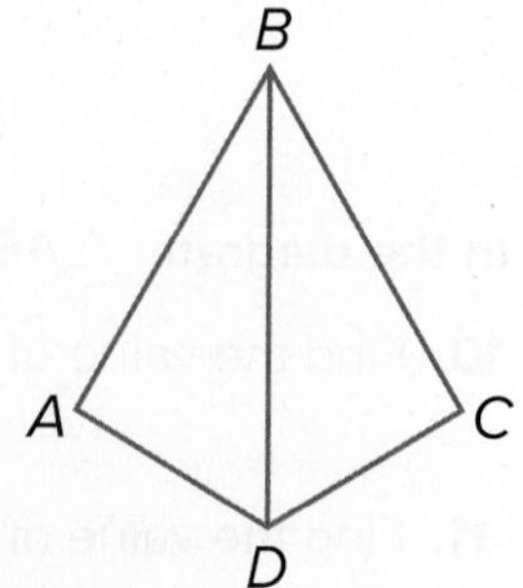

Mixed Exercises

18. PRECISION Beverly is using loyalty cards at her coffee shop. When a customer purchases a cup of coffee, he or she can present a loyalty card to be stamped with a star-shaped stamp that Beverly purchased specifically for this use. When the customer collects nine stamps, they receive their tenth cup of coffee for free. What property guarantees that the stamped designs are congruent?

Draw and label a figure to represent the congruent triangles. Then find the values of *x* and *y*.

19. $\triangle ABC \cong \triangle DEF$, $AB = 7$, $BC = 9$, $AC = 11 + x$, $DF = 3x - 13$, and $DE = 2y - 5$

20. $\triangle LMN \cong \triangle RST$, $m\angle L = 49°$, $m\angle M = 10y°$, $m\angle S = 70°$, and $m\angle T = (4x + 9)°$

21. $\triangle JKL \cong \triangle MNP$, $JK = 12$, $LJ = 5$, $PM = 2x - 3$, $m\angle L = 67°$, $m\angle K = (y + 4)°$ and $m\angle N = (2y - 15)°$

22. SIERPINSKI TRIANGLE The figure shown is a portion of the Sierpinski triangle. The triangle has the property that any triangle made from any combination of edges is equilateral. How many triangles in this portion are congruent to the black triangle at the bottom?

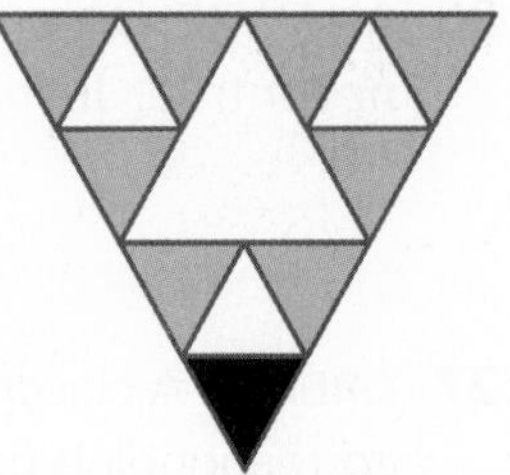

23. LOGO DESIGNS Refer to the design shown.

a. Indicate the triangles that appear to be congruent.

b. Name the congruent angles and congruent sides of a pair of congruent triangles.

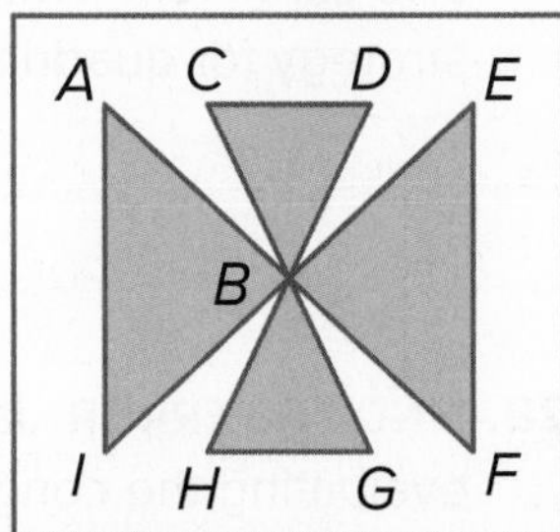

24. REASONING Igor noticed on a map that the triangle with vertices that are at the supermarket, the library, and the post office ($\triangle SLP$) is congruent to the triangle with vertices that are at Igor's home, Jasen's home, and Daran's home ($\triangle IJD$). That is, $\triangle SLP \cong \triangle IJD$.

a. The distance between the supermarket and the post office is 1 mile. Which path along the triangle $\triangle IJD$ is congruent to this?

b. The measure of $\angle LPS$ is 40°. Identify the angle that is congruent to this angle in $\triangle IJD$.

25. PROOF Copy and complete the two-column proof of the Third Angles Theorem by providing the reason for each statement.

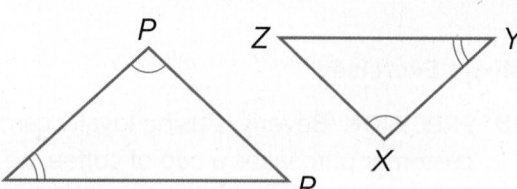

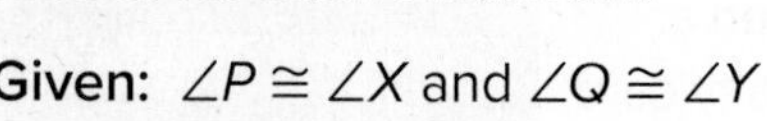

Given: $\angle P \cong \angle X$ and $\angle Q \cong \angle Y$

Prove: $\angle R \cong \angle Z$

Statements	Reasons
1. $\angle P \cong \angle X$, $\angle Q \cong \angle Y$	1. ?
2. $m\angle P = m\angle X$, $m\angle Q = m\angle Y$	2. ?
3. $m\angle P + m\angle Q + m\angle R = 180$ $180 = m\angle X + m\angle Y + m\angle Z$	3. ?
4. $m\angle P + m\angle Q + m\angle R = m\angle X + m\angle Y + m\angle Z$	4. ?
5. $m\angle X + m\angle Y + m\angle R = m\angle X + m\angle Y + m\angle Z$	5. ?
6. $m\angle R = m\angle Z$	6. ?
7. $\angle R \cong \angle Z$	7. ?

26. ANALYZE Determine whether the following statement is *sometimes*, *always*, or *never* true. Justify your argument.

Equilateral triangles are congruent.

27. CREATE A classmate is using the Third Angles Theorem to show that if two corresponding pairs of the angles of two triangles are congruent, then the third pair is also congruent. Write a question to help him decide whether he can use the same strategy for quadrilaterals.

28. FIND THE ERROR Jasmine and West are evaluating the congruent figures at right. Jasmine says that $\triangle CAB \cong \triangle ZYX$, and West says that $\triangle ABC \cong \triangle YXZ$. Is either of them correct? Explain your reasoning.

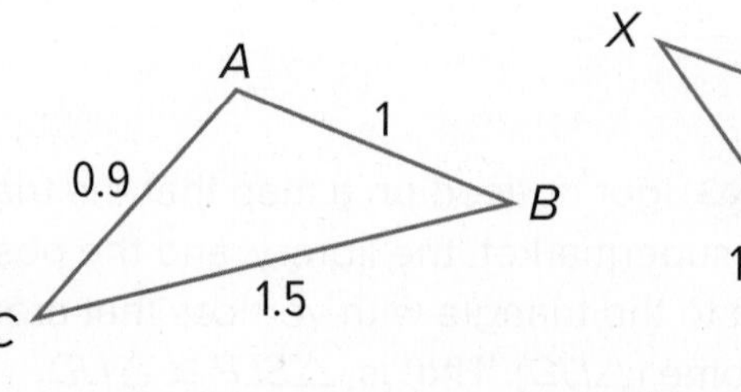

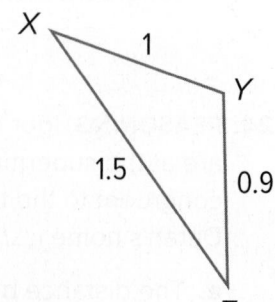

29. WRITE Justify why the order of the vertices is important when naming congruent triangles. Give an example to support your argument.

30. PERSEVERE Find the values of x and y if $\triangle PQS \cong \triangle RQS$.

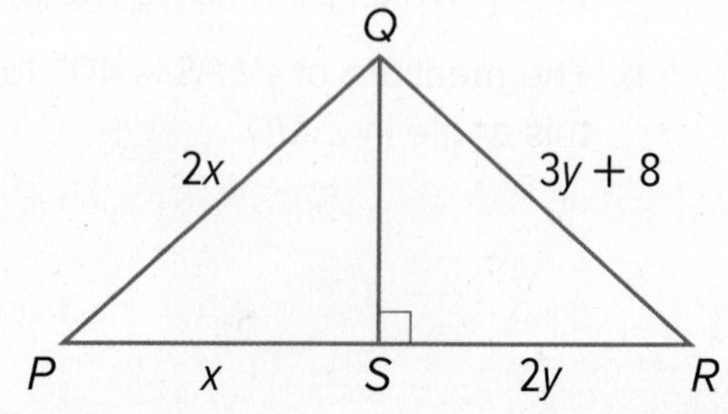

Proving Triangles Congruent: SSS, SAS

Explore Conditions That Prove Triangles Congruent

Online Activity Use dynamic geometry software to complete the Explore.

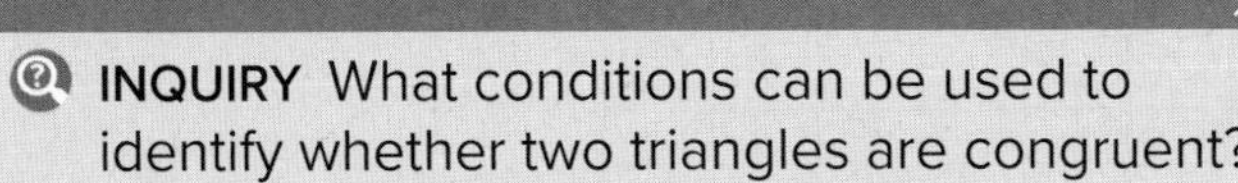

INQUIRY What conditions can be used to identify whether two triangles are congruent?

Today's Goals
- Use the SSS Congruence criterion for triangles to solve problems and prove relationships in geometric figures.
- Use the SAS Congruence criterion for triangles to solve problems and prove relationships in geometric figures.

Today's Vocabulary
included angle

Learn Proving Triangles Congruent: SSS

You can prove two triangles congruent by showing that all six pairs of corresponding parts are congruent. However, it is possible to prove two triangles congruent using fewer pairs of corresponding parts.

If two triangles have the same three side lengths, then there is a series of rigid motions that will show the two triangles congruent. This leads to the postulate below.

Postulate 14.1: Side-Side-Side (SSS) Congruence
If three sides of one triangle are congruent to three sides of a second triangle, then the triangles are congruent.

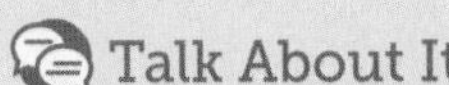

Talk About It!

Will two equilateral triangles always be congruent by SSS? Justify your argument.

Example 1 Use SSS to Prove Triangles Congruent

Write a flow proof to show that $\triangle QRT \cong \triangle SRT$.

Given: $\triangle QRS$ **is isosceles with** $\overline{QR} \cong \overline{SR}$. $\overrightarrow{RT}$ **bisects** $\overline{QS}$ **at point** T.

Prove: $\triangle QRT \cong \triangle SRT$

Proof:

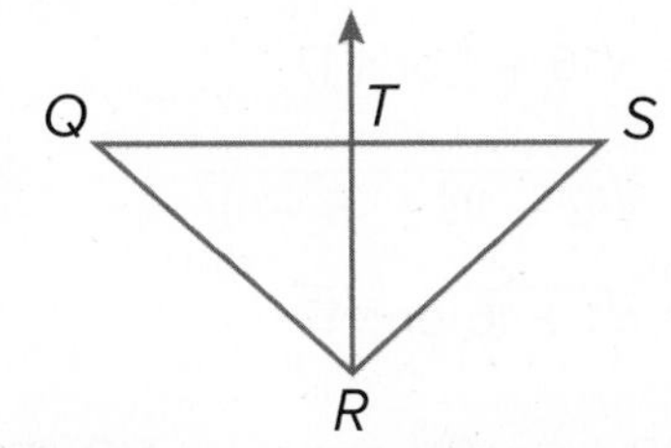

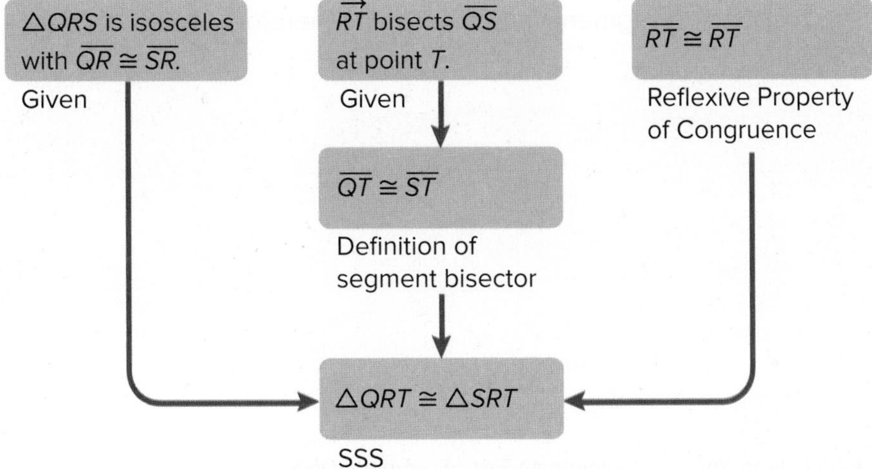

Go Online You can complete an Extra Example online.

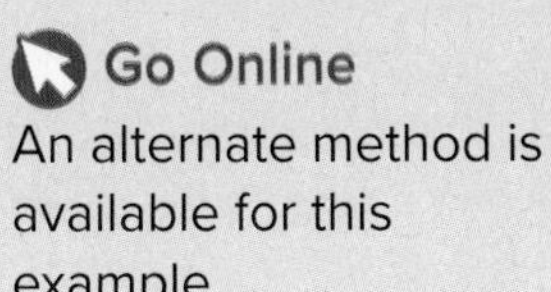

An alternate method is available for this example.

Example 2 Use SSS on the Coordinate Plane

Think About It!

Is the following statement *true* or *false*? Justify your argument.

If the congruent sides in one isosceles triangle have the same measure as the congruent sides in another isosceles triangle, then the triangles are congruent.

Triangle *JKL* has vertices *J*(2, 5), *K*(1, 1), and *L*(5, 2). Triangle *QNP* has vertices *Q*(−4, 4), *N*(−3, 0), and *P*(−7, 1). Is $\triangle JKL \cong \triangle QNP$?

Part A Graph the triangles.

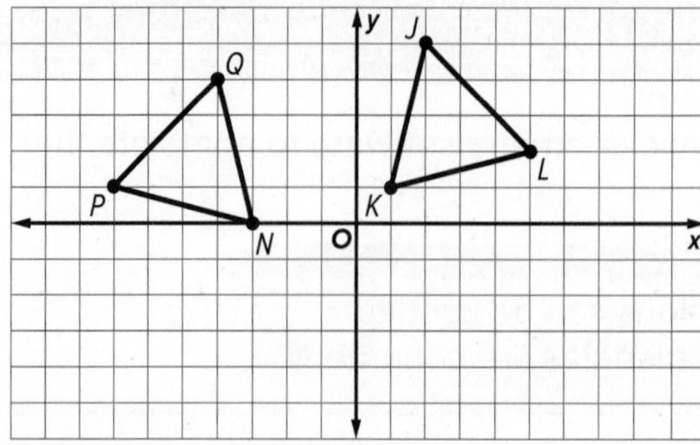

Part B Make a conjecture.

Use your graph to make a conjecture as to whether the triangles are congruent. Explain your reasoning.

From the graph, it appears that the triangles have the same shape and size, so we can conjecture that they are congruent.

Part C Support your conjecture.

Use the Distance Formula to show that all corresponding sides have the same measure.

$$JL = \sqrt{(5-2)^2 + (2-5)^2} = \sqrt{9+9} \text{ or } 3\sqrt{2}$$

$$QP = \sqrt{[-7-(-4)]^2 + (1-4)^2} = \sqrt{9+9} \text{ or } 3\sqrt{2}$$

$$LK = \sqrt{(1-5)^2 + (1-2)^2} = \sqrt{16+1} \text{ or } \sqrt{17}$$

$$PN = \sqrt{[-3-(-7)]^2 + (0-1)^2} = \sqrt{16+1} \text{ or } \sqrt{17}$$

$$KJ = \sqrt{(2-1)^2 + (5-1)^2} = \sqrt{1+16} \text{ or } \sqrt{17}$$

$$NQ = \sqrt{[-4-(-3)]^2 + (4-0)^2} = \sqrt{1+16} \text{ or } \sqrt{17}$$

$JL = QP$, $LK = PN$, and $KJ = NQ$. By the definition of congruent segments, all corresponding segments are congruent. Therefore, $\triangle JKL \cong \triangle QNP$ by SSS.

Go Online You can complete an Extra Example online.

Check

Triangle *ABC* has vertices *A*(1, 1), *B*(0, 3), and *C*(2, 5). Triangle *EFG* has vertices *E*(1, −1), *F*(2, −5), and *G*(4, −4). Is $\triangle ABC \cong \triangle EFG$?

Part A

Graph $\triangle ABC$ and $\triangle EFG$ on the same coordinate plane.

Part B

Find the side lengths of each triangle.

$AB =$ _?_; $BC =$ _?_; $AC =$ _?_; $EF =$ _?_; $FG =$ _?_; $EG =$ _?_

Part C

Is triangle *ABC* congruent to triangle *EFG*? Justify your argument.

A. No; $AC \neq FG$, so SSS congruence is not met.

B. No; $BC \neq FG$, so SSS congruence is not met.

C. Yes; all corresponding sides have the same measure, so SSS congruence is met.

D. Yes; all corresponding sides have the same measure, so by the definition of congruent figures, $\triangle ABC \cong \triangle EFG$.

Think About It!

Both legs of one right triangle are congruent to the legs of another right triangle. Are the triangles congruent? Justify your argument.

Learn Proving Triangles Congruent: SAS

The interior angle formed by two adjacent sides of a triangle is called an **included angle**.

If two triangles are formed using the same side lengths and included angle measure, then there is a series of rigid motions that will show that the two triangles are congruent. This leads to the postulate below.

Postulate 14.2: Side-Angle-Side (SAS) Congruence

If two sides and the included angle of one triangle are congruent to two sides and the included angle of a second triangle, then the triangles are congruent.

Go Online

You may want to complete the construction activities for this lesson.

Go Online You can complete an Extra Example online.

Study Tip

Assumptions Even though objects in the real world are not two-dimensional, you can still make some assumptions about their geometric relationships. For example, in three-dimensional objects you can apply the Reflexive Property of Congruence and the Vertical Angles Theorem.

Example 3 Use SAS to Prove Triangles Congruent

PLAYGROUND The playground equipment shown appears to be made of congruent triangles. If $\overline{KL} \cong \overline{LM}$ and $\angle JLK \cong \angle JLM$, write a two-column proof to prove that $\triangle JLK \cong \triangle JLM$. Complete the two-column proof by selecting the correct statements and reasons.

Statements/Reasons:

Given

Definition of congruent segments

SAS

SSS

Definition of angle bisector

Reflexive Property of Congruence

$\overline{JL} \cong \overline{JL}$

$\overline{JK} \cong \overline{JM}$

$\angle KJL \cong \angle MJL$

Statements	Reasons
1. $\overline{KL} \cong \overline{LM}$	1. Given
2. $\angle JLK \cong \angle JLM$	2. Given
3. $\overline{JL} \cong \overline{JL}$	3. Reflexive Property of Congruence
4. $\triangle JLK \cong \triangle JLM$	4. SAS

Check

KITES The kite shown appears to be made up of congruent triangles. If $\overline{DE} \cong \overline{FE}$ and $\overline{EG}$ bisects $\angle DEF$, prove that $\triangle DEG \cong \triangle FEG$. Copy and complete the two-column proof by selecting the correct statements and reasons.

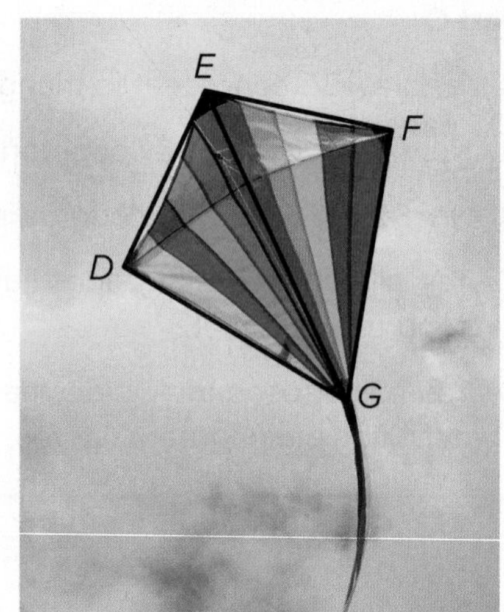

Statements/Reasons:

Given

Definition of congruent segments

SAS

SSS

Definition of angle bisector

Reflexive Property of Congruence

$\overline{DE} \cong \overline{FE}$

$\overline{EG} \cong \overline{EG}$

$\overline{DG} \cong \overline{FG}$

$\angle DEG \cong \angle FEG$

Given: $\overline{DE} \cong \overline{FE}$, $\overline{EG}$ bisects $\angle DEF$.
Prove: $\triangle DEG \cong \angle FEG$
Proof:

Statements	Reasons
1. ?	1. Given
2. $\overline{EG}$ bisects $\angle DEF$.	2. ?
3. ?	3. Definition of angle bisector
4. ?	4. ?
5. $\triangle DEG \cong \triangle FEG$	5. ?

 **Go Online** You can complete an Extra Example online.

Practice

Go Online You can complete your homework online.

Example 1

PROOF **Write the specified type of proof.**

1. two-column proof
Given: $\overline{AB} \cong \overline{XY}$, $\overline{AC} \cong \overline{XZ}$, $\overline{BC} \cong \overline{YZ}$
Prove: $\triangle ABC \cong \triangle XYZ$

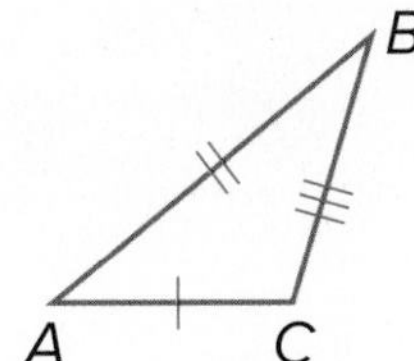

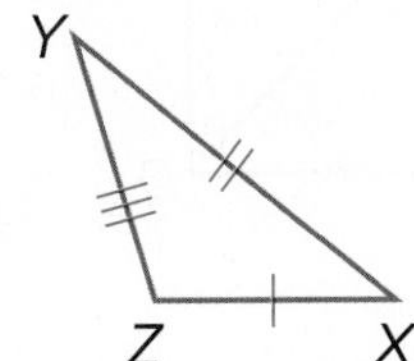

2. flow proof
Given: $\overline{RS} \cong \overline{UT}$, $\overline{RT} \cong \overline{US}$
Prove: $\triangle RST \cong \triangle UTS$

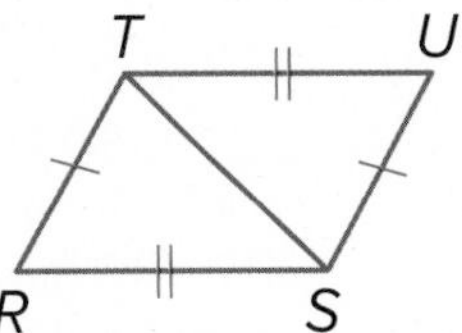

3. two-column proof
Given: $\overline{AB} \cong \overline{CB}$, D is the midpoint of $\overline{AC}$.
Prove: $\triangle ABD \cong \triangle CBD$

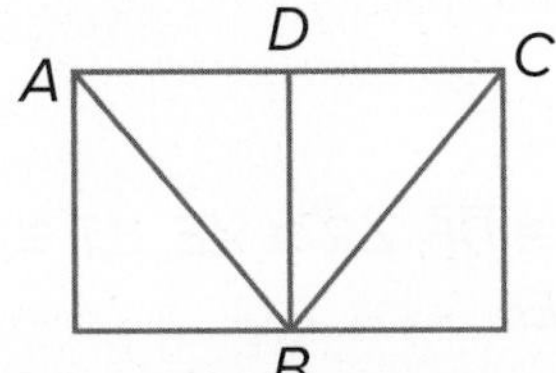

4. flow proof
Given: $\overline{RS} \cong \overline{TS}$, V is the midpoint of $\overline{RT}$.
Prove: $\triangle RSV \cong \triangle TSV$

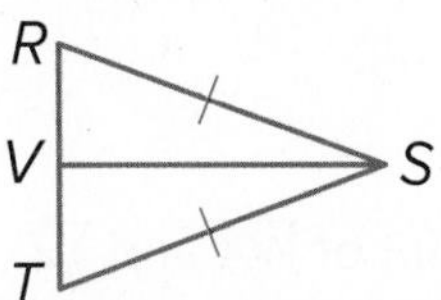

5. paragraph proof
Given: $\overline{QR} \cong \overline{SR}$, $\overline{ST} \cong \overline{QT}$
Prove: $\triangle QRT \cong \triangle SRT$

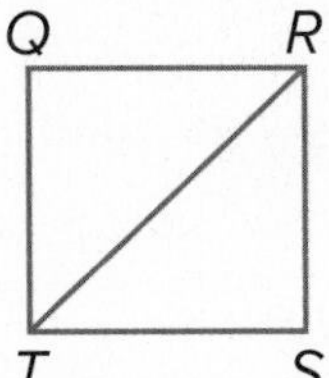

6. two-column proof
Given: $\overline{AB} \cong \overline{ED}$, $\overline{CA} \cong \overline{CE}$, $\overline{AC}$ bisects $\overline{BD}$
Prove: $\triangle ABC \cong \triangle EDC$

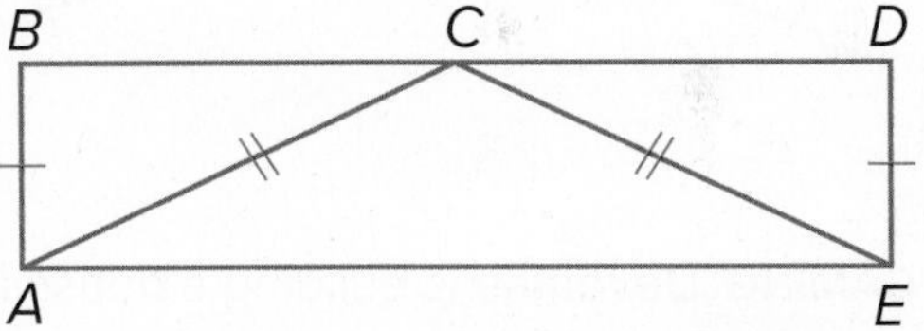

Example 2

REGULARITY **Determine whether $\triangle DEF \cong \triangle PQR$. Explain.**

7. $D(-6, 1)$, $E(1, 2)$, $F(-1, -4)$, $P(0, 5)$, $Q(7, 6)$, $R(5, 0)$

8. $D(-7, -3)$, $E(-4, -1)$, $F(-2, -5)$, $P(2, -2)$, $Q(5, -4)$, $R(0, -5)$

Determine whether $\triangle ABC \cong \triangle KLM$. Explain.

9. $A(-3, 3)$, $B(-1, 3)$, $C(-3, 1)$, $K(1, 4)$, $L(3, 4)$, $M(1, 6)$

10. $A(-4, -2)$, $B(-4, 1)$, $C(-1, -1)$, $K(0, -2)$, $L(0, 1)$, $M(4, 1)$

Example 3

PROOF **Write the specified type of proof.**

11. two-column proof
Given: $NP = PM$, $\overline{NP} \perp \overline{PL}$
Prove: $\triangle NPL \cong \triangle MPL$

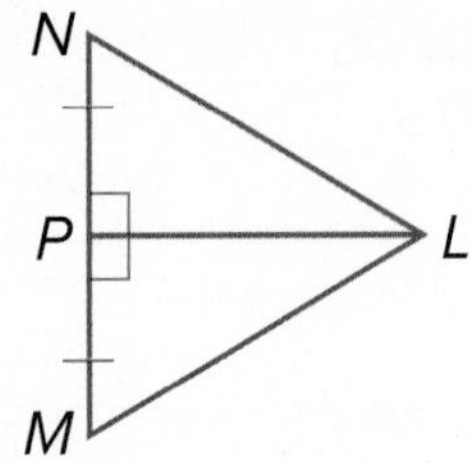

12. two-column proof
Given: $AB = CD$, $\overline{AB} \parallel \overline{CD}$
Prove: $\triangle ACD \cong \triangle DBA$

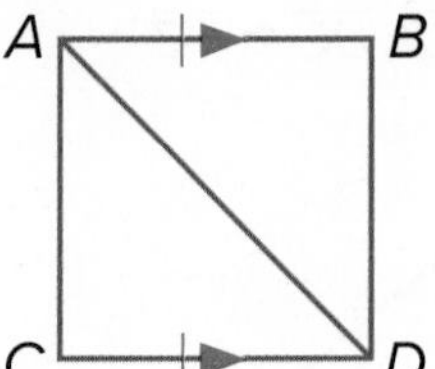

13. paragraph proof
Given: V is the midpoint of $\overline{WX}$ and $\overline{YZ}$.
Prove: $\triangle XVZ \cong \triangle WVY$

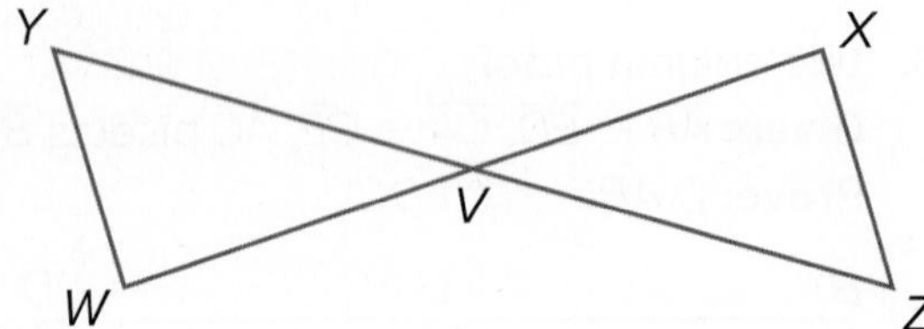

14. flow proof
Given: $\overline{PR} \cong \overline{DE}$, $\overline{PT} \cong \overline{DF}$, $\angle R \cong \angle E$, $\angle T \cong \angle F$
Prove: $\triangle PRT \cong \triangle DEF$

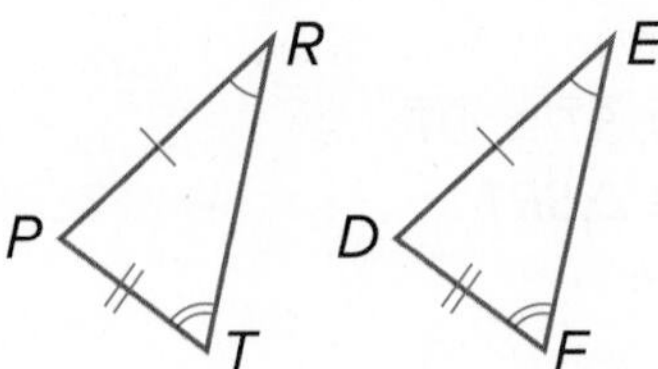

15. GAMING Devontae is building a house in a simulation video game. He wants the roof of the house and the main support beam to create congruent triangles. If $\overline{BD} \perp \overline{AC}$ and $\overline{BD}$ bisects $\overline{AC}$, write a two-column proof to prove $\triangle ABD \cong \triangle CBD$.

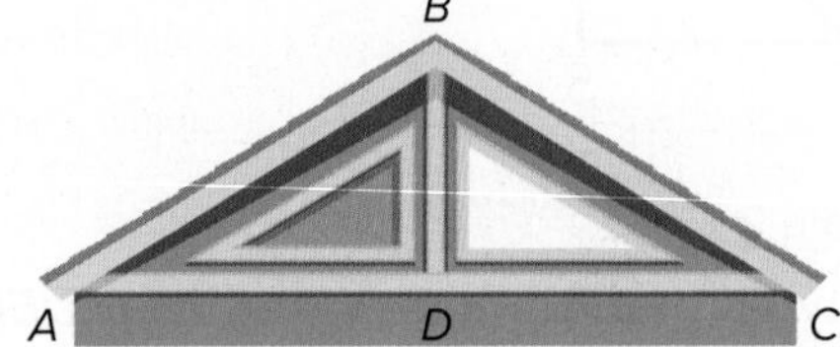

16. TECHNOLOGY Nevaeh has developed a new timer app. The icon for the app contains an hourglass that can be modeled by two triangles. If R is the midpoint of $\overline{QS}$ and $\overline{PT}$, write a paragraph proof to prove $\triangle PRQ \cong \triangle TRS$.

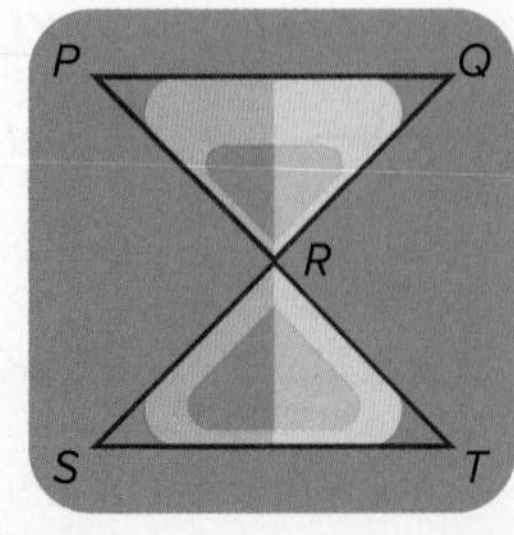

Mixed Exercises

Explain whether there is enough information given in each figure to prove that the triangles are congruent using SSS or SAS.

17.

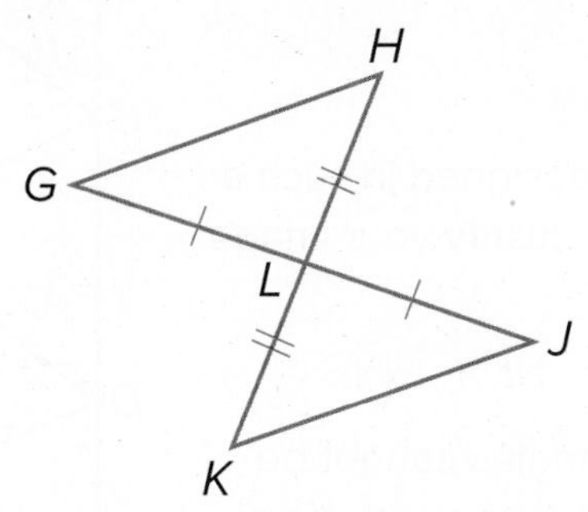

18.

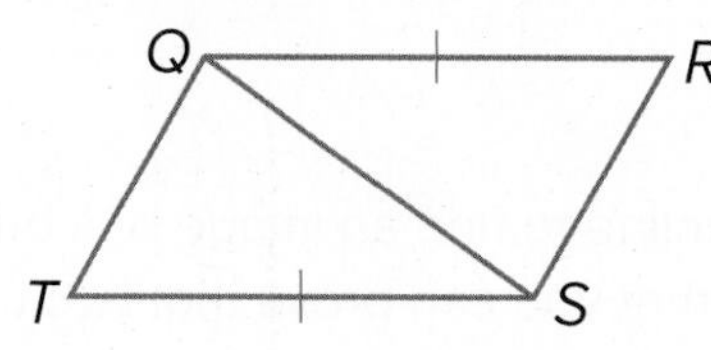

19.

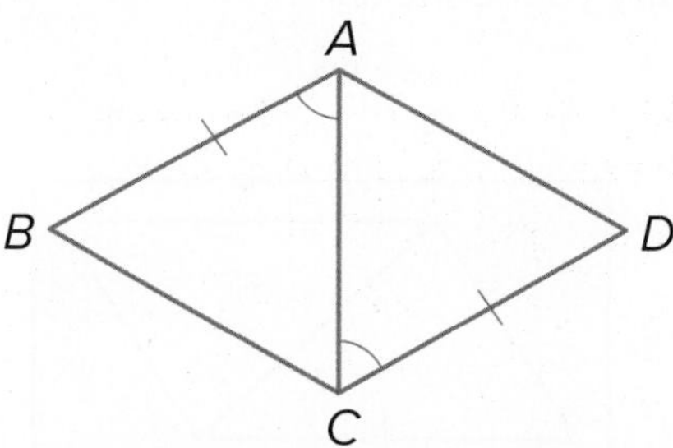

20.

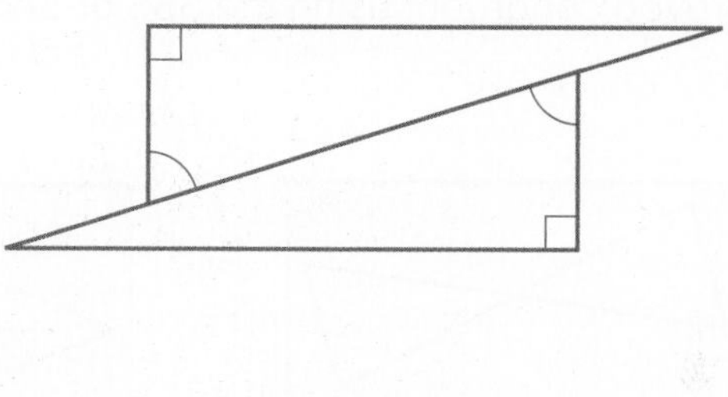

21. REASONING Tyson had three sticks of lengths 24 inches, 28 inches, and 30 inches. Is it possible to make two non-congruent triangles using the same three sticks? Explain.

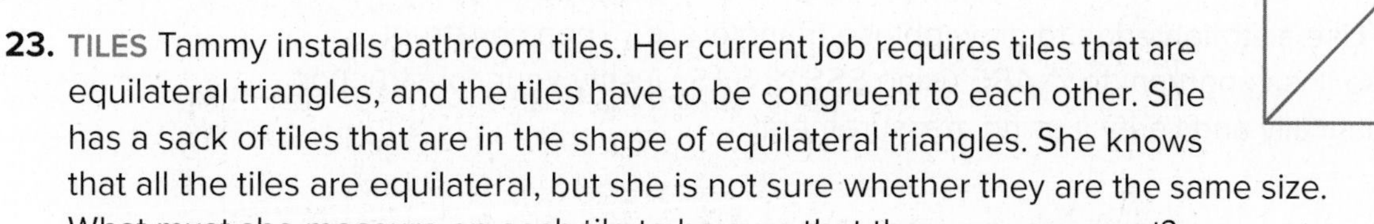

22. BAKERY Sonia made a sheet of baklava. She has markings on her pan so that she can cut them into large squares. After she cuts the pastry in squares, she cuts them diagonally to form two congruent triangles, as shown. Which postulate could you use to prove the two triangles congruent?

23. TILES Tammy installs bathroom tiles. Her current job requires tiles that are equilateral triangles, and the tiles have to be congruent to each other. She has a sack of tiles that are in the shape of equilateral triangles. She knows that all the tiles are equilateral, but she is not sure whether they are the same size. What must she measure on each tile to be sure that they are congruent?

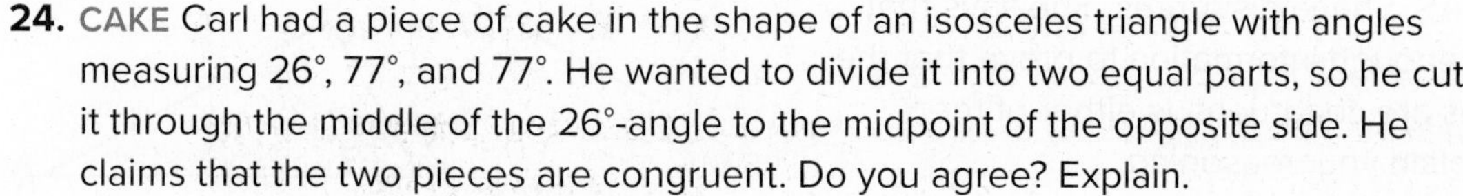

24. CAKE Carl had a piece of cake in the shape of an isosceles triangle with angles measuring 26°, 77°, and 77°. He wanted to divide it into two equal parts, so he cut it through the middle of the 26°-angle to the midpoint of the opposite side. He claims that the two pieces are congruent. Do you agree? Explain.

25. In the figure, $\overline{AC} \cong \overline{AD}$. Suppose you know $\angle C \cong \angle D$. Can you prove that $\triangle ABC \cong \triangle ABD$? Why or why not?

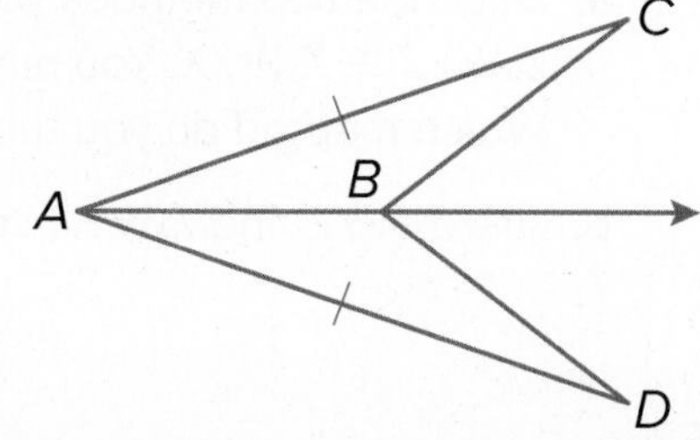

26. **USE A SOURCE** An engineer is designing a new cell phone tower. Part of the tower is shown in the figure. The engineer makes sure that line m is parallel to line n and that $\overline{AB} \cong \overline{CD}$.

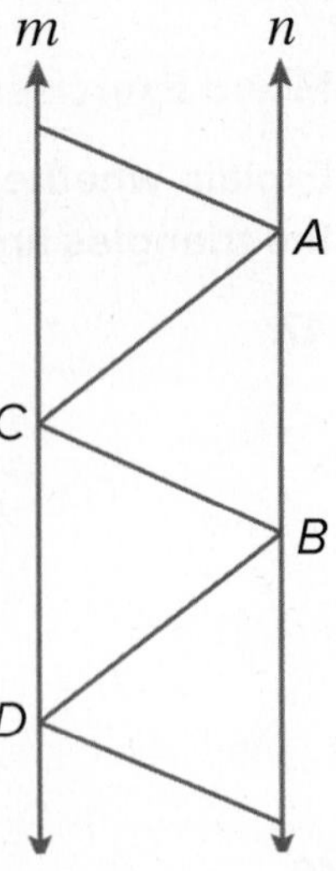

a. Can the engineer prove that $\triangle ABC \cong \triangle DCB$? Explain why or why not.

b. Go online to find an image of a bridge or a tower that is designed in such a way that you can prove that two triangles are congruent. Justify your image.

27. **WHICH ONE DOESN'T BELONG?** Determine which pair of triangles cannot be proved congruent using the SSS or SAS Postulates. Justify your conclusion.

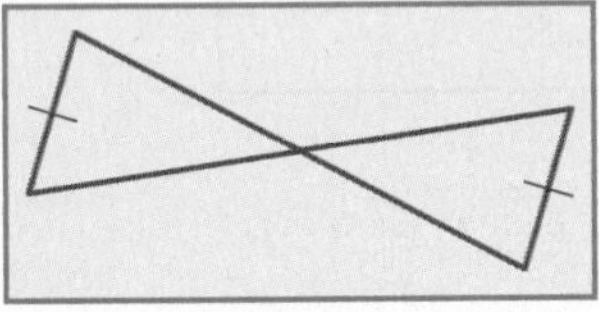

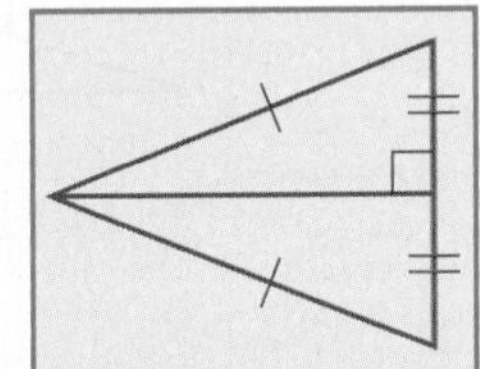

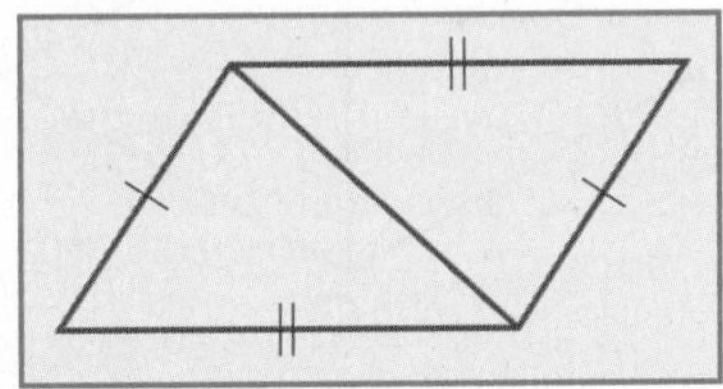

28. **ANALYZE** Determine whether the following statement is *true* or *false*. If true, justify your reasoning. If false, provide a counterexample.

If the congruent sides in one isosceles triangle have the same measure as the congruent sides in another isosceles triangle, then the triangles are congruent.

29. **WRITE** Two pairs of corresponding sides of two right triangles are congruent. Are the triangles congruent? Explain your reasoning.

30. **CREATE** Use a straightedge to draw obtuse triangle ABC. Then construct $\triangle XYZ$ so it is congruent to $\triangle ABC$ using SSS or SAS. Justify your construction mathematically and verify it using measurement.

31. **FIND THE ERROR** Bonnie says that $\triangle PQR \cong \triangle XYZ$ by SAS. Shada disagrees. She says that there is not enough information to prove that the two triangles are congruent. Is either of them correct? Explain your reasoning.

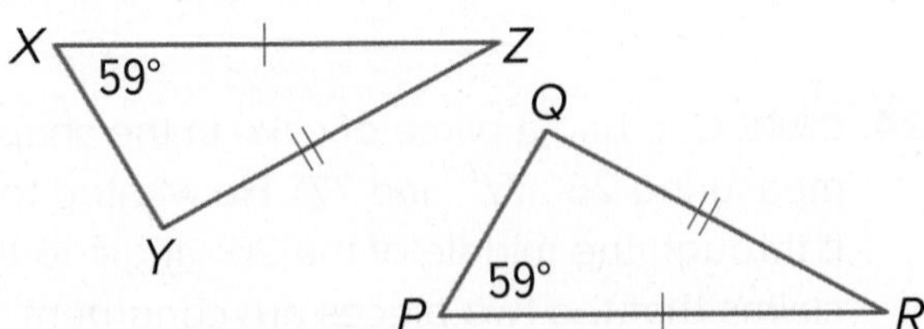

32. **PERSEVERE Refer to the graph shown.**

a. Describe two methods you could use to prove $\triangle WYZ \cong \triangle WYX$. You may not use a ruler or protractor. Which method do you think is more efficient? Explain.

b. Are $\triangle WYZ$ and $\triangle WYX$ congruent? Explain your reasoning.

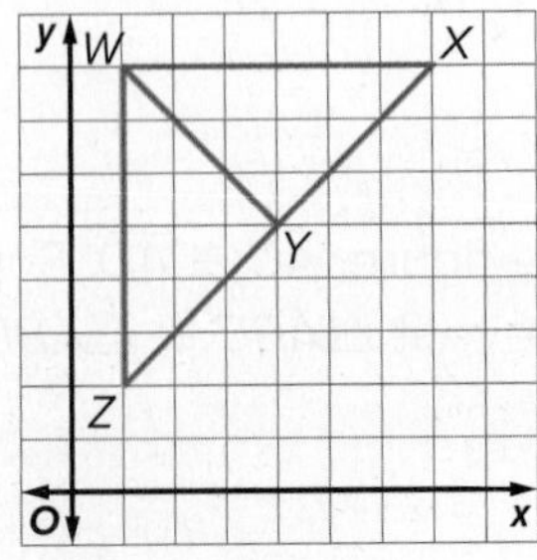

Proving Triangles Congruent: ASA, AAS

Learn Proving Triangles Congruent: ASA

An **included side** is the side of a triangle between two angles.

If two triangles are formed using the same two angle measures and included side length, then there is a series of rigid motions that will show the two triangles congruent. This leads to the postulate below.

Postulate 14.3: Angle-Side-Angle (ASA) Congruence

If two angles and the included side of one triangle are congruent to two angles and the included side of another triangle, then the triangles are congruent.

Example 1 Use ASA to Prove Triangles Congruent

Complete the two-column proof.

Given: $\angle BAC \cong \angle DEC$; $\overline{BD}$ bisects $\overline{AE}$.

Prove: $\triangle ACB \cong \triangle ECD$

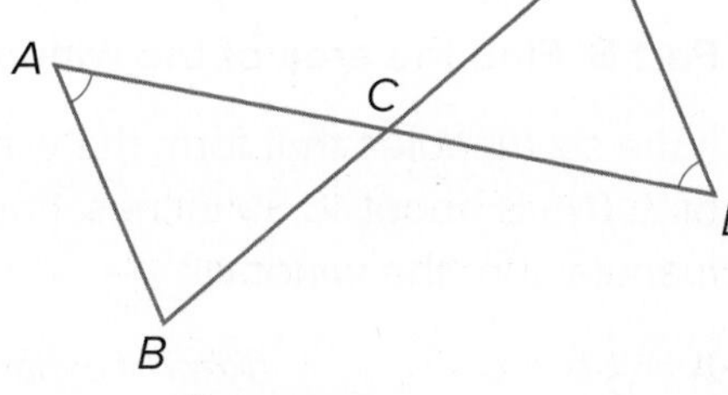

Statements	Reasons
1. $\angle BAC \cong \angle DEC$	1. Given
2. $\overline{BD}$ bisects $\overline{AE}$	2. Given
3. $\overline{AC} \cong \overline{EC}$	3. Definition of segment bisector
4. $\angle ACB \cong \angle ECD$	4. Vertical Angles Theorem
5. $\triangle ACB \cong \triangle ECD$	5. ASA

Check

Complete the two-column proof.

Given: $\overline{WX} \parallel \overline{YZ}$ and $\overline{WZ} \parallel \overline{YX}$

Prove: $\triangle WXZ \cong \triangle YZX$

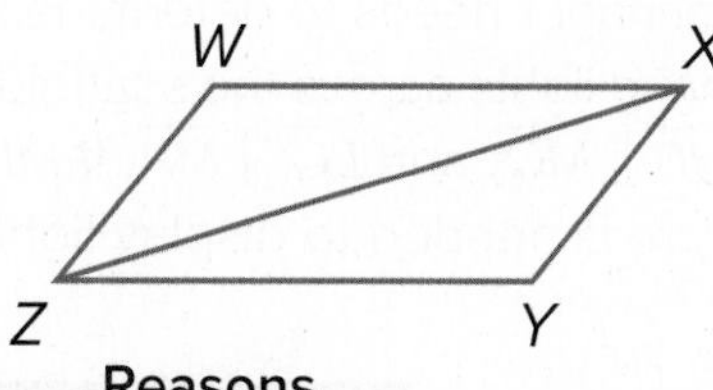

Statements	Reasons
1. ___?___	1. Given
2. $\overline{WZ} \parallel \overline{YX}$	2. ___?___
3. $\angle WXZ \cong \angle YZX$	3. ___?___
4. ___?___	4. Alternate Interior Angles Theorem
5. ___?___	5. Reflexive Property of Congruence
6. $\triangle WXZ \cong \triangle YZX$	6. ___?___

Go Online You can complete an Extra Example online.

Today's Goals

- Use the ASA congruence criterion for triangles to solve problems and prove relationships in geometric figures.
- Use the AAS congruence criterion for triangles to prove relationships in geometric figures.

Today's Vocabulary

included side

Think About It!

Can the Vertical Angles Theorem always be used to prove angles congruent in any two triangles? Justify your argument.

Example 2 Apply ASA Congruence

Use a Source

Navajo Native Americans often live in octagonal homes known as *hogans*. Use available resources to research the design of octagonal hogans. How can you use ASA and congruent triangles to find the area of the floor of a regular octagonal hogan?

PRODUCTION **A company that manufactures windows needs to determine the amount of glass required to make the hexagonal window shown. $\overline{PQ} \parallel \overline{TS}$, R is the midpoint of $\overline{PT}$, and ST is 12 inches.**

Part A Determine whether $\triangle PRQ$ is congruent to $\triangle TRS$.

Because $\overline{PQ}$ is parallel to $\overline{TS}$, $\angle RPQ \cong \angle RTS$ by the Alternate Interior Angles Theorem.

Because point R is the midpoint of $\overline{PT}$, $\overline{TR} \cong \overline{PR}$ by the Midpoint Theorem.
$\angle TRS$ and $\angle PRQ$ are vertical angles, so they are congruent by the Vertical Angles Theorem.

Therefore, by ASA, $\triangle PRQ \cong \triangle TRS$.

Part B Find the area of the window.

If the six triangles that form the window are congruent and the height of $\triangle TRS$ is about 10.39 inches, how much glass is required to manufacture the window?

$A = \frac{1}{2}bh$	Area of a triangle
$\approx \frac{1}{2}(12)(10.39)$	$b = 12$ and $h = 10.39$
≈ 62.34	Simplify.

The area of the window is approximately 62.34(6) or about 374.04 square inches.

Study Tip

Units of Measure Remember that when you are finding the area of a polygon, you are multiplying two dimensions. So, you should use *square units*.

Check

LIGHTING A theater uses scaffolding to hang stage lighting. The stage manager needs to determine how much electrical wire is needed to hang lights across the scaffolding from point L to N. $\overline{LN} \parallel \overline{QP}$, $\overline{NP} \parallel \overline{MQ}$, and $\overline{LQ} \parallel \overline{MP}$. If MN is 4 feet, how many feet of electrical wire is needed to display lights across the scaffolding?

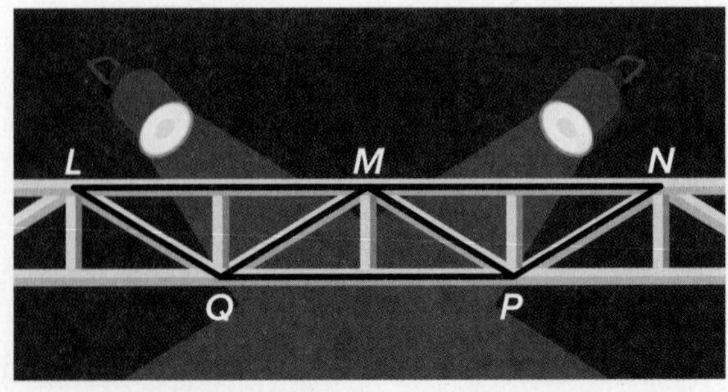

Go Online You can complete an Extra Example online.

Learn Proving Triangles Congruent: AAS

The congruence of two angles and a nonincluded side is also sufficient to prove two triangles congruent. This congruence relationship is a theorem because it can be proved using the Third Angles Theorem.

Theorem 14.5: Angle-Angle-Side (AAS) Congruence	
Words	If two angles and the nonincluded side of one triangle are congruent to the corresponding two angles and nonincluded side of a second triangle, then the two triangles are congruent.
Example	If $\angle A \cong \angle D$, $\angle B \cong \angle E$, and $\overline{BC} \cong \overline{EF}$, then $\triangle ABC \cong \triangle DEF$.

The proof of the AAS Congruence Theorem is below.

Given: $\angle L \cong \angle Q$
$\angle M \cong \angle R$
$\overline{MN} \cong \overline{RS}$

Prove: $\triangle LMN \cong \triangle QRS$

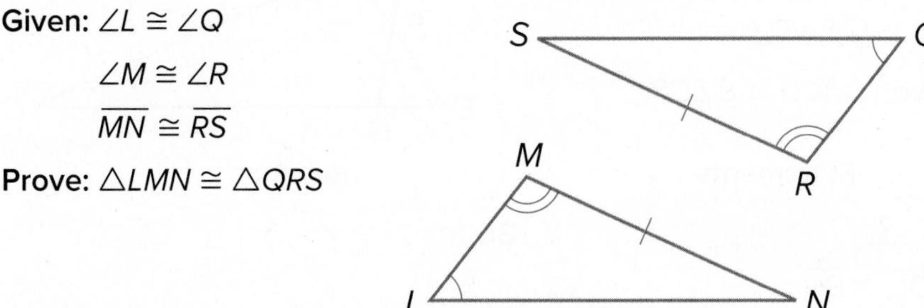

Proof:

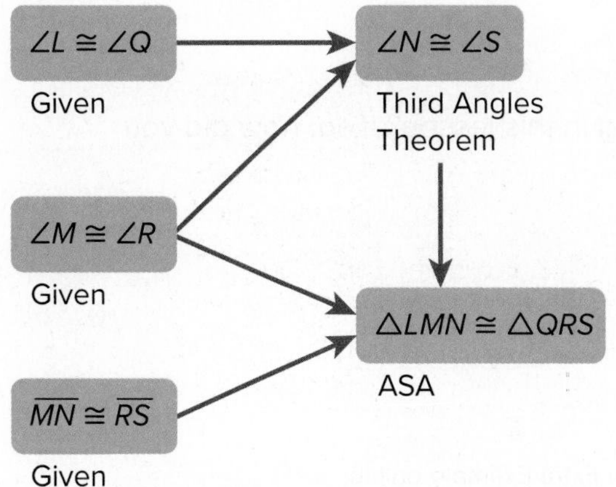

Talk About It!

Do you think angle-angle-angle, or AAA, could be used to prove triangles congruent? Provide an example to justify your reasoning.

Think About It!

Is hexagon *ABCDEF* congruent to hexagon *QRSTUV*? Use triangle congruence to justify your reasoning.

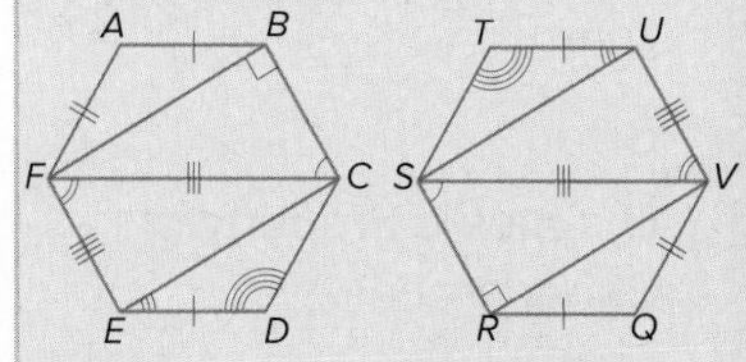

Go Online

You may want to complete the construction activities for this lesson.

Example 3 Use AAS to Prove Triangles Congruent

Choose the correct statements and reasons to complete the flow proof.

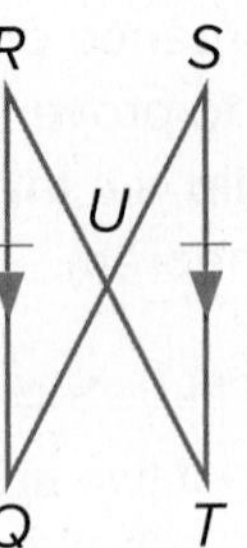

Given: $\overline{RQ} \cong \overline{ST}$ and $\overline{RQ} \parallel \overline{ST}$

Prove: $\triangle RUQ \cong \triangle TUS$

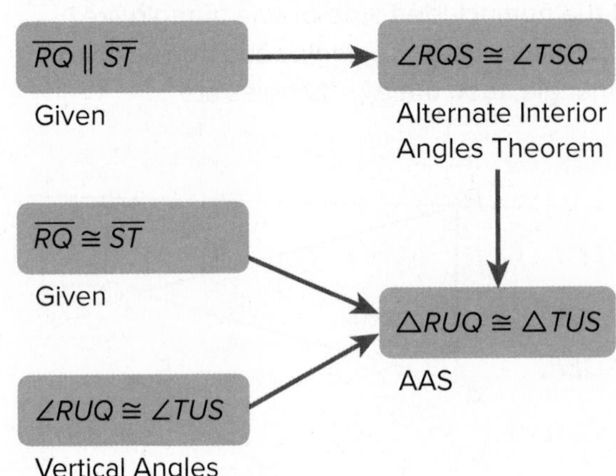

Statements/Reasons:

Alternate Exterior Angles Theorem

Alternate Interior Angles Theorem

Vertical Angles Theorem

$\angle RUQ \cong \angle TUS$

$\angle R \cong \angle T$

$\angle Q \cong \angle S$

$\triangle RUQ \cong \triangle SUT$

$\triangle RUS \cong \triangle QUT$

$\triangle RUQ \cong \triangle TUS$

Check

Choose the correct statements and reasons to complete the two-column proof.

Given: $\angle DAC \cong \angle BEC$ and $\overline{DC} \cong \overline{BC}$

Prove: $\triangle ACD \cong \triangle ECB$

Statements	Reasons
1. ___?___	1. Given
2. $\overline{DC} \cong \overline{BC}$	2. ___?___
3. ___?___	3. ___?___
4. $\triangle ACD \cong \triangle ECB$	4. ___?___

Statements/Reasons:

$\angle DAC \cong \angle BEC$

$\angle C \cong \angle C$

$\overline{AB} \cong \overline{ED}$

ASA

AAS

Given

Symmetric Property of Congruence

Reflexive Property of Congruence

Pause and Reflect

Did you struggle with anything in this lesson? If so, how did you deal with it?

Go Online to practice what you've learned about proving triangles congruent in the Put It All Together over Lessons 14-3 and 14-4.

Go Online You can complete an Extra Example online.

Practice

Go Online You can complete your homework online.

Example 1

PROOF **Write the specified type of proof.**

1. two-column proof

Given: $\overline{AB} \parallel \overline{CD}$, $\angle CBD \cong \angle ADB$

Prove: $\triangle ABD \cong \triangle CDB$

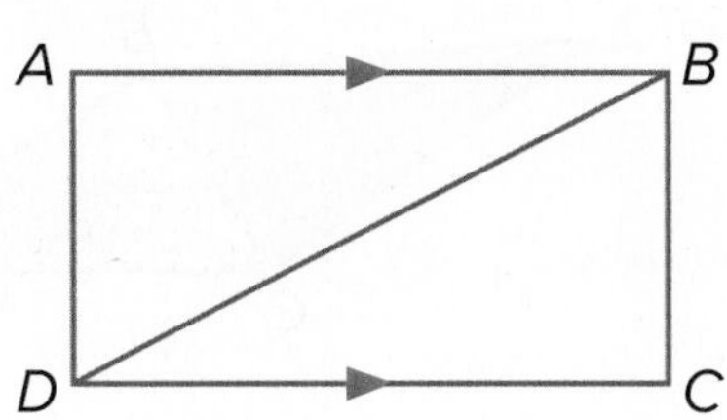

2. two-column proof

Given: $\angle S \cong \angle V$, and T is the midpoint of $\overline{SV}$.

Prove: $\triangle RTS \cong \triangle UTV$

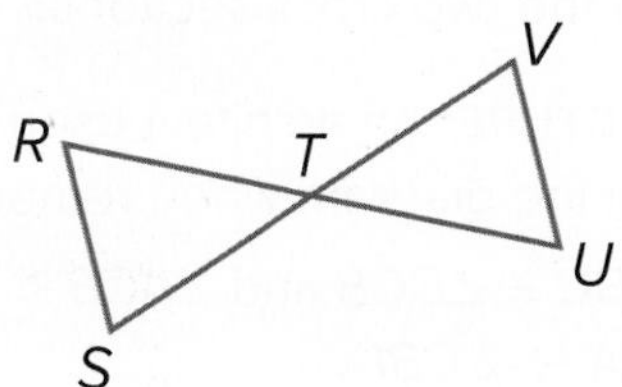

3. flow proof

Given: $\overline{AB} \cong \overline{CB}$, $\angle A \cong \angle C$, and $\overline{DB}$ bisects $\angle ABC$.

Prove: $\overline{AD} \cong \overline{CD}$

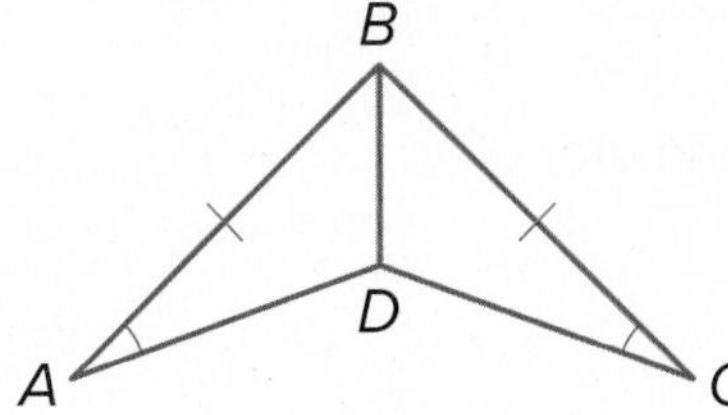

4. paragraph proof

Given: $\overline{CD}$ bisects $\overline{AE}$, $\overline{AB} \parallel \overline{CD}$, and $\angle E \cong \angle BCA$.

Prove: $\triangle ABC \cong \triangle CDE$

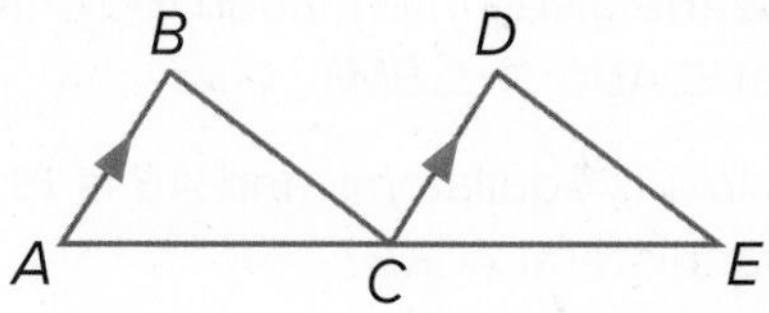

5. paragraph proof

Given: $\overline{CE}$ bisects $\angle BED$; $\angle BCE$ and $\angle ECD$ are right angles.

Prove: $\triangle ECB \cong \triangle ECD$

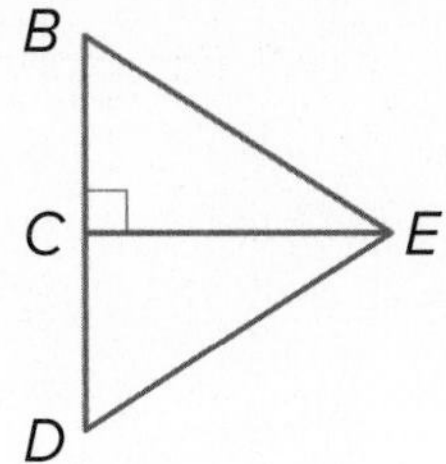

6. paragraph proof

Given: $\angle W \cong \angle Y$, $\overline{WZ} \cong \overline{YZ}$, and $\overline{XZ}$ bisects $\angle WZY$.

Prove: $\triangle XWZ \cong \triangle XYZ$

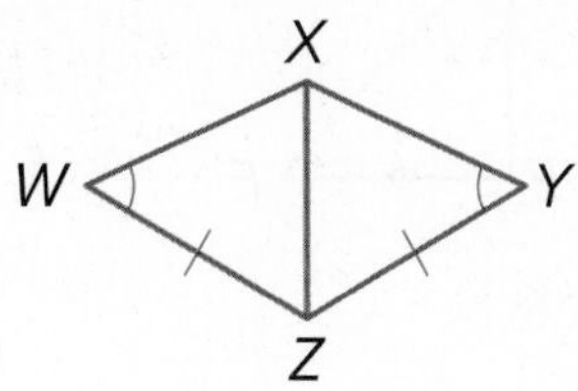

Example 2

7. **REASONING** Two doorstops have cross sections that are right triangles. Both have a 20° angle, and the lengths of the sides between the 90° and 20° angles are equal.

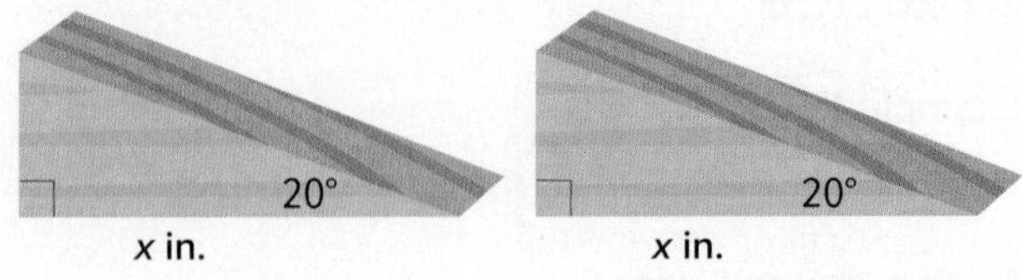

a. Are the cross sections congruent? Explain.

b. If each cross section has a height of 2 inches and $x = 5$, what is the combined area of the two cross sections?

8. **ARCHITECTURE** An architect used the stained-glass window design in the diagram when remodeling an art studio.

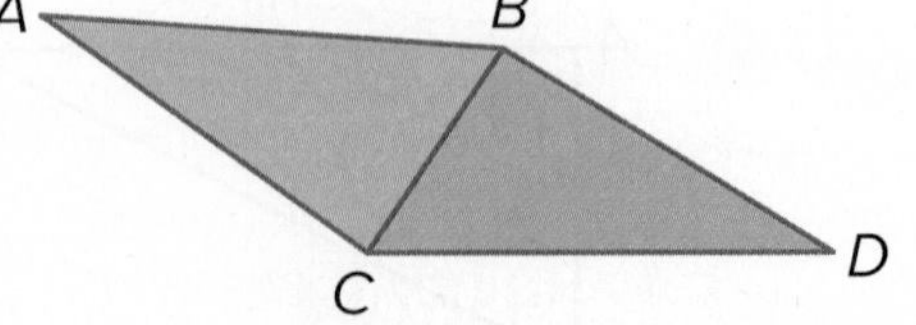

a. If $\angle ABC \cong \angle DCB$ and $\angle ACB \cong \angle DBC$, prove that $\triangle BCA \cong \triangle CBD$.

b. With CD as the base, if the height of $\triangle CBD$ is 1.4 meters and CD is 3.5 meters, how much glass is needed to make the entire window?

9. **BRIDGES** An engineering company that restores bridges needs to determine the amount of steel required to replace some trusses. $\overline{AC} \parallel \overline{BK}$, $\overline{CB} \parallel \overline{KM}$, and B is the midpoint of $\overline{AM}$.

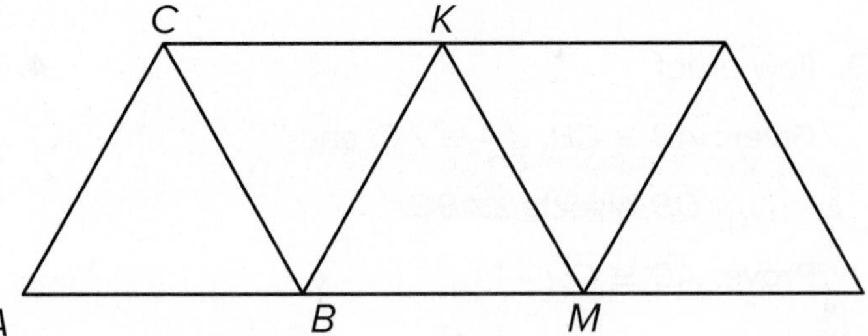

a. Use the given information to confirm that $\triangle ABC \cong \triangle BMK$.

b. $\triangle ABC$ is equilateral, and AB is 18.5 feet. What is the perimeter of quadrilateral $ACKB$?

PROOF **Write the specified type of proof.**

10. two-column proof

Given: $\overline{BC} \parallel \overline{EF}$, $\overline{AB} \cong \overline{DE}$, $\angle C \cong \angle F$

Prove: $\triangle ABC \cong \triangle DEF$

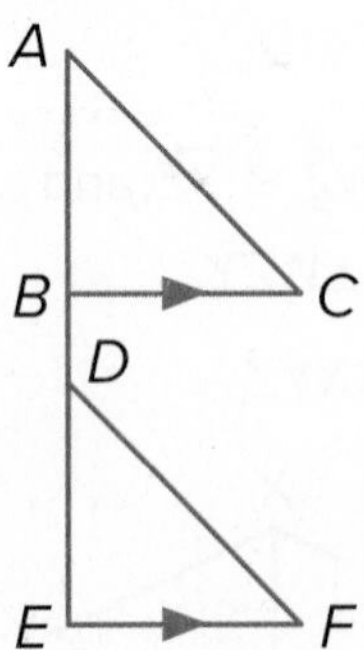

11. flow proof

Given: $\angle S \cong \angle U$, and $\overline{TR}$ bisects $\angle STU$.

Prove: $\triangle SRT \cong \triangle URT$

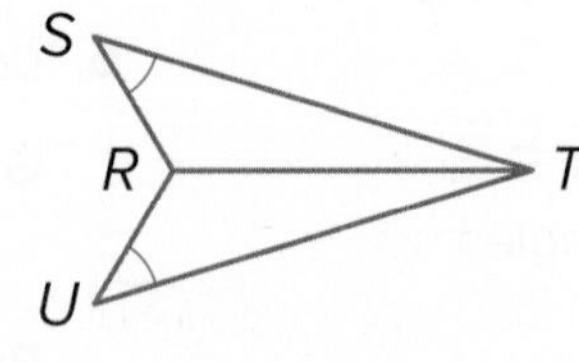

Example 3

Write the specified type of proof.

12. flow proof

Given: $\overline{JK} \cong \overline{MK}$, $\angle N \cong \angle L$

Prove: $\triangle JKN \cong \triangle MKL$

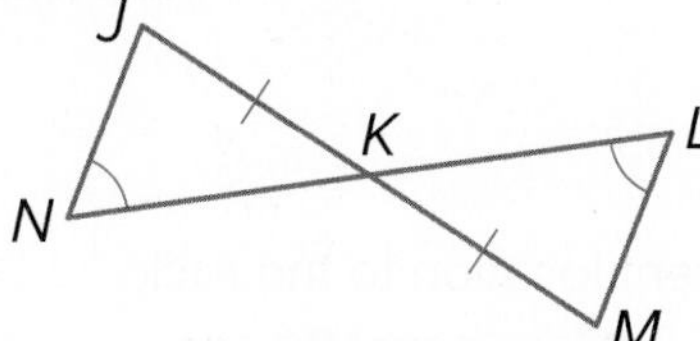

13. paragraph proof

Given: $\overline{DE} \parallel \overline{FG}$, $\angle E \cong \angle G$

Prove: $\triangle DFG \cong \triangle FDE$

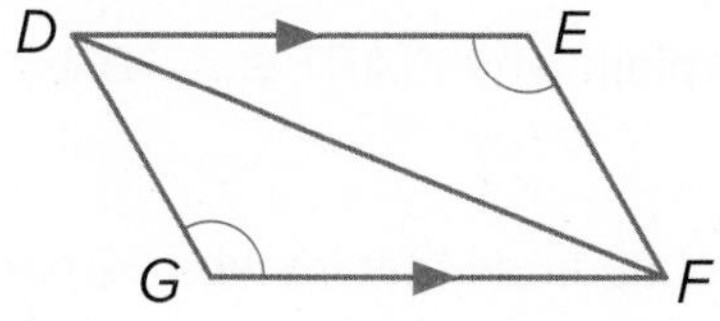

14. two-column proof

Given: V is the midpoint of $\overline{YW}$; $\overline{UY} \parallel \overline{XW}$,

Prove: $\triangle UVY \cong \triangle XVW$

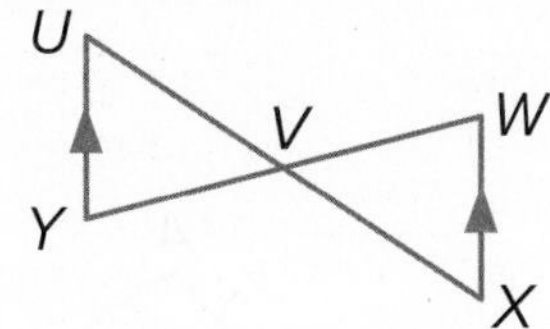

15. two-column proof

Given: $\overline{MS} \cong \overline{RQ}$,

$\overline{MS} \parallel \overline{RQ}$,

Prove: $\triangle MSP \cong \triangle RQP$

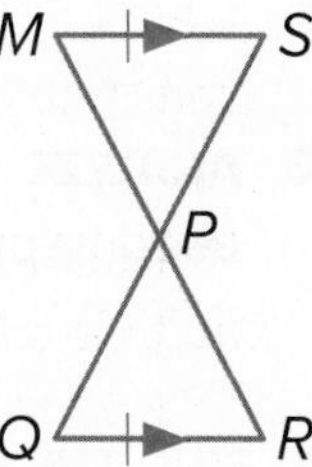

Mixed Exercises

16. USE TOOLS Use a compass and straightedge and the ASA Congruence Postulate to construct a triangle congruent to $\triangle PQR$.

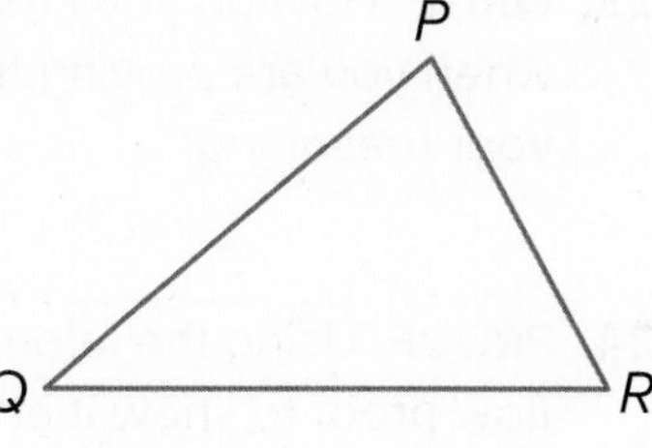

17. PRECISION Two people decide to take a walk. One person is in Bombay, and the other is in Milwaukee. They start by walking straight for 1 kilometer. Then both turn right at an angle of 110° and continue to walk straight again. After a while, both turn right again, but this time at an angle of 120°. Both walk straight for a while in this new direction until they end up where they started. Each person walked in a triangular path at their location. Are the two triangles they formed congruent? Explain.

18. **USE ESTIMATION** Delma came to a river during a hike, and she wanted to estimate the distance across it. She held her walking stick $\overline{AB}$ vertically on the ground at the edge of the river and sighted along the top of the stick across the river to the base of a tree T. Then she turned without changing the angle of her head and sighted along the top of the stick to a rock R, located on her side of the river.

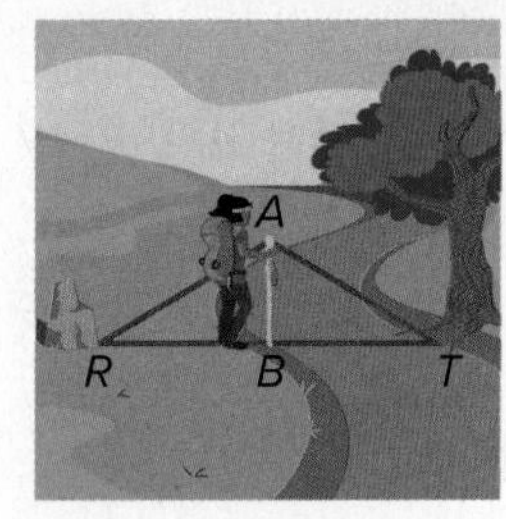

a. Explain why $\triangle ABT \cong \triangle ABR$.

b. Delma finds that it takes 27 paces to walk from her current location to the rock. She also knows that each of her paces is 14 inches long. Explain how she can use this information to estimate the distance across the river.

19. **PROOF** Write a paragraph proof.

Given: $\angle D \cong \angle F$

$\overline{GE}$ bisects $\angle DEF$.

Prove: $\overline{DG} \cong \overline{FG}$

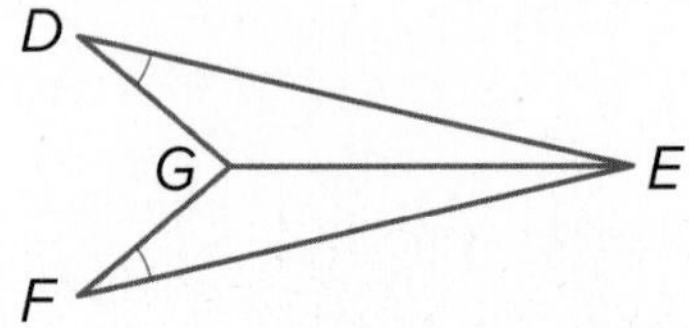

20. **ANALYZE** Find a counterexample to show why SSA (Side-Side-Angle) cannot be used to prove the congruence of two triangles.

21. **FIND THE ERROR** Tyrone says that it is not possible to show that $\triangle ADE \cong \triangle ACB$. Lorenzo disagrees, explaining that because $\angle ADE \cong \angle ACB$, $\angle AED \cong \angle ABC$, and $\angle A \cong \angle A$ by the Reflexive Property, $\triangle ADE \cong \triangle ACB$. Who is correct? Explain your reasoning.

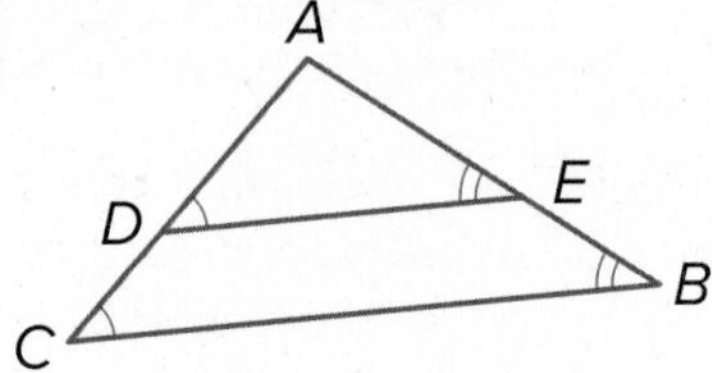

22. **CREATE** Draw and label two triangles that could be proved congruent by ASA.

23. **WRITE** How do you know which method (SSS, SAS, and so on) to use when you are proving triangle congruence? Use a table to explain your reasoning.

24. **PROOF** Using the information given in the diagram, write a flow proof to show that $\triangle PVQ \cong \triangle SVT$.

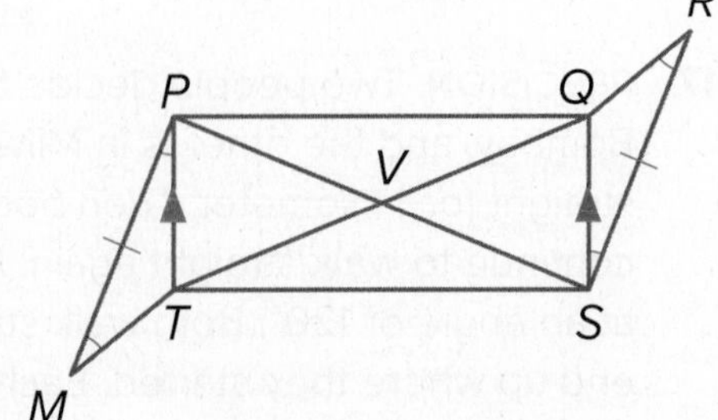

Proving Right Triangles Congruent

Explore Congruence Theorems and Right Triangles

 **Online Activity** Use graphing technology to complete the Explore.

INQUIRY What criteria can be used to prove right triangles congruent?

If two right triangles are formed using the criteria for leg-leg congruence, hypotenuse-angle congruence, leg-angle congruence, or hypotenuse-leg congruence, then there is a series of rigid motions that will show the two triangles congruent. This leads to the theorems below.

Learn Right Triangle Congruence

Theorem 14.6: Leg-Leg (LL) Congruence

Words	If the legs of one right triangle are congruent to the corresponding legs of another right triangle, then the triangles are congruent.
Example	Given right $\triangle ABC$ and right $\triangle DEF$, $\overline{AB} \cong \overline{DE}$ and $\overline{CB} \cong \overline{FE}$. So, $\triangle ABC \cong \triangle DEF$ by LL.

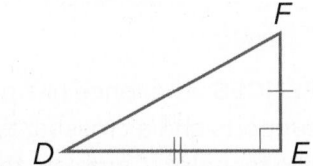

Theorem 14.7: Hypotenuse-Angle (HA) Congruence

Words	If the hypotenuse and an acute angle of one right triangle are congruent to the hypotenuse and the corresponding acute angle of another right triangle, then the triangles are congruent.
Example	Given right $\triangle ABC$ and right $\triangle DEF$, $\overline{AC} \cong \overline{DF}$ and $\angle C \cong \angle F$. So, $\triangle ABC \cong \triangle DEF$ by HA.

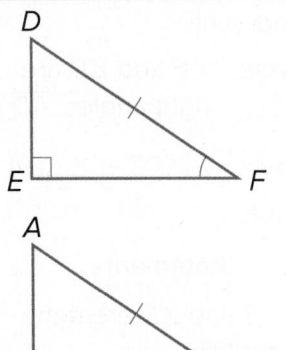

Theorem 14.8: Leg-Angle (LA) Congruence

If one leg and an acute angle of one right triangle are congruent to the corresponding leg and acute angle of another right triangle, then the triangles are congruent.

Theorem 14.9: Hypotenuse-Leg (HL) Congruence

If the hypotenuse and a leg of one right triangle are congruent to the hypotenuse and the corresponding leg of another right triangle, then the triangles are congruent.

Today's Goals

- Use the right triangle congruence theorems to prove relationships in geometric figures.

 **Talk About It!**

Can you declare that the given triangles are congruent by HA? Justify your argument.

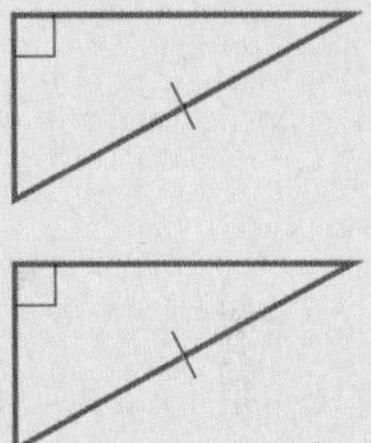

Go Online

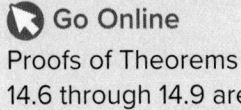

Proofs of Theorems 14.6 through 14.9 are available.

Example 1 Problem Solving with Right Triangles

HOME IMPROVEMENT **Craig and his brother are painting a house. The brothers use ladders that are the same length. If they place their ladders an equal distance from the house, will each ladder reach the same height on the house? Construct a logical argument.**

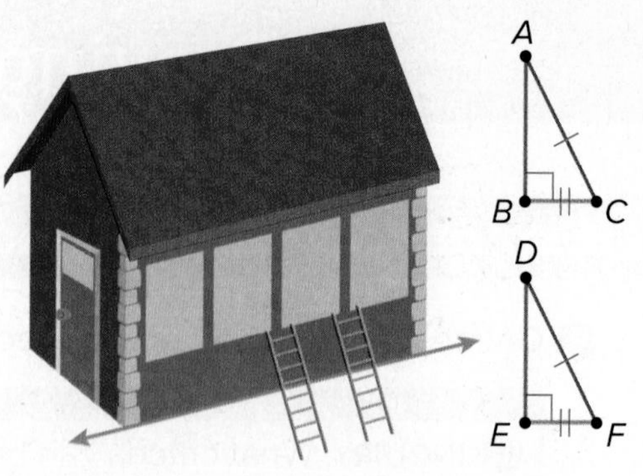

Draw a diagram to model this situation. It is given that the length of the ladders is the same and that they are placed the same distance from the house.

Because the wall of the house is perpendicular to the ground, the triangles formed by the house, the ground, and the ladders are right triangles. The hypotenuses are congruent because the ladders are the same length. The corresponding legs along the ground are congruent because the ladders are placed the same distance from the house. So the triangles are congruent by the Hypotenuse-Leg Congruence Theorem or HL. Thus, $\overline{AB}$ and $\overline{DE}$ are congruent by CPCTC. You can conclude that the ladders reach to the same height on the house.

Check

FENCES The fence has parallel supports and a crossbar that forms two triangles. Complete the proof to show that the triangles are congruent.

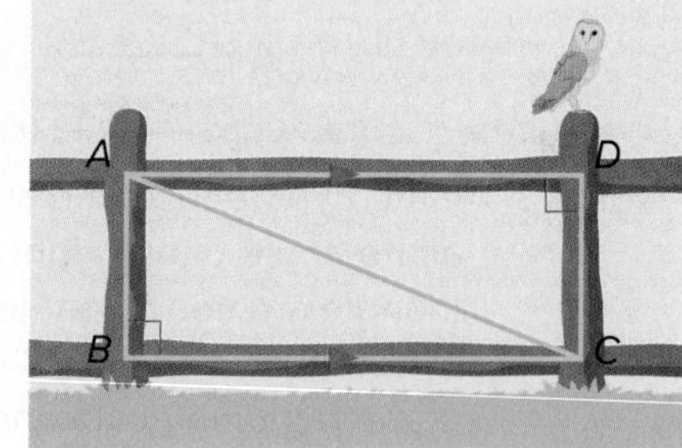

Given: $\angle B$ and $\angle D$ are right angles. $\overline{AD} \parallel \overline{BC}$.

Prove: $\triangle ABC \cong \triangle CDA$

Proof:

Statements	Reasons
1. $\angle B$ and $\angle D$ are right angles.	1. ?
2. $\triangle ABC$ and $\triangle CDA$ are right triangles.	2. ?
3. $\overline{AD} \parallel \overline{BC}$	3. ?
4. $\angle DAC \cong \angle BCA$	4. ?
5. $\overline{AC} \cong \overline{AC}$	5. ?
6. $\triangle ABC \cong \triangle CDA$	6. ?

Reasons:
- Alternate exterior angles are congruent.
- Alternate interior angles are congruent.
- Consecutive interior angles are congruent.
- Definition of right triangle
- Given
- HA
- HL
- LA
- Reflexive Property
- Symmetric Property

Go Online You can complete an Extra Example online.

Practice

Go Online You can complete your homework online.

Example 1

1. **CAMPING** In the diagram of the pup tent, the support pole is perpendicular to the ground. The base of the support pole is located at the midpoint of the segment connecting the bottom of the sides of the tent. Write a two-column proof to show that the triangles formed by the support pole are congruent.

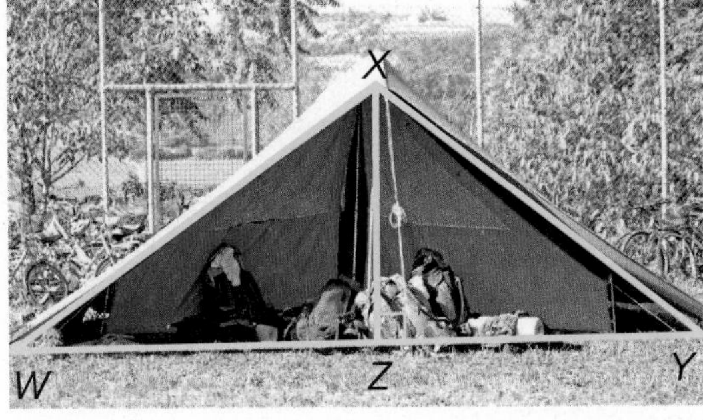

Given: $\overline{XZ} \perp \overline{WY}$; Z is the midpoint of $\overline{WY}$.

Prove: $\triangle WXZ \cong \triangle YXZ$

2. **TOWERS** The cell phone tower has parallel poles and diagonal support beams that form two triangles. Write a two-column proof to show that the triangles are congruent.

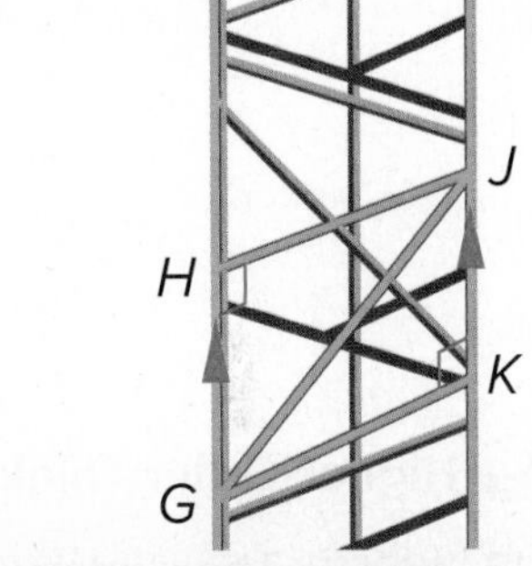

Given: $\angle H$ and $\angle K$ are right angles.

$\overline{GH} \parallel \overline{KJ}$

Prove: $\triangle GKJ \cong \triangle JHG$

3. **BRIDGES** In the diagram, the vertical support beam, $\overline{BX}$, is perpendicular to the deck of the bridge. The two diagonals, $\overline{AB}$ and $\overline{CB}$, are equal in length. Write a two-column proof to show that the triangles formed by the vertical support beam are congruent.

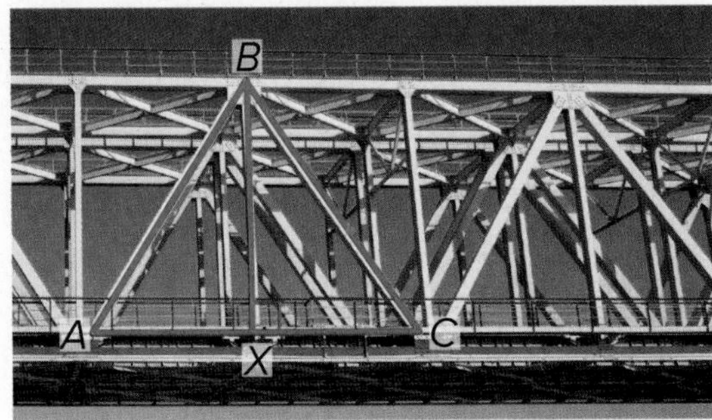

Given: $\overline{BX} \perp \overline{AC}$; $AB = CB$

Prove: $\triangle AXB \cong \triangle CXB$

Federicofoto/123RF, Alexey Baskakov/Hemera/Getty Images

Mixed Exercises

Determine whether each pair of triangles is congruent. If yes, include the theorem that applies.

4. 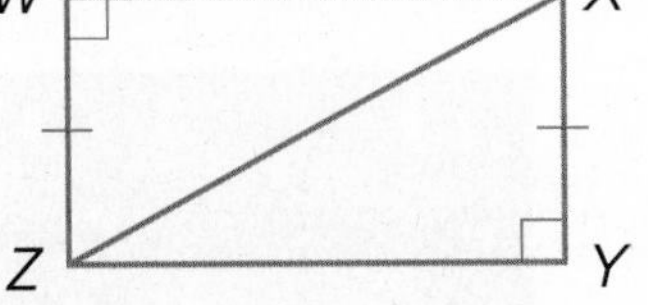

5. 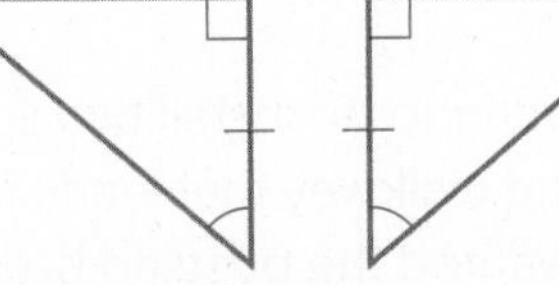

6. 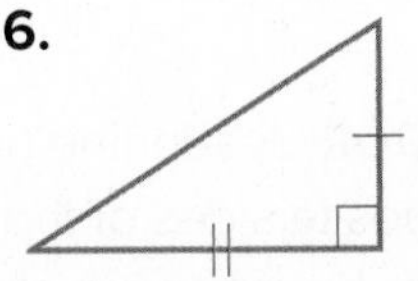 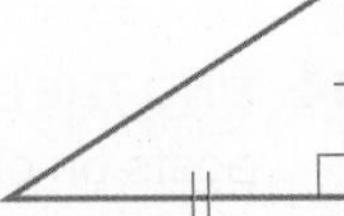

7. 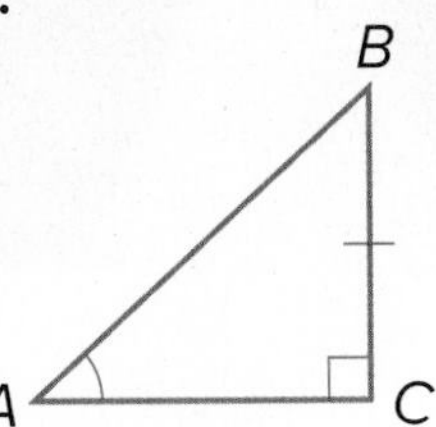 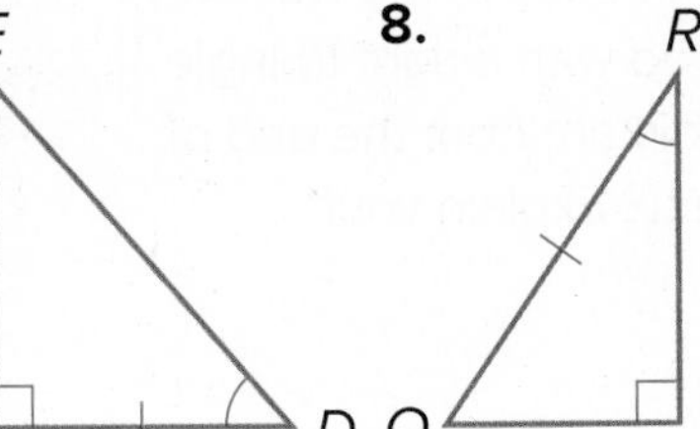

8. 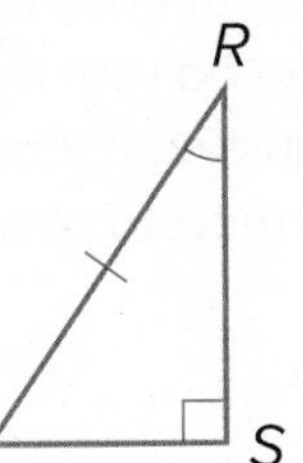 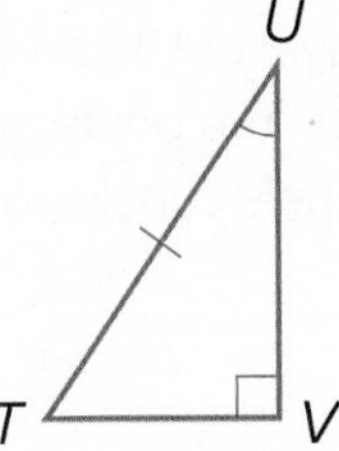

9. 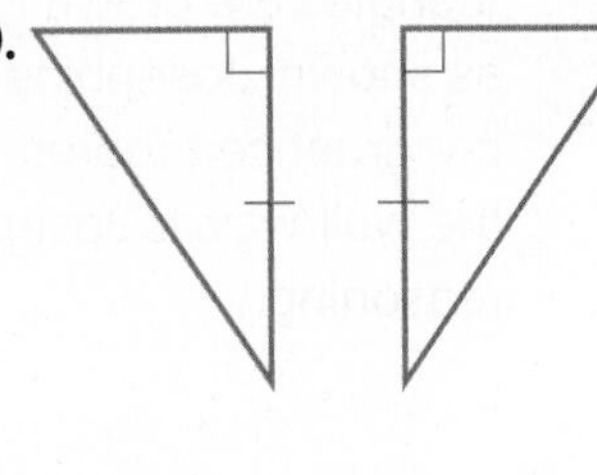

10. Which pairs of corresponding parts need to be congruent to prove that $\triangle ABC \cong \triangle XYZ$ using the indicated theorem?

a. HA

b. LL

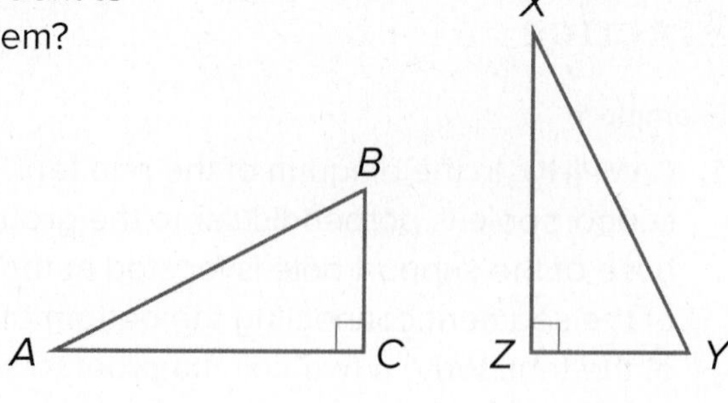

11. **PROOF** Write a two-column proof.

Given: $\overline{BX} \perp \overline{XA}$, $\overline{BY} \perp \overline{YA}$, and $\overline{XA} \cong \overline{YA}$

Prove: $\triangle BXA \cong \triangle BYA$

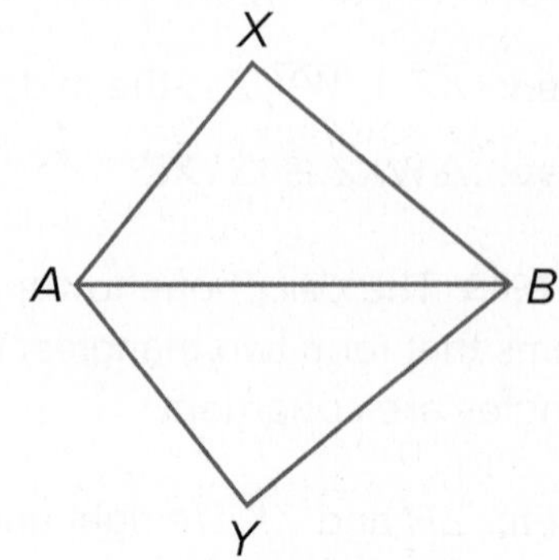

Higher-Order Thinking Skills

12. **WRITE** The sketch shows the side view of a sculpture that is being designed by an artist. Determine whether $\triangle ABC \cong \triangle DCA$. If yes, then provide a paragraph proof. If no, then explain your reasoning.

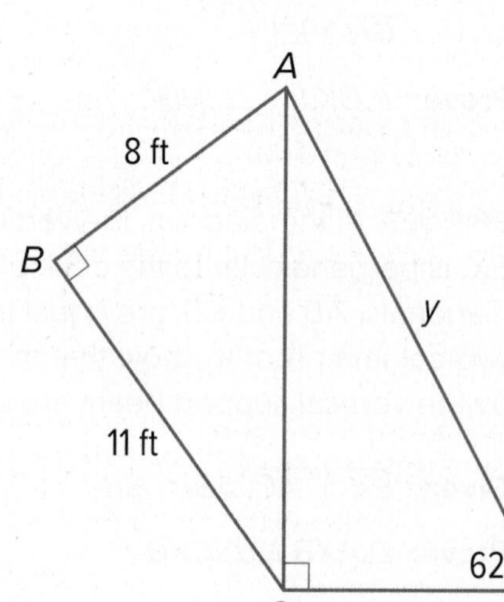

13. **PROOF** Write a paragraph proof.

Given: $\overline{BY} \perp \overline{AC}$; $\overline{CX} \perp \overline{AB}$; $AX = AY$

Prove: $\triangle ABY \cong \triangle ACX$

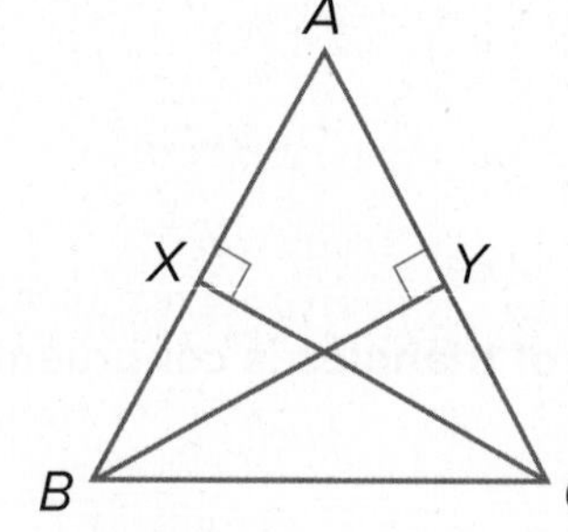

14. **FIND THE ERROR** Josephine hired a contractor to install two light posts on opposite sides of the end of the walkway that leads from the rear of her house to the alley. She wanted the posts to be equidistant from the end of the 40-foot walkway. Suppose two triangles are drawn from the light posts to both ends of the walkway as shown. Josephine says that it can be proved with a right triangle congruence theorem that the posts are equidistant from the end of the walkway. Is Josephine's conclusion correct? Explain your reasoning.

50 ft

50 ft

Isosceles and Equilateral Triangles

Explore Properties of Equilateral, Isosceles, and Scalene Triangles

Online Activity Use dynamic geometry software to complete the Explore.

> **INQUIRY** What are the differences between equilateral, isosceles, and scalene triangles?

Explore Isosceles and Equilateral Triangles

Online Activity Use dynamic geometry software to complete the Explore.

> **INQUIRY** What conjecture can you make about the relationship between the parts of isosceles and equilateral triangles?

Learn Isosceles Triangles

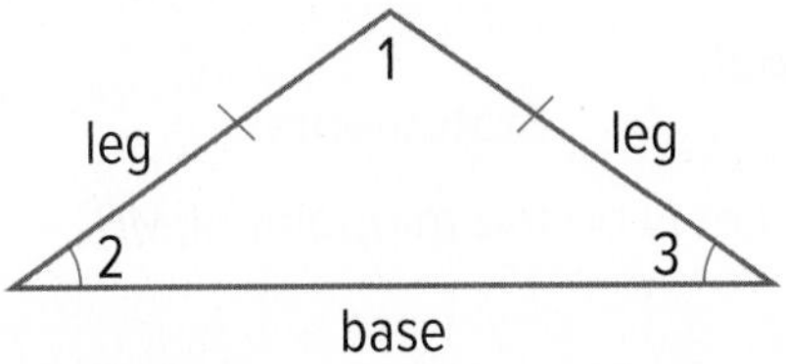

An **isosceles triangle** is a triangle with at least two sides congruent. The two congruent sides are called the **legs of an isosceles triangle**. The angle between the sides that are the legs is called the **vertex angle of an isosceles triangle**. $\angle 1$ is the vertex angle of the triangle. The side of the triangle opposite the vertex angle is called the *base*. The two angles formed by the base and the congruent sides are called the **base angles of an isosceles triangle**. $\angle 2$ and $\angle 3$ are the base angles.

Theorem 14.10: Isosceles Triangle Theorem	
Words	If two sides of a triangle are congruent, then the angles opposite those sides are congruent.
Example	If $\overline{AC} \cong \overline{BC}$, then $\angle 2 \cong \angle 1$. 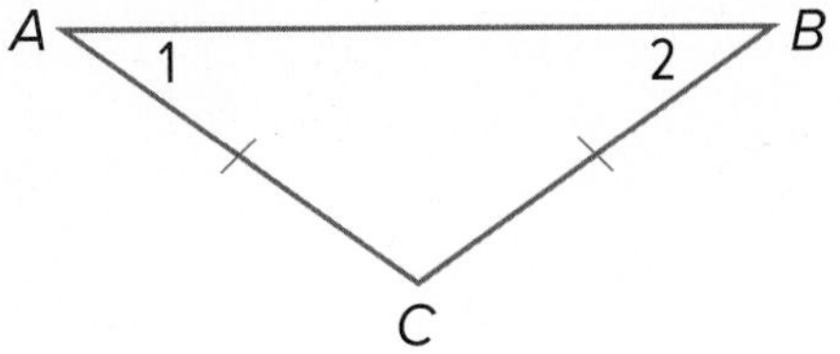

Today's Goals

- Solve problems involving isosceles triangles.
- Solve problems involving equilateral triangles.

Today's Vocabulary

isosceles triangle

legs of an isosceles triangle

vertex angle of an isosceles triangle

base angles of an isosceles triangle

auxiliary line

Math History Minute

Henry Dudeney (1857–1930) was a British government employee who enjoyed creating logic puzzles and mathematical games. One of Dudeney's greatest accomplishments was his success at solving a particular puzzle, the Haberdasher's Puzzle, that requires a person to cut an equilateral triangle into four pieces that can be rearranged to make a square.

Theorem 14.11: Converse of the Isosceles Triangle Theorem	
Words	If two angles of a triangle are congruent, then the sides opposite those angles are congruent.
Example	If $\angle 1 \cong \angle 2$, then $\overline{FE} \cong \overline{DE}$. 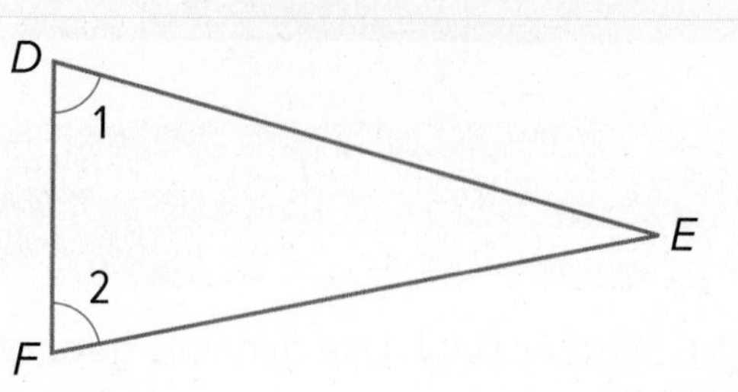

Go Online A proof of Theorem 5.11 is available.

Example 1 Prove the Isosceles Triangle Theorem

Prove the Isosceles Triangle Theorem.

To prove the Isosceles Triangle Theorem, draw an *auxiliary line* and use the two triangles that are formed. An **auxiliary line** is an extra line or segment drawn in a figure to help analyze geometric relationships.

Given: $\triangle LMP$, $\overline{LM} \cong \overline{LP}$

Prove: $\angle M \cong \angle P$

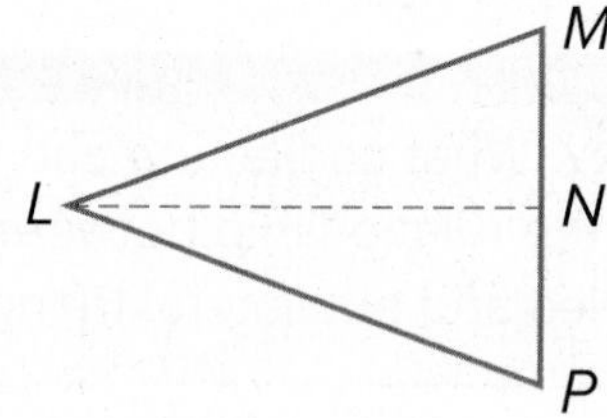

Proof:

Statements	Reasons
1. Let N be the midpoint of $\overline{MP}$.	1. Every segment has exactly one midpoint.
2. Draw an auxiliary segment $\overline{LN}$.	2. Two points determine a line.
3. $\overline{MN} \cong \overline{PN}$	3. Midpoint Theorem
4. $\overline{LN} \cong \overline{LN}$	4. Reflexive Property of Congruence
5. $\overline{LM} \cong \overline{LP}$	5. Given
6. $\triangle LMN \cong \triangle LPN$	6. SSS
7. $\angle M \cong \angle P$	7. CPCTC

Statements/Reasons:

Definition of congruence

Given

Midpoint Theorem

$\overline{LN} \cong \overline{LN}$

$\overline{MP} \cong \overline{LP}$

Go Online You can complete an Extra Example online.

Example 2 Find Missing Measures in Isosceles Triangles

Find $m\angle B$ and $m\angle C$.

Part A Determine side relationships.

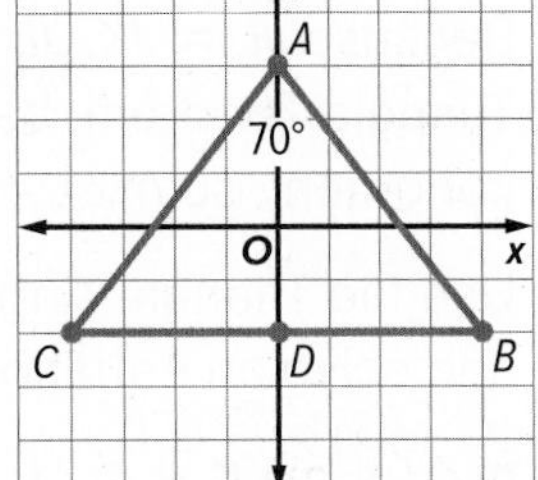

Use the Distance Formula to determine the measures of the sides of $\triangle ABC$. The coordinates of $\triangle ABC$ are $A(0, 3)$, $B(4, -2)$, and $C(-4, -2)$.

$AB = \sqrt{(0-4)^2 + [3-(-2)]^2}$ or $\sqrt{41}$ units

$AC = \sqrt{[0-(-4)]^2 + [3-(-2)]^2}$ or $\sqrt{41}$ units

$BC = \sqrt{[4-(-4)]^2 + [-2-(-2)]^2}$ or 8 units

So, $\triangle ABC$ is an isosceles triangle with $\overline{AB} \cong \overline{AC}$.

Part B Determine the angle measures.

Because $\overline{AB} \cong \overline{AC}$, we know that $\angle C \cong \angle B$ by the Isosceles Triangle Theorem.

$m\angle A + m\angle B + m\angle C = 180°$	Triangle Angle-Sum Theorem
$m\angle A + 2m\angle B = 180°$	Definition of congruent
$70° + 2m\angle B = 180°$	Substitute.
$m\angle B = m\angle C = 55°$	Solve.

Check

Find $m\angle XYZ$ and $m\angle YZX$.

$m\angle XYZ =$ _?_

$m\angle YZX =$ _?_

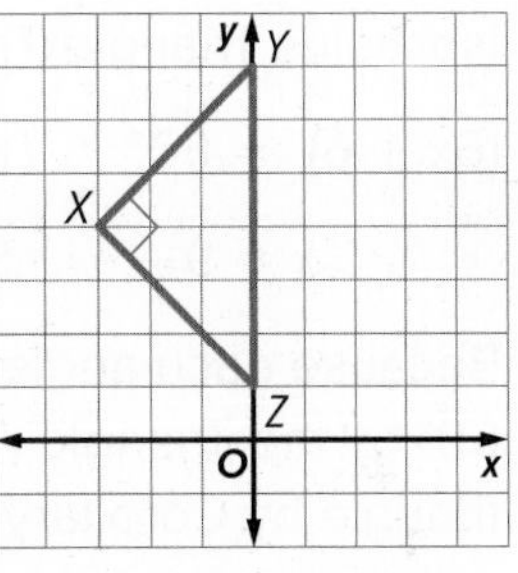

Learn Equilateral Triangles

The Isosceles Triangle Theorem leads to two corollaries about the angles of an equilateral triangle.

Corollary 14.3
A triangle is equilateral if and only if it is equiangular.
Corollary 14.4
Each angle of an equilateral triangle measures 60°.

You will prove Corollaries 14.3 and 14.4 in Exercises 18 and 19, respectively.

Watch Out!

Triangle Relationships We cannot use the Isosceles Triangle Theorem until we show that two sides of $\triangle ABC$ are congruent.

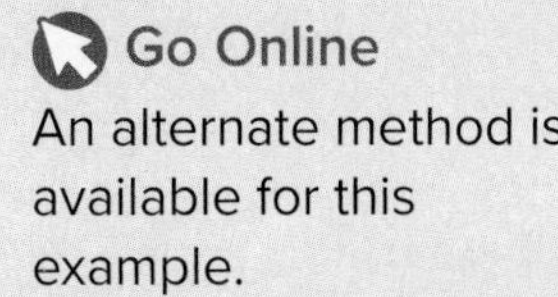

Go Online An alternate method is available for this example.

Example 3 Find Missing Measures in Equilateral Triangles

Study Tip

Isosceles Triangles Any isosceles triangle that has one 60° angle must be an equilateral triangle.

Find $m\angle J$.

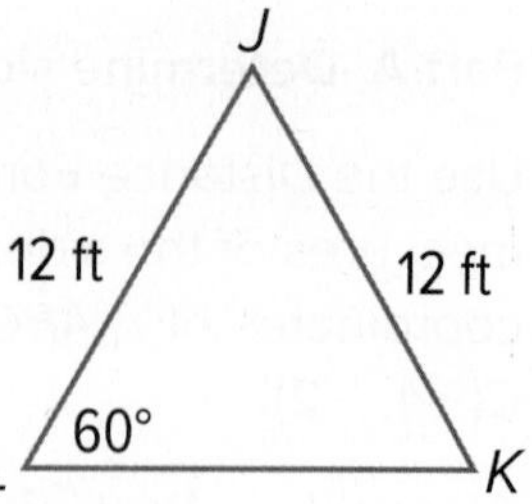

Because $JL = JK$, $\overline{JL} \cong \overline{JK}$. By the Isosceles Triangle Theorem, base angles L and K are congruent, so $m\angle L = m\angle K$.

Use the Triangle Angle-Sum Theorem to write and solve an equation to find $m\angle J$.

$m\angle J + m\angle K + m\angle L = 180°$	Triangle Angle-Sum Theorem
$m\angle J + 60° + 60° = 180°$	Isosceles Triangle Theorem
$m\angle J = 60°$	Solve.

Check

Find $m\angle R$ and PR.

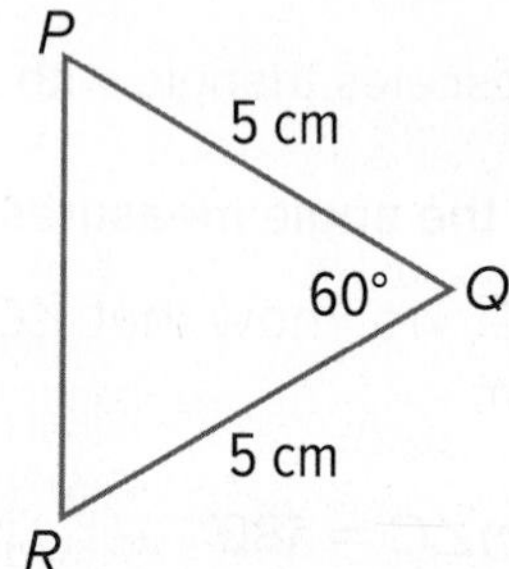

$m\angle R =$ __?__

$PR =$ __?__ cm

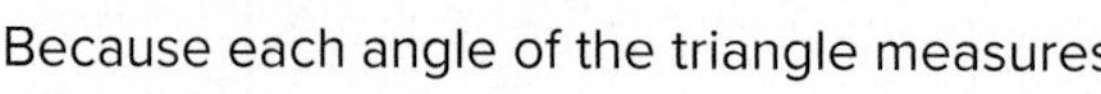

Talk About It!

Arturo argues that you do not have to use the properties of equilateral triangles to solve for y. Do you agree? Explain your reasoning.

Example 4 Find Missing Values

BILLIARDS **Find the value of each variable.**

Because $\overline{AB} \cong \overline{BC}$, $\angle ACB \cong \angle BAC$ by the Isosceles Triangle Theorem.

$(6x + 6)° = 60°$	Isosceles Triangle Theorem
$x = 9$	Solve.

Because each angle of the triangle measures 60° by the Triangle Angle-Sum Theorem, the triangle is an equilateral triangle by Corollary 5.3.

$4y - 2 = 2y + 2$	Corollary 14.3; definition of equilateral △
$y = 2$	Solve.

Check

ARCHITECTURE The main entrance to the Louvre Museum is a unique metal and glass pyramid. Find the value of each variable.

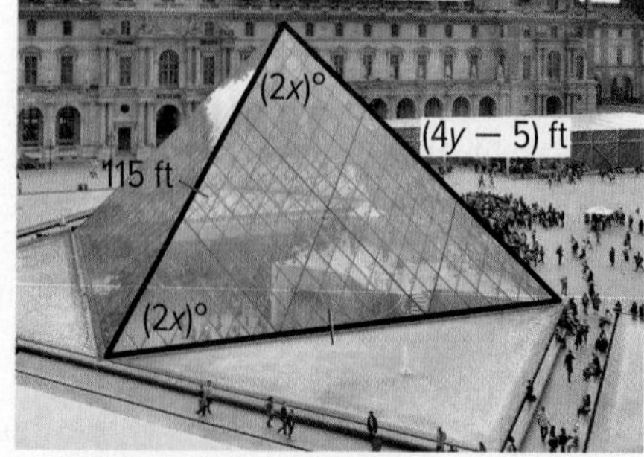

$x =$ __?__ and $y =$ __?__

Go Online

You may want to complete the construction activities for this lesson.

 **Go Online** You can complete an Extra Example online.

Kucher Serhii/Shutterstock; Sira Anamwong/Shutterstock

Practice

Go Online You can complete your homework online.

Example 1

1. PROOF Write a two-column proof.

Given: $\angle 1 \cong \angle 2$

Prove: $\overline{AB} \cong \overline{CB}$

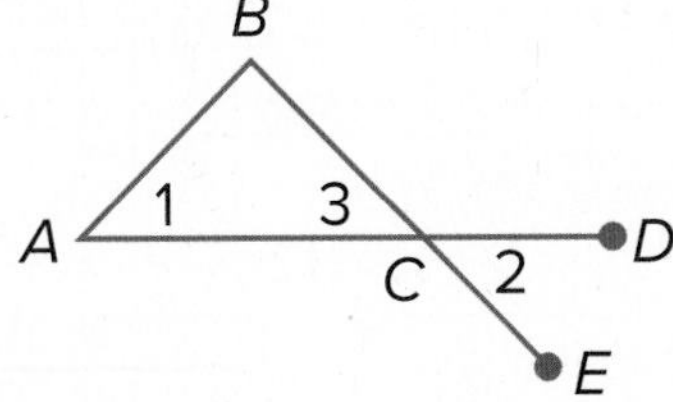

2. PROOF Write a two-column proof.

Given: $\overline{CD} \cong \overline{CG}$

$\overline{DE} \cong \overline{GF}$

Prove: $\overline{CE} \cong \overline{CF}$

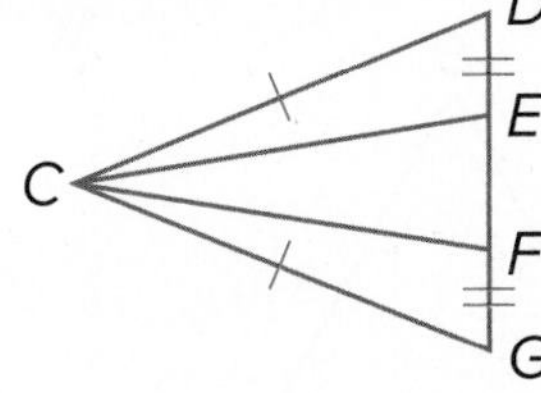

3. PROOF Write a two-column proof.

Given: $\overline{DE} \parallel \overline{BC}$

$\angle 1 \cong \angle 2$

Prove: $\overline{AB} \cong \overline{AC}$

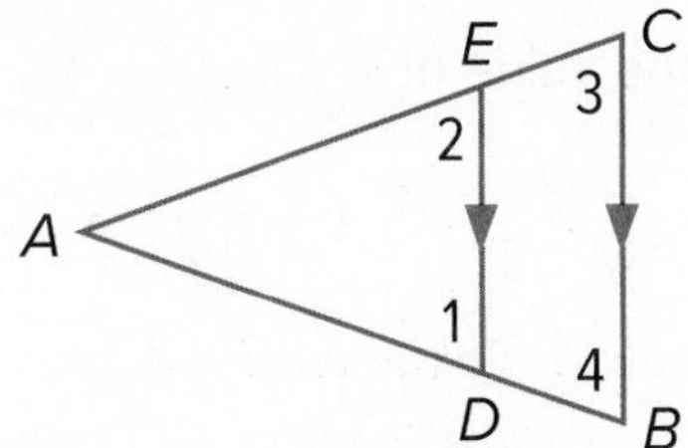

4. ROOFS In the picture, $\overline{BD} \perp \overline{AC}$ and $\triangle ABC$ is an isosceles triangle with base $\overline{AC}$. Write a two-column proof to prove that $\overline{BD}$ bisects the angle formed by the sloped sides of the roof, $\angle ABC$.

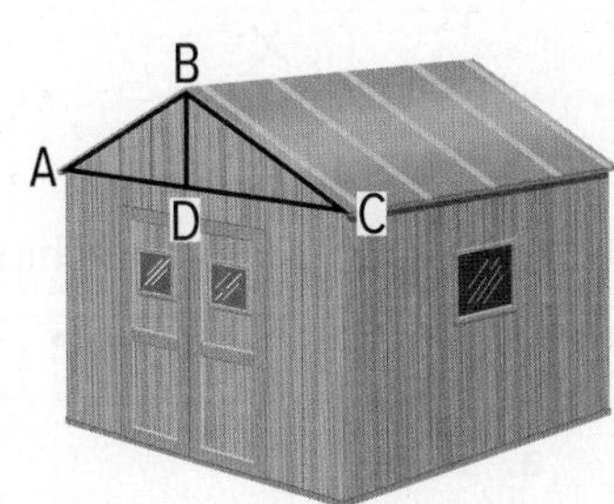

Example 2

5. Refer to the figure.

a. Find the measures of the sides of $\triangle ABC$. Show your work.

b. Find $m\angle A$. Show your work.

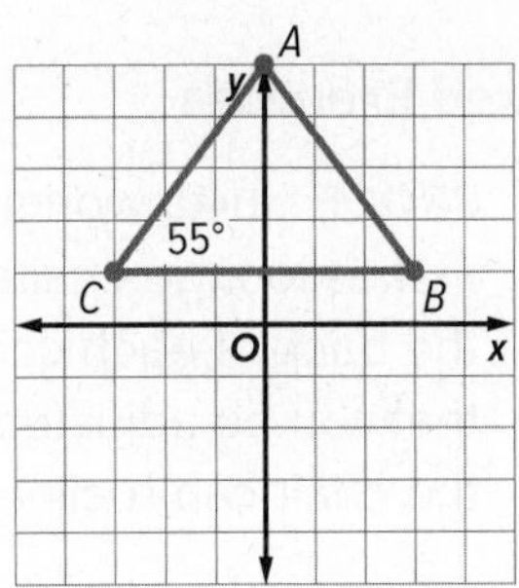

6. Find SR, ST, RT, $m\angle TRS$, and $m\angle RST$. Round to the nearest tenth, if necessary.

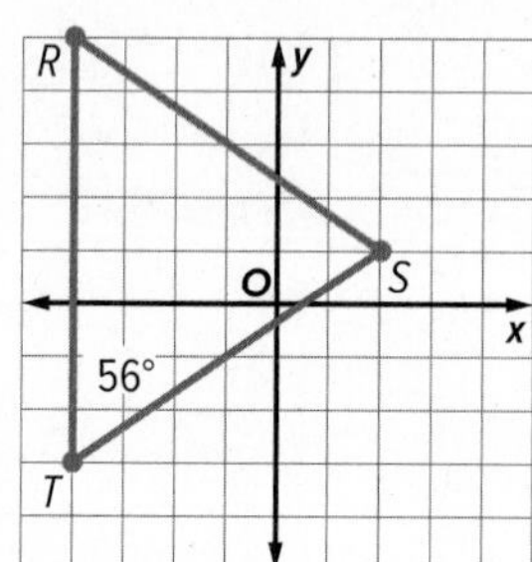

7. Find the measures of $\angle DEF$ and $\angle EFD$. Round to the nearest tenth, if necessary.

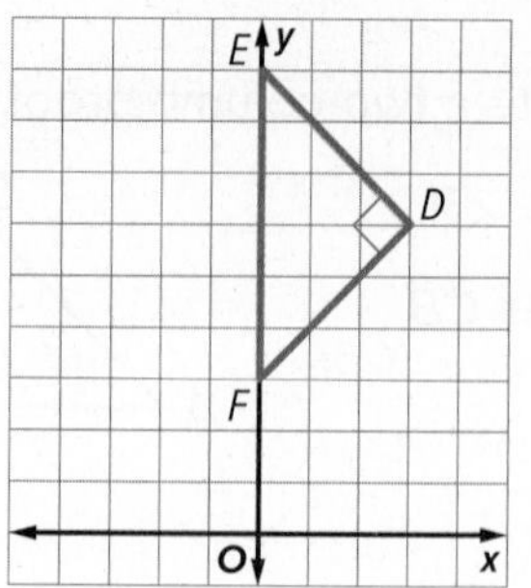

Examples 3 and 4

8. Find the value of x.

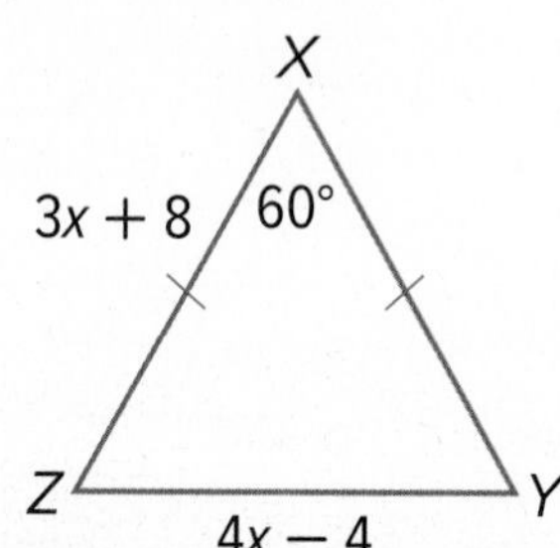

9. Find $m\angle B$ and AC.

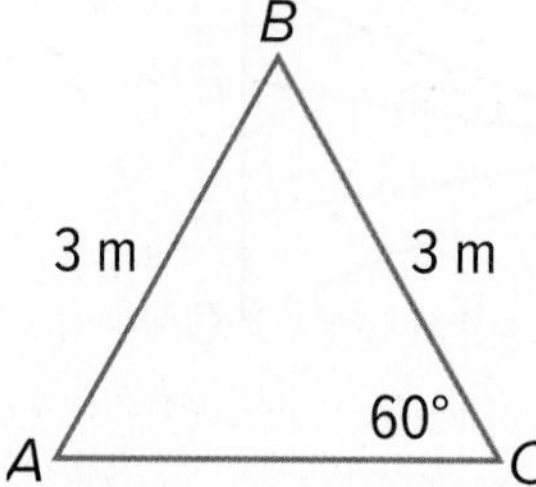

10. Find the value of x.

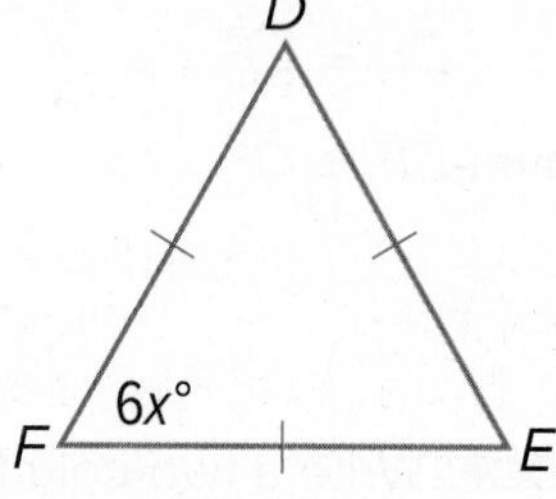

11. Find $m\angle Y$ and WY.

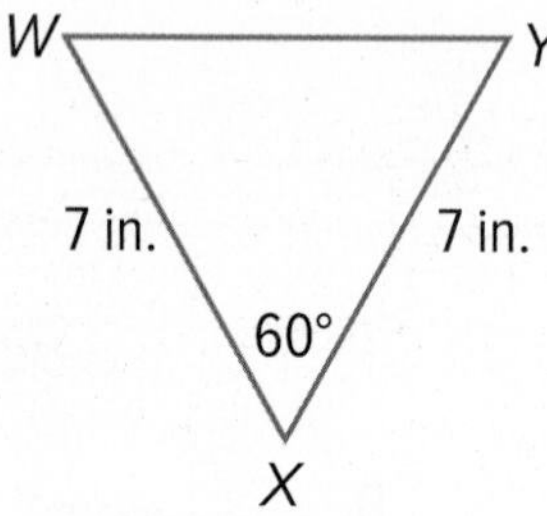

12. Find the value of x.

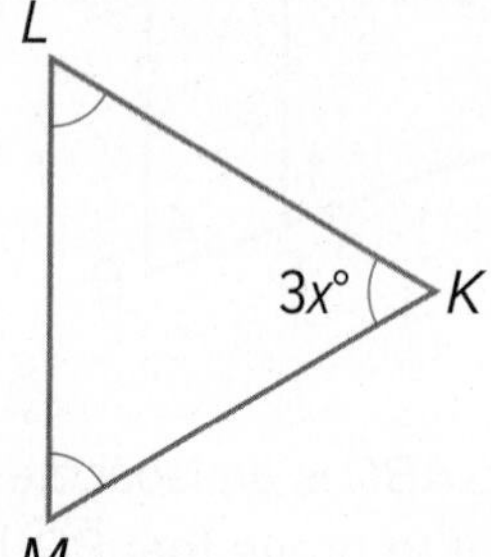

13. Find the value of x.

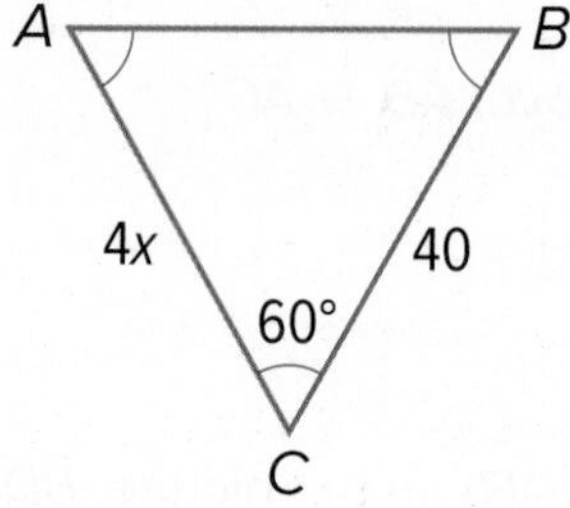

14. **CHIPS** Some tortilla chips can be modeled by a triangle.

a. Solve for x.

b. Solve for y.

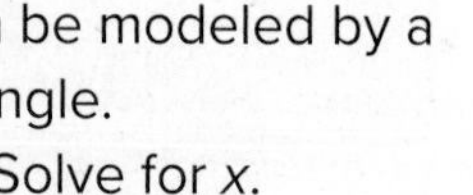

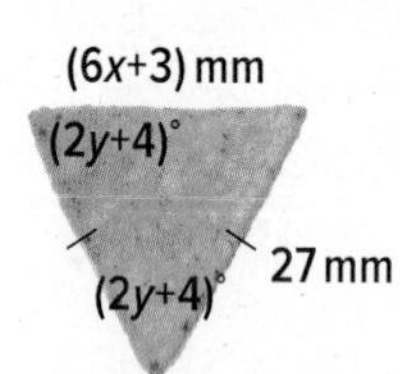

15. **SIGNS** Yield signs notify drivers to slow down and allow oncoming vehicles to proceed first.

a. Solve for x.

b. Solve for y.

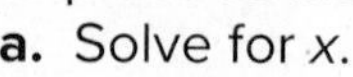

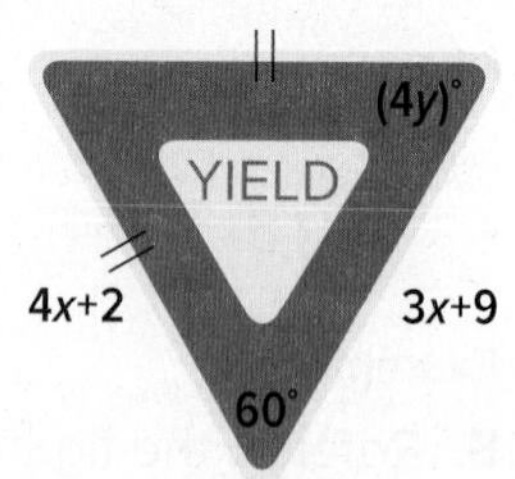

Maryna Kulchytska/Shutterstock

Mixed Exercises

16. **PROOF** Julita works for a company that makes lounge chairs. As shown in the figure, the back of each chair is an isosceles triangle that can be adjusted so the person sitting on the chair can recline.

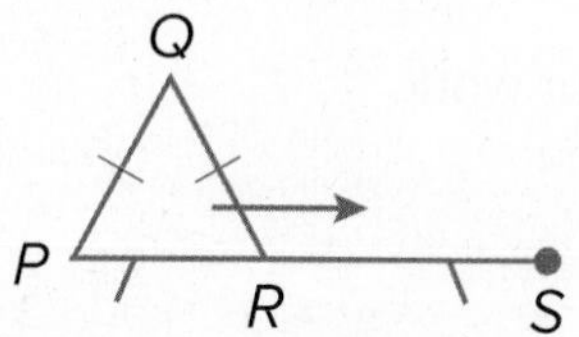

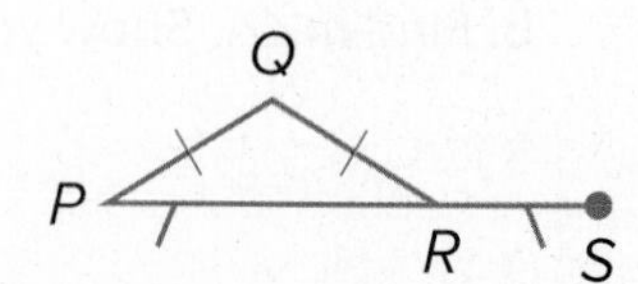

Suppose the chair is adjusted so $m\angle Q = 50°$. What is $m\angle QRS$? Write a paragraph proof to justify your argument.

Mixed Exercises

17. STRUCTURE Each of the triangles shown is isosceles.

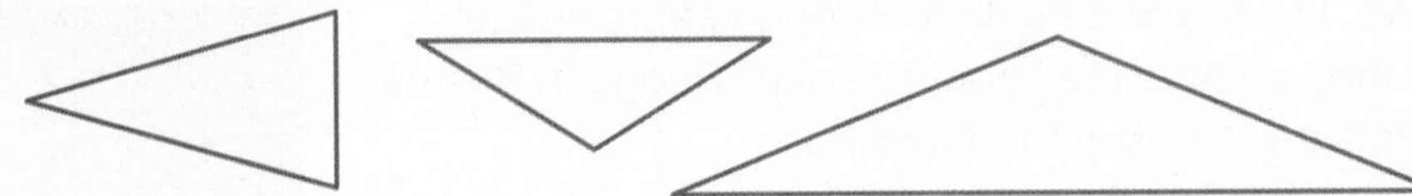

a. Use a ruler to find the midpoint of each side of each triangle. Copy and draw a triangle formed by connecting the midpoints of each side.

b. Look for patterns in your drawings. Make a conjecture about what you notice.

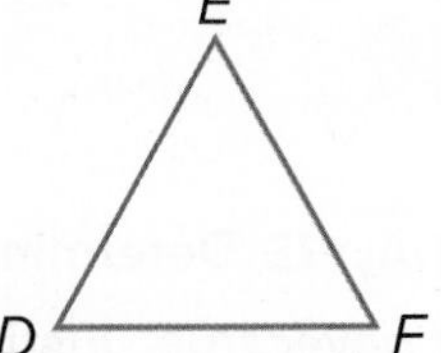

18. PROOF Write a two-column proof to prove each case of Corollary 14.3.

a. Case 1

Given: $\triangle DEF$ is an equilateral triangle.

Prove: $\triangle DEF$ is an equiangular triangle.

b. Case 2

Given: $\triangle DEF$ is an equiangular triangle.

Prove: $\triangle DEF$ is an equilateral triangle.

19. PROOF Write a two-column proof to prove Corollary 14.4.

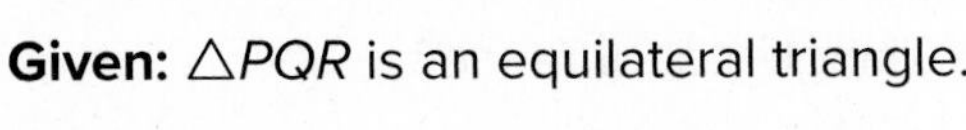

Given: $\triangle PQR$ is an equilateral triangle.

Prove: $m\angle P = m\angle Q = m\angle R = 60°$

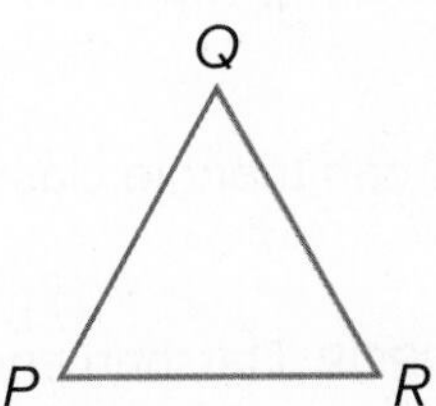

REGULARITY **Find each measure.**

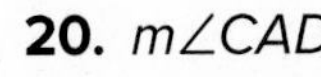

20. $m\angle CAD$

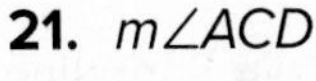

21. $m\angle ACD$

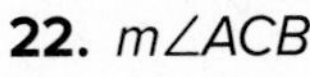

22. $m\angle ACB$

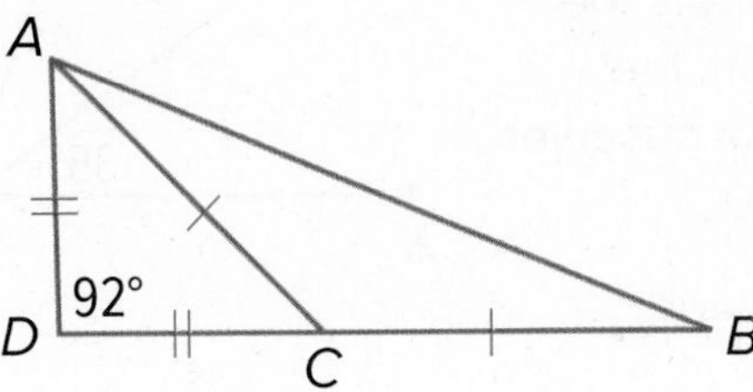

23. $m\angle ABC$

24. PATHS A marble path, as shown at the right, is constructed out of several congruent isosceles triangles. All the vertex angles measure 20°. What is the measure of angle 1 in the figure?

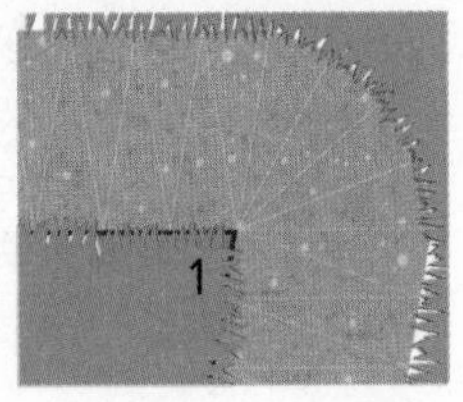

25. PRECISION Construct three different isosceles right triangles. Explain your method. Then verify your constructions using measurement and mathematics.

26. STATE YOUR ASSUMPTIONS Every day, cars drive through approximate isosceles triangles when they go over the Leonard Zakim Bridge in Boston. The ten-lane roadway forms the bases of the triangles.

a. If $m\angle A = 67°$, find $m\angle B$.

b. Find $m\angle C$.

c. What assumption is made when approximating that the bridge forms isosceles triangles?

ANALYZE **Determine whether the following statements are *sometimes*, *always*, or *never* true. Justify your argument.**

27. If the measure of the vertex angle of an isosceles triangle is an integer, then the measure of each base angle is an integer.

28. If the measures of the base angles of an isosceles triangle are integers, then the measure of its vertex angle is odd.

29. CREATE If possible, draw an isosceles triangle with base angles that are obtuse. If it is not possible, explain why not.

30. WRITE How can triangle classifications help you prove triangle congruence?

31. FIND THE ERROR Darshan and Miguela are finding $m\angle G$ in the figure shown. Darshan says that $m\angle G = 35°$, and Miguela says that $m\angle G = 60°$. Is either of them correct? Explain your reasoning.

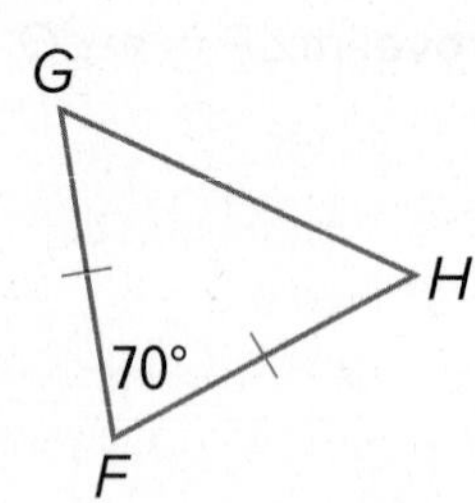

32. PERSEVERE A boat is traveling at 25 mi/h parallel to a straight section of the shoreline, $\overline{XY}$, as shown. An observer in a lighthouse L spots the boat when the angle formed by the boat, the lighthouse, and the shoreline is 35°. The observer spots the boat again when $m\angle CLX = 70°$.

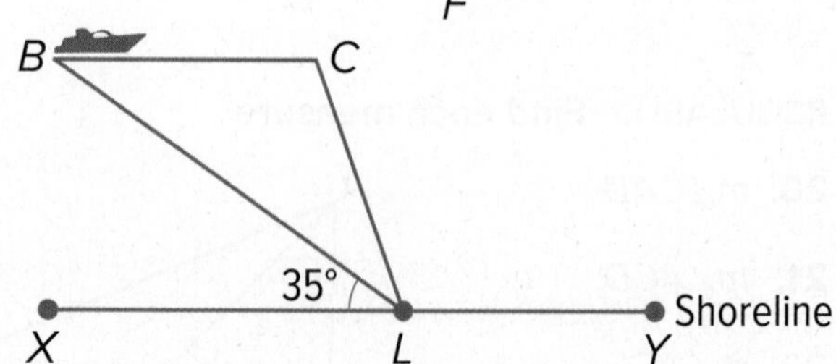

a. Explain how you can prove that $\triangle BCL$ is isosceles.

b. It takes the boat about 15 minutes to travel from point B to point C. When the boat is at point C, what is the distance to the lighthouse?

Triangles and Coordinate Proof

Learn Position and Label Triangles

Coordinate proofs use figures in the coordinate plane and algebra to prove geometric concepts. The first step in a coordinate proof is placing the figure on the coordinate plane.

Key Concept • Placing Triangles on the Coordinate Plane

1. Use the origin as a vertex, or the center of the triangle.
2. Place at least one side of the triangle on an axis.
3. Keep the triangle within the first quadrant if possible.
4. Use coordinates that make computations as simple as possible.

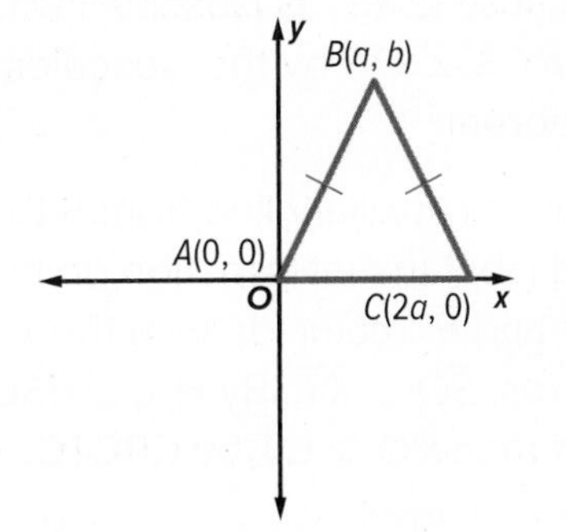

Example 1 Position and Label a Triangle

Position and label right $\triangle ABC$ with legs $\overline{AC}$ and $\overline{AB}$ so $\overline{AC}$ is $2a$ units long and $\overline{AB}$ is $2b$ units long.

Step 1 Position the triangle.

- Position the triangle in the first quadrant.
- Placing the right angle of the triangle, $\angle A$, at the origin will allow the two legs to be along the x- and y-axes.

Step 2 Determine the coordinates.

- Because C is on the y-axis, its x-coordinate is 0. Its y-coordinate is $2a$ because the leg is $2a$ units long.
- Because B is on the x-axis, its y-coordinate is 0. Its x-coordinate is $2b$ because the leg is $2b$ units long.

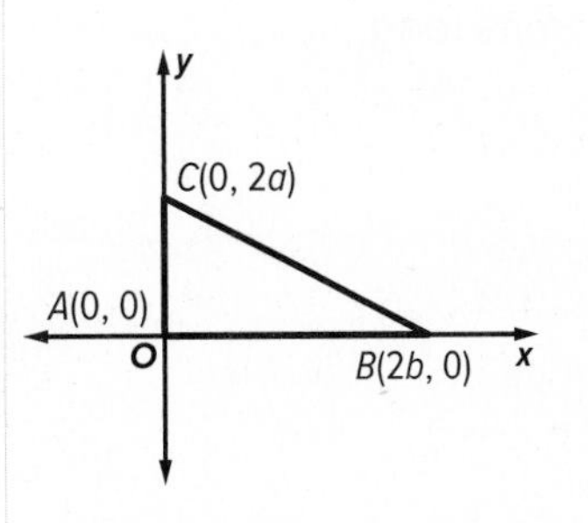

Check

Position and label isosceles triangle JKL on a coordinate plane such that the base $\overline{JL}$ is $2a$ units long, the vertex K is on the y-axis, and the height of the triangle is b units.

Go Online You can complete an Extra Example online.

Today's Goals

- Position a triangle on the coordinate plane and label the vertices.
- Write coordinate proofs to verify properties and to prove theorems about triangles.

Today's Vocabulary

coordinate proofs

Go Online

You can watch a video to see how to place figures on the plane for coordinate proofs.

 Think About It!

The coordinates of two vertices of an equilateral triangle are (0, 0) and ($2a$, 0). The height of the triangle is b units. The coordinates of the third vertex are in terms of a and b. What are the coordinates of the third vertex?

Example 2 Identify Missing Coordinates

Name the missing coordinates of isosceles $\triangle RST$.

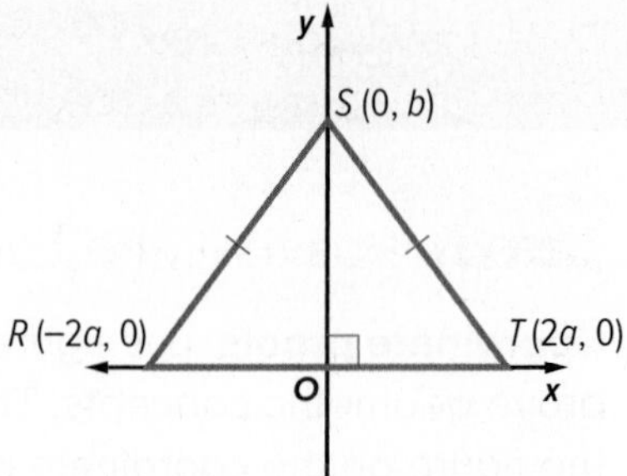

Step 1 Find y-coordinates of R and T.

The base of the triangle is positioned on the x-axis. So, the y-coordinate of R is 0, and the y-coordinate of T is 0.

Step 2 Use the properties of $\triangle RST$.

Because $\triangle RST$ is isosceles with $\overline{RS} \cong \overline{TS}$, $\angle SRT \cong \angle STR$ by the Isosceles Triangle Theorem.

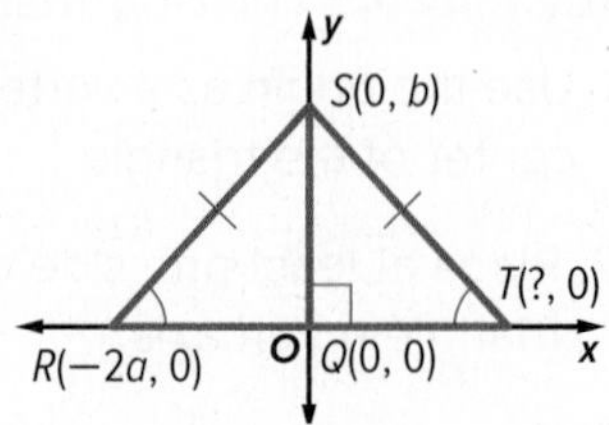

Draw an auxiliary line from S to the origin and label the intersection point Q. Because $\overline{SQ}$ and $\overline{RT}$ coincide with the x- and y-axes, $\overline{SQ} \perp \overline{RT}$. By HA, $\triangle RSQ \cong \triangle TSQ$, and then $\overline{RQ} \cong QT$ by CPCTC.

Step 3 Find the x-coordinate of T.

So, because the x-coordinate of R is $-2a$, the x-coordinate of T must be $2a$.

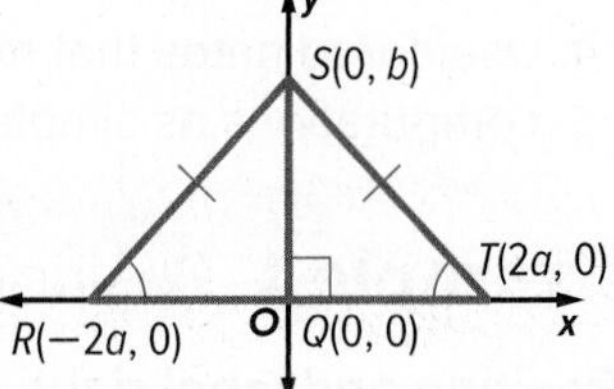

> **Study Tip**
>
> **Isosceles Triangles** You may want to place isosceles and equilateral triangles on the coordinate plane so that the uppermost vertex lies on the y-axis. Then you can use the properties of isosceles triangles to label the coordinates of the vertices, making the computations in coordinate proofs easier.

Check

Name the missing coordinates of isosceles right triangle ABC with $\overline{BC}$ a units long.

Explore Triangles and Coordinate Proofs

Online Activity Use the guiding exercises to complete the Explore.

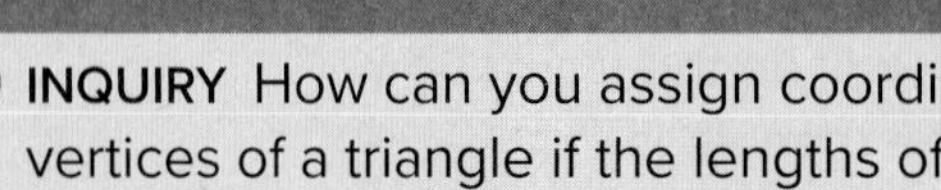

INQUIRY How can you assign coordinates to vertices of a triangle if the lengths of the sides are unknown?

Go Online You can complete an Extra Example online.

Learn Triangles and Coordinate Proof

Coordinate proofs use figures on the coordinate plane to prove geometric concepts and theorems.

Key Concept • Writing a Coordinate Proof
Step 1 Place the figure on the coordinate plane.
Step 2 Label the coordinates of the vertices of the figure.
Step 3 Use algebra to prove properties or theorems.

Study Tip

Coordinate Proofs These guidelines apply to all polygons, not just triangles.

Example 3 Write a Coordinate Proof

Write a coordinate proof to show that $\triangle FGH \cong \triangle FDC$.

Use the Distance Formula to find the length of each side of each triangle. If the sides of the triangles are congruent, then the triangles are congruent by SSS.

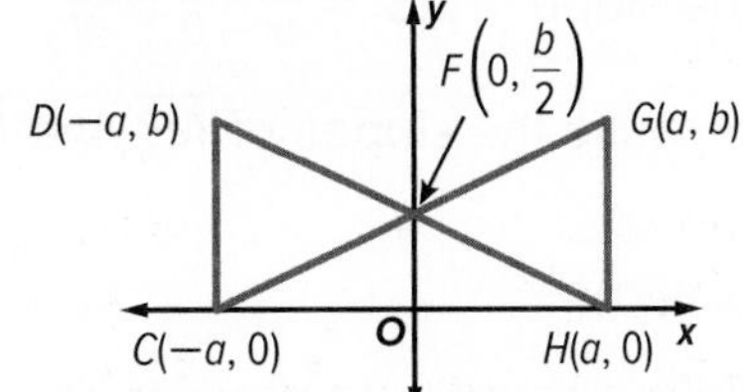

$DC = \sqrt{[-a - (-a)]^2 + (b - 0)^2}$ or b

$GH = \sqrt{(a - a)^2 + (b - 0)^2}$ or b

Because $DC = GH$, $\overline{DC} \cong \overline{GH}$ by the definition of congruence.

$DF = \sqrt{[0 - (-a)]^2 + \left(\frac{b}{2} - b\right)^2}$ or $\sqrt{a^2 + \frac{b^2}{4}}$

$GF = \sqrt{(a - 0)^2 + \left(b - \frac{b}{2}\right)^2}$ or $\sqrt{a^2 + \frac{b^2}{4}}$

$CF = \sqrt{[0 - (-a)]^2 + \left(\frac{b}{2} - 0\right)^2}$ or $\sqrt{a^2 + \frac{b^2}{4}}$

$HF = \sqrt{(a - 0)^2 + \left(0 - \frac{b}{2}\right)^2}$ or $\sqrt{a^2 + \frac{b^2}{4}}$

Because $DF = GF = CF = HF$, $\overline{DF} \cong \overline{GF} \cong \overline{CF} \cong \overline{HF}$, $\triangle FGH \cong \triangle FDC$ by SSS.

Go Online You can complete an Extra Example online.

Check

Write a coordinate proof to show that $\triangle ABX \cong \triangle CDX$.

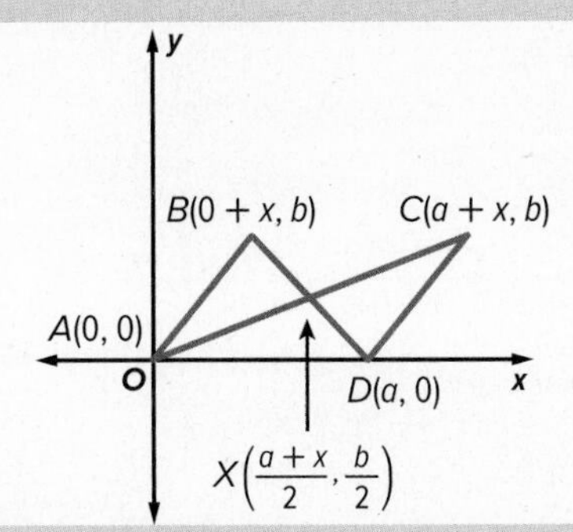

Proof:

The ___?___ of $\overline{AC}$ is $\left(\frac{0 + a + x}{2}, \frac{0 + b}{2}\right)$, or $\left(\frac{a + x}{2}, \frac{b}{2}\right)$. The midpoint of $\overline{BD}$ is $\left(\frac{0 + x + a}{2}, \frac{b + 0}{2}\right)$ or $\left(\frac{a + x}{2}, \frac{b}{2}\right)$. Because x is located at $\left(\frac{a + x}{2}, \frac{b}{2}\right)$, it is the midpoint of $\overline{AC}$ and $\overline{BD}$. By the definition of a segment bisector, $\overline{AC}$ bisects $\overline{BD}$ and $\overline{BD}$ bisects $\overline{AC}$.

Therefore, $\overline{BX} \cong \overline{XD}$ and $\overline{AX} \cong \overline{XC}$. From the ___?___, $CD = \sqrt{[(a + x) - a]^2 + (b - 0)^2}$ or $\sqrt{x^2 + b^2}$, and $AB = \sqrt{[(0 + x) - 0]^2 + (b - 0)^2}$ or $\sqrt{x^2 + b^2}$. Therefore, $\overline{CD} \cong \overline{AB}$ by the definition of ___?___, and $\triangle ABX \cong \triangle CDX$ by SSS.

Example 4 Prove a Theorem by Using Coordinate Geometry

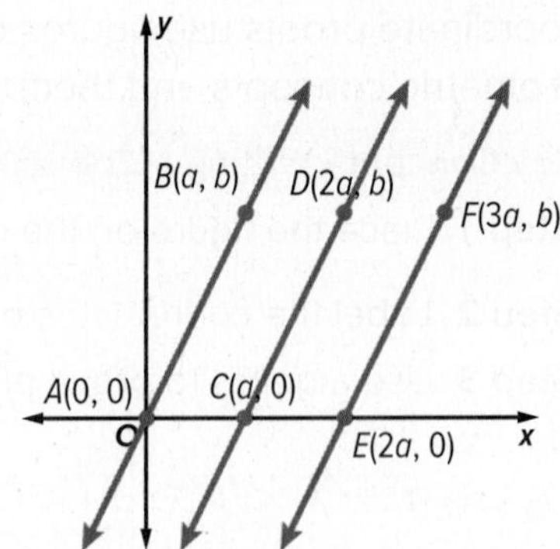

Write a coordinate proof to show that if two lines are each equidistant from a third line, then the two lines are parallel to each other.

Given: $\overleftrightarrow{AB}$ **and** $\overleftrightarrow{EF}$ **are equidistant from** $\overleftrightarrow{CD}$**.**

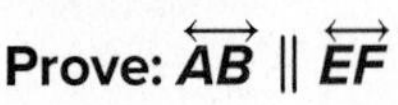

Prove: $\overleftrightarrow{AB} \parallel \overleftrightarrow{EF}$

Proof:

The slope of $\overleftrightarrow{AB} = \frac{b-0}{a-0} = \frac{b}{a}$. The slope of $\overleftrightarrow{EF} = \frac{b-0}{3a-2a} = \frac{b}{a}$.

Because the slopes of $\overleftrightarrow{AB}$ and $\overleftrightarrow{EF}$ are the same, $\overleftrightarrow{AB} \parallel \overleftrightarrow{EF}$.

Check

Write a coordinate proof to show that the three segments joining the midpoints of the sides of an isosceles triangle form another isosceles triangle.

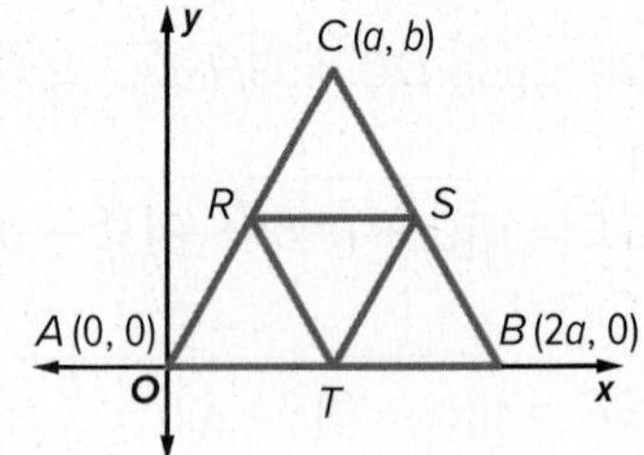

Given: Isosceles triangle ABC; $\overline{BC} \cong \overline{AC}$; R, S, and T are midpoints of their respective sides.

Prove: $\triangle RST$ is isosceles.

Go Online You can complete an Extra Example online.

Example 5 Classify a Triangle

NAVIGATION **The Polynesian Triangle is a triangle formed between the three Pacific island groups that form the South Pacific region known as Polynesia. The approximate coordinates in latitude and longitude of each vertex are Auckland, New Zealand (−40.9, 174.9), Honolulu, Hawaii (21.3, −157.9), and Easter Island (−27.1, −109.4).**

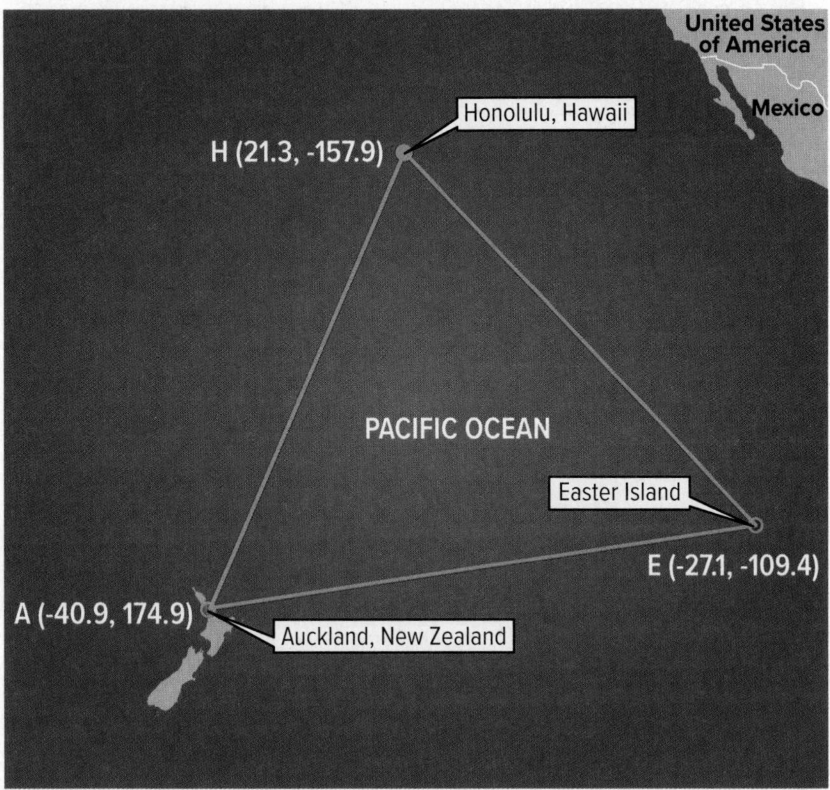

Study Tip

Units of Measure While the distance between cities is usually measured in miles or kilometers, latitude and longitude are measured in degrees relative to the Prime Meridian and the Equator.

Part A Estimate the type of triangle formed by the Polynesian islands.

The triangle appears to be a(n) acute scalene triangle.

Part B Use coordinate geometry to determine the type of triangle formed.

Use the Distance Formula to determine the length of each side of the triangle.

Round to the nearest tenth.

$$AE = \sqrt{[-40.9 - (-27.1)]^2 + [174.9 - (-109.4)]^2}$$
$$\approx 284.6$$

$$EH = \sqrt{(-27.1 - 21.3)^2 + [-109.4 - (-157.9)]^2}$$
$$\approx 68.5$$

$$AH = \sqrt{(-40.9 - 21.3)^2 + [174.9 - (-157.9)]^2}$$
$$\approx 338.6$$

Because the length of each side is different, the triangle is scalene.

Go Online You can complete an Extra Example online.

Check

GEOGRAPHY Eldora's family lives in New Mexico. She lives southwest of Rio Rancho, her uncle lives in Clines Corners, and her grandparents live in Rociada. Eldora has assigned coordinates to each location.

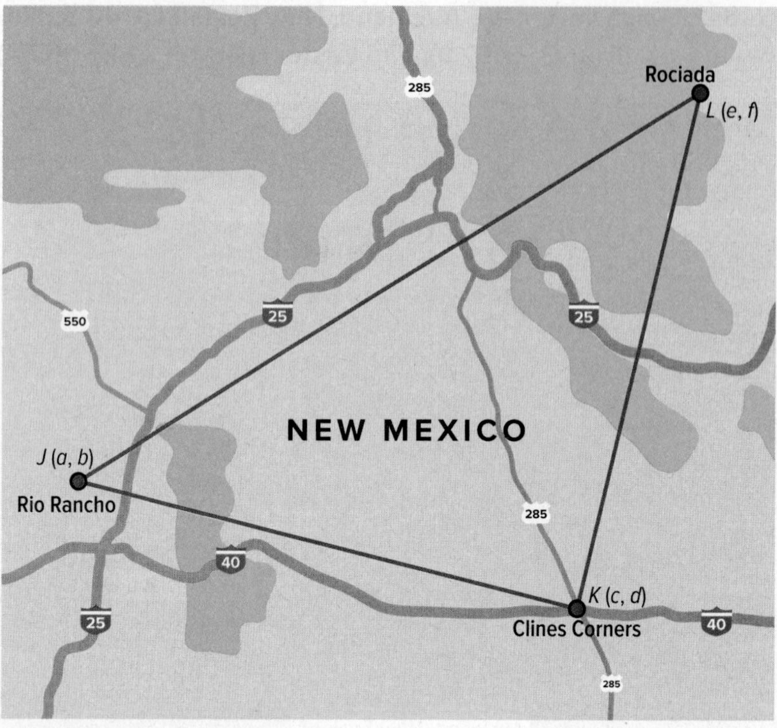

Part A Estimate the type of triangle formed.

A. acute scalene
B. obtuse scalene
C. right scalene
D. right isosceles
E. equilateral

Part B Which of the following can be used in a coordinate proof to show that the estimate chosen above is correct?

A. Use the Distance Formula to find the lengths of $\overline{JL}$, $\overline{JK}$, and $\overline{KL}$. If they are all equal, then the triangle is equilateral.

B. Use the Distance Formula to find the lengths of $\overline{JL}$, $\overline{JK}$, and $\overline{KL}$. If they are different, then the triangle is scalene.

C. Compare the slopes of $\overline{JK}$ and $\overline{KL}$. If the product of the slopes is -1, then the lines are perpendicular. Use the Distance Formula to find the lengths of $\overline{JK}$ and $\overline{KL}$. If the lengths are equal, then the triangle is a right isosceles triangle.

D. Compare the slopes of $\overline{JK}$ and $\overline{KL}$. If the product of the slopes is -1, then the lines are perpendicular. Use the Distance Formula to find the lengths of $\overline{JL}$, $\overline{JK}$, and $\overline{KL}$. If the lengths are different, then the triangle is a right scalene triangle.

Practice

Go Online You can complete your homework online.

Example 1

REGULARITY Position and label each triangle on the coordinate plane.

1. isosceles $\triangle ABC$ with base $\overline{AB}$ that is a units long and height that is b units

2. right $\triangle XYZ$ with hypotenuse $\overline{YZ}$, leg $\overline{XY}$ that is b units long, and leg $\overline{XZ}$ that is three times the length of $\overline{XY}$

3. isosceles right $\triangle RST$ with hypotenuse $\overline{RS}$ and legs $3a$ units long

4. right $\triangle JKL$ with legs $\overline{JK}$ and $\overline{KL}$ such that $\overline{JK}$ is a units long and leg $\overline{KL}$ is $4b$ units long

Example 2

Name the missing coordinate(s) of each triangle.

5.

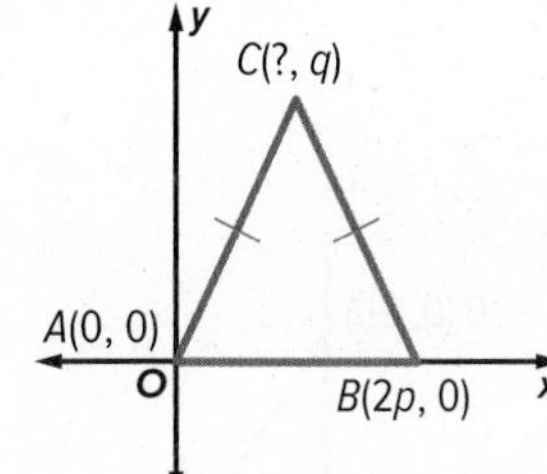

6.

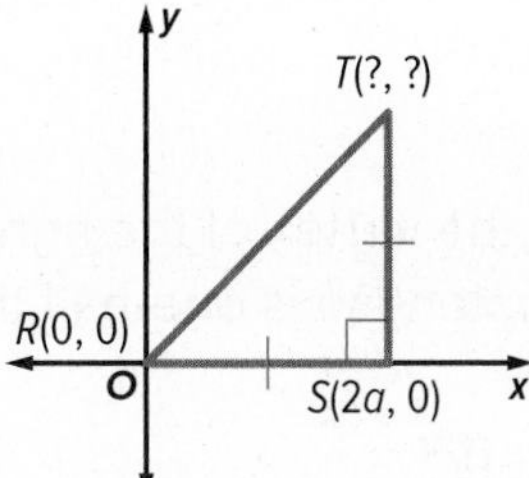

7.

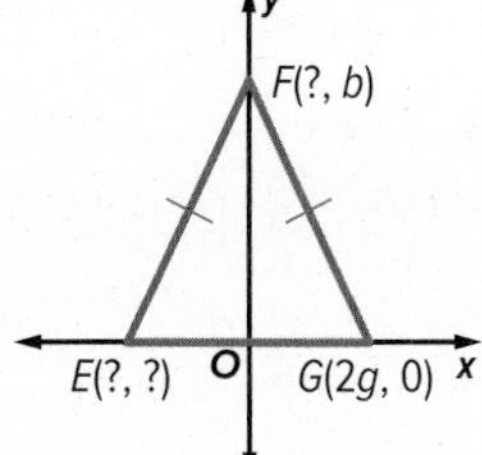

8.

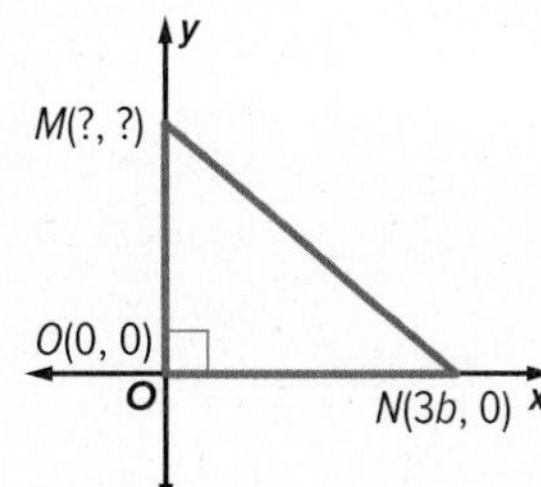

Examples 3 and 4

PROOF For Exercises 9–13, write a coordinate proof for each statement.

9. The segments joining the midpoints of the sides of a right triangle form a right triangle.

   Given: Point R is the midpoint of $\overline{AB}$.
   Point P is the midpoint of $\overline{BC}$.
   Point Q is the midpoint of $\overline{AC}$.

   Prove: $\triangle RPQ$ is a right triangle.

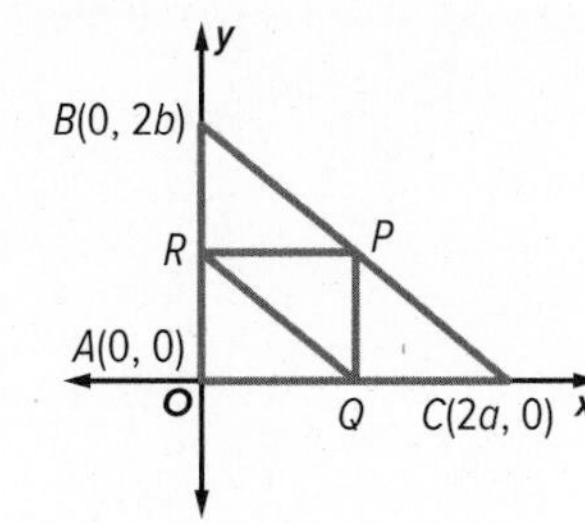

10. A segment from the vertex angle of an isosceles triangle to the midpoint of the base is perpendicular to the base.

Given: Isosceles $\triangle RST$; U is the midpoint of base $\overline{RT}$.

Prove: $\overline{SU} \perp \overline{RT}$

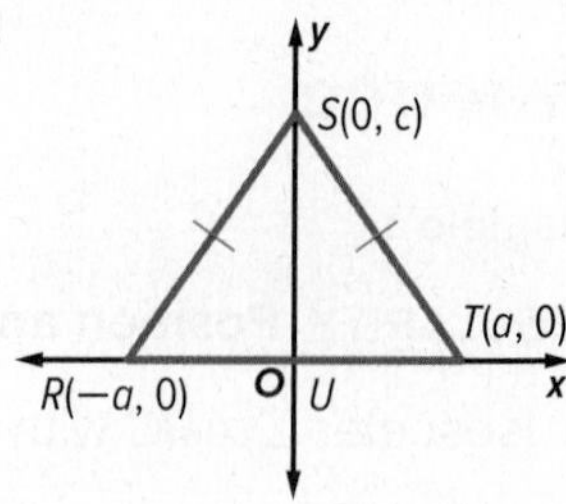

11. In an isosceles right triangle, the segment from the vertex of the right angle to the midpoint of the hypotenuse is perpendicular to the hypotenuse.

Given: isosceles right $\triangle ABC$ with right angle $\angle ABC$; M is the midpoint of $\overline{AC}$.

Prove: $\overline{BM} \perp \overline{AC}$

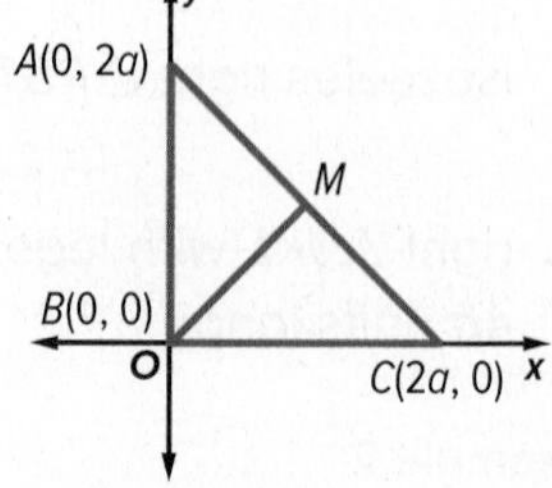

12. The measure of the segment that joins the vertex of the right angle in a right triangle to the midpoint of the hypotenuse is one-half the measure of the hypotenuse.

Given: right $\triangle ABC$; P is the midpoint of $\overline{BC}$.

Prove: $AP = \frac{1}{2}BC$

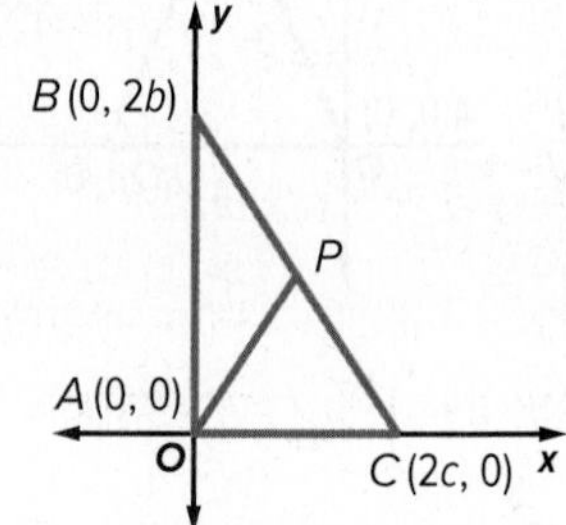

13. If a line segment joins the midpoints of two sides of a triangle, then its length is equal to one-half the length of the third side.

Given: S is the midpoint of $\overline{AC}$.

T is the midpoint of $\overline{BC}$.

Prove: $ST = \frac{1}{2}AB$

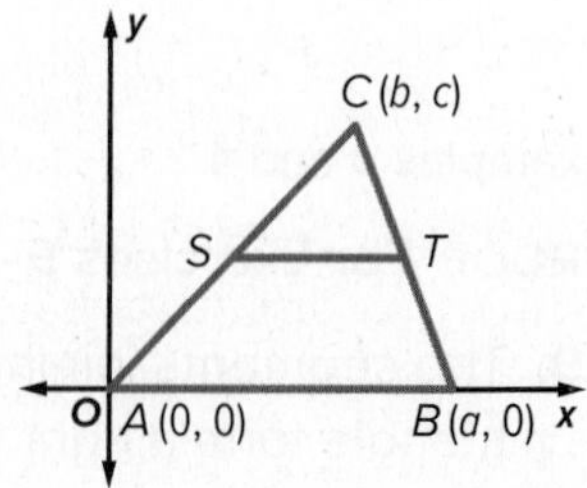

Example 5

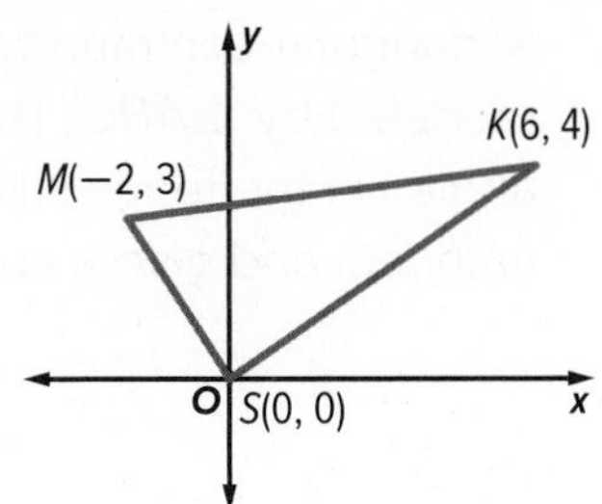

14. NEIGHBORHOODS Kalini lives 6 miles east and 4 miles north of her high school. After school, she works part time at the mall in a music store. The mall is 2 miles west and 3 miles north of the school. Use coordinate geometry to determine the type of triangle formed by Kalini's high school, her home, and the mall.

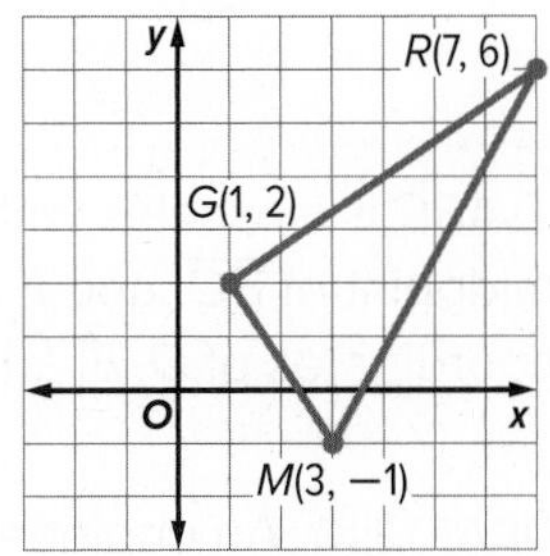

15. COUNTY FAIR The fair committee wants to print a map to distribute to vendors as they arrive to set up their booths at the fairgrounds. On a coordinate grid, the main gate is located at (3, −1), the grandstand is located at (1, 2), and the rides and games are located at (7, 6). Use coordinate geometry to determine the type of triangle formed by these locations.

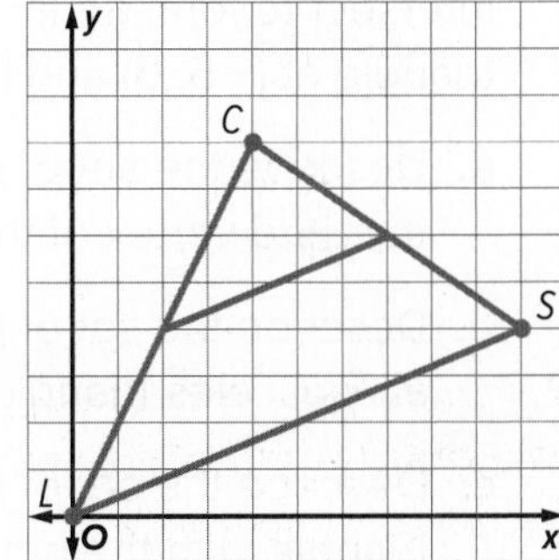

16. USE ESTIMATION A town is preparing for a 5K run. The race will start at city hall C. The course will take runners along straight streets to the library L, to the science museum S, and back to city hall for the finish.

a. Estimate the type of triangle formed by the course.

b. Use coordinate geometry to determine the type of triangle formed.

Mixed Exercises

REASONING **For Exercises 17 and 18, determine whether the triangle can be a right triangle. Explain.**

17. $X(0, 0)$, $Y(2h, 2h)$, $Z(4h, 0)$

18. $X(0, 0)$, $Y(1, h)$, $Z(2h, 0)$

19. SHELVES Martha has a shelf bracket shaped like a right isosceles triangle. She wants to know the length of the hypotenuse relative to the sides. She does not have a ruler but remembers the Distance Formula. She places the bracket in Quadrant I of a coordinate grid with the right angle at the origin. The length of each leg is a. What are the coordinates of the vertices that form the two acute angles?

20. FLAGS A flag is shaped like an isosceles triangle. A designer would like to make a drawing of the flag on a coordinate plane. She positions it so the base of the triangle is on the y-axis with one endpoint located at (0, 0). She locates the tip of the flag at $\left(a, \frac{b}{2}\right)$. What are the coordinates of the third vertex?

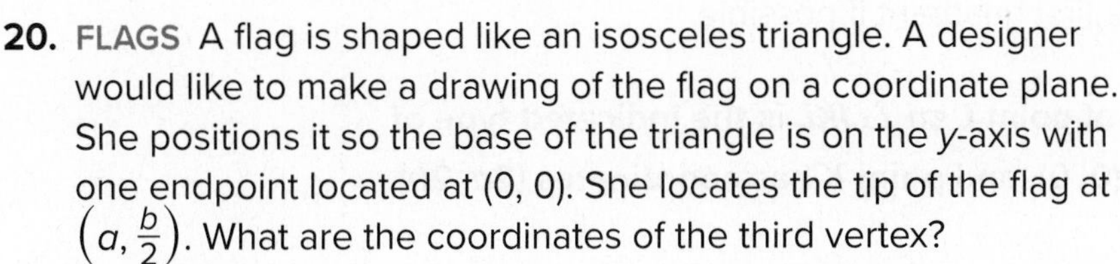

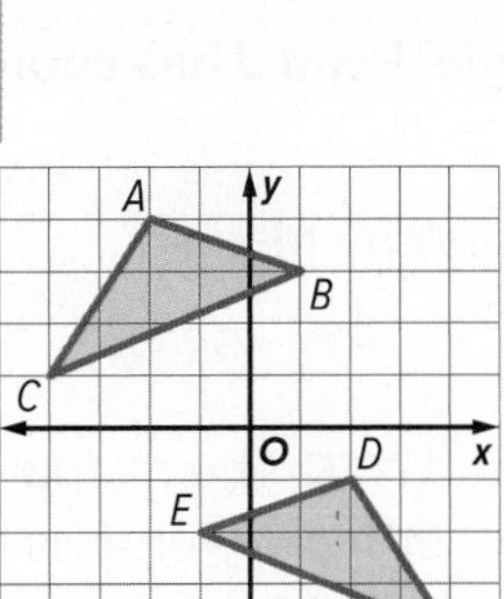

21. DESIGN Andrew is using a coordinate plane to design a quilt. Two of the triangular patches for the quilt are shown in the figure. Andrew wants to be sure that $\angle A$ and $\angle D$ have the same measure. Describe the main steps you can use to prove that $\angle A \cong \angle D$.

22. **COMMUNITY** A landscape architect is using a coordinate plane to design a triangular community garden. The fence that will surround the garden is modeled by $\triangle ABC$. The architect wants to know whether any of the three angles in the fence will be congruent. Determine the answer for the architect and give a coordinate proof to justify your response.

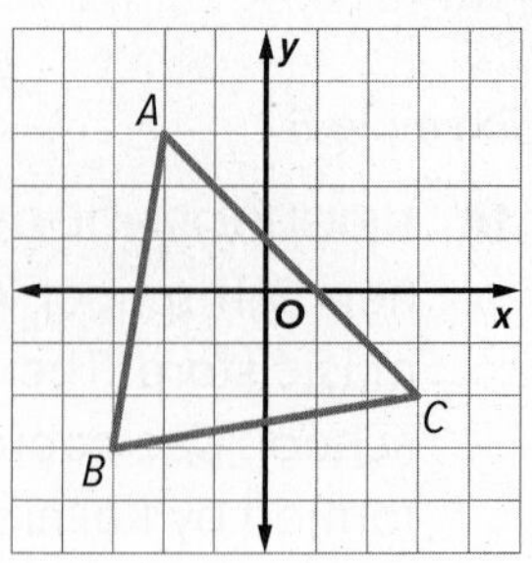

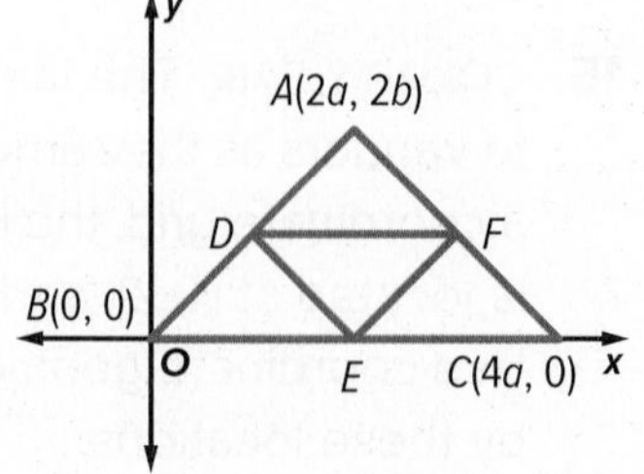

23. $\triangle ABC$ is isosceles with $\overline{AB} \cong \overline{AC}$. D is the midpoint of $\overline{AB}$, E is the midpoint of $\overline{BC}$, and F is the midpoint of $\overline{AC}$. What are the coordinates of D, E, and F?

24. **DRAFTING** An engineer is designing a roadway. Three roads intersect to form a triangle. The engineer marks two vertices of the triangle at $(-5, 0)$ and $(5, 0)$ on a coordinate plane.

a. Describe the set of points in the coordinate plane that could not be used as the third vertex of the triangle.

b. Describe the set of points in the coordinate plane that could be the vertex of an isosceles triangle.

c. Describe the set of points in the coordinate plane that would make a right triangle with the other two points if the right angle is located at $(-5, 0)$.

Higher-Order Thinking Skills

25. **CREATE** Draw an isosceles right triangle on the coordinate plane so the midpoint of its hypotenuse is the origin. Label the coordinates of the vertex.

26. **WRITE** Explain why following each guideline for placing a triangle on the coordinate plane is helpful in proving coordinate proofs.

a. Use the origin as a vertex of the triangle.

b. Place at least one side of the triangle on the x- or y-axis.

c. Keep the triangle within the first quadrant if possible.

PERSEVERE **Find the coordinates of point L so $\triangle JKL$ is the indicated type of triangle. Point J has coordinates $(0, 0)$, and point K has coordinates $(2a, 2b)$.**

27. scalene triangle

28. right triangle

29. isosceles triangle

30. **ANALYZE** The midpoints of the sides of a triangle are located at $(a, 0)$, $(2a, b)$ and (a, b). If one vertex is located at the origin, what are the coordinates of the other vertices? Explain your reasoning.

Review

Essential Question

How can you prove congruence and use congruent figures in real-world situations?

Showing combinations of angles and sides in two triangles congruent to one another results in the potential to show two triangles congruent. These congruent triangles can be used to represent objects used in the construction of buildings or mechanical objects.

Module Summary

Lesson 14-1 through 14-2

Angles and Sides

- The sum of the measures of the interior angles of a triangle is 180°.
- Two figures are congruent if and only if there is a rigid motion or series of rigid motions that maps one figure exactly onto the other
- In two congruent polygons, all the parts of one polygon are congruent to the corresponding parts of the other polygon.

Lesson 14-3 through 14-5

Ways to Prove Triangles Congruent

- Side-Side-Side (SSS) Congruence three sides of one triangle congruent to three sides of a second triangle
- Side-Angle-Side (SAS) Congruence two sides and the included angle of one triangle congruent to two sides and the included angle of a second triangle
- Angle-Side-Angle (ASA) Congruence two angles and the included side of one triangle congruent to two angles and the included side of a second triangle
- Angle-Angle-Side (AAS) Congruence two angles and the nonincluded side of one triangle congruent to two angles and the nonincluded side of a second triangle
- For right triangles, use the following ways to prove congruence.

  Leg-Leg Congruence (LL)

  Hypotenuse-Angle Congruence (HA)

  Leg-Angle Congruence (LA)

  Hypotenuse-Leg Congruence (HL)

Lesson 14-6

Isosceles and Equilateral Triangles

- If two sides of a triangle are congruent, then the angles opposite those sides are congruent.
- If two angles of a triangle are congruent, then the sides opposite those angles are congruent.
- Each angle of an equilateral triangle measures 60°.

Lesson 14-7

Coordinate Proof

- To write a coordinate proof: Place the figure on the coordinate plane. Label the vertices. Use algebra to prove properties or theorems.

Study Organizer

Use your Foldable to review this module. Working with a partner can be helpful. Ask for clarification of concepts as needed.

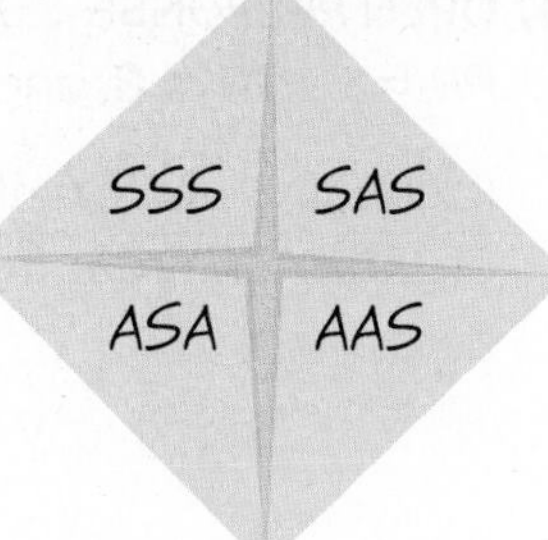

Test Practice

1. OPEN RESPONSE Find the measure of $\angle BCD$ in degrees. (Lesson 14-1)

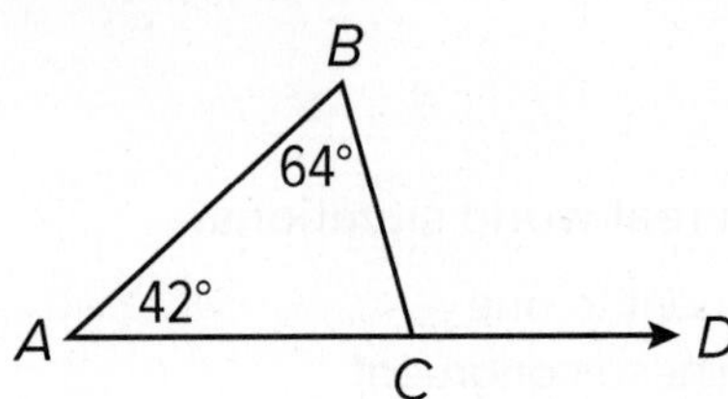

2. MULTIPLE CHOICE Find the value of x given the triangle below. (Lesson 14-1)

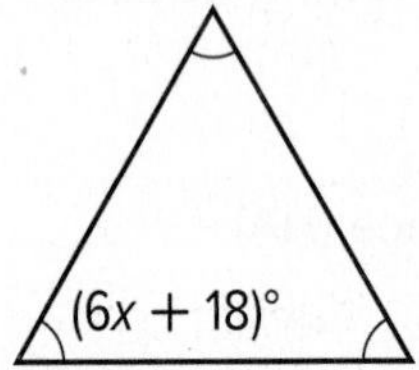

A. 7

B. 12

C. 60

D. 126

3. MULTI-SELECT In $\triangle PQR$, $\angle Q$ is a right angle.

Select all the statements about $\angle P$ and $\angle R$ that must be true. (Lesson 14-1)

A. $\angle P$ and $\angle R$ are complementary.

B. $\angle P$ and $\angle R$ are supplementary.

C. $\angle P$ and $\angle R$ are congruent.

D. $\angle P$ and $\angle R$ are acute.

E. $\angle P$ or $\angle R$ is obtuse.

4. OPEN RESPONSE $\triangle PRQ$ has side lengths $PR = 6$, $QR = 8$, and $PQ = 5$.

If $\triangle PRQ \cong \triangle CBA$, then put the side lengths of $\triangle CBA$ in order from shortest to longest. (Lesson 14-2)

5. MULTI-SELECT Given $\triangle DEF$ and $\triangle JLK$ where $\overline{DE} \cong \overline{JL}$, $\overline{FD} \cong \overline{KJ}$, $\overline{LK} \cong \overline{EF}$, $\angle D \cong \angle J$, $\angle E \cong \angle L$, and $\angle F \cong \angle K$, which of the following conclusions can be made? Select all that apply. (Lesson 14-2)

A. $\triangle DEF$ and $\triangle JLK$ are congruent.

B. $\triangle DEF$ and $\triangle JLK$ are not congruent.

C. A series of rigid motions will map $\triangle DEF$ onto $\triangle JLK$.

D. A series of rigid motions will not map $\triangle DEF$ onto $\triangle JLK$.

E. $\triangle FDE$ and $\triangle KJL$ are congruent.

F. $\triangle FDE$ and $\triangle KJL$ are not congruent.

6. MULTIPLE CHOICE Which postulate shows $\triangle ABC \cong \triangle DEF$? (Lesson 14-3)

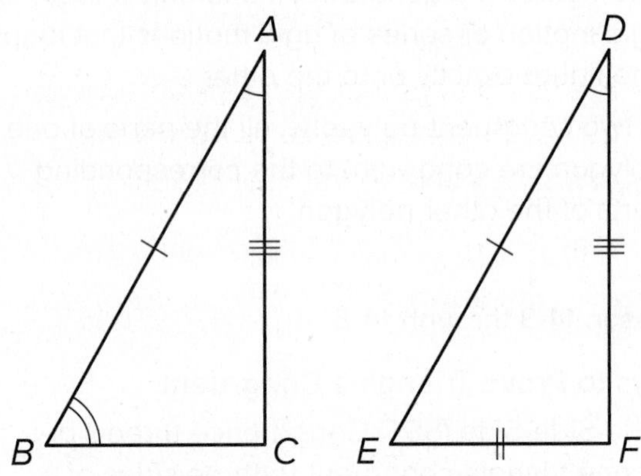

A. Side-Side-Side

B. Angle-Side-Angle

C. Side-Angle-Side

D. Angle-Angle-Side

7. **MULTIPLE CHOICE** In $\triangle JKL$ and $\triangle PQR$, $\overline{JK} \cong \overline{PQ}$ and $\overline{JL} \cong \overline{PR}$. Which additional statement would prove that $\triangle JKL \cong \triangle PQR$? (Lesson 14-3)

A. $\angle J \cong \angle P$

B. $\angle L \cong \angle R$

C. $\overline{JK} \cong \overline{PR}$

D. $\overline{JL} \cong \overline{KL}$

8. **OPEN RESPONSE** Stephanie and Fernando are building triangular prism birdhouses that have the same dimensions.

- Stephanie says that they should measure the lengths of two pairs of corresponding sides of the triangular bases and use a protractor to measure the included angles to be sure the bases are congruent.
- Fernando says that they can be sure the triangular bases are congruent if they measure the lengths of all three corresponding sides.

Which student is correct? (Lesson 14-3)

9. **MULTI-SELECT** In $\triangle ABC$ and $\triangle MNP$, $\angle A \cong \angle M$ and $\overline{BC} \cong \overline{NP}$. What additional piece(s) of information could be used to prove $\triangle ABC \cong \triangle MNP$ by AAS? Select all that apply. (Lesson 14-4)

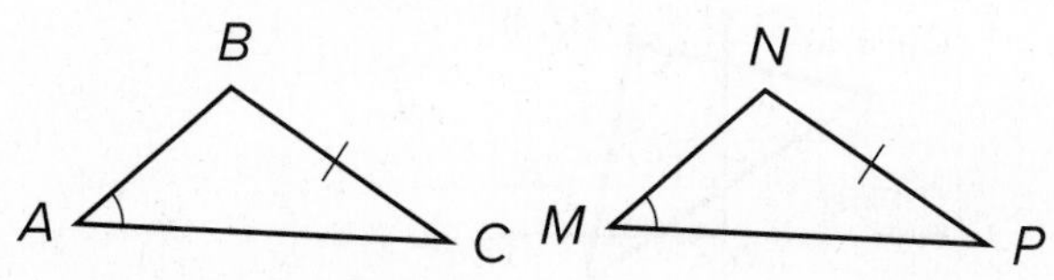

A. $\angle B \cong \angle N$

B. $\angle C \cong \angle P$

C. $\overline{AB} \cong \overline{MN}$

D. $\overline{AC} \cong \overline{MP}$

E. $\angle A \cong \angle N$

10. **OPEN RESPONSE** A technician is assembling parts for a radio antenna. He attaches two metal bars to 3-foot-long crosspieces so a triangle is formed, with each bar meeting the crosspiece at a 40° angle. Which postulate proves that all triangles formed this way are congruent? (Lesson 14-4)

11. **MULTIPLE CHOICE** In $\triangle RST$, $m\angle R = 85°$, $m\angle S = 33°$, and $RT = 17$.

Which set of measurements would make $\triangle RST \cong \triangle MNP$ by the AAS Theorem? (Lesson 14-4)

A. $m\angle M = 85°$, $m\angle N = 33°$, and $MP = 17$

B. $m\angle M = 85°$, $m\angle N = 33°$, and $MN = 17$

C. $m\angle M = 33°$, $m\angle N = 85°$, and $MP = 17$

D. $m\angle M = 33°$, $m\angle N = 85°$, and $MN = 17$

12. **MULTI-SELECT** Select all the pairs of triangles that must be congruent to each other. (Lesson 14-5)

A.

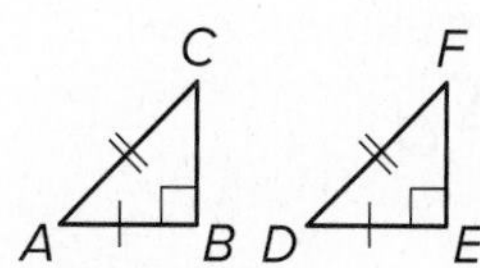

B.

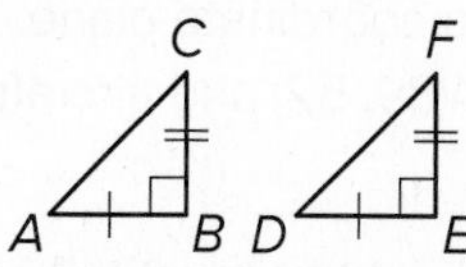

C.

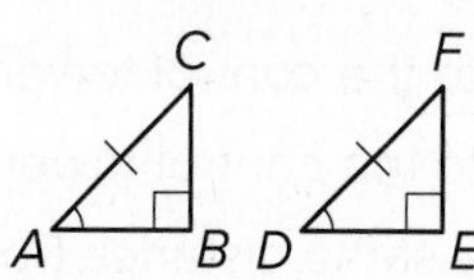

D.

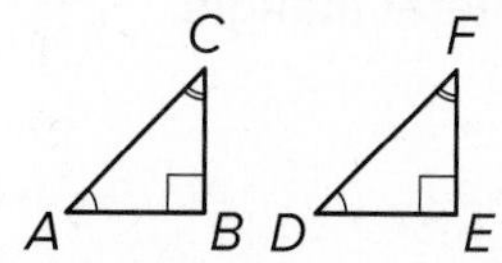

13. OPEN RESPONSE If the vertex angle of an isosceles triangle measures 86°, what is the angle measure in degrees of one of the base angles? (Lesson 14-6)

14. MULTIPLE CHOICE Find the value of x. (Lesson 14-6)

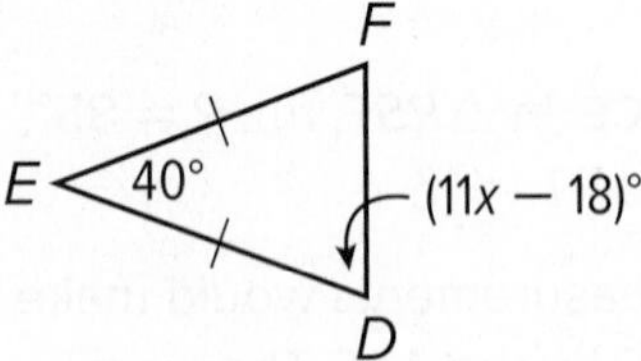

A. 5.3

B. 8

C. 14.4

D. 70

15. OPEN RESPONSE What is the length of segment QR? (Lesson 14-6)

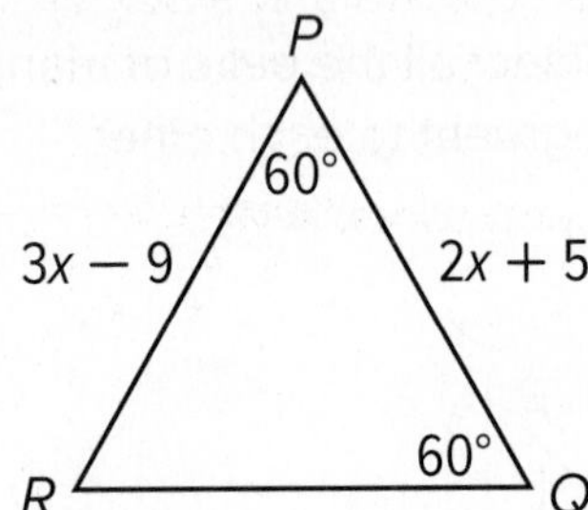

16. MULTIPLE CHOICE An air traffic control tower is located at $O(0, 0)$ on a coordinate plane. Aircraft A is located at $A(39, 52)$ and aircraft B is located at $B(25, 60)$.

What statement is true about this situation? (Lesson 14-7)

A. Aircraft A is closer to the control tower.

B. Aircraft B is closer to the control tower.

C. Both aircraft are the same distance from the control tower.

D. $\triangle OAB$ is an equilateral triangle.

17. MULTIPLE CHOICE A triangle drawn on a coordinate plane has vertices $A(0, 0)$, $B(0, 2b)$, and $C(2c, 0)$. Which expression represents the slope of $\overline{BC}$? (Lesson 14-7)

A. $-\frac{b}{c}$

B. $\frac{b}{c}$

C. $\frac{c}{b}$

D. $-\frac{c}{b}$

18. MULTIPLE CHOICE The given triangle will be used in a coordinate proof.

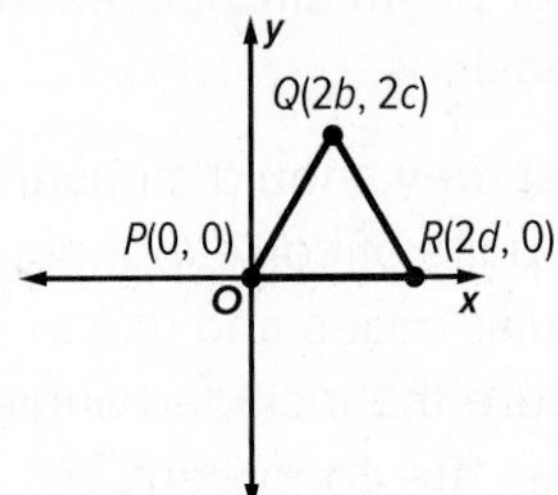

What are the coordinates of the midpoint of $\overline{QR}$? (Lesson 14-7)

A. $(b - d, 0)$

B. $(b + d, 0)$

C. $(b - d, c)$

D. $(b + d, c)$

19. MULTI-SELECT Use coordinate geometry to determine the type of triangle formed below. Select all that apply. (Lesson 14-7)

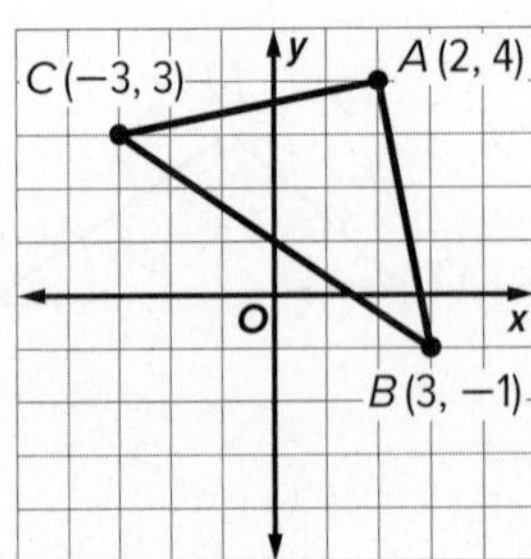

A. equilateral

B. isosceles

C. right

D. scalene

E. acute

Module 1

Quick Check

1. $\frac{2}{3}$ **3.** $\frac{1}{13}$ **5.** 4 **7.** $\frac{3}{2}$ or $1\frac{1}{2}$

Lesson 1-1

1. $2 + 12 \div 4$ **3.** $3 \times 11 + 7$ **5.** $6 - 3 - 1$ **7.** $24 \div 6 + 7$ **9.** $116 \div 4 + 28 - 33$ **11.** $\frac{3+7}{2}$ **13.** $(4 + 9) \times 3$ **15.** $\frac{10}{4 \times 5}$ **17.** $\frac{1+2}{20}$ **19.** $\frac{12+16}{3+4}$ **21.** $\frac{36+14}{2 \times 5}$ **23.** $\frac{6+15}{13-9}$ **25.** 365×85 **27.** $2 \times 35 - 5$ **29.** 12×7 **31.** 2744 **33.** 11 **35.** 42 **37.** 7 **39.** 56 **41.** 22 **43.** 8 **45.** 324 **47.** 29 **49a.** $20 \times 5 + 9$ **49b.** 109 flies **51a.** $2 \times \frac{1}{2}(30 + 50)24$ **51b.** 1920 in^2 **53.** $8^4 + 6$ **55.** Tamara; when evaluating, first perform the multiplication and division from left to right, and then the addition and subtraction from left to right. **57.** The cashier; Kelly should have entered the expression into her calculator as $3(18.95 - 2) + 2(11.50)$. **59.** $10 \times 18 + 8 \times 18$; You can also find the length of each side of the apartment, 18 and $10 + 8$, and then multiply: $18(10 + 8)$. **61.** The student should have added/subtracted from left to right. The correct value is 30.

Lesson 1-2

1. four times a number q **3.** 15 plus r **5.** 3 times x squared **7.** two times a plus six **9.** twenty-five plus six times a number squared **11.** three times a number raised to the fifth power divided by two **13.** 5 times g to the sixth power **15.** four minus five times h **17.** 1 less than 7 times x cubed **19.** 3 times n squared minus x **21.** 18 times the quantity p plus 5 **23.** $n - 35$ **25.** $\frac{1}{3}n$ **27.** $\frac{45}{r}$ **29.** $18 - 3d$ **31.** $\frac{20}{t^5}$ **33.** $(k + 2) - 15$ **35.** $2m + 6$ **37.** $\frac{6136}{y}$ **39.** $19.95 \times t - 10$ or $19.95t - 10$ **41.** $5f + 45h$ **43.** 1 **45.** 7 **47.** 149 **49.** $\frac{65}{4}$ **51.** 44 **53.** $\frac{1}{2}$ **55.** 18 **57.** 10 **59.** 16 **61.** 13 **63a.** $5t - 100$ **63b.** 1400 students **65a.** $1.75 + 3.45m$ **65b.** $29.35 **67a.** Sample answer: the quotient of $5x$ and 2 plus y cubed; $5x$ divided by 2 plus y to the third power **67b.** 18 **69.** 89 **71.** 52 **73.** Sample answer: the quotient of x minus 1 and 2; $\frac{x-1}{2}$ **75a.** $x - (36 \times 4)$ **75b.** $\frac{x - (36 \times 4)}{0.20}$ **75c.** 350 mi **77.** Sample answer: An algebraic expression is a math phrase that contains one or more numbers or variables. To write an algebraic expression from a real-world situation, first assign variables. Then determine arithmetic operations done on the variables. Finally, put the terms in order. **79.** Sample answer: Movie tickets cost $10 and a box of popcorn cost $5.25. You buy t movie tickets and a box of popcorn. What is the greatest number of movies tickets you can purchase with $50?

Lesson 1-3

1. Symmetric Property of Equality **3.** Symmetric Property of Equality **5.** 14 **7.** 34 **9a.** Exit 15 to Exit 8 **9b.** Symmetric Property of Equality

11. $= (3 \div 2)\frac{2}{3}$	Multiplicative Identify
$= \frac{3}{2} \cdot \frac{2}{3}$	Substitution
$= 1$	Multiplicative Identify
13. $= 2(5 - 5)$	Substitution
$= 2(0)$	Substitution
$= 0$	Multiplicative Property of Zero
15. $= 2(2 - 1) \cdot \frac{1}{2}$	Substitution
$= 2(1) \cdot \frac{1}{2}$	Substitution
$= 2 \cdot \frac{1}{2}$	Multiplicative Identity
$= 1$	Multiplicative Inverse

17. $= 4 + \frac{4}{9} + 7 + \frac{2}{9}$ Substitution
$= 4 + 7 + \frac{4}{9} + \frac{2}{9}$ Commutative (+)
$= 4 + 7 + \left(\frac{4}{9} + \frac{2}{9}\right)$ Associative (+)
$= 11 + \frac{6}{9}$ Substitution
$= 11\frac{6}{9}$ Substitution
$= 11\frac{2}{3}$ Substitution

19. $= (2 \cdot 8) \cdot (10 \cdot 2)$ Associative (×)
$= 16 \cdot 20$ Substitution
$= 320$ Substitution

21. $= \left(2\frac{3}{4} \cdot 1\frac{1}{8}\right) \cdot 32$ Associative (×)
$= \left(\frac{11}{4} \cdot \frac{9}{8}\right) \cdot 32$ Substitution
$= \frac{99}{32} \cdot 32$ Substitution
$= 99$ Substitution

23. $= 2 \cdot 5 \cdot 4 \cdot 3$ Commutative (×)
$= (2 \cdot 5) \cdot (4 \cdot 3)$ Associative (×)
$= 10 \cdot 12$ or 120 Substitution

25. $= \frac{4}{3} \cdot 3 \cdot 7 \cdot 10$ Commutative (×)
$= \left(\frac{4}{3} \cdot 3\right) \cdot (7 \cdot 10)$ Associative (×)
$= 4 \cdot 70$ or 280 Substitution

27. −64 **29.** −5 **31.** −9 **33.** Sample answer: Multiplicative Identity and Multiplicative Inverse **35.** 0; Additive Identity **37.** 1; Multiplicative Identity **39.** 5; Additive Identity **41.** 1; Multiplicative Inverse **43.** 3; Reflexive Property **45.** Yes; the Commutative and Associative Properties of Multiplication allow it to be rewritten. **47.** Sample answer: $126 + 28 + 52 = 126 + (28 + 52) = 126 + 80 = 206$ **49.** Sample answer: $5 = 3 + 2$ and $3 + 2 = 4 + 1$, so $5 = 4 + 1$; $5 + 7 = 8 + 4$ and $8 + 4 = 12$, so $5 + 7 = 12$.
51. $4 \div 8 \neq 8 \div 4$ because $4 \div 8 = \frac{1}{2}$ and $8 \div 4 = 2$, so there is no Commutative Property for division. $16 \div (8 \div 4) \neq (16 \div 8) \div 4$ because $16 \div (8 \div 4) = 16 \div 2 = 8$ and $(16 \div 8) \div 4 = 2 \div 4 = \frac{1}{2}$, so there is no Associative Property for division. As long as neither number is 0, when the order of division of two numbers is switched, the results are multiplicative inverses of each other.
53a. False; sample answer: $3 - 4 = -1$, which is not a whole number. **53b.** True **53c.** False; sample answer: $2 \div 3 = \frac{2}{3}$, which is not a whole number. **55.** $(2j)k = 2(jk)$; The other three equations illustrate the Commutative Property of Addition or Multiplication. This equation represents the Associative Property of Multiplication.

Lesson 1-4

1. 4(6) + 5(6); 54 **3.** 6(6) − 6(1); 30 **5.** 14(8) − 14(5); 42 **7a.** 39(23 + 2) **7b.** $975
9a. $10\left(3\frac{3}{5}\right)$ **9b.** $10\left(3\frac{3}{5}\right) = 10\left(3 + \frac{3}{5}\right) = 10(3) + 10\left(\frac{3}{5}\right) = 30 + 6 = 36$ yards of fabric
11. 7(500 − 3); 3479 **13.** $36\left(3 + \frac{1}{4}\right)$; 117
15. 5(90 − 1); 445 **17.** 15(100 + 4); 1560
19. 12(100 − 2); 1176 **21.** 3(10 + 0.2); 30.6
23. $2(x) + 2(4)$; $2x + 8$ **25.** $4(8) + (-3m)(8)$; $32 - 24m$ **27.** $2(17) + (-4n)(17)$; $34 - 68n$
29. $\frac{1}{3}(27) + (-2b)(27)$; $9 - 54b$ **31.** $6(2c) + 6(-cd^2) + 6(d)$; $12c - 6cd^2 + 6d$ **33.** $3(m) + 3(n)$; $3m + 3n$ **35.** $\frac{1}{2}(14) + (6a)(14)$; $7 + 84a$
37. $0.3(9) + (-6x)(9)$; $2.7 - 54x$ **39.** $18r$
41. $2m + 7$ **43.** $13m + 5p$ **45.** $14m + 11g$
47. $12k^3 + 12k$ **49.** $18g$ **51.** $5a^2$ **53.** $2q^2 + q$
55a. $3a + 5(a - b)$
55b. $3a + 5(a - b) = 3a + 5a - 5b$ Distributive Property
$= (3a + 5a) - 5b$ Associative (+)
$= 8a - 5b$ Substitution

57. $24x + 28$ **59.** $18d + 20$ **61.** $7y^3 + y^4$
63. $4b$ **65.** $20x + 37y$ **67.** $2n + 2m$ and $2(n + m)$ **69.** No; sample answer: 10 pounds 5 ounces is 10(16) + 5 = 165 ounces, but Ariana used the Distributive Property incorrectly. She should have written 8(20 + 2) = 8(20) + 8(2) = 160 + 16 = 176 ounces. **71.** Sample answer: Algebraic expressions are helpful because they are easier to interpret and apply than verbal expressions. They can also be written in a more simplified form.

Lesson 1-5

1. $|p - t|$ and $|t - p|$ **3.** $|r - w|$ and $|w - r|$
5. 15 **7.** 22 **9.** 37 **11.** 32 **13.** −62 **15.** 11
17. 10 **19.** 5 **21.** 5 **23.** 6 **25.** −7.4 **27.** 8.4
29. −15 **31.** 22 **33.** 14.5 **35a.** $|g - d|$ and $|d - g|$ **35b.** 5 meters **37.** Sample answer: A meteorologist says that the high temperature is going to be 89 degrees. If the actual high temperature that day is x, then

$|x - 89|$ represents the number of degrees the meteorologist is away from the actual high temperature.
39. False; sample answer: Suppose $a = 5$ and $b = -3$, then $|a + b| = |5 + (-3)| = |5 - 3| = |2| = 2$ and $|a| + |b| = |5| + |-3| = 5 + 3 = 8$. $2 \neq 8$, so Diaz's claim is not correct.

Lesson 1-6

1. 32 **3.** 11.4 seconds **5.** 0.512 **7.** Automatic Method: \$6000; Exact Method: \$5625
9. $2\frac{1}{3}$ snack bars **11.** 6 **13.** \$333.33
15. \$10,500 **17.** Because the number of students enrolled at Hartgrove High School can be counted, giving an exact enrollment is accurate. **19.** The map maker is probably accurate because the number of traffic lights in New York City is not very specific. **21.** Sample answer: Steve Nash would be selected as a free throw shooter. Michael Jordan would be selected as a free throw shooter. Shaquille O'Neal would not be selected as a free throw shooter. **23.** Light-years; sample answer: The distance from Earth to the star is very great so using the largest distance unit is appropriate in this situation. most accurate number of visitors at the zoo for a given temperature. **27.** Sample answer: The number of visitors does not increase at the same rate for each average daily temperature. **29.** Sample answer: An employer might consider the number of sick days an employee takes or the amount of sales an employee generates. **31.** Sample answer: \$28.43 because $3.299 \times 8.618 = 28.430782$. The answer could be accurate to the thousandths place, but it is only necessary to round to the nearest hundredths place because the penny is the smallest unit of money.

Module 1 Review

1. D **3.** $5(x + 7) - 4^3$ **5.** A **7.** B
9. A. True B. False C. True
11. B, D
13. C, E **15.** 67
17. No; sample answer: The manager rounded down, but actually spent much more than \$4000. It would have been better to report a greater amount so that it was clear her budget was not overspent.

Selected Answers

Module 2

Quick Check

1. $6n + 2$ **3.** $4b + 9$ **5.** 8 **7.** 32 **9.** 36

Lesson 2-1

1. $3m + 2 = 18$ **3.** $\frac{24}{x} = 14 - 2x$ **5.** $2 + 3h = 6$ **7.** $(48 + 33) + n = 107$ **9.** $2a + a^3 = b$ **11.** $x + x^2 = yz$ **13.** $A = \ell^2$ **15.** $P = 2\ell + 2w$ **17.** $I = prt$ **19.** The sum of *j* and sixteen is thirty-five. **21.** Seven times the sum of *p* and twenty-three is the same as one hundred two. **23.** Two-fifths of *v* plus three-fourths is identical to two-thirds of *x* squared. **25.** *g* plus 10 is the same as 3 times *g*. **27.** 4 times the sum of *a* and *b* is 9 times *a*. **29.** Half of the sum of *f* and *y* is *f* minus 5. **31.** Sample answer: The volume equals π times the radius squared times the height. The base is a circle so the expression πr^2 represents the area of the base. **33.** Sample answer: The interest equals the product of the principal, the rate, and the time. **35.** Sample answer: Force equals mass times acceleration. The expression *ma* represents the force on an object with mass *m* that is accelerating.
37. B **39.** A **41.** $y^2 - 12 = 5x$ **43.** $100 - 3b = 6b$ **45.** Four times *n* equals *x* times the difference of five and *n*. **47.** The sum of *y* and the product of 3 and the square of *x* is 5 times *x*. **49.** $V = \ell wh$ **51.** $m + 2m = 24$ or $3m = 24$ **53.** $c = 10w + 0.1(10w)$ or $c = 11w$
55a. It is correct. The product is squared, so parentheses are needed. **55b.** It is not correct. One-half of a number means to multiply, not divide, by one-half. It should be $\frac{1}{2}n + 3 = n - 2$. **57.** Sample answer: A teacher ordered 188 math books. The algebra books were packed in boxes of 12. The geometry books were packed in boxes of 10. He ordered one more box of algebra books than geometry books. How many books of each type book did he order? Let a = number of algebra books.
59. $S = 6\ell^2$ **61.** Sample answer: First, you should identify the unknown quantity or quantities for which you are trying to solve, and assign variables. Then, you should look for key words or phrases that can help you to determine operations that are being used. You can then write the equation using the numbers that you are given and the variables and operations that you assigned.

Lesson 2-2

1. 23 **3.** −43 **5.** −12 **7.** 73 **9.** −15 **11.** −54 **13.** $\frac{7}{20}$ **15.** $-\frac{7}{15}$ **17.** −937 **19.** −147 **21.** −25 **23.** $-\frac{9}{2}$ **25.** 15 **27.** 10 **29.** 64 **31.** 28 **33.** 18 **35.** 24 **37.** 27 **39.** 39 **41.** 64 **43.** 9 **45.** −12 **47.** 7 **49.** 64 **51.** −252 **53.** −52 **55.** $x + 33 = 2005$; $x = 1972$ **57.** $x - 21 = -9$; $x = 12$°C **59a.** Let p = the number of players who signed up for the soccer league. If 13% of the players who signed up for the soccer league dropped out, then 100% −13%, or 87% of the players finished the season. So, $0.87p$ represents the number of players who finished the season. **59b.** $0.87p = 174$ **59c.** $p = 200$; 200 players signed up for the soccer league
61. $\frac{2}{3} = -8n$; $-\frac{1}{12}$ **63.** $\frac{4}{5} = \frac{10}{16}n$; $\frac{32}{25}$ **65.** $4\frac{4}{5}n = 1\frac{1}{5}$; $\frac{1}{4}$ **67.** −77 **69.** $\frac{16}{3}$ **71.** −10 **73.** $-\frac{10}{7}$ or $-1\frac{3}{7}$ **75.** 18 **77.** 225 **79.** −14 **81.** 4 **83.** −49 **85.** 40 **87.** −15 **89.** $-\frac{8}{15}$ **91a.** $12x = 780$; $x = 65$ **91b.** $20 **93.** $x = 216$; Multiplication Property of Equality **95.** $y = -224$; Subtraction Property of Equality
97. $15 = b$; Division Property of Equality
99. $n - 16 = 29$ does not belong because for the other three, $n = 13$, and for this one $n = 45$.
101. Sample answer: $x - 4 = 10$ **103.** Sample answer: To solve $5x = 35$, I would divide each side by 5 to get $x = 7$. To solve $5 + x = 35$, I would subtract 5 from each side to get $x = 30$. In both equations I used properties of equality to isolate the variable. In the first equation I used the Division Property of Equality and I used the Subtraction Property of Equality in the second equation.

Lesson 2-3

1. -5 **3.** -5 **5.** 70 **7.** 27 **9.** 16
11. -61 **13.** $\frac{1}{2}a - 5.25 = 22.50$; \$55.50
15. $\frac{t-10}{15} = 4$; 70 treats
17. $71 = 2h - 1$; 36 inches
19. $\frac{18}{a}$ **21.** $\frac{-35}{a}$ **23.** $\frac{-24}{a}$ **25.** $\frac{-14}{a}$
27. 7 **29.** 10 **31.** -16 **33.** -2
35. 18 **37.** $(n - 2) \div 3 = 30$; 92 **39.** Sample answer: Both are correct. Dividing by a number and multiplying by that number's reciprocal are equivalent operations. **41a.** $x = \frac{-2}{a}$
41b. $x = 13a$ **41c.** $x = \frac{10}{a}$ **43.** Never; whenever three odd integers are added together, the sum is always odd.

Lesson 2-4

1. 6 **3.** 1 **5.** -2 **7.** -2 **9.** 14 **11.** 4
13. -5 **15.** 0 **17.** $7 + F = 4F + 1$; France won 2 gold medals and the U.S won 9 gold medals. **19.** $38 + 4x = 45.5 + 2.5x$; 5 years **21.** $180 - x = 10 + 2(90 - x)$; 10°
23. $9(5 + x) = 15\frac{3}{7}x$; 7 **25.** no solution
27. identity **29.** no solution **31.** one solution
33. identity **35.** no solution **37.** no solution
39. all numbers **41.** -25 **43.** 3
45. -2 **47.** 15 **49a.** Let n = the first odd integer; $2(n + 2) = 3n - 13$ **49b.** 17 and 19
51a. Let k = the number; $4k - 3 = 2k + 5$
51b. $k = 4$ **51c.** Substitute 4 for k in the expression for the perimeter of Figure 2, $2k + 5$. So the perimeter of Figure 2 is $2(4) + 5 = 8 + 5 = 13$. **51d.** Substitute 4 for k in the expression for the perimeter of Figure 1, $4k - 3$. So the perimeter of Figure 1 is $4(4) - 3 = 16 - 3 = 13$. The perimeter for Figure 1 and Figure 2 is the same, so the value of k is correct.
53. Anthony is correct. When Patty added m to each side, she subtracted the terms instead of adding them. **55.** Sample answer: $2(3x + 6) = 3(2x + 5)$ **57a.** Incorrect; the 2 must be distributed over both g and 5, then 10 must be subtracted from each side; 6.
57b. Correct; the Subtraction Property was used to combine the variable terms on the left side of the equation. The Division Property was used to isolate the variable on one side.
57c. Incorrect; to eliminate $-6z$ on the left side of the equal sign, $6z$ must be added to each side of the equation; 1. **59.** Sample answer: $2x + 1 = x + 9$

Lesson 2-5

1. $\{-2, 8\}$
−3 −2 −1 0 1 2 3 4 5 6 7 8 9

3. ∅
−5 −4 −3 −2 −1 0 1 2 3 4 5

5. $\{-3.25, 2\}$
−4 −3 −2 −1 0 1 2 3 4 5 6 7 8

7. ∅
−5 −4 −3 −2 −1 0 1 2 3 4 5

9. $\{-8, 16\}$
−8 −6 −4 −2 0 2 4 6 8 12 14 16

11. $\{2, 5\}$ **13.** $\{-6, 4\}$ **15.** $\{0, 4\}$
17. $\{-5, -1\}$ **19.** $|t - 400| = 15$; min = 385°F; max = 415°F **21a.** $|t - 20.9| = 5.3$ **21b.** 15.6 to 26.2; 10.3 to 31.5

23.

$|x - 35| = 0.5$	Absolute value equation
Case 1: $x - 35 = 0.5$	Definition of absolute value.
$x = 35.5$	Simplify
Case 2: $x - 35 = -0.5$	Definition of absolute value.
$x = 34.5$	Simplify

The bags of rock salt weigh no less than 34.5 pounds and no more than 35.5 pounds.
25. $|x| = 6$ **27.** $|x + 2| = 4$ **29.** $|x + 3| = 2$
31. $|x| = 4$
33. $\left\{-\frac{3}{2}, \frac{9}{2}\right\}$
−5 −4 −3 −2 −1 0 1 2 3 4 5

35. $\{5.5, -5.5\}$
−6 −4 −2 0 2 4 6

37. $\{2, -2\}$
−5 −4 −3 −2 −1 0 1 2 3 4 5

39. $|x| = 1\frac{1}{2}$ **41.** $|x - \frac{1}{4}| = \frac{1}{4}$ **43.** $|x + \frac{1}{3}| = 1$
45a. $|x - 38| = 2$ **45b.** 40°F, 36°F
47. $|x - 3| = 1$ **49.** $|a - b| + |b - c| = |a - c|$ **51.** Cami; The absolute value of a number cannot be negative. **53.** Sample answer: Let x = the temperature at night. Then the temperature is 4 ± 10 degrees.

Lesson 2-6

1. 40 **3.** 29.25 **5.** 9.8 **7.** 1.32 **9.** 0.84 **11.** 0.57 **13.** 6 **15.** 11 **17.** 18 **19.** 0.8 **21.** 11 **23.** −2 **25.** 1.44 **27.** −2.29 **29.** −2.2 **31.** 10 **33.** 3 **35.** −8.4 **37.** 12.5 gal **39.** $46.27 **41a.** 60 free throws **41b.** Sample answer: I assumed that Brent continues to make free throws at the same rate. **43.** 22.5 in. **45.** $3333.33 **47.** 204.55 mL **49.** 6 **51.** 10 **53.** 21 **55.** 8 **57.** 42 **59.** 27 **61.** 3 **63.** 15 **65.** −3 **67.** −0.4 **69.** −6 **71a.** Sample answer: 2.2 cm **71b.** Sample answer: about 6.6 miles **71c.** Sample answer: about 453.7 mi^2 **73a.** $4.50; because 8 potatoes cost $1.50, multiply by 3 to get a cost of $4.50 for 24 potatoes. **73b.** $4.13; Sample answer: $4.13 is slightly less than $4.50, which aligns with my estimate. **73c.** 37 **73d.** $0.19 **75.** $\frac{2}{4}$ or $\frac{1}{2}$ **77.** Ratios and rates each compare two numbers by using division. However, rates compare two measurements that involve different units of measure. **79.** $\frac{x}{100} = \frac{z}{y}$

Lesson 2-7

1. $y = \frac{x-1}{2}$ **3.** $f = \frac{5-g}{7}$ **5.** $t = \frac{x}{7}$ **7.** $r = \frac{q}{2}$ **9.** $a = -\frac{b}{8}$ **11.** $v = \frac{u-z}{w}$ **13.** $g = \frac{10j + 9h}{f}$ **15.** $t = \frac{3}{2}(r - v)$ **17.** $a = \frac{-33 + x}{10c}$ **19a.** $\ell = \frac{P - 2w}{2}$ **19b.** 14 m **21a.** $g = \frac{c-p}{13.50}$ **21b.** 6 games **25.** ≈ 12.96 trillion pounds **27.** 0.44 ft **29.** 24 miles **31.** 90.2 gallons **33.** 82 students **35.** $c = \frac{2k - 3g}{b}$ **37.** $c = \frac{5p - 6j}{8}$ **39.** $c = x - 2d$ **41.** $t = \frac{w - 11v}{31}$ **43.** $c = \frac{-13 + f}{10 - d}$ **45.** $r = \frac{A}{P} - 1 = \frac{2182.25}{2150} - 1 = 1.015 - 1 = 0.015$; The interest rate is 1.5%. **47a.** $y = \frac{mx + mt - z}{r}$ **47b.** Division by 0 is undefined, so in the original equation $m \neq 0$, and in the final equation $r \neq 0$. **49.** No; Sasha does not have a correct solution. When she multiplied F by $\frac{5}{9}$, she should have multiplied 32 by $\frac{5}{9}$.

Module 2 Review

1. A **3.** D **5.** D **7.** C **9.** B, D **11.** C **13.** Let d = age of dogwood tree (in years); $3d - 2 = \frac{1}{2}(d + 8 + 8)$; The gingko tree is 10 years old, and the dogwood tree is 4 years old. **15.** A **17.** $b = 8$ **19.** B **21.** D **23A.** $h = 3\frac{V}{B}$ **23B.** 12

Module 3

Quick Check

1, 3.

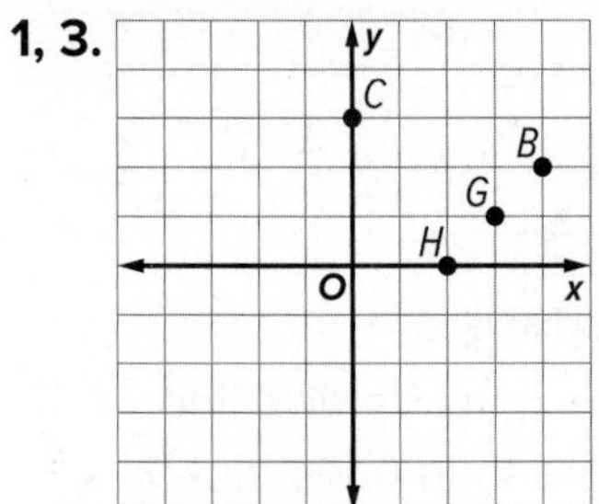

5. 20 **7.** −3

Lesson 3-1

1.

x	*y*
−1	−1
1	1
2	1
3	2

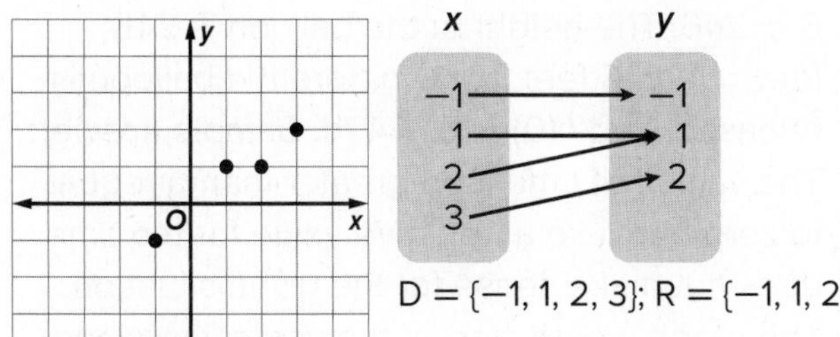

D = {−1, 1, 2, 3}; R = {−1, 1, 2}

3.

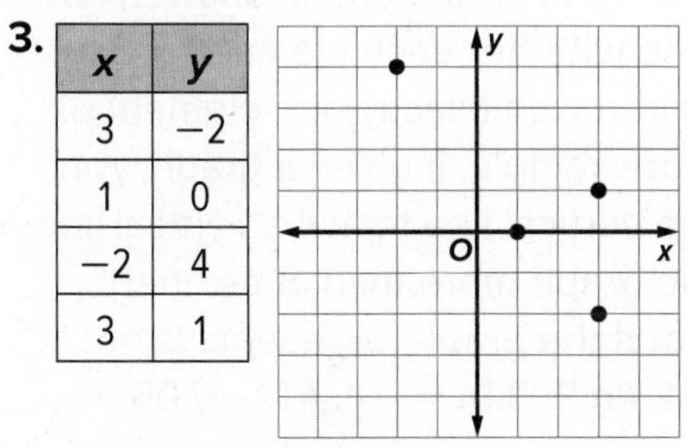

x	*y*
3	−2
1	0
−2	4
3	1

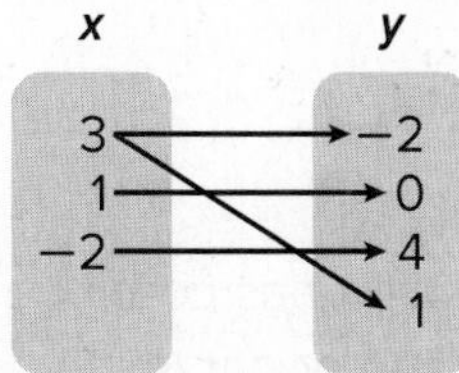

D = {−2, 1, 3}; R = {−2, 0, 1, 4}

5a. independent: price of item, dependent: number of items purchased **5b.** As the price of an item increases, the number of items purchased decreases.

7.

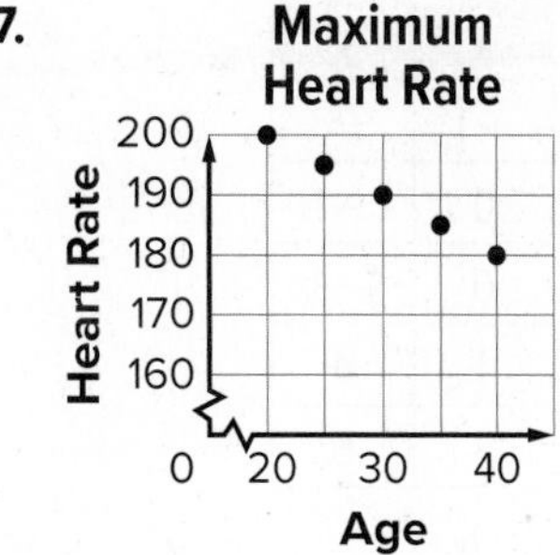

9. The *x*-axis represents the time in seconds. The *y*-axis represents the height of the elevator in feet. The *x*-axis has a scale of 1 mark = 1 second. The *y*-axis has a scale of 1 mark = 10 feet. The origin (0, 0) represents a height of 0 feet in 0 seconds. **11.** {(1, 7), (3, 45), (5, 11), (13, 15)} **13.** {(2, 5), (5, 0), (7, 8), (7, 10), (10, 2)} **15.** {(2, 80), (3, 120), (6, 240), (8, 320)}; The *x*-axis represents the number of gallons of syrup. The *y*-axis represents the number of gallons of sap. The *x*-axis has a scale of 1 mark = 1 gallon of syrup. The *y*-axis has a scale of 1 mark = 40 gallons of sap. The origin (0, 0) represents 0 gallons of sap makes 0 gallons of syrup.
17a. {(0, 12), (1, 8), (2, 23), (3, 28), (4, 11), (5, 11)}
17b. {0, 1, 2, 3, 4, 5}
17c. {8, 11, 12, 23, 28}
19. sample answer:

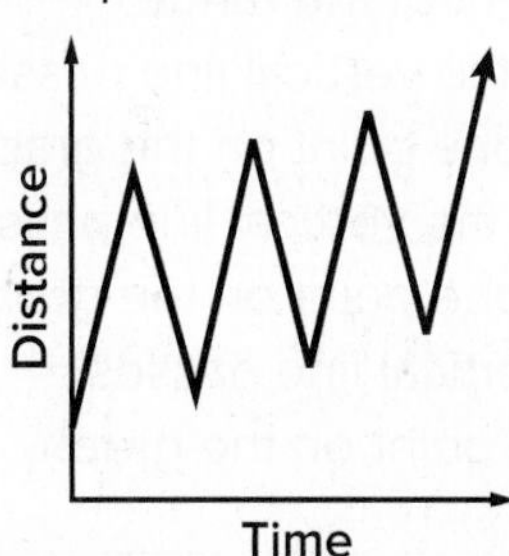

21. Tim drives away from the pizzeria, stops to make a delivery, continues to drive away from the pizzeria, stops to make another delivery, and then returns to the pizzeria. **23.** Disagree; The intersection point represents a time when Tim and Lauren were both at the same distance from the pizzeria. **25.** Disagree; sample counterexample: In the relation {(1, 2), (1, 3)}, the domain is {1}, so it has one element, while the range is {2, 3}, which has two elements. **27a.** Sample answer: {(−1, −3), (0, −3), (0, −1), (1, 4), (2, 5)}

27b. sample answer:

x	y
−1	−3
0	−1
0	−3
1	4
2	5

29. Sample answer: A dependent variable is determined by the independent variable for a given relation.

Lesson 3-2

1. Yes; for each element of the domain, there is only one element of the range. **3.** No; the element 4 in the domain is paired with both 2 and 5 in the range. **5.** No; the element 5 in the domain is paired with both −3 and 2 in the range. **7.** Yes; for each element of the domain, there is only one element of the range.
9. Yes; for any value *x*, the vertical line passes through no more than one point on the graph.
11. Yes; for any value *x*, the vertical line passes through no more than one point on the graph.
13. No; for $x > 0$, the vertical line passes through more than one point on the graph.
15a.

Year	2014	2015	2016	2017
Value ($)	254,000	293,000	338,000	372,000

15b. Domain: {2014, 2105, 2016, 2017}; Range: {254,000; 293,000; 338,000; 372,000}
15c. For each element of the domain, there is only one element of the range. So, this relation is a function. **17.** 26 **19.** 2 **21.** 42 **23.** 4 **25.** 6 **27.** $9b^2 - 3b$ **29.** $f(3.5) = 12.25$, which means the area of a square with a side of length 3.5 units is 12.25 square units.
31. $f(12) = \$435$, which is the cost of a gym membership for 12 months, or 1 year. **33.** −1 **35.** 14 **37.** −4 **39.** $-8y - 3$ **41.** $-2c - 8$ **43.** $-10d - 15$

45a.

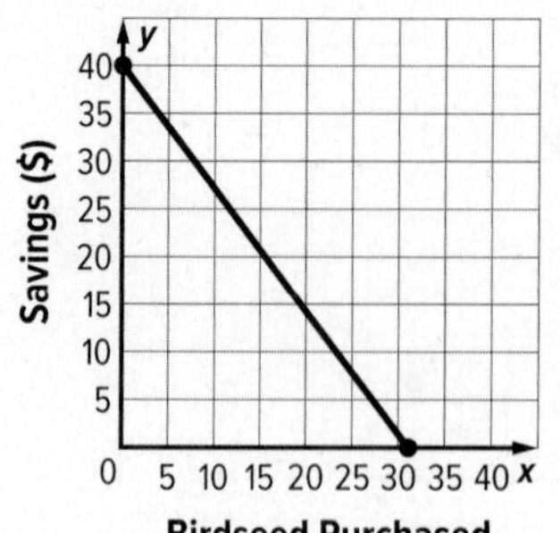

45b. Yes; for any value *x*, the vertical line passes through no more than one point on the graph. **45c.** $f(3) = 36.25$, which means if Aisha buys 3 pounds of birdseed, she saves \$36.25; $f(18) = 17.50$, which means if Aisha buys 18 pounds of birdseed, she saves \$17.50; $f(36) = -5$, which means if Aisha wants to buy 36 pounds of birdseed, she needs \$5 extra. **45d.** 8 pounds **47a.** $h(20) = 46$; The height of the balloon 20 seconds after it is released is 46 feet. **47b.** 2 minutes is $2(60) = 120$ seconds, so calculate $h(120)$ by substituting $t = 120$ in the equation; $h(120) = 2(120) + 6 = 246$; The height of the balloon is 246 feet. **47c.** 6 feet; $t = 0$ before the balloon is released, and $h(0) = 6$. **47d.** Sample answer: The values of *t* must be greater than or equal to zero because a negative value for the time does not make sense for the given situation. The graph would start at the vertical axis and go only to the right. **49.** Sample answer: You can determine whether each element of the domain is paired with exactly one element of the range. For example, if given a graph, you could use the vertical line test; if a vertical line intersects the graph more than once, then the relation that the graph represents is not a function. **51.** $f(g + 3.5) = -4.3g - 17.05$
53. Sample answer: $f(x) = 3x + 2$

Lesson 3-3

1. Neither; because the function has continuous sections but is not a single line or curve, it is neither continuous or discrete **3.** Discrete; because the function is made up entirely of individual points, it is discrete. **5.** Continuous; because the function is graphed with a single line, it is continuous. **7.** Discrete; because the function is made up entirely of individual points, it is discrete. **9.** discrete **11.** discrete

13. linear **15.** nonlinear **17.** nonlinear
19. nonlinear **21.** nonlinear **23.** linear
25a. linear

25b.

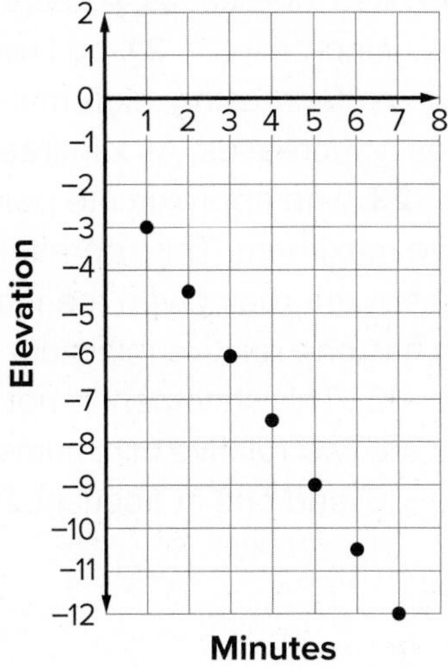

27a. nonlinear

27b.

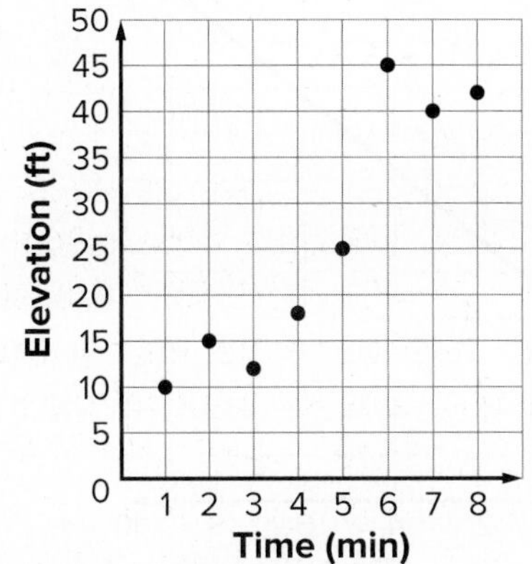

29. continuous; nonlinear **31.** neither; nonlinear **33.** discrete; nonlinear
35. Sample answer: A studio charges musicians to use the space and recording equipment by the hour, rounding a fraction of an hour up. So, for up to 1 hour, the studio charges \$100, but for up to 2 hours, the studio charges \$200, and so on. The function that models this situation is neither discrete nor continuous.

Lesson 3-4

1. x-intercept: $(-0.75, 0)$ y-intercept: $(0, 3)$ positive: when $x > -0.75$ negative: when $x < -0.75$ **3.** $x-$intercepts: $(0, 0)$ and $(2, 0)$ y-intercept: $(0, 0)$ positive: when $x < 0$ and when $x > 2$ negative: $0 < x < 2$
5. x-intercepts: $(-2, 0)$ y-intercept: $(0, 4)$ positive: when $x > -2$ negative: when $x < -2$

7. x-intercepts: $(-5, 0)$ and $(3, 0)$ y-intercept: $(0, 3)$ positive: $-5 < x < 3$ negative: $x < -5$ and when $x > 3$
9. x-intercepts: none y-intercept: $(0, -3)$ positive: never negative: always

11. The x-intercept is 0. The y-intercept is 0. This means that Ryan earns \$0 for working 0 hours. The function is positive when x is greater than 0, which means that Ryan earns money for working. No portion of the graph shows that the function is negative.
13a. The x-intercept is 6. The y-intercept is 1950. **13b.** The x-intercept means that after 6 months, Javier's remaining balance will be \$0, or it will take Javier 6 months to repay his parents. The y-intercept means that Javier owes his parents \$1950 after 0 months, or Javier initially borrowed \$1950 from his parents.

15. Sample graph; no solution

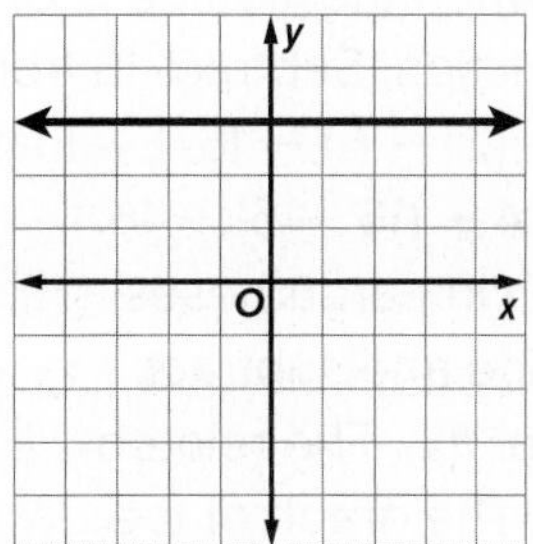

17. -2

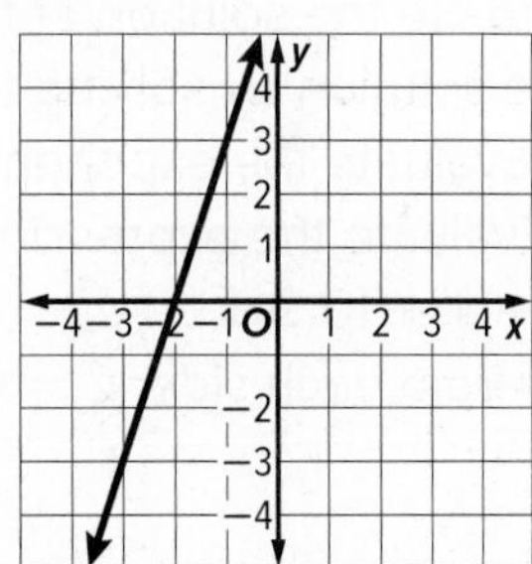

19. 1

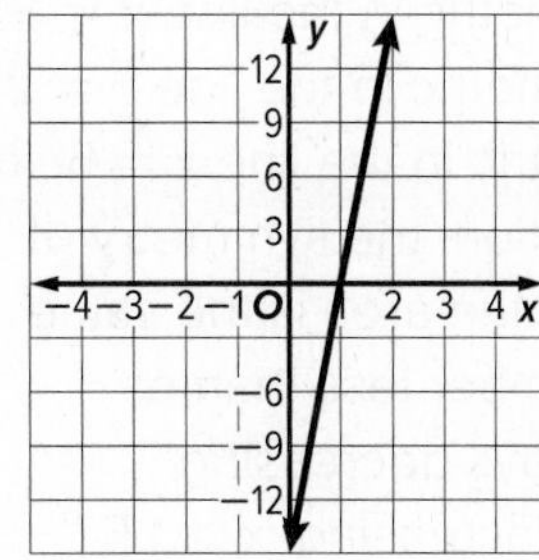

21. The zero of the function is at 32. This represents that after tying ribbon on 32 gift bags, Juanita will have no ribbon left.

23. x-intercepts: $(-2, 0)$ and $(2, 0)$ y-intercept: $(0, 4)$ positive: $-2 < x < 2$ negative: when $x < -2$ and when $x > 2$ **25.** The x-intercepts are 3 and 7. That means that the bird will be at sea level at 3 seconds and at 7 seconds. The y-intercept is 4.5. This means that at time 0, the bird was at a height of 4.5 feet. The function is positive when x is less than 3 and when x is greater than 7, which means that the bird is above sea level from 0 to 3 seconds and after 7 seconds. The function is negative when x is between 3 and 7, which means that the bird is below sea, or under water, for 4 seconds.

27. To find the x-intercept in a graph, find the place where the function crosses the x-axis. To find the y-intercept in a graph, find the place where the function crosses the y-axis. To find the x-intercept in a table, find the x-value when the y-value is 0. To find the y-intercept in a table, find the y-value when the x-value is 0. **29.** Find the related function. Subtract 16 from each side: $0 = x + 4 + (2^4 - 6) - 16$. Evaluate the exponent: $0 = x + 4 + (16 - 6) - 16$. Evaluate the expression in parentheses: $0 = x + 4 + 10 - 16$. Add and subtract: $0 = x - 2$. Replace 0 for $f(x)$. The related function is $f(x) = x - 2$. The graph of the related function intersects the x-axis at 2. This is the x-intercept, or zero. So the solution of the equation is 2. Check the solution by solving the equation algebraically. Evaluate the exponent: $16 = x + 4 + (16 - 6)$. Evaluate the expression in parentheses: $16 = x + 4 + 10$. Add: $16 = x + 14$. Subtract 14 from each side: $2 = x$.

Lesson 3-5

1. This function is symmetric in the line $x = -1$. **3.** This function is symmetric in the line $x = 2.5$. **5.** The graph is symmetric in the line $x = 5$. In the context of the situation, the symmetry of the graph tells you that the area is the same when the width is a number less than or greater than 5. **7.** always decreasing **9.** decreasing: $x < 1.5$; increasing: $x > 1.5$ **11.** extrema: B and D; rel min: D; rel max: B **13.** extrema: B and D; rel min: B; rel max: D **15.** Point A is a relative maximum. Point A represents the greatest height of the golf ball given the distance from the tee. **17.** As x decreases, y increases. As x increases, y increases. **19.** As x decreases, y decreases. As x increases, y increases. **21.** no line symmetry; always decreasing; extrema: none; As x decreases, y increases. As x increases, y decreases. **23.** The approximate point (2.5, 114) is a relative maximum. This represents the greatest height of the rock given the time. **25.** The graph has one relative minimum at about $(-2.25, -16)$; This statement is not true because there are two relative minimums: one at about $(-2.25, -16)$ and one at about $(2.25, -16)$.

Lesson 3-6

1.

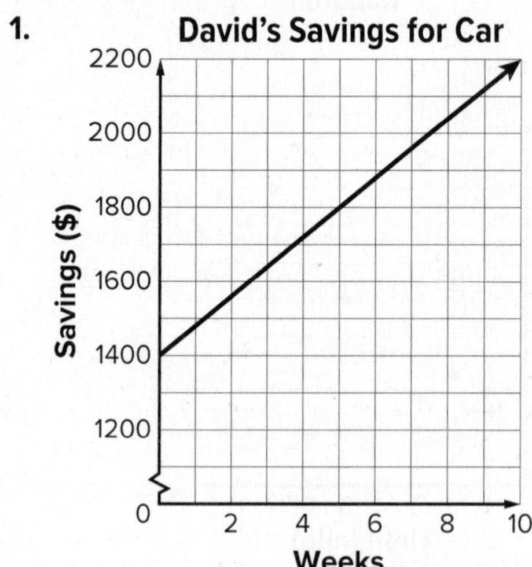

3.

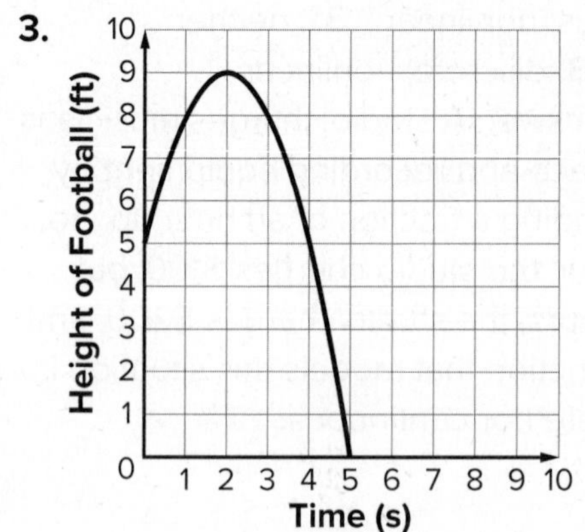

5.

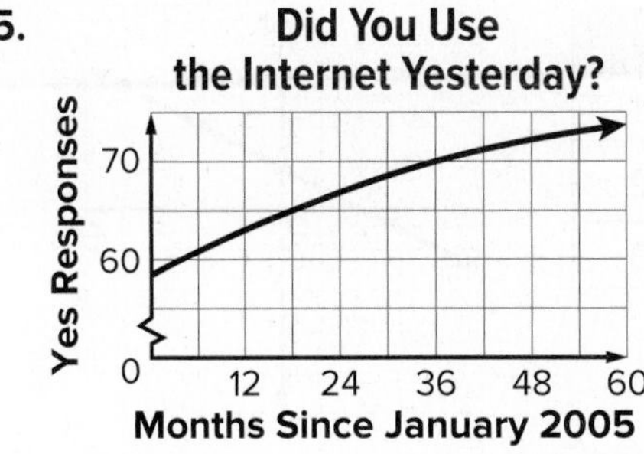

7. Sample answer: Internet use at home initially has a higher number of users than Internet use away from home. Both Internet use at home and Internet use away from home increase after 36 months since March 2004. Neither Internet use at home nor Internet use away from home reaches 0 users.

9.

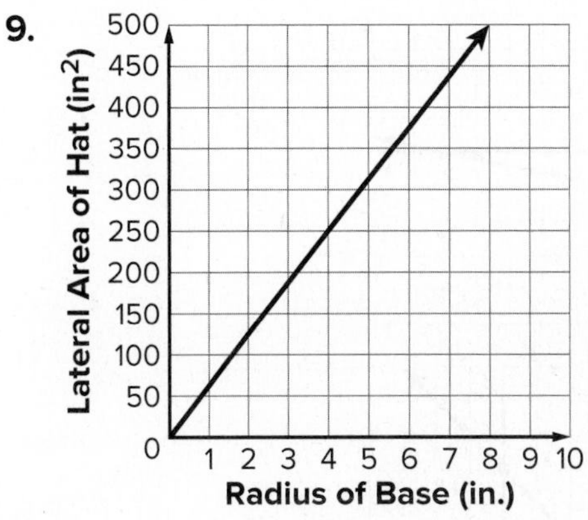

11. Sample answer: The graph on the calculator and the graph I sketched are both linear, increasing, and have an *x*- and *y*-intercept at 0. **13.** $P(x) = 28x - 840$; $P(x)$ is Aidan's profit from fixing and selling x bicycles. **15.** The *x*-intercept; To find the *x*-intercept, locate the point on the graph when $P(x) = 0$, which is 30. So, when 30 bicycles are bought and sold, Aidan makes a profit of $0.

Module 3 Review

1. B, C, F
3. C
5. Sample answer: The element –4 in the domain is paired with both 8 and 13 in the range. This relation is not a function. **7.** B **9.** (−1, 0) and (0, −1) **11.** 40; 60 **13.** Sample answer: In 1900 the population of Ohio was nearly 4 million more than the population of Florida. Both populations grew between 1900 and 1950. At this point, the population of Ohio exceeded that of Florida by approximately 5 million, indicating a greater growth rate for Ohio than Florida during those decades. Then from 1950 to 2000, the population of Ohio grew by about 3.4 million, whereas the population of Florida grew by about 13 million, indicating a significantly greater growth rate for Florida during those decades. In fact, by 2000, the population of Florida surpassed Ohio by more than 4 million people.

Module 4

Quick Check

1, 3, 5.

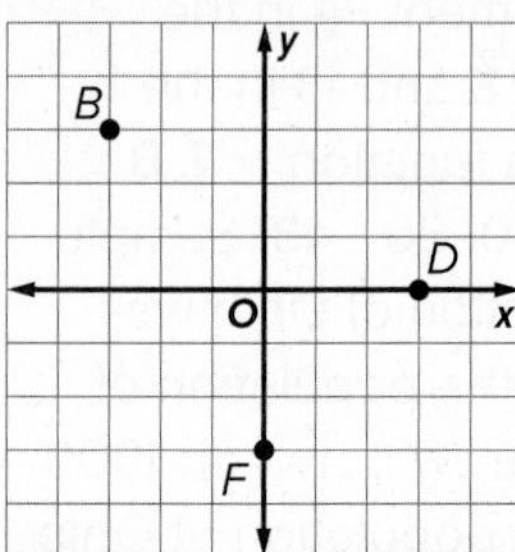

7. $y = -3x + 1$ **9.** $y = \frac{5}{2}x - 6$ **11.** $y = -10x + 6$

Lesson 4-1

1.

x	y
−2	0
−2	1
−2	2

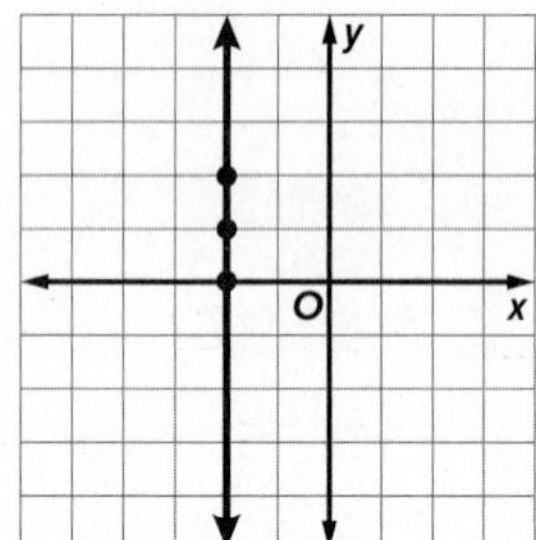

3.

x	y
−1	8
0	0
1	−8

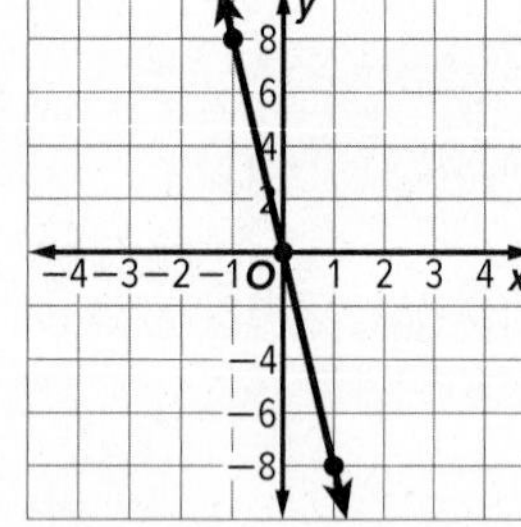

5.

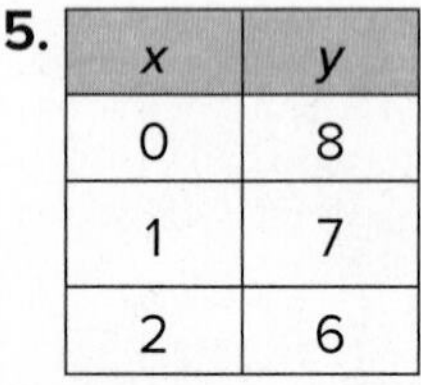

x	y
0	8
1	7
2	6

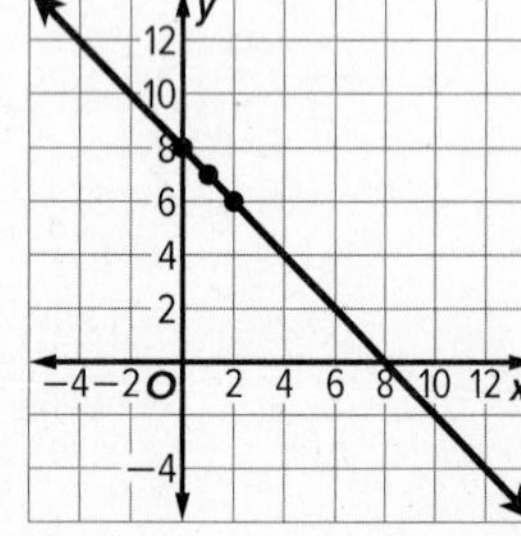

7.

x	y
0	1
2	2
4	3

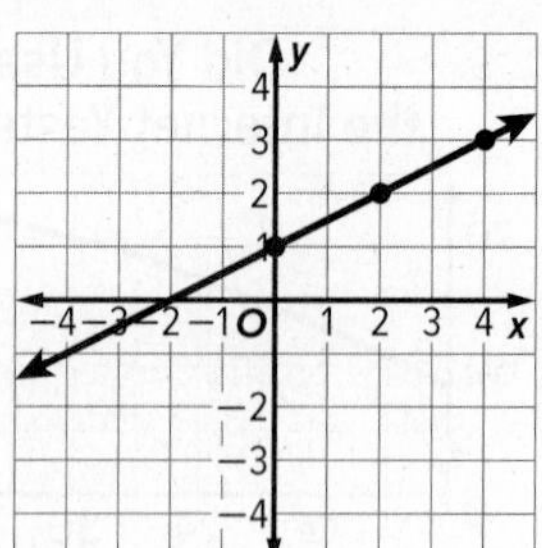

9.

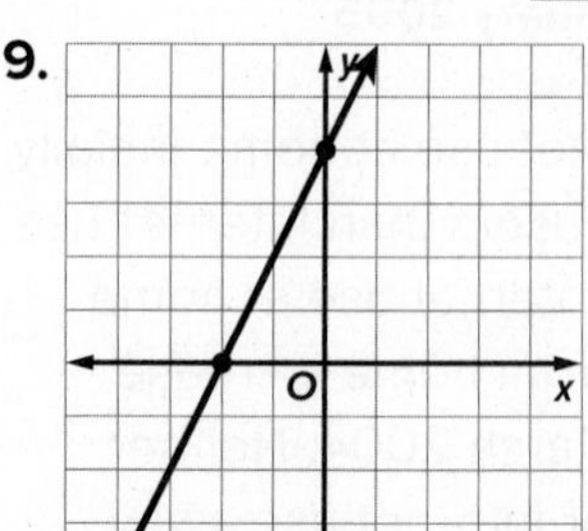

11.

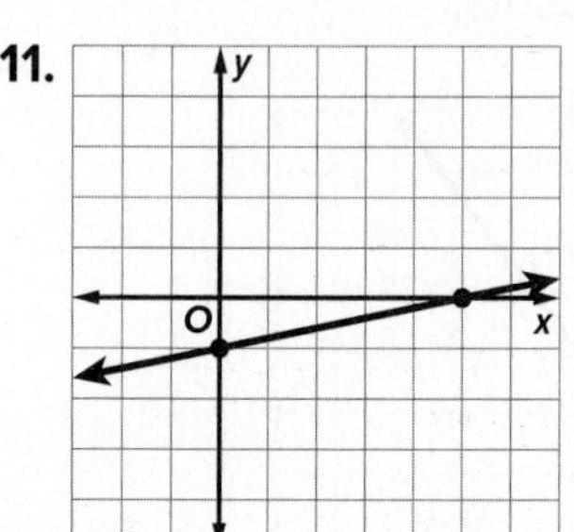

13.

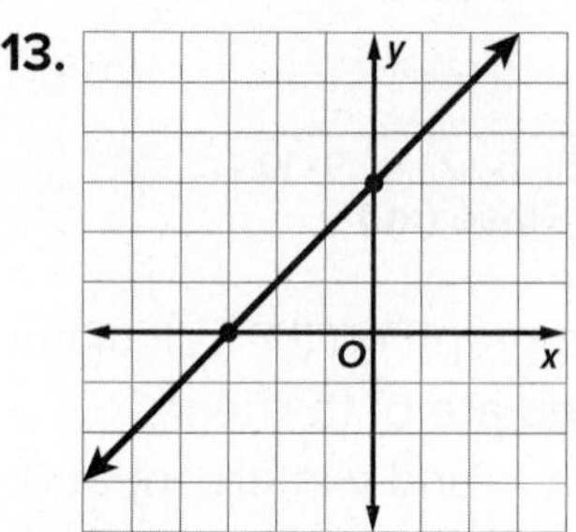

15a. The *x*-intercept is 6. This means that after 6 weeks, Amanda will have $0 in her school lunch account. The *y*-intercept is 210. This means that there was initially $210 in Amanda's school lunch account.

15b.

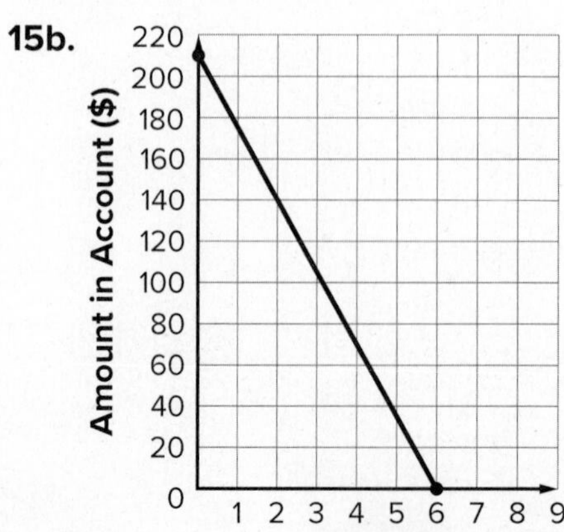

17.

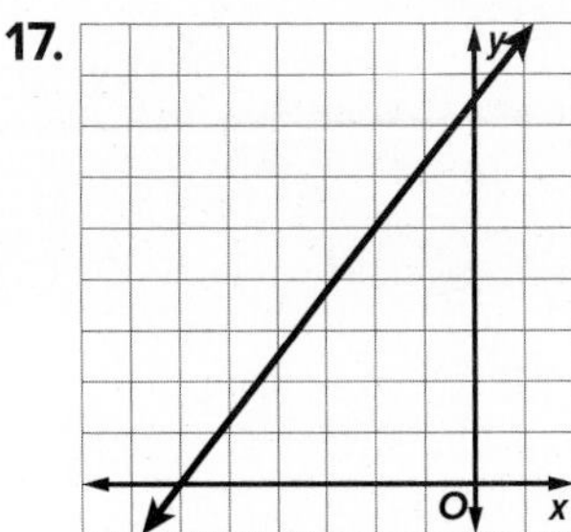

19.

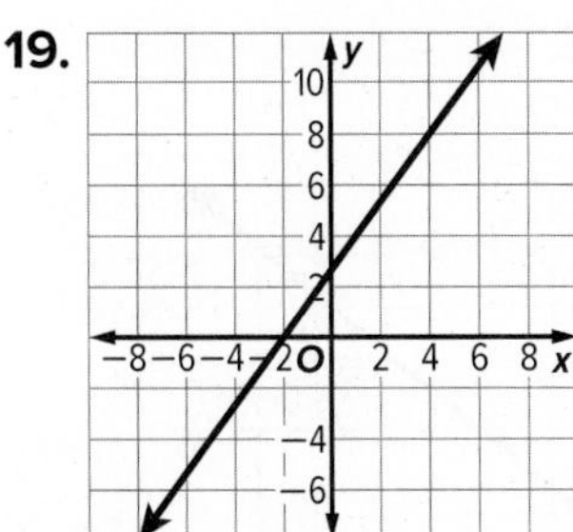

21. x-int: 7; y-int: −2
23. x-int: $1\frac{1}{3}$; y-int: 4
25. x-int: $-1\frac{1}{2}$; y-int: 1
27. $y = 1.7x + 40$; The y-intercept is 40. This means that it would cost $40 to hook up the car.

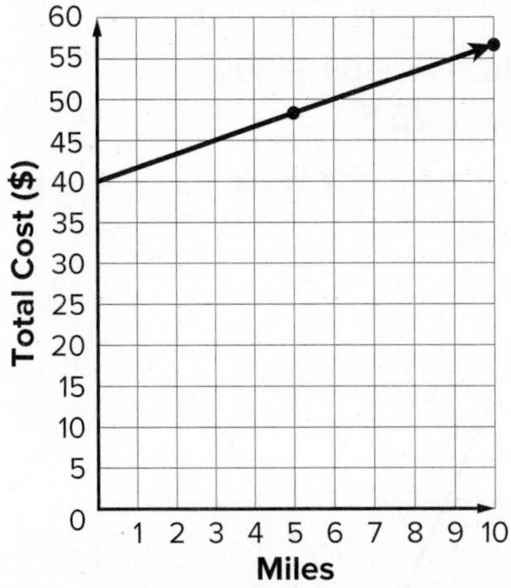

29. Sample answer: The x-intercept is −4. The x-intercept is not reasonable because the football team cannot lose −4 games. The y-intercept is 4. The y-intercept is reasonable because the y-intercept means that if the football team won 4 games, they lost 0 games. **31.** No; sample answer: A horizontal line only has a y-intercept and a vertical line only has an x-intercept. **33.** In the equation, let $y = 0$ to find the x-intercept: $2x + (0) = 4$. So the x-intercept is 2. In the equation, let $x = 0$ to find the y-intercept: $2(0) + y = 4$. So the y-intercept is 4. Robert graphed points at (2, 0) and (0, 4) and connected the points with a line. **35.** Sample answer: $y = 8$; horizontal line **37.** Sample answer: $x - y = 0$; line through (0, 0)

Lesson 4-2

1. $\frac{1}{5}$ **3.** increased about 1.9 people per square mile **5a.** −5; This means the temperature decreased 5°F per hour from 6 A.M. to 7 A.M. **5b.** −5; This means the temperature decreased 5°F per hour from 1 P.M. to 2 P.M. **7.** linear; $-\frac{1}{1}$ or −1 **9.** not linear **11.** $-\frac{3}{5}$ **13.** 1 **15.** 0 **17.** $\frac{1}{6}$ **19.** $\frac{4}{3}$ **21.** undefined **23.** 1 **25.** undefined **27.** 2 **29.** undefined **31.** −1 **33.** undefined **35.** $-\frac{7}{2}$ **37.** $\frac{5}{2}$ **39.** $\frac{3}{4}$ **41.** 6 **43.** 8 **45.** 11 **47.** $\frac{1}{20}$ **49.** $-\frac{1}{2}$ **51.** $\frac{1}{3}$ **53.** $\frac{1}{2}$ **55.** −1 **57.** $\frac{7}{4}$ **59.** 3 **61.** After drawing a graph, use the two points on the graph to determine the slope. This can be done by counting squares for the rise and run of the line or by using the coordinates of the points in the slope formula. **63.** The rate of change is $2\frac{1}{4}$ inches of growth per week. **65.** Step 1; she reversed the order of the x-coordinates in the formula. **67.** The difference in the x-values is always 0, and division by 0 is undefined.

Lesson 4-3

1. $y = 5x - 3$ **3.** $y = -6x - 2$ **5.** $y = 3x + 2$ **7.** $y = x - 12$ **9.** $y = 5x + 6$ **11.** $y = \frac{1}{3}x - 2$ **13.** $y = -0.25x - 3$ **15.** $y = 25x + 100$ **17.** $y = 0.12x + 9$

19.

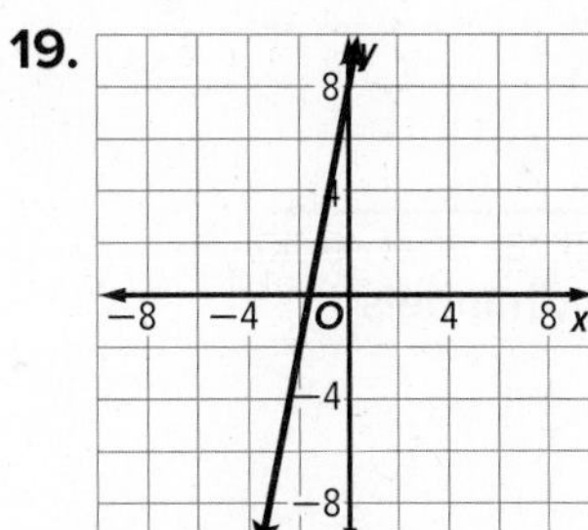

21.

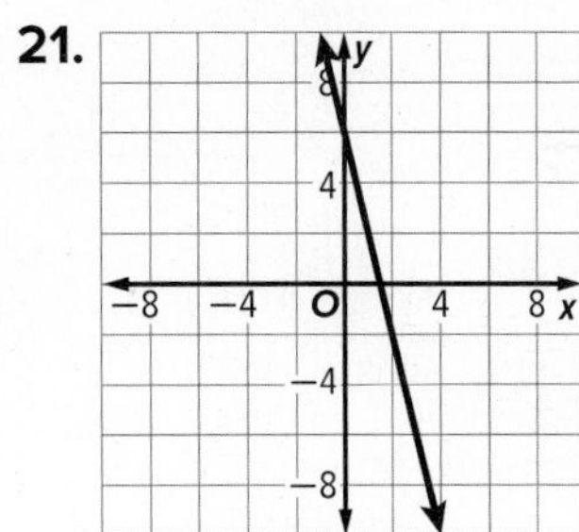

Selected Answers

23.

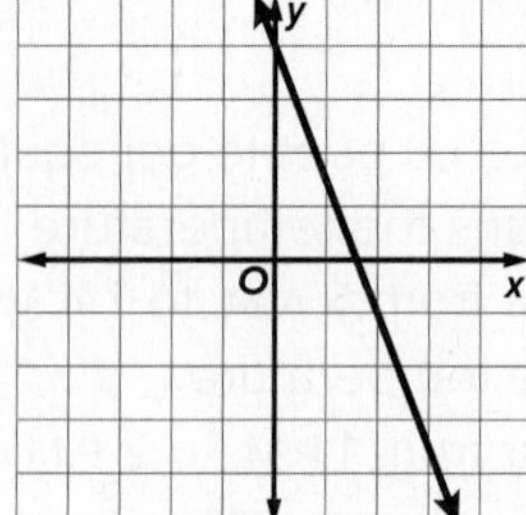

25.

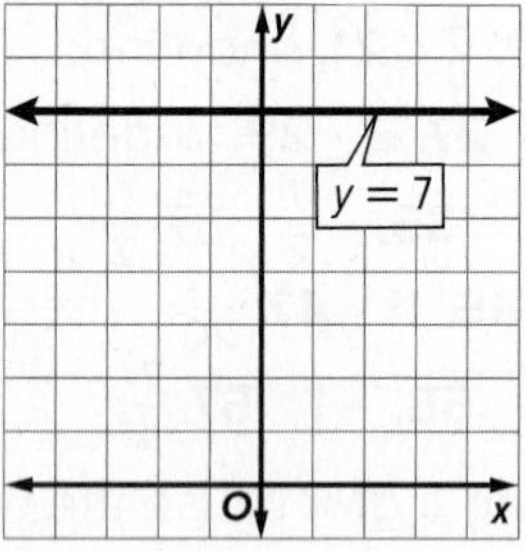

27.

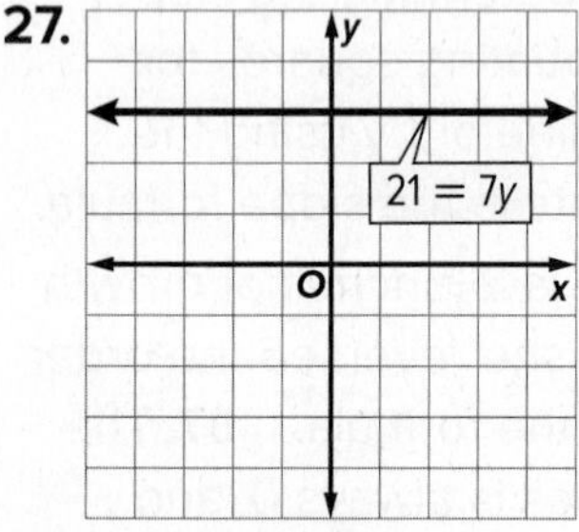

29a. $c = 13 + 8p$

29b.

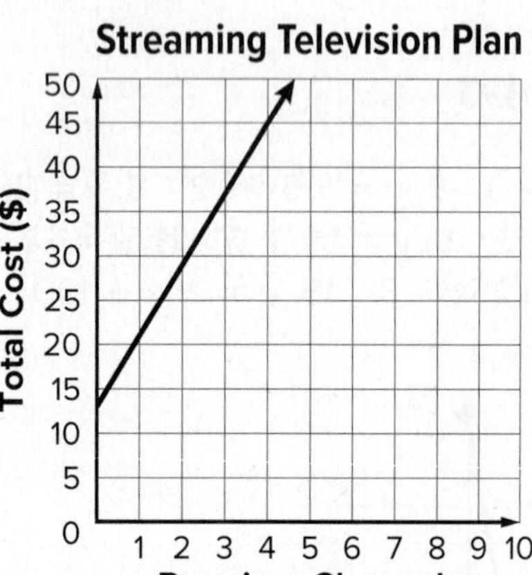

29c. \$37 **31.** $y = \frac{1}{2}x - 3$

33.

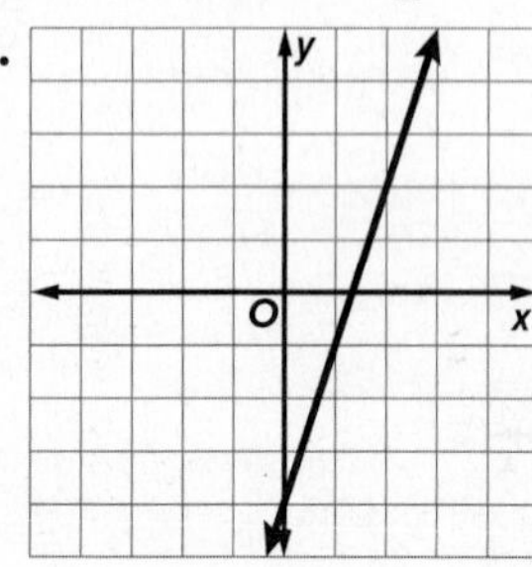

35.

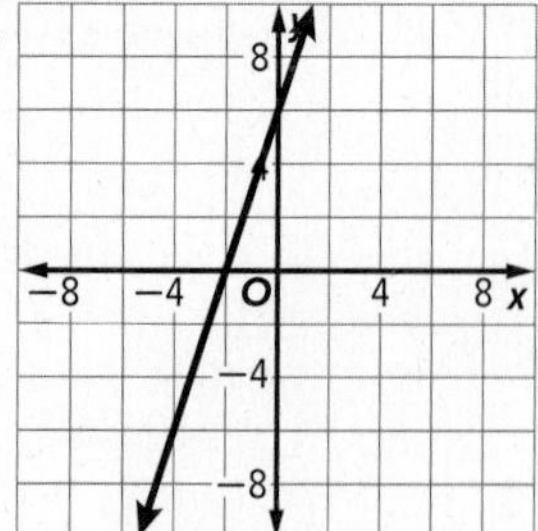

37. $y = 2x - 3$ **39.** $y = -x - 1$

41a. $T = 10x + 80$

41b.

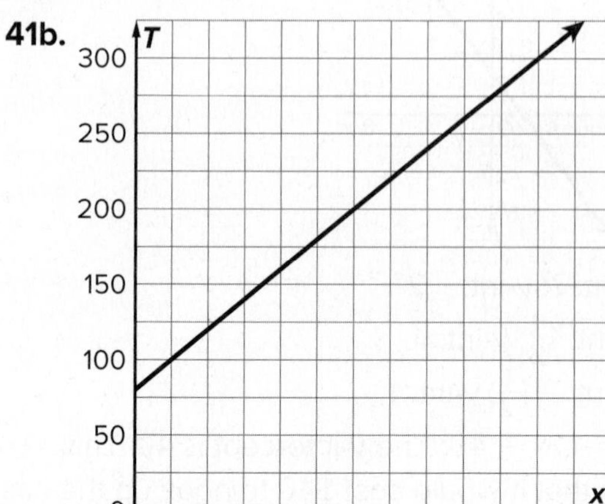

41c. 300°F **41d.** $T = -5x + 300$

43a. $y = -4.25x + 25.5$

43b.

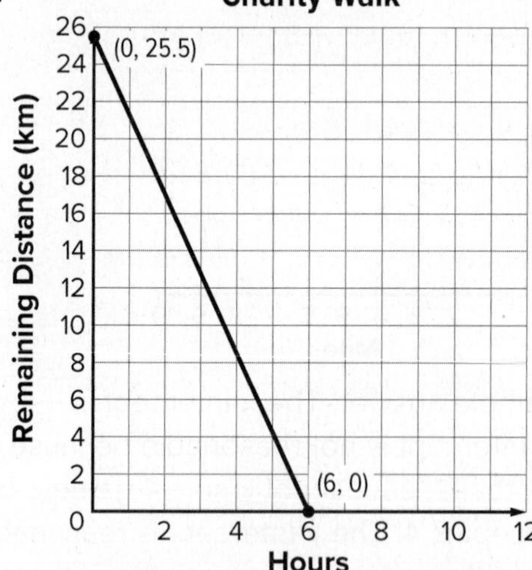

43c. The x-intercept (6) represents the number of hours it will take Jazmin to complete the walk. The y-intercept (25.5) represents the length of the walk. **43d.** Using the graph, I can determine the value of x when y equals $-17 + 25.5$ or 8.5 km, and use the value of the x-intercept. The value of x is 4 when $y = 8.5$ and the x-intercept is 6. Therefore, Jazmin has $6 - 4$ or 2 hours more to walk. **45.** Yes; you can find the value of x on the graph when $y = 0$; $x = \frac{1}{2}$.

47. Sample answer: $y = 25x + 200$; I have \$200 in savings and will save \$25 per week until I have enough money to buy a new phone. I can predict how much money I'll have after x number of weeks.

Lesson 4-4

1. $g(x)$ is a translation of the parent function 11 units up **3.** $g(x)$ is a translation of the parent function 7 units right **5.** $g(x)$ is a translation of the parent function 10 units left and 1 unit down **7.** $g(x) = 4x$; $g(x)$ is the translation of $f(x)$ 3.5 units down. **9.** $g(h) = 8h + 15$; $g(h)$ is the translation $f(h)$ of 5 units up. **11.** $g(x)$ is a vertical compression of the parent function by a factor of $\frac{1}{3}$ **13.** $g(x)$ is a horizontal compression of the parent function by a factor of $\frac{1}{3}$ **15.** $g(x)$ is a horizontal stretch of the parent function by a factor of 2.5 **17.** $g(x)$ is a vertical stretch of the parent function by a factor of 8 and a reflection across the x-axis **19.** $g(x)$ is a horizontal stretch of the parent function by a factor of $\frac{5}{4}$ and a reflection across the y-axis **21.** $g(x)$ is a horizontal compression of the parent function by a factor of $\frac{2}{3}$ and a reflection across the y-axis **23.** $g(x)$ is a translation of the parent function 2 units right and 8 units down **25.** $g(x)$ is a vertical compression of the parent function by a factor of $\frac{1}{5}$ **27.** $g(x)$ is a horizontal compression of the parent function by a factor of 0.4 **29.** $g(x) = x - 7$ **31.** $g(x) = 1.5x$; The graph of $g(x) = 1.5x$ is the graph of $f(x) = 0.50x$ stretched vertically by a factor of 3. **33a.** $g(x) = 1.29x$ **33b.** The graph of $g(x) = 1.29x$ is the graph of $f(x) = x$ stretched vertically by a factor of 1.29. **35.** $y = \frac{1}{a}x$; The function is horizontally stretched by a factor of a.

Lesson 4-5

1. This sequence has a common difference of 4 between its terms. This is an arithmetic sequence. **3.** This sequence does not have a common difference between its terms. This is not an arithmetic sequence. **5.** This sequence does not have a common difference between its terms. This is not an arithmetic sequence.

7. This sequence has a common difference of 3 between its terms. This is an arithmetic sequence. **9.** 1.06; 4.26, 5.32, 6.38 **11.** −2; 13, 11, 9 **13.** $\frac{1}{3}$; $3\frac{2}{3}$, 4, $4\frac{1}{3}$ **15.** 4; 19, 23, 27 **17.** 2; −5, −3, −1 **19.** $a_n = -5n + 2$; −33 **21.** $a_n = -4n - 7$; −35 **23a.** $f(n) = -4n + 128$

23b.

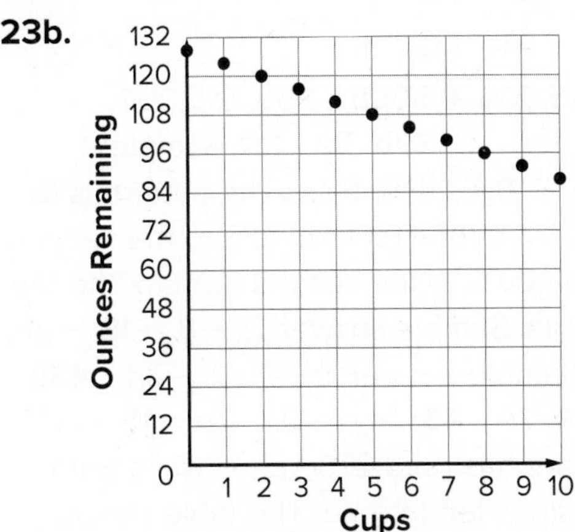

23c. 72 ounces **25a.** $f(n) = 0.17n + 0.71$

25b.

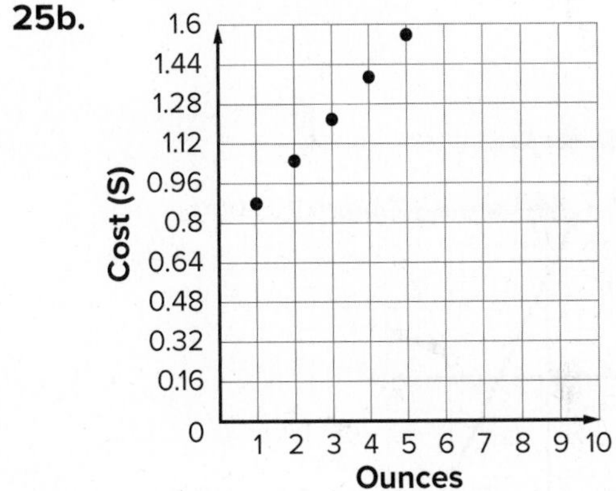

25c. 8 ounces **27a.** $f(n) = 2n + 28$

27b.

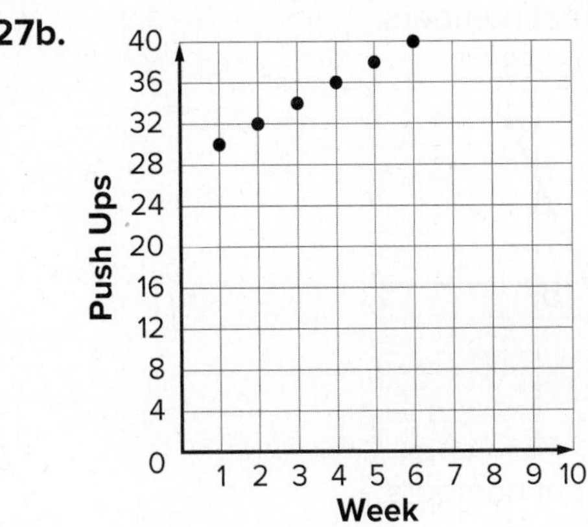

27c. 11th week **29.** This sequence does not have a common difference between its terms. This is not an arithmetic sequence. **31.** This sequence has a common difference of 2 between its terms. This is an arithmetic sequence.

Selected Answers

33. $a_n = -4n + 34$

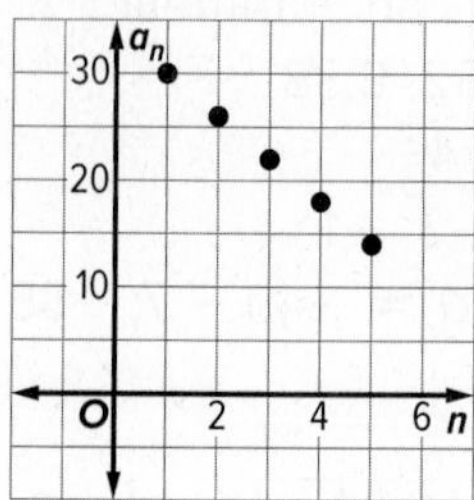

35a. $a_n = 3000 + 500n$ **35b.** $15,000 **37a.** $a_n = 3n - 1$ **37b.** 59 **39.** Sample answer: 5, 3, 8, 6, 11, 9, 14, ...; The pattern is to subtract 2 from the first term to find the second term, then add 5 to the second term to find the third term. **41.** Sample answer: 2, −8, −18, −28, ... **43a.** Sample answer: $a_n = -2 - 3n$ **43b.** $a_n = -19 + 7n$ **43c.** $a_n = 12 - 2n$ **45.** On day 9, Andre has read 270 pages while Sam has 270 pages left to read. The table shows that both functions have a value of 270 when $x = 9$.

Lesson 4-6

1.

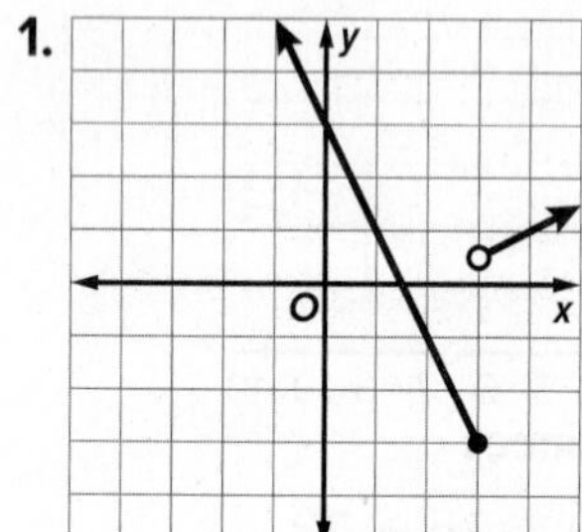

D = all real numbers,
R = $f(x) \geq -3$

3.

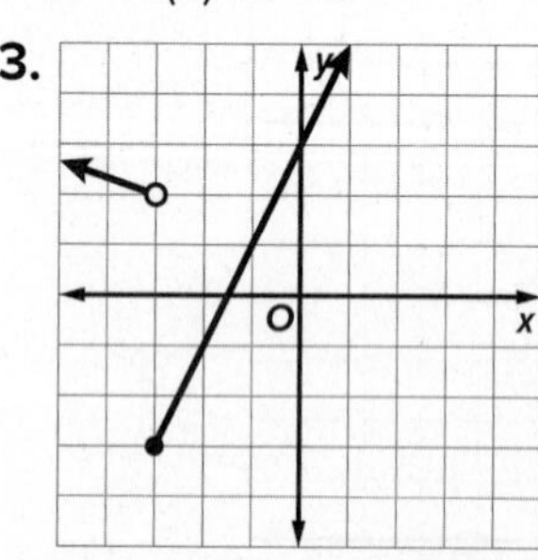

D = all real numbers,
R = $f(x) \geq -3$

5.

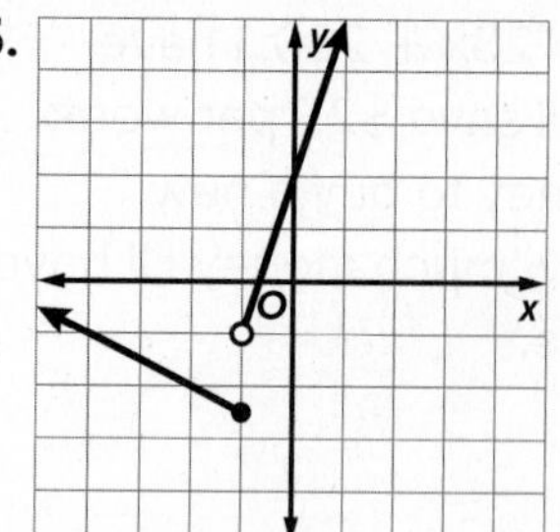

D = all real numbers,
R = $f(x) \geq -2.5$

7.

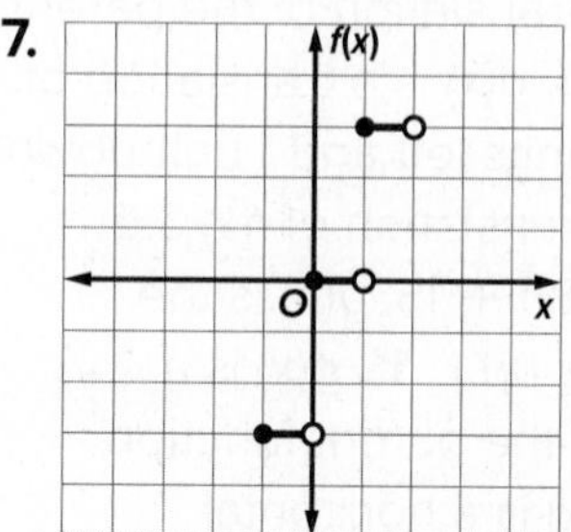

D = all real numbers,
R = all integer multiples of 3

9.

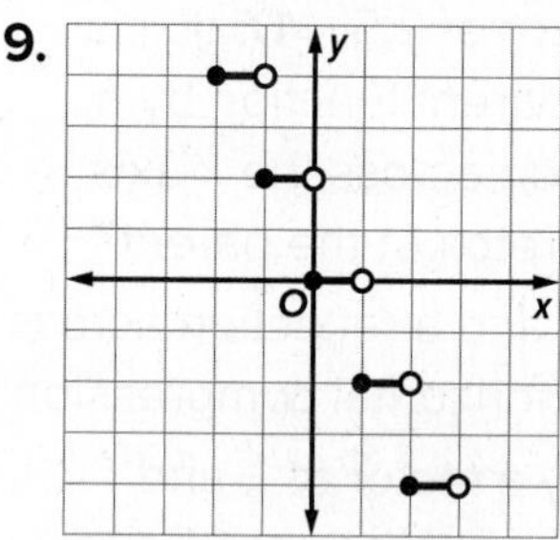

D = all real numbers,
R = all even integers

11.

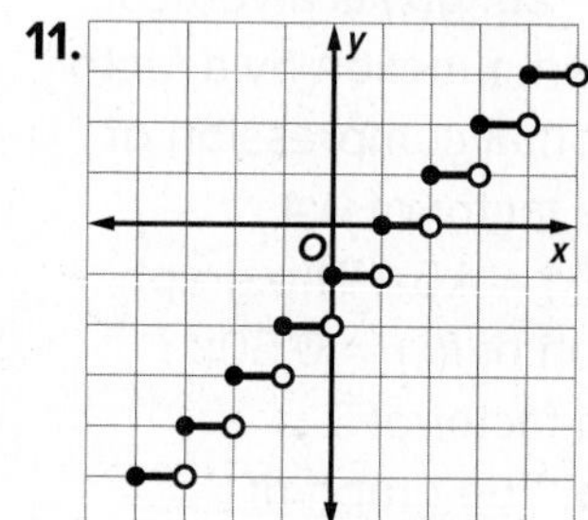

D = all real number,
R = all integers

13.

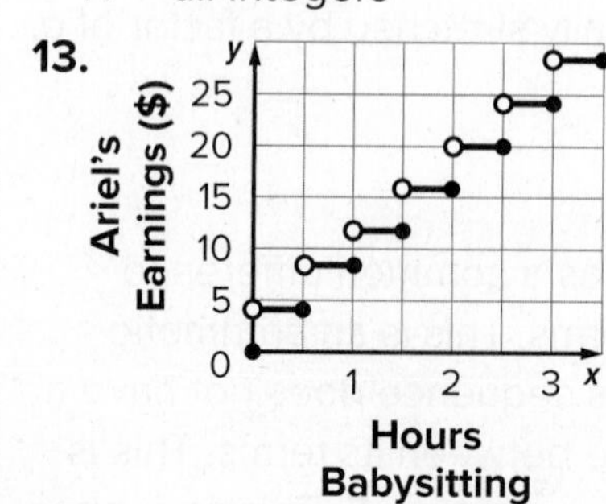

15. $f(x) = \begin{cases} 16.20 \text{ if } 0 < x \leq 1 \\ 19.30 \text{ if } 1 < x \leq 2 \\ 22.40 \text{ if } 2 < x \leq 3 \\ 25.50 \text{ if } 3 < x \leq 4 \\ 28.60 \text{ if } 4 < x \leq 5 \end{cases}$

$D = \{x \mid 0 < x < 5\}$;
$R = \{16.20, 19.30, 22.40, 25.50, 28.60\}$

17. $g(x) = \begin{cases} 2x + 1 & \text{if } x \leq 2 \\ x - 2 & \text{if } x > 2 \end{cases}$

19. $133.00

21a.

x	0	2	4	6	8
f(x)	0	75	175	275	375

21b. $f(x) = 25 + 50[[x]]$

21c.

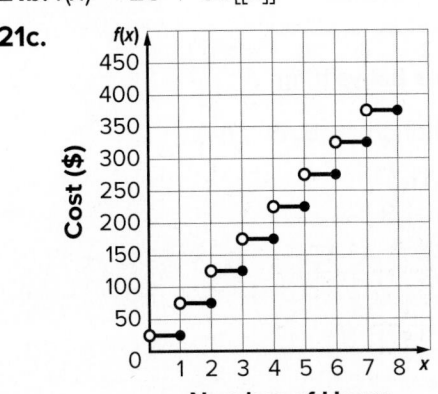

f(x)
450
400
350
300
250
200
150
100
50
0
1 2 3 4 5 6 7 8 x
Cost ($)
Number of Hours

21d. $2 < x \leq 3$

23. Sample answer:

$y = \begin{cases} -x & x < -4 \\ 2x & -4 \leq x \leq 2 \\ x & x > 2 \end{cases}$

25. A step function has different constants over different intervals of its domain. A piecewise-defined function can have different algebraic rules over different intervals of its domain.

27. $f(x) = \begin{cases} \frac{1}{2}x - 3 & x > 6 \\ -\frac{1}{2}x + 3 & x \leq 6 \end{cases}$

29. $R = f(x) \geq 0$

31. 2.4

Lesson 4-7

1. The graph of $g(x)$ is the parent function translated 5 units down. **3.** The graph of $g(x)$ is the parent function translated 2 units right and 7 units up. **5.** The graph of $g(x)$ is the parent function translated 1 unit up.

7. $f(x) = |x + 2|$ **9.** $f(x) = |x| - 3$ **11.** $f(x) = |x| + 1$ **13.** The graph of $g(x)$ is a horizontal compression of the parent function. **15.** The graph of $g(x)$ is a vertical stretch of the parent function. **17.** The graph of $g(x)$ is a horizontal stretch of a parent function. **19.** The graph of $g(x)$ is a reflection of the parent function across the x-axis and a vertical stretch. **21.** The graph of $g(x)$ is a reflection of the parent function across the y-axis and a horizontal stretch. **23.** The graph of $g(x)$ is a reflection of the parent function across the y-axis and a horizontal compression.

25.

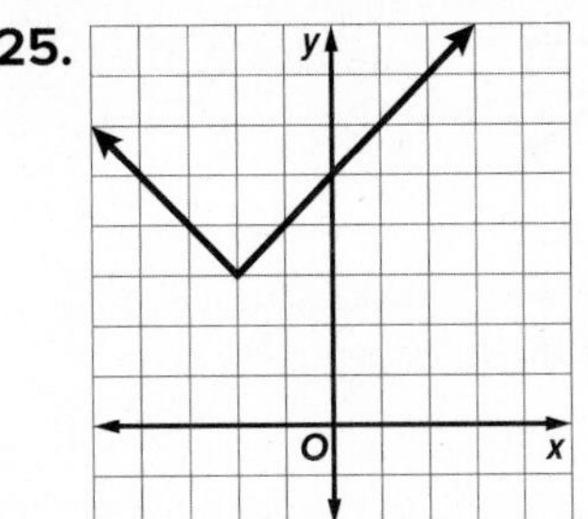

D = all real numbers,
$R = g(x) \geq 3$

27.

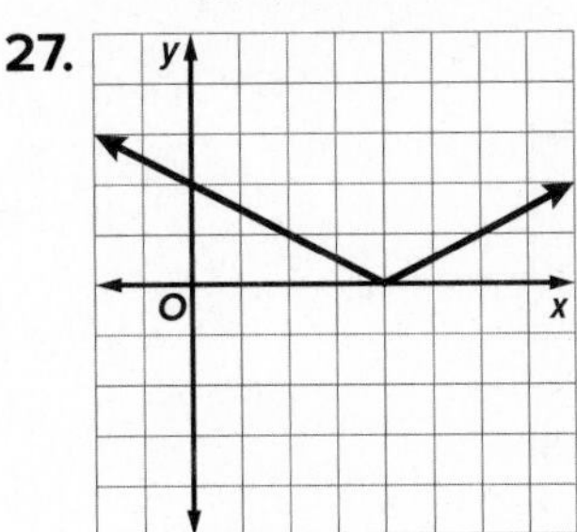

D = all real numbers,
$R = f(x) \geq 0$

29.

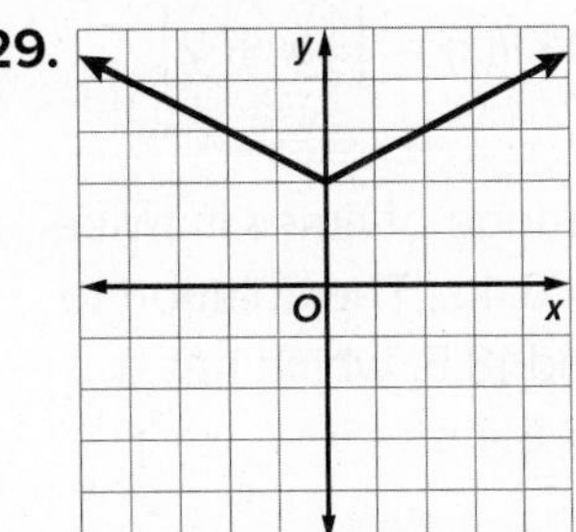

D = all real numbers,
$R = f(x) \geq 2$

Selected Answers

31.

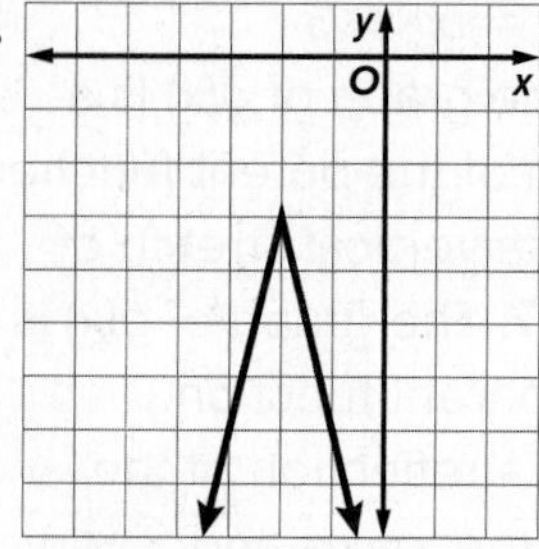

D = all real numbers,
R = $f(x) \leq -3$

33.

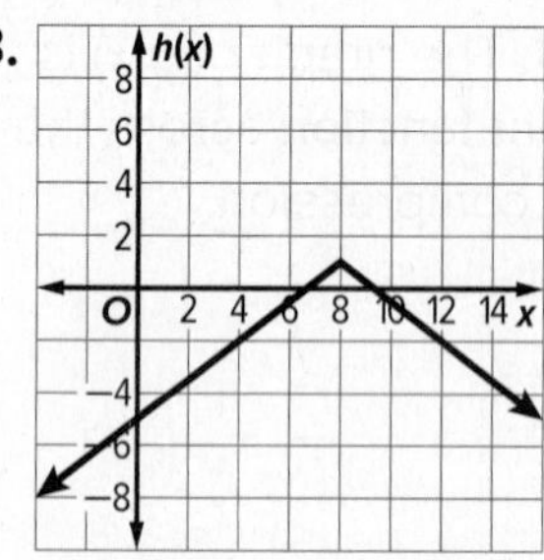

D = all real numbers,
R = $h(x) \leq 1$

35. $y = 65|10 - x|$

37.

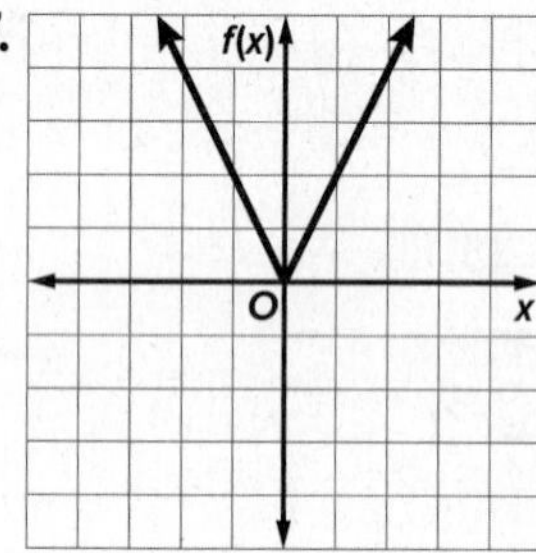

D = all real numbers,
R = $f(x) \geq 0$
The graph of $f(x)$ is the parent function horizontally compressed by a factor of $\frac{1}{2}$.

39. $f(x) = |-3x - 5|$ **41.** $f(x) = \left|\frac{1}{3}x + 2\right|$

43. $x = |s - 16|$

45. $x = |t - 21.7|$; The range of times is twice the value of x, 3.2(2) = 6.4 s; The solution to the equation is 24.9 and 18.5, which has a range of 24.9 − 18.5 = 6.4 s.

47. $x = |b - 12|$

49. To get the graph of $h(x)$, the parent absolute value function is reflected in the x-axis, then translated 2 units left and 3 units down.

51. $f(x) = \begin{cases} -x + 5 & \text{if } x < 3 \\ x - 1 & \text{if } x \geq 3 \end{cases}$

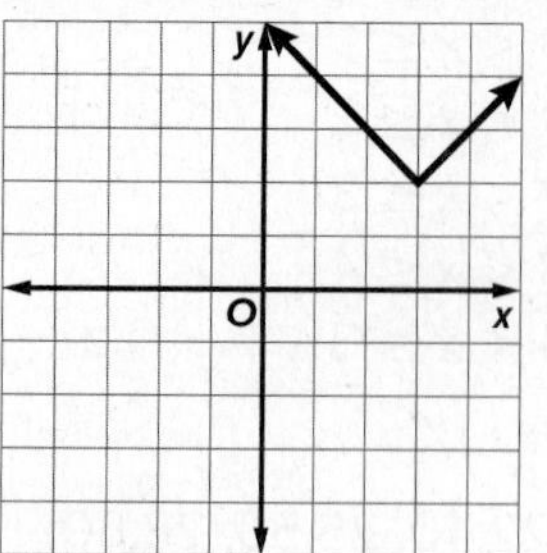

Module 4 Review

1.

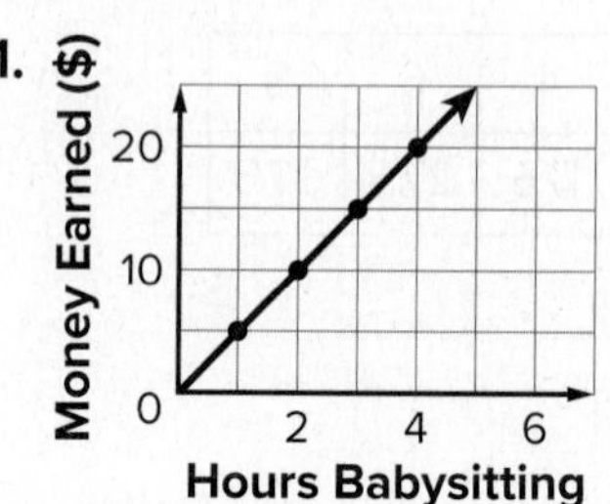

3. −500 gallons/hr **5.** B **7.** A

9. dilation **11.** C

13. $f(n) = 9n - 8$

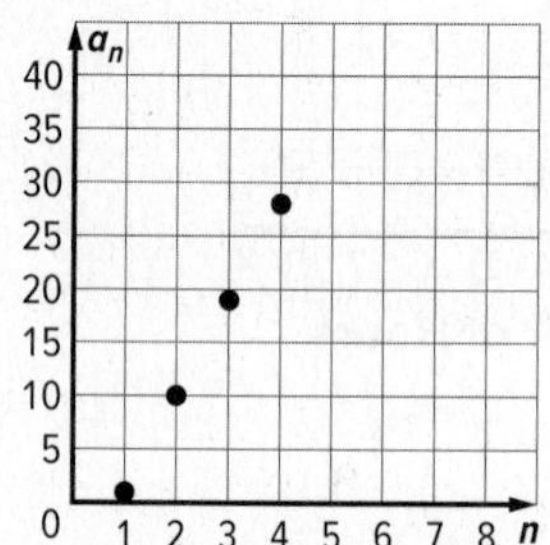

15.

Hours Worked, x	Money Earned, $f(x)$
30	270
35	315
40	360
45	427.5
50	495

17. B

19. Sample answer: It is translated 5 units up.

21. $f(x) = -|x - 4| + 3$

Module 5

Quick Check

1. $y = 5 - x$ **3.** $y = x + 5$ **5.** (4, 2) **7.** (2, −4) **9.** (−3, −3)

Lesson 5-1

1. $y = \frac{1}{2}x$ **3.** $-\frac{3}{4}x + \frac{17}{2}$ **5.** $y = \frac{1}{2}x + 1$ **7.** $d = 3t + 12$ **9.** $C = 2.54y + 62.38$ **11.** $y = -4$ **13.** $y = \frac{4}{3}x - \frac{1}{3}$ **15.** $y = -\frac{3}{2}x - \frac{9}{2}$ **17.** $y = -\frac{4}{11}x + \frac{58}{11}$ **19.** $y = -\frac{1}{2}x - \frac{9}{2}$ **21.** $y = \frac{1}{6}x + \frac{19}{24}$ **23.** $C = 10d + 12$ **25.** $T = -4.5x + 103$ **27.** $y = 3x - 1$ **29.** $y = -x - 4$ **31.** $y = -x + 3$ **33.** No; substituting 3 and −1 for x and y results in an equation that is not true. **35.** Yes; substituting 15 and −13 for x and y results in an equation that is true. **37.** Sample answer: (3, −3) **39.** Sample answer: (0, −5) **41.** Sample answer: (0, 4) **43.** C; x represents the number of plane tickets per order and y represents the total cost of an order. **45.** A; x represents the number of hours and y represents the oil level in the tank, in inches. **47a.** $y = x + 2.5$ **47b.** 10 **47c.** $y = x + 1.5$ **49a.** $y = 7.5x + 1$ **49b.** 1; Koby's puppy weighed 1 pound at birth (0 months) **49c.** 7.5; Koby's puppy gained 7.5 pounds a month for the first 6 months. **51.** Jacinta; Tess switched the x- and y-coordinates on the point that she entered in Step 3. **53.** Sample answer: Let y represent the number of quarts of water in a pitcher, and let x represent the time in seconds that water is pouring from the pitcher. As time increase by 1 second, the amount of water in the pitcher decrease by $\frac{1}{2}$ qt. An equation is $y = -\frac{1}{2}x + 4$. The slope is the rate at which the water is leaving the pitcher, $\frac{1}{2}$ quart per second. The y-intercept represents the amount of water in the pitcher when it is full, 4 qt.

Lesson 5-2

1. $y + 3 = -1(x + 6)$

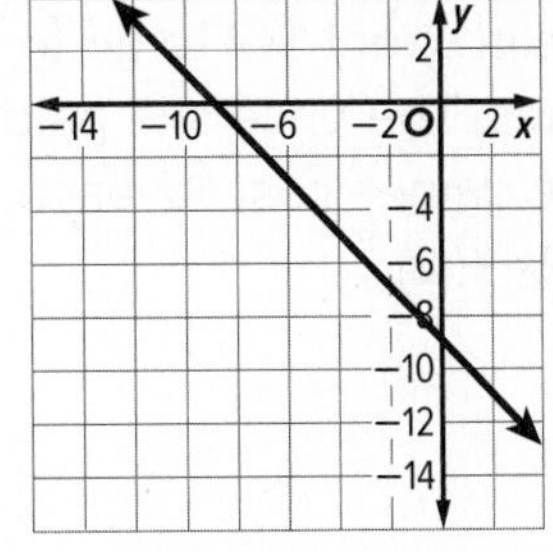

3. $y - 11 = \frac{4}{3}(x + 2)$

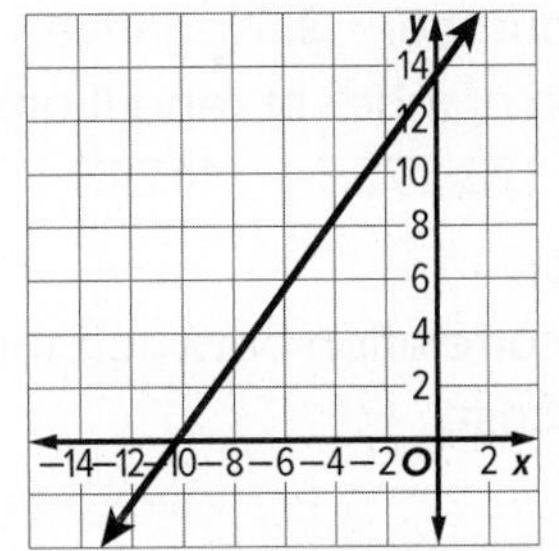

5. Sample answer: $y + 3 = -4(x - 1)$
7. Sample answer: $y - 3 = \frac{4}{3}(x - 3)$
9. $y = -6x - 47$ **11.** $y = \frac{1}{6}x - \frac{8}{3}$
13. $y - 18 = 3.5(x - 5)$ **15.** $2x - y = 6$
17. $x + 6y = -7$ **19.** $x - y = -1$
21. Sample answer: $2x + 3y = -13$
23. Sample answer: $3x + y = -3$
25. Sample answer: $y = x - 5$; $y = -x + 1$
27. Sample answer: $y = -5x + 2$; $y = \frac{1}{5}x + 2$
29. Sample answer:
$y = -\frac{3}{4}x + \frac{3}{2}$; $y = \frac{4}{3}x + \frac{17}{3}$
31. neither
33. perpendicular **35.** neither
37. Sample answer: $5x + 4y = 20$
39. $y = 9x + 5$; $9x - y = -5$
41. $y = -6x - 45$; $6x + y = -45$
43. $y = \frac{9}{10}x - 4\frac{3}{10}$; $9x - 10y = 43$
45. Yes; sample answer: The line that represents one of the ceiling walls has a slope of $-\frac{1}{4}$ and the line that represents the other ceiling wall has a slope of 4.
47a. Sample answer: $y - 0 = 0.5(x - 0)$
47b. $y = 0.5x$ **47c.** $x - 2y = 0$
49. Sample answer: You need to know the slope of the line and the y-intercept of the line, the slope and the coordinates of another point on the line, or the coordinates of two points on the line.

51. No; the line through (7, −10) and (3, −2) has a slope of −2 and $2x - y = -5$ has a slope of 2. **53.** Sample answer: $y - g = \frac{j - g}{h - f}(x - f)$ **55.** Sample answer: Jocari spent $18 to go to a carnival and play games. The price she paid included admission. The games cost $2 each; $y - 18 = 2(x - 5)$, $y = 2x + 8$.

Lesson 5-3

1. Positive; as time spent exercising increases, the more Calories are burned. **3.** Negative; as weight increases, the number of repetitions decreases. **5a.** $y = -328.275x + 3142.15$ **5b.** about 187.675 million **7a.** There is a positive correlation between the child's age and annual cost.

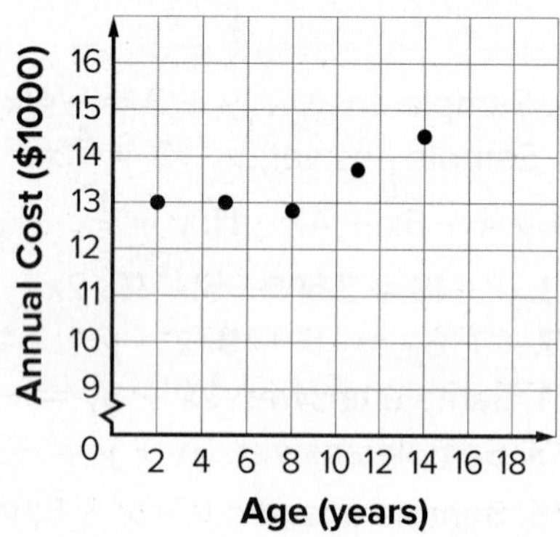

7b. $y = 270x + 10{,}640$ **7c.** about $15,230 **9.** no correlation **11a.**

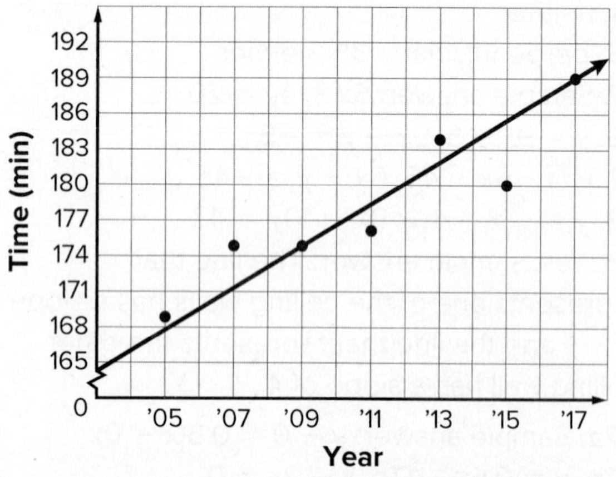

11b. Sample answer: x represents the number of years since 2005, so year 2005 is represented by $x = 0$ and year 2020 is represented by $x = 15$. Two points on the line of best fit are (4, 175) and (17, 189). Use these two points to find the slope to be 1.75 and the equation of the line of best fit to be $y = 1.75x + 168$. **11c.** Sample answer: about 196 minutes **11d.** Sample answer: Not all of the data points are close to the line of best fit, so there is not a consistent trend regarding the length of games. Therefore, the predicted game length may or may not be accurate. **13a.** positive correlation; As the number of years since 2007 increases, the price of a ticket increases. **13b.** $y = 2.87x + 64.24$ **13c.** about $133.12 **15a.**

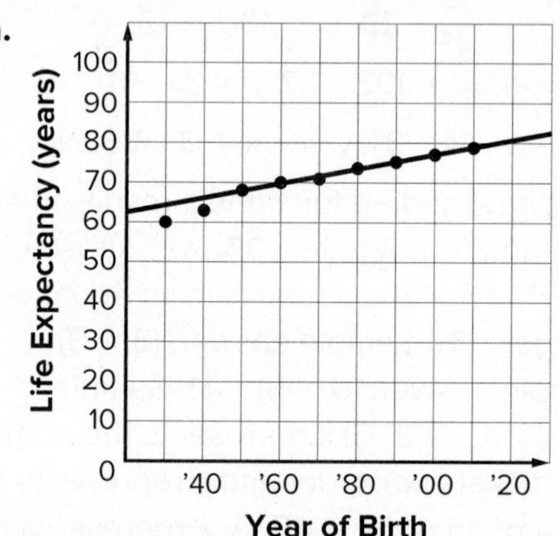

15b. Sample answer: About 81.8; The data show a positive correlation, so as the years increase, the life expectancy also increases. Therefore, the life expectancy should be higher than that of a baby born in 2010. **15c.** Sample answer: I assumed that the trend continues, so as the year increases, the life expectancy also increases. **17.** Sample answer: The salary of an individual and the years of experience that he or she has could be modeled using a scatter plot. This would be a positive correlation because the more experience an individual has, the higher the salary would likely be. **19.** Neither; line g has the same number of points above the line and below the line. Line f is close to 2 of the points; but for the rest of the data, there are 3 points above and 3 points below the line. **21.** Sample answer: You can visualize a line to determine whether the data has a positive or negative correlation. The graph shows the ages and heights of people. To predict a person's age given his or her height, write a linear equation for the line of fit. Then substitute the person's height and solve for the corresponding age. You can use the pattern in the scatter plot to make decisions.

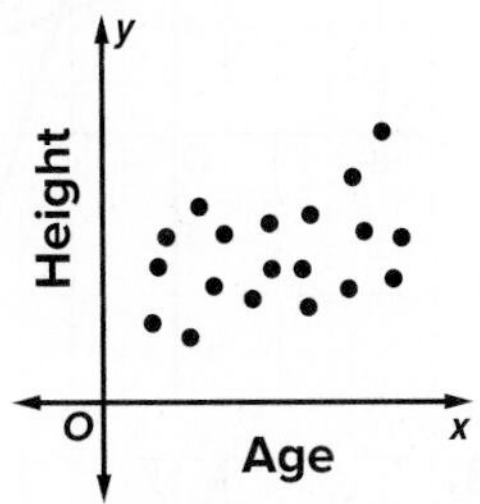

Lesson 5-4

1a.

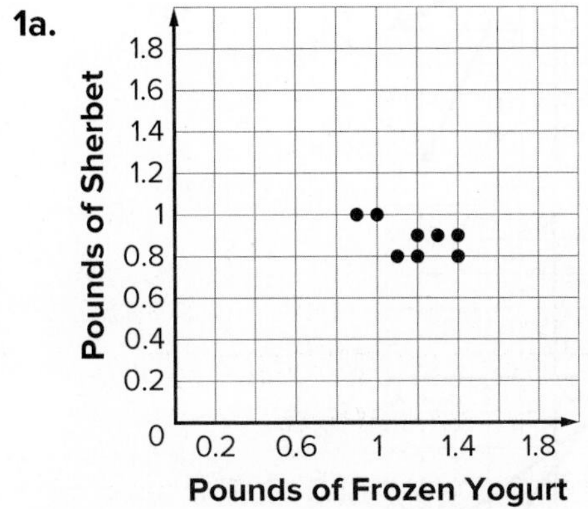

1b. Negative; as the number of pounds of frozen yogurt consumed increases, the number of pounds of sherbet consumed decreases.
1c. The relationship may be a causation. Since both are frozen desserts, eating more frozen yogurt may cause people to decrease the amount of sherbet they eat. Other things that might influence the data are an increase in frozen yogurt stores and a decrease in popularity or availability of sherbet.
3. Correlation, sample answer: Having a wider palm does not cause someone to watch less television. **5.** Causation; sample answer: An increase in the price of cereal likely causes customers to buy less cereal.

7a.

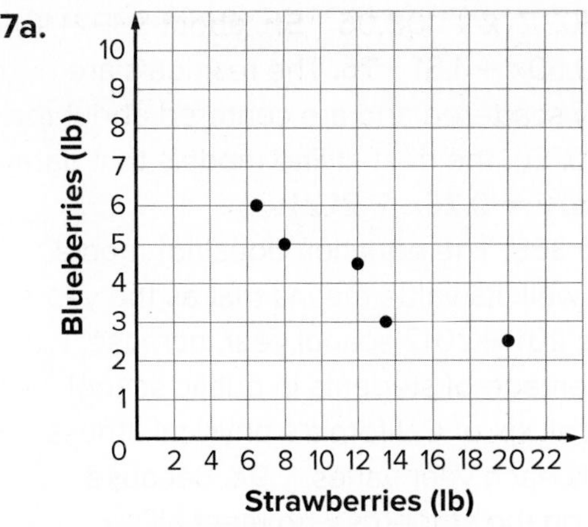

7b. Negative; as the number of pounds of strawberries produced increases, the number of pounds of blueberries produced decreases.
7c. The relationship is a correlation, but not a causation. A better yield of strawberries does not cause the blueberries to grow poorly. Other factors, such as temperature and rain, could be affecting the plants that week.

9. positive correlation and causation; Sample answer: Because pizzas are topped with cheese, an increase in the number of pizzas made cause more cheese to be used.
11. Sample answer: Two elements can have a strong correlation, but it does not mean that one causes the other. There could be an unknown factor affecting the elements.
13. Sample answer: Correlation does not mean causation. Even though there is a strong correlation that does not mean buying swimsuits causes the use of air conditioners. Another factor, like the temperature, could be affecting both swimsuit sales and use of air conditioners.

Lesson 5-5

1a. $y = -1.31x + 50.95$ **1b.** $r \approx -0.714$; The equation models the data fairly well. Its negative value means that as the years since 2010 increase, the total number of goals the soccer team scores each season decreases.
3a. $y = 8.52x + 3.18$
3b. $r \approx 0.999$; The equation models the data very well. Its positive value means that as the years since 2010 increase, sales, in millions of dollars, increase.

Selected Answers

5a. $y = 103.77x + 108.06$ **5b.** about $3221.16 **7a.** $y = 0.59x + 1.51$ **7b.** The residuals are randomly scattered and are centered about the line $y = 0$. So, the best-fit line models that data well. **9a.** $y = 0.26x + 21.21$
9b. $r \approx 0.359$; The equation does not model the data well. Its value means that as the years since the 2011–2012 school year increase, the percentage of students in public school who met all six of California's physical fitness standards each year varies. **9c.** Because the data on the students who meet all six standards is reported as a percentage, it cannot exceed 100. **11a.** $y = 140.4x + 13.8$
11b. $r \approx 0.999$; The equation models the data very well. Its positive value means that as the number of games increases, the cumulative number of yards increases.
11c. Sample answer: Because the data have a positive correlation, the total number of yards will increase and then number of games increases. So, the running back will have run for 950 yards between games 6 and 9.
11d. during game 7
13a. $y = 9619x + 443{,}918.8$ **13b.** $r \approx 0.999$; The equation models the data very well. Its positive value means that as the number of years since the 2010-2011 school year increases, the number of student athletes participating in college athletics each year increases.
13c. The residuals are randomly scattered and are centered about the line $y = 0$. So, the best-fit line models that data well.
13d. about 684,394
15. Apply a linear regression model to the data. Use the number of each test as the independent variable. If there is no correlation, the r-value will not be close enough to 1 or −1. If this is the case, the line of fit could not be used to predict the scores of the other students.
17a. $y = 84{,}345.0x + 5{,}003{,}868.3$
17b. about 7,365,528

Lesson 5-6

1. {(−1, −9), (−4, −7), (−7, −5), (−10, −3), (−13, −1)}
3. {(−2, −4), (−1, −2), (1, 0), (0, 2), (2, 4)}
5. {(−3, 5), (−9, 2), (−15, −1), (−21, −4)}
7. {(16, −1), (12, −2), (8, −3), (4, −4)}
9. {(−49, −4), (35, 8), (−28, −1), (7, 4)}

11. 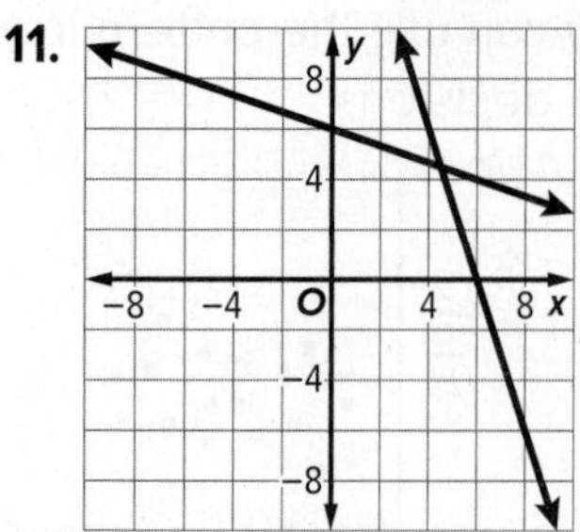

13. 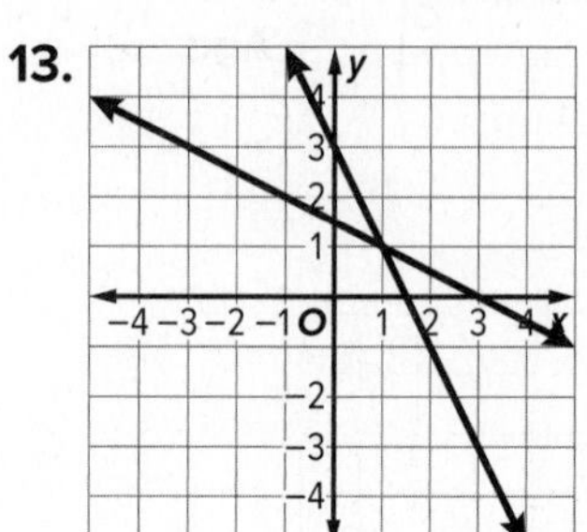

15. 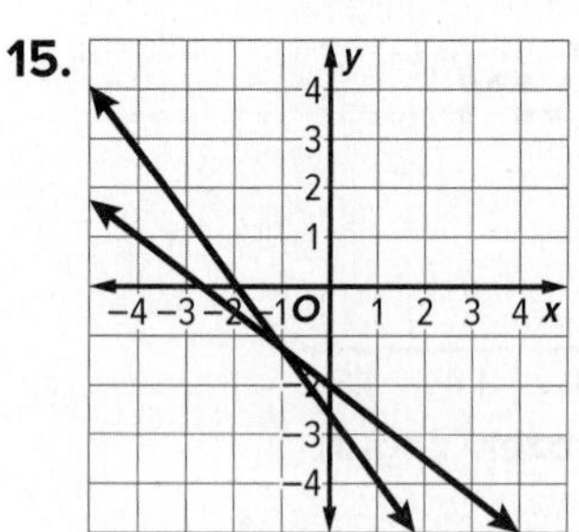

17. $f^{-1}(x) = \frac{x}{6} - 7$
19. $f^{-1}(x) = \frac{5}{2}(x + 16)$ **21.** $f^{-1}(x) = \frac{1 - 5x}{4}$
23a. $P^{-1}(x) = \frac{x + 36}{7.6}$ **23b.** x represents Alisha's profit and represents the number of dozens of brownies sold. **23c.** 5
25a. $C^{-1}(x) = \frac{x - 125}{16}$ **25b.** 108 feet
27. $f^{-1}(x) = \frac{1}{4}x + 6$ **29.** $f^{-1}(x) = 6x - 42$
31. $f^{-1}(x) = \frac{7}{2}x - 14$ **33.** $f^{-1}(x) = \frac{1}{7}x - \frac{6}{7}$
35. $f^{-1}(x) = 2x - 22$ **37.** B **39.** A
41. {(−k, b), (p, −g), (−m, −w), (q, r)} **43.** The slopes are reciprocals. For example, if the slope of one line is $\frac{2}{3}$, then the slope of the inverse function is $\frac{3}{2}$. **45.** Sample answer: This claim is incorrect. The −1 in the inverse function notation is not an exponent. As an example, the inverse function for $y = x + 1$ is found by switching x and y and solving for y, which gives $y = x - 1$. $y = x - 1$ is not the same as $y = \frac{1}{(x + 1)}$, which is not a line. This method does not work.

47. $a = 2$; $b = 14$

49. sometimes; Sample answer: $f(x)$ and $g(x)$ do not need to be inverse functions for $f(a) = b$ and $g(b) = a$. For example, if $f(x) = 2x + 10$, then $f(2) = 14$ and if $g(x) = x - 12$, then $g(14) = 2$, but $f(x)$ and $g(x)$ are not inverse functions. However, if $f(x)$ and $g(x)$ are inverse functions, then $f(a) = b$ and $g(b) = a$.

51. Sample answer: A situation may require substituting values for the dependent variable into a function. By finding the inverse of the function, the dependent variable becomes the independent variable. This makes the substitution an easier process.

Module 5 Review

1. A **3.** $y = 1.5x + 11$ **5.** A

7. $y - 4 = 2.5(x - 2)$ **9.** A

11. a positive correlation **13.** C

15. $f^{-1}(x) = -2x + 1$;

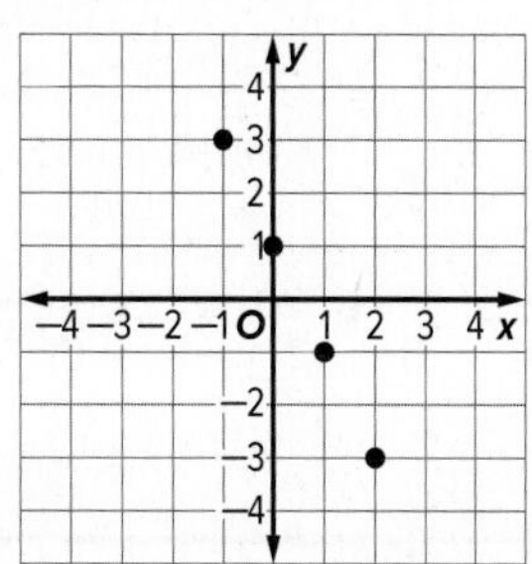

17. A

Selected Answers

Module 6

Quick Check

1. 4 **3.** −2 **5.** {−29, 7} **7.** {−1, 15}

Lesson 6-1

1.

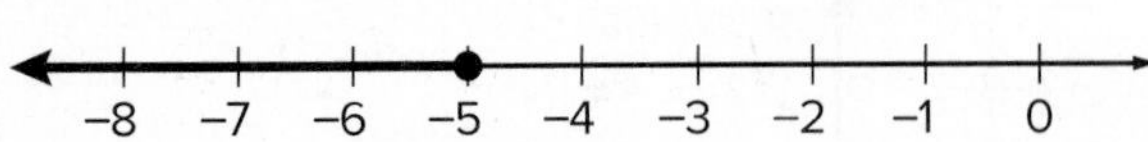

3.

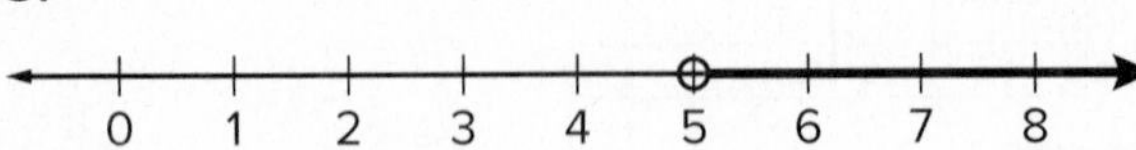

5.

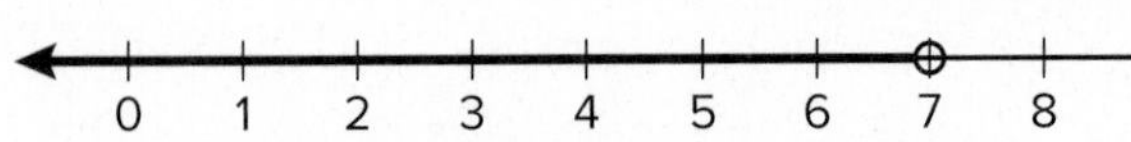

7. $t < -1$ **9.** $w < 5$ **11.** $b \geq -5$
13. $\{m \mid m < 7\}$ **15.** $\{r \mid r \leq 15\}$ **17.** $\{b \mid b \geq 2\}$
19. $\{c \mid c \leq -4\}$ **21.** $\{m \mid m \geq 4\}$
23. $\{r \mid r \geq 22\}$ **25.** $\{a \mid a \leq -4\}$
27. $\{w \mid w \geq -5\}$ **29.** $\{x \mid x \leq 5\}$
31. $\frac{3}{10}x \leq 4.50$, $x \leq \$15$ **33.** no more than 2.1 pounds per day **35.** at least 500 pieces
37. $\{m \mid m \leq -68\}$

−76 −72 −68 −64 −60

39. $\{c \mid c > 121\}$

106 112 118 124 130

41. $\{x \mid x \leq 20\}$

0 6 12 18 24

43. $\{h \mid h > 21\}$

10 14 18 22 26

45. $\{n \mid n \geq 108\}$

100 106 112 118 124

47. $\{r \mid r < 16\}$

0 6 12 18 24

49. $\{t \mid t > -1\}$

−8 −4 0 4 8

51. $\{z \mid z \geq 11\}$

0 6 12 18 24

53. $\{d \mid d > -2\frac{1}{2}\}$

−8 −4 0 4 8

55. d **57.** a **59.** b
61. Sample answer: Let n = the number. $n + 7 \leq -18$; $\{n \mid n \leq -25\}$
63. Sample answer: Let n = the number. $n + 2 \leq 1$; $\{n \mid n \leq -1\}$
65. Sample answer: Let n = the number. $-12n \leq 84$; $\{n \mid n \geq -7\}$
67. $\{g \mid g > 4\}$

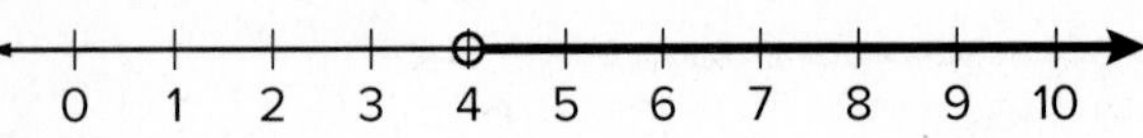

69. $\{x \mid x < 36\}$

30 31 32 33 34 35 36 37 38 39 40

71. $\{m \mid m < 5.4\}$

4 4.2 4.4 4.6 4.8 5 5.2 5.4 5.6 5.8 6

73. $\{c \mid c \geq 3.7\}$

3.5 3.6 3.7 3.8 3.9 4.0 4.1 4.2 4.3 4.4 4.5

75. \$22.23.
77. Sample answer: Let x represent the decibel level of the calls of a blue whale; $x - 83 \leq 105$; $x \leq 188$. The calls of a blue whale are less than or equal to 188 decibels.
79. $-\frac{x}{2} < 1$ **81a.** $x < \frac{7}{a}$ **81b.** $x \geq \frac{12}{a}$
81c. $x > 3$ **81d.** $x \geq \frac{1}{4}$

Lesson 6-2

1a. $15 + 2h \leq 35$ **1b.** $h \leq 10$; 10 hours
3a. $1.50 + 0.25(5x - 1) \leq 3.75$ **3b.** $x \leq 2$; 2 mi
3c. Because the service charges per $\frac{1}{a}$ mile, multiply a by the number of miles, x, to find the number of $\frac{1}{a}$ miles. Subtract 1 from the total number of $\frac{1}{a}$ miles, ax, to find the number of additional $\frac{1}{a}$ miles. Multiply the difference by the cost per additional $\frac{1}{a}$ mile, \$0.25, and add the cost for the first $\frac{1}{a}$ mile, \$1.50. This sum is

less than or equal to the total amount Demetri has in his pocket, \$3.75, so $1.50 + 0.25(ax - 1) \leq 3.75$. **5a.** $100 + 40x \leq 250$
5b. $x \leq 3.75$; 3 people

7. $21 > 15 + 2x$; $x < 3$

−10 −5 0 5 10

9. $\frac{x}{8} - 13 > -6$; $x > 56$

50 54 58 62 66

11. $37 < 7 - 10x$; $x < -3$

−10 −5 0 5 10

13. $-\frac{5}{4}x + 6 < 12$; $x > -\frac{24}{5}$

−10 −5 0 5 10

15. $15x + 30 < 10x - 45$; $x < -15$

−24 −16 −8 0 8

17. $\{a \mid a \leq 11\}$

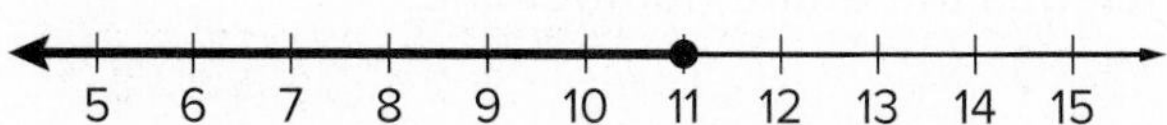

19. $\{b \mid b$ is a real number.$\}$

−5 −4 −3 −2 −1 0 1 2 3 4 5

21. $\{a \mid a \geq -9\}$

−15 −14 −13 −12 −11 −10 −9 −8 −7 −6 −5

23. $\{x \mid x \geq \frac{1}{2}\}$ **25.** $\{m | m \geq 18\}$
27. $\{w | w > -2\}$ **29.** $\{x \mid x \leq 8\}$
31. $\{x \mid x > -6\}$ **33.** $\{x \mid x \geq 1.5\}$
35. $\left\{p \mid p \leq 1\frac{1}{9}\right\}$

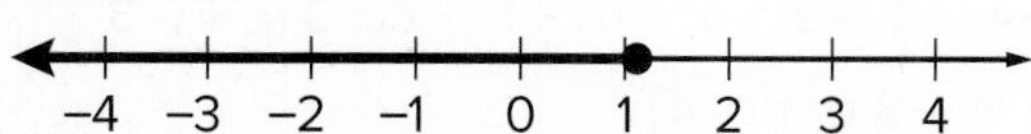

37a. $2x + 4 \leq 13$; $x \leq 4.5$
37b. 4.5 ft.
37c. 5ft.
39. Eric does not have any pencils. Based on his statement, the inequality is $6p + 15 < 20$, where p is the number of pencils. The solution of the inequality is $p < \frac{5}{6}$. However, the number of pencils must be a whole number, so $p = 0$.

41. $10n - 7(n + 2) > 5n - 12$
(Original inequality)
$10n - 7n - 14 > 5n - 12$
(Distributive Property)
$3n - 14 > 5n - 12$
(Combine like terms.)
$3n - 14 - 5n > 5n - 12 - 5n$
(Subtract $5n$ from each side.)
$-2n - 14 > -12$
(Simplify.)
$-2n - 14 + 14 > -12 + 14$
(Add 14 to each side.)
$-2n > 2$
(Simplify.)
$\frac{-2n}{-2} < \frac{2}{-2}$
Divide each side by −2. Change > to <.
$n < -1$
(Simplify.)
The solution set is $\{n \mid n < -1\}$.
43. $\frac{76 + 80 + 78 + x}{4} \geq 82$; $x \geq 94$; Mei needs a score of at least 94 on the next exam.
45. Sample answer: $2(2x - 1) < 10$
47. Let c = the number of baseball cards Ted has; $4c > 5c - 15$; $15 > c$; Ted has fewer than 15 cards. **49.** Ø; If the inequality is always true, the opposite inequality will always be false.
51. Sample answer: The solution set for the inequality that results in a false statement is the empty set, as in $12 \geq 15$. The solution set for an inequality in which any value of x results in a true statement is all real numbers, as in $12 \leq 12$.

Lesson 6-3

1. $\{f \mid 6 < f < 11\}$

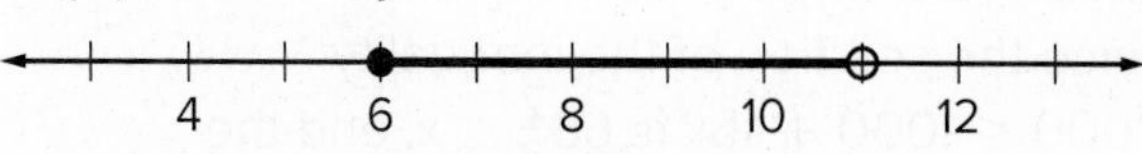

3. $\{y \mid y \geq 8$ or $y < -4\}$

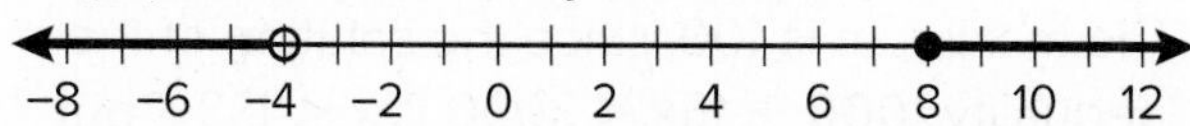

5. $\{p \mid -4 < p \leq 5\}$

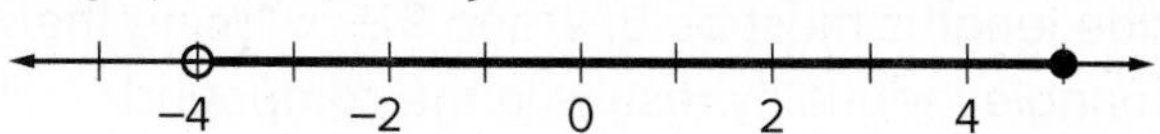

Selected Answers

7. $\{h \mid 2 \leq h < 3\}$

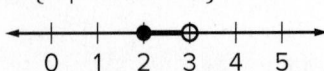

0 1 2 3 4 5

9. $\{y \mid y < -3\}$

−6 −5 −4 −3 −2 −1 0

11. $\{b \mid 4 < b \leq 5\}$

1 2 3 4 5 6 7

13. $\{m \mid m < -6 \text{ or } m > -1\}$

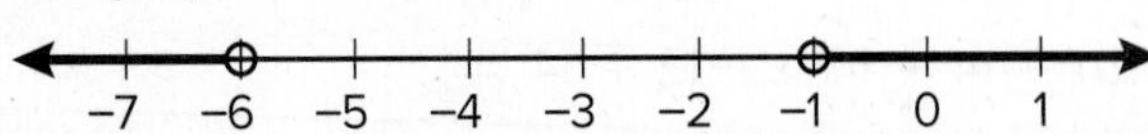

15. $\{m \mid 2 \leq m < 4\}$

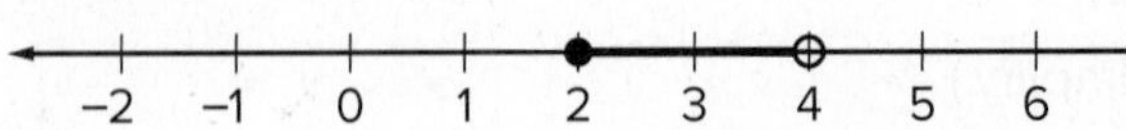

17a. $x + 8 < 20$ or $x + 8 > 35$ **17b.** $0 < x < 12$ or $x > 27$; Because the combined height of the sign and pole cannot be negative, the value of x must be greater than 0.

17c.

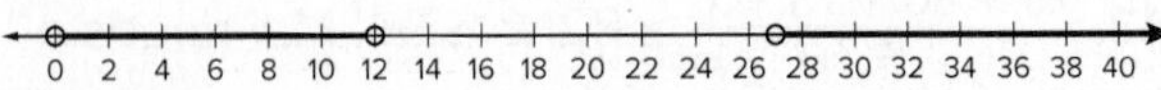

19. $-3 < x \leq 3$ **21.** $x < -2$ or $x \geq 1$
23. $b > 3$ or $b \leq 0$ **25.** $y < -1$ or $y \geq 1$
27. $f \mid -2 < f < -1\}$

−4 −3 −2 −1 0 1 2 3 4

29. $\{b \mid -2 < b < 6\}$

−2 −1 0 1 2 3 4 5 6

31. $\{a \mid -2 \leq a < 5\}$

−3 −2 −1 0 1 2 3 4 5

33. Sample answer: Let $n =$ the number. $n - 2 \leq 4$ or $n - 2 \geq 9$; $\{n \mid n \leq 6 \text{ or } n \geq 11\}$
35. $54° \leq x \leq 68°$ **37.** The minimum is 67, since the solution of the inequality $2000 \leq 1000 + 15x$ is $66\frac{2}{3} \leq x$, and the number of students must be a whole number. The maximum is 100, since the solution of the inequality $1000 + 15x \leq 3000$ is $x \leq 133\frac{1}{3}$, but the bus can only hold 100 students. **39a.** The side lengths must be 5, x, and $9 - x$. Using the Triangle Inequality results in the compound inequality $x + 5 > 9 - x$ and $14 - x > x$.

39b. The solution of the compound inequality is $2 < x < 7$, so each of the lengths must be greater than 2 m but less than 7 m. The sum of the two lengths must be 9 m.

41. $\{x \mid -2 < x < 5\}$

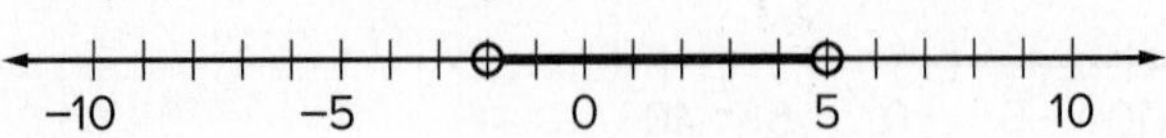

43a. $\$400 \leq x \leq \$800$ **43b.** $\$428 \leq x \leq \$856$
45. B
47. $x > -1$ or $x \geq 4$; This can be written as $x > -1$ because this is the union of two graphs.
49. The union of the two graphs is the graph on the left, so the graph on the left is the graph of the solution set for **Exercise 47.** The intersection of the two graphs is the graph on the right, so the graph on the right is the graph of the solution set for **Exercise 48.**
51a. $x > -\frac{4}{a}$ and $x \leq \frac{4}{a}$
51b. $x < -\frac{6}{a}$ or $x > 5a$
53. Sometimes; The graph of $x > 2$ or $x < 5$ includes the entire number line.

Lesson 6-4

1. $\{x \mid -24 < x < 8\}$

−28 −24 −20 −16 −12 −8 −4 0 4 8 12

3. $\{c \mid -3 \leq c \leq 4\}$

−5 −4 −3 −2 −1 0 1 2 3 4 5

5. $\{\varnothing\}$

−5 −4 −3 −2 −1 0 1 2 3 4 5

7. $\{r \mid r < -8 \text{ or } r > 4\}$

−10 −8 −6 −4 −2 0 2 4 6

9. $\{h \mid h \leq -3 \text{ or } h \geq 6\}$

−4 −2 0 2 4 6 8

11. $\{v \mid v$ is a real number.$\}$

−5 −4 −3 −2 −1 0 1 2 3 4 5

13. $\left\{n \mid n \leq -5\frac{1}{4} \text{ or } n \geq 3\frac{3}{4}\right\}$

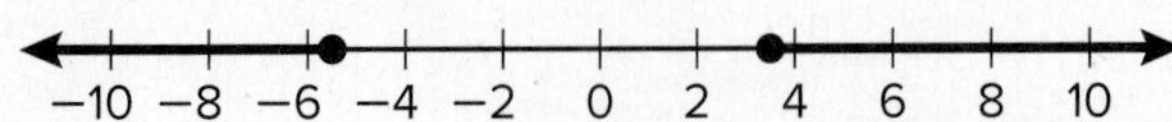

15. $\{h \mid -5\frac{2}{3} < h < 5\}$

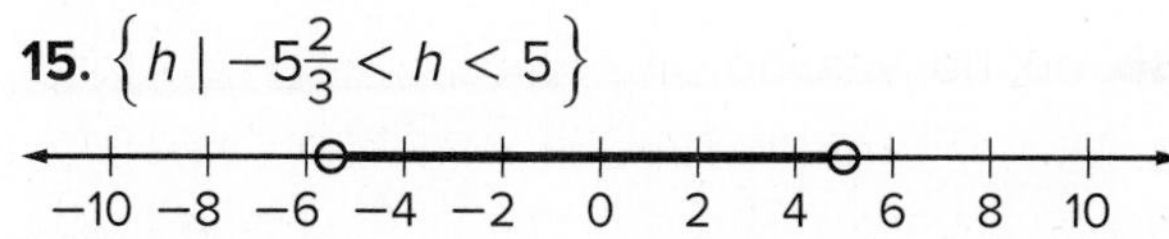

17. $\{\emptyset\}$

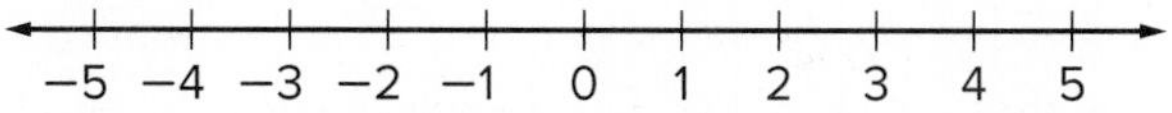

19. $\{r \mid -2 < r < \frac{2}{3}\}$

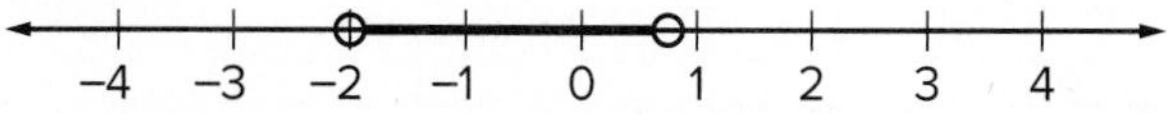

21. $\{x \mid 58.5 \leq x \leq 61.5\}$ **23a.** $|p - 130| \leq 3.05$

23b.

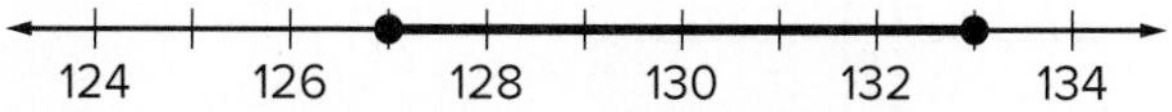

25a. $|x - 515| \leq 114$ **25b.** 287 to 743

27. $|n + 2| \geq 1$ **29.** $|w - 2| < 2$ **31.** $|x| > 1$

33. d **35.** c

37. $|x - 92| \leq 8$; $\{x \mid 84 \leq x \leq 100\}$

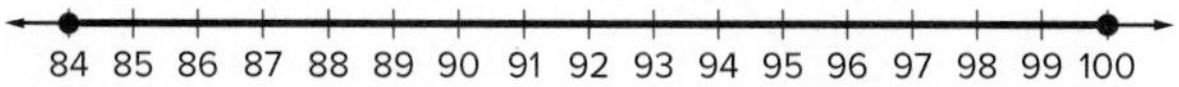

39. $\{x \mid -1 \leq x \leq 3\}$

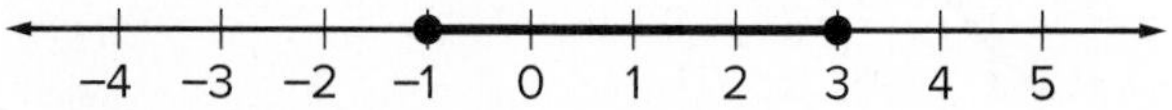

41. $\{x \mid -7 \leq x \leq 3\}$

43. $\{x \mid x > 18 \text{ or } x < -17\}$

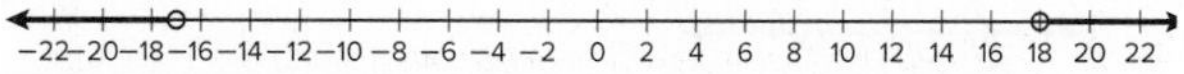

45. By definition, the absolute value is always greater than a negative number. Therefore, no matter what number is chosen, it will always be greater than −1 when evaluated in the absolute value inequality given. **47a.** Set the absolute value of an unknown variable, *x*, minus the recommended weight, 516, to be less than or equal to the variance of 4. So, the inequality $|x - 516| \leq 4$ represents the situation.

47b. Write two inequalities, one for each case: $x - 516 \leq 4$ and $-(x - 516) \leq 4$. For the first case, add 516 to both sides: $x \leq 520$. For the second case, distribute the negative on the left side, subtract 516 from both sides and divide by a negative 1 remembering to switch the inequality sign: $x \geq 512$. This means a box of cereal should have a minimum weight of 512 g and a maximum weight of 520 g.

49. The solution set for $|x - 2| > 4$ is $\{x \mid x < -2 \text{ or } x > 6\}$. The solution set for $-2x < 4$ or $x > 6$ is $\{x \mid x > -2\}$. One includes numbers greater than −2, and the other includes numbers less than −2 or greater than 6. These solution sets are not the same.

51. Jordan is correct. Chloe did not distribute the negative to both *x* and 3.

53. $(-8 \leq n < -3)$ or $(1 < n \leq 6)$. To solve this compound inequality, split it into two inequalities. The first one to solve is $|n + 1| > 2$ and the second one is $|n + 1| \leq 7$. The solution set of the entire problem is the overlap of the individual solutions.

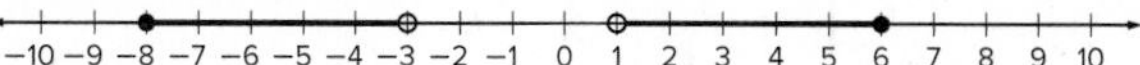

55. No; Sample answer: Lucita forgot to change the direction of the inequality sign for the negative case of the absolute value.

57. Sample answer: If $t = 0$, then the absolute value is equal to 0, not greater than 0.

59. Sample answer: When an absolute value is on the left and the inequality symbol < or ≤, the compound sentence uses *and*, and if the inequality symbol is > or ≥, the compound sentence uses *or*. To solve, if $|x| < n$, then set up and solve the inequalities $x < n$ and $x > -n$, and if $|x| > n$, then set up and solve the inequalities $x > n$ or $x < -n$.

Lesson 6-5

1.

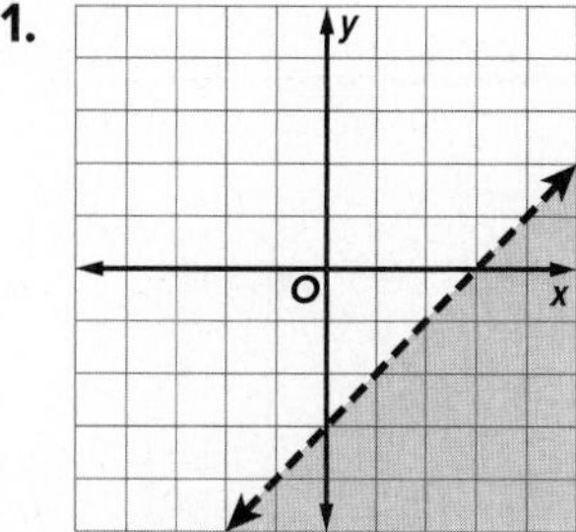

3.

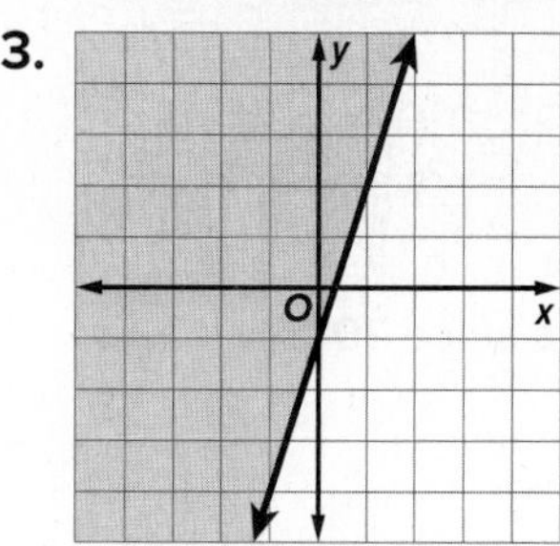

5.

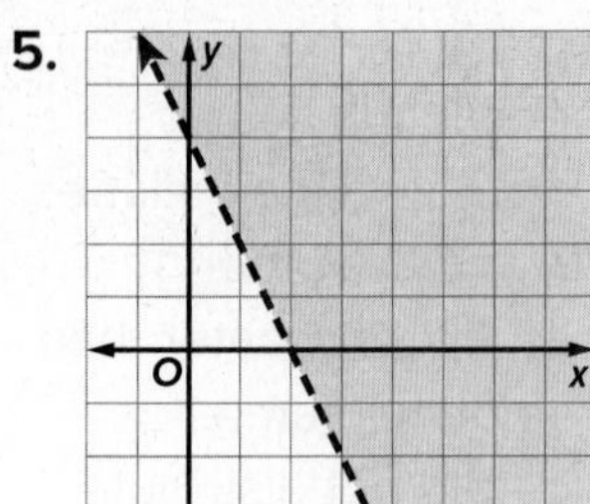

7.

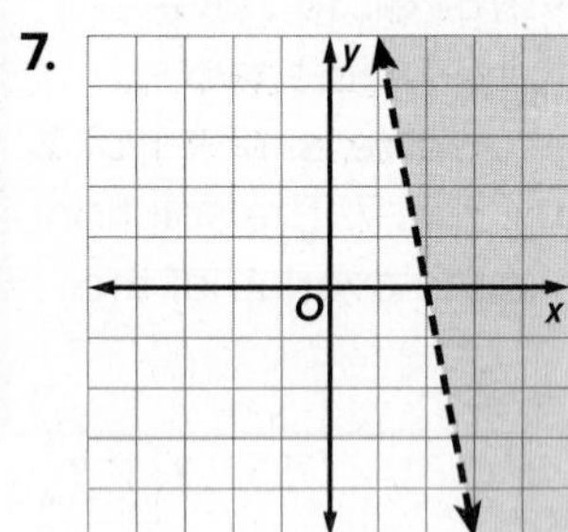

9.

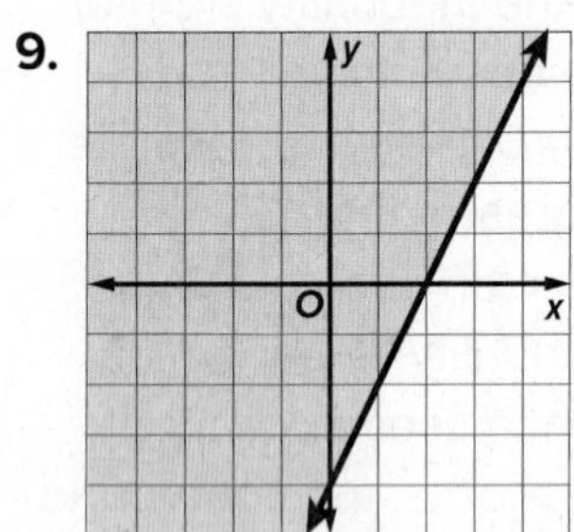

11.

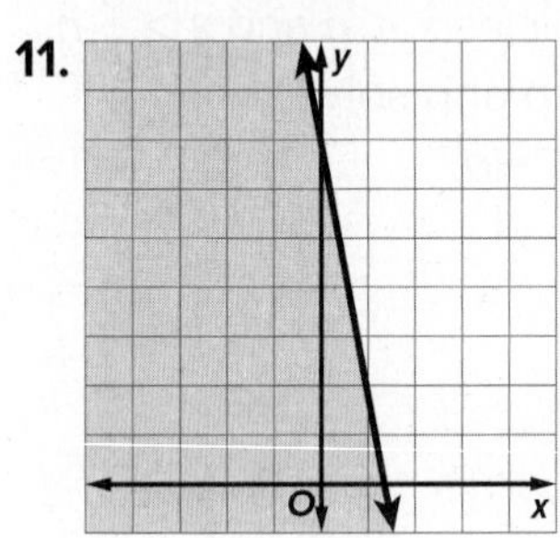

13a. $y < 1240x + 48{,}200$

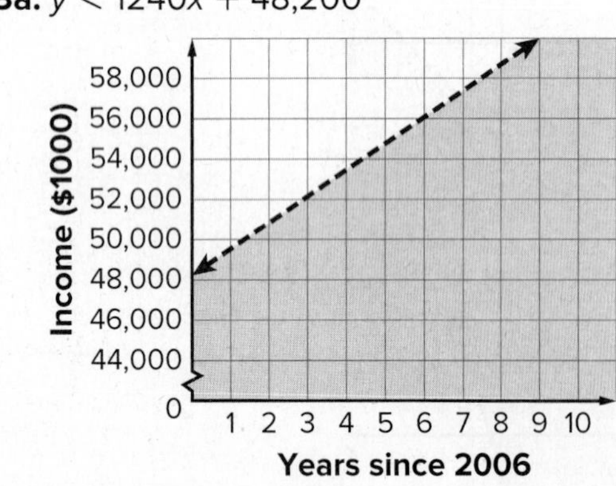

13b. no, no, yes, no

15.

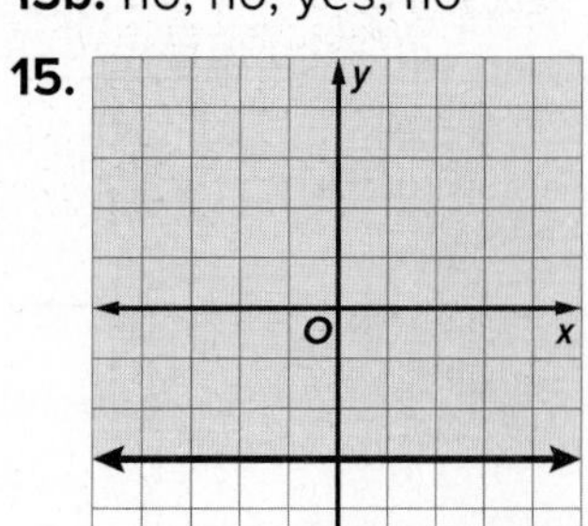

17.

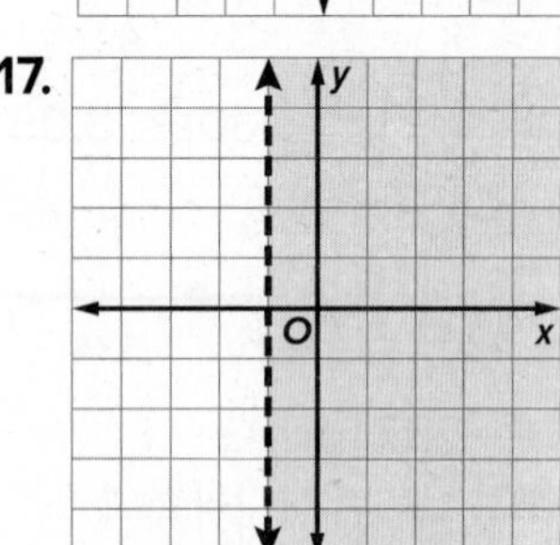

19.

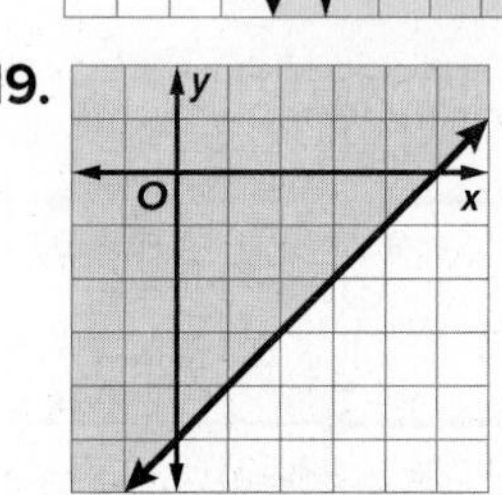

21.

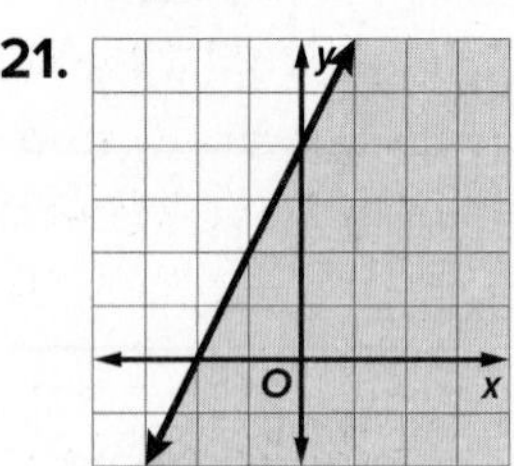

23.

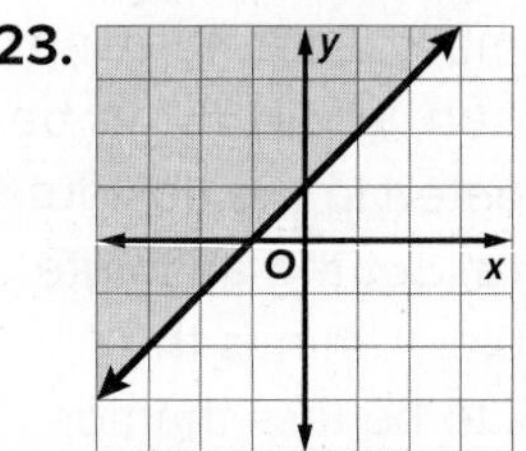

25a. $2.25p + 2b > 90$; $p \geq 0$ and $b \geq 0$

25b.

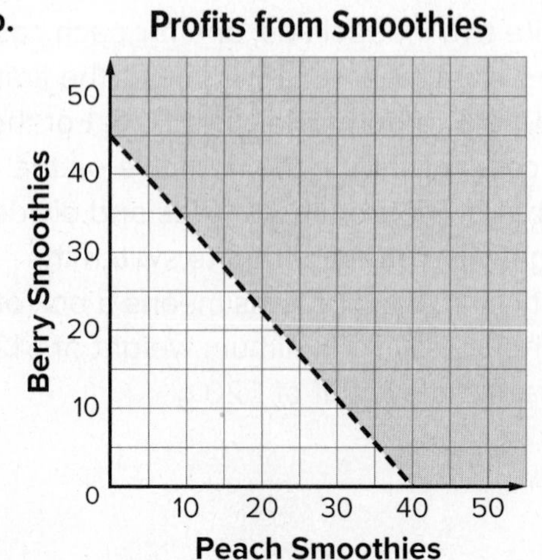

25c. The café sold more than 22 berry smoothies. **25d.** 41; Sample explanation: 40 peach smoothies results in a profit of exactly $90, so to make a profit of more than $90, the café must have sold 41 smoothies. **27.** The value of c must be positive. Since (0, 0) is a solution of the inequality, $a(0) + b(0) < c$ must be a true statement, so $0 < c$. **29.** Sample answer: $y < -x + 1$ **31.** Sample answer: The inequality $y > 10x + 45$ represents the cost of a monthly smartphone data plan with a one-time fee of $45, plus $10 per GB of data used. Both the domain and range are nonnegative real numbers because the GB used, and the total cost cannot be negative.

Module 6 Review

1. A

3. 8 rows

5. C

7. $\{t \mid t < -3\}$

9. A, B

11. $\{g \mid -5 \leq g\}$

13A. $\{h \mid -8 < h < 2\}$

13B. −10 −8 −6 −4 −2 0 2 4 6 8 10

15. B

17. C, D, B, A

Selected Answers

Module 7

Quick Check

1. (4, 0) **3.** (0, 0) **5.** $x = 6 - 2y$ **7.** $m = 2n + 6$

Lesson 7-1

1. 1; consistent; independent

3. 0; inconsistent

5. 1; consistent; independent

7. 1; consistent; independent

9. 1; consistent; independent

11. 1 solution; (0, −3)

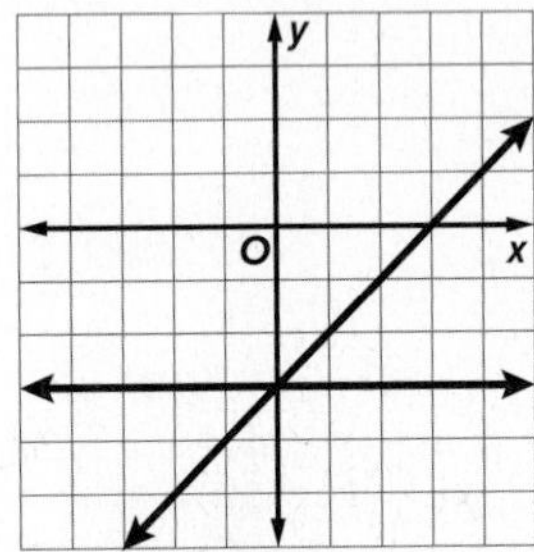

13. no solution

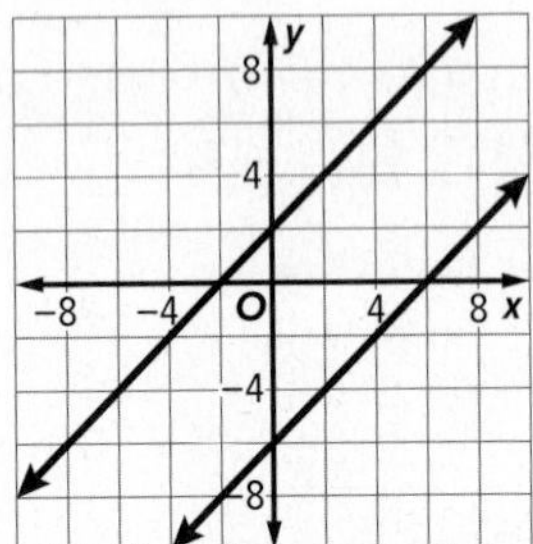

15. infinitely many solutions

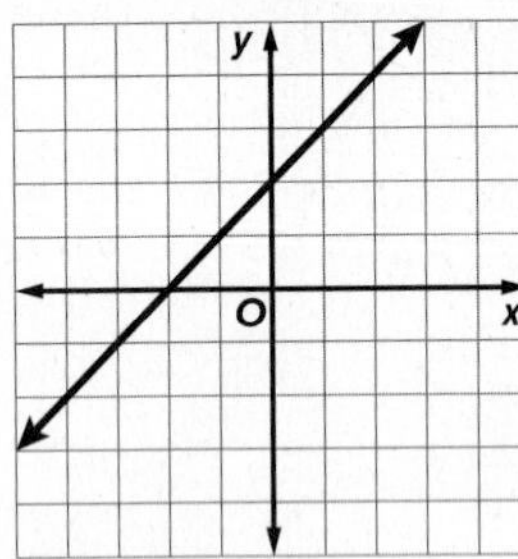

17a. $y = 400x + 1000$; $y = 5900 - 300x$

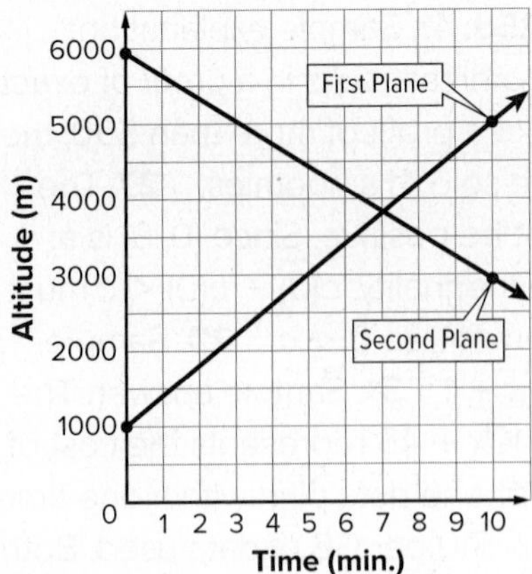

17b. After 7 minutes the planes will be at the same altitude.

19. $y = 3x + 6$ and $y = 6$; (0, 6)

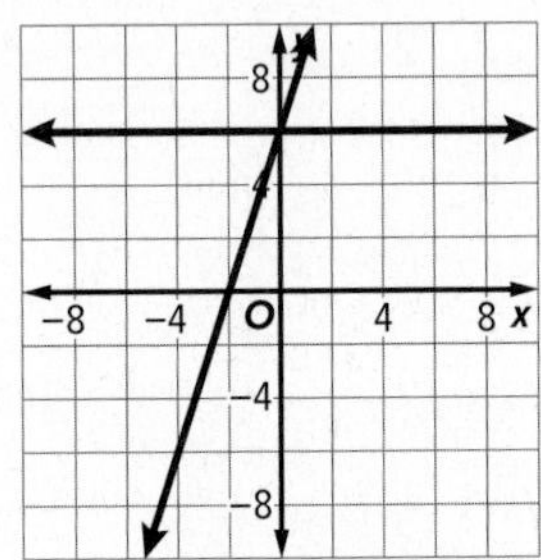

21. $y = -12x + 90$ and $y = 30$; (5, 30)

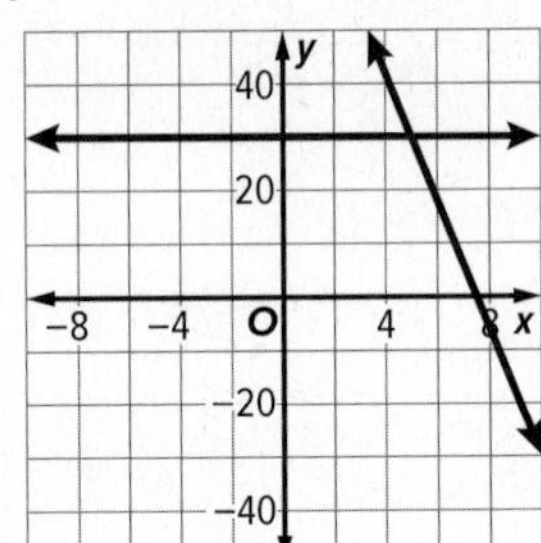

23. $y = 2x + 5$ and $y = 2x + 5$; infinitely many solutions

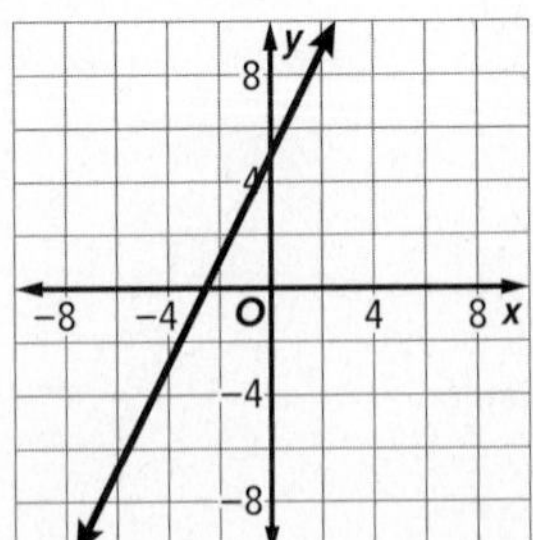

25. approximately (2.68, 1.01)

27. approximately (2.67, −0.88)

29. Sample answer: $x + y = 260$; $2.5x + 0.75y = 450$; approximately (145.71, 114.29); The bookstore will make a weekly profit of $450 with total weekly sales of 260 items when about 146 books and about 114 magazines are sold.

31. no solution; inconsistent

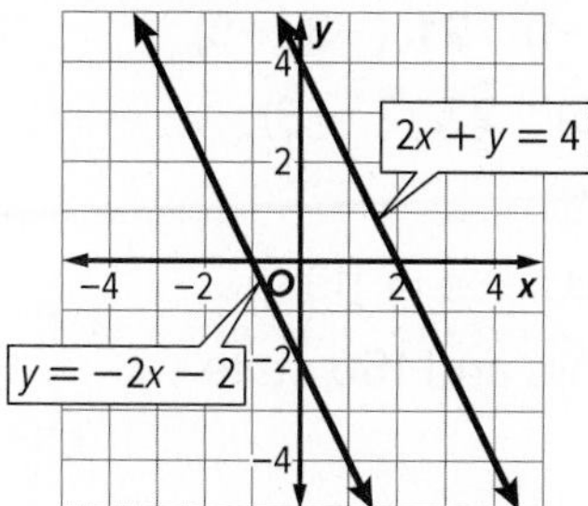

33. 1 solution; (1, −3); consistent; independent

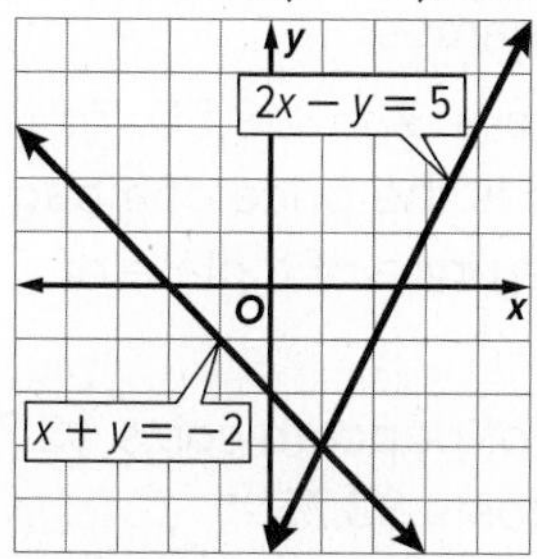

35. infinitely many solutions; consistent; dependent

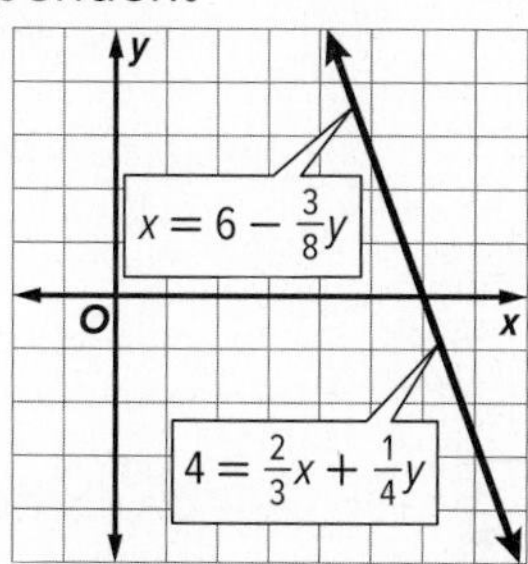

37. infinitely many solutions; consistent; dependent

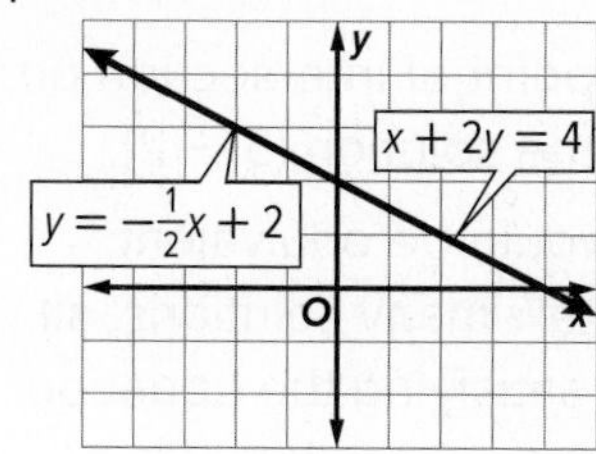

39. 1 solution; (2, 1); consistent; independent

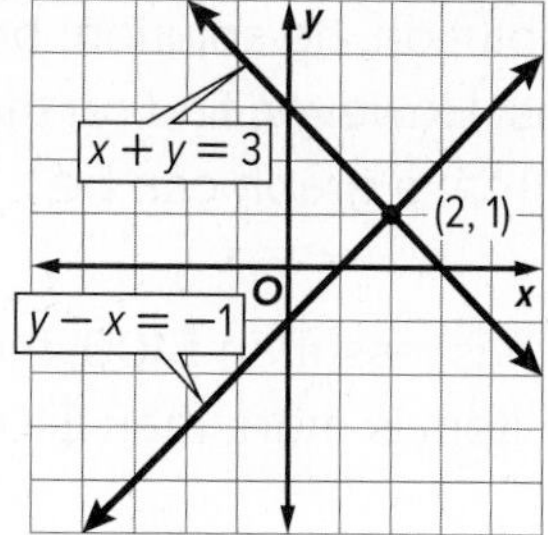

41a. x = time, in seconds, y = distance from where Olivia started to the finish line, in feet; $y = 20x$; $y = 15x + 150$

41b.

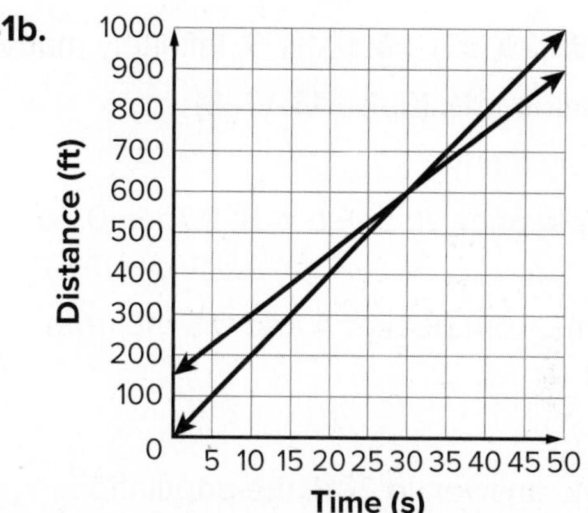

41c. 600 ft

43a. x = time walking in minutes, y = time on bike in minutes; $3x + 2y = 70$, $x = y + 15$

43b.

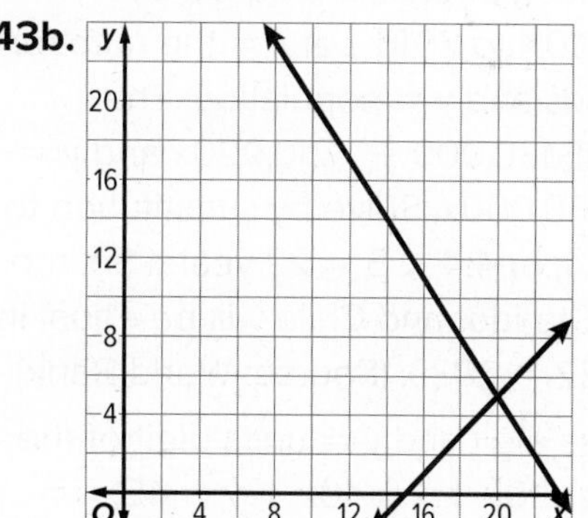

43c. 20 minutes

45. (–2, 3)

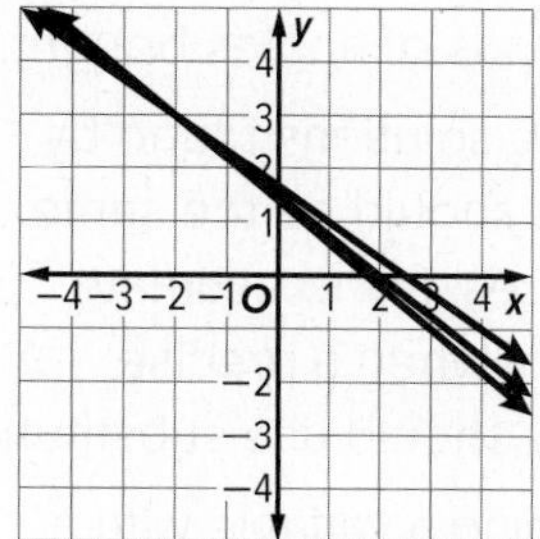

47. Sample answer: $4x + 2y = 14$, $12x + 6y = 18$; This system is inconsistent, while the others are consistent and independent.

49. Graphing clearly shows whether a system of equations has one solution, no solution, or infinitely many solutions. However, finding the exact value of x and y from a graph can be difficult.

51. Francisca; If the item is less than \$100, then \$10 off is better. Of the item is more than \$100, then the 10% is better.

Lesson 7-2

1. (1, 6) **3.** (29, 53) **5.** (1, 1) **7.** infinitely many

9. no solution **11.** (0, 1) **13.** (2, 5)

15. infinitely many

17a. Sample answer: $a + b = 5$; $0.7a + 0.2b = 0.65(5)$

17b. 4.5 mL from Beaker A and 0.5 mL from Beaker B

19. $\left(\frac{1}{2}, -\frac{3}{8}\right)$

21. Sample answer: In 2011, the population of Ecuador was about 15,180,000 and the population of Chile was about 17,150,000. The population of Ecuador increased by 1,210,000 and the population of Chile increased by 760,000 from 2011 to 2016. Let x = the number of 5-year periods and y = population. The system is $y = 15{,}180{,}000 + 1{,}210{,}000x$ and $y = 17{,}150{,}000 + 760{,}000x$. Solve by substitution to find that $x \approx 4.4$, or $4.4 \times 5 = 22$ years. So, the population of Ecuador and Chile will be equal in about $2011 + 22 = 2033$. (Source: World Bank)

23. Let x = tens digit and y = units digit of the original number; $10y + x = 10x + y - 45$; $x = 3y + 1$; (7, 2); The original number is 72.

25. Neither; Guillermo substituted incorrectly for b. Cara solved correctly for b, but misinterpreted the pounds of apples bought.

27. Sample answer: The solutions found by each of these methods should be the same. However, it may be necessary to estimate when using a graph. So, when a precise solution is needed, you should use substitution.

29. An equation containing a variable with a coefficient of 1 can easily be solved for the variable. That expression can then be substituted into the second equation for the variable.

Lesson 7-3

1. (−3, 4) **3.** (−3, 1) **5.** (4, −2) **7.** (8, −7)

9. (4, 7) **11.** (4, 1.5) **13.** (2, 1) **15.** (11, 0)

17. (−3, 7) **19.** (2, −1) **21.** (−3, −5)

23. (10, 4) **25.** (7, 5) **27.** (2, −3)

29. −2 and −4

31a. $r + s = 181$ and $r - s = 119$

31b. 31 state senators and 150 state representatives

33. (4, −1) **35.** $\left(-1, 3\frac{1}{3}\right)$

37. (−36, −4) **39.** 34 games

41a. Sample answer: $4p + 2n = 18.50$, $7p + 2n = 26.75$, where p is the price of a bag of popcorn and n is the price of a plate of nachos

41b. (2.75, 3.75); A bag of popcorn costs \$2.75 and a plate of nachos costs \$3.75.

43a. Add the equations because this will eliminate the variable y, and then you can solve for x.

43b. (5, −2)

43c.

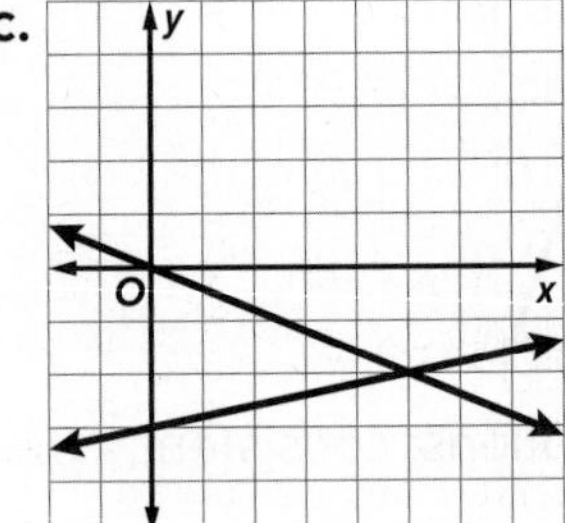

Sample answer: The point of intersection on the graph will match the solution (5, −2).

43d. The equations would be equivalent. There would be infinitely many solutions, all real numbers x and y satisfying the equation $x - 5y = 15$.

43e. There would be no solution because the lines would be parallel and would never intersect.

45. Sample answer: $x + y = 1$ and $-x - y = 1$; This system of equations has no solutions.

47. Sample answer: $-x + y = 5$, I used the solution to create another equation with the coefficient of the x-term being opposite of its corresponding coefficient.

49. Sample answer: It would be most beneficial to use elimination to solve a system of equations when one variable has either the same coefficient in both equations or one variable has coefficients that are additive inverses in the equations.

Lesson 7-4

1. $(-1, 3)$ **3.** $(-3, 4)$ **5.** $(-2, 3)$ **7.** $(3, 5)$

9. $(1, -5)$ **11.** $(0, 1)$

13a. $2x + y = 592.30$ and $x + 2y = 691.31$, where x is the number of MLB games and y is the number of NBA games

13b. MLB: \$164.43, NBA: \$263.44

15. 8 and -1

17. wash: \$6, vacuum: \$2

19a.

	Tropical Breeze	Kona Cooler	Total
Amount of Juice (qt)	t	k	10
Amount of Pineapple Juice (qt)	$0.2t$	$0.5k$	4

19b. $\left(3\frac{1}{3}, 6\frac{2}{3}\right)$; The owner should mix $3\frac{1}{3}$ qt of Tropical Breeze and $6\frac{2}{3}$ qt of Kona Cooler.

19c. $3\frac{1}{3}$ qt $+ 6\frac{2}{3}$ qt $= 10$ qt, so the total amount is correct, and $0.2\left(3\frac{1}{3}\text{ qt}\right) + 0.5\left(6\frac{2}{3}\text{ qt}\right) = 4$ qt, so the amount of pineapple juice in the new drink is correct.

21. Jason; In order to eliminate the t-terms, you can multiply the second equation by 2 and then subtract, or multiply the equation by -2 and then add. Daniela did not subtract the equations correctly.

23. Sample answer: $2x + 3y = 6$ and $4x + 9y = 5$

25. Sample answer: It is more helpful to use substitution when one of the variables has a coefficient of 1 or if a coefficient can be reduced to 1 without turning other coefficients into fractions. Otherwise, elimination is more helpful because it will avoid the use of fractions when solving the system.

Lesson 7-5

1.

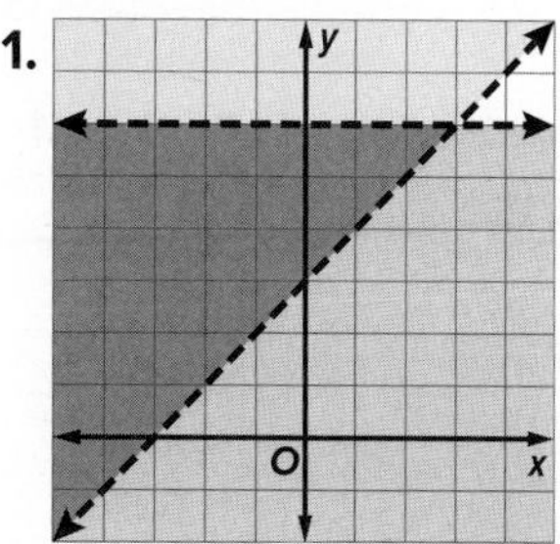

3.

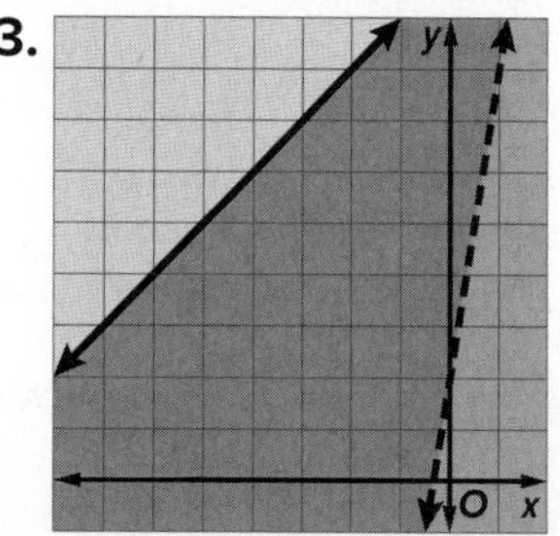

5. no solution

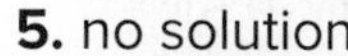

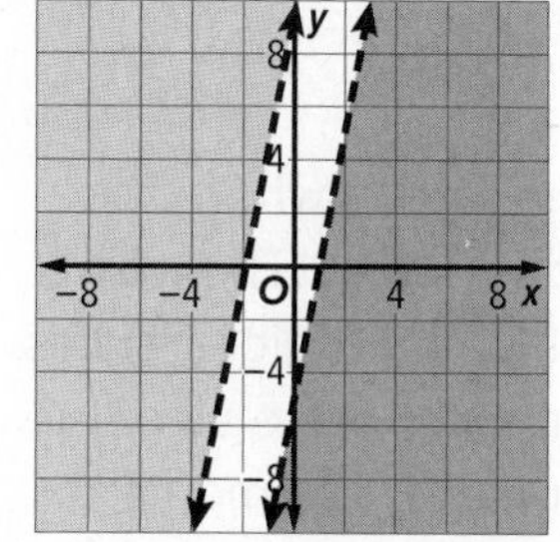

7.

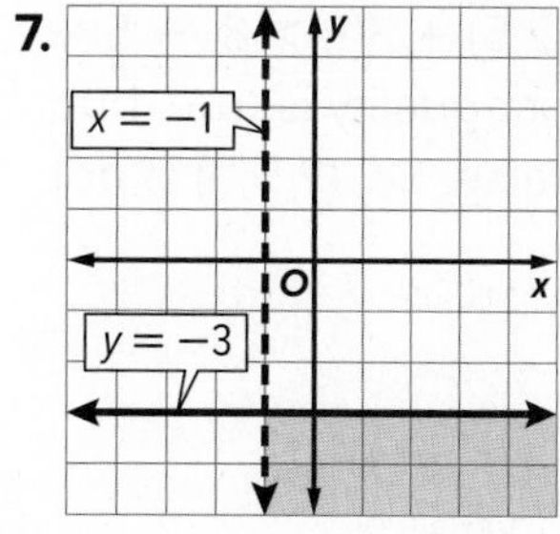

9.

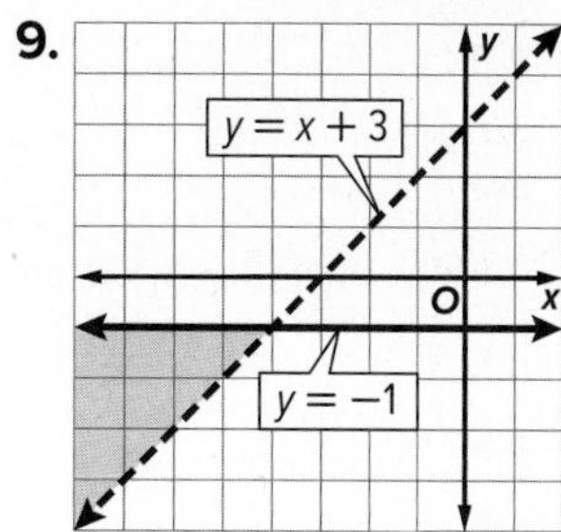

11. no solution

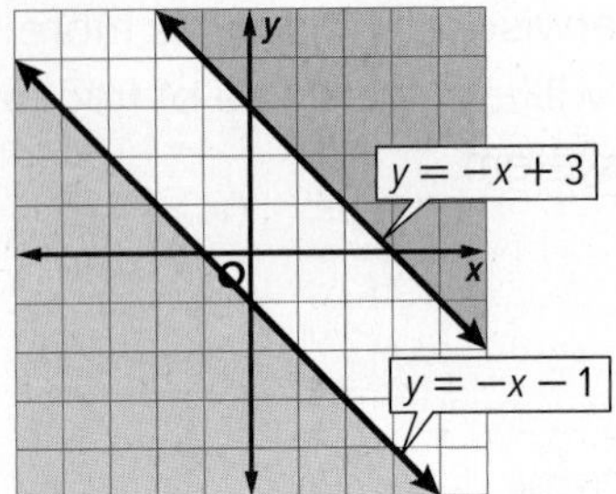

13a. Sample answer: Let x = hours at gym and y = miles of walking; $x \geq 9$, $x \leq 12$, $y \geq 4.5$, $y \leq 6$

13b.

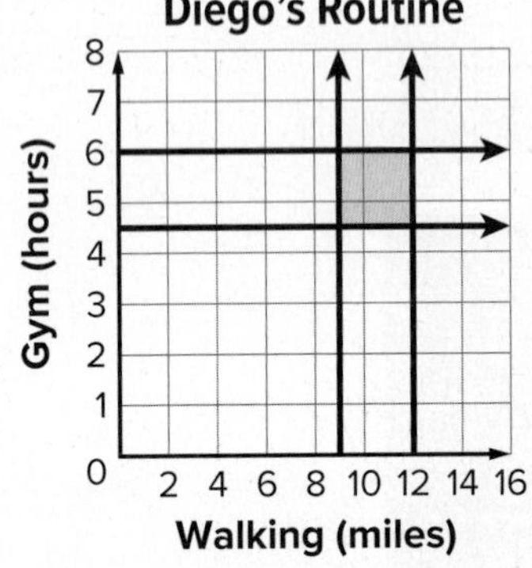

13c. Sample answers: gym 5 h, walk 9 mi; gym 6 h, walk 10 mi, gym 5.5 h, walk 11 mi

15. $y \leq x + 2$, $y \geq x - 3$

17. $y \geq x + 1$, $y < 1$

19. The solution set is the region where the graphs of the inequalities overlap. The point (2.5, 1) is not in the overlapping region, so it is not a solution. A solution must make all of the inequalities in the system true statements: $4x - 5y \geq 2 \rightarrow 4(2.5) - 5(1) \geq 2 \rightarrow 10 - 5 \geq 2 \rightarrow 5 \geq 2$; $2x + 3y > 8 \rightarrow 2(2.5) + 3(1) > 8 \rightarrow 5 + 3 > 8 \rightarrow 8 > 8$; The first inequality is true, but the second inequality is false. So, (2.5, 1) is not a solution.

21. Let x = tins of popcorn and y = tins of peanuts; $x + y \leq 200$; $x \geq y$; $3x + 4y \leq 900$; $x \geq 0$ and $y \geq 0$.

23. Sample answer: (3, 3)

25. Sometimes; sample answer: $y > 3$, $y < -3$ will have no solution, but $y < -3$, $y < 3$ will have solutions.

27. 9 units^2

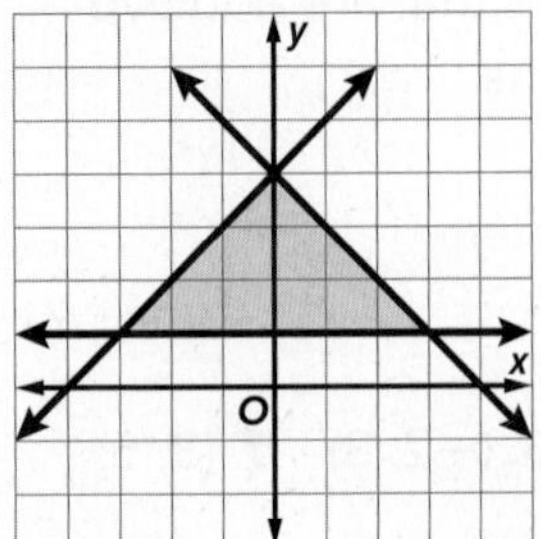

Module 7 Review

1. B, C, D **3.** B

5. 4 wooden frames and 3 plastic frames

7. one solution; (−2, 7)

9. (10, 4)

11. A, B **13.** $r = 6$, $t = 5$ **15.** D **17.** C

Module 8

Quick Check

1. −196

3. 0.25

5. 32

7. −3

Lesson 8-1

1. No; the domain values are at regular intervals and the range values have a common difference 3.

3. Yes; the domain values are at regular intervals and the range values have a common factor 2.

5. No; there is no common factor between the picture areas.

7.

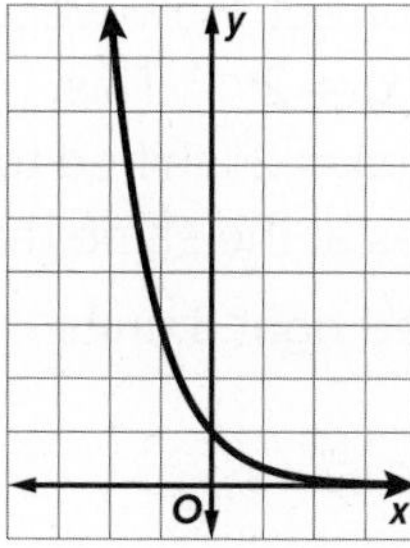

1; D = {all real numbers}, R = $\{y \mid y > 0\}$

9.

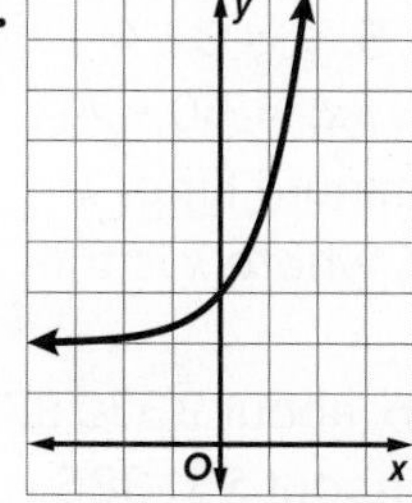

3; D = {all real numbers}, R = $\{y \mid y > 2\}$

11.

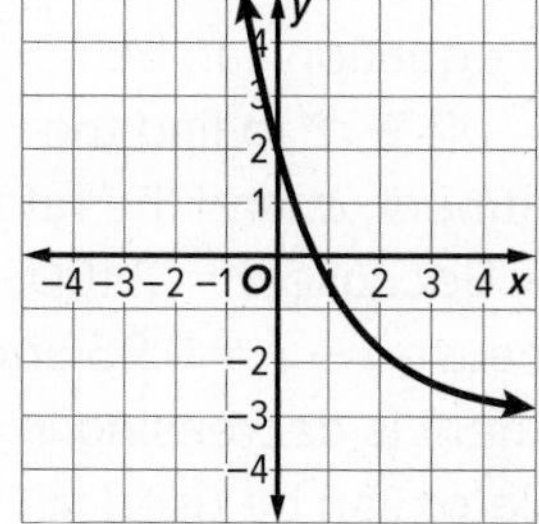

2; D = {all real numbers}, R = $\{y \mid y > -3\}$

13a. y-intercept = 50; D = {all real numbers}, R = $\{y \mid y > 0\}$

13b.

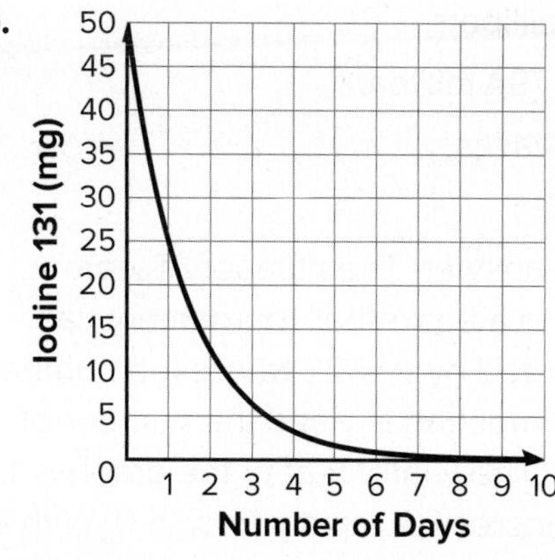

13c. Because time cannot be negative, the relevant domain is $\{x \mid x \geq 0\}$. Because the amount of Iodine 131 cannot be negative, and the amount when $x = 0$ is 50 mg, the relevant range is $\{y \mid 0 < y \leq 50\}$.

9.

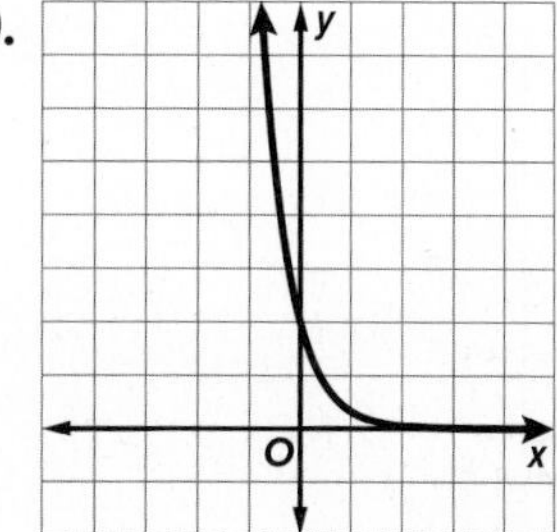

2; D = {all real numbers}, R = $\{y \mid y > 0\}$; $y = 0$

11.

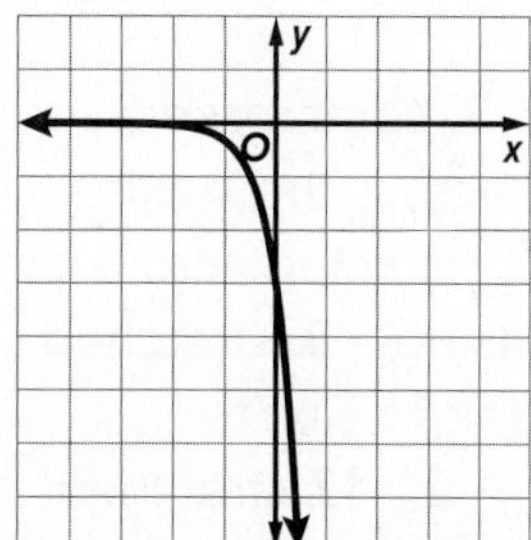

−3; D = {all real numbers}, R = $\{y \mid y < 0\}$; $y = 0$

13.

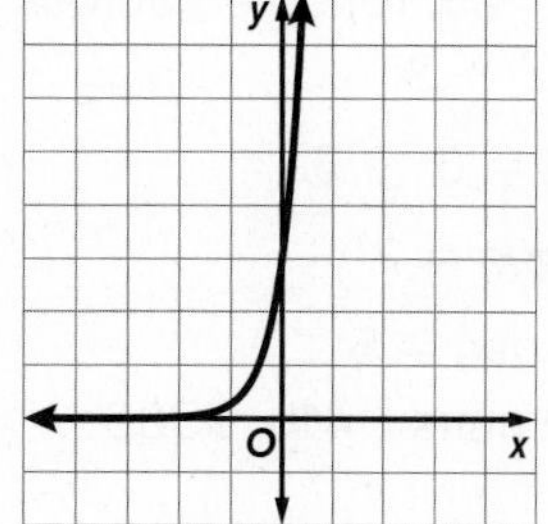

3; D = {all real numbers}, R = $\{y \mid y > 0\}$; $y = 0$

15a. 1038 millibars

15b. about 794 millibars

15c. It decreases.

17. $f(x) = 3(2^x)$

19. Sample answer: The number of teams competing in a basketball tournament can be represented by $y = 2^x$, where the number of teams competing is y and the number of rounds is x. The y-intercept of the graph is 1. The graph increases rapidly for $x > 0$. With an exponential model, each team that joins the tournament will play all of the other teams. If the scenario were modeled with a linear function, each team that joined would play a fixed number of teams.

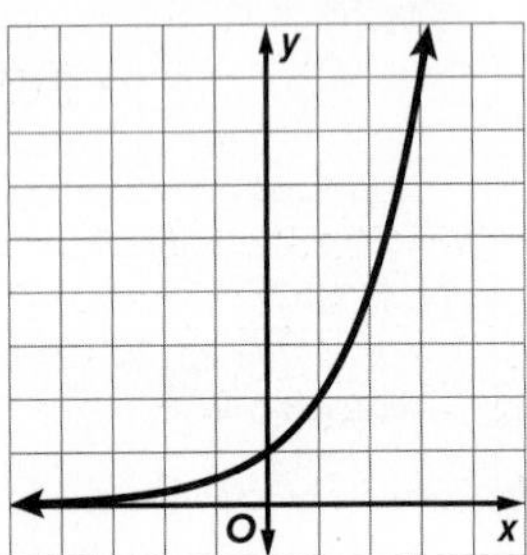

Lesson 8-2

1. translated up 8 units **3.** compressed horizontally **5.** reflected across the x-axis; translated 1 unit right **7.** reflected across the y-axis; translated 4 units up **9.** stretched vertically **11.** translated right 3 units **13.** $y = -2^x$ **15.** $y = 2^{-x} + 5$ **17.** stretched vertically by a factor of 2000 **19.** stretched vertically by a factor of 20 **21.** translated up 6 units **23.** reflected across the x-axis; compressed vertically **25.** reflected across the y-axis **27.** $g(x) = 2^x + 3$

29. $g(x) = 5^{x-2}$ **31.** $g(x) = 6^x + 5$

33. $g(x) = \frac{1}{2}(4^x)$ **35.** $g(x) = 2^{3x}$

37. $g(x) = 5^x - 2$ **39.** $g(x) = 5^{x-4}$

41a. translated up 500 units **41b.** $500

43. The graph has been reflected over the x-axis and reflected over the y-axis. It has been stretched vertically by a factor of 3 and shifted up 1 unit.

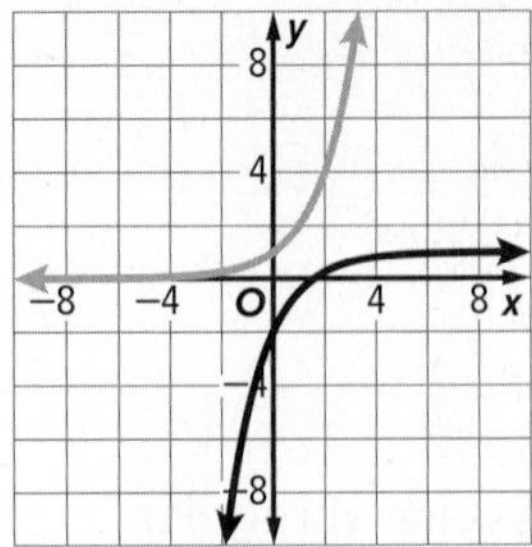

45. The graphs of these two exponential functions are the same. $f(x) = 4^{x+2} = 4^x \cdot 4^2 = 16 \cdot 4^x = g(x)$.

47. Jennifer is correct. Sample answer: As it is written, the function is multiplied by 2, which causes the graph to rise more rapidly than the parent graph, so Jennifer is correct. However, $g(x) = 2(2^x)$ is equivalent to $g(x) = 2^{x+1}$. This graph is the parent graph of $f(x) = 2^x$ shifted to the left one unit, but it still rises at the same rate.

49. The first pair; $g(x)$ is shifted right 3 units instead of left 3 units.

Lesson 8-3

1. $y = 4 \cdot 2^x$ **3.** $y = 10 \cdot 3^x$ **5.** $y = 3 \cdot 4^x$ **7.** $y = 3 \cdot 2^x$ **9.** $y = \left(\frac{1}{4}\right)^x$ **11.** $f(x) = 50 \cdot 2^x$, where x is the number of 30-minute time periods **13.** $f(x) = 43 \cdot (1.23)^x$, where x is the number of years since 2010.

15a. $P = 8{,}192{,}426(1.009)^t$ **15b.** about 9,370,872

17a. $Z = 60{,}000(0.90)^t$ **17b.** about $31,886

19. $2200 **21.** 360 million

23. $y = 2.6 \cdot 4^x$ **25.** about 77,529

27. Sample answer: The equation can be rewritten in the form $y = a(1 + r)^x$ to find the amount of original investment, a, and the rate of increase or decrease. Because $a = 2400$, he invested $2400. Because $1 + r = 0.95$ and is less than 1, his investment is decreasing in value. A graphing calculator can be used to find that the investment will be worth $1200 in about 13.5 years.

29a. $P(t) = 128(1.25)^t$

29b. an increase of approximately 41 deer per year **29c.** No; the amount of increase is exponential, not linear. **29d.** There is no common difference over equal intervals (differences are 32, 40 and 50). There is a common factor (factor is 1.25 in each case.)

31. about 9.2 years **33.** Sample answer: Exponential models can grow without bound, which is usually not the case for the situation that is being modeled. For instance, a population cannot grow without bound due to space and food constraints. Therefore, the situation that is being modeled should be carefully considered when used to make decisions.

35a. Sample answer: 5%; about 14.2 years

35b. Sample answer: 10%; about 6.6 years

35c. Sample answer: about 10.4 years; about $8320

Lesson 8-4

1a. $A(t) = (1.021)^t$; $A(t) = (1.0052)^{4t}$

1b. Bank B has the better plan because the effective quarterly interest rate is 0.8%, which is greater than the quarterly interest rate of about 0.52% for Bank A.

1c. About 3.2%; sample answer: This confirms the result of part **b** because 3.2% is greater than the annual interest rate at Bank A, so Bank B has the better plan.

3. Bank A; Bank A has a quarterly interest rate of 0.95%. Bank B has a quarterly interest rate of about 0.92%. Bank A's quarterly interest rate is higher.

5. Species B; the population of Species A is decreasing at a rate of about 0.25% per quarter. The population of Species B is decreasing at a rate of about 0.4% per quarter. The population of Species B is decreasing at a faster rate.

7. Plan A

9. Account A; Account A has a semi-annual interest rate of 2.3%. Account B has a semi-annual interest rate of about 2.1%. Account A's semi-annual interest rate is greater.

11. Account A; Account A has a monthly interest rate of 0.5%. Account B has a monthly interest rate of about 0.21%. Account A's monthly interest rate is greater.

13. $T(t) = 72 + 140(0.67)^t$

15. Sample answer: Bank A offers a savings account with a 0.6% interest rate compounded quarterly. Bank B offers a savings account with a 2% interest rate compounded annually. Bank A offers the better interest rate because it has a higher effective annual interest rate of about 2.4%.

Lesson 8-5

1. The ratios are not the same, so the sequence is not geometric.

3. Since the ratio is the same for all of the terms, 5, the sequence is geometric.

5. The ratios are not the same, so the sequence is not geometric.

7. Because the ratio is the same for all of the terms, $\frac{1}{2}$, the sequence is geometric.

9. The ratios are not the same, so the sequence is not geometric.

11. The ratios are not the same, so the sequence is not geometric.

13. −250, 1250, −6250

15. 108, 324, 972

17. −2058; −14,406; −100,842

19. 54, 162, 486

21. $\frac{1}{10}, \frac{1}{20}, \frac{1}{40}$

23. $\frac{1}{3}, \frac{1}{18}, \frac{1}{108}$

25. 387,420,489

27. 177,147

29. $a_n = 4 \cdot \left(\frac{3}{2}\right)^{n-1}$

31. $1310.72

33a. $a_n = P \cdot 1.005^n$

33b. $538.84

35. $a_n = \frac{9}{16}\left(\frac{2}{3}\right)^{n-1}$; $\frac{4}{81}$

37. $a_n = -8\left(\frac{1}{4}\right)^{n-1}$; $-\frac{1}{2048}$

Selected Answers

39. Sample answer: The average annual salary is about \$39,416, and the average annual rate of increase is about 3%. $a_n = 39{,}416(1.03)^{n-1}$; $\approx 69{,}116.19$; This means that after 20 years of employment the average annual salary will be about \$69,116.19.

41. −3, −12, −48

43a. The first method provides a starting salary of \$100 and an \$8 per month raise. The second method provides a starting salary of \$0.01 and doubles it each month.

43b. The first situation is linear because there is a common difference of \$8. The equation is $y = 8x + 92$. The second situation is exponential because it is a geometric sequence with a common ratio of 2. The equation is $y = 0.01(2)^{x-1}$.

43c. Sample answer: As long as I do not need money immediately, I would use the second method. In the last month, I would make $y = 0.01(2)^{23} = \$83{,}886.08$ due to the fact that the payment is growing exponentially. In the last month, in the first method I would make $y = 8(24) + 92 = \$284$.

45. If the values fit a geometric sequence, then $r = \sqrt{\frac{540}{180}} = \sqrt{3}$. This would mean that the interior angles of a square would have a sum of $180\sqrt{3} \approx 312°$. Since the sum of the angles in a square is 360°, this is not a geometric sequence.

47. Neither; Haro calculated the exponent incorrectly. Matthew did not calculate $(-2)^8$ correctly.

49. Sample answer: When graphed, the terms of a geometric sequence lie on a curve that can be represented by an exponential function. They are different in that the domain of a geometric sequence is the set of natural numbers, while the domain of an exponential function is all real numbers. Thus, geometric sequences are discrete, while exponential functions are continuous.

51. Sample answer: In the geometric sequence 6, 3, 1.5, ..., the value of r is 0.5 and the absolute value of a_{n+1} will be closer to zero than the value of a_n.

Lesson 8-6

1. 23, 30, 37, 44, 51

3. 8, 20, 50, 125, 312.5

5. 13, −29, 55, −113, 223

7. $a_1 = 12, a_n = a_{n-1} - 13, n \geq 2$

9. $a_1 = 2, a_n = a_{n-1} + 9, n \geq 2$

11. $a_1 = 40, a_n = -1.5a_{n-1}, n \geq 2$

13. $a_1 = 3, a_n = a_{n-1} - 1, n \geq 2$

15. $a_1 = 2, a_n = a_{n-1} + 1, n \geq 2$

17. $a_1 = \frac{5}{2}, a_n = a_{n-1} - \frac{1}{2}, n \geq 2$

19a. 875, 1050, 1225, 1400, 1575

19b. $a_1 = 175, a_n = a_{n-1} + 175, n \geq 2$

19c. $a_n = 175n$

21a. $a_1 = 6, a_n = 0.9a_{n-1}, n \geq 2$

21b. $a_n = 6(0.9)^{n-1}$

23. $a_n = -12n + 10$

25. $a_1 = 45, a_n = a_{n-1} - 7, n \geq 2$

27. $a_1 = -11, a_n = a_{n-1} + 5, n \geq 2$

29. $a_n = 16(4)^{n-1}$

31. $a_n = 500(1.05)^{n-1}$

33. Ramon has 2 parents, 4 grandparents, 8 great-grandparents, and so on. We can write a geometric sequence to count the number of ancestors in a given generation. The recursive formula is $a_1 = 2, a_n = 2_{n-1}, n >= 2$. The explicit formula is $a_n = 2^n$. Ramon's claim is about the 8th generation back: $a_8 = 2^8 = 256$. Ramon is correct.

35a. Sample answer: B3 = B2 + B1 and C2 = B2 ÷ B1

35b. The ratio approaches a constant value of 1.618034.... For larger values of n, the Fibonacci numbers behave like a geometric sequence with a common ratio of 1.618034....

37. Both; Sample answer: The sequence can be written as the recursive formula $a_1 = 2$, $a_n = (-1)a_{n-1}, n \geq 2$. The sequence can also be written as the explicit formula $a_n = 2(-1)^{n-1}$.

39. False; sample answer: A recursive formula for the sequence 1, 2, 3, ... can be written as $a_1 = 1, a_n = a_{n-1} + 1, n \geq 2$ or as $a_1 = 1, a_2 = 2$, $a_n = a_{n-2} + 2, n \geq 3$.

41. Sample answer: In an explicit formula, the nth term a_n is given as a function of n. In a recursive formula, the nth term a_n is found by performing operations to one or more of the terms that precede it.

Module 8 Review

1.

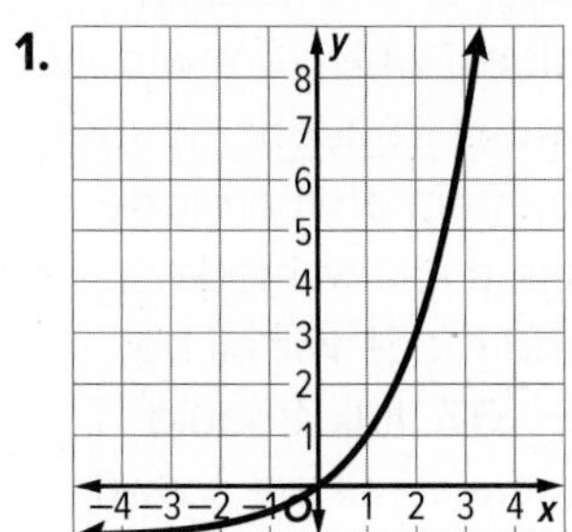

3. As x increases, y increases; and, as x decreases, y approaches 0.

5. A

7. A

9. D

11. The Local Credit Union offers Joey the better savings plan; sample answer: the monthly interest rate is 0.12% higher than at First & Loan, and the annual interest rate is 1.6% higher than at First & Loan.

13. 146 people

15. $a_1 = 20$, $a_n = a_{n-1} + 15$, $n \geq 2$

17. row 2: 12; row 3: 12, 48; row 4: 48, 192

Module 9

Quick Check

1. 45.88 **3.** $3\frac{3}{20}$ **5.** 82.4% **7.** 85.6%

Lesson 9-1

1. mean: 15.625, median: 15.5, mode: none
3. mean: 3.3, median: 2.5, mode: 2
5. mean: 5 students; median: 4 students; mode: 3 students **7.** mean: 54.75 mph; median: 54 mph; mode: 53 mph **9.** mean: about 2.8; median: 2.75; mode: 2 **11.** 25th percentile
13. Sample answer: The mean could be slightly higher because on a few of Saturday nights throughout the year, there were a very large number of people at the movies, which caused the mean to increase but did not affect the median. **15.** mean: 252; median: 245; mode: none **17.** 23 **19.** mean: 51.5, median: 51, mode: none **21.** 20 points **23.** 90th percentile **25a.** mean = 109,633; median = 66,556, no mode **25b.** Sample answer: The novels lower than the 50th percentile would be those consisting of words in the thirty-thousands and in the upper fifty-thousands. My prediction is correct because those three books are in the 47th percentile, which is just under the 50th percentile. **25c.** The median will change from 66,556 to 69,920, a difference of 3364 words. The mean will change from 109,633 to 111,065, a difference of 1432 words. **27.** Canada: 20th percentile; France: 50th percentile; Japan: 40th percentile; Russia: 60th percentile; Brazil: 10th percentile; Great Britain: 70th percentile

Olympic Medal Counts	
Country	**Total Medals**
Australia	29
Brazil	19
Canada	22
China	70
France	42
Great Britain	67
Japan	41
New Zealand	18
Russia	56
United States	121

29. Sample answer: I can assume that the data is tightly clustered around 37 because all three measures of center are close. **31.** 75th percentile
33. Because the mean is an average of all the numbers in the data set, it is most affected by outliers. An outlier on the high end will cause the mean to increase. The median is the middle value in the dataset, adding one high number should not have much effect on the median unless the dataset has values, which are widely spread. The mode is the most frequent number so the outlier will have no effect on the mode unless the outlier is the same as the mode.
35. The mean, median, and mode will all be multiplied by the number. **37.** Julio should have chosen the mean because all the growth values are close together. **39.** Sample answer: To find a percentile rank, order the data set in decreasing order. Count the number of items below the item you are ranking, and divide that by the total number of items. Multiply this answer by 100 to arrive at the percentile rank.

Lesson 9-2

1. Summer Reading Program

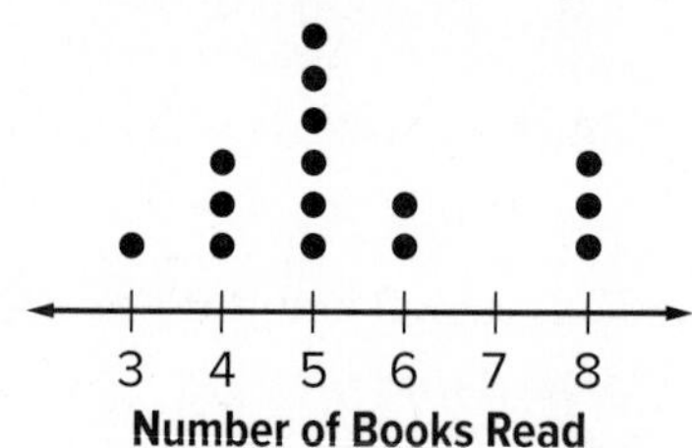

3.

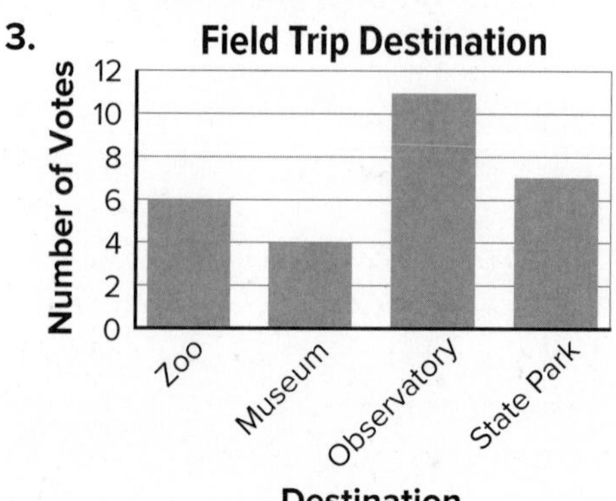

5. histogram

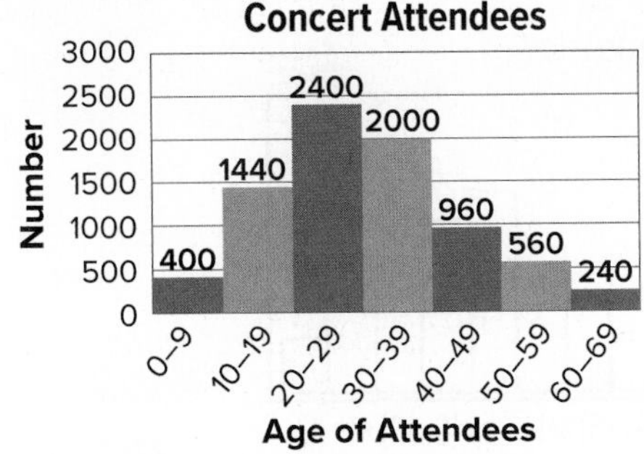

7a.

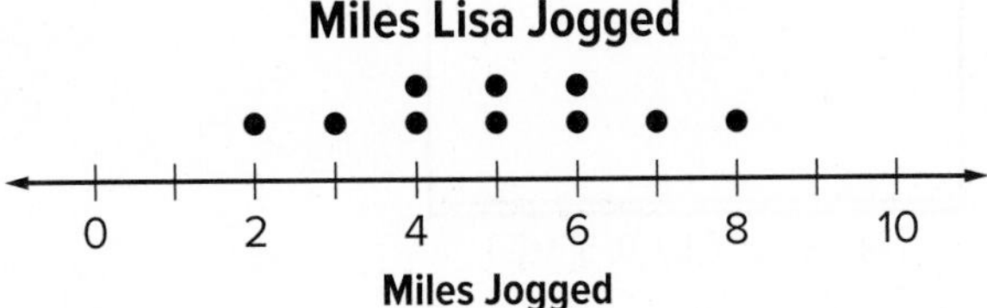

7b. 2 **7c.** 8

9a.

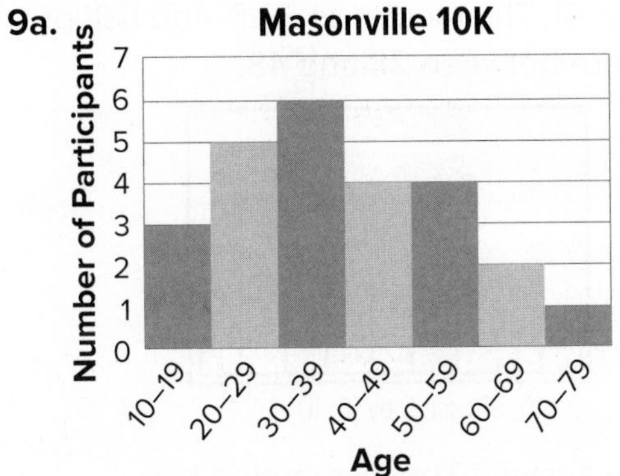

9b. 8 **9c.** 30–39

11. Sample answer: The scientist should break down the data into increments of two-tenths starting at 1 and going through 2.6.

13. Sample answer: 1) Because the data is clustered around ratings 7-10, it can be concluded that the product is well-liked by most customers and may have minor inconsistencies that certain people did not like. 2) Because there are only three low ratings, it can be concluded that dissatisfaction with the product could be a result of personal preference or manufacturer defect in a specific item.

15. Sample answer: If the range of the data is broad with specific, unrepeating values, then it makes the dot plot more meaningful if the range is divided up into equal intervals.

17. Sample answer: Bar graphs and histograms are similar because each displays data with bars. They are different because a bar graph is best used with data that are discrete and a histogram represents data that are continuous. For this reason, the bars in a bar graph do not touch and represent single values while the bars in a histogram touch and represent a range of values.

Lesson 9-3

1. Sample answer: The intended population is all students. By asking only students leaving basketball practice, Awan is not getting a representative example of the entire student body. **3.** Sample answer: The first sentence states a positive outcome of music education, which may bias the respondent toward support. This bias may serve people trying to keep music education in schools. **5.** Mean: 4, median: 4, mode: 2; The mean and median are appropriate measures to use to accurately summarize the data. **7.** Sample answer: The scale for vendor 1 starts at 70, and because of the size of the bars, it looks like their sales doubled in one year, when they increased about 50%. Vendor 2 had a larger increase in sales of approximately 67%.

9. Sample answer: The required class would be better because it is more likely to contain a representative sample of students. The elective class might not be representative of the whole student body because these courses are chosen for reasons such as personal preference or future career aspirations.

11. Median; sample answer: The two lowest weights are much lower than the others, so the mean will be affected by those outliers.

13. Sample answer: The original data are very close together, so it is likely that the measures of center will all be the same or very close. Adding an outlier of 24 to the data set will cause the mean to go up, but the median and mode would likely stay unchanged or very close to the original number. So, in this case the median or mode would best represent the center of data.

15. Sample answer: To assess a sample for bias, identify the intended population and sample method; then, based on this information assess whether there is potential sample bias. **21.** There are more extreme values in the lower end, which cause the mean to be lower than the median.

Lesson 9-4

1. 41 **3.** 62 **5.** 20

7. 62, 66, 73, 82, 99

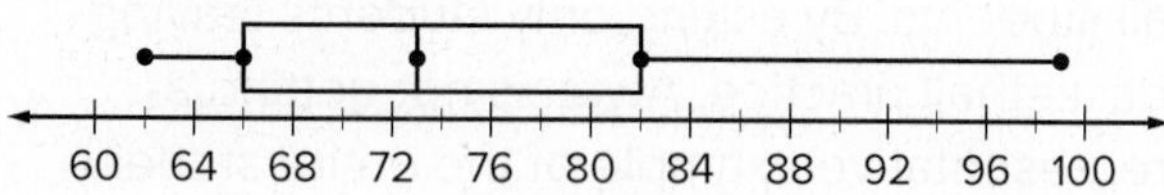

9. 35.2, 35.7, 35.9, 36.2, 36.5

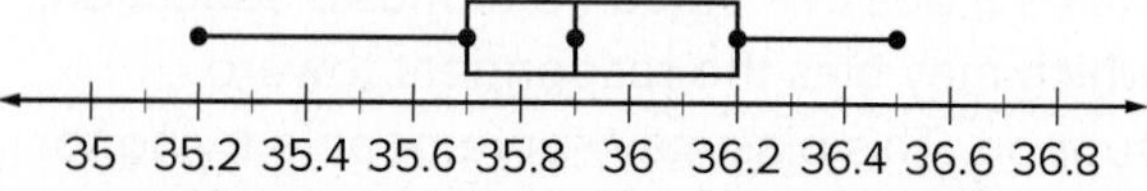

11. 25 **13.** 13 **15.** 20 **17.** 2.83 **19.** 2.97

21. 2.16; Since the standard deviation is large compared to the mean of 3, the number of goals scored each game is not relatively close to the mean. **23a.** 9, 36, 59, 67, 69 **23b.**

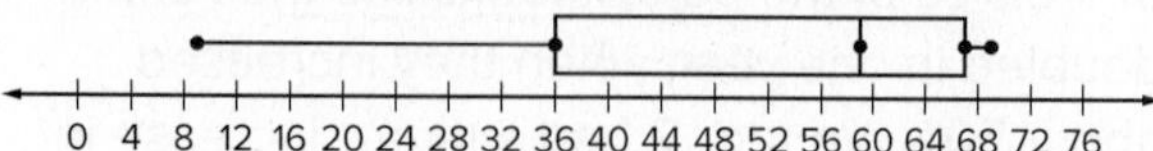

25. range: 14; minimum: 3; lower quartile: 6; median: 12; upper quartile: 14.5; maximum: 17; interquartile range: 8.5; standard deviation: 4.5

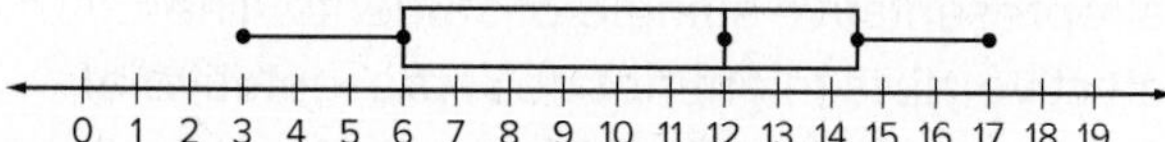

27. Both; sample answer; When an outlier is removed from a set of data, the spread and standard deviation of the data will decrease. When more values that are equal to the mean of a data set are added to the data set, the mean will be stronger and outliers will have less influence.

Lesson 9-5

1. symmetric

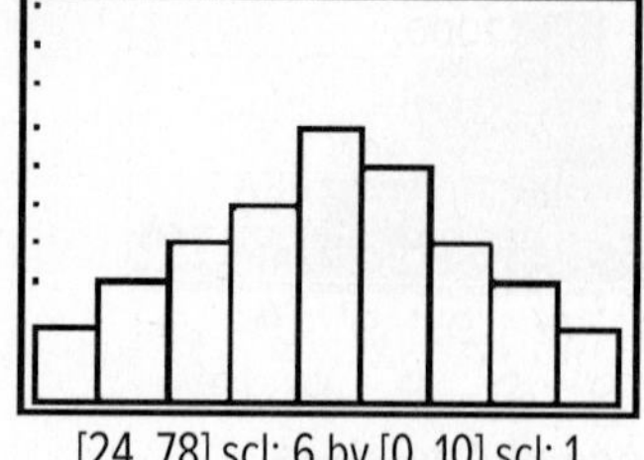

[24, 78] scl: 6 by [0, 10] scl: 1

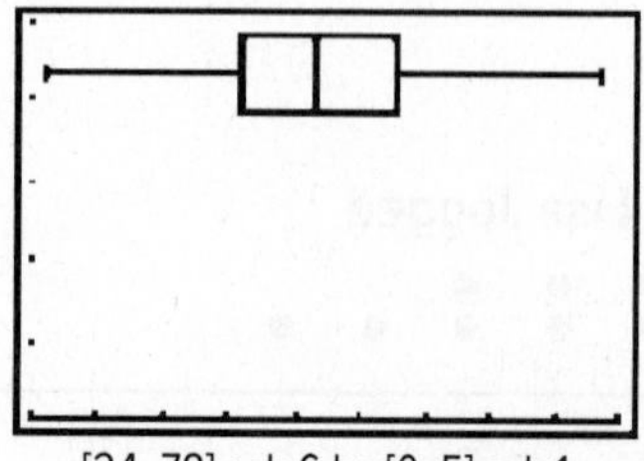

[24, 78] scl: 6 by [0, 5] scl: 1

3. Sample answer: The distribution is skewed, so use the five-number summary. The range is 53 – 12, or 41. The median is 39.5, and half of the data are between 28 and 48.

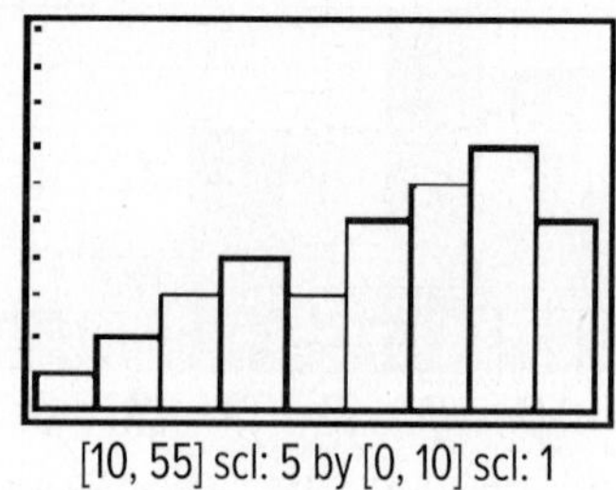

[10, 55] scl: 5 by [0, 10] scl: 1

5. Sample answer: The distribution is symmetric, so use the mean and standard deviation. The mean is about 58.7 with standard deviation of about 22.8.

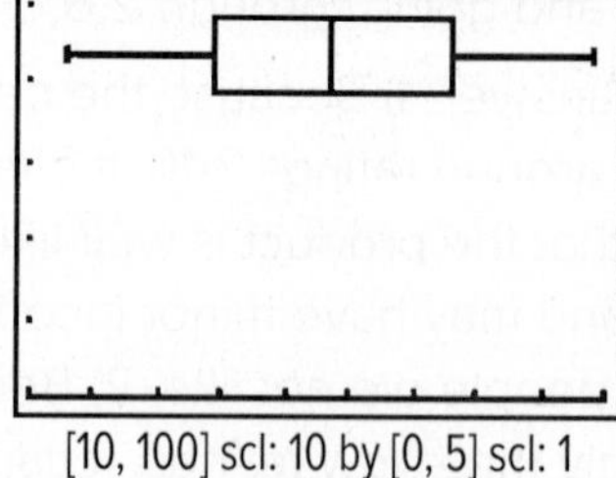

[10, 100] scl: 10 by [0, 5] scl: 1

7a.

75 100 125 150 175 200 225 250 275 300 325 350 375 400 425 450 475

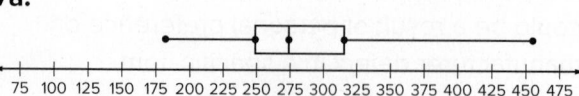

7b. min: 182, Q$_1$: 249, median: 274, Q$_3$: 315.5, max: 455 **7c.** The outlier mainly affects the mean. When the outlier is removed, the median decreases, but only $3 to $271. However, the mean changes from $289 to $266, which is more representative of the data as a whole.

9. negatively skewed

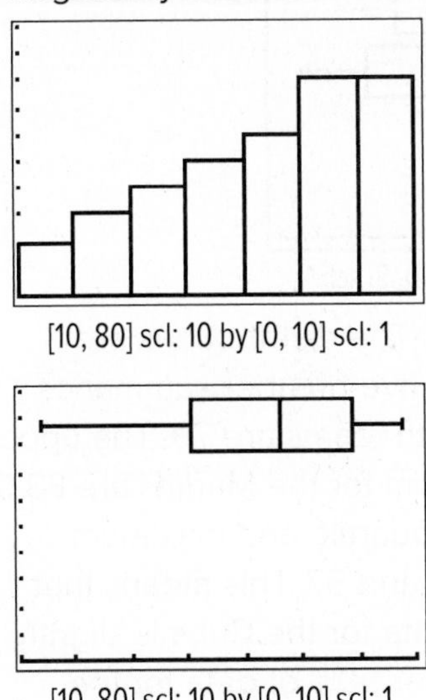

[10, 80] scl: 10 by [0, 10] scl: 1

[10, 80] scl: 10 by [0, 10] scl: 1

11. symmetric

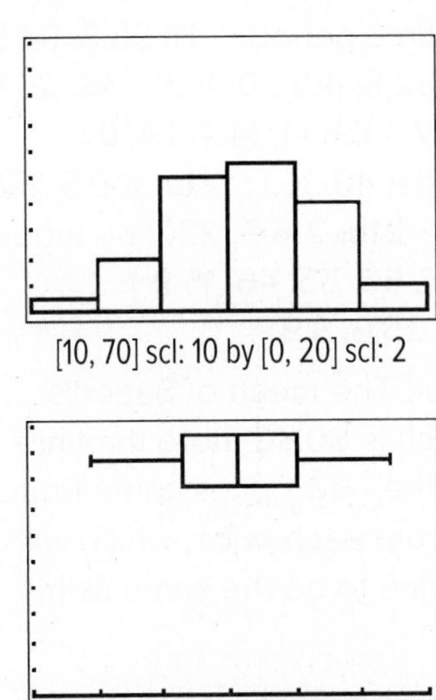

[10, 70] scl: 10 by [0, 20] scl: 2

[10, 70] scl: 10 by [0, 20] scl: 2

13. symmetric

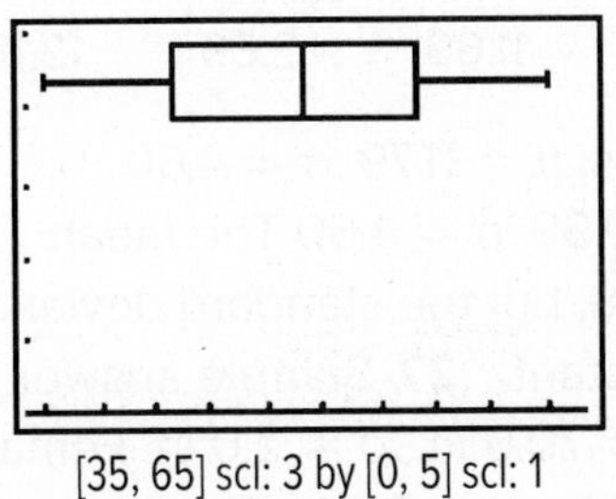

[35, 65] scl: 3 by [0, 5] scl: 1

15. Sample answer: The distribution is approximately symmetric, so use the mean and standard deviation. The mean is about 54.7 years with standard deviation of about 6.2 years.

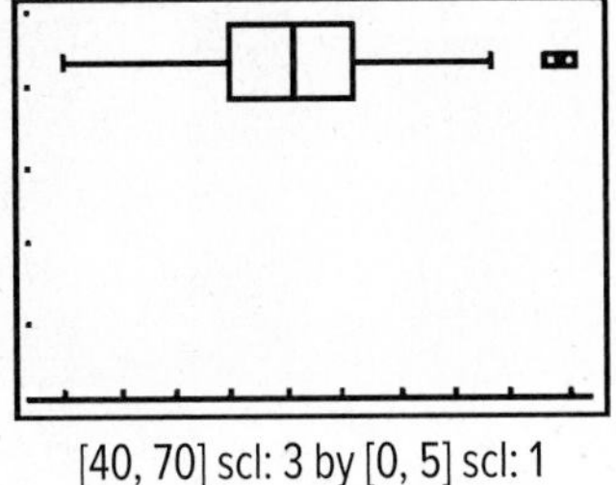

[40, 70] scl: 3 by [0, 5] scl: 1

17a. Sample answer: The distribution is skewed, so use the five-number summary. min: 62, max: 525, med: 103, Q1: 84, Q3: 290
17b. Sample answer: The distribution is symmetric, so use the mean and standard deviation. The mean is about 92.4 with standard deviation of about 18.4.

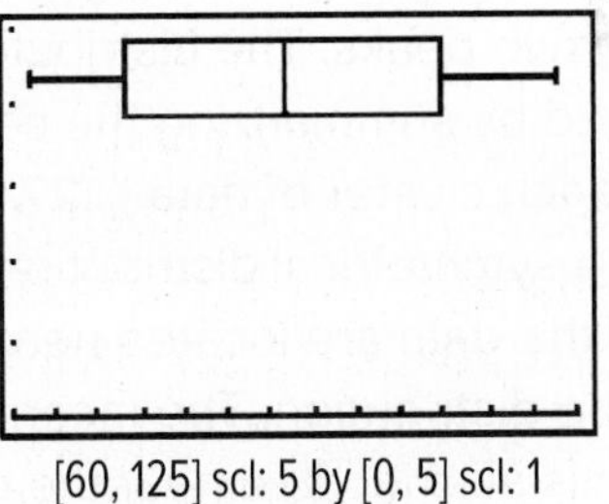

[60, 125] scl: 5 by [0, 5] scl: 1

17c. Original: mean 171.5, median 103; altered: mean about 92.4, median 92. The means differ by about 79.1, while the medians differ by 11.
19a. Sample answer: 225–230 g would be a reasonable advertised weight for either brand, so it is quite likely that they have the same advertised weight. Rafaello appears to have better control over the exact quantity in each package because its distribution is grouped more closely about the mean. **19b.** Sample answer: Both distributions have an inverted, symmetric U-shape with "tails" on either side. Leonardo's distribution is lower and wider.
21. Currently, Gerardo's distribution would be positively skewed. If he lost his longest streaks, the data would represent a symmetric distribution.

23a. negatively skewed

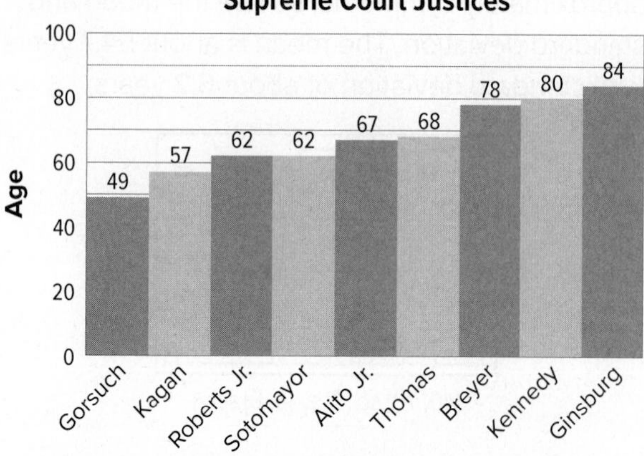

23b. The data is skewed, so use the five-number summary; min: 49, Q_1: 59.5, median: 67, Q_3: 79, max: 84. **23c.** There are no outliers in the data. **25.** Sample answer: A bimodal distribution is a distribution of data that is characterized by having data divided into two clusters, thus producing two modes and having two peaks. The distribution can be described by summarizing the center and spread of each cluster of data. **27.** Sample answer: In a symmetrical distribution, the majority of the data are located near the center of the distribution. The mean of the distribution is also located near the center of the distribution. Therefore, the mean and standard deviation should be used to describe the data. In a skewed distribution, the majority of the data lie either on the right or left side of the distribution. Because the distribution has a tail or may have outliers, the mean is pulled away from the majority of the data. The median is less affected. Therefore, the five-number summary should be used to describe the data.

Lesson 9-6

1. 60.9; 60; 60; 14; 4.7 **3.** 22.5; 21.5; no mode; 24; 7.4 **5.** 36.8; 38; 12; 56; 20.0 **7.** 26.8; 27.2; 29.6; 10.4; 3.5

9a. both negatively skewed

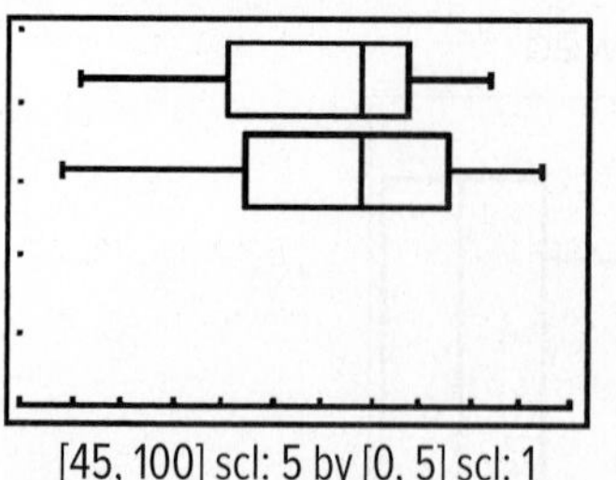

[45, 100] scl: 5 by [0, 5] scl: 1

9b. Sample answer: The distributions are skewed, so use the five-number summaries. The medians for both teams are 79. The upper quartile and maximum for the Marlins are 83.5 and 92. The upper quartile and maximum for the Cubs are 88 and 97. This means that the upper 50% of data for the Cubs is slightly higher than the upper 50% of data for the Marlins. Overall, we can conclude that the Cubs were slightly more successful than the Marlins during this time period. **11.** 93.5; 94.5; 97; 17; 5.1 **13.** 60; 62.5; 45; 50; 16.9 **15.** 25.9; 21.5; 17; 30; 10.3 **17.** 13.9; 14; 14.4; 1.4; 0.5 **19.** 75.8; 72; no mode; 48, 16.1 **21a.** 160.5; 166; no mode; 115; 33.9 **21b.** 216.5; 222; no mode; 115; 33.9 **23a.** 64.7, 66, 55, 46, 15.9 **23b.** 18.1, 18.9, 12.8, 25.6, 8.9

25a. Sample answer: The mean of Saeed's prices is $11.79, which is $0.80 more than his rival's mean price. The new prices come from subtracting $0.80 from each price, which will reduce the mean price to be the same as his rival's.

New Prices				
14.19	3.69	9.19	17.69	12.19
6.19	7.69	21.19	12.69	13.19
9.19	10.19	11.69	3.69	12.19

25b. Current prices: $\mu = 11.79$, $\sigma = 4.60$ New prices: $\mu = 10.99$, $\sigma = 4.60$ The mean has dropped by 0.8, but the standard deviation has remained constant **27.** Sample answer: Male students, $\bar{x} = 70.0$ in., $\sigma = 2.0$ in. Female students, $\bar{x} = 66.3$ in., $\sigma = 2.7$ in. Sample answer: For male students, the mean is 70.0 in., and the standard deviation is 2.0 in. For female students, the mean is 66.3 in., and the standard deviation is 2.7 in. On average, males are taller. However, because the standard deviation of males is smaller than that of females, the heights of females are more spread out.

29. Sample answer: Histograms show the frequency of values occurring within set intervals. This makes the shape of the distribution easy to recognize. However, no specific values of the data set can be identified from looking at the histogram, and the overall spread of the data can be difficult to determine. The box plot show the data divided into four sections. This aids when comparing the spread of one set of data to another. However, the box plots are limited because they cannot display the data any more specifically than showing it divided into four sections. **31.** $37,750
33. Sample answer: When two distributions are symmetric, determine how close the averages are and how spread out each set of data is. The mean and standard deviation are the best values to use for this comparison. When distributions are skewed, determine which direction the data is skewed and the degree to which the data is skewed. The mean and standard deviation cannot provide information in this regard, but get this information by comparing the range, quartiles, and medians found in the five-number summaries. So if one or both sets of data are skewed, it is best to compare their five-number summaries.

Lesson 9-7

1.

	Small	Large	Total
Cherry	35	20	55
Grape	25	15	40
Watermelon	15	15	30
Total	75	50	125

3. 30

5.

	Male	Female	Total
Spanish	22.5%	25%	47.5%
French	20%	15%	35%
German	7.5%	10%	17.5%
Total	50%	50%	100%

7. Sample answer: Most of the students are studying Spanish. **9.** Sample answer: Each conditional relative frequency represents the proportion of each candidate's support from each gender. **11.** 12

13.

	Male	Female	Total
Tree Swallow	5	7	12
Cardinal	5	10	15
Goldfinch	8	5	13
Total	18	22	40

15. 18

17.

	Sports or Clubs	No Sports or Clubs	Total
Freshmen	10%	12.5%	22.5%
Sophomores	12.5%	15%	27.5%
Juniors	10.6%	14.4%	25%
Seniors	11.9%	13.1%	25%
Total	45%	55%	100%

19. 55.6% **21.** 66 **23.** 100 **25.** 31 **27.** 38%

29.

Region	Apple	Sweet Potato	Pumpkin	Totals
West	77 ≈ 19.0%	4 ≈ 1.0%	13 ≈ 3.2%	94 ≈ 23.2%
Midwest	32 ≈ 7.9%	6 ≈ 1.5%	54 ≈ 13.3%	92 ≈ 22.7%
South	12 ≈ 3.0%	63 ≈ 15.6%	24 ≈ 5.9%	99 ≈ 24.4%
Northeast	92 ≈ 22.7%	2 ≈ 0.5%	26 ≈ 6.4%	120 ≈ 29.6%
Total	213 ≈ 52.6%	75 ≈ 18.5%	117 ≈ 28.9%	405 = 100%

31. Sample answer: The conditional relative frequencies based on pie preference give the probability of a person preferring a particular pie choice being from one of the U.S. regions. For example, there is an 84% probability that a person who prefers sweet potato pie is from the south.

Region	Apple	Sweet Potato	Pumpkin
West	36.2%	5.3%	11.1%
Midwest	15.0%	8.0%	46.2%
South	5.6%	84%	20.5%
Northeast	43.2%	2.7%	22.2%
Total	100%	100%	100%

33.

Vehicle Type	2WD	AWD	Totals
Hatchbacks	90	9	99
Sedans	60	13	73
SUVs	2	41	43
Total	152	63	215

35. Sample answer: Yes, there does appear to be an association. When the gasoline prices are higher, the distances traveled appear to be lower; when the gasoline prices are lower, the distances traveled appear to be higher.

37. Sample answer: A relative frequency is the ratio of the number in a category to the overall total of both categories. A conditional relative frequency is the ratio of the joint frequency to the marginal frequency. Therefore, it is important to understand what relationship is being analyzed because each two-way relative frequency table can provide two different conditional relative frequency tables.

Module 9 Review

1.

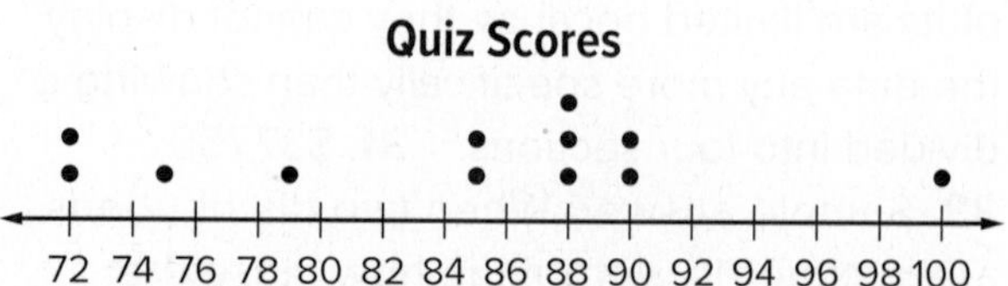

3. C, D **5.** The data could be separated into intervals of 10, from 0–9, 10–19, 20–29, and so on through 70–79. **7.** scaled dot plot; histogram

9.

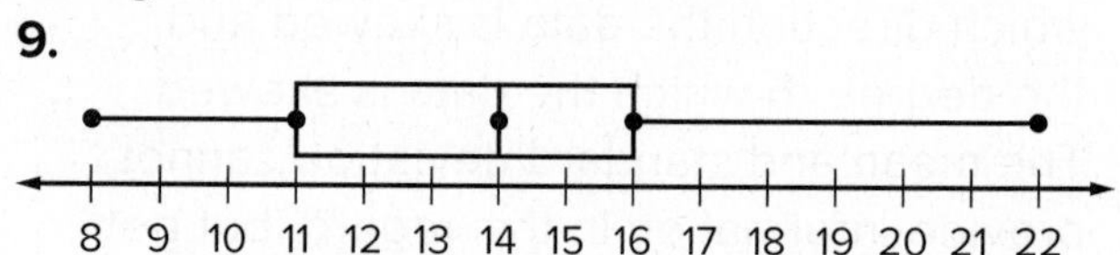

11. D **13.** C

Module 10
Quick Check

1. 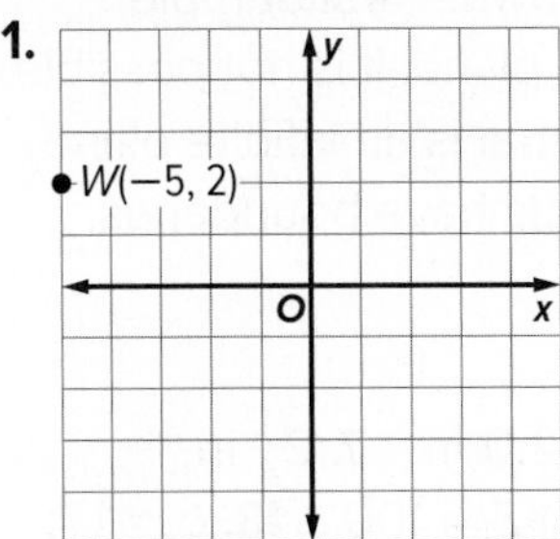

3. 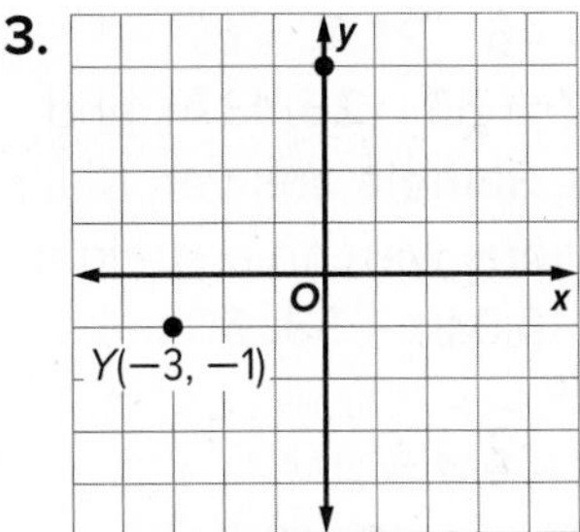

5. 20 **7.** 68

Lesson 10-1

1. Sample answer: Kelsey's jersey number is greater than 5 and less than 11. Marie and Kelsey are on the same team. Kylie's team scored 26 points. **3.** Sample answer: Tom mulched the yard of all three clients this week. Ms. Martinez paid Tom $115 this week. Mr. Hansen paid Tom to mow his lawn and mulch his yard. Mrs. Johnson used all of Tom's services this week. **5.** analytic geometry **7.** synthetic geometry **9.** synthetic geometry **11.** Sample answer: Pedro and Rafael ate the same type of salad. **13.** Sample answer: Theo is likely doing analytic geometry, because he is using a graph with points. **15.** Sample answer: Because Sydney's plan is on a grid, she is likely using analytic geometry; that is, assuming the grid is used as a coordinate system. **17.** Sample answer: The three routes are not a model for the axiomatic system. Axioms 1, 2, and 4 are satisfied. Axiom 3 is not satisfied because Route 3 visits Stadium District twice and it is not the first/last stop. Axiom 5 is not satisfied because all three routes visit Stadium District. **19.** Sample answer: The rules of a game are like the axioms of an axiomatic system. They establish what can happen within the game. Plays are like theorems. They are tested against the rules or axioms to see whether they are legal in the game. In basketball, it is a rule that during playing time 5 players from each team shall be on the playing court. A play in which 6 players are on the court is a violation because the rules allow exactly 5 players. **21.** Sample answer: The second figure does not satisfy all the axioms. The axioms do not specify that the line segments connecting the points need to be straight, so the first and third figures would work. **23.** Sample answer: The triangle on the coordinate grid does not belong because it illustrates analytic geometry, while the other two figures illustrate synthetic geometry.

Lesson 10-2

1. Sample answer: n and q **3.** plane R **5.** Sample answer: point P **7.** Yes; sample answer: Line n intersects line q when the lines are extended. **9.** plane **11.** plane **13.** point on a line **15.** line **17.** plane **19.** plane
21. Sample answer:

23. Sample answer:

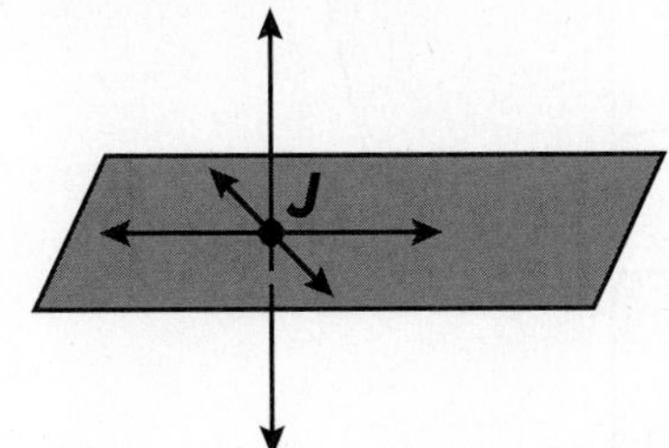

25. 5 **27.** *A*, *B*, *E*, *F* or *B*, *C*, *D*, *E* or *A*, *C*, *D*, *F* **29.** $\overleftrightarrow{AB}$, $\overleftrightarrow{AG}$, $\overleftrightarrow{AH}$, $\overleftrightarrow{BC}$, $\overleftrightarrow{BH}$, $\overleftrightarrow{CD}$, $\overleftrightarrow{DE}$, $\overleftrightarrow{DH}$, $\overleftrightarrow{EF}$, $\overleftrightarrow{FG}$, $\overleftrightarrow{FH}$ **31a.** The lines intersect at the vanishing points. **31b.** Sample answer: The walls of the building and the ground form planes.
33. Sample answer:

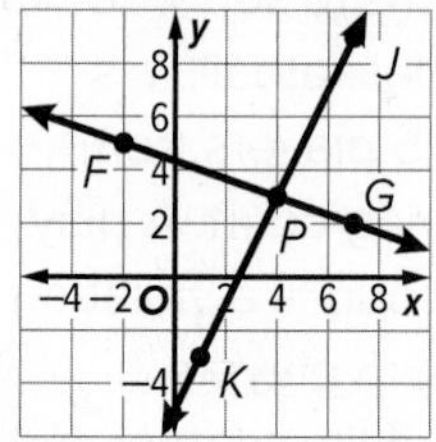

35. Sample answer: line q **37.** $\overleftrightarrow{CD}$ or $\overleftrightarrow{DC}$
39. Sample answer:

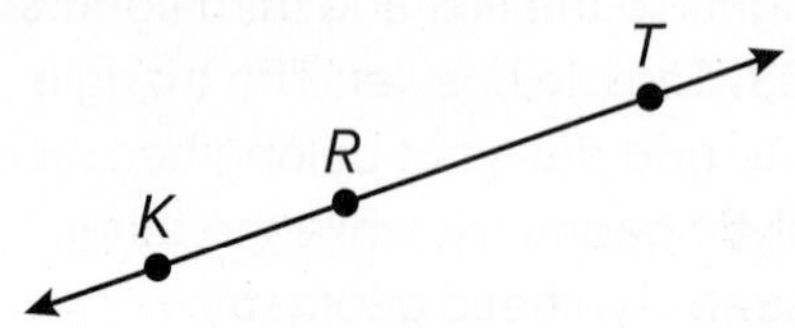

41. Sample answer:

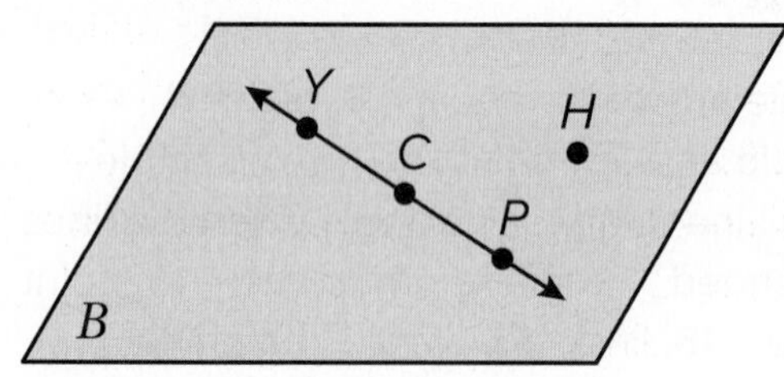

43. lines perpendicular to a plane
45. Sample answer:

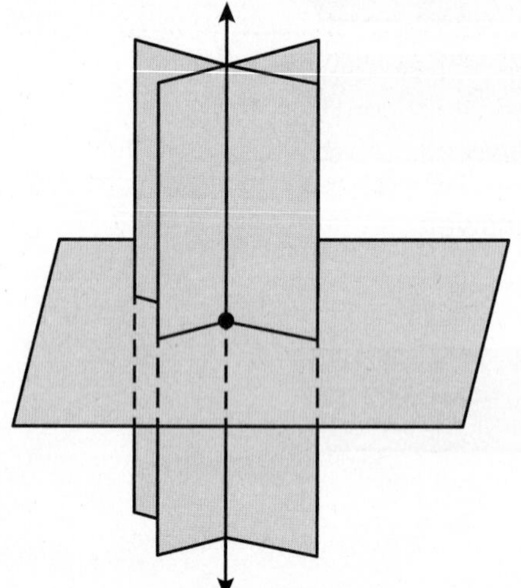

47. Sample answer: Hiroshi is correct. After you draw 3 lines from the first point to the other three points, one of the lines from each of the other three points is already drawn. **49.** Sample answer: A table is a finite plane. It is not possible to have a real-life object that is an infinite plane because all real-life objects have boundaries.

Lesson 10-3

1. 2.1 mm **3.** 1.1 cm **5.** 2.0 m **7.** $2\frac{1}{4}$ in. **9.** 5.3 mm **11.** $b = 12.5$; $YZ = 100$ **13.** $c = 1.7$; $YZ = 3.4$ **15.** $w = 4$; $YZ = 24$ **17.** $n = 4$; $YZ = 1$ **19.** $k = 6$; $YZ = 46$ **21.** $x = 6$; $YZ = 18$ **23.** $x = 10$; $YZ = 60$ **25.** 13 in. and 65 in. **27a.** 6 mi **27b.** Sample answer: I assumed the three locations were in a straight line. **29.** 4.4 mm **31.** 10.8 in. **33.** 66 units **35.** 3 **37.** 4
39. Sample answer: $AP + PM = AM$
41. 5184 ft **43.** $x = 3$; 13 mi **45.** 40 ft
47. $JK = 12$, $KL = 16$ **49.** $x = 3$; $y = 4$
51. Sample answer: $2.8 + BC = 5.3$; $BC = 2.5$ in.

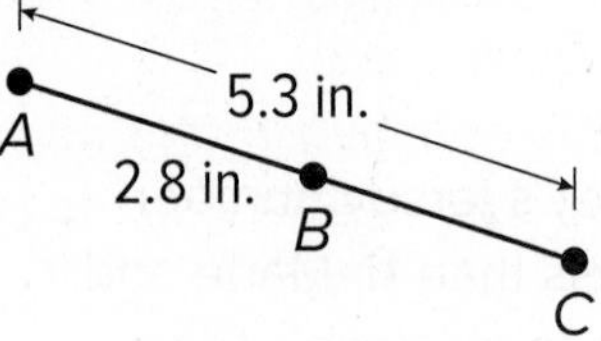

Lesson 10-4

1. 5 **3.** 9 **5.** 12 **7.** 3 **9.** 3 **11.** 9 **13.** 6 **15.** yes **17.** no **19.** no **21.** 10 units **23.** $\sqrt{89}$ or about 9.4 units **25.** $\sqrt{20}$ or about 4.5 units **27.** $\sqrt{20}$ or approximately 4.5 units **29.** Yes; sample answer: The distance between Mariah's house and the library is $\sqrt{74}$ or about 8.6 miles. Because $\frac{2}{3}$ of 12 miles is 8 miles, Mariah's bike ride is more than $\frac{2}{3}$ of the cycling portion of the triathlon. **31.** $\sqrt{37}$ or about 6.1 units **33.** $\sqrt{29}$ or about 5.4 units **35.** $\sqrt{13}$ or about 3.6 units **37.** $\sqrt{89}$ or about 9.4 units **39.** $3\sqrt{37}$ or about 18.2 units **41.** $\sqrt{52}$ or about 7.2 units **43.** no **45.** (0, −4), (0, 10) **47.** 10 in. **49.** No; sample answer: We know that $QU + UR = QR = 4$ and $QU = UR$, so $QU = 2$. Further, we know that $RV + VS = RS = 2$, and $RV = VS$, so $RV = 1$. Because QU is not equal to RV, we know that $\overline{QU}$ is not congruent to $\overline{RV}$. **51.** (5, 10)

53. Sample answer: Substitute 10 for d, (1, 3) for (x_1, y_1), and (9, y) for (x_2, y_2) in the Distance Formula: $10 = \sqrt{(9-1)^2 + (y-3)^2}$. Solve for y:

$$100 = (9-1)^2 + (y-3)^2$$
$$= 8^2 + (y-3)^2$$
$$= 64 + (y-3)^2$$
$$36 = (y-3)^2$$
$$6 = y - 3 \text{ or } -6 = y - 3$$
$$9 = y \text{ or } -3 = y$$

So, the y-coordinate of point B is 9 or −3.

Lesson 10-5

1. 6 **3.** 9 **5.** 8.4 **7.** −1 **9.** −5.5 **11.** −4 **13.** −1 **15.** −2 **17.** Y **19.** −4 **21.** −3 **23a.** 4.36 mi **23b.** 10 mi **25.** 720 mi **27.** $2\frac{2}{5}$ in. **29.** Sample answer: Draw $\overline{AB}$. Next, draw a construction line and place point C on it. From C, strike 6 arcs in succession of length AB. On the sixth segment of length AB, perform a segment bisector two times to create a $\frac{1}{4}AB$ length. Label the endpoint D. **31.** Sometimes; sample answer: If the coordinate of X is 0 and the coordinate of Y is negative, then the coordinate of W will be negative and less than the coordinate of X. If the coordinate of X is positive and the coordinate of Y is greater than the coordinate of X, then the coordinate of W will be greater than the coordinate of X.

Lesson 10-6

1. (−3.6, −2.2) **3.** $\left(1, 1\frac{2}{3}\right)$ **5.** $\left(\frac{14}{3}, 1\right)$ **7.** $\left(-\frac{7}{5}, 4\right)$ **9.** $\left(\frac{16}{7}, -3\right)$ **11.** $\left(1, \frac{5}{4}\right)$ **13.** $\left(\frac{20}{7}, -1\right)$ **15.** (−7, 11) and (−1, 1) **17.** (−3, −2) **19.** $\frac{3}{4}$ **21a.** Julianne substituted the wrong values for (x_1, y_1) and (x_2, y_2). **21b.** (1.6, 4.2) **23.** Sample answer: Because the distance from A to P is twice the distance from P to D, the distance from A to P could be 2 and the distance from P to D could be 1. Therefore, the fractional distance that P is from A to D is $\frac{2}{2+1}$ or $\frac{2}{3}$. The coordinates of point P are (5, 10).

25. Sample answer:

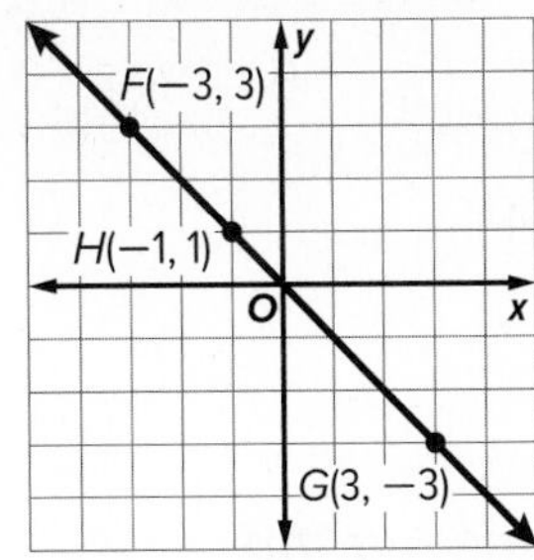

H is $\frac{1}{3}$ of the distance from F to G.

Lesson 10-7

1. −2 **3.** 0.5 **5.** 1.5 **7.** 3 **9.** −4.5 **11.** 8.5 **13.** 9 **15.** 3 **17.** 9 yd **19.** (4, 6) **21.** (−3, −3) **23.** (−1.5, 3.5) **25.** (5, 4) **27.** (18.5, 5.5) **29.** (−6.5, −3) **31.** (4.2, 10.4) **33.** A(1, 6) **35.** C(16, −4) **37.** C(−12, 13.25) **39.** 16 **41.** 2.5 **43.** 8 **45.** 7 **47.** 0.5 **49.** Q(8, 4) **51.** N(−5, 1) **53.** (1, 1) **55.** (6, 7.5) **57.** 48 m **59.** Sample answer: The midpoint of a segment is the average of the coordinates of the endpoints. Divide each coordinate of the endpoint that is not located at the origin by 2. For example, if the segment has coordinates (0, 0) and (−10, 6), the midpoint is located at $\left(-\frac{10}{2}, \frac{6}{2}\right)$ or (−5, 3). Using the Midpoint Formula, if the endpoints of the segment are (0, 0) and (a, b), the midpoint is $\left(\frac{a+0}{2}, \frac{b+0}{2}\right)$ or $\left(\frac{a}{2}, \frac{b}{2}\right)$.

61. Sample answer:

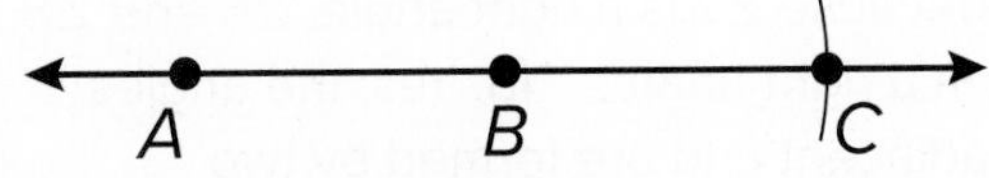

Module 10 Review

1. B, D

3. intersecting planes

5. 11

7. A

9. B

11. D

13. (3, 6)

15. B

17. C

Module 11

Quick Check

1. 63 **3.** 3 **5.** 1980 **7.** 344

Lesson 11-1

1. *A* **3.** $\angle ADC$, $\angle CDA$ **5.** 66° **7.** 78° **9.** 61° **11.** 56° **13.** Sample answer: $\angle SRQ$ and $\angle TRP$ **15.** 65 **17.** $x = 35$, $y = 85$ **19.** *R* **21.** *P* **23.** $\overrightarrow{NM}$ and $\overrightarrow{NV}$ **25.** $\overrightarrow{RP}$ and $\overrightarrow{RQ}$ or $\overrightarrow{RT}$ and $\overrightarrow{RQ}$ **27.** $\angle TPQ$ **29.** $\angle TPN$, $\angle NPT$, $\angle TPM$, $\angle MPT$ **31.** *S*, *Q* **33.** Sample answer: $\angle MPR$, $\angle PRQ$ **35.** $x = 48$, $y = 21$ **37.** Sample answer: $\angle HGE$, $\angle DGE$ **39.** Sample answer: $\angle BFC$, $\angle BFD$ **41.** Sample answer: $\angle STV$, $\angle VTW$, $\angle UTW$ **43a.** yes **43b.** no **43c.** no **43d.** no **45.** 45° **47a.** 45° **47b.** $\overrightarrow{EF}$ **49.** 168; sample answer: If $m\angle RMP = 21°$ and $\overrightarrow{MR}$ bisects $\angle QMP$, then $m\angle QMP = 2(21)$ or 42°. If $m\angle QMP = 42°$ and $\overrightarrow{MQ}$ bisects $\angle LMP$, then $m\angle LMP = 2(42)$ or 84°. If $m\angle LMP = 84°$ and $\overrightarrow{MP}$ bisects $\angle LMN$, then $m\angle LMN = 2(84) = 168°$.

Lesson 11-2

1. 72.5°, 107.5° **3.** 128°; 52° **5.** 45°; 135° **7.** $m\angle ABD = 47°$; $m\angle DBC = 43°$ **9.** $a = 20$ **11.** Yes; because $\angle 7$ is a right angle, $\angle 6$ and $\angle 8$ must form a right angle. **13.** Yes; the angles are nonadjacent and are formed by two intersecting lines. **15.** 40° **17.** 7 **19.** 92° **21.** No; the measures of the angles are unknown. **23.** No; the angles are not adjacent. **25.** $x = 25$ **27.** $x = 94$, $y = 79$, $z = 26$ **29.** Yes; sample answer: Angles that are right or obtuse do not have complements because their measures are greater than or equal to 90°. **31.** Sample answer: You can determine whether an angle is right if it is marked with a right angle symbol, if the angle is a vertical pair with a right angle, or if the angle forms a linear pair with a right angle.

33. Sample answer:

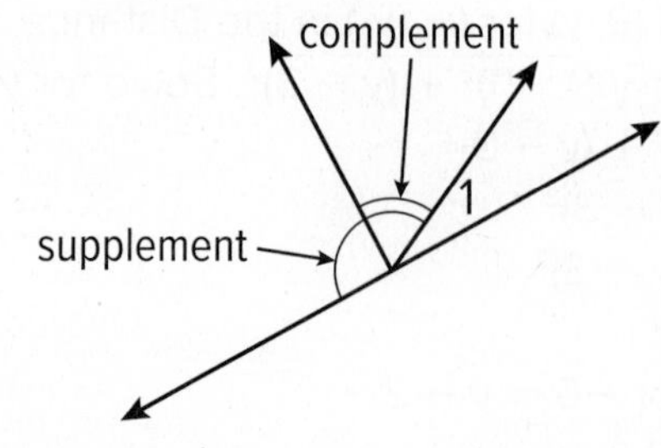

35. No; sample answer: Straight angles or angles that are greater than 180° do not have supplementary angles because their measures are greater than or equal to 180°.

Lesson 11-3

1. 20.9 cm; 16 cm^2 **3.** 17.8 cm; 25.1 cm^2 **5.** 11.3 cm; 8 cm^2 **7.** 15.7 in.; 19.6 in^2 **9.** 42.1 ft; 92.5 ft^2 **11.** 35.1 m^2 **13a.** 398 m **13b.** In part a, I assumed that there was no space between the field and the first lane of the track. I also assumed the athlete's body was centered on the border of the track. **13c.** 402 m **13d.** 30 cm **15.** quadrilateral; 20 units; 24 units^2 **17.** quadrilateral; 20 units; 25 units^2 **19.** \$650; sample answer: The side of the play area that is adjacent to the house does not need fencing. The remaining three sides of the play area on the grid have lengths 4, 5, and 4 units. The perimeter of the play area to be fenced on the grid is $P = 4 + 5 + 4 = 13$ units. Each unit on the grid represents 5 feet, so Derek will need 13(5 ft) or 65 ft of fencing. The cost of the fencing is \$10 per foot, so the total cost will be 65(\$10) = \$650. **21.** 78.5 square miles **23a.** $s = 4$, so $A = 4^2 = 16$ units^2 **23b.** $b = 2$, $h = \sqrt{3}$, so $A = \frac{1}{2}(2)(\sqrt{3}) = \sqrt{3}$ units^2 **23c.** $16 - \sqrt{3}$ or about 14.3 units^2; sample answer: The area not covered by the triangle is equal to the area of the square minus the area of the triangle. So, $A = (16 - \sqrt{3})$ units^2 or about 14.3 units^2. **25.** 290.93 units^2 **27.** 39 units; Sample answer: An equilateral triangle has congruent side lengths. Use the Distance Formula to find $KM = \sqrt{(10 - (-2))^2 + (6 - 1)^2} = 13$. So, $P = 3(KM) = 3(13)$ or 39 units.

Lesson 11-4

1. reflection **3.** rotation **5.** translation **7.** rotation **9.** $A'(2, 0)$, $B'(-1, -5)$, and $C'(4, -3)$ **11.** $A'(2, 2)$, $B'(-1, 7)$, and $C'(4, 5)$ **13.** $A'(-2, 0)$, $B'(1, -5)$, and $C'(-4, -3)$ **15.** $D'(4, 1)$, $E'(5, -2)$, and $F'(1, -2)$ **17.** $D'(5, -1)$, $E'(6, 2)$, and $F'(2, 2)$ **19.** $D'(-4,1)$, $E'(-5, -2)$, and $F'(-1,-2)$ **21.** translation along vector $\langle -10, 35\rangle$ **23a.** rotation **23b.** translation **25.** reflection; translation

27.

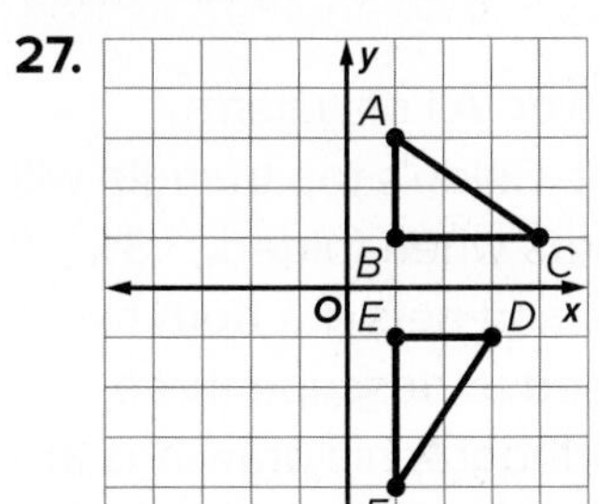

$D(3, -1)$, $E(1, -1)$, $F(1, -4)$

29. reflection in the x-axis **31.** translation 3 units right and 2 units down **33.** Antwan; sample answer: When you reflect a point across the x-axis, the reflected point is in the same place horizontally, but not vertically. When (2, 3) is reflected across the x-axis, the coordinates of the reflected point are (2, –3) because it is in the same location horizontally, but the other side of the x-axis vertically. **35.** Sample answer:

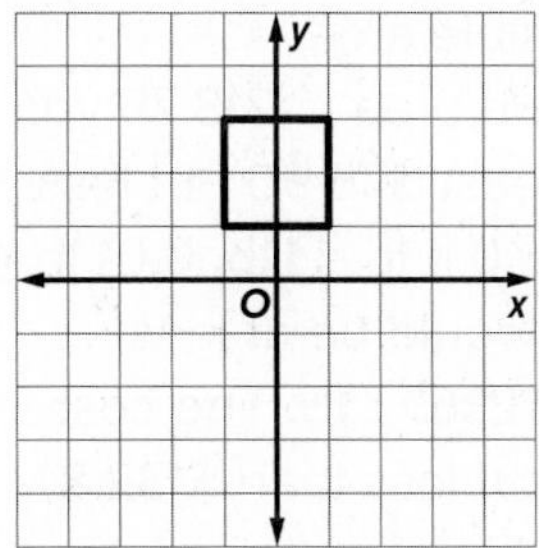

Lesson 11-5

1. not a polyhedron; cone **3.** polyhedron; rectangular pyramid; base $WXYZ$; faces $\square WXYZ$, $\triangle VWX$, $\triangle VXY$, $\triangle VYZ$, $\triangle VZW$; edges $\overline{WX}$, $\overline{XY}$, $\overline{YZ}$, $\overline{ZW}$, $\overline{WV}$, $\overline{XV}$, $\overline{YV}$, $\overline{ZV}$; vertices W, X, Y, Z, V **5.** sphere; not a polyhedron **7.** 26.9 cm^2; 7.7 cm^3 **9.** 800 ft^2; 1280 ft^3 **11.** 90π or about 282.7 yd^2; 100π or about 314.2 yd^3 **13a.** 35 ft^2 **13b.** 13 bags **15.** 2004 in^2 **17.** 3.9 cm **19.** 3 in. **21a.** 5027 ft^3 **21b.** 402 ft^3 **21c.** 5429 ft^3 **23a.** 0.012 m^3; $V = \frac{1}{3}Bh$, so $V = \frac{1}{3}(900)(40)$ or 12,000 cm^3. One cubic meter equals 1 million cubic centimeters, so the volume of one pyramid shaped lawn ornament is 0.012 m^3. **23b.** concrete: 28.5 kg; granite: 32.2 kg; marble 32.5 kg **23c.** If the volume of the lawn ornament stays the same, then the weight of the ornament increases as the density of the material used to make it increases. **25.** Neither; sample answer: The surface area is twice the sum of the areas of the top, front, and left side of the prism or $2(5 \cdot 3 + 5 \cdot 4 + 3 \cdot 4)$, which is 94 square inches. **27.** Sample answer: The cone and pyramid have nearly the same volumes. Cone: The area of the base is approximately 154 square centimeters, so $V \approx \frac{1}{3}(154)(28)$ or about 1437 cubic centimeters. The volume of the pyramid is greater by such a small amount that we can say the volumes are approximately equal. **29.** 27 mm^3

Lesson 11-6

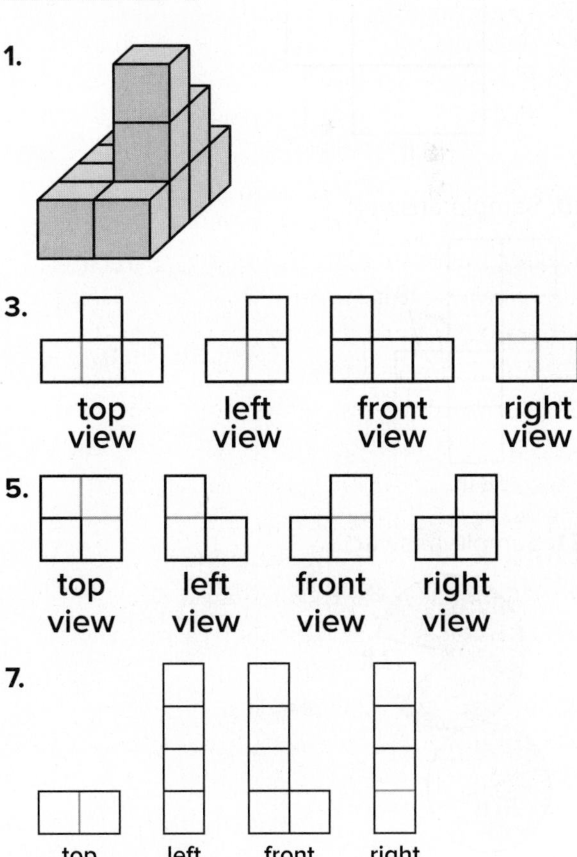

9. square pyramid, 144 cm^2

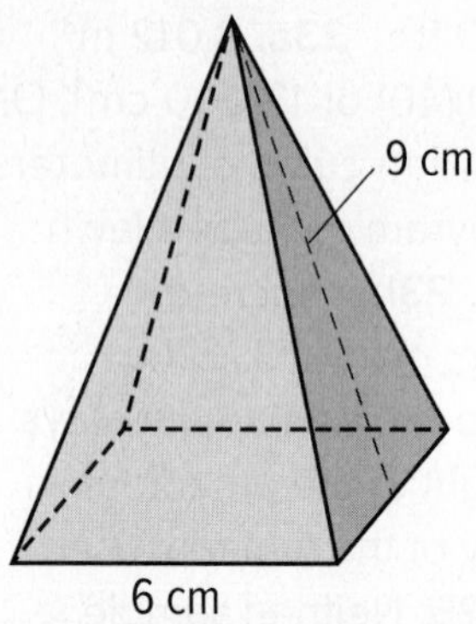

11. cylinder; 3.5π in^2 or 11.0 in^2

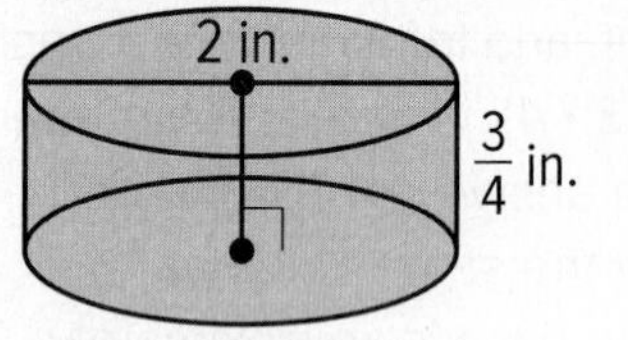

13. tetrahedron
15. octahedron
17. Sample answer:

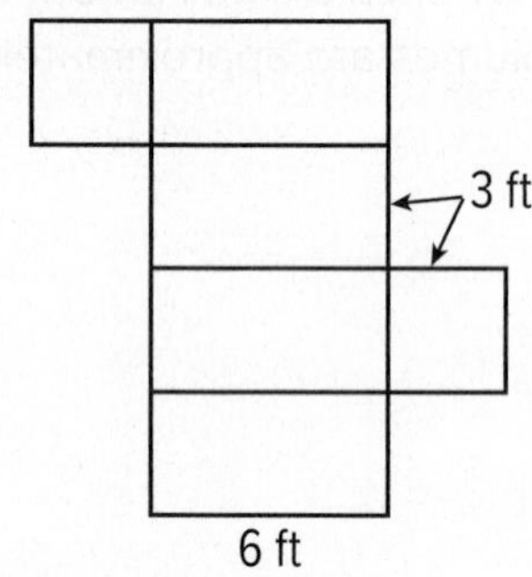

19. Sample answer:

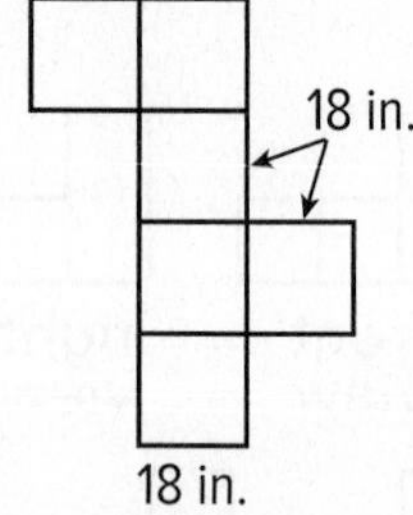

21. Sample answer:

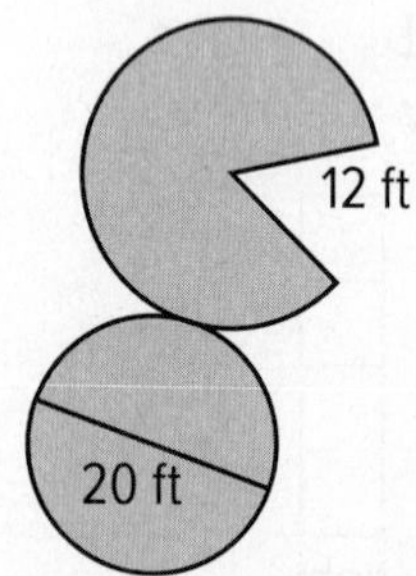

23.

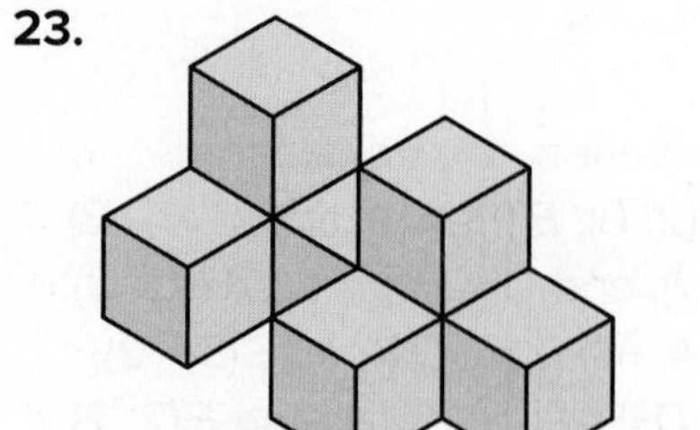

25. Sample answer: ice cream cone
27. No; the area of this net is 48 square centimeters.
29. Julian; sample answer: All of Julian's triangles are congruent. Caleb's top triangle will not match with the others when folded. **31.** Orthographic drawings and nets are both two-dimensional shapes used to describe three-dimensional figures. Orthographic drawings are views of the top, left, front, and right sides of an object, whereas nets can be folded to create a three-dimensional object.

Lesson 11-7

1. The sample is precise because there are consistently 17 or 18 rice cakes in each package. The sample indicates an innacurate claim of 20 rice cakes per package.
3. 0.05 milliamp **5.** 1.1°F **7a.** 16.5 yd **7b.** 12.5 yd **7c.** least possible area = 189.75 yd^2; greatest possible area = 218.75 yd^2 **7d.** The cost would be at least \$75.90 but less than \$87.50. **9.** 1 in. **11a.** 0.11 lb **11b.** 0.01 lb **11c.** 0.64 lb **13.** The cost would be at least \$955.94 but less than \$961.36. **15.** The cost would be at least \$107.81 but less than \$132.81. **17.** Accuracy is how well the information or data matches the true values. Precision is the repeatability of the measurement and level of measurement. Sample answer: 16.1 oz, 16.3 oz, 15.93 oz, 15.8 oz. **19a.** There are two faces that have an area of 111.4 in^2, two faces that have an area of 35.82 in^2, and two faces that have an area of 75.3 in^2. **19b.** 445.0 in^2 **19c.** The calculation of surface area is accurate to the nearest tenth. The true surface area falls between 443.31 in^2 and 446.60 in^2.
19d. 440.1 in^2

Lesson 11-8

1. 5 **3.** 2 **5.** 1 **7.** Michelle **9.** 4.5 to 5.5 in. **11.** 2.5 to 3.5 cm **13.** 4.375 to 4.625 in. **15.** 2 **17.** 4 **19.** 2 **21.** 7.2 mL **23.** 15.1 cm^2 **25.** The area of the circle is between 31.2 and 33.2 cm^2. **27.** 4 **29.** 5 **31.** 5 **33.** 0.663 gram **35.** Scale 1: 4; Scale 2: 3; Scale 3: 4 **37.** 79.40 yd^2 **39.** 8.3 g **41.** 624 in^3

43. Sample answer: The 0 in each dimension, 1.40 cm and 1.60 cm, is significant. The answer should be given with 3 significant figures as 2.24 square centimeters. **45.** Yes; sample answer: The zeros before and after the decimal are not significant because a nonzero number did not come before them. Therefore, both numbers have two significant figures. **47.** Sometimes; sample answer: A zero between two nonzero significant figures is always significant, a leading zero is never significant, and a zero at the end of a number is only significant when a decimal point is given in the number.

Module 11 Review

1. A, C **3.** Sample answer: Use paddy paper or wax paper. Draw an angle on the paper. Fold the paper so that one side of the angle is directly on top of the other side. Draw the angle bisector in the crease of the fold. **5.** B **7.** Perimeter: $\approx$ 22.4 units; Area: 30 square units **9.** B **11.** C **13.** 2144.66 in^3 **15.** B

Selected Answers

Module 12

Quick Check

1. 5 **3.** 2.3 **5.** $\angle BXD$, $\angle AXE$ **7.** 6

Lesson 12-1

1. Each term in the pattern is four more than the previous term; 24.

3. Each term is one half the previous term; $\frac{1}{16}$.

5. Each arrival time is 2 hours and 30 minutes prior to the previous arrival time; 7:30 A.M.

7. The shaded section in each circle has moved one section counterclockwise from its location in the previous circle.

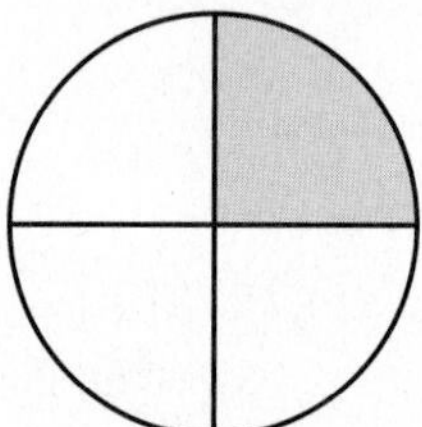

9. The product is an odd number.

11. They are equal. **13.** The lines are perpendicular.

15. $NP = PQ$ **17.** 180 cm

19. False; sample answer: Suppose $x = 2$, then $-x = -2$.

21. false;

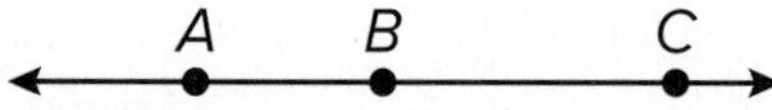

23. False; sample answer: The length could be 4 m, and the width could be 5 m.

25. Yes; sample answer: If no team got more than 5 medals, then the total number of medals could not be more than 5×6 or 30 medals.

27a. 1, 3, 5 **27b.** You get all the odd numbers.

27c. Their sum equals n^2.

29. Sample answer: When $n = 4$, $2^n - 1 = 15$ and $15 = 3 \times 5$.

31a. Sample answer: The number of regions doubles when you add a point on the circle.

31b. For six points, there should be 32 regions; however only 31 regions are formed. The conjecture is false.

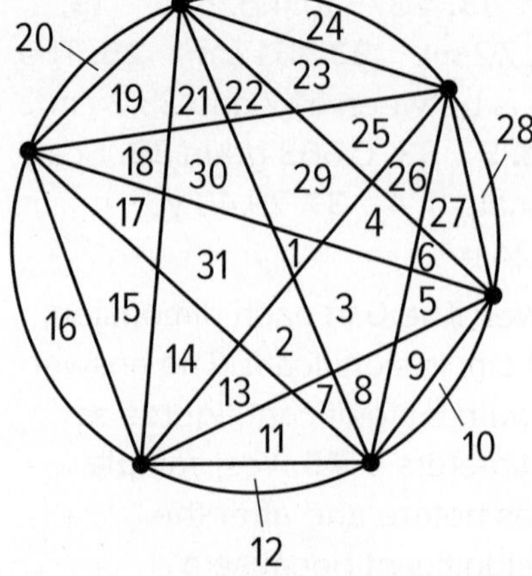

33. False; sample answer: If the two points create a straight angle that includes the third point, then the conjecture is true. If the two points do not create a straight angle with the third point, then the conjecture is false.

35. Sample answer: A postulate states that a line contains at least two points. These two points and all the points between them are line segments, by the definition of a line segment.

Lesson 12-2

1. $-3 - 2 = -5$, and vertical angles are congruent; p is true, and q is true, so p and q is true.

3. Vertical angles are congruent, or $2 + 8 \leq 10$; q is true, and $\sim r$ is true, so $q \vee \sim r$ is true.

5. $-3 - 2 \neq -5$, and not all vertical angles are congruent; $\sim p$ is false, and $\sim q$ is false, so $\sim p \wedge \sim q$ is false. **7.** H: there is no struggle; C: there is no progress **9.** H: you lead; C: I will follow **11.** H: two angles are vertical; C: they are congruent **13.** H: you were at the party; C: you received a gift; If you were at the party, then you received a gift. **15.** H: a figure is a circle; C: the area is πr^2; If a figure is a circle, then the area is πr^2. **17.** H: an angle is right; C: the angle measures 90°; If an angle is right, then the angle measures 90°. **19.** Converse: If you can buy five raffle tickets, then you have five dollars. The converse is true. Inverse: If you do not have five dollars, then you cannot buy five raffle tickets. The inverse is true.

Contrapositive: If you cannot buy five raffle tickets, then you do not have five dollars. The contrapositive is true.

21. Converse: If you do not take this medicine, then you can drive. The converse is true. Inverse: If you are not driving, then you can take this medicine. The inverse is true. Contrapositive: If you take this medicine, then you are not driving. The contrapositive is true.
23. Conditional: If it is raining, then the game will be cancelled. The conditional is true. Converse: If the game is cancelled, then it is raining. Counterexample: The game could be cancelled, and it is not raining. The converse is false. Because the converse is false, a biconditional statement cannot be written.

25. Conditional: If a polygon has four sides, then it is a quadrilateral. Converse: If a polygon is a quadrilateral, then it has four sides. The conditional and the converse are true, so the biconditional is true. **27.** true **29.** false

31. yes **33.** nothing **35.** If a ray does not bisect an angle, then it does not divide the angle into two congruent angles.

37. If two angles are right angles, then they are congruent.

39a. Sample answer: If you are in Houston, then you are in Texas. **39b.** Sample answer: Converse: If you are in Texas, then you are in Houston. Inverse: If you are not in Houston, then you are not in Texas. Contrapositive: If you are not in Texas, then you are not in Houston. **39c.** Converse: false; Inverse: false; Contrapositive: true

41. There exists at least one student at Hammond High school that does not have a locker.

43. For every real number x, $x^2 \neq x$. **45.** There exists a real number that does not have a real square root.

47. Truth table with the following columns:

p	q	$\sim p$	$\sim q$	$p \rightarrow q$	$q \rightarrow p$	$\sim p \rightarrow \sim q$	$\sim q \rightarrow \sim p$
T	T	F	F	T	T	T	T
T	F	F	T	F	T	T	F
F	T	T	F	T	F	F	T
F	F	T	T	T	T	T	T

Because column 5 is the same as column 8, the conditional is equivalent to its contrapositive. Because column 6 is the same as column 7, the converse and the inverse are equivalent.
49. Kiri; sample answer: When the hypothesis of a conditional is false, the conditional is always true.

Lesson 12-3

1. deductive **3.** deductive **5.** inductive
7. valid; Law of Detachment **9.** Invalid; your battery could be dead because it was old.
11. valid; Law of Detachment **13.** If Tina has a grade point average of 3.0 or greater, then she will have her name in the school paper.
15. If the measure of an angle is between 90° and 180°, then it is not acute. **17.** No valid conclusion; the conclusion of statement (1) is not the hypothesis of statement (2). **19.** The sum of the measures of ∠1 and ∠2 is 90°; Law of Detachment **21.** No valid conclusion; the conclusion of statement (1) is not the hypothesis of statement (2).

23. If Terryl completes a course with a grade of C, then he will have to take the course again; Law of Syllogism. **25.** Valid; Theo is inside the small and large circles, so the conclusion is valid.

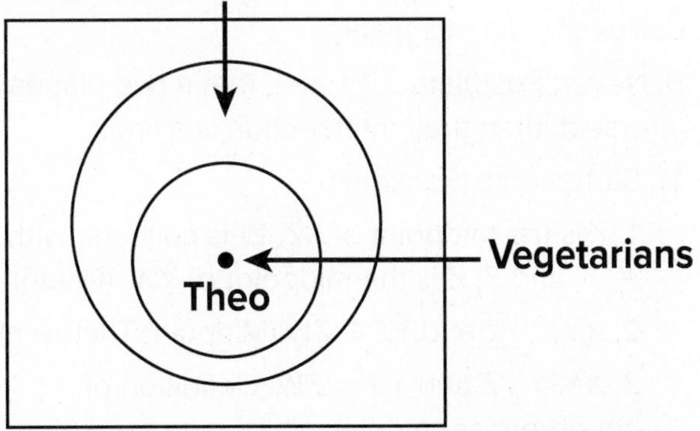

27. then Mozart did not live in Vienna in the early 1800s. **29.** The child is at least 5 years old. **31.** Energy costs will be higher in Florida.

33. Law of Detachment: $[(p \rightarrow q) \wedge p] \rightarrow q$. Law of Syllogism: $[(p \rightarrow q) \wedge (q \rightarrow r)] \rightarrow (p \rightarrow r)$.

35. Sample answer: Jonah's statement can be restated as, "Jonah is in Group B, and Janeka is in Group B." For this compound statement to be true, both parts of the statement must be true. If Jonah was in Group A, he would not be able to say that he is in Group B, because students in Group A must always tell the truth. Therefore, the statement that Jonah is in Group B is true. For the compound statement to be false, the statement that Janeka is in Group B must be false. Therefore, Jonah is in Group B, and Janeka is in Group A.

37. Sample answer: Given: If you are at the Willis Tower, then you are in Chicago. If you are in Chicago, then you are in Illinois. Conclusion: Therefore, if you are at the Willis Tower, then you are in Illinois.

39. D; Statement D follows logically from statements (1) and (2). Statements A, B, and C do not follow logically from statements (1) and (2).

Lesson 12-4

1. The two planes meet at the edge, which lies on line *t*. Postulate: If two planes intersect, then their intersection is a line.

3. Postulate 3.1; Through any two points, there is exactly one line.

5. Always; Postulate 3.2 states that through any three noncollinear points, there is exactly one plane.

7. Sometimes; the points do not have to be collinear to lie in a plane.

9. Never; Postulate 3.7 states that if two planes intersect, then their intersection is a line.

11. Statements (Reasons)

1. Y is the midpoint of $\overline{XZ}$. W is collinear with X, Y, and Z. Z is the midpoint of $\overline{YW}$. (Given)
2. $\overline{XY} \cong \overline{YZ}$ and $\overline{YZ} \cong \overline{ZW}$ (Midpoint Theorem)
3. $XY = YZ$ and $YZ = ZW$ (Definition of congruent segments)
4. $XY = ZW$ (Transitive Property of Equality)
5. $\overline{XY} \cong \overline{ZW}$ (Definition of congruent segments)

13. Statements (Reasons)

1. $SR = RT$ (Given)
2. R is the midpoint of $\overline{ST}$. (Definition of midpoint)
3. $SR = UR$ and $RT = RV$ (Given)
4. $SR = RT$, so $UR = RV$. (Substitution)
5. R is the midpoint of $\overline{UV}$. (Definition of midpoint)

15.

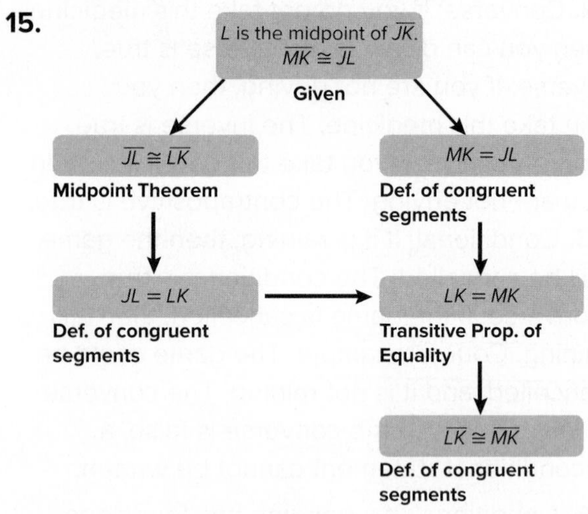

17. Given: B is the midpoint of $\overline{AC}$. C is the midpoint of $\overline{BD}$.
Prove: $AB = CD$
Proof: Because B is the midpoint of $\overline{AC}$ and C is the midpoint of $\overline{BD}$, we know by the Midpoint Theorem that $\overline{AB} \cong \overline{BC}$ and $\overline{BC} \cong \overline{CD}$. Because congruent segments have equal measures, $AB = BC$ and $BC = CD$. Thus, by the Transitive Property of Equality, $AB = CD$.

19. Postulate 3.3: A line contains at least two points. **21.** Iza is because a line can lie in a plane and intersect it in infinite points. **23.** 15 **25.** Ana is correct. Sample answer: The proof should begin with the given, which is that $\overline{AB}$ is congruent to $\overline{BD}$ and A, B, and D are collinear. Therefore, Ana began the proof correctly. **27a.** Plane Q is perpendicular to plane P. **27b.** Line a is perpendicular to plane P. **29.** Sometimes; if the points were noncollinear, then there would be exactly one plane by Postulate 3.2 shown by Figure 1. If the points were collinear, then there would be infinitely many planes. Figure 2 shows what two planes through collinear points would look like. More planes would rotate around the three points.

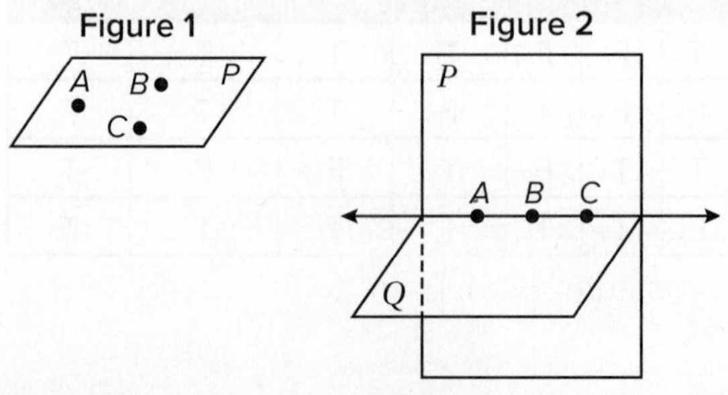

Lesson 12-5

1.

1. C is the midpoint of $\overline{AE}$.
   C is the midpoint of $\overline{BD}$.
   $\overline{AE} \cong \overline{BD}$
2. Definition of midpoint
3. Definition of $\cong$ segments
4. $AE = AC + CE$
   $BD = BC + CD$
5. Substitution Property
6. Substitution Property
7. $2AC = 2CD$
8. $AC = CD$
9. Definition of $\cong$ segments

3. Given: $\overline{VZ} \cong \overline{VY}$ and $\overline{WY} \cong \overline{XZ}$

Prove: $\overline{VW} \cong \overline{VX}$

Proof

Statements (Reasons)

1. $\overline{VZ} \cong \overline{VY}$ and $\overline{WY} \cong \overline{XZ}$ (Given)
2. $VZ = VY$ and $WY = XZ$ (Def. of $\cong$ segments)
3. $VZ = VX + XZ$ and $VY = VW + WY$ (Segment Addition Postulate)
4. $VX + XZ = VW + WY$ (Substitution Prop.)
5. $VX + WY = VW + WY$ (Substitution Prop.)
6. $VX = VW$ (Sub. Prop. of =)
7. $VW = VX$ (Symmetric Property)
8. $\overline{VW} \cong \overline{VX}$ (Def. of $\cong$ segments)

5. Clara is shorter than Chad when Maria is shorter than Luna; Clara and Chad are the same height when Maria is the same height as Luna.

7. No, it's not possible. Lola's house must be less than a mile from each house because she lives between them.

9a. Given: $\overline{AC} \cong \overline{GI}$, $\overline{FE} \cong \overline{LK}$, $AC + CF + FE = GI + IL + LK$

Prove: $\overline{CF} \cong \overline{IL}$

Proof

Statements (Reasons)

1. $\overline{AC} \cong \overline{GI}$, $\overline{FE} \cong \overline{LK}$, $AC + CF + FE = GI + IL + LK$ (Given)
2. $AC = GI$ and $FE = LK$ (Def. of $\cong$ segments)
3. $AC + CF + FE = AC + IL + LK$ (Substitution Property)
4. $AC - AC + CF + FE = AC - AC + IL + LK$ (Subtraction Property of Equality)
5. $CF + FE = IL + LK$ (Substitution Property)
6. $CF + FE = IL + FE$ (Substitution Property)
7. $CF + FE - FE = IL + FE - FE$ (Subtraction Property of Equality)
8. $CF = IL$ (Substitution Property)
9. $\overline{CF} \cong \overline{IL}$ (Def. of $\cong$ segments)

11a. Both segments are half the length of two congruent segments, so the lengths of the shorter segments must be the same.

11b.

1. $\overline{AB} \cong \overline{CD}$; M is the midpoint of $\overline{AB}$ and $\overline{CD}$.
2. Congruent segments have equal lengths.
3. Definition of midpoint
5. Segment Addition Postulate
6. Substitution Property of Equality
8. Substitution Property of Equality
9. $AM = CM$
10. $\overline{AM} \cong \overline{CM}$

13. Neither; because $\overline{AB} \cong \overline{CD}$ and $\overline{CD} \cong \overline{BF}$, then $\overline{AB} \cong \overline{BF}$ by the Transitive Property of Congruence.

15. Sample diagram:

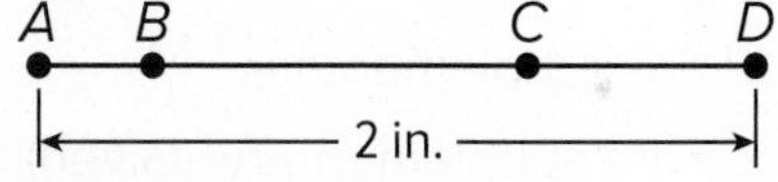

17. No; The Segment Addition Postulate only applies to points that are collinear, but points P, Q, and R are not collinear.

19. Because $\overline{PQ} \cong \overline{RS}$ and congruent segments have equal lengths, $PQ = RS$. Because Q is the midpoint of $\overline{PR}$, $PQ = QR$. By the Substitution Property of Equality, $QR = RS$ so R is the midpoint of $\overline{QS}$.

Lesson 12-6

1. 113° **3.** 74° **5.** 46°

7. Statements (Reasons)

1. $\angle 1$ and $\angle 2$ form a linear pair.
   $\angle 3$ and $\angle 4$ form a linear pair. (Def. of linear pair)
2. $\angle 1$ and $\angle 2$ are supplementary.
   $\angle 3$ and $\angle 4$ are supplementary. (Supp. Thm)
3. $\angle 1 \cong \angle 3$ (Given)

Selected Answers

4. $\angle 2 \cong \angle 4$ ($\cong$ Supp. Thm)

9. Statements (Reasons)

1. $m\angle ABC = m\angle DEF$ (Given)
2. $\angle ABC \cong \angle DEF$ (Def. of $\cong$ angles)
3. $\angle ABC$ and $\angle DEF$ are supplementary. (Given)
4. $\angle ABC$ and $\angle DEF$ are rt. angles. (If two $\angle$s are $\cong$ and supp., then each $\angle$ is a rt. $\angle$.)

11. 36°; 84°

13. $m\angle 6 = m\angle 8 = 73°$, $m\angle 7 = 107°$ ($\cong$ Supp. Thm. and Vert. $\angle$s Thm.)

15. $m\angle 3 = 62°$, $m\angle 1 = m\angle 4 = 45°$ ($\cong$ Comp. and Supp. Thm.)

17. $m\angle 9 = 156°$, $m\angle 10 = 24°$ ($\cong$ Supp. Thm.)

19. Statements (Reasons)

1. $\angle PQT$ and $\angle TQR$ form a linear pair. (Given)
2. $\angle PQR$ is a straight angle. (Given from figure)
3. $m\angle PQR = 180°$ (Def. of straight angle)
4. $m\angle PQT + m\angle TQR = m\angle PQR$ (Angle Add. Post.)
5. $m\angle PQT + m\angle TQR = 180°$ (Subs.)
6. $\angle PQT$ and $\angle TQR$ are supplementary. (Def. of supp. angles)

21. Statements (Reasons)

1. $\angle 1$ and $\angle 2$ are supplementary.
   $\angle 2$ and $\angle 3$ are supplementary. (Given)
2. $m\angle 1 + m\angle 2 = 180°$
   $m\angle 2 + m\angle 3 = 180°$ (Def. of supp. angles)
3. $m\angle 1 + m\angle 2 = m\angle 2 + m\angle 3$ (Subs.)
4. $m\angle 1 = m\angle 3$ (Subtraction Prop. of =)
5. $\angle 1 \cong \angle 3$ (Def. of $\cong$ angles)

23. Statements (Reasons)

1. $\ell \perp m$ (Given)
2. $\angle 1$ is a right angle. (Def. of $\perp$)
3. $m\angle 1 = 90°$ (Def. of rt. angles)
4. $\angle 1 \cong \angle 4$ (Vert. Angles Thm)
5. $m\angle 1 = m\angle 4$ (Def. of $\cong$ angles)
6. $m\angle 4 = 90°$ (Subs.)
7. $\angle 1$ and $\angle 2$ form a linear pair.
   $\angle 3$ and $\angle 4$ form a linear pair. (Def. of linear pairs)
8. $m\angle 1 + m\angle 2 = 180°$, $m\angle 4 + m\angle 3 = 180°$ (Linear pairs are supp.)
9. $90° + m\angle 2 = 180°$, $90° + m\angle 3 = 180°$ (Subs.)
10. $m\angle 2 = 90°$, $m\angle 3 = 90°$ (Subtraction)
11. $\angle 2$, $\angle 3$, and $\angle 4$ are rt. angles. (Def. of rt. angles (Steps 6, 10))

25. Statements (Reasons)

1. $\angle 1 \cong \angle 2$, $\angle 1$ and $\angle 2$ are supplementary. (Given)
2. $m\angle 1 + m\angle 2 = 180°$ (Def. of supp. angles)
3. $m\angle 1 = m\angle 2$ (Def. of $\cong$ angles)
4. $m\angle 1 + m\angle 1 = 180°$ (Subs.)
5. $2(m\angle 1) = 180°$ (Subs.)
6. $m\angle 1 = 90°$ (Div. Prop.)
7. $m\angle 2 = 90°$ (Subs. (steps 3, 6))
8. $\angle 1$ and $\angle 2$ are rt. angles. (Def. of rt. angles)

27. By the Transitive Property, if any two angles are equal to the angle of the template, then they must be equal to each other.

29. Sample answer: $m\angle WXY = 90°$

Given: $m\angle WXZ = 45°$, $\angle WXZ \cong \angle YXZ$

Prove: $m\angle WXY = 90°$

Proof:

Statements (Reasons)

1. $m\angle WXZ = 45°$, $\angle WXZ \cong \angle YXZ$ (Given)
2. $m\angle WXZ = m\angle YXZ$ (Def. of $\cong$ $\angle$s)
3. $m\angle YXZ = 45°$ (Substitution)
4. $m\angle WXY = m\angle WXZ + m\angle YXZ$ (Angle Add. Post.)
5. $m\angle WXY = 45° + 45°$ (Substitution)
6. $m\angle WXY = 90°$ (Substitution)

31. Each of these theorems uses the words *or to congruent angles* indicating that this case of the theorem must also be proved true. The first proof of each theorem only addressed the *to the same angle* case of the theorem.

Proof of the Congruent Complements Theorem (Case 2: Congruent Angles)

Given: $\angle ABC \cong \angle DEF$, $\angle GHI$ is complementary to $\angle ABC$, $\angle JKL$ is complementary to $\angle DEF$.

Prove: $\angle GHI \cong \angle JKL$

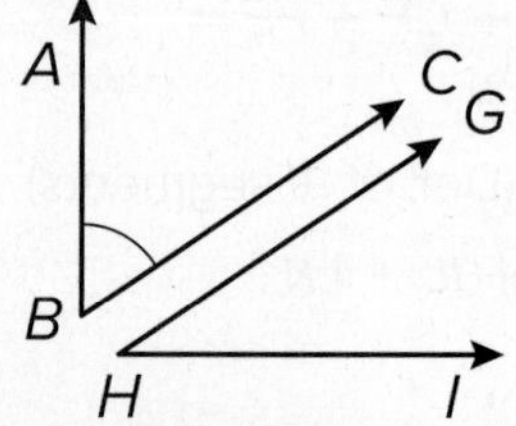

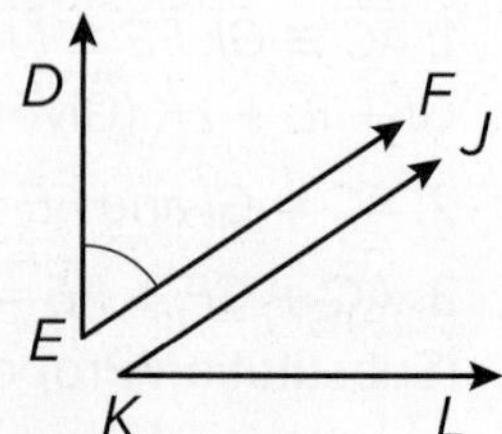

Proof:

Statements (Reasons)

1. $\angle ABC \cong \angle DEF$, $\angle GHI$ is complementary to $\angle ABC$, $\angle JKL$ is complementary to $\angle DEF$. (Given)
2. $m\angle ABC + m\angle GHI = 90°$, $m\angle DEF + m\angle JKL = 90°$ (Def. of compl. angles)
3. $m\angle ABC = m\angle DEF$ (Def. of $\cong$ angles)
4. $m\angle ABC + m\angle JKL = 90°$(Subs.)
5. $90° = m\angle ABC + m\angle JKL$ (Symm. Prop.)
6. $m\angle ABC + m\angle GHI = m\angle ABC + m\angle JKL$ (Trans. Prop.)
7. $m\angle ABC - m\angle ABC + m\angle GHI = m\angle ABC - m\angle ABC + m\angle JKL$ (Subt. Prop.)
8. $m\angle GHI = m\angle JKL$ (Subs.)
9. $\angle GHI \cong \angle JKL$ (Def. of $\cong$ angles)

Proof of the Congruent Supplements Theorem (Case 2: Congruent Angles)

Given: $\angle ABC \cong \angle DEF$, $\angle GHI$ is supplementary to $\angle ABC$, $\angle JKL$ is supplementary to $\angle DEF$.

Prove: $\angle GHI \cong \angle JKL$

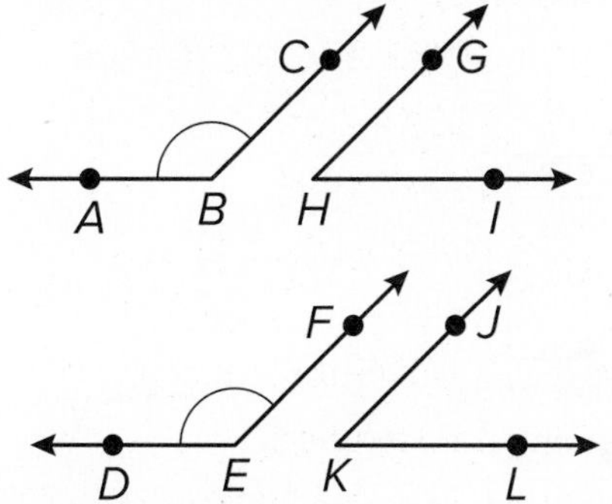

Proof:

Statements (Reasons)

1. $\angle ABC \cong \angle DEF$, $\angle GHI$ is suppl. to $\angle ABC$, $\angle JKL$ is suppl. to $\angle DEF$. (Given)
2. $m\angle ABC + m\angle GHI = 180°$, $m\angle DEF + m\angle JKL = 180°$ (Def. of suppl. angles)
3. $m\angle ABC = m\angle DEF$ (Def. of $\cong$ angles)
4. $m\angle ABC + m\angle JKL = 180°$ (Subs.)
5. $180° = m\angle ABC + m\angle JKL$ (Symm. Property)
6. $m\angle ABC + m\angle GHI = m\angle ABC + m\angle JKL$ (Trans. Prop.)
7. $m\angle ABC - m\angle ABC + m\angle GHI = m\angle ABC - m\angle ABC + m\angle JKL$ (Subt. Prop.)
8. $m\angle GHI = m\angle JKL$ (Subs.)
9. $\angle GHI \cong \angle JKL$ (Def. of $\cong$ angles)

Lesson 12-7

1. $\overline{BF}$, $\overline{CG}$, and $\overline{DH}$ **3.** *ABCD* and *EFGH* or *ABFE* and *CDHG* **5.** $\overline{EF}$, $\overline{AB}$, and $\overline{DC}$
7. Sample answer: *ABCD* and *DCGH* could be characterized as perpendiculars, because *DCGH* contains segment $\overline{CG}$, which is perpendicular to *ABCD*. **9.** line d; alternate exterior **11.** line d; corresponding **13.** line a; consecutive interior **15.** line c; alternate interior **17.** line d; corresponding **19.** line p
21. 100° **23.** 80° **25.** 100° **27.** 170°
29. $x = 28$, $y = 47$; Use supplementary angles to find x. Then use alternate exterior angles to find y. **31.** $x = 12$, $y = 31$ **33.** 105°
35. 105° **37.** 75° **39.** 94°

41. 64°; sample answer: Opposite sides of a rectangle are parallel. So, the top and bottom lines on the side panel are parallel and cut by a transversal, which is the dashed line. Therefore, $\angle 1$ and the 116°-angle are consecutive interior angles, so their sum is 180°. $m\angle 1 + 116° = 180°$, so $m\angle 1 = 64°$.

43. Sample answer:

$\angle 1$ and $\angle 4$ are alternate interior angles. $\angle 2$ and $\angle 3$ are also alternate interior angles. $\angle 1$ and $\angle 2$ are complementary angles, and $\angle 3$ and $\angle 4$ are complementary angles.

45. By the Corresponding Angles Postulate, $\angle 1 \cong \angle 13$ and $\angle 13 \cong \angle 9$. By the Transitive Property, $\angle 1 \cong \angle 9$. So, $m\angle 1 = m\angle 9$. By the Corresponding Angles Postulate, $\angle 4 \cong \angle 8$ and $\angle 8 \cong \angle 12$. By the Transitive Property, $\angle 4 \cong \angle 12$. So, $m\angle 4 = m\angle 12$. It is given that $m\angle 1 - m\angle 4 = 25°$. By the Substitution Property, $m\angle 9 - m\angle 12 = 25°$.

47. By the Vertical Angles Theorem, $\angle 7 \cong \angle 5$. By the Corresponding Angles Theorem $\angle 5 \cong \angle 1$. By the Transitive Property, $\angle 1 \cong \angle 7$.

49.

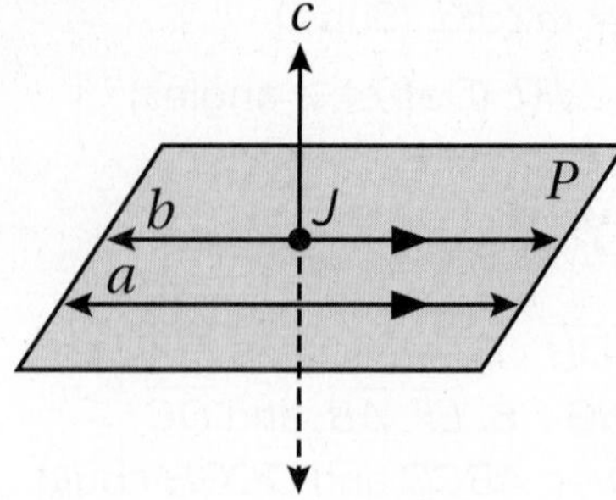

51. Sometimes; sample answer: $\overleftrightarrow{AB}$ intersects $\overleftrightarrow{EF}$ depending on where the planes intersect.

53. $x = 171$ or $x = 155$; $y = 3$ or $y = 5$

55. No; sample answer: From the definition of skew lines, the lines must not intersect and cannot be coplanar. Different planes cannot be coplanar, but they are always parallel or intersecting. Therefore, planes cannot be skew.

Lesson 12-8

1. parallel

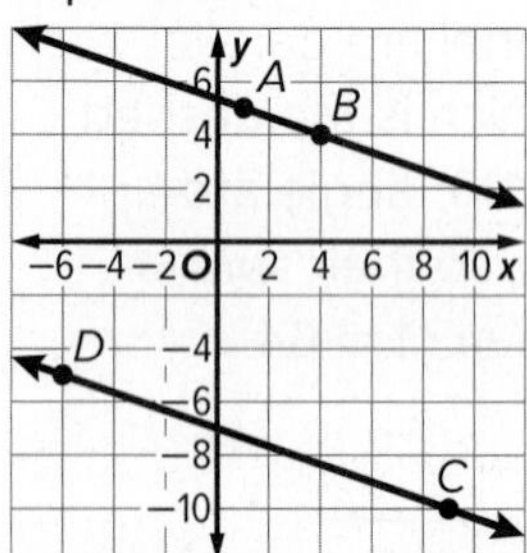

3. neither

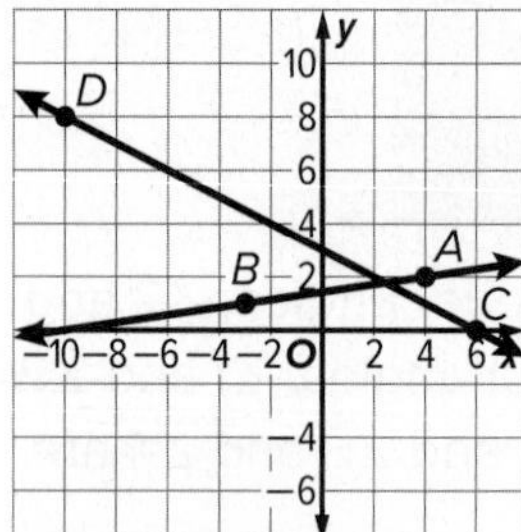

5. perpendicular

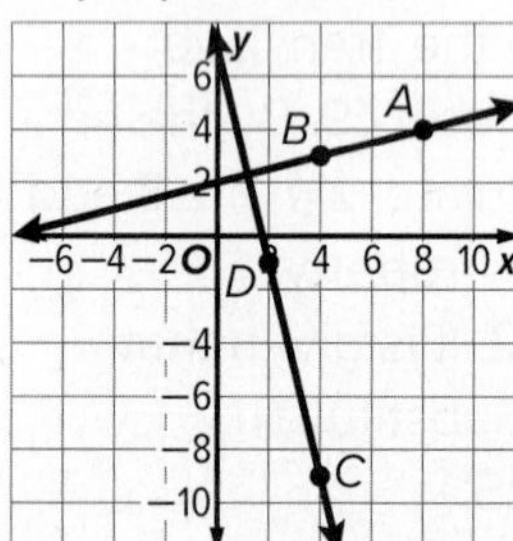

7. perpendicular 9. parallel 11. perpendicular 13. neither 15. parallel

17.

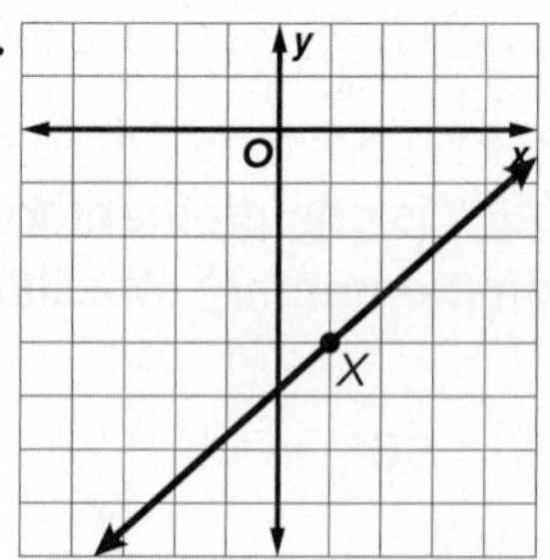

19.

21a. $y = -\frac{2}{3}x + \frac{11}{3}$ 21b. When the drummer crosses the x-axis, the y-coordinate will be 0, so solve $0 = -\frac{2}{3}x + \frac{11}{3}$ to find the x-coordinate; solving shows that $x = \frac{11}{2}$ or $5\frac{1}{2}$, so the x-coordinate will be greater than 5.

23. $y = -10$ 25. $y = \frac{1}{5}x + \frac{12}{5}$

27. $y = -8$

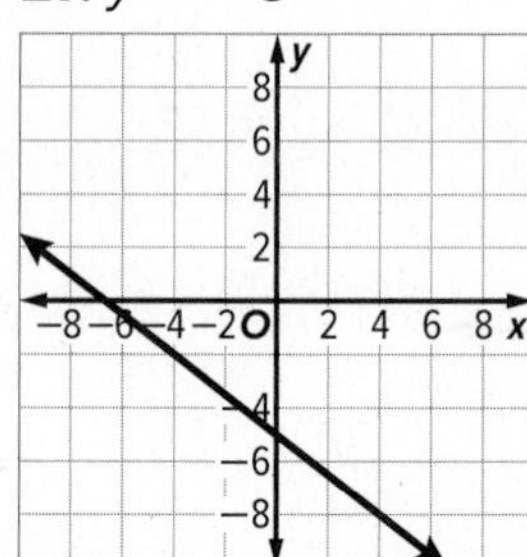

29. $y = 0$

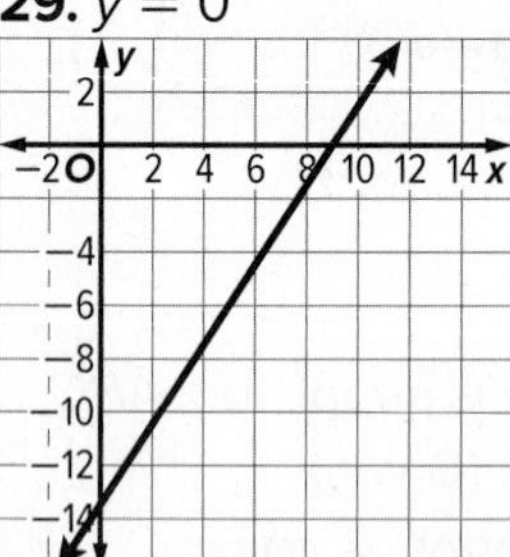

31. $y = -\frac{3}{2}x - 7$; $y = \frac{2}{3}x - \frac{8}{3}$

33. No; none of the slopes are equal, and no two of the slopes have a product of −1.

35a. 450 yd 35b. −3; Ford Street and 6th Street are parallel, so they have the same slope. 35c. Both have a slope of $\frac{1}{3}$ because both are perpendicular to Ford and 6th, and the slope of a perpendicular is given by the negative reciprocal. 35d. 200 yd

37. $S\left(0, -5\frac{1}{2}\right)$; The slope of $\overline{QR}$ is $\frac{2-4}{3-(-2)} = -\frac{2}{5}$, so the slope of $\overline{RS}$ is $\frac{5}{2}$. Let the coordinates of S be $(0, y)$ because S must be on the y-axis. Solve $\frac{5}{2} = \frac{y-2}{0-3}$ for y. $y = -5\frac{1}{2}$, so the coordinates of S are $\left(0, -5\frac{1}{2}\right)$.

39a. $y - 1 = a(x - 5)$; the line must have slope a to be parallel to line p.

39b. $y = -\frac{1}{a}x + \frac{2}{a} + 3$; the line must have slope $-\frac{1}{a}$ to be perpendicular to line p.

41a. $B(2, 4)$ and $D(10, -4)$

41b. Sample answer: The slopes of $\overline{AB}$ and $\overline{DC}$ are undefined, so they are parallel to each other. The slopes of $\overline{AD}$ and $\overline{BC}$ are 0, so they are parallel to each other.

41c. Sample answer: Because the slope of $\overline{AB}$ is undefined and the slope of $\overline{BC}$ is zero, the lines are perpendicular to each other. Therefore, they form a right angle, which measures 90°. The same logic applies to all the sides.

43. Yes; the slope of the line through the points $(-2, 2)$ and $(2, 5)$ is $\frac{3}{4}$. The slope of the line through the points $(2, 5)$ and $(6, 8)$ is $\frac{3}{4}$. Because these lines have the same slope and have a point in common, their equations would be the same. Therefore, all the points are on the same line, and all the points are collinear.

45. Two nonvertical lines are parallel if and only if they have the same slope. Two nonvertical lines are perpendicular if and only if the product of their slopes is −1.

Lesson 12-9

1. $a \parallel b$; Alternate Interior Angles Converse

3. $\ell \parallel m$; Alternate Exterior Angles Converse

5. $g \parallel h$; Converse of Corresponding Angles Thm.

7. 22 **9.** 6 **11.** 13 **13.** 20

15. Parallelogram; sample answer: The top edges are perpendicular to the vertical line, so they are a single line. The bottom edge is also a single line and perpendicular to the same line as the top, so it is parallel to the top. The top edge is a transversal to the left and right slanted edges, and the angles are supplementary. So, the left and right edges are parallel.

17a. 108°; sample answer: To ensure that the horizontal part of the A is truly horizontal, it should be parallel to the dashed line. Therefore, $\angle 2$ and the 108°-angle are alternate interior angles, and $m\angle 2 = 108°$. $\angle 1$ and $\angle 2$ are congruent angles, so $m\angle 1 = 108°$.

17b. Sample answer: One side of the A is longer than the other.

19. Sample answer: It is given that $\angle 1 \cong \angle 2$. Also, $\angle 1 \cong \angle 3$, because these are vertical angles. Therefore, $\angle 2 \cong \angle 3$ by the Transitive Property of Congruence. This shows that $\ell \parallel m$ by the Converse of Corresponding Angles Theorem.

21.

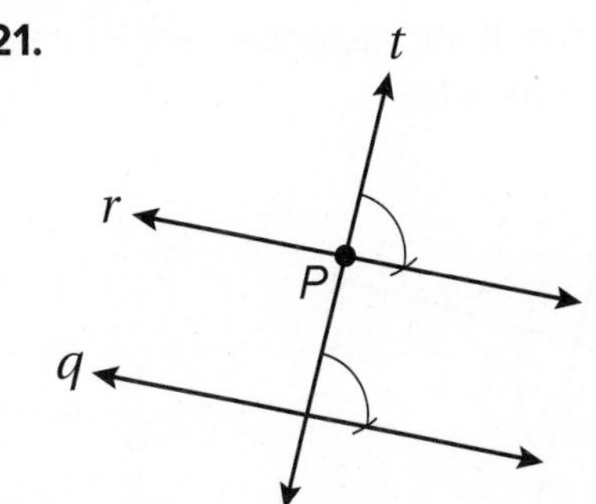

23. Sample answer: Because the corners are right angles, each pair of opposite sides is perpendicular to the same line. Therefore, each pair of opposite sides is parallel.

25. Daniela is correct. $\angle 1$ and $\angle 2$ are alternate interior angles for $\overline{WX}$ and $\overline{YZ}$. So, if alternate interior angles are congruent, then the lines are parallel.

27a.

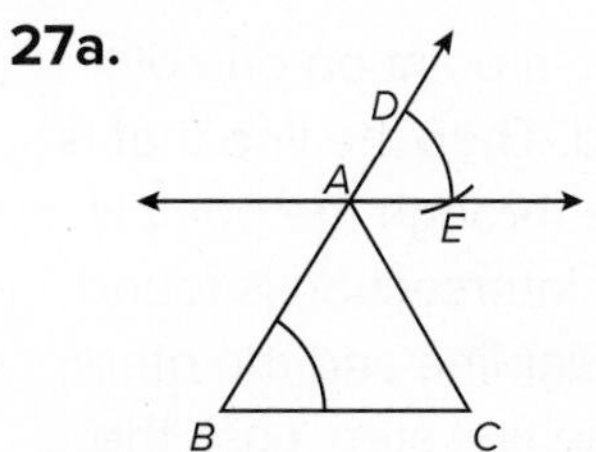

27b. Sample answer: Using a straightedge, the lines are equidistant. So, they are parallel.

27c. Sample answer: $\angle ABC$ was copied to construct $\angle DAE$. So, $\angle ABC \cong \angle DAE$. $\angle ABC$ and $\angle DAE$ are corresponding angles, so by the Converse of the Corresponding Angles Theorem, $\overleftrightarrow{AE} \parallel \overleftrightarrow{BC}$.

29. Yes; sample answer: A pair of angles can be supplementary and congruent if the measure of both angles is 90°, because the sum of the angle measures would be 180°.

Lesson 12-10

1. $\sqrt{2}$ or about 1.41 units **3.** 6 units **5.** $\sqrt{10}$ or about 3.16 units

7. yes; 4.24 in.

9. 8 units **11.** $\sqrt{10}$ or about 3.16 units **13.** $3\sqrt{2}$ or about 4.24 units

15. $4\sqrt{17}$ or about 16.49 units **17.** $\sqrt{14.76}$ or about 3.84 units **19.** 0 units

21. Sample answer: Isaiah can measure the perpendicular distance between the wires in two different places. If the distances are equal, then the wires are parallel.

23.

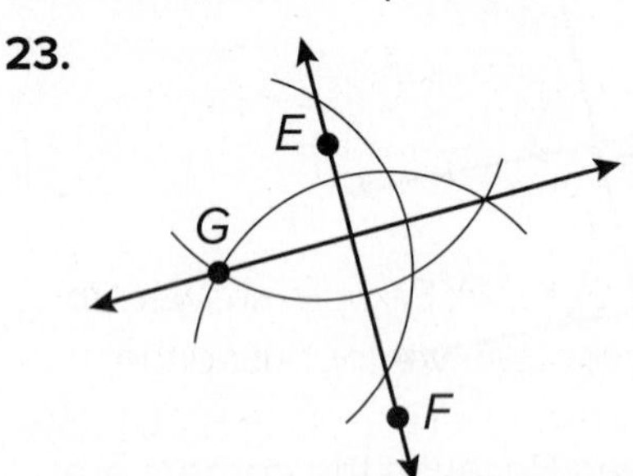

25. No; Sample answer: A path that is perpendicular to the tee box would be the shortest. The angle that the tee box makes with the path that Mark walked is less than 90°, so it is not the shortest possible path.

27. Sample answer: The point is on the line. The two lines are the same line.

29. Sample answer: First, a point on one of the parallel lines is found. Then the line that is perpendicular to the line through the point is found. Then the point of intersection is found between the perpendicular line and the other line that is not used in the first step. Last, the Distance Formula is used to determine the distance between the pair of intersection points. This value is the distance between the pair of parallel lines.

31. Sometimes; sample answer: The distance can only be found if the line is parallel to the plane.

33. If two planes are each equidistant from a third plane, then the two planes are parallel to each other.

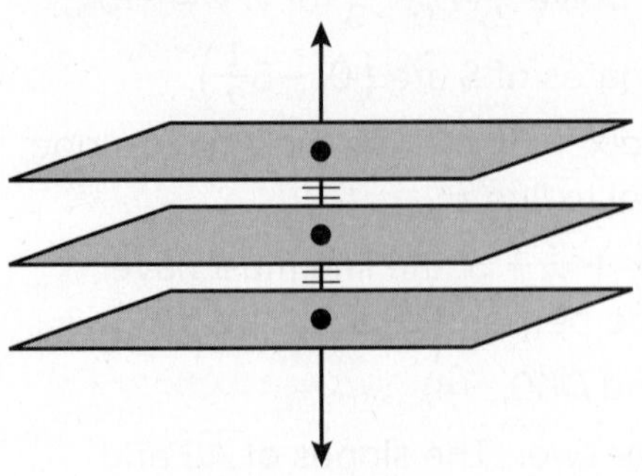

Module 12 Review

1. 6

3. B

5. $x = 29$, $m\angle ABD = 66°$, $m\angle DBC = 24°$

7. A, D, E

9. Lines *m* and *p*

11. 102° **13.** A, B, C, D **15.** B

Module 13

Quick Check

1. −14 **3.** −2 **5.** (3, 0) **7.** (1, −1)

Lesson 13-1

1. $A'(2, -3)$, $B'(1, 0)$, $C'(-3, -2)$

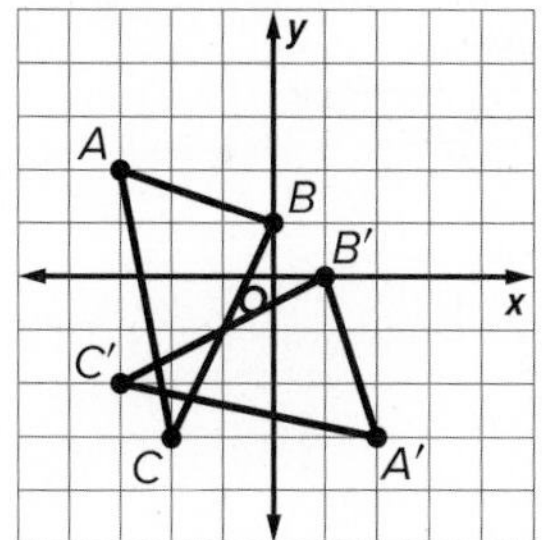

3. $R'(3, -2)$, $S'(4, 2)$, $T'(-3, 2)$, $U'(-4, -2)$

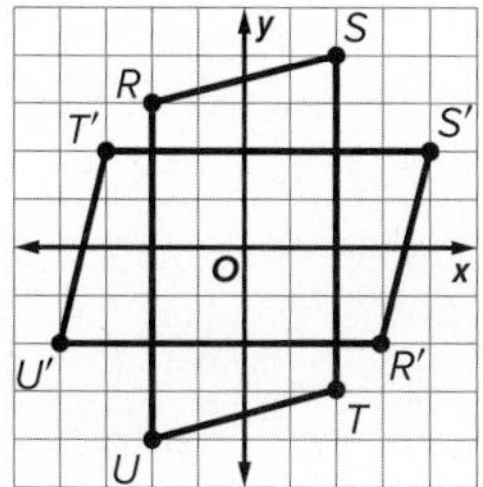

5. $S'(-7, 5)$

7.

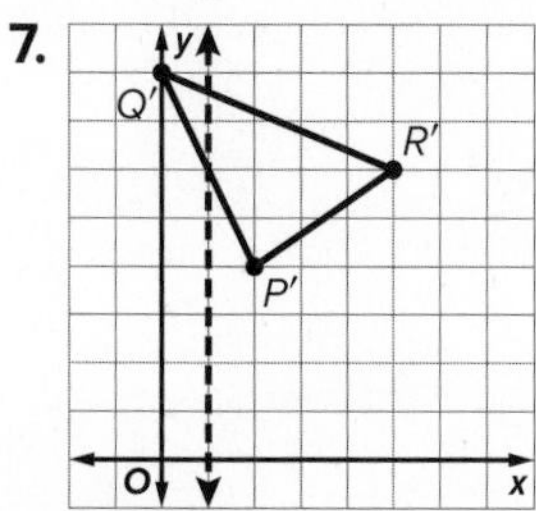

9. $W'(-7, 14)$

11.

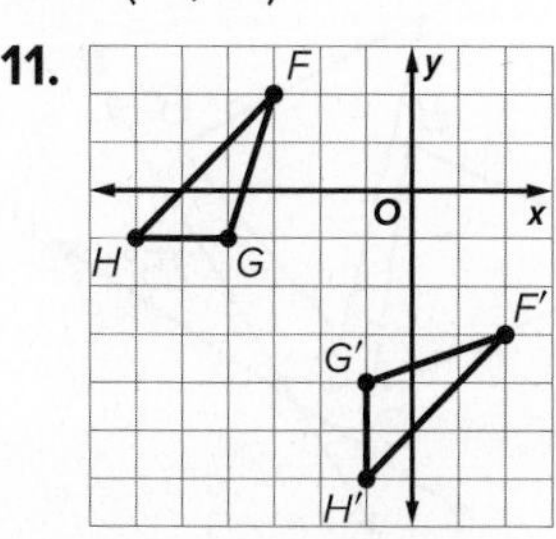

13.

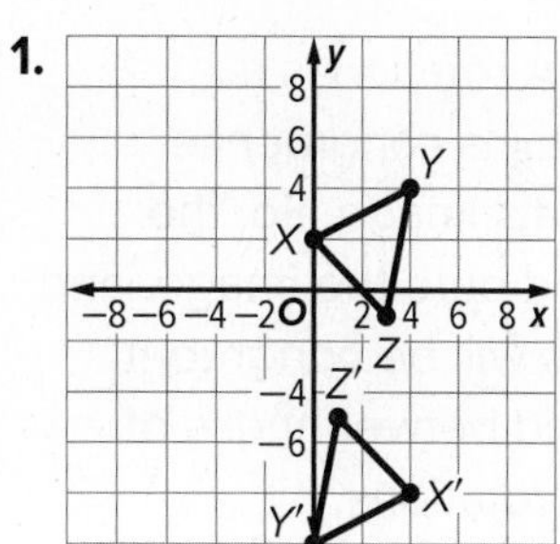

15. reflection in the line $y = -1$

17. Sample answer: The M can be represented with the points (0, 0), (0, 3), (1, 1), (2, 0), and (2, 3). Reflecting in the line $y = x$ gives (0, 0), (3, 0), (1, 1), (0, 2), and (3, 2). **19.** $(-b, a)$

Lesson 13-2

1. $\triangle J'K'L'$ is a translation of $\triangle JKL$. This translation vector can be represented as $\langle 2, 5 \rangle$.

3. (4, 3), (5, 5), (6, 3)

5.

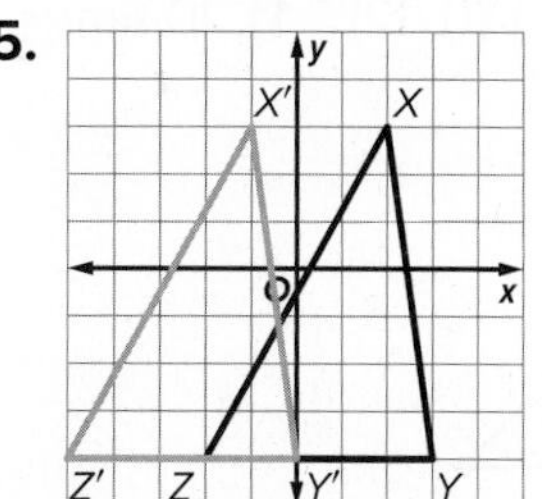

7.

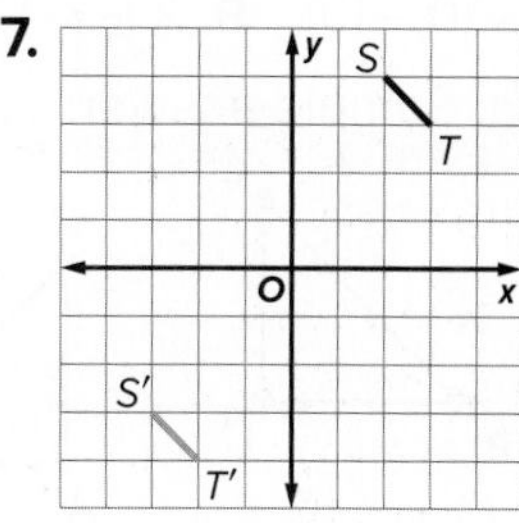

9. $Q'(2, -7)$ **11.** $B'(12, 2)$

13. No; sample answer: The size has been changed.

15. $(x, y) \rightarrow (x, y + 4)$

17. (0, 2), (2, 2), (0, 0), (2, 0); Sample answer: To minimize the length of the vector, I used the vertex closest to the origin, (2, 1), as the preimage for the point translated to the origin.

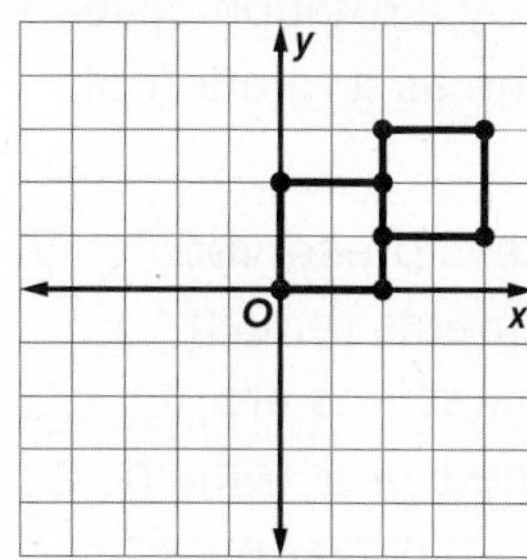

Lesson 13-3

1.

3.

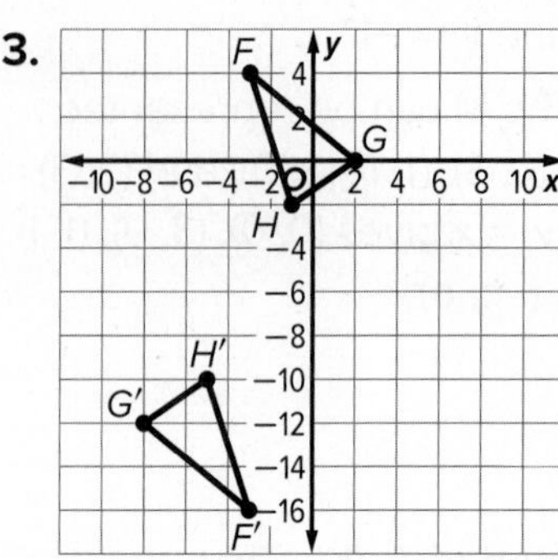

5. home plate: (3, 23), first base: (3, 13), second base: (13, 13), third base: (13 , 23)

7. $K'(0, -10)$ 9. $X'(-3, 13)$, $Y'(0, 12)$ 11. 10 times

13a. Sample answer:

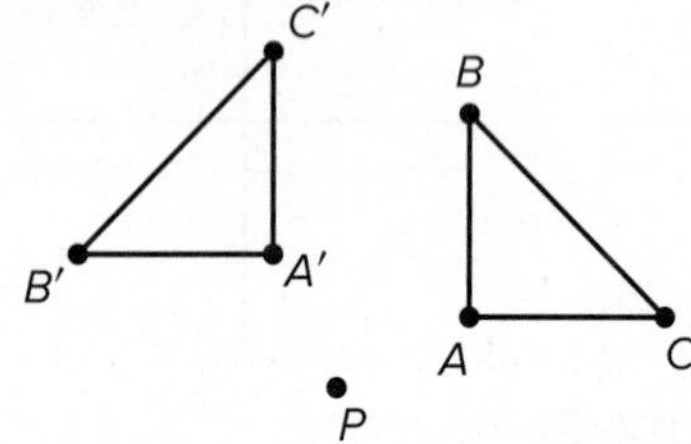

13b. Sample answer: 270° clockwise rotation about point P

15. Sample answer: A rotation followed by another rotation is still a rotation. For example, a rotation of 30° clockwise followed by a rotation of 20° counterclockwise is the same as a rotation of 10° clockwise. A rotation of 30° counterclockwise followed by a rotation of 15° counterclockwise is the same as a rotation of 45° counterclockwise.

17. Sample answer: Distance is preserved because the lengths of segments remain the same measure. Angle measures are preserved because angle measures remain the same measure. Parallelism is preserved because parallel lines remain parallel. Collinearity is preserved because points remain on the same lines.

19. Yes; sample answer: A rotation is a transformation that maintains congruence of the original figure and its image. So, the preimage can be mapped onto the image, and corresponding segments will be congruent. Therefore, collinearity and betweenness of points are maintained in rotations.

Lesson 13-4

1.

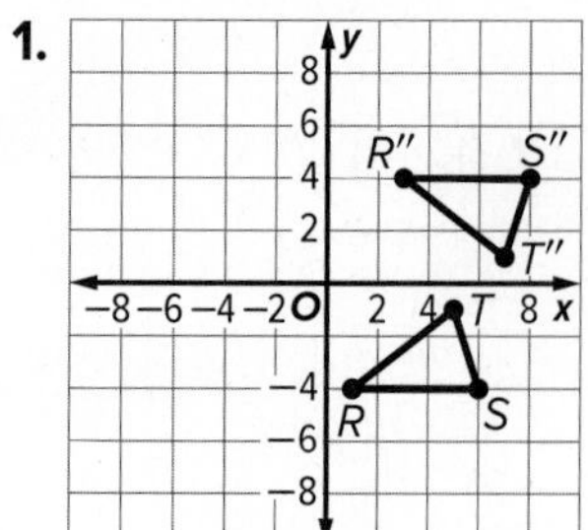

3.

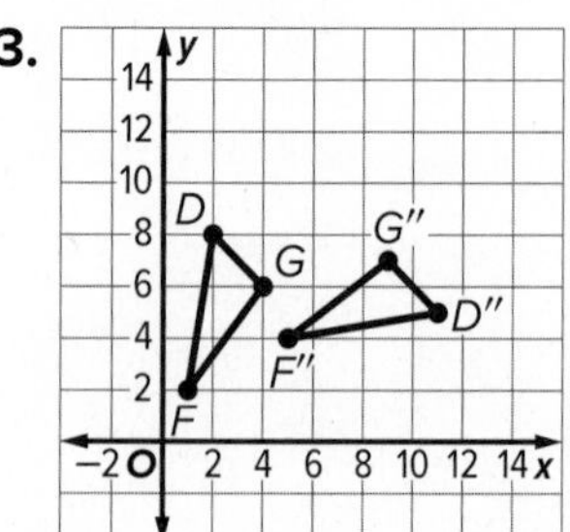

5.

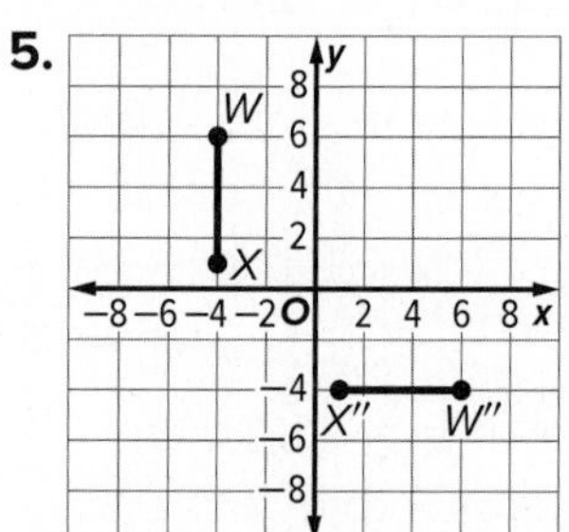

7.

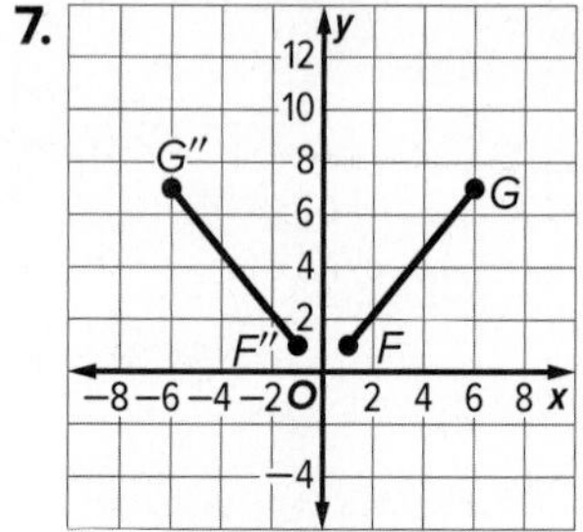

9. a 50° clockwise rotation about the point where lines u and v intersect

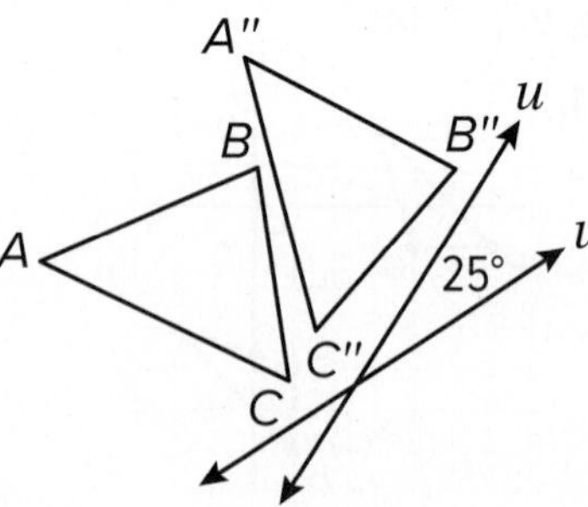

11. a 90° clockwise rotation about the point where lines u and v intersect

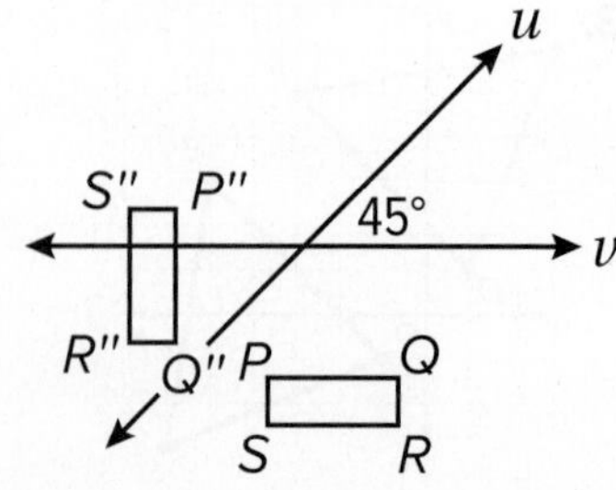

13. $\triangle JKL \cong \triangle MNP$; reflection in y-axis followed by translation along $\langle -1, 2\rangle$

15. $\triangle JKL \cong \triangle MNP$; translation along $\langle 2, 0\rangle$ followed by 180° rotation about the origin

17. Sample answer: Reflect the first two shapes in a horizontal line through the midpoints of the vertical segments, and then translate to the right and repeat.

19.

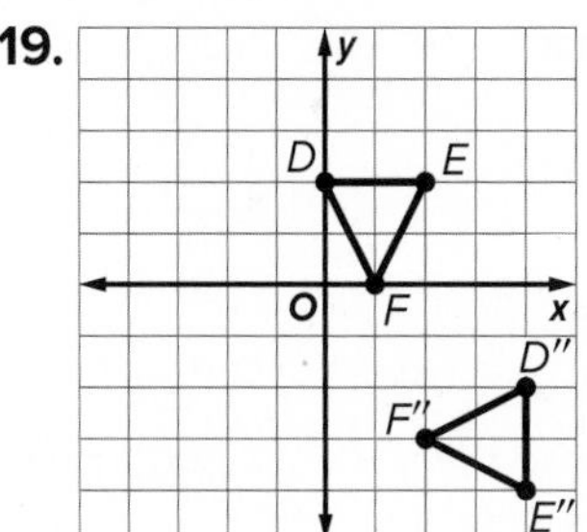

21. (3, −1), (2, −5), (−5, −2)

23. (1, 1), (0, 5), (−7, 2)

25. Sometimes; sample answer: If the lines of reflection intersect, the composition is a rotation.

27. Sometimes; sample answer: This is true if the point is the origin.

29. Sample answer: Proof: It is given that a translation along $\langle a, b\rangle$ maps R to R' and S to S'. Using the definition of a translation, points R and S move the same distance in the same direction, therefore $\overline{RS} \cong \overline{R'S'}$. It is also given that a reflection in a maps R' to R'' and S' to S''. Using the definition of a reflection, points R' and R'' and points S' and S'' are the same distance from line a, so $\overline{R'S'} \cong \overline{R''S''}$. By the Transitive Property of Congruence, $\overline{RS} \cong \overline{R''S''}$.

31. Sometimes; sample answer: The order of rotating by 180° about the origin and reflecting in the line $y = x$ does not change the location of the final image.

33. $P'''(1, -2)$, $Q'''(2, 1)$, $R'''(-1, 3)$, $S'''(-2, 0)$

Lesson 13-5

1. $x = \frac{180(5-2)}{5} = 108°$; Because 108° is not a factor of 360°, a regular pentagon will not tessellate the plane.

3. $x = \frac{180(9-2)}{9} = 140°$; Because 140° is not a factor of 360°, a regular 9-gon will not tessellate the plane.

5. yes; 2 regular hexagons, 2 equilateral triangles

7. tessellation; uniform, semi-regular

9. Yes; sample answer: reflection, rotation, translation

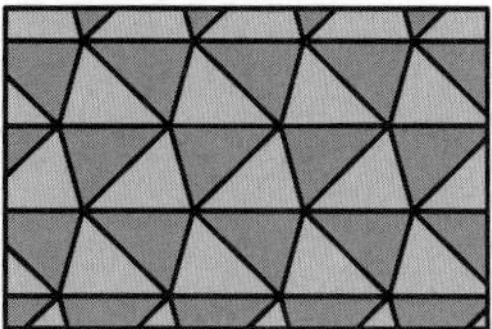

11. Never; sample answer: Each interior angle of a regular dodecagon is $\frac{180°(12-2)}{12} = 150°$. Because 150° is not a factor of 360°, a regular dodecagon will not tessellate the plane.

13. Never; sample answer: Each interior angle of a regular 15-gon is $\frac{180°(15-2)}{15} = 156°$. Because 156° is not a factor of 360°, a regular 15-gon will not tessellate the plane.

15. translation **17.** rotation **19.** Sample answer: Translations can be performed because the pieces slide. Rotations can be performed because each piece can be turned. Reflections cannot be performed because the back of a piece cannot be used to create the puzzle.

21.

Lesson 13-6

1. yes; 3

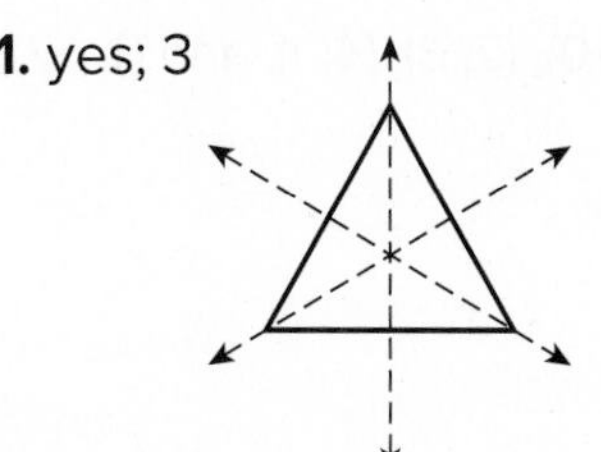

3. no

5. yes; 5

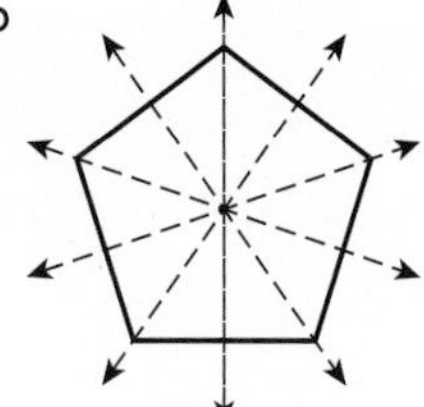

7a. Yes; the hubcap can map onto itself with a rotation that is less than 360°.

7b. Yes; the hubcap can map onto itself with a rotation that is less than 360°.

7c. Yes; the hubcap can map onto itself with a rotation that is less than 360°.

9. Yes; the symbol can map onto itself with a rotation that is less than 360°.

11. yes; 3; 120°

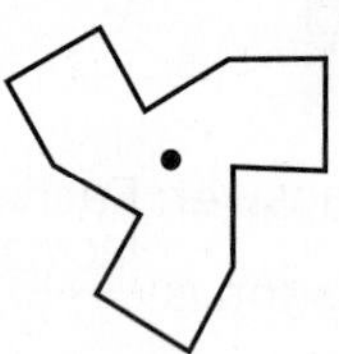

13. yes; 2; 180°

15.

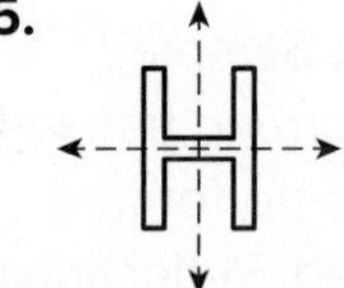

17. 2; 180° **19.** pentagon **21.** yes; order of symmetry: 3; magnitude of symmetry: 120° **23.** yes; order of symmetry: 6; magnitude of symmetry: 60°

25. Sample answer: A rectangular mirror with two lines of symmetry, one vertical and one horizontal, through the middle or a spoon with one line of symmetry down the middle.

27. 24; 360° ÷ 15° = 24, so the order of symmetry is 24. This means there are 24 sides.

29. Sample answer: (–1, 0), (2, 3), (4, 1), and (1, –2);

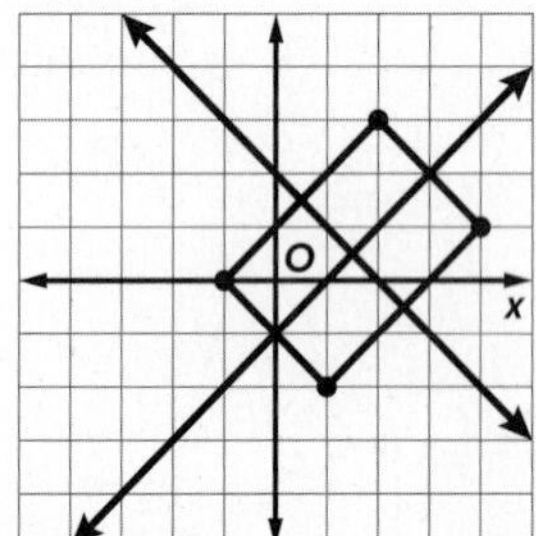

31. Sample answer: In both rotational and line symmetry a figure is mapped onto itself. However, in line symmetry the figure is mapped onto itself by a reflection, and in rotational symmetry a figure is mapped onto itself by a rotation. A figure can have line symmetry and rotational symmetry.

Module 13 Review

1.

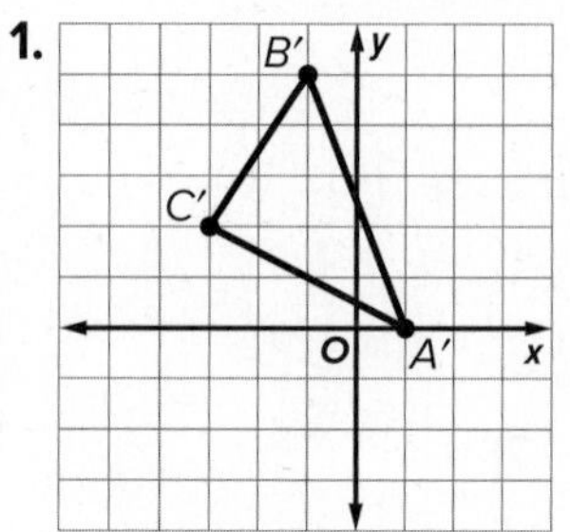

3. (–9, 6)

5. Sample answer: The length of $\overline{AC}$ is not the same as the length of $\overline{EG}$.

7. C **9.** Quadrant II **11.** false **13.** A, D

15. C

17. order = 24; magnitude = 15°

Module 14

Quick Check

1. right **3.** obtuse **5.** 10.8 **7.** 18.0

Lesson 14-1

1. $m\angle 1 = 30°$, $m\angle 2 = 60°$
3. $m\angle 1 = 109°$, $m\angle 2 = 29°$, $m\angle 3 = 71°$
5. 50° **7.** 98° **9.** 62° **11.** 26° **13.** 55°
15. $x = 20$; 40°, 60°, 80° **17.** $x = 11$; 80°, 117°
19.

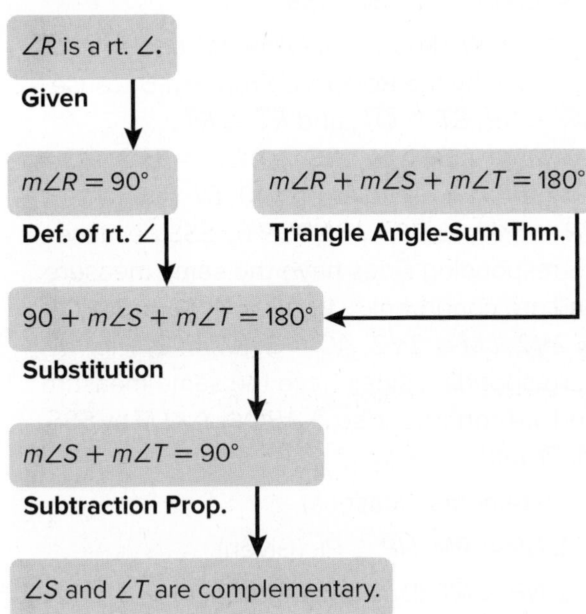

21. $m\angle D = 27°$, $m\angle E = 81°$, $m\angle F = 72°$
23. $m\angle J = 18°$, $m\angle K = 72°$, $m\angle L = 90°$
25. $m\angle Z < 23°$; Sample answer: Because the sum of the measures of the angles of a triangle is 180° and $m\angle X = 157°$, $157° + m\angle Y + m\angle Z = 180°$, so $m\angle Y + m\angle Z = 23°$. If $m\angle Y$ was 0°, then $m\angle Z$ would equal 23°. But because an angle must have a measure greater than 0°, $m\angle Z$ must be less than 23°, so $m\angle Z < 23°$.
27. 90° **29.** $m\angle 1 = 55°$, $m\angle 2 = 75°$, $m\angle 3 = 55°$, $m\angle 4 = 15°$ **31.** Sample answer: Draw a triangle and then tear the corners off the triangle. Arrange the three corners so the angles are adjacent. The angles now form a straight angle. Because a straight angle measures 180°, the sum of the measures of the angles of a triangle is 180°.

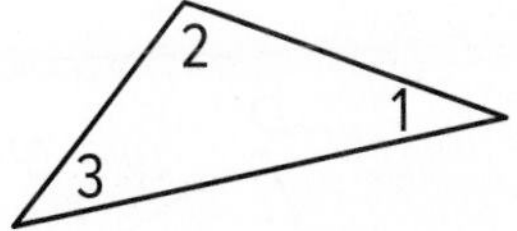

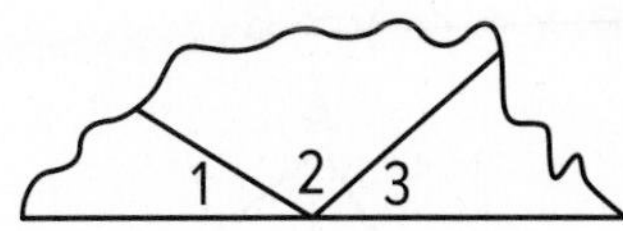

33. Sample answer: Because an exterior angle is acute, the adjacent angle must be obtuse. Because another exterior angle is right, the adjacent angle must be right. A triangle cannot contain a right angle and an obtuse angle because the sum would be greater than 180°. Therefore, a triangle cannot have an obtuse, an acute, and a right exterior angle.
35. Sample answer: I found the measure of the second angle by subtracting the first angle from 90° because the acute angles of a right triangle are complementary.

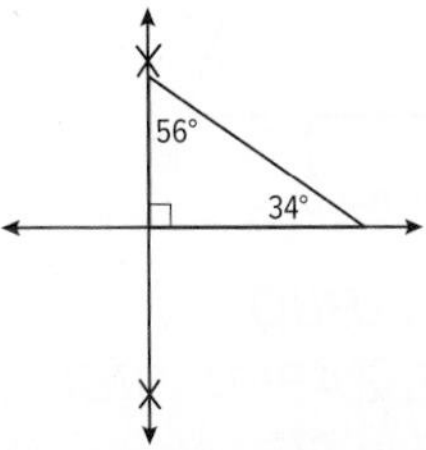

Lesson 14-2

1. $\angle A \cong \angle D$; $\angle ABC \cong \angle DCB$; $\angle ACB \cong \angle DBC$; $\overline{AC} \cong \overline{DB}$; $\overline{AB} \cong \overline{DC}$; $\triangle ABC \cong \triangle DCB$

3. $\angle X \cong \angle A$; $\angle Y \cong \angle B$; $\angle Z \cong \angle C$; $\overline{XY} \cong \overline{AB}$; $\overline{XZ} \cong \overline{AC}$; $\overline{YZ} \cong \overline{BC}$; $\triangle XYZ \cong \triangle ABC$

5. $\angle R \cong \angle J$; $\angle T \cong \angle K$; $\angle S \cong \angle L$; $\overline{RT} \cong \overline{JK}$; $\overline{TS} \cong \overline{KL}$; $\overline{RS} \cong \overline{JL}$; $\triangle RTS \cong \triangle JKL$

7. 48 **9.** 5 **11.** 35 **13.** 41°

15. Proof:

Statements (Reasons)

1. $\overline{AB} \cong \overline{CB}$, $\overline{AD} \cong \overline{CD}$ (Given)
2. $\overline{BD} \cong \overline{BD}$ (Reflexive Prop. of Congruence)
3. $\angle ABD \cong \angle CBD$, $\angle ADB \cong \angle CDB$ (Given)
4. $\angle A \cong \angle C$ (Third Angles Theorem)
5. $\triangle ABD \cong \triangle CBD$ (Def. of congruent triangles)

17. Proof: It is given that $\overline{BD}$ bisects $\angle ABC$ and $\angle ADC$. Therefore, $\angle ABD \cong \angle CBD$ and $\angle ADB \cong \angle CDB$ by the definition of angle bisector. By the Third Angles Theorem, $\angle A \cong \angle C$. It is given that $\overline{AB} \cong \overline{CB}$ and $\overline{AD} \cong \overline{CD}$. By the Reflexive Property of Congruence, $\overline{BD} \cong \overline{BD}$. Therefore, $\triangle ABD \cong \triangle CBD$ by Def. of congruent triangles.

19. $x = 12; y = 6$

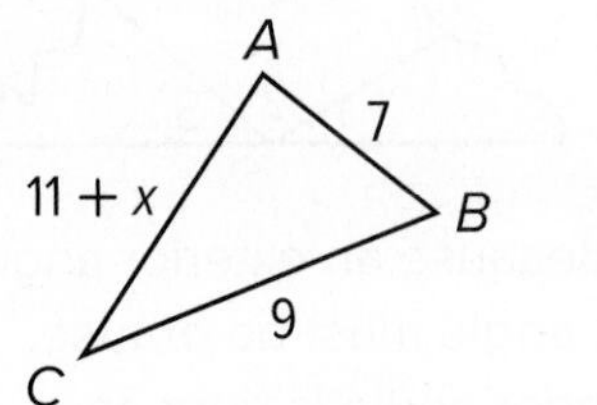

21. $x = 4; y = 19$

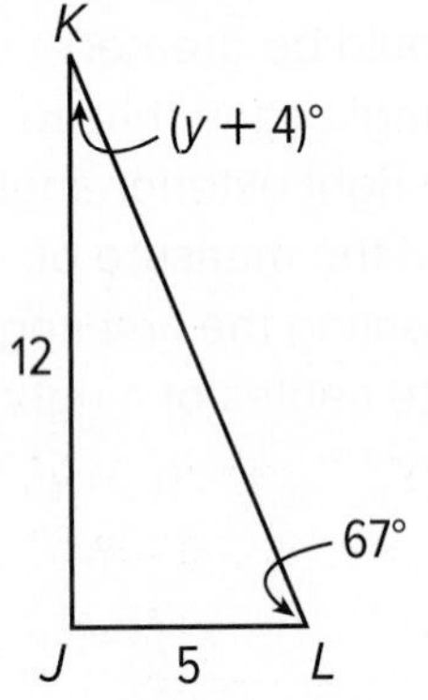

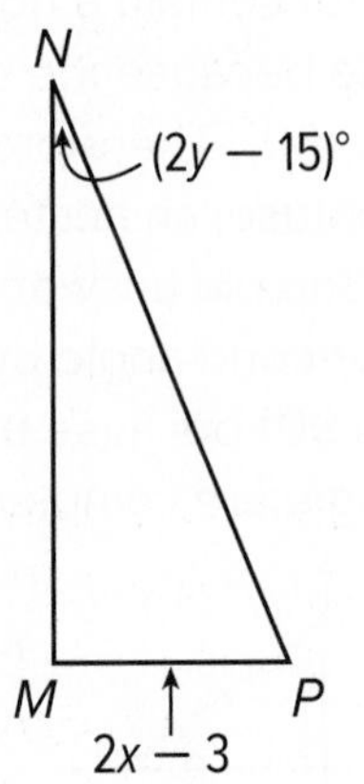

23a. $\triangle ABI \cong \triangle EBF$, $\triangle CBD \cong \triangle HBG$

23b. Sample answer: $\angle A \cong \angle E$, $\angle ABI \cong \angle EBF$, $\angle I \cong \angle F$; $\overline{AB} \cong \overline{EB}$, $\overline{BI} \cong \overline{BF}$, $\overline{AI} \cong \overline{EF}$

25. Statements (Reasons)

1. $\angle P \cong \angle X$, $\angle Q \cong \angle Y$ (Given)
2. $m\angle P = m\angle X$, $m\angle Q = m\angle Y$ (Def. of congruent angles)
3. $m\angle P + m\angle Q + m\angle R = 180$
   $180 = m\angle X + m\angle Y + m\angle Z$ (Triangle Angle-Sum Thm.)
4. $m\angle P + m\angle Q + m\angle R = m\angle X + m\angle Y + m\angle Z$ (Transitive Property)
5. $m\angle X + m\angle Y + m\angle R = m\angle X + m\angle Y + m\angle Z$ (Substitution Property)
6. $m\angle R = m\angle Z$ (Subtraction Prop. of Eq.)
7. $\angle R \cong \angle Z$ (Def. of congruent angles)

27. Sample answer: Do you think that the sum of the angles of a quadrilateral is constant? If so, do you think that the final pair of corresponding angles will be congruent if three other pairs of corresponding angles are congruent for a pair of quadrilaterals?

29. Sample answer: When naming congruent triangles, it is important that the corresponding vertices be in the same location for both triangles because the location indicates congruence. For example, if $\triangle ABC$ is congruent to $\triangle DEF$, then $\angle A \cong \angle D$, $\angle B \cong \angle E$, and $\angle C \cong \angle F$.

Lesson 14-3

1. Statements (Reasons)

1. $\overline{AB} \cong \overline{XY}$
   $\overline{AC} \cong \overline{XZ}$
   $\overline{BC} \cong \overline{YZ}$ (Given)
2. $\triangle ABC \cong \triangle XYZ$ (SSS Post.)

3. Proof:

Statements (Reasons)

1. $\overline{AB} \cong \overline{CB}$, D is the midpoint of $\overline{AC}$. (Given)
2. $\overline{AD} \cong \overline{DC}$ (Definition of midpoint)
3. $\overline{BD} \cong \overline{BD}$ (Reflexive Property of Congruence)
4. $\triangle ABD \cong \triangle CBD$ (SSS)

5. Proof: We know that $\overline{QR} \cong \overline{SR}$ and $\overline{ST} \cong \overline{QT}$. $\overline{RT} \cong \overline{RT}$ by the Reflexive Property. Because $\overline{QR} \cong \overline{SR}$, $\overline{ST} \cong \overline{QT}$, and $\overline{RT} \cong \overline{RT}$, $\triangle QRT \cong \triangle SRT$ by SSS. **7.** $DE = 5\sqrt{2}$, $PQ = 5\sqrt{2}$, $EF = 2\sqrt{10}$, $QR = 2\sqrt{10}$, $DF = 5\sqrt{2}$, $PR = 5\sqrt{2}$; $\triangle DEF \cong \triangle PQR$ by SSS because corresponding sides have the same measure and are congruent. **9.** $AB = 2$, $KL = 2$, $BC = 2\sqrt{2}$, $LM = 2\sqrt{2}$, $AC = 2$, $KM = 2$; The corresponding sides have the same measure and are congruent, so $\triangle ABC \cong \triangle KLM$ by SSS.

11. Proof:

Statements (Reasons)

1. $NP = PM$, $\overline{NP} \perp \overline{PL}$ (Given)
2. $\overline{NP} \cong \overline{MP}$ (Def. of congruence)
3. $\angle MPL$ and $\angle NPL$ are rt. angles. ($\perp$ lines form rt. angles.)
4. $\angle MPL \cong \angle NPL$ (All right angles are congruent.)
5. $\overline{PL} \cong \overline{PL}$ (Reflexive Property of $\cong$)
6. $\triangle NPL \cong \triangle MPL$ (SAS)

13. Proof: Because V is the midpoint of $\overline{YZ}$ and the midpoint of $\overline{WX}$, by the Midpoint Theorem, $\overline{YV} \cong \overline{VZ}$ and $\overline{WV} \cong \overline{XV}$. Because $\angle YVW$ and $\angle ZVX$ are vertical angles, by the Vertical Angles Theorem, the angles are congruent. Therefore, by SAS, $\triangle XVZ \cong \triangle WVY$.

15. Proof:

Statements (Reasons)

1. $\overline{BD} \perp \overline{AC}$; $\overline{BD}$ bisects $\overline{AC}$. (Given)
2. $\angle BDA$ and $\angle BDC$ are rt. angles. ($\perp$ lines form rt. angles.)

3. $\angle BDA \cong \angle BDC$ (All right angles are congruent.)
4. $\overline{AD} \cong \overline{DC}$ (Def. of segment bisector)
5. $\overline{BD} \cong \overline{BD}$ (Reflexive Property of $\cong$)
6. $\triangle ABD \cong \triangle CBD$ (SAS)

17. Yes; sample answer: $\angle GLH$ and $\angle JLK$ are vertical angles, so they are congruent. Therefore, $\triangle GLH \cong \triangle JLK$ by the SAS Congruence Postulate. **19.** Yes; sample answer: The triangles share the side $\overline{AC}$, so they have two pairs of congruent sides. The given congruent angles are included angles, so $\triangle ABC \cong \triangle CDA$ by SAS. **21.** No; sample answer: The sticks do not change size, so any arrangement will yield a congruent triangle. **23.** Sample answer: She needs to measure one side of each tile because all the tiles are equilateral triangles. **25.** No; sample answer: You cannot use SAS because the angle congruence that we are given is not an included angle between two sides that are known to be congruent, and SSS cannot be used because only 2 sides of each triangle are known to be congruent. **27.** First pair; sample answer: The second pair can be shown congruent by SAS or SSS, and the third pair can be shown congruent by SSS. **29.** Case 1: You know that the hypotenuses are congruent and that one pair of legs are congruent. Then the Pythagorean Theorem says that the other pair of legs are congruent, so the triangles are congruent by SSS. Case 2: You know that the pairs of legs are congruent and that the right angles are congruent, so the triangles are congruent by SAS. **31.** Shada; to use SAS, the angle must be the included angle.

Lesson 14-4

1. Proof:

Statements (Reasons)

1. $\overline{AB} \parallel \overline{CD}$ (Given)
2. $\angle CBD \cong \angle ADB$ (Given)
3. $\angle ABD \cong \angle CDB$ (Alternate Interior Angles Theorem)
4. $\overline{BD} \cong \overline{BD}$ (Reflexive Property of Congruence)
5. $\triangle ABD \cong \triangle CDB$ (ASA)

3. Proof:

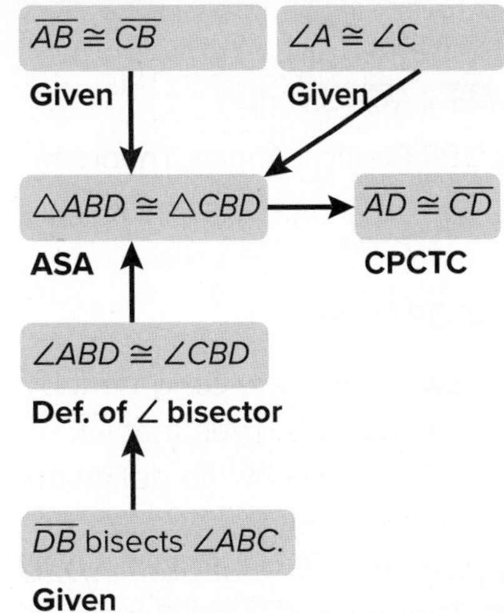

5. Proof: We are given that $\overline{CE}$ bisects $\angle BED$ and that $\angle BCE$ and $\angle ECD$ are right angles. Because all right angles are congruent, $\angle BCE \cong \angle ECD$. By the definition of angle bisector, $\angle BEC \cong \angle DEC$. The Reflexive Property tells us that $\overline{EC} \cong \overline{EC}$. By the Angle-Side-Angle Congruence Postulate, $\triangle ECB \cong \triangle ECD$.
7a. yes; by the ASA Congruence Postulate
7b. 10 in^2 **9a.** Sample answer: Because $\overline{AC} \parallel \overline{BK}$, $\angle CAB \cong \angle KBM$ by the Corresponding Angles Theorem. Because $\overline{CB} \parallel \overline{KM}$, $\angle ABC \cong \angle BMK$ by the Corresponding Angles Theorem. Because B is the midpoint of $\overline{AM}$, $\overline{AB} \cong \overline{BM}$ by the Midpoint Theorem. Therefore, by the ASA Congruence Postulate, $\triangle ABC \cong \triangle BMK$.
9b. 74 ft

11. Proof:

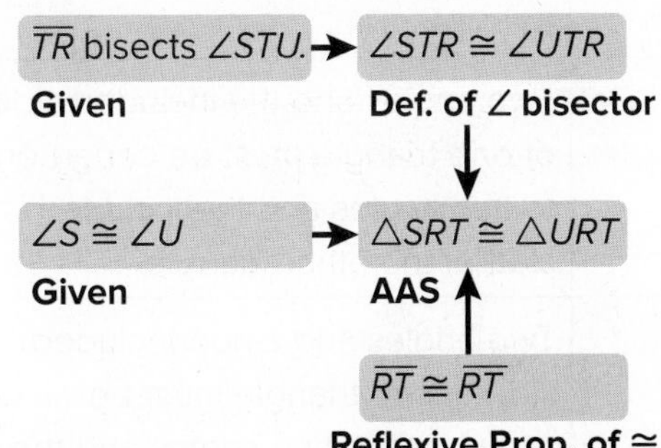

13. Proof: It is given that $\angle E \cong \angle G$ and $\overline{DE} \parallel \overline{FG}$. By the Alternate Interior Angles Theorem, $\angle DFG \cong \angle FDE$. $\overline{DF} \cong \overline{DF}$ by the Reflexive Property of Congruence. Therefore, $\triangle DFG \cong \triangle FDE$ by AAS.

15. Proof:

Statements (Reasons)

1. $\overline{MS} \cong \overline{RQ}$, $\overline{MS} \parallel \overline{RQ}$ (Given)
2. $\angle SPM \cong \angle QPR$ (Vertical Angles Theorem)
3. $\angle SMP \cong \angle QRP$ (Alternate Interior Angles Theorem)
5. $\triangle MSP \cong \triangle RQP$ (AAS)

17. Yes; sample answer: They are congruent by AAS. **19.** Proof: Because it is given that $\overline{GE}$ bisects $\angle DEF$, $\angle DEG \cong \angle FEG$ by the definition of an angle bisector. It is given that $\angle D \cong \angle F$. By the Reflexive Property, $\overline{GE} \cong \overline{GE}$. So, $\triangle DEG \cong \triangle FEG$ by AAS. Therefore, $\overline{DG} \cong \overline{FG}$ by CPCTC. **21.** Tyrone; Lorenzo showed that all three corresponding angles were congruent, but AAA is not a proof of triangle congruence.
23.

Method	Use when...
Definition of Congruent Triangles	All corresponding parts of one triangle are congruent to the corresponding parts of the other triangle.
SSS	The three sides on one triangle must be congruent to the three sides of the other triangle.
SAS	Two sides and the included angle of one triangle must be congruent to two sides and the included angle of the other triangle.
ASA	Two angles and the included side of one triangle must be congruent to two angles and the included side of the other triangle.
AAS	Two angles and a non-included side of one triangle must be congruent to two angles and the corresponding non-included side of the other triangle.

Lesson 14-5

1. Statements (Reasons)

1. $\overline{XZ} \perp \overline{WY}$ (Given)
2. $\angle XZW$ and $\angle XZY$ are right angles. ($\perp$ lines form right angles.)
3. $\triangle WXZ$ and $\triangle YXZ$ are right triangles. (Definition of right triangle)
4. Z is the midpoint of $\overline{WY}$. (Given)
5. $\overline{WZ} \cong \overline{ZY}$ (Definition of midpoint)
6. $\overline{XZ} \cong \overline{XZ}$ (Reflexive Property of Congruence)
7. $\triangle WXZ \cong \triangle YXZ$ (LL Congruence Theorem)

3. Proof:

Statements (Reasons)

1. $\overline{BX} \perp \overline{AC}$ (Given)
2. $\angle AXB$ and $\angle CXB$ are rt. $\angle$s. (Definition of $\perp$ lines)
3. $\triangle AXB$ and $\triangle CXB$ are rt. $\triangle$s. (Definition of right $\triangle$s)
4. $\overline{XB} \cong \overline{XB}$ (Reflexive Property of Congruence)
5. $AB = CB$ (Given)
6. $\overline{AB} \cong \overline{CB}$ (Definition of congruent)
7. $\triangle AXB \cong \triangle CXB$ (HL Congruence Thm.)

5. Yes; LA **7.** No; not enough information
9. No; not enough information
11. Proof:

Statements (Reasons)

1. $\overline{BX} \perp \overline{XA}$, $\overline{BY} \perp \overline{YA}$ (Given)
2. $\angle BXA$ and $\angle BYA$ are rt. $\angle$s. (Definition of $\perp$ lines)
3. $\triangle BXA$ and $\triangle BYA$ are rt. $\triangle$s. (Definition of right $\triangle$s)
4. $\overline{XA} \cong \overline{YA}$ (Given)
5. $\overline{BA} \cong \overline{BA}$ (Reflexive Property of Congruence)
6. $\triangle BXA \cong \triangle BYA$ (HL Congruence Theorem)

13. Proof: By the definition of $\perp$ segments, $\angle AYB$ and $\angle AXC$ are right angles. By the definition of right triangles, $\triangle AYB$ and $\triangle AXC$ are right triangles. By the definition of congruent segments, $\overline{AX}$ is congruent to $\overline{AY}$.

By the Reflexive Property of Congruence, $\angle BAY$ is congruent to $\angle CAX$. Therefore by LA, $\triangle ABY$ is congruent to $\triangle ACX$.

Lesson 14-6

1. Proof:

Statements (Reasons)

1. $\angle 1 \cong \angle 2$ (Given)

2. $\angle 2 \cong \angle 3$ (Vertical Angles Thm.)

3. $\angle 1 \cong \angle 3$ (Transitive Prop. of $\cong$)

4. $\overline{AB} \cong \overline{CB}$ (Conv. of Isos. Triangle Thm.)

3. Proof:

Statements (Reasons)

1. $\overline{DE} \parallel \overline{BC}$ (Given)
2. $\angle 1 \cong \angle 4$, $\angle 2 \cong \angle 3$ (Corresponding angles are $\cong$.)
3. $\angle 1 \cong \angle 2$ (Given)
4. $\angle 1 \cong \angle 3$ (Transitive Property of $\cong$.)
5. $\angle 3 \cong \angle 4$ (Substitution.)
6. $\overline{AB} \cong \overline{AC}$ (Converse of Isosceles Triangle Theorem.)

5a. The coordinates of $\triangle ABC$ are $A(0, 5)$, $B(3, 1)$, and $C(-3, 1)$.

$AC = \sqrt{[0-(-3)]^2 + (5-1)^2}$ or 5 units

$AB = \sqrt{(0-3)^2 + (5-1)^2}$ or 5 units

$BC = 6$ units

So, $\triangle ABC$ is an isosceles triangle with $\overline{AB} \cong \overline{AC}$.

5b. Because $\overline{AB} \cong \overline{AC}$, we know that $\angle C \cong \angle B$ by the Isosceles Triangle Theorem.

$m\angle A + m\angle B + m\angle C = 180°$	Triangle Angle-Sum Theorem
$m\angle A + 2m\angle C = 180°$	Definition of congruent
$m\angle A + 2(55) = 180°$	Substitute.
$m\angle A + 110 = 180°$	Multiply.
$m\angle A = 70°$	Solve.

7. $m\angle DEF = 45°$ and $m\angle EFD = 45°$ **9.** 60°; 3 m **11.** 60°; 7 in. **13.** 10 **15a.** 7 **15b.** 15

17a.

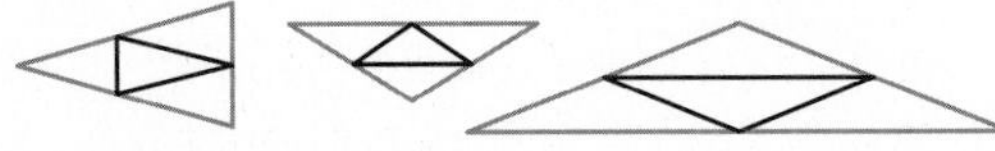

17b. Sample answer: The triangle formed by connecting the midpoints of the sides of an isosceles triangle is an isosceles triangle.

19. Proof:

Statements (Reasons)

1. $\triangle PQR$ is an equilateral triangle. (Given)
2. $\overline{PQ} \cong \overline{QR} \cong \overline{PR}$ (Def. of equilateral triangle)
3. $\angle P \cong \angle Q \cong \angle R$ (Isosceles Triangle Theorem)
4. $m\angle P = m\angle Q = m\angle R$ (Def. of congruence)
5. $m\angle P + m\angle Q + m\angle R = 180°$ (Triangle Angle-Sum Thm.)
6. $3m\angle P = 180°$ (Substitution)
7. $m\angle P = 60°$ (Division Property)
8. $m\angle P = m\angle Q = m\angle R = 60°$ (Substitution)

21. 44° **23.** 22°

25.

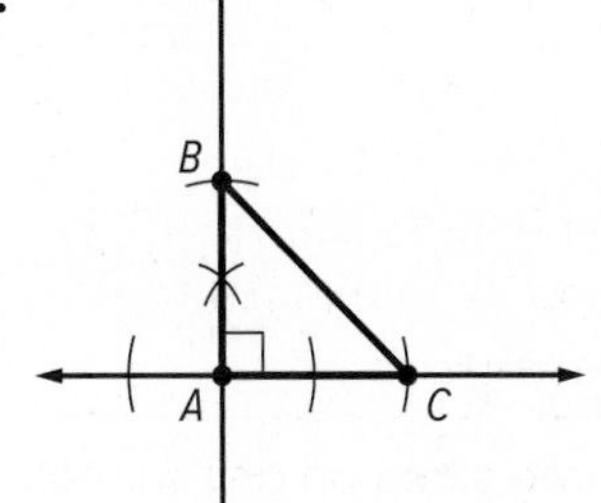

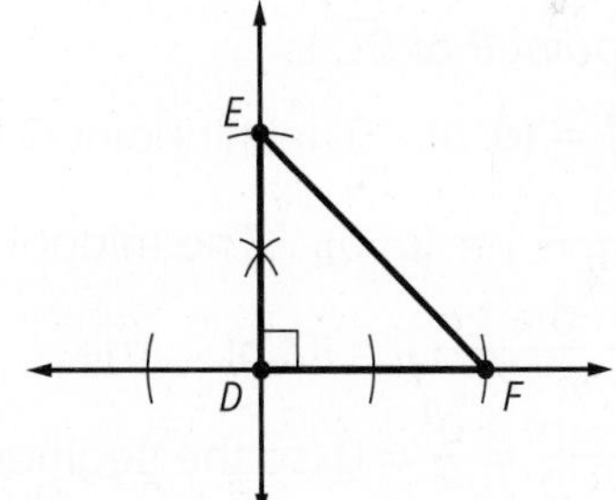

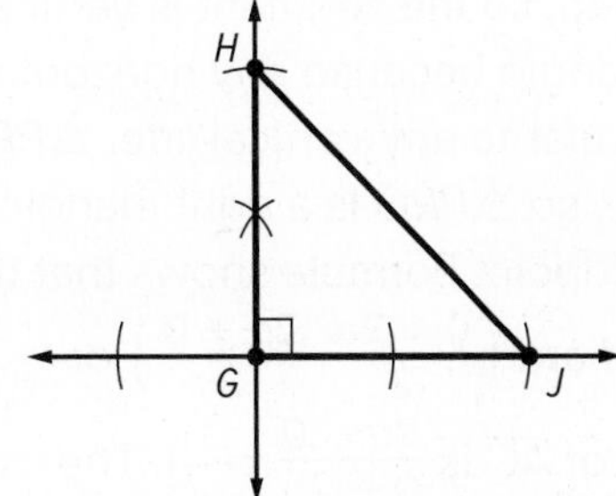

Sample answer: I constructed a pair of perpendicular segments and then used the same compass setting to mark points that are equidistant from their intersection. I measured both legs for each triangle. Because $AB = AC = 1.3$ cm, $DE = DF = 1.9$ cm, and

$GH = GJ = 2.3$ cm, the triangles are isosceles. I used a protractor to confirm that $\angle A$, $\angle D$, and $\angle G$ are all right angles.
27. Sometimes; sample answer: Only if the measure of the vertex angle is even.
29. Sample answer: It is not possible because a triangle cannot have more than one obtuse angle. **31.** No; $m\angle G = \frac{180-70}{2}$ or 55°.

Lesson 14-7

1.

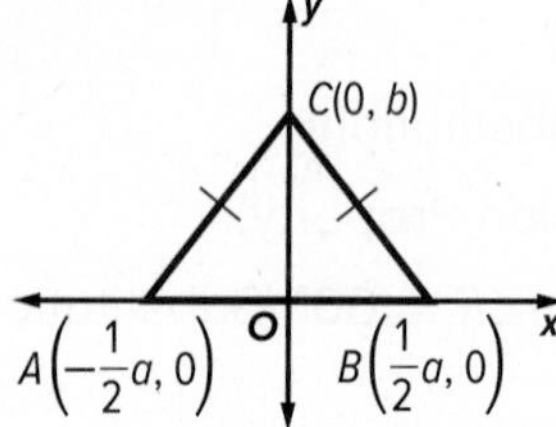

3.

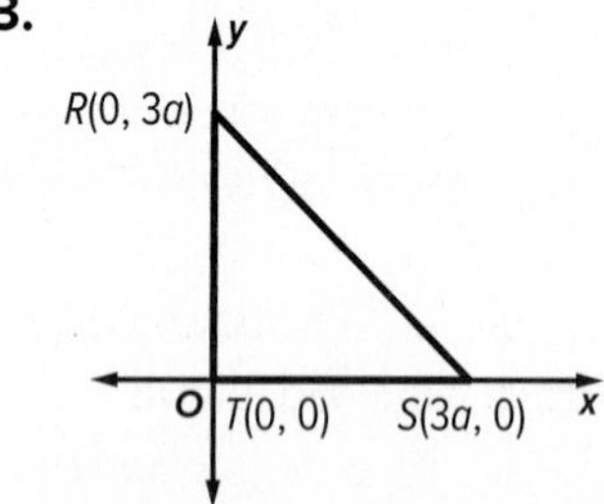

5. $C(p, q)$ **7.** $E(-2g, 0)$; $F(0, b)$ **9.** Sample answer: The midpoint P of $\overline{BC}$ is $\left(\frac{0+2a}{2}, \frac{2b+0}{2}\right) = (a, b)$. The midpoint Q of $\overline{AC}$ is $\left(\frac{0+2a}{2}, \frac{0+0}{2}\right) = (a, 0)$. The midpoint R of $\overline{AB}$ is $\left(\frac{0+0}{2}, \frac{0+2b}{2}\right) = (0, b)$. The slope of $\overline{RP}$ is $\frac{b-b}{a-0} = \frac{0}{a} = 0$, so the segment is horizontal. The slope of $\overline{PQ}$ is $\frac{b-0}{a-a} = \frac{0}{a}$, which is undefined, so the segment is vertical. $\angle RPQ$ is a right angle because any horizontal line is perpendicular to any vertical line. $\triangle PRQ$ has a right angle, so $\triangle PRQ$ is a right triangle.
11. Proof: The Midpoint Formula shows that the coordinates of M are $\left(\frac{0+2a}{2}, \frac{2a+0}{2}\right)$ or (a, a). The slope of $\overline{AC}$ is $\frac{2a-0}{0-2a} = -1$. The slope of $\overline{BM}$ is $\frac{a-0}{a-0} = 1$. The product of the slopes is -1, so $\overline{BM} \perp \overline{AC}$. **13.** Proof: The coordinates of S are $\left(\frac{b}{2}, \frac{c}{2}\right)$, and the coordinates of T are $\left(\frac{a+b}{2}, \frac{c}{2}\right)$.

$ST = \sqrt{\left(\frac{a+b}{2} - \frac{b}{2}\right)^2 + \left(\frac{c}{2} - \frac{c}{2}\right)^2}$ or $\frac{a}{2}$

$AB = \sqrt{(a-0)^2 + (0-0)^2}$ or a

$ST = \frac{1}{2}AB$

15. The slope between the grandstand and the rides and games is $\frac{2}{3}$. The slope between the grandstand and the main gate is $-\frac{3}{2}$. Because $\frac{2}{3} \cdot -\frac{3}{2} = -1$, the triangle formed by these three locations is a right triangle.

17. Slope of $\overline{XY} = 1$, slope of $\overline{YZ} = -1$, slope of $\overline{ZX} = 0$; because $1(-1) = -1$, $\overline{XY} \perp \overline{YZ}$. Therefore, $\triangle XYZ$ is a right triangle.

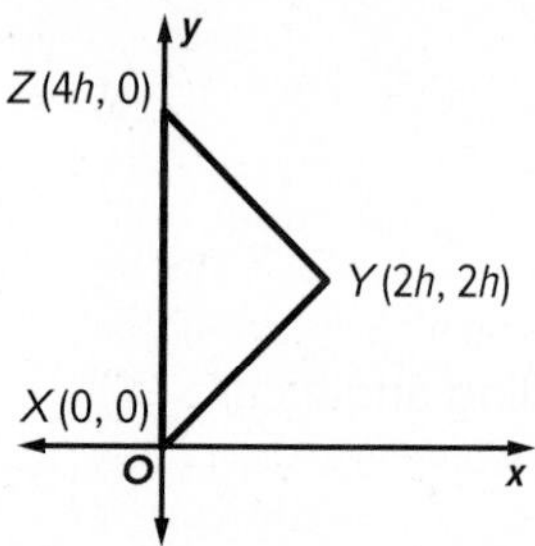

19. $(a, 0)$ and $(0, a)$ **21.** Sample answer: Use the Distance Formula to find the length of each side of each triangle. Show that the triangles are congruent by SSS. Conclude that $\angle A \cong \angle D$ using CPCTC. **23.** $D(a, b)$, $E(2a, 0)$, $F(3a, b)$
25. Sample answer:

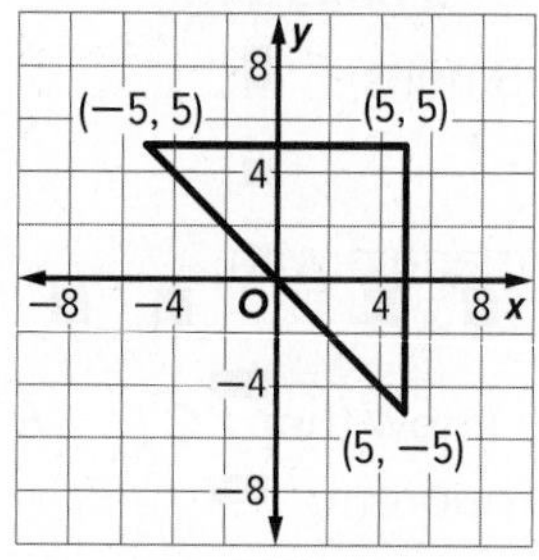

27. Sample answer: $(a, 0)$
29. Sample answer: $(4a, 0)$ or $(0, 4b)$

Module 14 Review

1. 106 **3.** A, D **5.** A, C, E **7.** A **9.** A, B
11. A **13.** 47 **15.** 33 units **17.** A **19.** B

Glossary

English	Español

A

30°-60°-90° triangle A right triangle with two acute angles that measure 30° and 60°.

triángulo 30°-60°-90° Un triángulo rectángulo con dos ángulos agudos que miden 30° y 60°.

45°-45°-90° triangle A right triangle with two acute angles that measure 45°.

triángulo 45°-45°-90° Un triángulo rectángulo con dos ángulos agudos que miden 45°.

absolute value The distance a number is from zero on the number line.

valor absoluto La distancia que un número es de cero en la línea numérica.

absolute value function A function written as $f(x) = |x|$, in which $f(x) \geq 0$ for all values of x.

función del valor absoluto Una función que se escribe $f(x) = |x|$, donde $f(x) \geq 0$, para todos los valores de x.

accuracy The nearness of a measurement to the true value of the measure.

exactitud La proximidad de una medida al valor verdadero de la medida.

additive identity Because the sum of any number a and 0 is equal to a, 0 is the additive identity.

identidad aditiva Debido a que la suma de cualquier número a y 0 es igual a, 0 es la identidad aditiva.

additive inverses Two numbers with a sum of 0.

inverso aditivos Dos números con una suma de 0.

adjacent angles Two angles that lie in the same plane and have a common vertex and a common side but have no common interior points.

ángulos adyacentes Dos ángulos que se encuentran en el mismo plano y tienen un vértice común y un lado común, pero no tienen puntos comunes en el interior.

adjacent arcs Arcs in a circle that have exactly one point in common.

arcos adyacentes Arcos en un circulo que tienen un solo punto en común.

algebraic expression A mathematical expression that contains at least one variable.

expresión algebraica Una expresión matemática que contiene al menos una variable.

algebraic notation Mathematical notation that describes a set by using algebraic expressions.

notación algebraica Notación matemática que describe un conjunto usando expresiones algebraicas.

alternate exterior angles When two lines are cut by a transversal, nonadjacent exterior angles that lie on opposite sides of the transversal.

ángulos alternos externos Cuando dos líneas son cortadas por un ángulo transversal, no adyacente exterior que se encuentran en lados opuestos de la transversal.

alternate interior angles When two lines are cut by a transversal, nonadjacent interior angles that lie on opposite sides of the transversal.

ángulos alternos internos Cuando dos líneas son cortadas por un ángulo transversal, no adyacente interior que se encuentran en lados opuestos de la transversal.

altitude of a parallelogram A perpendicular segment between any two parallel bases.

altitud de un paralelogramo Un segmento perpendicular entre dos bases paralelas.

altitude of a prism or cylinder A segment perpendicular to the bases that joins the planes of the bases.

altitude of a pyramid or cone A segment perpendicular to the base that has the vertex as one endpoint and a point in the plane of the base as the other endpoint.

altitude of a triangle A segment from a vertex of the triangle to the line containing the opposite side and perpendicular to that side.

ambiguous case When two different triangles could be created or described using the given information.

amplitude For functions of the form $y = a \sin b\theta$ or $y = a \cos b\theta$, the amplitude is $|a|$.

analytic geometry The study of geometry that uses the coordinate system.

angle The intersection of two noncollinear rays at a common endpoint.

angle bisector A ray or segment that divides an angle into two congruent angles.

angle of depression The angle formed by a horizontal line and an observer's line of sight to an object below the horizontal line.

angle of elevation The angle formed by a horizontal line and an observer's line of sight to an object above the horizontal line.

angle of rotation The angle through which a figure rotates.

apothem A perpendicular segment between the center of a regular polygon and a side of the polygon or the length of that line segment.

approximate error The positive difference between an actual measurement and an approximate or estimated measurement.

arc Part of a circle that is defined by two endpoints.

arc length The distance between the endpoints of an arc measured along the arc in linear units.

altitud de un prisma o cilindro Un segmento perpendicular a las bases que une los planos de las bases.

altitud de una pirámide o cono Un segmento perpendicular a la base que tiene el vértice como un punto final y un punto en el plano de la base como el otro punto final.

altitud de triángulo Un segmento de un vértice del triángulo a la línea que contiene el lado opuesto y perpendicular a ese lado.

caso ambiguo Cuando dos triángulos diferentes pueden ser creados o descritos usando la información dada.

amplitud Para funciones de la forma $y = a$ sen $b\theta$ o $y = a \cos b\theta$, la amplitud es $|a|$.

geometría analítica El estudio de la geometría que utiliza el sistema de coordenadas.

ángulo La intersección de dos rayos no colineales en un extremo común.

bisectriz de un ángulo Un rayo o segmento que divide un ángulo en dos ángulos congruentes.

ángulo de depresión El ángulo formado por una línea horizontal y la línea de visión de un observador a un objeto por debajo de la línea horizontal.

ángulo de elevación El ángulo formado por una línea horizontal y la línea de visión de un observador a un objeto por encima de la línea horizontal.

ángulo de rotación El ángulo a través del cual gira una figura.

apotema Un segmento perpendicular entre el centro de un polígono regular y un lado del polígono o la longitud de ese segmento de línea.

error aproximado La diferencia positiva entre una medida real y una medida aproximada o estimada.

arco Parte de un círculo que se define por dos puntos finales.

longitude de arco La distancia entre los extremos de un arco medido a lo largo del arco en unidades lineales.

area The number of square units needed to cover a surface.

área El número de unidades cuadradas para cubrir una superficie.

arithmetic sequence A pattern in which each term after the first is found by adding a constant, the common difference *d*, to the previous term.

secuencia aritmética Un patrón en el cual cada término después del primero se encuentra añadiendo una constante, la diferencia común *d*, al término anterior.

asymptote A line that a graph approaches.

asíntota Una línea que se aproxima a un gráfico.

auxiliary line An extra line or segment drawn in a figure to help analyze geometric relationships.

línea auxiliar Una línea o segmento extra dibujado en una figura para ayudar a analizar las relaciones geométricas.

average rate of change The change in the value of the dependent variable divided by the change in the value of the independent variable.

tasa media de cambio El cambio en el valor de la variable dependiente dividido por el cambio en el valor de la variable independiente.

axiom A statement that is accepted as true without proof.

axioma Una declaración que se acepta como verdadera sin prueba.

axiomatic system A set of axioms from which theorems can be derived.

sistema axiomático Un conjunto de axiomas de los cuales se pueden derivar teoremas.

axis of symmetry A line about which a graph is symmetric.

eje de simetría Una línea sobre la cual un gráfica es simétrico.

axis symmetry If a figure can be mapped onto itself by a rotation between 0° and 360° in a line.

eje simetría Si una figura puede ser asignada sobre sí misma por una rotación entre 0° y 360° en una línea.

B

bar graph A graphical display that compares categories of data using bars of different heights.

gráfico de barra Una pantalla gráfica que compara las categorías de datos usando barras de diferentes alturas.

base In a power, the number being multiplied by itself.

base En un poder, el número se multiplica por sí mismo.

base angles of a trapezoid The two angles formed by the bases and legs of a trapezoid.

ángulos de base de un trapecio Los dos ángulos formados por las bases y patas de un trapecio.

base angles of an isosceles triangle The two angles formed by the base and the congruent sides of an isosceles triangle.

ángulo de la base de un triángulo isosceles Los dos ángulos formados por la base y los lados congruentes de un triángulo isosceles.

base edge The intersection of a lateral face and a base in a solid figure.

arista de la base La intersección de una cara lateral y una base en una figura sólida.

base of a parallelogram Any side of a parallelogram.

base de un paralelogramo Cualquier lado de un paralelogramo.

base of a pyramid or cone The face of the solid opposite the vertex of the solid.

base de una pirámide o cono La cara del sólido opuesta al vértice del sólido.

bases of a prism or cylinder The two parallel congruent faces of the solid.

bases de un prisma o cilindro Las dos caras congruentes paralelas de la figura sólida.

bases of a trapezoid The parallel sides in a trapezoid.

bases de un trapecio Los lados paralelos en un trapecio.

best-fit line The line that most closely approximates the data in a scatter plot.

línea de ajuste óptimo La línea que más se aproxima a los datos en un diagrama de dispersión.

betweenness of points Point C is between A and B if and only if A, B, and C are collinear and $AC + CB = AB$.

intermediación de puntos El punto C está entre A y B si y sólo si A, B, y C son colineales y $AC + CB = AB$.

bias An error that results in a misrepresentation of a population.

sesgo Un error que resulta en una tergiversación de una población.

biconditional statement The conjunction of a conditional and its converse.

declaración bicondicional La conjunción de un condicional y su inverso.

binomial The sum of two monomials.

binomio La suma de dos monomios.

bisect To separate a line segment into two congruent segments.

bisecar Separe un segmento de línea en dos segmentos congruentes.

bivariate data Data that consists of pairs of values.

datos bivariate Datos que constan de pares de valores.

boundary The edge of the graph of an inequality that separates the coordinate plane into regions.

frontera El borde de la gráfica de una desigualdad que separa el plano de coordenadas en regiones.

bounded When the graph of a system of constraints is a polygonal region.

acotada Cuando la gráfica de un sistema de restricciones es una región poligonal.

box plot A graphical representation of the five-number summary of a data set.

diagram de caja Una representación gráfica del resumen de cinco números de un conjunto de datos.

C

categorical data Data that can be organized into different categories.

datos categóricos Datos que pueden organizarse en diferentes categorías.

causation When a change in one variable produces a change in another variable.

causalidad Cuando un cambio en una variable produce un cambio en otra variable.

center of a circle The point from which all points on a circle are the same distance.

centro de un círculo El punto desde el cual todos los puntos de un círculo están a la misma distancia.

center of a regular polygon The center of the circle circumscribed about a regular polygon.

centro de un polígono regular El centro del círculo circunscrito alrededor de un polígono regular.

center of dilation The center point from which dilations are performed.

centro de dilatación Punto fijo en torno al cual se realizan las homotecias.

center of rotation The fixed point about which a figure rotates.

centro de rotación El punto fijo sobre el que gira una figura.

center of symmetry A point in which a figure can be rotated onto itself.

centro de la simetría Un punto en el que una figura se puede girar sobre sí misma.

central angle of a circle An angle with a vertex at the center of a circle and sides that are radii.

ángulo central de un círculo Un ángulo con un vértice en el centro de un círculo y los lados que son radios.

central angle of a regular polygon An angle with its vertex at the center of a regular polygon and sides that pass through consecutive vertices of the polygon.

ángulo central de un polígono regular Un ángulo con su vértice en el centro de un polígono regular y lados que pasan a través de vértices consecutivos del polígono.

centroid The point of concurrency of the medians of a triangle.

baricentro El punto de intersección de las medianas de un triángulo.

chord of a circle or sphere A segment with endpoints on the circle or sphere.

cuerda de un círculo o esfera Un segmento con extremos en el círculo o esfera.

circle The set of all points in a plane that are the same distance from a given point called the center.

círculo El conjunto de todos los puntos en un plano que están a la misma distancia de un punto dado llamado centro.

circular function A function that describes a point on a circle as the function of an angle defined in radians.

función circular Función que describe un punto en un círculo como la función de un ángulo definido en radianes.

circumcenter The point of concurrency of the perpendicular bisectors of the sides of a triangle.

circuncentro El punto de concurrencia de las bisectrices perpendiculares de los lados de un triángulo.

circumference The distance around a circle.

circunferencia La distancia alrededor de un círculo.

circumscribed angle An angle with sides that are tangent to a circle.

ángulo circunscrito Un ángulo con lados que son tangentes a un círculo.

circumscribed polygon A polygon with vertices outside the circle and sides that are tangent to the circle.

poligono circunscrito Un polígono con vértices fuera del círculo y lados que son tangentes al círculo.

closed If for any members in a set, the result of an operation is also in the set.

cerrado Si para cualquier número en el conjunto, el resultado de la operación es también en el conjunto.

closed half-plane The solution of a linear inequality that includes the boundary line.

semi-plano cerrado La solución de una desigualdad linear que incluye la línea de limite.

codomain The set of all the *y*-values that could possibly result from the evaluation of the function.

codominar El conjunto de todos los valores y que podrían resultar de la evaluación de la función.

coefficient The numerical factor of a term.

coeficiente El factor numérico de un término.

coefficient of determination An indicator of how well a function fits a set of data.

coeficiente de determinación Un indicador de lo bien que una función se ajusta a un conjunto de datos.

cofunction identities Identities that show the relationships between sine and cosine, tangent and cotangent, and secant and cosecant.

collinear Lying on the same line.

combination A selection of objects in which order is not important.

combined variation When one quantity varies directly and/or inversely as two or more other quantities.

common difference The difference between consecutive terms in an arithmetic sequence.

common logarithms Logarithms of base 10.

common ratio The ratio of consecutive terms of a geometric sequence.

common tangent A line or segment that is tangent to two circles in the same plane.

complement of *A* All of the outcomes in the sample space that are not included as outcomes of event *A*.

complementary angles Two angles with measures that have a sum of 90°.

completing the square A process used to make a quadratic expression into a perfect square trinomial.

complex conjugates Two complex numbers of the form $a + bi$ and $a - bi$.

complex fraction A rational expression with a numerator and/or denominator that is also a rational expression.

complex number Any number that can be written in the form $a + bi$, where a and b are real numbers and i is the imaginary unit.

component form A vector written as $<x, y>$, which describes the vector in terms of its horizontal component x and vertical component y.

identidades de cofunción Identidades que muestran las relaciones entre seno y coseno, tangente y cotangente, y secante y cosecante.

colineal Acostado en la misma línea.

combinación Una selección de objetos en los que el orden no es importante.

variación combinada Cuando una cantidad varía directamente y / o inversamente como dos o más cantidades.

diferencia común La diferencia entre términos consecutivos de una secuencia aritmética.

logaritmos comunes Logaritmos de base 10.

razón común El razón de términos consecutivos de una secuencia geométrica.

tangente común Una línea o segmento que es tangente a dos círculos en el mismo plano.

complemento de *A* Todos los resultados en el espacio muestral que no se incluyen como resultados del evento *A*.

ángulo complementarios Dos ángulos con medidas que tienen una suma de 90°.

completar el cuadrado Un proceso usado para hacer una expresión cuadrática en un trinomio cuadrado perfecto.

conjugados complejos Dos números complejos de la forma $a + bi$ y $a - bi$.

fracción compleja Una expresión racional con un numerador y / o denominador que también es una expresión racional.

número complejo Cualquier número que se puede escribir en la forma $a + bi$, donde a y b son números reales e i es la unidad imaginaria.

forma de componente Un vector escrito como $<x, y>$, que describe el vector en términos de su componente horizontal x y componente vertical y.

composite figure A figure that can be separated into regions that are basic figures, such as triangles, rectangles, trapezoids, and circles.

figura compuesta Una figura que se puede separar en regiones que son figuras básicas, tales como triángulos, rectángulos, trapezoides, y círculos.

composite solid A three-dimensional figure that is composed of simpler solids.

solido compuesta Una figura tridimensional que se compone de figuras más simples.

composition of functions An operation that uses the results of one function to evaluate a second function.

composición de funciones Operación que utiliza los resultados de una función para evaluar una segunda función.

composition of transformations When a transformation is applied to a figure and then another transformation is applied to its image.

composición de transformaciones Cuando una transformación se aplica a una figura y luego se aplica otra transformación a su imagen.

compound event Two or more simple events.

evento compuesto Dos o más eventos simples.

compound inequality Two or more inequalities that are connected by the words *and* or *or*.

desigualdad compuesta Dos o más desigualdades que están unidas por las palabras *y* u *o*.

compound interest Interest calculated on the principal and on the accumulated interest from previous periods.

interés compuesto Intereses calculados sobre el principal y sobre el interés acumulado de períodos anteriores.

compound statement Two or more statements joined by the word *and* or *or*.

enunciado compuesto Dos o más declaraciones unidas por la palabra *y* o *o*.

concave polygon A polygon with one or more interior angles with measures greater than 180°.

polígono cóncavo Un polígono con uno o más ángulos interiores con medidas superiores a 180°.

concentric circles Coplanar circles that have the same center.

círculos concéntricos Círculos coplanarios que tienen el mismo centro.

conclusion The statement that immediately follows the word *then* in a conditional.

conclusión La declaración que inmediatamente sigue la palabra *entonces* en un condicional.

concurrent lines Three or more lines that intersect at a common point.

líneas concurrentes Tres o más líneas que se intersecan en un punto común.

conditional probability The probability that an event will occur given that another event has already occurred.

probabilidad condicional La probabilidad de que un evento ocurra dado que otro evento ya ha ocurrido.

conditional relative frequency The ratio of the joint frequency to the marginal frequency.

frecuencia relativa condicional La relación entre la frecuencia de la articulación y la frecuencia marginal.

conditional statement A compound statement that consists of a premise, or hypothesis, and a conclusion, which is false only when its premise is true and its conclusion is false.

enunciado condicional Una declaración compuesta que consiste en una premisa, o hipótesis, y una conclusión, que es falsa solo cuando su premisa es verdadera y su conclusión es falsa.

cone A solid figure with a circular base connected by a curved surface to a single vertex.

cono Una figura sólida con una base circular conectada por una superficie curvada a un solo vértice.

confidence interval An estimate of the population parameter stated as a range with a specific degree of certainty.

intervalo de confianza Una estimación del parámetro de población se indica como un rango con un grado específico de certeza.

congruent Having the same size and shape.

congruente Tener el mismo tamaño y forma.

congruent angles Two angles that have the same measure.

ángulo congruentes Dos ángulos que tienen la misma medida.

congruent arcs Arcs in the same or congruent circles that have the same measure.

arcos congruentes Arcos en los mismos círculos o congruentes que tienen la misma medida.

congruent polygons All of the parts of one polygon are congruent to the corresponding parts or matching parts of another polygon.

poligonos congruentes Todas las partes de un polígono son congruentes con las partes correspondientes o partes coincidentes de otro polígono.

congruent segments Line segments that are the same length.

segmentos congruentes Línea segmentos que son la misma longitud.

congruent solids Solid figures that have exactly the same shape, size, and a scale factor of 1:1.

sólidos congruentes Figuras sólidas que tienen exactamente la misma forma, tamaño y un factor de escala de 1:1.

conic sections Cross sections of a right circular cone.

secciones cónicas Secciones transversales de un cono circular derecho.

conjecture An educated guess based on known information and specific examples.

conjetura Una suposición educada basada en información conocida y ejemplos específicos.

conjugates Two expressions, each with two terms, in which the second terms are opposites.

conjugados Dos expresiones, cada una con dos términos, en la que los segundos términos son opuestos.

conjunction A compound statement using the word *and*.

conjunción Una declaración compuesta usando la palabra *y*.

consecutive interior angles When two lines are cut by a transversal, interior angles that lie on the same side of the transversal.

ángulos internos consecutivos Cuando dos líneas se cortan por un ángulo transversal, interior que se encuentran en el mismo lado de la transversal.

consistent A system of equations with at least one ordered pair that satisfies both equations.

consistente Una sistema de ecuaciones para el cual existe al menos un par ordenado que satisfice ambas ecuaciones.

constant function A linear function of the form $y = b$; The function $f(x) = a$, where a is any number.

función constante Una función lineal de la forma $y = b$; La función $f(x) = a$, donde a es cualquier número.

constant of variation The constant in a variation function.

constante de variación La constante en una función de variación.

constant term A term that does not contain a variable.

término constante Un término que no contiene una variable.

constraint A condition that a solution must satisfy.

restricción Una condición que una solución debe satisfacer.

constructions Methods of creating figures without the use of measuring tools.

construcciones Métodos de creación de figuras sin el uso de herramientas de medición.

continuous function A function that can be graphed with a line or an unbroken curve.

función continua Una función que se puede representar gráficamente con una línea o una curva ininterrumpida.

continuous random variable The numerical outcome of a random event that can take on any value.

variable aleatoria continua El resultado numérico de un evento aleatorio que puede tomar cualquier valor.

contrapositive A statement formed by negating both the hypothesis and the conclusion of the converse of a conditional.

antítesis Una afirmación formada negando tanto la hipótesis como la conclusión del inverso del condicional.

convenience sample Members that are readily available or easy to reach are selected.

muestra conveniente Se seleccionan los miembros que están fácilmente disponibles o de fácil acceso.

converse A statement formed by exchanging the hypothesis and conclusion of a conditional statement.

recíproco Una declaración formada por el intercambio de la hipótesis y la conclusión de la declaración condicional.

convex polygon A polygon with all interior angles measuring less than 180°.

polígono convexo Un polígono con todos los ángulos interiores que miden menos de 180°.

coordinate proofs Proofs that use figures in the coordinate plane and algebra to prove geometric concepts.

pruebas de coordenadas Pruebas que utilizan figuras en el plano de coordenadas y álgebra para probar conceptos geométricos.

coplanar Lying in the same plane.

coplanar Acostado en el mismo plano.

corollary A theorem with a proof that follows as a direct result of another theorem.

corolario Un teorema con una prueba que sigue como un resultado directo de otro teorema.

correlation coefficient A measure that shows how well data are modeled by a regression function.

coeficiente de correlación Una medida que muestra cómo los datos son modelados por una función de regresión.

corresponding angles When two lines are cut by a transversal, angles that lie on the same side of a transversal and on the same side of the two lines.

ángulos correspondientes Cuando dos líneas se cortan transversalmente, los ángulos que se encuentran en el mismo lado de una transversal y en el mismo lado de las dos líneas.

corresponding parts Corresponding angles and corresponding sides of two polygons.

partes correspondientes Ángulos correspondientes y lados correspondientes.

cosecant The ratio of the length of a hypotenuse to the length of the leg opposite the angle.

cosecante Relación entre la longitud de la hipotenusa y la longitud de la pierna opuesta al ángulo.

cosine The ratio of the length of the leg adjacent to an angle to the length of the hypotenuse.

coseno Relación entre la longitud de la pierna adyacente a un ángulo y la longitud de la hipotenusa.

cotangent The ratio of the length of the leg adjacent to an angle to the length of the leg opposite the angle.

cotangente La relación entre la longitud de la pata adyacente a un ángulo y la longitud de la pata opuesta al ángulo.

coterminal angles Angles in standard position that have the same terminal side.

ángulos coterminales Ángulos en posición estándar que tienen el mismo lado terminal.

counterexample An example that contradicts the conjecture showing that the conjecture is not always true.

contraejemplo Un ejemplo que contradice la conjetura que muestra que la conjetura no siempre es cierta.

critical values The *z*-values corresponding to the most common degrees of certainty.

valores críticos Los valores z correspondientes a los grados de certeza más comunes.

cross section The intersection of a solid and a plane.

sección transversal Intersección de un sólido con un plano.

cube root One of three equal factors of a number.

raíz cúbica Uno de los tres factores iguales de un número.

cube root function A radical function that contains the cube root of a variable expression.

función de la raíz del cubo Función radical que contiene la raíz cúbica de una expresión variable.

curve fitting Finding a regression equation for a set of data that is approximated by a function.

ajuste de curvas Encontrar una ecuación de regresión para un conjunto de datos que es aproximado por una función.

cycle One complete pattern of a periodic function.

ciclo Un patron completo de una función periódica.

cylinder A solid figure with two congruent and parallel circular bases connected by a curved surface.

cilindro Una figura sólida con dos bases circulares congruentes y paralelas conectadas por una superficie curvada.

D

decay factor The base of an exponential expression, or $1 - r$.

factor de decaimiento La base de una expresión exponencial, o $1 - r$.

decomposition Separating a figure into two or more nonoverlapping parts.

descomposición Separar una figura en dos o más partes que no se solapan.

decreasing Where the graph of a function goes down when viewed from left to right.

decreciente Donde la gráfica de una función disminuye cuando se ve de izquierda a derecha.

deductive argument An argument that guarantees the truth of the conclusion provided that its premises are true.

argumento deductivo Un argumento que garantiza la verdad de la conclusión siempre que sus premisas sean verdaderas.

deductive reasoning The process of reaching a specific valid conclusion based on general facts, rules, definitions, or properties.

razonamiento deductivo El proceso de alcanzar una conclusión válida específica basada en hechos generales, reglas, definiciones, o propiedades.

define a variable To choose a variable to represent an unknown value.

definir una variable Para elegir una variable que represente un valor desconocido.

defined term A term that has a definition and can be explained.

término definido Un término que tiene una definición y se puede explicar.

definitions An explanation that assigns properties to a mathematical object.

definiciones Una explicación que asigna propiedades a un objeto matemático.

degree The value of the exponent in a power function; $\frac{1}{360}$ of the circular rotation about a point.

grado Valor del exponente en una función de potencia. $\frac{1}{360}$ de la rotación circular alrededor de un punto.

degree of a monomial The sum of the exponents of all its variables.

grado de un monomio La suma de los exponents de todas sus variables.

degree of a polynomial The greatest degree of any term in the polynomial.

grado de un polinomio El grado mayor de cualquier término del polinomio.

density A measure of the quantity of some physical property per unit of length, area, or volume.

densidad Una medida de la cantidad de alguna propiedad física por unidad de longitud, área o volumen.

dependent A consistent system of equations with an infinite number of solutions.

dependiente Una sistema consistente de ecuaciones con un número infinito de soluciones.

dependent events Two or more events in which the outcome of one event affects the outcome of the other events.

eventos dependientes Dos o más eventos en que el resultado de un evento afecta el resultado de los otros eventos.

dependent variable The variable in a relation, usually y, with values that depend on x.

variable dependiente La variable de una relación, generalmente y, con los valores que depende de x.

depressed polynomial A polynomial resulting from division with a degree one less than the original polynomial.

polinomio reducido Un polinomio resultante de la división con un grado uno menos que el polinomio original.

descriptive modeling A way to mathematically describe real-world situations and the factors that cause them.

modelado descriptivo Una forma de describir matemáticamente las situaciones del mundo real y los factores que las causan.

descriptive statistics The branch of statistics that focuses on collecting, summarizing, and displaying data.

estadística descriptiva Rama de la estadística cuyo enfoque es la recopilación, resumen y demostración de los datos.

diagonal A segment that connects any two nonconsecutive vertices within a polygon.

diagonal Un segmento que conecta cualquier dos vértices no consecutivos dentro de un polígono.

diameter of a circle or sphere A chord that passes through the center of a circle or sphere.

diámetro de un círculo o esfera Un acorde que pasa por el centro de un círculo o esfera.

difference of squares A binomial in which the first and last terms are perfect squares.

diferencia de cuadrados Un binomio en el que los términos primero y último son cuadrados perfectos.

difference of two squares The square of one quantity minus the square of another quantity.

diferencia de dos cuadrados El cuadrado de una cantidad menos el cuadrado de otra cantidad.

dilation A nonrigid motion that enlarges or reduces a geometric figure; A transformation that stretches or compresses the graph of a function.

dilatación Un movimiento no rígido que agranda o reduce una figura geométrica; Una transformación que estira o comprime el gráfico de una función.

dimensional analysis The process of performing operations with units.

análisis dimensional El proceso de realizar operaciones con unidades.

direct variation When one quantity is equal to a constant times another quantity.

variación directa Cuando una cantidad es igual a una constante multiplicada por otra cantidad.

directed line segment A line segment with an initial endpoint and a terminal endpoint.

segment de línea dirigido Un segmento de línea con un punto final inicial y un punto final terminal.

directrix An exterior line perpendicular to the line containing the foci of a curve.

directriz Una línea exterior perpendicular a la línea que contiene los focos de una curva.

discontinuous function A function that is not continuous.

función discontinua Una función que no es continua.

discrete function A function in which the points on the graph are not connected.

función discreta Una función en la que los puntos del gráfico no están conectados.

discrete random variable The numerical outcome of a random event that is finite and can be counted.

variable aleatoria discreta El resultado numérico de un evento aleatorio que es finito y puede ser contado.

discriminant In the Quadratic Formula, the expression under the radical sign that provides information about the roots of the quadratic equation.

discriminante En la Fórmula cuadrática, la expresión bajo el signo radical que proporciona información sobre las raíces de la ecuación cuadrática.

disjunction A compound statement using the word *or*.

disyunción Una declaración compuesta usando la palabra *o*.

distance The length of the line segment between two points.

distancia La longitud del segmento de línea entre dos puntos.

distribution A graph or table that shows the theoretical frequency of each possible data value.

distribución Un gráfico o una table que muestra la frecuencia teórica de cada valor de datos posible.

domain The set of the first numbers of the ordered pairs in a relation; The set of x-values to be evaluated by a function.

dominio El conjunto de los primeros números de los pares ordenados en una relación; El conjunto de valores x para ser evaluados por una función.

dot plot A diagram that shows the frequency of data on a number line.

gráfica de puntos Una diagrama que muestra la frecuencia de los datos en una línea numérica.

double root Two roots of a quadratic equation that are the same number.

raíces dobles Dos raíces de una función cuadrática que son el mismo número.

E

e An irrational number that approximately equals 2.7182818....

e Un número irracional que es aproximadamente igual a 2.7182818 ....

edge of a polyhedron A line segment where the faces of the polyhedron intersect.

arista de un poliedro Un segmento de línea donde las caras del poliedro se cruzan.

elimination A method that involves eliminating a variable by combining the individual equations within a system of equations.

eliminación Un método que consiste en eliminar una variable combinando las ecuaciones individuales dentro de un sistema de ecuaciones.

empty set The set that contains no elements, symbolized by { } or ∅.

conjunto vacio El conjunto que no contiene elementos, simbolizado por { } o ∅.

end behavior The behavior of a graph at the positive and negative extremes in its domain.

comportamiento extremo El comportamiento de un gráfico en los extremos positivo y negativo en su dominio.

enlargement A dilation with a scale factor greater than 1.

ampliación Una dilatación con un factor de escala mayor que 1.

equation A mathematical statement that contains two expressions and an equal sign, =.

ecuación Un enunciado matemático que contiene dos expresiones y un signo igual, =.

equiangular polygon A polygon with all angles congruent.

polígono equiangular Un polígono con todos los ángulos congruentes.

equidistant A point is equidistant from other points if it is the same distance from them.

equidistante Un punto es equidistante de otros puntos si está a la misma distancia de ellos.

equidistant lines Two lines for which the distance between the two lines, measured along a perpendicular line or segment to the two lines, is always the same.

líneas equidistantes Dos líneas para las cuales la distancia entre las dos líneas, medida a lo largo de una línea o segmento perpendicular a las dos líneas, es siempre la misma.

equilateral polygon A polygon with all sides congruent.

polígono equilátero Un polígono con todos los lados congruentes.

equivalent equations Two equations with the same solution.

ecuaciones equivalentes Dos ecuaciones con la misma solución.

equivalent expressions Expressions that represent the same value.

expresiones equivalentes Expresiones que representan el mismo valor.

evaluate To find the value of an expression.

evaluar Calcular el valor de una expresión.

even functions Functions that are symmetric in the y-axis.

incluso funciones Funciones que son simétricas en el eje y.

event A subset of the sample space.

evento Un subconjunto del espacio de muestra.

excluded values Values for which a function is not defined.

valores excluidos Valores para los que no se ha definido una función.

experiment A sample is divided into two groups. The experimental group undergoes a change, while there is no change to the control group. The effects on the groups are then compared; A situation involving chance.

experimento Una muestra se divide en dos grupos. El grupo experimental experimenta un cambio, mientras que no hay cambio en el grupo de control. A continuación se comparan los efectos sobre los grupos; Una situación de riesgo.

experimental probability Probability calculated by using data from an actual experiment.

probabilidad experimental Probabilidad calculada utilizando datos de un experimento real.

exponent When n is a positive integer in the expression x^n, n indicates the number of times x is multiplied by itself.

exponente Cuando n es un entero positivo en la expresión x^n, n indica el número de veces que x se multiplica por sí mismo.

exponential decay Change that occurs when an initial amount decreases by the same percent over a given period of time.

desintegración exponencial Cambio que ocurre cuando una cantidad inicial disminuye en el mismo porcentaje durante un período de tiempo dado.

exponential decay function A function in which the independent variable is an exponent, where $a > 0$ and $0 < b < 1$.

función exponenciales de decaimiento Una ecuación en la que la variable independiente es un exponente, donde $a > 0$ y $0 < b < 1$.

explicit formula A formula that allows you to find any term a_n of a sequence by using a formula written in terms of n.

fórmula explícita Una fórmula que le permite encontrar cualquier término a_n de una secuencia usando una fórmula escrita en términos de n.

exponential equation An equation in which the independent variable is an exponent.

ecuación exponencial Una ecuación en la que la variable independiente es un exponente.

exponential form When an expression is in the form x^n.

forma exponencial Cuando una expresión está en la forma x^n.

exponential function A function in which the independent variable is an exponent.

función exponencial Una función en la que la variable independiente es el exponente.

exponential growth Change that occurs when an initial amount increases by the same percent over a given period of time.

crecimiento exponencial Cambio que ocurre cuando una cantidad inicial aumenta por el mismo porcentaje durante un período de tiempo dado.

exponential growth function A function in which the independent variable is an exponent, where $a > 0$ and $b > 1$.

función de crecimiento exponencial Una función en la que la variable independiente es el exponente, donde $a > 0$ y $b > 1$.

exponential inequality An inequality in which the independent variable is an exponent.

desigualdad exponencial Una desigualdad en la que la variable independiente es un exponente.

exterior angle of a triangle An angle formed by one side of the triangle and the extension of an adjacent side.

ángulo exterior de un triángulo Un ángulo formado por un lado del triángulo y la extensión de un lado adyacente.

exterior angles When two lines are cut by a transversal, any of the four angles that lie outside the region between the two intersected lines.

ángulos externos Cuando dos líneas son cortadas por una transversal, cualquiera de los cuatro ángulos que se encuentran fuera de la región entre las dos líneas intersectadas.

exterior of an angle The area outside of the two rays of an angle.

exterior de un ángulo El área fuera de los dos rayos de un ángulo.

extraneous solution A solution of a simplified form of an equation that does not satisfy the original equation.

solución extraña Una solución de una forma simplificada de una ecuación que no satisface la ecuación original.

extrema Points that are the locations of relatively high or low function values.

extrema Puntos que son las ubicaciones de valores de función relativamente alta o baja.

extreme values The least and greatest values in a set of data.

valores extremos Los valores mínimo y máximo en un conjunto de datos.

F

face of a polyhedron A flat surface of a polyhedron.

cara de un poliedro Superficie plana de un poliedro.

factored form A form of quadratic equation, $0 = a(x - p)(x - q)$, where $a \neq 0$, in which p and q are the x-intercepts of the graph of the related function.

forma factorizada Una forma de ecuación cuadrática, $0 = a(x - p)(x - q)$, donde $a \neq 0$, en la que p y q son las intercepciones x de la gráfica de la función relacionada.

factorial of n The product of the positive integers less than or equal to n.

factorial de n El producto de los enteros positivos inferiores o iguales a n.

factoring The process of expressing a polynomial as the product of monomials and polynomials.

factorización por agrupamiento Utilizando la Propiedad distributiva para factorizar polinomios que possen cuatro o más términos.

factoring by grouping Using the Distributive Property to factor some polynomials having four or more terms.

factorización El proceso de expresar un polinomio como el producto de monomios y polinomios.

family of graphs Graphs and equations of graphs that have at least one characteristic in common.

familia de gráficas Gráficas y ecuaciones de gráficas que tienen al menos una característica común.

feasible region The intersection of the graphs in a system of constraints.

región factible La intersección de los gráficos en un sistema de restricciones.

finite sample space A sample space that contains a countable number of outcomes.

espacio de muestra finito Un espacio de muestra que contiene un número contable de resultados.

finite sequence A sequence that contains a limited number of terms.

secuencia finita Una secuencia que contiene un número limitado de términos.

five-number summary The minimum, quartiles, and maximum of a data set.

resumen de cinco números El mínimo, cuartiles *y* máximo de un conjunto de datos.

flow proof A proof that uses boxes and arrows to show the logical progression of an argument.

demostración de flujo Una prueba que usa cajas y flechas para mostrar la progresión lógica de un argumento.

focus A point inside a parabola having the property that the distances from any point on the parabola to them and to a fixed line have a constant ratio for any points on the parabola.

foco Un punto dentro de una parábola que tiene la propiedad de que las distancias desde cualquier punto de la parábola a ellos ya una línea fija tienen una relación constante para cualquier punto de la parábola.

formula An equation that expresses a relationship between certain quantities.

fórmula Una ecuación que expresa una relación entre ciertas cantidades.

fractional distance An intermediary point some fraction of the length of a line segment.

distancia fraccionaria Un punto intermediario de alguna fracción de la longitud de un segmento de línea.

frequency The number of cycles in a given unit of time.

frecuencia El número de ciclos en una unidad del tiempo dada.

function A relation in which each element of the domain is paired with exactly one element of the range.

función Una relación en que a cada elemento del dominio de corresponde un único elemento del rango.

function notation A way of writing an equation so that $y = f(x)$.

notación functional Una forma de escribir una ecuación para que $y = f(x)$.

G

geometric means The terms between two nonconsecutive terms of a geometric sequence; The nth root, where n is the number of elements in a set of numbers, of the product of the numbers.

medios geométricos Los términos entre dos términos no consecutivos de una secuencia geométrica; La enésima raíz, donde n es el número de elementos de un conjunto de números, del producto de los números.

geometric model A geometric figure that represents a real-life object.

modelo geométrico Una figura geométrica que representa un objeto de la vida real.

geometric probability Probability that involves a geometric measure such as length or area.

probabilidad geométrica Probabilidad que implica una medida geométrica como longitud o área.

geometric sequence A pattern of numbers that begins with a nonzero term and each term after is found by multiplying the previous term by a nonzero constant r.

secuencia geométrica Un patrón de números que comienza con un término distinto de cero y cada término después se encuentra multiplicando el término anterior por una constante no nula r.

geometric series The indicated sum of the terms in a geometric sequence.

series geométricas La suma indicada de los términos en una secuencia geométrica.

glide reflection The composition of a translation followed by a reflection in a line parallel to the translation vector.

reflexión del deslizamiento La composición de una traducción seguida de una reflexión en una línea paralela al vector de traslación.

greatest integer function A step function in which $f(x)$ is the greatest integer less than or equal to x.

función entera más grande Una función del paso en que $f(x)$ es el número más grande menos que o igual a x.

growth factor The base of an exponential expression, or $1 + r$.

factor de crecimiento La base de una expresión exponencial, o $1 + r$.

H

half-plane A region of the graph of an inequality on one side of a boundary.

semi-plano Una región de la gráfica de una desigualdad en un lado de un límite.

height of a parallelogram The length of an altitude of the parallelogram.

altura de un paralelogramo La longitud de la altitud del paralelogramo.

height of a solid The length of the altitude of a solid figure.

altura de un sólido La longitud de la altitud de una figura sólida.

height of a trapezoid The perpendicular distance between the bases of a trapezoid.

altura de un trapecio La distancia perpendicular entre las bases de un trapecio.

histogram A graphical display that uses bars to display numerical data that have been organized in equal intervals.

histograma Una exhibición gráfica que utiliza barras para exhibir los datos numéricos que se han organizado en intervalos iguales.

horizontal asymptote A horizontal line that a graph approaches.

asíntota horizontal Una línea horizontal que se aproxima a un gráfico.

hyperbola The graph of a reciprocal function.

hipérbola La gráfica de una función recíproca.

hypothesis The statement that immediately follows the word *if* in a conditional.

hipótesis La declaración que sigue inmediatamente a la palabra *si* en un condicional.

I

identity An equation that is true for every value of the variable.

identidad Una ecuación que es verdad para cada valor de la variable.

identity function The function $f(x) = x$.

función identidad La función $f(x) = x$.

if-then statement A compound statement of the form *if p, then q*, where p and q are statements.

enunciado si-entonces Enunciado compuesto de la forma *si p, entonces q*, donde p y q son enunciados.

image The new figure in a transformation.

imagen La nueva figura en una transformación.

imaginary unit i The principal square root of -1.

unidad imaginaria i La raíz cuadrada principal de -1.

incenter The point of concurrency of the angle bisectors of a triangle.

incentro El punto de intersección de las bisectrices interiores de un triángulo.

included angle The interior angle formed by two adjacent sides of a triangle.

ángulo incluido El ángulo interior formado por dos lados adyacentes de un triángulo.

included side The side of a triangle between two angles.

lado incluido El lado de un triángulo entre dos ángulos.

inconsistent A system of equations with no ordered pair that satisfies both equations.

inconsistente Una sistema de ecuaciones para el cual no existe par ordenado alguno que satisfaga ambas ecuaciones.

increasing Where the graph of a function goes up when viewed from left to right.

crecciente Donde la gráfica de una función sube cuando se ve de izquierda a derecha.

independent A consistent system of equations with exactly one solution.

independiente Un sistema consistente de ecuaciones con exactamente una solución.

independent events Two or more events in which the outcome of one event does not affect the outcome of the other events.

eventos independientes Dos o más eventos en los que el resultado de un evento no afecta el resultado de los otros eventos.

independent variable The variable in a relation, usually x, with a value that is subject to choice.

variable independiente La variable de una relación, generalmente x, con el valor que sujeta a elección.

index In nth roots, the value that indicates to what root the value under the radicand is being taken.

índice En enésimas raíces, el valor que indica a qué raíz está el valor bajo la radicand.

indirect measurement Using similar figures and proportions to measure an object.

medición indirecta Usando figuras y proporciones similares para medir un objeto.

indirect proof One assumes that the statement to be proven is false and then uses logical reasoning to deduce that a statement contradicts a postulate, theorem, or one of the assumptions.

demostración indirecta Se supone que la afirmación a ser probada es falsa y luego utiliza el razonamiento lógico para deducir que una afirmación contradice un postulado, teorema o uno de los supuestos.

indirect reasoning Reasoning that eliminates all possible conclusions but one so that the one remaining conclusion must be true.

razonamiento indirecto Razonamiento que elimina todas las posibles conclusiones, pero una de manera que la conclusión que queda una debe ser verdad.

inductive reasoning The process of reaching a conclusion based on a pattern of examples.

razonamiento inductive El proceso de llegar a una conclusión basada en un patrón de ejemplos.

inequality A mathematical sentence that contains $<, >, \leq, \geq$, or $\neq$.

desigualdad Una oración matemática que contiene uno o más de $<, >, \leq, \geq$, o $\neq$.

inferential statistics When the data from a sample is used to make inferences about the corresponding population.

estadísticas inferencial Cuando los datos de una muestra se utilizan para hacer inferencias sobre la población correspondiente.

infinite sample space A sample space with outcomes that cannot be counted.

espacio de muestra infinito Un espacio de muestra con resultados que no pueden ser contados.

infinite sequence A sequence that continues without end.

secuencia infinita Una secuencia que continúa sin fin.

informal proof A paragraph that explains why the conjecture for a given situation is true.

prueba informal Un párrafo que explica por qué la conjetura para una situación dada es verdadera.

initial side The part of an angle that is fixed on the *x*-axis.

lado inicial La parte de un ángulo que se fija en el eje *x*.

inscribed angle An angle with its vertex on a circle and sides that contain chords of the circle.

ángulo inscrito Un ángulo con su vértice en un círculo y lados que contienen acordes del círculo.

inscribed polygon A polygon inside a circle in which all of the vertices of the polygon lie on the circle.

polígono inscrito Un polígono dentro de un círculo en el que todos los vértices del polígono se encuentran en el círculo.

intercept A point at which the graph of a function intersects an axis.

interceptar Un punto en el que la gráfica de una función corta un eje.

intercepted arc The part of a circle that lies between the two lines intersecting it.

arco intersecado La parte de un círculo que se encuentra entre las dos líneas que se cruzan.

interior angle of a triangle An angle at the vertex of a triangle.

ángulo interior de un triángulo Un ángulo en el vértice de un triángulo.

interior angles When two lines are cut by a transversal, any of the four angles that lie inside the region between the two intersected lines.

ángulos interiores Cuando dos líneas son cortadas por una transversal, cualquiera de los cuatro ángulos que se encuentran dentro de la región entre las dos líneas intersectadas.

interior of an angle The area between the two rays of an angle.

interior de un ángulo El área entre los dos rayos de un ángulo.

interquartile range The difference between the upper and lower quartiles of a data set.

rango intercuartil La diferencia entre el cuartil superior *y* el cuartil inferior de un conjunto de datos.

intersection A set of points common to two or more geometric figures; intersection The graph of a compound inequality containing *and*.

intersección Un conjunto de puntos communes a dos o más figuras geométricas; intersección La gráfica de una desigualdad compuesta que contiene la palabra *y*.

intersection of *A* and *B* The set of all outcomes in the sample space of event *A* that are also in the sample space of event *B*.

intersección de *A* y *B* El conjunto de todos los resultados en el espacio muestral del evento *A* que también se encuentran en el espacio muestral del evento *B*.

interval The distance between two numbers on the scale of a graph.

intervalo La distancia entre dos números en la escala de un gráfico.

interval notation Mathematical notation that describes a set by using endpoints with parentheses or brackets.

notación de intervalo Notación matemática que describe un conjunto utilizando puntos finales con paréntesis o soportes.

inverse A statement formed by negating both the hypothesis and conclusion of a conditional statement.

inverso Una declaración formada negando tanto la hipótesis como la conclusión de la declaración condicional.

inverse cosine The ratio of the length of the hypotenuse to the length of the leg adjacent to an angle.

inverso del coseno Relación de la longitud de la hipotenusa con la longitud de la pierna adyacente a un ángulo.

inverse functions Two functions, one of which contains points of the form (a, b) while the other contains points of the form (b, a).

funciones inversas Dos funciones, una de las cuales contiene puntos de la forma (a, b) mientras que la otra contiene puntos de la forma (b, a).

inverse relations Two relations, one of which contains points of the form (a, b) while the other contains points of the form (b, a).

relaciones inversas Dos relaciones, una de las cuales contiene puntos de la forma (a, b) mientras que la otra contiene puntos de la forma (b, a).

inverse sine The ratio of the length of the hypotenuse to the length of the leg opposite an angle.

inverso del seno Relación de la longitud de la hipotenusa con la longitud de la pierna opuesta a un ángulo.

inverse tangent The ratio of the length of the leg adjacent to an angle to the length of the leg opposite the angle.

inverso del tangente Relación de la longitud de la pierna adyacente a un ángulo con la longitud de la pierna opuesta a un ángulo.

inverse trigonometric functions Arcsine, Arccosine, and Arctangent.

funciones trigonométricas inversas Arcsine, Arccosine y Arctangent.

inverse variation When the product of two quantities is equal to a constant k.

variación inversa Cuando el producto de dos cantidades es igual a una constante k.

isosceles trapezoid A quadrilateral in which two sides are parallel and the legs are congruent.

trapecio isósceles Un cuadrilátero en el que dos lados son paralelos y las patas son congruentes.

isosceles triangle A triangle with at least two sides congruent.

triángulo isósceles Un triángulo con al menos dos lados congruentes.

J

joint frequencies Entries in the body of a two-way frequency table. In a two-way frequency table, the frequencies in the interior of the table.

frecuencias articulares Entradas en el cuerpo de una tabla de frecuencias de dos vías. En una tabla de frecuencia bidireccional, las frecuencias en el interior de la tabla.

joint variation When one quantity varies directly as the product of two or more other quantities.

variación conjunta Cuando una cantidad varía directamente como el producto de dos o más cantidades.

K

kite A convex quadrilateral with exactly two distinct pairs of adjacent congruent sides.

cometa Un cuadrilátero convexo con exactamente dos pares distintos de lados congruentes adyacentes.

L

lateral area The sum of the areas of the lateral faces of the figure.

área lateral La suma de las áreas de las caras laterales de la figura.

lateral edges The intersection of two lateral faces.

aristas laterales La intersección de dos caras laterales.

lateral faces The faces that join the bases of a solid.

caras laterales Las caras que unen las bases de un sólido.

lateral surface of a cone The curved surface that joins the base of a cone to the vertex.

superficie lateral de un cono La superficie curvada que une la base de un cono con el vértice.

lateral surface of a cylinder The curved surface that joins the bases of a cylinder.

superficie lateral de un cilindro La superficie curvada que une las bases de un cilindro.

leading coefficient The coefficient of the first term when a polynomial is in standard form.

coeficiente líder El coeficiente del primer término cuando un polinomio está en forma estándar.

legs of a trapezoid The nonparallel sides in a trapezoid.

patas de un trapecio Los lados no paralelos en un trapezoide.

legs of an isosceles triangle The two congruent sides of an isosceles triangle.

patas de un triángulo isósceles Los dos lados congruentes de un triángulo isósceles.

like radical expressions Radicals in which both the index and the radicand are the same.

expresiones radicales semejantes Radicales en los que tanto el índice como el radicand son iguales.

like terms Terms with the same variables, with corresponding variables having the same exponent.

términos semejantes Términos con las mismas variables, con las variables correspondientes que tienen el mismo exponente.

line A line is made up of points, has no thickness or width, and extends indefinitely in both directions.

línea Una línea está formada por puntos, no tiene espesor ni anchura, y se extiende indefinidamente en ambas direcciones.

line of fit A line used to describe the trend of the data in a scatter plot.

línea de ajuste Una línea usada para describir la tendencia de los datos en un diagrama de dispersión.

line of reflection A line midway between a preimage and an image; The line in which a reflection flips the graph of a function.

línea de reflexión Una línea a medio camino entre una preimagen y una imagen; La línea en la que una reflectión voltea la gráfica de una función.

line of symmetry An imaginary line that separates a figure into two congruent parts.

línea de simetría Una línea imaginaria que separa una figura en dos partes congruentes.

line segment A measurable part of a line that consists of two points, called endpoints, and all of the points between them.

segmento de línea Una parte medible de una línea que consta de dos puntos, llamados extremos, y todos los puntos entre ellos.

line symmetry A graph has line symmetry if it can be reflected in a vertical line so that each half of the graph maps exactly to the other half.

simetría de línea Un gráfico tiene simetría de línea si puede reflejarse en una línea vertical, de modo que cada mitad del gráfico se asigna exactamente a la otra mitad.

linear equation An equation that can be written in the form $Ax + By = C$ with a graph that is a straight line.

ecuación lineal Una ecuación que puede escribirse de la forma $Ax + By = C$ con un gráfico que es una línea recta.

linear extrapolation The use of a linear equation to predict values that are outside the range of data.

extrapolación lineal El uso de una ecuación lineal para predecir valores que están fuera del rango de datos.

linear function A function in which no independent variable is raised to a power greater than 1; A function with a graph that is a line.

función lineal Una función en la que ninguna variable independiente se eleva a una potencia mayor que 1; Una función con un gráfico que es una línea.

linear inequality A half-plane with a boundary that is a straight line.

desigualdad lineal Un medio plano con un límite que es una línea recta.

linear interpolation The use of a linear equation to predict values that are inside the range of data.

interpolación lineal El uso de una ecuación lineal para predecir valores que están dentro del rango de datos.

linear pair A pair of adjacent angles with noncommon sides that are opposite rays.

par lineal Un par de ángulos adyacentes con lados no comunes que son rayos opuestos.

linear programming The process of finding the maximum or minimum values of a function for a region defined by a system of inequalities.

programación lineal El proceso de encontrar los valores máximos o mínimos de una función para una región definida por un sistema de desigualdades.

linear regression An algorithm used to find a precise line of fit for a set of data.

regresión lineal Un algoritmo utilizado para encontrar una línea precisa de ajuste para un conjunto de datos.

linear transformation One or more operations performed on a set of data that can be written as a linear function.

transformación lineal Una o más operaciones realizadas en un conjunto de datos que se pueden escribir como una función lineal.

literal equation A formula or equation with several variables.

ecuación literal Un formula o ecuación con varias variables.

logarithm In $x = b^y$, y is called the logarithm, base b, of x.

logaritmo En $x = b^y$, y se denomina logaritmo, base b, de x.

logarithmic equation An equation that contains one or more logarithms.

ecuación logarítmica Una ecuación que contiene uno o más logaritmos.

logarithmic function A function of the form $f(x) =$ log base b of x, where $b > 0$ and $b \neq 1$.

función logarítmica Una función de la forma $f(x) =$ base log b de x, donde $b > 0$ y $b \neq 1$.

logically equivalent Statements with the same truth value.

lógicamente equivalentes Declaraciones con el mismo valor de verdad.

lower quartile The median of the lower half of a set of data.

cuartil inferior La mediana de la mitad inferior de un conjunto de datos.

M

magnitude The length of a vector from the initial point to the terminal point.

magnitud La longitud de un vector desde el punto inicial hasta el punto terminal.

magnitude of symmetry The smallest angle through which a figure can be rotated so that it maps onto itself.

magnitud de la simetria El ángulo más pequeño a través del cual una figura se puede girar para que se cargue sobre sí mismo.

major arc An arc with measure greater than 180°.

arco mayor Un arco con una medida superior a 180°.

mapping An illustration that shows how each element of the domain is paired with an element in the range.

cartografía Una ilustración que muestra cómo cada elemento del dominio está emparejado con un elemento del rango.

marginal frequencies In a two-way frequency table, the frequencies in the totals row and column; The totals of each subcategory in a two-way frequency table.

frecuencias marginales En una tabla de frecuencias de dos vías, las frecuencias en los totales de fila y columna; Los totales de cada subcategoría en una tabla de frecuencia bidireccional.

maximum The highest point on the graph of a function.

máximo El punto más alto en la gráfica de una función.

maximum error of the estimate The maximum difference between the estimate of the population mean and its actual value.

error máximo de la estimación La diferencia máxima entre la estimación de la media de la población y su valor real.

measurement data Data that have units and can be measured.

medicion de datos Datos que tienen unidades y que pueden medirse.

measures of center Measures of what is average.

medidas del centro Medidas de lo que es promedio.

measures of spread Measures of how spread out the data are.

medidas de propagación Medidas de cómo se extienden los datos son.

median The beginning of the second quartile that separates the data into upper and lower halves.

mediana El comienzo del segundo cuartil que separa los datos en mitades superior e inferior.

median of a triangle A line segment with endpoints that are a vertex of the triangle and the midpoint of the side opposite the vertex.

mediana de un triángulo Un segmento de línea con extremos que son un vértice del triángulo y el punto medio del lado opuesto al vértice.

metric A rule for assigning a number to some characteristic or attribute.

métrico Una regla para asignar un número a alguna caracteristica o atribuye.

midline The line about which the graph of a function oscillates.

linea media La línea sobre la cual oscila la gráfica de una función periódica.

midpoint The point on a line segment halfway between the endpoints of the segment.

punto medio El punto en un segmento de línea a medio camino entre los extremos del segmento.

midsegment of a trapezoid The segment that connects the midpoints of the legs of a trapezoid.

segment medio de un trapecio El segmento que conecta los puntos medios de las patas de un trapecio.

midsegment of a triangle The segment that connects the midpoints of the legs of a triangle.

segment medio de un triángulo El segmento que conecta los puntos medios de las patas de un triángulo.

minimum The lowest point on the graph of a function.

mínimo El punto más bajo en la gráfica de una función.

minor arc An arc with measure less than 180°.

arco menor Un arco con una medida inferior a 180°.

mixture problems Problems that involve creating a mixture of two or more kinds of things and then determining some quantity of the resulting mixture.

problemas de mezcla Problemas que implican crear una mezcla de dos o más tipos de cosas y luego determinar una cierta cantidad de la mezcla resultante.

monomial A number, a variable, or a product of a number and one or more variables.

monomio Un número, una variable, o un producto de un número y una o más variables.

monomial function A function of the form $f(x) = ax^n$, for which a is a nonzero real number and n is a positive integer.

función monomial Una función de la forma $f(x) = ax^n$, para la cual a es un número real no nulo y n es un entero positivo.

multi-step equation An equation that uses more than one operation to solve it.

ecuaciones de varios pasos Una ecuación que utiliza más de una operación para resolverla.

multiplicative identity Because the product of any number a and 1 is equal to a, 1 is the multiplicative identity.

identidad multiplicativa Dado que el producto de cualquier número a y 1 es igual a, 1 es la identidad multiplicativa.

multiplicative inverses Two numbers with a product of 1.

inversos multiplicativos Dos números con un producto es igual a 1.

multiplicity The number of times a number is a zero for a given polynomial.

multiplicidad El número de veces que un número es cero para un polinomio dado.

mutually exclusive Events that cannot occur at the same time.

mutuamente exclusivos Eventos que no pueden ocurrir al mismo tiempo.

N

natural base exponential function An exponential function with base e, written as $y = e^x$.

función exponencial de base natural Una función exponencial con base e, escrita como $y = e^x$.

natural logarithm The inverse of the natural base exponential function, most often abbreviated as $\ln x$.

logaritmo natural La inversa de la función exponencial de base natural, más a menudo abreviada como $\ln x$.

negation A statement that has the opposite meaning, as well as the opposite truth value, of an original statement.

negación Una declaración que tiene el significado opuesto, así como el valor de verdad opuesto, de una declaración original.

negative Where the graph of a function lies below the x-axis.

negativo Donde la gráfica de una función se encuentra debajo del eje x.

negative correlation Bivariate data in which y decreases as x increases.

correlación negativa Datos bivariate en el cual y disminuye a x aumenta.

negative exponent An exponent that is a negative number.

exponente negativo Un exponente que es un número negativo.

negatively skewed distribution A distribution that typically has a median greater than the mean and less data on the left side of the graph.

distribución negativamente sesgada Una distribución que típicamente tiene una mediana mayor que la media y menos datos en el lado izquierdo del gráfico.

net A two-dimensional figure that forms the surfaces of a three-dimensional object when folded.

red Una figura bidimensional que forma las superficies de un objeto tridimensional cuando se dobla.

no correlation Bivariate data in which x and y are not related.

sin correlación Datos bivariados en los que x e y no están relacionados.

nonlinear function A function in which a set of points cannot all lie on the same line

función no lineal Una función en la que un conjunto de puntos no puede estar en la misma línea

nonrigid motion A transformation that changes the dimensions of a given figure.

movimiento no rígida Una transformación que cambia las dimensiones de una figura dada.

normal distribution A continuous, symmetric, bell-shaped distribution of a random variable.

distribución normal Distribución con forma de campana, simétrica y continua de una variable aleatoria.

*n*th root If $a^n = b$ for a positive integer n, then a is the nth root of b.

raíz enésima Si $a^n = b$ para cualquier entero positive n, entonces a se llama una raíz enésima de b.

*n*th term of an arithmetic sequence The nth term of an arithmetic sequence with first term a_1 and common difference d is given by $a_n = a_1 + (n - 1)d$, where n is a positive integer.

enésimo término de una secuencia aritmética El enésimo término de una secuencia aritmética con el primer término a_1 y la diferencia común d viene dado por $a_n = a_1 + (n - 1)d$, donde n es un número entero positivo.

numerical expression A mathematical phrase involving only numbers and mathematical operations.

expresión numérica Una frase matemática que implica sólo números y operaciones matemáticas.

O

oblique asymptote An asymptote that is neither horizontal nor vertical.

asíntota oblicua Una asíntota que no es ni horizontal ni vertical.

observational study Members of a sample are measured or observed without being affected by the study.

estudio de observación Los miembros de una muestra son medidos o observados sin ser afectados por el estudio.

octant One of the eight divisions of three-dimensional space.

octante Una de las ocho divisiones del espacio tridimensional.

odd functions Functions that are symmetric in the origin.

funciones extrañas Funciones que son simétricas en el origen.

one-to-one function A function for which each element of the range is paired with exactly one element of the domain.

función biunívoca Función para la cual cada elemento del rango está emparejado con exactamente un elemento del dominio.

onto function A function for which the codomain is the same as the range.

sobre la función Función para la cual el codomain es el mismo que el rango.

open half-plane The solution of a linear inequality that does not include the boundary line.

medio plano abierto La solución de una desigualdad linear que no incluye la línea de limite.

opposite rays Two collinear rays with a common endpoint.

rayos opuestos Dos rayos colineales con un punto final común.

optimization The process of seeking the optimal value of a function subject to given constraints.

optimización El proceso de buscar el valor óptimo de una función sujeto a restricciones dadas.

order of symmetry The number of times a figure maps onto itself.

orden de la simetría El número de veces que una figura se asigna a sí misma.

ordered triple Three numbers given in a specific order used to locate points in space.

triple ordenado Tres números dados en un orden específico usado para localizar puntos en el espacio.

orthocenter The point of concurrency of the altitudes of a triangle.

ortocentro El punto de concurrencia de las altitudes de un triángulo.

orthographic drawing The two-dimensional views of the top, left, front, and right sides of an object.

dibujo ortográfico Las vistas bidimensiónales de los lados superior, izquierdo, frontal y derecho de un objeto.

oscillation How much the graph of a function varies between its extreme values as it approaches positive or negative infinity.

oscilación Cuánto la gráfica de una función varía entre sus valores extremos cuando se acerca al infinito positivo o negativo.

outcome The result of a single event; The result of a single performance or trial of an experiment.

resultado El resultado de un solo evento; El resultado de un solo rendimiento o ensayo de un experimento.

outlier A value that is more than 1.5 times the interquartile range above the third quartile or below the first quartile.

parte aislada Un valor que es más de 1,5 veces el rango intercuartílico por encima del tercer cuartil o por debajo del primer cuartil.

P

parabola A curved shape that results when a cone is cut at an angle by a plane that intersects the base; The graph of a quadratic function.

parábola Forma curvada que resulta cuando un cono es cortado en un ángulo por un plano que interseca la base; La gráfica de una función cuadrática.

paragraph proof A paragraph that explains why the conjecture for a given situation is true.

prueba de párrafo Un párrafo que explica por qué la conjetura para una situación dada es verdadera.

parallel lines Coplanar lines that do not intersect; Nonvertical lines in the same plane that have the same slope.

líneas paralelas Líneas coplanares que no se intersecan; Líneas no verticales en el mismo plano que tienen pendientes iguales.

parallel planes Planes that do not intersect.

planos paralelas Planos que no se intersecan.

parallelogram A quadrilateral with both pairs of opposite sides parallel.

paralelogramo Un cuadrilátero con ambos pares de lados opuestos paralelos.

parameter A measure that describes a characteristic of a population; A value in the equation of a function that can be varied to yield a family of functions.

parámetro Una medida que describe una característica de una población; Un valor en la ecuación de una función que se puede variar para producir una familia de funciones.

parent function The simplest of functions in a family.

función basica La función más fundamental de un familia de funciones.

Pascal's triangle A triangle of numbers in which a row represents the coefficients of an expanded binomial $(a + b)^n$.

triángulo de Pascal Un triángulo de números en el que una fila representa los coeficientes de un binomio expandido $(a + b)^n$.

percent rate of change The percent of increase per time period.

por ciento tasa de cambio El porcentaje de aumento por período de tiempo.

percentile A measure that tells what percent of the total scores were below a given score.

percentil Una medida que indica qué porcentaje de las puntuaciones totales estaban por debajo de una puntuación determinada.

perfect cube A rational number with a cube root that is a rational number.

cubo perfecto Un número racional con un raíz cúbica que es un número racional.

perfect square A rational number with a square root that is a rational number.

cuadrado perfecto Un número racional con un raíz cuadrada que es un número racional.

perfect square trinomials Squares of binomials.

trinomio cuadrado perfecto Cuadrados de los binomios.

perimeter The sum of the lengths of the sides of a polygon.

perimetro La suma de las longitudes de los lados de un polígono.

period The horizontal length of one cycle.

periodo La longitud horizontal de un ciclo.

periodic function A function with y-values that repeat at regular intervals.

función periódica Una función con y-valores aquella repetición con regularidad.

permutation An arrangement of objects in which order is important.

permutación Un arreglo de objetos en el que el orden es importante.

perpendicular Intersecting at right angles.

perpendicular Intersección en ángulo recto.

perpendicular bisector Any line, segment, or ray that passes through the midpoint of a segment and is perpendicular to that segment.

mediatriz Cualquier línea, segmento o rayo que pasa por el punto medio de un segmento y es perpendicular a ese segmento.

perpendicular lines Nonvertical lines in the same plane for which the product of the slopes is –1.

líneas perpendiculares Líneas no verticales en el mismo plano para las que el producto de las pendientes es -1.

phase shift A horizontal translation of the graph of a trigonometric function.

cambio de fase Una traducción horizontal de la gráfica de una función trigonométrica.

pi The ratio $\frac{\text{cricumference}}{\text{diameter}}$.

pi Relación $\frac{\text{circunferencia}}{\text{diámetro}}$.

piecewise-defined function A function defined by at least two subfunctions, each of which is defined differently depending on the interval of the domain.

piecewise-linear function A function defined by at least two linear subfunctions, each of which is defined differently depending on the interval of the domain.

plane A flat surface made up of points that has no depth and extends indefinitely in all directions.

plane symmetry When a plane intersects a three-dimensional figure so one half is the reflected image of the other half.

Platonic solid One of five regular polyhedra.

point A location with no size, only position.

point discontinuity An area that appears to be a hole in a graph.

point of concurrency The point of intersection of concurrent lines.

point of symmetry The point about which a figure is rotated.

point of tangency For a line that intersects a circle in one point, the point at which they intersect.

point symmetry A figure or graph has this when a figure is rotated 180° about a point and maps exactly onto the other part.

polygon A closed plane figure with at least three straight sides.

polyhedron A closed three-dimensional figure made up of flat polygonal regions.

polynomial A monomial or the sum of two or more monomials.

polynomial function A continuous function that can be described by a polynomial equation in one variable.

función definida por piezas Una función definida por al menos dos subfunciones, cada una de las cuales se define de manera diferente dependiendo del intervalo del dominio.

función lineal por piezas Una función definida por al menos dos subfunciones lineal, cada una de las cuales se define de manera diferente dependiendo del intervalo del dominio.

plano Una superficie plana compuesta de puntos que no tiene profundidad y se extiende indefinidamente en todas las direcciones.

simetría plana Cuando un plano cruza una figura tridimensional, una mitad es la imagen reflejada de la otra mitad.

sólido platónico Uno de cinco poliedros regulares.

punto Una ubicación sin tamaño, solo posición.

discontinuidad de punto Un área que parece ser un agujero en un gráfico.

punto de concurrencia El punto de intersección de líneas concurrentes.

punto de simetría El punto sobre el que se gira una figura.

punto de tangencia Para una línea que cruza un círculo en un punto, el punto en el que se cruzan.

simetría de punto Una figura o gráfica tiene esto cuando una figura se gira 180° alrededor de un punto y se mapea exactamente sobre la otra parte.

polígono Una figura plana cerrada con al menos tres lados rectos.

poliedros Una figura tridimensional cerrada formada por regiones poligonales planas.

polinomio Un monomio o la suma de dos o más monomios.

función polinómica Función continua que puede describirse mediante una ecuación polinómica en una variable.

polynomial identity A polynomial equation that is true for any values that are substituted for the variables.

population All of the members of a group of interest about which data will be collected.

population proportion The number of members in the population with a particular characteristic divided by the total number of members in the population.

positive Where the graph of a function lies above the x-axis.

positive correlation Bivariate data in which y increases as x increases.

positively skewed distribution A distribution that typically has a mean greater than the median.

postulate A statement that is accepted as true without proof.

power function A function of the form $f(x) = ax^n$, where a and n are nonzero real numbers.

precision The repeatability, or reproducibility, of a measurement.

preimage The original figure in a transformation.

prime polynomial A polynomial that cannot be written as a product of two polynomials with integer coefficients.

principal root The nonnegative root of a number.

principal square root The nonnegative square root of a number.

principal values The values in the restricted domains of trigonometric functions.

principle of superposition Two figures are congruent if and only if there is a rigid motion or series of rigid motions that maps one figure exactly onto the other.

prism A polyhedron with two parallel congruent bases connected by parallelogram faces.

identidad polinomial Una ecuación polinómica que es verdadera para cualquier valor que se sustituya por las variables.

población Todos los miembros de un grupo de interés sobre cuáles datos serán recopilados.

proporción de la población El número de miembros en la población con una característica particular dividida por el número total de miembros en la población.

positiva Donde la gráfica de una función se encuentra por encima del eje x.

correlación positiva Datos bivariate en el cual y aumenta a x disminuye.

distribución positivamente sesgada Una distribución que típicamente tiene una media mayor que la mediana.

postulado Una declaración que se acepta como verdadera sin prueba.

función de potencia Una ecuación polinomial que es verdadera para una función de la forma $f(x) = ax^n$, donde a y n son números reales no nulos.

precisión La repetibilidad, o reproducibilidad, de una medida.

preimagen La figura original en una transformación.

polinomio primo Un polinomio que no puede escribirse como producto de dos polinomios con coeficientes enteros.

raíz principal La raíz no negativa de un número.

raíz cuadrada principal La raíz cuadrada no negativa de un número.

valores principales Valores de los dominios restringidos de las functiones trigonométricas.

principio de superposición Dos figuras son congruentes si y sólo si hay un movimiento rígido o una serie de movimientos rígidos que traza una figura exactamente sobre la otra.

prisma Un poliedro con dos bases congruentes paralelas conectadas por caras de paralelogramo.

probability The number of outcomes in which a specified event occurs to the total number of trials.

probabilidad El número de resultados en los que se produce un evento especificado al número total de ensayos.

probability distribution A function that maps the sample space to the probabilities of the outcomes in the sample space for a particular random variable.

distribución de probabilidad Una función que mapea el espacio de muestra a las probabilidades de los resultados en el espacio de muestra para una variable aleatoria particular.

probability model A mathematical representation of a random event that consists of the sample space and the probability of each outcome.

modelo de probabilidad Una representación matemática de un evento aleatorio que consiste en el espacio muestral y la probabilidad de cada resultado.

projectile motion problems Problems that involve objects being thrown or dropped.

problemas de movimiento del proyectil Problemas que involucran objetos que se lanzan o caen.

proof A logical argument in which each statement is supported by a statement that is accepted as true.

prueba Un argumento lógico en el que cada sentencia está respaldada por una sentencia aceptada como verdadera.

proof by contradiction One assumes that the statement to be proven is false and then uses logical reasoning to deduce that a statement contradicts a postulate, theorem, or one of the assumptions.

prueba por contradicción Se supone que la afirmación a ser probada es falsa y luego utiliza el razonamiento lógico para deducir que una afirmación contradice un postulado, teorema o uno de los supuestos.

proportion A statement that two ratios are equivalent.

proporción Una declaración de que dos proporciones son equivalentes.

pure imaginary number A number of the form bi, where b is a real number and i is the imaginary unit.

número imaginario puro Un número de la forma bi, donde b es un número real e i es la unidad imaginaria.

pyramid A polyhedron with a polygonal base and three or more triangular faces that meet at a common vertex.

pirámide Poliedro con una base poligonal y tres o más caras triangulares que se encuentran en un vértice común.

Pythagorean identities Identities that express the Pythagorean Theorem in terms of the trigonometric functions.

identidades pitagóricas Identidades que expresan el Teorema de Pitágoras en términos de las funciones trigonométricas.

Pythagorean triple A set of three nonzero whole numbers that make the Pythagorean Theorem true.

triplete Pitágorico Un conjunto de tres números enteros distintos de cero que hacen que el Teorema de Pitágoras sea verdadero.

Q

quadrantal angle An angle in standard position with a terminal side that coincides with one of the axes.

ángulo de cuadrante Un ángulo en posición estándar con un lado terminal que coincide con uno de los ejes.

quadratic equation An equation that includes a quadratic expression.

ecuación cuadrática Una ecuación que incluye una expresión cuadrática.

quadratic expression An expression in one variable with a degree of 2.

expresión cuadrática Una expresión en una variable con un grado de 2.

quadratic form A form of polynomial equation, $au^2 + bu + c$, where u is an algebraic expression in x.

forma cuadrática Una forma de ecuación polinomial, $au^2 + bu + c$, donde u es una expresión algebraica en x.

quadratic function A function with an equation of the form $y = ax^2 + bx + c$, where $a \neq 0$.

función cuadrática Una función con una ecuación de la forma $y = ax^2 + bx + c$, donde $a \neq 0$.

quadratic inequality An inequality that includes a quadratic expression.

desigualdad cuadrática Una desigualdad que incluye una expresión cuadrática.

quadratic relations Equations of parabolas with horizontal axes of symmetry that are not functions.

relaciones cuadráticas Ecuaciones de parábolas con ejes horizontales de simetría que no son funciones.

quartic function A fourth-degree function.

función cuartica Una función de cuarto grado.

quartiles Measures of position that divide a data set arranged in ascending order into four groups, each containing about one fourth or 25% of the data.

cuartiles Medidas de posición que dividen un conjunto de datos dispuestos en orden ascendente en cuatro grupos, cada uno de los cuales contiene aproximadamente un cuarto o el 25% de los datos.

quintic function A fifth-degree function.

función quíntica Una función de quinto grado.

R

radian A unit of angular measurement equal to $\frac{180°}{\pi}$ or about 57.296°.

radián Una unidad de medida angular igual o $\frac{180°}{\pi}$ alrededor de 57.296°.

radical equation An equation with a variable in a radicand.

ecuación radical Una ecuación con una variable en un radicand.

radical expression An expression that contains a radical symbol, such as a square root.

expresión radicales Una expresión que contiene un símbolo radical, tal como una raíz cuadrada.

radical form When an expression contains a radical symbol.

forma radical Cuando una expresión contiene un símbolo radical.

radical function A function that contains radicals with variables in the radicand.

función radical Función que contiene radicales con variables en el radicand.

radicand The expression under a radical sign.

radicando La expresión debajo del signo radical.

radius of a circle or sphere A line segment from the center to a point on a circle or sphere.

radio de un círculo o esfera Un segmento de línea desde el centro hasta un punto en un círculo o esfera.

radius of a regular polygon The radius of the circle circumscribed about a regular polygon.

radio de un polígono regular El radio del círculo circunscrito alrededor de un polígono regular.

range The difference between the greatest and least values in a set of data; The set of second numbers of the ordered pairs in a relation; The set of y-values that actually result from the evaluation of the function.

rango La diferencia entre los valores de datos más grande or menos en un sistema de datos; El conjunto de los segundos números de los pares ordenados de una relación; El conjunto de valores y que realmente resultan de la evaluación de la función.

rate of change How a quantity is changing with respect to a change in another quantity.

tasa de cambio Cómo cambia una cantidad con respecto a un cambio en otra cantidad.

rational equation An equation that contains at least one rational expression.

ecuación racional Una ecuación que contiene al menos una expresión racional.

rational exponent An exponent that is expressed as a fraction.

exponente racional Un exponente que se expresa como una fracción.

rational expression A ratio of two polynomial expressions.

expresión racional Una relación de dos expresiones polinomiales.

rational function An equation of the form $f(x) = \frac{a(x)}{b(x)}$, where $a(x)$ and $b(x)$ are polynomial expressions and $b(x) \neq 0$.

función racional Una ecuación de la forma $f(x) = \frac{a(x)}{b(x)}$, donde $a(x)$ y $b(x)$ son expresiones polinomiales y $b(x) \neq 0$.

rational inequality An inequality that contains at least one rational expression.

desigualdad racional Una desigualdad que contiene al menos una expresión racional.

rationalizing the denominator A method used to eliminate radicals from the denominator of a fraction or fractions from a radicand.

racionalizando el denominador Método utilizado para eliminar radicales del denominador de una fracción o fracciones de una radicand.

ray Part of a line that starts at a point and extends to infinity.

rayo Parte de una línea que comienza en un punto y se extiende hasta el infinito.

reciprocal function An equation of the form $f(x) = \frac{n}{b(x)}$, where n is a real number and $b(x)$ is a linear expression that cannot equal 0.

función recíproca Una ecuación de la forma $f(x) = \frac{n}{b(x)}$, donde n es un número real y $b(x)$ es una expresión lineal que no puede ser igual a 0.

reciprocal trigonometric functions Trigonometric functions that are reciprocals of each other.

funciones trigonométricas recíprocas Funciones trigonométricas que son reciprocales entre sí.

reciprocals Two numbers with a product of 1.

recíprocos Dos números con un producto de 1.

rectangle A parallelogram with four right angles.

rectángulo Un paralelogramo con cuatro ángulos rectos.

recursive formula A formula that gives the value of the first term in the sequence and then defines the next term by using the preceding term.

formula recursiva Una fórmula que da el valor del primer término en la secuencia y luego define el siguiente término usando el término anterior.

reduction A dilation with a scale factor between 0 and 1.

reducción Una dilatación con un factor de escala entre 0 y 1.

reference angle The acute angle formed by the terminal side of an angle and the x-axis.

ángulo de referencia El ángulo agudo formado por el lado terminal de un ángulo en posición estándar y el eje x.

reflection A function in which the preimage is reflected in the line of reflection; A transformation in which a figure, line, or curve is flipped across a line.

reflexión Función en la que la preimagen se refleja en la línea de reflexión; Una transformación en la que una figura, línea o curva se voltea a través de una línea.

regression function A function generated by an algorithm to find a line or curve that fits a set of data.

función de regresión Función generada por un algoritmo para encontrar una línea o curva que se ajuste a un conjunto de datos.

regular polygon A convex polygon that is both equilateral and equiangular.

polígono regular Un polígono convexo que es a la vez equilátero y equiangular.

regular polyhedron A polyhedron in which all of its faces are regular congruent polygons and all of the edges are congruent.

poliedro regular Un poliedro en el que todas sus caras son polígonos congruentes regulares y todos los bordes son congruentes.

regular pyramid A pyramid with a base that is a regular polygon.

pirámide regular Una pirámide con una base que es un polígono regular.

regular tessellation A tessellation formed by only one type of regular polygon.

teselado regular Un teselado formado por un solo tipo de polígono regular.

relation A set of ordered pairs.

relación Un conjunto de pares ordenados.

relative frequency In a two-way frequency table, the ratios of the number of observations in a category to the total number of observations; The ratio of the number of observations in a category to the total number of observations.

frecuencia relativa En una tabla de frecuencia bidireccional, las relaciones entre el número de observaciones en una categoría y el número total de observaciones; La relación entre el número de observaciones en una categoría y el número total de observaciones.

relative maximum A point on the graph of a function where no other nearby points have a greater y-coordinate.

máximo relativo Un punto en la gráfica de una función donde ningún otro punto cercano tiene una coordenada y mayor.

relative minimum A point on the graph of a function where no other nearby points have a lesser y-coordinate.

mínimo relativo Un punto en la gráfica de una función donde ningún otro punto cercano tiene una coordenada y menor.

remote interior angles Interior angles of a triangle that are not adjacent to an exterior angle.

ángulos internos no adyacentes Ángulos interiores de un triángulo que no están adyacentes a un ángulo exterior.

residual The difference between an observed y-value and its predicted y-value on a regression line.

residual La diferencia entre un valor de y observado y su valor de y predicho en una línea de regresión.

rhombus A parallelogram with all four sides congruent.

rombo Un paralelogramo con los cuatro lados congruentes.

rigid motion A transformation that preserves distance and angle measure.

movimiento rígido Una transformación que preserva la distancia y la medida del ángulo.

root A solution of an equation.

raíz Una solución de una ecuación.

rotation A function that moves every point of a preimage through a specified angle and direction about a fixed point.

rotación Función que mueve cada punto de una preimagen a través de un ángulo y una dirección especificados alrededor de un punto fijo.

rotational symmetry A figure can be rotated less than 360° about a point so that the image and the preimage are indistinguishable.

simetría rotacional Una figura puede girar menos de 360° alrededor de un punto para que la imagen y la preimagen sean indistinguibles.

S

sample A subset of a population.

muestra Un subconjunto de una población.

sample space The set of all possible outcomes.

espacio muestral El conjunto de todos los resultados posibles.

sampling error The variation between samples taken from the same population.

error de muestreo La variación entre muestras tomadas de la misma población.

scale The distance between tick marks on the *x*- and *y*-axes.

escala La distancia entre las marcas en los ejes *x* e *y*.

scale factor of a dilation The ratio of a length on an image to a corresponding length on the preimage.

factor de escala de una dilatación Relación de una longitud en una imagen con una longitud correspondiente en la preimagen.

scatter plot A graph of bivariate data that consists of ordered pairs on a coordinate plane.

gráfica de dispersión Una gráfica de datos bivariados que consiste en pares ordenados en un plano de coordenadas.

secant Any line or ray that intersects a circle in exactly two points; The ratio of the length of the hypotenuse to the length of the leg adjacent to the angle.

secante Cualquier línea o rayo que cruce un círculo en exactamente dos puntos; Relación entre la longitud de la hipotenusa y la longitud de la pierna adyacente al ángulo.

sector A region of a circle bounded by a central angle and its intercepted arc.

sector Una región de un círculo delimitada por un ángulo central y su arco interceptado.

segment bisector Any segment, line, plane, or point that intersects a line segment at its midpoint.

bisectriz del segmento Cualquier segmento, línea, plano o punto que interseca un segmento de línea en su punto medio.

self-selected sample Members volunteer to be included in the sample.

muestra auto-seleccionada Los miembros se ofrecen como voluntarios para ser incluidos en la muestra.

semicircle An arc that measures exactly 180°.

semicírculo Un arco que mide exactamente 180°.

semiregular tessellation A tessellation formed by two or more regular polygons.

teselado semiregular Un teselado formado por dos o más polígonos regulares.

sequence A list of numbers in a specific order.

secuencia Una lista de números en un orden específico.

series The indicated sum of the terms in a sequence.

serie La suma indicada de los términos en una secuencia.

set-builder notation Mathematical notation that describes a set by stating the properties that its members must satisfy.

notación de construción de conjuntos Notación matemática que describe un conjunto al declarar las propiedades que sus miembros deben satisfacer.

sides of an angle The rays that form an angle.

lados de un ángulo Los rayos que forman un ángulo.

sigma notation A notation that uses the Greek uppercase letter S to indicate that a sum should be found.

notación de sigma Una notación que utiliza la letra mayúscula griega S para indicar que debe encontrarse una suma.

significant figures The digits of a number that are used to express a measure to an appropriate degree of accuracy.

dígitos significantes Los dígitos de un número que se utilizan para expresar una medida con un grado apropiado de precisión.

similar polygons Two figures are similar polygons if one can be obtained from the other by a dilation or a dilation with one or more rigid motions.

polígonos similares Dos figuras son polígonos similares si uno puede ser obtenido del otro por una dilatación o una dilatación con uno o más movimientos rígidos.

similar solids Solid figures with the same shape but not necessarily the same size.

sólidos similares Figuras sólidas con la misma forma pero no necesariamente del mismo tamaño.

similar triangles Triangles in which all of the corresponding angles are congruent and all of the corresponding sides are proportional.

triángulos similares Triángulos en los cuales todos los ángulos correspondientes son congruentes y todos los lados correspondientes son proporcionales.

similarity ratio The scale factor between two similar polygons.

relación de similitud El factor de escala entre dos polígonos similares.

similarity transformation A transformation composed of a dilation or a dilation and one or more rigid motions.

transformación de similitud Una transformación compuesto por una dilatación o una dilatación y uno o más movimientos rígidos.

simple random sample Each member of the population has an equal chance of being selected as part of the sample.

muestra aleatoria simple Cada miembro de la población tiene la misma posibilidad de ser seleccionado como parte de la muestra.

simplest form An expression is in simplest form when it is replaced by an equivalent expression having no like terms or parentheses.

forma reducida Una expresión está reducida cuando se puede sustituir por una expresión equivalente que no tiene ni términos semejantes ni paréntesis.

simulation The use of a probability model to imitate a process or situation so it can be studied.

simulación El uso de un modelo de probabilidad para imitar un proceso o situación para que pueda ser estudiado.

sine The ratio of the length of the leg opposite an angle to the length of the hypotenuse.

seno La relación entre la longitud de la pierna opuesta a un ángulo y la longitud de la hipotenusa.

sinusoidal function A function that can be produced by translating, reflecting, or dilating the sine function.

función sinusoidal Función que puede producirse traduciendo, reflejando o dilatando la función sinusoidal.

skew lines Noncoplanar lines that do not intersect.

líneas alabeadas Líneas no coplanares que no se cruzan.

slant height of a pyramid or right cone The length of a segment with one endpoint on the base edge of the figure and the other at the vertex.

altura inclinada de una pirámide o cono derecho La longitud de un segmento con un punto final en el borde base de la figura y el otro en el vértice.

slope The rate of change in the *y*-coordinates (rise) to the corresponding change in the *x*-coordinates (run) for points on a line.

pendiente La tasa de cambio en las coordenadas *y* (subida) al cambio correspondiente en las coordenadas *x* (carrera) para puntos en una línea.

slope criteria Outlines a method for proving the relationship between lines based on a comparison of the slopes of the lines.

criterios de pendiente Describe un método para probar la relación entre líneas basado en una comparación de las pendientes de las líneas.

solid of revolution A solid figure obtained by rotating a shape around an axis.

sólido de revolución Una figura sólida obtenida girando una forma alrededor de un eje.

solution A value that makes an equation true.

solución Un valor que hace que una ecuación sea verdadera.

solve an equation The process of finding all values of the variable that make the equation a true statement.

resolver una ecuación El proceso en que se hallan todos los valores de la variable que hacen verdadera la ecuación.

solving a triangle When you are given measurements to find the unknown angle and side measures of a triangle.

resolver un triángulo Cuando se le dan mediciones para encontrar el ángulo desconocido y las medidas laterales de un triángulo.

space A boundless three-dimensional set of all points.

espacio Un conjunto tridimensional ilimitado de todos los puntos.

sphere A set of all points in space equidistant from a given point called the center of the sphere.

esfera Un conjunto de todos los puntos del espacio equidistantes de un punto dado llamado centro de la esfera.

square A parallelogram with all four sides and all four angles congruent.

cuadrado Un paralelogramo con los cuatro lados y los cuatro ángulos congruentes.

square root One of two equal factors of a number.

raíz cuadrada Uno de dos factores iguales de un número.

square root function A radical function that contains the square root of a variable expression.

función raíz cuadrada Función radical que contiene la raíz cuadrada de una expresión variable.

square root inequality An inequality that contains the square root of a variable expression.

square root inequality Una desigualdad que contiene la raíz cuadrada de una expresión variable.

standard deviation A measure that shows how data deviate from the mean.

desviación tipica Una medida que muestra cómo los datos se desvían de la media.

standard error of the mean The standard deviation of the distribution of sample means taken from a population.

error estandar de la media La desviación estándar de la distribución de los medios de muestra se toma de una población.

standard form of a linear equation Any linear equation can be written in this form, $Ax + By = C$, where $A \geq 0$, A and B are not both 0, and A, B, and C are integers with a greatest common factor of 1.

forma estándar de una ecuación lineal Cualquier ecuación lineal se puede escribir de esta forma, $Ax + By = C$, donde $A \geq 0$, A y B no son ambos 0, y A, B y C son enteros con el mayor factor común de 1.

standard form of a polynomial A polynomial that is written with the terms in order from greatest degree to least degree.

forma estándar de un polinomio Un polinomio que se escribe con los términos en orden del grado más grande a menos grado.

standard form of a quadratic equation A quadratic equation can be written in the form $ax^2 + bx + c = 0$, where $a \neq 0$ and a, b, and c are integers.

forma estándar de una ecuación cuadrática Una ecuación cuadrática puede escribirse en la forma $ax^2 + bx + c = 0$, donde $a \neq 0$ y a, b, y c son enteros.

standard normal distribution A normal distribution with a mean of 0 and a standard deviation of 1.

distribución normal estándar Distribución normal con una media de 0 y una desviación estándar de 1.

standard position An angle positioned so that the vertex is at the origin and the initial side is on the positive x-axis.

posición estándar Un ángulo colocado de manera que el vértice está en el origen y el lado inicial está en el eje x positivo.

statement Any sentence that is either true or false, but not both.

enunciado Cualquier oración que sea verdadera o falsa, pero no ambas.

statistic A measure that describes a characteristic of a sample.

estadística Una medida que describe una característica de una muestra.

statistics An area of mathematics that deals with collecting, analyzing, and interpreting data.

estadísticas El proceso de recolección, análisis e interpretación de datos.

step function A type of piecewise-linear function with a graph that is a series of horizontal line segments.

función escalonada Un tipo de función lineal por piezas con un gráfico que es una serie de segmentos de línea horizontal.

straight angle An angle that measures 180°.

ángulo recto Un ángulo que mide 180°.

stratified sample The population is first divided into similar, nonoverlapping groups. Then members are randomly selected from each group.

muestra estratificada La población se divide primero en grupos similares, sin superposición. A continuación, los miembros se seleccionan aleatoriamente de cada grupo.

substitution A process of solving a system of equations in which one equation is solved for one variable in terms of the other.

sustitución Un proceso de resolución de un sistema de ecuaciones en el que una ecuación se resuelve para una variable en términos de la otra.

supplementary angles Two angles with measures that have a sum of 180°.

ángulos suplementarios Dos ángulos con medidas que tienen una suma de 180°.

surface area The sum of the areas of all faces and side surfaces of a three-dimensional figure.

área de superficie La suma de las áreas de todas las caras y superficies laterales de una figura tridimensional.

survey Data are collected from responses given by members of a group regarding their characteristics, behaviors, or opinions.

encuesta Los datos se recogen de las respuestas dadas por los miembros de un grupo con respecto a sus características, comportamientos u opiniones.

symmetric distribution A distribution in which the mean and median are approximately equal.

distribución simétrica Un distribución en la que la media y la mediana son aproximadamente iguales.

symmetry A figure has this if there exists a rigid motion—reflection, translation, rotation, or glide reflection—that maps the figure onto itself.

simetría Una figura tiene esto si existe una reflexión-reflexión, una traducción, una rotación o una reflexión de deslizamiento rígida-que mapea la figura sobre sí misma.

synthetic division An alternate method used to divide a polynomial by a binomial of degree 1.

división sintética Un método alternativo utilizado para dividir un polinomio por un binomio de grado 1.

synthetic geometry The study of geometric figures without the use of coordinates.

geometría sintética El estudio de figuras geométricas sin el uso de coordenadas.

synthetic substitution The process of using synthetic division to find a value of a polynomial function.

sustitución sintética El proceso de utilizar la división sintética para encontrar un valor de una función polynomial.

system of equations A set of two or more equations with the same variables.

sistema de ecuaciones Un conjunto de dos o más ecuaciones con las mismas variables.

system of inequalities A set of two or more inequalities with the same variables.

sistema de desigualdades Un conjunto de dos o más desigualdades con las mismas variables.

systematic sample Members are selected according to a specified interval from a random starting point.

muestra sistemática Los miembros se seleccionan de acuerdo con un intervalo especificado desde un punto de partida aleatorio.

T

tangent The ratio of the length of the leg opposite an angle to the length of the leg adjacent to the angle.

tangente La relación entre la longitud de la pata opuesta a un ángulo y la longitud de la pata adyacente al ángulo.

tangent to a circle A line or segment in the plane of a circle that intersects the circle in exactly one point and does not contain any points in the interior of the circle.

tangente a un círculo Una línea o segmento en el plano de un círculo que interseca el círculo en exactamente un punto y no contiene ningún punto en el interior del círculo.

tangent to a sphere A line that intersects the sphere in exactly one point.

tangente a una esfera Una línea que interseca la esfera exactamente en un punto.

term A number, a variable, or a product or quotient of numbers and variables.

término Un número, una variable, o un producto o cociente de números y variables.

term of a sequence A number in a sequence.

terminal side The part of an angle that rotates about the center.

tessellation A repeating pattern of one or more figures that covers a plane with no overlapping or empty spaces.

theorem A statement that can be proven true using undefined terms, definitions, and postulates.

theoretical probability Probability based on what is expected to happen.

transformation A function that takes points in the plane as inputs and gives other points as outputs. The movement of a graph on the coordinate plane.

translation A function in which all of the points of a figure move the same distance in the same direction; A transformation in which a figure is slid from one position to another without being turned.

translation vector A directed line segment that describes both the magnitude and direction of the slide if the magnitude is the length of the vector from its initial point to its terminal point.

transversal A line that intersects two or more lines in a plane at different points.

trapezoid A quadrilateral with exactly one pair of parallel sides.

trend A general pattern in the data.

trigonometric equation An equation that includes at least one trigonometric function.

trigonometric function A function that relates the measure of one nonright angle of a right triangle to the ratios of the lengths of any two sides of the triangle.

trigonometric identity An equation involving trigonometric functions that is true for all values for which every expression in the equation is defined.

término de una sucesión Un número en una secuencia.

lado terminal La parte de un ángulo que gira alrededor de un centro.

teselado Patrón repetitivo de una o más figuras que cubre un plano sin espacios superpuestos o vacíos.

teorema Una afirmación o conjetura que se puede probar verdad utilizando términos, definiciones y postulados indefinidos.

probabilidad teórica Probabilidad basada en lo que se espera que suceda.

transformación Función que toma puntos en el plano como entradas y da otros puntos como salidas. El movimiento de un gráfico en el plano de coordenadas.

traslación Función en la que todos los puntos de una figura se mueven en la misma dirección; El movimiento de un gráfico en el plano de coordenadas.

vector de traslación Un segmento de línea dirigido que describe tanto la magnitud como la dirección de la diapositiva si la magnitud es la longitud del vector desde su punto inicial hasta su punto terminal.

transversal Una línea que interseca dos o más líneas en un plano en diferentes puntos.

trapecio Un cuadrilátero con exactamente un par de lados paralelos.

tendencia Un patrón general en los datos.

ecuación trigonométrica Una ecuación que incluye al menos una función trigonométrica.

función trigonométrica Función que relaciona la medida de un ángulo no recto de un triángulo rectángulo con las relaciones de las longitudes de cualquiera de los dos lados del triángulo.

identidad trigonométrica Una ecuación que implica funciones trigonométricas que es verdadera para todos los valores para los cuales se define cada expresión en la ecuación.

trigonometric ratio A ratio of the lengths of two sides of a right triangle.

relación trigonométrica Una relación de las longitudes de dos lados de un triángulo rectángulo.

trigonometry The study of the relationships between the sides and angles of triangles.

trigonometría El estudio de las relaciones entre los lados y los ángulos de los triángulos.

trinomial The sum of three monomials.

trinomio La suma de tres monomios.

truth value The truth or falsity of a statement.

valor de verdad La verdad o la falsedad de una declaración.

two-column proof A proof that contains statements and reasons organized in a two-column format.

prueba de dos columnas Una prueba que contiene declaraciones y razones organizadas en un formato de dos columnas.

two-way frequency table A table used to show frequencies of data classified according to two categories, with the rows indicating one category and the columns indicating the other.

tabla de frecuencia bidireccional Una tabla utilizada para mostrar las frecuencias de los datos clasificados de acuerdo con dos categorías, con las filas que indican una categoría y las columnas que indican la otra.

two-way relative frequency table A table used to show frequencies of data based on a percentage of the total number of observations.

tabla de frecuencia relativa bidireccional Una tabla usada para mostrar las frecuencias de datos basadas en un porcentaje del número total de observaciones.

U

unbounded When the graph of a system of constraints is open.

no acotado Cuando la gráfica de un sistema de restricciones está abierta.

undefined terms Words that are not formally explained by means of more basic words and concepts.

términos indefinidos Palabras que no se explican formalmente mediante palabras y conceptos más básicos.

uniform motion problems Problems that use the formula $d = rt$, where d is the distance, r is the rate, and t is the time.

problemas de movimiento uniforme Problemas que utilizan la fórmula $d = rt$, donde d es la distancia, r es la velocidad y t es el tiempo.

uniform tessellation A tessellation that contains the same arrangement of shapes and angles at each vertex.

teselado uniforme Un teselado que contiene la misma disposición de formas y ángulos en cada vértice.

union The graph of a compound inequality containing *or*.

unión La gráfica de una desigualdad compuesta que contiene la palabra *o*.

union of A and B The set of all outcomes in the sample space of event A combined with all outcomes in the sample space of event B.

unión de A y B El conjunto de todos los resultados en el espacio muestral del evento A combinado con todos los resultados en el espacio muestral del evento B.

unit circle A circle with a radius of 1 unit centered at the origin on the coordinate plane.

círculo unitario Un círculo con un radio de 1 unidad centrado en el origen en el plano de coordenadas.

univariate data Measurement data in one variable.

datos univariate Datos de medición en una variable.

upper quartile The median of the upper half of a set of data.

cuartil superior La mediana de la mitad superior de un conjunto de datos.

valid argument An argument is valid if it is impossible for all of the premises, or supporting statements, of the argument to be true and its conclusion false.

argumento válido Un argumento es válido si es imposible que todas las premisas o argumentos de apoyo del argumento sean verdaderos y su conclusión sea falsa.

variable A letter used to represent an unspecified number or value; Any characteristic, number, or quantity that can be counted or measured.

variable Una letra utilizada para representar un número o valor no especificado; Cualquier característica, número, o cantidad que pueda ser contada o medida.

variable term A term that contains a variable.

término variable Un término que contiene una variable.

variance The square of the standard deviation.

varianza El cuadrado de la desviación estándar.

vertex Either the lowest point or the highest point of a function.

vértice El punto más bajo o el punto más alto en una función.

vertex angle of an isosceles triangle The angle between the sides that are the legs of an isosceles triangle.

ángulo del vértice de un triángulo isósceles El ángulo entre los lados que son las patas de un triángulo isósceles.

vertex form A quadratic function written in the form $f(x) = a(x - h)^2 + k$.

forma de vértice Una función cuadrática escribirse de la forma $f(x) = a(x - h)^2 + k$.

vertex of a polyhedron The intersection of three edges of a polyhedron.

vértice de un polígono La intersección de tres bordes de un poliedro.

vertex of an angle The common endpoint of the two rays that form an angle.

vértice de un ángulo El punto final común de los dos rayos que forman un ángulo.

vertical angles Two nonadjacent angles formed by two intersecting lines.

ángulos verticales Dos ángulos no adyacentes formados por dos líneas de intersección.

vertical asymptote A vertical line that a graph approaches.

asíntota vertical Una línea vertical que se aproxima a un gráfico.

vertical shift A vertical translation of the graph of a trigonometric function.

cambio vertical Una traducción vertical de la gráfica de una función trigonométrica.

volume The measure of the amount of space enclosed by a three-dimensional figure.

volumen La medida de la cantidad de espacio encerrada por una figura tridimensional.

work problems Problems that involve two people working at different rates who are trying to complete a single job.

problemas de trabajo Problemas que involucran a dos personas trabajando a diferentes ritmos que están tratando de completar un solo trabajo.

X

x-intercept The *x*-coordinate of a point where a graph crosses the *x*-axis.

intercepción *x* La coordenada *x* de un punto donde la gráfica corte al eje de *x*.

Y

y-intercept The *y*-coordinate of a point where a graph crosses the *y*-axis.

intercepción *y* La coordenada *y* de un punto donde la gráfica corte al eje de *y*.

Z

z-value The number of standard deviations that a given data value is from the mean.

valor *z* El número de variaciones estándar que separa un valor dado de la media.

zero An *x*-intercept of the graph of a function; a value of *x* for which $f(x) = 0$.

cero Una intercepción *x* de la gráfica de una función; un punto *x* para los que $f(x) = 0$.

Index

F

G

S

T

U

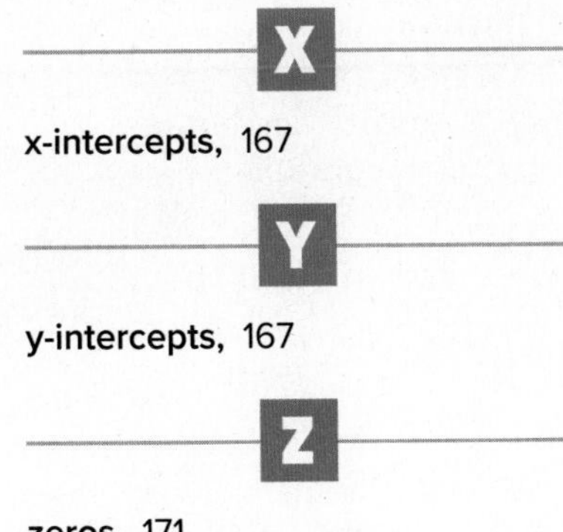